CRC Handbook of Phosphorus-31 Nuclear Magnetic Resonance Data

Editor
John C. Tebby
Head, Division of Chemistry
Staffordshire Polytechnic
Stoke-on-Trent, U.K.

CRC Press is an imprint of the
Taylor & Francis Group, an informa business

First published 1991 by CRC Press
Taylor & Francis Group
6000 Broken Sound Parkway NW, Suite 300
Boca Raton, FL 33487-2742

Reissued 2018 by CRC Press

A Library of Congress record exists under LC control number: 90032631

ISBN 13: 978-1-138-10589-8 (hbk)
ISBN 13: 978-1-138-56002-4 (pbk)
ISBN 13: 978-0-203-71212-2 (ebk)

Visit the Taylor & Francis Web site at http://www.taylorandfrancis.com and the
CRC Press Web site at http://www.crcpress.com

PREFACE

This Handbook of Phosphorus-31 NMR Data represents the culmination of several years of collaboration by phosphorus chemists from all parts of the world. Only the contributors will know the considerable painstaking effort required to follow exacting presentation and sequencing rules. Their perseverance is gratefully acknowledged. The editor thanks contributors for their patience while awaiting publication. Many other chemists have generously donated NMR data related to their own research fields. Accordingly, the authors wish to acknowledge and thank John Verkade, Reinhard Schmutzler, Louis Quin, E. Nieke, Dennis Hall, Jean P. Dutasta, Sheldon Cremer, and Hans J. Bestmann for their valuable contributions. Ekkehard Fluck and Stephan Kleemann donated data covering all phosphorus compounds appearing in the literature for the year 1983.

The bar charts in Chapter 1 were meticulously produced by Miss Jenny Keeling, Department of Applied Sciences, Staffordshire Polytechnic.

Every effort has been made to ensure the accurate transfer and organization of data. Information regarding errors which readers may encounter would be gratefully received.

THE EDITOR

John Tebby, D.Sc., is Head of the Division of Chemistry at Staffordshire Polytechnic, Stoke-on-Trent, U.K.

Professor Tebby graduated in 1958 from Brunel College of Technology, London, following eight years of evening and part-time study while working at Glaxo Laboratories, Greenford. He studied for his Ph.D. degree in organic chemistry at Nottingham University, and from 1961 to 1964 he was a research associate at Stanford University and research fellow at University of California Medical Center, San Francisco.

He was elected Fellow of the Royal Society of Chemistry in 1974 and he has been an active officer of the Staffordshire Local Section of the Royal Society of Chemistry since 1978, including two years as Chairman. Professor Tebby is a senior referee for the Chemical Society in the field of phosphorus chemistry, spectroscopy and reaction mechanisms.

He was appointed Research Associate at the University of Keele from 1979 and was Visiting Professor at Queen's University Belfast in 1986. He was on secondment to the Molecular Modelling Section of Smith Kline and French Research in 1986-1987.

In 1987, he was awarded a D.Sc. degree by Nottingham University and appointed to a personal professorship at Staffordshire Polytechnic.

Professor Tebby has presented many papers to National and International Conferences and lectured widely in the U.S. and Europe. He is well known for his regular chapter on Physical Properties in the Specialist Periodical Reports on Phosphorus Chemistry which is published annually by the Chemical Society.

He has current major research interests in reaction mechanisms, structure elucidation and spectroscopic studies of organophosphorus compounds including conformational studies of phosphonium ylides; he is also working on syntheses of phosphonate monomers and phosphorus compounds for metal extraction, water soluble catalysts and superoxide detection. Other research interests include molecular modeling, molecular recognition, synthesis of compounds with biological activity and computer applications in NMR spectroscopy. These interests have been supported by grants from SERC, various industries, the National Advisory Board and Staffordshire Polytechnic.

CONTRIBUTORS

Jean-François Brazier
Maître de Conférence
Department of Chemistry
Université Paul Sabatier
Toulouse, France

Ramon Burgada
Laboratoire de Chimie des
Organo-éléments
Université Paris-Nord
Paris, France

Peter J. Diel
Wissenschaftliche Spezialist
Agrochemical Division
Ciba-Geigy AG
Basel, Switzerland

Keith B. Dillon
Senior Lecturer
Department of Chemistry
University of Durham
Durham, England

Ronald S. Edmundson
Lecturer (retired)
University of Bradford
Bradford, England

Ekkehard Fluck
Professor
Gmelin Institüt
Max-Planck-Gesellschaft
Frankfurt, Germany

Michael J. Gallagher
Associate Professor
School of Chemistry
University of New South Wales
Kensington, New South Wales, Australia

Hélène Germa
Maître de Conférence
Department of Chemistry
Enseignement Supérieur
Toulouse, France

David Vaughan Griffiths
Senior Lecturer
Department of Chemistry
University of Keele
Staffordshire, England

Penelope A. Griffiths
Research Fellow
Department of Chemistry
University of Keele
Staffordshire, England

Harry R. Hudson
Professor
Department of Applied Chemistry
The Polytechnic of North London
London, England

Konstantin Karaghiosoff
Institute of Inorganic Chemistry
University of Munich
Munich, Germany

Stephan G. Kleemann
Professor
Fachhochschule Muenchen
Munich, Germany

Max Koenig
Director of Research
Department of Chemistry
CNRS
Toulouse, France

S. S. Krishnamurthy
Professor
Department of Inorganic and Physical
Chemistry
Indian Institute of Science
Bangalore, India

Lydia Lamandé
Head of Research
Department of Chemistry
CNRS
Toulouse, France

Yves Leroux
Laboratoire de Chimie Structurale
Biomoléculaire
Université de Paris-Nord
Bobigny, France

Ludwig Maier
Wissenschaftlicher Experte
Agrochemical Division
Ciba-Geigy AG
Basel, Switzerland

Jacques Navech
Director of Research
Department of Chemistry
CNRS
Toulouse, France

Andrew Platt
Department of Applied Sciences
Staffordshire Polytechnic
Stoke-on-Trent, U.K.

Jean G. Riess
Professor
Unité de Chimie Moleculaire
Université de Nice
Nice, France

Alfred Schmidpeter
Professor
Institute for Inorganic Chemistry
University of Munich
Munich, Germany

Jacqueline Seyden-Penne
Université de Paris-Sud
Orsay, France

Marie-Paule Simonnin
École Nationale Supérieure de Chemie
Paris, France

Tekla Strzalko
Université de Paris-Sud
Orsay, France

Brian J. Walker
Head, Department of Chemistry
The Queens University of Belfast
Belfast, Northern Ireland

Robert Wolf
Director of Research
Department of Chemistry
CNRS
Toulouse, France

TABLE OF CONTENTS

Chapter 1

USE OF THE HANDBOOK AND GENERAL TRENDS

John C. Tebby

SCOPE AND OBJECTIVES

The handbook aims to provide a compilation of phosphorus-31 chemical shifts of organophosphorus compounds which is as comprehensive as possible in the available space. Important coupling constants involving phosphorus are given but references to NMR data relating specifically to other nuclei are not included. Inevitably there has had to be some selection of the data since in 1987 it was estimated that there were over 40,000 items of ^{31}P NMR data in the literature. Nevertheless, the Handbook contains all classes of nonmetallic compounds known at the time of compilation, covered with as large a variety of structure and stereochemistry as possible. Some metallic compounds have been included, notably when the metal is stabilizing an interesting structure and those compounds used as organic synthons. It has been our goal that each entry provides additional information not provided by other entries. Significant differences in chemical shift for otherwise similar structures was also a valid reason for inclusion of data in the Handbook. In most cases where the solvent and temperature were recorded they are also shown.

The data is organized systematically, primarily according to the atomic environment of the phosphorus nuclei and then according to molecular formula. This allows the retrieval of NMR data for a specific compound and, where that is not available, to locate data for a phosphorus atom with as similar an environment as possible. It is also possible to estimate the effects of any remaining structural differences by comparing other related data. In the first place, the data are categorized according to the coordination number of the phosphorus atom, and then, within each coordination state, they are organized according to recognized or well-defined classes. For example, the four coordinate compounds are divided into phosphonium salts, phosphates, phosphonates, phosphinates, phosphine oxides, etc. The sequence within each class is determined primarily by the atoms bound to phosphorus: the presence and size of rings and then the connectivities of the phosphorus-bound atoms follow. The final prioritizing factor is molecular formula. Details and examples of the sequencing rules are given in the following section.

The coordination number is taken as the number of strongly covalently bound atoms to phosphorus. The bonds may be single, double or treble. Ionic or weakly coordinated atoms are not counted.

Occasionally contributors to the Handbook have deviated slightly from these strict rules in order to collect together other important structural features. For example, the polyphosphines and their oxides are of particular interest to coordination chemists and therefore they have been placed in separate subsections throughout Tables E and L, respectively. For these tables it may be necessary to search two sections in order to be sure that the nearest structure about phosphorus has been identified. When two or more different phosphorus atoms are part of one molecule, the NMR data of each phosphorus environment appear at the appropriate position in the tables. In order to assist the identification of a structure for a recorded chemical shift the data are also presented in the form of a series of bar charts in the final section of this chapter. There is a comparison of the range of chemical shifts for compounds in each of the main classes of compounds right across the six coordination states. The data are also further analyzed according to each table and for some cases for the main subsections. In these tables certain structural features are identified on the bar lines.

The tables may also be used to estimate and study specific structural effects. For example, the change in chemical shift caused by incorporating a phosphorus atom in a small ring may be estimated by seeking matching cyclic and acyclic compounds having identical P-bound atoms and connectivities. The effect of replacing one group X by another group Y may be estimated by seeking matches differing only in X and Y.

Several measures were taken in order to make the most efficient use of space. An extensive list of abbreviations was used for functional groups and commonly occurring substructures, as well as for stereochemistry and solvents. Data from references quoted for other tables are not always duplicated; such references are prefixed by the letter of the relevant table.

STRUCTURE DEFINITION

The atoms covalently bound to phosphorus are defined either as a subheading, i.e., **P-bound atoms** = XXX at the start of a set of data, or they are listed in the first column of the table depending on the frequency with which they are changing. The P-bound atoms are presented as a formula arranged in the usual Chemical Abstracts manner (i.e., C before H, and H before other elements listed in alphabetical order). The atoms are in their normal coordination state with respect to their group position in the periodic table, e.g., four coordinate for carbon and silicon, three coordinate for nitrogen and phosphorus, two coordinate for oxygen and sulfur, etc. except where indicated otherwise. Coordination states one lower than normal are indicated by a prime, e.g., C′, N′, and O′ for carbon, nitrogen, and oxygen, respectively; whereas coordination states two lower than normal are identified by double primes, e.g., C″ and N″ for two and one coordinate carbon and nitrogen, respectively. Higher than normal valencies are indicated by the valency number shown as a superscript, e.g., N^4, P^4, and S^4 for four coordinate nitrogen, phosphorus, and sulfur, respectively. Two or more P-bound atoms of the same element with different coordination states are placed in the order of decreasing coordination number. Thus the P-bound atom formula for methylphenylcyanophosphine is CC′C″. The coordination differences are also used in the definitions of the connectivities. This is discussed later.

The incorporation of the phosphorus atom in a cyclic structure is indicated by adding the size of the **Ring** or **Rings** either immediately after the P-bound atom formula, e.g., $CC'_2/5$ for 2-methylphosphole or immediately in front of the connectivities — in both cases separated by a slash. Phosphorus atoms in bicyclic and tricyclic structures are indicated in a similar manner, e.g., $C'_3/6,6$ for 1-phosphabarrelene P(CH=CH)3CH. The number of rings incorporating the phosphorus atom is determined by the number of bonds that must be cleaved in order to create an acyclic phosphorus group. When there is a choice of ring size the smallest is stated and the larger omitted, e.g., $C_3/5$ and not $C_3/6$ for P-methyl 2-phosphanorbornane.

The next stage of structure definition concerns the second sphere of atoms surrounding the phosphorus atom. The atoms connected to each P-bound atom are stated in sets of **Connectivities** which are listed in the same order as the sequence of atoms in the P-bound atom formula. For example, methylphenylcyanophosphine with P-bound atoms = CC′C″ has connectivities $H_3;C'_2;N''$ since there are three hydrogen atoms bound to the methyl carbon, two ortho three coordinate carbons connected to the quaternary carbon of the phenyl group and a one coordinate nitrogen atom connected to the cyano carbon atom. When, as in this example, the P-bound atoms are nonidentical the sets of connectivities are separated by semicolons.

On the other hand, when there are two or more identical P-bound atoms, e.g., C_3 for methylethyl(t-butyl)phosphine, the connectivities are stated in their priority order (see next section), with each set of connectivities separated by a comma, i.e., H_3,CH_2,C_3. If two or more of the connectivities are the same they are collected in parentheses, e.g., trimethylphosphine with P-bound atoms = C_3 has connectivities $(H_3)_3$.

The next stage of structure definition is the **Molecular formula,** whether or not it is listed. A table will contain a column of molecular formula only when the P-bound atoms are presented as subheadings, i.e., when there are substantial numbers of compounds with similar structures. The molecular formulae are presented in the usual manner, i.e., C before H and H before other elements placed in alphabetical order. Differences in hybridization/coordination are ignored.

Finally the **Molecular structure** is given using an extended range of abbreviations for functional groups and common substructures. Most structural ambiguities that arise can be resolved by reference to the connectivities. Usually the functional groups bound to phosphorus are given in the same order as the P-bound atoms except where this might lead to structural ambiguity. Cyclic structures are usually indicated by placing a superscript dot or an asterisk beside atoms which are directly bound so that a ring is formed. Alternatively atoms involved in a ring involving a phosphorus atom are simply enclosed in parentheses, e.g., P-methylphosphole Me-P(CH=CHCH=CH). The stereochemistry is defined either immediately after the structure or as superscripted abbreviations following the chemical shift, e.g., 16^{C} indicates a *cis* isomer with a phosphorus chemical shift of 16 ppm.

THE SEQUENCE RULES

The sequence of the compounds within each table (or within the main subsections of a table) is determined by applying the following rules to the above structural features:

a. *First* according to the priorities of the **P-bound atoms.**
b. *Secondly* according to the priorities of any **Ring** or **Rings** which incorporate the phosphorus atom.
c. *Thirdly* according to the priorities of the **Connectivities**.
d. *Fourthly* according to the priorities of the **Molecular formula**.

Details are as follows.

PRIORITIZATION USING P-BOUND ATOMS

The P-bound atoms are the first factor which determines the sequence of the compounds within each table and they are as follows.

(i) The sequence follows that used by the Hill formula index, i.e., the Chemical Abstracts system modified so that all inorganic formulae appear first. The Molecular Formula Index in the Aldrich Chemical catalogue is an example. Thus compounds with *no* P-C or P-H bonds precede those *with* P-H bonds but *no* P-C bonds which in turn precede those *with* P-C bonds. The sequence within each of these sections follows the usual rules, i.e., the formulae appear in order to increasing numbers of the first quoted atom, e.g., $AB_3 > A_2B_2 > A_3B$ then similarly for the next atom, etc. Any differences in coordination number are ignored at this stage, e.g., formula C′C″H is considered as C_2H.*

(ii) For those compounds which differ *only* with respect to the coordination/hybridization of one or more of the P-bound atoms, priority is given to atoms with highest coordination number, e.g., for C 4 > 3 > 2 > and for N 3 > 2 > 1. The same principle is applied to other elements. With respect to the combinations of two elements with different coordination numbers, the coordination of the first atom is considered before the second atom appearing in the formula, e.g., CN > CN′ > C′N > C′N′. See Example Set A, next page.

PRIORITIZATION ACCORDING TO P-HETEROCYCLE RINGS

This rule applies to compounds with P-bound formulae that are identical in all respects, including identical coordination numbers. Any acyclic compounds are followed by monocyclic P-heterocycles — given in order of increasing ring size 3 > 4 > 5 > 6, etc; these

* The sequence is based on the concept that $C_oH_oZ_k$ has the highest priority, and that $C_oH_nZ_{o-n}$ has next highest priority, with $C_mH_{o-n}Z_{o-n}$ having lowest priority, with the priorities falling within each section as m, n, and x increase. Thus the order of P-bound atoms formulae is $C_oH_oZ_{o-n} > C_oH_nZ_{o-n} > C_mH_{o-n}Z_{o-n}$, where Z represents all atoms other than C and H given alphabetical order.

are followed by bicyclic and then tricyclic heterocycles — each group also being placed in order of increasing ring size, the smallest ring appearing first and having the highest priority, e.g., 3,5 > 3,6 > 5,5 > 5,6 > 6,6 etc., > 3,6,6 > 5,5,5 > 5,6,6 > 6,6,6 etc. Note that only four or higher coordination states can form tricyclic structures incorporating phosphorus.

PRIORITIZATION USING THE CONNECTIVITIES

The connectivities are the next factor for determining the sequence. The compounds must have the same P-bound atoms in the same hybridization and with same sized heterocyclic rings. The connectivities of the atoms in the P-bound atom formula are considered consecutively until a difference is found in comparison with an adjacent entry. The prioritization of the connectivities follow the same Hill system extended to include differences in hybridization/coordination as used for the sequence of the P-bound atom formulae. The order is such that for compounds with one simple alkyl group, i.e., one sp^3 carbon bound to phosphorus the sequence is methyl > ethyl > i-propyl > t-butyl.

As stated above in an earlier section, when two or more identical P-bound atoms have different connectivities, the sets of connectivities are presented in order of their own priority, e.g., $H_3 > CH_2 > C_2H > C_3$. Since this order also determines the sequence of the entries, the group with the highest priority dominates the sequence. See Example Sets B and C.

PRIORITIZATION USING MOLECULAR FORMULAE

When two or more compounds have identical P-bound atoms, heterocyclic ring size and connectivities, then the molecular formulae controls the sequence in the usual manner. In most tables the ordering of the entries at this stage is determined by the number of carbon atoms in the compound. However the Hill system of prioritization also applies to the molecular formulae and thus formulae with no carbon and hydrogen have priority over formulae with hydrogen but no carbon, and formulae which contain most carbon have the lowest priority. The sequence within each of these sections follows the usual rules, i.e., the formulae appear in order of increasing numbers of the first quoted atom, e.g., $AB_3 > A_2B_2 > A_3B$ then similarly for the next atom, etc.

EXAMPLES OF PRIORITIZATION

SET A: Prioritization of Three Coordinate Compounds with Variation of Three P-Bound Atoms — Namely A, B, X, (Three Elements Other Than C and H), H (Hydrogen) and C (Carbon)

The order is as follows:

$ABX > AB_2 > A_2B > A_2X > A_3 > BX_2 > B_2X > B_3 > X_3 > HAB > HAX >$
$HA_2 > HB_2 > HX_2 > H_2A > H_2B > H_2X > H_3 > CAB > CAX > CA_2 > CBX >$
$CB_2 > CX_2 > CHA > CHB > CHX > CH_2 > C_2H > C_3$

Each of these formulae is subdivided when one or more of the atoms have different coordination states. For example, if each of the atoms in the P-bound atom formula CAB existed in normal and one less than normal coordination state then these compounds would appear in the following sequence:

$$CAB > CAB' > CA'B > CA'B' > C'AB > C'AB' > C'A'B > C'A'B'$$

The existence of lower or higher coordination numbers for any of these atoms would further extend the number of divisions.

SET B: Prioritization of Compounds when Connectivities of a P-Bound Four Coordinate Carbon Atom Are a Controlling Factor

The connectivities to a four coordinate carbon are probably the most diverse and therefore it is illustrated below. The example applies to all classes of compound and it also usefully illustrates the principles for prioritizing. Note that Br has been used as the first non C or H element. However less common elements such as arsenic would preceed Br. The symbols X, Y, and Z represent atoms higher in the alphabet than Br, but not H, nor C. Not all possible combinations are included.

First in the sequence are compounds with connectivities XYZ: BrXY (e.g., BrClF > $BrCl_2$ > BrF_2) > Br_2X > Br_3 > ClXY > Cl_2X > Cl_3 > FXY > F_2N > F_2N' > F_2N'' > etc.

Next in the sequence are compounds with connectivities involving hydrogen but not carbon, e.g., HXY: HBrX > HBr_2 > HXY > HX_2 > etc. to HTe_2 followed by H_2X: H_2Br > H_2Cl > etc. followed by H_3: (i.e., methyl). Note that isotopes, e.g., D_3 are not distinguished.

Finally, connectivities involving carbon have the lowest priority and therefore appear last in the sequence: CBrX > CBr_2 > CXY > CX_2 > CHBr > CHX > CHX' etc. > CH_2 (representing ethyl and other n-alkyl and beta substituted ethyl groups) > $C'H_2$ (allyl or benzyl or $RCOCH_2$ etc.) > $C''H_2$ (propargyl or cyanomethyl) > C_2Br > C'CBr (e.g., 1-bromo-1-phenylethyl) etc. > C_2X > CC'X >> C''_2X (e.g., 1-bromo-1,1-dicyanomethyl) > C_2H (isopropyl) > CC'H (e.g., 1-arylethyl or 1-vinylethyl or 1-acylethyl) > CC''H (e.g., 1-cyanoethyl or 1-ethynylethyl) > C_3 (t-butyl etc.) > C_2C' (e.g., 1-methyl-1-phenylethyl) > C_2C'' (e.g., 1-cyano-1-methylethyl) > CC'_2 (e.g., 1,1-diphenylethyl) > CC'C'' (e.g., 1-cyano-1-phenylethyl) > CC''_2 (e.g., 1,1-dicyanoethyl) > C'_3 (e.g., 1,1,1-triphenylmethyl) > etc. to > C''_3 (e.g., 1;1,1-tricyanomethyl).

SET C: Prioritizing Compounds Using the Connectivity Formula of a P-Bound Three Coordinate Carbon Atom

This could apply to a table of acyl, aryl and vinyl primary phosphines all with P-bound atom formula = $C'H_2$.

First priority is given to connectivities XY': BrO' (bromoacetyl) > ClO' > NN' (amidinyl) > NO' (amido), OO' (alkoxycarbonyl) > O'S etc.

Next connectivities HX': HN' > HO' (formyl) > HS' etc.

Finally connectivities CX, CH and C_2: C'Br (1-bromovinyl) > C'Cl etc. > CN' > CO' (acetyl) > CS' etc. > C'H (vinyl) > CC' (1-alkylvinyl) > C'_2 (e.g., 1-phenylvinyl) > C'C'' (1-cyanovinyl).

SELECTION AND PRESENTATION OF THE NMR DATA

The NMR data is given in the fourth column of each table. The phosphorus chemical shift(s), quoted to one decimal place, are presented first. Positive shifts are downfield of 85% phosphoric acid. In some cases where the sign designation has not been stated in the literature it may not be possible to confidently state whether the shift is upfield or downfield of the reference. A query precedes such parameters. Normally no more than two literature values of the chemical shift are quoted for each compound. A range of shifts (e.g., −3.2/−4.7) is given where more than two have been reported. Solvent, temperature and pH (if appropriate) are given in parentheses immediately following the chemical shift. The list of abbreviations for the solvents are given on page 6. Mixed solvents are indicated using a plus sign, e.g., (c + r). Important coupling constants involving phosphorus follow the chemical shift data. Unless otherwise specified the coupling path is that given at the head of the table. Data on compounds with different stereochemistry are separated by semicolons whereas data for the same compound from different literature sources are separated by commas.

DATA SOURCES AND REFERENCES

A large number of phosphorus chemists throughout the world have been involved in providing the data. While the main authors of each chapter have been responsible for sorting, selecting, and presenting the data, many others have provided extensive lists of data recorded in their laboratories and other data relating to their specialist areas of research.

During the process of placing the compounds in the correct sequence many entries have been moved about within the tables. As a consequence the order of appearance of references in the tables are frequently not in numerical sequence. A letter preceding a reference number relates to the table in which the reference may be found.

ABBREVIATIONS

The commonly accepted abbreviations for alkyl, aryl, functional groups, substructures, and stereochemistry definers have been extended in order to make as efficient use of the space available as possible.

Abbreviations for Alkyl, Aryl, and Heteroaryl Groups

Alkyl Groups

Me = Methyl	Et = Ethyl	Pr = Propyl
Bu = Butyl	Pe = Pentyl	neoPe = neoPentyl
Hex = Hexyl	Oct = Octyl	Non = Nonyl
Dec = Decyl	Nb = Norbornyl	Nbe = Norbornenyl
Cp = C_5H_5	Cpm = C_5Me_5	Bz = Benzyl

Aryl Groups

Ph = Phenyl	Ph′ = C_6H_4 (substituted phenyl)	
Tol = Tolyl	Xyl = Xylyl	Mes = Mesityl
Sms = Supermesityl (2,4,6-tri-t-butylphenyl)		

Heteroaryl

Fur = Furyl	Pyr = Pyridyl	Quin = Quinoline-8-hydroxy
Bipyr = Bipyridine	7-Me-Quin = 7-Methyl-8-hydroxyquinoline	

Abbreviations for Functional Groups

Bis = Tms_2CH	Lp = Lone pair of electrons	Pip = N-Piperidyl
Mpip = Me_4-piperidyl	Morph = Morpholinyl	Tf = Trifluoromethyl
Tms = Trimethylsilyl	Tos = Tosylate	Tris = Tms_3C

Abbreviations for Substructures

Ad = Adamantyl	Cat = Catechol	Dop = Di-orthophenylene
Glc = Glycolyl	Oam = o-Aminophenol	Nox = $OC^*HCH_2CMe_2N(O)CMe_2C^*$
Pfp = Perfluoropinacol	Pnc = Pinacol	Don = Dioxynaphthalene

Abbreviations for Stereochemical Symbols

Stereochemical Prefixes for Alkyl Groups

c = cyclo
o = ortho
t = tertiary
i = iso
p = para
m = meta
s = sec

Stereochemical Superscripts to Follow NMR Shift Data

C = Cis
E = E-geometry
R_P = Rectus configuration at P
T = Trans
D = Diastereoisomer
I = Isomer
S_P = Sinister configuration at P
Z = Z-geometry
dl = Racemate
M = Meso

Abbreviations for Solvent and Conditions

The symbols appear in brackets together with temperature, if known, after the chemical shift.

a = acetone
d = dimethyl sulfoxide
g = carbon tetrachloride
j = ether
m = MeCN (acetonitrile)
p = pyridine
s = carbon disulfide
v = dimethylformamide
y = MeOH
ae = Et_3N
de = 1,2-dichlorobenzene
he = C_6F_6
ne = nitrobenzene
ze = $MeOCH_2CH_2OMe$
b = benzene
e = ethanol
h = hexane
k = tetrahydrofuran
n = neat
q = petroleum ether
t = toluene
w = water
z = $MeOCH_2CH_2OH$
be = 1-bromonaphthalene
ee = ethyl acetate
ke = dioxan
pe = diphenyl ether
CP/MAS = Cross Polarization with Magic Angle Spinning
c = chloroform
f = Freon 11, $CFCl_3$
i = isopropanol
l = trifluoroacetic acid
o = cyclohexane
r = dichloromethane
u = acetic acid
x = hexachloroacetone
ce = chloroethane
fe = nitromethane
me = mesitylene
re = dichloroethane

BAR CHARTS OF ^{31}P CHEMICAL SHIFTS

Given a chemical shift of an unknown compound the bar charts provide an aid to identify possible structures.

The first bar chart (Figure 1.1) shows the ranges of chemical shifts right across all the compounds in the Handbook. All six possible coordination states are differentiated and the three and four coordinate compounds are subdivided into their primary structural classes.

The second set of bar charts (Figures 1.2 to 1.4 and 1.6 to 1.9) provide further breakdown of the ranges according to the nature of the P-bound atoms. Due to the very large number of four coordinated compounds, Figure 1.5 gives the chemical shift ranges for the well-defined classes of compounds and the associated set of Figures 1.5a to 1.5i provide the breakdown according to the nature of the P-bound atoms.

At each inspection stage other structural information may be used to narrow the field of possible structures. The tables are so constructed that further detail of trends can be obtained by direct inspection.

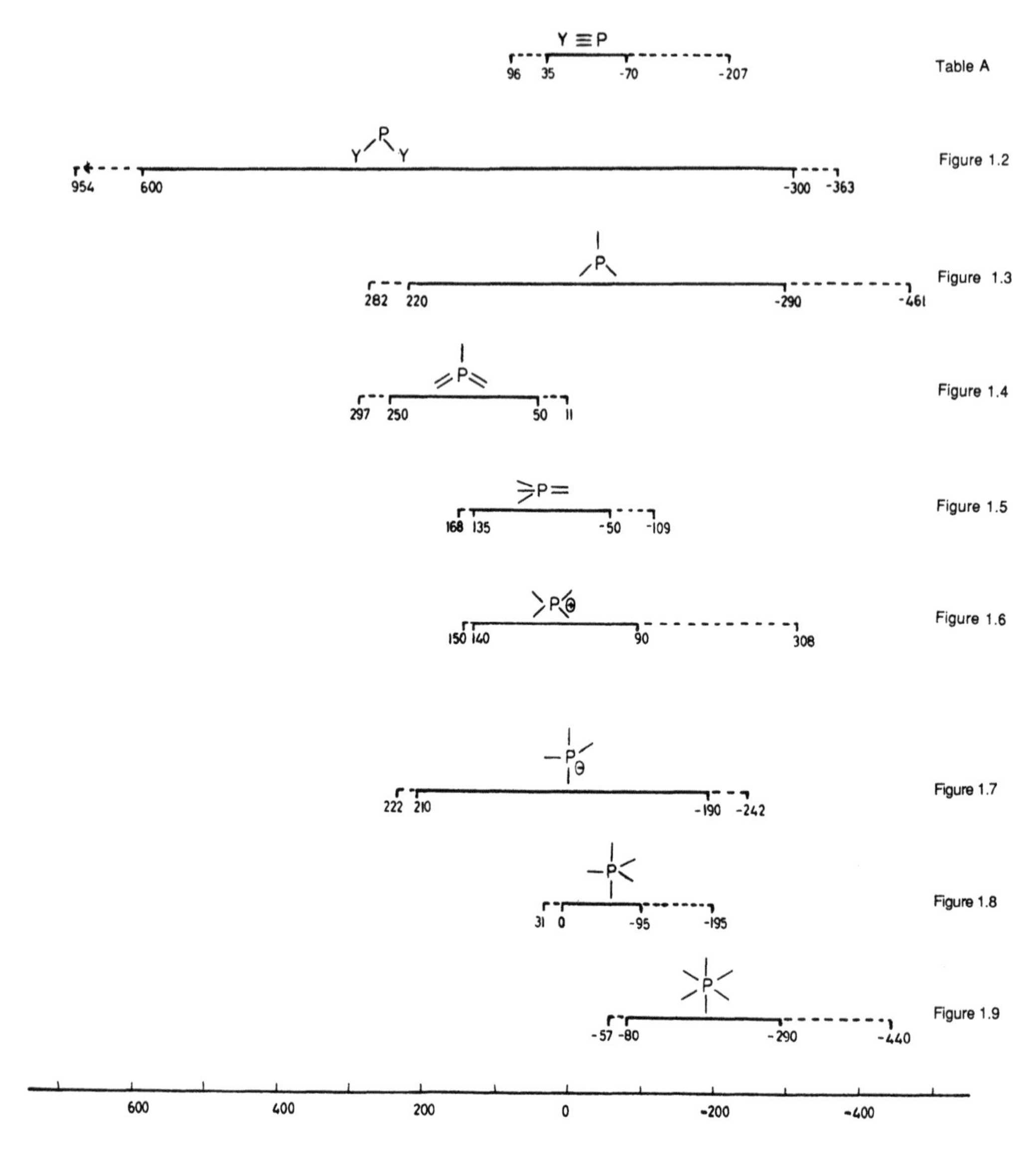

FIGURE 1.1. ^{31}P NMR chemical shifts for all compounds. Dotted lines show the extensions of the ranges to extreme values and 5% or less compounds falls into these outer regions.

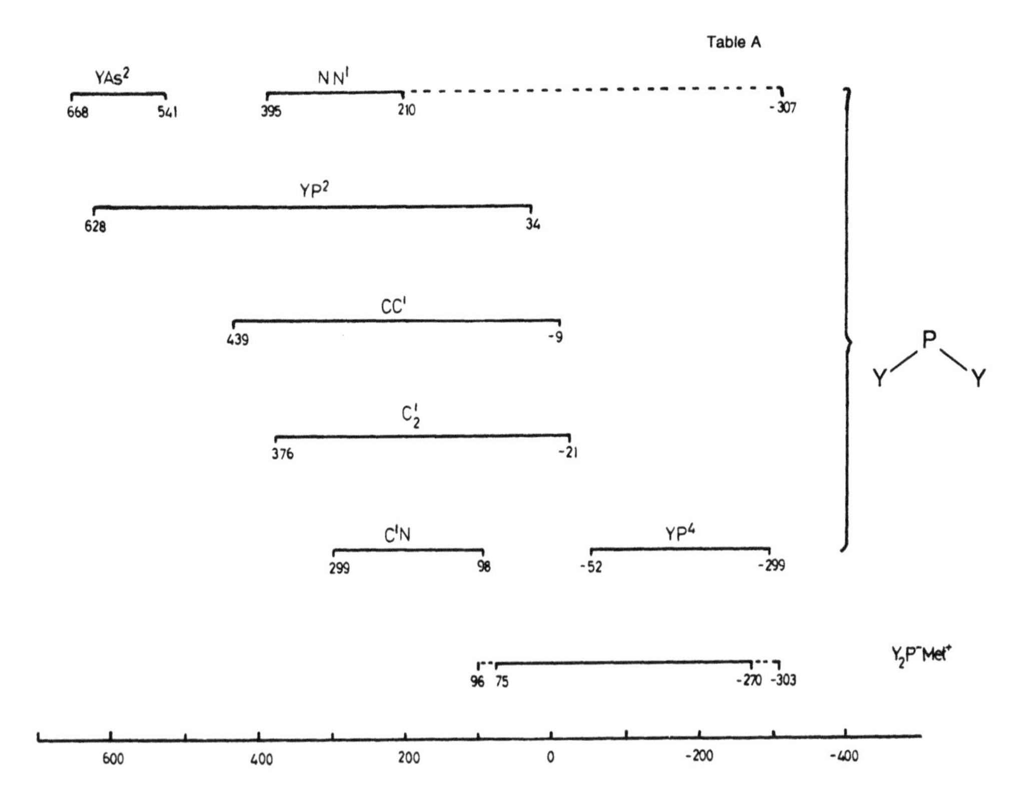

FIGURE 1.2. ^{31}P NMR chemical shifts for two coordinate compounds and alkali metal phosphides. P-bound atoms are given above each bar line; Y = variety of elements. (Table A.)

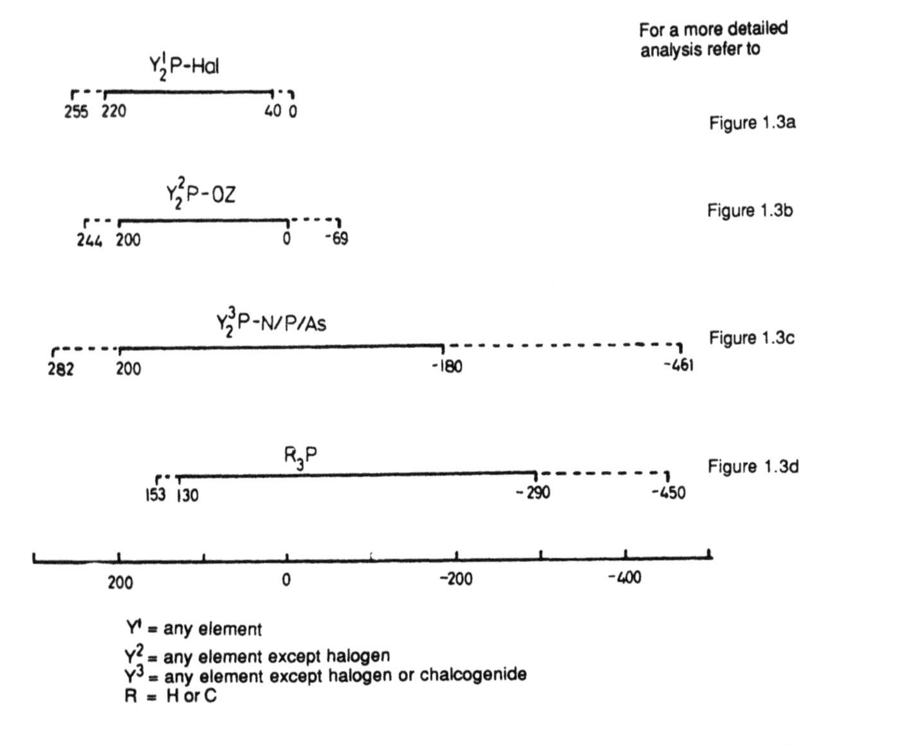

FIGURE 1.3. ^{31}P NMR chemical shifts of three coordinate (λ3 σ3) phosphorus compounds. (Tables B to E.)

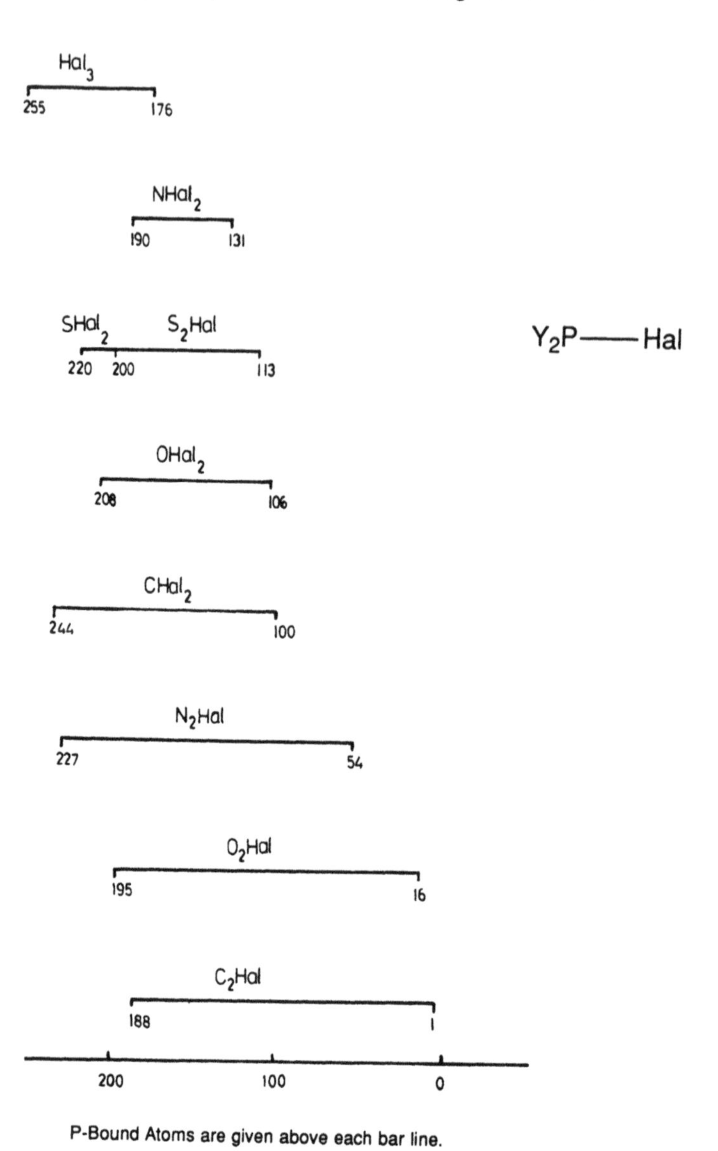

FIGURE 1.3a. ^{31}P NMR chemical shifts of three coordinate ($\lambda 3$ $\sigma 3$) phosphorus compounds containing a P-halogen bond. (Table B.)

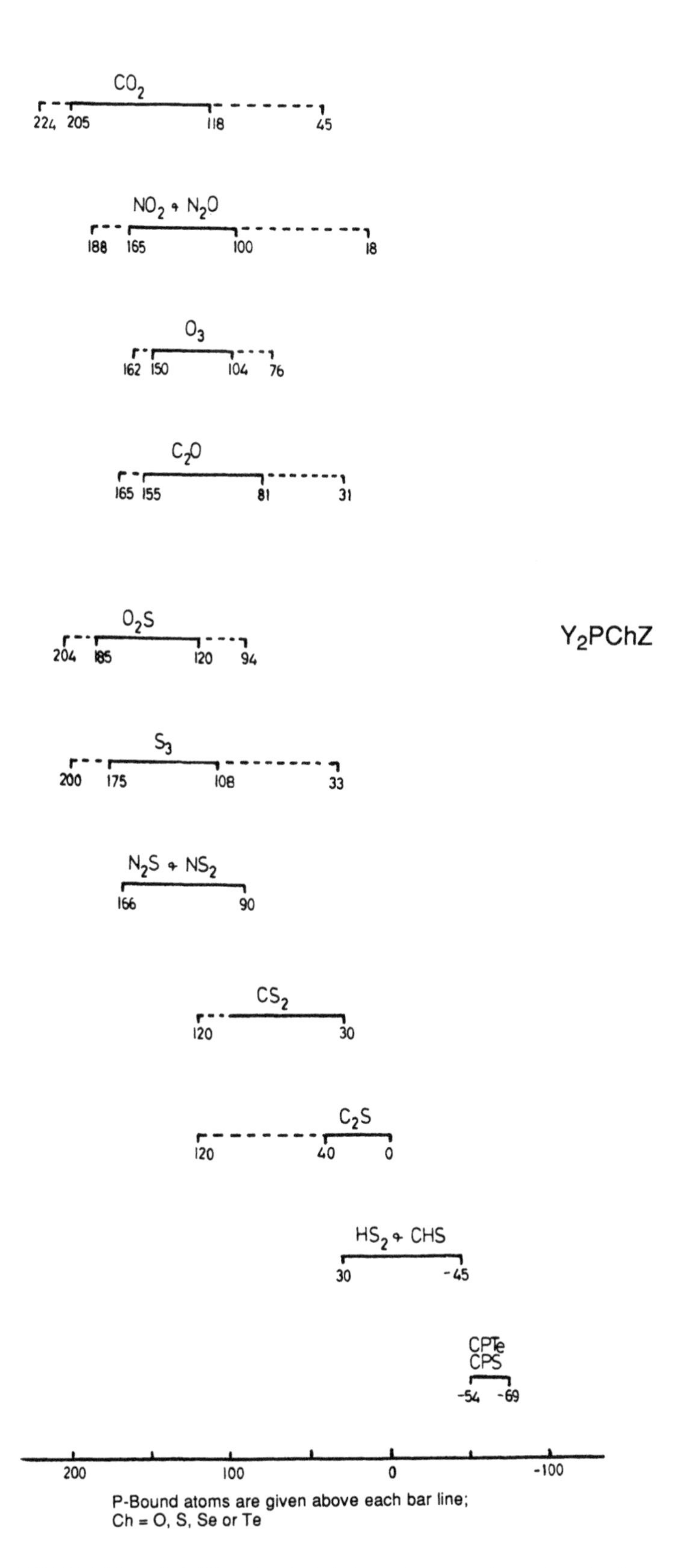

FIGURE 1.3b. [31]**P NMR chemical shifts of three coordinate (λ3 σ3) phosphorus compounds containing P-chalcogenide bonds but no bonds to halogens. (Table C; includes phosphites.)**

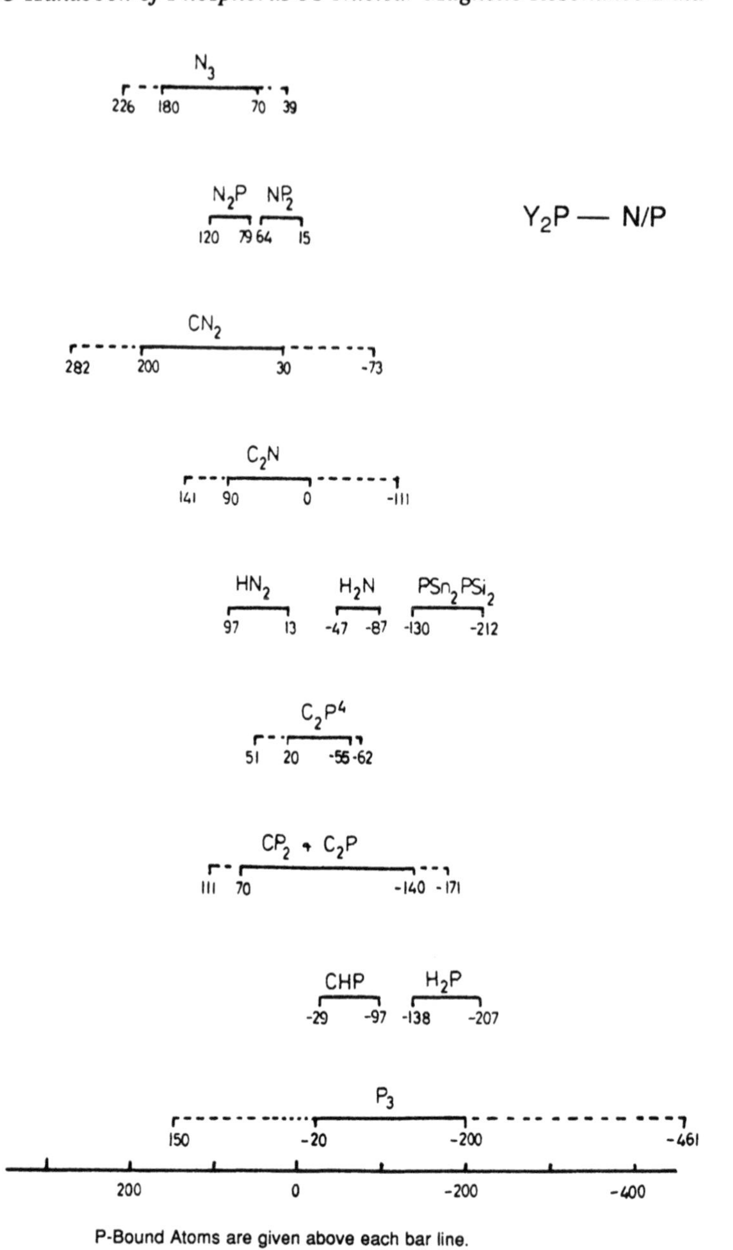

FIGURE 1.3c. ^{31}P NMR chemical shifts of three coordinate (λ3 σ3) phosphorus compounds containing a phosphorus bond to a Group V element but no bonds to halogen or chalcogenide. (Table D; includes aminophosphines and polyphosphines.)

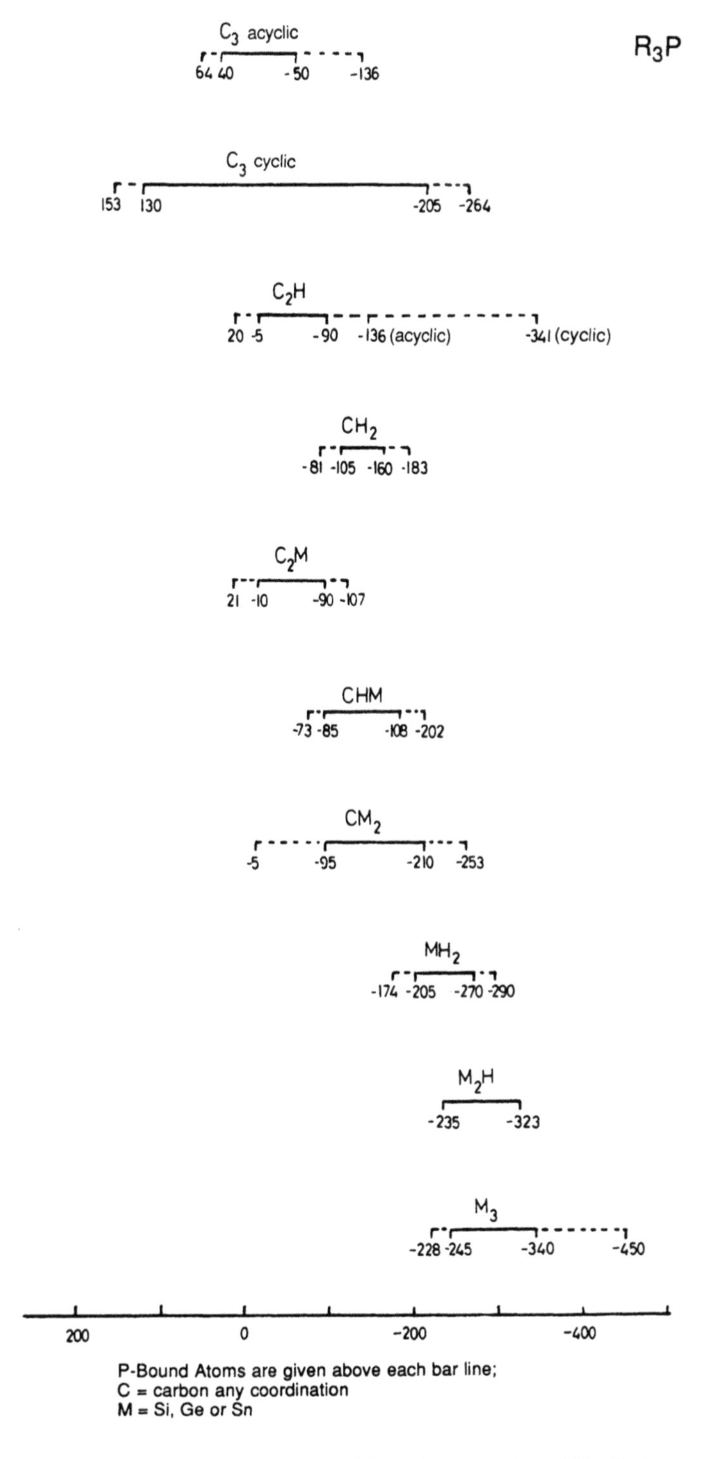

FIGURE 1.3d. ^{31}P NMR chemical shifts of three coordinate (λ3 σ3) phosphorus compounds containing phosphorus bonds to Group IV elements and hydrogen. (Table E: Phosphines.)

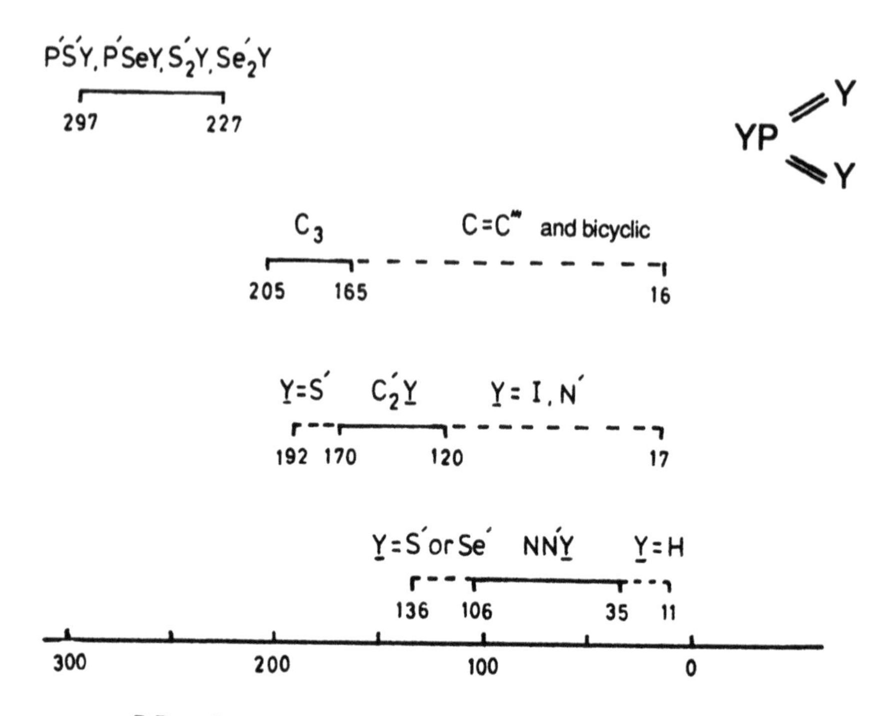

FIGURE 1.4. ^{31}P NMR chemical shifts of three coordinate five valent ($\lambda 5$ $\sigma 3$) compounds. (Table F; includes metaphosphoryl compounds.)

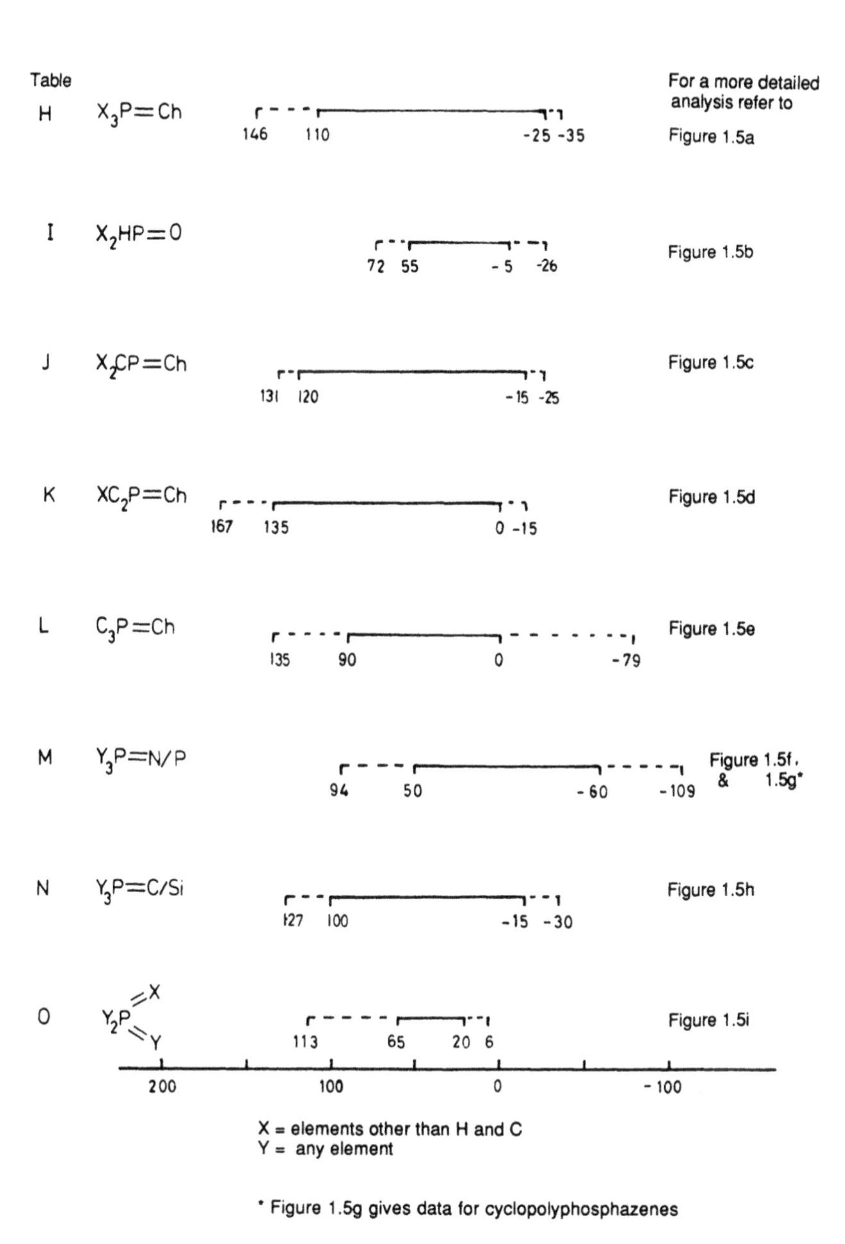

FIGURE 1.5. ^{31}P NMR chemical shifts of four coordinate phosphorus compounds containing one of two formal multiple bonds.

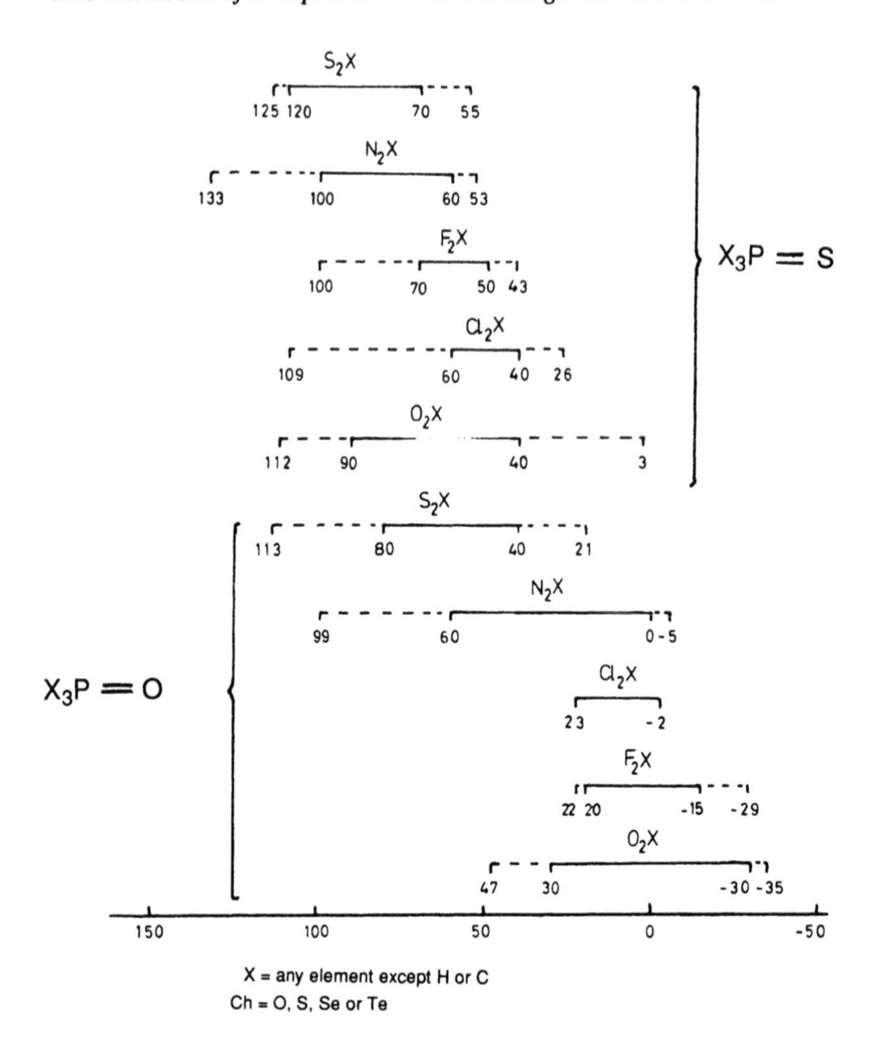

FIGURE 1.5a. ^{31}P NMR chemical shifts of four coordinate phosphorus compounds containing a P=Ch bond but no bonds to hydrogen or Group IV elements. (Table H; includes phosphoric acids and their derivatives.)

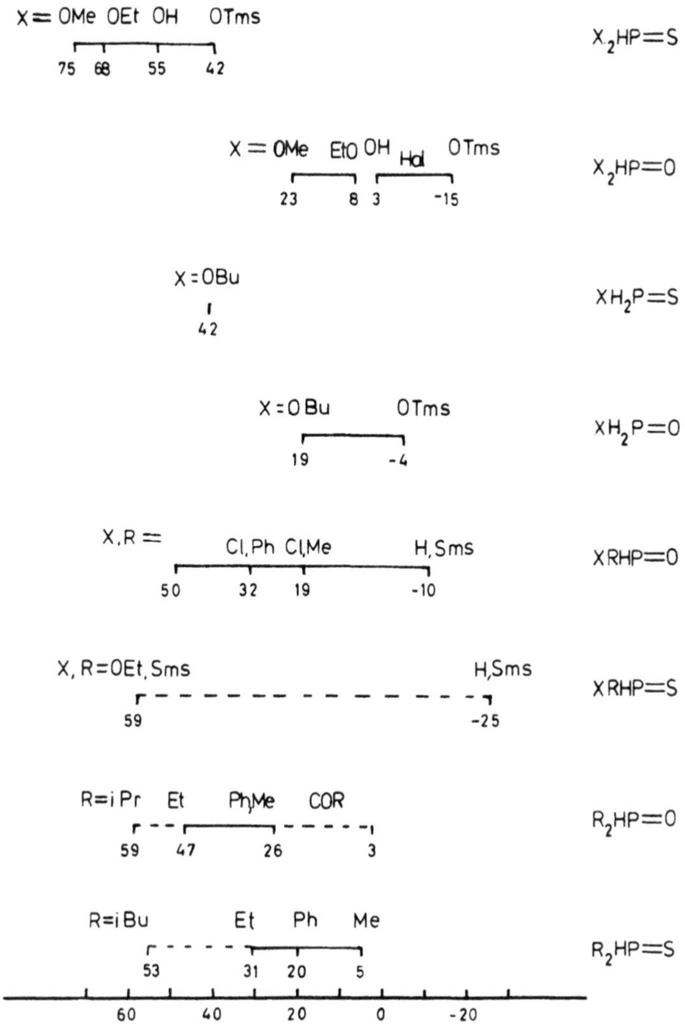

The identity of X and for R are given above the bar line;

R = carbon group

X = P-Bound Groups other than H or C

FIGURE 1.5b. ^{31}P NMR chemical shifts of four coordinate phosphorus compounds containing P=Ch and PH bonds. (Table I: Phosphorous and phosphinous acids and their derivatives.)

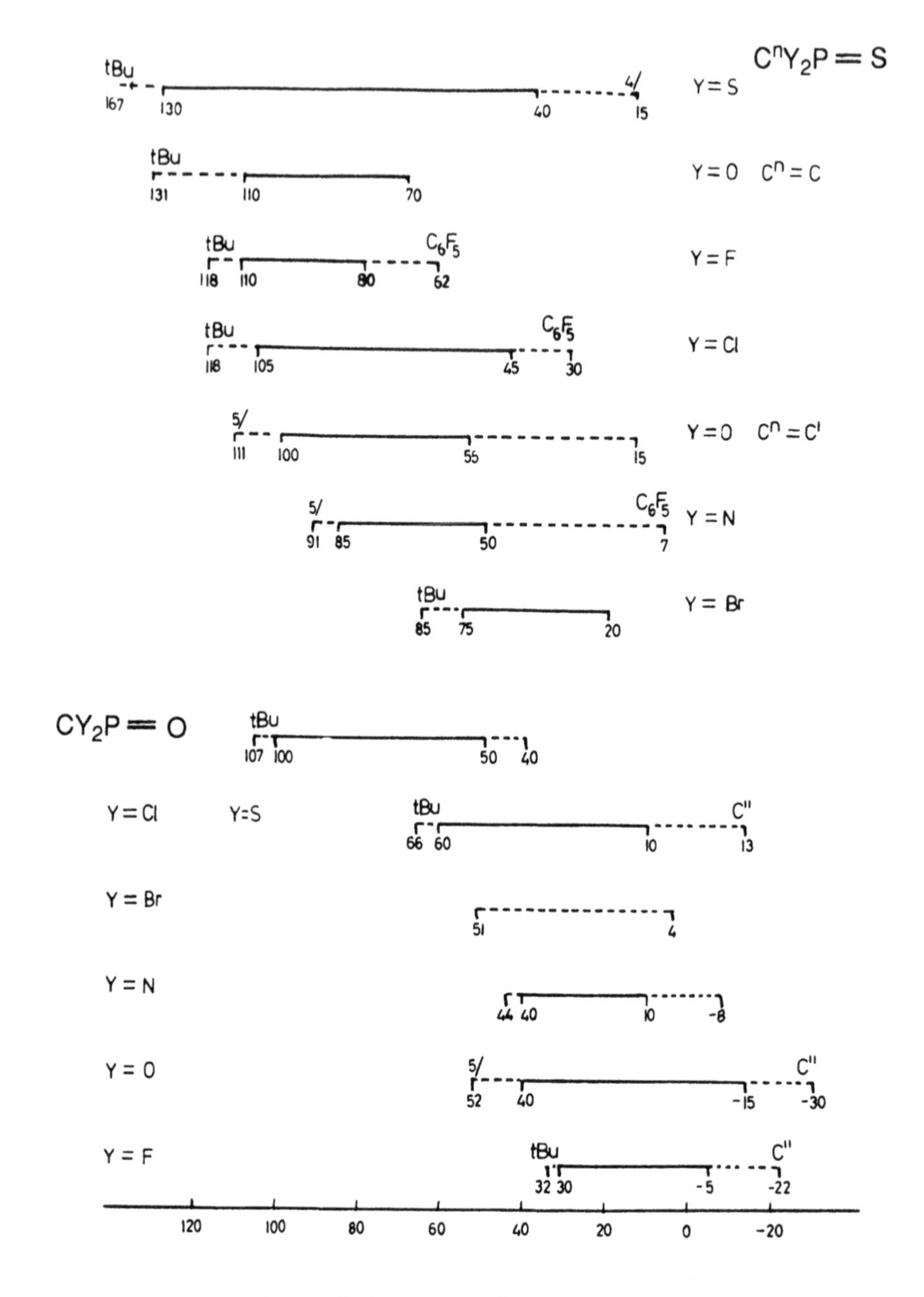

The coordination state of the P-Bound carbon atoms and the size of any rings incorporating phosphorus are indicated at the side, the identity of the organic group or ring size are indicated for the extreme values.

FIGURE 1.5c. ^{31}P NMR chemical shifts of four coordinate phosphorus compounds containing one P=Ch bond and one P–C bond. (Table J: Phosphonic acids and their derivatives.)

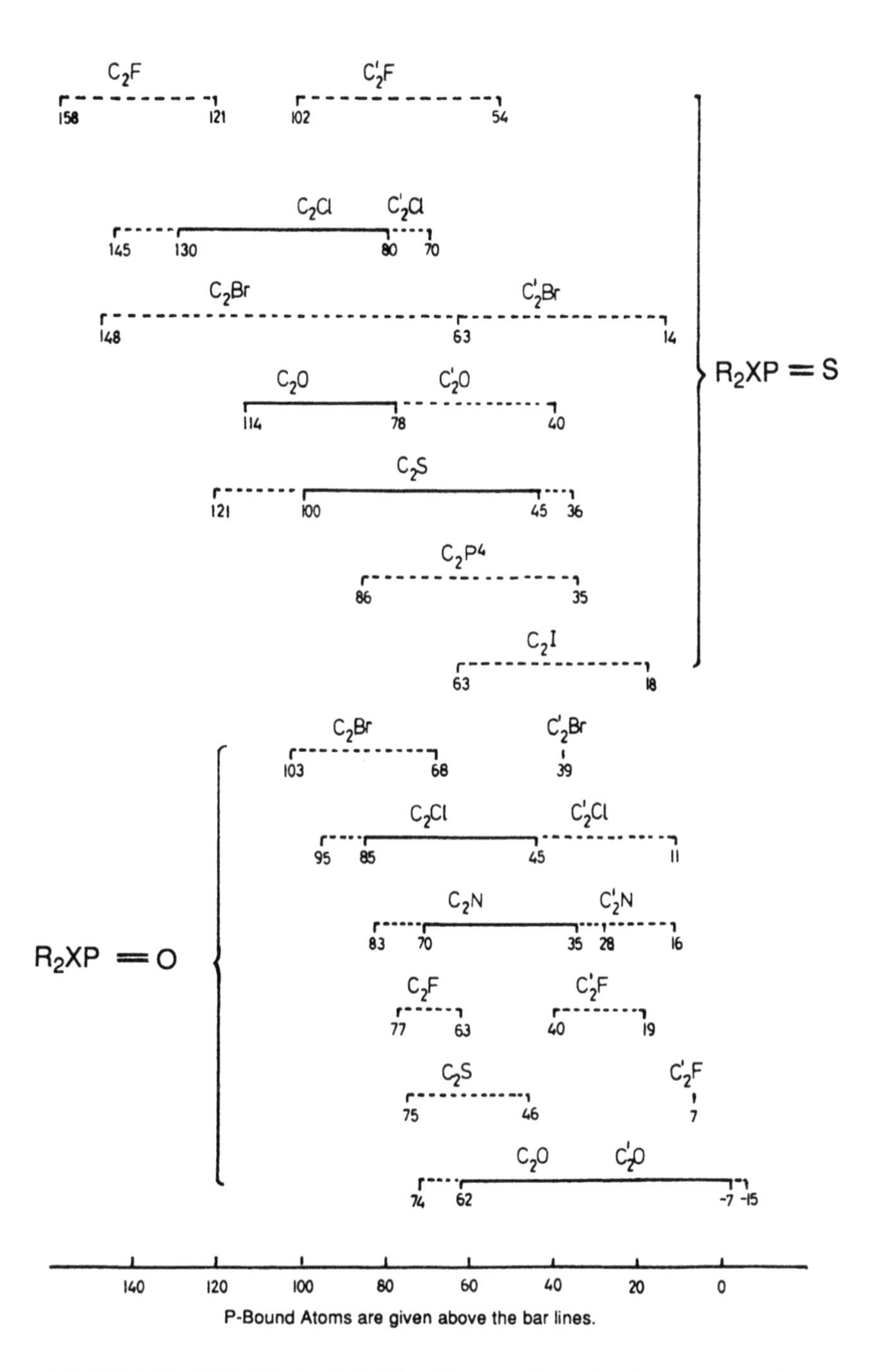

FIGURE 1.5d. ^{31}P NMR chemical shifts of four coordinate phosphorus compounds containing one P=Ch bond and two P–C bonds. (Table K: Phosphinic acids and their derivatives.)

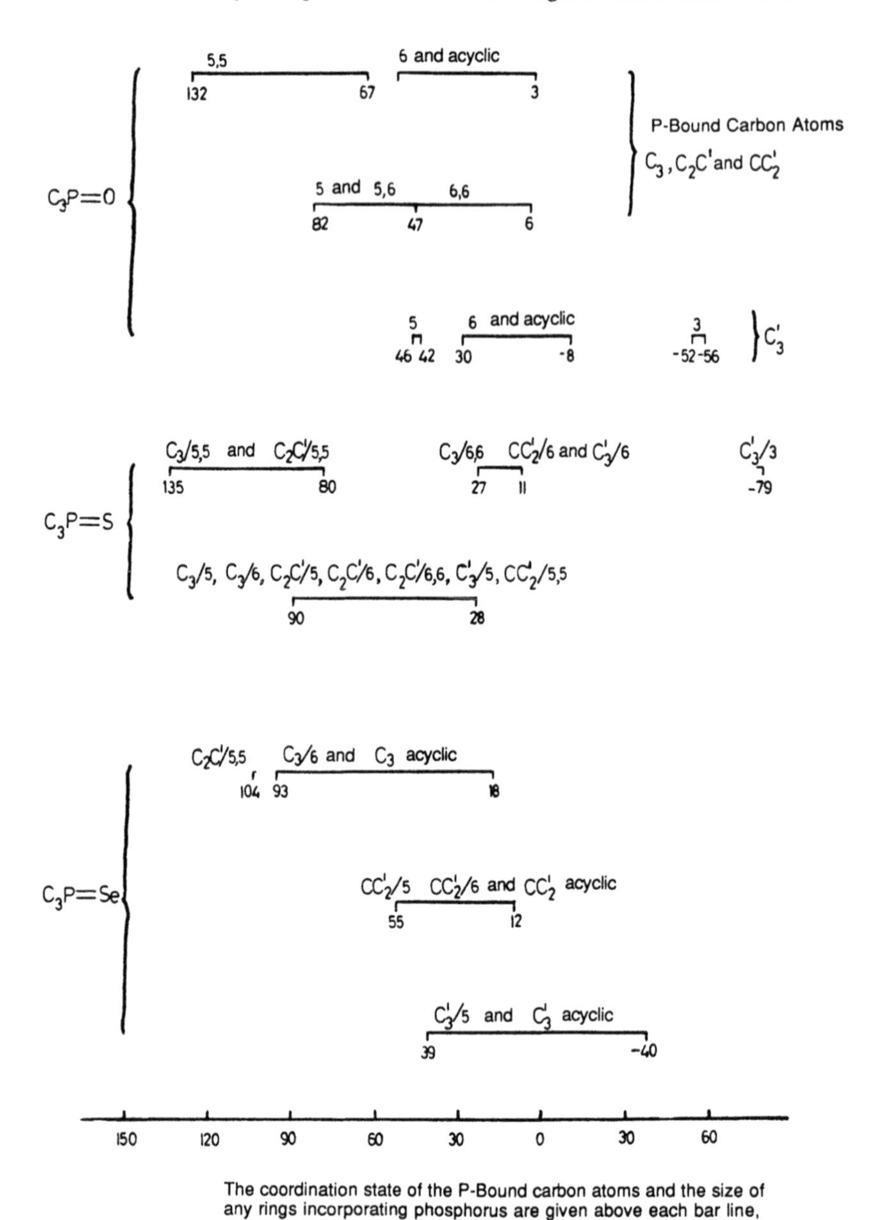

FIGURE 1.5e. ^{31}P NMR chemical shifts of four coordinate phosphorus compounds containing one P=Ch bond and four P–C bonds. (Table L: Phosphine chalcogenides.)

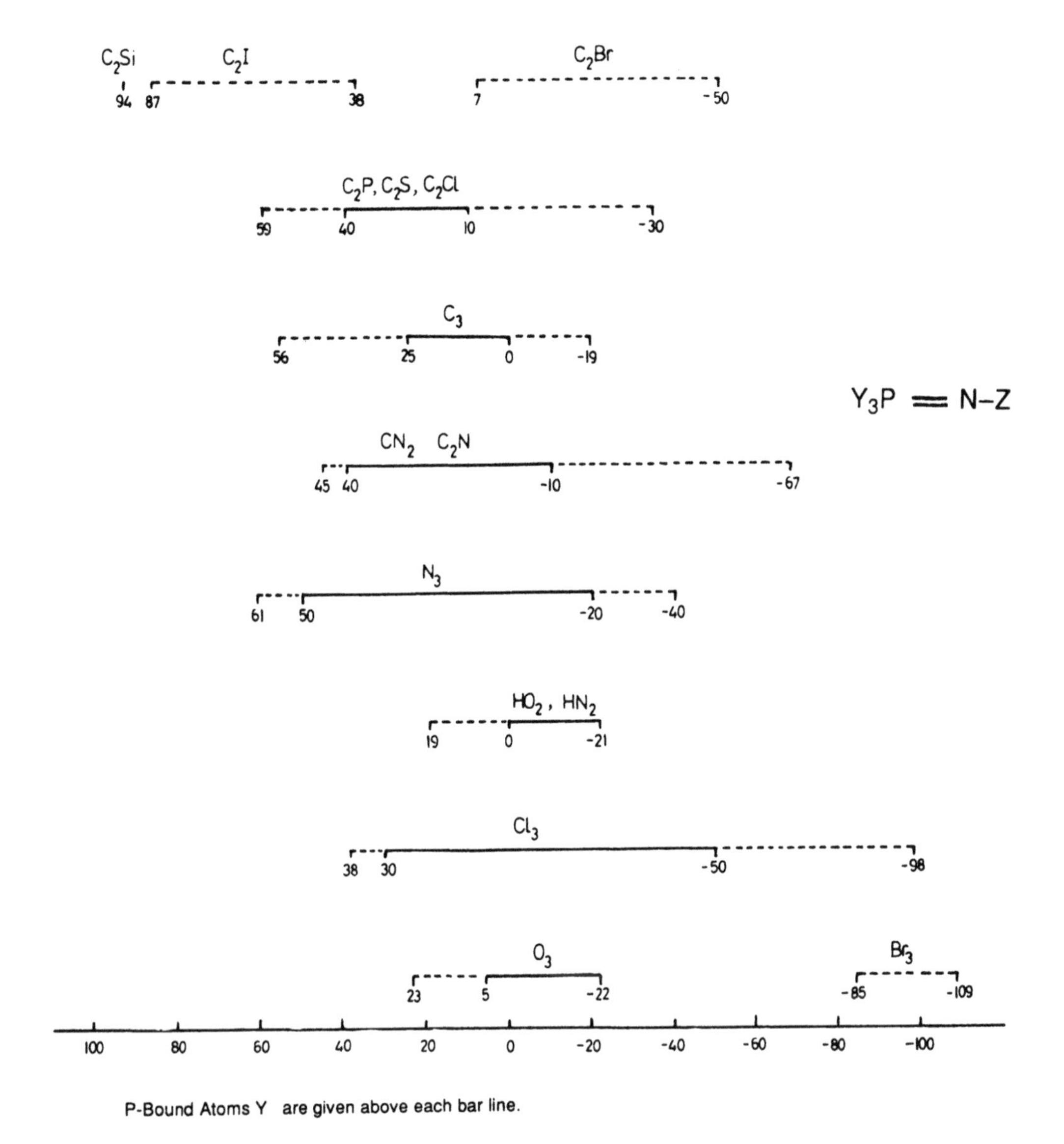

FIGURE 1.5f. ^{31}P NMR chemical shifts of four coordinate phosphorus compounds containing a formal multiple bond from phosphorus to a Group V atom. (Table M1; includes all phosphazenes except cyclopolyphosphazenes.)

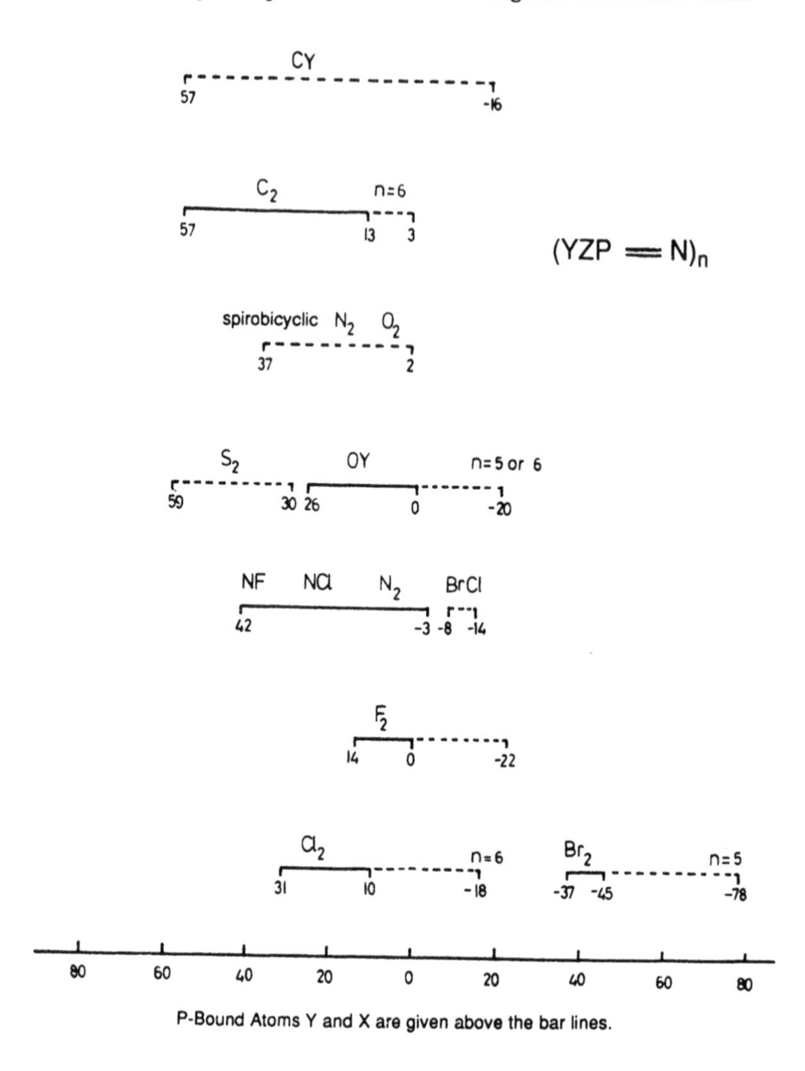

FIGURE 1.5g. ^{31}P NMR chemical shifts of four coordinate phosphorus compounds in the cyclopolyphosphazene class. (Table M2.)

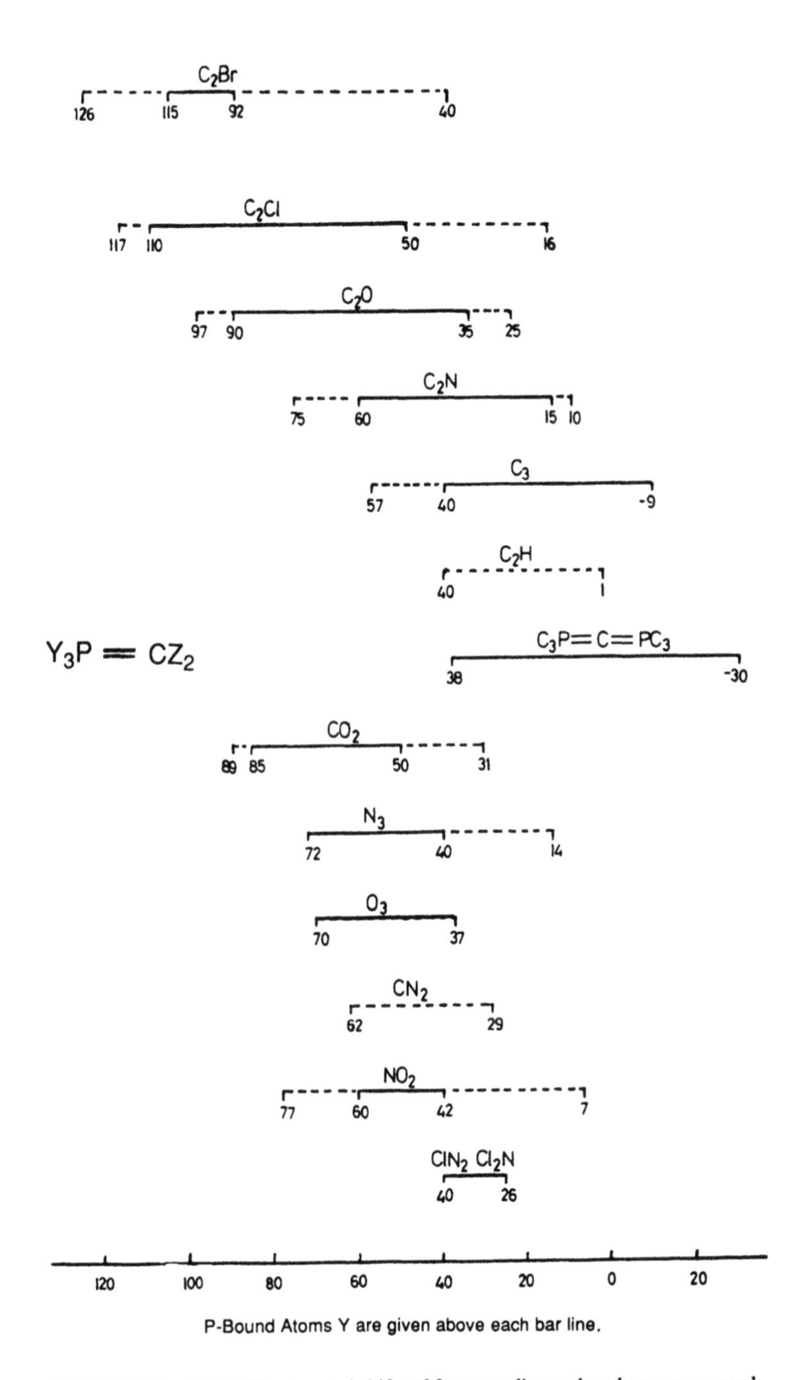

FIGURE 1.5h. ^{31}P NMR chemical shifts of four coordinate phosphorus compounds containing a formal multiple bond to a Group IV element. (Table N; includes phosphonium ylides.)

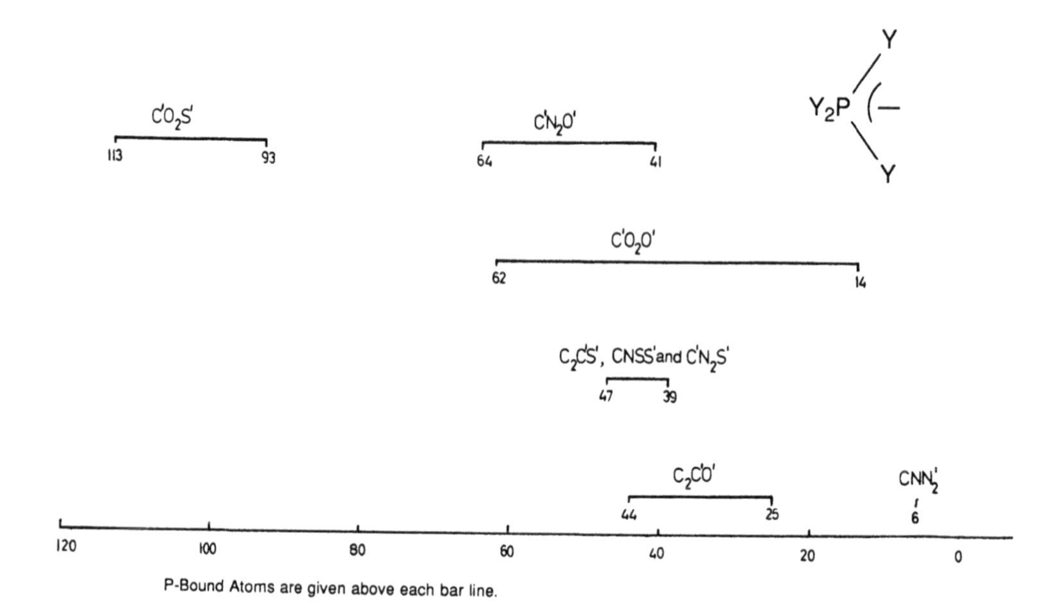

FIGURE 1.5i. ^{31}P NMR chemical shifts of four coordinate phosphorus compounds containing two formal multiple bonds at phosphorus. (Table O; includes phosphoryl stabilized carbanions.)

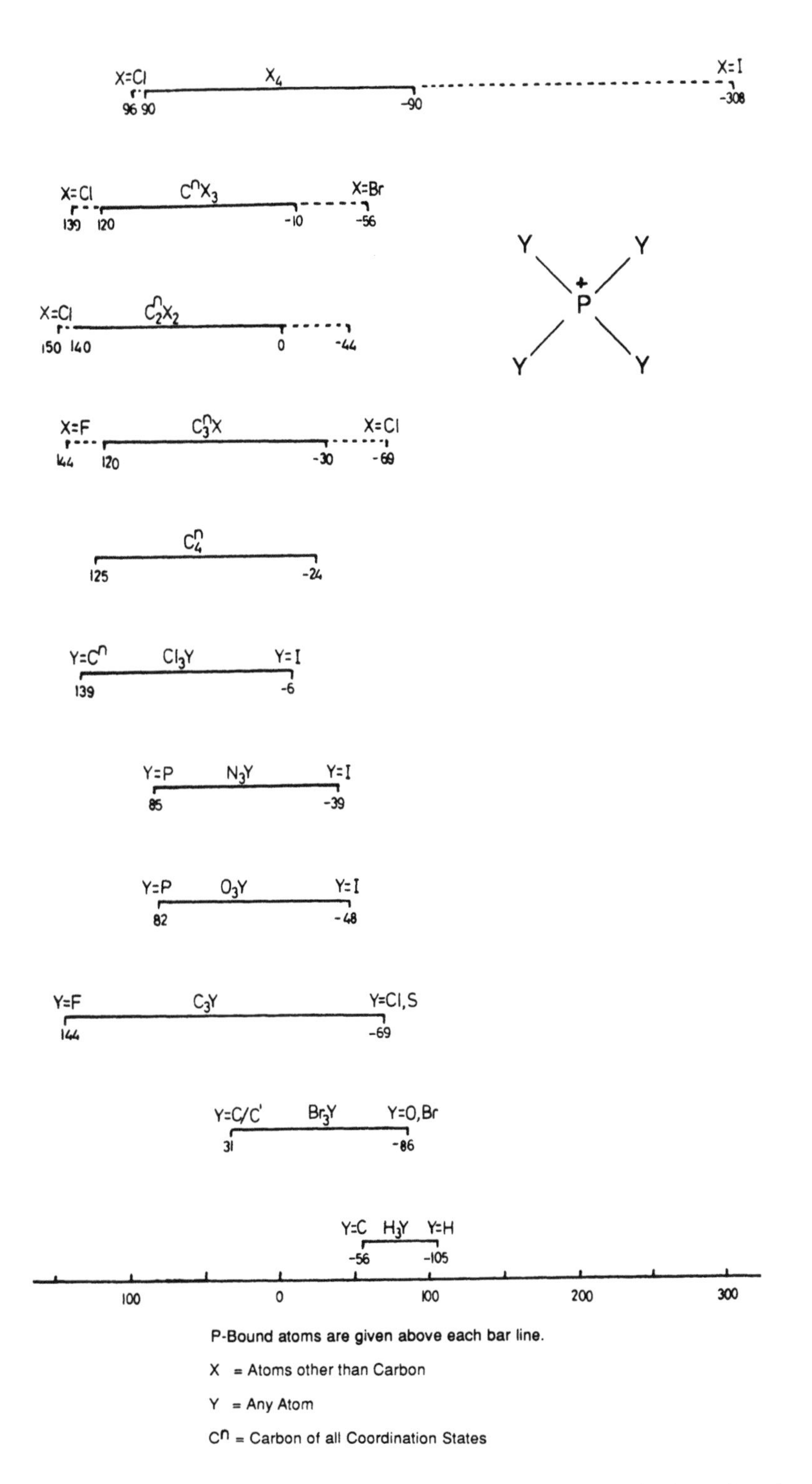

FIGURE 1.6. ^{31}P NMR chemical shifts of four coordinate (λ5 σ4) phosphonium salts and betaines. (Table G.)

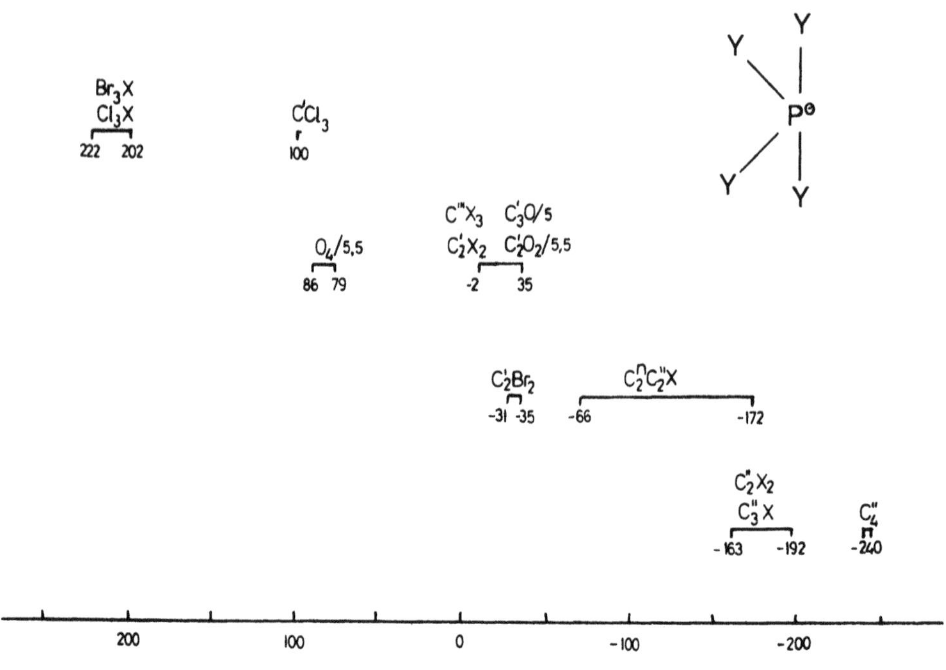

P-Bound Atoms are given above the bar line;

X = atoms other than C and H

FIGURE 1.7. ^{31}P NMR chemical shifts of four coordinates phosphorus compounds in which the phosphorus atoms formally bear a negative charge. (Table P: Phosphoranides.)

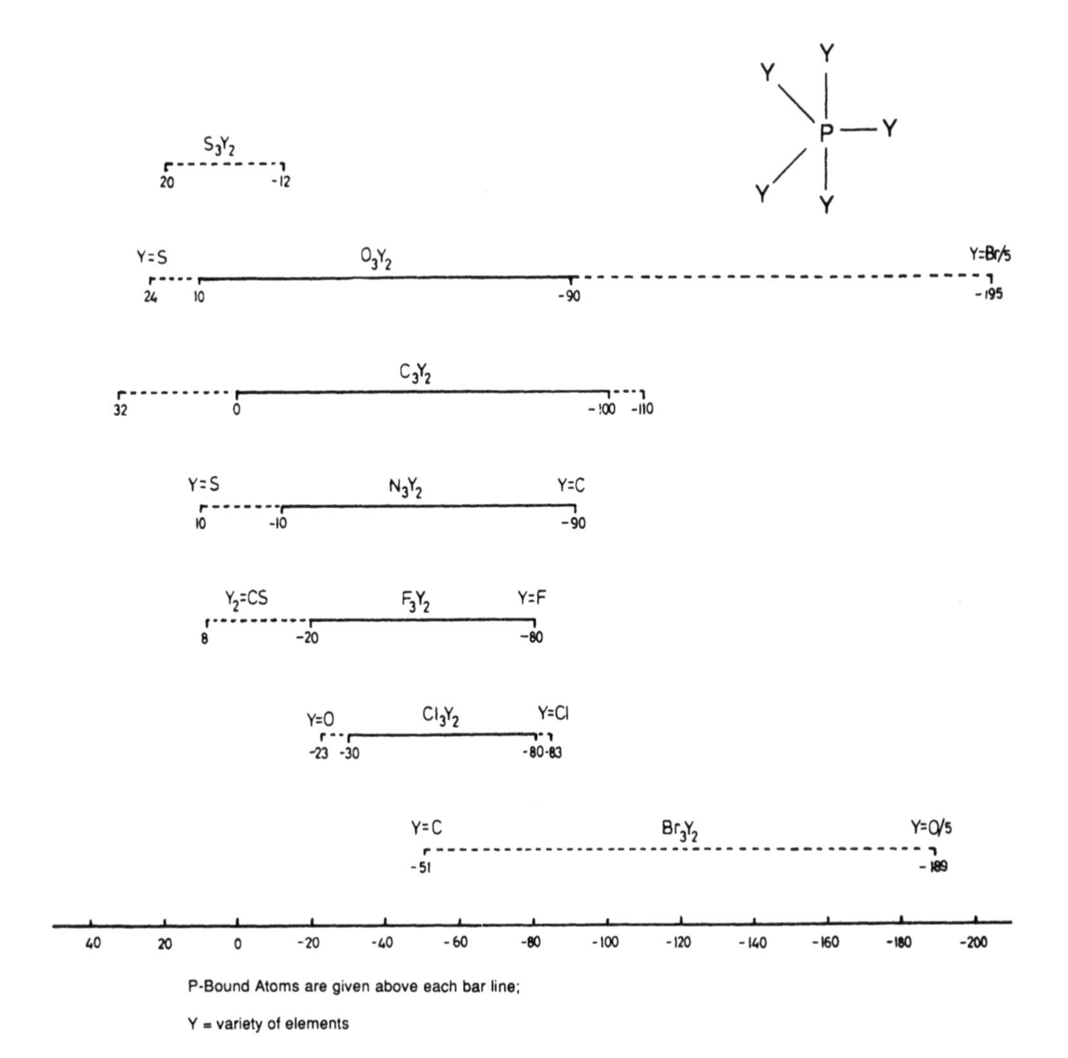

FIGURE 1.8. ^{31}P NMR chemical shifts of five coordinate (λ5 σ5) phosphorus compounds. (Table Q: Phosphoranes.)

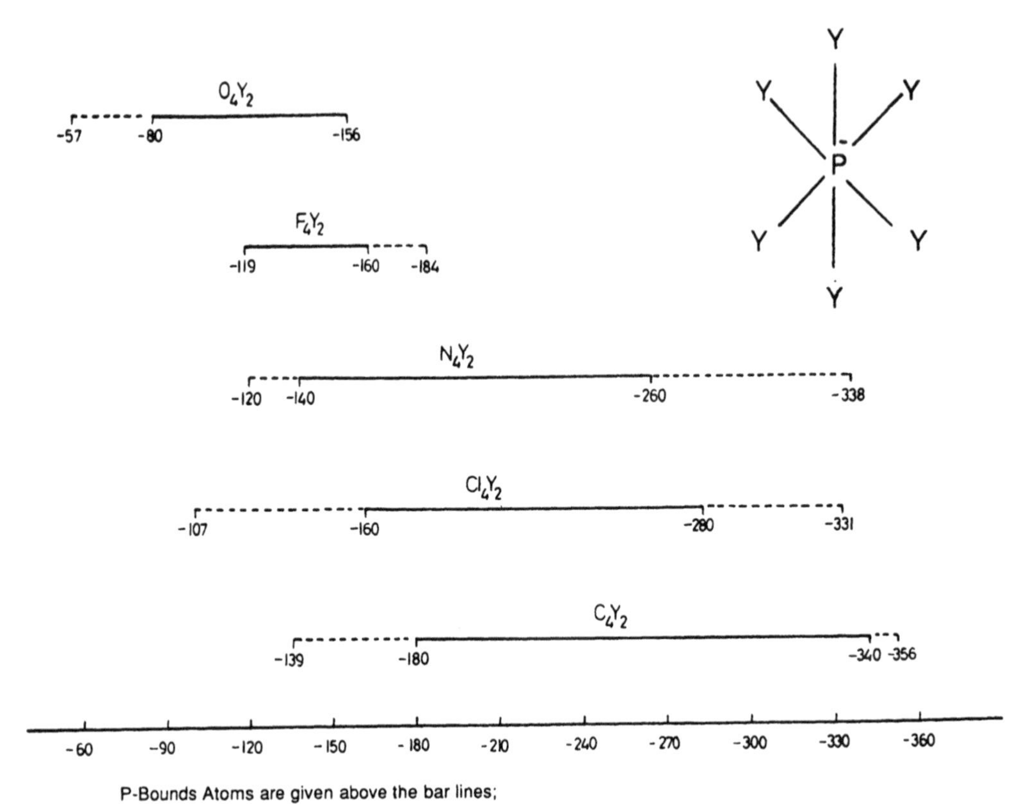

FIGURE 1.9. ^{31}P NMR chemical shifts of six coordinate (λ5 σ6) compounds. (Table R.)

Chapter 2

^{31}P NMR DATA OF ONE AND TWO COORDINATE PHOSPHORUS COMPOUNDS

Compiled and presented by
Alfred Schmidpeter and Konstantin Karaghiosoff

NMR data are presented in Table A. It is divided into two sections: (A1) One coordinate compounds; (A2) Two coordinate compounds. The latter section includes data on phosphenium salts, phosphaalkenes and phosphabenzenes.

The phosphaalkynes R–C≡P are the only stable representatives of one coordinate phosphorus compounds. Their ^{31}P chemical shift ranges from 96 to -207 and moves to higher field with increasing electronegativity of R in the order of

$$R = SiMe_3, \text{aryl}, H, \text{alkyl}, F$$

For the large number of two-coordinate phosphorus compounds two recent $\delta^{31}P$-compilations are available.[8,21] The first one covers the acyclic and the more or less saturated cyclic compounds up to mid 1985; the second one covers the fully unsaturated heterocyclic systems up to mid 1987. Their $\delta^{31}P$-values extend from 954 to -363 with most of them between 600 and -300. This wide shift range corresponds to a variation of phosphorus character from a phosphenium type with a low field chemical shift to a phosphide type with a high field chemical shift. (The overall charge of the molecule is, however, not important.)

$\delta^{31}P$ of acyclic compounds X–P=Y moves from low to high field in the order of

$$Y = NR_2^+, PR, NR, CR_2, PR_2S^-, PR_2O^-, CNR$$

and much less pronounced in the order of

$$X = tBu, Me, Sms, Mes, Ph, \text{alkenyl}, \text{alkinyl}, CN$$

Consequently, and as shown in Figure 1, individual shift ranges can be given for

λ^3,λ^3-Diphosphenes and their homologues, their oxides, sulfides and η^1-complexes	$-P{=}(P^2,As^2,Sb^2)$ $-P{=}(P^3,As^3)$
λ^3-Phosphazenes (iminophosphanes) and aminophosphenium ions	$-P{=}N^{2,3}$
Phosphaalkenes and 2-phospha-allylic ions	$-P{=}C^3$
Sila- and germaphosphenes	$-P{=}Si^3,Ge^3$
1-Phospha-cumulenes, cyanophosphides	$-P{=}C^2$
λ^3,λ^5-Diphosphenes, triphosphenium ions, and phosphoryl phosphides	$-P{=}P^4$

In phospha-alkenes ($Y = CR_2$) the substituents at carbon cause in the order of

$$R = SiMe_3, SSiMe_3, H, Ph, tBu, NMe_2, OSiMe_3$$

a phosphorus shift to higher field.

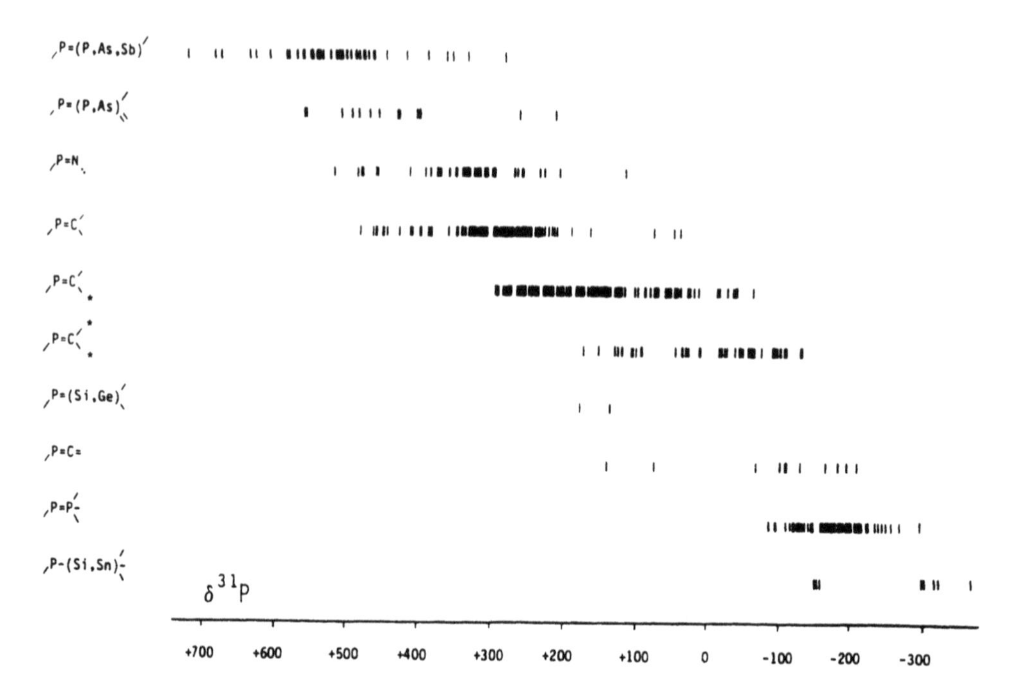

FIGURE 1. ³¹P-shift ranges of acyclic two-coordinate phosphorus compounds: every line represents a compound. The three ranges of phosphaalkenes refer to representatives with no, one and both carbon substituents being F or O- or N-bonded (indicated by *), respectively.

For unsaturated five- and six-membered two-coordinate phosphorus heterocycles $\delta^{31}P$ is mostly found in a range from 300 to 50. The compounds may be classified according to the ring size and to the P-bound ring members.

The five-membered heterocycles are

S,N: 1,3,2-thiazaphospholes
O,N: 1,3,2-oxazaphospholes
P,N: 1,2,3-azadiphospholes
N,N: 1,3,2-diazaphospholes, 1,2,4,3-triazaphospholes, 1,2,4,3,5-triazaphosphaboroles
S,C: 1,2,5-thiadiphospholes
N,C: 1,2-azaphospholes, 1,2,3-diazaphospholes, 1,3,4-diazaphospholes, 1,3,4-thiazaphospholes, 1,2,3,4-triazaphospholes, 1,2,3,5-diazadiphospholes
C,C: 1,3-azaphospholes, 1,3-diphospholes, 1,3-oxaphospholes, 1,3-thiaphospholes, 1,2,4-diazaphospholes, 1,2,4-oxazaphospholes, 1,2,4-thiazaphospholes

The six-membered heterocycles are

N,N: 1,3,5,2,4,6-triazaphosphadiborines
N,C: 1,2-azaphosphinines
C,C: phosphinines, 1,3-azaphosphinines, 1,4-azaphosphinines, 1,4-diphosphinines, 1,3,5-triphosphinines

As can be seen from Figure 2 only broad and overlapping ranges result from this subdivision. Four of them are sufficiently populated to make the following conclusions. (a) Phosphorus is less shielded in six-membered rings than in five-membered rings, as shown for the CPC case. (b) A nitrogen instead of a carbon as neighbor also causes a phosphorus shift to lower field, as shown by the five-membered rings with CPC, CPN, and NPN

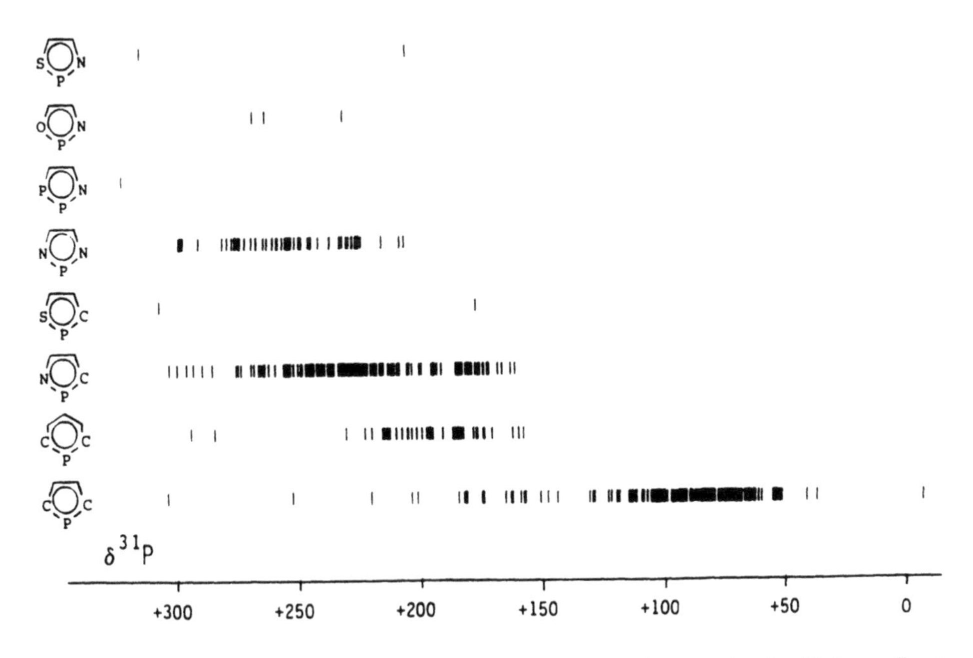

FIGURE 2. ^{31}P-shift ranges of unsaturated two-coordinate phosphorus heterocycles classified according to the ring size and to the P-bound ring members.

fragments. (c) The effect of an α-substituent on $\delta^{31}P$ is similar to that in the acyclic case; for 2,5-dimethyl-1,2,3-diazaphospholes depending on the 4-substituent it moves to higher field in the order of

4–R = $PSHal_2$, $PS(OMe)_2$, $POPh_2$, $PO(OMe)_2$, $P(CN)_2$, $PHal_2$, PPhHal, PPh_2, $P(OMe)_2$, SPh, H, Me, Hal, NHPh

from δ = 275 (4–R = PSF_2) to 172 (4–R = NHPh). In the same sense protonation of the 1,2,4-diazaphosphole anion results in a downfield chemical shift:

	$C_2H_2PN_2^-$	$C_2H_2PN_2H$	$C_2H_2PN_2H_2^+$
$\delta^{31}P$ =	68	83	95

TABLE A
Compilation of ^{31}P NMR Data of One and Two Coordinate Phosphorus Compounds

Section A1: One Coordinate Phosphorus Compounds (Phosphaacetylenes)

P-bound atoms/ rings	Connectivities	Structure	NMR data (δP [solv. temp] $^nJ_{PZ}$ Hz)	Ref.
C	F	F-C≡P	−207	167
	Si	Tms-C≡P	+96	168
	H	H-C≡P	−32(n, t, r, −80°)	42, 169, 170
	C	Me-C≡P	−60	170
		$tBuCH_2$-C≡P	−51(b)	171
		iPr-C≡P	−64(b)	171
		tBu-C≡P	−69	172
		1-MeC_5H_8-C≡P	−66(b)	171
		1-MeC_6H_{10}-C≡P	−57(b)	171
		1-Ad-C≡P	−67(b)	173
		$ClC(C_6H_4)_3C$-C≡P	−15(b)	174
		$HC(C_6H_4)_3C$-C≡P	−18(b)	174
		$MeC(C_6H_4)_3C$-C≡P	−17(b)	174
	C′	Ph-C≡P	−32(r)	175
		Sms-C≡P	+34(c)	134

Section A2: Two Coordinate Phosphorus Compounds

P-bound atoms/ rings	Connectivities	Structure	NMR data (δP [solv. temp] $^nJ_{PZ}$ Hz)	Ref.
ClN	C_2	$ClP{=}NMe_2^+\ AlCl_4^-$	+325(r)	1
		$ClP{=}NEt_2^+\ AlCl_4^-$	+317(r)	2
ClP^2	C′	Cl**P**=PSms	+523(k) $^1J_{PP}$598	3
Fe^8N	$C'_5C''_2;C_2H$	$Cpm(CO)_2FeP{=}N(iPr)_2^+\ PF_6^-$	+954(t)	4
Fe^8P^2	$C'_5C''_2;C'$	$Cpm(CO)_2Fe$**P**=PSms	+715(b) $^1J_{PP}$594	5
N′Ni	$C'_5P^4;C$	$Cpm(Ph_3P)NiP{=}NtBu$	+752(t)	29
NO	$C;C_2$	$tBuCH_2OP{=}N(iPr)_2^+\ AlCl_4^-$	+297(r)	20
NO/5	C′B;C′	**1**	+270(b)	32
N′O	C′;C′	SmsOP=NSms	+139(b)	176
NP^2	$C_2;N$	TmpP=PTmp	+471(b)	30
	$C_2;C'$	Tmp**P**=PSms	+461(k) $^1J_{PP}$580	31
$NP^2/5$	CC′;C′	**2**; P-2	+323 $^1J_{PP}$480	21
NS/5	C′H;C′	**3**; X = $AlCl_4$	+317(r)	21
N_2	$(Si_2)_2$	$Tms_2NP{=}NTms_2^+\ AlCl_4^-$	+450(r)	6
	$Si_2;CAl$	$Tms_2NP{=}N(tBu)AlCl_3$	+445(r)	7
	$(C_2)_2$	$Me_2NP{=}NMe_2^+\ AlCl_4^-$	+264(r)	1
$N_2/4$	AlSi;CAl	**4**; R = tBu, R′ = Tms	+369(r)	7
	$(CAs)_2$	**5**; R = tBu, X = $AlCl_4$	+357	8
		6; R = tBu, X = $AlCl_4$	+412	8
	$(C'Li)_2$	**7**; R = Sms	+337(b)	9
	$(CP)_2$	**8**; R = tBu, R′ = Me, X = $AlCl_4$	+252(m) $^2J_{PP}$36	10
	$(CSn^5)_2$	**9**; R = tBu	+274(r) +259(b)	11
$N_2/5$	CB′;CN	**10**; R = Me, X = $AlCl_4$	+228(c)	12
$N_2/6$	$(CB')_2$	**11**; R = Me, X = $AlCl_4$	+302(r)	12
NN′	PSi;P	$tBu_2P(Tms)N$**P**$=NP(tBu)_2$	+395 $^2J_{PP}$229	13
	$Si_2;P^4$	$Tms_2NP{=}NP(tBu)_2Et^+\ Br^-$	+244(b) $^2J_{PP}$49	14
	$Si_2;Si$	$Tms_2NP{=}NTms$	+326(o)	15
	C′H;C′	SmsHNP=NSms	+268(b)	16
	$C_2;N'$	$TmpP{=}NNTms_2$	+364	17
	$C_2;N''$	$N_3P{=}N(iPr)_2^+\ AlCl_4^-$	+280(r)	2
	$C_2;P^4$	$iPr_2NP{=}NPBu_3^+\ AlCl_4^-$	−307(r) $^2J_{PP}$50	18
	$C_2;P$	$Me_2NP{=}NPNMe_2Cl^{2+}\ (AlCl_4^-)_2$	+294 $^2J_{PP}$91	19
	$C_2;C$	$tBu_2NP{=}N$-tBu	+316	15

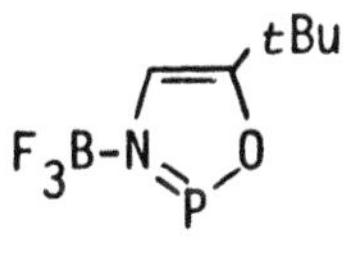

1 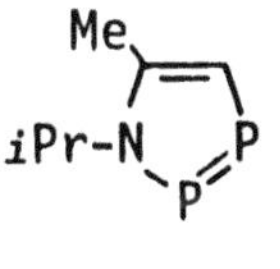 2 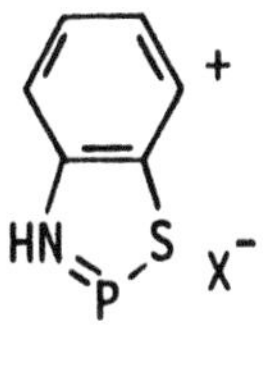 3 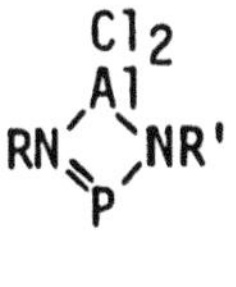 4 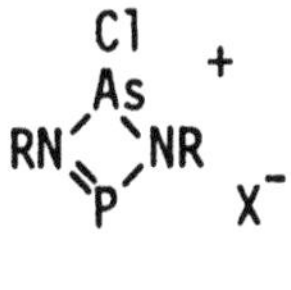 5

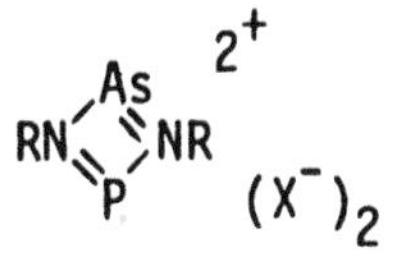

6 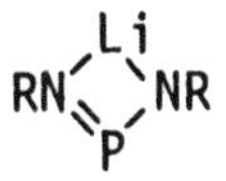 7 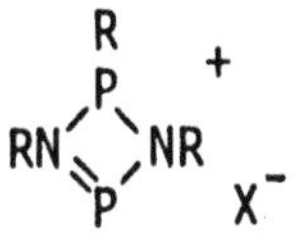 8 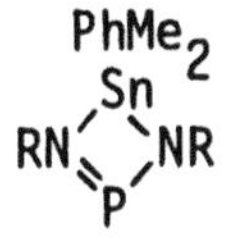 9 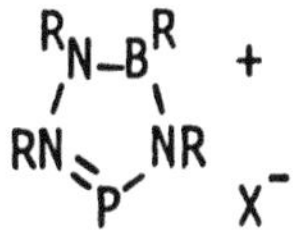 10

P-bound atoms/ rings	Connectivities	Structure	NMR data (δP [solv. temp] $^nJ_{PZ}$ Hz)	Ref.
	C_2;C″	SCNP=NEt_2^+ $AlCl_4^-$	+265(r)	20
NN′/5	N′Si;C′	**12**; R = Tms, R′ = SMe	+281(c)	21
	CN′;C′	**12**; R = R′ = Me	+254(c)	22
	C′N′;C′	**12**; R = R′ = Ph	+245(b)	23
	C′H;N′	**13**	+249(t)	24
	C′H;C′	**14**; R = H, R′ = R″ = CN	+255(m)	25
	CC′;C′	**14**; R = Me, R′ = NMe_2, R″ = H	+210(c)	21
		14; R = Me, R′ = R″ = CN	+234(c)	26
		15; X = $MeSO_4$	+278(c)	27
N'_2	$(P^4)_2$	$(Me_2N)_3$PNP=NP$(NMe_2)_3^+$ BF_4^-	+301(f)	28
N'_2/5	N′;C′	**16**	+257(c)	22
		17	+291(k)	21
	C'_2	**18**; M = H_2NMe_2	+262(m)	25
		19	+300(c)	27
OP^2	C;C′	tBuOP=PSms	+524 $^1J_{PP}$574	3
OsP^2	$C'_5C''_2$;C′	Cpm$(CO)_2$Os**P**=PSms	+632(k) $^1J_{PP}$584	33
P^2Ru	$C'_5C''_2$;C′	Cpm$(CO)_2$RuP=PSms	+677(b) $^1J_{PP}$597	5
P^2Si	C′;C_3	TmsP=PSms	+628(c) $^1J_{PP}$575	38
P^4Sn	N_3;C'_3	Ph_3Sn**P**=P$(NMe_2)_3$	−299(k) $^1J_{PP}$590	8
	C'_2O;C'_3	Ph_3Sn**P**=PPh_2OLi	−220(k) $^1J_{PP}$489	8
P^4_2	$(N_3)_2$	$(Me_2N)_3$P**P**=P$(NMe_2)_3^+$ $AlCl_4^-$	−194(r) $^1J_{PP}$518	8
	$(N_2O)_2$	O=$(Me_2N)_2$P**P**=P$(NMe_2)_2$OLi	−200(k) $^1J_{PP}$411	8
	N'_3;C'_3	$(Me_2N)_3$P**P**=PPh_3^+ $AlCl_4^-$	−180(r) $^1J_{PP}$493,523	34
	O_2O';C_3	O=$(EtO)_2$P**P**=PEt_3	−218 $^1J_{PP}$444,365	15
	$(C'_2N')_2$	PhN=Ph_2P**P**=PPh_2−NPhLi	−120(k) $^1J_{PP}$410	35
	$(C'_2O)_2$	O=Ph_2P**P**=PPh_2ONa	−116(k) $^1J_{PP}$398	36
	$(C_3)_2$	Me_3P**P**=PMe_3^+ $AlCl_4^-$	−156(r) $^1J_{PP}$452	8
		Bu_3P**P**=PBu_3^+ $AlCl_4^-$	−229(r) $^1J_{PP}$473	34
	$(C'_3)_2$	Ph_3P**P**=PPh_3^+ $AlCl_4^-$	−174(r) $^1J_{PP}$502	34
P^4_2/4	$(CN_2)_2$	**20**; R = NEt_2, X = $AlCl_4$	−90(r) $^1J_{PP}$347	8
P^4_2/5	$(CC'_2)_2$	**21**; R = Ph, X = $AlCl_4$	−232(r) $^1J_{PP}$452	37
P^4_2/8	$(C'_2N')_2$	**22**; R = Ph	−142(k) $^1J_{PP}$429	35
	$(C'_3)_2$	**23**; R = Ph, R′ = CN	−141(k) $^1J_{PP}$425	35
P^4P	N_3;C'_2	Ph_2P**P**=P$(NMe_2)_3$	−163(k) $^1J_{PP}$550,241	8
	C'_2O;C'_2	Ph_2P**P**=PPh_2OLi	−121(k) $^1J_{PP}$464,253	8
P^4P^2	C'_2O;P^4	$LiOPh_2$P=**P**-P=PPh_2OLi	−183(k) $^1J_{PP}$573,268	36
PP^2	C_2;C′	tBu_2P**P**=PSms	+601 $^1J_{PP}$586	3
HP^4	C'_2O	H**P**=PPh_2ONa	−190(k) $^1J_{PP}$401	8

TABLE A (continued)
Compilation of ^{31}P NMR Data of One and Two Coordinate Phosphorus Compounds

Section A2: Two Coordinate Phosphorus Compounds

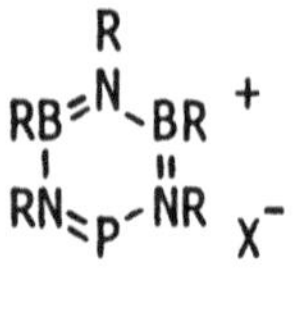

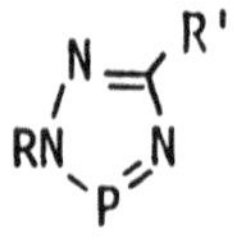

P-bound atoms/ rings	Connectivities	Structure	NMR data (δP [solv. temp] $^nJ_{PZ}$ Hz)	Ref.
CAs^2	Si_3;C	$Tms_3CP{=}AsCTms_3$	+668	38
C′As	Si_2;C_2	$tBu_2AsP{=}CTms_2$	+471	39
$C'As^2$	C'_2;N	$SmsP{=}AsNTms_2$	+541(h)	40
	C'_2;C	$SmsP{=}AsCHTms_2$	+575	38
C′Br	Si_2	$BrP{=}CTms_2$	+342	41
C′Cl	Si_2	$ClP{=}CTms_2$	+343	44
	H_2	$ClP{=}CH_2$	+300(r)	42
	CO	ClP=C(tBu)OTms	+182(b)	43
C′F	Si_2	$FP{=}CTms_2$	+367	45
C′Fe	Si_2;$C'_5C''_2$	$Cpm(CO)_2FeP{=}CTms_2$	+642(h)	46
	$C'_5C''_2$;C′O	$Cp(CO)_2FeP{=}CMes(OTms)$	+208	5
C′Ge	N_2;C'_3	$Ph_3GeP{=}C(NEt_2)_2$	−53(b)	47
C′Ge′	C'_2;C'_2	$SmsP{=}GeMes_2$	+175(b)	48
C′I	Si_2	$IP{=}CTms_2$	+361	45
C′Mo	C'_2;C'_{10}	$SmsP{=}MoCp_2$	+800(b)	49
	Si_2;$C'_5C''_3$	$Cp(CO)_3MoP{=}CTms_2$	+529(b)	50
	Si_2;$C'_5C''_2$	$Cp(CO)_2MoP{=}CTms_2$	+497(r)	51
C″Mo	C;$P^4_2P^2$	$[(Et_2P)_2C_2H_4]_2Mo(P{\equiv}CtBu)_2$	+10(k) $^2J_{PP}$40	52
CN	C_3;C_2	$tBuP{=}NMe_2^+\ AlCl_4^-$	+513(r)	1
CN′	C'_2H;C	CpP=N−tBu	+283(b)	177
CN′	C_3;C	tBuP=N−tBu	+472(t)	80
	C_3;C′	tBuP=NSms	+490(b)	81
C′N	Cl_2;Si_2	$Tms_2NP{=}CCl_2$	+252	53
	N_2;Si_2	$Tms_2NP{=}C(NMe_2)_2$	+98(n)	54
	S_2;C_2	$Et_2NP{=}C(SMe)_2$	+233(k)	60
	Si_2;C_2	$TmpP{=}CTms_2$	+407(b)	61
	HCl;Si_2	$Tms_2NP{=}CHCl$	+272(r)	55
	HP;Si_2	$Tms_2NP{=}CHPtBu_2$	+299 $^2J_{PP}$319	56
	HSi;C_2	$iPr_2NP{=}CHTms$	+273(k)	62
	C′Si;CN	$(MeNP{=}CPhTms)_2$	+245 (pentane)	58
	C′Si;C′N	PhHNPhNP=CPhTms	+240(r)	58

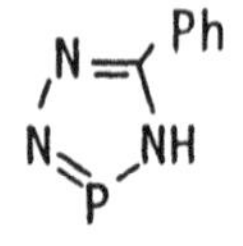

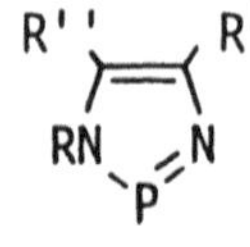

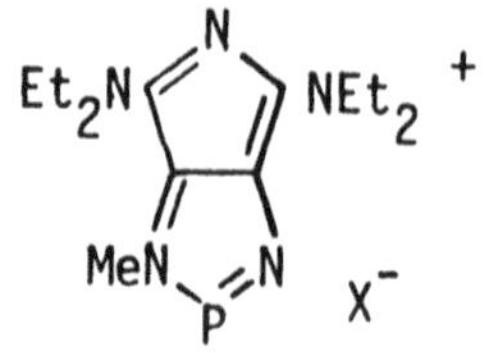

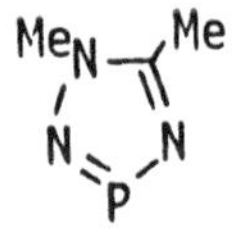

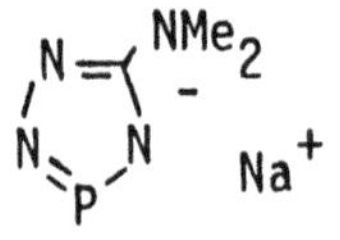

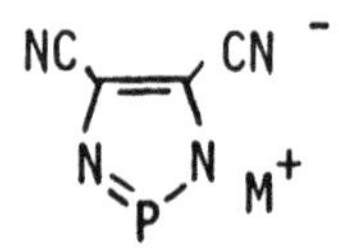

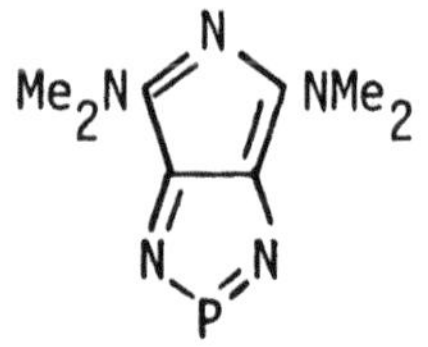

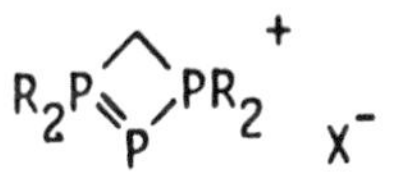

20 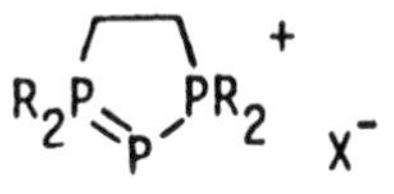 **21** 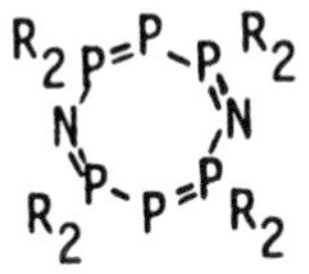 **22** 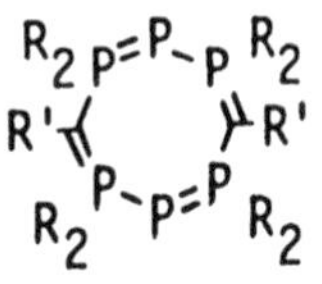 **23**

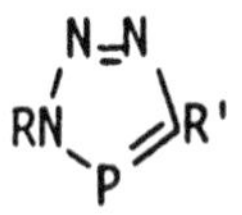

24 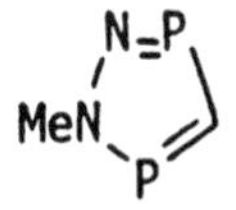 **25** 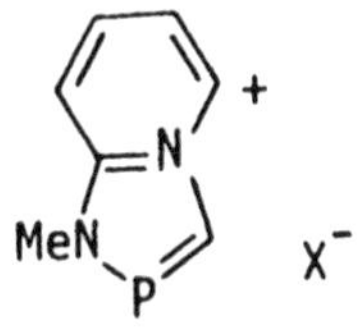 **26** 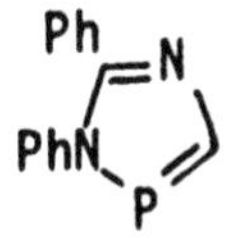 **27** 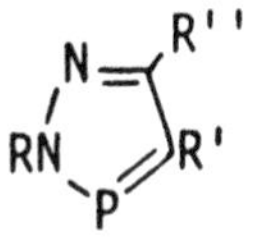 **28**

P-bound atoms/ rings	Connectivities	Structure	NMR data (δP [solv. temp] $^nJ_{PZ}$ Hz)	Ref.
	C′Si;CP^2	iPrN(P=CPhTms)$_2$	+266(c)	59
	C′Si;C$_2$	Me$_2$NP=CPhTms	+248	44
	CH;Si$_2$	Tms$_2$NP=CHtBu	+269(c)	57
CN	C′$_2$H;C$_2$	CpmP=NMe$_2^+$ AlCl$_4^-$	+111(r)	1
C′N/5	N′Si;CN′	**24**; R = CH$_2$Ph, R′ = Tms	+212(r)	64
	HP^2;CN′	**25**	+280(c) $^2J_{PP}$23	66
	HN′;CN′	**24**; R = Me, R′ = H	+188	65
	CN′;HN′	**24**; R = H, R′ = tBu	+161(c)	63
	HN;CC′	**26**; X = MeSO$_4$	+187(m)	78
	HN′;C′$_2$	**27**	+212(c)	21
	C′Cl;CN′	**28**; R = Me, R′ = Cl, R″ = Me	+212(c)	73
	(CN′)$_2$	**24**; R = Me, R′ = tBu	+173(c)	63
	C′N;CN′	**28**; R = R″ = Me, R′ = NHPh	+172(c)	74
	C′P^4;CN′	**28**; R = R″ = Me, R′ = PSF$_2$	+275(c)	69
	C′P;CN′	**28**; R = Me, R′ = PCl$_2$, R″ = Et	+248 $^1J_{PP}$61	75
	C′S;CN	**29**; R = H, R′ = SMe, X = Br	+220(c)	67
	C′S;CN′	**28**; R = R″ = Me, R′ = SPh	+236(c)	67
	C′S;CC′	**30**; R = CO$_2$Et, X = BF$_4$	+290(c)	21
	C′H;HN′	**28**; R = R′ = H, R″ = Me	+220(g)	71
	C′H;CN	**29**; R = R′ = H, X = Cl	+231(c)	68
		29; R = Me, R′ = H, X = BF$_4$	+230(r)	69
	C′H;CN′	**28**; R = R″ = Me, R′ = H	+229(c)	68
	C′H;C′N′	**28**; R = Ph, R′ = H, R″ = Me	+225	77
	C′H;CC′	**31**; R = Me, R′ = H, R″ = CH$_2$NMe$_2$	+220(c)	79
	CC′;CN	**29**; R = H, R′ = Me, X = Cl	+238(c)	70
	CC′;CN′	**28**; R = Me, R′ = tBu, R″ = Ph	+224(c)	76
	(CC′)$_2$	**31**; R = Me, R′ = CH$_2$NMe$_2$, R″ = H	+217(c)	79
	C′$_2$;HN′	**28**; R = H, R′ = R″ = CO$_2$Me	+255(c)	72
	C′$_2$;C′Si	**31**; R = Tms, R′ = R″ = CO$_2$Me	+238(c)	79
C′N′	C′Si;C′	Ph$_2$C=NP=CPhTms	+299^Z; +258^E	83
	C′$_2$;Si	SmsP=NTms	+476(k)	82
	C′$_2$;C′	MesP=NSms	+456(b)	81
C′N′/5	HN;C′	**32**	+182(c)	21
		33; R = H	+195(c)	78
	CN′;N	**34**; R = tBu, R′ = Tms	+183(b)	84
	C′N;C′	**33**; R = CO$_2$Et	+233(c)	21

TABLE A (continued)
Compilation of ^{31}P NMR Data of One and Two Coordinate Phosphorus Compounds

Section A2: Two Coordinate Phosphorus Compounds

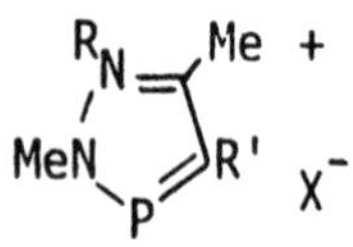

29 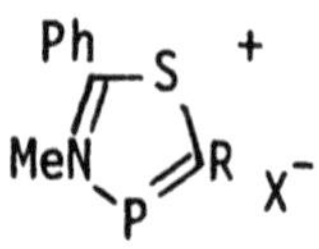 30 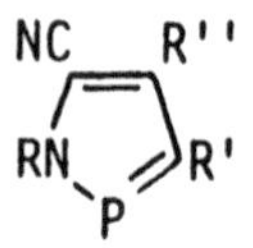 31 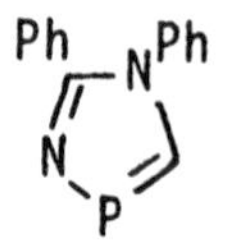 32 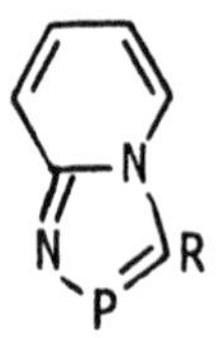 33 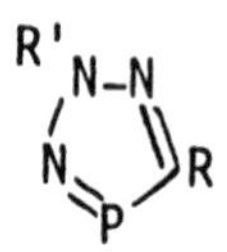 34

P-bound atoms/ rings	Connectivities	Structure	NMR data (δP [solv. temp] $^nJ_{PZ}$ Hz)	Ref.
	C'S;C'	**35**; R = NMe_2, R' = Ph	+249(c)	85
	C'S;C'	**35**; R = Ph, R' = CO_2Et	+300(c)	85
	C'H;N	**36**	+223(c)	68
C'N'/6	CC';C'	**37**	+263(c)	86
C'Ni	Si_2;C'_5P^4	$Cpm(Ph_3P)Ni$**P**$=CTms_2$	+723(t) $^2J_{PP}$35	29
C'O	C'Si;C	MeOP=CPhTms	+305	44
		tBuOP=CPhTms	+291	44
C'O/5	C'_2;C	**38**; X = CF_3SO_3	+196	87
CP^4	F_3;C_3	F_3C**P**$=PMe_3$	−81(k)	88
CP^2	C_3;N	tBu**P**=PTmp	+383 $^1J_{PP}$611	94
	Si_3;C	Tms_3C**P**=**P**$CTms_3$	+599(r)	38
	C_3;C'	tBu**P**=PSms	+532 $^1J_{PP}$576	8
$C'P^4$	N_2;C_2S'	$(Me_2N)_2C$=**P**$-P(tBu)_2$=S	−54(j) $^1J_{PP}$439	90
	C'_2;C_2O	Mes**P**=$P(OTms)tBu_2$	−213(b) $^1J_{PP}$429	89
	C'_2;C'_2O	Ph**P**=PPh_2ONa	−81(k) $^1J_{PP}$442	8
C'P	N_2;C_2	tBu_2P**P**$=C(NMe_2)_2$	−11(b) $^1J_{PP}$256	90
	S_2;C_2	tBu_2P**P**$=C(SMe)_2$	+175(c) $^1J_{PP}$223	60
	Si_2;CSi	tBuTmsP**P**$=CTms_2$	+478(b) $^1J_{PP}$244	92
	C'Si;CP^2	tBuP(**P**$=CPhTms)_2$	+319(c) $^1J_{PP}$262	59
	C'Si;C_2	tBu_2P**P**=CPhTms	+331(c) $^1J_{PP}$232	59
	C'_2;C'O'	Sms**P**=P(O)Sms	+207 $^1J_{PP}$684	91
	$(C'_2)_2$	Sms**P**$=P(CPh_2)Sms$	+208 $^1J_{PP}$676	93
$C'P^2$	N_2;C'	$[$**P**$=C(NMe_2)_2]_2$	+34(k)	95
	CO;C'	$[$**P**$=C(OTms)tBu]_2$	+132(b)	43
	C'_2;Cl	Sms**P**=PCl	+473(k) $^1J_{PP}$598	3
	C'_2;Fe^8	Sms**P**$=PFe(CO)_2Cpm$	+554(b) $^1J_{PP}$594	5
	C'_2;N	Sms**P**=PTmp	+336(k) $^1J_{PP}$580	31
	C'_2;Os^8	Sms**P**$=POs(CO)_2Cpm$	+543(k) $^1J_{PP}$584	33
	C'_2;P	Sms**P**$=PPtBu_2$	+487 $^1J_{PP}$586	3
	C'_2;Ru^8	Sms**P**$=PRu(CO)_2Cpm$	+552(b) $^1J_{PP}$597	5
	C'_2;C	Sms**P**=P-tBu	+525 $^1J_{PP}$576	8
$C'P^2/5$	N;C'H	**2**; P-3	+244 $^1J_{PP}$480	21
C"P	N";C'_2	Ph_2P**P**$=CN^-$ [18]-Crown-6-K^+	−169(k) $^1J_{PP}$302	8
$C"P^4$	N";C'_2N	NC**P**$=PPh_2NPhLi$	−180(k) $^1J_{PP}$380	8
	N";C'_2O	NC**P**$=PPh_2ONa$	−161(k) $^1J_{PP}$363	8
	N";C_3	NC**P**$=PBu_3$	−216(k) $^1J_{PP}$412	8
C'Pt/3	CPt^4;$C'P^4_2$	**39**; R = Ph, R' = tBu	+57 $^2J_{PP}$24,15	96
$C'S^3$	C'O;C'_2	MesP=S(OTms)Mes	+140^E; +139^Z(b)	89
C'S	C'Si;C	BuSP=CPhTms	+301	44
C'S/5	P^2S;C'	**40**; R = Tms	+310(j) $^2J_{PP}$59	97
C'Sb	C'_2;C	$SmsP=SbCHTms_2$	+620	38
C'Si	N_2;C_3	$TmsP=C(NMe_2)_2$	−47	98

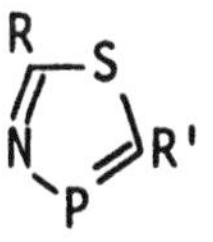

35 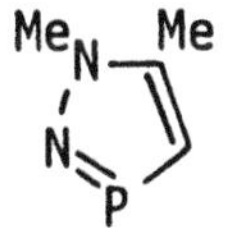 36 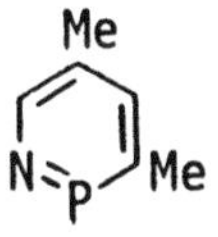 37 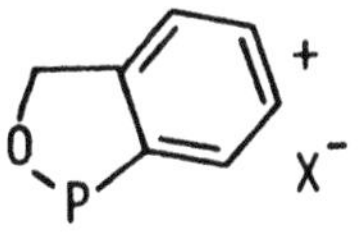 38 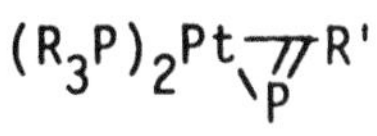 39 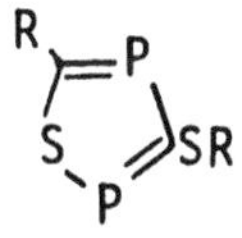 40

P-bound atoms/ rings	Connectivities	Structure	NMR data (δP [solv. temp] $^{n}J_{PZ}$ Hz)	Ref.
	CO;C_3	TmsP=C(OTms)tBu	$+120^{Z}$; $+124^{E}$	99
	C'_2;C_3	TmsP=CPh_2	+286(j)	100
C'Si/3	CSi;C_2C'	**41**; R = tBu	+274(b)	101
C'Si'	C'_2;C'_2	SmsP=$SiMes_2$	+136(k)	102
C″Si	N';C_3	tBu_3SiP=CNNa	−294(k)	8
C'Sn	N_2;C'_3	Ph_3SnP=C$(NEt_2)_2$	−65(b)	47
C″Sn	N';C'_3	Ph_3SnP=CNLi	−313(k)	8
C'Sn'	C'_2;C_2	SmsP=Sn$(CHTms_2)_2$	+205	103
CW	HSi_2;C'_{10}	Tms_2CHP=WCp_2	+680(b)	49
C'W	Si_2;C'_5,C''_2	Cp$(CO)_2$WP=$CTms_2$	+446(b)	50
	Si_2;C'_5,C''_3	Cp$(CO)_3$WP=$CTms_2$	+505(b)	50
C″W	C;P^4_4,P^2	$[(Ph_2P)_2C_2H_4]_2$W(P≡C-tBu$)_2$	−17(t) $^2J_{PP}$27	52
C'H	CN	HP=CMe(NMe_2)	-30^{E}, $+54^{Z}$(b)	104
	CO	HP=CtBu(OTms)	$+38^{E}$, $+54^{Z}$(b)	44
C_2	C_3,CC'_2	Cpm-P-tBu^+ $AlCl_4^-$	+168(r)	1
CC'	F_3;FN	CF_3P=CF(NMe_2)	-9^{Z}(t)	105
	F_3;F_2	CF_3P=CF_2	+18	106
	F_3;HF	CF_3P=CHF	+150(c)	107
	H_2P^2;CO	CH_2(P=C(OTms)tBu$)_2$	+124(n)	108
	H_3;O_2	MeP=C$(OTms)_2$	−18(c)	109
	H_3;S_2	MeP=C$(STms)_2$	+276(ze)	110
	CH_2;C'Si	BuP=C(Tms)Ph	+304(c)	59
	C_3;NS	tBuP=C(NMe_2)STms	+219(t)	110
	C_3;N_2	tBuP=C$(NMe_2)_2$	+92(b)	104
	C_3;OP	tBu**P**=C(OTms)P(tBu)Cl	+250(t) $^2J_{PP}$381	111
	C_3;OP	tBu**P**=C(OTms)P(Tms)tBu	+232(b) $^2J_{PP}$10	111
	C_3;ORe^8	tBuP=C(OTms)ReCpm(CO)(NO)	$+240^{E}$, $+272^{Z}$(b)	112
	C_3;Si_2	tBuP=$CTms_2$	+439(b)	44
	C_3;HN	tBuP=C(NMe_2)H	+119(b)	100
	C_3;CO	tBuP=C(OTms)tBu	+176(cyclopentane)	44
	C_3;C'_2	tBuP=C(CO_2Me)Ph	+160	113
CC'/4	COP;OP	**42**; R = OTms, R' = COR″, R″ = tBu	+279(b) $^2J_{PP}$16	43
CC'/4,6	C_2C';C_2	**43**; R = R' = tBu	+316	8
		43; R = tBu, R' = 1-Ad	+317	8
CC″	C_3;N'	tBuP=CN-tBu	−65	159
	C_3;O'	tBuP=CO	−180	160
C'_2	BrLi;C'_2	SmsP=CLiBr	+257(b)	114
	BrSi;C'_2	SmsP=CTmsBr	+303(b)	115
	BrSn;C'_2	SmsP=C$(SnMe_3)$Br	+298(b)	114
	ClLi;C'_2	SmsP=CLiCl	+179(b)	114
	Fe^8O;C'_2	SmsP=C(OTms)FeCpm$(CO)_2$	+269	5
	$(N_2)_2$	$(Me_2N)_2$CP=C$(NMe_2)_2^+$ Cl^-	−21(c)	116
	N_2;C'_2	PhP=C$(NMe_2)_2$	+28(b)	117
	OP;C'_2	Sms**P**=C(OTms)PSmsH	+171(b) $^2J_{PP}$66	118
	O_2;C'_2	PhP=C$(OTms)_2$	+27(c)	109
	PS;C'_2	Sms**P**=C(SH)PSmsH	+241(b) $^2J_{PP}$235	119
	S_2;C'_2	PhP=C$(STms)_2$	+262(ze)	110
CC'	CC'_2;Si_2	CpmP=$CTms_2$	+371(r)	120

TABLE A (continued)
Compilation of ^{31}P NMR Data of One and Two Coordinate Phosphorus Compounds

Section A2: Two Coordinate Phosphorus Compounds

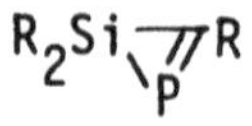

41 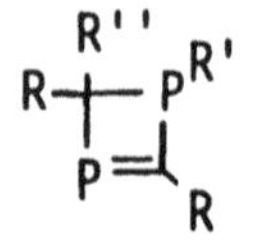 42 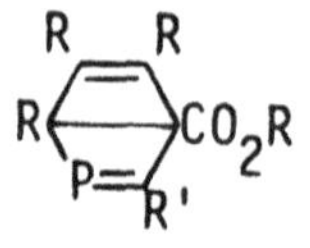 43 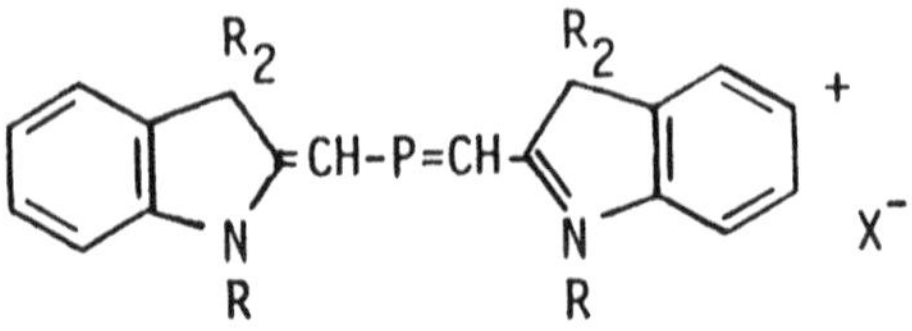 44

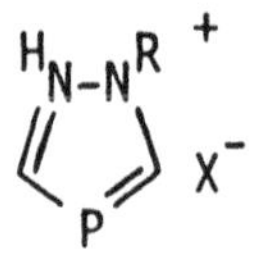

45 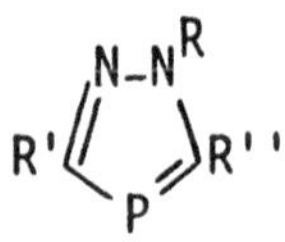 46 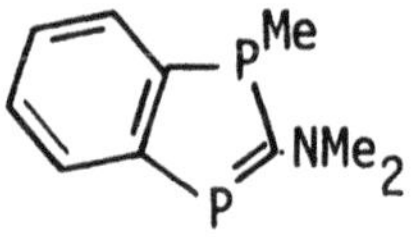 47 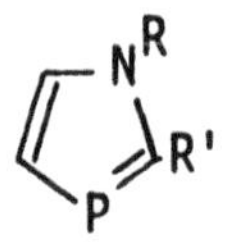 48 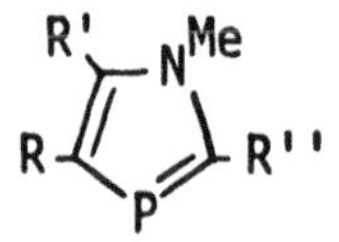 49

P-bound atoms/ rings	Connectivities	Structure	NMR data (δP [solv. temp] $^nJ_{PZ}$ Hz)	Ref.
C'_2	$Si_2;C'_2$	$PhP{=}CTms_2$	+376(b)	44
	$HBr;C'_2$	SmsP=CHBr	+262, +269(b)	115
	$HCl;C'_2$	SmsP=CHCl	+250(b)	121
	$(HN)_2$	$(Me_2N)HCP{=}CH(NMe_2)^+\ Cl^-$	+42(c)	122
	$HN;C'_2$	$PhP{=}CH(NMe_2)$	+70^E; +68^Z (cyclopentane)	44
	$HP;C'_2$	SmsP=CH(PSmsCl)	+322(b) $^2J_{PP}$357	123
		$SmsP{=}CH(PSmsMoCp(CO)_2)$	+324(b) $^2J_{PP}$55	124
		SmsP=CH(PSmsH)	+278(b) $^2J_{PP}$98	123
	$HSi;C'_2$	SmsP=CHTms	+328	125
	$H_2;C'_2$	$SmsP{=}CH_2$	+290	125
	$(C'N)_2$	$(Ph_2N)PhCP{=}CPh(NPh_2)^+Cl^-$	+172(c)	8
	CO;CO'	MeOCP=C(OH)Me	+73(b)	128
		tBuOCP=C(OLi)tBu	+51(b)	126
		tBuOCP=C(OTms)tBu	+124(b)	44
	C'O;C'O'	SmsOCP=C(OH)Sms	+123(c)	129
	$CO;C'_2$	PhP=C(OLi)tBu	+67	127
	$CO;C'_2$	PhP=C(OTms)tBu	+137 (cyclopentane)	44
	$C'Si;C'_2$	PhP=CPhTms	+277	44
	$(C'H)_2$	**44**; R = Me, X = BF_4	+356(c)	130
	$C'H;C'_2$	SmsP=CHPh	+259^E, +242^Z(b)	132
		$(SmsP{=}CH)_2$	+270^{EE}, +246^{ZZ}, +243^{EZ}, +285^{EZ}(c)	131
	$(C'_2H)_2$	$Fc\text{-}P\text{-}Fc^+\ AlCl_4^-$	+184(r)	1
		$PhP{=}CPh_2$	+232(j)	100
$C'_2/5$	$(HN)_2$	**45**; R = H, X = Cl	+95(c)	21
	BN;HN'	**46**; R = Me, R' = H, R'' = BMe_2	+104(c)	21
	$NP;C_2$	**47**	+104 $^2J_{PP}$27	143
	NSi;N'Si	**46**; R = R' = R'' = Tms	+178(c)	21
		46; R = H, R' = R'' = Tms	+151(c)	21
	NSn;HN'	**46**; R = Me, R' = H, R'' = $SnMe_3$	+131(c)	133

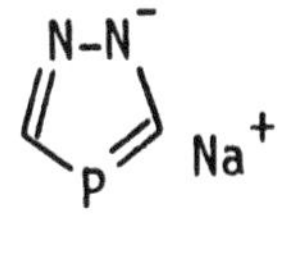

50 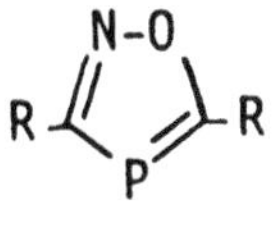 51 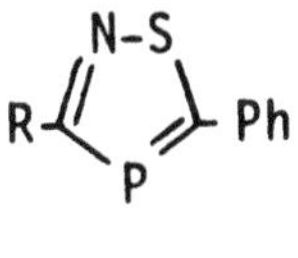 52 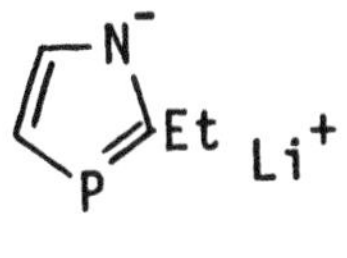 53 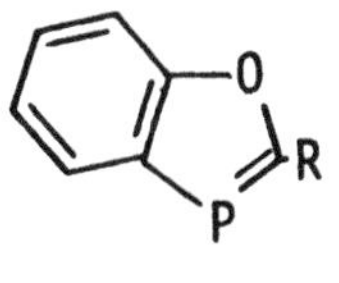 54

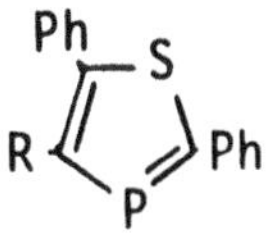

55 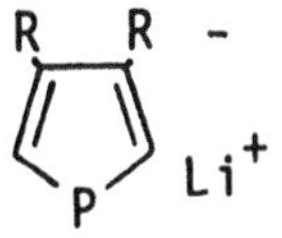 56 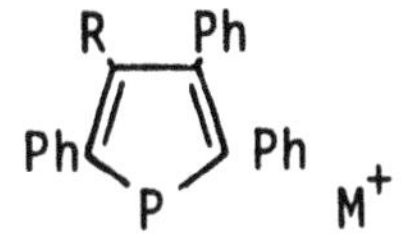 57 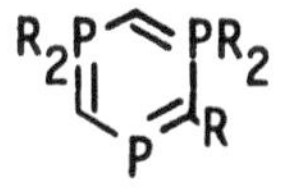 58 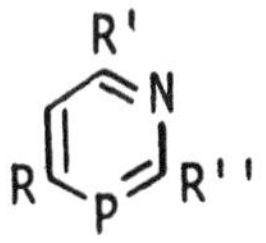 59

P-bound atoms/ rings	Connectivities	Structure	NMR data (δP [solv. temp] $^nJ_{PZ}$ Hz)	Ref.
	$(HN)_2$	**45**; R = Ph, X = BF_4	+102(c)	21
	HN;HN′	**46**; R = R′ = R″ = H	+83(c) +84(r) +86(t)	122
	HN;CN′	**46**; R = R″ = H, R′ = Me	+84(c)	21
	HN;C′N′	**46**; R = R″ = H, R′ = Ph	+81(c)	135
	HN;HN′	**46**; R = Me, R′ = R″ = H	+92(c)	21
		46; R = COMe, R′ = R″ = H	+107(c)	21
		46; R = Ph, R′ = R″ = H	+99(c)	122
	HN;C′N′	**46**; R = Me, R′ = Ph, R″ = H	+87(c)	136
		46; R = R′ = Ph, R″ = H	+93(c)	136
	HN;C′H	**48**; R = R′ = H	+98	139
		48; R = Tms, R′ = H	+114	139
	HN;CC′	**49**; R = Me, R′ = Ph, R″ = H	+99(c)	140
	$(HN')_2$	**50**	+68(w)	21
	HN′;HO	**51**; R = H	+84(c)	21
	HN′;CN	**46**; R = R″ = Me, R′ = H	+94(c)	133
		46; R = Tms, R′ = H, R″ = tBu	+86(b)	137
	CN;CN′	**46**; R = H, R′ = R″ = Me	+79(c)	21
		46; R = H, R′ = R″ = tBu	+65(c)	63
		46; R = R″ = Me, R′ = tBu	+84(c)	137
	CN;C′N′	**46**; R = Et, R′ = Ph, R″ = Me	+88(c)	136
	CN;C′H	**48**; R = Tms, R′ = Me	+122	139
	CN′;C′N	**46**; R = H, R′ = tBu, R″ = Ph	+74(c)	137
	CN′;CO	**51**; R = tBu	+69(c) +69(t) +68(b)	21
	CN′;C′S	**52**; R = tBu	+178(c)	138
	CN′;C′H	**53**	+78	139
	C′N;C′N′	**46**; R = H, R′ = R″ = Ph	+74(c) +76(c)	122
		46; R = R′ = R″ = Ph	+95(c)	122
	$C'N;C'_2$	**49**; R = R″ = Ph, R′ = OH	+63(c)	141
	C′N′;C′O	**51**; R = Ph	+70(r) +72(c)	21
	$CO;C'_2$	**54**; R = tBu	+78(c)	142
	$C'O;C'_2$	**54**; R = Ph	+86(c)	142
	C′S;CC′	**55**; R = tBu	+221(c)	138
	$C'S;C'_2$	**55**; R = Ph	+217(c)	144
	$(C'H)_2$	**56**; R = H	+77(k)	145
		56; R = Me	+59(k)	145
	$(C'_2)_2$	**57**; R = H, M = Li	+103(k)	146
		57; R = Ph, M = K	+99(k)	145
$C'_2/6$	$HP^4;CP^4$	**58**; R = NMe_2, R′ = tBu	+296(b) $^2J_{PP}$49,45	148
	CN′;CC′	**59**; R = tBu, R′ = Me, R″ = CF_3	+163	21
	$C'N';C'_2$	**59**; R = R′ = R″ = Ph	+163(c)	147
	C′Si,C′H	**60**; R = R′ = Me, R″ = Tms	+215(o)-dichlorobenzene)	64

TABLE A (continued)
Compilation of ^{31}P NMR Data of One and Two Coordinate Phosphorus Compounds

Section A2: Two Coordinate Phosphorus Compounds

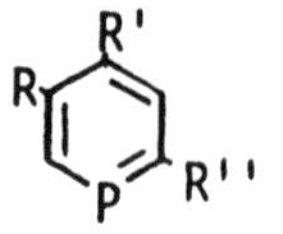

60 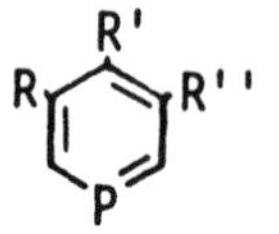 **61** 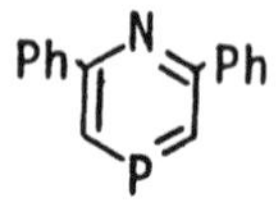 **62** 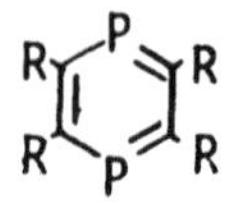 **63** 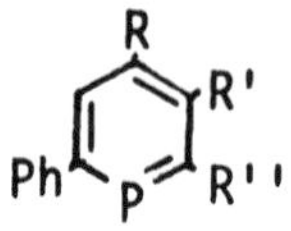 **64**

P-bound atoms/ rings	Connectivities	Structure	NMR data (δP [solv. temp] $^nJ_{PZ}$ Hz)	Ref.
	$(C'H)_2$	**61**; R = R′ = R″ = H	+211	151
		61; R = R′ = Me, R″ = H	+188(c)	152
		61; R = Ph, R′ = H, R″ = OH	+213(c)	153
		61; R = Ph, R′ = R″ = H	+209	154
		62	+245	149
	$C'H,C'_2$	**60**; R = OH, R′ = H, R″ = CO_2H	+231(w)	156
		60; R = R′ = H, R″ = Ph	+201(t)	157
		60; R = R′ = Me, R″ = Ph	+189(t)	157
	$(CC')_2$	**63**; R = CF_3	+287(c)	150
	CC',C'_2	**64**; R = H, R′ = R″ = Et	+198(c)	155
	$(C'_2)_2$	**64**; R = R″ = Ph, R′ = H	+178(pyridine)	158
C′C″	C'_2,N'	PhP=CNtBu	−99	159
	C'_2,N''	$PhP{=}CN^-$ [18]-Crown-6-K^+	−109(k)	161
	C'_2,O'	SmsP=CO	−207(b)	178
	C'_2,P^2	SmsP=C=PSms	+141(b) +143(c)	118, 179
	C'_2,C'	$SmsP{=}CCH_2$	+62(b)	162
		SmsP=CCHPh	+76(c)	163
		$SmsP{=}CCPh_2$	+72(c)	164
		$SmsP{=}CCCPh_2$	+157(c)	165
C''_2	$(N'')_2$	$NCP{=}CN^-$ [18]-Crown-6-K^+	−193(k)	166

REFERENCES

1. **Cowley, A. H. and Kemp, R. A.,** *Chem. Rev.*, 85, 367, 1985.
2. **Marre-Mazieres, M. R., Sanchez, M., and Wolf, R.,** *Nouv. J. Chim.*, 9, 605, 1985.
3. **Markovskii, L. N., Romanenko, V. D., and Ruban, A. V.,** *Phosphorus Sulfur*, 30, 447, 1987.
4. **Gladysz, J. A., Nakazama, H., Bahro, W. E., and Bertrand, G.,** *Inorg. Chem.*, 23, 3431, 1984.
5. **Weber, L., Reizig, K., Boese, R., and Polk, M.,** *Angew. Chem.*, 97, 583, 1985; *Angew. Chem. Int. Ed. Engl.*, 24, 604, 1985; **Weber, L. and Reizig, K.,** *Angew. Chem.*, 97, 868, 1985; *Angew. Chem. Int. Ed. Engl.*, 24, 865, 1985.
6. **Cowley, A. H., Lattmann, M., and Wilburn, J. C.,** *Inorg. Chem.*, 20, 2916, 1981.
7. **Niecke, E. and Kröher, R.,** *Z. Naturforsch.*, B34, 837, 1979.
8. **Lochschmidt, S. and Schmidpeter, A.,** *Phosphorus Sulfur*, 29, 73, 1986.
9. **Hitchcock, P. B., Jasim, H. A., Lappert, M. F., and Williams, H. D.,** *J. Chem. Soc. Chem. Commun.*, 1634, 1986.

10. **Scherer, O. J. and Schnabl, G.,** *Chem. Ber.*, 109, 2996, 1976.
11. **Bürklin, M., Hanecker, E., Nöth, H., and Storch, W.,** *Angew. Chem.*, 97, 980, 1985; *Angew. Chem. Int. Ed. Engl.*, 24, 999, 1985.
12. **Barlos, K., Nöth, H., and Wrackmeyer, B.,** *J. Chem. Soc. Dalton Trans.*, 601, 1979.
13. **Markovskii, L. N., Romanenko, V. D., Klebanskii, E. O., Povolotskii, M. I., Chernega, A. N., Antipin, M. Yu., and Struchkov, Yu. T.,** *Zh. Obshch. Khim.*, 56, 1721, 1986; *J. Gen. Chem. USSR*, 56, 1524, 1986.
14. **Markovskii, L. N., Romanenko, V. D., Klebanskii, E. O., Povolotskii, M. I., Chernega, A. N., Antipin, M. Yu., and Struchkov, Yu. T.,** *Zh. Obshch. Khim.*, 57, 1020, 1987; *J. Gen. Chem. USSR*, 57, 909, 1987.
15. **Fluck, E.,** *Topics Phosphorus Chem.*, 10, 193, 1980.
16. **Hitchcock, P. B., Lappert, M. F., Rai, A. K., and Williams, H. D.,** *J. Chem. Soc. Chem. Commun.*, 1633, 1986.
17. **Dressler, U., Niecke, E., Pohl, S., Saak, W., Schoeller, W. W., and Schäfer, H. G.,** *J. Chem. Soc. Chem. Commun.*, 1086, 1986.
18. **Mazieres, M. R., Sanchez, M., Bellan, J., and Wolf, R.,** *Phosphorus Sulfur*, 26, 97, 1986.
19. **Marre, M. R., Sanchez, M., and Wolf, R.,** *J. Chem. Soc. Chem. Commun.*, 566, 1984.
20. **Mazieres, M. R., Roques, L., Sanchez, M., Majoral, J.-P., and Wolf, R.,** *Tetrahedron*, 43, 2109, 1987.
21. **Karaghiosoff, K. and Schmidpeter, A.,** *Phosphorus Sulfur*, 36, 217, 1988.
22. **Schmidpeter, A., Luber, J., and Tautz, H.,** *Angew. Chem.*, 89, 554, 1977; *Angew. Chem. Int. Ed. Engl.*, 16, 546, 1977.
23. **Charbonnel, Y. and Barrans, J.,** *Tetrahedron*, 32, 2039, 1976.
24. **Lopez, L., Majoral, J.-P., Meriem, A., N'Gando M'Pondo, Th., Navech, J., and Barrans, J.,** *J. Chem. Soc. Chem. Commun.*, 183, 1984.
25. **Karaghiosoff, K., Majoral, J.-P., Meriem, A., Navech, J., and Schmidpeter, A.,** *Tetrahedron Lett.*, 24, 2137, 1983.
26. **Schmidpeter, A. and Karaghiosoff, K.,** *Z. Naturforsch.*, B36, 1273, 1981.
27. **Karaghiosoff, K., Sheldrick, W. S., and Schmidpeter, A.,** *Chem. Ber.*, 119, 3213, 1986.
28. **Marchenko, A. P., Koidan, G. N., Pinchuk, A. M., and Kirsanov, A. V.,** *Zh. Obshch. Khim.*, 54, 1774, 1984; *J. Gen. Chem. USSR*, 54, 1581, 1984.
29. **Gudat, D. and Niecke, E.,** *J. Chem. Soc. Chem. Commun.*, 10, 1987.
30. **Romanenko, V. D., Klebanskii, E. O., Shulgin, V. F., and Markovskii, L. N.,** *Zh. Obshch. Khim.*, 54, 465, 1984; *J. Gen. Chem. USSR*, 54, 415, 1984.
31. **Markovskii, L. N., Romanenko, V. D., Klebanskii, E. O., and Iksanova, S. V.,** *Zh. Obshch. Khim.*, 55, 1867, 1985.
32. **Balitskii, Yu. V., Negrebetskii, V. V., and Gololobov, Yu. G.,** *Zh. Obshch. Khim.*, 51, 475, 1981.
33. **Weber, L., Reizig, K., Bungardt, D., and Boese, R.,** *Organometallics*, 6, 110, 1987.
34. **Schmidpeter, A., Lochschmidt, S., and Sheldrick, W. S.,** *Angew. Chem.*, 97, 214, 1985; *Angew. Chem. Int. Ed. Engl.*, 24, 226, 1985.
35. **Schmidpeter, A. and Burget, G.,** *Angew. Chem.*, 97, 602, 1985; *Angew. Chem. Int. Ed. Engl.*, 24, 580, 1985.
36. **Schmidpeter, A., Burget, G., von Schnering, H.-G., and Weber, D.,** *Angew. Chem.*, 96, 795, 1984; *Angew. Chem. Int. Ed. Engl.*, 23, 816, 1984.
37. **Schmidpeter, A., Lochschmidt, S., and Sheldrick, W. S.,** *Angew. Chem.*, 94, 72, 1982; *Angew. Chem. Int. Ed. Engl.*, 21, 63, 1982.
38. **Cowley, A. H.,** *Polyhedron*, 3, 389, 1984.
39. **Kolodyazhnyi, O. I., Shevchenko, I. V., and Kukhar, V. P.,** *Zh. Obshch. Khim.*, 55, 1862, 1985; *J. Gen. Chem. USSR*, 55, 1655, 1985.
40. **Romanenko, V. D., Klebanskii, E. O., and Markovskii, L. N.,** *Zh. Obshch. Khim.*, 55, 2141, 1985; *J. Gen. Chem. USSR*, 55, 1899, 1985.
41. **Appel, R., Peters, J., and Westerhaus, A.,** *Tetrahedron Lett.*, 22, 4957, 1981.
42. **Pellerin, B., Denis, J.-M., Perrocheau, J., and Carrie, R.,** *Tetrahedron Lett.*, 27, 5723, 1986.
43. **Appel, R., Barth, V., and Knoch, F.,** *Chem. Ber.*, 116, 938, 1983.
44. **Appel, R., Knoll, F., and Ruppert, I.,** *Angew. Chem.*, 93, 771, 1981; *Angew. Chem. Int. Ed. Engl.*, 20, 731, 1981.
45. **Markovskii, L. N., Romanenko, V. D., and Kachkovskaya, L. S.,** *Zh. Obshch. Khim.*, 54, 2800, 1984; *J. Gen. Chem. USSR*, 54, 2511, 1984.
46. **Gudat, D., Niecke, E., Arif, A. M., Cowley, A. H., and Quashie, S.,** *Organometallics*, 5, 593, 1986.
47. **Markovskii, L. N., Romanenko, V. D., Sarina-Pidvarko, T. V., and Povolotskii, M. I.,** *Zh. Obshch. Khim.*, 55, 221, 1985; *J. Gen. Chem. USSR*, 55, 194, 1985.
48. **Escudie, J., Couret, C., Satge, J., Andrianarison, M., and Andriamizaka, J.-D.,** *J. Am. Chem. Soc.*, 107, 3378, 1985.

49. **Hitchcock, P. B., Lappert, M. F., and Leung, W.-P.,** *J. Chem. Soc. Chem. Commun.*, 1282, 1987.
50. **Gudat, D., Niecke, E., Malisch, W., Hofmockel, U., Quashie, S., Cowley, A. H., Arif, A. M., Krebs, B., and Dartmann, M.,** *J. Chem. Soc. Chem. Commun.*, 1687, 1985.
51. **Cowley, A. H., Norman, N. C., and Quashie, S.,** *J. Am. Chem. Soc.*, 106, 5007, 1984.
52. **Hitchcock, P. B., Maah, M. J., Nixon, J. F., Zora, J. A., Leigh, G. J., and Bahar, M. A.,** *Angew. Chem.*, 99, 497, 1987; *Angew. Chem. Int. Ed. Engl.*, 26, 474, 1987.
53. **Prishchenko, A. A. and Lutsenko, I. F.,** *Zh. Obshch. Khim.*, 51, 2630, 1981; *J. Gen. Chem. USSR*, 51, 2268, 1981.
54. **Markovskii, L. N., Romanenko, V. D., and Pidvarko, T. I.,** *Zh. Obshch. Khim.*, 52, 1925, 1982; *J. Gen. Chem. USSR*, 52, 1707, 1982.
55. **Prishchenko, A. A., Gramor, A. V., Luzikov, Yu. N., Borisenko, A. A., Lazhko, E. I., Klaus, K., and Lutsenko, I. F.,** *Zh. Obshch. Khim.*, 54, 1520, 1984; *J. Gen. Chem. USSR*, 54, 1354, 1984.
56. **Lutsenko, I. F., Prishchenko, A. A., Borisenko, A. A., and Novikova, E. S.,** *Dokl. Akad. Nauk SSSR*, 256, 1401, 1981.
57. **Niecke, E., Böske, J., Krebs, B., and Dartmann, M.,** *Chem. Ber.*, 118, 3227, 1985.
58. **Appel, R., Korte, S., Halstenberg, M., and Knoch, F.,** *Chem. Ber.*, 115, 3610, 1982.
59. **Appel, R., Kündgen, U., and Knoch, F.,** *Chem. Ber.*, 118, 1352, 1985.
60. **Kukhar, V. P., Shevchenko, I. V., and Kolodyazhnyi, O. I.,** *Zh. Obshch. Khim.*, 55, 264, 1985; *J. Gen. Chem. USSR*, 55, 230, 1985.
61. **Romanenko, V. D., Polyachenko, L. K., and Markovskii, L. N.,** *Phosphorus Sulfur*, 19, 189, 1984.
62. **Markovskii, L. N., Romanenko, V. D., Ruban, A. V., and Polyachenko, L. K.,** *Zh. Obshch. Khim.*, 56, 1907, 1986; *J. Gen. Chem. USSR*, 56, 1685, 1986.
63. **Rösch, W. and Regitz, M.,** *Angew. Chem.*, 96, 898, 1984; *Angew. Chem. Int. Ed. Engl.*, 23, 900, 1984.
64. **Yeung Lam Ko, Y. Y. C. and Carrie, R.,** *J. Chem. Soc. Chem. Commun.*, 1984, 1640.
65. **Blatter, K., Rösch, W., Vogelbacher, U.-J., Fink, J., and Regitz, M.,** *Angew. Chem.*, 99, 67, 1987; *Angew. Chem. Int. Ed. Engl.*, 26, 85, 1987.
66. **Schmidpeter, A., Leyh, Ch., and Karaghiosoff, K.,** *Angew. Chem.*, 97, 127, 1985; *Angew. Chem. Int. Ed. Engl.*, 24, 124, 1985.
67. **Högel, J. and Schmidpeter, A.,** *Chem. Ber.*, 118, 1621, 1985.
68. **Weinmaier, J. H., Luber, J., Schmidpeter, A., and Pohl, S.,** *Angew. Chem.*, 91, 442, 1979; *Angew. Chem. Int. Ed. Engl.*, 18, 412, 1979.
69. **Weinmaier, J. H., Brunnhuber, G., and Schmidpeter, A.,** *Chem. Ber.*, 113, 2278, 1980.
70. **Friedrich, P., Huttner, G., Luber, J., and Schmidpeter, A.,** *Chem. Ber.*, 111, 1558, 1978.
71. **Negrebetskii, V. V., Bogelfer, L. Ya., Bobkova, R. G., Ignatova, N. P., and Shvetsov-Shilovskii, N. I.,** *Zh. Strukt. Khim.*, 19, 64, 1978; *J. Struct. Chem. USSR*, 19, 52, 1978.
72. **Schmidpeter, A. and Klehr, H.,** *Z. Naturforsch*, B38, 1484, 1983.
73. **Rösch, W. and Regitz, M.,** *Synthesis*, 591, 1984.
74. **Högel, J., Schmidpeter, A., and Sheldrick, W. S.,** *Chem. Ber.*, 116, 549, 1983.
75. **Luber, J. and Schmidpeter, A.,** *J. Chem. Soc. Chem. Commun.*, 887, 1976.
76. **Rösch, W. and Regitz, M.,** *Synthesis*, 689, 1987.
77. **Vasilev, A. V., Vilkov, L. V., Ignatova, N. P., Melnikov, N. N., Negrebetskii, V. V., Shvetsov-Shilovskii, N. I., and Khaikin, L. S.,** *J. Prakt. Chem.*, 314, 806, 1972.
78. **Karaghiosoff, K., Cleve, C., and Schmidpeter, A.,** *Phosphorus Sulfur*, 28, 289, 1986.
79. **Karaghiosoff, K., Klehr, H., and Schmidpeter, A.,** *Chem. Ber.*, 119, 410, 1986.
80. **Niecke, E., Rüger, R., and Schoeller, W. W.,** *Angew. Chem.*, 93, 1110, 1981; *Angew. Chem. Int. Ed. Engl.*, 20, 1034, 1981.
81. **Romanenko, V. D., Drapailo, A. B., Ruban, A. V., and Markovskii, L. N.,** *Zh. Obshch. Khim.*, 56, 714, 1986; *J. Gen. Chem. USSR*, 56, 635, 1986.
82. **Romanenko, V. D., Ruban, A. V., and Markovskii, L. N.,** *J. Chem. Soc. Chem. Commun.*, 187, 1983.
83. **Appel, R., Knoch, F., and Zimmerman, R.,** *Chem. Ber.*, 118, 814, 1985.
84. **Rösch, W., Facklam, Th., and Regitz, M.,** *Tetrahedron*, 43, 3247, 1987.
85. **Schmidpeter, A., Karaghiosoff, K., Cleve, C., and Schomburg, D.,** *Angew. Chem.*, 97, 125, 1985; *Angew. Chem. Int. Ed. Engl.*, 24, 123, 1985.
86. **Bordieu, C. and Foucaud, A.,** *Tetrahedron Lett.*, 28, 4673, 1987.
87. **Dahl, O.,** *Tetrahedron Lett.*, 23, 1493, 1982.
88. **Burg, A. B.,** *J. Inorg. Nucl. Chem.*, 33, 1575, 1971.
89. **Zurmühlen, F. and Regitz, M.,** *Angew. Chem.*, 99, 65, 1987; *Angew. Chem. Int. Ed. Engl.*, 26, 83, 1987.
90. **Romanenko, V. D., Sarina, T. V., Povolotskii, M. I., and Markovskii, L. N.,** *Zh. Obshch, Khim.*, 55, 1437, 1985; *J. Gen. Chem. USSR*, 55, 1280, 1985.
91. **Yoshifuji, M., Ando, K., Toyota, K., Shima, I., and Inamoto, M.,** *J. Chem. Soc. Chem. Commun.*, 419, 1983.

92. **Romanenko, V. D., Ruban, A. V., Iksanova, S. V., Polyachenko, L. K., and Markovskii, L. N.,** *Phosphorus Sulfur*, 22, 365, 1985.
93. **Etemad-Moghadam, G., Bellan, J., Tachon, Ch., and Koenig, M.,** *Tetrahedron*, 43, 1793, 1987.
94. **Markovskii, L. N., Romanenko, V. D., and Kirsanov, A. V.,** *Phosphorus Sulfur*, 18, 31, 1983.
95. **Romanenko, V. D., Kachkovskaya, L. S., and Markovskii, L. N.,** *Zh. Obshch. Khim.*, 55, 2140, 1985; *J. Gen. Chem. USSR*, 55, 1898, 1985.
96. **Burckett-St. Laurent, J. C. T. R., Hitchcock, P. B., Kroto, H. W., and Nixon, J. F.,** *J. Chem. Soc. Chem. Commun.*, 1141, 1981.
97. **Appel, R. and Moors, R.,** *Angew. Chem.*, 98, 570, 1986; *Angew. Chem. Int. Ed. Engl.*, 25, 567, 1986.
98. **Markovskii, L. N., Romanenko, V. D., and Pidvarko, T. V.,** *Zh. Obshch. Khim.*, 52, 1925, 1982; *J. Gen. Chem. USSR*, 52, 1707, 1982.
99. **Barron, A. R., Cowley, A. H., and Hall, S. W.,** *J. Chem. Soc. Chem. Commun.*, 980, 1987.
100. **Becker, G., Uhl, W., and Wessely, H.-J.,** *Z. Allg. Anorg. Chem.*, 479, 41, 1981.
101. **Schäfer, A., Weidenbruch, M., Saak, W., and Pohl, S.,** *Angew. Chem.*, 99, 806, 1987; *Angew. Chem. Int. Ed. Engl.*, 26, 776, 1987.
102. **Schmit, C. N., Lock, F. M., and Bickelhaupt, F.,** *Tetrahedron Lett.*, 25, 3011, 1984.
103. **Escudie, J., Couret, C., Adrianarison, M., Raharinirina, A., and Satge, J.,** *Phosphorus Sulfur*, 30, 377, 1987.
104. **Issleib, K., Leissring, E., Riemer, M., and Oehme, H.,** *Z. Chem.*, 23, 99, 1983.
105. **Grobe, J., le Van, D., and Nientiedt, J.,** *Z. Naturforsch.*, B41, 149, 1986.
106. **Eshtiagh-Hosseni, H., Kroto, H., and Nixon, J. F.,** *J. Organomet. Chem.*, 181, C1, 1979.
107. **Grobe, J., le Van, D., and Nientiedt, J.,** *Z. Naturforsch.*, B42, 984, 1987.
108. **Becker, G. and Mundt, O.,** *Z. Allg. Anorg. Chem.*, 443, 53, 1978.
109. **Appel, R.,** *Tetrahedron Lett.*, 25, 4447, 1984.
110. **Becker, G., Becker, W., and Uhl, G.,** *Z. Allg. Anorg. Chem.*, 518, 21, 1984.
111. **Appel, R. and Paulen, W.,** *Chem. Ber.*, 116, 109, 1983.
112. **Weber, L. and Reizig, K.,** *Angew. Chem.*, 97, 53, 1985; *Angew. Chem. Int. Ed. Engl.*, 24, 53, 1985.
113. **Kolodyazhnyi, O. I.,** *Zh. Obshch. Khim.*, 50, 230, 1980.
114. **Appel, R., Casser, C., and Immenkeppel, M.,** *Tetrahedron Lett.*, 26, 3551, 1985.
115. **Appel, R. and Casser, C.,** *Tetrahedron Lett.*, 25, 4109, 1984.
116. **Schmidpeter, A., Lochschmidt, S., and Willhalm, A.,** *Angew. Chem.*, 95, 561, 1983; *Angew. Chem. Int. Ed. Engl.*, 22, 545, 1983.
117. **Oehme, H., Leissring, E., and Meyer, H.,** *Z. Chem.*, 21, 407, 1981.
118. **Appel, R., Fölling, P., Josten, B., Siray, M., Winkhaus, V., and Knoch, F.,** *Angew. Chem.*, 96, 620, 1984; *Angew. Chem. Int. Ed. Engl.*, 23, 619, 1984.
119. **Appel, R., Fölling, P., Krieger, L., Siray, M., and Knoch, F.,** *Angew. Chem.*, 96, 981, 1984; *Angew. Chem. Int. Ed. Engl.*, 23, 970, 1984.
120. **Gudat, D., Niecke, E., Krebs, B., and Dartmann, M.,** *Chimia*, 39, 277, 1985.
121. **Appel, R., Casser, C., Immenkeppel, M., and Knoch, F.,** *Angew. Chem.*, 96, 905, 1984; *Angew. Chem. Int. Ed. Engl.*, 23, 895, 1984.
122. **Schmidpeter, A. and Willhalm, A.,** *Angew. Chem.*, 96, 901, 1984; *Angew. Chem. Int. Ed. Engl.*, 23, 903, 1984.
123. **Karsch, H. H.,** *Tetrahedron Lett.*, 25, 3687, 1984.
124. **Karsch, H. H., Reisacher, H. U., Huber, B., Müller, G., Malisch, W., and Jörg, K.,** *Angew. Chem.*, 98, 468, 1986; *Angew. Chem. Int. Ed. Engl.*, 25, 455, 1986.
125. **Issleib, K., Schmidt, H., and Wirkner, Ch.,** *Z. Allg. Anorg. Chem.*, 488, 75, 1982.
126. **Becker, G., Rössler, M., and Uhl, G.,** *Z. Allg. Anorg. Chem.*, 495, 73, 1982.
127. **Becker, G., Rössler, M., and Schneider, E.,** *Z. Allg. Anorg. Chem.*, 439, 121, 1978.
128. **Becker, G.,** *Z. Allg. Anorg. Chem.*, 480, 38, 1981.
129. **Märkl, G. and Sejpka, H.,** *Tetrahedron Lett.*, 27, 1771, 1986.
130. **Gamon, N. and Reichardt, C.,** *Angew. Chem.*, 89, 418, 1977; *Angew. Chem. Int. Ed. Engl.*, 16, 404, 1977; *Liebigs Ann. Chem.*, 2072, 1980.
131. **Appel, R., Hünerbein, J., and Siabalis, N.,** *Angew. Chem.*, 99, 810, 1987; *Angew. Chem. Int. Ed. Engl.*, 26, 779, 1987.
132. **Romanenko, V. D., Ruban, A. V., Povolotskii, M. I., Polyachenko, L. K., and Markovskii, L. N.,** *Zh. Obshch. Khim.*, 56, 1186, 1986; *J. Gen. Chem. USSR*, 56, 1044, 1986.
133. **Kerschl, S., Wrackmeyer, B., Willhalm, A., and Schmidpeter, A.,** *J. Organomet. Chem.*, 319, 49, 1987.
134. **Märkl, G. and Sejpka, H.,** *Tetrahedron Lett.*, 27, 171, 1986.
135. **Märkl, G. and Trötsch, I.,** *Angew. Chem.*, 96, 899, 1984; *Angew. Chem. Int. Ed. Engl.*, 23, 901, 1984.
136. **Märkl, G. and Pflaum, S.,** *Tetrahedron Lett.*, 27, 4415, 1986.

137. **Rösch, W., Hees, U., and Regitz, M.,** *Chem. Ber.*, 120, 1645, 1987.
138. **Rösch, W., Richter, H., and Regitz, M.,** *Chem. Ber.*, 120, 1809, 1987.
139. **Heinicke, J.,** *Tetrahedron Lett.*, 27, 5699, 1986.
140. **Märkl, G. and Dorfmeister, G.,** *Tetrahedron Lett.*, 27, 4419, 1986.
141. **Märkl, G. and Dorfmeister, G.,** *Tetrahedron Lett.*, 28, 1089, 1987.
142. **Heinicke, J. and Tzschach, A.,** *Phosphorus Sulfur*, 25, 345, 1985.
143. **Schmidt, H. and Issleib, K.,** *Wiss. Z. Univ. Halle*, 32, 41, 1983.
144. **Märkl, G., Eckl, E., Jakobs, U., Ziegler, M. L., and Nuber, B.,** *Tetrahedron Lett.*, 28, 2119, 1987.
145. **Charrier, C., Bonnard, H., de Lauzon, G., and Mathey, F.,** *J. Am. Chem. Soc.*, 105, 6871, 1983.
146. **Charrier, C., Bonnard, H., Mathey, F., and Neibecker, D.,** *J. Organomet. Chem.*, 231, 361, 1982.
147. **Märkl, G. and Dorfmeister, G.,** *Tetrahedron Lett.*, 28, 1093, 1987.
148. **Fluck, E., Becker, G., Neumüller, B., Knebl, R., Heckmann, G., and Riffel, R.,** *Z. Naturforsch.*, B42, 1213, 1987.
149. **Märkl, G. and Matthews, D.,** *Angew. Chem.*, 84, 1069, 1972; *Angew. Chem. Int. Ed. Engl.*, 11, 1019, 1972.
150. **Kobayashi, J., Hamana, H., Fujino, Sh., Ohsawa, A., and Kumadaki, I.,** *J. Am. Chem. Soc.*, 102, 252, 1980.
151. **Ashe, A. J.,** *J. Am. Chem. Soc.*, 93, 3293, 6690, 1971; **Ashe, A. J., Sharp, R. R., and Tolan, J. W.,** *J. Am. Chem. Soc.*, 98, 5451, 1976.
152. **Alcaraz, J.-M. and Mathey, F.,** *Tetrahedron Lett.*, 25, 4659, 1984.
153. **Märkl, G., Adolin, G., Kees, F., and Zander, G.,** *Tetrahedron Lett.*, 18, 3445, 1977.
154. **Märkl, G. and Hock, K.,** *Tetrahedron Lett.*, 24, 2645, 1983.
155. **Charrier, C., Bonnard, H., and Mathey, F.,** *J. Org. Chem.*, 47, 2376, 1982.
156. **Pellon, P. and Hamelin, J.,** *Tetrahedron Lett.*, 27, 5611, 1986.
157. **Alcaraz, J.-M., Deschamps, B., and Mathey, F.,** *Phosphorus Sulfur*, 19, 45, 1984.
158. **Märkl, G.,** *Angew. Chem.*, 78, 907, 1966; *Angew. Chem. Int. Ed. Engl.*, 5, 846, 1966.
159. **Kolodyazhnyi, O. I.,** *Zh. Obshch. Khim.*, 53, 1226, 1983; *J. Gen. Chem. USSR*, 53, 1093, 1983.
160. **Appel, R. and Paulen, W.,** *Tetrahedron Lett.*, 24, 2639, 1983.
161. **Schmidpeter, A., Zirzow, K.-H., Burget, G., Huttner, G., and Jibril, I.,** *Chem. Ber.*, 117, 1695, 1984.
162. **Märkl, G. and Reitinger, S.,** *Tetrahedron Lett.*, 29, 463, 1988.
163. **Appel, R., Winkhaus, V., and Knoch, F.,** *Chem. Ber.*, 119, 2466, 1986.
164. **Yoshifuji, M., Toyota, K., Shibayama, K., and Inamoto, M.,** *Tetrahedron Lett.*, 25, 1809, 1984.
165. **Märkl, G., Sejpka, H., Dietl, S., Nuber, B., and Ziegler, M. L.,** *Angew. Chem.*, 98, 1020, 1986; *Angew. Chem. Int. Ed. Engl.*, 25, 1003, 1986.
166. **Schmidpeter, A. and Zwaschka, F.,** *Angew. Chem.*, 89, 747, 1977; *Angew. Chem. Int. Ed. Engl.*, 16, 704, 1977.
167. **Eshtiagh-Hosseini, H. E., Kroto, H. W., and Nixon, J. F.,** *J. Chem. Soc. Chem. Commun.*, 653, 1979.
168. **Appel, R. and Westerhaus, A.,** *Tetrahedron Lett.*, 22, 2159, 1981.
169. **Anderson, S. P., Goldwhite, H., Ko, D., and Letsou, A.,** *J. Chem. Soc. Chem. Commun.*, 744, 1975.
170. **Fuchs, E. P. O., Hermesdorf, M., Schnurr, W., Rösch, W., Heydt, H., Regitz, M., and Binger, P.,** *J. Organomet. Chem.*, 338, 329, 1988.
171. **Rösch, W., Vogelbacher, U., Allspach, Th., and Regitz, M.,** *J. Organomet. Chem.*, 306, 39, 1986.
172. **Becker, G., Gresser, G., and Uhl, W.,** *Z. Naturforsch.*, B36, 16, 1981.
173. **Allspach, Th., Regitz, M., Becker, G., and Becker, W.,** *Synthesis*, 31, 1986.
174. **Märkl, G. and Sejpka, H.,** *Tetrahedron Lett.*, 26, 5507, 1985.
175. **Appel, R., Maier, G., Reisenauer, H. P., and Westerhaus, A.,** *Angew. Chem.*, 93, 215, 1981; *Angew. Chem. Int. Ed. Engl.*, 20, 197, 1981.
176. **Markovskii, L. N., Romanenko, V. D., Ruban, A. V., Drapailo, A. B., Chernega, A. N., Antipin, M. Yu., and Struchkov, Yu. T.,** *Zh. Obshch. Khim.*, 58, 291, 1988.
177. **Gudat, D., Niecke, E., Krebs, B., and Dartmann, M.,** *Organometallics*, 5, 2376, 1986.
178. **Appel, R. and Paulen, W.,** *Angew. Chem.*, 95, 807, 1983; *Angew. Chem. Int. Ed. Engl.*, 22, 785, 1983.
179. **Yoshifuji, M., Toyota, K., and Inamoto, N.,** *J. Chem. Soc. Chem. Commun.*, 689, 1984.

Chapter 2a

^{31}P NMR DATA OF ALKALI METAL PHOSPHIDES

Compiled and presented by
Michael J. Gallagher

The NMR data for alkali metal phosphides are given in Table A3. While most phosphides are best described as two coordinate, the phosphorus atom often has additional loose coordination to the metal atoms. As a consequence, the metal atom has been included as one of the P-bound atoms and the phosphides put into a separate table. These compounds constitute a relatively small group whose chemistry has been largely developed and explored by Issleib and his co-workers, though in recent years there has been a quickening of interest, particularly in structure and spectroscopy. The structures of the phosphides appear to be a function of solvent, temperature, and concentration and considerable differences in shift values are observed when these parameters are varied.[3,6] It would be expected that with the high electron density at phosphorus in these substances the chemical shifts would be sensitive to steric and electronic effects. This is certainly the case but neither the direction nor magnitude of these shifts is easily predictable. In general $\Delta\delta$ for the formal process

$$R_1R_2PH \rightarrow R_1R_2PM$$

is upfield when R_1 and R_2 are alkyl or hydrogen but downfield if either group is aryl, though the latter effect may be very small (<1 ppm). Phosphines with sterically demanding substituents also show downfield shifts (2 to 20 ppm) on conversion to their phosphides. When the phosphorus atom forms part of an aromatic system, e.g., in a phosphole, then the downfield shift is very large (ca. 200 ppm).

Unexpectedly, complexation of the metal, e.g., with crown ethers, also results in a downfield shift, though such a process might reasonably be expected to result in increased electron density at phosphorus. X-ray studies of crystalline crown ether complexes confirm that the phosphorus is no longer bonded to the metal.[20,23] Chemical shifts also vary considerably with the metal.

So unpredictable are shift values in this class of substances that literature comparisons should only be made when solvent, temperature, and concentration are approximately the same and wherever possible chemical evidence, e.g., protonation, should be used to confirm assignments. The use of formulae based on summing α and β substituent effects should be restricted to the solvent systems used for determining them.[4] Here, as elsewhere, α effects are positive and large, β effects are also positive but smaller, and λ effects are negative.

Coupling Constants — The phosphides of phosphine and primary phosphines have $^1J_{PH}$ smaller than in the parent compound, usually by about 30 Hz and for primary phosphines most fall in the range 160 to 170 Hz. Lithium phosphides without hydrogen attached will show ^1J_P Li at low temperatures and this is usually 30 to 40 Hz (for ^{7}Li). Variations occur with degree of association in solution[183] and also very crowded phosphides seem to show larger values. Phosphides from polyphosphines may show $^1J_{PP}$ (ca. 200 to 400 Hz). Few $^1J_{PC}$ have been measured; those that have fall mainly in the ranges 16 to 29 Hz (aliphatic) and 43 to 54 Hz (aromatic). There is an increase in $^1J_{PC}$ for the sequence

$$RR'PH \rightarrow RR'PLi.\ solvent \rightarrow RR'PLi\ Crown\ ether$$

and the effect is larger for aromatic (33-43;1-10) than for aliphatic (4-18; 1-10) carbons.[3]

TABLE A3
Alkali Metal Phosphides

P-bound atoms/ rings	Connectivities	Structure	NMR data (δP [solv. temp] $^nJ_{PZ}$ Hz)	Ref.
$P_2Li/5$	$(C'_2)_2$	**1**	−174(k) J369	18
HLiSi	C_3	TmsPHLi	−108.4, −110(k), −120(j), J171, 172, 189	6,7
H_2Cs	None	H_2PCs	−243(v)	1
H_2K	—	H_2PK	−256(k) J134	2
			−252(v)	1
			−272(Me_3N) J139	1
			−280(NH_3)	1
H_2Li	—	H_2PLi	−251(v)	1
H_2Na		H_2PNa	−303.4(k) J147	2
CCsP	C_2';C'P	$Cs_2(PPh)_3$	−42.5(k)	17
CKN	C_3;CP	$(t\text{-}BuKP)_2N\text{-}i\text{-}Pr$	20.3(k)	19
CKP	CH_2;P	EtP(K)P(K)P(K)Et	−78.5 J306	1
	CH_2;CK	$(EtPK)_2$	−79.6(k)	1
	C_3;CK	$(t\text{-}BuPK)_2$	−24.3(k)	19
	C_3;C_2	t-BuPKP(t-Bu)i-Pr	−71.4(k) J350	19
C'KP	C'_2;C'P	$K_2(PPh)_3$	−49.8(k) −50(k)	1, 17
		$K_2(PPh)_4$	−70.0(k)	17
CLiP	H_3;CP	$(MePLi)_2(PMe)_2$	−155.6(k) J274	1
	CH_2;CC'_2	EtPhPPEtLi	−112.8(k) J396	1
C'LiP	C'_2;C'P	$Li_2(PPh)_4$	−83.4(k), −86(k), J216	1, 17
C'LiP4	CC';C'_2	**2**	−53.2(k)	15
C'NaP	C'_2;C'P	$Na_2(PPh)_4$	−91.4(k)	17
CHK	C_3	t-BuPHK	−60(ze)	19
CHLi	Si_3	Tms_3CPHLi	−92.1(k) J168	7
	C_3	t-BuPHLi	−73.3 J165	3
	C_3	Et_3CPHLi	−115 J169	3
C'HLi	C'_2	PhPHLi	−111.9(k), −120.9(k)	3, 4
		1,2-$C_6H_4(PHLi)_2$	−127.4(j) −127.7(k)	5
			−118.3(ae) J163	20
	C_2'	MesPHLi	−154.3(k)/−163.8(ae) J167/170	3, 4, 20
C'HNa	C'_2	PhPHNa	−120.9(k) J163	2
C'_2Cs	$(C'_2)_2$	Ph_2PCs	0.0(ne)	6
C_2K	C_2H; C_3	t-BuPK-iPr	22(z/l)	19
C'_2K	$(C'O')_2$	1,2-$C_6H_4(CO)_2PK$	43.3	24
$C'_2K/5$	$(C'H)_2$	**3**; R_1 = H, R_2 = Me M = K	55.8(k)	12
		3; R_2 = H, R_2 = Ph M = K	78.7(k), 76.7(k)	9, 12
$C'_2K/5$	$(C'_2)_2$	**3**; R_1 = R_2 = Ph M = K	96.1(k)	11
	$(C'_2)_2$	Ph_2PK	−12.4	6
C_2Li	$(HSi_2)_2$	$(Tms_2CH)_2PLi$	−254 J80, −115($(Me_2NCH_2)_2$) J122	22
	$(H_3)_2$	Me_2PLi	−141.2(k)	8
	$(CH_2)_2$	Et_2PLi	−64.2(k), −72.5(j) J49	4
		Pr_2PLi	−81.8(k) −91.4(ae) −93.7(j) J40	4, 8
		Bu_2PLi	−82.8(k) −89.3(j) −91.3(ae) J36, 38	4, 8
		i-Bu_2PLi	−95.4(k)	8
	$(C_2H)_2$	i-Pr_2PLi	−12.9(k), −4.7(ae) J67	4
		s-Bu_2PLi	−17.8(k), −25.7(j), −24.7(ae) J63, 66,	4, 8

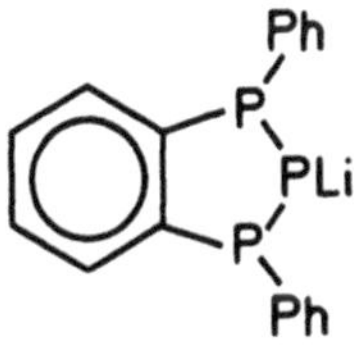

1 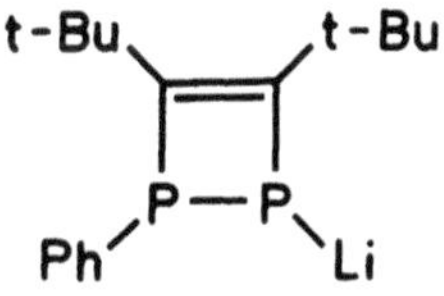 2 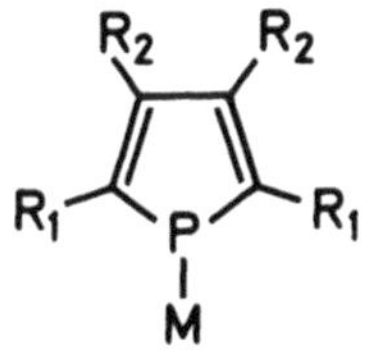 3

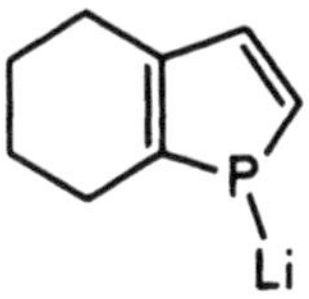

4 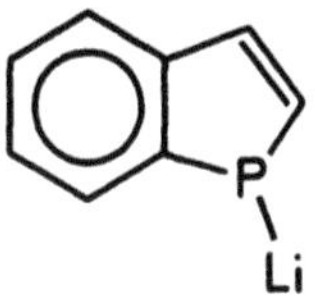 5 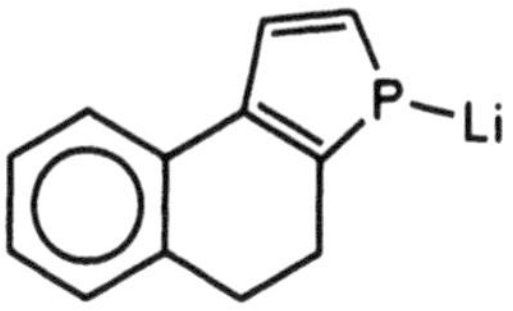 6

P-bound atoms/ rings	Connectivities	Structure	NMR data (δP [solv. temp] $^nJ_{PZ}$ Hz)	Ref.
		$(c\text{-}C_6H_{11})_2PLi$	−7.1, −9.5	20
	$(C_3)_2$	t-Bu_2PLi	29.9/38.5(j)	4, 21
			30.9(ae) J66, 67	
CC′Li	$H_2P;C'_2$	$(PhPLi)_2CH_2$	−79.9(b/k)	10
	$H_3;C'_2$	PhMePLi	−78.4(k), −80(k)	4, 8, 9
			−95.5(j) J45	
	$H_3;C'_2$	1,2-$(MePLi)_2C_6H_4$	−101.8(k), −106(j) J36, 39	4
	$CH_2;C'_2$	PhEtPLi	−41.9(k), −42.7(k),	3,9
			−54.9(j) J47, −69(ae)	4
				4
		PhPrPLi	−52.8(k)	8
			−54.9(j) J47 −69(ae)	4
		PhPrPLi	−52.8(k)	8
		PhBuPLi	−51.8(K)	8
		i-BuPhPLi	−57.9(k)	8
		1,2-$(EtPLi)_2C_6H_4$	−61.7(k) −64.5(j) J38	4
		1,2-$(BuPLi)_2C_6H_4$	−78.4(k) −82.2(j) J40,41	4
		$(PhPLiCH_2)_2$	−61.9(k)	9
		$(PhPLi_2(CH_2)_3$	−63.0(k)	9
		$(PhPLi)_2(CH_2)_4$	−54.9(k)	9
		$(PhPLi)_2(CH_2)_5$	−55.6(k)	9
		$Ph_2P(CH_2)_2PhPLi$	−41.9(k)	9
		$Ph_2P(CH_2)_3PhPLi$	−54.6(k)	9
		$Ph_2P(CH_2)_4PhPLi$	−53.1(k)	9
		$Ph_2P(CH_2)_5PhPLi$	−55.7(k)	9
		i-PrPhPLi	−11.8(k), −40.1(j)	3, 4
			−35.5(ae) J44, 33	
		s-BuPhPLi	−17.5(k)	8
		$c\text{-}C_6H_{11}PhPLi$	−23.2(k)	8
		1,2-$(iPrPLi)_2C_6H_4$	−37.6(k), −41.2(j) J39	4
	$C_3;C'_2$	t-BuPhPLi	11.2(j),7(k) J44,50	3, 4
CC′Na	$C'_2;N''$	$Ph(CN)P^- Me_4N^+$	−109(k)	13
C'_2Li	$(CO')_2$	$(t\text{-}BuCO)_2PLi$	52.6(ze)	14
	$(C'H)_2$	**3**; $R_1 = R_2 = H$ M = Li	76(k)	11

TABLE A3 (continued)
Alkali Metal Phosphides

P-bound atoms/ rings	Connectivities	Structure	NMR data (δP [solv. temp] $^nJ_{PZ}$ Hz)	Ref.
	C'H, CC'	**4**	73.3(k)	16
		5	81.7	16
	C'H, C'_2	**6**	40	11
	$(C'_2)_2$	Ph_2PLi	−36(j), −20.8/−26.1(k)	3, 19
			−27.4(b/$(Me_2NCH_2)_2$)	17, 23
			−13.65(b/$(Me_2N)_3PO$)	
			−6.9(k/crown ether)	
		$(4\text{-}CH_3C_6H_4)_2PLi$	−26.4(k)	9
		$(2\text{-}CH_3C_6H_4)_2PLi$	−59.6	3
		$(2,4,6\text{-}Me_3C_6H_2)_2PLi$	−89.3(j), −63.7(k)	20
			−42.8 (12-crown-4)	
C'_2Na	$(C\ O')_2$	$(t\text{-}BuCO)_2PNa$	42.2(ze)	14
	$(C'_2)_2$	Ph_2PNa	−22.5(k), −24.5(k)	1, 2, 6
			−29.9(j), 3.3(v)	

REFERENCES

1. **Maier, L.**, *Organic Phosphorus Compounds*, Kosolapoff, G. M. and Maier, L., Eds., Wiley-Interscience, New York, 1972, chap. 2.
2. **Batchelor, R. and Birchall, T.**, *J. Am. Chem. Soc.*, 104, 674, 1982.
3. **Zschunke, A., Reimer, M., Schmidt, H., and Issleib, K.**, *Phosphorus Sulphur*, 17, 237, 1983.
4. **Zschunke, A., Reimer, M., Krech, F., and Issleib, K.**, *Phosphorus Sulfur*, 349, 1985.
5. **Zschunke, A., Reimer, M., Lessering, E., and Issleib, K.**, *Z. Anorg. Allg. Chem.*, 525, 35, 1985.
6. **Zschunke, A., Bauer, E., Schmidt, H., and Issleib, K.**, *Z. Anorg. Allg. Chem.*, 495, 115, 1982.
7. **Cowley, A., Kilduff, J. E., Newman, T. H., and Pakulski, M.**, *J. Am. Chem. Soc.*, 104, 5820, 1982.
8. **Grim, S. O. and Molenda, R. P.**, *Phosphorus*, 4, 189, 1974.
9. **Gallagher, M. J., Brooks, P., Sarroff, A., and Brayan, J.**, unpublished data.
10. **Schmidbaur, H. and Schnatterer, S.**, *Chem. Ber.*, 119, 2832, 1986.
11. **Nief, F., Charrier, C., Mathey, F., and Simalty, M.**, *Phosphorus Sulfur*, 259, 1982.
12. **deLauzon, G., Charrier, C., Bonnard, H., and Mathey, F.**, *Tetrahedron Lett.*, 23, 511, 1982.
13. **Deng, R. M. K. and Dillon, K. B.**, *J. Chem. Soc. Chem. Commun.*, 1170, 1981.
14. **Becker, G., Rössler, M., and Uhl, G.**, *Z. Anorg. Allg. Chem.*, 495, 73, 1982.
15. **Charrier, C., Mathey, F., Robert, F., and Jeannin, V.**, *J. Chem. Soc. Chem. Commun.*, 1707, 1984.
16. **Quin, L. D. and Orton, W. L.**, *J. Chem. Soc. Chem. Commun.*, 401, 1979.
17. **Hoffmann, P. R. and Caulton, K. G.**, *J. Am. Chem. Soc.*, 97, 6470, 1975.
18. **Schmidpeter, A., Burget, G., and Sheldrick, W. S.**, *Chem. Ber.*, 118, 3849, 1985.
19. **Baudler, M. and Kupprat**, *Z. Anorg. Allg. Chem.*, 533, 153, 1986.
20. **Bartlett, R. A., Olmstead, M. M., and Power, P. P.**, *Inorg. Chem.*, 26, 1941, 1987.
21. **Jones, R. A., Stuart, A. L., and Wright, T. C.**, *J. Am. Chem. Soc.*, 103, 7459, 1983.
22. **Hitchcock, P. B., Lappert, M. F., Power, P. P., and Smith, S. J.**, *J. Chem. Soc. Chem. Commun.*, 1669, 1984.
23. **Hope, H., Olmstead, M. M., Power, P. P., and Xu, X.**, *J. Am. Chem. Soc.*, 106, 1984.
24. **Liotta, C. L., McLaughlin, M. L., Van der Veer, D. G., and O'Brien, B.**, *Tetrahedron Lett.*, 25, 1665, 1984.

Chapter 3

^{31}P NMR DATA OF THREE COORDINATE ($\lambda 3$ $\sigma 3$) PHOSPHORUS COMPOUNDS CONTAINING PHOSPHORUS BONDS TO HALOGEN

Compiled and presented by
Stephan G. Kleemann, Ekkehard Fluck, and John C. Tebby

NMR data was also donated by
H. W. Roesky, Universitäet Göettingen, FRG
E. Nieke, Universitäet Bielefeld, FRG
C. D. Hall, King's College London, U.K.
J. P. Dutasta, Université Scientifique et Médicale de Grenoble, France

The NMR data is presented in Table B with no subsections. The compilation consists of compounds with one, two, and three phosphorus halogen bonds and thus includes phosphonous and phosphinous halides. Interestingly, only two compounds with a phosphorus bond to hydrogen and halogen were found.

A bar chart summary of the range of chemical shifts is given in Figures 1.3 and 1.3a in Chapter 1. While the total chemical shift range for this class of three coordinate compounds is only slightly higher frequency (lower field) than the corresponding phosphites and amides, it is clear that the presence of a phosphorus halogen bond lowers the screening of the three valent three coordinate phosphorus atom. Thus the phosphorus trihalides (except for PF_3) are all at the high frequency end of the 255 to 1 ppm range for this group of compounds. While it is well recognized that fluorine produces a unique bond to phosphorus, once one of the three fluorine atoms is replaced by another halogen most of the deshielding influence of the P-F bonds on the phosphorus chemical shift is removed.

Replacement of only one halogen (Cl, Br, or I) by another group causes the signal to move to lower frequency; the extent increases in the order $S < C \approx N \leqslant O$. Thus, while sulfur has a marginal effect of ca. 10 ppm only, the introduction of carbon or nitrogen causes movement of the order 30 to 50 ppm and oxygen by 30 to 80 ppm. Although the difluorides resonate across a similar frequency range as the rest of the dihalogeno compounds, the movement relative to PF_3 is in the opposite direction toward higher frequency due to the anomalously low frequency chemical shift (97 ppm) for the latter.

Compounds in which two halogen atoms (I, Br, and Cl) have been replaced by carbon resonate at lower frequencies still and occupy the low frequency end of the total range (usually to δ_P 120/60 ppm). Replacement of a second halogen by another nitrogen group or replacement of a second bromine by oxygen produces a small additional shielding effect of ca. 10/40 ppm; whereas replacement of a second chloro or fluoro group by another oxygen atom is virtually without effect. Thus the order of the effect of second halogen replacements is $O < S < N < C$. The large influence of carbon and the negligible influence of a second oxygen may reflect the fact that tertiary phosphines generally appear at much lower frequencies whereas phosphites resonate at quite high frequencies.

The effect of replacing halides by other atoms depends on the halide being replaced. Thus, the relative mean chemical shift frequencies of the halogen/oxygen and the halogen/nitrogen compounds are Br (200/150) > Cl (160/140) > F (140/110) whereas the relative mean chemical shift frequencies of the organohalides, i.e., P-bound atoms $=CHal_2$ and C_2Hal, are in the reverse order, i.e., F (190/160) > Cl (190/100) > Br (180/100) > I (130/70 ppm).

Among the halogens, bromine mixed with other hetero atoms has the most consistent high frequency effect. Exceptions occur only when the strong anisotropic influence of a directly bound triple bond is operating.

An analysis of the effects of introducing organic groups shows that all the iodides, including PI_3, resonate to low frequency of the bromides and chlorides. There is relatively little difference between the bromides and the chlorides. For the fluorides the presence of organic groups narrows the range of chemical shifts, e.g.,

P-Bound Atoms	FX_2	CFX	C_2F
δ_P	255/47	245/121	192/123

The chloro compounds exhibit properties intermediate of fluorine and bromine, i.e., the introduction of one organic group narrows the range but the introduction of two organic groups moves the signals to much lower frequencies.

The screening of phosphorus is dependent on the hybridization of the P-bound carbon. Relative to an sp^3 carbon, an sp^2 carbon has signals at lower frequencies for all the organophosphorus halides. The difference is least for fluorine. In common with nearly all phosphorus compounds the presence of P-fluoroalkyl groups cause resonance at frequencies 30/60 ppm lower than compounds containing the corresponding alkyl group. This also appears to apply to fluoroalkoxy groups relative to alkoxy in this class of compounds.

The incorporation of phosphorus in a three-membered ring strongly reduces the screening of phosphorus which is another general feature of phosphorus compounds.

TABLE B
Compilation of ^{31}P NMR Data of Three Coordinate (λ3 σ3) Phosphorus Compounds Containing Phosphorus Bonds to Halogen

P-bound atoms/ rings	Connectivities	Structure	NMR data (δP[solv,temp]$^{n}J_{PA}$ Hz)	Ref.
BrClO	C′	BrClPOPh	ca. 190	108
$BrCl_2$		$BrPCl_2$	228	108
BrF_2		$BrPF_2$	218 $^{1}J_{PF}$1395	128
BrI_2		$BrPI_2$	208	107
BrN_2	$(C_2)_2$	$BrP(NMe_2)_2$	184.2(n)	136
$BrN_2/5$	CB,CN	BrP*N(Me)B(Me)N(Me)N*(Me)	186.0	160
	$(C_2)_2$	$BrP*(NMeCH_2)_2$	190.2(r)	136
$BrN_2/6$	$(CB)_2$	BrP*N(Me)B(Me)N(Me)B(Me)N*(Me)	178.0	150
BrN'_2	$(C'')_2$	$BrP(NCO)_2$	126.7	83
		$BrP(NCS)_2$	111.5	83
BrO_2	$(C')_2$	$BrP(OPh)_2$	176.0	108
$BrO_2/5$	$(C')_2$	BrP-Cat	195.6, 174	145, 164
BrS_2	$(C)_2$	$BrP(SEt)_2$	198	35
		$BrP(SPr)_2$	198	23
	$(C')_2$	$BrP(SPh)_2$	184.2	56
Br_2Cl		Br_2PCl	228	108
Br_2F		Br_2PF	255 $^{1}J_{PF}$1292	128
Br_2I		Br_2PI	224	107
Br_2N	C_2	Br_2PNMe_2	190(n), 176	136, 164
Br_2O	C′	Br_2POPh	ca. 200	108
Br_2S	C	Br_2PSMe	203.5	56
	C′	Br_2PSPh	203.5	56
Br_3		Br_3P	227	22
ClF_2		$ClPF_2$	175.7(34°) $^{1}J_{PF}$1369	7
ClI_2		$ClPI_2$	208	90
ClNO	C_2;C	$ClP(OMe)N(iPr)_2$	183.9	20
		$ClP(OMe)N*(CH_2)_3C*H_2$	182.7	20
		$ClP(OMe)N*CMe_2(CH_2)_3C*Me_2$	194.4	20
		$ClP(OMe)N*(CH_2)_2OCH_2C*H_2$	172.1	20
		$ClP(NEt_2)OCHTf_2$	172.8	47
ClNO/5	CP;C	$ClP*N(PCl_2)CH_2CH_2O*$	154;158	37
	C_2;C	$ClP*N(Me)CH_2CH_2O*$	168.3(b)	46
ClNP	C_2;ClN	$(ClP*N\text{-}iPr_2)_2$	127.7	140
ClNSi	C_2;Si_3	$ClP(Mpip)SiTms_3$	142.1	163
ClN_2	$(C_2)_2$	$ClP(NMe_2)_2$	160(n)	136
		$ClP(NEt_2)_2$	154(n)	136
		$ClP(N\text{-}iPr_2)_2$	140.8(r)	59
$ClN_2/4$	$(Si_2)_2$	**1**; R = Tms, R′ = Cl, R″ = NHEt	173.1/170.5	117
		1; R = Tms, R′ = Cl, R″ = NEt_2	173.2/169.9	117
		1; R = Tms, R′ = Cl, R″ = NH-tBu	173.2/172.2	117
		1; R = Tms, R′ = Cl, R″ = NMeTms	173.4/171.4	117
		1; R = Tms, R′ = Cl, R″ = Ph	186.3/202.0	117
		1; R = Tms, R′ = Cl, R″ = NHPh	174.1/169.9	117
		1; R = Tms, R′ = Cl, R″ = NPr_2	175.2/172.9	117
	Si_2,CSi	**1**; R = tBu, R′ = R″ = Cl	170.2	118
		1; R = tBu, R′ = Me, R″ = Cl	185.3/198.5	118
		1; R = tBu, R′ = Et, R″ = Cl	186.3/199.4	118
		1; R = tBu, R′ = R″ = Me	211.7	118
		1; R = tBu, R′ = Ph, R″ = Cl	191.2	118
	$(CP^4)_2$	ClP*-N(tBu)-P(S)(Cl)-N*tBu	145.5/157.3	148
	$(CP)_2$	**2**; R = R′ = Et, X = Y = Cl	227.3	152
	$(C'P)_2$	**2**; R = R′ = Ph, X = Y = Cl (cis)	208.7	149

TABLE B (continued)
Compilation of ³¹P NMR Data of Three Coordinate (λ3 σ3) Phosphorus Compounds Containing Phosphorus Bonds to Halogen

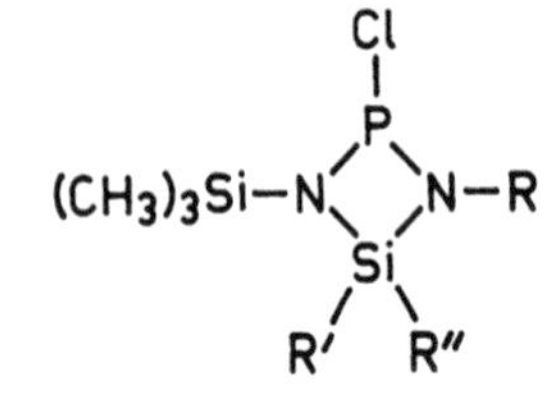

1 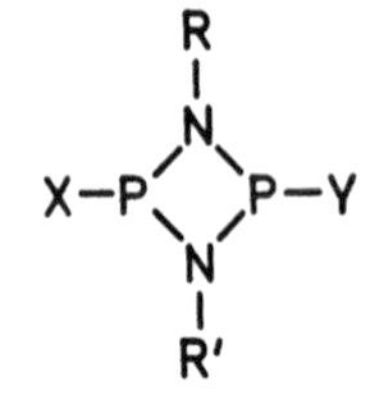 **2** 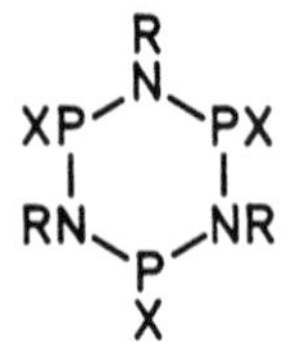 **3**  **4**

P-bound atoms/ rings	Connectivities	Structure	NMR data (δP[solv,temp]ⁿJ$_{PA}$ Hz)	Ref.
		2; R = R′ = Ph, X = MeO, Y = Cl	176.4	149
		2; R = R′ = Ph, X = EtO, Y = Cl	175.9	149
	$(CS^4)_2$	ClP*-N(tBu)-SO_2-N*(tBu)	107.5	124
ClN_2/5	CB,CN	ClP*N(Me)-B(Me)-N(Me)-N*(Me)	166.0;161.0	81, 160
	$(C'H)_2$	ClP*NHC(CN)=C(CN)N*H	136	19
	$(C_2)_2$	ClP(NMe-CH_2)$_2$	166.5(r) $^1J_{PN}$63.3	136
ClN_2/6	$(CB)_2$	ClP*N(Me)B(Me)N(Me)B(Me)N*(Me)	160.2	150
	$(CP)_2$	**3**; R = Me, X = Cl (cis/trans)	101.4/127.8/132.4	153
		3; R = Et, X = Cl (cis/trans)	104.1/129.0;135.4	151
		3; R = nPr, X = Cl (cis/trans)	105.3/130.6/136.5	151
		4; R = Et, X = Cl (A)	54.4/129.7	151
		4; R = Et, X = Cl (B)	119.4/36.9/130.8	151
	C_2,CC′	ClP*N(Me)C(O)CH_2CH_2N*(Me)	145.2	45
	CC′,C'_2	ClP*N(Ph)C(O)CH_2CH_2N*(Ph)	131.2	45
ClN'_2	$(C'')_2$	ClP(NCO)$_2$	128	83
		ClP(NCS)$_2$	114.0	83
ClN'_2/5	$(C')_2$	ClP(N=C*OMe)$_2$	143.6	127
		ClP(N=C*OCH_2Tf)$_2$	107.6	123
ClOS/5	C;C	ClP*SCH_2CH_2O*	104.2(b)	46
ClO_2	$(C)_2$	ClP(OMe)$_2$	169	100
		ClP(OEt)$_2$	165	111
		ClP(O$CH_2CF_2CHF_2$)$_2$	168	25
		ClP(O-iPr)$_2$	165.4	93
		ClP(O-tBu)$_2$	170.3	101
		ClP(1-Ad)	174.8	165
	$(C')_2$	ClP(OCH_2CH_2Cl)O-2,4,-$Cl_2C_6H_3$	15.9	146
		ClP(OPh)$_2$	157.3	108
		5	161.8	93
ClO_2/5	$(C)_2$	ClP-Glc	167(n)	110
	$(C')_2$	ClP-Cat	173.0	145
ClO_2/6	$(C)_2$	ClP*O$(CH_2)_3$O*	153(n)	110
		6	148.1	94
		ClP*O$CH_2CMe_2CH_2$O*	146.7	147
	$(C')_2$	**7**	151.2	144
ClO_2/8	$(C)_2$	ClP*O$(CH_2)_2$O$(CH_2)_2$O*	155(r)	131
	$(C')_2$	**8**	136.3	162
ClP_2/3	PSi,P_2	ClP*-P(Tms)-P*-$PTms_2$	10 $^1J_{PP}$-232	29
ClS_2	$(C)_2$	ClP(SMe)$_2$	188.2	56
		ClP)SEt)$_2$	188	35
		ClP(SCH_2CH=CH_2)$_2$	185.1	56
		ClP(S-iPr)$_2$	188	33

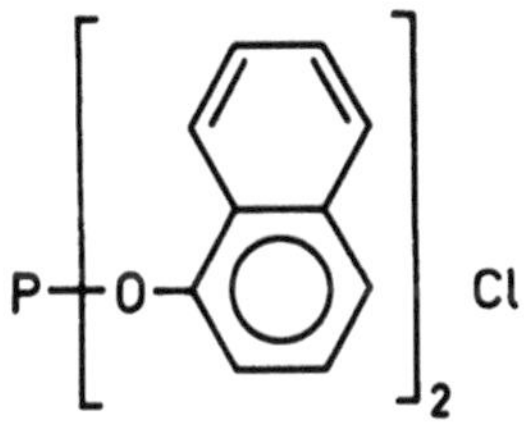

5  6 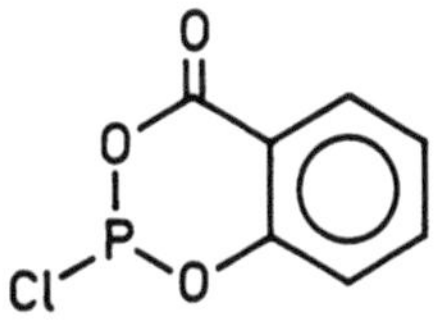 7

P-bound atoms/ rings	Connectivities	Structure	NMR data (δP[solv,temp]$^{n}J_{PA}$ Hz)	Ref.
		$ClP(SBu)_2$	185.5	96
		$ClP(S\text{-}tBu)_2$	173.2	93
		$ClP(SC_5H_{11})_2$	187.2	56
		$ClP(SC_7H_{15})_2$	187.7	56
		$ClP(SCH_2Ph)_2$	180.3	56
	$(C')_2$	$ClP(SPh)_2$	182.7	56
$ClS_2/5$	$(C)_2$	$ClP^*SCH_2CH_2S^*$	168.1(b)	46
Cl_2F		Cl_2PF	217.6(34°) $^1J_{PF}$1329	7
Cl_2I		Cl_2PI	211	90
Cl_2N	CP^4	$(Cl_2PNMe)_3PO$	162.9	14
		$(Cl_2PNMe)_2P(O)Ph$	167.1(cd)	13
		$(Cl_2PNMe)_2P(O)NHMe$	169.7	14
	CP	$(Cl_2P)_2NMe$	159.5(n) $^1J_{PN}$86.0	136
		$Cl_2PN^*CH_2CH_2OP^*(Cl)$	154;158	37
	C_2	Cl_2PNMe_2	166.0(n)	136
		Cl_2PNEt_2	162.0(n)	136
		$Cl_2PN^*(CH_2)_3C^*H_2$	164.5	113
	CC′	$Cl_2PN(Me)Ph$	158.3	91
	C'_2	Cl_2PNPh_2	151.3	92
Cl_2N'	C″	Cl_2PNCO	165.7	83
		Cl_2PNCS	155.3	83
Cl_2O	C	Cl_2POMe	180.5	110
		$Cl_2POCH_2CH_2Cl$	177.6	94
		Cl_2POEt	177.0	111
		$Cl_2PO(CH_2)_3OH$	178.6	93
		$Cl_2PO\text{-}iPr$	174.4	93
		$Cl_2POCH_2CH{=}CHCCl_3$	177.6	134
		Cl_2POBu	178.8	88
		$Cl_2PO(1\text{-}Ad)$	166.1	165
	C′	$Cl_2POC(tBu){=}CHCl$	201	38
		Cl_2POPh	173	108
		$Cl_2PO(2,4\text{-}Cl_2C_6H_3)$	182.8	146
		$Cl_2PO(4\text{-}MeC_6H_4)$	175.9	88
		9	184	93
Cl_2P	Cl_2	Cl_2PPCl_2	155	95
Cl_2S	C	Cl_2PSMe	206.0	56
		Cl_2PSCH_2STf	220.2	2
		Cl_2PSEt	210.7	56
		$Cl_2PSCH_2CH{=}CH_2$	210.1	56
		Cl_2PSPr	212.6	56
		$Cl_2PS\text{-}iPr$	211.0	56
		Cl_2PSBu	209.7	96
		$Cl_2PSC_5H_{11}$	210.4	56
		Cl_2PSCH_2Ph	205.5	56
		$Cl_2PSC_7H_{15}$	211.0	56
	C′	Cl_2PSPh	204.2	56

TABLE B (continued)
Compilation of ^{31}P NMR Data of Three Coordinate (λ3 σ3) Phosphorus Compounds Containing Phosphorus Bonds to Halogen

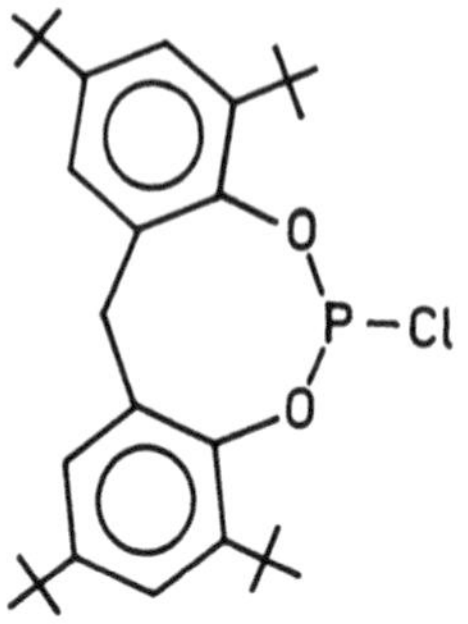

8 **9**

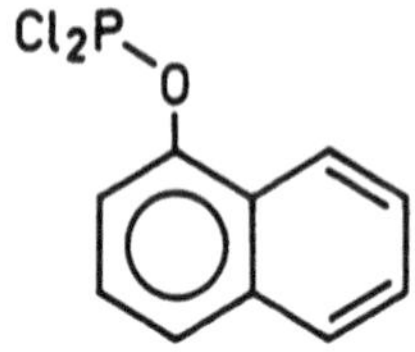

P-bound atoms/ rings	Connectivities	Structure	NMR data (δP[solv,temp]^aJ_{PA} Hz)	Ref.
		$Cl_2P(4\text{-}SC_6H_4Cl)$	198.2	94
Cl_3		PCl_3	219.5	38
FNO	C_2;C	$FP(NEt_2)OCH(Tf)_2$	153.0	47
		$Fp(NMe_2)OBu$	149.7 $^1J_{PF}$1117	15
	C_2;C′	$FP(NEt_2)OPh$	144.1 $^1J_{PF}$1135	15
FNO/6	C_2;C	$FP{*}N(CH_2CH_2Cl)(CH_2)_3O{*}$	158.7	126
FNSi	C_2;Si_3	FP(Mpip) Si Tms_3	215.3 $^1J_{PF}$870	163
FN_2	$(C_2)_2$	$FP(NMe_2)_2$	150.9(n) $^1J_{PF}$1046	15
		$FP(NEt_2)_2$	150.4 $^1J_{PF}$1038	15
FN_2/4	$(CS^4)_2$	$FP(N\text{-}tBu)_2SO_2$	89.5 $^1J_{PF}$1227.5	124
FN_2/6	$(CP)_2$	**2**; R = Et, X = Y = F (cis/trans)	110.4/111.2;117.8	151
		4; R = Et, X = F (A)	46.6/116.0	151
		4; R = Et, X = F (B)	106.8/35.4/119.3	151
FN'_2	$(C'')_2$	$FP(NCO)_2$	127.9 $^1J_{PF}$1226	83
		$FP(NCS)_2$	126.5 $^1J_{PF}$1270	70
FO_2	$(C)_2$	$FP(OBu)_2$	131.1 $^1J_{PF}$1204	15
		$FP(OCTf_2CN)_2$	90.1 $^1J_{PF}$1350	16
		$FP(OCTf_2CN)_2$ x $Fe(CO)_4$	16.7	16
	C,C′	FP(OBu)OPh	123.1 $^1J_{PF}$1230	15
	$(C')_2$	$FP(OPh)_2$	121.3 $^1J_{PF}$1264	15
FO_2/5	$(C)_2$	FP-Glc	123.8 $^1J_{PF}$1225	15
	$(C')_2$	FP-Cat	123.1 $^1J_{PF}$1305	103
FO_2/6	$(C)_2$	$FP{*}OCH_2CMe_2CH_2O{*}$	111.5	125
		$FP{*}OCH_2CMe(CH_2Cl)CH_2O{*}$	112.2/112.8	125
F_2I		F_2PI	242.2	82
F_2N	$(P)_2$	$(F_2P)_3N$	150.3(bd) $^1J_{PF}$1224	161
	HP	$(F_2P)_2NH$	144.4(bd) $^1J_{PF}$-1253	161
	HSi	$F_2PNHSiH_3$	150.4(bd) $^1J_{PF}$-1215	142
	H_2	F_2PNH_2	147.5(bd) $^1J_{PF}$-1200	161
	CO	$F_2PN{*}(CH_2)_4O{*}$	138.5 $^1J_{PF}$1199	15
	CP^4	$(F_2PNMe)_3PO$	134.3 $^1J_{PF}$1287	14
	CP	$(F_2P)_2NMe$	141.3	158
		$(F_2P)_2NMe$ x $Fe(CO)_4$	189.1/136.8 $^1J_{PF}$1177/1268	158
	CH	F_2PNHMe x $Fe(CO)_4$	193.7 $^1J_{PF}$1146	158
	C_2	F_2PNMe_2	143.2 $^1J_{PF}$1198	15
		$F_2PN{*}CH_2C{*}H_2$	140.5 $^1J_{PF}$1192	15
		$F_2PN{*}(CH_2)_3C{*}H_2$	146.2 $^1J_{PF}$1198	113

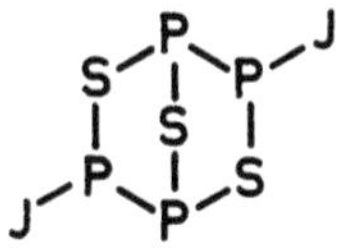

10

P-bound atoms/ rings	Connectivities	Structure	NMR data (δP[solv,temp]^{n}J$_{PA}$ Hz)	Ref.
		$F_2PN^*(CH_2)_4C^*H_2$	140.5 $^1J_{PF}$1199	113
		F_2PNEt_2	144.0 $^1J_{PF}$1195	106
	CC′	$F_2PNMePh$	138.1 $^1J_{PF}$1217	15
		$F_2PNEtPh$	137.1 $^1J_{PF}$1213	15
		$F_2PNBuPh$	136.5 $^1J_{PF}$1214	15
	C″	F_2PNCO	130.6 $^1J_{PF}$1310	83
		F_2PNCS	132 $^1J_{PF}$1340	70
F_2O	P^4	$(F_2PO)_2P(F)O$	106.8 $^1J_{PF}$1390	6
		$(F_2PO)_3PO$	107.9 $^1J_{PF}$1385	6
		$(F_2PO)_2P(O)Ph$	110.7 $^1J_{PF}$1363	6
		$F_2POP(O)Ph_2$	113.7 $^1J_{PF}$1358	6
	P	F_2POPF_2	114.6 $^1J_{PF}$1358	86
		$(F_2PO)_2POH$	108(27°) $^1J_{PF}$1375	9
		$(F_2PO)_3P$	112.3(27°) $^1J_{PF}$1349	9
		$F_2POP(OMe)_2$	104.8(27°) $^1J_{PF}$1322	9
		$F_2POP(OPh)_2$	114.1(27°) $^1J_{PF}$1335	9
	C	F_2POMe	111 $^1J_{PF}$1275	110
		$F_2POCH_2CH_2OPF_2$	112.0(n) $^1J_{PF}$1295	84
		$F_2POCH_2CH{=}CH_2$	111.9 $^1J_{PF}$1290	103
		F_2POPr	111.5 $^1J_{PF}$1287	84
		F_2POCTf_2CN x $Fe(CO)_4$	54.3 (vs. P_4O_6)	16
		F_2POBu	112.3 $^1J_{PF}$1288	15
		F_2PO-iBu	111.5 $^1J_{PF}$1280	137
		$F_2POC_5H_{11}$	111.6 $^1J_{PF}$1284	15
		$F_2POC_6H_{13}$	111.1 $^1J_{PF}$1285	15
		F_2POCH_2Ph	111.3 $^1J_{PF}$1288	137
		F_2POCH_2-4-ClC_6H_4	111.8 $^1J_{PF}$1290	137
		F_2POCH_2-4-$NO_2C_6H_4$	111.5 $^1J_{PF}$1293	137
		F_2POCH_2-4-MeC_6H_4	112.1 $^1J_{PF}$1282	137
		F_2POCH_2-4-$MeOC_6H_4$	111.7 $^1J_{PF}$1281	137
		F_2PO-1-Ad	122.7 $^1J_{PF}$1296	137
	C′	F_2POPh	110.1(n) $^1J_{PF}$1326	103
		$F_2POC_6H_4OPF_2$	109.8(n) $^1J_{PF}$1328	103
		$F_2P(O$-4-$MeC_6H_4)$	208 $^1J_{PF}$1154	15
F_2P	F_2	F_2PPF_2	226	104
	H_2	F_2PPH_2	293.7	82
F_3		F_3P	97 $^1J_{PF}$1400	110
IPS	$P;S_2$	**10**	128.9(s) $^1J_{PP}$243.6	41
I_2P	I_2	I_4P_2	105.6(s)	41
I_3		I_3P	178(s)	110
CBrCl	H_3	BrClPMe	190.0	56
	C_2H	BrClP-2-Nb(endo/exo)	191.3/183.9(c)	120
C′BrCl	C'_2	BrClPPh	158	69
CBrN	$F_3;C_2$	$BrP(Tf)NEt_2$	95.2	21
	$H_3;C_2$	$BrP(Me)NMe_2$	161.1(n)	56
C′BrP	$C'_2;BrC'$	BrP(Ph)P(Br)Ph	74.9;62.2(c,rac/meso)	5
	$C'_2;BrC'_2$	$(BrP^*(4\text{-}MeOC_6H_4))_2$	74.1;62.9(c,rac/meso)	5
		$(BrP^*(4\text{-}ClC_6H_4))_2$	70.8;59.8(c,rac/meso)	5
		$(BrP(4\text{-}FC_6H_4))_2$	71.1;60.0(c,rac/meso)	5

TABLE B (continued)
Compilation of ^{31}P NMR Data of Three Coordinate ($\lambda 3\ \sigma 3$) Phosphorus Compounds Containing Phosphorus Bonds to Halogen

P-bound atoms/ rings	Connectivities	Structure	NMR data ($\delta P[solv,temp]^nJ_{PA}$ Hz)	Ref.
CBr_2	HCl_2	Br_2PCHCl_2	142.5	26
	H_2Cl	Br_2PCH_2Cl	150.6	26
	H_3	Br_2PMe	184.0	56
	CBrCl	$Br_2PCBrClMe$	144.2(g)	78
	CBr_2	Br_2PCBr_2Me	144.0(g)	78
	CH_2	Br_2PEt	194.7(n)	26
		$Br_2PCH_2CH(Br)Me$	183	71
	C_2H	$Br_2P(iPr)$	201.5	26
		$Br_2P(C_6H_{11})$	190.5	26
		Br_2P(-)menthyl	214.0	26
	C'_2H	$Br_2P(cC_5H_5)$	145.3	28
	C_3	$Br_2P(tBu)$	203.8	26
	CC'_2	$Br_2P(cMe_5C_5)$	120.5	28
$C'Br_2$	C'H	$Br_2PCH{=}C(Et)OEt$	158.7	24
		$Br_2PCH{=}C(iPr)OEt$	159.1	24
		$Br_2PCH{=}C(Bu)OBu$	157.6	24
	C'_2	$Br_2PC_6F_5$	113.5	64
		$Br_2P(2\text{-}BrC_6H_4)$	146.7	26
		$Br_2P(3\text{-}BrC_6H_4)$	147.2;157(n)	26;53
		$Br_2P(4\text{-}BrC_6H_4)$	148.9	26
		$Br_2P(2\text{-}ClC_6H_4)$	144.0	26
		$Br_2P(3\text{-}ClC_6H_4)$	146.7	26
		$Br_2P(4\text{-}ClC_6H_4)$	147.2(c)	5
		$Br_2P(2\text{-}FC_6H_4)$	135.9	26
		$Br_2P(3\text{-}FC_6H_4)$	141.7	26
		$Br_2P(4\text{-}FC_6H_4)$	147.7(c)	5
		Br_2PPh	150.7(c)	5
		$Br_2P(4\text{-}MeOC_6H_4)$	151.7(c)	5
		$Br_2P(4\text{-}Me_2NC_6H_4)$	156.1	26
CClF	H_3	ClFPMe	240	55
CClN	$ClSi_2;C_2$	$ClP(NMe_2)C(Cl)Tms_2$	145.2(c)	4
	$F_3;C_2$	$ClP(Tf)NEt_2$	95.7	21
	$H_2Cl;C_2$	$ClP(CH_2Cl)NMe_2$	123	98
	$H_3;CP^4$	$(ClPMeNMe)_2P(O)Ph$	132.1(c)	13
	$H_3;C_2$	$ClP(Me)NMe_2$	150.7(c)	136
		$ClP(Me)NEt_2$	143.0(n)	56
	$C_2H;Si_2$	$ClP(iPr)N(Tms)_2$	162.6(c)	12
	$C_3;C_2$	$ClP(tBu)NEt_2$	156.7	47
C'ClN	$C'_2;CP^4$	$(ClPPhNMe)_2P(O)NHMe$	128.2 $^1J_{PP}$76	132
		$(ClPPhNMe)_2P(O)Ph$	120.1(c)	13
		$(ClPPhNMe)_3P(O)$	125.2 $^1J_{PP}$80	132
	$C'_2;CH$	ClP(Ph)NH(tBu)	118	99
	$C'_2;C_2$	$ClP(Ph)NMe_2$	141.0(n)	98
		$ClP(Ph)NEt_2$	140.4(n)	98
CClO	$C_3;C$	ClP(OEt)tBu	207.2(b)	1
		ClP(OiPr)tBu	203.5(b)	1
		$(ClP(tBu)OCH_2CH_2)_2$	207.9	130
CClP	$Si_3;CCl$	$ClP(CTms_3)P(Cl)CTms_3$	144.5	27
	$Si_3;CH$	$ClP(CTms_3)P(H)CTms_3$	173.6;154.1	11
	$C_3;CCl$	$ClP(tBu)P(Cl)CMe_3$	106.7;113.8	141
CClS	$C_3;C$	$(ClP(tBu)SCH_2)_2CH_2$	182.3/182.6(b)	129
CCl_2	$ClSi_2$	$Cl_2PC(Cl)Tms_2$	180.6(c)	4
	Cl_3	Cl_2PCCl_3	147	87

P-bound atoms/ rings	Connectivities	Structure	NMR data (δP[solv,temp]$^nJ_{PA}$ Hz)	Ref.
	Si_3	Cl_2PCTms_3	233(b)	8
	H_2Cl	Cl_2PCH_2Cl	158.9	56
	H_2P	$Cl_2PCH_2PCl_2$	147.7	138
		Cl_2PCH_2PCl-tBu	185.3 $^1J_{PP}$101	138
	H_3	Cl_2PMe	192.0	57
	CHCl	$Cl_2PCH(Cl)Me$	161.6	58
	CH_2	$Cl_2PCH_2CH_2Cl$	185.0	59
		Cl_2PEt	196.3(n)	56
		$Cl_2P(CH_2)_3Cl$	182(n)	110
		Cl_2PPr	201	63
		$Cl_2P(CH_2)_3PCl_2$	192	68
		$Cl_2P(CH_2)_4PCl_2$	192	68
		Cl_2PBu	194.9(n)	60
		Cl_2P-IBu	200	89
		$Cl_2PC_6H_{11}$	204.0	88
		$Cl_2PC_{10}H_{11}$	163.4	67
		$Cl_2PCH_2CH(Cl)Ph$	182.8	30
	C_2H	Cl_2P-2-Nb(endo/exo)	196.8/187.5(c)	120
		Cl_2P-7-Nb	199.7(c)	119
		Cl_2P-7-Nbe	190.9/199.7(c)	119
	C'_2H	$Cl_2P(cC_5H_5)$	155	28
		$Cl_2P(cC_5H_5)$ x $W(CO)_5$	125	28
	CC'_2	$Cl_2P(cMe_5C_5)$	125.5	28
$C'Cl_2$	C'H	$Cl_2PCH{=}CHOEt$	170	32
	C'H	$Cl_2PCH{=}CHOBu$	167	24
		$Cl_2PCH{=}CHPh$	163.5	30
		$Cl_2PCH{=}CHOPh$	163.1	30
		$Cl_2PCH{=}C(iPr)OEt$	169.2	24
	CC'	**11**	155.1	30
	C'_2	$Cl_2PC_6F_5$	137.0	64
		$Cl_2P(2\text{-}BrC_6H_4)$	153.6	26
	C'_2	$Cl_2P(3\text{-}BrC_6H_4)$	157.0	26
	C'_2	$Cl_2P(4\text{-}BrC_6H_4)$	158.4	26
	C'_2	$Cl_2P(2\text{-}ClC_6H_4)$	151.8	26
	C'_2	$Cl_2P(3\text{-}ClC_6H_4)$	156.9	26
	C'_2	$Cl_2P(4\text{-}ClC_6H_4)$	158.3	26
	C'_2	$Cl_2P(2\text{-}FC_6H_4)$	147.9	26
	C'_2	$Cl_2P(3\text{-}FC_6H_4)$	152	65
	C'_2	$Cl_2P(4\text{-}FC_6H_4)$	158.8	80
	C'_2	Cl_2PPh	166	66
	C'_2	$Cl_2P(2,5\text{-}Me_2C_6H_3)$	169	88
	C'_2	$Cl_2P(4\text{-}NMe_2C_6H_4)$	164.5	80
	C'_2	**12**	163.4/165.0	88
$C''Cl_2$	C''	$Cl_2PC{\equiv}CPCl_2$	108.6	133
CFN	$Si_3;Si_2$	$FP(CTms_3)NTms_2$	245.5	62
	Si_3;CSi	$FP(CTms_3)N$-tBuTms	207.7	62
		$FP(CTms_3)NMesTms$	236.0	62
		$FP(CTms_3)NAdTms$	208.5	62
	$H_3;C_2$	$FP(Me)NMe_2$	168.9 $^1J_{PF}$917	72
	$CHSi;Si_2$	**13**; R = $NTms_2$, R' = H, X = F	187.5(b)	154
	CC'Si;CSi	**13**; R = N(tBu)Tms,R' = Me, X = F	178.1(b) $^1J_{PF}$868	154
	$C_3;C_2$	$FP(NEt_2)tBu$	181.0	47
	$C'_2:C_2$	$FP(NMe_2)Ph$	159.8 $^1J_{PF}$985	109
		$FP(NEt_2)Ph$	156.0 $^1J_{PF}$987	98
C''FN	$C'';C_2$	$F(NEt_2)PC{\equiv}CP(NEt_2)F$	123.2(b) $^1J_{PF}$1000.2	40
		F(morph)PC≡CP(morph)F	120.8(r) $^1J_{PF}$996	40
CFO	C'_2;C	$FP(Ph)OCHTf_2$	229 $^1J_{PF}$1116	135
CF_2	Cl_3	F_2PCCl_3	131 $^1J_{PF}$1290	113

TABLE B (continued)
Compilation of ^{31}P NMR Data of Three Coordinate (λ3 σ3) Phosphorus Compounds Containing Phosphorus Bonds to Halogen

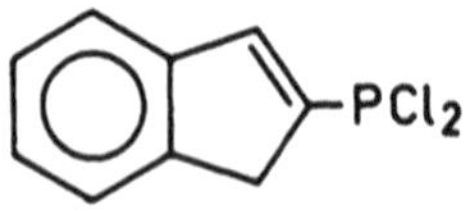

PCl_2 Cl_2P t-C_4H_9, N—N, $(CH_3)_2Si$, CH_3, R′ PXR   P—Cl

11 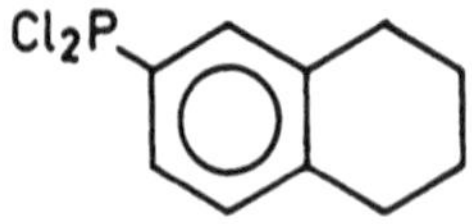   **12** 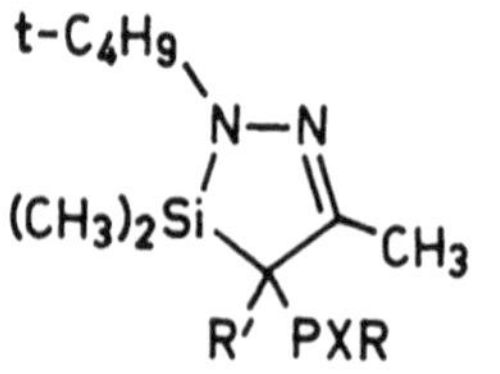   **13** 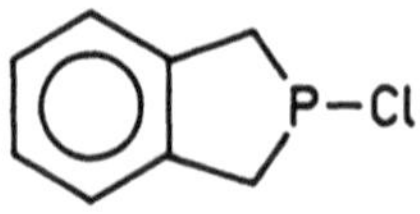   **14**

P-bound atoms/ rings	Connectivities	Structure	NMR data (δP[solv,temp]^{n}J$_{PA}$ Hz)	Ref.
	F_3	F_2PTf	158.3 $^1J_{PF}$1245	102
	H_2Cl	F_2PCH_2Cl	201.8 $^1J_{PF}$1203	109
	H_3	F_2PMe	244.2	50
	CH_2	F_2PEt	234.1	51
		F_2PPr	167.8 $^1J_{PF}$1250	102
		F_2P-nBu	242.3 $^1J_{PF}$1162	15
	$C'H_2$	F_2PCH_2Ph	223.8	54
	C'_2H	$F_2P(cC_5H_5)$	195	28
	CC'_2	$F_2P(cMe_5C_5)$	159.5	28
$C'F_2$	C'_2	$F_2PC_6F_5$	193.4	52
		$F_2P(4\text{-}ClC_6H_4)$	196.8	51
		F_2PPh	208.3 $^1J_{PF}$1174	109
		F_2PTol (o/p mixt.)	208.3/207.8 $^1J_{PF}$1161	15
		$F_2P(4\text{-}Tol)$	206.4	54
$C''F_2$	$C''P_3$	$F_2PC{\equiv}CPF_2$	165.0 $^1J_{PF}$1222.2	40
ClN	$F_3;C_2$	$IP(NEt_2)Tf$	86.6	21
CI_2	H_3	I_2PMe	130.6	105
	CH_2	I_2PEt	147.0(n)	39
		I_2PBu	140.3(n)	39
$C'I_2$	C'_2	I_2PPh	100.1(n)	39
CHCl	Si_3	$ClP(H)C(Tms)_3$	70.5 $^1J_{PH}$160.6	11
C'HCl	C'_2	$ClP(H)2,4,6\text{-}(tBu)_3C_6H_2$	26.4 $^1J_{PH}$215.0	11
C_2Br	$(F_3)_2$	$BrP(Tf)_2$	33.7	85
	$(H_3)_2$	$BrPMe_2$	87.9(n)	56
	H_3,CH_2	$BrP(Me)Et$	98.5	79
	$(CH_2)_2$	$BrPEt_2$	116.2	56
	$(C_2H)_2$	$BrP(iPr)_2$	126.9	26
$C_2Br'5$	$(CH_2)_2$	$BrP^*CH_2CH{=}CHC^*H_2$	111.4(n)	36
		$BrP^*CH_2C(Me){=}C(Me)C^*H_2$	104.8(n)	48
	C_2H,CH_2	$BrP^*CH{=}C(Me)CH_2C^*H_2$	130.6(n)	48
CC'Br	$H_3;C'_2$	$BrP(Me)Ph$	77.0	56
C'_2Br	$(C'_2)_2$	$BrP(C_6F_5)_2$	13.0	64
		$BrP(Ph)C_6F_5$	39.3	64
		$BrPPh_2$	72.9	26
		$BrP(2,6\text{-}Me_2C_6H_3)_2$	72.2	31
C_2Cl	$(F_3)_2$	$ClP(Tf)_2$	50.0	76
	HS_2,C_3	$ClP(tBu)CH(S\text{-}iBu)_2$	119.4	34
	HSi_2,CC'_2	$ClP(cC_5Me_5)CHTms_2$	171.0	10
	H_2Cl,H_3	$ClP(CH_2Cl)Me$	99.4	97
	H_2P,C_3	$ClP(tBu)CH_2PCl_2$	107.2 $^1{}_{PP}$101	138
		$ClP(tBu)CH_2PCl(tBu)$	114.8	138
		$ClP(tBu)CH_2P(tBu)_2$	125.4 $^2J_{PP}$137.3	138

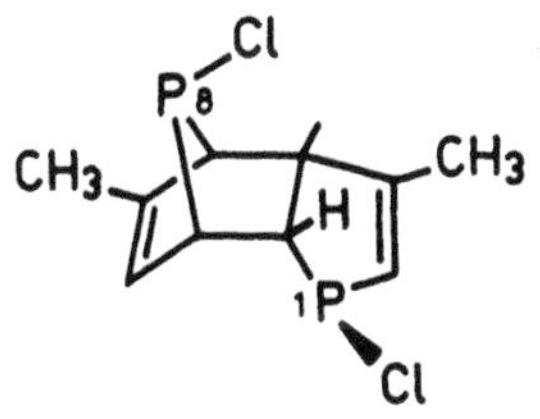

15 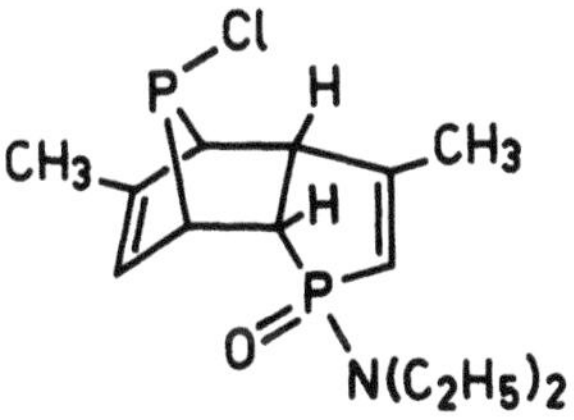 **16** 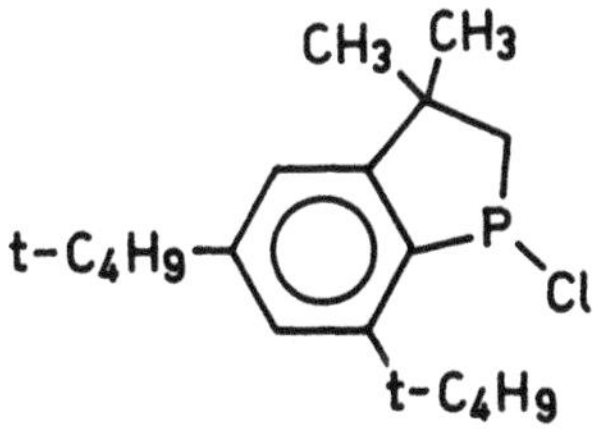 **17**

P-bound atoms/ rings	Connectivities	Structure	NMR data (δP[solv,temp]^{a}J$_{PA}$ Hz)	Ref.
	$(H_3)_2$	$ClPMe_2$	96	77
	H_3,CH_2	ClP(Me)Et	105.2	56
	H_3,C_2H	ClP(Me)iBu	95.5(c)	44
	H_3,C_3	ClP(Me)tBu	117.5	43
		$ClP(Me)C(Me)_2CHMe_2)$	109.2(r)	49
	$C'HSi,C_3$	**3**; R = tBu, R′ = H, X = Cl	58.7(b)	
	$(CH_2)_2$	$ClPEt_2$	119.0	56
	CH_2,C_2H	$ClP(iPr)(CH_2)_2P(Cl)iPr$	121.4(b)	1
	CH_2,C_3	ClP(Et)tBu	132.7(c)	44
	$(C_2H)_2$	$ClP(cPr)_2$	123.3	17
		ClP(iPr)iBu	118.9(c)	44
	C_2H,C_3	ClP(iPr)tBu	140.8(c)	44
	C_2H,C_3	$ClP(tBu)(CH_2)_2P(Cl)tBu$	128.1;130.6(b,rac/meso)	1
	C_3,CC'_2	$ClP(tBu)(cC_5Me_5)$	168	10
$C_2Cl/5$	$(CH_2)_2$	$ClP{*}CH_2C(Me){=}CHC{*}H_2$	127.5(n)	48
		$ClP{*}CH_2C(Me){=}C(Me)C{*}H_2$	111.6(c)	18
	$CH_2,C'H$	$ClP{*}CH{=}C(Me)CH_2C{*}H_2$	132.5(n)	48
	$(C'H_2)_2$	**14**	125.0	123
	$(CC'H)_2$	**15**	54.5(P-8) $^3J_{PP}$7.3	121
		16	43.6	121
CC′Cl	$ClSi_2;C'_2$	$ClP(Ph)C(Cl)Tms_2$	93.9(c)	4
	$H_3;C'_2$	ClP(Me)Ph	83.4(n)	61
	$CH_2;C'_2$	ClP(Et)Ph	97.0	56
	$C_2H;C'_2$	ClP(iBu)Ph	83.7(c)	44
	$C'_2H;C'_2$	$ClP(cC_5H_5)Ph$	85.5	28
		$ClP(cC_5H_5)Ph \times W(CO)_5$	92.6	28
	$C_3;OP$	ClP(tBu)C(OTms)=P(tBu)	106.3	3
	$C_3;C'_2$	ClP(tBu)Ph	107	42
CC′Cl/5	$CH_2;C'_2$	**17**	111.5(c)	155
	$C_2H;C'H$	**15**	124.3(P-1) $^3J_{PP}$7.3	121
CC″Cl	$C_3;C''$	ClP(tBu)C≡CPh	82(c)	166
C'_2Cl	$(C'_2)_2$	$ClP(C_6F_5)_2$	37	64
		$ClP(C_6F_5)Ph$	57.1	64
		$ClP(3\text{-}FC_6H_4)_2$	74.7	65
		$ClP(4\text{-}FC_6H_4)_2$	79	65
		$ClPPh_2$	81.5	66
		$ClPPh_2 \times Fe(CO)_4$	164.7	139
C_2F	$(F_3)_2$	$FPTf_2$	123.9 $^1J_{PF}$1013	85
	$(H_3)_2$	$FPMe_2$	187	73
	H_3,CH_2	FP(Me)Et	182	74
	$(CH_2)_2$	$FPPr_2$	138.8 $^1J_{PF}$1020	102
	$(C'_2H)_2$	$FP(cC_5H_5)_2$	192	28
C'_2F	$(C'_2)_2$	$FP(C_6F_5)_2$	136.0	75
		$FPPh_2$	162.3 $^1J_{PF}$883	15
C_2I	$(F_3)_2$	$IPTf_2$	0.8 $^2J_{PF}$73.2	76
	$(H_3)_2$	$IPMe_2$	60.6(re)	39
	$(CH_2)_2$	$IPEt_2$	77.5(re)	39

REFERENCES

1. **Weisheit, R., Stendel, R., Messbauer, B., Langer, C., and Walther, B.,** *Z. Anorg. Allg. Chem.*, 504, 147, 1983.
2. **Mason, M. G., Swepston, P. N., and Ibers, J. A.,** *Inorg. Chem.*, 22, 411, 1983.
3. **Appel, R. and Paulen, W.,** *Chem. Ber.*, 116, 109, 1983.
4. **Appel, R., Huppertz, M., and Westerhaus, A.,** *Chem. Ber.*, 116, 114, 1983.
5. **Hinke, A. and Kuchen, W.,** *Chem. Ber.*, 116, 3003, 1983.
6. **Ebsworth, E. A. V., Hunter, G. M., and Rankin, D. W. H.,** *J. Chem. Soc. Dalton Trans.*, 245, 1983.
7. **Dillon, K. B. and Platt, A. W. G.,** *J. Chem. Soc. Dalton Trans.*, 1159, 1983.
8. **Eaborn, C., Retta, N., and Smith, J. D.,** *J. Chem. Soc. Dalton Trans.*, 905, 1983.
9. **Ebsworth, E. A. V., Hunter, G. M., and Rankin, D. W. H.,** *J. Chem. Soc. Dalton Trans.*, 245, 1983.
10. **Cowley, A. H. and Mehrota, S. K.,** *J. Am. Chem. Soc.*, 105, 2074, 1983.
11. **Cowley, A. H., Kilduff, J. E., Norman, N. C., and Pakulski, M.,** *J. Am. Chem. Soc.*, 105, 4845, 1983.
12. **O'Neal, H. R. and Neilson, R. H.,** *Inorg. Chem.*, 22, 814, 1983.
13. **Kleemann, S. and Fluck, E.,** *Z. Anorg. Allg. Chem.*, 516, 79, 1984.
14. **Fluck, E. and Kleemann, S.,** *Z. Anorg. Allg. Chem.*, 461, 187, 1980.
15. **Riesel, L., Sturm, D., Nagel, A., Taudien, S., Beuster, A., and Karwatzki, A.,** *Z. Anorg. Allg. Chem.*, 542, 157, 1986.
16. **Bauer, D. P. and Ruff, J. K.,** *Inorg. Chem.*, 22, 1686, 1983.
17. **Schmidbaur, H. and Schier, A.,** *Synthesis*, 372, 1983.
18. **Quin, L. D. and Szewczyk, J.,** *Phosphorus Sulfur*, 21, 161, 1984.
19. **Karaghiosoff, K., Majoral, J. P., Meriem, A., Navech, J., and Schmidpeter, A.,** *Tetrahedron Lett.*, 24, 2137, 1983.
20. **McBride, L. J. and Caruthers, M. H.,** *Tetrahedron Lett.*, 24, 245, 1983.
21. **Volbach, W. and Ruppert, I.,** *Tetrahedron Lett.*, 24, 5509, 1983.
22. **Aleinikov, S. F., Krutikov, V. I., Golovanov, A. V., Maslennikov, I. G., and Lavrent'ev,** *Zh. Obshch. Khim.*, 53, 1678, 1983.
23. **Sinyashin, O. G., Kostin, V. P., Batyeva, E. S., Pudovik, A. N., and Ivasyuk, N. V.,** *Zh. Obshch. Khim.*, 53, 1706, 1983.
24. **Trostyanskaya, I. G., Efimova, I. V., Kazankova, M. A., and Lutsenko, I. F.,** *Zh. Obshch. Khim.*, 53, 236, 1983.
25. **Konovalova, I. V., Mikhailova, N. V., Burnaeva, L. A., and Pudovik, A. N.,** *Zh. Obshch. Khim.*, 53, 1715, 1983.
26. **Hinke, A. and Kuchen, W.,** *Phosphorus Sulfur*, 15, 93, 1983.
27. **Escudie, J., Couret, C., Ranaivonjatovo, H., Satge, J., and Jaud, J.,** *Phosphorus Sulfur*, 17, 221, 1983.
28. **Deschamps, B. and Mathey, F.,** *Phosphorus Sulfur*, 17, 317, 1983.
29. **Fritz, G. and Haerer, J.,** *Z. Anorg. Allg. Chem.*, 500, 14, 1983.
30. **Timokhin, B. V., Dmitriev, V. I., Vengel'nikova, V. N., Donskikh, V. I., and Kalabina, A. V.,** *Zh. Obshch. Khim.*, 53, 291, 1983.
31. **Schmidbaur, H., Schnatterer, S., Dash, K. C., and Aly, A. A. M.,** *Z. Naturforsch.*, 38b, 62, 1983.
32. **Gazizov, M. B., Moskva, V. V., and Khairullin, R. A.,** *Zh. Obshch. Khim.*, 53, 2142, 1983.
33. **Sinyashin, O. G., Kostin, V. P., Batyeva, E. S., and Pudovik, A. N.,** *Zh. Obshch. Khim.*, 53, 472, 1983.
34. **Kolodyazhnyi, O. I., Shevchenko, I. V., and Kukhar, V. P.,** *Zh. Obshch. Khim.*, 53, 473, 1983.
35. **Sinyashin, O. G., Kostin, V. P., Batyeva, E. S., and Pudovik, A. N.,** *Zh. Obshch. Khim.*, 53, 502, 1983.
36. **King, R. B. and Sadanani, N. D.,** *J. Org. Chem.*, 50, 1719, 1985.
37. **Pudovik, M. A., Mironova, T. A., and Pudovik, A. N.,** *Zh. Obshch. Khim.*, 53, 2464, 1983.
38. **Kolodka, T. V., Loktionova, R. A., and Gololobov, Yu. G.,** *Zh. Obshch. Khim.*, 53, 2476, 1983.
39. **Kudryavtseva, L. I., Feshchenko, N. G., and Povolotskii, M. I.,** *Zh. Obshch. Khim.*, 53, 2684, 1983.
40. **Svara, J., Fluck, E., Stezowski, J. J., and Maier, A.,** *Z. Anorg. Allg. Chem.*, 545, 47, 1987.
41. **Buadler, M., Kloth, B., Koch, D., and Tolls, E.,** *Z. Naturforsch.*, 30b, 340, 1975.
42. **Foss, V. L., Solodenko, V. A., Veits, Yu. A., and Lutsenko, I. F.,** *Zh. Obshch. Khim.*, 49, 1724, 1979.
43. **Haegele, G., Rossing, G., Kueckelhaus, W., and Seega, J.,** *Z. Naturforsch.*, 39b, 1574, 1984.
44. **Wolfsberger, W.,** *Chemiker-Ztg.*, 110, 449, 1986.
45. **Nifantiev, E. E., Zavalishina, A. I., and Smirnova, E. I.,** *Phosphorus Sulfur*, 10, 261, 1981.
46. **Denney, D. B., Denney, D. Z., and Liu, Lun-Tsu,** *Phosphorus Sulfur*, 22, 71, 1985.

47. **Dakternieks, D. and Giacomo, R. Di.,** *Phosphorus Sulfur,* 24, 217, 1985.
48. **Quin, L. D. and Myers, D. K.,** *J. Org. Chem.,* 36, 1285, 1971.
49. **Symmes Jr., C. and Quin, L. D.,** *J. Org. Chem.,* 43, 1250, 1978.
50. **Drozd, G. I., Ivin, S. Z., and Sheluchenko, V. V.,** *Zh. Vses. Khim. Obshch.,* 12, 474, 1967, *Chem. Abstr.,* 67, 108705f, 1967.
51. **Drozd, G. I., Ivin, S. Z., Sheluchenko, V. V., Tetel'baum, B. I., Luganskii, G. M., and Varshavskii, A. D.,** *Zh. Obshch. Khim.,* 37, 1343, 1967.
52. **Fild, M. and Schmutzler, R.,** *J. Chem. Soc. (A),* 840, 1969.
53. **Quin, L. D., Gratz, J. P., and Barket, T. P.,** *J. Org. Chem.,* 33, 1034, 1968.
54. **Drozd, G. I., Ivin, S. Z., Sheluchenko, V. V., and Tetel'baum, B. I.,** *Zh. Obshch. Khim.,* 37, 958, 1967. *Chem. Abstr.,* 68, 39735x, 1968.
55. **Sheluchenko, V. V., Dubov, S. S., Drozd, G. I., and Ivin, S. Z.,** *Zh. Strukt. Khim.,* 9, 909, *Chem. Abstr.,* 70, 24524v, 1969.
56. **Moedritzer, K., Maier, L., and Groenweghe, L. C. D.,** *J. Chem. Eng. Data,* 7, 307, 1962.
57. **Maier, L.,** *Helv. Chim. Acta,* 47, 2137, 1964.
58. **Maier, L.,** *Helv. Chim. Acta,* 52, 1337, 1969.
59. **King, R. B. and Sundaram, P. M.,** *J. Org. Chem.,* 49, 1784, 1984.
60. **Quin, L. D., Gordon, M. D., and Lee, S. O.,** *Org. Magn. Reson.,* 6, 503, 1974.
61. **Maier, L.,** *J. Inorg. Nucl. Chem.,* 24, 1073, 1962.
62. **Haase, M., Klingebiel, U., and Skoda, L.,** *Z. Naturforsch.,* 39b, 1500, 1984.
63. **Fields, R., Hazeldine, R. N., and Wood, N. F.,** *J. Chem. Soc. (C),* 1370, 1970.
64. **Fild, M., Glemser, O., and Hollenberg, I.,** *Z. Naturforsch.,* 21b, 920, 1966.
65. **De Ketelaere, R., Muylle, E., Vanermen, W., Claeys, E., and Van der Kelen, G. P.,** *Bull. Soc. Chim. Belg.,* 78, 219, 1969.
66. **Fluck, E. and Binder, H.,** *Z. Anorg. Allg. Chem.,* 354, 139, 1967.
67. **Houalla, D. R. and Wolf, R.,** *Bull. Soc. Chem.,* 1152, 1963.
68. **Sommer, K.,** *Z. Anorg. Allg. Chem.,* 376, 37, 1970.
69. **Finch, A., Gardner, P. J., and Sen Gupta, K. K.,** *J. Chem. Soc. (B),* 1162, 1966.
70. **Roesky, H. W.,** *Chem. Ber.,* 100, 2142, 1967.
71. **Fontal, B. and Goldwhite, H.,** *J. Org. Chem.,* 31, 3804, 1966.
72. **Schmutzler, R.,** *Angew. Chem.,* 76, 570, 1964.
73. **Seel, F. and Rudolph, K. H.,** *Z. Anorg. Allg. Chem.,* 363, 233, 1968.
74. **Sheluchenko, V. V., Dubov, S. S., Drozd, G. I., and Ivin, S. Z.,** *Zh. Strukt. Khim.,* 9, 909, 1968; *Chem. Abstr.,* 70, 24524v, 1969.
75. **Fild, M. and Schmutzler, R.,** *J. Chem. Soc. (A),* 840, 1969.
76. **Packer, K. J.,** *J. Chem. Soc.,* 960, 1963.
77. **Seel, F., Gombler, W., and Rudolph, K. H.,** *Z. Naturforsch.,* 23b, 387, 1968.
78. **Bauer, G., Haegele, G., and Sartori, P.,** *Phosphorus and Sulfur,* 8, 95, 1980.
79. **Maier, L.,** *Helv, Chim. Acta,* 47, 2137, 1964.
80. **Grabiak, R. C., Miles, J. A., and Schwenzer, G. M.,** *Phosphorus Sulfur,* 9, 197, 1980.
81. **Barlos, K., Noeth, H., Wrackmeyer, B., and McFarlane, W.,** *J. Chem. Soc. Dalton Trans.,* 801, 1979.
82. **Rudolph, R. W. and Schiller, H. W.,** *J. Am. Chem. Soc.,* 90, 3581, 1968.
83. **Fluck, E.,** *Z. Naturforsch.,* 19b, 869, 1964.
84. **Schmutzler, R.,** *Chem. Ber.,* 96, 2435, 1963.
85. **Packer, K. J.,** *J. Chem. Soc.,* 960, 1963.
86. **Riess, J. G. and Van Wazer, J. R.,** *J. Am. Chem. Soc.,* 88, 2341, 1966.
87. **Nixon, J. F.,** *J. Inorg. Nucl. Chem.,* 27, 1281, 1965.
88. **Crutchfield, M. M., Dungan, C. H., Letcher, J. H., Mark, V., and Van Wazer, J. R., Eds.,** *P^{31} Nuclear Magnetic Resonance,* Topics in Phosphorus Chemistry, Vol. 5, John Wiley & Sons, New York, 1967, Ref. 341.
89. **Crutchfield, M. M., Dungan, C. H., Letcher, J. H., Mark, V., and Van Wazer, J. R., Eds.,** *P^{31} Nuclear Magnetic Resonance,* Topics in Phosphorus Chemistry, Vol. 5, John Wiley & Sons, New York, 1967, Ref. 283.
90. **Crutchfield, M. M., Dungan, C. H., Letcher, J. H., Mark, V., and Van Wazer, J. R., Eds.,** *P^{31} Nuclear Magnetic Resonance,* Topics in Phosphorus Chemistry, Vol. 5, John Wiley & Sons, New York, 1967, Ref. 335.
91. **Crutchfield, M. M., Dungan, C. H., Letcher, J. H., Mark, V., and Van Wazer, J. R., Eds.,** P^{31} Nuclear Magnetic Resonance, *Topics in Phosphorus Chemistry, Vol. 5, John Wiley & Sons, New York,*
92. **Nielsen, M. L., Pustinger, J. V., and Strobel, J.,** *J. Chem. Eng. Data,* 9, 167, 1964.
93. **Crutchfield, M. M., Dungan, C. H., Letcher, J. H., Mark, V., and Van Wazer, J. R., Eds.,** *P^{31} Nuclear Magnetic Resonance,* Topics in Phosphorus Chemistry, Vol. 5, John Wiley & Sons, New York, 1967, Ref. 316.

94. **Crutchfield, M. M., Dungan, C. H., Letcher, J. H., Mark, V., and Van Wazer, J. R., Eds.,** *P^{31} Nuclear Magnetic Resonance,* Topics in Phosphorus Chemistry, Vol. 5, John Wiley & Sons, New York, 1967, Ref. 288.
95. **Sandoval, A. A. and Moser, H. C.,** *Inorg. Chem.,* 2, 27, 1963.
96. **Crutchfield, M. M., Dungan, C. H., Letcher, J. H., Mark, V., and Van Wazer, J. R., Eds.,** *P^{31} Nuclear Magnetic Resonance,* Topics in Phosphorus Chemistry, Vol. 5, John Wiley & Sons, New York, London, Sidney, 1967, Ref. 290.
97. **Crutchfield, M. M., Dungan, C. H., Letcher, J. H., Mark, V., and Van Wazer, J. R., Eds.,** *P^{31} Nuclear Magnetic Resonance,* Topics in Phosphorus Chemistry, Vol. 5, John Wiley & Sons, New York, 1967, Ref. 308.
98. **Schmutzler, R.,** *J. Chem. Soc.,* 5630, 1965.
99. **Hart, W. A. and Sisler, H. H.,** *Inorg. Chem.,* 3, 617, 1964.
100. **Mark, V., Dungan, M., Crutchfield, M. M., and Van Wazer, J. R.,** *P^{31} Nuclear Magnetic Resonance,* Topics in Phosphorus Chemistry, Vol. 5, John Wiley & Sons, New York, 1967, Ref. 314.
101. **Mark, V. and Van Wazer, J. R.,** *J. Org. Chem.,* 29, 1006, 1964.
102. **Nixon, J. F. and Schmutzler, R.,** *Spectrochim. Acta,* 20, 1835, 1964.
103. **Reddy, G. S. and Schmutzler, R.,** *Z. Naturforsch.,* 20b, 104, 1965.
104. **Rudolf, R. W., Taylor, R. C., and Parry, R. W.,** *J. Am. Chem. Soc.,* 88, 3729, 1966.
105. **Maier, L.,** *Helv. Chim. Acta,* 46, 2026, 1963.
106. **Schmutzler, R.,** *Inorg. Chem.,* 3, 415, 1964.
107. **Cowley, A. H. and Cohen, S. T.,** *Inorg. Chem.,* 4, 1221, 1965.
108. **Fluck, E., Van Wazer, J. R., and Groengweghe, L. C. D.,** *J. Am. Chem. Soc.,* 81, 6363, 1959.
109. **Schmutzler, R.,** *Angew. Chem. Int. Ed. Engl.,* 3, 513, 1964.
110. **Jones, R. A. Y. and Katritzky, A. R.,** *Angew. Chem.,* 74, 60, 1962.
111. **Fluck, E. and Van Wazer, J. R.,** *Z. Anorg. Allg. Chem.,* 307, 113, 1961.
112. **Tetel'baum, B. I., Sheluchenko, V. V., Dubov, S. S., Drozd, G. I., and Ivin, S. Z.,** *Zh. Vses. Khim. Obshch.,* 12, 351, 1967, *Chem. Abstr.,* 68, 73847v, 1968.
113. **Barlow, C. G. and Nixon, J. F.,** *J. Chem. Soc.,* 228, 1966.
114. **Nixon, J. F.,** *Chem. Ind. London,* 1555, 1963.
115. **King, R. B. and Lee, T. W.,** *Inorg. Chem.,* 21, 319, 1982.
116. **King, R. B. and Gimeno, J.,** *Inorg. Chem.,* 17, 2390, 1978.
117. **Klingebiel, U., Werner, P., and Meller, A.,** *Chem. Ber.,* 110, 2905, 1977.
118. **Klingebiel, U., Werner, P., and Meller, A.,** *Monatshef. Chem.,* 107, 939, 1976.
119. **Littlefiel, L. B. and Quin, L. D.,** *Org. Magn. Reson.,* 12, 199, 1979.
120. **Quin, L. D., Gallagher, M. J., Cunkle, G. T., and Chesnut, D. B.,** *J. Am. Chem. Soc.,* 102, 3136, 1980.
121. **Quin, L. D. and Szewczyk, J.,** *J. Chem. Soc. Chem. Commun.,* 1551, 1984.
122. **Quin, L. D. and Bernhardt, F. C.,** *Magn. Reson. Chem.,* 23, 929, 1985.
123. **Roesky, H. W. and Hofmann, H.,** *Z. Naturforsch.,* 39b, 1315, 1984.
124. **Cowley, A. H., Mehrotra, S. K., and Roesky, H. W.,** *Inorg. Chem.,* 20, 712, 1981.
125. **White, D. W., Bertrand, R. D., McEwen, G. K., and Verkade, J. G.,** *J. Am. Chem. Soc.,* 92, 7125, 1970.
126. **Okruszek, A. and Verkade, J. G.,** *Phosphorus Sulfur,* 7, 235, 1979.
127. **Hofmann, H.,** *Ph.D. thesis,* Frankfurt, 1984.
128. **Mueller, A., Niecke, E., and Glemser, O.,** *Z. Anorg. Allg. Chem.,* 350, 256, 1967.
129. **Dutasta, J. P., Martin, J., and Robert, J. B.,** *J. Org. Chem.,* 42, 1662, 1977.
130. **Dutasta, J. P., Guimaraes, A. C., and Robert, J. B.,** *Tetrahedron Lett.,* 9, 801, 1977.
131. **Dutasta, J. P., Robert, J. B., and Vincens, M.,** *Tetrahedron Lett.,* 11, 933, 1979.
132. **Fluck, E., Kleemann, S., Hess, H., and Riffel, H.,** *Z. Anorg. Allg. Chem.,* 486, 187, 1982.
133. **Kleemann, S. and Fluck, E.,** *Int. Conf. on Phosphorus Chem.,* ICPC, Durham, 1981; *Chemiker-Ztg.,* 105, 326, 1981.
134. **Seidel, P. and Ugi, I.,** *Z. Naturforsch.,* 35b, 1584, 1980.
135. **Dakternieks, D., Roeschenthaler, G.-V., and Schmutzler, R.,** *Z. Naturforsch.,* 33b, 507, 1978.
136. **Barlos, K., Noeth, H., and Wrackenmeyer, B.,** *Z. Naturforsch.,* 33b, 515, 1978.
137. **Krueger, W., Sell, M., and Schmutzler, R.,** *Z. Naturforsch.,* 38b, 1074, 1984.
138. **Karsch, H. H.,** *Z. Naturforsch.,* 38b, 1027, 1983.
139. **Knoll, L.,** *Z. Naturforsch.,* 33b, 396, 1978.
140. **King, R. B., Sadanani, N. D., and Sundaram, P. M.,** *J. Chem. Soc. Chem. Commun.,* 477, 1983.
141. **Baudler, M., Hellmann, J., and Hahn, J.,** *Z. Anorg. Allg. Chem.,* 489, 11, 1982.
142. **Anderson, D. W. W., Bentham, J. E., and Rankin, D. W. H.,** *J. Chem. Soc. Dalton,* 1215, 1973.
143. **Colqhoun, I. J. and McFarlane, W.,** *J. Chem. Soc. Faraday II,* 73, 722, 1977.

144. **Crutchfield, M. M., Dungan, C. H., Letcher, J. H., Mark, V., and Van Wazer, J. R.,** in *P^{31} Nuclear Magnetic Resonance,* Topics in Phosphorus Chemistry, Vol. 5, John Wiley & Sons, New York, 1967, Ref. 325.
145. **Fluck, E., Gross, H., Binder, H., and Gloede, J.,** *Z. Naturforsch.*, 20b, 1125, 1966.
146. **Crutchfield, M. M., Dungan, C. H., Letcher, J. H., Mark, V., and Van Wazer, J. R.,** in *P^{31} Nuclear Magnetic Resonance,* Topics in Phosphorus Chemistry, Vol. 5, John Wiley & Sons, New York, 1967, Ref. 294.
147. **Crutchfield, M. M., Dungan, C. H., Letcher, J. H., Mark, V., and Van Wazer, J. R.,** in *P^{31} Nuclear Magnetic Resonance,* Topics in Phosphorus Chemistry, Vol. 5, John Wiley & Sons, New York, 1967, Ref. 298.
148. **Scherer, O. J., Kulbach, N. T., and Glaessel, W.,** *Z. Naturforsch.*, 33b, 652, 1978.
149. **Zeiss, W. and Weis, J.,** *Z. Naturforsch.*, 32b, 485, 1977.
150. **Barlos, K. and Noeth, H.,** *Z. Naturforsch.*, 35b, 415, 1980.
151. **Harvey, D. A., Keat, R., and Rycroft, D. S.,** *J. Chem. Soc. Dalton Trans.*, 425, 1983.
152. **Bulloch, G. and Keat, R.,** *J. Chem. Soc. Dalton Trans.*, 2010, 1974.
153. **Zeiss, W. and Barlos, K.,** *Z. Naturforsch.*, 34b, 423, 1979.
154. **Hesse, M. and Klingebiel, U.,** *Z. Anorg. Allg. Chem.*, 501, 57, 1983.
155. **Yoshifuji, M., Shima, I., Ando, K., and Inamoto, N.,** *Tetrahedron Lett.*, 24, 933, 1983.
156. **Schaefer, H., Zipfel, J., Migula, B., and Binder, D.,** *Z. Anorg. Allg. Chem.*, 501, 111, 1983.
157. **Baxter, S. G., Collins, R. L., Cowley, A. H., and Sena, S. F.,** *Inorg. Chem.*, 22, 3475, 1983.
158. **King, R. B., Lee, T. W., and Kim, J. H.,** *Inorg. Chem.*, 22, 3112, 1983.
159. **Severson, S. J., Cymbaluk, T. H., Ernst, R. D., Higashi, J. M., and Parry, R. W.,** *Inorg. Chem.*, 22, 3834, 1983.
160. **Barlos, K. and Noeth, H.,** *Z. Naturforsch.*, 35b, 407, 1980.
161. **Arnold, D. E. J. and Rankin, D. W. H.,** *J. Chem. Soc. Dalton,* 889, 1975.
162. **Odorisio, P. A., Pastor, S. D., Spivack, J. D., Steinhuebel, L., and Rodebaugh, R. K.,** *Phosphorus Sulfur,* 15, 9, 1983.
163. **Haase, M. and Klingbiel, U.,** *Z. Naturforsch.*, 41B, 697, 1986.
164. **Gloede, J.,** *Z. Anorg. Allg. Chem.*, 531, 17, 1985.
165. **Yurchenko, R. I. and Klepa, T. I.,** *Zh. Obshch. Khim.*, 56, 1044, 1986.
166. **Mathieu, R., Caminade, A. M., Majoral, J. P., and Daran, J. C.,** *J. Am. Chem. Soc.*, 108, 8007, 1986.

Chapter 4

^{31}P NMR DATA OF THREE COORDINATE ($\lambda 3$ $\sigma 3$) PHOSPHORUS COMPOUNDS CONTAINING BONDS TO CHALCOGENIDES (O, S, Se, Te) BUT NO BONDS TO HALOGEN

Compiled and presented by
Andrew W. G. Platt and Stephan G. Kleemann

NMR data was also donated by
E. Fluck, Gmelin-Institut, Frankfurt/Main, FRG
J. Verkade, Iowa State University, Ames, Iowa
C. D. Hall, King's College London, U.K.
S. E. Cremer, Marquette University, Wisconsin
J. P. Dutasta, Université Scientifique et Médicale de Grenoble, France
H. W. Roesky, Universität Göttingen, FRG

The NMR data are presented with no subdivisions in Table C. The Table includes data on phosphites, phosphonites, phosphinites, and their thio, seleno and telluro analogues.

TRENDS IN ^{31}P NMR CHEMICAL SHIFTS

EFFECT OF THE P-BOUND ATOM

There appear to be no regular trends in the chemical shifts due to P-bound atoms. Thus simple changes due to electronegativity of the P-bound atoms or heavy atom effects do not dominate the shift values. Thus in general the resonances of O_3 species do not appear at higher frequencies than, say, S_3 compounds.

Some weak trends are discernible on successive substitution of one P-bound atom by another. For instance, along the series OS_2, O_2S, O_3 there is a general decrease in the shift. Similar effects are observed in the S_3, S_2C, SC_2 sequence where the trend is a more pronounced decrease in the resonance frequency. The O_3, O_2C, OC_2 series does not show the same trend in shifts. The decrease between O_2C and OC_2 (and on to the lower frequency shifts of phosphines) is as expected; however, in general, $\delta O_3 > \delta O_2C$ reversing this trend.

It must be stressed that these are, for the most part, weak trends with many shift ranges overlapping and consequently it must be expected that there will be many exceptions to the above generalizations.

Resonances of P(III) atoms directly bound to P(V) are not shifted significantly from those observed in, say, CO_2 compounds. However, phosphorus atoms directly bonded to P(III) are shifted to lower frequency by over 100 ppm.

EFFECT OF RING SIZE

The incorporation of the P-bound atom into a ring system containing 5 or more atoms does not cause any large changes in the shift values compared with those of acyclic compounds. A notable exception to this is the case of compounds with O_2S as P-bound atoms. Here, cyclic compounds have shifts up to 60 ppm higher frequency of acyclic derivatives.

Where data are available (e.g., with O_3 compounds) it seems that involving the P atom in a bicyclic system contracts the range of chemical shifts, in this case to about 10 ppm.

STEREOCHEMICAL EFFECTS

There have been many studies concerning the case of cyclic compounds where *cis* and *trans* isomers are usually apparent by differences in chemical shift ranging from 0.5 ppm to 20 ppm or more. In many cases where isomers are expected, and observed in the NMR spectrum, assignments to *cis* and *trans* species have not been made.

In unsymmetrically substituted 5-membered rings it is observed that the *cis* isomer invariably gives a signal of slightly higher frequency than that of the *trans* isomer by about 5 ppm. This appears to be independent of the P-bound atoms. In some cases these results have been confirmed by stereospecific oxidation of the P(III) compound followed by crystal structure determination of the P(V) derivative formed. The same situation is apparent in 6-membered ring compounds where the ring adopts the chair conformation. Here $\delta cis > \delta trans$ by up to 5 ppm (exceptionally by 40 ppm). These results have also been confirmed by stereochemical oxidation and appear to be independent of the P-bound atoms.

COUPLING CONSTANTS

$^1J_{PH}$ values vary from 640 to 150 Hz, with values usually in the region of 200 Hz. From the limited data available there appears to be some dependence on the P-bound atoms; for example, $^1J_{PH}$ is greater in compounds with P-bound atoms HS_2 than with HO_2.

One bond couplings to selenium range from 200 to 550 Hz while those to tellurium fall in the slightly narrower range of 350 to 500 Hz.

Range of Chemical Shift According to P-bound Atoms (Extreme Values in Parentheses)

P-bound atoms		Range
NOS		150—120
NS_2		170—90
NO_2	(186)	165—120
N_2O		190—100 (18)
N_2S		160—110
OS_2		180—150
O_2S		185—120 (94)
O_3 acyclic		150—120
O_3 5-membered rings		150—120
O_3 6-membered rings		130—100
S_3	(198)	125—110 (33)
HNO		120—10
HO_2		165—140 (105)
CO_2	(224)	190—130 (45)
CS_2	(120)	105—25
C_2O		165—90 (35)
C_2S	(120)	30—0
C_2Te		100—60 (5)

TABLE C

Compilation of ^{31}P NMR Data of Three Coordinate (λ3 σ3) Phosphorus Compounds Containing Phosphorus Bonds to Chalcogenides (O, S, Se, Te) But No Bonds to Halogen

Rings/connectivities	Formula	Structure	NMR Data (δP[solv]^{a}J$_{PZ}$ Hz)	Ref.
	P-bound atoms = NOS			
C_2;C;C′	$C_{14}H_{21}N_2OPS$	P(Et$_2$N)(OEt)(SC=CHNDop)	134	9
5/C_2;C;C	$C_4H_{10}NOPS$	Me$_2$NP(OC$_2$H$_4$S)	149	46
	$C_7H_{14}NOPS$	PipP(OC$_2$H$_4$S)	129.5	46
5/CC′;C;C′	$C_9H_{12}NOPS$	(EtO)P(N(Me)1,2-C$_6$H$_4$S)	125	64
	$C_{10}H_{14}NOPS$	(EtO)P(N(Et)1,2-C$_6$H$_4$S)	125	64
	P-bound atoms = NO$_2$			
HP;(C)$_2$	$C_{10}H_{23}NO_2P_2$	(EtO)$_2$PNHP(iPr)$_2$	137.5 J$_{PP}$20	221
PSi;(C)$_2$	$C_{11}H_{29}NO_2P_2Si$	(EtO)$_2$PN(Tms)PEt$_2$	161.0	221
	$C_{11}H_{33}NO_2P_2Si$	(EtO)$_2$PN(Tms)P(iPr)$_2$	164.0 J$_{PP}$19	221
CP;(C)$_2$	$C_{12}H_{29}NO_2P_2$	(EtO)$_2$PN(tBu)PEt$_2$	155.0 J$_{PP}$23	221
	$C_{13}H_{31}NO_4P_2Se$	(iPrO)$_2$PN(Me)P(Se)(O-iPr)$_2$	138.3 J$_{PSe}$901 J$_{PP}$90	222
	$C_{14}H_{31}NO_2P_2$	(EtO)$_2$PN(tBu)P(O-iPr)$_2$	160.0 J$_{PP}$30	221
	$C_{17}H_{39}NO_4P_2Se$	(iBuO)$_2$PN(Me)P(Se)(O-iBu)$_2$	139.8 J$_{PSe}$915 J$_{PP}$83	222
		(BuO)$_2$PN(Me)P(Se)(OBu)$_2$	140 J$_{PSe}$914 J$_{PP}$82	222
C′P;(C)$_2$	$C_{14}H_{25}NO_4P_2$	(EtO)$_2$PN(Ph)P(OEt)$_2$	142	221
	$C_{16}H_{29}NO_2P_2$	(EtO)$_2$PN(Ph)P(iPr)$_2$	142.0 J$_{PP}$185	221
	$C_{17}H_{30}NO_4P_2S$	(EtO)$_2$PN(O$_2$SC$_6$H$_4$Me-4)P(iPr)$_2$	148.6	221
C′H;(C)$_2$	$C_{11}H_{19}N_2O_2P$	P(NHPh)(OC$_2$H$_4$NHMe)(OEt)	140	8
	$C_{14}H_{25}NO_4P_2S$	(EtO)$_2$PNH(C$_6$H$_4$SP(OEt)$_2$-2)	186	64
C_2;C,C′	$C_{10}H_{20}NO_3P$	Me$_2$NP(OiPr)OC(Me)=CHC(O)Me	140,142	205
		Et$_2$NP(OMe)OC(Me)=CHC(O)Me	136,140	205
	$C_{13}H_{22}NO_3P$	Et$_2$NP(OEt)OC(Me)=CHC(O)Me	141,143	205
	$C_{15}H_{26}NO_3P$	Et$_2$NP(OiBu)OC(Me)=CHC(O)Me	142,145	205
C_2;(C)$_2$	$C_4H_{12}NO_2P$	P(NMe$_2$)(OMe)$_2$	148	1
	$C_6H_{16}NO_2P$	P(NMe$_2$)(OEt)	144.7	2
	$C_8H_{16}NO_2P$	P(NMeCH$_2$C≡CH)(OEt)$_2$	143.6	223
	$C_8H_{20}NO_2P$	P(NEt$_2$)(OEt)	154.6	3
		P(NMe$_2$)(OPr)$_2$	144.9	3
	$C_{10}H_{24}NO_2P$	P(NEt$_2$)(OBu)$_2$	144.6	3
(C)$_2$;C;C′	$C_{15}H_{21}NO_4P$	PN(CH$_2$)$_3$CHCOOEt(OEt)(OPh)	141.2(c)	11
CC′;C_2	$C_{15}H_{27}NO_4P_2S$	P(OEt)$_2$N(Me)DopSP(OEt)$_2$	126	64
CC″;(C)$_2$	$C_{13}H_{18}NO_2P$	PN(C≡CH)(CH$_2$Ph)(OEt)$_2$	141.4	3
C′$_2$;(C)$_2$	$C_{12}H_{16}NO_2P$	P(NDopCH=CH)(OEt)$_2$	128	9
	$C_{12}H_{15}NO_2PS$	P(NDopC(SH)=CH)(OEt)$_2$	128	10
	$C_{14}H_{17}NO_4P_2$	P(NC(Me)=CHC(Me)=C(P(OEt)$_2$)(OEt)$_2$	124	224
	$C_{14}H_{19}NO_2P$	P(NDopCH=CH)(OPr)$_2$	128	9
	$C_{14}H_{19}NO_2PS$	P(NDopC(SH)=CH)(OPr)$_2$	128	10
	$C_{15}H_{21}NO_2P$	P(NDopC(Me)=CH)(OPr)$_2$	108	9
	$C_{16}H_{29}NO_4P_2$	P(NC(Me)=C(Me)C(Me)=C(P(OEt)$_2$)(OEt)$_2$	124	224
	$C_{18}H_{25}NO_4P_2$	P(NC(Me)=CHC(Me)=C(P(OPr)$_2$)(OPr)$_2$	125	224
5/Si$_2$;(C′)$_2$	$C_{10}H_{24}NO_2PSi_2$	PN(Tms)$_2$OCMe=CMeO	162	20

TABLE C (continued)
Compilation of ^{31}P NMR Data of Three Coordinate (λ3 σ3) Phosphorus Compounds Containing Phosphorus Bonds to Chalcogenides (O, S, Se, Te) But No Bonds to Halogen

Rings/connectivities	Formula	Structure	NMR Data (δP[solv]^{a}J$_{PZ}$ Hz)	Ref.
5/C′H;(C′)$_2$	$C_{10}H_{15}N_4O_2P$	P(NHC(CN)=C(CN)-NH$_2$)(OC$_2$Me$_4$O)	140	17
	$C_{12}H_{10}NO_2P$	(PhO)PNHDopO	127	29
5/C$_2$;(C)$_2$	$C_4H_8NO_2P$	(CH$_2$CH$_2$N)PGlc	150.4	3
	$C_4H_{10}NO_2P$	Me$_2$NPGlc	140.6	1
	$C_5H_9F_3NO_2P$	MeNC$_2$H$_4$OP(OCH$_2$Tf)	140.8	23
	$C_5H_{12}NO_2P$	P(NMe$_2$)OCHMeCH$_2$O	137.3^{C}, 142.5^{T}	12
		EtNC$_2$H$_4$OPOEt	134	24
	$C_6H_8F_6NO_2P$	Tf$_2$CHNC$_2$H$_4$OPOMe	147.2	23
	$C_8H_{16}NO_2P$	P(NMe$_2$)O(1,2-cHex)O	144^{c}, 148^{T}	13
	$C_8H_{18}NO_2P$	Me$_2$NPPnc	150, 142.5	1
	$C_9H_{16}NO_2P$	PN(sBu)(CH$_2$C≡CH)Glc	140	14
		P(NEtCH$_2$C≡CH)-(OCHMeCHMeO)	143, 132	14
	$C_{10}H_{14}NO_2P$	P(NMe$_2$)(OCH$_2$CHPhO)	151, 146.6	13
	$C_{10}H_{18}NO_6P$	P(NEt$_2$)(OCH(COOMe)-CH(COMe)O)	158b	15
	$C_{11}H_{23}NO_2P$	tBuNCH(tBu)CH$_2$OPOMe	122.5	25
	$C_{12}H_{25}NO_2P$	tBuNCH(tBu)CH$_2$OPOEt	123.0	25
	$C_{16}H_{18}NO_2P$	P(NMe$_2$)(OCH(Ph)CH(Ph)O)	145^{M}	1
5/C$_2$;C,C′	$C_9H_{12}NO_2P$	MeNC$_2$H$_4$OPOPh	134	26
5/C$_2$;(C′)$_2$	$C_8H_{10}NO_2P$	Me$_2$N(PCat)	147.5	1
	$C_8H_{16}NO_2$	Et$_2$NPOCMe=CMeO	140.5	19
	$C_{10}H_{18}NO_2P$	Et$_2$NPO(1,2-cHexene)O	142.0	19
	$C_{11}H_{14}NO_2P$	Pip(PCat)	126	21
	$C_{12}H_{12}NO_2P$	Me$_2$P(2,3-Don)	148.2	22
	$C_{12}H_{18}NO_2P$	iPr$_2$N(BCat)	154.8	226
	$C_{14}H_{16}NO_2P$	Et$_2$NP(2,3-Don)	150.1	2
5/CC′;(C)$_2$	$C_{10}H_{14}NO_2P$	PhNC$_2$H$_4$OPOEt	122	26
	$C_{11}H_{14}NO_3P$	PhNC(O)CH$_2$OPOPr	121	27
	$C_{11}H_{16}NO_2P$	PhNCHMeCHOP(OEt)	124, 128	28
5/CC′;C,C′	$C_{14}H_{14}NO_2P$	PhNC$_2$H$_4$OPOPh	126	28
5/CC′;C′$_2$	$C_{18}H_{16}NO_2P$	PN(Et)(Ph)(2,3-Don)	142.2	22
5/C′$_2$;(C)$_2$	$C_6H_8NO_2P$	P(NCH=CHCH=CH)Glc	124	1
	$C_{14}H_{14}NO_2P$	P(NPh$_2$)Glc	134	18
5/C′$_2$;CC′	$C_{14}H_{12}NO_3P$	PhNC(O)CH$_2$OPOPh	115	27
6/CH;(C)$_2$	$C_5H_{12}NO_2P$	P(NHC$_3$H$_6$O)(OEt)	126	37
6/C′H;(C)$_2$	$C_9H_{14}NO_2P$	P(NHPh)(OC$_2$H$_4$CHMeO)	136.0^{T}, 121.0^{C}	35
6/C$_2$(P)$_2$	$C_{18}H_{42}N_3O_3P_3$	[OP(N-iPr$_2$)]$_3$	140.3, 131.3 J13.5	40
6/C$_2$;(C)$_2$	$C_5H_{12}NO_2P$	P(NMe$_2$)(OC$_3$H$_6$O)	143.0	30
	$C_6H_{14}NO_2P$	P(NMe$_2$)(OC$_2$H$_4$CHMeO)	138.5^{T}, 142.7^{C}	30
	$C_7H_{10}NO_2P$	P(NC$_3$H$_6$CH$_2$)(OC$_3$H$_6$O)	122.0	30
	$C_7H_{16}NO_2P$	P(NMe$_2$)(OCHMeCH$_2$CHMeO)	137^{T}, 141^{C}(b)	31
		P(N(iPr)C$_3$H$_6$O)(OMe)	138.5	38
	$C_8H_{12}NO_2P$	P(NC$_3$H$_6$CH$_2$)(OC$_2$H$_4$CHMeO)	120.5^{T}	30
	$C_8H_{18}NO_2P$	P(NMe$_2$)(OCHMeCH$_2$CMe$_2$O)	137.9^{C}, 137.5^{T}	30
		P(N(iPr)C$_3$H$_6$O)(OEt)	137	38
	$C_8H_{19}N_2O_2P$	P(NMeC$_3$H$_6$O)(OC$_3$H$_6$NHMe)	134.7	39
	$C_9H_{20}NO_2P$	P(NMe$_2$)(OCH$_2$CH(tBu)CH$_2$O)	142.6 (b)	32
	$C_9H_{21}N_2O_2P$	P(N(iPr)C$_3$H$_6$O)(OC$_3$H$_6$NH$_2$)	134.0	39

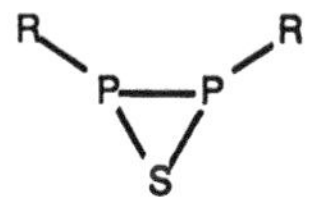

1 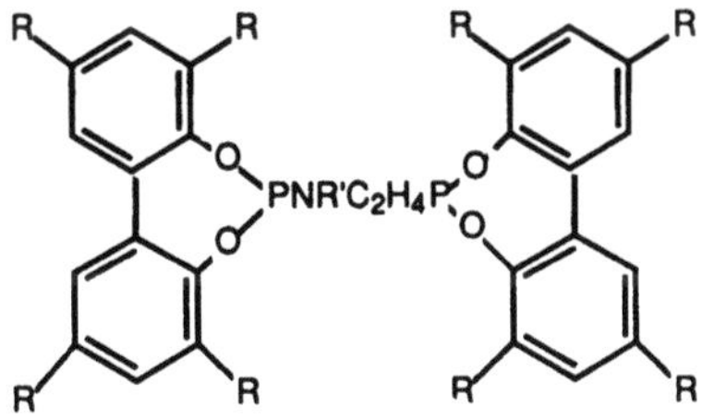 2 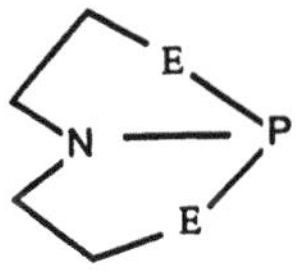 3

Rings/connectivities	Formula	Structure	NMR Data (δP[solv]ᵃJ$_{PZ}$ Hz)	Ref.
	$C_{10}H_{20}NO_3P$	$P(Et_2)(OCH(OMe)CH_2O)CHCH_2O)$	142, 131^{I}	33
	$C_{10}H_{20}NO_2P$	$P(NMe_2)(OCH_2CMe_2CH(iPr)O)$	144.9^{C}, 136.4^{T}	34
		$P(NEt_2)(OCHMeCH_2CMe_2O)$	139.8^{C}, 137.5^{T}	30
	$C_{11}H_{25}N_2O_2P$	$PN(iPr)C_3H_6O)(OC_3H_6NMe_2)$	135.2	39
	$C_{12}H_{27}N_2O_2P$	$P(N(iPr)C_3H_6O)(OC_3H_6NH(iPr))$	134.0	38
	$C_{14}H_{26}NO_2P$	$P(NEt_2)$(1,8-Don)	127.6	22
$6/C_2;(C')_2$	$C_9H_{16}NO_2P$	$P(NEt_2)OC(=CH_2)CH=CMeO$	132	205
	$C_{12}H_{11}NO_2P$	$P(NMe_2)$(1,8-Don)	126	22
$6/CC';(C)_2$	$C_{11}H_{16}NO_2P$	$P(NEtPh)(OC_3H_6O)$	139.0	30
$6/C'_2;(C)_2$	$C_{15}H_{16}NO_2P$	$P(NPh_2)(OC_3H_6O)$	125.3	30
	$C_{16}H_{18}NO_2P$	$P(NPh_2)(OC_2H_4CHMeO)$	120.3^{T}	30
$7/C_2;(C')_2$	$C_{43}H_{87}NO_5P_2$	**2**; R^1 = Me, R = tBu	132.2	42
	$C_{44}H_{89}NO_5P_2$	**2**; R^1 = Et, R = tBu	132.0	42
	$C_{46}H_{93}NO_5P_2$	**2**; R^1 = nBu, R = tBu	132.5	42
$8/C_2;(C)_2$	$C_6H_{14}NO_3P$	$OC_2H_4OC_2H_4OPNMe_2$	150.6	41
$5,5/C_2;(C)_2$	$C_4H_8NO_2P$	**3**; E = O	139.0(c)	43
	$C_{18}H_{28}NO_2P$	**4**	158.0(t)	44
$5,5/C_2;(C')_2$	$C_8H_{16}NO_2P$	**4**; RR = Me	163	225
	$C_{12}H_{20}NO_2P$	**5**; R = tBu	187(r)	45
$5,10/C_3;(C')_2$	$C_{16}H_{32}N_2O_4P_2$	**6**	140.8	225
P-bound atoms = N'O$_2$				
$C';(C)_2$	$C_3H_6NO_3P$	$P(N=C=O)(OMe)_2$	125	5
	$C_5H_{10}NO_2PS$	$P(N=C=S)(OEt)_2$	121	6
	$C_7H_6F_8NO_3P$	$P(N=C=O)(OCH_2CF_2CHF_2)_2$	127	227
	$C_9H_{20}NO_3P$	$P(N=C(iPr)(OMe))(OEt)_2$	138.2	7
$5/C';(C)_2$	$C_7H_{12}NO_2PS$	$P(N=C=S)(OCMe_2CMe_2O)$	175.2(c)	16
	$C_7H_{12}NO_3P$	$P(N=C=O)(OCMe_2CMe_2O)$	127.6(c)	16
$6/C';(C)_2$	$C_6H_{10}NO_2PSe$	$P(N=C=Se)(OC_2H_4CHMeO)$	121.0^{C}, 136.0^{T}	35
	$C_{16}H_{16}NO_2P$	$P(N=CPh_2)(OC_3H_6O)$	121.0	36
P-bound atoms = NPS				
$3/Si_2;NS;P$	$C_{12}H_{36}N_2P_2SSi$	**1**; R = $NTms_2$	−21.9	134
P-bound atoms = NS$_2$				
$C_2;(C_2)$	$C_6H_{12}NPS_2$	$Me_2NP(SMe)_2$	166	209
$6/CH;(C)_2$	$C_7H_{16}NPS_2$	$BuNHP(SC_3H_6S)$	87.5	47
$6/C_2;(C)_2$	$C_5H_{10}NPS_2$	$(CH_2CH_2N)P(SC_3H_6S)$	117.5	47
	$C_8H_{16}NPS_2$	$PipP(SC_3H_6S)$	122.9	47
	$C_9H_{20}NPS_2$	$iPr_2NP(SC_3H_6S$	121.0	47
	$C_{10}H_{22}NPS_2$	$iPr(tBu)NP(SC_3H_6S)$	125.0	47
$5,5/C_2;(C)_2$	$C_4H_8PS_2$	**2**; E = S	162.3	48

TABLE C (continued)
Compilation of ^{31}P NMR Data of Three Coordinate ($\lambda 3\ \sigma 3$) Phosphorus Compounds Containing Phosphorus Bonds to Chalcogenides (O, S, Se, Te) But No Bonds to Halogen

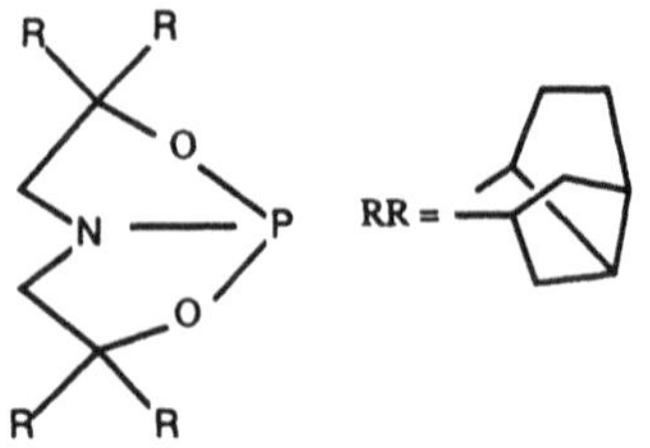

4 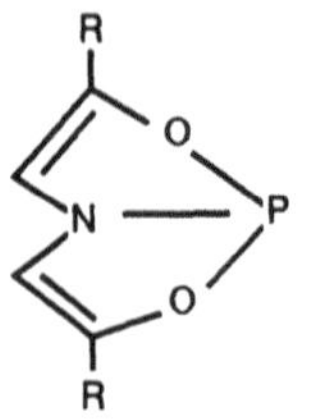 5 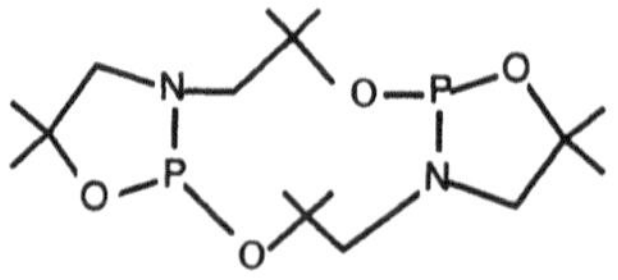 6

Rings/connectivities	Formula	Structure	NMR Data ($\delta P[solv]^a J_{PZ}$ Hz)	Ref.
P-bound atoms = N^4NO				
$C_2H;C_2;C$	$C_5H_{16}N_2OP$	$P(Me_2NH^+)(Me_2N)(OMe)$	130.8	49
P-bound atoms = N_2O				
$(C_2)_2;P$	$C_{16}H_{40}N_4O_2P_2$	$[P(NEt_2)_2]_2O$	123.5	51
$(C_2)_2;C$	$C_5H_{15}N_2OP$	$P(NMe_2)_2(OMe)$	138.4	49
	$C_6H_{17}N_2OP$	$P(NMe_2)_2(OEt)$	135.7	2
	$C_{10}H_{25}N_2OP$	$P(NEt_2)_2(OEt)$	134	3
	$C_{12}H_{29}N_2OP$	$P(NEt_2)_2(OBu)$	132.1	2
	$C_{13}H_{31}N_2OP$	$P(N\text{-}iPr_2)_2(OMe)$	130.9	49
$(C_2)_2;C'$	$C_{14}H_{25}N_2OP$	$P(NEt_2)_2(OPh)$	131	3
	$C_{13}H_{27}N_2O_2P$	$P(Et_2N)_2OC(Me){=}CHC(O)Me$	128	205
	$C_{19}H_{44}N_4O_2P_2$	$P(NEt_2)_2(OC(OMe){=}CHP(NEt_2)_2$	140	50
$C_2,C'_2;C$	$C_6H_{10}Cl_2N_3OP$	$P(NMe_2)(NCH{=}NC(Cl){=}C(Cl))(OMe)$	130.9	49
	$C_{14}H_{21}N_2OP$	$P(NEt_2)(NCH{=}CHDop)(OEt)$	122	9
$4/(C'P)_2;C$	$C_{14}H_{16}N_2O_2P_2$	$[N(Ph)P(OMe)]_2$	184.8, 136.5^I	52
	$C_{15}H_{18}N_2O_2P_2$	$N(Ph)P(OMe)N(Ph)P(OEt)$	188.3^C, 141.9^T	206
	$C_{16}H_{20}N_2O_2P_2$	$[N(Ph)P(OEt)]_2$	182.0, 134.9^I	52
	$C_{32}H_{28}N_2O_2P_2$	$[N(Ph)P(O\text{-}tBu)]_2$	136.8^T	206
$5/CB,CN;C$	$C_5H_{15}BN_3OP$	$MeNP(OMe)NMeNMeBMe$	126.0	207
$5/CSi,CC';C$	$C_{12}H_{21}OPSi$	$PhNC_2H_4OPMeTms$	120	26
$5/CP,C_2;C$	$C_4H_{14}Cl_2N_2OP_2$	$Et_2NPN(PCl_2)C_2H_4O$	137 $J_{PP}60$	208
$5/C'P,C_2:C$	$C_{22}H_{44}N_4OP_2$	$[P(NEt_2)_2]DopOP(NEt_2)_2$	102	56
$5/C'P,CC';C'$	$C_{22}H_{23}N_3O_2P_2$	$(PhNC_2H_4OP)_2NPh$	114	8
$5/(CH)_2;C$	$C_9H_5N_8OP$	$P(NHC(CN)N{=}C(CN)C(CN{=}NC(CN)NH)OMe$	90	17
$5/C_2,C'H;C$	$C_{16}H_{25}N_2OP$	$tBuNCH(tBu)CH_2OPNHPh$	123.0	25
$5/C_2,C'H;C'$	$C_{10}H_{15}N_2OP$	Et_2NPOam	128.0	29
$5/(C_2)_2;C$	$C_5H_{13}N_2OP$	$P(NMeC_2H_4NMe)(OMe)$	123.2	5
	$C_5H_{13}N_2OP$	$(MeNC_2H_4O)PNMe_2$	132.5	1
	$C_6H_{12}F_3N_2OP$	$TfP(MeNC_2H_4NMe)(OCH_2Tf)$	140.7	23
	$C_6H_{15}N_2OP$	$MeNCH_2CH(Me)OPNMe_2$	131, 133^I	54
	$C_7H_{11}F_6N_2OP$	$Tf_2P(N(Me)C_2H_4NMe)(OCHTf_2)$	145.4	2
	$C_7H_{17}N_2OP$	$(MeNCMe_2CH_2O)PNMe_2$	138, 185^I	54
		$(MeNCHMeCHMeO)PNMe_2$	128, 133^I	54
	$C_8H_{17}N_2OP$	$(N\text{-}iPrNC_2H_4O)PNC_2H_4N(iPr)_2$	149.8	38

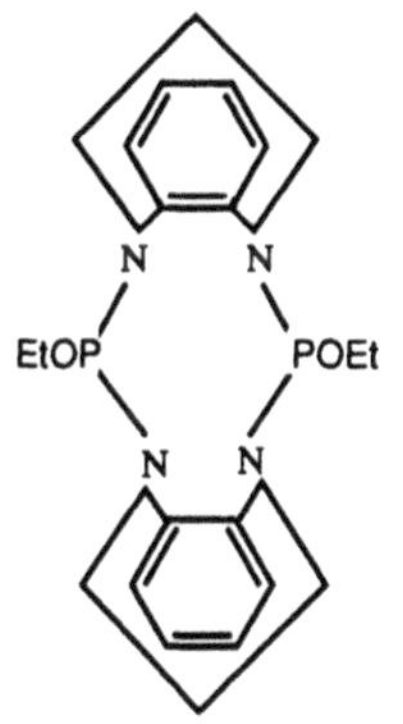

7

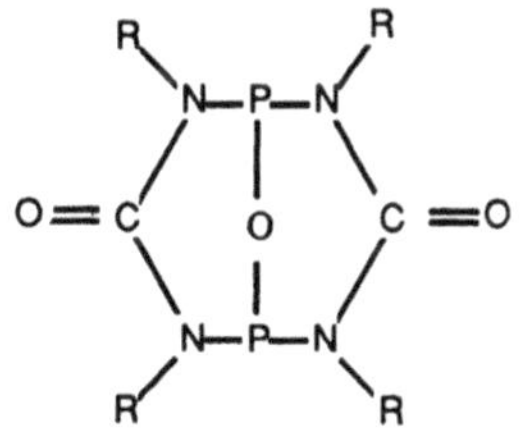

8

Rings/connectivities	Formula	Structure	NMR Data (δP[solv]ᵃJ$_{PZ}$ Hz)	Ref.
	$C_9H_{19}N_2OP$	(1,2-cHexN(Me)O)POMe	137	38
	$C_{10}H_{23}N_2OP$	(iPrNC$_2$H$_4$O)PNEt$_2$	137	38
	$C_{11}H_{17}N_2OP$	MeNCH$_2$CH(Ph)OPNMe$_2$	129, 132[I]	54
	$C_{12}H_{24}N_2OP$	tBuNCH(tBu)CH$_2$OPNMe$_2$	136.2	25
	$C_{12}H_{24}N_2O_5P$	MeN*CHOCH(C(OMe)$_2$ CH$_2$OMeCH(OMe)*CHOPNMe$_2$	149.4, 132.8[I]	55
	$C_{13}H_{20}N_3O_2P$	MeNCH(Ph)CH-(CONHMe)OPNMe$_2$	129	54
	$C_{14}H_{28}N_2O_5P$	MeNCH(Ph)CH-(CONHMe)OPNMe$_2$	152.0, 134.8[I]	55
	$C_{17}H_{21}N_2O_5P$	MeNCH(Ph)CHOPNMe$_2$	127	54
5/C$_2$,CC′;C	$C_6H_{13}N_2O_2P$	MeNCC(O)CHMeOPNMe$_2$	129.5	1
	$C_{16}H_{15}N_2OP$	PhNC$_2$H$_4$OPNMe$_2$	119	1
5/CC′,C′H;C	$C_{14}H_{15}N_2OP$	PhNC$_2$H$_4$OPNHPh	106	8
6/(C′H)$_2$; C	$C_5H_5N_4OP$	MeOPNHC(CN)=C(CN)NH	90	17
6/C′H,C$_2$;C′	$C_{11}H_{15}N_2O_2P$	OC(O)DopNHPNEt$_2$	154	57
6/(C$_2$)$_2$;Si	$C_6H_{15}N_2OP$	MeNC$_3$H$_6$OPNMe$_2$	139.4	39
	$C_{35}H_{35}N_2OP$	PhCH$_2$NC$_3$H$_6$N(CH$_2$Ph)P(OSiPh$_3$)	116.5	60
6/(C$_2$)$_2$;C	$C_7H_{15}N_2OP$	MeNCH(Me)C$_2$H$_4$N(Me)POMe	130.1	58
		MeNCH$_2$CHMeCH$_2$POMe	131.6, 122.6[I](b)	59
	$C_8H_{19}N_2OP$	MeNC$_3$H$_6$OPNEt$_2$	144.1	39
	$C_{12}H_{19}N_2OP$	PhCH$_2$NC$_3$H$_3$OPNMe$_2$	137.5	39
	$C_{14}H_{21}N_2O_2P$	BzNC$_3$H$_6$OP*NC$_2$H$_4$OCH$_2$*CH$_2$	133.2	39
	$C_{19}H_{25}N_2OP$	PhCH$_2$NC$_3$H$_6$N(CH$_2$Ph)P(OEt)	128.3	60
10/(CC′)$_2$;P	$C_{22}H_{30}N_4O_2P_2$	**7**	132.0	61
6,6/(CC′)$_2$;C	$C_{18}H_{14}F_6N_4O_3P_2$	**8**	17.8	62

P-bound atoms = NN′O

Rings/connectivities	Formula	Structure	NMR Data	Ref.
C$_2$;C′;C	$C_{14}H_{21}N_2OPS$	P(NEt$_2$)(NCH=C(SH)Dop)(OEt)	122	9
	$C_{14}H_{21}N_2OP$	P(NEt$_2$)(NCH=CHDop)(OEt)	122	9

P-bound atoms = N$_2$S

Rings/connectivities	Formula	Structure	NMR Data	Ref.
(C$_2$)$_2$;C′	$C_9H_{21}N_3PS$	P(NMe$_2$)$_2$(SC(SMe)=CMe$_2$)	162	63
	$C_{13}H_{21}N_3PS$	P(SC=NHDop)(NMe)$_2$	136	9

TABLE C (continued)
Compilation of ^{31}P NMR Data of Three Coordinate (λ3 σ3) Phosphorus Compounds Containing Phosphorus Bonds to Chalcogenides (O, S, Se, Te) But No Bonds to Halogen

Rings/connectivities	Formula	Structure	NMR Data (δP[solv][a] J_{PZ} Hz)	Ref.
5/CB,CN;C	$C_5H_{15}BN_3PS$	MeSPN(Me)B(Me)N(Me)N(Me)	158.0	207
5/C'P,C$_2$;C'	$C_{22}H_{44}N_5PS$	SDopN(P(NEt$_2$)$_2$)P(NEt$_2$)	114	64
5/(C$_2$)$_2$;C	$C_5H_{13}N_2PS$	MeSPN(Me)C$_2$H$_4$N(Me)	160.1	209
5/CC';C'	$C_{11}H_{17}N_2PS$	SDopN(Me)P(NEt$_2$)	108	64
P-bound atoms = OS$_2$				
C;(C)$_2$	$C_6H_{15}OPS_2$	P(OEt)(SEt)$_2$	153.5	3
	$C_9H_{21}OPS_2$	P(OMe)(SBu)$_2$	162.1	3
C';(C)$_2$	$C_{6\text{-}8}H_{14\text{-}18}O_2PS_2$	P(OC(O)Me)(SR)$_2$	176.0	65, 210
	$C_{11}H_{15}O_2PS_2$	P(OC(O)Ph)(SEt)$_2$	180	65
5/C;(C)$_2$	$C_4H_6F_3OPS_2$	P(SC$_2$H$_4$S)(OCH$_2$Tf)	153.4	23
	$C_5H_5F_6OPS_2$	P(SC$_2$H$_4$S)(OCHTf$_2$)	177.7	23
6/C;(C)$_2$	$C_4H_9OPS_2$	P(SC$_3$H$_6$S)(OMe)	153.0	47
P-bound atoms = O$_2$S				
(C)$_2$;C'	$C_6H_{14}NO_2PS$	P(OEt)$_2$SC(Me)=NH	124	66
	$C_8H_{18}NO_2PS$	P(OPr)$_2$SC(Me)=NH	124	66
	$C_{10}H_{15}O_2PS$	P(OEt)$_2$SPh	186	64
	$C_{12}H_{16}NO_2PS$	P(OEt)$_2$(SC=CHNHDop)	126	9
	$C_{14}H_{20}NO_2PS$	P(OPr)$_2$(SC=CHNHDop)	125	9
	$C_{15}H_{27}NO_4P_2S$	(EtO)$_2$PSDop-N(Me)P(OEt)$_2$	186	64
5/(C)$_2$;C	$C_2H_5O_2PS$	P(OC$_2$H$_4$S)(OH)	176	67
	$C_3H_7O_2PS$	P(OC$_2$H$_4$S)(OMe)	170.3	66
	$C_4H_9O_2PS_2$	P(OC$_2$H$_4$S)(OC$_2$H$_4$SH)	176.0	67
	$C_4H_6F_3O_2PS$	P(OC$_2$H$_4$S)(OCH$_2$Tf)	173.4	23
	$C_5H_5F_6O_2PS$	P(OC$_2$H$_4$S)(OCHTf$_2$)	185.7	23
5/(C')$_2$;C	$C_7H_7O_2PS$	CatP(SMe)	204(n)	68
	$C_8H_9O_2PS$	CatP(SEt)	204(n)	68
6/(C)$_2$;C	$C_5H_{11}O_2PS$	P(OC$_3$H$_6$S)(OMe)	179.6, 185.0[l]	69
	$C_{11}H_{15}O_2PS$	P(OC$_2$H$_4$CHMeS)(OCH$_2$Ph)	93.5	70
P-bound atoms = O$_3$				
K,(C)$_2$	$C_4H_{10}KO_3P$	P(OMe)$_2$(OK)	152.3	3
Li,(C)$_2$	$C_4H_{10}LiO_3P$	P(OMe)$_2$(OLi)	141	3
	$C_8H_{18}LiO_3P$	P(OBu)$_2$(OLi)	145	3
Na,(C)$_2$	$C_4H_{10}NaO_3P$	P(OMe)$_2$(ONa)	153.0	3
	$C_8H_{18}NaO_3P$	P(OB)$_2$(ONa)	153.5	3
P$_3$	$F_6O_3P_4$	P(OPF$_2$)$_3$	110.0 J_{PF}18	211
P,(C)$_2$	$C_2H_6F_2O_3P_2$	P(OMe)$_2$(OPF$_2$)	120.7 J_{PF}12	211
P,(C')$_2$	$C_{12}H_{10}F_2O_3P_2$	P(OPh)$_2$(OPF$_2$)	121.3, J_{PF}11	211
S,(C)$_2$	$C_5H_{13}O_5PS$	P(OEt)$_2$(OSO$_2$Me)	134.6	80
	$C_7H_{17}O_5PS$	P(OPr)$_2$(OSO$_2$Me)	132.9	80
(Si)$_2$,C	$C_7H_{21}O_3PSi_3$	P(OMe)(OTms)$_2$	117	82
Si,(C)$_2$	$C_5H_{15}O_3PSi$	P(OMe)$_2$(OTms)	126.2	81
	$C_7H_{19}O_3PSi$	P(OEt)$_2$(OTms)	128.0	3
		P(OEt)$_2$(OTms)	126.4	3
	$C_{11}H_{17}O_3PSi$	P(OBu)$_2$(OTms)	119.0	81

Rings/connectivities	Formula	Structure	NMR Data (δP[solv][a]J_{PZ} Hz)	Ref.
$(C)_3$	$C_3H_9O_3P$	$P(OMe)_3$	141-139.6	3
	$C_4H_{11}O_4P$	$P(OMe)_2(OC_2H_4OH)$	135	71
	$C_5H_{13}O_3PS$	$P(OMe)_2(OC_2H_4SMe)$	145.0	72
	$C_6H_{15}Cl_3O_3P$	$P(OC_2H_4Cl)_3$	138.7	3
	$C_6H_{15}O_3P$	$P(OEt)_3$	140-137.1	3
	$C_6H_{16}NO_3P$	$P(OMe)_2(OC_2H_4NMe_2)$	145.0	72
	$C_7H_{13}O_3P$	$P(OEt)_2(OCH_2CCH)$	137	3
	$C_7H_{14}NO_3P$	$P(OEt)_2OCHMeCN$	138.2	3
	$C_7H_{15}O_3P$	$P(OEt)_2(OCH_2CH{=}CH_2)$	141.1	3
	$C_7H_{17}O_3P$	$P(OEt)_2(O\text{-}iPr)$	137.2	3
	$C_7H_{18}NO_3P$	$P(OEt)_2OC_2H_4NHMe$	139	26
	$C_8H_{17}O_4P$	$P(OMe)_2OCMe_2CH_2C(O)Me$	135.3	214
	$C_8H_{19}O_3P$	$P(OEt)_2(O\text{-}tBu)$	135-130.5	3
		$P(OEt)(O\text{-}iPr)_2$	137	3
	$C_8H_{20}O_3P$	$(P(OEt)_2)_2O$	127	79
	$C_9F_{18}N_3O_3P$	$P(OCTf_2CN)_3$	75.9	213
	$C_9H_3F_{18}O_3P$	$P(OCHTf_2)_3$	141.0	212
	$C_9H_9O_3P$	$P(OCH_2C{=}CH)_3$	135	3
	$C_9H_{15}Br_3Cl_3O_3P$	$P(OCH_2CHBrCH_2Cl)_3$	141.5	3
	$C_9H_{15}Cl_6O_3P$	$P(OCH_2CHClCH_2Cl)_3$	141.7	3
	$C_9H_{17}O_3P$	$P(O\text{-}iPr)_2(OCH_2CCH)$	137.3	3
	$C_9H_{18}Cl_3O_3P$	$P(OCH_2CHClMe)_3$	142.8	3
	$C_9H_{21}O_3P$	$P(O\text{-}iPr)_3$	138-136.9	3
		$P(OEt)_2(ONp)$	137.2	3
	$C_{10}H_{23}O_3P$	$P(OEt)(OtBu)_2$	130.8	3
	$C_{11}H_{17}O_3P$	$P(OEt)_2(OBz)$	138	3
	$C_{12}H_{23}O_3P$	$P(OEt)(OC(Me)_2CHCH_2$	132.1	3
		$P(OBu)_3$	137.7-142.6	3
		$P(O\text{-}iBu)_3$	138.2-139.8	3
		$P(O\text{-}tBu)_3$	138.2	3
	$C_{12}H_{27}O_3PS_3$	$P(OC_2H_4SEt)_3$	139.1	3
	$C_{12}H_{30}N_3O_3P$	$P(OC_2H_4NMe_2)_3$	139.1	3
	$C_{15}H_3F_{24}O_3P$	$P(OCF(CH_2)_4)_3$	139.8	73
	$C_{15}H_{33}O_3P$	$P(OCH(Me)(Pr))_3$	140	74
		$P(OCH(Me)(iPr))_3$	141	74
		$P(ONp)_3$	137-139	74
	$C_{21}H_{21}O_3P$	$P(OBz)_3$	138.8	3
	$C_{27}H_{51}O_6N_3P$	$P(OCHCH_2CMe_2NCMe_2CH_2)_3$	132 b	12
	$C_{7\text{-}13}H_{13\text{-}25}NO_2P$	$P(OR)_2(OCH(CCl_3)$ $NCH_2CH_2CH_2O)$	138-140	76
$(C)_2,C'$	$C_8H_{14}ClO_4P$	$P(OEt)_2(OCH{=}C(Cl)COMe)$	131.8	77
	$C_9H_{17}O_4P$	$P(OEt)_2(OC(Me){=}CHC(O)Me)$	132	205
	$C_{10}H_{16}NOP$	$P(OEt)_2(OC_6H_4NH_2\text{-}2)$	136	64
	$C_{10}H_{15}O_2P$	$P(OEt)_2(OC_6H_4OH\text{-}2)$	136	64
	$C_{10}H_{18}ClO_4P$	$P(OPr)_2(OCHC(Cl)COMe)$	131.8	77
	$C_{10}H_{21}O_4P$	$P(OPr)_2(OC(Me){=}CHC(O)Me)$	130	205
	$C_{12}H_{11}O_3P$	$P(OMe)_2(OPh)$	135.2	3
	$C_{12}H_{14}O_6P_2$	$P(OEt)_2(OC_6H_4OP(OEt_2)\text{-}2)$	132	64
	$C_{14}H_{30}O_3P_2$	$P(OEt)_2(OC(OEt)CHP(iPr)_2)$	133	50
	$C_{16}H_{34}O_3P_2$	$P(OiPr)_2(OC(OEt)CHP(iPr)_2)$	147.6	50
$C,(C')_2$	$C_{22}H_{23}O_3P$	$P(OEt)(OC(CH_2)CH{=}CHPh)_2$	131.3	78
$(C')_3$	$C_{18}H_{15}O_6P$	$P(OC(O)Ph)_3$	130	215
	$C_{18}H_{15}O_3P$	$P(OPh)_3$	125-129	3
	$C_{19}H_{16}ClO_3P$	$P(OPh)(OC_6H_4Cl\text{-}4)(OC_6H_4Me\text{-}4)$	127.8(a)	11
	$C_{21}H_{21}O_3P$	$P(OC_6H_4Me\text{-}4)_3$	127.6	3

TABLE C (continued)
Compilation of ^{31}P NMR Data of Three Coordinate (λ3 σ3) Phosphorus Compounds Containing Phosphorus Bonds to Chalcogenides (O, S, Se, Te) But No Bonds to Halogen

Rings/connectivities	Formula	Structure	NMR Data (δP[solv]^{a}J_{PZ} Hz)	Ref.
P-bound atoms = O_3				
5/P,$(C)_2$	$C_4H_8O_5P_2$	PGlc(OPGlc)	121.4	3
5/P,$(C')_2$	$C_8H_{12}O_5P$	(POCMe=CMeO)$_2$O	116	20
5/Si,$(C')_2$	$C_9H_{21}O_3PSi$	P(Pnc)(OTms)	126.9(d)	16
5/$(C)_3$	$C_3H_7O_3P$	P(Glc)(OMe)	132.6	12
	$C_4H_9O_3P$	P(Glc)(OEt)	133.5	2, 83
		P(OCHMeCH$_2$O)(OMe)	139.6^{C}, 135.3^{T}	12
	$C_4H_9O_4P$	P(Glc)(OC$_2$H$_4$OH)	134	84
	$C_5H_{11}O_3P$	P(OCHMeCHMeO)(OMe)	135, 150^{M}, 140^{dl}	85
		P(OCMe$_2$CH$_2$O)(OMe)	145	85
		P(Glc)(OPr)	134.4	3
		P(OCH$_2$CHMeO)(OEt)	134	3
	$C_6H_{11}O_3P$	P(OCH(CH$_2$OCH$_2$)CHO)(OEt)	134	86
	$C_6H_{12}O_6P_2$	PGlc(OC$_2$H$_4$OPGlc)	138	87
	$C_6H_{13}O_3P$	P(Glc)(OBu)	134	87
		P(Glc)(O-iBu)	134	3
		P(OC$_2$H$_4$O)(O-tBu)	134	87
		P(OCHMeCMe$_2$O)(OMe)	137, 150^{I}	85
		P(OCHMeCH$_2$O)(O-iPr)	139, 142^{I}	85
	$C_7H_{11}O_7P$	P(OCH(COOMe)CH-(COOMe)O)(OMe)	139, 140^{I}	15
	$C_7H_{15}O_3P$	P(OCH(tBu)CH$_2$O)(OMe)	134, 142^{I}	85
		P(Pnc)(OMe)	147	85
		P(OCHMeCH$_2$O)(O-tBu)	140.2^{C}, 137.8^{T}	
	$C_8H_{15}O_3P$	P(O-1,2-cHexO)(OEt)	136-139	86
	$C_{10}H_{17}O_7P$	P(OCH(COOEt)CH-(COOEt)O)(OEt)	139	15
	$C_{10}H_{13}O_4P$	P(Glc)(OCH$_2$C$_6$H$_4$OMe-4)	135.4	3
	$C_{10}H_{13}O_4P$	P(Glc)(OCH$_2$C$_6$H$_4$OMe-4)	135.4	3
	$C_{10}H_{20}O_6P_2$	P$_a$(Glc)OC$_2$H$_4$OP$_b$(Pnc)	P$_a$ = 138.5-140	87
			P$_b$ = 153-154.4	87
		P(Glc)OC$_2$Me$_4$OP(Glc)	154.5	87
	$C_{10}H_{21}O_3P$	P(Pnc)(OBu)	149	87
		P(Pnc)(O-tBu)	143-144	87
	$C_{11}H_{14}NO_6P$	P(Glc)(OC$_2$H$_4$OC(O)NHPh)	132	84
	$C_{12}H_{24}NaO_4P$	P(Pnc)(OC$_2$Me$_4$ONa)	140.0	88
	$C_{13}H_{19}N_2O_4P$	P(Glc)(OCHMeCH$_2$N-(Me)C(O)NHPh	141.0	84
	$C_{20}H_{25}O_3P$	P(Pnc)(OCHPhCH$_2$Ph)	147	84
	$C_{20}H_{25}O_4P$	P(Pnc)(OCHPhCH(OH)Ph)	147.0	84
5/$(C)_2$,C'	$C_4H_7O_4P$	P(Glc)(OC(O)Me)	127.2	3
	$C_6H_9O_3P$	P(Glc)(OC(=CH$_2$)CH=CH$_2$)	127.0	78
	$C_8H_9O_3P$	P(Glc)(OPh)	120-128.1	3
	$C_8H_{13}O_3P$	P(Glc)(OC(=CH$_2$)CH=CMe$_2$)	126.5	78
	$C_{10}H_{11}O_4P$	P(OCMe$_2$C(O)O)(OPh)	127.0	27
	$C_{12}H_{13}O_3P$	P(OC$_2$H$_4$O)(OC(=CH$_2$)CH=CHPh)	127	78
5/C,$(C')_2$	$C_6H_{11}O_3P$	P(OCH=CHO)(OBu)	121.5(g)	19
	$C_8H_{15}O_3P$	P(OCMe=CMeO)(OBu)	120.0(g)	19

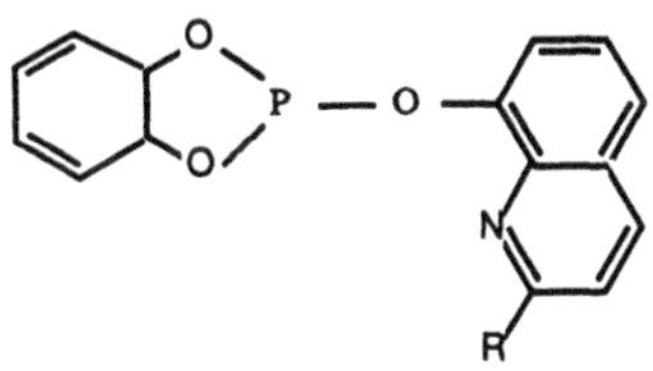

9

Rings/connectivities	Formula	Structure	NMR Data (δP[solv]*J_{PZ} Hz)	Ref.
	$C_8H_8ClO_3P$	PCat(OC_2H_4Cl)	139.1	89
	$C_8H_9O_4P$	PCat(OC_2H_4OH)	133.4	89
	$C_8H_9O_3P$	PCat(OEt)	127.3	217
	$C_9H_{10}O_4ClO_3P$	PCat(OCH_2CHClMe)	127.5	3
	$C_{10}H_{12}ClO_3P$	PCat(OCHClCHMe_2)	130.0	90
	$C_{10}H_{14}NO_3P$	PCat($OC_2H_4NMe_2$)	137.5	89
	$C_{10}H_{17}NO_3P$	Ṗ(O-1,2-c-HexeneȮ)(OBu)	122.0(g)	19
	$C_{10}H_{19}O_3P$	Ṗ(OCEt=CEtȮ)(OBu)	122.0(g)	19
	$C_{11}H_9O_3P$	2,3-Don(OMe)	127.8	22
	$C_{12}H_{11}O_3P$	2,3-Don(OEt)	128.9	22
5/(C′)$_3$	$C_8H_4F_3O_4P$	PCatO(C(O)Tf)	130	218
	$C_{12}H_{13}Cl_2O_3P$	PCatO(C(tBu)=CCl_2)	134.9	219
	$C_6H_9O_3P$	Ṗ(OCMe=CMeȮ)(OCH=CH_2)	119.0	71
	$C_7H_{11}O_3P$	Ṗ(OCMe=CMeȮ)(OCMe=CH_2)	122	20
	$C_8H_{11}O_3P$	Ṗ(O-1,2-c-HexeneȮ)(OCH=CH_2)	120	20
	$C_9H_{11}O_3P$	Ṗ(O-1,2-c-HexeneȮ)(OCMe=CH_2)	124	20
	$C_{10}H_9Cl_2O_4P$	PCatO(C(C=CH_2)CCl_2C(O)Me)	130.2	91
	$C_{12}H_9O_3P$	PCat(OPh)	128	92
	$C_{12}H_{11}O_3P$	1,2-Don(OEt)	123.0	22
	$C_{12}H_{18}O_5P_2$	PCat(OCH(iPr)PO$(OMe)_2$)	130	90
	$C_{13}H_{13}O_3P$	1,2-Don(OPr)	123.5	22
		2,3-Don(OPr)	127.9	22
	$C_{14}H_{20}NO_4P$	PCat(OCHCH_2CMe_2N-(O)CMe_2CH_2)	129.0(b)	12
	C_{10}—C_{18}	PCat(OCR^1R^2PO$(OR^3)_2$	127-142	3
	$C_{13}H_{10}BrO_3P$	2,3-Don(OC_6H_4Br-4)	128.1	22
	$C_{15}H_{10}NO_3P$	**9**; R = H	126	92
	$C_{16}H_{11}O_3P$	2,3-Don(OPh)	127.5	22
		1,2-Don(OPh)	129.8	22
	$C_{16}H_{10}BrO_3P$	1,2-Don(OC_6H_4Br-4)	127.3	94
	$C_{16}H_{12}NO_3$	**9**; R = CH_3	125	93
6/(P)$_2$,C′	$C_{42}H_{63}O_6P_3$	[OP$OC_6H_3$$(tBu)_2$-2,6]$_3$	127.5(t), 119.7(d)	101
	$C_{42}H_{69}O_9P_3$	[OP$OC_6H_2$$(tBu)_2$-2,6-OMe-4]$_3$	128.3(t), 120.3(d)	101
		[OP$OC_6H_2$$(tBu)_2$-2,6-Me-4]$_3$	127.9(t), 119.0(d)	101
	$C_{54}H_{87}O_6P_3$	[OP$OC_6H_2$$(tBu)_2$-2,6-Bu-4]$_3$	127.8(t), 119.6(d)	101
	$C_{54}H_{97}O_6P_3$	[OPOC_6H_2(tBu)]$_3$	128.1(t), 119.4(d)	101
6/P, (C)$_2$	$C_5H_{12}O_5P$	ȮC_2H_4P(O)(OMe)OṖ(OMe)	132.5	100
6/Si,(C)$_2$	$C_7H_{17}O_3PSi$	(TmsO)Ṗ(OC_2H_4CHMeȮ)	120.1^C, 114.0^T	98
6/(C)$_3$	$C_4H_9O_3P$	(MeO)ṖOC_3H_6Ȯ)	130.1	95
	$C_5H_{11}O_3P$	(MeO)Ṗ(OCHMeC_2H_4Ȯ)	129.8^C, 123.5^T	95

TABLE C (continued)
Compilation of ^{31}P NMR Data of Three Coordinate (λ3 σ3) Phosphorus Compounds Containing Phosphorus Bonds to Chalcogenides (O, S, Se, Te) But No Bonds to Halogen

Rings/connectivities	Formula	Structure	NMR Data (δP[solv][a]J_{PZ} Hz)	Ref.
		$(EtO)\dot{P}(OC_3H_6\dot{O})$	128-132	3
	$C_6H_{13}O_3P$	$(MeO)\dot{P}(OCH_2CMe_2CH_2\dot{O})$	122.6	95
		$(MeO)\dot{P}(OCHMeCH_2CHMe\dot{O})$	131.5^C, 127.2^T	95
	$C_6H_{14}O_5P$	$\dot{O}CH_2P(O)(OEt)CH_2O\dot{P}(OEt)$	124.3, 113.7	99
	$C_7H_{13}O_3P$	$(MeO)\dot{P}(OCH(1,2\text{-}cHex)CHCH_2\dot{O})$	132.0^C, 129.3^T	95
	$C_7H_{14}ClO_3P$	$(ClC_2H_4O)\dot{P}(OCH_2CMe_2CH_2\dot{O})$	122.6	3
	$C_7H_{15}O_3P$	$(tBuO)\dot{P}(OC_3H_6\dot{O})$	126.2	30
		$(EtO)\dot{P}(OCH_2CMe_2CH_2\dot{O})$	121.6	30
		$(MeO)\dot{P}(OCHMeCH_2CMe_2\dot{O})$	130.4^C,128.8^T	30
	$C_7H_{17}O_3P$	$(tBuO)\dot{P}(OC_2H_4CHMe\dot{O})$	129.3^C, 123.3^T	30
	$C_8H_{15}O_3P$	$(EtO)\dot{P}O\text{-}1,3\text{-}cHex\dot{O}$	124.0	3
	$C_8H_5O_3P$	$(EtO)\dot{P}(OCH_2\dot{C}H(CH_2CH_2CH_2)\text{-}\dot{C}H\dot{O})$	125.0	3
	$C_8H_{16}ClO_3P$	$(MeCHCl\text{-}CH_2O\dot{P}(OCH_2CMe_2CH_2\dot{O})$	121.3	3
	$C_8H_{17}O_3P$	$(MeO)\dot{P}(OCH_2CH(tBu)CH_2\dot{O})$	124.6^C, 131.4^T	96a
		$(PrO)\dot{P}(OCH_2CMe_2CH_2\dot{O})$	121.3	3
	$C_9H_{17}O_3P$	$(EtO)\dot{P}(OCH_2\dot{C}H(CH_2CH_2CH_2)\text{-}\dot{C}H\dot{O})$	125.0	3
	$C_9H_{19}O_3P$	$(MeO)\dot{P}(OCH_2CMe_2CH(iPr)\dot{O})$	132.8^C, 127.1^T	30
		$(MeO)\dot{P}(OCH(iPr)CMe_2CH_2\dot{O})$	134.0^C, 127.4^T	97
	$C_{10}H_{21}O_3P$	$(EtO)\dot{P}(OCH_2CMe_2CH(iPr)\dot{O})$	133.1^c, 127.1^T	30, 34
	$C_{11}H_{23}O_3P$	$(iPrO)\dot{P}(OCH_2CMe_2CH(iPr)\dot{O})$	132.8^C, 124.9^T	30
	$C_{12}H_{24}ClO_3P$	$(PeCHCl\text{-}CH_2O)\dot{P}(OCH_2CMe_2CH_2\dot{O})$	123.1	3
	$C_{17}H_{31}NO_8P$	$EtO\dot{P}OCH(CO_2Et,NH)CH[O\dot{C}H\text{-}(OCMe_2O\dot{C}H]CH\dot{O}$	137	97
6/(C)$_2$,C′	$C_6H_{11}O_4P$	$(MeC(O)O)\dot{P}(OC_2H_4CHMe\dot{O})$	117.5^T	30
	$C_{10}H_9Cl_2O_3P$	$\dot{O}CH_2CHClCHClCH_2O\dot{P}(OPh)$	128	107
	$C_{10}H_{11}O_3P$	$\dot{O}CH_2CH{=}CHCH_2O\dot{P}(OPh)$	130.4	108
	$C_{10}H_{13}O_3P$	$(PhO)\dot{P}(OC_2H_4CHMe\dot{O})$	124.7^C, 122.3^T	30
	$C_{11}H_{15}O_3P$	$(PhO)\dot{P}(OCH_2CMe_2CH_2O)$	114.8	3
	$C_{15}H_{15}O_4P$	$PhC(O)C_6H_4O\dot{P}OC_3H_6\dot{O}$	122.8	220
	$C_{16}H_{17}O_4P$	$PhC(O)C_6H_4O\dot{P}OCHMeC_2H_4\dot{O}$	119.8	220
	$C_{17}H_{19}O_4P$	$PhC(O)C_6H_4O\dot{P}OCH_2CMe_2CH_2\dot{O}$	114.7	220
	$C_{13\text{-}21}H_{26\text{-}41}O_7P_2$	$(RO)_2P(O)CH{=}CHO\dot{P}\text{-}(OCH_2CMe_2CH_2\dot{O})$	118-119	3
6/C,(C′)$_2$	$C_7H_9Cl_2O_3P$	$(EtO)\dot{P}(OC({=}CH_2)CCl_2C({=}CH_2)(\dot{O})$	118.0	132
	$C_7H_{11}O_3P$	$(EtO)\dot{P}(OC({=}CH_2)CH{=}C(Me)\dot{O})$	116.0	205
	$C_8H_{13}O_3P$	$(iPr)\dot{P}(OC({=}CH_2)CH{=}C(Me)\dot{O})$	114	205
	$C_{11}H_9O_3P$	1,8-Don(OMe)	112.5	22
	$C_{12}H_{11}O_3P$	1,8-Don(OEt)	112.0	94
	$C_{13}H_{13}O_3P$	1,8-Don(OPr)	114.0	94
6/(C′)$_3$	$C_{16}H_{10}BrO_3P$	1,8-Don(OC_6H_4Br-4)	104.6	94
	$C_{16}H_{11}O_3P$	1,8-Don(OPh)	104.0	94
7/(C$_2$),C′	$C_{14}H_{13}O_3P$	**10**	124.5	108
7/C,(C′)$_2$	$C_{29}H_{43}O_3P$	**11**; R = CH_2CHMe_2, R^1 = Me	153.6	109

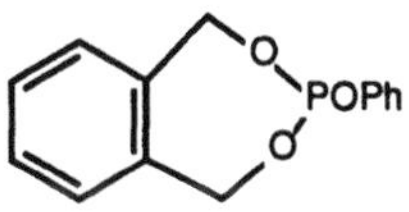

10

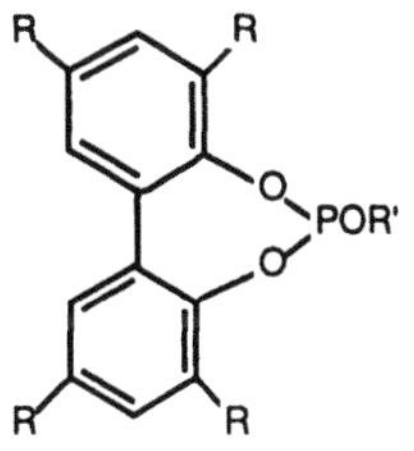

11

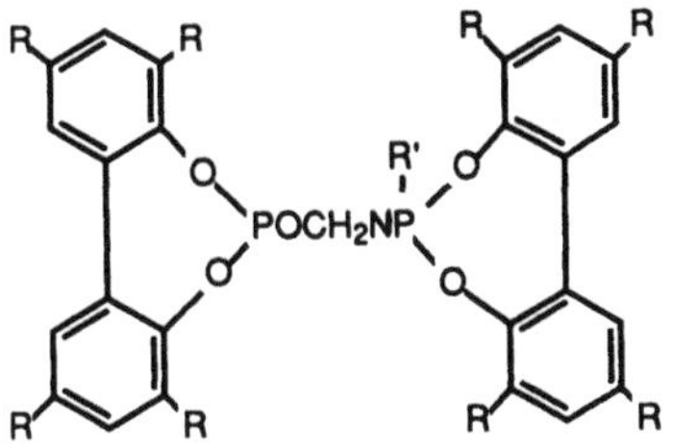

12

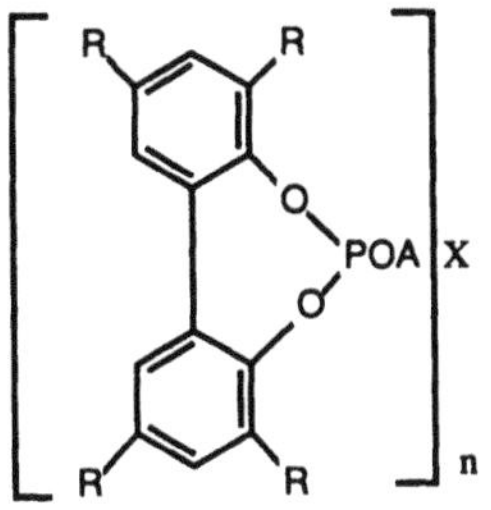

13

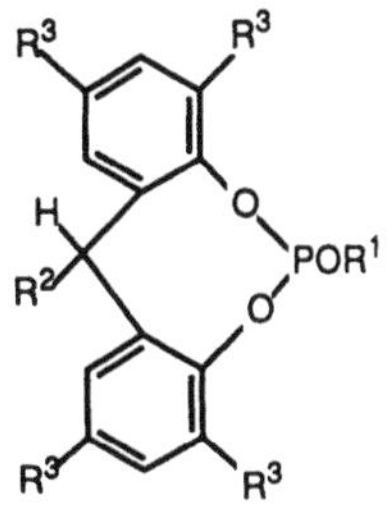

14

Rings/connectivities	Formula	Structure	NMR Data (δP[solv][a]J_{PZ} Hz)	Ref.
	$C_{32}H_{49}O_3P$	**11**; R = CH_2CHMe_2, R^1 = CH_2CHMe_2	162.2	109
	$C_{59}H_{87}NO_5P_2$	**12**; R = tBu, R^1 = Me	146.9	42
	$C_{60}H_{89}NO_5P_2$	**12**; R = tBu, R^1 = Et	147.8	42
	$C_{62}H_{93}NO_5P_2$	**12**; R = tBu, R^1 = Bu	146.7	42
	$C_{64}H_{97}NO_5P_2$	**13**; A = C_2H_4, N = 2, X = NtBu	136.0[C]	42
	$C_{117}H_{168}O_{12}P_4$	**13**; A = CH_2, n = 4, X = C	138.4	110
8/(C)$_3$	$C_5H_1O_4P$	$\dot{O}C_2H_4OC_2H_4O\dot{P}(OMe)$	138.6(b)	41
	$C_{24}H_{51}N_3O_6P_2$	tBuN-($C_2H_4O\dot{P}OC_2H_4N(tBu)C_2H_4\dot{O})_2$	137.8	106
11/(C)$_3$	$C_9H_{21}N_2O_3P$	$\dot{O}[C_2H_4N(Me)]_2C_2H_4O\dot{P}(OMe)$	139	105
8/C,(C′)$_2$	$C_{30}H_{45}O_3P$	**14**; R^1 = Me, R^2 = H, R^3 = tBu	129.5(b)	111
	$C_{31}H_{47}NO_3P$	**14**; R^1 = C_2H_4NHMe, R^2 = H, R^3 = tBu	129.2	42
	$C_{31}H_{47}O_3P$	**14**; R^1 = Me, R^2 = Me, R^3 = tBu	129.0[C]	111
	$C_{31}H_{52}F_3O_3P$	**14**; R^1 = CH_2Tf, R^2 = H, R^3 = tBu	125.9	112
	$C_{31}H_{55}O_3P$	**14**; R^1 = Et, R^2 = H, R^3 = tBu	128.6	112
	$C_{32}H_{49}NO_3P$	**14**; R^1 = C_2H_4NHMe, R^2 = H, R^3 = tBu	129.5	42
	$C_{33}H_{51}O_3P$	**14**; R^1 = Me R^2 = nPr, R^3 = tBu	129.0[C]	111
	$C_{35}H_{55}NO_3P$	**14**; R^1 = C_2H_4NH-tBu, R^2 = H, R^3 = tBu	130.0	42
	$C_{36}H_{57}NO_3P$	**14**; R^1 = C_2H_4NH-tBu, R^2 = Me, R^3 = tBu	129.5	42
	$C_{40}H_{67}O_3P$	**14**; R^1 = Oct, R^2 = Pr, R^3 = tBu	129.1[C]	111

TABLE C (continued)
Compilation of ^{31}P NMR Data of Three Coordinate (λ3 σ3) Phosphorus Compounds Containing Phosphorus Bonds to Chalcogenides (O, S, Se, Te) But No Bonds to Halogen

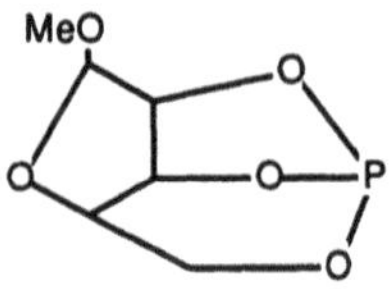

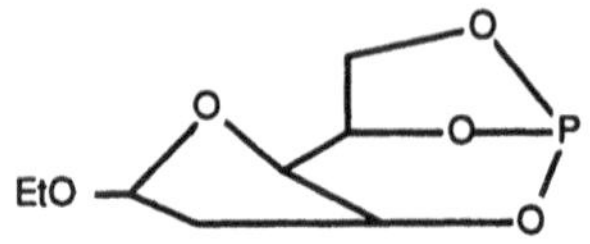

Rings/connectivities	Formula	Structure	NMR Data (δP[solv][a]J_{PZ} Hz)	Ref.
8/(C′)$_3$	$C_{36}H_{49}O_3P$	**14**; R[1] = Ph, R[2] = H, R[3] = tBu	128.7	113
5,6/(C)$_3$	$C_7H_9O_4P$	**15**	127.6	102
	$C_8H_{13}O_5P$	**16**	115.0	103
	$C_8H_{13}O_3PS$	**17**	117.2	103
	$C_9H_{13}O_7P$	**18**	117.0	102
	$C_9H_{13}O_6P$	**19**	117.0	104
		20	117.0	30
	$C_{24}H_{21}O_5PSi$	**21**; X = $SiPh_3$	124.2	102
5,6/(C)$_2$,C′	$C_9H_{11}O_7P$	**22**	126	97
<u>P-bound atoms = PS_2</u>				
4,5/P[4]S;P[4]P	P_4S_5	**23**; P_b	120	3
<u>P-bound atoms = P_2S</u>				
3,5/(PS)$_2$;P	P_4S_3	**24**; E = S, P_b	−120; −103	3
4,5/(S$_2$)$_2$;P	P_4S_5	**23**; P_c	20	3
<u>P-bound atoms = P_2Se</u>				
3,5/(PSe)$_2$;P	P_4Se_3	**24**; E = Se, P_b	−106	3

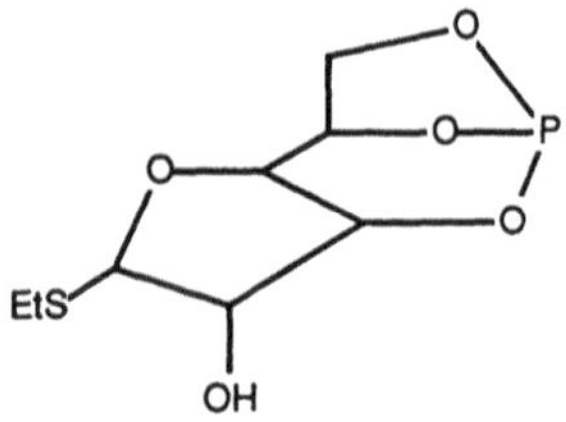

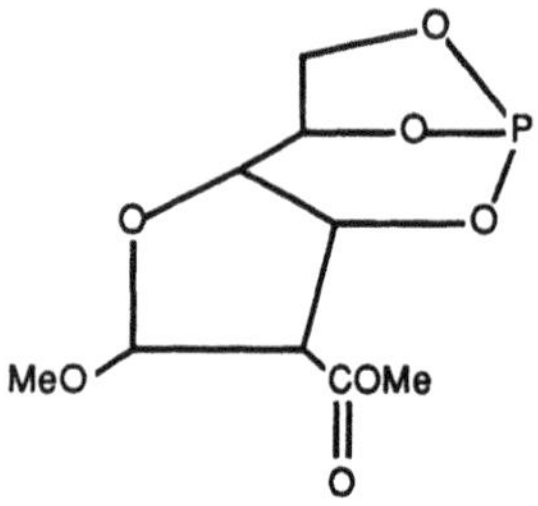

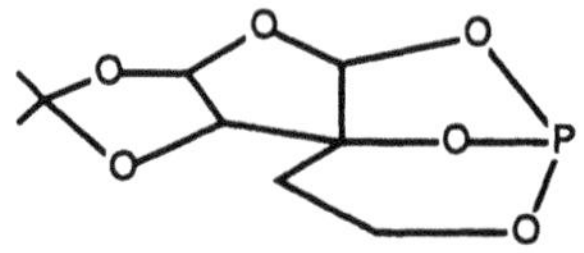

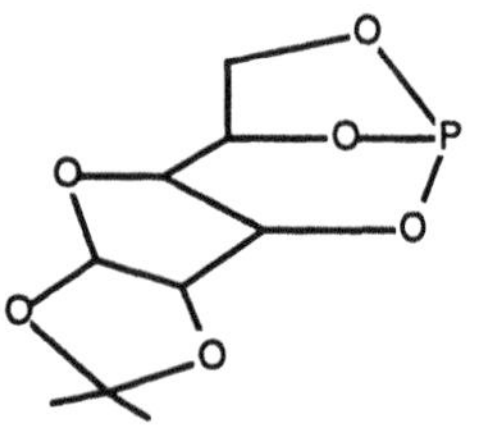

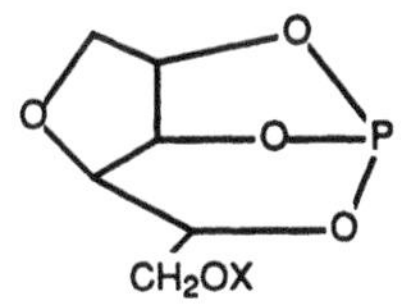

21 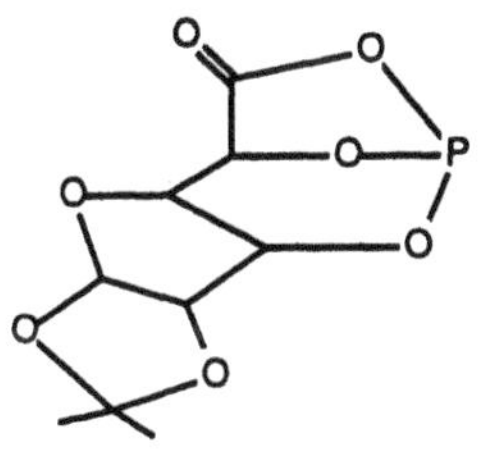 **22** 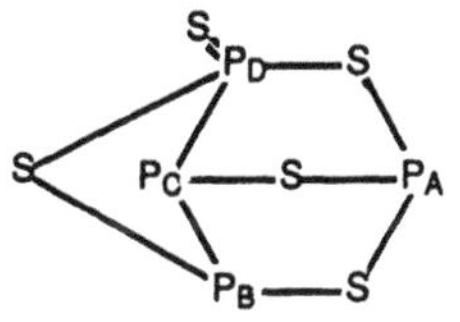 **23**

Rings/connectivities	Formula	Structure	NMR Data (δP[solv][a]J_{PZ} Hz)	Ref.
P-bound atoms = S_3				
P^4,(C)$_2$	$C_8H_{20}O_2P_2S_4$	$(EtS)_2PSP(S)(OEt)_2$	121 J_{PP}51	228
(C)$_3$	$C_3H_9PS_3$	$P(SMe)_3$	124.5, 125.0	114, 115
	$C_6H_{15}PS_3$	$P(SMe)_2(SBu)$	121.7	3
		$P(SEt)_3$	114.7, 115.5	116, 115
	$C_9H_{21}PS_3$	$P(SMe)(SBu)_2$	118.0	3
	$C_{12}H_{27}PS_3$	$P(SBu)_3$	116-117	3
	$C_{21}H_{21}PS_3$	$P(SCH_2Ph)_3$	198	117
(C′)$_3$	$C_{18}H_{15}PS_3$	$P(SPh)_3$	130.5, 180.0	3, 115
5/(C)$_3$	$C_6H_{12}P_2S_3$	$(\dot{S}C_2H_4S\dot{P}SCH_2)_2$	110b	67
	$C_6H_{13}PS_3$	$\dot{S}C_2H_4S\dot{P}SBu$	108.0	26
5,5/(C)$_3$	$C_9H_{17}PS_3$	$P(SCH_2)_3CC_5H_{11}$	32.8	117a
5/C,(C′)$_2$	$C_{11}H_{13}PS_3$	$\dot{S}$-(1,2-C_6H_3Me-4)$S\dot{P}SBu$	108.0	23
5,5/P^4,(P)$_2$	P_4S_3	23, P_a	72	3
	P_4S_3	**24**; E = S, P_a	200	3
5,5/(P)$_3$	$C_2H_6P_4S_3$	**25**; X = SMe	176, 187.4[I]	118
	$C_4H_{19}P_4S_3$	**25**; X = SEt	172.4, 186.2[I]	118
	$C_{12}H_{10}P_4S_5$	**25**; X = SPh	175.7, 187.4[I]	118
	$I_2P_4S_3$	**25**; X = I	195.9	118
P-bound atoms = Se_3				
5/5(P)$_3$	P_4Se_3	**24**; E = S, P_a	38	3
(C)$_3$	$C_3H_9PSe_3$	$(MeSe)_3P$	107	114
P-bound atoms = HNO				
Si;C	$C_7H_{20}NOPSi_2$	HP(NTms$_2$)(OEt)	109.5	119
	$C_9H_{24}NOPSi_2$	HP(NTms$_2$)(OPr)	87	119
	$C_{11}H_{28}NOPSi_2$	HP(NTms$_2$)(OBu)	111.7	119
Si;C$_2$	$C_5H_{16}NOPSi$	HP(NMe$_2$)(OTms)	121.9 J_{PH}216.6	120
C$_2$;C	$C_4H_{12}NOP$	HP(NMe$_2$)(OEt)	13 J_{PH}640	205
	$C_5H_{14}NOP$	HP(NMe$_2$)(O-iPr)	13.5	205
5/CC′;C′	$C_6H_{14}NOP$	$\dot{O}C_3H_6N$(iPr)$\dot{P}$h	112	34
P-bound atoms = HO_2				
(Si)$_2$	$C_6H_{19}O_2PSi_2$	HP(OTms)$_2$	140 J_{PH}176	119
(C)$_2$	$C_2H_7O_2P$	HP(OMe)$_2$	168, 171.5	121, 122
	$C_4H_{12}O_2P$	HP(OEt)$_2$	160	123
	$C_{10}H_{23}O_2P$	HP(O-neoPe)$_2$	162	121

TABLE C (continued)
Compilation of ^{31}P NMR Data of Three Coordinate (λ3 σ3) Phosphorus Compounds Containing Phosphorus Bonds to Chalcogenides (O, S, Se, Te) But No Bonds to Halogen

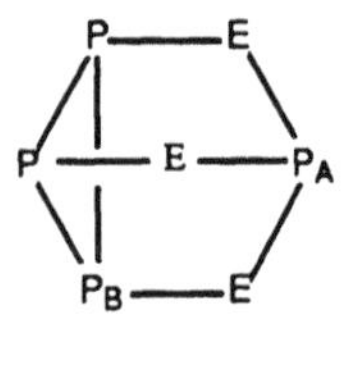

24 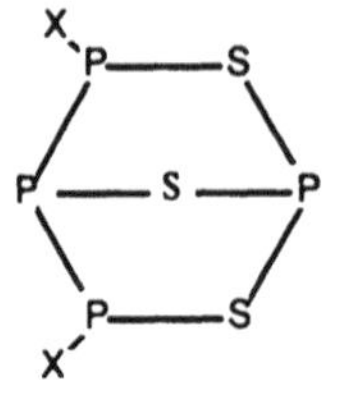 **25** 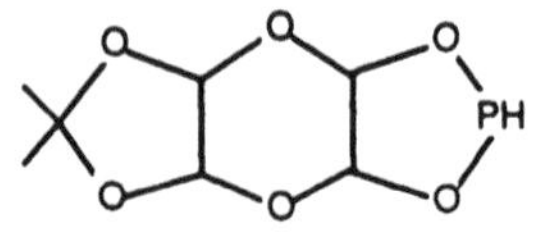 **26**

Rings/connectivities	Formula	Structure	NMR Data (δP[solv]ⁿJ$_{PZ}$ Hz)	Ref.
5/(C)$_2$	$C_7H_{11}O_6P$	**26**	139	30
5/(C')$_2$	$C_4H_7O_2P$	OCMe=CMeOPH	165 J_{PH}160	20
	$C_6H_5O_2P$	CatPH	164 J_{PH}157	20
6/(C)$_2$	$C_3H_7O_2P$	HP(OC$_3$H$_6$O)	154 J_{PH}169	124
	$C_4H_9O_2P$	HP(OC$_2$H$_4$CHMeO)	149.2	34
	$C_5H_{11}O_2P$	HPOCH$_2$CMe$_2$CH$_2$O	147.3	34
	$C_6H_{13}O_2P$	HPOCHMeCH$_2$CMe$_2$O	119.1 J_{PH}164, 105.4 J_{PH}158[1]	34
	$C_8H_{16}O_2P$	HPOCH$_2$CMe$_2$CH(iPr)O	151.9 J_{PH}166, 131.8 J_{PH}164[1]	34
P-bound atoms = HS$_2$				
(C)$_2$	$C_2H_7PS_2$	HP(SMe)$_2$	29 J_{PH}214	124a
	$C_4H_{11}PS_2$	HP(SEt)$_2$	7 J_{PH}214	124a
	$C_6H_{15}PS_2$	HP(SPr)$_2$	10 J_{PH}215	124a
6/(C)$_2$	$C_3H_7PS_2$	HPSC$_3$H$_6$S	−1.5	76
	$C_4H_9PS_2$	HPSC$_2$H$_4$CHMeS	−4.5, −45.3[1]	76
P-bound atoms = COP				
CH$_2$;C;C$_2$O'	$C_{16}H_{36}O_2P_2$	BuP(OBu)P(O)tBu$_2$	135 J_{PP}288	231
C$_2$H;C;C$_2$O'	$C_{13}H_{30}O_2P_2$	iPrP(OBu)P(O)iPr$_2$	127 J_{PP}265	231
	$C_{15}H_{34}O_2P_2$	iPrP(OBu)P(O)tBu$_2$	134 J_{PP}324	231
C$_3$;C;C$_2$O'	$C_{14}H_{32}O_2P_2$	tBuP(OBu)P(O)iPr$_2$	140 J_{PP}290	231
	$C_{16}H_{36}O_2P_2$	tBuP(OBu)P(O)tBu$_2$	155 J_{PP}335	231
C$_3$;C';C$_2$O'	$C_{18}H_{33}O_2P_2$	tBuP(OPh)P(O)tBu$_2$	135 J_{PP}308	231
P-bound atoms = COS				
5/CH$_2$;C;C'	$C_9H_{11}OPS$	OC$_3$H$_6$PSPh	79.0	1
P-bound atoms = CO$_2$				
Cl$_3$;(C)$_2$	$C_5H_4Cl_3F_6O_2P$	Cl$_3$CP(OCH$_2$Tf)$_2$	137.2	229
	$C_7H_6Cl_3F_8O_2P$	Cl$_3$CP(OCH(CF$_2$H)$_2$)$_2$	138.9	229
	$C_{11}H_{18}F_3OP$	TfP(O-iBu)$_2$	133.3	230
H$_2$Cl;(C')$_2$	$C_{13}H_{12}ClO_2P$	ClCH$_2$P(OPh)$_2$	153.6	3

Rings/connectivities	Formula	Structure	NMR Data (δP[solv][a]J_{PZ} Hz)	Ref.
$H_2P^4;(C)_2$	$C_9H_{22}O_5P$	$(EtO)_2P(O)CH_2P(OEt)_2$	165.5 J_{PP}41	151
	$C_{13}H_{28}O_5P_2$	$iPr_2P(OC)CH_2P(O\text{-}iPr)_2$	163.1	151
	C_{11}—C_{15}	$R^1COP(O)(OR)CH_2P(OR)_2$	163-166 J_{PP}35-45	169
$H_2P;(C)_2$	$C_{11}H_{24}O_6P$	$[(EtO)_2P]_2CHCOOMe$	168.3	50
$H_3;(Si)_3$	$C_7H_{21}O_2PSi_2$	$MeP(OTms)_2$	171.9	153
H_3:C,Si	$C_5H_{15}O_2PSi$	MeP(OMe)(OTms)	152.0	153
$H_3;(C)_2$	$C_3H_9O_2P$	$MeP(OMe)_2$	182.5, 200.8	147, 3
	$C_5H_{13}O_2P$	MeP(OMe)(O-iPr)	176.6	148
	$C_7H_5F_{12}O_2P$	$MeP(OCHTf_2)_2$	224	149
	$C_7H_{17}O_2P$	$MeP(O\text{-}iPr)_2$	173.0	150
	$C_9H_{21}O_2P$	$MeP(O\text{-}iBu)_2$	177.9	232
$H_3;(C')_2$	$C_{13}H_{13}O_2P$	$MeP(OPh)_2$	178.5	152
	$C_9H_{17}O_4P$	$MeP(OC(O)Pr)_2$	159.7	215
	$C_{15}H_{17}O_4P$	$MeP(OC(O)CH_2Ph)_2$	176.0	215
$CF_2;(C)_2$	$C_{10}H_{10}F_{13}O_2P$	$C_6F_{13}P(OEt)_2$	144.5	162
	$C_{11}H_8F_7O_2P$	$C_3F_7P(O\text{-}iBu)_2$	142.6	230
CH_2;P,C	$C_{16}H_{36}O_2P_2$	$BuP(OBu)(OPBu_2)$	182 J_{PP}93	231
$CH_2;(C)_2$	$C_4H_{11}O_2P$	$EtP(OMe)_2$	188.3(n)	154
	$C_5H_{13}O_2P$	$iPrP(OMe)_2$	157.9	122
	$C_5H_{13}O_3P$	$McOC_2H_4P(OMe)_2$	145.2(n)	155
	$C_5H_{13}O_2P$	$PrP(OMe)_2$	187.6(n)	154
	$C_6H_{15}O_2P$	$BuP(OMe)_2$	187.6(n)	154
	$C_8H_{11}F_8O_2P$	$EtP(OCH_2CF_2CF_2H)_2$	201.8	229
	$C_{12}H_{11}F_{16}O_2P$	$EtP(OCH_2C_3F_6CF_2H)_2$	204.2	229
C_2H;P,C	$C_{13}H_{30}O_2P_2$	$iPrP(OBu)(OP\text{-}iPr_2)$	182 J_{PP}82	231
	$C_{16}H_{36}O_2P_2$	$BuP(OBu)(OP\text{-}tBu_2)$	189 J_{PP}105	231
$C_2H;(C)_2$	$C_9H_{17}O_2P$	$NbP(OMe)_2$	185.3exo, 192.4endo	156
	$C_9H_{13}O_2P$	$(c\text{-}Hex\text{-}4Me)P(OMe)_2$	191.9ax, 189.9eq	158
	$C_{12}H_{25}O_2P$	$c\text{-}HexP(O\text{-}iPr)_2$	179.6	3
	$C_{14}H_{27}O_2P$	$c\text{-}HexP(OBu)_2$	184.2	3
C_3;P,C	$C_{14}H_{32}O_2P_2$	$tBuP(OBu)(OP\text{-}iPr_2)$	180 J_{PP}88	231
	$C_{16}H_{36}O_2P_2$	$tBuP(OBu)(OPBu_2)$	177 J_{PP}98	231
		$tBuP(OBu)(OP\text{-}tBu_2)$	186 J_{PP}90	231
C_3;P,C';	$C_{16}H_{28}O_2P_2$	$tBuP(OPh)(OP\text{-}iPr_2)$	179 J_{PP}90	231
	$C_{18}H_{32}O_2P_2$	$tBuP(OPh)(OPBu_2)$	180 J_{PP}81	231
		$tBuP(OPh)(OP\text{-}tBu_2)$	182 J_{PP}82	231
C_3;S,C	$C_6H_{15}O_4PS$	$tBuP(OSO_2Me)(OMe)$	141.4	80
$C_3;(C)_2$	$C_{10}H_{23}O_2P$	$tBuP(O\text{-}iPr)_2$	178.8	233
$5/C'H_2;(C)_2$	$C_{10}H_{13}O_2P$	$PhCH_2\dot{P}OCHMeC_2\dot{O}$	169^C, 165.2^T	12
$5/C_2H;(C)_2$	$C_8H_{15}O_2P$	c-HexPGlc	187.4(n)	170
$5/C_3;(C)_2$	$C_7H_{15}O_2P$	$tBu\dot{P}OCHMeCH_2\dot{O}$	207.4^C, 201.5^T	12
$6/H_3;(C)_2$	$C_6H_{13}O_2P$	$Me\dot{P}OCH_2CMe_2CH_2\dot{O}$	145.1(c), 163(b)	175, 173
	$C_8H_{17}O_2P$	$Me\dot{P}OCH_2CH(tBu)CH_2\dot{O}$	161.8, 185.2	174
$6/CH_2;(C')_2$	$C_7H_9Cl_2O_2P$	$Et\dot{P}OC(=CH_2)CCl_2C(=CH_2)\dot{O}$	142	91
	$C_8H_{11}Cl_2OP$	$Pr\dot{P}OC(=CH_2)CCl_2C(=CH_2)\dot{O}$	118	91
$7/C_3;(C)_2$	$C_8H_{17}O_2P$	$tBu\dot{P}OC_4H_8\dot{O}$	180.4	179
$8/H_3;(C)_2$	$C_5H_7O_3P$	$Me\dot{P}OCH_2CH(tBu)CH_2\dot{O}$	161.8, 185.2	174
	$C_5H_{11}O_2PS$	$Me\dot{P}OC_2H_4SC_2H_4\dot{O}$	182.8(b)	177
	$C_6H_{13}O_2P$	$Me\dot{P}OC_5H_{10}\dot{O}$	173.2(b)	177
	$C_6H_{14}O_2P_2$	$Me\dot{P}OC_2H_4P(Me)C_2H_4\dot{O}$	175.2, 183.0(b)	178

TABLE C (continued)
Compilation of ^{31}P NMR Data of Three Coordinate ($\lambda 3$ $\sigma 3$) Phosphorus Compounds Containing Phosphorus Bonds to Chalcogenides (O, S, Se, Te) But No Bonds to Halogen

Rings/connectivities	Formula	Structure	NMR Data ($\delta P[solv]^{a}J_{PZ}$ Hz)	Ref.
	$C_8H_{17}O_2P$	$MePOC_2H_4CMe_2C_2H_4O$	174.7(b)	177
	$C_9H_{15}NO_2P$	$MePOC_2H_4N(Me)C_2H_4O$	182.0(b)	177
$12/H_3;(C)_2$	$C_{12}H_{26}O_4P_2$	$[MePOCH_2CMe_2CH_2O]_2$	165.5, 169	180
$14/C_3;(C)_2$	$C_{16}H_{34}O_4P_2$	$tBuPOC_4H_8OP(tBu)OC_4H_8O$	187.1, 186.6	179
$16/H_3;(C)_2$	$C_{10}H_{22}O_6P_2$	$[MePO(C_2H_4O)_2]_2$	176.3(b)	177
	$C_{12}H_{28}N_2O_4P_2$	$[MePOC_2H_4N(Me)C_2H_4O]_2$	176.4, 176.3(b)	177
P-bound atoms = $C'O_2$				
$C'S;(C)_2$	$C_8H_{11}Cl_2O_2PS$	$(SC{=}CHCH{=}CH)P(OC_2H_4Cl)_2$	151	168
$C'H;(C)_2$	$C_8H_{17}O_2P$	$Me_2C{=}CHP(OEt)_2$	158	144a
	$C_{10}H_{20}BrO_3P$	$EtOC(Br){=}CHP(O\text{-}iPr)_2$	147	150
$CC';(C)_2$	$C_{10}H_{20}BrO_3P$	$EtOC(Br){=}C(Et)P(OEt)_2$	159	159
	$C_{10}H_{22}BrO_3$	$EtOC(Br){=}C(iPr)P(OEt)_2$	170	159
	$C_{12}H_{28}O_6P_2$	$(EtO)_2P(O)C(Et){=}C(Me)P(OEt)_2$	159	161
	$C_{13}H_{26}BrO_3P$	$BuOC(Br){=}C(Me)P(O\text{-}iPr)_2$	146.1	159
		$MeOC(Br){=}C(iPr)P(OEt)_2$	170	159
$C'_2;(C)_2$	$C_8H_{11}O_2P$	$PhP(OMe)_2$	159	3
	$C_8H_{16}Br_2O_3P$	$EtOC(Br){=}CHP(OEt)_2$	145	159
	$C_{10}H_{19}F_6O_2P$	$PhP(OCH_2Tf)_2$	167.4	229
	$C_{10}H_{14}NO_4P$	$P(C_6H_4NO_2\text{-}4)(OEt)_2$	133	206
	$C_{10}H_{15}O_2P$	$PhP(OEt)_2$	153.5	160
	$C_{11}H_7F_{12}O_2P$	$PhP(OCHTf_2)_2$	190.0	212
	$C_{11}H_{17}O_2P$	$P(C_6H_4Me\text{-}4)(OEt)_2$	178	3
	$C_{12}H_{16}BrO_2P$	$BrCH{=}C(Ph)P(OEt)_2$	135.0	163
	$C_{12}H_{19}O_2P$	$PhP(O\text{-}iPr)_2$	151.2	3
	$C_{13}H_{23}O_2P$	$PhP(OBu)_2$	157.3, 178	164, 3
	$C_{14}H_{23}O_2P$	$P(C_6H_3Me_2\text{-}2,4)(O\text{-}iPr)_2$	146.9	3
	$C_{16}H_{23}O_4P$	$PhP(OCH_2CHC_3H_6O)_2$	158.8, 157.8, 156.7	1
$C'_2;C,C'$	$C_{14}H_{16}NO_3P$	$PhP(OC_2H_4NH_2)(OC_6H_4OH\text{-}4)$	160	144
$C'_2;(C')_2$	$C_{14}H_{19}O_4P$	$PhP(OC(O)Pr)_2$	142.9	215
	$C_{16}H_{24}NO_2P$	$1,2\text{-}C_6H_4NHCH{=}CNP(OBu)_2$	129	16
	$C_{18}H_{15}O_2P$	$PhP(OPh)_2$	164.9	3
	$C_{18}H_{15}O_4P$	$PhP(OC_6H_4CHO\text{-}4)_2$	158.3	166
	$C_{19}H_{19}O_4P$	$PhP(OC(O)CH_2Ph)_2$	143.2	215
	$C_{20}H_{19}O_4P$	$PhP(OC_6H_4C(O)Me\text{-}4)_2$	164.9	166
	$C_{34}H_{47}O_2P$	$PhP(OC_6H_3(tBu)_2\text{-}2,4)_2$	155.0	167
$5/C'_2;(C)_2$	$C_8H_9O_2P$	PhPGlc	162	169a
	$C_9H_{11}O_2P$	$PhPOCHMeCH_2O$	190.4^{C}, 185.4^{T}	32
$5/C'_2;C,C'$	$C_{13}H_{11}O_2P$	**27**; R = H	162.7(c)	140
	$C_{14}H_{13}O_2P$	**27**; R = Me	160.2^{C}, 157.2^{T}	141
$5/C'_2;(C')_2$	$C_{12}H_8BrO_2P$	$CatPC_6H_4Br\text{-}4$	175.1	171
	$C_{14}H_{13}O_2P$	$CatPC_6H_4Et\text{-}4$	179.4	171
		$CatPC_6H_4Me_2\text{-}3,5$	180.5	171
	$C_{15}H_{15}O_2P$	$CatPC_6H_4iPr\text{-}4$	179.6	171
$6/CO';(C)_2$	$C_6H_{11}O_3P$	$MeC(O)POC_2H_4CHMeO$	44.5	70
$6/C'O';(C)_2$	$C_{11}H_{13}O_3P$	$PhC(O)POC_2H_4CHMeO$	53.4	70
$6/C'_2;(C)_2$	$C_9H_{11}O_2P$	$PhPOCH_2CMe_2CH_2O$	152.2(c)	172

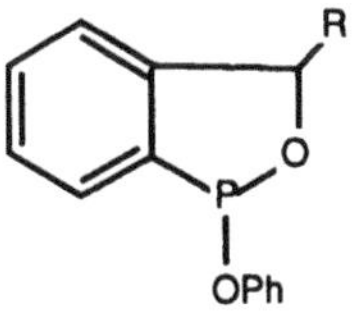

27

Rings/connectivities	Formula	Structure	NMR Data ($\delta P[solv]^{a} J_{PZ}$ Hz)	Ref.
$8/C'_2;(C)_2$	$C_{10}H_{13}O_3P$	$PhPOC_2H_4OC_2H_4O$	168.0(b)	41
	$C_{14}H_{17}NO_2P$	$PhPOC_2H_4N(tBu)C_2H_4O$	161.6(b)	177
$8/C'_2;(C)_2$	$C_{14}H_{18}O_4P_2$	$PhPOC_2H_4OP(Ph)OC_2H_4O$	155.8, 155.2(t)	169a
P-bound atoms = C″O₂				
$5/N'';(C)_2$	$C_7H_{12}NO_2P$	$N{\equiv}CPOC_2Me_4O$	174.6(c)	16
$6/N'';(C)_2$	$C_6H_{10}NO_2P$	$N{\equiv}CPOCH_2CMe_2CH_2O$	100.7	175
P-bound atoms = CPS				
$3/Si_3;SCS;P$	$C_{18}H_{54}P_2Si_3$	1; R = $CTms_3$	−53.9	135
P-bound atoms = C′PS				
$3/C'_2;C'S;P$	$C_{36}H_{58}P_2S$	1; R = SMS	−65.1(c,d)	136
P-bound atoms = CPTe				
$3/C_3;CTe;P$	$C_8H_{18}P_2Te$	tBuPTeP-TBu	−69	137
$4/C_3;CP;P$	$C_{12}H_{27}P_3Te$	tBuPP(tBu)P(tBu)Te	−60.2	137
P-bound atoms = CS₂				
$H_3;(C)_2$	$C_5H_{13}PS_2$	$MeP(SEt)_2$	71.0	115
	$C_7H_{17}PS_2$	$PrP(SEt)_2$	81.0	115
	$C_8H_{19}PS_2$	$BuP(SEt)_2$	80.5	115
$H_3;C,C'$	$C_9H_{13}PS_2$	MeP(SEt)(SPh)	86.5	115
$CH_2;(C)_2$	$C_6H_{15}PS_2$	$EtP(SEt)_2$	83.5	115
$CH_2;(C')_2$	$C_{14}H_{15}PS_2$	$EtP(SPh)_2$	100.0	115
$C_3;(C)_2$	$C_8H_{19}PS_2$	$tBuP(SEt)_2$	11.6	181
$5/H_3;(C)_2$	$C_3H_7PS_2$	$MePSC_2H_4S$	41	87
$6/H_3;(C)_2$	$C_4H_9PS_2$	$MePSC_3H_6S$	23.4	47
$6/C_3;(C)_2$	$C_6H_{15}PS_2$	$tBuPSC_3H_6S$	97.3b	183
	$C_9H_{19}PS_2$	$tBuPSCH_2CMe_2CH_2S$	95.2	181
$8/H_3;(P)_2$	$C_4H_{12}P_4S_4$	$(SPMe)_4$	85.9	184
$13/C_3;(C)_2$	$C_{14}H_{30}P_2S_4$	$tBuPSC_3H_6SP(tBu)SC_3H_3S$	104.0^{C}, 121.0^{T}	183
P-bound atoms = C′S₂				
$C'_2;(C)_2$	$C_{10}H_{15}PS_2$	$PhP(SEt)_2$	75.0	118
	$C_{12}H_{19}P_2$	$PhP(SPr)_2$	78.0	115
$C'_2;(C')_2$	$C_{18}H_{15}PS_2$	$PhP(SPh)_2$	90.8	182
$6/C'_2;(C)_2$	$C_9H_{11}PS_2$	$PhPSC_3H_6S$	38.3	181
$6/C'_2;(C)_2$	$C_{10}H_{12}PS_2$	$PhSPCH_2CH(Me)CH_2S$	25.5^{C}, 39.1^{T}	181
	$C_{11}H_{15}PS_2$	$PhPSCH_2CMe_2CH_2S$	25.2	181
		$PhPSCH(Me)CH_2CH(Me)S$	37.0^{C}, 63.5^{T}	181

TABLE C (continued)
Compilation of ^{31}P NMR Data of Three Coordinate (λ3 σ3) Phosphorus Compounds Containing Phosphorus Bonds to Chalcogenides (O, S, Se, Te) But No Bonds to Halogen

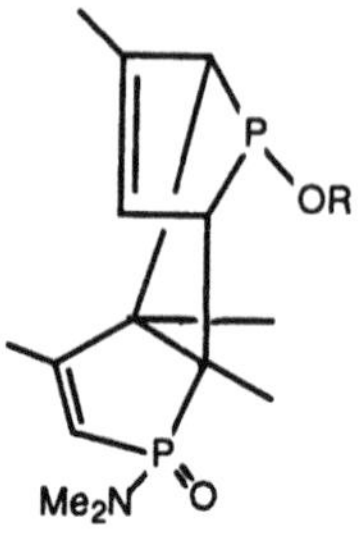

28

Rings/connectivities	Formula	Structure	NMR Data (δP[solv][a]J_{PZ} Hz)	Ref.
	$C_{13}H_{19}PS_2$	$PhPSC(tBu)C_2H_4S$	28.8(c)	184
		$PhPSCH_2CH(tBu)CH_2S$	28.8	181
P-bound atoms = CHO				
CH_2;Si	$C_7H_{19}OPSi$	HP(Bu)(OTms)	88.1	11
P-bound atoms = CHS				
H_3;C	C_3H_9PS	HP(Me(SEt)	−45 J_{PH}210	12
CH_2;C	$C_4H_{11}PS$	HP(Et)(SEt)	−22 J_{PH}200	12
C_3:C	$C_6H_{15}PS$	HP(tBu)(SEt)	13 J_{PH}200	12
P-bound atoms = C'HS				
C'_2;C	C_7H_9PS	HPSMe(Ph)	−21 J_{PH}199	12
P-bound atoms = C_2O				
$(F_3)_2$;B	$C_4H_4F_6BOP$	Tf_2POBMe_2	31.2	194
	$C_5H_3BF_{12}O_2P_2$	$[Tf_2PO]_2BMe$	81.3	194
	$C_6BF_{18}O_3P_3$	$[Tf_2PO]_3B$	36.3	194
F_3,H_3;C	$C_3H_6F_3OP$	TfMePOMe	91.8	193
$(F_3)_2$;C	$C_4H_5F_6OP$	Tf_2POEt	92.3	3
HCl_2,CH_2;C	$C_8H_{17}CL_2OP$	$Bu(CHCl_2)PO$-iPr	121(c)(b)	128
HS_2,C_3;C	$C_{14}H_{31}OPS_2$	$tBuCH(SiBu)_2POMe$	139.8(c)	191
$(H_3)_2$;C	$C_5H_7F_6OP$	$Me_2POCHTf_2$	163	83
	$C_6H_{15}OP$	Me_2PO-tBu	90.9	3
H_3,C_3;C	$C_8H_{13}F_6OP$	$Me(tBu)P(OCHTf_2)$	172	3
$(CH_2)_2$;C	$C_{12}H_{36}OP$	Bu_2POBu	126	1
	$C_{16}H_{35}OP$	$Et_2POC_{12}H_{35}$	133.9	3
$(C_2H)_2$;P	$C_{12}H_{28}OP_2$	$[iPr_2P]_2O$	156	186
C_2H,C_3;P	$C_{14}H_{32}OP_2$	$[(tBu)(iPr)P]_2O$	156(c)	187
5/(CC'H)$_2$;C	$C_{16}H_{24}Cl_3NO_2P_2$	**28**; R = CH_2CCl_3	104.8 J_{PP}6.4	186
5/(CC'H)$_2$;C'	$C_{20}H_{27}NO_2P_2$	**28**; R = Ph	88.9 J_{PP}4.9	186
6/CH_2,CC'O;C	$C_{21}H_{25}O_4PS$	$MeOPCH_2$ CMe=$CMeCH_2$CPhOTos	120.9, 121.1[I]	197
6/C'H_2,CC'O;C	$C_{26}H_{27}O_4PS$	$PhOPCH_2$ CMe=$CMeCH_2$CPhOTos	111.9, 118.9[I]	197

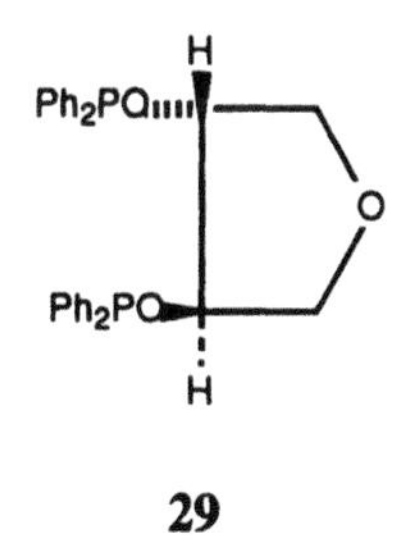

29

Rings/connectivities	Formula	Structure	NMR Data (δP[solv][a]J_{PZ} Hz)	Ref.
P-bound atoms = CC′O				
C_3;C'_2;P^4	$C_{20}H_{28}O_2P_2$	(tBu)(Ph)POP(O)(Ph)(tBu)	126.7, 129.2J_{PP}18.8	189
C_3;C'_2;S^4	$C_{11}H_{17}O_3PS$	(tBu)(Ph)POSO$_2$Me	145.2	190
5/CH_2,C′H;C	$C_7H_{12}ClOP$	$\dot{P}$(OC$_2$H$_4$Cl)C$_2$H$_4$CMe=$\dot{C}$H	150	195
	$C_9H_{17}OP$	$\dot{P}$(OtBu)C$_2$H$_4$CMe=$\dot{C}$H	127	195
5/CH_2,C'_2;C	$C_9H_{11}OP$	Ph$\dot{P}$C$_3$H$_6$$\dot{O}$	110.2	196
	$C_{10}H_{13}OP$	Ph$\dot{P}$C$_4$H$_8$$\dot{O}$	111.1	196
P-bound atoms = C′$_2$O				
(C′H)$_2$;C	C_5H_9OP	(CH$_2$=CH)$_2$POMe	164.5	188
(C'_2)$_2$;S^4	$C_{13}H_{13}O_3PS$	Ph$_2$POSO$_2$Me	123.9	190
(C'_2)$_2$;C	$C_{13}H_{13}OP$	Ph$_2$POMe	115.6	3
	$C_{14}H_{12}F_3OP$	Ph$_2$POCH$_2$Tf	142	215
	$C_{14}H_{15}OP$	Ph$_2$POEt	109.8	1
	$C_{15}H_{15}BrOP$	(4-BrC$_6$H$_4$)$_2$P(O-iPr)	106	87
	$C_{15}H_{15}Cl_2OP$	(4-ClC$_6$H$_4$)$_2$P(O-iPr)	102	87
	$C_{15}H_{15}F_2OP$	(4-FC$_6$H$_4$)$_2$P(O-iPr)	103	87
	$C_{15}H_{15}OP$	Ph$_2$P(OCH$_2$CH=CH$_2$)	113.4	3
	$C_{16}H_{19}OP$	Ph$_2$POBu	111.1	3
		(4-MeC$_6$H$_4$)(Ph)P(O-iPr)	107	87
	$C_{16}H_{19}O_2P$	(2-MeOC$_6$H$_4$PO-iPr	105	87
	$C_{17}H_{15}F_6O_2P$	(4-TfC$_6$H$_4$)$_2$P(O-iPr)	100	87
	$C_{17}H_{19}O_2P$	Ph$_2$P(OCH$_2$CHC$_3$H$_6$O)	119.5(a)	1
	$C_{27}H_{21}OP$	(4-MeC$_6$H$_4$)$_2$PO-iPr	106	87
	$C_{17}H_{21}OP$	(3-MeC$_6$H$_4$)$_2$PO-iPr	106	87
	$C_{17}H_{21}O_3P$	(4-MeOC$_6$H$_4$)$_2$PO-iPr	106	87
		(2-MeOC$_6$H$_4$)$_2$PO-iPr	90	87
	$C_{17}H_{21}OPS_2$	(4-MeSC$_6$H$_4$)$_2$PO-iPr	104	87
	$C_{18}H_{21}OP$	Ph$_2$P(O-cHex)	104	87
	$C_{18}H_{21}O_2P$	Ph$_2$POCH$_2$cPe	119.4(a)	1
	$C_{18}H_{23}O_2P$	(4-iPrOC$_6$H$_4$)(Ph)PO-iPr	105	87
	$C_{19}H_{26}NOP$	(4-Et$_2$NC$_6$H$_4$)(Ph)PO-iPr	106	87
	$C_{19}H_{28}OP$	Ph$_2$POCH$_2$-cHex)	117.0(a)	1
	$C_{28}H_{26}O_3P_2$	**29**	111.25(c)	75
	$C_{29}H_{30}O_2P_2$	Ph$_2$POCHMeCH$_2$CHMeOPPh$_2$	106.1	192
(C'_2)$_2$;C′	$C_{22}H_{21}O_2P$	Ph$_2$POC(O)(Mes)	103.2(r)	16
P-bound atoms = C$_2$S				
(H$_3$)$_2$;C	C_3H_9PS	Me$_2$PSMe	7.8	202
(F$_3$)$_2$;C	C_3F_9PS	Tf$_2$PSTf	12.8	3
	$C_3H_3F_6PS$	Tf$_2$PSMe	37.1	202
F$_3$,H$_3$;C	$C_3H_3F_6PS$	MeTfPSTf	27	193
	$C_3H_6F_3PS$	MeTfPSMe	27	193

TABLE C (continued)
Compilation of ^{31}P NMR Data of Three Coordinate (λ3 σ3) Phosphorus Compounds Containing Phosphorus Bonds to Chalcogenides (O, S, Se, Te) But No Bonds to Halogen

Rings/connectivities	Formula	Structure	NMR Data (δP[solv]ᵃJ$_{PZ}$ Hz)	Ref.
$(CH_2)_2$;C′	$C_6H_{15}PS$	Et_2PSEt	37.0	115
	$C_9H_{11}PS_2$	$Et_2PSC(SMe)$=CMe_2	120	115
	$C_{10}H_{21}PS_2$	$Et_2PSC(SEt)$=CMe_2	120	63
	P-bound atoms = CC′S			
CH_2,C′$_2$;C	$C_{10}H_{15}PS$	EtPhPSEt	12.8	3
5/CH_2,C′$_2$;C	$C_9H_{13}PS$	$Ph\dot{P}C_3H_6\dot{S}$	4.8	196
6/CH_2;C′$_2$;C	$C_{10}H_{15}PS$	$Ph\dot{P}C_4H_8\dot{S}$	22.3	196
	P-bound atoms = C′$_2$S			
(C′$_2$)$_2$;C	$C_{14}H_{15}PS$	Ph_2PSEt	26.5	115
	P-bound atoms = C_2Se			
$(H_3)_2$;C	C_3H_9PSe	Me_2PSeMe	0.8 J_{Se}218	202
$(F_3)_2$;C	C_3F_9PSe	Tf_2PSeTf	14.1	1
$(F_3)_2$;C	$C_3H_3F_6PSe$	Tf_2PSeMe	27.9 J_{PSe}294	202
	P-bound atoms = C_2Te			
$(F_3)_2$;C	$C_3H_3F_6PTe$	Tf_2PTeMe	4.3 J_{PF}64.5	184
$(C_2H)_2$;C′	$C_{13}H_{22}PTe$	$iPr_2P(TeC_6H_4Me$-4)	62 J_{PTe}495	203
$(C_3)_2$;Ge	$C_{11}H_{27}PGeTe$	$tBu_2PteGeMe_3$	68 J_{PF}384	204
$(C_3)_2$;P	$C_{16}H_{36}P_2Te$	$[tBu_2P]_2Te$	85.5 J_{PTe}451	204
$(C_3)_2$;Si	$C_{11}H_{27}PSiTeTms$	$tBu_2PTeTms$	68.2 J_{PTe}384	204
$(C_3)_2$;Sn	$C_{11}H_{27}PSnTe$	$tBu_2PTeSnMe_3$	64.5 J_{PTe}348	204
$(C_3)_2$;Te	$C_{18}H_{45}PTe_2Si_3$	$tBu_2PTeTeC(Tms)_3$	102.7 J_{PTe}501	201
$(C_3)_2$;C	$C_{18}H_{45}TeSi_3$	$tBu_2P(TeC(Tms)_3)$	93.8 J_{PTe}561	205
$(C_3)_2$;C′	$C_{15}H_{25}PTe$	$tBu_2P(TeC_6H_4Me$-4)	93.8 J_{PTe}532	203

REFERENCES

1. **Burgada, R.,** *Bull. Soc. Chim. Fr.*, 137, 1971.
2. **Houalla, D., Sanchez, M., and Wolf, R.,** *Bull. Soc. Chim. Fr.*, 2368, 1965.
3. **Mark, V., Duncan, C., Crutchfield, M., and Van Wazer, J.,** *Topics in Phosphorus Chemistry,* Vol. 5, John Wiley, New York, 1967.
4. **Benworth, W.,** *Helv. Chim. Acta,* 68, 1907, 1985.
5. **Konovalova, I. V., Gareev, R. D., Burnaeva, L. A., Faskhudtinova, T. A., and Pudovik, A. N.,** *J. Gen. Chem. USSR,* 46, 2283, 1976.
6. **Oftiserov, E. N., Zyabilikova, T. A., Batyeva, E. S., and Pudovik, A. N.,** *Bull. Akad. Sci. USSR Ser. Khim.*, 25, 1325, 1976.
7. **Charbonnel, Y., Barrans, J., and Burgada, R.,** *Bull. Soc. Chim. Fr.*, 1868, 1970.
8. **Pudovik, M. A., Vorob'eva, L. A., and Pudovik, A. N.,** *J. Gen. Chem. USSR,* 51, 420, 1981.
9. **Gurevich, P. A., Razumov, I. A., Komina, T. V., Klimentova, G. U., and Zykova, T. V.,** *Zh. Obshch. Khim.*, 55, 1257, 1985.

10. **Komina, T. V.,** *Zh. Obshch. Khim.*, 55, 1327, 1985.
11. **Abicht, H. P., Spencer, J. T., and Verkade, J. G.,** *Inorg. Chem.*, 24, 2132, 1985.
12. **Roschenthaler, G. V.,** *Phosphorus Sulphur*, 4, 373, 1978.
13. **Gema, H., Sanchez, M., Burgada, R., and Wolf, R.,** *Bull. Soc. Chim. Fr.*, 612, 1970.
14. **Shtynilina, A. A., Bayadina, E. V., and Nurekdinov, I. A.,** *Zh. Obshch. Khim.*, 55, 1287, 1985.
15. **Samitov, Yu. Yu., Musina, A. A., Gurrarri, L. I., Mukmenev, E. T., and Arbuzov, B. A.,** *Izv. Akad. Nauk. SSSR Ser. Khim.*, 1518, 1975.
16. **Brierley, J., Dickstein, J. I., and Trippett, S.,** *Phosphorus Sulphur*, 7, 167, 1979.
17. **Karaghiosoff, K., Majoral, J. P., Merim, A., Navech, J., and Schmidpeter, A.,** *Tetrahedron Lett.*, 24, 2137, 1983.
18. **Pudovik, A. N., Batyeva, E. S., Al'fonsov, V. A., and Mizhivitskii, M. D.,** *Bull. Akad. Sci. USSR Ser. Khim.*, 27, 189, 1978.
19. **Kudryavstena, T. N., Karlstedt, N. B., Proskurnina, M. V., Boganova, M. V., Shestakova, T. G., and Lutsenko, I. F.,** *J. Gen. Chem. USSR*, 52, 912, 1982.
20. **Karlstedt, N. B., Kudryavstena, T. N., Proskurnina, M. V., and Lutsenko, I. F.,** *J. Gen. Chem. USSR*, 52, 1754, 1982.
21. **Burgada, R., Lafaille, L., and Mathis, F.,** *Bull. Soc. Chim. Fr.*, 2, 341, 1974.
22. **Voropai, L. M., Ruchinka, N. G., Milliaressi, E. E., and Nifant'ev, E. E.,** *Zh. Obshch. Khim.*, 55, 55, 1985.
23. **Denney, D. B., Denney, D. Z., and Liu, L-T.,** *Phosphorus Sulphur*, 22, 71, 1985.
24. **Pudovik, M. A., Kibardina, L. K., Pudovik, A. N.,** *J. Gen. Chem. USSR*, 51, 420, 1981.
25. **Balitski, Yu. V.,** *J. Gen. Chem. USSR*, 50, 231, 1980.
26. **Pudovik, N. A., Kibardina, L. K., and Pudovik, A. N.,** *J. Gen. Chem. USSR*, 52, 1310, 1982.
27. **Munoz, A., Garrigues, R., and Wolf, R.,** *Phosphorus Sulphur*, 4, 47, 1978.
28. **Pudovik, M. A., Kibardina, L. K., and Pudovik, A. N.,** *J. Gen. Chem. USSR*, 52, 677, 1982.
29. **Pudovik, M. A., Mikhailov, Yu. B., and Pudovik, A. N.,** *Zh. Obshch. Khim.*, 55, 1184, 1985.
30. **Nifant'ev, E. E., Sorokina, S. F., and Borisenko, A. A.,** *J. Gen. Chem. USSR*, 55, 1481, 1985.
31. **Mosbo, J. A. and Verkade, J. G.,** *J. Am. Chem. Soc.*, 95, 4659, 1973.
32. **Bentrude, W. G. and Tan, H-W.,** *J. Am. Chem. Soc.*, 98, 1850, 1976.
33. **Nifant'ev, E. E., Elepina, L. T., Borisenko, A. A., and Koroteev, M. P.,** *Phosphorus Sulphur*, 5, 315, 1979.
34. **Nifant'ev, E. E., Sorokina, S. F., Borisenko, A. A., Zavalishina, A. I., and Vorobjeva, L. A.,** *Tetrahedron*, 37, 3183, 1981.
35. **Stec, W. J. and Okruszek, A.,** *J. Chem. Soc. Perkin (I)*, 1828, 1975.
36. **Borisov, E. V., Akhlebinim, A. K., and Nifnat'ev, E. E.,** *Zh. Obshch. Khim.*, 54, 2345, 1984.
37. **Pudovik, M. A., Shaludina, O. S., Ivanova, L. K., Terent'eva, S. N., and Pudovik, A. N.,** *J. Gen. Chem. USSR*, 44, 482, 1974.
38. **Nifant'ev, E. E., Predvolitelev, D. A., and Grachev, M. K.,** *J. Gen. Chem. USSR*, 46, 475, 1976.
39. **Nifant'ev, E. E., Teleshev, A. T., and Shikovets, T. A.,** *J. Gen. Chem. USSR*, 56, 258, 1986.
40. **Niecke, E., Zorn, H., Krebs, B., and Heskel, G.,** *Angew. Chem. Int. Ed.*, 19, 709, 1980.
41. **Dutasta, J. P., Robert, J. B., and Vincens, M.,** *Tetrahedron Lett.*, 933, 1979.
42. **Odoriso, P. A., Pastor, S. D., and Spivak, J. D.,** *Phosphorus Sulphur*, 19, 1, 1984.
43. **Denney, D. B., Denney, D. Z., Hammond, P. J., Huang, C., Liu, L-T., and Tseng, K-S.,** *Phosphorus Sulphur*, 15, 281, 1983.
44. **Arduengo, A. J., Dixon, D. A., and Roe, D. C.,** *J. Am. Chem. Soc.*, 108, 6821, 1986.
45. **Culley, S. A. and Arduengo, A. J.,** *J. Am. Chem. Soc.*, 106, 1164, 1984.
46. **Bernard, D., Javignac, P., and Burgada, R.,** *Bull. Soc. Chim. Fr.*, 1657, 1972.
47. **Martin, J., Robert, J. B., and Taieb, C.,** *J. Phys. Chem.*, 80, 2417, 1976.
48. **Jurkschat, K., Mugge, C., Tzschach, A., Ohlig, W., and Zschunke, A.,** *Tetrahedron Lett.*, 23, 1345, 1982.
49. **Moore, F. M. and Beaucage, S. L.,** *J. Org. Chem.*, 50, 2019, 1985.
50. **Novikova, Z. S., Skorobogatova, S. Ya., Prischenko, A. A., and Lutsenko, I. F.,** *J. Gen. Chem. USSR*, 46, 568, 1976.
51. **Foss, V. L., Veits, Yu. A., Lutashev, N. V., and Lutsenko, I. F.,** *J. Organomet. Chem.*, 121, c27, 1976.
52. **Kawashima, T. and Inamoto, N.,** *Bull. Chem. Soc. Jpn.*, 49, 1924, 1976; **Niecke, E. and Ringel, G.,** *Angew. Chem. Int. Ed.*, 16, 486, 1977.
53. **Ramirez, F., Patwardhan, A. V., Kugler, H. J., and Smith, C. P.,** *Tetrahedron Lett.*, 3055, 1963.
54. **Burgada, R. and Launenco, C.,** *C.R. Acad. Sci. Fr.*, 273, 1536, 1971.
55. **Rumyantseva, S. A., Sisneros, Ks., Koroteev, M. P., and Nifant'ev, E. E.,** *J. Gen. Chem. USSR*, 54, 181, 1984.
56. **Pudovik, M. A., Mikhailov, Yu. B., and Pudovik, A. N.,** *Zh. Obshch. Khim.*, 54, 2794, 1984.

57. **Kuliev, A. K., Moskva, V. V., and Akhmedzade, D. A.,** *Zh. Obshch. Khim.*, 55, 457, 1985.
58. **Mosbo, J. A.,** *Tetrahedron Lett.*, 4789, 1976.
59. **Hutchins, R. O., Maryanoff, B. E., Albrand, J. P., Cogne, A., Gagnaire, D., and Robert, J. B.,** *J. Am. Chem. Soc.*, 91, 9151, 1972.
60. **Nifant'ev, E. E., Sorokova, S. F., and Vorob'eva, L. A.,** *Zh. Obshch. Khim.*, 55, 308, 1985.
61. **Nifant'ev, E. E., Zavalishina, A. I., Smirnova, E. I., and Filimonov, V. F.,** *J. Gen. Chem. USSR*, 55, 2498, 1985.
62. **Roesky, H. W., Ambrosius, K., Banek, M., and Sheldrich, G. M.,** *J. Chem. Soc. Dalton Trans.*, 2855, 1984.
63. **Danchenko, M. N., Budilova, I. Yu., and Sinitsa, A. D.,** *J. Gen. Chem. USSR*, 55, 838, 1985.
64. **Pudovik, M. A., Mikhailov, Yu. B., and Pudovik, A. N.,** *Zh. Obshch. Khim.*, 53, 2468, 1983.
65. **Sinyashin, O. G., Kostin, V. P., Batyeva, E. S., Pudovik, A. N., and Ivasyuk, N. V.,** *Zh. Obshch. Khim.*, 53, 1706, 1983.
66. **Al'Fonsov, V. A., Pudovik, M. A., and Batyeva, E. S.,** *Zh. Obshch. Khim.*, 55, 559, 1985.
67. **Nifant'ev, E. E., Zavalishina, A. I., Sorokina, S. F., and Borisenko, A. A.,** *J. Gen. Chem. USSR*, 46, 469, 1976.
68. **Danchenko, M. N. and Sinitsa, A. D.,** *J. Gen. Chem. USSR*, 54, 422, 1984.
69. **Okruszek, A. and Stec, W. J.,** *Z. Naturforsch.*, 30B, 430, 1975.
70. **Nifant'ev, E. E., Komlev, I. V., Konyeva, I. P., Zavalishina, A. I., and Tul'chinskii, V. M.,** *J. Gen. Chem. USSR*, 43, 2353, 1973.
71. **Burgada, R., Lafaille, L., and Mathis, F.,** *Bull. Soc. Chim. Fr.*, 2, 341, 1974.
72. **Van Gendera, M. H. P.,** *Phosphorus Sulphur*, in press.
73. **Kirolevets, A. A. and Ragulin, L. I.,** *Zh. Obshch. Khim.*, 55, 461, 1985.
74. **Chaudri, B. A., Goodurin, D. G., and Hudson, H. R.,** *J. Chem. Soc. B*, 1290, 1970.
75. **Jackson, W. R. and Lovel, C. G.,** *Aust. J. Chem.*, 35, 2069, 1982.
76. **Grechkin, N. P. and Grishina, L. N.,** *J. Gen. Chem. USSR*, 48, 2002, 1978.
77. **Malenko, D. M., Repina, L. A., and Sinitsa, A. D.,** *Zh. Obshch. Khim.*, 54, 2402, 1984.
78. **Malenko, D. M. and Sinitsa, A. D.,** *J. Gen. Chem. USSR*, 56, 195, 1986.
79. **Onys'ko, P. P., Kiseleva, E. I., and Kim, T. V.,** *Zh. Obshch. Khim.*, 55, 454, 1985.
80. **Dabkowski, W., Michalski, J. and Skrzypczriski, Z.,** *J. Chem. Soc. Chem. Commun.*, 1260, 1982.
81. **Chernystev, E. A., Burgerenko, E. F., Akat'eva, A. S., and Naumov, A. D.,** *J. Gen. Chem. USSR*, 45, 231, 1975.
82. **Nesterov, L. V. and Alexandrova, N. A.,** *Zh. Obshch. Khim.*, 55, 1742, 1985.
83. **Livenkov, I. V.,** *Zh. Obshch. Khim.*, 54, 2504, 1984.
84. **Burgada, R.,** *Phosphorus Sulphur*, 2, 237, 1976.
85. **Denney, D. Z., Chen, G. Y., and Denney, D. B.,** *J. Am. Chem. Soc.*, 91, 6838, 1969.
86. **Hall, C. D.,** personal communication.
87. **Blackburn, G. M., Cohen, J. S., and Todd, A. R.,** *Tetrahedron Lett.*, 2837, 1964.
88. **Granoth, I. and Martin, J. C.,** *J. Am. Chem. Soc.*, 7434, 1978.
89. **Coates, P. C.,** unpublished work.
90. **Tsivanin,** *J. Gen. Chem. USSR*, 54, 2357, 1984.
91. **Malenko, D. M., Repina, L. A., and Sinitsa, A. D.,** *Zh. Obshch. Khim.*, 55, 698, 1985.
92. **Cong, C. B., Munoz, A., Koenig, M., and Wolf, R.,** *Tetrahedron Lett.*, 2297, 1977.
93. **Koenig, M., El Khatib, F., Munoz, A., and Wolf, R.,** *Tetrahedron Lett.*, 23, 421, 1982.
94. **Nifant'ev, E. E., Milliaressi, E. E., Ruchtina, N. G., Druyan-Poleshchuk, L. M., Vasyanina, L. K., and Tyan, E. A.,** *J. Gen. Chem. USSR*, 51, 1295, 1981.
95. **Haemers, M., Ottinger, R., Zimmerman, D., and Riesse, J.,** *Tetrahedron*, 29, 3539, 1973.
96. **Eliel, E. L., Chandrasekvan, S., Carpenter, L. E., and Verkade, J. G.,** *J. Am. Chem. Soc.*, 108, 6651, 1986.

96a. **Bajuva, G. S. and Bentrude, W. G.,** *Tetrahedron Lett.*, 421, 1978.

97. **Koroteev, M. P., Pugushova, N. M., Lidak, M. Yu., Abbasov, E. M., and Nifant'ev, E. E.,** *J. Gen. Chem. USSR*, 55, 2453, 1985.

97a. **Edmunson, R. S., Johnson, O., Jones, D. W., and King, J. T.,** *J. Chem. Soc. Perkin (II)*, 69, 1985.

98. **Albrand, J. P. and Dutasta, J. P.,** *Tetrahedron Lett.*, 4584, 1974.
99. **Nifant'ev, E. E., Legin, G. Ya., Sorkina, S. F., and Borisenko, A. A.,** *J. Gen. Chem. USSR*, 49, 2331, 1979.
100. **Gazizov, M. B., Zakharov, V. M., Khairullin, R. A., Moskva, V. V., Muzin, R. Z., Efremov, Yu. Ya., and Savel'eva, E. I.,** *J. Gen. Chem. USSR*, 54, 200, 1984.
101. **Chaser, D. W., Fackler, J. P., Mazany, A. M., Komoroski, R. A., and Kroenke, W. J.,** *J. Am. Chem. Soc.*, 108, 5956, 1986.
102. **Koroteev, M. P., Lutsenko, I. F. and Nifant'ev, E. E.,** *J. Gen. Chem. USSR*, 54, 245, 1984.
103. **Nifant'ev, E. E., Koroteev, M. P., Zhane, Z. K., and Borisenko, A. A.,** *Tetrahedron Lett.*, 4125, 1977.

104. **Nifant'ev, E. E., Teleshev, A. T., and Koroteev, M. P.,** *J. Gen. Chem. USSR,* 53, 1476, 1983.
105. **Grandjean, J., Laszlo, P., Picavet, J. P., and Silwa, H.,** *Tetrahedron Lett.,* 1861, 1978.
106. **Devillers, J., Houalla, D., and Bonnet, J. J.,** *Nouv. J. Chem.,* 4, 179, 1980.
107. **Arbuzov, B. A., Kalyrov, R. A., Arshinova, R. P., Sharikov, I. Kh., and Shagidullin, R. R.,** *J. Gen. Chem. USSR,* 55, 1755, 1985.
108. **Guimeraes, A. C. and Robert, J. B.,** *Tetrahedron Lett.,* 473, 1976.
109. **Pastor, S. D., Spivak, J. D., Steinhuetel, L., and Matzura, C.,** *Phosphorus Sulphur,* 15, 253, 1983.
110. **Odoriso, P. A., Pastor, S. D., Spivak, J. D., Bini, D., and Rodenbaugh, R. K.,** *Phosphorus Sulphur,* 19, 285, 1984.
111. **Odoriso, P. A., Pastor, S. D., Spivak, J. D., and Steinhuetel, L.,** *Phosphorus Sulphur,* 15, 9, 1983.
112. **Abdon, W. M., Denney, D. B., Denney, D. Z., and Pastor, S. D.,** *Phosphorus Sulphur,* 22, 99, 1985.
113. **Odoriso, P. A., Pastor, S. D., and Spivak, J. D.,** *Phosphorus Sulphur,* 20, 273, 1984.
114. **Maier, L.,** *Helv. Chim. Acta,* 59, 252, 1976.
115. **Krasil'nikova, E. A.,** *Russ. Chem. Rev.,* 46, 861, 1977.
116. **Olah, G. A. and McFarland, C. W.,** *J. Org. Chem.,* 40, 2582, 1975.
117. **Khokhlov, P. S., Bereseneva, L. S., and Savenkov, N. F.,** *J. Gen. Chem. USSR,* 54, 867, 1984.
118. **Tattershall, B. W.,** *J. Chem. Soc. Dalton Trans.,* 1707, 1985.
119. **Litvansov, M. V., Prishchenko, A. A., and Lutsenko, I. F.,** *J. Gen. Chem. USSR,* 55, 1461, 1985.
120. **Niecke, E. and Ringel, G.,** *Angew. Chem. Int. Ed.,* 16, 486, 1977.
121. **Stec, W., Uznonski, B., Houalla, D., and Wolf, R. C. R.,** *Heb. Seances Acad. Sci. Serc.,* 281, 727, 1975.
122. **Gallagher, M. J. and Honegger, H.,** *Tetrahedron Lett.,* 2987, 1977.
123. **Nifant'ev, E. E., Zavalishina, A. I., and Sorkina, S. F.,** *Zh. Obshch. Khim.,* 46, 1184, 1976.
124. **Nifant'ev, E. E., Sorokina, S. F., and Borisenko, A. A.,** *J. Gen. Chem. USSR,* 48, 2158, 1978.
125. **Dahl, O.,** *Tetrahedron Lett.,* 23, 1493, 1982.
126. **Romanenko, V. D., Ruben, A. V., and Markoviski, L. N.,** *J. Chem. Soc. Chem. Commun.,* 187, 1983.
127. **Roesky, H. W., Ambrosius, K., Banek, M., and Sheldrick, G. M.,** *J. Chem. Soc. Dalton Trans.,* 2855, 1984.
128. **Prishenko, A. A., Gromov, A. V., Kadyko, M. I., and Lutsenko, I. F.,** *J. Gen. Chem. USSR,* 53, 2250, 1983.
129. **Novikova, Z. S., Kabachnik, M. M., Monin, E. A., Boirsenko, A. A., and Nifant'ev, E. E.,** *J. Gen. Chem. USSR,* 55, 290, 1985.
130. **Schmidpeter, A., Laber, J., Riedl, H., and Volz, M.,** *Phosphorus Sulphur,* 3, 171, 1977.
131. **Monin, E. A., Borisenko, A. A., Lutsenko, A. I., Kabachnik, M. M., Novikova, Z. S., and Lutsenko, I. F.,** *J. Gen. Chem. USSR,* 56, 235, 1986.
132. **Konovalova, I. V., Mironov, V. F., Ofitserov, E. N., and Pudovik, A. N.,** *Zh. Obshch. Khim.,* 53, 470, 1983.
133. **Kolodka, T. V., Loktionova, R. A., and Gololobov, Yu. G.,** *Zh. Obshch. Khim.,* 53, 2476, 1983.
134. **Niecke, E. and Rveger, R.,** *Angew. Chem.,* 95, 154, 1983.
135. **Escudie, J., Couret, C., Ranaivonjatovo, H., Satage, J., and Jaud, J.,** *Phosphorus Sulphur,* 17, 221, 1983.
136. **Yoshifuji, M., Ando, K., Shibayama, K., Inamoto, N., Hirotou, K., and Higuchi, T.,** *Angew. Chem.,* 95, 416, 1983.
137. **Du Mont, W. W., Severengiz, T., and Meyer, B.,** *Angew. Chem.,* 95, 1025, 1983.
138. **Shibaev, V. I., Garabadziu, A. V., and Rodin, A. A.,** *Zh. Obshch. Khim.,* 53, 1743, 1983.
139. **Komina, T. V.,** *Zh. Obshch. Khim.,* 55, 1327, 1985.
140. **Dahl, B. M., Dahl, O., and Trippett, S.,** *J. Chem. Soc. Perkin (II),* 2239, 1981.
141. **Dahl, O.,** *Tetrahedron Lett.,* 22, 3281, 1981.
142. **Schmidpeter, A., Legh, C., and Karaghiosoff, K.,** *Angew. Chem.,* 97, 127, 1985.
143. **Robert, J. B. and Weichman, H.,** *J. Org. Chem.,* 43, 3031, 1978.
144. **Pastor, S. D., Spivak, J. D., and Steinhuebel, L. P.,** *Phosphorus Sulphur,* 22, 169, 1985.
144a. **Pudovik, A. N.,** *J. Gen. Chem. USSR,* 55, 1316, 1985.
145. **Balitski, Yu. V., Kasukhin, L. F., Ponomarchuk, M. D., and Golobov, Yu. G.,** *J. Gen. Chem. USSR,* 50, 34, 1980.
146. **Appel, R., Kuendga, A., and Knock, F.,** *Chem. Ber.,* 118, 1352, 1985.
147. **Abraham, K. M. and Van Wazer, J. R.,** *J. Inorg. Nucl. Chem.,* 37, 541, 1975.
148. **Szafraniec, L. J., Szafraniec, L., and Aaron, H. S.,** *J. Org. Chem.,* 47, 1937, 1982.
149. **Roschenthaler, G. V.,** *Z. Naturforsch.,* 33B, 131, 1978.
150. **Laurent, J-P., Jugie, G., and Wolf, R.,** *J. Chim. Phys. Physiochim. Biol.,* 66, 409, 1969.
151. **Odinets, I. L., Novikova, Z. S., and Lutsenko, I. F.,** *J. Gen. Chem. USSR,* 55, 488, 1985.
152. **Schmidpeter, A. and Ebeling, J.,** *Angew. Chem. Int. Ed.,* 7, 209, 1968.
153. **Ruchinka, N. G., Druyan-Poleshchek, L. M., Vasynina, L. M. and Ayan, E. A.,** *Zh. Obshch. Khim.,* 51, 1528, 1981.

154. **Quinn, L. D., Gordon, M. D., and Lee, S. D.,** *Org. Magn. Reson.*, 6, 503, 1974.
155. **Van Genderen, M. H. P., Koole, L. H., Bernd, C. C. M., Scheper, O., Ven der Vers, L. J. M., and Buck, H. M.,** *Phosphorus Sulphur,* in press.
156. **Quinn, L. D., Gallagher, M. J., Cunkle, G. T., and Chesnut, D. B.,** *J. Am. Chem. Soc.*, 102, 3136, 1980.
157. **Kozolov, E. S., Tolmachev, A. A., and Chernega, A. N.,** *Zh. Obshch. Khim.*, 55, 1262, 1985.
158. **Gordon, M. D. and Quinn, L. D.,** *J. Am. Chem. Soc.*, 98, 15, 1976.
159. **Rodionov, I. L., Kazankova, M. A., and Lutsenko, I. F.,** *J. Gen. Chem. USSR,* 54, 417, 1984.
160. **Tolman, C. A.,** *J. Am. Chem. Soc.*, 92, 2956, 1970.
161. **Novikova, Z. S., Kurin, A. N., and Lutsenko, I. F.,** *J. Gen. Chem. USSR,* 55, 2405, 1985.
162. **Kato, M. and Yambe, M.,** *J. Chem. Soc. Chem. Commun.*, 1173, 1981.
163. **Kruglov, A. S., Oskot-skii, E. L., Dogadina, A. V., Ionin, B. I., and Petrov, A. A.,** *J. Gen. Chem. USSR,* 48, 1371, 1978.
164. **Maier, L.,** *Helv. Chim. Acta,* 52, 858, 1969.
165. **Razumov, A. I., Gurevich, P. A., Baigil'dina, S. Yu., Zykova, T. V., and Salakhutdinova, R. A.,** *J. Gen. Chem. USSR,* 44, 2547, 1974.
166. **Harper, S. D. and Arduego, A. J.,** *J. Am. Chem. Soc.*, 104, 2497, 1982.
167. **Pastor, S. D., Spivak, J. D., and Steinhuebel, L. P.,** *Phosphorus Sulphur,* 22, 169, 1985.
168. **Aliev, R. Z., Khairullin, V. K., and Makhmutova, S. F.,** *J. Gen. Chem. USSR,* 46, 59, 1976.
169. **Novikova, Z. S., Prisenko, A. A., and Lutsenko, I. F.,** *J. Gen. Chem. USSR,* 49, 412, 1979.
169a. **Dutasta, J. P., Guimaraes, A. G., Martin, J., and Robert, J. B.,** *Tetrahedron Lett.*, 1519, 1975.
170. **Gordon, M. D. and Quinn, L. D.,** *J. Magn. Reson.*, 22, 149, 1976.
171. **Vasil'ev, V. V., Razumova, N. A., and Dodgadeava, L. V.,** *J. Gen. Chem. USSR,* 46, 461, 1976.
172. **Singh, G.,** *J. Org. Chem.*, 44, 1060, 1979.
173. **Albrand, J. P., Dutasta, J. P., and Robert, J. B.,** *J. Am. Chem. Soc.*, 96, 4584, 1974.
174. **Bentrude, W. G., Yee, K. C., Bertrand, R. D., and Grant, D. M.,** *J. Am. Chem. Soc.*, 93, 797, 1971.
175. **Stec, W., Sudol, T., and Uzanski, B.,** *J. Chem. Soc. Chem. Commun.*, 467, 1975.
176. **Tzschach, A., Jurkschat, K., Zschunke, A., Mugge, C., Altenbrunn, B., Piccinni-Leopardi, C., Germain, G., Declerq, J. P., and Van Meerssche, M.,** *J. Cryst. Spectros. Res.*, 15, 423, 1985.
177. **Dutasta, J. P.,** *J. Chem. Res.*, 22(m), 0361, 1986.
178. **Dutasta, J. P., Jurkstat, K., and Robert, J. B.,** *Tetrahedron Lett.*, 22, 2549, 1981.
179. **Dutasta, J. P., Robert, J. B., and Guimares, A. C.,** *Tetrahedron Lett.*, 9, 801, 1977.
180. **Albrand, J. P. and Dutasta, J. P.,** *Tetrahedron Lett.*, 4584, 1974.
181. **Peake, S. C. and Schmultzler, R.,** *J. Chem. Soc. A,* 1049, 1970.
182. **Lucas, J.,** Ph.D. thesis, Frankfurt, 1984.
183. **Dutasta, J. P., Martin, J., and Robert, J. B.,** *J. Org. Chem.*, 42, 1662, 1977.
184. **Hutchins, R. O. and Maryanoff, B. E.,** *J. Am. Chem. Soc.*, 101, 1600, 1979.
184a. **Lensch, C., Clegg, W., and Sheldrick, G. M.,** *J. Chem. Soc. Dalton Trans.*, 723, 1984.
185. **Grayson, M. and Farley, C. E.,** *J. Chem. Soc. Chem. Commun.*, 830, 1967.
186. **Foss, V. L., Veits, Yu. A., Kukhmisterov, P. L., Solodenko, V. A., and Lutsenko, I. F.,** *J. Gen. Chem. USSR,* 47, 437, 1977.
187. **Foss, V. L., Solodenko, V. A., Veits, Yu. A., and Lutsenko, I. F.,** *J. Gen. Chem. USSR,* 49, 1510, 1979.
188. **King, R. B. and Masker, W. F.,** *J. Am. Chem. Soc.*, 99, 4001, 1977.
189. **Krawiecka, B., Michalski, J., and Wojna-Tadeusiak, E.,** *J. Org. Chem.*, 51, 4201, 1986.
190. **Dabkowski, W., Michalski, J., and Skrzypczriski, Z.,** *J. Chem. Soc. Chem. Commun.*, 1260, 1982.
191. **Kolodyazhnyi, O. I., Shevchenko, and Kukhar, V. P.,** *J. Gen. Chem. USSR,* 53, 416, 1983.
192. **Bakos, J., Toth, I., and Marko, L.,** *J. Org. Chem.*, 46, 5427, 1981.
193. **Burg, A. B. and Kang, D. K.,** *J. Am. Chem. Soc.*, 92, 1901, 1970.
194. **Burg, A. B. and Basi, J. S.,** *J. Am. Chem. Soc.*, 91, 1937, 1969.
195. **Vizel, A. U., Krupov, V. K., Zyryanova, L. I., and Arbuzov, B. A.,** *J. Gen. Chem. USSR,* 46, 1536, 1976.
196. **Kobayashi, S., Suzaki, M., and Saregusa, T.,** *Bull. Chem. Soc. Jpn.*, 58, 2153, 1985.
197. **Appel, R. and Zimmermann, R.,** *Tetrahedron Lett.*, 24, 3591, 1983.
198. **Szewczyk, J. and Quin, L. D.,** *J. Organomet. Chem.*, 52, 1190, 1987.
199. **Quin, L. D., Kislaus, J. C., and Merch, K. A.,** *J. Org. Chem.*, 48, 4466, 1983.
200. **Granoth, I. and Martin, J. C.,** *J. Am. Chem. Soc.*, 103, 2711, 1981.
201. **Seel, F. and Velleman, K. D.,** *Chem. Ber.*, 105, 406, 1972.
202. **Grobe, J., Kohne-Wachter, M., and Le Van, D.,** *Z. Anorg. Allg. Chem.*, 519, 67, 1984.
203. **DuMont, W. W., Kubiniok, S., and Severingiz, T.,** *Z. Anorg. Allg. Chem.*, 531, 21, 1985.
204. **DuMont, W. W.,** *Angew. Chem. Int. Ed.*, 19, 554, 1980.
205. **Mukhametov, F. S., Korshin, E. E., Shagidullin, R. R., and Rizpolozhenskii, N. I.,** *Zh. Obshch. Khim.*, 53, 2452, 1983.

206. **Zeiss, W. and Weis, J.,** *Z. Naturforsch.*, 32b, 485, 1977.
207. **Barlos, K. and Noeth, H.,** *Z. Naturforsch.*, 35b, 407, 1980.
208. **Pudovik, M. A., Mironova, T. A., and Pudovik, A. N.,** *Zh. Obshch. Khim.*, 53, 2464, 1983.
209. **Barlos, K. and Noeth, H.,** *Z. Naturforsch.*, 33b, 515, 1978.
210. **Sinyashin, O. G., Kostin, V. P., Batyeva, E. S., Pudovik, A. N., and Ivasyuk, N. V.,** *Zh. Obshch. Khim.*, 53, 302, 1983.
211. **Ebsworth, E. A. V., Hunter, G. M., and Rankin, D. W. H.,** *J. Chem. Soc. Dalton Trans.*, 1983, 1983.
212. **Denney, D. B., Denney, D. Z., Hammond, P. J., Liu, L-T., and Wang, Y-P.,** *J. Org. Chem.*, 48, 2159, 1983.
213. **Bauer, D. P. and Ruff, J. K.,** *Inorg. Chem.*, 22, 1686, 1983.
214. **Wroblewski, A. E.,** *Tetrahedron*, 39, 1809, 1983.
215. **Azizian, H. and Morris, R. H.,** *Inorg. Chem.*, 22, 6, 1983.
216. **Bollmacher, H. and Sartori, P.,** *Chem. Ztg.*, 107, 121, 1983.
217. **Budilova, I. Yu., Gusar, N. I., and Golobov, Yu. G.,** *Zh. Obshch. Khim.*, 53, 285, 1983.
218. **Konovalova, I. V., Mironov, V. F., Ofitserov, E. N., and Pudovik, A. N.,** *Zh. Obshch. Khim.*, 53, 470, 1983.
219. **Kolodka, T. V., Loktionova, R. A., and Gololobov, Yu. G.,** *Zh. Obshch. Khim.*, 53, 2476, 1983.
220. **Chekmacheva, O. I., Suerbaev, H. A., Komlev, I. V., Dakhonv, P. P., Korolova, O. R., and Mastryukova, T. A.,** *Zh. Obshch. Khim.*, 53, 281, 1983.
221. **Foss, V. L., Veits, Yu., Chernykh, T. E., Staroverova, I. N., and Lutsenko, I. F.,** *Zh. Obshch. Khim.*, 53, 2489, 1983.
222. **Nikoronova, L. K., Enikeev, K. M., Grechkin, N. P., Ismaev, I. E., Il'yasov, A. V., and Nuretdinov, I. A.,** *Zh. Obshch. Khim.*, 53, 2454, 1983.
223. **Angelov, C. M. and Dahl, O.,** *Tetrahedron Lett.*, 24, 4323, 1983.
224. **Gurevch, P. A., Kiselev, V. V., Moskva, V. V., Zykova, T. V., and Salakhutinov, R. A.,** *Zh. Obshch. Khim.*, 53, 2145, 1983.
225. **Bonningue, C., Houalla, D., Wolf, R., and Jand, J.,** *J. Chem. Soc. Perkin Trans. II*, 773, 1983.
226. **Chaus, M. P., Gusar, N. I., and Gololobov, Yu. G.,** *Zh. Obshch. Khim.*, 53, 244, 1983.
227. **Konovalova, I. V., Burnaeva, L. A., Gareev, R. D., Lyakhova, A. S., Moshikina, T. M., and Pudovik, A. N.,** *Zh. Obshch. Khim.*, 53, 1471, 1983.
228. **Sinyashin, O. G., Kostin, V. P., Batyeva, E. S., and Pudovik, A. N.,** *Zh. Obshch. Khim.*, 53, 472, 1983.
229. **Krutikov, V. I., Semenova, E. S., Maslennikov, I. G., and Lavrentev, A. N.,** *Zh. Obshch. Khim.*, 53, 1557, 1983.
230. **Shibaev, V. I., Garabadziu, A. V., and Rodin, A. A.,** *Zh. Obshch. Khim.*, 53, 1743, 1983.
231. **Foss, V. L., Chernykh, T. E., Staroverova, I. N., Veits, Yu. A., and Lutsenko, I. F.,** *Zh. Obshch. Khim.*, 53, 2184, 1983.
232. **Boenigk, W., Fischer, U., and Haegele, G.,** *Phosphorus Sulphur*, 16, 263, 1983.
233. **Weisheit, R., Stendel, R., Messbauer, B., Langer, C., and Walther, B.,** *Z. Anorg. Allg. Chem.*, 504, 147, 1983.

Chapter 5

^{31}P NMR DATA OF THREE COORDINATE ($\lambda 3\ \sigma 3$) PHOSPHORUS COMPOUNDS CONTAINING PHOSPHORUS BONDS TO GROUP V ELEMENTS (N, P, As, Sb) BUT NO BONDS TO HALOGENS NOR CHALCOGENIDES

Compiled and presented by
Yves Leroux, Ramon Burgada, Stephan G. Kleemann, and Ekkehard Fluck

NMR data was also donated by
H. W. Roesky, Universität Göttingen, FRG
C. D. Hall, King's College, London, U.K.

The NMR data is presented in Table D and it is divided into two sections. The first part, Section D1, contains compounds with P-N bonds and it was compiled by Y. Leroux and R. Burgada. A brief discussion of the trends of the chemical shifts with structure is given below.

Section D2 contains compounds with P-P, P-As, and P-Sb bonds. It was compiled by S. G. Kleemann and E. Fluck and it consists mainly of polyphosphines. An analysis, by J. C. Tebby, of the trends in the NMR data of the polyphosphines and related compounds follows that relating to D1.

Bar chart summaries of the ranges of chemical shifts for the aminophosphines and polyphosphines are given in Figure 1.3 and 1.3c in Chapter 1.

TRENDS IN CHEMICAL SHIFTS FOR AMINOPHOSPHINES

Most of the chemical shifts of the acyclic triamides ($P(NR_2)_3$ occur within the narrow range of 100 to 130 ppm including the cases when two of the nitrogen atoms belong to a saturated 5-membered heterocyclic ring. Unsaturated groups bound to nitrogen induce the signal to move upfield to 40 to 70 ppm. When the phosphorus atom is part of a 4-membered ring (e.g., compounds *9*) the effect of altering the substituents is small. Conversely, the chemical shift is very dependent on geometric and steric factors. The *trans* isomers of compound *8* appear to much lower field (165 to 226) of their *cis* isomers (76 to 98 ppm).

The secondary amides (P-bound atoms HN_2) resonate between 12.7 and 82 ppm, e.g., 42.1 for the di-isopropylamide $(iPr_2N)_2PH$ while the primary amides (P-bound atoms H_2N) are strongly shifted to high fields, e.g., −82.9 ppm for the iso-propylamide iPr_2NPH_2.

The inclusion of the phosphorus atom in a 3-membered ring usually produces a large shielding effect. Thus in comparison with $MeP(NEt_2)_2$ 80.4 or $iPr_2NP(NTms_2)_2$ 116.6 the 3-membered heterocycle $iPr_2NP^*N\text{-}tBuC^*H\text{-}iPr$ resonates at −73.3 ppm. The bicyclic compound *39* similarly resonates at −74, −77 ppm. Inclusion of phosphorus in 4-, 5-, and 6-membered rings does not have such a dramatic effect and substituent effects are usually more dominant. The series $Tms_2NP(R)TmsNH$ which has resonances at 59.1 R = tBu, 84 R = iPr, and 106 R = Ph shows a trend opposite to that usually observed.

TRENDS IN CHEMICAL SHIFTS FOR POLYPHOSPHINES AND RELATED COMPOUNDS

The chemical shift range for most of the three coordinate compounds in Section D2 is 200 to −180 ppm. Although the ranges overlap, this is in general to lower frequency (higher field) than phosphites and the halides (Tables B and C) but to higher frequency than the phosphines (Table E). Compared to nitrogen a P-bound phosphorus atom has a greater shielding effect. It also differs from nitrogen in that compounds with three P-bound phosphorus atoms are the most shielded and replacing phosphorus by carbon moves the signals to higher frequency which is the reverse of the trends for the aminophosphines. Compounds which have both nitrogen and phosphorus P-bound atoms have intermediate chemical shift ranges, shielding increasing in the order $N_3 < N_2P < NP_2 < P_3$.

Analysis of the trends within the polyphosphines i.e., P-bound atoms = P_3 shows that as with most phosphorus compounds those with phosphorus as part of 3-membered rings resonate at the lowest frequencies. For the polyphosphines this range is from −142 to −461 ppm. In contrast, when phosphorus participates in 4-membered rings the opposite occurs and signals appear at the higher frequency end (150 to 52 ppm) of the polyphosphine range. The 5-membered heterocycles resonate at intermediate values (−22 to −65 ppm). Most bicyclic polyphosphines resonate in the range −100 to −180 although the extremes of the range extend to +85 and −17 ppm. The bicyclic compounds with fused 3- and 5-membered rings appear to low frequency (−98 to −217) whereas bicyclic compounds which do not involve a 3-membered ring resonate at the high frequency region (85 to −63 ppm). Hydrogen bound to phosphorus exerts its usual deshielding effect relative to carbon. Thus the chemical shift ranges for compounds with P-bound atoms C_2P, CHP, and H_2P are 40 to −60, −29 to −97, and −138 to −207, respectively. In contrast to most other phosphorus compounds, the trifluoromethyl group appears to have a deshielding influence relative to the methyl analogues. This tendency also applies to the arsenic (III) compounds.

Variable temperature studies of one bond PP coupling constants of diphosphines have shown that the coupling is strongly dependent on the conformation about acyclic P-P bonds. The phosphorus chemical shifts of conformationally mobile di⁻ and polyphosphines are also likely to be temperature dependent.

The relatively few arsenic compounds have P-bound atoms C_2As and they resonate (11 to −45 ppm) within the range for C_2P compounds. There are only four antimony compounds all of which appear at lower frequencies, the lowest — Ph_2PSbPh_2 — resonating at −156 ppm.

TABLE D
Three Coordinate (λ3 σ3) Compounds Containing A Phosphorus Bond to a Group V Element But No Bonds to Halogen or Chalcogenide

Section D1: Three Coordinate Compounds with P-N Bonds

P-bound atoms/ rings	Connectivities	Formula	Structure	NMR data (δp[solv]J_{PZ} Hz)	Ref.
Ge_2P	$(H_3)_2;F_2$	$H_6F_2Ge_2P_2$	$F_2PP(GeH_3)_2$	−192.8(r) $^1J_{PP}$ −295.7 $^2J_{PPF}$68.6	52
$NP_2/4$	$Si_2;(NP)_2$	$C_{24}H_{72}N_4P_4Si_8$	$(Tms_2NP^{\cdot})_4$	63.8	1, 20
	CSi;(C Si)$_2$	$C_{23}H_{54}NP_3Si_2$	TmsN-tBuPP˙ tBuSi-tBu$_2$P˙ tBu	14.5 $^1J_{PP}$144	2
	CSi;(NP)$_2$	$C_{28}H_{72}N_4P_4Si_4$	(TmsN-tBuP˙)4	47.5	3
	$C_2;(NP)_2$	$C_{24}H_{56}N_4P_4$	(˙P N-iPr$_2$)$_4$	18.7	4
		$C_{48}H_{88}N_4P_4$	(cHex$_2$NP˙)$_4$	25.1	5
	$(C_2)_2;(N_2O')$	$C_8H_{24}N_4OP_2$	(Me$_2$N)$_2$P(=O)P(NMe$_2$)$_2$	78.7	6
$N_2P^4/5,5$	(CC')$_2$;N$_3$	$C_{11}H_{22}F_3N_5O_5P_2S$	**1**	42.7(r) $^1J_{PP}$170.6	27
$N_2P/5$	$(C_2)_2;N_2$	$C_{16}H_{36}N_4P_2$	(iPr˙N(CH$_2$)$_2$N-iPr˙P)$_2$	120	7
		$C_{20}H_{44}N_4P_2$	**2**; R = (CH$_2$)$_2$, R^1 = iBu, X = NiBu	120	8
$N_2P/6$	$(C_2)_2;N_2$	$C_{24}H_{52}N_4P_2$	**2**; R = CHMeCH$_2$CH$_2$, R^1 = iBu, X = NiBu	120.4	8
			2; R = CHMeCH$_2$CH$_2$, R^1 = tBu, X = NtBu	121	8
N_3	N'P;(C$_2$)$_2$	$C_{13}H_{22}N_6P_2S$	**3**; R = Me	P^{a}71.9 $^2J_{PNP}$165	10
	N'P;(C$_2$)$_2$	$C_{19}H_{34}N_6P_2S$	**3**; R = Et	P^{a}106.6 $^2J_{PNP}$188	10
	N'P;(C$_2$)$_2$	$C_{25}H_{46}N_6P_2S$	**3**; R = Pr	P^{a}111 $^2J_{PNP}$195	10
		$C_{31}H_{58}N_6P_2S$	**3**; R = Bu	P^{a}108.7 $^2J_{PNP}$204	10
	Si$_2$;CH,C$_2$	$C_{14}H_{38}N_3PSi_2$	Et$_2$N-P(NTms$_2$)NH-tBu	108.6	9
	(CN)$_3$	$C_9H_{27}N_6P$	P(NMeNMe$_2$)$_3$	101.5 $^3J_{PNCH}$4.5	11
	C'N';(C$_2$)$_2$	$C_{11}H_{17}N_3P$	**4**; R = Me	107 $^3J_{PNCH}$9.7	12
	CP;(C$_2$)$_2$	$C_9H_{27}N_5P_2$	MeN(P(NMe$_2$)$_2$)$_2$	118.1(b) $^2J_{PCP}$315 $^1J_{PH}$3.3	13
		$C_{10}H_{29}N_5P_2$	(Me$_2$N)$_2$P N EtP(N Me$_2$)$_2$	121.7	14
	C'P;(C'H)$_2$	$C_{36}H_{33}N_6P_3$	**5**; R = Ph	P^{a}62	16
	C'P;(C$_2$)$_2$	$C_{26}H_{54}N_7P_3$	**6**; R = NEt$_2$	P^{a}106.5 $^2J_{PP}$43.5	15
	(C$_2$)$_3$	$C_6H_{12}N_3P$	P(˙NCH$_2$C˙H$_2$)$_3$	129	101
		$C_6H_{14}N_3P$	Me$_2$N P(N˙CH$_2$C˙H$_2$)$_2$	119.2	18
		$C_6H_{18}N_3P$	P(N Me$_2$)$_3$	121.5/123 $^3J_{PNCH}$9/8	19, 28
		$C_8H_{22}N_3P$	(Me$_2$N)$_2$P N Et$_2$	121.6	18
		$C_{10}H_{26}N_3P$	(Me$_2$N)$_2$ P N Pr$_2$	122.5	18
		$C_{12}H_{30}N_3P$	(Me$_2$N)$_2$ P N Bu$_2$	122.2	18
			P(N Et$_2$)$_3$	118/119	19
		$C_{18}H_{42}N_3P$	P(N Pr$_2$)$_3$	121.5	19
			P(N iPr$_2$)$_3$	133.6	19
	(C')$_3$	$C_3N_3O_3P$	P(NCO)$_3$	96.4/97	19
		$C_3N_3PS_3$	P(NCS)$_3$	85.6	19, 21
		$C_{21}F_{36}N_3PS_6$	**7**	64.2	22
$N_3/4$	BSi;Si$_2$;CB	$C_{17}H_{45}BN_3PSi_3$	Tms$_2$NP˙NtBuBBuN˙Tms	114.5	25
	(NP)$_2$;C$_2$	$C_{14}H_{36}N_6P_2$	**8**; R^1 = R^2 = NME$_2$, X = Y = NMe-tBu	79.2	23
	(PSi)$_2$;Si$_2$	$C_{18}H_{54}N_4P_2Si_6$	**8**; R^1 = R^2 = Tms, X = Y = NTms$_2$	225.9	26
	(PSi)$_2$;C$_2$	$C_{10}H_{30}N_4P_2Si_2$	**8**; R^1 = R^2 = Tms, X = Y = NMe$_2$	106.1^c; 198^T	24

TABLE D (continued)
Three Coordinate (λ3 σ3) Compounds Containing A Phosphorus Bond to a Group V Element But No Bonds to Halogen or Chalcogenide

Section D1: Three Coordinate Compounds with P-N Bonds

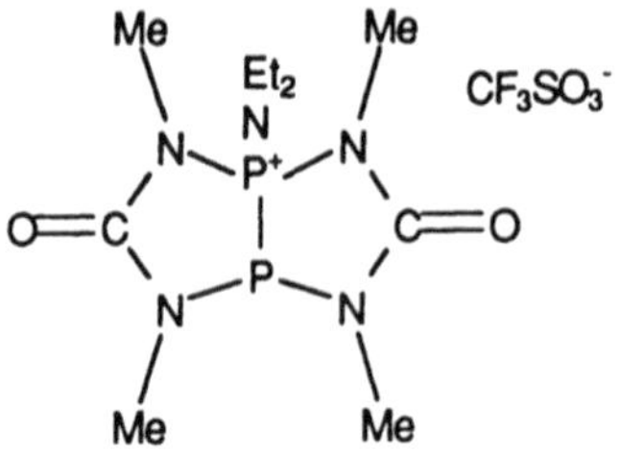

1 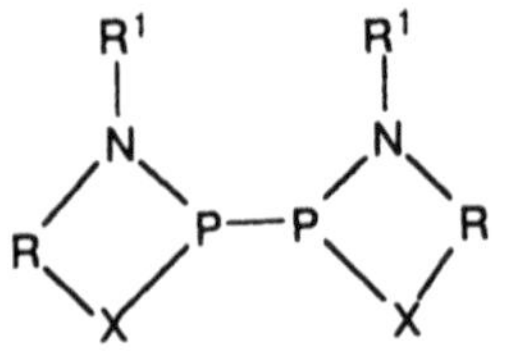 2

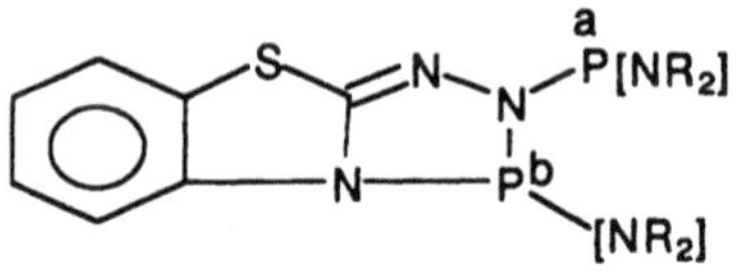

3 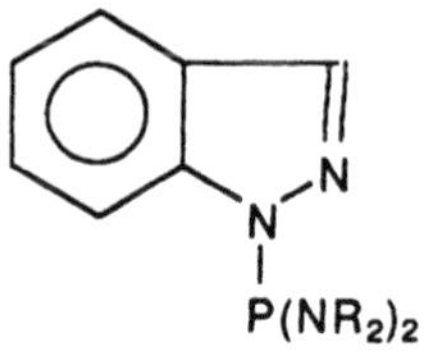 4

P-bound atoms/ rings	Connectivities	Formula	Structure	NMR data (δp[solv]J_{PZ} Hz)	Ref.
		$C_{14}H_{38}N_4P_2Si_2$	**8**; $R^1 = R^2$ = Tms, X = Y = NEt_2	104.2^c; 201.4^T	24
		$C_{18}H_{46}N_4P_2Si_2$	**8**; $R^1 = R^2$ = Tms, X = Y = $N(iPr)_2$	95.5^c	24
	$(PSi)_2$;CSi	$C_{18}H_{50}N_4P_2Si_4$	**8**; $R^1 = R^2$ = Tms, X = Y = NTmsiPr	99.7	26
	$(PSi)_2$;C_2	$C_{14}H_{34}N_4P_2Si_2$	$(TmsNPN^{\cdot}CH_2CH_2CH_2C^{\cdot}H_2)_2$	98^c; 183^T	24
		$C_{16}H_{38}N_4P_2Si_2$	$(TmsNPN^{\cdot}CH_2CH_2CH_2CH_2C^{\cdot}H_2)_2$	102.7^c; 196.5^T	24
		$C_{18}H_{46}N_4P_2Si_2$	**8**; $R^1 = R^2$ = Tms, X = Y = $NiPr_2$	95.0	26
	$(PSi)_2$;CC′	$C_{20}H_{34}N_4P_2Si_2$	**8**; $R^1 = R^2$ = Tms, X = Y = NMePh	106.8^c; 208.7^T	24
	$(Si_2)_3$	$C_{12}H_{36}Cl_2N_3PSi_5$	**9**; R = R^1 = Tms, X = Cl	122.0	29
	$(Si_2)_2$;HSi	$C_9H_{28}Cl_2N_3PSi_3$	**9**; R = H, R^1 = Tms, X = Cl	105.6	29
	$(Si_2)_2$;CSi	$C_{10}H_{30}Cl_2N_3PSi_4$	**9**; R = Me, R^1 = Tms, X = Cl	105.4	29
	$(Si_2)_2$;C_2	$C_{14}H_{36}Cl_2N_3PSi_3$	**9**; R = R^1 = tBu, X = Cl	108.4	29
		$C_{14}H_{38}ClN_4PSi_3$	**9**; R = R^1 = Et, X = NEt_2	102.0/95.5	29
		$C_8H_{24}Cl_2N_3PSi_3$	**9**; R = R^1 = Me, X = Cl	109.9	29
		$C_{10}H_{28}Cl_2N_3PSi_3$	**9**; R = R^1 = Et, X = Cl	108.2	29
	$(CB)_2$;C_2	$C_{18}H_{41}BN_3P$	$iPr_2NP^{\cdot}$ N-tBuBBuN$^{\cdot}$tBu	85.0	25
	$(CP)_2$;CSi	$C_{16}H_{42}N_4P_2Si_2$	**8**; $R^1 = R^2$ = tBu, X = Y = NTmsMe	90.7/89.9^c(c)	29
		$C_{18}H_{46}N_4P_2Si_2$	**8**; $R^1 = R^2$ = iPr, X = Y = NTmsiPr	186.4	29
	$(CP)_2$;CH	$C_{10}H_{26}N_4P_2$	**8**; $R^1 = R^2$ = tBu, X = Y = NMeH	98.1^c; 172.4^T(c)	14

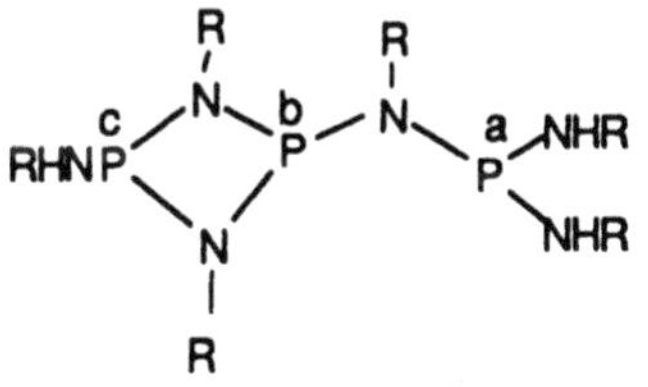

5

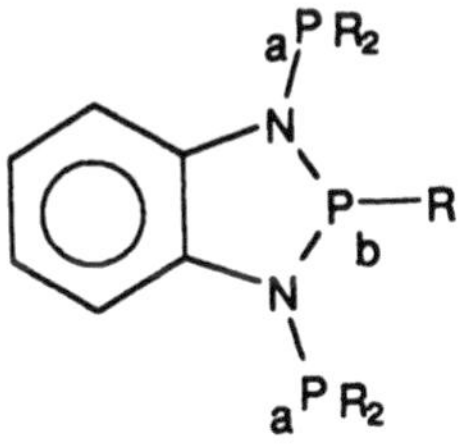

6

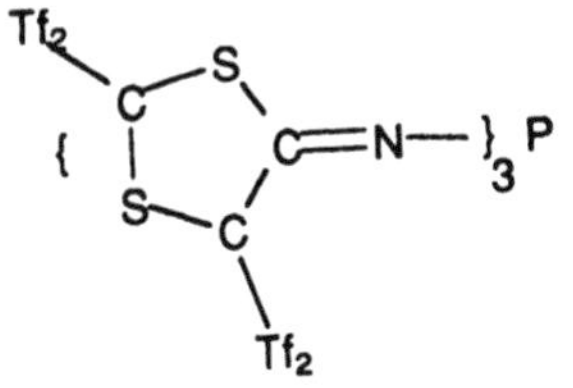

7

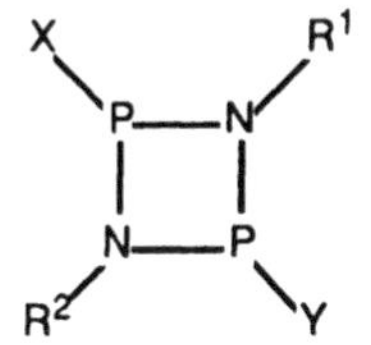

8

P-bound atoms/ rings	Connectivities	Formula	Structure	NMR data (δp[solv]J_{PZ} Hz)	Ref.
		$C_{11}H_{28}N_4P_2$	8; $R^1 = R^2$ = tBu, X = NMeH, Y = NMe_2	99.2[c]; 170.7/ 184.3[T](c)	14
		$C_{12}H_{30}N_4P_2$	8; $R^1 = R^2$ = tBu, X = Y = NEtH	94.7[c](c)	14
	$(CP)_2$;CH	$C_{16}H_{38}N_4P_2$	8;$R^1 = R^2$ = tBu, X = Y = NHtBu	89.4/88	30, 31, 32
	$(CP)_2$;C_2	$C_6H_{18}N_4P_2$	8; $R^1 = R^2$ = Me, X = Y = NMe_2	114.8[c]; 200.9[T](c)	14
		$C_7H_{19}N_4OP_2$	HNMe O=P* NMe $PNEt_2N^*Me$	87.6 $^1J_{PP}$7.5	33
		$C_8H_{22}N_4P_2$	8; $R^1 = R^2$ = Et, X = Y = NMe_2	107.9[c]; 191.8[T](c)	14
		$C_{12}H_{30}N_4P_2$	8; $R^1 = R^2$ = tBu, X = Y = NMe_2	95.0[c]; 184.7[T](c)	14
		$C_9H_{24}N_4P_2$	8; R^1 = Me, R^2 = tBu, X = Y = NMe_2	105.6/192.3(c)	32
		$C_{10}H_{26}N_4P_2$	8; R^1 = Et, R^1 = tBu, X = Y = NMe_2	101.4/186.4(c)	32
		$C_{10}H_{26}N_4P_2$	8; $R^1 = R^2$ = Me, X = Y = NEt_2	198.4[T]; 111.5[c]	34
		$C_{12}H_{30}N_4P_2$	8; $R^1 = R^2$ = tBu, X = Y = NMe_2	95.0/184.7(c)	32
		$C_{12}H_{30}N_4P_2$	8; X = Y = NMe_2, $R^1 = R_2$ = tBu	95.0/184.7(c)	32
		$C_{16}H_{38}N_4P_2$	8; X = Y = NEt_2, $R^1 = R^2$ = tBu	91.3(c)	32
		$C_{16}H_{34}N_4O_2P_2$	8; $R^1 = R^2$ = tBu, X = Y = Morp	94.8[c](c)	14
		$C_{16}H_{36}N_4P_2$	8; $R^1 = R^2$ = tBu, X = Y = NBu	76.7[c]/165.1(c)	14
	$(CP)_2$;CH	$C_{16}H_{38}N_4P_2$	8; $R^1 = R^2$ = tBu, X = Y = NHtBu	89.4(c)/88	30, 31
	$(CP)_2$;C_2	$C_{16}H_{38}N_4P_2$	8; $R^1 = R^2$ = tBu, X = Y = NEt_2	91.3[c](c)	32

TABLE D (continued)
Three Coordinate (λ3 σ3) Compounds Containing A Phosphorus Bond to a Group V Element But No Bonds to Halogen or Chalcogenide

Section D1: Three Coordinate Compounds with P-N Bonds

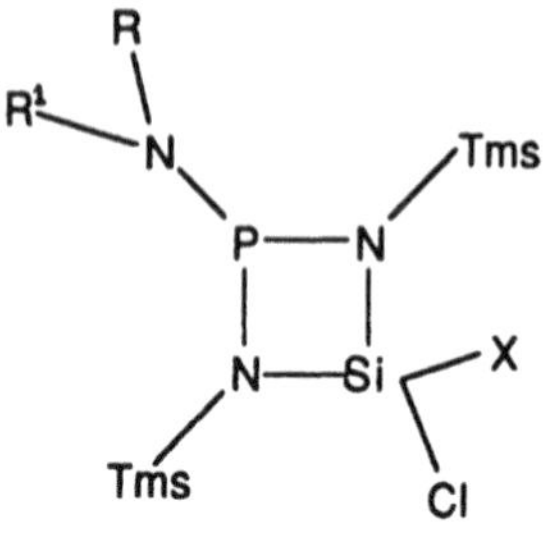

9

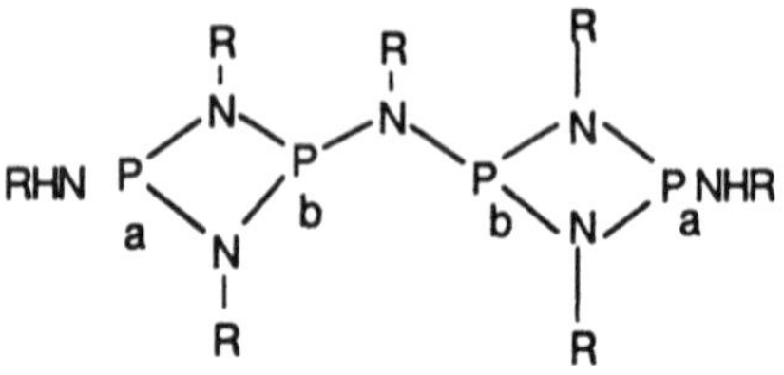

10

P-bound atoms/ rings	Connectivities	Formula	Structure	NMR data (δp[solv]J_{PZ} Hz)	Ref.
		$C_{18}H_{38}N_4P_2$	8; $R^1 = R^2$ = tBu, X = Y = Pip	91.9^c; 182.3^T(c)	14
		$C_{18}H_{42}N_4P_2$	8; $R^1 = R^2$ = tBu, X = Y = NMe-tBu	180.0/120.9; $^2J_{PNP}$38	26
		$C_{20}H_{46}N_4P_2$	8; $R^1 = R^2$ = tBu, X = Y = $NiPr_2$	85.4	26
	CP;C'P;C_2	$C_{12}H_{19}F_3N_4P_2$	8; R^1 = $C_6H_4CF_3$-o, R^2 = Me, X = Y = NMe_2	178.4	35
		$C_{16}H_{28}F_2N_4P_2$	8; R^1 = $C_6H_3F_2$-o.p. R^2 = Et. X = Y = NEt_2	170.0	35
		$C_{17}H_{28}ClF_3N_4P_2$	8; R^1 = $C_6H_3CF_3$-o,Cl-p, R^2 = Et, X = Y = NEt_2	178.0	35
		$C_{17}H_{29}F_3N_4P_2$	8; R^1 = $C_6H_4CF_3$-o, R^2 = Et. X = Y = NEt_2	194.4	35
		$C_{17}H_{31}FN_4P_2$	8; R^1 = $C_6H_3CH_3$-o,F-m, R^2 = Et, X = Y = NEt_2	177	35
	$(C'P^4)_2,C_2$	$C_{22}H_{25}N_3O_1P_2$	$Et_2NP^{\cdot}NPhP(=O)PhN^{\cdot}Ph$	80.0; $^3J_{PH}$9.6	38
	$(C'P)_3$	$C_{42}H_{37}N_7P_4$	10; R = Ph	b 112.8/111.8(t)	16
	$(C'P)_2$;C'H	$C_{42}H_{37}N_7P_4$	10; R = Ph	a 107/107.6(t)	36
		$C_{24}H_{22}N_4P_2$	8; $R^1 = R^2$ = Ph, X = Y = NHPh	104/113.6	16, 19
		$C_{36}H_{33}N_6P_3$	5; R = Ph	P^b106.9	16
	$(C'P)_2$;C_2	$C_8H_{18}N_4O_2P_2$	8; $R^1 = R^2$ = COMe, X = Y = NMe_2	154	37
		$C_{16}H_{22}N_4P_2$	8; $R^1 = R^2$ = Ph, X = Y = NMe_2	101.0/166.5(c)	32
		$C_{16}H_{20}Cl_2N_4P_2$	8; $R^1 = R^2$ = C_6H_4 Cl-p, X = Y = NMe_2	100.8/166.1(c)	32
		$C_{18}H_{26}N_4P_2$	8; $R^1 = R^2$ = C_6H_4 Me-p, X = Y = NMe_2	166.8(c)	32
		$C_{18}H_{26}N_4O_2P_2$	8; $R^1 = R^2$ = C_6H_4 OMe-p, X = Y = NMe_2	101.5/168.9(c)	32
		$C_{20}H_{30}N_4P_2$	8; $R^1 = R^2$ = Ph, X = Y = NEt_2	162.2(c)	32
	CP;C'P;C_2	$C_{12}H_{19}F_3N_4P_2$	8; R^2 = Me, R^1 = $C_6H_4CF_3$-o, X = Y = NMe_2	178.4	35

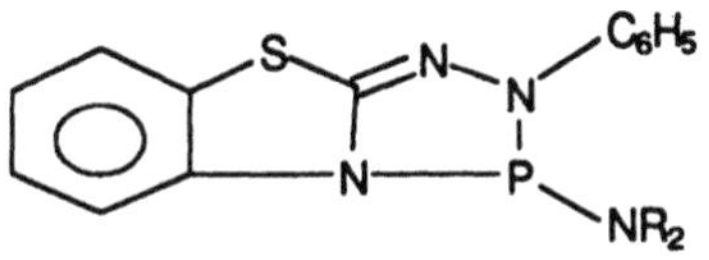

11

P-bound atoms/ rings	Connectivities	Formula	Structure	NMR data (δp[solv]J_{PZ} Hz)	Ref.
		$C_{16}H_{28}F_2N_4P_2$	**8**; R^2 = Et, R^1 = $C_6H_3F_2$-o,p, X = Y = NEt_2	170.0	35
		$C_{17}H_{28}ClF_3N_4P_2$	**8**; R^2 = Et, R^1 = $C_6H_3CF_3$-o, Cl-p X = Y = NEt_2	178	35
		$C_{17}H_{29}F_3N_4P_2$	**8**; R^2 = Et, R^1 = $C_6H_4CF_3$-o, X = Y = NEt_2	194.4	35
		$C_{17}H_{31}FN_4P_2$	**8**; R^2 = Et, R^1 = C_6H_3 CH_3-o,F-m, X = Y = NEt_2	177	35
	$(CP)_2$;CH	$C_{16}H_{38}N_4P_2$	$(tBuN{\cdot}HP{\cdot}NtBu)_2$	89.4/88	30, 31, 32
	$(C,S^4)_2;C_2$	$C_8H_{20}N_3O_2PS$	$Et_2NP{\cdot}NEtSO_2N{\cdot}Et$	79.5	40
		$C_{10}H_{24}N_3O_2PS$	$Me_2NP{\cdot}NtBuSO_2N{\cdot}tBu$	68.4, $^2J_{PNC}$8.2 $3J_{PNCC}$4.4	39, 102
	$(C'S^4)_2;C_2$	$C_{16}H_{20}N_3O_2PS$	$Et_2NP{\cdot}NPhS(O_2)N{\cdot}Ph$	67.7	38
	$C_2,(CC')_2$	$C_5H_{12}N_3OP$	$Me_2NP{\cdot}NMeCON{\cdot}Me$	93.6 $^3J_{PNCH}$12(acy) $^3J_{PNCH}$4.3(cy)	37
		$C_7H_{16}N_3OP$	$Et_2NP{\cdot}NMeC({=}O)N{\cdot}Me$	85.1	38
N_3/5	N'P;C_2;CC'	$C_{13}H_{22}N_6P_2S$	**3**; R = Me	39.1 $^2J_{PNP}$165	10
		$C_{15}H_{15}N_4PS$	**11** R = Me	74.1(c)	10
		$C_{17}H_{19}N_4PS$	**11**; R = Et	71.1(c)	10
		$C_{19}H_{23}N_4PS$	**11**; R = Pr	72.4(c)	10
		$C_{21}H_{27}N_4PS$	**11**; R = Bu	72.4(c)	10
		$C_{19}H_{34}N_6P_2S$	**3**; R = Et	76.9 $^3J_{PNP}$188	10
		$C_{25}H_{46}N_6P_2S$	**3**; R = Pr	77.3 $^2J_{PNP}$195	10
		$C_{31}H_{58}N_6P_2S$	**3**; R = Bu	77.1 $^2J_{PNP}$204	10
	$(C'P)_2;C_2$	$C_{26}H_{54}N_7P_3$	**5**; R = NEt_2	p^b 100,5. $^2J_{PNP}$43.5	15
	$(C_2)_3$	$C_6H_{16}N_3P$	$Me_2NP{\cdot}NMe(CH_2)_2N{\cdot}Me$	114/116J_{PNCH}9.1 J_{PNCH}23.8	19, 28
		$C_8H_{18}N_3P$	$C(CH_2)_3N{*}PN{\cdot}Me$ $(CH_2)_2N{\cdot}Me$	104.8	19
		$C_{12}H_{30}N_6P_2$	$(-CH_2NMeP{\cdot}$ $NMe(CH_2)_2N{\cdot}Me)_2$	114 $^3J_{PNCH}$12.5(cy) $^3J_{PNCH}$5.4(acy)	40
N_3/6	$(CN)_2;C_2$	$C_8H_{24}N_6P_2$	$(Me_2NP{\cdot}NMeN{\cdot}Me)_2$	130(b)	41
	$(C_2)_3$	$C_8H_{18}N_3P$	$CH_2{\cdot}CH_2N{\cdot}$ $P{*}NMeCHMeCH_2CH_2N{*}Me$	133.9	42
		$C_{10}H_{24}N_3P$	$Et_2NP{\cdot}NMeCHMe$- $CH_2CH_2N{\cdot}Me$	129.7/107.3	42
	$(C_2)_2$,CC'	$C_9H_{20}N_3OP$	$Et_2NP{\cdot}NMeCOCH_2CH_2N{\cdot}Me$	107	43
	$C_2,(CC')_2$	$C_{17}H_{20}N_3P$	$CH_2{\cdot}CH_2N{\cdot}P{*}NPh$- $CH_2CH_2CH_2N{*}Ph$	121.9	44
		$C_{17}H_{22}N_3P$	$Me_2NP{\cdot}NPhCH_2CH_2CH_2N{\cdot}Ph$	105.2	44
		$C_{18}H_{24}N_3P$	$Me_2NP{\cdot}NPhCHMe$- $CH_2CH_2N{\cdot}Ph$	97.3	44

TABLE D (continued)
Three Coordinate (λ3 σ3) Compounds Containing A Phosphorus Bond to a Group V Element But No Bonds to Halogen or Chalcogenide

Section D1: Three Coordinate Compounds with P-N Bonds

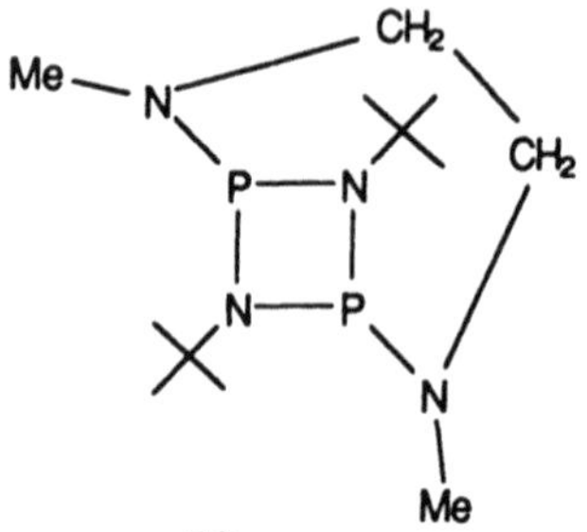

12 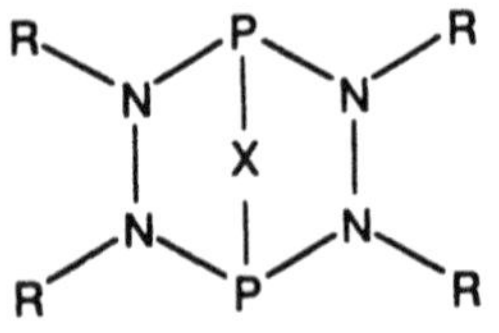 **13** 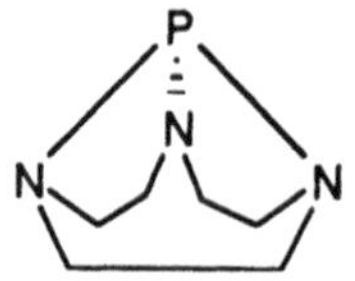 **14**

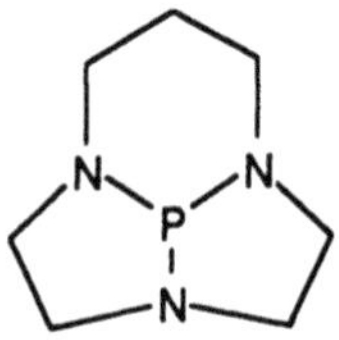

15 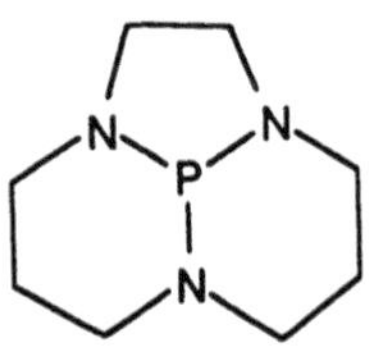 **16** 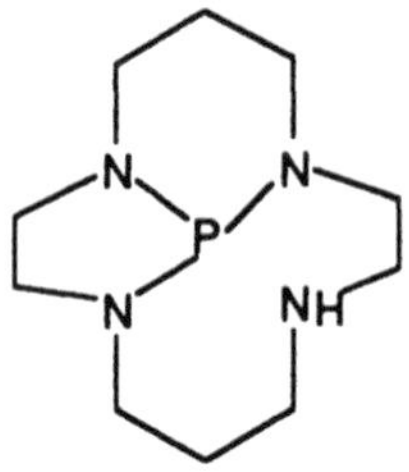 **17**

P-bound atoms/ rings	Connectivities	Formula	Structure	NMR data (δp[solv]J_{PZ} Hz)	Ref.
		$C_{19}H_{26}N_3P$	$Et_2NP^{\cdot}NPhCH_2CH_2CH_2N^{\cdot}Ph$	106.9	44
		$C_{20}H_{28}N_3P$	$Et_2NP^{\cdot}NPhCHMe$-$CH_2CH_2N^{\cdot}Ph$	94.3	44
		$C_{23}H_{34}N_3P$	$Bu_2NP^{\cdot}NPhCH_2CH_2CH_2N^{\cdot}Ph$	105.5	44
N_3/4,7	$(CP)_2,C_2$	$C_{12}H_{28}N_4P_2$	**12**	155(c)	45
N_3/5,5	$(CN)_2$;CP	$C_5H_{15}N_5P_2$	**13**; X = MeN, R = Me	101.8 $^3J_{PNCH}$12.8	41
	$(C_2)_3$	$C_6H_{12}N_3P$	**14**	142/155	101
		$C_7H_{14}N_3P$	**15**	111.6(b)	49
N_3/5.6	$(C_2)_3$	$C_8H_{16}N_3P$	**16**	122.7(b)	49
		$C_{10}H_{21}N_4P$	**17**	111.9(b)	50
N_3/6,6	NP;$(C'C)_2$	$C_7H_{15}N_5O_2P_2$	**18**	106.9	46
	$(CN)_3$	$C_5H_{16}N_6P_2$	**13**; X = MeNNH, R = Me	106.7/103.0(b) $^3J_{PNNP}$32 $^3J_{PNCH}$15	41
		$C_6H_{18}N_6P_2$	**13**; X = MeN-NMe, R = Me	109(b) $^3J_{PNCH}$15.2	41
		$C_{12}H_{30}N_6P_2$	**13**; X = EtNNEt. R = Et	98.4(b)	41
		$C_6H_{18}N_6P_2$	**19**	107.4/109 $^3J_{PNNP}$32	47
	$(CP)_3$	$C_6H_{18}N_6P_4$	**20**; R = Me	78.4 $^3J_{PNCH}$16	19
		$C_{18}H_{42}N_6P_4$	**20**; R = iPr	84(b)	48
	$(CN)_3$	$C_6H_{15}N_6P$	**21**	102	51
	$(C_2)_2$	$C_8H_{18}N_3P$	**22**; R = Me	83.5/84.0(m)	93, 84
	$(C_2)_3$	$C_9H_{18}N_3P$	**23**	109.2(b)	49
		$C_{10}H_{21}N_4P$	**24**	111.6/115.5(b)	50

18

19

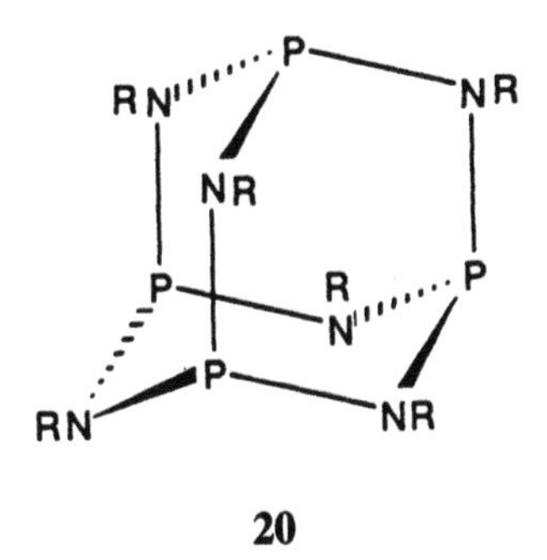

20

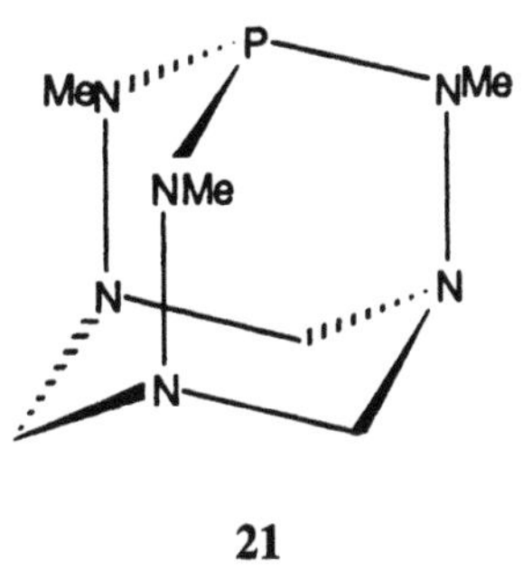

21

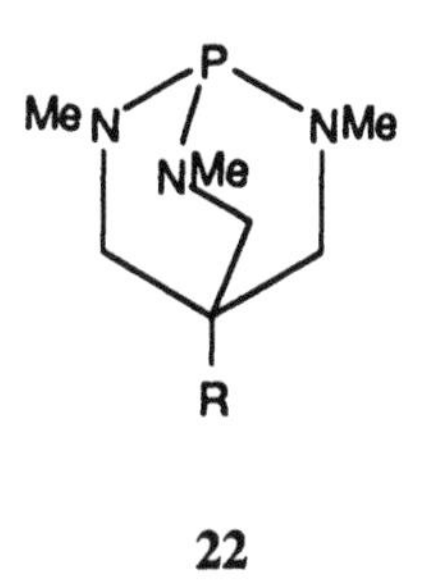

22

P-bound atoms/ rings	Connectivities	Formula	Structure	NMR data (δp[solv]J_{PZ} Hz)	Ref.
		$C_{12}H_{25}N_4P$	**25**	104.8(b)	50
	$(C_2)_2$	$C_{12}H_{26}N_3P$	**22**; R = Pent	86.6	73
N_3/6.7	$(CP)_2;C_2$	$C_{13}H_{33}N_6P_3$	**26**	101(ax); 95.7(eq)	34
N_3/8,8	$(CP)_3$	$C_{18}H_{42}N_6P_4$	**27**	145.8 (r)	48
N_2N'	$(C_2)_2;C'$	$C_9H_{22}N_3OP$	$(Me_2N)_2P\text{-}N{=}COMePr$	100.4	17
PSi_2	$F_2;(H_3)_2$	$H_6F_2P_2Si_2$	$F_2PP(SiH_3)_2$	−211.8(r) $^1J_{PP}$301	52
	$F_2;(C_3)_2$	$C_6H_{18}F_2P_2Si_2$	$F_2P\text{-}PTms_2$	−128.7(r) $^1J_{PP}$381	52
PSn_2	$F_2;(C_3)_2$	$C_6H_{18}F_2P_2Sn_2$	$F_2PP(SnMe_3)_2$	−160(r) $^1J_{PP}$380 $^2J_{PPF}$48	52
		$C_{24}H_{54}F_2P_2Sn_2$	$F_2PP(SnBu_3)_2$	−169(r) $^1J_{PP}$382.6 $^2J_{PPF}$45.3	52
P_2Sn	$(F_2)_2;C_3$	$C_{12}H_{27}F_4P_3Sn$	$(F_2P)_2PSnBu_3$	−33(r) $^1J_{PP}$360 $^2J_{PPF}$59	52
HN_2	$(Si_2)_2$	$C_{12}H_{37}N_2PSi_4$	$(Tms_2N)_2PH$	58.2(b.)58.8 (c)59.6 $^1J_{PH}$207	53, 54, 80
	$(C_2)_2$	$C_4H_{13}N_2P$	$(Me_2N)_2PH$	95 $^1J_{PH}$250	56
		$C_8H_{21}N_2P$	$(Et_2N)_2PH$	77(b) $^1J_{PH}$259/249	56
		$C_{12}H_{29}N_2P$	$(iPr_2N)_2PH$	42.1(c) $^1J_{PH}$254	56, 57
HN_2/5	$(C_2)_2$	$C_4H_{11}N_2P$	$HP^{\cdot}NMe(CH_2)_2N^{\cdot}Me$	97.4 $^1J_{PH}$150	57, 58
		$C_{10}H_{23}N_2P$	$HP^{\cdot}NtBuCH_2CH_2N^{\cdot}tBu$	57, 9(h) $^1J_{PH}$156	56, 57
HN_2/6	$(C_2)_2$	$C_6H_{15}N_2P$	$HP^{\cdot}NMeCHMeCH_2CH_2N^{\cdot}Me$	82.2/47.3 $^1J_{PH}$192/202	42
		$C_{11}H_{25}N_2P$	$HP^{\cdot}NtBuCH_2CH_2CH_2N^{\cdot}tBu$	49.6 $^1J_{PH}$214	56
		$C_{12}H_{27}N_2P$	$HP\text{-}{}^{\cdot}NtBuCH_2CH_2CHMeN^{\cdot}tBu$	12.7 $^1J_{PH}$229	56
H_2N	Si_2	$C_{12}H_{32}NPSi_2$	$H_2PN(SiMe_2tBu)_2$	−47.3 $^1J_{PH}$191	3
	C_2	$C_9H_{20}NP$	$H_2PN^{\cdot}CMe_2CH_2\text{-}CH_2CH_2C^{\cdot}Me_2$	−86.6(k) $^1J_{PH}$202	59

TABLE D (continued)
Three Coordinate (λ3 σ3) Compounds Containing A Phosphorus Bond to a Group V Element But No Bonds to Halogen or Chalcogenide

Section D1: Three Coordinate Compounds with P-N Bonds

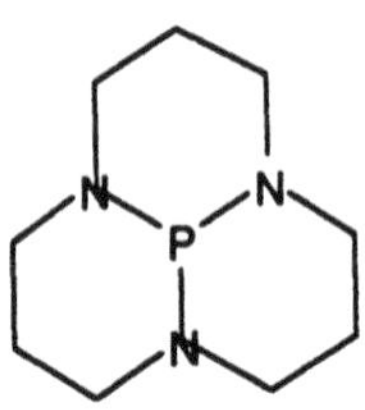

23 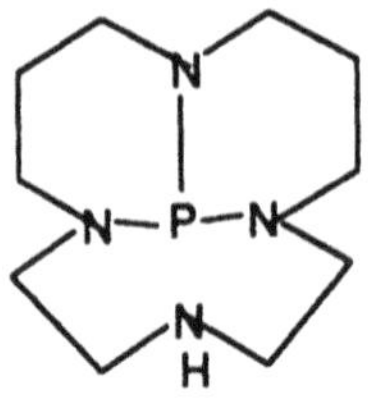 24 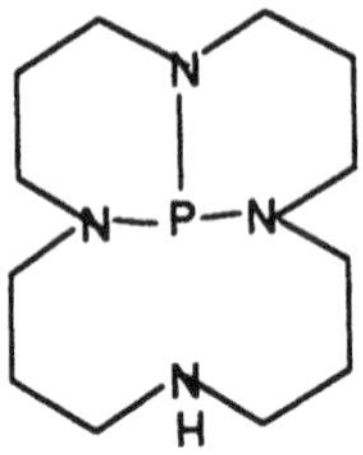 25

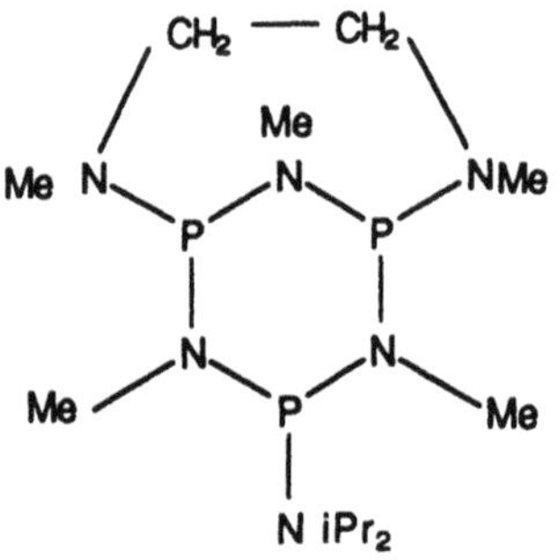

26 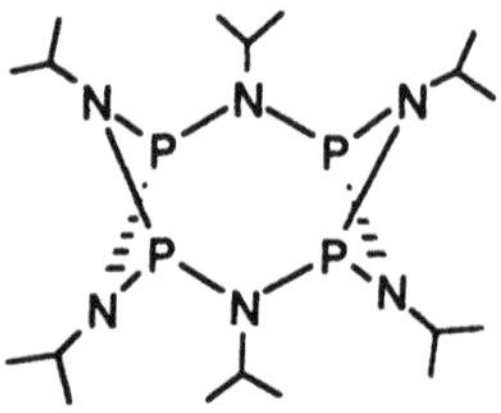 27

P-bound atoms/ rings	Connectivities	Formula	Structure	NMR data (δp[solv]J_{PZ} Hz)	Ref.
		$C_6H_{16}NP$	H_2PNiPr_2	−82.9(k) $^1J_{PH}$196	59
CNP^4	$H_2Si;Si_2;CNN'$	$C_{21}H_{39}N_3P_2Si_5$	**28**	38(r) $^1J_{PP}$290	60
CNP/5	$CC'H;Si_2;CN$	$C_{17}H_{42}N_2P_2Si_4$	**29**	57.4/64.2 J_{PP}330.2	20
$C'NP^5$/4,5	$C'_2;CC';C'N_3$	$C_{18}H_{22}N_4O_2P_2$	**30**; R = Ph, R^1 = Me	1.9 J_{PP}156	62
	$C'_2;C'_2;C'N_3$	$C_{28}H_{26}N_4O_2P_2$	**30**; R = R^1 = Ph	6.3 J_{PP}190	62
		$C_{30}H_{14}F_{16}N_4O_2P_2$	**30**; R = C_6F_5, R^1 = C_6H_4-Tf-p	−8.6 J_{PP}181.5	63
$C'NP^4$/5	$C'_2;CC';C'NO'$	$C_{15}H_{16}N_2O_2P_2$	**31**; R = Me	5.7 J_{PP}180	61
	$C'_2;C'_2;C'NO'$	$C_{20}H_{18}N_2O_2P_2$	**31**; R = Ph	10.2 J_{PP}185	61
CN_2	$ClSi_2;(C_2)_2$	$C_{11}H_{30}ClN_2PSi_2$	$(Me_2N)_2PCClTms_2$	86.1	64
	$H_2Si;(Si_2)_2$	$C_{16}H_{47}N_2PSi_5$	$(Tms_2N)_2PCH_2Tms$	107.1 J_{PCH}3.6	65, 66
	$H_2Si;Si_2;C_2$	$C_{14}H_{39}N_2PSi_3$	$Tms_2P(NEt_2)CH_2Tms$	92	60
	$H_2Si;HSi;C_2$	$C_{11}H_{31}N_2PSi_2$	$TmsNHP(NEt_2)CH_2Tms$	68.5(r)	60
	$H_3;(Si_2)_2$	$C_{13}H_{39}N_2PSi_4$	$(Tms_2N)_2PMe$	105.3	66
	$H_3;(C_2)_2$	$C_5H_{15}N_2P$	$MeP(NMe_2)_2$	86.4	19
		$C_9H_{19}N_2O_2P$	$MeP(N^{\cdot}CH_2CH_2OCH_2CH_2^{\cdot})_2$	82.5	19
		$C_9H_{23}N_2P$	$MeP(NEt_2)_2$	80.4	19
		$C_{15}H_{31}N_2P$	$MeP(NMeC_6H_{11})_2$	79.6	19
	$H_3;(CC')_2$	$C_{15}H_{19}N_2P$	$MeP(N\ Me\ Ph)_2$	76.2	19
	$H_3;(C')_2$	$C_3H_3N_2PS_2$	$MeP(N{=}C{=}S)_2$	37.9 $^2J_{PCH}$15.8	19
	$C'HSi;(Si_2)_2$	$C_{22}H_{51}N_2PSi_5$	$(Tms_2N)_2P\ CH\ Ph\ Tms$	111.5 $^2J_{PCH}$6.0	65
	$CH_2;(Si_2)_2$	$C_{14}H_{41}N_2PSi_4$	$(Tms_2N)_2P\ Et$	114.3 $^2J_{PCH}$18.6	65, 66

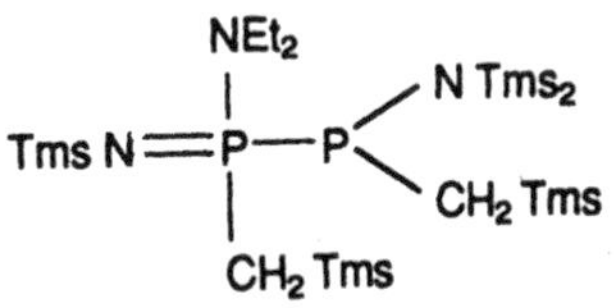

28 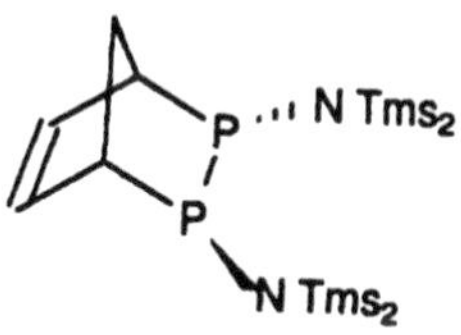 **29**

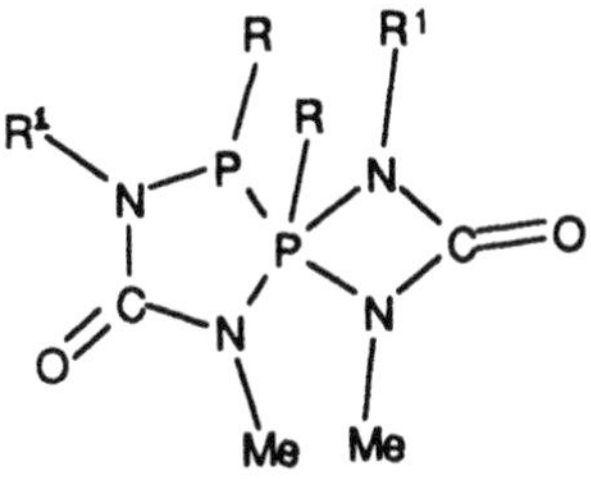

30 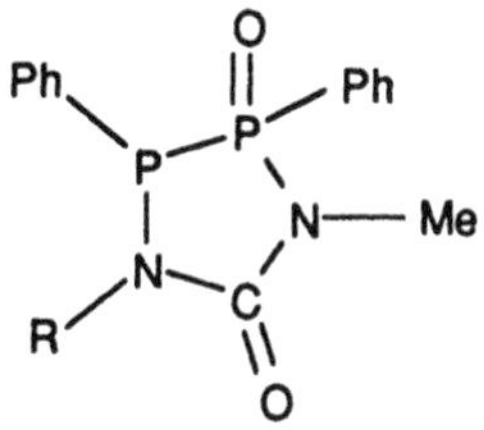 **31**

P-bound atoms/ rings	Connectivities	Formula	Structure	NMR data (δp[solv]J_{PZ} Hz)	Ref.
	$CH_2;(C_2)_2$	$C_6H_{17}N_2P$	Et P (N $Me_2)_2$	99.9	19
	$CH_2;(C_2)_2$	$C_7H_{20}N_2P_2$	MeP(H) $CH_2CH_2P(N\ Me_2)_2$	95(r) $^3J_{PCCP}$29	68
	$CH_2;(C_2)_2$	$C_8H_{21}N_2P$	iBu P(N $Me_2)_2$	92.4	19
		$C_8H_{22}N_2P_2$	Me_2 P $CH_2CH_2P(NMe_2)_2$	95.3(r) $^3J_{PCCP}$29	68
		$C_{12}H_{22}N_2P_2$	$PhP(H)CH_2CH_2P(NMe_2)_2$	93.6(r) $^3J_{PCCP}$25	68
		$C_{12}H_{29}N_2P$	$BuP(NEt_2)_2$	99.8	19
		$C_{13}H_{35}N_4P_3$	$MeP(CH_2CH_2P(NMe_2)_2)_2$	95.2(r) $^3J_{PCCP}$29	68
		$C_{14}H_{31}N_2P$	$C_6H_{11}P(NEt_2)_2$	107.1	19
		$C_{18}H_{26}N_2P_2$	$Ph_2PCH_2CH_2P(NMe_2)_2$	94.8(r) $^3J_{PCCP}$37	68
		$C_{18}H_{37}N_4P_3$	$PhP(CH_2CH_2P(NMe_2)_2)_2$	95.0(r) $^3J_{PCCP}$25	68
		$C_{18}H_{44}N_4P_2$	$((Et_2N)_2PCH_2\text{-})_2$	90.7/91.1(b)	57, 69
		$C_{18}H_{48}N_6P_4$	$P(CH_2CH_2P(NMe_2)_2)_3$	95.6(r) $^3J_{PCCP}$28	68
		$C_{19}H_{46}N_4P_2$	$((Et_2N)_2P\ CH_2)_2CH_2$	87.8	69
		$C_{20}H_{48}N_4P_2$	$((Et_2N)_2P\ CH_2CH_2\text{-})_2$	87.9	69
		$C_{21}H_{50}N_4P_2$	$((Et_2N)_2P\ CH_2CH_2)_2CH_2$	87.9	69
		$C_{22}H_{52}N_4P_2$	$((Et_2N)_2P\ CH_2CH_2CH_2\text{-})_2$	88.0	69
		$C_{23}H_{54}N_4P_2$	$((Et_2N)_2P\ CH_2CH_2CH_2)_2CH_2$	88.1	69
		$C_{24}H_{56}N_4P_2$	$((Et_2N)_2P\ CH_2CH_2CH_2CH_2\text{-})_2$	88.1	69
		$C_{25}H_{58}N_4P_2$	$((Et_2N)_2P\text{-}$ $CH_2CH_2CH_2CH_2)_2CH_2$	88.2	69
		$C_{26}H_{60}N_4P_2$	$((Et_2N)_2P\text{-}$ $CH_2CH_2CH_2CH_2CH_2\text{-})_2$	88.2	69
	$C'H_2;(Si_2)_2$	$C_{19}H_{43}N_2PSi_4$	$(Tms_2N)_2P\ CH_2$ Ph	112.3 $^2J_{PCH}$6.9	65, 66
	$C_2H;Si_2;HSi$	$C_{12}H_{35}N_2PSi_3$	(Tms_2N) (TmsNH)P iPr	84	67
	$C_3;Si_2;HSi$	$C_{13}H_{37}N_2PSi_3$	(Tms_2N) (Tms NH) PtBu	59	67
	$C_2H;(Si_2)_2$	$C_{15}H_{43}N_2PSi_4$	$(Tms_2N)_2P$ iPr	116.6(c)	54
$CN_2/3$	$CHN;(C_2)_2$	$C_{14}H_{31}N_2P$	iPr_2 N P'NtBuC'HiPr	− 73.3(b)	70
	$CHN;(C_2)_2$	$C_{15}H_{33}N_2P$	iPr_2 N P'NtBuC'HtBu	− 73.3(b) J_{PCH}2.3, $^2J_{PNCH}$3.1	70
$CN_2/4$	$H_3;(CP)_2$	$CH_{21}ClN_2P_2$	**8**; $R^1 = R^2 =$ tBu, X = Me, Y = Cl	200.4(m) $^2J_{PNP}$33	71
		$C_{10}H_{24}N_2P_2$	**8**; $R^1 = R^2 =$ tBu, X = Y = Me	171.4(b)	71
	$CF_2;CC',C'_2$	$C_{11}H_7F_8N_2OP$	C_2F_5P'NMe(CO)N'C $_6H_4$-4Tf	84.6 $^1J_{JPF}$69/74	62

TABLE D (continued)
Three Coordinate (λ3 σ3) Compounds Containing A Phosphorus Bond to a Group V Element But No Bonds to Halogen or Chalcogenide

Section D1: Three Coordinate Compounds with P-N Bonds

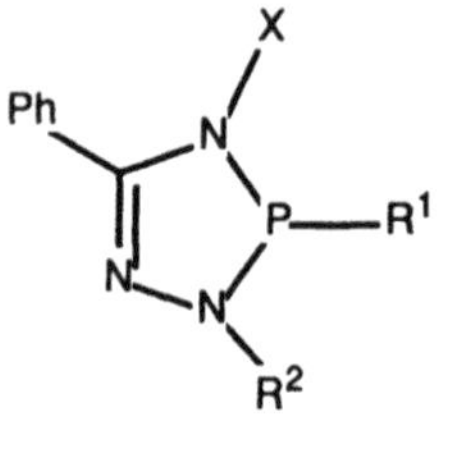

32 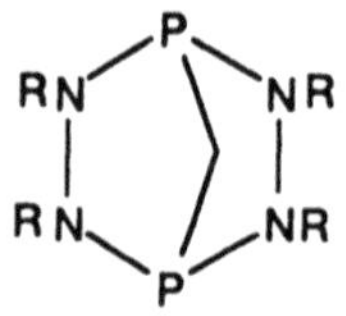 **33**

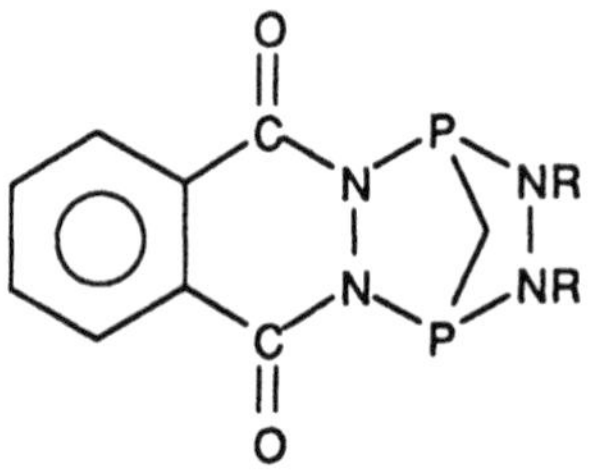

34 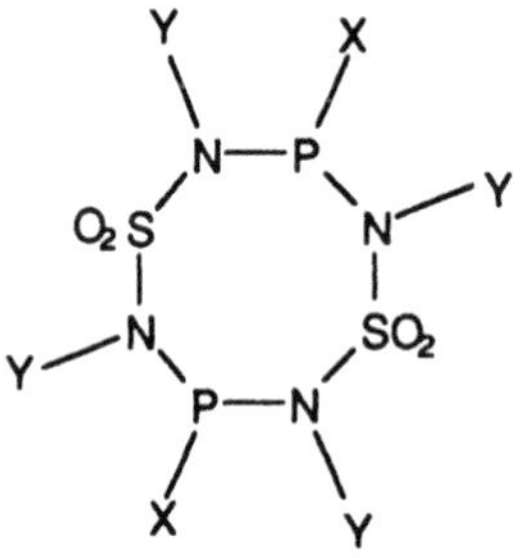 **35**

P-bound atoms/ rings	Connectivities	Formula	Structure	NMR data (δp[solv]J_{PZ} Hz)	Ref.
	C_3;$(CP)_2$	$C_{10}H_{24}N_2P_2$	**8**; $R^1=R^2=Me$, X = Y = tBu	201.7^c(r)	72
			8; $R^1=R^2=Me$, X = Y = tBu	281.7^t(r)	72
CN_2/5	H_3;CN,CP	$C_5H_{15}N_3P$	MeP˙NMeNMePMeN˙Me	117.8(b) $^3J_{PNCH}$14.3	41
	H_3;CN′,C′H	$C_8H_{12}N_3P$	**32**; R^1 = Me, R^2 = Me, X = H	95.8 $^2J_{PNH}$32	74
	H_3;C′N′,C′H	$C_{13}H_{14}N_3P$	**32**; R^1 = Me, R^2 = Ph, X = H	85 $^2J_{PNH}$33	74
	CHN′;CN′,C $_2$	$C_{15}H_{33}N_4P$	iPr_2N P˙NtBu N=N C˙HtBu	17.8(b) $^2J_{PCH}$4.3, $^3J_{PNCH}$2.6	70
	CH_2;CN′;C′H	$C_9H_{14}N_3P$	**32**; R^1 = Et, R^2 = Me, X = H	100.1	74
		$C_{11}H_{18}N_3P$	**32**; R^1 = Bu, R^2 = Me, X = H	97.4 $^2J_{PNH}$24	74
	CH_2;CN′,CC′	$C_{10}H_{16}N_3P$	**32**; R^1 = Et, R^2 = Me, X = Me	88	74
	CH_2;C′N′;C′H	$C_{14}H_{16}N_3P$	**32**; R^1 = Et, R^2 = Ph, X = H	97.6	74
		$C_{16}H_{20}N_3P$	**32**; R^1 = Bu, R^2 = Ph, X = H	76.5	74
CN_2/6	H_3;$(CN)_2$	$C_6H_{18}N_4P_2$	$(MeP˙NMeN˙Me)_2$	91.1(b)	41
	CH_2;$(C_2)_2$	$C_{24}H_{52}N_4P_2$	(-CH_2P˙NtBuCH_2CH_2-CH_2N˙tBu$)_2$	61.8(c)	57
CN_2/8	H_3;(C $S^4)_2$	$C_{10}H_{26}N_4O_4P_2S_2$	**35**; X = Me, Y = Et	86.3/88.5	103
CN_2/5,5	H_2P;$(CN)_2$	$C_5H_{14}N_4P_2$	**33**; R = Me	86(r)	75
	H_2P;CN,C′N	$C_{11}H_{12}N_4O_2P_2$	**34**; R = Me	77.5(b)	75
C′N_2	C′H;$(C_2)_2$	$C_6H_{15}N_2P$	$(Me_2N)_2$ P CH=CH_2	93.6(r)	68
		$C_{10}H_{23}N_2P$	$(Et_2N)_2$ P CH=CH_2	89.9(k)	57
		$C_{14}H_{31}N_2P$	$(iPr_2N)_2$ P CH=CH_2	53.1(c)	57
	C′$_2$;Si_2,HSi	$C_{15}H_{33}N_2PSi_3$	(Tms_2 N)(Tms NH) P Ph	106	67
	C′$_2$;$(CN)_2$	$C_{12}H_{23}N_4P$	Ph P(N Me N $Me_2)_2$	80.2 $^3J_{PNCH}$5	11
	C′$_2$;CN′;C′H	$C_{13}H_{14}N_3P$	**32**; R^1 = Ph, R^2 = Me, X = H	77 $^2J_{PNH}$34	74

P-bound atoms/ rings	Connectivities	Formula	Structure	NMR data (δp[solv]J_{PZ} Hz)	Ref.
	C'$_2$;C'N';C'H	$C_{18}H_{16}N_3P$	**32**; R[1] = R[2] = Ph, X = H	71 $^2J_{PNH}$37	74
	C'$_2$;(CP)$_2$	$C_{13}H_{20}N P$	-(P Mes NtBu)-$_n$	102/97	78
	C'$_2$;(C'P)$_2$	$C_{15}H_{15}Cl N P$	-(P Mes N C_6H_4Cl-p)-$_n$	72/70.7	78
	C'$_2$;(C$_2$)$_2$	$C_{10}H_{17}N_2P$	PhP(NMe$_2$)$_2$	100.3	19
		$C_{14}H_{20}F_5N_2P$	C_6F_5P(NEt$_2$)$_2$	79.2	79
		$C_{14}H_{25}N_2P$	Ph P(N Et$_2$)$_2$	98/95.6	19
		$C_{16}H_{30}N_2P$	2,6-di-MeC_6H_4P(N Et$_2$)$_2$	92.8	19
		$C_{18}H_{33}N_2P$	Ph P (N Pr$_2$)$_2$	98	19
C'N$_2$/4	C'$_2$;CP,C'P	$C_{44}H_{76}N_2P_2$	**36**; X = 2,4,6-tBu$_3C_6H_2$, Y = tBu	250.7	81
		$C_{54}H_{80}N_2P_2$	**36**; X = 2,4,6-tBu$_3C_6H_2$, Y = Mes	250.7	81
		$C_{66}H_{104}N_2P_2$	**36**; X = 2,4,6tBu$_3C_6H_2$, Y = 2,4tBu$_2$, 6MeC_6H_2	251.7	81
		$C_{72}H_{116}N_2P_2$	**36**; X = Y = 2,4,6-tBu$_3C_6H_2$	252.4(b)	81
	C'$_2$;(C'$_2$)$_2$	$C_{19}H_{15}N_2OP$	PhP$^\cdot$N(Ph)CON$^\cdot$Ph	91.8	73
C'N$_2$/6	C'H;(C$_2$)$_2$	$C_{13}H_{27}N_2P$	CH$_2$ = CHP$^\cdot$NtBuCH$_2$CH$_2$-CH$_2$N$^\cdot$tBu	56.4(h)	57
	CC';(C$_2$)$_2$	$C_{13}H_{21}N_2P$	PhP$^\cdot$NEt(CH$_2$)$_3$N$^\cdot$Et	86.8	82
C'N$_2$/8	C'$_2$;(CS^4)$_2$	$C_{20}H_{30}N_4O_4P_2S_2$	**35**; X = Ph, Y = Et	91.0	103
C'N'$_2$	C'$_2$;(C''')$_2$	$C_8F_5N_2O_2P$	C_6F_5P(N=C=O)$_2$	70.5	76
		$C_8F_5N_2PS_2$	C_6F_5P(N=C=S)$_2$	62.2	76
			Ph P(N=C=S)$_2$	33	77
C''N$_2$/6	N'';(CN)$_2$	$C_6H_{12}N_6P_2$	(N≡C-P$^\cdot$NMeN$^\cdot$Me)$_2$	50.4(b)	41
CHN	HSi$_2$;Si$_2$	$C_{13}H_{38}N P Si_4$	Tms$_2$ N P(H)CH Tms$_2$	5.9(c) J_{PH}210	87
	H$_2$Si;Si$_2$	$C_{10}H_{30}N P Si_3$	Tms$_2$N PH CH$_2$Tms	− 3.2(c)	54
	C$_2$H;Si$_2$	$C_9H_{26}N PSi_2$	Tms$_2$N PH iPr	17.6(c) J_{PH}206 $^2J_{PCH}$2.4	86
	C$_3$;Si$_2$	$C_{10}H_{28}N P Si_2$	Tms$_2$ N PH tBu	33.4(c)	54
C'HN	C'$_2$;Si$_2$	$C_{12}H_{24}N P_2Si_2$	Tms$_2$ N PH Ph	9.0(c) J_{PH}215	54, 86
C$_2$N	(F$_3$)$_2$;C$_2$	$C_4H_6F_6N P$	Tf$_2$ P N Me$_2$	46.3 $^2J_{PCF}$85.6	19
	HSi$_2$,H$_2$Si;Si$_2$	$C_{17}H_{48}N P Si_5$	Tms$_2$ N P(CH$_2$Tms)CH Tms$_2$	60.6(c)	87
	HSi$_2$,H$_3$;Si$_2$	$C_{14}H_{40}N P Si_4$	Tms$_2$ N P(CH$_3$)CH Tms$_2$	48,3(c)	87
		$C_{11}H_{32}N P Si_3$	Tms$_2$ N P(CH$_2$ Tms)Me	36,2(c)	88
	H$_2$Si,C$_2$H;Si$_2$	$C_{13}H_{36}N P Si_3$	Tms$_2$ N P iPr CH$_2$ Tms	59.8, $^2J_{PCH}$44	86
	H$_2$Si,C$_3$,Si$_2$	$C_{14}H_{36}N P Si_3$	tBu(CH$_2$ Tms) P N$^\cdot$Si Me$_2$(CH$_2$)$_2$Si$^\cdot$ Me$_2$	56.6(c)	88
	(H$_3$)$_2$,Si$_2$	$C_{14}H_{36}N P Si_2$	(tBuMe$_2$Si)$_2$NPMe$_2$	31,2(c)	88
	(H$_3$)$_2$,CP	$C_5H_{15}N P_2$	Me$_2$N(PMe$_2$)$_2$	38.9(b) $^2J_{PNP}$168 $^3J_{PNCH}$5.4	13
	(H$_3$)$_2$;C$_2$	$C_4H_{12} N P$	Me$_2$PNMe$_2$	38.2/39	19
	(H$_3$)$_2$,CC'	$C_9H_{14}N P$	Me$_2$P N Ph Me	29.8	19
	H$_3$,CH$_2$,C$_2$	$C_{14}H_{34}N_2P_2$	(Et$_2$N P Me CH$_2$CH$_2$-)$_2$	42,6	69
	(CF$_2$)$_2$,Si$_2$	$C_{10}H_{18}F_{10}NP Si_2$	(C_2F_5)$_2$ P N Tms$_2$	48,2	89
	(CH$_2$)$_2$;PSi	$C_{11}H_{29}N O_2P_2 Si$	Et$_2$ P N Tms P(O Et)$_2$	47.5	92
	(CH$_2$)$_2$;CP	$C_{12}H_{29}N O_2P_2$	Et$_2$ P N tBu P(OEt)$_2$	44,0 $^2J_{PNP}$23	92
	(CH$_2$)$_2$;CH	$C_{10}H_{24}N P$	Bu$_2$ P N HEt	37	19
	(CH$_2$)$_2$;C$_2$	$C_6H_{16} N P$	Et$_2$P N Me$_2$	63.4 $^3J_{PNCH}$8	19
		$C_8H_{16} N P$	Et$_2$ P N Me CH$_2$C≡CH	67.6	90
		$C_8H_{19} N P_2$	Et$_2$NP$^\cdot$CH$_2$CH$_2$PHCH$_2$CH$_2^\cdot$	43 $^3J_{PCCP}$4.4	91
		$C_8H_{20} N P$	Et$_2$PNEt$_2$	61.6	19
		$C_{12}H_{30}N P_3$	Et$_2$NP(CH$_2$CH$_2$PMe$_2$)$_2$	59, 9 $^3J_{PCCP}$26	68
		$C_{16}H_{38}N_2P_2$	(Et$_2$NPEtCH$_2$CH$_2$-)$_2$	57.2	68
		$C_{28}H_{30}N P_3$	Et$_2$NP(CH$_2$CH$_2$PPh$_2$)	60.5 $^3J_{PCCP}$31	68
	(CH$_2$)$_2$;CC'	$C_{11}H_{18} N P$	Et$_2$PNPhMe	52.1	19
	CH$_2$;C$_2$H;C$_2$	$C_{16}H_{38}N_2P_2$	(Et$_2$NPiPrCH$_2$-)$_2$	70.0/72.8	69
		$C_{17}H_{40}N_2P_2$	(Et$_2$NPiPrCH$_2$)$_2$CH$_2$	67.2/67.6	69
		$C_{18}H_{42}N_2P_2$	(Et$_2$NPiPrCH$_2$CH$_2$-)$_2$	67.9	69

TABLE D (continued)
Three Coordinate (λ3 σ3) Compounds Containing A Phosphorus Bond to a Group V Element But No Bonds to Halogen or Chalcogenide

Section D1: Three Coordinate Compounds with P-N Bonds

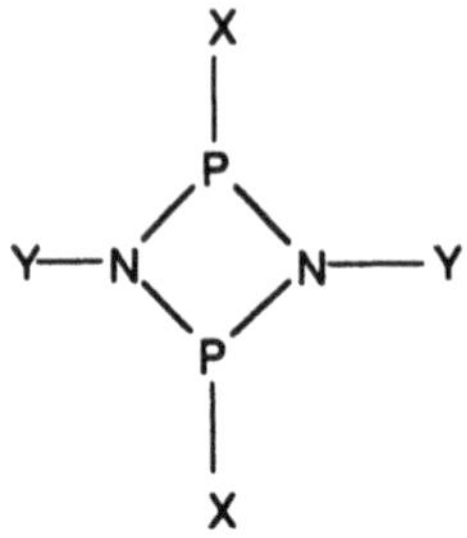

36 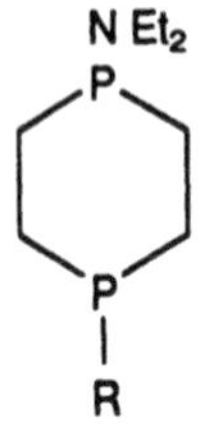 37 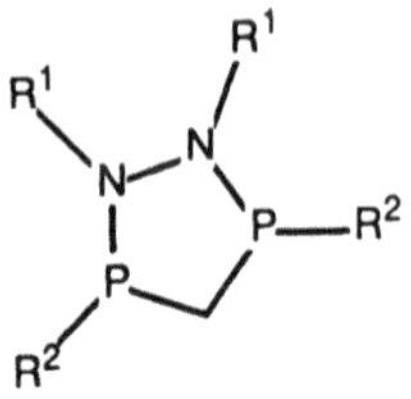 38 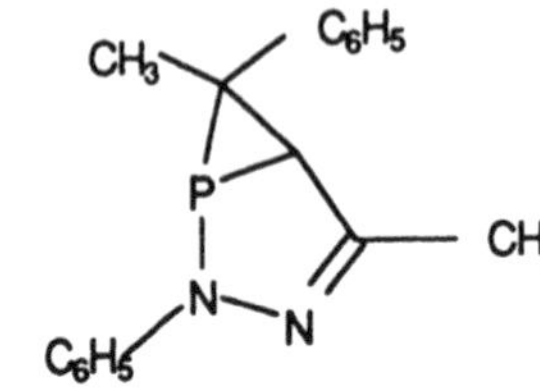 39

P-bound atoms/ rings	Connectivities	Formula	Structure	NMR data ($\delta p[solv]J_{PZ}$ Hz)	Ref.
	$(C'H_2)_2;Si_2$	$C_{12}H_{28}N\ P\ Si_2$	$Tms_2NP(CH_2CH{=}CH_2)_2$	49.6(c)	88
		$C_{20}H_{32}N\ P\ Si_2$	$Tms_2NP(CH_2Ph)_2$	62.5(c)	88
	$(C_2H)_2;PSi$	$C_{13}H_{33}N\ O_2P_2Si$	$iPr_2PNTmsP(OEt)_2$	64 $^2J_{PNP}$19	92
	$(C_2H)_2;Si_2$	$C_{12}H_{32}N\ PSi_2$	Tms_2NPiPr_2	79.3(c)	54
	$(C_2H)_2;HP$	$C_{10}H_{25}N\ O_2P_2$	$iPr_2PNHP(OEt)_2$	48.5 $^2J_{PNP}$200	92
	$(C_2H)_2;CP$	$C_{14}H_{33}\ N\ O_2\ P_2$	$iPr_2PNtBuP(OEt)_2$	65.0 $^2J_{PNP}$30	92
	$C_2H,C_3;Si_2$	$C_{13}H_{34}N\ P\ Si_2$	$Tms_2NPiPrtBu$	87.2(c)	86
$C_2N/3$	$(CHSi)_2;Si_2$	$C_{14}H_{38}NPSi_4$	$Tms_2NP^{\cdot}CHTmsC^{\cdot}HTms$	−110.9	94
$C_2N/5$	$(HN'Si)_2;Si_2$	$C_{14}H_{38}N_3P\ Si_4$	$Tms_2NP^{\cdot}CHTmsN{=}NC^{\cdot}HTms$	−0.7	94
	$CH_2,C_2H;HN$	$C_6H_{16}N_2P_2$	**37**; R^1 = H, R^2 = iPr	54.8(j)	75
	$CH_2,C_3;HN$	$C_8H_{20}N_2P_2$	**37**; R_1 = H, R_2 = tBu	64.3(r)	75
	$CH_2,C_3;CN$	$C_{10}H_{24}N_2P_2$	**37**; R_1 = Me, R_2 = tBu	92.7(j)	75
	$(C'H_2)_2;C_2$	$C_8H_{16}N\ P$	$Et_2NP^{\cdot}CH_2CH = CHCH_2^{\cdot}$	61.6(c)	95
		$C_8H_{16}\ N\ P$	$Me_2N\ P^{\cdot}CH_2\ C(Me){=}C(Me)$-$CH_2^{\cdot}$	52.6/51(c)	95, 96
		$C_{10}H_{20}N\ P$	$Et_2N\ P^{\cdot}CH_2$-$C(Me){=}C(Me)CH_2^{\cdot}$	47.0(c)	95
$C_2N/6$	$(CH_2)_2;C_2$	$C_{11}H_{27}\ N\ P_2\ Si$	$Et_2N\ P^{\cdot}CH_2CH_2\ P\ Tms$-$CH_2CH_2^{\cdot}$	49.2 $^3J_{PCCP}$20.5	91
		$C_{14}H_{23}N\ P_2$	**38**; R = Ph	47.5/48.2 $^3J_{PCCP}$17	68
		$C_{15}H_{25}N\ P_2$	**38**; R = CH_2 Ph	45.2/48.1 $^3J_{PCCP}$17	68
$C_2N/3,5$	$CC'H;C_2C';C'N'$	$C_{17}H_{17}N_2P$	**39**	−74/−77	104
$C_2N/5,5$	$C_2H,CC'H;C'N'$	$C_{10}H_{13}N_2O\ P$	**40**	50.5(endo); 47.1(exo)	97
C_2N'	$(F_3)_2;C'$	$C_3F_6NO\ P$	$Tf_2\ P\ N{=}C{=}O$	34.5 $^2J_{PCF}$88.0	19
	$(F_3)_2;C'$	$C_3F_6NP\ S$	$Tf_2\ P\ N{=}C{=}S$	34.5 $^2J_{PCF}$87.3	19
$CC'N$	$HSi_2;C'_2H;Si_2$	$C_{15}H_{40}N\ P\ Si_4$	$Tms_2\ N\ P\ (CH{=}CH_2)CH$-$Tms_2$	51.5(c)	87
	$HSi_2;C'_3;Si_2$	$C_{19}H_{42}N\ P\ Si_4$	$Tms_2N\ P\ (Ph)\ CH\ Tms_2$	51.1	87
	$H_2Si;C'_3;Si_2$	$C_{16}H_{34}N\ PSi_3$	$Tms_2N\ P\ (CH_2\ Tms)Ph$	41.9(c)	88
	$CH_2;C'H;(CH_2)_2$	$C_{10}H_{23}N\ P_2$	$Et_2N\ P\ (CH{=}CH_2)CH_2CH_2P\ Me_2$	55.4 $^3J_{PCCP}$28	68
		$C_{20}H_{27}N\ P_2$	$Et_2N\ P(CH{=}CH_2)CH_2CH_2\ P\ Ph_2$	55.4 $^3J_{PCCP}$34	68
		$C_{24}H_{38}N_2P_2$	$(Et_2N\ P\ Ph\ CH_2CH_2$-$)_2$	45.0	69
	$C'H_2;C'_2;Si_2$	$C_{15}H_{28}N\ P\ Si_2$	$Tms_2\ N\ P\ (CH_2CH{=}CH_2)\ Ph$	46.9(c)	73

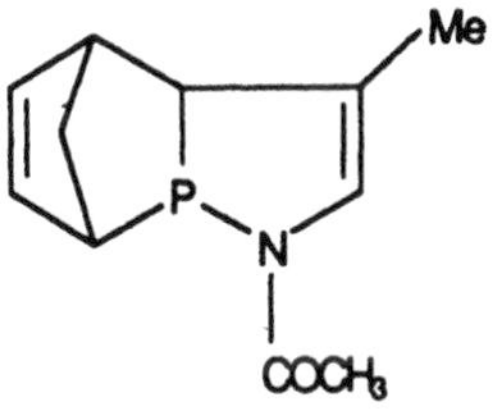

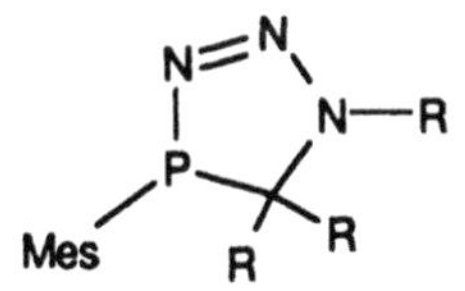

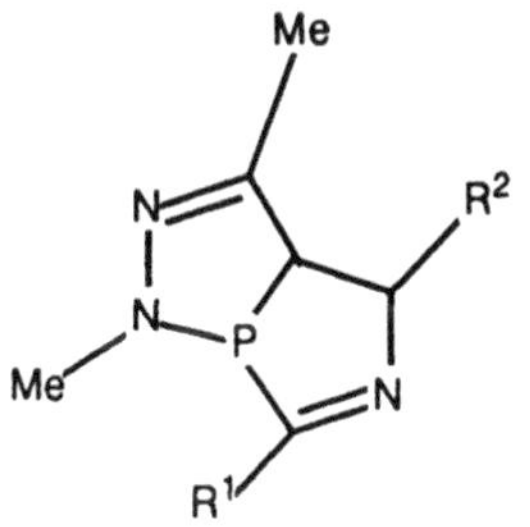

40 41 42

P-bound atoms/ rings	Connectivities	Formula	Structure	NMR data (δp[solv]J_{PZ} Hz)	Ref.
		$C_{19}H_{30}N P Si_2$	$Tms_2N P (CH_2Ph)Ph$	54.3(c)	88
	$C_2H;C'_2;Si_2$	$C_{15}H_{30}N P Si_2$	Tms_2 N P-iPr-Ph	57.7(c)	54
CC'N/5	$C'_2N;C'_2;N'_2$	$C_{28}H_{26}N_3P$	**41**, R = Ph	140.8(b)	98
CC'N/5,5	C'HN;C'N;CN'	$C_9H_{14}N_5O_2P$	**41**; R^1 = $C_6H_4NO_2$-p, R^2 = CH_3	61.5	99
		$C_9H_{14}N_5O_2P$	**42**; R^1 = CH_3, R^2 = $C_6H_4NO_2$-p	57.5	99
		$C_{14}H_{17}N_4P$	**42**; R^1 = R^2 = Ph	52.9	99
C'_2N	$(C'_2H)_2;C_2$	$C_8H_{16}N P$	$Et_2N P (CH{=}CH_2)_2$	52.9	68
	$(C'_3)_2$;AlP	$C_{26}H_{26}Al N P_2$	Me_2 Al $N(PPh_2)_2$	51/42	100
	$(C'_2)_2$;GaP	$C_{26}H_{26}GaN P_2$	Me_2 Ga $N(PPh_2)_2$	51/56	100
		$C_{28}H_{30}GaN P_2$	Et_2 Ga $N((PPh_2)_2)_2$	66	100
	$(C'_2)_2$;HN	$C_{14}H_{17}N_2P$	Ph_2P NH NMe_2	37.6	19
	$(C'_2)_2$;CP	$C_{25}H_{23}N P_2$	Me $N(PPh_2)_2$	72.8 $^2J_{PNP}$280? $^3J_{PNCH}$3?	13
		$C_{26}H_{25}N P_2$	Ph_2 P NEt P Ph_2	61	19
	$(C'_2)_2$;CH	$C_{16}H_{20}N P$	Ph_2 P NH tBu	22.4 $^2J_{PNH}$11.5	19
	$(C'_2)_2;C_2$	$C_{14}H_{16}N P$	Ph_2 P N Me_2	63.9	19
		$C_{16}H_{10}F_{10}N P$	$(C_6F_5)_2$ P NEt_2	21.9	79
		$C_{16}H_{15}F_5N P$	C_6F_5 P Ph NEt_2	47	79
		$C_{16}H_{16}N P$	Ph_2 P N Me CH_2 C≡CH	66.9	90
C'_2N'	$(C'_2)_2$;C″	$C_{13}F_{10}N O P$	$(C_6F_5)_2$ PN=C=O	17.0	76
		$C_{13}F_{10}N P S$	$(C_6F_5)_2$ PN=C=S	12.4	76
		$C_{13}H_{10}N P S$	Ph_2 P N=C=S	55	77

Section D2: Three Coordinate Compounds with P-P, P-As, and P-Sb Bonds

P-bound atoms/ rings	Connectivities	Structures	NMR data (δP[solv,temp]$^nJ_{PZ}$ Hz)	Ref.
P_3	$(CH)_3$	$P(PH\text{-}tBu)_3$	−145.2 $^1J_{PP}$ −226	106
$P_3/3$	ClP,$(PSi)_2$	TmsP*-P(Cl)-P*-$PTms_2$	−142 $^1J_{PP}$232 and 196	107
	$(PSi)_2$,CP	TmsP*-P(tBu)-P*-$PTms_2$	−174	107
	$(P_2)_3$	P_4	−450, −461 (n)	108
$P_3/4$	$P_2,(CP)_2$	**43**; R = Me	ca. 150	111
		43; R = Et	ca. 102	111
		43; R = iPr	ca. 52	111
$P_3/5$	$(LiP)_3$	**44**; R = R^1 = R_2 = Li	−50(k)	109
	$(LiP)_2$,PSi	**44**; R = R^1 = Li, R^2 = Tms	−50(k) and −91.9	109, 110
	$(LiP)_2$,CP	**44**; R = R^1 = Li, R^2 = Me	−40(k)	109
		44; R = R^1 = Li, R^2 = tBu	−61(k)	109
	LiP,$(PSi)_2$	**44**; R = Li, R^1 = R^2 = Tms	−58(k)	109

TABLE D (continued)
Three Coordinate (λ3 σ3) Compounds Containing A Phosphorus Bond to a Group V Element But No Bonds to Halogen or Chalcogenide

Section D2: Three Coordinate Compounds with P-P, P-As, and P-Sb Bonds

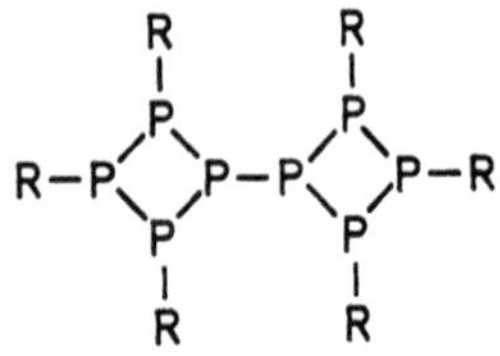

43 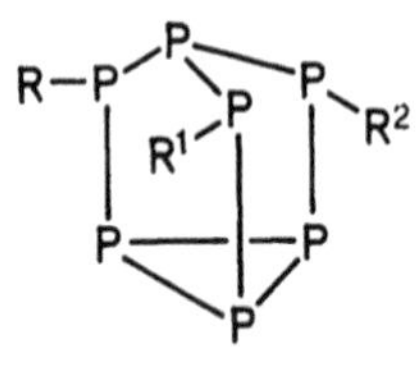 **44** 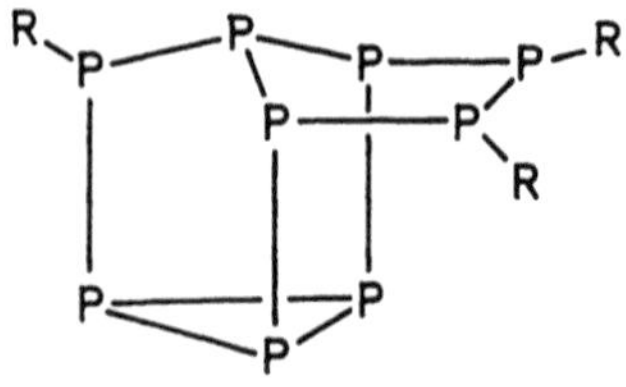 **45**

P-bound atoms/ rings	Connectivities	Formula	Structure	NMR data (δp[solv]J_{PZ} Hz)	Ref.
	LiP,$(CP)_2$	**44**; R = Li, R^1 = R_2 = Me		−22(k)	109
		44; R = Li, R^1 = R^2 = tBu		−50(k)	109
	$(P_2)_2$,CP	**45**; R = iPr		−48.7	112
		45; R = tBu		−64.9	112
	$(CP)_3$	**44**-sym; R = R^1 = R^2 = Me		−65.2	110
		44; R = R^1 = R^2 = Me		−57.0	110
P_3/3,5	LiP,$(P_2)_2$	**44**; R = R^1 = R^2 = Li		−153(k), −154	110, 111
		44; R = R^1 = Li, R^2 = Me		−98 and −145(k)	109
		44; R = Li, R^1 = R^2 = Me		−123(k)	109
		44; R = R^1 = Li, R^2 = tBu		−137 and −109(k)	109
		44; R = Li, R^1 = R^2 = Tms		−106(k)	109
P_3/3,5	LiP,$(P_2)_2$	**44**; R = Li, R^1 = R^2 = tBu		−106(k)	109
		44; R = R^1 = Li, R^2 = Tms		−118 and −135(k)	109
	PSi,$(P_2)_2$	**44**; R = R^1 = Li, R^2 = Tms		−196(k)	109
		44; R = Li, R^1 = R^2 = Tms		−169(k)	109
	$(P_2)_3$	**45**; R = iPr		−149.0/−177.1	112
		45; R = tBu		−182.3, −141.4/ −182.3[I]	112
	$(P_2)_2$,CP	**44**; R = R^1 = Li, R^2 = Me		−194(k)	109
		44; R = Li, R^1 = R^2 = Me		−149(k)	109
		44-sym; R = R^1 = R^2 = Me		−110.7, −151.1, −169.2	110
		44; R = R^1 = R^2 = Me		−161.7	110
		44; R = R^1 = Li, R^2 = tBu		−217(k)	109
		44; R = Li, R^1 = R^2 = tBu		−154(k)	109
P_3/5,6	$(P_2)_3$	**46**		multiplett +85/−30	113
		47		multiplett +85/−30	113
HP_2	H_2	P_3H_5		−179.1 $^1J_{PP}$146.7	114
H_2P	F_2	H_2P-PF_2		−137.6	115
	HP	P_3H_5		−162.6 $^1J_{PP}$146.7	114
	H_2	P_2H_4		−204.0 $^1J_{PP}$108.2	114
D_2P	D_2	P_2D_4		−206.8	114
CPSi/4	C_3;CN;C_2 P	tBuTmsNP*P-tBuSi-tBu$_2$P*-tBu		−60.3	84
	C_3;CP;C_2 P	PhP*P-tBuSiBu$_2$P*-tBu		−43.6	83
CPSb/3	C_3;CP;CSb	tBuP'-Sb(tBu)-P'tBu		−42.5, −40.0, −69.3[J] $^1J_{PP}$245	134
CPSb/4	C_3;$(CP)_2$	**48**		−82.1 $^1J_{PP}$ −167.1	134
	C_3;$(CSb)_2$	**49**		−54.5	134
CP_2	H_3;CP,C_2	Me_2P-P(Me)-P(Me)-PMe_2 x $Mo(CO)_4$		−56.6 $^1J_{PP}$196 and 297	119
	CH_2;$(CK)_2$	Et(K)P-P(Et)-P(K)Et		−23.9 $^1J_{PP}$306	120

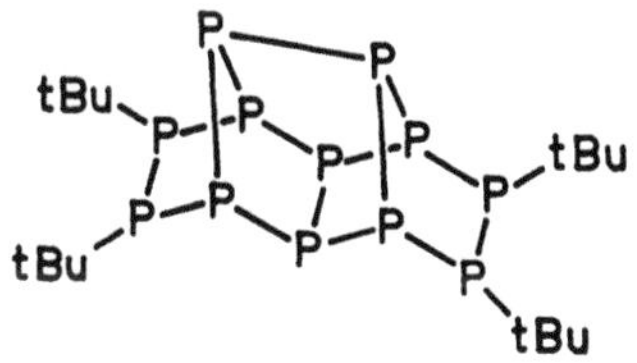

46

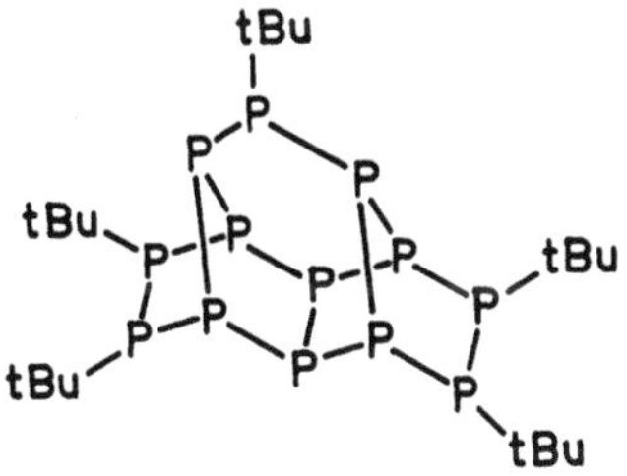

47

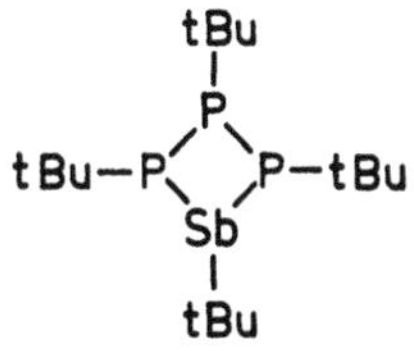

48

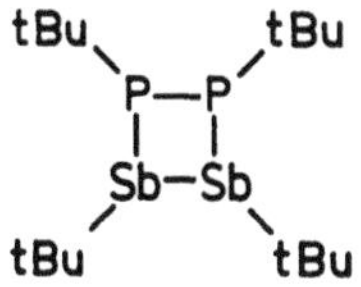

49

P-bound atoms/ rings	Connectivities	Formula	Structure	NMR data (δp[solv]J_{PZ} Hz)	Ref.
	C_3;$(CBr)_2$	Br(tBu)P-P(tBu)-P(tBu)Br		$-6.0/-16.7^l$(k)	175
	C_3;$(CCl)_2$	Cl(tBu)P-P(tBu)-P(tBu)Cl		$-6.2/-16.9^l$(k)	175
	C_3;$(CI)_2$	I(tBu)P-P(tBu)-P(tBu)I		$-13.3/-21.5^l$(k)	175
	C_3;$(CLi)_2$	Li(tBu)P-P(tBu)-P(tBu)Li		-47.9(k) $^1J_{PP}274$	176
	C_3;CP,CSi	Tms(tBu)P-P(tBu)-P(tBu)-P(tBu)Tms		1.1 AA′BB′	121
	C_3;CP,CH	tBuP(H)-P(tBu)-P(tBu)-P(H)tBu		-30.9, -24	118
	C_3;CP,C_2	Me_2P-P(tBu)-P(tBu)-PMe_2 x $Mo(CO)_4$		-28.8 $^1J_{PP}251$ and 338	119
	C_3;$(CSi)_2$	Tms(tBu)P-P(tBu)-P(tBu)Tms		-9.3, -11.8^l	176
	C_3;$(CH)_2$	H(tBu)P-P(tBu)-P(tBu)H		-48.5(b) $^1J_{PP}230$	176
$CP_2/3$	H_3;$(CP)_2$	$(P^*\text{-Me})_3$		-157.3, -171.0 $^1J_{PP}170$	177
	CH_2;$(CP)_2$	$(P^*\text{-Et})_3$		-157.3, -145.0 $^1J_{PP}178$	177, 178
	C_2H;$(CP)_2$	$(P^*\text{-iPr})_3$		-132.3, -128.6 $^1J_{PP}184$	122, 178
		$(P^*\text{-sBu})_3$		-135.7, -141.0, -136.5	122
		$(P^*\text{-cHex})_3$		-141.7, -138.9^l(k) $^1J_{PP}185$	180
	C_3;$(PSi)_2$	$TmsP^*$-P(tBu)-P^*-$PTms_2$		-105	107
	C_3;$(CP)_2$	$(P^*\text{-tBu})_3$		-71.9, -108.1 $^1J_{PP}201$	179
$CP_2/4$	H_3;$(CP)_2$	**43**; R = Me		ca. 50	111
	H_3,$(CC')_2$	$tBuP^*$-P(Me)-P(tBu)-C^*=O		-153.8 AX_2 $J_{AX}148$	126
	CH_2;$(CP)_2$	$(P^*\text{-Et})_4$		17	123, 188
		$(P^*\text{-}CH_2CH_2CN)_4$		2	123
		$(P^*\text{-Pr})_4$		13	123
		$(P^*\text{-iPr})_4$		-66	123
		$(P^*\text{-Bu})_4$		10, 14	123, 129
		$(P^*\text{-iBu})_4$		13	123, 130
		$(P^*\text{-}CHEt_2)_4$		-70	123
		$(P^*\text{-cHex})_4$		-69.1	123
		$(P^*\text{-}C_8H_{17})_4$		14, 13	123, 130

TABLE D (continued)
Three Coordinate (λ3 σ3) Compounds Containing A Phosphorus Bond to a Group V Element But No Bonds to Halogen or Chalcogenide

Section D2: Three Coordinate Compounds with P-P, P-As, and P-Sb Bonds

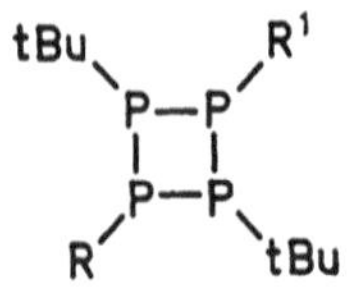

50 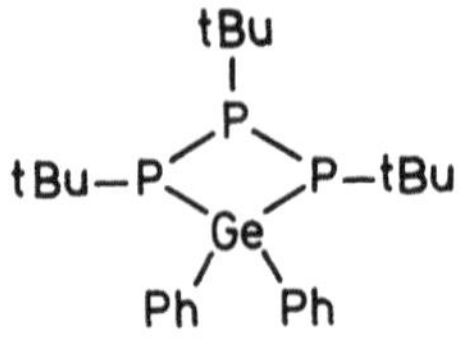 **51** 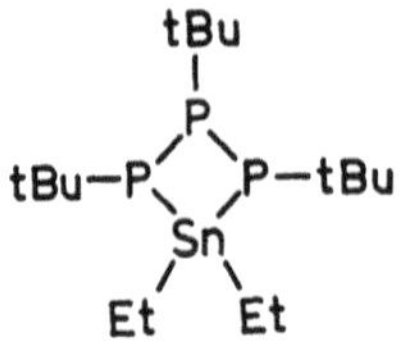 **52**

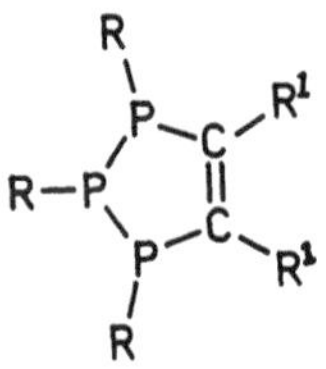

53 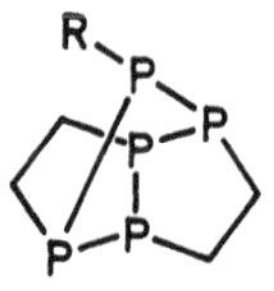 **54** 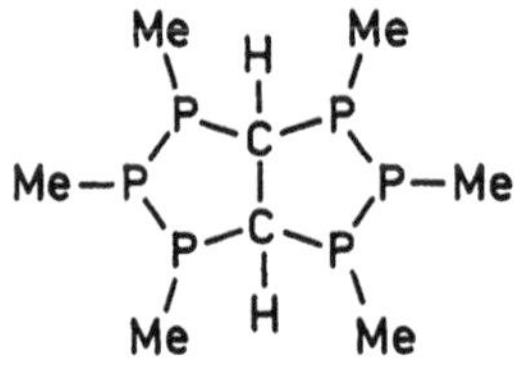 **55**

P-bound atoms/ rings	Connectivities	Formula	Structure	NMR data (δp[solv]J_{PZ} Hz)	Ref.
		43; R = Et		ca. 50	111
		48; R = iPr		ca. 52	111
	C_3;LiP;PSi	**50**; R = Li, R^1 = Tms		−52 $^1J_{PP}$189 and 123	125
	C_3;LiP;CP	**50**; R = Li, R^1 = tBu		−54 $^1J_{PP}$192 and 164	125
	C_3;PSi;CP	**50**; R = Tms, R^1 = tBu		−66 $^1J_{PP}$108 and 146	125
	C_3;HP,CP	**50**; R = H, R^1 = tBu		−74 $^1J_{PP}$93 and 137 $^2J_{PH}$31	125
	C_3;$(CGe)_2$	**51**		−17.7 $^1J_{PP}$ −152.6	132
	C_3;$(CNN')_2$	TmsN=P*(tBu)P(tBu)P(tBu)-(=NTms)N*Tms		14.2 $^1J_{PP}$174	127
	C_3;$(CP)_2$	$(P^*\text{-tBu})_4$		−57.8	131
		50; R = Li, R^1 = tBu		9 $^1J_{PP}$164	125
		50; R = Li, R^1 = Tms		−128 $^1J_{PP}$123	125
		50; R = H, R^1 = tBu		−19 $^1J_{PP}$137 $^3J_{PH}$13	125
		50; R = Tms, R^1 = tBu		−15 $^1J_{PP}$146	125
	C_3;$(CSb)_2$	**48**;		−35.7 $^1J_{PP}$ −167.1	134
	C_3;$(CSn)_2$	**52**;		−5.9 $^1J_{PP}$ −182.2	133
	C_3;$(CTe)_2$	tBuP*-P(tBu)-P(tBu)-Te		−4	136
	C_3;$(CC')_2$	tBuP*-P(tBu)-P(tBu)-C*=O		−97.6 AX_2 J_{AX}164	126
CP_2/5	F_3;$(CC')_2$	**53**; R = R^1 = Tf		−55	140
	H_3;$(CC')_2$	**53**; R = Me, R^1 = H		−68.0 $^1J_{PP}$236	160
	H_3;$(P_2)_2$	**44**; R = R^1 = Li, R^2 = Me		31	109
		44; R = Li, R^1 = R^2 = Me		15	109
		44-sym; R = R^1 = R^2 = Me		4.6, 73.5, 101.7	110
		44; R = R^1 = R^2 = Me		−103.0	110
	H_3;$(CP)_2$	$(P^*\text{-Me})_5$		20, 19	123, 130
		54; R = Me		multiplett +45/−60	141
	H_3;$(C_2)_2$	**55**		20^T, -45^C	142
	H_3;$(CC')_2$	MeP*-P(Me)-CH=CH-P*Me		−68 $^1J_{PP}$236.2	142
	$(CSi)_2$;C_3	**56**		52.1, 37.9(t) $^1J_{PP}$ −306/−317	146

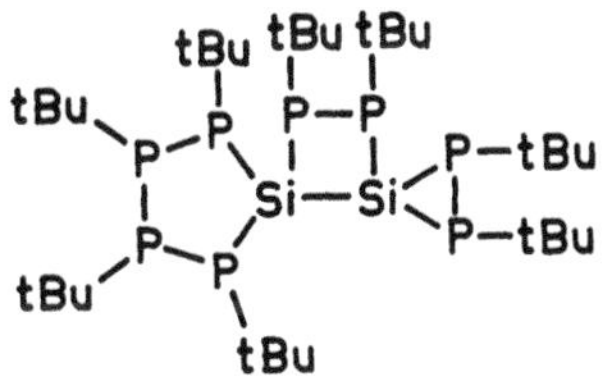

56 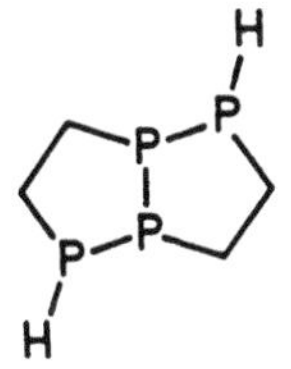 **57** 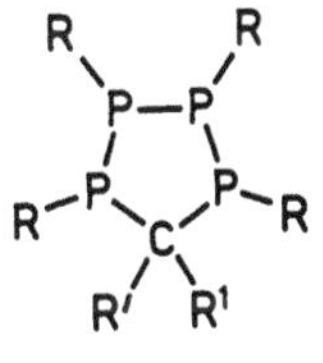 **58**

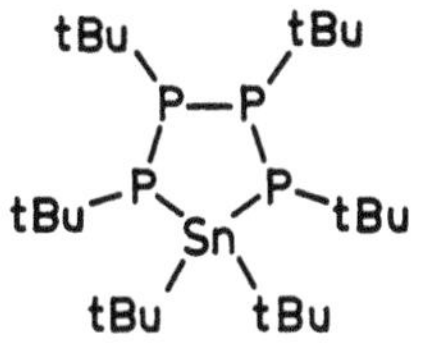

59 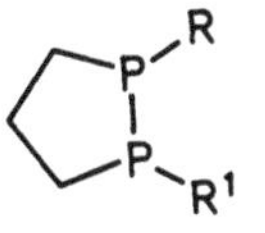 **60**

P-bound atoms/ rings	Connectivities	Formula	Structure	NMR data (δp[solv]J_{PZ} Hz)	Ref.
	CH$_2$;(CP)$_2$	**54**; R = Et		multiplett +33/−48	141
	CH$_2$;CP,CH	**57**		−23.3 $^1J_{PP}$224, 237	158
	C$_2$H;(CP)$_2$	(P*-cHex)$_5$		7.2	123
	C$_2$H;(C$_2$)$_2$	**58**; R = cHex, R^1 = H		−4.2 $^1J_{PP}$253	123
	C$_3$;(P$_2$)$_2$	**44**; R = R^1 = Li, R^2 = tBu		91	109
		44; R = Li, R^1 = R^2 = tBu		54	109
	C$_3$;(CP)$_2$	**59**		18.8(k)	133
	C$_3$;CP;CC′	**58**; R = tBu, R^1 = O		−45.3 AA′XX′	144
CP$_2$/5,6	C$_3$;(CP)$_2$	**46**		multiplett +85/−30	113
		47		multiplett +85/−30	113
C′P$_2$	C′$_2$;CP,(C′)$_2$	Me$_2$P-P(Ph)-P(Ph)-PMe$_2$ x Mo(CO)$_4$		30.1 $^1J_{PP}$111 and 234	119
	C′$_2$;(C$_2$)$_2$	Me$_2$P-P(Ph)-PMe$_2$ x Mo(CO)$_4$		−28.1 $^1J_{PP}$190	119
	C′$_2$;(C′)$_2$	Ph$_2$P-P(Ph)-PPh$_2$		−3.9	167
		Ph$_2$P-P(Ph)-PPh$_2$ x Mo(CO)$_4$		−68.2 $^1J_{PP}$172	119
C′P$_2$/3	C′$_2$;(C′P)$_2$	(P*-Ph)$_3$		−131.7, −147.2 $^1J_{PP}$186	153, 160, 181
		(P*-Mes)$_3$		−109.2, −143.3 $^1J_{PP}$184	124
		(P*-2,4,6-iPr$_3$C$_6$H$_2$)$_3$		−99.3, −133 $^1J_{PP}$180	124
C′P$_2$/4	C′$_2$;(C′P)$_2$	(P*-C$_6$F$_5$)$_4$		−67	137
		(P*-Ph)$_4$		4.5, −4.6, −9	123, 138, 139
	C′$_2$;(CSi)$_2$	PhP*-P(tBu)-Si(tBu)$_2$-P*tBu		−66.2 $^1J_{PP}$117	135
	C′$_2$;(CC′)$_2$	tBuP*-P(Ph)-P(tBu)-C*=O		−133.4 $^1J_{PP}$148	126
C′P$_2$/5	C′$_2$;(C′P)$_2$	(P*-4-FC$_6$H$_4$)$_5$		ca. −9	145
		(P*-4-ClC$_6$H$_4$)$_5$		ca. −3	145
	C′$_2$;(C′P)$_2$	(P*-Ph)$_5$		ca. −4	145
		(P*-4-MeOC$_6$H$_4$)$_5$		ca. −12	145
	C′$_2$;C′P,CC′	**58**; R = Ph, R^1 = Me		−30.7 $^1J_{PP}$235, 217	144
CHP	Si$_3$;CCl	Tms$_3$C-P(H)-P(Cl)CTms$_3$		−36.7, −29.4 $^1J_{PP}$224, 380	116, 117
	C$_3$;CP	tBuP(H)-P(tBu)-P(tBu)-P(H)tBu		−44.6, −55.1	118
	C$_3$;P$_2$	P(PH-tBu)$_3$		−8.7 $^1J_{PP}$ −226	106
CHP/5	CH$_2$;CLi	**60**; R = H, R^1 = Li		−68 $^1J_{PP}$160, 298	159
	CH$_2$;CH	**60**; R = R^1 = H		−97 $^1J_{PP}$167, 190	159
	CH$_2$;CSi	**60**; R = H, R^1 = Tms		−89 $^1J_{PP}$188	159

TABLE D (continued)
Three Coordinate (λ3 σ3) Compounds Containing A Phosphorus Bond to a Group V Element But No Bonds to Halogen or Chalcogenide

Section D2: Three Coordinate Compounds with P-P, P-As, and P-Sb Bonds

P-bound atoms/ rings	Connectivities	Formula	Structure	NMR data (δp[solv]J_{PZ} Hz)	Ref.
	CH_2;C_2	**60**; R = H, R[1] = nBu		−86 $^1J_{PP}$194	159
	CH_2;CP	**57**		−65.4 $^1J_{PP}$224	158
C_2As	$(F_3)_2$;C_2	Tf_2P-$AsMe_2$		9.0, 10.7	147, 148
		Tf_2P-$AsMe_2$ x $Mo(CO)_5$		68.7	148
	$(H_3)_2$;C_2	Me_2P-$AsTf_2$		−44.5, −43.5	147, 148
		Me_2P-$AsTf_2$ x $Mo(CO)_5$		−2.2	148
C_2P^4	$(F_3)_2$;O_3	Tf_2P-P(O)$(OMe)_2$		−23.7 $^1J_{PP}$126 $^2J_{PCF}$72	149
		Tf_2P-P(O)$(OPr)_2$		−22.0 $^1J_{PP}$116 $^2J_{PCF}$71	149
		Tf_2P-P(O)$(O\text{-}iPr)_2$		−20.2 $^1J_{PP}$114 $^2J_{PCF}$71	149
		Tf_2P-P(O)$(OBu)_2$		−22.0 $^1J_{PP}$115 $^2J_{PCF}$71	149
		Tf_2P-P(O)$(O\text{-}iBu)_2$		−20.5 $^1J_{PP}$115 $^2J_{PCF}$73	149
	$(H_3)_2$;C_2S	Me_2P-P(S)Me_2		−62 $^1J_{PP}$330 and 360	155
		(Me_2P-P(S)Me_2)(cC_5Me_5)Rh(C_2H_4)		8.3 $^1J_{PP}$132	152
	H_3,CH_2;C_2S	Me(Et)P-P(S)(Me)Et		−48 $^1J_{PP}$332, 362	155
	$(CH_2)_2$;CO_2	Bu_2P-P(O)(OPh)tBu		−50 $^1J_{PP}$310	163
		Bu_2P-P(O)(OBu)Bu		−52 $^1J_{PP}$222	163
		Bu_2P-P(O)(OBu)iPr		−55 $^1J_{PP}$240	163
	$(CH_2)_2$;C_2N	Et_2P-PEt_2(=NSO_2-1,4-MeC_6H_4)		−36.0 $^1J_{PP}$270	164
	$(CH_2)_2$;C_2S	Et_2P-P(S)Et_2		−37 $^1J_{PP}$364, 366	155
	$(CH_2)_2$;N′O_2	Et_2P-P(=N-Tms)$(OEt)_2$		−42.5 $^1J_{PP}$200	150
	$(C_2H)_2$;CO_2	iPr_2P-P(O)(OBu)tBu)		−10 $^1J_{PP}$330	163
		iPr_2P-P(O)(OPh)tBu		−8 $^1J_{PP}$330	163
	$(C_2H)_2$;C_2N	iPr_2P(=NSO_2-1,4-MeC_6H_4)-P$(iPr)_2$		−7.5 $^1J_{PP}$325(50.5)	164
	$(C_2H)_2$;N′O_2	$(iPr)_2P$-P(=NH)$(OEt)_2$		−7.5 $^1J_{PP}$235	150
		$(iPr)_2P$-P(=N-Tms)$(OEt)_2$		−11.5 $^1J_{PP}$215	150
		$(iPr)_2P$-P(=N-tBu)$(OEt)_2$		−10.0 $^1J_{PP}$210	150
		$(iPr)_2P$-P(=N-Ph)$(OEt)_2$		−8.0 $^1J_{PP}$250	150
		$(iPr)_2P$-P(=N-Tos)$(OEt)_2$		−6.3 $^1J_{PP}$286	150
	$(C_3)_2$;CO_2	tBu_2P-P(O)(OBu)iPr		20 $^1J_{PP}$328	163
		tBu_2P-P(O)(OBu)Bu		20 $^1J_{PP}$290	163
		tBu_2P-P(O)(OPh)tBu		−7 $^1J_{PP}$325	163
C_2P	$(F_3)_2$;C_2	Tf_2P-PMe_2		10.8 $^1J_{PP}$266	148
		Tf_2P-PMe_2 x $Mo(CO)_5$		4.7	148
	$(HSi_2)_2$;C_2	$(Tms_2CH)_2P$-P$(CHTms_2)_2$		19.0	151
	$(H_3)_2$;C_2	((cMe_5C_5)Rh(C_2H_4)P*$Me_2)_2$		19.8 $^1J_{PP}$127 $^1J_{PRh}$198	152
		(Me_2P-PMe_2)(cMe_5C_5)Rh(C_2H_4)		−4.4/−64.2 $^1J_{PP}$241	152
		(Me_2P-PMe_2)RhI_2(cC_5Me_5)		−15.9/−58.1 $^1J_{PP}$290	152
		Me_2P-PTf_2		−57.5 $^1J_{PP}$266	148
		Me_2P-PTf_2 x $Mo(CO)_5$		−8.4	148
		Me_2P-PMe_2		−59.5	155, 156
	H_3,CH_2;C_2	Me(Et)P-P(Me)Et		−44.7	155, 156
		Me(Et)P-P(Et)iPr		−52, −53 $^1J_{PP}$213, 226	157
		Me(Et)P-$PiPr_2$		−58.9 $^1J_{PP}$259	157
		Me(Et)P-P(Et)tBu		−54.7 $^1J_{PP}$229, 257	157
		Me(Et)P-P(iPr)tBu		−54, −52.6 $^1J_{PP}$251, 265	157
		Me(Et)P-P(Me)tBu		−58.6, −58 $^1J_{PP}$233, 196	157

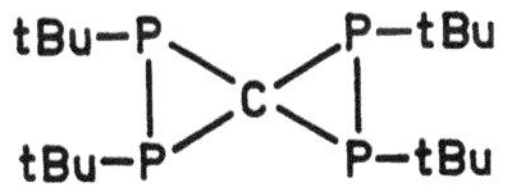

61 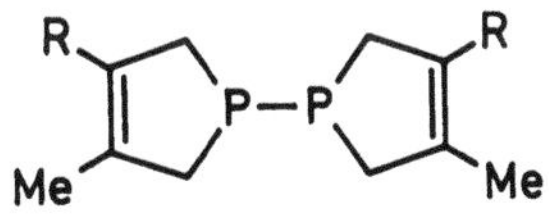 **62** 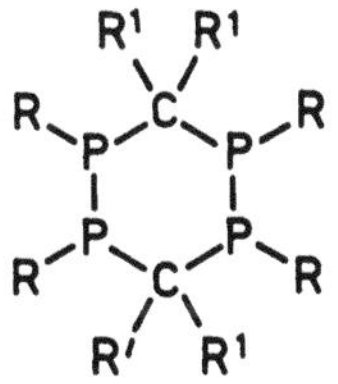 **63**

64 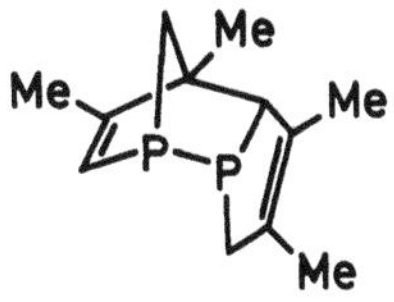 **65** 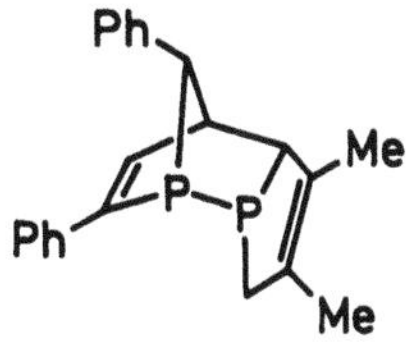 **66**

P-bound atoms/ rings	Connectivities	Formula	Structure	NMR data (δp[solv]J_{PZ} Hz)	Ref.
	$H_3,C_3;C_2$	Me(tBu)P-P(Me)Et		−20, −11.0 $^1J_{PP}$233, 196	157
	$(CH_2)_2;C_2$	Et_2P-PEt_2		−34.3	155, 156
		Et(nBu)P-P(nBu)Et		−37.5	162
		Et_2P-P-$cHex_2$		−42.2 $^1J_{PP}$282	162
	$CH_2,C_2H;C_2$	Et(iPr)P-P(Me)Et		−8, −12 $^1J_{PP}$213, 226	157
	$CH_2,C_3;C_2$	Et(tBu)P-P(Me)Et		10.2, 0.3 $^1J_{PP}$229, 257	157
	$(C_2H)_2;C_2$	iPr_2P-P(Me)Et		3.0 $^1J_{PP}$259	157
		$cHex_2P$-PEt_2		−13.8 $^1J_{PP}$282	162
		$cHex_2P$-PPh_2		−8.2 $^1J_{PP}$224	162
		$cHex_2P$-$PcHex_2$		−21.5	165
	$(C_3)_2;C_2$	tBu_2P-$P(tBu)_2$		40.0	131
	$C_3,C_2H;C_2$	tBu(iPr)P*-P(Me)Et		25.5, 18.6 $^1J_{PP}$251, 265	157
$C_2P/3$	$Cl_2P,C_3;C_2$	tBuP*-CCl_2-P*tBu		−76.1	168
	$P_3,C_3;C_2$	**61**		−132.0, −109.7^I(i)	168
	$H_2P,C_3;C_2$	tBuP*-CH_2-P*tBu		−168.8	169, 154
	$CHP,C_3;C_2$	tBuP*-CH(Me)-P*tBu		−132.2	154
	$C_2P,C_3;C_2$	tBuP*-CMe_2-P*tBu		−91.7	169, 154
$C_2P/4$	$OP,C_3;CP$	tBuP*-P(Me)-P(tBu)-C*=O		108.4 AX_2 J_{AX}148	126
		tBuP*-P(tBu)-P(tBu)-C*=O		91.4 AX_2 J_{AX}164	126
	$OP,C_3;C'P$	tBuP*-P(Ph)-P(tBu)-C*=O		111.1 $^1J_{PP}$148	126
$C_2P/5$	$H_3,CHP;CP$	**55**		−20^T, 5^C	142
	$H_2P,C_2H;CP$	**58**; R = cHex, R^1 = H		28.9 $^1J_{PP}$253	123
	$(CH_2)_2;CLi$	**60**; R = Li, R^1 = nBu		−25 $^1J_{PP}$135	159
	$(CH_2)_2;C_2$	**60**; R, R^1 = C*H_2-CH_2-C*H_2		−27	159
	$(CH_2)_2;CH$	**60**; R = H, R^1 = nBu		−39 $^1J_{PP}$194	159
	$(CH_2)_2;C_2$	**60**; R = R^1 = nBu		−27	159
	$(C'H_2)_2;C_2$	**62**; R = H		−26.1, 26.3^I(c)	172
		62; R = Me		−42.6(c)	172
$C_2P/6$	$H_2P,C_3;C_2$	**63**; R = tBu, R^1 = H		−26.2	169
	$C_2P,C'_2;CC'$	**63**; R = Ph, R^1 = Me		multiplett 0/−20	144
$C_2P/5,6$	$C'H_2,C_2H;CC'$	**64**		−35.5(c) $^1J_{PP}$193	171
		65		−24.1(c) $^1J_{PP}$185	171
		66		−28.6(c) $^1J_{PP}$200	171

TABLE D (continued)
Three Coordinate (λ3 σ3) Compounds Containing A Phosphorus Bond to a Group V Element But No Bonds to Halogen or Chalcogenide

Section D2: Three Coordinate Compounds with P-P, P-As, and P-Sb Bonds

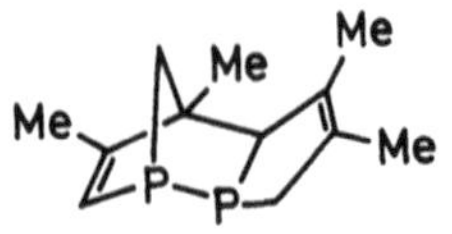

67 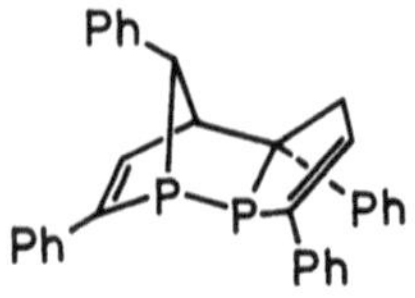 **68** 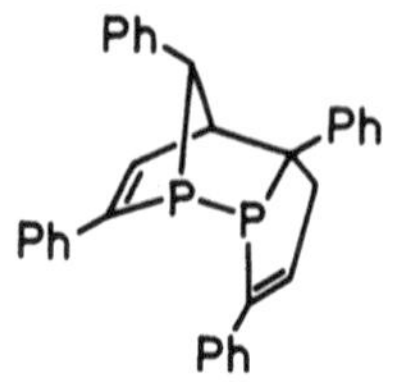 **69** 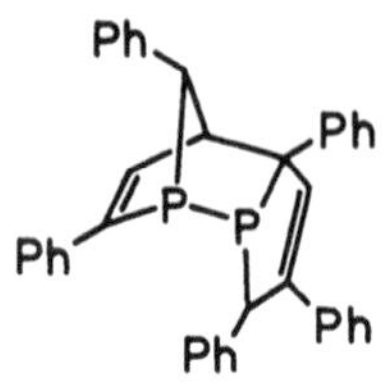 **70**

P-bound atoms/ rings	Connectivities	Formula.	Structure	NMR data (δp[solv]J_{PZ} Hz)	Ref.
	CH_2,CC'H;CC'	**67**		−17.8(c) $^1J_{PP}$185	171
C_2P/5,6	C'_2H,CC'_2;CC'	**68**		18.3(c) $^1J_{PP}$213	171
CC'P/5,6	C_2C';C'_2;CC'	**69**		22.8(c) $^1J_{PP}$211	171
		70		17.1(c) $^1J_{PP}$200	171
$CC'P^4$	H_3;C'_2;CC'S	Me(Ph)P-P(S)(Me)Ph		−42 $^1J_{PP}$238, 240	155
		Me(Sms)P-P(S)(Bu)Sms		−50.1(k) $^1J_{PP}$310	174
	CH_2;C'_2;CC'S	Et(Sms)P-P(S)(Bu)Sms		−31.4(k) $^1J_{PP}$322	174
		Pr(Sms)P-P(S)(Bu)Sms		−37.4(k) $^1J_{PP}$319	174
		Bu(Sms)P-P(S)(Bu)Sms		−36.6(k) $^1J_{PP}$319	174
		$tBuCH_2$(Sms)P-P(S)(Bu)Sms		−37.5(k) $^1J_{PP}$334	174
		C_8H_{17}(Sms)P-P(S)(Bu)Sms		−36.6(k) $^1J_{PP}$321	174
		$PhCH_2$(Sms)P-P(S)(Bu)Sms		−28.5(k) $^1J_{PP}$329	174
		$PhCH_2CH_2$(Sms)P-P(S)(Bu)Sms		−37.1(k) $^1J_{PP}$317	174
$CC'P^4$/5,5	CH_2;C'H;C_2S	**71**		−55.1(c) $^1J_{PP}$234	171
CC'P	H_3;C'_2;CC'	Me(Ph)P-P(Me)Ph		−38.2^C, −41.7^T	155
	CH_2;C'_2;CLi	Et(Ph)P-P(Li)Et		−17.2 $^1J_{PP}$396	120
	CH_2;C'_2;CC'	Et(Ph)P-P(Ph)Et		−21.5^C, −28.3^T	166
		Et(Sms)P-P(Bu)Sms		−19.5, −23.6(k) $^1J_{PP}$170	174
		Pr(Sms)P-P(Bu)Sms		−24.1(k)	174
		Bu(Sms)P-P(Bu)Sms		−23.1(k)	174
		$tBuCH_2$(Sms)P-P(Bu)Sms		−18.7, −34.8(k) $^1J_{PP}$222	174
		C_8H_{17}(Sms)P-P(Bu)Sms		−23.2(k)	174
		$PhCH_2$(Sms)P-P(Bu)Sms		−22.9(k)	174
CC'P/3	C_2P;C'_2;CC'	**58**; R = Ph, R^1 = Me		−122	144
	C_3;C'P;CC'	tBuP*-C(=CH_2)-P*tBu		−143.2	154
		tBuP*-C(=CMe_2)-P*tBu		−139.1	154
		tBuP*-C(=C(4-ClC_6H_4)$_2$)-P*tBu		−149.7	154
CC'P/4	F_3;CC';CC'	TfP*-C(Tf)=C(Tf)-P*Tf		40 $^1J_{PP}$55	140
	CC';C'_2;C'_2	PhP*-CPh=CPh-P*Ph		−37.5	170
CC'P/5	F_3;CC';CP	**53**; R = R^1 = Tf		41 $^1J_{PP}$220	140
	H_3;C'H;H_3P	**53**; R = Me, R^1 = H		11.3 $^1J_{PP}$236	160
	OP,C_3;CP	**58**; R = tBu, R^1 = O		67.9 AA'X'	143
	H_3;C'H;CP	Me*P-CH=CH-P(Me)-P*Me		11.3 $^1J_{PP}$236.2	142
	C_2P;C'_2;C'P	**58**; R = Ph, R^1 = Me		40.6 $^1J_{PP}$217.6	144
	CC'H;C'_2;C'_2	**72**		−1.7(c) $^1J_{PP}$370	171
CC'P/4,5	C_2;CC'H,C_2C'	**73**		−13.6(c) $^1J_{PP}$73.2	171
	C_2;(C_2C')$_2$	**73**		−4.5(c) $^1J_{PP}$73.2	171
CC'P/5,6	CH_2;C'H;C_2	**64**		−49.1(c) $^1J_{PP}$193	171
		67		−56.4(c) $^1J_{PP}$185	171
		65		−63.1(c) $^1J_{PP}$185	171
	CC'H;C'_2;C_2	**66**		−46.0(c) $^1J_{PP}$200	171

71 72 73 74

P-bound atoms/ rings	Connectivities	Formula	Structure	NMR data (δp[solv]J_{PZ} Hz)	Ref.
		70		-20.3(c) $^1J_{PP}$213	171
	CC'H;C'$_2$;CC'	**68**		-22.6(c) $^1J_{PP}$211	171
	C_2C',C'$_2$;CC'	**69**		-25.9(c) $^1J_{PP}$200	171
C'$_2$P	(C'$_2$)$_2$;C'P	Ph_2P-P(Ph)-PPh_2		-15.9	167
	(C'$_2$)$_2$;(C')$_2$	Ph_2P-PPh_2		-15.6, -15.2	145
	(C'$_2$)$_2$;C_2	Ph_2P-$PcHex_2$		-28.8 $^1J_{PP}$224	162
C'$_2$P/5	(C'$_2$)$_2$;CC'	**72**		28.7(c) $^1J_{PP}$370	171
	(C'$_2$)$_2$;C'$_2$	**74**		-14.6(c)	171
C'$_2$Sb	(C'$_2$)$_2$;C'$_2$	Ph_2P-$SbPh_2$		-155.6	173

REFERENCES

1. **Niecke, E. and Rüger, R.,** *Angew. Chem. Int. Ed.*, 22, 155, 1983.
2. **Klingebiel, U., Vater, N., Clegg, W., Haase, M., and Sheldrick, G. M.,** *Z. Naturforsch.*, 38b, 1557, 1983.
3. **Niecke, E., Rueger, R., Lysek, M., Pohl, S., and Schoeller, W.,** *Angew. Chem.*, 95(6), 495, 1983.
4. **King, R. B., Sadanani, N. D., and Sundaram, P. M.,** *J. Chem. Soc. Chem. Commun.*, 477, 1983.
5. **King, R. B. and Sadanani, N. D.,** *J. Org. Chem.*, 50, 1719, 1985.
6. **Foss, V. L., Lukashev, N. V., and Lutsenko, I. F.,** *Zh. Obshch. Khim.*, 50, 1236, 1980.
7. **Komlev, I. V., Zavalishina, A. I., Chernikevitch, I. P., Predvoditelev, D. A., and Nifantev, E. E.,** *Zh. Obsch. Khim.*, 42, 802, 1971.
8. **Komlev, I. V., Zavalishina, A. I., Chernikevitch, I. P., Predvoditelev, D. A., and Nifantev, E. E.,** *Zhur. Obsch. Khim.*, 42, 802, 1972.
9. **Markovski, L. N., Romanenko, V. D., and Ruban, A. V.,** *Phosphorus Sulfur*, 9, 221, 1980.
10. **Zhang, J. and Cao, Z.,** *Synthesis*, 1068, 1985.
11. **Kanamüller, J. H. and Sisler, H. H.,** *Inorg. Chem.*, 6, 1765, 1967.
12. **Elguero, J. and Wolf, R.,** *C.R. Acad. Sci. Fr. Ser. C*, 265, 1507, 1967.
13. **Colquhoun, I. J. and McFarlane, W.,** *J. Chem. Soc. Dalton Trans.*, 1674, 1977.
14. **Keat, R., Rycroft, D. S., and Thompson, D. G.,** *J. Chem. Soc. Dalton Trans.*, 321, 1980.
15. **Barendt, J. M., Haltiwanger, C. R., and Norman, A. D.,** *J. Am. Chem. Soc.*, 108, 3127, 1986.
16. **Thompson, M. L., Tarassoli, A., Haltiwanger, R. C., and Norman, A. D.,** *Inorg. Chem.*, 26(5), 684, 1987.
17. **Charbonnel, Y., Barrans, J., and Burgada, R.,** *Bull. Soc. Chim. Fr.*, 1366, 1970.
18. **Marchenko, A. P., Koidan, G. N., Povolotskii, M. I., and Pinchuk, A. M.,** *Zh. Obshch. Khim.*, 53, 1513, 1983.
19. **Crutchfield, M. M., Dungan, C. H., Letcher, J. H., Mark, V., and Van Wazer, J. R.,** P^{31}Nuclear Magnetic Resonance, in *Topics in Phosphorus Chemistry*, Vol. 5, Wiley Interscience, New York, 1967.
20. **Niecke, E. and Rueger, R.,** *Angew. Chem.*, 95(2), 154, 1983.
21. **Fluck, E., Goldmann, F. L., and Rümpler, K. D.,** *Z. Anorg. Allg. Chem.*, 338, 52, 1965.

22. **Roesky, H. W., Dhathathreyan, K. S., Noltemeyer, M., and Sheldrick, G. M.,** *Z. Naturforsch.*, 40b, 240, 1985.
23. **Scherer, O. J. and Gläbel, W.,** *Angew. Chem.*, 87, 667, 1975.
24. **Zeib, W., Feldt, C., Weis, J., and Dunkel, G.,** *Chem. Ber.*, 111, 1180, 1978.
25. **Paetzold, P., Plotho, C., Niecke, E., and Rueger, R.,** *Chem. Ber.*, 116, 1678, 1983.
26. **Scherer, O. J. and Gläbel, W.,** *Chem. Ber.*, 110, 3874, 1977.
27. **Shomburg, D., Bettermann, G., Ernst, L., Schmutzler, R.,** *Angew. Chem.*, 97(11), 971, 1985.
28. **Von Weinmaier, J. H., Luber, J., Schmidpeter, A., and Pohl, S.,** *Angew. Chem.*, 91, 442, 1979.
29. **Klingebiel, U., Werner, P., and Meller, A.,** *Chem. Ber.*, 110, 2905, 1977.
30. **Rietzel, M., Katti, K. V., and Roesky, H. W.,** unpublished.
31. **Holmes, R. R. and Forstner, J. A.,** *Inorg. Chem.*, 2, 380, 1963.
32. **Bulloch, G., Keat, R., and Thompson, D. G.,** *J. Chem. Soc., Dalton Trans.*, 99, 1977.
33. **Kleemann, S., Fluck, E., and Schwartz, W.,** *Phosphorus Sulfur*, 12, 19, 1981.
34. **Zeiss, W., Pointner, A., Engelhardt, C., Klehr, H.,** *Z. Anorg. Allg. Chem.*, 475, 256, 1981.
35. **Fluck, E. and Wachtler, D.,** *Justus Liebigs Ann. Chem.*, 1125, 1979.
36. **Thompson, M. L., Haltiwanger, R. C., and Norman, A. D.,** *J. Chem. Soc. Chem. Commun.*, 647, 1979.
37. **Devillers, J., Willson, M., and Burgada, R.,** *Bull. Soc. Chim.*, 4670, 1968.
38. **Fluck, E. and Richter, H.,** *Chem. Ber.*, 116, 610, 1983.
39. **Cowley, A. H., Mehrotra, S. K., and Roesky, H. W.,** *Inorg. Chem.*, 20, 712, 1981.
40. **Scherer, O. J. and Wokulat, J.,** *Z. Natur. F.*, 22, 474, 1967; **Burgada R.,** *Bull. Soc. Chim. Fr.*, 136, 1971.
41. **Nöth, H. and Ullmann, R.,** *Chem. Ber.*, 109, 1942, 1976.
42. **Nifantev, E. E., Zavalishina, A. I., Sorokina, S. F., Borisenko, A. A., Smirnova, E. I., Kurochkin, V. V., and Moiseeva, L. I.,** *Zh. Obshch. Khim.*, 49, 64, 1979.
43. **Nifantev, E. E., Zavalishina, A. I., and Smirnova, E. I.,** *Phosphorus Sulfur*, 10, 261, 1981.
44. **Smirnova, E. I., Zavalishina, A. I., Borisenko, A. A., Rybina, M. N., and Nifantev, E. E.,** *Zh. Obshch. Khim.*, 51, 1956, 1981.
45. **Keat, R. and Thompson, D. G.,** *Angew. Chem.*, 89, 829, 1977.
46. **Roesky, H. W. and Amirzadeh-Asl, D.,** *Z. Naturforsch.*, 38b, 460, 1983.
47. **Goetze, R., Nöth, H., and Payne, D. S.,** *Chem. Ber.*, 105, 2637, 1972; **Kroshefsky, R. D. and Verkade, J. G.,** *Phosphorus Sulfur*, 6, 397, 1979.
48. **Scherer, O. J., Andres, K., Krüger, C., Tsay, Y. H., and Wolmershäuser, G.,** *Angew. Chem.*, 92, 563, 1980.
49. **Atkins, T. J.,** *Tetrahedron Lett.*, 45, 4331, 1978.
50. **Atkins, T. J. and Richman, J. E.,** *Tetrahedron Lett.*, 52, 5149, 1978.
51. **Benhammou, M., Kraemer, R., Germa, H., Majoral, J. P., and Navech, J.,** *Phosphorus Sulfur*, 14, 105, 1982.
52. **Ebsworth, E. A. V., Hutchinson, D. J., Mac Donald, E. K., and Runkin, D. W. H.,** *Inorg. Nucl. Chem. Lett.*, 17, 19, 1980.
53. **Romanenko, V. D., Shulın, V. F., Scopenko, V. V., and Markovski, L. N.,** *J. Chem. Soc. Chem. Commun.*, 15, 809, 1983.
54. **O'Neil, L. R. and Neilson, R. H.,** *Inorg. Chem.*, 22(5), 815, 1983.
55. **Kaul, H., Greissinger, D., Malisch, W., Klein, H., and Thewalt, U.,** *Angew. Chem. Suppl.*, 47, 1983.
56. **Snow, S. S., Jiang, D. X., and Parry, R. W.,** *Inorg. Chem.*, 24(10), 1460, 1985.
57. **King, R. B. and Sundaram, P. M.,** *J. Org. Chem.*, 49, 1784, 1984.
58. **Snow, S. S., Jiang, D., and Parry, R. W.,** *Inorg. Chem.*, 23, 2063, 1984.
59. **King, R. B. and Sadanani, N. D.,** *J. Chem. Soc. Chem. Commun.*, 955, 1984.
60. **Li, B. L. and Neilson, R. H.,** *Inorg. Chem.*, 25(3), 358, 1986.
61. **Banek, M., Ph.D. thesis,** Frankfurt, 1980; **Sheldrick, W. S., Pohl, S., Zamankhan, H., Banek, M., Amirzadeh-Asl, D., and Roesky, H. W.,** *Chem. Ber.*, 114, 2132, 1981.
62. **Roesky, H. W., Ambrosius, K., Banek, M., and Sheldrick, W. S.,** *Chem. Ber.*, 113, 1847, 1980.
63. **Roesky, H. W., Ambrosius, K., and Sheldrick, W. S.,** *Chem. Ber.*, 112, 1365, 1979.
64. **Appel, R., Huppertz, M., and Westerhaus, A.,** *Chem. Ber.*, 116, 114, 1983.
65. **Li, B. L., Engenito, J. S., Neilson, R. H., and Wisian-Neilson, P.,** *Inorg. Chem.*, 22, 575, 1983.
66. **Li, B. L., Engenito, J. S., Neilson, R. H., Wisian-Neilson, P.,** *Inorg. Chem.*, 22(4), 577, 1983.
67. **O'Neal, H. R. and Neilson, R. H.,** *Inorg. Chem.*, 23, 1372, 1984.
68. **King, R. B. and Masler, W. F.,** *J. Am. Chem. Soc.*, 99(12), 4001, 1977.
69. **Diemert, K., Kuchen, W., and Kutter, J.,** *Phosphorus Sulphur*, 15, 155, 1983.
70. **Niecke, E., Seyer, A., and Wildbredt, D. A.,** *Angew. Chem.*, 93, 119, 1981. (Eng., 20, 674—679).
71. **Sherer, O. J. and Schnabl, G.,** *Chem. Ber.*, 109, 2996, 1976.
72. **Scherer, O. J. and Günter, S.,** *Angew. Chem. Int. Ed.*, 15, 772, 1976.

73. **Verkade, J.,** *Inorg. Chem.*, 8, 2115, 1969.
74. **Haddad, M., M'Pondo, T. N., Malavaud, C., Lopez, L., Barrans, J.,** *Phosphorus Sulphur*, 20, 337, 1984.
75. **Brauer, D. J., Gol, F., Hietkamp, S., Stelzer, O.,** *Chem. Ber.*, 119(9), 2767, 1986.
76. **Fild, M.,** *Z. Naturforsch.*, 23b, 604, 1968.
77. **Pudovik, A. N., Romanov, G. V., and Stepanova, T. Y.,** *Zh. Obshch. Khim.*, 49, 1425, 1979.
78. **Lehousse, C., Haddad, M., and Barrans, J.,** *Tetrahedron Lett.*, 23, 4171, 1982.
79. **Fild, M., Glemser, O., and Hollenberg, I.,** *Z. Naturforsch.*, 21b, 920, 1966.
80. **Cowley, A. H. and Kemp, R. A.,** *Inorg. Chem.*, 22, 547, 1983.
81. **Yoshifuji, M., Shibayama, K., Toyota, K., Inamoto, N., and Nagase, S.,** *Chem. Lett.*, 237, 1985.
82. **Hitchins, R. O., Maryanoff, B. E., Alhond, J. P., Cogne, A., Gagnaire, D., Robert, J. B.,** *J. Am. Chem. Soc.*, 94, 9151, 1972.
83. **Clegg, W., Haase, M., Klingebiel, U., and Sheldrick, G. M.,** *Chem. Ber.*, 116, 146, 1983.
84. **Verkade, J.,** *Phosphorus Sulphur*, 6, 397, 1979.
85. **Fild, M., Hollenberg, I., and Glemser, O.,** *Naturwisschschaften*, 4, 89, 1967.
86. **O'Neal, H. R. and Neilson, R. H.,** *Inorg. Chem.*, 22, 814, 1983.
87. **Ford, R. R. and Neilson, R. H.,** *Polyhedron*, 643, 1986.
88. **Ford, R. R., Goodman, M. A., Neilson, R. H., Roy, A. K., Wettermark, U. G., and Wisian-Neilson, P.,** *Inorg. Chem.*, 23(14), 2067, 1984.
89. **Lucas, J., Ph.D. thesis,** Frankfurt, 1984.
90. **Angelov, C. M. and Dahl, O.,** *Tetrahedron Lett.*, 24, 1643, 1983.
91. **Hackney, M. L. J. and Norman, A. D.,** *J. Chem. Soc. Chem. Commun.*, 11, 850, 1986.
92. **Foss, V. L., Veits, Yu. A., Chernykh, T. E., Staroverova, I. N., and Lutsenko, I. F.,** *Zh. Obshch. Khim.*, 53, 2489, 1983.
93. **Verkade, J.,** *Inorg. Chem.*, 14, 3090, 1975.
94. **Niecke, E., Leuer, M., Wildbredt, D. A., and Schoeller, W. N.,** *J. Chem. Soc. Chem. Commun.*, 1171, 1983.
95. **Quinn, L. D. and Szewczyk, J.,** *Phosphorus Sulfur*, 21, 161, 1984.
96. **Hammond, P. J., Scott, G., and Hall, C. D.,** *J. Chem. Soc. Perkin Trans. II*, 205, 1982.
97. **Yeung Lam Ko, Y. Y. C., Carrié, R., Toupet, L., and De Sarlo, F.,** *Bull. Soc. Chim. Fr.*, 1, 115, 1986 (Eng).
98. **Van der Knaap, T. A., Klebach, T. C., Visser, F., Lourens, R., Bickelhaupt, F.,** *Tetrahedron*, 40(6), 991, 1984.
99. **Högel, J., Schmidpeter, A., and Sheldrick, S.,** *Chem. Ber.*, 116(2), 553, 1983.
100. **Schmidbaur, H., Lauteschlaeger, S., and Milewski-Mahrla, B.,** *Chem. Ber.*, 116, 1403, 1983.
101. **Verkade, J.,** *J. Am. Chem. Soc.*, 101, 4921, 1979; **Verkade, J.,** *Phosphorus Sulphur*, 6, 397, 1979.
102. **Cowley, A. H., Mehrotra, S. K., and Roesky, H. W.,** *Inorg. Chem.*, 20, 712, 1981.
103. **Roesky, H. W., Mehrotra, S. K., Platte, C., Amirzadeh-Asl, D., and Roth, B.,** *Z. Naturforsch.*, 35b, 1130, 1980.
104. **Litvinov, L. A., Stuchkov, B. A., Dianova, E. N., and Zabotina, E. Y.,** *Izv. Akad. Nauk. SSSR Ser. Khim.*, 10, 2275, 1983.
106. **Baudler, M., Hellmann, J., and Schmidt, Th.,** *Z. Naturforsch.*, B38, 537, 1983.
107. **Fritz, G. and Haerer, J.,** *Z. Anorg. Allg. Chem.*, 500, 14, 1983.
108. **Fluck, E. in** *P^{31} Nuclear Magnetic Resonance*, Crutchfield, M. M., Dungan, C. H., Letcher, J. H., Mark, V., and Van Wazer, J. R., Eds., Topics in Phosphorus Chemistry, Vol. 5, John Wiley & Sons, New York, 1967, Ref. 304.
109. **Fritz, G., Haerer, J., and Matern, E.,** *Z. Anorg. Allg. Chem.*, 504, 38, 1983.
110. **Baudler, M. and Pontzen, Th.,** *Z. Naturforsch.*, B38, 955, 1983.
111. **Baudler, M. and Aktalay, Y.,** *Z. Anorg. Allg. Chem.*, 496, 29, 1983.
112. **Baudler, M., Aktalay, Y., Kazmierczak, K., and Hahn, J.,** *Z. Naturforsch.*, B38, 428, 1983.
113. **Baudler, M., Aktalay, Y., Arndt, V., Tebbe, K.-F., and Feher, M.,** *Angew. Chem.*, 95, 1005, 1983.
114. **Junkes, P., Baudler, M., Dobbers, J., and Rackwitz, D.,** *Z. Naturforsch.*, B27, 1451, 1972.
115. **Rudolph, R. W. and Schiller, H. W.,** *J. Am. Chem. Soc.*, 90, 3581, 1968.
116. **Cowley, A. H., Kilduff, J. E., Norman, N. C., and Pakulski, M.,** *J. Am. Chem. Soc.*, 105, 4845, 1983.
117. **Escudie, J., Couret, C., Ranaivonjatovo, H., Satge, J., and Jaud, J.,** *Phosphorus Sulfur*, 17, 221, 1983.
118. **Baudler, M., Reuschenbach, G., Hellmann, J., and Hahn, J.,** *Z. Anorg. Allg. Chem.*, 499, 89, 1983.
119. **Stelzer, O. and Weferling, N.,** *Z. Naturforsch.*, B35, 74, 1980.
120. **Fluck, E. and Issleib, K.,** *Z. Anorg. Allg. Chem.*, 339, 274, 1965.
121. **Baudler, M., Reuschenbach, G., and Hahn, J.,** *Chem. Ber.*, 116, 847, 1983.

122. **Baudler, M., Fuerstenberg, G., Suchomel, H., and Hahn, J.,** *Z. Anorg. Allg. Chem.*, 498, 57, 1983.
123. **Henderson, W. A., Jr., Epstein, M., and Seichter, F. S.,** *J. Am. Chem. Soc.*, 85, 2462, 1963.
124. **Smit, C. N., van der Knaap, Th.A., and Bickelhaupt, F.,** *Tetrahedron Lett.*, 24, 2031, 1983.
125. **Fritz, G., Haerer, J., and Stoll, K.,** *Z. Anorg. Allg. Chem.*, 504, 47, 1983.
126. **Appel, R. and Paulen, W.,** *Chem. Ber.*, 116, 2371, 1983.
127. **Niecke, E., Rueger, R., Krebs, B., and Dartmann, M.,** *Angew. Chem.*, 95, 570, 1983.
128. **Issleib, K. and Mitscherling, B.,** *Z. Naturforsch.*, B15, 267, 1960.
129. **Rauhut, M. M. and Semsel, A. M.,** *J. Org. Chem.*, 28, 1822, 1963.
130. **Smith, L. R. and Mills, J. L.,** *J. Chem. Soc. Chem. Commun.*, 808, 1974.
131. **Issleib, K. and Hoffmann, M.,** *Chem. Ber.*, 99, 1320, 1966.
132. **Baudler, M. and Suchomel, H.,** *Z. Anorg. Allg. Chem.*, 506, 22, 1983.
133. **Baudler, M. and Suchomel, H.,** *Z. Anorg. Allg. Chem.*, 505, 39, 1983.
134. **Baudler, M. and Klautke, S.,** *Z. Naturforsch.*, B38, 121, 1983.
135. **Clegg, W., Haase, M., Klingebiel, U., and Sheldrick, G. M.,** *Chem. Ber.*, 116, 146, 1983.
136. **Du Mont, W. W., Severengiz, T., and Meyer, B.,** *Angew. Chem.*, 95, 1018, 1983.
137. **Fild, M., Hollenberg, I., and Glemser, O.,** *Naturwissenschaften*, 4, 89, 1967.
138. **Nielsen, M. L., Pustinger, J. V., and Strobel, J.,** *J. Chem. Eng. Data*, 9, 167, 1964.
139. **Crutchfield, M. M., Dungan, C. H., Letcher, J. H., Mark, V., and Van Wazer, J. R., Eds.,** *P^{31} Nuclear Magnetic Resonance*, Topics in Phosphorus Chemistry, Vol. 5, John Wiley & Sons, New York, 1967, Ref. 319.
140. **Mahler, W.,** *J. Am. Chem. Soc.*, 86, 2306, 1964.
141. **Baudler, M. and Esat, S.,** *Chem. Ber.*, 116, 2711, 1983.
142. **Baudler, M. and Tolls, E.,** *Z. Naturforsch.*, B33, 691, 1978.
143. **Appel, R. and Paulen, W.,** *Chem. Ber.*, 116, 109, 1983.
144. **Baudler, M. and Carlsohn, B.,** *Z. Naturforsch.*, B32, 1490, 1977.
145. **Hinke, A. and Kuchen, W.,** *Chem. Ber.*, 116, 3141, 1983.
146. **Baudler, M., Pontzen, T., Schings, U., Tebbe, K.-F., and Feher, M.,** *Angew. Chem.*, 95, 803, 1983.
147. **Cavell, R. G. and Dobbie, R. C.,** *J. Chem. Soc. A*, 1406, 1968.
148. **Grobe, J. and Duc le Vam,** *Z. Naturforsch.*, B35, 694, 1980.
149. **Maslennikov, I. G., Lyubimova, M. V., Krutikov, I. G., and Lavrent'ev, A. N.,** *Zh. Obshch. Khim.*, 53, 2489, 1983.
150. **Foss, V. L., Veits, Yu. A., Chernykh, T. E., Staroverova, I. N., and Lutsenko, I. F.,** *Zh. Obshch. Khim.*, 53, 2489, 1983.
151. **Cowley, A. H. and Kemp, R. A.,** *Inorg. Chem.*, 22, 547, 1983.
152. **Klingert, B. and Werner, H.,** *Chem. Ber.*, 116, 1450, 1983.
153. **Baudler, M., Koch, D., Tolls, E., Diedrich, K., and Kloth, B.,** *Z. Anorg. Allg. Chem.*, 420, 146, 1976.
154. **Baudler, M.,** *Z. Chem.*, 24, 356, 1984.
155. **Maier, L.,** *J. Inorg. Nucl. Chem.*, 24, 275, 1962.
156. **Moedritzer, K., Maier, L., and Groenweghe, L. C. D.,** *J. Chem. Eng. Data*, 7, 307, 1962.
157. **Moedritzer, K.,** *J. Am. Chem. Soc.*, 83, 4381, 1961.
158. **Baudler, M., Warnau, M., and Koch, D.,** *Chem. Ber.*, 111, 2828, 1978.
159. **Issleib, K. and Thorausch, P.,** *Phosphorus Sulfur*, 4, 137, 1978.
160. **Baudler, M., Carlsohn, B., Kloth, B., and Koch, D.,** *Z. Anorg. Allg. Chem.*, 432, 67, 1977.
161. **Baudler, M. and Tolls, E.,** *Z. Chem.*, 19, 418, 1979.
162. **Issleib, K. and Krech, K.,** *Chem. Ber.*, 98, 2086, 1965.
163. **Foss, V. L., Kudinova, V. V., and Lutsenko, I. F.,** *Zh. Obshch. Khim.*, 53, 2193, 1983.
164. **Foss, V. L., Chernykh, T. E., Staroverova, I. N., Veits, Yu. A., and Lutsenko, I. F.,** *Zh. Obshch. Khim.*, 53, 2184, 1983.
165. **Issleib, K. and Seidel, W.,** *Chem. Ber.*, 92, 2681, 1959.
166. **Krech, F., in** *P^{31} Nuclear Magnetic Resonance*, Crutchfield, M. M., Dungan, C. H., Letcher, J. H., Mark, V., and Van Wazer, J. R., Eds., Topics in Phosphorus Chemistry, Vol. 5, John Wiley & Sons, New York, 1967, Ref. 313.
167. **Maier, L.** in *P^{31} Nuclear Magnetic Resonance*, Crutchfield, M. M., Dungan, C. H., Letcher, J. H., Mark, V., and Van Wazer, J. R., Eds., Topics in Phosphorus Chemistry, Vol. 5, John Wiley & Sons, New York, 1967, Ref. 315.
168. **Baudler, M. and Leonhardt, W.,** *Angew. Chem.*, 95, 632, 1983.
169. **Baudler, M. and Saykowski, F.,** *Z. Naturforsch.*, B33, 1208, 1978.
170. **Meriem, A., Majoral, J.-P., Revel, M., and Navech, J.,** *Tetrahedron Lett.*, 24, 1975, 1983.
171. **Charrier, C., Bonnard, H., de Lauzon, G., and Mathey, F.,** *J. Am. Chem. Soc.*, 105, 6871, 1983.
172. **Quin, L. D. and Szewczyk, J.,** *Phosphorus Sulfur*, 21, 161, 1984.
173. **Kuhn, N. and Sartori, P.,** *Chem. Ztg.*, 107, 257, 1983.
174. **Yoshifuji, M., Shibayama, K., and Inamoto, N.,** *Chem. Lett.*, 115, 1984.

175. **Baudler, M. and Hellmann, J.**, *Z. Anorg. Allg. Chem.*, 509, 53, 1984.
176. **Baudler, M., Hellmann, J., and Reuschenbach, G.**, *Z. Anorg. Allg. Chem.*, 509, 38, 1984.
177. **Baudler, M., Hahn, J., and Clef, E.**, *Z. Naturforsch.*, B39, 438, 1984.
178. **Baudler, M.**, *Pure Appl. Chem.*, 52, 755, 1980.
179. **Baudler, M., Hahn, J., Dietsch, H., and Fuerstenberg, G.**, *Z. Naturforsch.*, B31, 1305, 1976.
180. **Baudler, M., Pinner, Ch., Gruner, Ch., Hellmann, J., Schwamborn, M., and Kloth, B.**, *Z. Naturforsch.*, B32, 1244, 1977.
181. **Baudler, M., Carlsohn, B., Boehm, W., and Reuschenbach, G.**, *Z. Naturforsch.*, B31, 558, 1976.

Chapter 6

^{31}P NMR DATA OF THREE COORDINATE ($\gamma 3$ $\sigma 3$) PHOSPHORUS COMPOUNDS CONTAINING PHOSPHORUS BONDS TO GROUP IV ELEMENTS AND HYDROGEN ONLY

Compiled and presented by
Ludwig Maier, Peter J. Diel, and John C. Tebby

NMR data was also donated by
L. D. Quin, University of Massachusetts, Amherst, MA
S. E. Cremer, Marquette University, Milwaukee, Wisconsin,
E. Nieke, Universität Bielefeld, FRG
H. W. Roesky, Universität Göttingen, FRG
C. D. Hall, King's College London, U.K.
J. Verkade, Iowa State University, Ames, Iowa

The NMR data is presented in Table E. It is divided into four main sections each with further subdivisions as shown below:

E1: Primary phosphines
- E1a: Primary alkylphosphines
- E1b: Primary alkyl di- and triphosphines
- E1c: Primary alkenyl/aryl mono- and diphosphines

E2: Secondary phosphines
- E2a: Secondary dialkylphosphines
- E2b: Secondary dialkyl diphosphines
- E2c: Secondary cyclic dialkylphosphines
- E2d: Secondary alkyl alkenyl/aryl phosphines
- E2e: Secondary alkyl alkenyl/aryl di- and triphosphines

E3: Tertiary phosphines
- E3a: Tertiary trialkylphosphines, saturated
- E3b: Tertiary trialkyl polyphosphines
- E3c: Tertiary cyclic trialkylphosphines
- E3d: Tertiary dialkyl alkenyl/aryl phosphines
- E3e: Tertiary dialkyl alkenyl/aryl polyphosphines
- E3f: Tertiary cyclic dialkyl alkenyl/aryl phosphines
- E3g: Tertiary dialkyl alkynylphosphines
- E3h: Tertiary alkyl dialkenyl/aryl phosphines
- E3i: Tertiary alkyl dialkenyl/aryl polyphosphines
- E3j: Tertiary cyclic alkyl dialkenyl/aryl phosphines
- E3k: Tertiary alkyl dialkynyl phosphines
- E3l: Tertiary trialkenyl/aryl phosphines
- E3m: Tertiary trialkenyl/aryl polyphosphines
- E3n: Tertiary cyclic trialkenyl/aryl phosphines
- E3o: Tertiary alkenyl/aryl alkynyl and trialkynyl phosphines

E4: Phosphines containing P-Si, P-Ge, and P-Sn bonds
- E4a: Phosphines with no P-C bonds
- E4b: Phosphines with one or two P-C bonds

While the data in the following Table E is organized according to the general sequencing rules of the handbook the larger sections have also been separated into two additional classes, i.e., monophosphines and polyphosphines. The latter polyphosphine groups consist of compounds which have two or more phosphine groups. The NMR data relate to the phosphorus atom in boldface. (Note polyphosphines with P-P bonds are covered in Table D of Chapter 5). The division caters to the special interests of chemists who are working on coordination chemistry. The sections on mono- and polyphosphines overlap with respect to the nature of the spheres of atoms surrounding the phosphorus atoms. Thus, to find structures with the closest fit of atoms surrounding a particular phosphorus atom it is advisable to search both sections.

TRENDS IN THE ^{31}P NMR CHEMICAL SHIFTS OF PHOSPHINES

The hybridization of phosphorus, which may involve mixed s, p, and d orbital participations, makes the interpretation of ^{31}P-chemical shifts a very difficult task. For phosphines with only s and p orbital participation the electronegativity of substituents on phosphorus and the angles between them are the two most important variables determining ^{31}P-chemical shifts and coupling constants. As the groups on P increase in steric size the C-P-C- angle opens and the ^{31}P-chemical shift moves to lower field. Increase of the electronegativity of the groups on phosphorus also moves the ^{31}P-chemical shift to lower field. On this basis the calculation of ^{31}P-chemical shifts of phosphines by the equations given below, using additive constants σ^P for the various groups, can be understood.

Primary phosphines[281] $$\delta(\text{ppm}) = -163.5 + 2.5\,\sigma^{P\ a}$$

Secondary phosphines[282] $$\delta(\text{ppm}) = -99 + 1.5 \sum_{n=1}^{2} \sigma_n^P$$

Tertiary phosphines[282] $$\delta(\text{ppm}) = -62 + \sum_{n=1}^{3} \sigma_n^P$$

^{31}P-chemical shifts to high field can be anticipated if structural constraints require small CPC angles as in

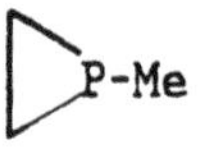

P-Me (^{31}P -251.0) or N P Me (^{31}P -264)

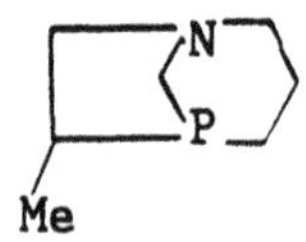

In P_4 where the angles are constrained to 60°, the chemical shift is −450 ppm, the highest $\delta(^{31}P)$ known. It follows that sizable differences in ^{31}P-shifts have been observed when phosphorus functions occupy axial or equatorial positions on 6-membered rings. ^{31}P-nmr has been used in the determination of conformational free energies of phosphorus groups on the cyclohexane ring.[15,16] However, the contracted C-P-C- angle in bridged saturated phosphines does not seem to cause any special effect.[220] The most unusual effect is that replacement of a CH_2CH_2- unit in a 7-phosphanorbornane with a $-CH{=}CH-$ unit causes very strong downfield shifting of the ^{31}P-resonance. In the anti-phosphole dimer series, values in the δ range + 30 to + 50 are observed, representing a 50 to 70 ppm displacement from the saturated model. The effect is far more pronounced in the syn than in the anti isomers, giving the syn-7-phosphanorbornene series the most downfield values ever recorded for tertiary phosphines (see Table E).[220]

In the propenyl- and styryl- (olefinic?) substituted phosphines the ^{31}P shifts show a simple and useful correlation: the Z isomer absorbs to higher field than the E. Thus the effect of steric compression which is commonly to shift the affected nucleus to higher field, is found here also for phosphorus.[172]

σ^P-Values (Calculated from the Measured Shifts of Tertiary Phosphines, R_3P)

R-group	σ^P	R-group	σ^P	R-group	σ^P
N≡C	−24.8	Me	0	$PhCH_2$	17
$CH_3C{\equiv}C$	−11	neopentyl	3	Ph	18
HC≡C	−10	i-Bu	7	c-C_5H_9	21
3-furyl	−7	Pr, Bu	10	c-C_6H_{11}	23
2,6-$F_2C_6H_3$	−5.5	n-C_nH_{2n+1} ($n \geqq 3$)	10	sec-Bu	23
C_6F_5	−5	$NCCH_2CH_2$	13	i-Pr	27
2,6-$(RO)_2C_6H_3$	−1	Et	14	t-Bu	41
Et_2NCH_2	−1	$CH_2{=}CH$	14	t-Am	42
				CH_3CO	42

[a] Owing to the new convention all signs of the originally published equations changed.

Furthermore, the ^{31}P-chemical shift is somewhat dependent on the solvent and the temperature. For Ph_3P the dependence of the ^{31}P-chemical shift in toluene from the temperature was observed to be 1.3 Hz/degree between −90° to 70°C (chemical shift thermometer).[72] The chemical shift of solids, measured by CP/MAS (= ^{31}P-cross-polarization spectra using magic-angle-spinning techniques), is at somewhat higher field than the solution shifts.

In the following table is given a survey of the ^{31}P-chemical shift range of different types of compounds (the second values in each section exclude 2.5% of the shifts at highest field and 2.5% of the shifts at lowest field).

Survey of the ^{31}P-chemical shift range of different types of compounds

Type	Section	Highest shift	Compound	Lowest shift	Compound
PH_3	E1	−239	PH_3	−230	PD_3
Aliph. RPH_2	E2	−175.8	$(CH_2{=}CHCH_2)_2NCH_2PH_2$	−81.2	$tBuPH_2$
		(−163.6	CH_3PH_2	−106	$iPrPH_2$)[a]
Aryl, alkenyl, acyl-PH_2	E3	−183.1	$C_6F_5PH_2$	−123.8	$PhPH_2$
		(−160.2	2,4,6-$Me_3C_6H_2PH_2$	−124.0	4-$BrC_6H_4PH_2$)[a]
$H_2P(CH_2)_xPH_2$	E4	−138.6	$H_2P(CH_2)_3PH_2$	−91	$H_2PC(Bu)_2PH_2$
		(−137.0	$H_2P(CH_2)_4PH_2$	−105	$H_2PCH(C_{10}H_{21}$-n)PH_2)[a]
Aliph. R_2PH	E5	−101.9	$(Et_2NCH_2)_2PH$	+20.1	tBu_2PH
		(−99.5	Me_2PH	−26.4	iPr_2PH)[a]
Aryl-R_2PH	E6	−143	$(C_6F_5)_2PH$	−40.2	(3-$MeC_6H_4)_2PH$
		(−59.1	(2-$MeC_6H_4)_2PH$	−40.9	(3-$FC_6H_4)_2PH$)[a]
RR^1PH	E7	−93.4	$Me_2PCH_2(Me)PH$	−37.2	$(H_2NCH_2CH_2)BuPH$
		(−87	MePrPH	−77	MeEtPH)[a]
Aryl, Alkenyl, acyl-R^1R^2PH	E8	−92.2	$Ph(C_6F_5)PH$	−4.0	3,5-$tBu_2C_6H_3(tBu)PH$
		(−72.3	MePhPH	−26.4	iPrPhPH)[a]
Cyclic R_2PH	E9	−341	$C'H_2CH_2P'H$	−0.1	5: R^1 = Ph, R^2 = H, R^3 = Me
		(−79	8: X = O	−3.5	5: R^1, R^2 = $-(CH_2)_4-$, R^3 = Me)[a]
$RHP(CH_2)PHR$	E10	−88.9	$MeHPCH_2PHMe$	−9.8	2-$(tBuHP)C_6H_4(PHtBu)$(isomer)
		(−84.3	$MeHP(CH_2)_3PMe(CH_2)_3PHMe$	−26.1	2-$(iPrHP)C_6H_4(PHiPr)$(isomer)[a]
R_3P	E11	−135.7	$(NC)_3P$	+63.5	$(CH_3CO)_3P$
		(−88.4	$(PhC{\equiv}C)_3P$	+19.3	iPr_3P)[a]
R_2R^1P	E12	−121.3	$(NC)_2PC_6F_5$	+60.8	tBu_2PCOCF_3
		(−50.6	$nBu_2PC{\equiv}CEt$	+40.2	tBu_2PPh)[a]
$R^1R^2R^3P$	E13	−40.25	MePh(2-$Me_2NCHMeC_6H_4$)P	+40.8	$CH_3CO(Ph)tBuP$
		(−37.9	MePrPhP	+34.1	$(CF_3CHFCO)(Bz)iPrP$)[a]
Polytert. Polyphosphines	E14	−91.9(P_A)	$Me_2P_ACH_2CF_2PMe_2$	+30	Ph(PhCO)PC[=C(CN)$_2$]P(COPh)Ph
		(−52.9	$(Et_2NCH_2)_2P(CH_2)_3P(CH_2NEt_2)_2$	+20.4	$Ph_2PC(O)CHMe(O)CPPh_2$)[a]
Cyclic R_3P	E15	−251	$C'H_2CH_2P'Me$	+58.09	49
		(−204.5	$C'H_2CH(CH{=}CH_2)P'Ph$	+35.1	$C'H_2CHMeSCHPhP'Ph$(*trans/cis*)[a]
Polycyclic R_3P	E16	−264		+152.5	R = t'Bu
		(−89		+133.7	R = Ph
Metal-PH_2	E17	−290	F_3SiPH_2	−174	Br_3SiPH_2
		(−274	H_3SiPH_2	−207	FBr_2SiPH_2)[a]
$Metal_2PH$	E18	−323	$(H_3Si)_2PH$	−235	$(Me_3Si)_2PH$
		(−286	$(MeH_2Si)_2PH$	−258.5	$(Me_2HSi)_2PH$)[a]
$Metal_3P$	E19	−450	$(H_3Sn)_3P$	−228	$(Me_3Ge)_3P$
		(−289	$Me_3Ge(Me_3Sn)_2P$	−245	$Me_3Ge(Me_3Si)_2P$)[a]

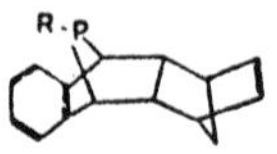

[a] Range of shifts excluding 2.5% at highest and 2.5% at lowest field.

TABLE E

^{31}P NMR Data of Three Coordinate (λ3σ3) Compounds Containing Phosphorus Bonds to Hydrogen and/or Group IV Atoms but Not to Halogen, Chalcogenide or Group V Atoms

Section E1: Primary Phosphines

E1a: Primary Alkylphosphines

Rings/ Connectivities	Formula	Structure	^{31}P NMR data (δP [solv, temp] J_{PH} Hz)	Ref.
P-bound atoms = H3				
	H_3P	PH_3	−239 J189.2, −235	1, 2
	D_3P	PD_3	−230(−80°)	2
P-bound atoms = CH$_2$				
F_3	CH_2F_3P	$TfPH_2$	−129 J201	3
NN′	$C_{16}H_{21}N_2PSi$	TmsPhNC(=NPh)PH_2	−148 J220	4
Si_3	$C_{10}H_{29}PSi$	$TrisPH_2$	−132.6(e+t) J167.8	5
HSi_2	$C_{27}H_{21}PSi$	$BisPH_2$	−149 J185	6
H_2N	$C_7H_{14}NP$	$(CH_2{=}CHCH_2)_2NCH_2PH_2$	−175.8 J203	7
H_2Si	$C_8H_{17}PSi$	$(CH_2{=}CHCH_2)_2MeSi$-CH_2PH_2	−158.5 J188	9
H_3	CH_5P	$MePH_2$	−163.5 J201	3, 10
CF_2	$C_2H_3ClF_3P$	$CHClFCF_2PH_2$	−140.3	3
CH_2	C_2H_7P	$EtPH_2$	−128 J185	3
	C_2H_8NP	$H_2NCH_2CH_2PH_2$	−150.4 J194	3
	C_3H_6NP	$NCCH_2CH_2PH_2$	−135 J195	3
	$C_3H_7O_2P$	$HO_2CCH_2CH_2PH_2$	−136.9 J194	3
	$C_4H_9O_2P$	$HO_2CCH_2CH_2CH_2PH_2$	−138 J195	3
	$C_4H_{11}P$	$BuPH_2$	−140, −135	3
		$iBuPH_2$	−151	3
	$C_5H_{13}P$	$PePH_2$	−139 J192	3
		$tBuCH_2PH_2$	−162.4(r)	273
	$C_8H_{19}P$	$OctPH_2$	−128.5 J183	3
	$C_{10}H_{23}P$	$DecPH_2$	−139 J190	3
C′H_2	$C_2H_5O_2P$	$HO_2CCH_2PH_2$	−142.6 J198	3
	C_7H_9P	$BzPH_2$	−120.9 J184.5	3
C_2H	C_3H_9P	$iPrPH_2$	−106 J195	3
	$C_4H_{11}P$	$sBuPH_2$	−120.5 J182	3
	$C_6H_{13}P$	$cHexPH_2$	−110 J162, −111.7 J186, −111.8	3, 14, 15
	$C_7H_{15}P$	4-Me-c$HexPH_2$ (trans)	−111.8(27°), −111(−100°,C_2H_3Cl)	15, 16
		4-Me-c$HexPH_2$ (cis)	−122.5(27°), −122.5	15, 16
		1 (cis) (eqMe-axP)	−133.4(−100°,C_2H_3Cl)	15
		2 (cis) (axMe-eqP)	−110.2(−100°,C_2H_3Cl)	15
	$C_9H_{21}P$	$sNonPH_2$	−116 J190	3
	$C_{10}H_{21}P$	4-tBu-c$HexPH_2$	−111.6T(27°), −131.3^{C}(27°)	15, 16
	$C_{16}H_{15}P$	**3**	−115.4(g)	13
	$C_{22}H_{19}OP$	**4**	−123.9, −126.4^{t}(c)	13
CC′H	$C_4H_9O_2P$	$HO_2CCH(Et)PH_2$	−120.8 J194	3
C_3	$C_4H_{11}P$	$tBuPH_2$	−82, −81.2(k) J180	3, 17
	$C_7H_{17}P$	Et_3CPH_2	−111.6(k) J190	17

TABLE E (continued)
^{31}P NMR Data of Three Coordinate (λ3σ3) Compounds Containing Phosphorus Bonds to Hydrogen and/or Group IV Atoms but Not to Halogen, Chalcogenide or Group V Atoms

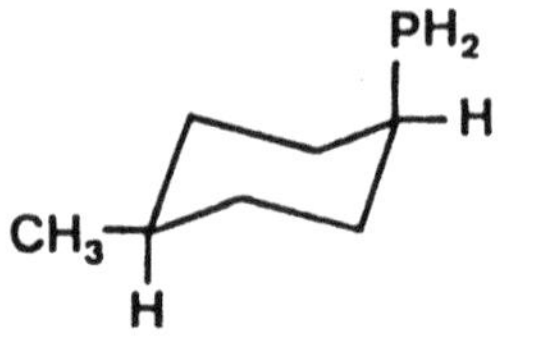

1 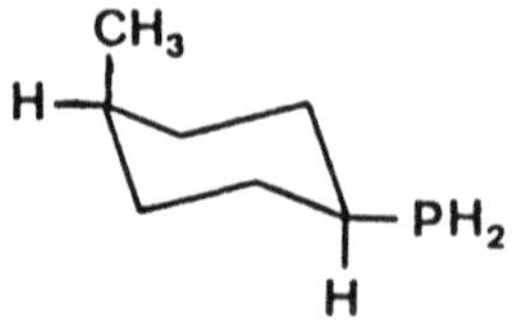 2 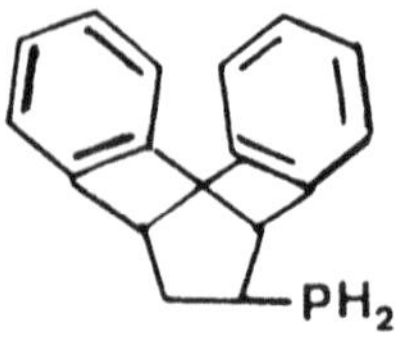 3 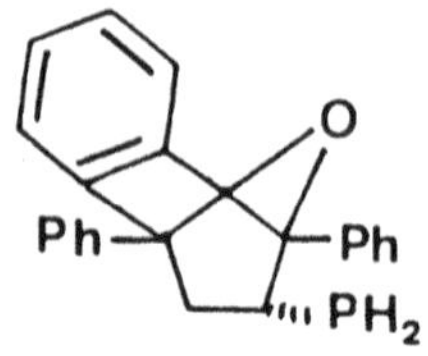 4

Rings/ Connectivities	Formula	Structure	^{31}P NMR data (δP [solv, temp] J_{PH} Hz)	Ref.
		E1b: Primary Alkyl Di- and Triphosphines		
	P-bound atoms = CH_2			
H_2P	CH_6P_2	$H_2PCH_2PH_2$	−121.8 J190, −126 J199	3
	$C_2H_8P_2$	$MeHPCH_2PH_2$	−138.9(r) J201	8
	$C_3H_{10}P_2$	$Me_2PCH_2PH_2$	−152.3(r) J189.6	8
	$C_4H_{12}P_2$	$iPrHPCH_2PH_2$	−138(r)	8
	$C_8H_{11}P_2$	$BzHPCH_2PH_2$	−139 J193.3	8
CHP	C_5H_{14}	$H_2PCH(Bu)PH_2$	−109 J194	3
	$C_6H_{24}P_2$	$H_2PCHNonPH_2$	−105 J191	3
	$C_{11}H_{26}P_2$	$H_2PCHDecPH_2$	−108 J192	3
	$C_{12}H_{28}P_2$	$H_2PCH(n\text{-}C_{11}H_{23})PH_2$	−107 J193	3
CH_2	$C_2H_8P_2$	$H_2PCH_2CH_2PH_2$	−130.8 J193	3
	$C_3H_{10}P_2$	$H_2PCH_2CH_2CH_2PH_2$	−138.6 J190	3
	$C_4H_{12}OP_2$	$H_2PCH_2CH_2OCH_2CH_2PH_2$	?? J193	3
	$C_4H_{12}P_2$	$H_2P(CH_2)_4PH_2$	−137.0 J192	3
	$C_5H_{14}P_2$	$H_2P(CH_2)_5PH_2$	−136.5 J191	3
	$C_6H_{14}P_2$	$CH_2{=}CHCH_2PH(CH_2)_3PH_2$	−135.3(c) J190.4	11
	$C_6H_{18}P_4$	$P(CH_2CH_2PH_2)_3$	−125.3	12
	$C_8H_{12}P_2$	$PhHPCH_2CH_2PH_2$	−129.1	12
	$C_{10}H_{17}P_3$	$(H_2PCH_2CH_2)_2P(Ph)$	−126.7	12
	$C_{11}H_{28}P_4$	$H_2P(CH_2)_3PMe(CH_2)_3PMe(CH_2)_3PH_2$	−139.3 J196	26
	$C_{12}H_{21}P_3$	$H_2P(CH_2)_3PPh(CH_2)_3PH_2$	−135.6 J194	27
	$C_{14}H_{16}P_2$	$Ph_2PCH_2CH_2PH_2$	−124.6	12
	$C_{18}H_{26}P_4$	$H_2P(CH_2)_2PPh(CH_2)_2PPh(CH_2)_2PH_2$	−126.7	12
	$C_{21}H_{32}P_4$	$H_2P(CH_2)_3PPh(CH_2)_3PPh(CH_2)_3PH_2$	−141.6 J194	26
C_2P	$C_9H_{22}P_2$	$H_2PC(Bu)_2PH_2$	−91 J194	3
		E1c: Primary Alkenyl/aryl Mono- and Di- Phosphines		
	P-bound atoms = $C'H_2$			
C'H	C_2H_5P	$CH_2{=}CHPH_2$	−135.7 J197	13
C'_2	C_6DH_6P	PhPHD	−125.0 J203 J_{PD}31	3, 18a

Rings/ Connectivities	Formula	Structure	^{31}P NMR data (δP [solv, temp] J_{PH} Hz)	Ref.
	$C_6D_2H_5P$	$PhPD_2$	−127 J_{PD}31, −124.7	3, 18a
		$C_6D_5PH_2$	−122.9	18a
	C_6D_7P	$C_6D_5PD_2$	−125.2	18a
	$C_6H_2F_5P$	$C_6F_5PH_2$	−183.1 J216	3
	C_6H_6BrP	2-$BrC_6H_4PH_2$ 4-$BrC_6H_4PH_2$	−130.6 J204; −124.0 J201	19
	C_6H_6ClP	2-$ClC_6H_4PH_2$ 4-$ClC_6H_4PH_2$	−127.5 J204; −124.1 J201	19
	C_6H_6FP	4-$FC_6H_4PH_2$	−125.4 J199	19
	C_6H_7OP	2-$HOC_6H_4PH_2$	−147.4(b) J204	20
		$PhPH_2$	−118.7/−125.7 J195, −122.3±0.3(k) J198	3, 18a, 21
	C_6H_7PS	2-$HSC_6H_4PH_2$	−131.7(b) J210	20
	C_6H_8NP	$H_2NDopPH_2$	−149.9(b) J199	20
	$C_6H_8P_2$	$(H_2P)_2Dop$	−124.3(c)	23
	C_7H_9OP	4-$MeOC_6H_4PH_2$	−125.8 J198	19
	C_7H_9P	2-$TolPH_2$ 4-$TolPH_2$	−130.9 J200; −124.5 J200	19
	$C_8H_{11}ClP_2$	$ClCH_2CH_2PHDopPH_2$	−124.9 J200	269
	$C_8H_{11}OP$	4-$EtOC_6H_4PH_2$	−125.5 J199	19
	$C_8H_{12}NP$	4-$Me_2NC_6H_4PH_2$	−126.1 J202	19
	$C_9H_{12}NO_2P$	$EtO_2CNHDopPH_2$	−143.1 J203	22
	$C_9H_{13}P$	$MesPH_2$	−160.2 J203	17
	$C_9H_{14}P_2$	iPrHPDop**P**H_2	−122.6	23
	$C_{10}H_{16}P_2$	tBuHPDop**P**H_2	−124.3	24
	$C_{12}H_{12}P_2$	PhHPDop**P**H_2	−124.4	24
	$C_{13}H_{13}N_2OP$	$PhNHCONHDopPH_2$	−144.7 J199	22
	$C_{18}H_{31}P$	$SmsPH_2$	−132.4(t) J206	5

Section E2: Secondary Phosphines

E2a: Secondary Dialkylphosphines

P-bound atoms = C_2H

Rings/ Connectivities	Formula	Structure	^{31}P NMR data	Ref.
$(F_3)_2$	C_2HF_6P	Tf_2PH	−50.7 J218, −49.8 J217	3
$(H_2N)_2$	$C_{10}H_{25}N_2P$	$(Et_2NCH_2)_2PH$	−101.9 J194	3
$(H_3)_2$	C_2H_7P	Me_2PH	−99.5 J188, −99.5 J191	3
H_3,CH_2	C_3H_9P	MeEtPH	−77.0 J191	3
	$C_4H_{11}P$	MePrPH	−87.0 J196	3
	$C_5H_{13}P$	MeBuPH	−86.0 J202	3
	$C_{11}H_{25}P$	MeDecPH	−85.3 J194	3
	$C_{12}H_{29}O_2P$ 3	Me(iPrO)P(O) $(CH_2)_3$(Me) P$(CH_2)_3$**PH**Me	−84.1 J189	27
$(CHO)_2$	$C_4H_5Cl_6O_2P$	$(CCl_3CHOH)_2PH$	−43 J226	3
$(CH_2)_2$	$C_4H_{11}P$	Et_2PH	−55.5 J190	3
	$C_6H_9N_2P$	$(NCCH_2CH_2)_2PH$	−75 J195	3
	$C_6H_{15}P$	Pr_2PH	−73 J178, −72 J192	3, 28
	$C_6H_{16}NP$	$(H_2NCH_2CH_2)BuPH$	−37.4	3
	$C_8H_{19}P$	Bu_2PH	−69.5 J180, −69.3 J189	3, 28
		iBu_2PH	−82.5 J194, −84.6 J197	3, 28

TABLE E (continued)
^{31}P NMR Data of Three Coordinate (λ3σ3) Compounds Containing Phosphorus Bonds to Hydrogen and/or Group IV Atoms but Not to Halogen, Chalcogenide or Group V Atoms

Rings/ Connectivities	Formula	Structure	^{31}P NMR data (δP [solv, temp] J_{PH} Hz)	Ref.
	$C_{10}H_{23}P$	Pe_2PH	−69.4 J189	28
		neoPe_2PH	−98.1(r) J189	273
	$C_{16}H_{35}P$	Oct_2PH	−71.5 J196	3
	$C_{24}H_{51}P$	$(nC_{12}H_{25})_2PH$	−71 J193	3
CH_2,C_2H	$C_8H_{18}NP$	$(H_2NCH_2CH_2)$cHexPH	−37.4	3
$(C_2H)_2$	$C_6H_{15}P$	iPr_2PH	−26.4 J198	28
$(C_3)_2$	$C_8H_{19}P$	tBu_2PH	+20.1 J197, +19.9(z) J200.6	3, 29

E2b: Secondary Dialkyl Diphosphines

<u>**P-bound atoms = C2H**</u>

Rings/ Connectivities	Formula	Structure	^{31}P NMR data	Ref.
H_2P,H_3	$C_2H_8P_2$	H_2PCH_2(Me)**PH**	−69.1(r)	8
	$C_3H_{10}P_2$	$MeHPCH_2PHMe$	−85.5 and −88.9$^{dl+meso}$ J210	8
	$C_4H_{12}P_2$	Me_2PCH_2(Me)**PH**	−93.4(r)	8
$H_2P,C'H_2$	$C_8H_{12}P_2$	$H_2PCH_2(PhCH_2)$**PH**	−44.8 J193.2(r)	8
	$C_{15}H_{18}P_2$	$BzHPCH_2PHBz$	−59.5 and −60.4$^{dl+meso}$(r)	8
	$C_{22}H_{24}P_2$	Bz_2PCH_2**PH**Bz	−68.6 J198.5(r)	8
H_2P,C_2H	$C_4H_{12}P_2$	H_2PCH_2(iPr)**PH**	−29.4 J191.2(r)	8
	$C_7H_{18}P_2$	$iPrHPCH_2PHPr$-i	−44.4 J184(r)	8
	$C_{10}H_{24}P_2$	$(iPr)_2PCH_2$(iPr)**PH**	−46.6 J194.3(r)	8
H_2P,C_3	$C_9H_{22}P_2$	$tBuHPCH_2PHBu$-t	−26.1 and −27.4$^{dl+meso}$(r)	8
H_3,CH_2	$C_9H_{23}P_3$	$MeHP(CH_2)_3PMe(CH_2)_3PHMe$	−84.3(r)	27
	$C_{12}H_{29}O_2P_3$	Me(iPrO)P(O)$(CH_2)_3$(Me)**PH**	−84.1 J189	27
	$C_{13}H_{32}P.$	$MeHP(CH_2)_3PMe(CH_2)_3PMe(CH_2)_3PHMe$	−87.3 J196	26
	$C_{14}H_{25}P_3$	$MeHP(CH_2)_3PPh(CH_2)_3PHMe$	−84.65(r)	27
$(CH_2)_2$	$C_6H_{14}P_2$	$H_2P(CH_2)_3(CH_2{=}CHCH_2)PH$	−70.1 J195	12
	$C_6H_{16}P_2$	$EtHPCH_2CH_2PHEt$	−57.3 J191	3
	$C_6H_{16}P_2S_2$	$YHPCH_2CH_2PHY$; Y = CH_2CH_2SH	−70.0 J189.2	45
		$YHP(CH_2)_3PHY$; Y = CH_2CH_2SH	−73.67 J175.2	45
	$C_8H_{20}P_2S_2$	$YHP(CH_2)_2PHY$; Y = $CH_2CHMeSH$	−68.4 J187.2	45
	$C_8H_{20}P_2S_2$	$YHP(CH_2)_4PHY$; Y = CH_2CH_2SH	−78.2 J179.4	45
$CH_2,C'H_2$	$C_6H_{14}P_2$	$CH_2{=}CHCH_2PH(CH_2)_3PH_2$	−70.1(c)	11

Rings/ Connectivities	Formula	Structure	^{31}P NMR data (δP [solv, temp] J_{PH} Hz)	Ref.

E2c: Secondary Cyclic Dialkylphosphines

P-bound atoms = C2H

Rings/ Connectivities	Formula	Structure	^{31}P NMR data (δP [solv, temp] J_{PH} Hz)	Ref.
3/$(CH_2)_2$	C_2H_5P	C*H_2CH_2P*H	−341 J155	3
5/H_2N,C_2H	$C_7H_{14}NP$	MeC*HCH_2NHCH_2P*H	−48.7, −62.2 J187	7
5/C'HS,CH_2	$C_9H_{11}PS$	5; R_1 = Ph, R_2 = R_3 = H	−15.6T; −5.6C	40
	$C_{10}H_{13}OPS$	5; R_1 = MeOPh, R_2 = R_3 = H	−19.9T; −9.6C	40
	$C_{10}H_{13}PS$	5; R_1 = Ph, R_2 = H, R_3 = Me	$-21.9^{C,T}$; $-0.1^{C,C}$ $-32.0^{T,T}$; $-12.5^{T,C}$	40
5/CH_2,C_2S	$C_7H_{13}PS$	5; R_1,R_2 = $-(CH_2)_4-$, R_3 = H	−52.0	40
	$C_8H_{15}PS$	5; R_1,R_2 = $-(CH_2)_4-$, R_3 = Me	−19.6T; −3.5C	40
5/CH_2,C.H	$C_5H_{11}P$	MeC*$HCH_2CH_2CH_2$P*H	-56.6^{E} J196.6; -46.6^{Z} J190.4	41
5/C'H_2,CC'H	C_5H_9P	MeC*HCH=CHCH_2P*H	-62^{E} J191.2; -47.7^{Z} J190.2	41
6/H_2N,CH_2	$C_7H_{14}NP$	6; X = (NCH_2CH=CH_2)	−76.8 J194	7
6/H_2Si,CH_2	$C_8H_{17}PSi$	6; X = (Si(Me)CH_2CH=CH_2)	−73.6 J187	9
6/$(CH_2)_2$	C_4H_9OP	7; X = O	−79.0 J185	3
	$C_5H_{12}OP_2$	7; X = POMe	-59.9; -66.3^{I}	42
	$C_8H_{19}NP_2$	7; X = NEt	−58.2 J193.2	42
	$C_8H_{19}OPSi$	8; X = SiMe(OMe)	−74.1 J195(b)	43
	$C_{13}H_{21}OPSi$	8; SiPh(OMe)	−73.6 J194(b)	43
	$C_{14}H_{32}OP_2Si_2$	9; R = Me	−74.9 J202(b)	43
	$C_{24}H_{36}OP_2Si_2$	9; R = Ph	−74.6 J202(b)	43
6/$(C_2O)_2$	$C_{10}H_{17}O_3P$	10	−51.3 J178	3

E2d: Secondary Alkyl Alkenyl/Aryl Phosphines

P-bound atoms = CC'H

Rings/ Connectivities	Formula	Structure	^{31}P NMR data (δP [solv, temp] J_{PH} Hz)	Ref.
OO';C_2H	$C_8H_{15}O_2P$	MeO_2CPH-cHex	−45 J217	14
	$C_9H_{17}O_2P$	EtO_2CPH-cHex	−44.8 J217	14
	$C_{11}H_{21}O_2P$	BuO_2CPH-cHex	−44.7 J217	14
H_3;C'_2	C_7H_9P	MePhPH	−72.3 J222, −70.6 J201	3, 28
CH_2;C'_2	$C_8H_{11}P$	EtPhPH	−43.7 J200.6, −44.9 J203.6(k)	3, 17
	$C_8H_{12}NP$	($H_2NCH_2CH_2$)PhPH	−52 J204(b)	33
		EtPHDopNH_2	−64.9 J223	22
	$C_9H_{13}P$	PrPhPH	−53.4 J202	28
	$C_{10}H_{15}P$	iBuPhPH	−60.2 J188	28
		BuPhPH	−52.4 J200	28
C_2H;C'_2	$C_9H_{13}P$	iPrPhPH	−26.4 J203.8(k), −25.7 J201	17, 28
	$C_{10}H_{15}P$	sBuPhPH	−30.1 J204; −34.4 J202	28
	$C_{11}H_{15}P$	cPePhPH	−37.1 J207	28
C_3;C'_2	$C_{10}H_{15}P$	tBuPhPH	−5.7 J200; −4.6 J206	3, 17
	$C_{13}H_{21}P$	tBuPHMes	−72.0 J218(k)	36

TABLE E (continued)
^{31}P NMR Data of Three Coordinate (λ3σ3) Compounds Containing Phosphorus Bonds to Hydrogen and/or Group IV Atoms but Not to Halogen, Chalcogenide or Group V Atoms

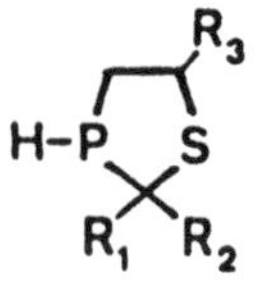

Rings/ Connectivities	Formula	Structure	^{31}P NMR data (δP [solv, temp] J_{PH} Hz)	Ref.
	$C_{18}H_{31}P$	$tBuPHC_6H_3tBu_2$-3,5	−4.0 J199; −4.6 J196	17
	$C_{22}H_{39}P$	tBuPHSms	−72 J218(c)	38

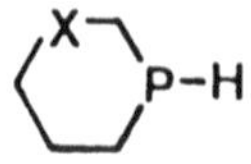

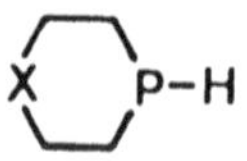

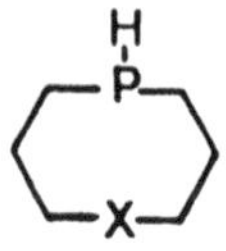

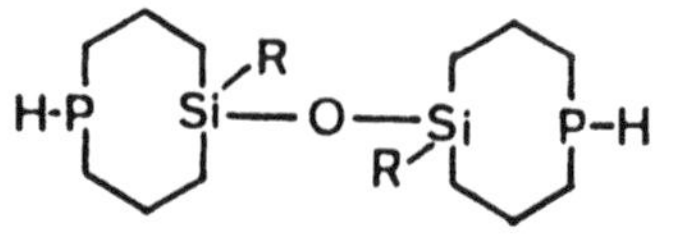

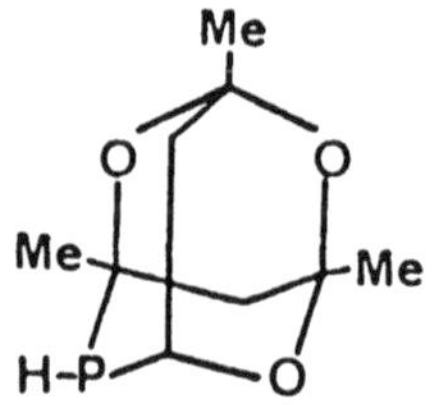

E2e: Secondary Alkyl Alkenyl/Aryl Polyphosphines

P-bound atoms = CC′H

Rings/ Connectivities	Formula	Structure	^{31}P NMR data (δP [solv, temp] J_{PH} Hz)	Ref.
$H_2P;C'_2$	$C_{13}H_{14}P_2$	$PhHPCH_2PHPh$	−54.1 J232, −55.5(c) J223	46, 47
$H_3;C'_2$	$C_8H_{12}P_2$	MeHPDopPHMe	−75.5, −76.4^{I}(c)	23, 48
$CH_2;C'_2$	$C_8H_{11}ClP_2$	$(ClCH_2CH_2)$PHDopPH2	−47.9 J195	269
	$C_8H_{12}P_2$	$PhHPCH_2CH_2PH_2$	−48.1	12
	$C_{14}H_{16}P_2$	$PhHPCH_2CH_2PHPh$	−50.8	26
	$C_{20}H_{20}P_2$	$Ph_2PCH_2CH_2PHPh$	−46.6(r)	11
	$C_{21}H_{22}P_2$	$Ph_2P(CH_2)_3PHPh$	−54.5 J203(b)	33
	$C_{21}H_{30}P_2S_2$	$PhHP[(CH_2)_3S]_2PHPh$	−53.4 J202	33
	$C_{21}H_{34}P_2$	$cHex_2P(CH_2)_3PhPH$	−54.9 J204.5(b)	34
	$C_{24}H_{29}P_3$	$PhHP(CH_2)_3PPh$-$(CH_2)_3PHPh$	−54.4, −54.8^{I}(r)	27
	$C_{32}H_{29}P_3$	$(Ph_2P)_2CHCH_2PHPh$	−54.8	35
	$C_{33}H_{40}P$ 4	$PhHP(CH_2)_3PPh$-$(CH_2)_3PPh(CH_2)_3PHPh$	−53.6; −54.0^{I}	26
$C_2H;C'_2$	$C_9H_{14}P_2$	$iPrHPDopPH_2$	−29.7	23
	$C_{12}H_{20}P_2$	iPrHPDopPHPr-i	−26.1/−31.5, −32.8(t,b)	23, 48
	$C_{15}H_{26}P_2$	$iPr_2PDopPH(iPr)$	−23.6	23

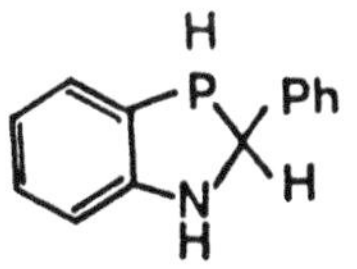

11 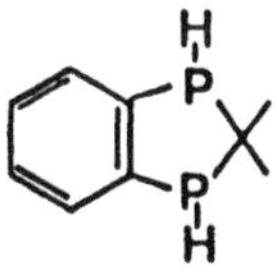 12 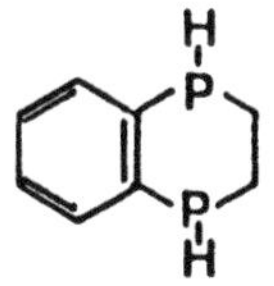 13 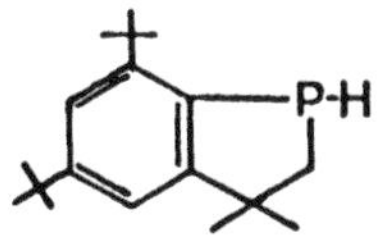 14 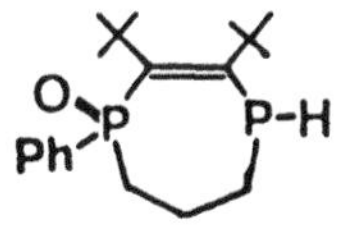 15

Rings/ Connectivities	Formula	Structure	^{31}P NMR data (δP [solv, temp] J_{PH} Hz)	Ref.
$C_3;C'_2$	$C_{10}H_{16}P$	tBuPHDopPH$_2$	−14.4(c)	24
	$C_{12}H_{20}P_2$	(tBuHP)$_2$Dop	−9.8, −18.6[I](c)	24
	$C_{16}H_{20}P_2$	tBuHPDopPHPh	−15.2, −17.6	24

E2f: Secondary Cyclic Alkyl Alkenyl/Aryl Phosphines

P-bound atoms = CC'H

Rings/ Connectivities	Formula	Structure	^{31}P NMR data	Ref.
5/C'HC;C'$_2$	$C_{13}H_{12}NP$	**11**	−50 J198, −59.8 J180[I]	22
5/C$_2$P;C'$_2$	$C_9H_{12}P_2$	**12**	+17.25, −4.67[I]	48
6/CH$_2$;C'$_2$	$C_8H_{10}P_2$	**13**	−78.2 J197, −79.7 J194	269
	$C_{18}H_{29}P$	**14**	−77(c)	38
7/CH$_2$;CC'	$C_{19}H_{32}O_2P_2$	**15**	−38.1 J244, −41.5[I] J288	44

E2g: Secondary Dialkenyl/Aryl Phosphines (Includes Diphosphines)

P-bound atoms = C'$_2$H

Rings/ Connectivities	Formula	Structure	^{31}P NMR data	Ref.
OO',C'$_2$	$C_8H_9O_2P$	MeO$_2$CPHPh	−52.3 J235	14
	$C_9H_{11}O_2P$	EtO$_2$CPHPh	−52.1 J235	14
	$C_{11}H_{15}O_2P$	BuO$_2$CPHPh	−52.0 J235	14
	$C_{12}H_{16}NO_4P$	EtO$_2$CPHDopNHCO$_2$Et	−58.8 J240	22
	$C_{13}H_{11}O_2P$	PhO$_2$CPHPh	−50.6 J236	14
CO',C'$_2$	$C_9H_{11}OP$	EtCOPhPH	−29.1 J231	3
	$C_{11}H_{15}OP$	Me$_2$CHCH$_2$COPhPH	−39.1 J218	3
C'N,C'$_2$	$C_{10}H_8N_3P$	(NC)$_2$C=C(NH$_2$)PHPh	−46.0 J230	39
C'H,C'$_2$	$C_{14}H_{13}P$	PhCH=CHPHPh	−64.4(c)	272
C'P;C'$_2$	$C_{16}H_{12}N_2P_2$	(PhHP)$_2$C=C(CH)$_2$	−34.5, −36.0[I]	39
(C'$_2$)$_2$	$C_{12}HF_{10}P$	(C$_6$F$_5$)$_2$PH	−143 J238	3
	$C_{12}H_6F_5P$	C$_6$F$_5$PHPh	−92.2 J225	3
	$C_{12}H_9Cl_2P$	(3-ClC$_6$H$_4$)2PH	−41.5(y)	30
	$C_{12}H_9F_2P$	(3-FC$_6$H$_4$)$_2$PH	−40.9 J220.4(c)	31
		(4-FC$_6$H$_4$)$_2$PH	−44.01 J219.4(c), −44.9 J216.3(y)	30, 31
	$C_{12}DH_{10}P$	Ph$_2$PD	−42.4 J_{PD}34	3
	$C_{12}H_{11}P$	Ph$_2$PH	−41 J220; −43.8 J239	3
	$C_{12}H_{12}NP$	H$_2$NDopPHPh	−45.8 J203	22
	$C_{12}H_{12}P_2$	H$_2$PDopPHPh	−43.3(c)	24
	$C_{13}H_9ClF_3P$	(3-ClC$_6$H$_4$)(4-TfC$_6$H$_4$)PH	−42.0(y)	30
	$C_{14}H_9F_6P$	(4-TfC$_6$H$_4$)$_2$PH	−42.7(y)	30
	$C_{14}H_{15}OP$	4-MeOC$_6$H$_4$PHTol-1	−44.4(y)	30
	$C_{14}H_{15}O_2P$	(4-MeOC$_6$H$_4$)$_2$PH	−45.4(y)	30

TABLE E (continued)
^{31}P NMR Data of Three Coordinate (λ3σ3) Compounds Containing Phosphorus Bonds to Hydrogen and/or Group IV Atoms but Not to Halogen, Chalcogenide or Group V Atoms

Rings/ Connectivities	Formula	Structure	^{31}P NMR data (δP [solv, temp] J_{PH} Hz)	Ref.
	$C_{14}H_{15}P$	2-Tol_2PH	−59.10 J219;	3, 32
		3-Tol_2PH	−40.2 J214;	
		4-Tol_2PH	−42.9 J212	
	$C_{16}H_{20}P_2$	tBuHPDopPHPh	−42.5, −45.7	24
	$C_{18}H_{16}P_2$	PhHPDopPHPh	−43.0, −44.0[l](c)	24
	$C_{20}H_{27}P$	$(4\text{-}tBuC_6H_4)_2PH$	−43.9 J212	3
	$C_{24}H_{28}P_2$	$(SmsHP)_2Dop$	−83.2, −83.7[l](c)	24

Section E3: Tertiary Phosphines

E3a: Tertiary Trialkylphosphines

P-bound atoms = C3

Rings/ Connectivities	Formula	Structure	^{31}P NMR data (δP [solv, temp] J_{PH} Hz)	Ref.
$(F_3)_3$	C_3F_9P	Tf_3P	−2.5	49
$(F_3)_2,H_2P$	$C_4H_2ClF_9P_2$	Tf_2PCH_2PClTf	?−9.5	88
	$C_4H_2F_{10}P_2$	Tf_2PCH_2PFTf	?−9.5	88
$(F_3);2,H_3$	$C_3H_3F_6P$	Tf_2PMe	−5.76	3
$(F_3)_2,CHF$	$C_4H_2F_9P$	$Tf_2PCHFCHF_2$	−7.8	3
$F_3,(H_3)_2$	$C_3H_6F_3P$	$TfPMe_2$	−26.9	3
$HCoP,(H_3)_2$	$C_{23}H_{41}CoP_4$	$(Me_3P)_2Me_2Co^*$-$PMeC^*HPMe_2$	−53.8	90
$HP^4{}_2,(H_3)_2$	$C_{17}H_{23}P_3S_2$	$Me_2PCH[P(S)Ph_2]$-$[P(S)Me_2]$	−25.1(c,v)	86
$HSi_2,(C_3)_2$	$C_{15}H_{37}PSi_2$	$tBu_2PCHTms_2$	47.4	101
$(H_2Cl)_2,H_3$	$C_3H_7Cl_2P$	$(ClCH_2)_2PMe$	−26.9(t)	89
$H_2Cl,(H_3)_2$	C_3H_8ClP	Me_2PCH_2Cl	−42.0	89
$(H_2N)_3$	$C_{15}H_{36}N_3P$	$(EtNCH_2)_3P$	−65.3	3
$(H_2O)_3$	$C_3H_9O_3P$	$(HOCH_2)_3)P$	−24.8, −29.1, −31	3, 51
$(H_2O)_2,H_3$	$C_3H_9O_2P$	$(HOCH_2)_2PMe$	−36.9(a)	89
$(H_2O)_2,C_3$	$C_{14}H_{23}O_2P$	$(HOCH_2)_2PAda$-1	0.6(b)	283
$(H_2P_4)_3$	$C_{39}H_{36}P_4S_3$	$[Ph_2(S)PCH_2]_3P$	−44.7(c+t)	86
$H_2P_4,(C'H_2)_2$	$C_{22}H_{24}OP_2$	$Bz_2PCH_2P(O)HBz$	−27.2(r)	8
$H2P^4,(C_2H)_2$	$C_{10}H_{24}OP_2$	$iPr_2PCH_2P(O)HPr$-i	−3.3(r)	8
$H_2P,(C_2H)_2$	$C_{10}H_{23}LiP_2$	iPr_2PCH_2PLiPr-i	−0.25(r)	99
$(H_3)_3$	C_3H_9P	Me_3P	−62, −63.3, −61.5 (k,30o)	3, 53
			−67(absorbed on h-y zeolite)	52
			−55 to −60(physisorbed to r-alumina)	271
$(H_3)_2,CF_2$	$C_4H_7ClF_3P$	Me_2PCF_2CHClF	−26.8	3
	$C_4H_8F_3P$	$Me_2PCF_2CH_2F$	−32.0	3
	$C_8H_6F_{13}P$	$Me_2PC_6F_{13}$	−22.7(b)	91
$(H_3)_2,CHF$	$C_4H_8F_3P$	$Me_2PCHFCHF_2$	−61.0	3
$(H_3)_2,CH_2$	$C_4H_{11}P$	Me_2PEt	−48.5	3

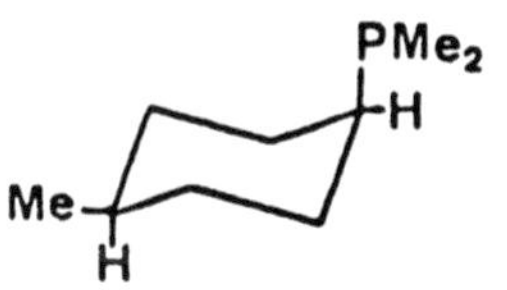

16

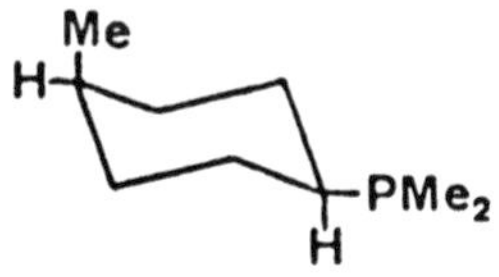

17

Rings/ Connectivities	Formula	Structure	^{31}P NMR data (δP [solv, temp] J_{PH} Hz)	Ref.
	$C_5H_{13}Cl_2PSi$	$Me_2PCH_2CH_2SiMeCl_2$	−45.0(r)	92
	$C_5H_{15}PSi$	$Me_2PCH_2CH_2SiMeH_2$	−46.0(r)	92
	$C_6H_{16}ClPSi$	$Me_2PCH_2CH_2SiMe_2Cl$	−46.0(r)	92
	$C_6H_{17}PSi$	$Me_2PCH_2CH_2SiMe_2H$	−46.0(r)	92
	$C_9H_{23}OPSi$	$Me_2P(CH_2)_4OTms$	−53.6	3
	$C_{11}H_{18}ClPSi$	$Me_2PCH_2CH_2SiMePhCl$	−45.0(r)	92
	$C_{11}H_{19}PSi$	$Me_2PCH_2CH_2SiMePhH$	−46.0(r)	92
	$C_{12}H_{27}P$	Me_2PDec	−52.5	3
	$C_{14}H_{31}P$	$Me_2P(nC_{12}H_{25})$	−52.5	3
$(H_3)_2,C'H_2$	$C_9H_{12}BrP$	$Me_2PCH_2Ph'Br$-2	−43.85	93
$(H_3)_2,C_2H$	$C_8H_{17}P$	Me_2PcHex	−43.1(j)	94
	$C_9H_{19}P$	$Me_2PcHexMe$-4)	−42.5^T (27°), −43.7^T (−100°); −49.0C(27)	15
		16	−57.2(−100°)	15
		17	−44.2(−100°)	15
	$C_{12}H_{25}P$	Me_2P(cHex-tBu-4)	−42.5^T(27°); −54.8(27°)	15
$(H_3)_2,C_3$	$C_6H_{15}P$	Me_2P-tBu	−28.7	95
	$C_{14}H_{23}P$	Me_2PAda-1	−31.5	283
$H_3,(CF_2)_2$	$C_{13}H_3F_{26}P$	$MeP(nC_6F_{13})_2$	5.5(b)	91
$H_3,(CH_2)_2$	$C_5H_{13}P$	$MePEt_2$	−34.0	3
	$C_5H_{13}PS_2$	$MeP(CH_2CH_2SH)_2$	−42.3(r)	96
	$C_7H_{17}P$	$MePPr_2$	−45.0(j)	94
	$C_{25}H_{53}P$	$MeP(nC_{12}H_{25})_2$	−34.0	3
$H_3,(C'H_2)_2$	$C_{15}H_{17}P$	$MePBz_2$	−30.0(j)	94
	$C_{11}H_{13}O_2P$	$MeP(CH_2Fur$-2$)_2$	−35.2(c)	60
$H_3,(C_2H)_2$	$C_7H_{13}P$	$MeP(C^*HCH_2C^*H_2)_2$	−8.37(b)	62
H_3,C_2H,C_3	$C_8H_{19}P$	Me(iPr)(tBu)P	1.5	188
$H_3,(C_3)_2$	$C_9H_{19}P$	MeP-tBu2	10.8(z)	29
$(CHO)_3$	$C_6H_6Cl_9O_3P$	$P(CHOHCCl_3)_3$	18(b)	54
$(CH_2)_3$	$C_6H_{15}P$	Et_3P	−20.4, −19.7, −21, −21* (*physisorbed to silica-alumina)	3, 271
$(CH_2)_3$	$C_9H_{12}N_3P$	$(NCCH_2CH_2)_3P$	−23.0, −25.5	3, 55
	$C_9H_{21}P$	Pr_3P	−33.0, −32.4(r)	3, 56
	$C_{12}H_{21}P$	$(CH_2{=}CHCH_2CH_2)_3P$	−30.2	57
	$C_{12}H_{27}P$	Bu_3P	−32.3, −33.3, −32* (*physisorbed to silica-alumina)	3, 271
		iBu_3P	−40.0, −45.3, −48.4	3, 58
	$C_{15}H_{33}P$	$(tBuCH_2)_3P$	−57.5(r)	273
		Pe_3P	−34.0	3
	$C_{24}H_{51}P$	Oct_3P	−31.8, −32.8	3
		$(BuEtCHCH_2)_3P$	−48.5	3
	$C_{36}H_{75}P$	$(nC_{12}H_{25})_3P$	−36.0	3
$(CH_2)_2,C_2O$	$C_9H_{20}OP$	$Et_2PC^*OHCH_2CH_2$-$CH_2C^*H_2$	−53.0	3

TABLE E (continued)
^{31}P NMR Data of Three Coordinate (λ3σ3) Compounds Containing Phosphorus Bonds to Hydrogen and/or Group IV Atoms but Not to Halogen, Chalcogenide or Group V Atoms

Rings/ Connectivities	Formula	Structure	^{31}P NMR data (δP [solv, temp] J_{PH} Hz)	Ref.
$(CH_2)_2,C'_2H$	$C_{21}H_{31}N_2P$	$Et_2PCH(C_6H_4NMe_2\text{-}4)_2$	−7.4(c)	97
$(CH_2)_2,C'_2$	$C_{24}H_{29}P_3$	$(Me_2PCH_2CH_2)PPh\text{-}(CH_2CH_2PPh_2)$	−17.9(r)	274
$(CH_2)_2,C_3$	$C_8H_{19}P$	$Et_2P\text{-tBu}$	−6.6(b); −6.86	98, 95
	$C_8H_{19}PS_2$	$(HSCH_2CH_2)_2P\text{-tBu}$	−1.4(r)	96
	$C_{10}H_{23}P$	$Pr_2P\text{-tBu}$	−8.7	95
	$C_{12}H_{27}P$	$Bu_2P\text{-tBu}$	−4.6, −4.4	95
$(CH_2)_3$	$C_{14}H_{25}N_2P$	$OctP(CH_2CH_2CN)_2$	−25.7	3
	$C_{16}H_{35}P$	$Et_2P(nC_{12}H_{25})$	−26.0	3
	$C_{20}H_{43}P$	$Bu_2P(nC_{12}H_{25})$	−32.4	3
		Oct_2PBu	−33.3	3
$CH_2,(C_2H)_2$	$C_9H_{20}ClP$	$iPr_2PCH_2CH_2CH_2Cl$	1.4(r)	99
	$C_{14}H_{27}P$	$cHex_2PEt$	−0.56(j)	94
$CH_2,(C_2H)_2$	$C_{15}H_{27}ClP$	$cHex_2PCH_2CH_2CH_2Cl$	−6.2(b)	33
	$C_{15}H_{29}P$	$cHex_2PPr$	−6.96(j)	94
$CH_2,(C_3)_2$	$C_{10}H_{23}P$	$EtP\text{-tBu}_2$	33.7, 33.6(b)(z)	29, 98
	$C_{11}H_{25}P$	$PrP\text{-tBu}_2$	26.3	95
	$C_{12}H_{27}P$	$BuP\text{-tBu}_2$	26.6	3
$(C'H_2)_3$	$C_9H_{15}P$	$(CH_2{=}CHCH_2)_3P$	−34.5	3
	$C_{15}H_{15}O_3P$	$(2\text{-FurCH}_2)_3P$	−24.8(c)	60
$(C'H_2)_3$	$C_{21}H_{21}P$	Bz_3P	−10.4, −12.9	3, 79
$(C'H_2)_2,C_3$	$C_{10}H_{19}P$	$(CH_2{=}CHCH_2)_2P\text{-tBu}$	−6.0	3
	$C_{14}H_{19}O_2P$	$(2\text{-FurCH}_2)_2P\text{-tBu}$	−0.2(c)	60
$C'H_2,C_2H,C_3$	$C_{10}H_{21}P$	$(CH_2{=}CHCH_2)(iPr)(tBu)P$	14.4	3
$C'H_2,(C_3)_2$	$C_{14}H_{29}OP$	$tBuCOCH_2P(tBu)_2$	24.1(b)	102
	$C_{15}H_{25}P$	$BzP(tBu)_2$	33.7(z)	29
$C'H_2;(C_3)_2$	$C_{16}H_{25}OP$	$PhCOCH_2P(tBu)_2$	30.6	102
$(C''H_2)_3$	$C_6H_6N_3P$	$(NCCH_2)_3P$	−33.1	61
$(C_2H)_3$	$C_9H_{15}P$	$(C^*H_2CH_2C^*H)_3P$	16.3	62
	$C_9H_{21}P$	iPr_3P	19.3	3
	$C_{12}H_{27}P$	iBu_3P	7.8,8.0	3
	$C_{15}H_{27}P$	cPe_3P	4.7,0.8	3
	$C_{18}H_{33}P$	$cHex_3P$	7,9.4(b),7.5(CP/MAS)	3, 70
$C_2H,(C_3)_2$	$C_{11}H_{25}P$	$iPrP(tBu)_2$	45.9(z)	29
$C'_2H,(C_3)_2$	$C_{25}H_{39}N_2P$	$tBu_2PCH(C_6H_4NMe_2\text{-}4)2$	44.4(c)	97
$(C_3)_3$	$C_{12}H_{27}P$	tBu_3P	61.1,63.0,61.5(z, −13°)	3, 29

E3b: Tertiary Trialkyl Polyphosphines

P-bound atoms = C_3

Rings/ Connectivities	Formula	Structure	^{31}P NMR data (δP [solv, temp] J_{PH} Hz)	Ref.
$(F_3)_2,H_2P$	$C_6H_2F_{15}P_3$	$Tf_2PCH_2PTfPTf_2$	?−1.52, −4.2/−8.2, +25.37	88
	$C_6H_8F_9NP_2$	$Tf_2PCH_2PTfNMe_2$	?−4.8	88
	$C_8H_4F_{18}P_4$	TfYP-PYTf; Y = CH_2PTf_2	2.63 and 10.46 (not assigned)	88
$HCoP,(H_3)_3$	$C_{23}H_{41}CoP_4$	$(Me_3P)_2Me_2Co^*PMeC^*HPMe_2$	−53.8	90

Rings/ Connectivities	Formula	Structure	^{31}P NMR data (δP [solv, temp] J_{PH} Hz)	Ref.
HPSi,HSi_2,H_3	$C_{14}H_{38}P_2Si_3$	Me_2PCHTms**P**Me-($CHTms_2$)	−23.3 or −38.6 J_{PP} 97	101
HPSi,$(H_3)_2$	$C_{14}H_{38}P_2Si_3$	Me_2**P**CHTmsPMe-($CHTms_2$)	−23.3 or −38.6 J_{PP} 97	101
H_2Cl,H_2P,C_2H	$C_{11}H_{25}ClP_2$	iPr_2PCH_2**P**(iPr)CH_2Cl	−28.4 JPP 93.3	99
$(H_2N)_2$,H_2P	$C_{21}H_{50}N_4P_2$	$Y_2PCH_2PY_2$; Y = CH_2NEt_2	−56.5	3
$(H_2N)_2$,CH_2	$C_{22}H_{52}N_4P_2$	$Y_2PCH_2CH_2PY_2$; Y = CH_2NEt_2	−48.7	3
	$C_{23}H_{54}N_4P_2$	$Y_2P(CH_2)_3PY_2$; Y = CH_2NEt_2	−52.9	3
	$C_{24}H_{56}N_4P_2$	$Y_2P(CH_2)_4PY_2$; Y = CH_2NEt_2	−53.2	3
$(H_2P)_3$	$C_9H_{24}P_4$	**P**$(CH_2PMe_2)_3$	−56.7	227
$(H_2P)_2$,H_3	$C_7H_{19}P_3$	$(Me_2PCH_2)_2$**P**Me	−48.9	248
$(H_2P)_2$,C_2H	$C_{21}H_{48}P_4$	Y_2**P**-iPr; Y = $CH_2P(iPr)_2$	−24.5(r); -24.8^{l}(r)	99
H_2P,$(H_3)_2$	$C_3H_{10}P_2$	Me_2**P**CH_2PH_2	−40.9(r)	8
	$C_4H_{12}P_2$	Me_2**P**CH_2PHMe	−51.4(r)	8
	$C_5H_{14}P_2$	$Me_2PCH_2PMe_2$	−41.6(c), −55.6(k,30°)	53, 89, 226
	$C_7H_{19}P_3$	$(Me_2P)_3CH$	−48.1(b), −48.5 (k, −90°)	53, 247
H_2P,$(H_3)_2$	$C_7H_{19}P_3$	Me_2**P**CH_2PMeCH_2**P**Me_2	−55.8	248
	$C_9H_{24}P_4$	$P(CH_2$**P**$Me_2)_3$	−43.0	227
	$C_{11}H_{18}P_3$	$(Me_2$**P**$)_2CHPPh_2$	−42.7(b)	247
H_2P,$(H_3)_2$	$C_{11}H_{26}P_2$	Me_2**P**CH_2P-tBu	−48.7(b)	227
	$C_{15}H_{18}P_2$	Me_2**P**CH_2PPh_2	−53.0	53, 227
	$C_{27}H_{27}P_3$	Me_2**P**$CH(PPh_2)_2$	−37.0(b)	247
H_2P_4,$(H_3)_2$	$C_{10}H_{26}P_4$	Me_2**P**CH2P-(Me_2)=C$(PCMe_2)_2$	−57.6(b)	103
H_2P,H_3,C_2H	$C_9H_{22}P_2$	iPrMePCH$_2$PMe-iPr	−31.0; $-31.3^{dl+meso}$(r)	99
H_2P,$(CH_2)_2$	$C_{11}H_{26}P_2$	Me_2PCH_2**P**-tBu_2	17.2(b)	227
	$C_{25}H_{58}P_6$	$Y_2PCH_2PY_2$; Y = $CH_2CH_2PEt_2$	−28.2(a)	250
H_2P,CH_2,C_2H	$C_{22}H_{50}P_4$	iPrYPCH$_2$CH$_2$PY-iPr; Y = $CH_2P(iPr)_2$	−20.8(r); -21.0^{l}(r)	99
	$C_{23}H_{52}P_4$	iPrYP$(CH_2)_3$PY-iPr; Y = $CH_2P(iPr)_2$	−19.25(r), −19.3(r)	99
	$C_{25}H_{56}P_4$	iPrYPCH$_2$PY-iPr; Y = $(CH_2)_3P(iPr)_2$	−22.6(r); -22.7^{l}(r)	99
	$C_{26}H_{58}P_4$	iPrYP$(CH_2)_6$PY-iPr; Y = $CH_2P(iPr)_2$	−19.0(r)	99
	$C_{30}H_{66}P_4$	iPrYP$(CH_2)_{10}$PY-iPr; Y = $CH_2P(iPr)_2$	−19.0(r)	99
H_2P,$(C'H_2)_2$	$C_{22}H_{24}P_2$	Bz_2**P**CH_2PHBz	−19.5(r)	8
	$C_{29}H_{30}P_2$	$Bz_2PCH_2PBz_2$	−30.6(r)	8
H_2P,$(C_2H)_2$	$C_{10}H_{23}ClP_2$	iPr_2**P**CH_2PCl-iPr	−5.2(r)	8
	$C_{10}H_{24}P_2$	iPr_2**P**CH_2PH-iPr	−3.2(r)	8
	$C_{11}H_{25}ClP_2$	iPr_2**P**$CH_2P(iPr)CH_2Cl$	−5.6 J_{PP} 93.3	99
	$C_{20}H_{46}P_4$	iPrYP-PY-iPr; Y = $CH_2P(iPr)_2$	−3.7; -2.7^{l}	99
	$C_{21}H_{48}P_4$	iPrYPCH$_2$PY-iPr; Y = $CH_2P(iPr)_2$	−5.9(r); -6.2^{l}(r)	99
	$C_{22}H_{50}P.$	iPrYPCH$_2$CH$_2$PY-iPr; Y = CH_2**P**$(iPr)_2$	−12.1(r); -12.1^{l}(r)	99

TABLE E (continued)
^{31}P NMR Data of Three Coordinate (λ3σ3) Compounds Containing Phosphorus Bonds to Hydrogen and/or Group IV Atoms but Not to Halogen, Chalcogenide or Group V Atoms

Rings/ Connectivities	Formula	Structure	^{31}P NMR data (δP [solv, temp] J_{PH} Hz)	Ref.
	$C_{23}H_{52}P_4$	iPrYP$(CH_2)_3$PY-iPr; Y = $CH_2\mathbf{P}(iPr)_2$	−5.13(r); −5.14[I](r)	99
	$C_{26}H_{58}P_4$	iPrYP$(CH_2)_6$PY-iPr; Y = $CH_2\mathbf{P}(iPr)_2$	−5.3(r)	99
	$C_{30}H_{66}P_4$	iPrYP$(CH_2)_{10}$PY-iPr; Y = $CH_2\mathbf{P}(iPr)_2$	−5.3(r)	99
$(H_3)_2,CF_2$	$C_6H_{14}F_2P_2$	$Me\mathbf{P}CF_2CH_2PMe_2$	−58.8	233
$(H_3)_2,CH_2$	$C_6H_{14}F_2P_2$	$MePCF_2CH_2\mathbf{P}Me_2$	−91.9	233
	$C_6H_{16}P_2$	$Me_2PCH_2CH_2PMe_2$	−49.4; −42.4(c)	3, 226
	$C_9H_{23}P_3$	$MeP(CH_2CH_2\mathbf{P}Me_2)_2$	−49.0(b)	249
	$C_{12}H_{30}P_4$	$P(CH_2CH_2\mathbf{P}Me_2)_3$	−48.9(r)	274
	$C_{14}H_{25}P_3$	$PhP(CH_2CH_2\mathbf{P}Me_2)_2$	−48.8(b)	249
	$C_{14}H_{34}P_4$	$YMePCH_2CH_2PMeY$; Y = $(CH_2)_3\mathbf{P}Me_2$	−61	254
	$C_{15}H_{36}P_4$	$YMeP(CH_2)_3PMeY$; Y = $(CH_2)_3\mathbf{P}Me_2$	−62.3	254
	$C_{16}H_{29}P_3$	$PhP[(CH_2)_3\mathbf{P}Me_2]_2$	−53.4(r)	27
	$C_{17}H_{22}P_2$	$Me_2\mathbf{P}(CH_2)_3PPh_2$	−52.7(c)	240
	$C_{18}H_{24}P_2$	$Me_2\mathbf{P}(CH_2)_4PPh_2$	−50.1(c)	240
	$C_{18}H_{42}P_4$	$P(CH_2CH_2CH_2CH_2\mathbf{P}$-$Me_2)_3$	−53.6(t)	252
	$C_{24}H_{29}P_3$	$(Me_2\mathbf{P}CH_2CH_2)PPh$-$(CH_2CH_2PPh_2)$	−49.0(r)	274
$(H_3)_2,C'H_2$	$C_{12}H_{20}P_2$	1,3-$(Me_2PCH_2)_2C_6H_4$	−46.5(b)	243
$H_3,(CH_2)_2$	$C_9H_{23}P_3$	$(MeHP(CH_2)_3)_2\mathbf{P}Me$	−43.0(r)	27
	$C_9H_{23}P_3$	$(Me_2PCH_2CH_2)_2\mathbf{P}Me$	−34.8(b)	249
	$C_{11}H_{28}P_4$	$YMe\mathbf{P}(CH_2)_3\mathbf{P}MeY$; Y = $(CH_2)_3PH_2$	−43.8; −44.8[I]	26
	$C_{13}H_{32}P_4$	$YMe\mathbf{P}(CH_2)_3\mathbf{P}MeY$; Y = $(CH_2)_3PHMe$	−43.6; −44.6[I]	26
	$C_{14}H_{28}O_4P_2$	$YMePCH_2CH_2PMeY$; Y = CH_2CH_2CHGlc	−37.23, −37.32(r)	230
	$C_{14}H_{34}P_4$	$YMe\mathbf{P}CH_2CH_2\mathbf{P}MeY$; Y = $(CH_2)_3PMe_2$	−46.3; −46.4[I]	254
	$C_{15}H_{30}O_4P_2$	$YMeP(CH_2)_3PMeY$; Y = CH_2CH_2CHGlc	−42.23(r)	230
	$C_{15}H_{36}P_4$	$YMe\mathbf{P}(CH_2)_3\mathbf{P}MeY$; Y = $(CH_2)_3PMe_2$	−49.2	254
	$C_{18}H_{40}O_4P_2$	$MeYP(CH_2)_2PYMe$; Y = $(CH_2)_3(OEt)_2$	−38.2(r)	230
	$C_{19}H_{44}O_4P_4$	$MeYP(CH_2)_3PYMe$; Y = $(CH_2)_3P(O)$ (O-iPr)Me	−43.4; −44.3[I]	26
	$C_{23}H_{52}O_6P_4$	$MeYP(CH_2)_3PYMe$; Y = $(CH_2)_3P(O)$-$(O\text{-}iPr)_2$	−43.6; −44.8[I]	26
	$C_{29}H_{31}P_3$	$Me\mathbf{P}(CH_2CH_2PPh_2)_2$	−33.6	249
$(CH_2)_3$	$C_{10}H_{24}P_2$	$Et_2PCH_2CH_2PEt_2$	−19.3	3
	$C_{12}H_{30}P_4$	$\mathbf{P}(CH_2CH_2PMe_2)_3$	−20.1(r)	274

Rings/ Connectivities	Formula	Structure	^{31}P NMR data (δP [solv, temp] J_{PH} Hz)	Ref.
	$C_{14}H_{20}N_4P_2$	$(NCCH_2CH_2)_2PCH_2CH_2$-$P(CH_2CH_2CN)_2$	−21.4	3
	$C_{18}H_{24}P_2$	Et_2**P**$CH_2CH_2PPh_2$	−20.7(k)	58
	$C_{18}H_{33}P_3$	$PhP(CH_2CH_2$**P**$Et_2)_2$	−17.3	250
	$C_{18}H_{42}P_4$	**P**$(CH_2CH_2CH_2$-$CH_2PMe_2)_3$	−32.7(t)	252
	$C_{19}H_{26}P_2$	Et**P**$(CH_2)_3PPh_2$	−23.1 (c)	240
	$C_{20}H_{28}P_2$	Et_2**P**$(CH_2)_4PPh_2$	−21.7 (c)	240
	$C_{22}H_{32}P_2$	Bu_2**P**$CH_2CH_2PPh_2$	−26.1	231
	$C_{25}H_{58}P_6$	$Y_2PCH_2PY_2$; Y = CH_2CH_2**P**Et_2	−19.3(a)	250
	$C_{31}H_{32}NP_3$	$NCCH_2CH_2$**P**-$(CH_2CH_2PPh_2)_2$	−21.0(b)	33
	$C_{31}H_{36}NP_3$	$H_2N(CH_2)_3$**P**-$(CH_2CH_2PPh_2)_2$	−22.6(b)	33
	$C_{35}H_{74}P_2$	$Oct_2P(CH_2)_3POct_2$	−33.7	3
	$C_{42}H_{42}P_4$	**P**$(CH_2CH_2PPh_2)_3$ (Tetraphos II)	−13.5; −14.5(r)	117, 253
	$C_{66}H_{69}P_7$	$PhP[CH_2CH_2$**P**-$(CH_2CH_2PPh_2)_2]_2$	−18.7	255
	$C_{78}H_{69}P_7$	$P[CH_2CH_2P$ $(CH_2CH_2PPh_2)_2]_3$	$-18.2P^a$, $-18.2P^b$	255
$(CH_2)_2,C_3$	$C_{34}H_{41}P_3$	$tBuP(CH_2CH_2CH_2PPh_2)_2$	−7.2(t)	252
$CH_2,(C_2H)_2$	$C_{21}H_{34}P_2$	$cHex_2$**P**$(CH_2)_3PHPh$	−7.7(b)	34
	$C_{25}H_{56}P_4$	$iPrYPCH_2PY$-iPr; Y = $(CH_2)_3$**P**$(iPr)_2$	0.6(r);0.5^I(r)	99
	$C_{26}H_{48}P_2$	$cHex_2PCH_2CH_2P(cHex)_2$	−0.6(k)	58
$(CH_2)_2,C_2H$	$C_{36}H_{61}P_3$	$PhP[(CH_2)3$**P**-$cHex_2]_2$	−7.8(b)	33

E3c: Tertiary Cyclic Trialkyl Phosphines

<u>P-bound atoms = C_3</u>

Rings/ Connectivities	Formula	Structure	^{31}P NMR data	Ref.
$3/H_3,(CH_2)_2$	C_3H_7P	$MeP{*}CH_2C{*}H_2$	−251.0	3
$3/H_3,CH_2,CC'H$	C_7H_9P	$MeF{*}CH(CH{=}CH_2)C{*}H_2$	-216.1^C; -216.7^T(b)	194
$3/CH_2,C_2H,CC'H$	$C_{10}H_{17}P$	cH-exP*CH(CH=CH$_2$)C*H$_2$	-190.2^C; -190.3^T(b)	194
	$C_{14}H_{25}P$	(−)Menthyl-$PCH(CH{=}CH_2)CH_2$	-190.1^C; -191.0^T(b)	194
$3/CH_2,CC'H,C_3$	$C_8H_{15}P$	$tBuP{*}CH(CH{=}CH_2)C{*}H_2$	-160.1^C; -171.0^T(b)	194
$3/(CC'H)_2,C_3$	$C_{12}H_{17}P$	**18**; R = tBu	−141.6	195
$4/H_3,(C_2H)_2$	$C_8H_{11}P$	**19**	47.4(c)	220
		20	39.5(c)	220
	$C_{11}H_{20}P$	**21**	25.5(c)	220
$5/H_3,(CH_2)_2$	C_5H_8OP	**22**	−36.4(b)	201
	$C_5H_9Br_2P$	$MeP{*}CH_2CHBr$-$CHBrC{*}H_2$	−41.5	3
	$C_5H_{11}P$	$MeP{*}Ch_2CH_2CH_2C{*}H_2$	−32.0	202
	$C_6H_{13}P$	$MeP{*}CH_2CHMe$-$CH_2C{*}H_2$	$-33.4; -33.8^I$	3, 202
	$C_7H_{15}P$	$MeP{*}CH_2CHMeCHMe$-$C{*}H_2$(cis)	$-41.6; -55.4^I$	202
$5/H_3,CH_2,C_2H$	$C_{10}H_{18}P_2$	**23**; P^2	−21.7(c)	220
$5/H_3,(C'H_2)_2$	C_5H_9P	$MeP{*}CH_2CHCHC{*}H_2$	−41.3, −41.6	3, 202

TABLE E (continued)
^{31}P NMR Data of Three Coordinate (λ3σ3) Compounds Containing Phosphorus Bonds to Hydrogen and/or Group IV Atoms but Not to Halogen, Chalcogenide or Group V Atoms

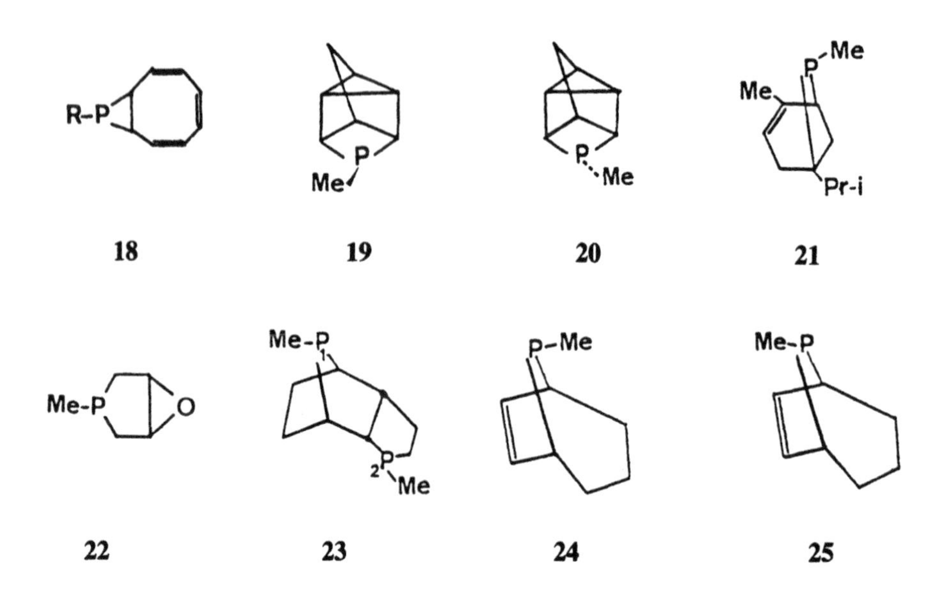

Rings/ Connectivities	Formula	Structure	^{31}P NMR data (δP [solv, temp] J_{PH} Hz)	Ref.
	$C_6H_{11}P$	MeP*CH₂CMeCHC*H₂	−33.5	202
	$C_7H_{13}P$	MeP*CH₂CMeCMeC*H₂	−50.7	202
5/H₃,C'H₂,CC'H	$C_6H_{11}P$	MeP*CHMeCHCHC*H₂	−16.7^C; −28.2^T	202
5/H₃,(C₂H)₂	$C_8H_{13}P$	**24**	−18.2(c)	220
	$C_8H_{13}P$	**25**	49.7(c)	220
	$C_9H_{15}P$	**26**	−8.2(c)	220
	$C_{10}H_{18}P_2$	**23**; P^1	−18.0(c)	220
	$C_{12}H_{20}P_2$	**27**	−1.8(c)	220
		28	P^1 −15.1(c);P^2 −17.2(c)	220
5/H₃,(CC'H)₂	$C_7H_{13}P$	**29**	−8.9	202
	$C_{10}H_{14}P_2$	**30**; R^1 = Me, R^2 = H,	P^1 30.2; P^2 7.9	220
		31; R^1 = Me, R^2 = R^3 = H,	P^1 96.5	220, 221
	$C_{12}H_{18}P_2$	**30**; R^1 = R^2 = Me,	P^1 26.5	220
5/H₃;(CC'H)₂	$C_{12}H_{18}P_2$	**31**; R^1 = R^2 = Me, R^3 = H,	P^1 100.8	220
	$C_{13}H_{20}P_2$	**31**; R^1 = R^2 = R^3 = Me,	P^1 98.8	220
5/(CH₂)₂,C'H₂	$C_{11}H_{14}BrP$	BrDopCH₂P*(CH₂)₃C*H₂	−15.6	190
	$C_{11}H_{15}P$	BzP*(CH₂)₃C*H₂	−14.4	202
5/(C'H₂)₃	$C_{11}H_{13}P$	BzP*CH₂CH=CHC*H₂	−23.5	202
	$C_{12}H_{15}P$	BzP*CH₂CMe=CHC*H₂	−18.7	202
	$C_{13}H_{17}P$	BzP*CH₂CMe= CMeC*H₂	−29.9	202
5/(CC'H)₂,C₃	$C_{19}H_{23}P$	**32**; R = tBu	152.5	222
6/(CHO)₃	$C_{12}H_9BCl_9O_2P$	**33**; R^1 = CH(OH)CCl₃, R^2 = CCl₃	−10.0(m), −4.0(v)	54

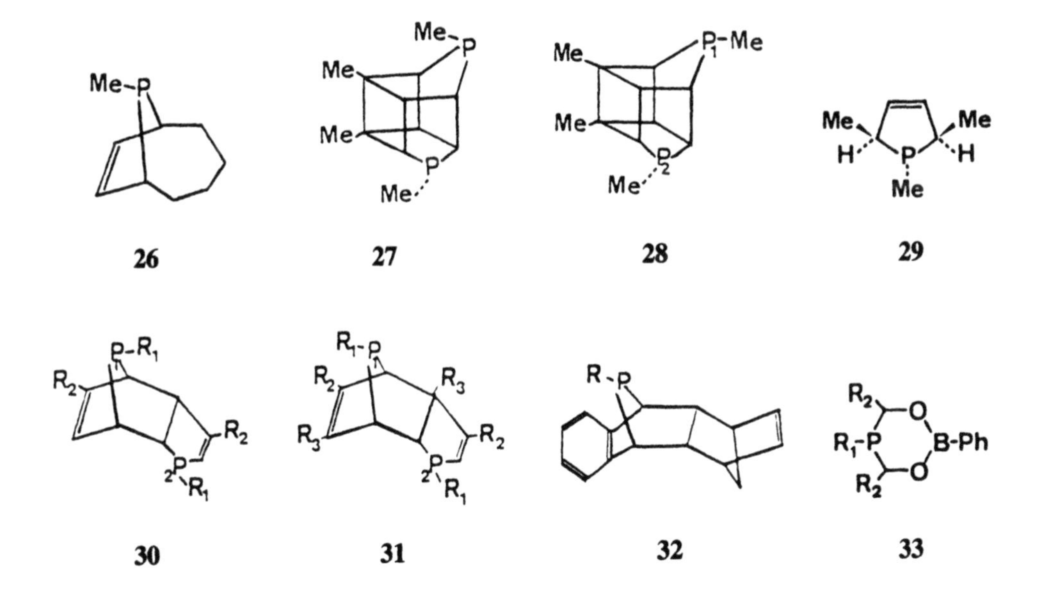

Rings/ Connectivities	Formula	Structure	^{31}P NMR data (δP [solv, temp] J_{PH} Hz)	Ref.
6/(CHO)$_3$	$C_{12}H_8B_2Cl_{18}O_7P_2$	**34**; R^1 = CH(OH)CCl$_3$, R^2 = CCl$_3$	−4(v)	54
6/C′HO,CH$_2$,C$_3$	$C_{14}H_{21}OP$	tBuP*CHPhOCH$_2$-CH$_2$C*H$_2$	−13.7^T	214
6/(CH$_2$)$_3$	$C_{12}H_{25}OP$	OctP*CH$_2$CH$_2$OCH$_2$-C*H$_2$ ·	−52.0	3
	$C_{16}H_{33}OP$	nC$_{12}$H$_{25}$P*CH$_2$CH$_2$-OCH$_2$C*H$_2$	−52.0	3
7/H$_2$O,(CH$_2$)$_2$	$C_{14}H_{23}OPSi$	HOCH$_2$P*(CH$_2$)$_3$-SiPhOMe(CH$_2$)$_2$C*H$_2$	−18.2	43
8/H$_3$,(CH$_2$)$_2$	$C_5H_{11}PS_2Sn$	**35**; R^1 = Me, X = S, Y = Sn	−60.1(r)	96
	$C_6H_{14}O_2P$	**35**; R^1 = Me, X = O, Y = PMe	−42.1, −38.5$^{C/T}$	217
8/(CH$_2$)$_2$,C$_3$	$C_8H_{17}PS_2Sn$	**35**; R^1 = tBu, X = S, Y = Sn	−30.3(r)	96
5,5/(CC′O)$_3$	$C_6H_9H_6P$	**36**	−21.0	3
5,6/(CH$_2$)$_3$	$C_5H_{10}N_3P$	**37**	−89.3	223
5,6/(CH$_2$)$_2$,C$_2$H	$C_7H_{14}NP$	**38**	−26.4	7
6,6/(H$_2$O)$_3$	$C_3H_6O_3P_2$	P(OCH$_2$)$_3$**P**	−67.0	50
6,6/(CH$_2$)$_3$	$C_3H_6O_3P_2$	**39**; n = O	−67.0	3
	$C_3H_6O_3P_2S$	**39**; n = 1	−70.0	3
	$C_6H_{12}N_3P$	**40**; R = Lp	−101.6(w)	224
	$C_7H_{14}NP$	**41**; R = N	−77.0	7
	$C_7H_{14}N_3P$	**42**; n = 2	−88.0(c)	223
	$C_7H_{15}IN_3P$	**40**; R = Me, X = I	−88.4(w)	224
	$C_8H_{15}O_3P$	**43**	−81.0	3
	$C_8H_{17}PSi$	**41**; R = SiMe	−37.4	9
	$C_{11}H_{22}N_3P$	**42**; n = 6	−67.0(c)	223
	$C_{12}H_{19}ClN_3P$	**40**; R = Bz, X = Cl	−86.0(w)	224
8,8/(CH$_2$)$_3$	$C_9H_{18}P_2$	**44**	−28.2(b)	11

TABLE E (continued)
^{31}P NMR Data of Three Coordinate ($\lambda 3\sigma 3$) Compounds Containing Phosphorus Bonds to Hydrogen and/or Group IV Atoms but Not to Halogen, Chalcogenide or Group V Atoms

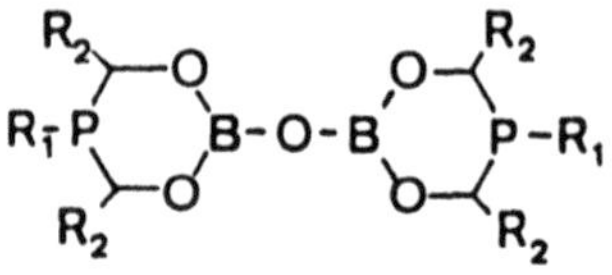

34 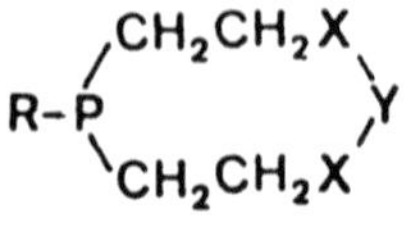 35 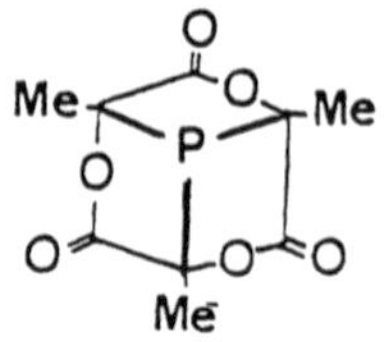 36 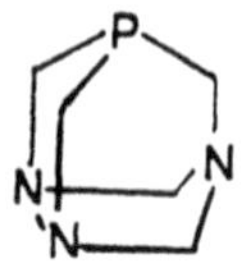 37

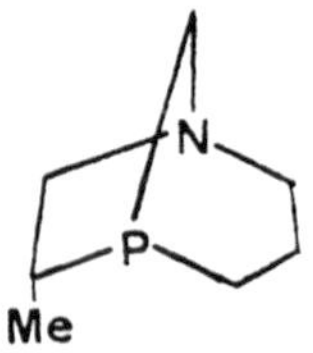

38 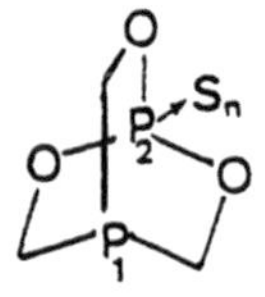 39 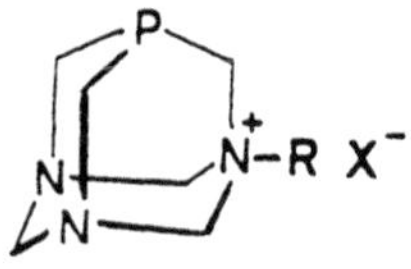 40 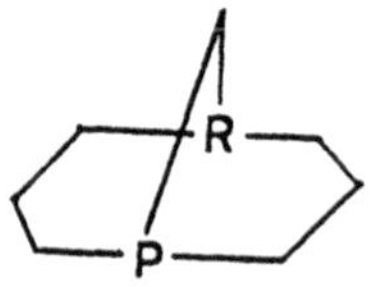 41 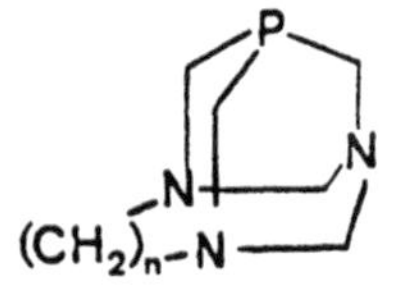 42

Rings/ Connectivities	Formula	Structure	^{31}P NMR data (δP [solv, temp] J_{PH} Hz)	Ref.
		E3d: Tertiary Dialkyl Alkenyl/Aryl Phosphines		
P-bound atoms = C2C′				
$HSi_2,H_3;C'_2$	$C_{17}H_{24}PSi_2$	$(Tms_2CH)MePMes$	−42.9	101
$H_2Li,H_3;C'_2$	$C_8H_{10}LiP$	$(LiCH_2)MePPh$	−22.9	3
$(H_2N)_2;C'_2$	$C_{12}H_{19}N_2P$	$(Et_2NCH_2)_2PPh$	−51.3	3
$H_2N,H_3;C'_2$	$C_{11}H_{16}NO_2P$	$(HO_2CCHMeNHCH_2)$-MePPh	−22.31(y)	129
$H_2N,C_3;C'_2$	$C_{19}H_{24}NO_2P$	$(HO_2CCHPhNHCH_2)$ tBuPPh	−19.04(y)	129
$(H_2P^4)_2;C'_2$	$C_{38}H_{29}P_3S_2$	$(Ph_2P(S)CH_2)_2PPh$	−42.2(c,r)	86
$(H_3)_2;P^4Si$	$C_9H_{24}P_2Si$	Me_2**P**$C(Tms){=}PMe_3$	−44.2	103
$(H_3)_2;P^4P$	$C_{10}H_{26}P_3$	$Me_2PCH_2P(Me_2)$ $=C($**P**$Me_2)$.	−39.3(b)	103
$(H_3)_2;HP^4$	$C_6H_{15}LiP_2$	Me_2**P**$CH{=}PMe_2CH_2Li$	−59.1(b)	103
	$C_6H_{16}P_2$	Me_2**P**$CH{=}PMe_3$	−61.0(b)	103
$(H_3)_2;C'N'$	$C_7H_{10}NP$	Me_2P(Pyr-2)	−41.6(b)	104
$(H_3)_2;CO'$	C_4H_9OP	Me_2PCOMe	−19.7(pe)	64
$(H_3)_2;C'_2$	$C_8H_6F_5P$	$Me_2PC_6F_5$	−47.8(a)	3, 105
	$C_8H_{10}ClP$	$Me_2PPh'Cl$	-49.4^{ORTHO}(b); -37.8^{PARA}	106
	$C_8H_{11}P$	Me_2PPh	−46.9, −47.5, −46.2	3, 10
	$C_9H_{13}OP$	$Me_2PPh'OMe$-4	−39.1	107
	$C_9H_{13}P$	Me_2PTol-4	−38.6	107
	$C_{10}H_{16}NP$	$Me_2PPh'(NMe2)$-2	−52.2(b)	106

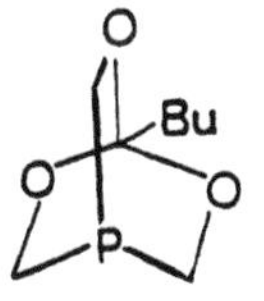

43 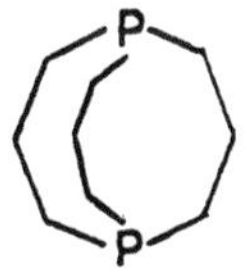 44 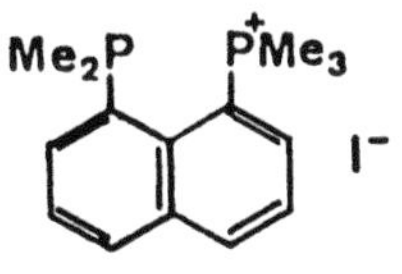 45

Rings/ Connectivities	Formula	Structure	^{31}P NMR data (δP [solv, temp] J_{PH} Hz)	Ref.
	$C_{11}H_{19}OPSi$	$Me_2PPh'OTms$-2	−54.0, −54.3	108, 109
	$C_{15}H_{21}IP_2$	**45**	−50.6(c)	110
$H_3,CH_2;C'_2$	$C_9H_{12}ClP$	$Me(ClCH_2CH_2)PPh$	−37.9(c)	131
	$C_9H_{13}P$	MeEtPPh	−32.8(j)	94
	$C_{10}H_{15}P$	MePrPPh	−37.9(j)	94
	$C_{10}H_{19}ClP$	$Me(ClCH_2CH_2CH_2)PPh$	−36.3(c)	131
$H_3,C'H_2;C'_2$	$C_{10}H_{13}P$	$Me(CH_2{=}CHCH_2)PPh$	−18.2	3
	$C_{14}H_{14}BrP$	$Me(2\text{-}BrPh'CH_2)PPh$	−29.8(b)	190
	$C_{14}H_{15}P$	MeBzPPh	−30.9(j)	94
$H_3,C_2H;C'_2$	$C_{13}H_{19}P$	Me(cHex)PPh	−25.4(j)	94
	$C_{17}H_{27}P$	Me(2-Menthyl)PPh	$-34.7^R; -31.9^S$(b)	191
		Me(2-neoMenthyl)PPh	$-38.9^R; -36.4^S$(b)	191
$H_3,CC'H;C'_2$	$C_{24}H_{23}OP$	$Me(PhCH{=}CHCOCH_2$-CHPh)PPh	−17	3
$H_3,C_3;C'_2$	$C_{11}H_{17}P$	Me(tBu)PPh	−11.2	188
	$C_{14}H_{25}OPSi$	Me(tBu)PPh'OTms-2	−27.3	109
$(CHO)_2;C'_2$	$C_{10}H_9Cl_6O_2P$	$(CCl_3CHOH)_2PPh$	$-10.0;0.0^I$(b)	54
	$C_{10}H_{15}O_2P$	$(CH_3CHOH)_2PPh$	$-12.0;2.0^I$(b)	54
$(C'HO)_2;C'_2$	$C_{20}H_{17}N_2O_6P$	$(4\text{-}NO_2Ph'CHOH)_2PPh$	$-14.0;2.0^I$(b)	54
	$C_{20}H_{19}O_2P$	$(PhCHOH)_2PPh$	$-6.0;10.0^I$(b)	54
$(CH_2)_2;C'_2$	$C_{10}H_{10}F_5P$	$Et_2PC_6F_5$	−23.4	3
	$C_{10}H_{14}BrP$	$Et_2PPh'Br$-4	−16.0	111
	$C_{10}H_{14}ClP$	$Et_2PPh'Cl$-4	−15.9	111
	$C_{10}H_{14}FP$	$Et_2PPh'F$-4	−16.1	111
	$C_{10}H_{15}P$	Et_2PPh	−15.1, −16.0, −17.1	3
	$C_{10}H_{17}N_2P$	$(H_2NCH_2CH_2)_2PPh$	−25.6(b); (HCl) −35.8	3, 33
	$C_{11}H_{17}OP$	$Et_2PPh'OMe$-4	−17.6	111
	$C_{12}H_{13}N_2P$	$(NCCH_2CH_2)_2PPH$	−23.8	55
	$C_{12}H_{19}P$	Pr_2PPh	−27.7, −27.3, −25.4(c)	3, 95, 112
	$C_{12}H_{20}NP$	$Et_2PPh'(NMe_2)$-4	−19.1	111
		$Et_2PPh'(NHEt)$-4	−17.79(c)	113
	$C_{13}H_{21}P$	Et_2PMes	−21.9	84
	$C_{13}H_{23}OPSi$	$Et_2PDopOTms$	−25.3; −26.4(c)	109, 111
	$C_{14}H_{19}P$	$(CH_2{=}CHCH_2CH_2)_2PPh$	−25.4	3
	$C_{14}H_{23}P$	Bu_2PPh	−25.8, −26.2	3
		iBu_2PPh	−34.2	3
	$C_{16}H_{23}O_2P$	$(O^*CH_2CH_2CH_2$-$C^*HCH_2)_2PPh$	−33.4, −34.4, −35.7(c)	115
	$C_{16}H_{23}P$	$(CH_2{=}CHCH_2$-$CH_2CH_2)_2PPh$	−26.3	57
	$C_{16}H_{27}P$	$(tBuCH_2)_2PPh$	−42.0	116
	$C_{20}H_{37}O_4P_3$	(Me(iPrO)(O)P-$(CH_2)_3)_2PPh$	−26.9(r)	27
	$C_{22}H_{25}N_2PO_2$	$(H_2NDopSCH_2CH_2)_2PPh$	−27.3(b)	33
	$C_{23}H_{45}O_6P_3$	$[(iPrO)_2(O)P(CH_2)_3]_2PPh$	−27.8(r)	27

TABLE E (continued)
^{31}P NMR Data of Three Coordinate (λ3σ3) Compounds Containing Phosphorus Bonds to Hydrogen and/or Group IV Atoms but Not to Halogen, Chalcogenide or Group V Atoms

Rings/ Connectivities	Formula	Structure	^{31}P NMR data (δP [solv, temp] J_{PH} Hz)	Ref.
	$C_{30}H_{41}O_4P_3$	[Ph(iPrO)(O)P-(CH$_2$)$_3$]$_2$PPh	−27.4(r)	27
	$C_{34}H_{33}As_2P$	(Ph$_2$AsCH$_2$CH$_2$)$_2$PPh	−16.6	117
CH$_2$,C$_2$H;C'$_2$	$C_{14}H_{21}P$	Et(cHex)PPh	−8.1(j)	94
	$C_{15}H_{23}P$	Pr(cHex)PPh	−13.5(j)	94
(C'H$_2$)$_2$;C'$_2$	$C_{12}H_{15}P$	(MeCH=CH)$_2$PPh	−57.9^{ZZ}; −21.4^{EE}; −39.8^{EZ}	118
	$C_{12}H_{15}P$	(CH$_2$=CHCH$_2$)$_2$PPh	−27.5	3
	$C_{16}H_{15}O_2P$	(2-FurCH$_2$)$_2$PPh	−22.7(c)	60
	$C_{18}H_{17}N_2P$	(2-PyrCH$_2$)$_2$PPh	−13.7(c)	119
	$C_{19}H_{21}O_2P$	(2-FurCH$_2$)$_2$PMes	−27.9(c)	60
	$C_{20}H_{19}P$	Bz$_2$PPh	−12.1	3
	$C_{16}H_{17}P$	CH$_2$=CHCH$_2$(Bz)PPh	−20.7	3
C'H$_2$,C$_2$H;CO'	$C_{12}H_{17}OP$	iPrBzPCOMe	27.9(pe)	64
	$C_{13}H_{15}F_4OP$	iPrBzPCOCHFTf	34.1(pe)	64
C'H$_2$,C$_3$;OO'	$C_{13}H_{19}O_2P$	tBuBzPCO$_2$Me	19.5(pe)	64
	$C_{16}H_{23}O_4P$	tBuBzPCO$_2$-CHMeCO$_2$Me	19.1(pe)	64
C'H$_2$,C$_2$;CO'	$C_{13}H_{16}F_3OP$	tBuBzPCOTf	36.1(pe)	64
(C"H$_2$)$_2$;C'$_2$	$C_{10}H_9N_2P$	(NCCH$_2$)$_{2P}$Ph	−28.9	61
(C$_2$H)$_2$;NO'	$C_9H_{20}NOP$	iPr$_2$PCONMe$_2$	8.0(pe)	64
	$C_{13}H_{20}NOP$	iPr$_2$PCONHPh	21.5(pe)	64
(C$_2$H)$_2$;NS'	$C_{19}H_{28}NPS$	cHex$_2$PCSNHPh	44.3(r)	120
(C$_2$H)$_2$;OO'	$C_8H_{17}O_2P$	iPr$_2$PCO$_2$Me	16.9(pe)	64
(C$_2$H)$_2$;CO'	$C_8H_{14}F_3OP$	iPr$_2$PCOTf	41.2(r)	64
	$C_8H_{16}FOP$	iPr$_2$PCOCH$_2$F	32(q)	64
	$C_8H_{17}OP$	iPr$_2$PCOMe	38.7(pe)	64
	$C_9H_{15}F_4OP$	iPr$_2$PCOCHFTf	40.9(q)	64
	$C_{10}H_{15}F_6OP$	iPr$_2$PCOCHTf$_2$	47.1(r)	64
	$C_{11}H_{23}OP$	iPr$_2$PCO-tBu	15.1(pe)	64
	$C_{13}H_{18}BrOP$	iPr$_2$PCOPh'Br-4	26.1	64
	$C_{14}H_{22}Cl_3OP$	cHex$_2$PCOCCl$_3$	20.5(c)	121
	$C_{14}H_{23}CL_2OP$	cHex$_2$PCOCHCl$_2$	26.4(c)	121
	$C_{14}H_{24}ClOP$	cHex$_2$PCOCH$_2$Cl	29.1(c)	121
(C$_2$H);C'O'	$C_{13}H_{19}OP$	iPr$_2$PCOPh	26.5(pe)	64
(C$_2$H)$_2$;C'$_2$	$C_{12}H_{19}P$	iPr$_2$PPh	6.8(c),9.3	3, 112
	$C_{14}H_{23}P$	2-Bu$_2$PPh	1.8	3
	$C_{15}H_{27}OPSi$	iPr$_2$PC$_6$H$_4$OTms-2	−2.7	108
	$C_{16}H_{23}P$	cPe$_2$PPh	1.6	3
	$C_{18}H_{27}P$	cHex$_2$PPh	2.5	3
C$_2$H,C$_3$;CO'	$C_{12}H_{25}OP$	iPr-tBuPCO-tBu	21.5(pe)	64
(C$_3$)$_2$;CO'	$C_{10}H_{18}F_3OP$	tBu$_2$PCOTf	60.8(pe)	64
	$C_{13}H_{27}OP$	tBu$_2$PCO-tBu	31.7(pe)	64
(C$_3$)$_2$;C'$_2$	$C_{14}H_{23}P$	tBu$_2$PPh	40.2(k),38.0	29, 95
	$C_{15}H_{25}P$	tBu$_2$PTol-4	37.02	95

Rings/ Connectivities	Formula	Structure	^{31}P NMR data (δP [solv, temp] J_{PH} Hz)	Ref.

E3e: Tertiary Dialkyl Alkenyl/Aryl Polyphosphines

P-bound atoms = C_2C'

Rings/ Connectivities	Formula	Structure	^{31}P NMR data (δP [solv, temp] J_{PH} Hz)	Ref.
$H_2P,C'H_2;C'_2$	$C_{19}H_{22}P_2$	$(CH_2{=}CHCH_2)PhPCH_2$-$PPh(CH_2CH{=}CH_2$	−31.6(c); −32.2$^{dl+MESO}$	46
$(H_3)_2;P^4P$	$C_8H_{20}LiP_3$	$(Me_2P)_2C{=}PMe_2CH_2Li$	−48.8(b)	103
	$C_8H_{21}P_3$	$(Me_2P)_2C{=}PMe_3$	−43.2(b)	103
	$C_9H_{24}P_4$	Me_2**P**$C({=}PMe_2$-$PMe_2)$**P**Me_2	−34.0(b)	103
$(H_3)_2;C'_2$	$C_{10}H_{16}P_2$	$(Me_2P)_2Dop$	−55.3(c)	23
	$C_{10}H_{20}P_2$	2,2′-$(Me_2P)_2$Diphenyl	−55.9(c)	245
	$C_{14}H_{18}P_2$	1,8-$(Me_2P)_2$Naphthalene	−54.5(c)	110
$H_3,CH_2;C'_2$	$C_{15}H_{17}P_2$	MePh**P**$CH_2CH_2PPh_2$	−31.7	3
	$C_{23}H_{26}P_2$	MePh**P**$(CH_2)_4PPh_2$	−38.2(c)	240
	$C_{25}H_{30}P_2$	MePh**P**$(CH_2)_6PPh_2$	−32.1(c)	240
$H_3,C_3;C'_2$	$C_{16}H_{30}P_2$	$(tBuMeP)_2Dop$	−23.3, −25.8(c)	24
	$C_{18}H_{24}P_2$	tBuMe**P**DopPMePh	−24.3, −24.0(c)	24
$(CH_2)_2,CC'$	$C_{21}H_{19}F_3N_2P_2$	$Ph_2PCH{=}CTf$**P**-$(CH_2CH_2CN)_2$	−32.9^C(c) J_{PP}131	173
$(CH_2)_2;C'_2$	$C_{10}H_{17}P_3$	$(H_2PCH_2CH_2)_2$**P**Ph	−20.7	12
	$C_{10}H_{25}P_3$	$(Me_2PCH_2CH_2)_2$**P**Ph	−18.1(b)	249
	$C_{12}H_{21}P_3$	$(H_2P(CH_2)_3)_2$**P**Ph	−26.5(r)	27
	$C_{14}H_{25}P_3$	$[MeHP(CH_2)_3]_2$**P**Ph	−24.8(r)	27
	$C_{15}H_{27}NP_2$	Et_2**P**Dop$(NMePEt_2)$	−32.26	114
	$C_{16}H_{29}P_3$	$[Me_2P(CH_2)_3]_2$**P**Ph	−26.5(r)	27
	$C_{18}H_{26}P_4$	YPPh$(CH_2)_2$**P**PhY; Y = $(CH_2)_2PH_2$	−19.2	12
	$C_{18}H_{33}P_3$	$(Et_2PCH_2CH_2)_2$**P**Ph	−19.6	250
	$C_{20}H_{30}N_2P_2$	YPhP$(CH_2)_4$PPhY; Y = $CH_2Ch_2NH_2$	−29.0	3
	$C_{21}H_{30}O_2P_2$	YPhP$(CH_2)_3$PPhY; Y = $(CH_2)_3OH$	−25.3; −25.6^I	26
	$C_{21}H_{32}N_2P_2$	YPhP$(CH_2)_3$PPhY; Y = $(CH_2)3NH_2$	−25.9(b)	33
	$C_{21}H_{32}P_4$	YPh**P**$(CH_2)_3$**P**PhY; Y = $(CH_2)_3PH_2$	−28.5	26
	$C_{22}H_{24}P_2$	EtPh**P**$CH_2CH_2PPh_2$	−17.0, −15.9	3, 231
	$C_{23}H_{26}P_2$	PrPh**P**$CH_2CH_2PPh_2$	−21.7	3
		EtPh**P**$(CH_2)_3PPh_2$	−21.2(c)	240
	$C_{24}H_{28}P_2$	iBuPh**P**$CH_2CH_2PPh_2$	−25.4	3
		EtPh**P**$(CH_2)_4PPh_2$	−19.8(c)	240
	$C_{24}H_{29}NP$	$H_2N(CH_2)_3$Ph**P**-$(CH_2)_3PPh_2$	−27.2(b)	33
	$C_{24}H_{29}P_3$	$(PhHP(CH_2)_3)_2$**P**Ph	−28.1(r)	27
	$C_{25}H_{30}P_2$	cPePh**P**$CH_2CH_2PPh_2$	−21.3	3
	$C_{26}H_{32}P_2$	EtPh**P**$(CH_2)_6PPh_2$	−17.4	240
	$C_{26}H_{24}N_2P_2$	$Ph_2PCH{=}CPhP$-$(CH_2CH_2CN)_2$	−28.2^C(c)	173
	$C_{26}H_{33}NP_2$	$Me_2N(CH_2)_3$**P**-$Ph(CH_2)_3PPh_2$	−27.2(b)	33

TABLE E (continued)
^{31}P NMR Data of Three Coordinate (λ3σ3) Compounds Containing Phosphorus Bonds to Hydrogen and/or Group IV Atoms but Not to Halogen, Chalcogenide or Group V Atoms

Rings/ Connectivities	Formula	Structure	^{31}P NMR data (δP [solv, temp] J_{PH} Hz)	Ref.
	$C_{30}H_{32}OP_2$	$PhO(CH_2)_3PhP$-$(CH_2)_3PPh_2$	−27.5(b)	33
	$C_{33}H_{40}P_4$	$YPhP(CH_2)_3PPhY$; Y = $(CH_2)_3PHPh$	−27.4	26
	$C_{33}H_{56}O_6P_4$	$YPhP(CH_2)_3PPhY$; Y = $(CH_2)_3P(O)(O$-iPr).	−27.2	26
	$C_{34}H_{33}P_3$	$Ph\mathbf{P}(CH_2CH_2PPh_2)_2$	−16.6	250
	$C_{36}H_{61}P_3$	$Ph\mathbf{P}(CH_2CH_2CH_2P$-$cHex_2)_2$	−28.1(b)	33
	$C_{36}H_{37}P_3$	$Ph\mathbf{P}[(CH_2)_3PPh_2]_2$	−28.2(b)	33
	$C_{39}H_{52}O_4P$	$YPhP(CH_2)_3PPhY$; Y = $(CH_2)_3P(O)(OPr)Ph$	−27.9	26
	$C_{42}H_{42}P_4$	$YPh\mathbf{P}(CH_2)_2\mathbf{P}PhY$; Y = $(CH_2)_2PPh_2$ Tetraphos I	−18.1(r)	253
	$C_{46}H_{42}P_4$	$Ph_2P(CH_2)_2\mathbf{P}Ph$-$CH_2CH(PPh_2)_2$	−23.8(r)	35
	$C_{50}H_{51}P_5$	$YPh\mathbf{P}CH_2CH_2\mathbf{P}Y_2$; Y = $CH_2CH_2PPh_2$	−18.0(broad)	255
		$Ph\mathbf{P}(CH_2CH_2\mathbf{P}Ph$-$CH_2CH_2PPh_2)_2$	−18.2	255
	$C_{58}H_{51}P_5$	$PhP[Ch_2CH(PPh_2)_2]_2$	−29.7	35
	$C_{66}H_{69}P_7$	$PhP[CH_2CH_2P$-$(CH_2CH_2PPh_2)_2]_2$	−18.1	255
	$C_{74}H_{78}P_8$	$[(Ph_2PCH_2CH_2)_2PCH_2$-$CH_2\mathbf{P}PhCH_2]_2$	−18.1(broad)	255
$CH_2,C_2H;C'_2$	$C_{23}H_{26}P_2$	$iPrPh\mathbf{P}CH_2CH_2PPh_2$	−4.1, −2.6	3, 231
	$C_{24}H_{28}P_2$	2-$BuPh\mathbf{P}CH_2CH_2PPh_2$	−7.1	3
$CHS,C_2H;C'_2$	$C_{28}H_{40}P_2$	$cHexPhP(CH_2)_4PPh$-$cHex$	−13.2(j)	94
$CH_2,C_2H;C'_2$	$C_{33}O_{38}P_2$	$cHexPh\mathbf{P}(CH_2)_6PPh_2$	−7.6	240
$(C_2H)_2;CO'$	$C_{14}H_{22}Cl_3OP$	$(cHex)_2PCOCCl_3$	20.5	284
	$C_{14}H_{23}CL_2OP$	$(cHex)_2PCOCHCl_2$	26.4	284
	$C_{14}H_{24}ClOP$	$(cHex)_2PCOCH_2Cl$	29.1	284
$(C_2H)_2;CC'$	$C_{29}H_{44}F_6P_2$	**46**	−17.0	3
$(C_2H)_2;C'_2$	$C_{15}H_{26}P_2$	$iPr_2\mathbf{P}DopPH$-iPr	−1.7	23
$(C_3)_2;C'_2$	$C_{22}H_{40}P_2$	$(tBu_2P)_2Dop$	6(c)	24

E3f: Tertiary Cyclic Dialkyl Alkenyl/Aryl Phosphines

<u>**P-bound atoms = C_2C'**</u>

Rings/ Connectivities	Formula	Structure	^{31}P NMR data	Ref.
$3/(CH_2)_2;C'_2$	C_8H_9P	$PhP^*CH_2C^*H_2$	−234.0	3
$3/CH_2,CC'H;C'_2$	$C_{10}H_{11}P$	$PhP^*CH(CH{=}CH_2)C^*H_2$	−204.5	194
$3/(CC'H)_2;C'_2$	$C_{14}H_{13}P$	**18**; R = Ph	−181.0	3
$4/(HPN)_2;C'_2$	$C_{14}H_{18}Cl_2N_2O_2P_2$	**47**; R = NH_3^+	−51.2(y)	199
	$C_{18}H_{26}Cl_2N_2O_2P_2$	**47**; R = $NHMe_2^+$	−10.7(y)	199

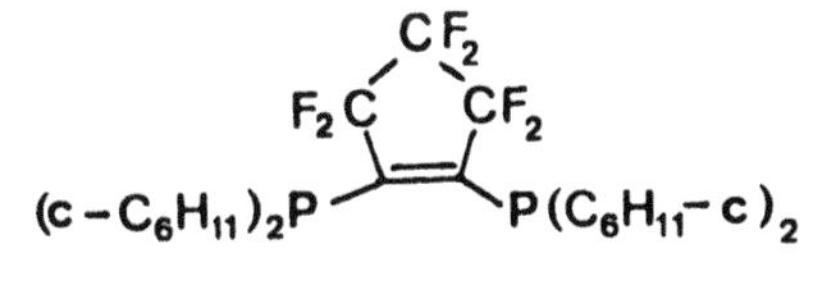

46

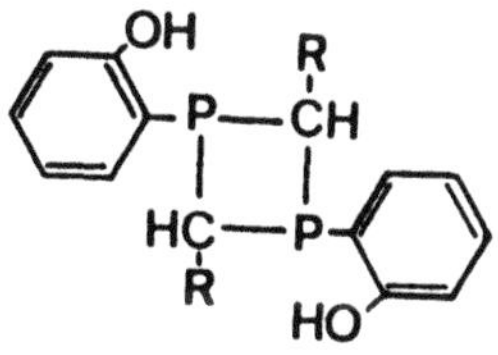

47

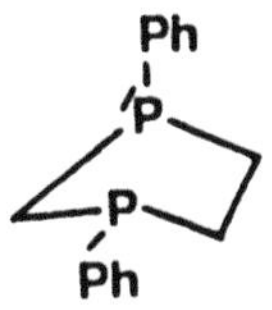

48

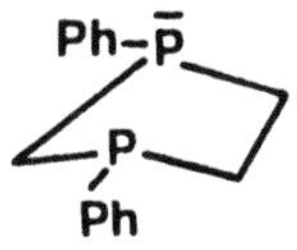

49

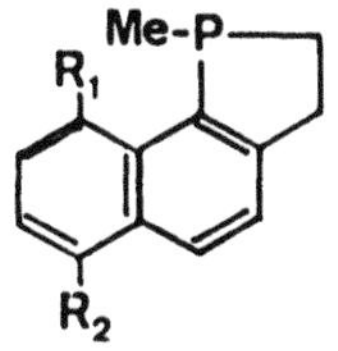

50

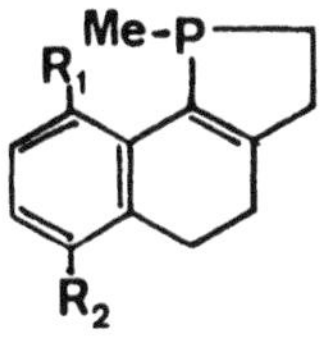

51

Rings/ Connectivities	Formula	Structure	^{31}P NMR data (δP [solv, temp] J_{PH} Hz)	Ref.
$4/(C_3)_2;C'_2$	$C_{13}H_{19}P$	$PhP^*CMe_2CH_2C^*Me_2$	49.1(b)	200
$5/H_2P,CH_2;C'_2$	$C_{15}H_{16}P_2$	**48** and **49**	−0.7; −1.5^{I}	4, 46
$5/H_3,CH_2;C'H$	$C_6H_{11}P$	$MeP^*CHCMeCH_2C^*H_2$	−15.2	202
$5/H_3,CH_2;C'_2$	$C_{13}H_{13}P$	**50**; $R^1 = R^2 = H$	−23.3(c)	205
	$C_{13}H_{15}P$	**51**; $R^1 = R^2 = H$	−17.3(c)	205
	$C_{15}H_{17}P$	**50**; $R^1 = R^2 = Me$	−10.3(c)	205
	$C_{15}H_{19}P$	**51**; $R^1 = R^2 = Me$	−6.6(c)	205
$5/H_3,C_2H;C'H$	$C_{10}H_{14}P_2$	**30**; $R^1 = Me$, $R^2 = H$ P^2	7.9	220
$5/H_3,C_2H;C'H$	$C_{10}H_{14}P_2$	**31**; $R^1 = Me$, $R^2 = R^3 = H$ P^1	2.3	220, 221
	$C_{12}H_{18}P_2$	**30**; $R^1 = R^2 = Me$ P^2	−9.6	220
		31; $R^1 = R^2 = Me$, $R^3 = H$ P^1	−2.6, −1.9	220, 221
	$C_{13}H_{20}P_2$	**31**; $R^1 = R^2 = R^3 = Me$ P^1	−14.4	220, 221
$5/H_3,C_2H;CC'$	$C_7H_{12}P$	$MeP^*CMeCHC^*H_2$-CHMe	−8.2^{C};4.7^{T}	204
$5/H_3,CC'H;C'H$	$C_9H_{11}P$	**52**; $R^1 = R^2 = H$, $R^3 = Me$	11.7	203
	$C_{11}H_{15}P$	**52**; $R^1 = R^2 = R^3 = Me$	11.9	203
		53	6.7	203
$5/C'HN,CH_2;C'_2$	$C_{15}H_{16}NP$	$PhP^*CHPhNHCH_2C^*H_2$	−2.9	3
$5/C'HN,CC'H;C'_2$	$C_{17}H_{18}NOP$	PhP*CHPhNMe-COC*HMe	−11.7^{CC}; −12.0^{CT}; −9.5^{TT}	40
	$C_{18}H_{20}NOP$	PhP*CHPhNEt-COC*HMe	−12.6^{CT}; −13.2^{TC}; −10.0^{TT}	40
$5/CHO,C'H_2;C'_2$	$C_{12}H_{15}O_2P$	$PhP^*CHPrOCOC^*H_2$	−23.5^{C}; −27.0^{T}	40
	$C_{13}H_{17}O_2P$	$PhP^*CH(tBu)OCOC^*H_2$	−20.2^{CT}; −20.7^{TC}; −24.5^{TT}	40
$5/CHO,CC'H;C'_2$	$C_{12}H_{15}O_2P$	PhP*CHEtOCOC*HMe	−7.0^{CC}; −9.5^{CT}; −13.5^{TC}; −15.5^{TT}	40

TABLE E (continued)
^{31}P NMR Data of Three Coordinate (λ3σ3) Compounds Containing Phosphorus Bonds to Hydrogen and/or Group IV Atoms but Not to Halogen, Chalcogenide or Group V Atoms

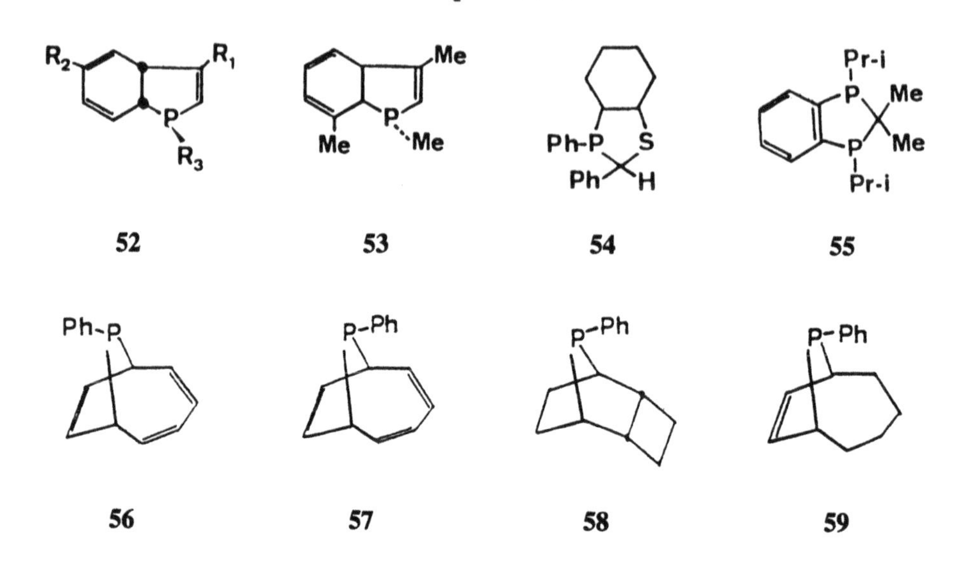

Rings/ Connectivities	Formula	Structure	^{31}P NMR data (δP [solv, temp] J_{PH} Hz)	Ref.
5/C'HO,CH$_2$;C'$_2$	$C_{15}H_{15}OP$	PhP*CHPhOCH$_2$C*H$_2$	-7.5^{C}; -4.0^{T}	40
	$C_{16}H_{17}OP$	PhP*CHPhOCHMeC*H$_2$	-1.0^{CC}; -0.0^{CT}; 2.5^{TC}; 2.0^{TT}	40
5/C'HS,CH$_2$;C'$_2$	$C_{16}H_{17}PS$	PhP*CHPhSCHMeC*H$_2$	27.6^{CC}; 21.5^{CT}; 35.1^{TC}; 24.4^{TT}	40
5/C'HS,C$_2$H;C'$_2$	$C_{19}H_{21}PS$	**54**	19.4^{C}; 30.6^{T}	40
5/CH$_2$,C$_2$N;C'$_2$	$C_{13}H_{20}NP$	PhP*CEt$_2$NHCH$_2$C*H$_2$	0.0	3
	$C_{14}H_{20}NP$	PhP*C(cPe)NHCH$_2$C*H$_2$	3.5	3
5/(C'H$_2$)$_2$;C'$_2$	$C_{10}H_{11}P$	PhP*CH$_2$CH=CHC*H$_2$	−25.3	202
	$C_{11}H_{13}P$	PhP*CH$_2$CMe=CHC*H$_2$	−18.6	202
	$C_{12}H_{15}P$	PhP*CH$_2$CMe=CMe-C*H$_2$	−34.0, −34.5	3, 202
5/C'H$_2$,CC'H;C'$_2$	$C_{11}H_{13}P$	PhP*CHMeCH=CHC*H$_2$	-9.2^{T}	202
5/C$_2$P,C$_2$H;C'$_2$	$C_{15}H_{24}P_2$	**55**	58.1;46.4^{1}(t,b)	48
5/(C$_2$H)$_2$,C'$_2$	$C_{14}H_{13}P$	**56**	−14.0(c)	3, 220
	$C_{14}H_{13}P$	**57**	−79.0(c)	3, 220
	$C_{14}H_{17}P$	**58**	22.0(c)	220
		59	−8.1(c)	220
5/(CC'H)$_2$;C'$_2$	$C_{14}H_{13}P$	**60**	147.0(c)	195, 220
		61	98.8(c)	195, 220
	$C_{14}H_{18}P_2$	**31**; R^1 = Ph, R^2 = R^3 = H P^1	114.2	220, 221
	$C_{15}H_{20}P_2$	**31**; R^1 = Ph, R^2 = Me, R^3 = H P^1	119.8	220, 221
	$C_{20}H_{18}P_2$	**30**; R^1 = Ph, R^2 = H P^1	44.9	220
	$C_{21}H_{19}P$	**32**; R = Ph	133.7	222
	$C_{22}H_{22}P_2$	**30**; R^1 = Ph, R^2 = Me P^1	48.5	220

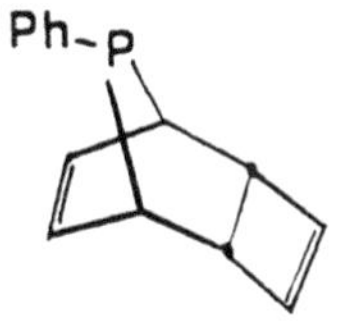

60 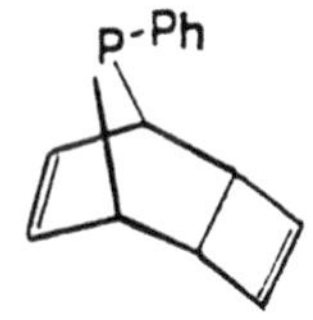 61 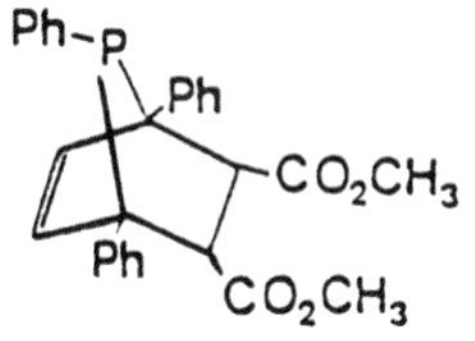 62

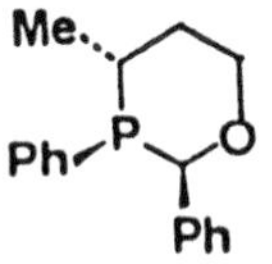

63 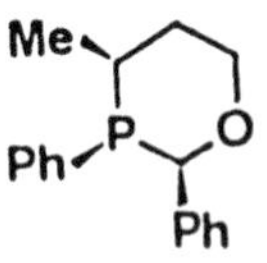 64 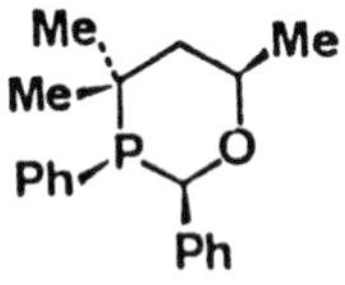 65 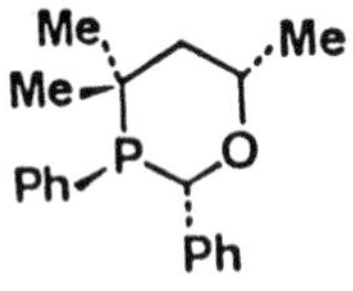 66

Rings/ Connectivities	Formula	Structure	^{31}P NMR data (δP [solv, temp] J_{PH} Hz)	Ref.
5/$(CC'_2)_2$;C'_2	$C_{28}H_{25}P$	**62**	137.0(c)	220
6/H_2P,CH.,C'_2	$C_{16}H_{18}P_2$	PhP*CH_2PPhCH_2 CH_2C*H_2	-34.5^{C}; -33.3^{T}; -23.7, $-45.9^{C}(-92^{O})^{RR/SS}$	46
6/CHN.$C'H_2$;C'_2	$C_{15}H_{22}NP$	PhP*CH_2CMe=CMe CH_2C*$HNMe_2$	-46	216
6/$(CHO)_2$;C'_2	$C_{16}H_{12}BClO_2P$	**33**; R^1 = Ph, R^2 = CCl_3	-32; -19^{I}	54
	$C_{20}H_{14}B_2Cl_{12}O_{52}P_2$	**34**; R^1 = Ph, R^2 = CCl_3	-32(b)	54
6/CHO,CH_2;C'_2	$C_{14}H_{21}OP$	PhP*CH-tBu- OCH_2CH_2C*H_2	-9.1^{C}; -22.4^{T}	214
6/$(C'HO)_2$;C'_2	$C_{26}H_{20}BN_2O_6P$	**33**; R^1 = Ph, R^2 = Ph'NO_2-4	-12.0; -24.0^{I}	54
	$C_{40}H_{28}B_2N_4O_5P_2$	**34**; R^1 = Ph, R^2 = Ph'NO_2-4	-12.0(b)	54
6/C'HO,CH_2;C'_2	$C_{16}H_{17}OP$	PhP*CHPhOCH_2 CH_2C*H_2	-22.6^{C}; -33.7^{T}	214
6/C'HO,C_2H;C'_2	$C_{17}H_{19}OP$	**53**	-11.7	214
		64	-11.8	214
6/C'HO,C_3;C'_2	$C_{19}H_{24}OP$	**65**	-8.8	214
		66	-8.1	214
6/$(CC'H)_2$;C'_2	$C_{27}H_{21}OP$	PhP*CHPhCH_2COCH_2- C*HPh	-6.0	215
6/$(C_3)_2$;C'_2	$C_{15}H_{21}OP$	PhP*CMe_2CH_2COCH_2- C*Me	-16.1	215
7/H_2P,CH_2;C'_2	$C_{17}H_{20}P_2$	PhP*CH_2PPhCH_2CH_2 CH_2C*H_2	-27.6	46, 47
10/$(CH_2)_2$;C'_2	$C_{20}H_{26}P_2$	PhP*$(CH_2)_4$PPH- $(CH_2)_3$C*H_2	-16.0, -19.1	218
5,7/$(CH_2)_2$;C'_2	$C_{10}H_{12}P_2$	**67**	-75.6	269
7,7/(CH_2);NO'	$C_{18}H_{22}NOPSi$	**68**; R = Ph, X = O, Y = NPh	-21.3(k)	43
	$C_{13}H_{20}NOPSi$	**68**; R = Me, X = O, Y = NPh	-20.8(k)	43
7,7/(CH_2);N'O	$C_{13}H_{20}$:NOPSi	**68**; R = Me, X = NPh, Y = O	-4.3(k)	43

TABLE E (continued)
^{31}P NMR Data of Three Coordinate (λ3σ3) Compounds Containing Phosphorus Bonds to Hydrogen and/or Group IV Atoms but Not to Halogen, Chalcogenide or Group V Atoms

Rings/ Connectivities	Formula	Structure	^{31}P NMR data (δP [solv, temp] J_{PH} Hz)	Ref.
	$C_{18}H_{22}NOPSi$	**68**; R = Ph, X = NPh, Y = O	−4.9(k)	43
7,7/(CH_2);NS′	$C_{13}H_{20}NPSSi$	**68**; R = Me, X = S, Y = NPh	11.5(k)	43
	$C_{18}H_{22}NPSSi$	**68**; R = Ph, X = S, Y = NPh	11.8(k)	43

E3g: Tertiary Dialkyl Alkynylphosphines

P-bound atoms = C_2C''

Rings/ Connectivities	Formula	Structure	^{31}P NMR data	Ref.
$(F_3)_2$;N″	C_3F_6NP	Tf_2PCN	−40.7	186
$(H_3)_2$;N″	C_3H_6NP	Me_2PCN	−63.0(pe)	64
$(C_2H)_2$;N″	$C_7H_{14}NP$	iPr_2PCN	−7.5(pe)	64
$(CH_2)_2$;C″	$C_{10}H_{19}P$	$iBu_2PC{\equiv}CH$	−56.0	87
	$C_{12}H_{23}P$	$nBu_2PC{\equiv}CEt$	−50.6	87
	$C_{16}H_{23}P$	$nBu_2PC{\equiv}CPh$	−49.8	87
$(C_2H)_2$;C″	$C_9H_{17}P$	$iPr_2PC{\equiv}CMe$	−13.2	87
	$C_{12}H_{23}P$	$iPr_2PC{\equiv}CBu$	−13.9	87
$(C_3)_2$;C″	$C_{18}H_{36}P_2$	$(tBu)_2PC{\equiv}CP(tBu)_2$	14.5	87

E3h: Tertiary Alkyl Dialkenyl/Aryl Phosphines

P-bound atoms = CC'_2

Rings/ Connectivities	Formula	Structure	^{31}P NMR data	Ref.
HPSi;$(C'_2)_2$	$C_{25}H_{38}ClNP_2Si_3$	$Ph_2PCHTmsPCl(NTms_2)$	−5.9	101
HP^4_2;$(C'_2)_2$	$C_{27}H_{27}P_3S_2$	$Ph_2PCH(P(S)Ph_2)$-$(P(S)Me_2$	−7.8(c,r)	86
	$C_{37}H_{31}O_2P_3$	$Ph_2PCH[P(O)Ph_2]_2$	−15.2(r,c)	128
	$C_{37}H_{31}P_3SSe$	$Ph_2PCH(P(S)Ph_2)$-$(P(Se)Ph_2$	−8.7(r,c)	127
	$C_{37}H_{31}P_3S_2$	$Ph_2PCH(P(S)Ph_2)_2$	−9.7(r,c), −10.0(r,c)	127, 128
	$C_{41}H_{31}O_4P_3Cr$	$Ph_2PC^*HPPh_2Cr$-$(CO)_4P^*Ph_2$	−25.0(r,c)	128
	$C_{41}H_{31}O_4P_3W$	$Ph_2PC^*HPPh_2W$ $(CO)_4P^*Ph_2$	−23.0(r,c)	128
H_2N;$(C'_2)_2$	$C_{15}H_{16}NO_2P$	$Ph_2PCH_2NHCH_2COOH$	−10.4	129
	$C_{16}H_{18}NO_2P$	$Ph_2PCH_2NHCHMe$-COOH(L)	−21.6(y)	129
	$C_{16}H_{18}NO_3P$	Ph_2PCH_2NHCH-$(CH_2OH)COOH$(L)	−21.5(y)	129
	$C_{17}H_{22}NP$	$Ph_2PCH_2NEt_2$	−27.8, −27.3	3
	$C_{18}H_{18}NO_2P$	Ph_2PCH_2NHCH-(iPr)COOH(L)	−21.3(y)	129
	$C_{18}H_{25}N_2P$	$Ph_2PCH_2NMeCH_2$-CH_2NMe_2	−26.6	130
	$C_{22}H_{22}NO_2P$	$Ph_2PCH_2NHCHBzCO_2H$ (L)	−19.8(y)	129

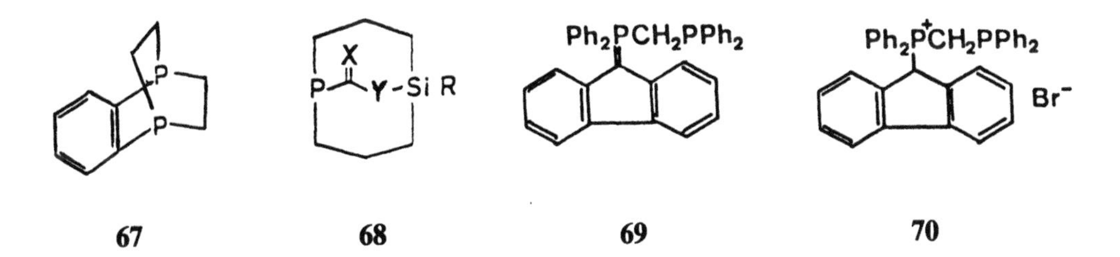

Rings/ Connectivities	Formula	Structure	^{31}P NMR data (δP [solv, temp] J_{PH} Hz)	Ref.
$H_2O;(C'_2)_2$	$C_{23}H_{31}OP$	$Ph_2PCH_2OMenthyl$	−18.5(c)	131
	$C_{25}H_{22}OP_2$	$Ph_2PCH_2OPPh_2$	−14.1	132
$H_2P;(C'_2)_2$	$C_{25}H_{22}OP_2$	$Ph_2PCH_2P(O)Ph_2$	−28.4(c,r)	133
	$C_{25}H_{22}P_2S$	$Ph_2PCH_2P(S)Ph_2$	−28.0(c,r)	133
	$C_{26}H_{25}IP_2$	$Ph_2PCH_2P^+Ph_2Me\ I^-$	−30.6(l+w); −26.5(c,r)	125, 133
	$C_{31}H_{27}NP_2$	$Ph_2PCH_2PPh_2$(=CHPyr-2)	−28.1(b)	134
	$C_{31}H_{28}ClNP_2$	$Ph_2PCH_2P^+Ph_2$-(CH_2Pyr-2) Cl^-	−31.1	134
	$C_{32}H_{29}BrP_2$	$Ph_2PCH_2P^+Ph_2Bz\ Br^-$	−31.8(l+w)	135
	$C_{34}H_{33}BrP_2$	$Ph_2PCH_2P^+Ph_2Mes\ Br^-$	−32.3(l+w)	134
	$C_{38}H_{30}P_2$	**69**	−29.4(b)	134
	$C_{38}H_{31}BrP_2$	**70**	−31.6(l+w)	134
$H_2S;(C'_2)_2$	$C_{14}H_{15}PS$	$MeSCH_2PPh_2$	−21.0	3
	$C_{19}H_{17}PS$	$PhSCH_2PPh_2$	−18.4(c)	25
$H_2Se;(C'_2)_2$	$C_{19}H_{17}PSe$	$PhSeCH_2PPh_2$	−16.5(c)	136
$H_2Si;(C'_2)_2$	$C_{15}H_{18}ClPSi$	$ClMe_2SiCH_2PPh_2$	−25.0(r)	92
	$C_{15}H_{19}PSi$	$Me_2SiHCH_2PPh_2$	−22.0(r)	92
	$C_{16}H_{21}PSi$	$TmsCH_2PPh_2$	−23.0(r), −30.7	92, 137
	$C_{19}H_{17}Cl_2PSi$	$Cl_2PhSiCH_2PPh_2$	−27.0(r)	92
	$C_{19}H_{19}PSi$	$PhSiH2CH_2PPh_2$	−20.0(r)	92
	$C_{20}H_{20}ClPSi$	$ClPhMeSiCH_2PPh_2$	−25.0(r)	92
	$C_{20}H_{21}PSi$	$PhMeSiHCH_2PPh_2$	−22.0(r)	92
	$C_{26}H_{22}ClPSi$	$ClPh_2SiCH_2PPh_2$	−26.0(r)	92
	$C_{25}H_{23}PSi$	$Pb_2SiHCH_2PPh_2$	−22.0(r)	92
	$C_{31}H_{27}PSi$	$Ph_3SiCH_2PPh_2$	−24.0(r)	92
$H_3;(C'N')_2$	$C_{11}H_{11}N_2P$	$MeP(Pyr\text{-}2)_2$	−19.9(c)	104
$H_3;(CO')_2$	$C_5H_9O_2P$	$MeP(COMe)_2$	31.0(pe)	64
$H_3;(C'O)_2$	$C_9H_9O_2P$	$MeP(Fur\text{-}2)_2$	−69.2(c)	115
$H_3;CO',C'_2$	$C_9H_8Cl_3OP$	$MeP(COCCl_3)Ph$	−5.5(c)	189
	$C_9H_9Cl_2OP$	$MePPh(COCHCl_2)$	−3.3(c)	189
	$C_9H_{10}ClOP$	$MePPh(COCH_2ClCO)$	−2.7(c)	189
$H_3;(C'_2)_2$	$C_{13}H_3F_{10}P$	$MeP(C_6F_5)_2$	−52.2	3
	$C_{13}H_{11}Cl_2P$	$MeP(Ph'Cl\text{-}4)_2$	−20.2	107
	$C_{13}H_{13}P$	$MePPh_2$	−28, −27.5(c)	3
	$C_{15}H_{17}P$	$MeP(Tol\text{-}4)_2$	−21.0	107
	$C_{17}H_{22}NP$	MePPh(Ph'CHMe-NMe_2-2)	−40.3; −40.9[l]	192
	$C_{28}H_{35}P$	MePMes(C_6H_2-2-Mes-4,6-Me_2)	−37.8(b)	193
$C'HO;(C'_2)_2$	$C_{19}H_{12}T_5OP$	$C_6T_5CH(OH)PPh_2$	1.6(t)	152
	$C_{19}H_{17}OP$	$PhCH(OH)PPh_2$	−3.6(t)	152
$CHP;(C'_2)_2$	$C_{26}H_{24}P_2S$	$Ph_2P(S)CHMePPh_2$	−13.0(c,r)	86
	$C_{30}H_{32}P_2S$	$Ph_2P(S)CH(neoPe)$**P**Ph_2	4.1(c,r)	86
	$C_{38}H_{33}P_3Se.$	$Ph_2P(Se)CH_2[P(Se)$-$Ph_2]CHPPh_2$	−3.9	138

TABLE E (continued)
^{31}P NMR Data of Three Coordinate (λ3σ3) Compounds Containing Phosphorus Bonds to Hydrogen and/or Group IV Atoms but Not to Halogen, Chalcogenide or Group V Atoms

Rings/ Connectivities	Formula	Structure	^{31}P NMR data (δP [solv, temp] J_{PH} Hz)	Ref.
CH_2;CO′,C′$_2$	$C_{10}H_{10}Cl_3OP$	EtP(COCCl$_3$)Ph	6.0(c)	189
	$C_{10}H_{11}Cl_2OP$	EtP(COCHCl$_2$)Ph	10.3(c)	189
	$C_{10}H_{12}ClOP$	EtP(COCH$_2$Cl)Ph	12(c)	189
	$C_{11}H_{12}Cl_3OP$	PrP(COCCl$_3$)Ph	1.3(c)	189
	$C_{11}H_{13}Cl_2OP$	EtP(COCHCl$_2$)Ph	6.1(c)	189
	$C_{11}H_{14}ClOP$	EtP(COCH$_2$Cl)Ph	7.5(c)	189
CH_2;(C′H)$_2$	$C_6H_{11}P$	EtP(CH=CH$_2$)$_2$	−20.8	3
CH_2;(C′$_2$)$_2$	$C_{14}H_5F_{10}P$	EtP(C$_6$F$_5$)$_2$	−44.0	3
	$C_{14}H_{14}Cl_3PSi$	SiCl$_3$CH$_2$CH$_2$PPh$_2$	−11.0	34
	$C_{14}H_{14}P(SiO_3)_x$	Silica gel-CH$_2$CH$_2$PPh$_2$	−10.5(CP/MAS)	70
	$C_{14}H_{15}P$	EtPPh$_2$	−12.0(c), −12.5, −13.5	1, 3
	$C_{14}H_{17}PSi$	SiH$_3$CH$_2$CH$_2$PPh$_2$	−12(r)	92
	$C_{15}H_{14}NP$	NCCH$_2$CH$_2$PPh$_2$	−17.6, −16.3(r)	55, 139
	$C_{15}H_{15}Cl_2P$	PrP(Ph′Cl-4)	−17.5	111
	$C_{15}H_{16}ClP$	ClCH$_2$CH$_2$CH$_2$PPh$_2$	−17.9(b)	33
	$C_{15}H_{16}Cl_3PSi$	SiCl$_3$(CH$_2$)$_2$PPh$_2$	−18.0(r)	92
	$C_{15}H_{17}P$	PrPPh$_2$	−17.6, −16.4	3, 111
	$C_{15}H_{19}PSi$	MeSiH$_2$CH$_2$CH$_2$PPh$_2$	−11.0(r)	92
	$C_{15}H_{19}PSi$	SiH$_3$(CH$_2$)$_2$PPh$_2$	−18.0(r)	92
	$C_{16}H_{16}NP$	NCCHMeCH$_2$PPh$_2$	−18.0(r)	139
	$C_{16}H_{17}P$	CH$_2$=CHCH$_2$CH$_2$PPh$_2$	−16.5	57
	$C_{16}H_{18}ClP$	Cl(CH$_2$)$_4$PPh$_2$	−17.1(c)	140
	$C_{16}H_{18}PSi$	Me$_2$SiHCH$_2$CH$_2$PPh$_2$	−11.0(r)	92
	$C_{16}H_{19}P$	BuPPh$_2$	−17.1;17.3(r)	3, 141
	$C_{16}H_{19}P$	iBuPPh$_2$	−21	3
	$C_{16}H_{20}ClPSi$	Me$_2$SiClCH$_2$CH$_2$PPh$_2$	−10.0(r)	92
	$C_{17}H_{19}OP$	O*(CH$_2$)$_3$CH*CH$_2$PPh$_2$	−21.8(c)	115
	$C_{17}H_{19}P$	CH$_2$=CH(CH$_2$)$_3$PPh$_2$	−17.6	57
	$C_{17}H_{21}P$	neoPePPh$_2$	−23.9(c)	1, 3
	$C_{17}H_{22}ClPSi$	MeSiCl$_2$(CH$_2$)$_3$PPh$_2$	−17.0(r)	92
	$C_{17}H_{23}N_2O_3P$	Me$_3$N$^+$CH$_2$-CH$_2$PPh$_2$ NO$_3^-$	−21.3(y)	142
	$C_{17}H_{23}PSi$	TmsCH$_2$CH$_2$PPh$_2$	−18.6	137
	$C_{17}H_{23}PSi$	MeSiH$_2$CH$_2$CH$_2$CH2PPh$_2$	−17.0(r)	92
	$C_{18}H_{25}PSi$	TmsCH$_2$CH$_2$CH$_2$PPh$_2$	−12.0(r), −26.1	92, 137
	$C_{19}H_{19}OP$	O*CMe=CH-CH=C*CH$_2$CH$_2$PPh$_2$	−15.6(c)	60
	$C_{19}H_{27}PSi$	Tms(CH$_2$)$_4$PPh$_2$	−24.9	137
	$C_{20}H_{19}Cl_2PSi$	PhSiCl$_2$CH$_2$CH$_2$PPh$_2$	−11.0(r)	92
	$C_{20}H_{21}PSi$	PhSiH$_2$CH$_2$CH$_2$PPh$_2$	−11.0(r)	92
	$C_{20}H_{26}P(SiO_2)x$	Silica gel-(CH$_2$)$_8$PPh$_2$	−15.0	143
	$C_{20}H_{27}P$	EtPMes$_2$	−18.1	84
	$C_{20}H_{29}O_3PSi$	Si(OEt)$_3$CH$_2$CH$_2$PPh$_2$	−9.9(b), −10.2(c)	33, 70
	$C_{21}H_{22}ClPSi$	MePhSiClCH$_2$CH$_2$PPh$_2$	−10.0(r)	92
	$C_{21}H_{23}PSi$	MePhSiCHC$_2$CH$_2$PPh$_2$	−11.0(r)	92
	$C_{26}H_{24}AsP$	Ph$_2$AsCH$_2$CH$_2$PPh$_2$	−12.0(k), −14.9(k), −11.5(CP/MAS)	58, 70, 144, 145
	$C_{26}H_{24}ClPSi$	Ph$_2$SiClCH$_2$CH$_2$PPh$_2$	−10.0(r)	92
	$C_{26}H_{24}OP_2$	Ph$_2$POCH$_2$CH$_2$PPh$_2$	−22.8	132

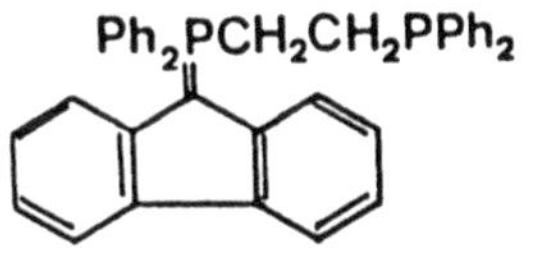

71

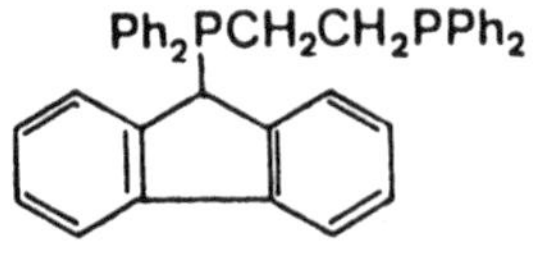

72

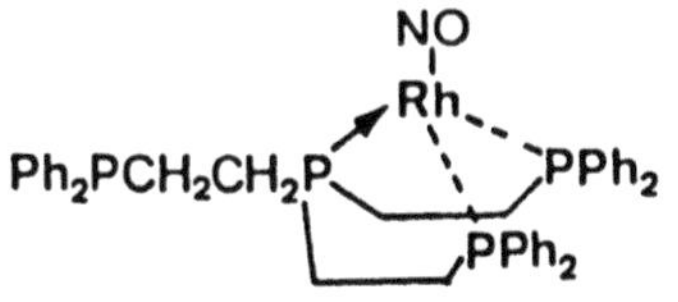

73

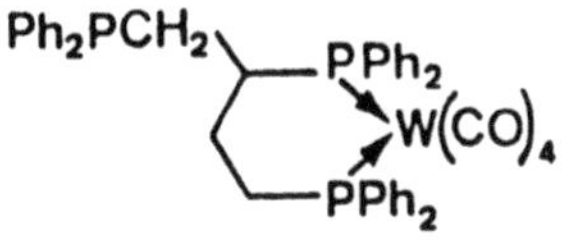

74

Rings/ Connectivities	Formula	Structure	³¹P NMR data (δP [solv, temp] J_{PH} Hz)	Ref.
	$C_{26}H_{24}P_2Se$	$Ph_2P(Se)CH_2CH_2PPh_2$	−14.0(r), −15.2(r,35°)	146, 147
	$C_{26}H_{25}PSi$	$Ph_2SiHCH_2CH_2PPh_2$	−10.0(r)	92
	$C_{27}H_{26}P_2Se$	$Ph_2P(Se)(CH_2)_3PPh_2$	−18.9, −20.2(r,35°)	146, 147
	$C_{28-30}H_{28-32}P_2Se$	$Ph_2P(Se)(CH_2)_{4-6}PPh.$	−18.0/−18.3(r,35°)	147
	$C_{31}H_{28}P_2$	C*H=CHCH=CH- C*=$PPh_2CH_2CH_2PPh_2$	−13.1(c)	148
	$C_{33}H_{31}BF_4P_2$	$BzPh_2P^+CH_2CH_2PPh_2$ BF_4^-	−12.1(r)	149
	$C_{39}H_{32}P_2$	**71**	−11.2(c)	148
	$C_{39}H_{33}BrP_2$	**72**	−12.7	148
	$C_{42}H_{42}NOP_4Rh$	**73**	−16.3(t)	150
	$C_{44}H_{37}O_4P_3W$	**74**	−20.4	151
C′H₂;CO′,C′₂	$C_{15}H_{12}Cl_3OP$	$BzPPhCOCCl_3$	3.9(c)	189
	$C_{15}H_{12}F_3OP$	Bz(TfCO)PPh	12.8(c)	165
	$C_{15}H_{13}Cl_2OP$	$Bz(CHCl_2CO)PPh$	9.6(c)	189
	$C_{15}H_{14}ClOP$	$Bz(CH_2ClCO)PPh$	11.6(c)	189
	$C_{15}H_{15}OP$	$Bz(MeCO)PPh_2$	15.8(c)	166
C′H₂;(C′₂)₂	$C_{15}H_{15}P$	CH_2=$CHCH_2PPh_2$	−17.1	3
	$C_{17}H_{15}OP$	2-FurCH_2PPh_2	−16.1(c)	60
	$C_{19}H_{16}P$	Polymer-4-Ph′CH_2PPh_2	−5.3(CP/MAS)	153
	$C_{19}H_{17}P$	$BzPPh_2$	−10.4	3, 79, 107
C″H₂;(C′₂)₂	$C_{14}H_{12}NP$	$NCCH_2PPh_2$	−18.0	61
C₂O;(C′₂)₂	$C_{15}H_{11}F_6OP$	$Tf_2C(OH)PPh_2$	−5.4	152
CC′O;(C′₂)₂	$C_{20}H_{16}F_3OP$	$TfPhC(OH)PPh_2$	−10.5	152
C₂P;(C′₂)₂	$C_{28}H_{29}P_2Hal$	$MePh_2P^+CMe_2PPh_2$ Hal^-	3.5	154
	$C_{33}H_{31}ClP_2$	$Ph_3P^+CMe_2PPh_2\ Cl^-$	5.5,2.8	3
C₂H;(OO′)₂	$C_{10}H_{17}O_4P$	$cHexP(CO_2Me)_2$	4.4	14
	$C_{12}H_{21}O_4P$	$cHexP(CO_2Et)_2$	3.6	14
	$C_{16}H_{29}O_4P$	$cHexP(CO_2Bu)_2$	3.4	14
C₂H;CO′,C′₂	$C_{11}H_{12}Cl_3OP$	$iPrP(COCCl_3)Ph$	12.4(c)	189
	$C_{11}H_{13}Cl_2OP$	$iPrP(COCHCl_2)Ph$	20.4(c)	189
	$C_{11}H_{14}ClOP$	$iPrP(COCH_2Cl)Ph$	22.4(c)	189
	$C_{14}H_{16}F_3OP$	cHexP(COTf)Ph	20.0(c)	165
	$C_{14}H_{19}OP$	cHexP(COMe)Ph	23.2(c)	166
C₂H;(C′₂)₂	$C_{15}H_{19}P$	$iPrPPh_2$	0.7(c),0.2,1.5(r)	1, 3, 156

TABLE E (continued)
^{31}P NMR Data of Three Coordinate (λ3σ3) Compounds Containing Phosphorus Bonds to Hydrogen and/or Group IV Atoms but Not to Halogen, Chalcogenide or Group V Atoms

Rings/ Connectivities	Formula	Structure	^{31}P NMR data (δP [solv, temp] J_{PH} Hz)	Ref.
	$C_{16}H_{19}P$	$sBuPPh_2$ (−)R	−2.8	155
	$C_{17}H_{19}P$	$cPePPh_2$	−3.9	3
	$C_{17}H_{21}P$	$iPrCHMePPh_2$ (+)R	−7.7	155
	$C_{18}H_{21}P$	$cHexPPh_2$	−4.4	3
	$C_{28}H_{29}ClP$	$MePh_2P^+CMe_2PPh_2\ Cl^-$	3.7(y)	157
CC'H;(C'$_2$)$_2$	$C_{24}H_{21}N_2OP$	C*H=CHNCH=N*-$COCH_2CHPhPPh_2$	−0.6	158
C'$_2$H;(C'$_2$)$_2$	$C_{29}H_{31}N_2P$	$(4\text{-}Me_2NPh')_2CHPPh_2$	−4.7(c)	97
C_3;(C'O)$_2$	$C_{12}H_{15}O_2P$	$tBuP(2\text{-}Fur)_2$	−28.9(c)	115
C_3;C'O,C'$_2$	$C_{12}H_{14}F_3OP$	tBuP(COTf)Ph	39.2(c)	165
	$C_{12}H_{17}OP$	tBuP(COMe)Ph	40.8(c)	166
C_3;(C'$_2$)$_2$	$C_{16}H_{19}P$	$tBuPPh_2$	17.2(c),17.1(c)	1, 3
	$C_{17}H_{21}P$	$EtMe_2CPPh_2$	15.4	3
	$C_{18}H_{23}P$	$Et_2MeCPPh_2$	11.4	3
		$tBuP(Tol\text{-}4)_2$	15.2	159

E3i: Tertiary Alkyl Dialkenyl/aryl Polyphosphines

P-bound atoms = CC'$_2$

Rings/ Connectivities	Formula	Structure	^{31}P NMR data (δP [solv, temp] J_{PH} Hz)	Ref.
HPSi;(C'$_2$)$_2$	$C_{28}H_{30}P_2Si$	$TmsCH(PPh_2)_2$	−12.4(c+r)	225
HP^{4}P;(C'$_2$)$_2$	$C_{37}H_{31}P_3S$	$Ph_2P(S)CH(\mathbf{P}Ph_2)_2$	−13.2(c+r)	86, 128
	$C_{37}H_{31}P_3Se$	$Ph_2P(Se)CH(\mathbf{P}Ph_2)_2$	−12.4(c+r)	127, 128
HP$_2$;(C'$_2$)$_2$	$C_{11}H_{18}P_3$	$(Me_2P)_2CH\mathbf{P}Ph_2$	−17.9(b)	247
HP$_2$;(C'$_2$)$_2$	$C_{19}H_{31}P_3$	$CH(PPh_2)_3$	−9.3(b), −10.4(c,r)	127, 247
HP$_2$;(C'$_2$)$_2$	$C_{27}H_{27}P_3$	$Me_2PCH(\mathbf{P}Ph_2)_2$	−13.9(b)	247
H$_2$As;(C'$_2$)$_2$	$C_{32}H_{29}AsP_2$	$PhAs(CH_2PPh_2)_2$	−20.9	256
H$_2$N;(C'$_2$)$_2$	$C_{29}H_{29}NO_2P_2$	$MeOOCCH_2N(CH_2PPh_2)_2$	−26.6	129
	$C_{30}H_{31}NO_3P_2$	$MeOOCCH(CH_2OH)$ $N(CH_2PPh_2)_2$ (L)	−26.3	129
	$C_{30}H_{34}N_2P_2$	$Me_2NCH_2CH_2N$ $(CH_2PPh_2)_2$	−27.2	130
	$C_{30}H_{34}N_2P_2$	$Ph_2PCH_2NMeCH_2CH_2$ $NMeCh_2PPh_2$	−26.9	130
	$C_{32}H_{33}NO_4P_2$	MeOOCCH (CH_2COOMe) $N(CH_2PPh_2)$ (L)	−26.4	129
	$C_{33}H_{37}NO_2P_2$	MeOOCCH(tBu) $N(CH_2PPh_2)_2$	−24.2	129
	$C_{42}H_{43}N_2P_3$	$Ph_2P^1CH_2NMeCH_2CH_2N$ $(CH_2P^2Ph_2)_2$	P^1 −26.9; P^2 −28.4	130
	$C_{54}H_{52}N_2P_4$	$Y_2NCH_2CH_2NY_2$; Y = CH_2PPh_2	−28.1	130
H$_2$P^4;(C'$_2$)$_2$	$C_{38}H_{34}P_3$	$(Ph_2PCH_2)_2P^+Ph_2Cl$	−27.1(c) J_{PP}62.6	125
H$_2$P^4;(C'$_2$)$_2$	$C_{50}H_{42}P_4$	$Ph_2PCH_2PPh_2{=}C(PPh_2)_2$	−26.8(b,t)	125
H$_2$P^4;(C'$_2$)$_2$	$C_{54}H_{52}Br_2P_4$	$YPh_2P^+(CH_2)_4P^+Ph_2Y$; Y = $CH_2PPh_2\ 2Br^-$	−27.6(c)	257
H$_2$P;(C'$_2$)$_2$	$C_{15}H_{18}P_2$	$Me_2PCH_2\mathbf{P}Ph_2$	−23.1(b)	53, 227

Rings/ Connectivities	Formula	Structure	^{31}P NMR data (δP [solv, temp] J_{PH} Hz)	Ref.
$H_2P;(C'_2)_2$	$C_{25}H_{22}P_2$	$CH_2(PPh_2)_2$	−23.6(a), −22.9(20°), −24.8(−90°), −23.2(CP/MAS)	3, 70, 73
$H_2P;(C'_2)_2$	$C_{29}H_{30}P_2$	CH_2(PTol-2)$_2$	−44.1	228
	$C_{33}H_{38}P_2$	CH_2(PXyl-2,6)$_2$	−27.8	229
	$C_{37}H_{46}P_2$	CH_2(PMes)$_2$	−30.8(b)	193
	$C_{38}H_{33}P_3$	$Ph_2PCH_2PPh_2$=CHPPh$_2$	−28.8,J_{PP}154.1 and 57.5	125
$H_3;(C'_2)_2$	$C_{18}H_{24}P_2$	(PhMe**P**)Dop(PMe-tBu)	−33.5,36.9(c)	24
	$C_{20}H_{20}P_2$	(MePhP)$_2$Dop	−35.5, −36.1[t](c)	24
$CHP^4;(C'_2)_2$	$C_{38}H_{33}P_3Se$	$Ph_2P(Se)CH$-(PPh_2)CH_2PPh_2	−9.1	138
$CHP;(C'_2)_2$	$C_{26}H_{24}P_2$	CHMe(**P**Ph$_2$)$_2$	−6.3(c)	225
	$C_{32}H_{29}P_2$	$PhPHCH_2CH$(**P**Ph$_2$)$_2$ nonequiv.	−3.9 and −5.8(r)	35
	$C_{38}H_{33}P_3$	Ph_2PCH_2CH(**P**Ph$_2$)$_2$	−4.0(r)	35
	$C_{38}H_{33}P_3Se$	$Ph_2P(Se)CH_2CH(PPh_2)_2$	−2.9	138
	$C_{46}H_{42}P_4$	$Ph_2P(CH_2)_2$ $PPhCH_2CH(PPh_2)_2$	−4.0(r)	35
	$C_{58}H_{51}AsP_4$	PhAs[($CH_2CH(PPh_2)$]$_2$	−3.2 and −4.6(r)	35
	$C_{58}H_{51}P_5$	PhP[CH_2CH(**P**Ph$_2$)$_2$]$_2$ nonequiv.	−2.9 and −4.8	35
$CH_2;CO',C'_2$	$C_{18}H_{20}O_2P_2$	(MeCO)PhPCH$_2$CH$_2$P-(COMe)Ph	11.6(c)	232
	$C_{28}H_{24}O_2P_2$	(PhCO)PhPCH$_2$CH$_2$P-(COPh)Ph	4.3,4.9(c)	232
$CH_2;CC',C'_2$	$C_{23}H_{21}F_3P_2$	Ph_2PCH=CTf**P**EtPh	−25.8[C](c) J_{PP}131	173
	$C_{26}H_{30}P_2$	Ph_2PCH=C(tBu)**P**EtPh	−15.8(c)	173
$CH_2;(C'_2)_2$	$C_{14}H_{16}P_2$	$PH_2CH_2CH_2$**P**PH$_2$	−14.7	12
	$C_{15}H_{17}P_2$	MePhPCH$_2$CH$_2$**P**Ph$_2$	−13.2	3
	$C_{17}H_{22}P_2$	$Me_2P(CH_2)_3$**P**Ph$_2$	−16.4(c)	240
	$C_{18}H_{24}P_2$	$Et_2PCH_2CH_2$**P**Ph$_2$	−14.4(k)	58
		$Me_2P(CH_2)_4$**P**Ph$_2$	−15.47(c)	240
	$C_{19}H_{26}P_2$	$Et_2P(CH_2)_3$**P**Ph$_2$	−16.33(c)	240
	$C_{20}H_{20}P_2$	PhPHCH$_2$CH$_2$**P**Ph$_2$	−14.1(r)	11
	$C_{20}H_{28}P_2$	$Et_2P(CH_2)_4$**P**Ph$_2$	−15.84(c)	240
	$C_{21}H_{22}P$	PhPH(CH$_2$)2**P**Ph$_2$	−18.0(b)	33
	$C_{22}H_{24}P_2$	EtPhPCH$_2$CH$_2$**P**Ph$_2$	−13.7, −12.9	3, 231
	$C_{22}H_{32}P_2$	$Bu_2PCH_2CH_2$**P**Ph$_2$	−12.8	231
	$C_{23}H_{26}P_2$	PrPhPCH$_2$CH$_2$**P**Ph$_2$	−13.2	3
		iPrPhPCH$_2$CH$_2$**P**Ph$_2$	−13.3, −12.8	3, 231
		EtPhP(CH$_2$)$_3$**P**Ph$_2$	−17.1(c)	240
		MePhP(CH$_2$)$_4$**P**Ph$_2$	−18.8(c)	240
	$C_{24}H_{28}P_2$	EtPhP(CH$_2$)$_4$**P**Ph$_2$	−16.1(c)	240
		sBuPhPCH$_2$CH$_2$**P**Ph$_2$	−12.9	3
		iBuPhPCH$_2$CH$_2$**P**Ph$_2$	−13.5	3
	$C_{24}H_{29}NP_2$	$H_2N(CH_2)_3$PhP-(CH$_2$)$_3$**P**Ph$_2$	−18.2(b)	33
	$C_{24}H_{29}P_3$	(Ph$_2$**P**CH$_2$CH$_2$) PPh(CH$_2$CH$_2$PMe$_2$)	−13.5(r)	274
	$C_{25}H_{30}P_2$	cPePhPCH$_2$CH$_2$**P**Ph$_2$	−13.7	3
	$C_{25}H_{30}P_2$	MePhP(CH$_2$)$_6$**P**Ph$_2$	−18.5(c)	240
	$C_{26}H_4F_{20}P_2$	$Y_2PCH_2CH_2PY_2$; Y = C_6F_5	−44.5(a)	105
	$C_{26}H_{22}F_2P_2$	$Ph_2PCH_2CH_2P^a$(Ph'F-3)$_2$	−12.1, −11.1[a](b)	31

TABLE E (continued)
^{31}P NMR Data of Three Coordinate ($\lambda 3\sigma 3$) Compounds Containing Phosphorus Bonds to Hydrogen and/or Group IV Atoms but Not to Halogen, Chalcogenide or Group V Atoms

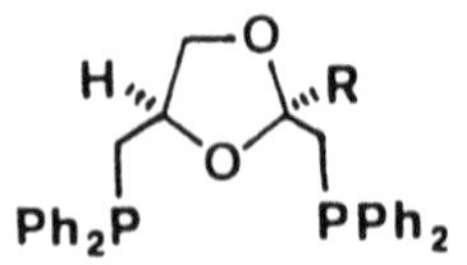

75 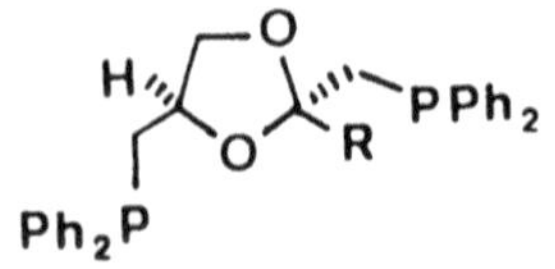 76 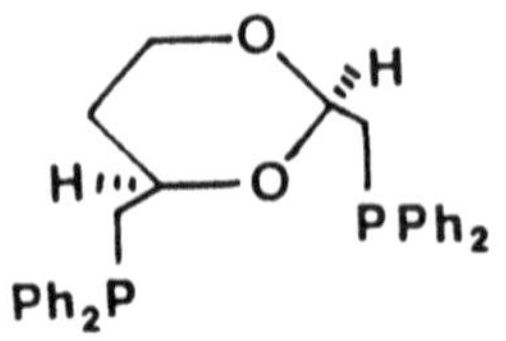 77

Rings/ Connectivities	Formula	Structure	^{31}P NMR data (δP [solv, temp] J_{PH} Hz)	Ref.
	$C_{26}H_{22}F_2P_1$	$Ph_2PCH_2CH_2P^a(Ph'F\text{-}4)_2$	−12.2 and −14.5~(b) J_{PP}35	31
	$C_{26}H_{24}P_2$	$Ph_2PCH_2CH_2PPh_2$	−13.2, −14.8(−90°), −12.6(CP/MAS)	3, 70, 73
	$C_{26}H_{32}P_2$	$EtPhP(CH_2)_6PPh_2$	−15.5	240
	$C_{26}H_{33}NP_2$	$Me_2N(CH_2)_3PhP\text{-}(CH_2)_3PPh_2$	−18.2(b)	33
	$C_{27}H_{26}P_2$	$Ph_2PCH_2CHMeP^aPh_2$	−20.8	234
		R-prophos	−20.6 J_{PP}20.6	235
		S-prophos	−23.3	236
	$C_{27}H_{26}P_2$	$Ph_2P(CH_2)_3PPh_2$	−17.5(r), −17.2(c), −18.4(b)	146, 241
	$C_{28}H_{22}F_6P_2$	$Ph_2PCH_2CH_2P^a(Ph'Tf\text{-}3)_2$	−12.2 and −11.2~(b) J_{PP}35	31
	$C_{28}H_{26}P_2$	$Ph_2PCH{=}CPhPEtPh$	−21.1^C(c) J_{PP}146	173
		$Ph_2P(CH_2)_2CHMePPh_2$	−16.8(c)	234
		$Ph_2P(CH_2)_4PPh_2$	−17.8(t), −16.3(c), −16.1(b)	58, 241
	$C_{29}H_{28}O_2P_2$	**75**; R = H	−23.2 and −25	258
	$C_{29}H_{30}P_2$	$Ph_2P(CH_2)_5PPh_2$	−16.4(c), −18.3(k)	58, 241
		$Ph_2PCH_2CMe_2CH_2PPh_2$	−25.3(c)	242
	$C_{29}H_{31}P_3$	$MeP(CH_2CH_2PPh_2)_2$	−13.6	249
	$C_{30}H_{30}O_2P_2$	**75**; R = Me	−21.2 and −23.2	258
	$C_{30}H_{30}O_2P_3$	**76**; R = Me	−21.9 and −22.5(c)	258
		77	−23.7(c), −22.9 and −23.5(y)	258
	$C_{30}H_{32}OP$	$PhO(CH_2)_3PhP\text{-}(CH_2)_3PPh_2$	−18.2(b)	33
	$C_{30}H_{32}P_2$	$Ph_2P(CH_2)_6PPh_2$	−18.2(k), −16(c), −17.2(b)	58, 228, 241
		$2\text{-}Tol_2PCH_2CH_2P(Tol\text{-}2)_2$	−33.5	228
	$C_{30}H_{32}P_2$	$4\text{-}Tol_2PCH_2CH_2P(Tol\text{-}4)_2$	−10.1(k)	58
	$C_{31}H_{32}NP_3$	$NCCH_2CH_2P(CH_2\text{-}CH_2PPh_2)_2$	−13.9(b)	33
	$C_{31}H_{34}P_2$	$2\text{-}Tol_2P(CH_2)_3P(Tol\text{-}2)_2$	−40.0	228
	$C_{31}H_{36}NP_3$	$H_2N(CH_2)_3P\text{-}(CH_2CH_2PPh_2)$.	−13.5(b)	33
	$C_{31\text{-}40}H_{34\text{-}52}P_2$	$Ph_2P(CH_2)_{7\text{-}16}PPh_2$	−15.5/−15.7(c), −17.1/ −17.7(r, −90°)	141, 244

Rings/ Connectivities	Formula	Structure	^{31}P NMR data (δP [solv, temp] J_{PH} Hz)	Ref.
	$C_{31}H_{32}O_2P_2$	YC*HCOMe$_2$OC*HY; Y = CH$_2$PPh$_2$	−25.0^{+diop}; −24.1(y)$^{-diop}$	235, 236
	$C_{32}H_{28}P_2$	Ph$_2$PCH$_2$CHPhPPh$_2$	−15.4(c)	237
	$C_{32}H_{36}P_2$	2-Tol$_2$P(CH$_2$)$_4$P(Tol-2)$_2$	−38.4	228
	$C_{33}H_{38}P_2$	cHexPhP(CH$_2$)$_6$**P**Ph$_2$	−15.5	240
	$C_{34}H_{33}P_3$	PhP(CH$_2$CH$_2$**P**Ph$_2$)$_2$	−12.8	250
	$C_{34}H_{40}P_2$	2-Tol$_2$P(CH$_2$)$_6$P(Tol-2)$_2$	−37.4	228
	$C_{34}H_{41}P_3$	tBuP[(CH$_2$)$_3$**P**Ph$_2$]$_2$	−17.6(t)	252
	$C_{35}H_{32}O_2P_2$	**75**; R = Ph	−20.0 and −22.7	258
	$C_{35}H_{32}O_2P_3$	**76**; R = Ph	−23.0 and −25.5(c)	258
	$C_{36}H_{37}P_3$	PhP[(CH$_2$)$_3$PPh$_2$]$_2$	−18.1(b)	75
	$C_{36}H_{44}P_2$	2-Tol$_2$P(CH$_2$)$_8$P(Tol-2)$_2$	−38.0	228
	$C_{37}H_{33}O_2P_2$	**78**	−23.1	213
	$C_{38}H_{33}P_3$	Ph$_2$PCH$_2$CH(PPh$_2$)$_2$	−19.6(r)	35
	$C_{38}H_{33}P_3Se$	Ph$_2$P(Se)CH-(PPh$_2$)CH$_2$PPh$_2$	−18.0	138
	$C_{42}H_{42}P_4$	YPhP(CH$_2$)$_2$PPhY; Y = (CH$_2$)$_2$PPh$_2$ Tetraphos I	−14.6(r)	253
		P(CH$_2$CH$_2$PPh$_2$)$_3$ Tetraphos II	−17.8; −18.8(r)	117, 253
	$C_{46}H_{42}P_4$	Ph$_2$P(CH$_2$)$_2$ PPhCH$_2$CH(PPh$_2$)$_2$	−11.2(r)	35
	$C_{50}H_{51}P_5$	Y$_2$PPhCH$_2$CH$_2$PY$_2$; Y = CH$_2$CH$_2$**P**Ph$_2$	−13.1	255
		PhP(CH$_2$CH$_2$PPhCH$_2$-CH$_2$**P**Ph$_2$)$_2$	−13.4	255
	$C_{66}H_{69}P_7$	PhP[CH$_2$CH$_2$P-(CH$_2$CH$_2$**P**Ph$_2$)$_2$]$_2$	−13.2	255
	$C_{74}H_{78}P_8$	[(Ph$_2$PCH$_2$CH$_2$)$_2$-PCH$_2$CH$_2$**P**PhCH$_2$]$_2$	−13.3	255
	$C_{78}H_{69}P_7$	P[CH$_2$CH$_2$P(CH$_2$CH$_2$-**P**Ph$_2$)$_2$]$_2$	−13.4	255
C$_2$H;(C'$_2$)$_2$	$C_{27}H_{26}P_2$	Ph$_2$PCH$_2$CHMe**P** Ph$_2$ S-prophos	−1.1	234
		R-prophos	−0.9;1.7 J_{PP}20.6	
	$C_{28}H_{27}NP_2$	**79**; R = H (3R,4R)	−8.3(d,y)	238
	$C_{28}H_{28}ClNP_2$	**79**; R = H.HCl (3R,4R)	−15.3(y)	238
	$C_{28}H_{28}P_2$	Ph$_2$P(CH$_2$)$_2$CHMe**P**Ph2	−2.6(c)	234
	$C_{28}H_{28}P_2$	Ph$_2$PCHMeCHMePPh$_2$ (S,S-chiraphos)	−10.7(y), −11	235, 236
	$C_{33}H_{33}NO_3P_2$	**79**; R = CO(CH$_2$)$_3$-COOH (3R,4R)	−11.2 and −12.0(d) (AB sig)	238
	$C_{33}H_{35}NO_3P_2$	**79**; R = COCH$_2$OCH$_2$-CH$_2$OMe (3R,4R)	−10.7(r)	238
	$C_{35}H_{31}NOP_2$	**79**; R = COPh (3R,4R)	−16.3(b)	238
	$C_{35}H_{33}NP_2$	**79**; R = CH$_2$Ph (3R,4R)	−2.9(b)	238
	$C_{35}H_{34}ClNP_2$	**79**; R = CH$_2$Ph.HCl (3R,4R)	−12.5(y), −11.3, −14.3(r)	238
	$C_{35}H_{35}NO_2P_2$	**79**; R = CO$_2$tBu (3R,4R)	−11.6(r)	238
	$C_{37}H_{43}NO_5P_2$	**79**; R = COCH$_2$-(OCH$_2$CH$_2$)$_3$-OMe(3R,4R)	−10.2(r)	238

TABLE E (continued)
^{31}P NMR Data of Three Coordinate ($\lambda 3\sigma 3$) Compounds Containing Phosphorus Bonds to Hydrogen and/or Group IV Atoms but Not to Halogen, Chalcogenide or Group V Atoms

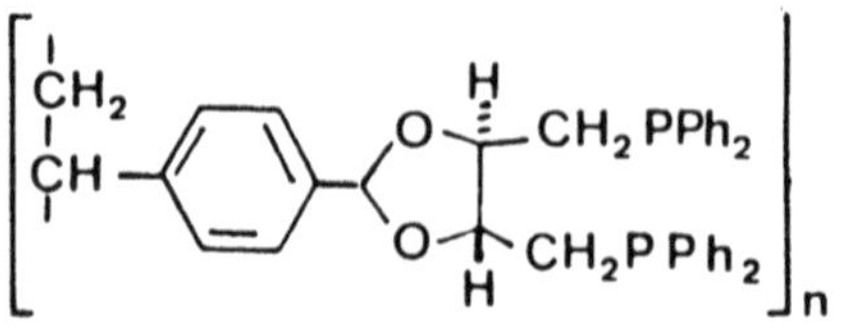

78 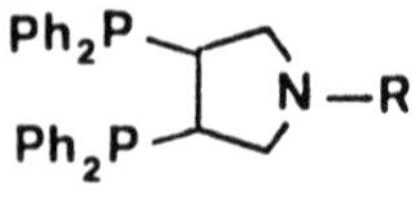 **79**

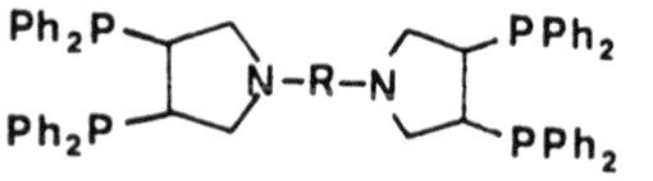

80 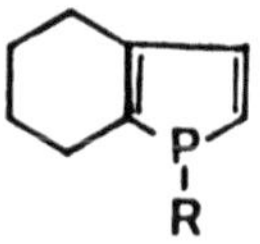 **81** 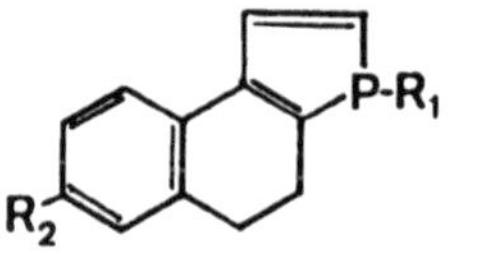 **82** 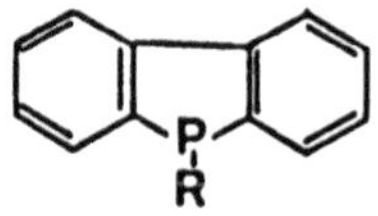 **83**

Rings/ Connectivities	Formula	Structure	^{31}P NMR data (δP [solv, temp] J_{PH} Hz)	Ref.
	$C_{58}H_{52}N_2O_2P_4$	**80**; R = –COCO– (3R,4R)	–8.5 and –9.2(d)	238
	$C_{64}H_{56}N_2O_2P_4$	**80**; R = –COPh′-CO(p)– (3R,4R)	–11.6(d)	238
CC′H;(C′$_2$)$_2$	$C_{32}H_{28}P_2$	$Ph_2PCH_2CHPhPPh_2$	9.2(c)	237
	$C_{38}H_{32}P_2$	$Ph_2PCHPhCHPhPPh_2$	22.4(c)	237
	$C_{39}H_{32}P_2$	$Ph_2PCHPhCH_2CHPhPPh_2$	4.4(r)	237
	$C_{40}H_{36}P_2$	$Ph_2PCHPhCH_2CH_2CH$-$PhPPh_2$	5.9(c)	237

E3j: Tertiary Cyclic Alkyl Dialkenyl/Aryl Phosphines

P-bound atoms = CC′$_2$

Rings/ Connectivities	Formula	Structure	^{31}P NMR data (δP [solv, temp] J_{PH} Hz)	Ref.
3/H$_3$;(C′$_2$)$_2$	$C_{15}H_{13}P$	MeP*CPh=C*Ph	–191.6(r)	196
3/C$_3$;(C′$_2$)$_2$	$C_{18}H_{19}P$	tBuP*CPh=C*Ph	–149.6(r)	196
5/H$_3$;(C′H)$_2$	C_5H_7P	MeP*CH=CHCH=C*H	–8.0. –8.7, –8.7(b)	3, 203
	C_6H_9P	MeP*CH=CMeCH=C*H	–6.6(c)	203
5/H$_3$;C′H,C′$_2$	$C_9H_{13}P$	**81**; R = Me	–7.6(c)	205
	$C_{13}H_{13}P$	**82**; R^1 = Me, R^2 = H	–4.1(c)	205
	$C_{14}H_{15}OP$	**82**; R^1 = Me, R^2 = OMe	–4.9(c)	205
5/H$_3$;(C′$_2$)$_2$	$C_{13}H_{11}P$	**83**; R = Me	–24.8(r)	156
5/CH$_2$;C′H,C′$_2$	$C_{11}H_{13}P$	PhP*CH=CMeCH$_2$C*H$_2$	4.0(r)	202
5/CH$_2$;(C′$_2$)$_2$	$C_{14}H_{13}P$	**83**; R = Et	–9.4(r)	156
	$C_{31}H_{28}O_2P$	**84**; R^1 = R^2 = Me	–20.8	212
	$C_{37}H_{30}O_2P$	**84**; R^1 = H, R^2 = Ph′(CH=CH$_2$)-4	–20.9;polymer –21.1	212

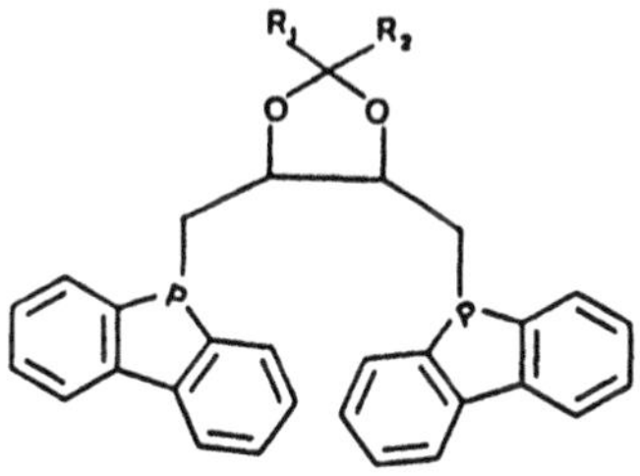

84 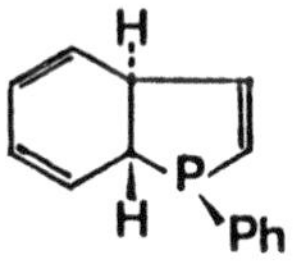 85 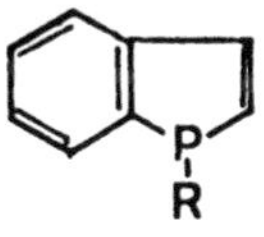 86 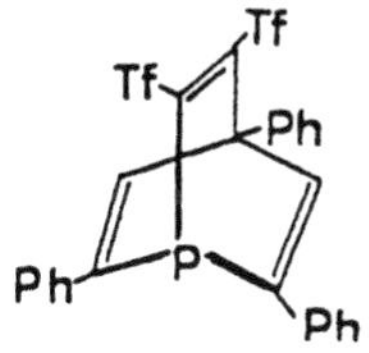 87

Rings/ Connectivities	Formula	Structure	^{31}P NMR data (δP [solv, temp] J_{PH} Hz)	Ref.
5/C'H_2;(C'H)$_2$	$C_{11}H_{11}P$	PhCH$_2$P*CH=CH-CH=C*H	?5.7(c)	207
5/C"H_2;(C'H)$_2$	$C_8H_{10}NP$	NCCH$_2$P*CH=CMe-CMe=C*H	−23.4(c)	206
5/C_2H;C'H,C'$_2$	$C_{20}H_{18}P_2$	**30**; R[1] = Ph, R[2] = H P[2]	9.9	220
	$C_{20}H_{18}P_2$	**31**; R[1] = Ph, R[2] = R[3] = H P[2]	15.7	220, 221
	$C_{22}H_{20}P_2$	**30**; R[1] = Ph, R[2] = Me P[2]	11.4	220
		31: R[1] = Ph, R[2] = Me, R[3] = H P[2]	15.0	220, 221
5/C_2H;CC',C'$_2$	$C_{12}H_{14}P$	PhP*CMe=CHCH$_2$-C*HMe	9.9[C];22.6[T]	204
5/C_2H;(C'$_2$)$_2$	$C_{15}H_{15}P$	**83**; R = iPr	2.0(r)	156
5/CC'H;C'H,C'$_2$	$C_{14}H_{13}P$	**85**	−16.4(c)	195
	$C_{16}H_{17}P$	**52**; R[1] = R[2] = Me, R[3] = Ph	25.6(c)	203
5/C_3;(C'H)$_2$	$C_9H_{15}P$	tBuP*CH=CMeCH=C*H	40.5	208
	$C_{10}H_{17}P$	tBuP*CH=CMeCMe=C*H	27.5	208
5/C_3;C'H,C'$_2$	$C_{12}H_{15}P$	**86**; R = tBu	23.5(c)	210
6,6/CC',(C'$_2$)$_2$	$C_{27}H_{17}F_6P$	**87**	−65.0	3

E3k: Tertiary Alkyl Di-alkynyl Phosphines

P-bound atoms = CC"$_2$

Rings/ Connectivities	Formula	Structure	^{31}P NMR data	Ref.
H_3;(N")$_2$	$C_3H_3N_2P$	MeP(CN)$_2$	−81.4	187
C_2H;(C")$_2$	$C_8H_{11}P$	iBuP(C-CH)$_2$	−53.0	87

E3l: Tertiary Trialkenyl/Arylphosphines

P-bound atoms = C'$_3$

Rings/ Connectivities	Formula	Structure	^{31}P NMR data	Ref.
NO',(C'$_2$)$_2$	$C_{19}H_{16}NOP$	PhNHCOPPh$_2$	10.8(pe)	64
NS',(C'$_2$)$_2$	$C_{15}H_{16}NPS$	Me$_2$NCSPPh$_2$	17.6(k)	160
	$C_{19}H_{16}NPS$	PhNHCSPPh$_2$	19.2(r)	120
	$C_{20}H_{18}NPS$	MePhNCSPPh$_2$	18.7(k)	160
	$C_{22}H_{24}NPSSi$	TmsNPhCSPPh$_2$	19.3(r)	120

TABLE E (continued)
^{31}P NMR Data of Three Coordinate (λ3σ3) Compounds Containing Phosphorus Bonds to Hydrogen and/or Group IV Atoms but Not to Halogen, Chalcogenide or Group V Atoms

Rings/ Connectivities	Formula	Structure	^{31}P NMR data (δP [solv, temp] J_{PH} Hz)	Ref.
N'S,(C'$_2$)$_2$	$C_{15}H_{16}NPS$	MeSC(=NMe)PPh$_2$	−7.9(k)	160
	$C_{20}H_{18}NPS$	MeSC(=NPh)PPh$_2$	−3.4(k)	160
(P^4)$_2$,(C'$_2$)$_2$	$C_{49}H_{40}ClP_3$	(Ph$_3$P)$_2$CP$^+$Ph$_2$ Cl$^-$	−1.3(20°), −0.18 (r−70°)	123
P^{4}P,(C'$_2$)$_2$	$C_{42}H_{35}AuClP_3$	Ph$_2$PC(=PPh$_3$)PPh$_2$.AuCl	−0.7	122
HP^4,(C'$_2$)$_2$	$C_{25}H_{21}LiP_2S$	Ph$_2$P(S)CH$^-$PPh$_2$ Li$^+$	−17.4(j-43°)	124
	$C_{26}H_{23}LiP_2$	Ph$_2$P(=CH$_2$)CH$^-$PPh$_2$ Li$^+$	−17.7(b,t)	125
	$C_{31}H_{26}NNaP_2$	2-PyrCH$^-$PPh$_2$=CHPPh$_2$ Na$^+$	−14.3(k)	126
	$C_{26}H_{24}P_2$	MePh$_2$P=CHPPh$_2$	−20.9(b)	125
	$C_{28}H_{28}P_2$	iPrPh$_2$P=CHPPh$_2$	−20.0(b)	154, 157
	$C_{31}H_{27}NP_2$	2-PyrCH$_2$Ph$_2$P=CHPPh$_2$	−20.5(b)	134
	$C_{32}H_{27}NaP_2$	PhCH$^-$Ph$_2$P=CHPPh$_2$ Na$^+$	−16.0(b)	135
	$C_{32}H_{28}P_2$	BzPh$_2$P=CHPPh$_2$	−20.1(b)	135
C'Cl,(C'$_2$)$_2$	$C_{16}H_{10}ClN_2P$	(NC)$_2$C=CClPPh$_2$	10.0	39
C'N,(C'$_2$)$_2$	$C_{16}H_{12}N_3P$	(NC)$_2$C=C(NH$_2$)PPh$_2$	−3.5	39
	$C_{17}H_{16}NP$	N-Me-PyrrolylPPh$_2$	−29.7	63
(C'N')$_3$	$C_{15}H_{12}N_3P$	2-Pyr$_3$P	−1.3(c)	59
(C'N')$_2$,C'$_2$	$C_{16}H_{13}N_2P$	2-Pyr$_2$PPh$_2$	−2.9(c)	59
C'N',(C'$_2$)$_2$	$C_{17}H_{14}NP$	2-PyrPPh$_2$	−4.1	59
(CO')$_3$	$C_6H_9O_3P$	(MeCO)$_3$P	63.5	64
(C'O)$_3$	$C_{12}H_9O_3P$	2-Fur$_3$P	−76.4(r)	63
(C'O)$_2$,C'$_2$	$C_{14}H_{11}O_2P$	(2-Fur)$_2$PPh	−50.8(r), −46.5(c)	63, 115
CO',(C'$_2$)$_2$	$C_{14}H_{10}ClF_2OP$	CClF$_2$COPPh$_2$	15.8(c)	161
	$C_{14}H_{10}Cl_3OP$	CCl$_3$COPPh$_2$	8.6(c)	162
	$C_{14}H_{10}F_3OP$	TfCOPPh$_2$	17.9(pe),18.2(c)	64, 163
	$C_{14}H_{11}Cl_2OP$	CHCl$_2$COPPh$_2$	14.0(c)	162
	$C_{14}H_{11}F_2OP$	CHF$_2$COPPh$_2$	15.6(c)	163
	$C_{14}H_{12}BrOP$	CH$_2$BrCOPPh$_2$	16.5(c)	164
	$C_{14}H_{12}ClOP$	CH$_2$ClCOPPh$_2$	15.1(c)	162, 163
	$C_{14}H_{13}OP$	MeCOPPh$_2$	15.0(pe),17.5(c)	64, 163
	$C_{15}H_{10}F_5OP$	C$_2$F$_5$COPPh$_2$	22.2(c)	163
	$C_{15}H_{13}Cl_2OP$	MeCCl$_2$COPPh$_2$	8.0(c)	162
	$C_{15}H_{14}ClOP$	CH$_2$ClCH$_2$COPPh$_2$	18.5(c)	161
		MeCHClCOPPh$_2$	14.5(c)	161
	$C_{16}H_8F_9OP$	TfCOP(Ph'Tf-3)$_2$	15.5(c)	165
	$C_{16}H_{10}F_7OP$	C$_3$F$_7$COPPh$_2$	23.2(c)	163
	$C_{16}H_{11}F_6OP$	MeCOP(Ph'Tf-3)$_2$	15.6(c)	166
	$C_{16}H_{14}Cl_3OP$	CCl$_3$COP(Tol-4)$_2$	7.7(c)	121
	$C_{16}H_{15}Cl_2OP$	CHCl$_2$COP(Tol-4)$_2$	12.7(c)	121
	$C_{16}H_{16}ClOP$	CH$_2$ClCOP(Tol-4)$_2$	14.2(c)	121
		CH$_2$ClCH$_2$CH$_2$COPPh$_2$	17.4(c)	161
		MeCHClCH$_2$COPPh$_2$	19.4(c)	161
	$C_{20}H_{25}OP$	MeCOPMes$_2$	1.7(c)	166
	$C_{22\text{-}24}H_{29\text{-}33}OP$	RCOPPh$_2$; R = $C_{9\text{-}11}H_{19\text{-}23}$	16.6 to 16.5(c)	167

Rings/ Connectivities	Formula	Structure	^{31}P NMR data (δP [solv, temp] J_{PH} Hz)	Ref.
	$C_{23}H_{29}OP$	CH_2=CH$(CH_2)_8$COPPh$_2$	16.4(c)	167
C'O,(C'$_2$)$_2$	$C_{16}H_{13}OP$	2-FurPPh$_2$	−26.6(r)	63
(C'O')$_3$	$C_{21}H_{15}O_3P$	(PhCO)$_3$P	60.9(liq. cryst,40°)	76
CO',(C'$_2$)$_2$	$C_{15}H_{10}F_5OP$	C_2F_5COPPh$_2$	22.2(c)	163
C'O',(C'$_2$)$_2$	$C_{18}H_{14}NOP$	3-PyrCOPPh$_2$	15.2(c)	167
	$C_{19}H_{10}F_5OP$	C_6F_5COPPh$_2$	27.8(c)	163
	$C_{19}H_{13}Cl_2OP$	(3,5-$C_6H_3Cl_2$)COPPh$_2$	14.4(c)	167
		(3,4-$C_6H_3Cl_2$)COPPh$_2$	14.0(c)	167
	$C_{19}H_{14}BrOP$	(2-BrPh')COPPh$_2$	19.7(c)	64, 168
	$C_{19}H_{14}ClOP$	(4-ClPh')COPPh$_2$	13.5(c)	167
	$C_{19}H_{14}NO_3P$	(2-NO2Ph')COPPh$_2$	21.5(c)	169
	$C_{19}H_{15}OP$	PhCOPPh$_2$	13.4,12.7(c)	169, 170
	$C_{20}H_{16}ClOP$	(4-ClCH2Ph')COPPh$_2$	19.8(c)	167
	$C_{20}H_{17}OPS$	4-MeSPh')COPPh$_2$	14.7(c)	170
	$C_{20}H_{17}O_2P$	(4-MeOPh')COPPh$_2$	11.3(c)	167
	$C_{20}H_{17}O_3PS$	(2-MeSO2Ph')COPPh$_2$	23.6(c)	169, 170
	$C_{21}H_{17}OP$	PhCH=CHCOPPh$_2$	17.3(c)	158
	$C_{24}H_{20}ClN_2OP$	C*H=CHN=CHN-*CPhC=CHCOPPh$_2$	12.5(c)	158
C'P^4,(C'$_2$)$_2$	$C_{26}H_{22}P_2Se$	Ph$_2$P(Se)C(=CH$_2$)PPh$_2$	−14.6(r)	146
	$C_{27}H_{25}IP_2$	MePh$_2$P$^+$C(=CH2)PPh$_2$ I$^-$	−6.2(c)	171
(C'S)$_3$	$C_{12}H_9PS$	(2-Thienyl)$_3$P	−45.8(r)	63
(C'S)$_2$,C'$_2$	$C_{14}H_{11}PS_2$	(2-Thienyl)$_2$PPh	−33.6(r)	63
C'S,(C'$_2$)$_2$	$C_{16}H_{13}PS$	2-ThienylPPh$_2$	−19.3(r)	63, 156
(C'H)$_3$	C_6H_9P	(CH$_2$=CH)$_3$P	−20.7	3
C'H,(C'$_2$)$_2$	$C_{14}H_{13}P$	CH$_2$=CHPPh$_2$	−13.8, −11.7	3
	$C_{15}H_{15}P$	MeCH=CHPPh$_2$	−32.7^E, −32.1^E(c); −14.2^Z	118, 172
	$C_{16}H_{11}ClF_4P$	C*F$_2$CF$_2$CCl=C*PPh$_2$	−25.1	3
	$C_{16}H_{17}P$	Me$_2$C=CHPPh$_2$	−28.3	118
	$C_{17}H_{11}ClF_6P$	C*F$_2$CF$_2$CF$_2$CCl=C*PPh$_2$	−19.6	6
	$C_{20}H_{17}P$	PhCH=CHPPh$_2$	−20.1^Z; −11.6^E(c)	172
	$C_{26}H_{22}P_2Se$	Ph$_2$P(Se)CH=CHPPh$_2$	−29.0C; −7.1T(r)	146
	$C_{33}H_{29}BrP_2$	BzP$^+$Ph$_2$CH=CHPPh$_2$ Br$^-$	−24.7^C; −3.2^T	3, 173
(CC')$_2$,C'$_2$	$C_{16}H_{11}N_4P$	[(NC)$_2$C=CMe]$_2$PPh	23.0	3
CC',(C'$_2$)$_2$	$C_{15}H_{15}P$	CH$_2$=CMePPh$_2$	−0.9	118
	$C_{16}H_{17}P$	MeCH=CMePPh$_2$	5.9^E; −18.5^Z	118
	$C_{17}H_{13}N_2P$	(NC)$_2$C=CMePPh$_2$	4.0	39
	$C_{18}H_{15}N_2P$	(NC)$_2$C=CEtPPh$_2$	7.0	39
(C'$_2$)$_3$	$C_{12}H_9O_3P$	(3-Fur)$_3$P	−81.9(r)	63
	$C_{12}H_9S_3P$	(3-Thienyl)$_3$P	−43.2(r)	63
	$C_{18}H_5F_{10}P$	(C$_6$F$_5$)$_2$PPh	−48.7, −42.5	3, 80
	$C_{18}H_6F_9P$	P(C$_6$H$_2$F$_3$-2,3,6)$_3$	−78.5	65
		P(C$_6$H$_2$F$_2$-2,6)$_3$	−78.5	65
	$C_{18}H_{10}F_5P$	C$_6$F$_5$PPh$_2$	−26.3	3
	$C_{18}H_{12}Br_3P$	P(Ph'Br-4)$_3$	−8.2	66
	$C_{18}H_{12}Cl_3P$	P(Ph'Cl-2)$_3$	−9.2	66
		P(Ph'Cl-3)$_3$	−4.4(c), −5.3(t)	3, 30
		P(Ph'Cl-4)$_3$	−9.2, −8.5(r), −9(c60°)	3, 67
	$C_{18}H_{12}F_3P$	P(Ph'F-2)$_3$	−44.6	67
		P(Ph'F-3)$_3$	−6.5, −5.2	3, 69
	$C_{18}H_{12}N_3O_6P$	P(Ph'NO2-4)$_3$	5.1(c)	71
	$C_{18}H_{13}Cl_2P$	PhP(Ph'Cl-2)$_2$	−18.8	66

TABLE E (continued)
^{31}P NMR Data of Three Coordinate (λ3σ3) Compounds Containing Phosphorus Bonds to Hydrogen and/or Group IV Atoms but Not to Halogen, Chalcogenide or Group V Atoms

Rings/ Connectivities	Formula	Structure	^{31}P NMR data (δP [solv, temp] J_{PH} Hz)	Ref.
(C'F$_2$)$_3$	$C_{18}H_{13}Cl_2P$	PhP(Ph'Cl-4)$_2$	−8.0(t)	30
(C'$_2$)$_3$	$C_{18}H_{13}F_2P$	PhP(Ph'F-2)$_2$	−32.5(t)	68
	$C_{18}H_{13}N_2O_4P$	Ph$_2$P(Ph'NO2-4)	4.6(c)	71
	$C_{18}H_{14}BrP$	Ph$_2$P(Ph'Br-4)	−4.8	111
	$C_{18}H_{14}ClP$	Ph$_2$P(Ph'Cl-2)	−10.9	66
		Ph$_2$P(Ph'Cl-4)	−6.9(t),6.6	30, 111
	$C_{18}H_{14}Cl_2OP_2$	Ph$_2$P(Ph'OPCl$_2$-2)	−16.9(r)	174
	$C_{18}H_{14}FP$	Ph$_2$P(Ph'F-2)	−19.8	68
	$C_{18}H_{14}NaOP$	Ph$_2$P(Ph'ONa-2)	−22.1(r)	174
	$C_{18}H_{15}OP$	Ph$_2$P(Ph'OH-2)	−18.1(r)	174
	$C_{18}H_{15}P$	Ph$_3$P	−4.7(r)/−8(j); −7.2(CP/MAS), −6.0(10°), −7.7 (−90°),1.3Hz/°C	3, 63, 74, 70, 72, 73
	$C_{19}H_{12}Cl_2F_3P$	(3-ClPh')$_2$P(Ph'Tf-4)	−7.5(t)	30
	$C_{19}H_{14}F_3P$	Ph$_2$P(Ph'F-3)	−10.9. −12.0	3, 66
	$C_{19}H_{14}NP$	Ph$_2$P(Ph'CN-3)	−8.3(r)	139
	$C_{19}H_{17}OP$	Ph$_2$P(Ph'OMe-2)	−16.0, −16.8(r), −13.5(c)	77, 174, 175
		Ph$_2$P(Ph'OMe-3)	−4.0(c)	175
		Ph$_2$P(Ph'OMe-4)	−6.0(c)	175
	$C_{19}H_{17}P$	Ph$_2$PTol-2	−13.0(r)	3
	$C_{19}H_{18}ClP$	4-Tol$_2$P(Ph'Cl-4)	−8.2	66
	$C_{20}H_{12}ClF_6P$	(4-TfPh')2P(Ph'Cl-3)	−6.3(t)	30
	$C_{20}H_{13}F_6P$	PhP(Ph'Tf-2)$_2$	−14.5	3
	$C_{20}H_{17}P$	Ph$_2$P(Ph'CH=CH2-4) polymer	−5.8(c), −7.0 (CP/MAS); −4.3/−6.2	176, 177
	$C_{20}H_{18}BrO_2P$	(4-BrPh')P(Ph'OMe-4)$_2$	−8.6	111
	$C_{20}H_{18}ClO_2P$	(4-ClPh')P(Ph'OMe-4)$_2$	−8.4	111
	$C_{20}H_{19}OP$	Ph$_2$P(Ph'OEt-2)	−15.2	77
	$C_{20}H_{19}O_2P$	PhP(Ph'OMe-2)$_2$	−27.2, −26.3	32, 66, 77
		Ph$_2$P(C$_6$H$_3$(OMe)$_2$-2,6	−24.9	77
	$C_{20}H_{19}P$	PhP(Tol-2)$_2$	−21.4	66
	$C_{20}H_{20}NP$	Ph$_2$P(Ph'NMe$_2$-2)	−13.5	178
		Ph$_2$P(Ph'NMe$_2$-4)	−7.0	111
		Ph$_2$P(Ph'NHEt-4)	−7.0(c)	113
	$C_{21}H_{12}F_9P$	P(Ph'Tf-2)$_3$	−18.4(r)	3, 66
		P(Ph'Tf-3)$_3$	−4.3(r), −5(r)	63, 66
		P(Ph'Tf-4)$_3$	−7(t), −6	30, 75
	$C_{21}H_{21}OP$	4-Tol$_2$P(Ph'OMe-4)	−9.2(t)	30
		Ph$_2$P(Ph'OPr-2)	−15.8	77
		Ph$_2$P(Ph'O-iPr-2)	−15.1	77
	$C_{21}H_{21}O_2P$	4-TolP(Ph'OMe-4)$_2$	−10.0	30
	$C_{21}H_{21}O_3P$	P(Ph'OMe-2)$_3$	−37.1, −38.5(r)	77, 78
		P(Ph'OMe-3)$_3$	−2.1(r)	3
		P(Ph'OMe-4)$_3$	−10.8(t), −10.5(c60°), −13.7(CP/MAS)	30, 67, 70

Rings/ Connectivities	Formula	Structure	^{31}P NMR data (δP [solv, temp] J_{PH} Hz)	Ref.
	$C_{21}H_{21}P$	P(Tol-2)$_3$	−30.5(c60°), −29.9(r)	3, 67, 78
		4-TolP(Tol-2)$_2$	−22.1	66
		P(Tol-3)$_2$	−5.7(c60°), −5.2(r)	67, 78
		P(Tol-4)$_2$	−8.0(r), −8.4(c60°), −10.2(CP/MAS)	3, 32, 67, 70
	$C_{21}H_{21}PS_3$	P(Ph'SMe-4)$_3$	−8.3(r)	3
	$C_{21}H_{23}OPSi$	Ph$_2$P(Ph'OTms-2)	−16.2(r)	174
	$C_{21}H_{23}O_2P$	PhP(Ph'OEt-2)$_2$	−24.1	77
	$C_{22}H_{19}N_2P$	Ph$_2$PC$_6$H$_3$(NMe$_2$)$_2$-2,5	−12.7	178
		Ph$_2$PC$_6$H$_3$(NMe$_2$)$_2$-2,6	−32.2	178
	$C_{22}H_{23}O_2P$	Ph$_2$PC$_6$H$_3$(OEt)$_2$-2,3	−16.2	77
		Ph$_2$PC$_6$H$_3$(OEt)$_2$-2,5	−14.7	77
		Ph$_2$PC$_6$H$_3$(OEt)$_2$-2,6	−24.1	77
	$C_{22}H_{23}O_4P$	PhP(C$_6$H$_3$(OMe)$_2$-2,3	−28.5	77
		PhP(C$_6$H$_3$(OMe)$_2$-2,5	−24.8	77
		PhP(C$_6$H$_3$(OMe)$_2$-2,6	−43.1	77
	$C_{22}H_{23}P$	Ph$_2$P(Ph'Bu-4)	−6.5, −6.5(r)	85, 179
	$C_{22}H_{24}NO_2P$	(4-MeOPh')$_2$P-(Ph'NMe$_2$-4)	−9.5	111
	$C_{22}H_{24}NP$	Ph$_2$P[Ph'CHMe-(NMe$_2$)-2]	−17.0(−)^S; −20.5^R	178, 180
		Ph$_2$P(Ph'NEt$_2$-2)	−14.1	178
	$C_{22}H_{25}ClP$	Ph$_2$P[Ph'(CHMeNMe$_2$)-2].HCl	−20.0(r)	180
	$C_{22}H_{25}N_2P$	PhP(Ph'NMe$_2$-3)$_2$.HCl	−2.1(c)	181
	$C_{23}H_{23}N_2O_2P$	PhP(Ph'NO$_2$-4)-(Ph'NC$_5$H$_{10}$-4)	5.3(c)	183
	$C_{23}H_{27}O_2P$	PhP(Ph'OPr-2)$_2$	−24.6	77
		PhP(Ph'O-iPr-2)$_2$	−23.3	77
	$C_{24}H_9F_{18}P$	P(C$_6$H$_3$Tf-2,6)$_3$	−42.8(a)	81
	$C_{24}H_{19}ClOP_2$	Ph$_2$P(Ph'(OPPhCl)-2	−15.8(r)	174
	$C_{24}H_{19}F_3OP_2$	Ph$_2$P[Ph'(OPPhF$_3$)-2]	−20.3(r)	174
	$C_{24}H_{22}NP$	**88**	−17.4	178
	$C_{24}H_{26}NP$	**89**	−16.1	178
		Ph$_2$PC*HNMeCH$_2$-CH$_2$CH$_2$C*H.	−18.2	178
	$C_{24}H_{27}O_2P$	Ph$_2$PC$_6$H$_3$(OPr)$_2$-2,3	−16.8	77
		Ph$_2$PC$_6$H$_3$(O-iPr)$_2$-2,3	−16.0	77
		Ph$_2$PC$_6$H$_3$(OPr)$_2$-2,5	−15.1	77
		Ph$_2$PC$_6$H$_3$(O-iPr)$_2$-2,5	−14.4	77
		Ph$_2$PC$_6$H$_3$(OPr)$_2$-2,6	−24.7	77
		Ph$_2$PC$_6$H$_3$(O-iPr)$_2$-2,6	−23.9	77
	$C_{24}H_{27}O_6P$	P[(C$_6$H$_3$(OMe)$_2$-2,3]$_3$	−40.8	77
		[P(C$_6$H$_3$(OMe)$_2$-2,6]$_3$	−65.7	77
	$C_{24}H_{27}P$	Ph$_2$P(Ph'Hex-4)	−6.2(r)	85, 179
	$C_{24}H_{28}NP$	Ph$_2$P(Ph'CH-iPrNMe$_2$-2)	−5.1	178
	$C_{24}H_{30}As_3P$	P(C$_6$H$_4$AsMe$_2$-2)$_3$	−20.2(r)	83
	$C_{24}H_{30}N_3P$	P(Ph'NMe$_2$-4)$_3$	−11.5(r)	3
	$C_{25}H_{30}NP$	Ph$_2$PPh'(CH-tBuNMe$_2$)-2	−6.0	178
	$C_{26}H_{26}NP$	**90**	−19.4	178
		91	−15.2	178
		92	−15.0	178
	$C_{26}H_{31}O_2P$	Ph$_2$PC$_6$H$_3$(OBu)$_2$-2,3	−16.8	77
		Ph$_2$PC$_6$H$_3$(O-iBu)$_2$-2,3	−16.8	77
		Ph$_2$PC$_6$H$_3$(O-sBu)$_2$-2,3	−16.1	77

TABLE E (continued)
^{31}P NMR Data of Three Coordinate (λ3σ3) Compounds Containing Phosphorus Bonds to Hydrogen and/or Group IV Atoms but Not to Halogen, Chalcogenide or Group V Atoms

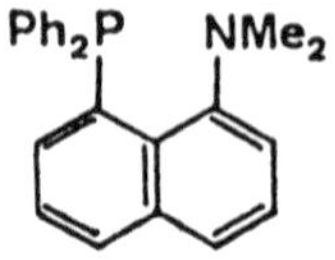

Ph_2P NMe_2 — **88**

Ph_2P NMe_2 — **89**

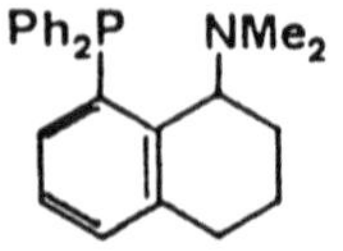

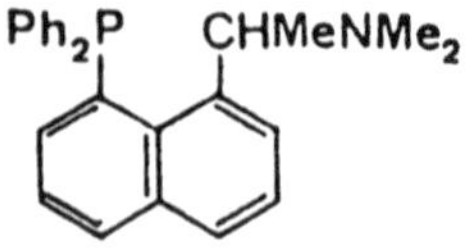

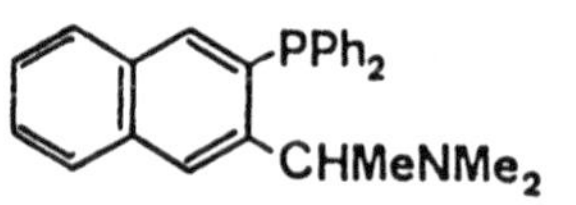

Ph_2P $CHMeNMe_2$ — **90**

PPh_2 $CHMeNMe_2$ — **91**

Ph_2P $CHMeNMe_2$ — **92**

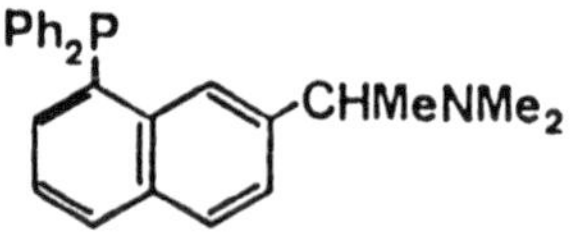

Rings/ Connectivities	Formula	Structure	^{31}P NMR data (δP [solv, temp] J_{PH} Hz)	Ref.
		$Ph_2PC_6H_3(OBu)_2$-2,5	−15.0	77
		$Ph_2PC_6H_3(O\text{-}iBu)_2$-2,5	−15.1	77
		$Ph_2PC_6H_3(O\text{-}sBu)_2$-2,5	−14.4	77
		$Ph_2PC_6H_3(OBu)_2$-2,6	−24.7	77
		$Ph_2PC_6H_3(O\text{-}iBu)_2$-2,6	−24.7	77
		$Ph_2PC_6H_3(O\text{-}sBu)_2$-2,6	−24.0	77
	$C_{26}H_{31}O_4P$	$PhPC_6H_3(OEt)_2$-2,3	−27.8	77
		$PhPC_6H_3(OEt)_2$-2,5	−24.0	77
		$PhPC_6H_3(OEt)_2$-2,6	−40.8	77
	$C_{26}H_{31}P$	$Ph_2PPh'Oct$-4	−6.5(r)	85, 179
	$C_{27}H_{33}P$	Mes_3P	−39.5(r), −36.2	74, 84
		$P(Ph'Pr\text{-}4)_3$	−8.2(r)	85
		$P(Ph'\text{-}iPr\text{-}4)_3$	−7.8(r)	3, 66
	$C_{27}H_{39}PSi_3$	$P(Ph'Tms\text{-}3)_3$	−5.6(r)	3
		$P(Ph'Tms\text{-}4)_3$	−5.8(r)	3
	$C_{28}H_{27}OP$	$Ph_2PPh'(CMe_2\text{-}COCH_2Ph)\text{-}_2$	−10.3	182
	$C_{28}H_{32}N_3O_2P$	$(4\text{-}C_5H_{10}NPh')_2P\text{-}(Ph'NO_2\text{-}4)$	6.8(c)	183
	$C_{30}H_{24}AsP$	$Ph_2P[Ph'(AsPh_2)\text{-}2]$	−14.2(k)	58
	$C_{30}H_{24}F_2OP$	$Ph_2PPh'(OPF_2Ph_2)$-2	−20.3(r)	174
	$C_{30}H_{39}O_4P$	$PhP[C_6H_3(OPr)_2\text{-}2,3]_3$	−27.8	77
		$PhP[C_6H_3(O\text{-}iPr)_2\text{-}2,3]_3$	−26.2	77
		$PhP[C_6H_3(OPr)_2\text{-}2,5]_3$	−24.2	77
		$PhP[C_6H_3(O\text{-}iPr)_2\text{-}2,5]_3$	−23.3	77
		$PhP[C_6H_3(OPr)_2\text{-}2,6]_3$	−41.1	77
		$PhP[C_6H_3(O\text{-}iPr)_2\text{-}2,6]_2$	−39.0	77
	$C_{30}H_{39}O_6P$	$P[C_6H_3(OEt)_2\text{-}2,3]_3$	−64.6	77
	$C_{30}H_{39}P$	$P(Ph'Bu\text{-}4)_3$	−7.6	85
		$P(Ph'tBu\text{-}4)_3$	−9.1(r)	3
		$PhP(Ph'Hex\text{-}4)_2$	−7.0(r)	85, 179

Rings/ Connectivities	Formula	Structure	^{31}P NMR data (δP [solv, temp] J_{PH} Hz)	Ref.
	$C_{30}H_{41}BrOP_2$	$Ph_2PPh'(OP^+Bu_3)$-2 Br^-	−16.9(r)	174
	$C_{31}H_{26}P_2$	$Ph_2PPh'(PPh_2{=}CH_2)$-2	−15.1(t)	184
	$C_{31}H_{27}IP_2$	$Ph_2PPh'(P^+MePh_2)$-2 I^-	−16.0(c)	184
	$C_{34}H_{47}O_4P$	$PhP[C_6H_3(OBu)_2$-2,6$]_2$	−41.0	77
		$PhP[C_6H_3(O$-iBu$)_2$-2,6$]_2$	−41.0	77
		$PhP[C_6H_3(O$-sBu$)_2$-2,6$]_2$	−39.2	77
	$C_{34}H_{47}P$	$PhP(Ph'Oct$-4$)_2$	−7.0(r)	85, 179
	$C_{35}H_{34}ClN_2O_4P$	$Ph_2PPh'[C^+(Ph'NMe_2$-4$)_2]$-4 ClO_4^-	−3.8	185
	$C_{36}H_{27}P$	$P(Ph'Ph$-2$)_3$	−27.0	3
	$C_{36}H_{51}O_6P$	$P[C_6H_3(OPr)_2$-2,3$]_3$	−40.2	77
		$P[C_6H_3(O$-iPr$)_2$-2,3$]_3$	−36.8	77
		$P[C_6H_3(OPr)_2$-2,5$]_3$	−32	77
		$P[C_6H_3(OPr)_2$-2,6$]_3$	−65.4	77
		$P[C_6H_3(O$-iPr$)_2$-2,6$]_3$	−63.7	77
	$C_{36}H_{51}P$	$P(Ph'Hex$-3$)_3$	−7.7(r)	85
	$C_{38}H_{39}O_2P$	$PhP[Ph'(CMe_2OCH_2Ph)$-2$]_2$	−7.7(r)	85
	$C_{42}H_{63}O_6P$	$P[C_6H_3(OBu)_2$-2,6$]_3$	−65.3	77
		$P[C_6H_3(O$-iBu$)_2$-2,6$]_3$	−65.6	77
		$P[C_6H_3(O$-sBu$)_2$-2,6$]_3$	−64.9	77
	$C_{42}H_{63}P$	$P(Ph'Oct$-4$)_3$	−8.2	85
	$C_{54}H_{42}AsP_3$	$As[Ph'(PPh_2)$-2$]_3$	−11.4	3
	$C_{54}H_{42}As_3P$	$P[Ph'(AsPh_2)$-2$]_3$	−5.3	3

E3m: Tertiary Trialkenyl/Aryl Polyphosphines

P-bound atoms = C'_3

Rings/ Connectivities	Formula	Structure	^{31}P NMR data	Ref.
$P^4P,(C'_2)_2$	$C_{38}H_{33}P_3$	$MePh_2P{=}C(\mathbf{P}Ph_2)_2$	−2.6	123
	$C_{43}H_{35}P_3$	$Ph_3P{=}C(\mathbf{P}Ph_2)_2$	−7.2,4.2(c, −60°)	125
	$C_{50}H_{41}AgP_4$	$Ph_2PCH^-PPh_2{=}C(\mathbf{P}Ph_2)_2$ Ag^+	3.3	125
	$C_{50}H_{41}KP_4$	$Ph_2PCH^-PPh_2{=}C(\mathbf{P}Ph_2)_2$ K^+	3.6	125
$HP^4,(C'_2)_2$	$C_{38}H_{32}LiP_3$	$Ph_2\mathbf{P}CH^-PPh_2{=}CHPPh_2$ Li^+	−17.4(b+t)	125
	$C_{50}H_{41}AgP_4$	$Ph_2\mathbf{P}CH^-PPh_2{=}C(PPh_2)_2$ Ag^+	−2.9	125
	$C_{50}H_{41}KP_4$	$Ph_2\mathbf{P}CH^-PPh_2{=}C(PPh_2)_2$ K^+	−14.2	125
	$C_{50}H_{41}NaP_3$	$Ph_2P^aCH^-PPh_2{=}CHP^bPh$ Na^+	−16.7P^a,3.2P^b	125
	$C_{54}H_{50}P_4$	$Ph_2PCH{=}PPh_2$ $(CH_2)_4PPh_2{=}CHPPh_2$	−20.6	257
$HP,(C'_2)_2$	$C_{25}H_{21}LiP_2$	$CHLi(PPh_2)_2$	−3.7 J_{PH}46	124
$CO',(C'_2)_2$	$C_{16}H_{14}Cl_3OP$	$Tol_2PCOCCl_3$	7.7	284
	$C_{16}H_{15}Cl_2OP$	$Tol_2PCOCHCl_2$	12.7	284
	$C_{16}H_{16}ClOP$	Tol_2PCOCH_2Cl	14.2	284
	$C_{27}H_{22}O_2P_2$	$Ph_2PCOCH_2COPPh_2$	21.9(c)	239
	$C_{28}H_{24}O_2P_2$	$Ph_2PCOCHMeCOPPh_2$	20.4(c)	239
	$C_{28\text{-}32}H_{24\text{-}32}O_2P_2$	$Ph_2PCO(CH_2)_{2\text{-}6}COPPh_2$	16.7/16.9(c)	239
	$C_{29}H_{26}O_2P_2$	$Ph_2PCOCMe_2COPPh_2$	12.4(c)	239

TABLE E (continued)
^{31}P NMR Data of Three Coordinate (λ3σ3) Compounds Containing Phosphorus Bonds to Hydrogen and/or Group IV Atoms but Not to Halogen, Chalcogenide or Group V Atoms

Rings/ Connectivities	Formula	Structure	^{31}P NMR data (δP [solv, temp] J_{PH} Hz)	Ref.
	$C_{31}H_{30}O_2P.$	$Ph_2PCOCH_2CMe_2CH_2$-$COPPh_2$	20.5(c)	239
	$C_{27}H_{22}O_2P_2$	$Ph_2PC(OH){=}CHCOPPh_2$ (Enol)	2.0(c)	239
C′O′,C′P,C′$_2$	$C_{30}H_{20}N_2O_2P_2$	$C(CN)_2{=}C(PPhCOPh)_2$	30	39
C′O′,(C′$_2$)$_2$	$C_{38}H_{28}O_2P_2$	$Dop(COPPh_2)_2$	17.4	169
C′P,(C′$_2$)$_2$	$C_{26}H_{22}P_2$	$CH_2{=}C(PPh_2)_2$	−3.9(r)	146
	$C_{28}H_{20}N_2P_2$	$C(CN)_2{=}C(PPh_2)_2$	10	39
C′H,(C′$_2$)$_2$	$C_{21}H_{19}F_3N_2P_2$	Ph_2**P**CH=CTfP-$(CH_2CH_2CN)_2$	−23^C(c) J_{PP}131	173
	$C_{23}H_{21}F_3P_2$	Ph_2**P**CH=CTfPEtPh	−25.8^C(c) J_{PP}131	173
	$C_{24}H_{18}N_2P_2$	Ph_2**P**CH=C(tBu)-$P(CH_2CH_2CN)_2$	−35.5^C(c) J_{PP}18	173, 245
	$C_{26}H_{22}P_2$	$Ph_2PCH{=}CHPPh_2$	−7.6^T(c), −8.4(r); −19.9(c); −23.1^C (c), −14.2(k)	58, 146 172, 234
	$C_{26}H_{24}N_2P_2$	Ph_2**P**CH=CPhP-$(CH_2CH_2CN)_2$	−28.2^C(c)	173
	$C_{26}H_{30}P_2$	Ph_2**P**CH=C(tBu)PEtPh	−37.5^C(c)	173
	$C_{27}H_{21}F_3P_2$	Ph_2**P**CH=CTfPPh$_2$	−26.3^C(c) J_{PP}131	173
	$C_{28}H_{26}P_2$	Ph_2**P**CH=CPhPEtPh	−27.8^C(c) J_{PP}146	173
	$C_{30}H_{30}P_2$	Ph_2**P**CH=C(tBu)PPh$_2$	−32.2^C(c) J_{PP}37	173
	$C_{32}H_{26}P_2$	Ph_2**P**CH=CPhPPh$_2$	−28.2^C(c) J_{PP}146	173
CC′,(C′$_2$)$_2$	$C_{27}H_{21}F_3P_2$	$Ph_2PCH{=}CTf$**P**Ph$_2$	−18^C(c) J_{PP}131	173
	$C_{28}H_{20}F_4P_2$	**93**	−22.7	3
	$C_{30}H_{30}P_2$	$Ph_2PCH{=}C(tBu)$**P**Ph$_2$	−32.7^C(c) J_{PP}37	173
(C′$_2$)$_3$	$C_{30}H_{24}P_2$	$Dop(PPh_2)_2$	−15.2(k), −14.3(b)	58, 184
	$C_{32}H_{26}P_2$	$Ph_2PCH{=}CPh$**P**Ph$_2$	−7.0C(c) J_{PP}146	173
	$C_{44}H_{32}P_2$	2,2′-$(PPh_2)_2$-dinaphthalene	−12.8(b + y)	246
	$C_{36}H_{28}ClO_2P_3$	ClP(OPh′**P**Ph$_2$)$_2$	−18.2(r)	174
	$C_{42}H_{33}O_2P_3$	PhP(OPh′**P**Ph$_2$)$_2$	−17.7(r)	174
	$C_{42}H_{33}F_2OP_3$	PhPF$_2$(OPh′**P**Ph$_2$)$_2$	−20.3	174
	$C_{42}H_{33}P_3$	PhP^a(Ph′P^{b}Ph$_2$-2)$_2$	P^a −16.6 and P^b −14.5	3, 251
	$C_{54}H_{42}P_4$	P^a(Ph′P^{b}Ph$_2$-2)$_3$	P^a −19.5 and P^b −13.1	251

E3n: Tertiary Cyclic Trialkenyl/Aryl Phosphines

<u>**P-bound atoms = C′$_3$**</u>

Rings/ Connectivities	Formula	Structure	^{31}P NMR data	Ref.
3/(CC′)$_3$	$C_{12}H_{20}ClP$	EtC*=CEtP*(CEt=CClEt)	−181.7(r)	196
3/(CC′)$_2$,C′$_2$	$C_{12}H_{15}P$	EtC*=CEtP*Ph	−188.4(r), −188.2(h)	196, 197
3/(C′$_2$)$_3$	$C_{20}H_{15}P$	PhC*=CPhP*Ph	−189.8(r), −190.3(b)	196, 198
5/C′O,(C′$_2$)$_2$	$C_{14}H_8F_3OP$	**83**; R = COTf	10.9(c)	211
	$C_{14}H_{11}OP$	**83**; R = COMe	13.1(c)	211
5/C′S,(C′$_2$)$_2$	$C_{16}H_{11}PS$	**83**; R = 2-Thienyl	−26.3(r)	156
5/(C′H)$_2$,C′$_2$	$C_{11}H_{11}P$	C*H=CMeCH=CHP*Ph	10.0	3
	$C_{12}H_{13}P$	C*H=CMeCMe=CHP*Ph	−2.0	3

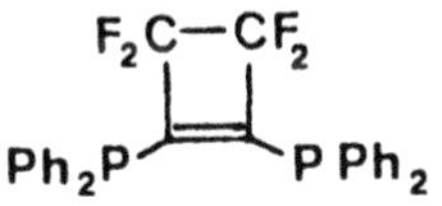

93 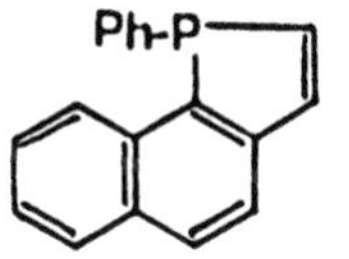 94 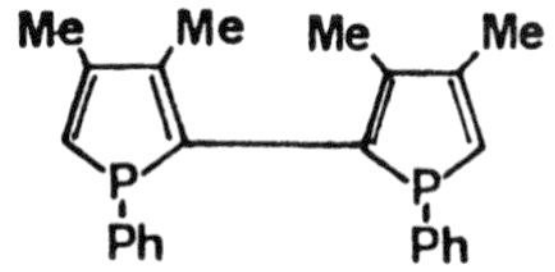 95

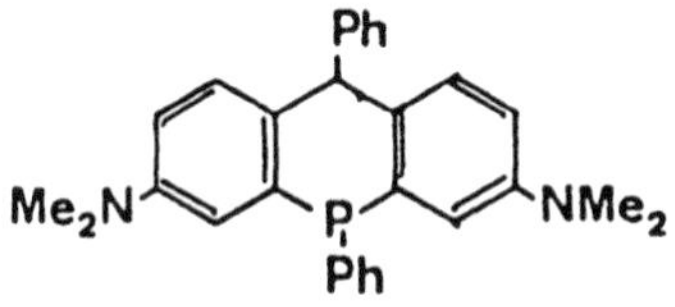

96 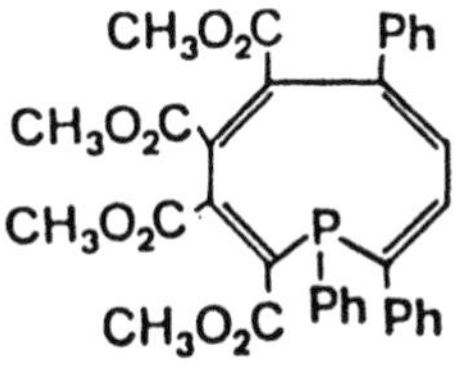 97 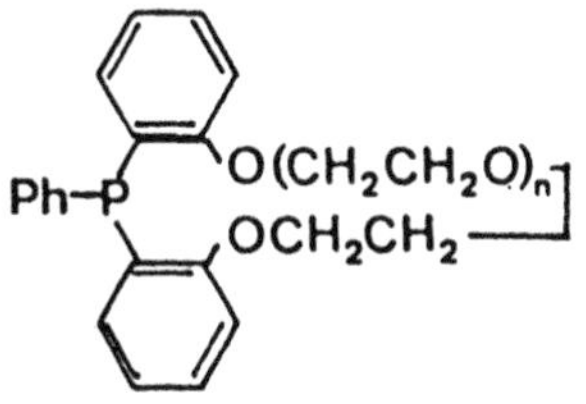 98

Rings/ Connectivities	Formula	Structure	^{31}P NMR data (δP [solv, temp] J_{PH} Hz)	Ref.
5/C'H,CC',C'$_2$	$C_{14}H_{15}P$	**81**; R = Ph	10.5(c)	205
	$C_{18}H_{15}P$	**82**; R^1 = Ph, R^2 = H	12.8(c)	205
5/C'H,(C'$_2$)$_2$	$C_{14}H_{11}P$	**86**; R = Ph	−2.0(c)	210
	$C_{18}H_{13}P$	**94**	0.1(c)	205
	$C_{24}H_{24}P_2$	**95**	12.5(c)	209
5/(C'$_2$)$_3$	$C_{18}H_{13}P$	**83**	−10.2	3
	$C_{22}H_{17}P$	PhP*CPh=CHCH=C*Ph	3.0,2.9(r)	3, 156
6/(C'$_2$)$_3$	$C_{29}H_{29}N_2P$	**96**	−20.4(c)	181
9/(C'$_2$)$_3$	$C_{38}H_{29}O_8P$	**97**	−35.0(c)	3
18/(C'$_2$)$_3$	$C_{26}H_{29}O_5P$	**98**, n = 3	−29(c)	219
21/(C'$_2$)$_3$	$C_{28}H_{33}O_6P$	**98**; n = 4	−27(c)	219

E3o: Tertiary Alkenyl/Aryl Alkynyl and Trialkynyl Phosphines

P-bound atoms = C'$_2$C"

(C'$_2$)$_2$;C"	$C_{14}H_{11}P$	$Ph_2PC{\equiv}CH$	−33.3	87
	$C_{15}H_{13}P$	$Ph_2PC{\equiv}CMe$	−34.3, −32.7	87, 270
	$C_{20}F_{15}P$	$(C_6F_5)PC{\equiv}CC_6F_5$	−74.7	80
	$C_{20}H_{15}P$	$Ph_2PC{\equiv}CPh$	−33.6(r)	82
	$C_{26}H_{20}P_2$	$Ph_2PC{\equiv}CPPh_2$	−32.3	146
	$C_{26}H_{20}P_2Se$	$Ph_2PC{\equiv}CP(Se)Ph_2$	−33.4(r)	146

P-bound atoms = C'C"$_2$

C'$_2$;(N")$_2$	$C_8F_5N_2P$	$C_6F_5P(CN)_2$	−121.3	80
C'$_2$;(C")$_2$	$C_{22}F_{15}P$	$C_6F_5P(C{\equiv}CC_6F_5)_2$	−29.4	80
	$C_{22}H_{15}P$	$PhP(C{\equiv}CPh)_2$	−61.2, −60.9(r)	80, 82

TABLE E (continued)
^{31}P NMR Data of Three Coordinate (λ3σ3) Compounds Containing Phosphorus Bonds to Hydrogen and/or Group IV Atoms but Not to Halogen, Chalcogenide or Group V Atoms

Rings/ Connectivities	Formula	Structure	^{31}P NMR data (δP [solv, temp] J_{PH} Hz)	Ref.
P-bound atoms = C″$_3$				
$(N'')_3$	C_3N_3P	$P(CN)_3$	−135.7	64
$(C'')_3$	C_6H_3P	$P(C{\equiv}CH)_3$	−91	87
	C_9H_9P	$P(C{\equiv}CHMe)_3$	87.0, −88.3(c,31°)	87, 270
	$C_{24}F_{15}P$	$P(C{\equiv}CC_6F_5)_3$	−90.0	80
	$C_{24}H_{15}P$	$P(C{\equiv}CPh)_3$	−88.4	82

Section E4: Phosphines Containing P-Si, P-Ge and P-Sn Bonds

E4a: Phosphines with No P-C Bonds

P-bound atoms	Connectivities	Structure	NMR data (δ_P J_{PH})	Ref.
GeSiSn	$C_3;C_3;C_3$	$Me_3GePTmsSnMe_3$	−264, −263.8(b) J_{PSn}731.6	275, 277
$GeSi_2$	$C_3;(C_3)_2$	$Me_3GePTms_2$	−245, −244.9(b)	275, 277
$GeSn_2$	$C_3;(C_3)_2$	$Me_3GeP(SnMe_3)_2$	−289, −289.4(b)	275, 277
Ge_2Si	$(C_3)_3;C_3$	$(Me_3Ge)_2PTms$	−237.1, −237.3(b)	275, 277
Ge_2Sn	$(C_3)_2;C_3$	$(Me_3Ge)_2PSnMe_3$	−256, −256.3(b) J_{PSn}770	275, 277
Ge_3	$(H'_3)_3$	$P(GeH_3)_3$	−338	260
	$(C_3)_3$	$P(GeMe_3)_3$	−228, −229	260, 275, 277
$SiSn_2$	$C_3;(C_3)_2$	$TmsP(SnMe_3)_2$	−296, −296(b) J_{PSn}761	275, 277
Si_2Sn	$(C_3)_3;C_3$	Tms_2PSnMe_3	−271, −270.4(b) J_{PSn}649	275, 277
Si_3	$(H_3)_3$	$P(SiH_3)_3$	−378(n), −372 J_{PSi}42.2	259, 260
	$(CH_2)_3$	$P(SiH_2Me)_3$	−328(n) J_{PSi}37	259
	$(C_2H)_3$	$P(SiHMe_2)_3$	−284(n) J_{PSi}32	259
	$(C_3)_3$	$PTms_3$	−252.6(b), −251(n) J_{PSi}27.5, −252(b)	98, 226, 275
	$(C_3)_2,C'_3$	Tms_2PSiPH_3	−254.3(b)	268
	$C_3,(C'_3)_2$	$TmsP(SiPh_3)_2$	−255(b)	268
	$(C'_3)_3$	$P(SiPh_3)_3$	−264.5(b)	268
Sn_3	$(H_3)_3$	$P(SnH_3)_3$	−450	260
	$(C_3)_3$	$P(SnMe_3)_3$	−330, −328, −328.8(b) J_{PSn}832	98, 275, 277
	$(C'_3)_3$	$P(SnPh_3)_3$	−323	260
HSi_2	$(H_3)_2$	$HP(SiH_3)_2$	−323(n) J_{PSi}34.8	259
	$(CH_2)_2$	$HP(SiH_2Me)_2$	−286(n) J_{PSi}30.2	259
	$(C_2H)_2$	$HP(SiHMe_2)_2$	−258.5(n) J_{PSi}27.0	259
	$(C_3)_2$	$HPTms_2$	−235(n), −237(b) J_{PSi}25.0	259, 260
H_2Ge	C_2P	$(H_2P)_2GeMe_2$	−222 ± 3 J180	276
	C_2H	$H_2PGeHMe_2$	−243 ± 3 J176	276

Rings/ Connectivities	Formula	Structure	^{31}P NMR data (δP [solv, temp] J_{PH} Hz)	Ref.
	C_3	H_2PGeMe_3	−236±3 J177	276
H_2Si	BrF_2	$H_2PSiBrF_2$	−243	259
	BrP_2	$(H_2P)_3SiBr$	−192 J_{PSi}219	259
	Br_2F	H_2PSiBr_2F	−207	259
	Br_2P	$(H_2P)_2SiBr_2$	−181	259
	Br_3	H_2PSiBr_3	−174	259
	Cl_3	H_2PSiCl_3	−205	259
	F_3	H_2PSiF_3	−290	259
	HP_2	$(H_2P)_3SiH$	−216	259, 276
	H_2P	$(H_2P)_2SiH_2$	−252	259, 276
	H_3	H_2PSiH_3	−274 J_{PSi}26.3	259
	CClP	$(H_2P)_2SiClMe$	−209	259
	CCl_2	H_2PSiCl_2Me	−231.5	259
	CP_2	$(H_2P)_3SiMe$	−211	259
	CHBr	$H_2PSiHBrMe$	−226	259
	CHCl	$H_2PSiHClMe$	−229.5	259
	CHP	$(H_2P)_2SiHMe$	−231	259
	CH_2	H_2PSiH_2Me	−256 J_{PSi}21.6	259
	C_2P	$(H_2P)_2SiMe_2$	−220 J187	259, 276
	C_2H	$H_2PSiHMe_2$	−242, −244 J182 J_{PSi}18.0	259, 276
	C_3	H_2PTms	−235,239 J_{PSi}16.2	259, 276
		$H_2PSiMe(CH_2CH{=}CH_2)_2$	−253.1(b) J186	43
	C_2C'	$H_2PSiPh(CH_2CH{=}CH_2)_2$	−257.2(b) J186	43
H_2Sn	C_3	H_2PSnMe_3	−269(b) J_{PSn}463	276, 277

E4b: Phosphines with One or Two P-C Bonds

Rings/ Connectivities	Formula	Structure	^{31}P NMR data (δP [solv, temp] J_{PH} Hz)	Ref.
C′GeP/5	C'_2;C_3;CC′	PhP*$GeMe_2CH_2CH_2$P*Ph	−101.2(b)	6
		PhP*$GeEt_2CH_2CH_2$P*Ph	−103.7(b)	6
CGeSi	C_3;C_3;C_3	tBuPTms($GeMe_3$)	−95.9	277
C′GeSi	C'_2;C_3;C_3	PhPTms($GeMe_3$)	−130.1(b)	277
CGeSn	C_3;C_3;C_3	tBuP($GeMe_3$)($SnMe_3$)	−93.6(b) J_{PSn}782	277
C′GeSn	C'_2;C_3;C_3	PhP($GeMe_3$)($SnMe_3$)	−140.0(b) J_{PSn}699	277
CGe_2	H_3;$(C_3)_2$	MeP$(GeMe_3)_2$	−177.0	260
	C_3;$(C_3)_2$	tBuP$(GeMe_3)_2$	−82.3(b)	277
CGe_2/5	C_3;$(C_3)_2$	tBuP*$GeMe_2CH_2$-CH_2Ge*Me_2	−127(b)	266
$C'Ge_2$	C'_2;$(C_3)_2$	PhP$(GeMe_3)_2$	−127, −122.3(b)	260, 277
$C'Ge_2$/5	C'_2;$(C_3)_2$	PhP*$GeMe_2CH_2CH_2$-Ge*Me_2	−171(b)	266
$C'Ge_2$/6	C'_2;C_2P,C_3	PhP*$GeMe_2CH_2CH_2$PPh-Ge*Me_2	−126.2; −126.0[l]	6
C′PSi/5	C'_2;CC′;C_3	PhP*$SiMe_2CH_2CH_2$P*Ph	−105	6
CSiSn	C_3;C_3;C_3	tBuPTms($SnMe_3$)	−109.4(b)	277
C′SiSn	C'_2;C_3;C_3	PhPTms($SnMe_3$)	−147.1(b) J_{PSn}657.5	277
CSi_2	NN′;$(C_3)_2$	PhN(Tms)C(=NPh)$PTms_2$	−137	4
	H_3;$(H_3)_2$	MeP$(SiH_3)_2$	−253(n) J_{PSi}39.7	259
	H_3;$(CH_2)_2$	MeP$(SiH_2Me)_2$	−232(n) J_{PSi}33.7	259
	H_3;$(C_2H)_2$	MeP$(SiHMe_2)_2$	−213(n) J_{PSi}28.8	259
	H_3;$(C_3)_2$	$MePTms_2$	−197.0(n), −195.5 J_{PSi}25.7	259, 261
	CH_2;$(C_3)_2$	$EtPTms_2$	−168.2(b)	98
	C_2H;$(C_3)_2$	$cHexPTms_2$	−139.0 J_{PSi}27.3	261
	C_3;$SiCl_2$,C_3	tBuPTmsSiClTms_2	−95.2(k)	267
	C_3;$(C_3)_2$	$tBuPTms_2$	−109.5, −108.4(b)	261, 277

TABLE E (continued)
^{31}P NMR Data of Three Coordinate (λ3σ3) Compounds Containing Phosphorus Bonds to Hydrogen and/or Group IV Atoms but Not to Halogen, Chalcogenide or Group V Atoms

Rings/ Connectivities	Formula	Structure	^{31}P NMR data (δP [solv, temp] J_{PH} Hz)	Ref.
$CSi_2/5$	$C_3;(C_3)_2$	$tBuP*SiMe_2CH_2$-CH_2Si*Me_2	−131.2(b)	266
$C'Si_2$	$CO';(C_3)_2$	$tBuCOPTms_2$	−34.0(b)	262
	$C'_2;Si_2Cl,C_3$	$MesPTmsSiClTms_2$	−148.4(k)	267
	$C'_2;(C_3)_2$	$PhPTms_2$	−137.4(b), −132.8(m)	261, 278
$C'Si_2/5$	$C'_2;(C_3)_2$	$PhP*SiMe_2CH_2$-CH_2Si*Me_2	−170.5(b)	266
$C'Si_2/6$	$C'_2;(C_2Si)_2$	$PhP*SiMe_2SiMe_2PPh$-$SiMe_2Si*Me_2$	−139.4(b)	280
CSn_2	$CH_2;(C_3)_2$	$EtP(SnMe_3)_2$	−188.3(b)	98
	$C_3;(C_3)_2$	$tBuP(SnMe_3)_2$	−111.1(b) J_{PSn}812	277
$CSn_2/5$	$C_3;(C_3)_2$	$tBuP*SnMe_2CH_2CH_2$-$Sn*Me_2$	−161(b)	266
$C'Sn_2$	$C'_2;(C_3)_2$	$PhP(SnMe_3)_2$	−164.9(b), −163.9(b)	98, 277
	$C'_2;(C_3)_2$	$PhP(SnBu_3)_2$	−170(b)	260
	$C'_2;(C'_3)_2$	$PhP(SnPh_3)_2$	−163	260
$C'Sn_2/5$	$C'_2;(C_3)_2$	$PhP*SnMe_2CH_2CH_2$-$Sn*Me_2$	−190(b)	266
	$C'_2;(CP)_2$	**99**	−5.2(b)	279
CHGe	$HSi_2;Cl_3$	$Tms_2CHPHGeCl_3$	−72.5	6
C'HGe	$C'_2;C'_2Cl$	$SmsPHGeClR_2$; R = Mes or Sms	−90.9 J215	6
	$C'_2;C'_2F$	$SmsPHGeFMes_2$	−110.6 J209	6
	$C'_2;C'_2N$	$SmsPHGeNHPhMes_2$	−94.9 J212	6
	$C'_2;C'_2O$	$SmsPHGeOHMes_2$	−105.6 J209	6
		$SmsPHGeOMeMes_2$	−107 J212	6
		$SmsPHGeOCOMeMes_2$	−98.6 J226	6
	$C'_2;C'_2S$	$SmsPHGe(S\text{-}iPr)Mes_2$	−86.3 J215	6
	$C'_2;C_3$	$PhPHGeMe_3$	−119, −117(b)	260, 277
	$C'_2;C'_3$	Sms-$PHGe(CH{=}PMe_3)Mes_2$	−86.3 J215	6
	$C'_2;C'_2C''$	$SmsPHGe(C{\equiv}CPh)Mes_2$	−106.3 J212	6
C'HGe/5	$C'_2;C_2P$	**100**; R^1 = H, R^2 = Et, M = Ge	−117.5 J175.9, −123.2 J179.6(b)	48
CHSi	$H_3;H_3$	$HPMeSiH_3$	−202 J_{PSi}31	259
	$H_3;CH_2$	$HPMeSiH_2Me$	−190 J_{PSi}27.5	259
	$H_3;C_2H$	$HPMeSiHMe_2$	−181 J_{PSi}19.8	259
	$H_3;C_3$	HPMeTms	−175 J_{PSi}17.4	259
C'HSi	$NN';C_3$	PhN(Tms)C(=NPh) PHTms	−141.7 J216	4
	$C'_2;C_3$	$PhPHTms_2$	−121.1(b), −56.2(b)	277, 278
		$Dop(PHTms_2)_2$	−121.0(b)	48
C'HSi/5	$C'_2;C_2P$	**100**; R^1 = H, R^2 = Me, M = Si.	−119.3 J183.1, −110.4 J181.1(b)	48
C'HSn	$C'_2;C_3$	$PhPHSnMe_3$	−141.7(b) J_{PSn}537	277
C_2Ge	$(C_3)_2;C_3$	tBu_2PGeMe_3	14.3(b)	277
$C_2Ge/5$	$CH_2,C_2H;C_3$	$C*H_2CH_2CH_2CH$-$MeP*GeMe_3$	-66.4^E; -59.4^Z(b)	41
	$C'H_2,CC'H;C_3$	$C*H_2CH{=}CHCHMe$-$P*GeMe_3$	-73.0^E; -58.6^Z(b)	41

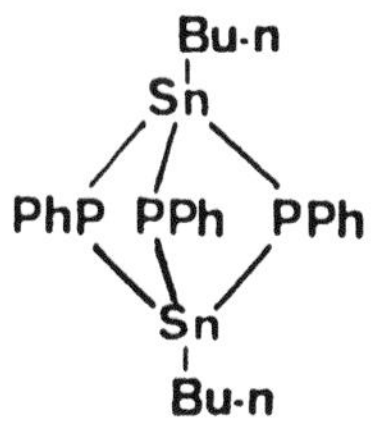

99 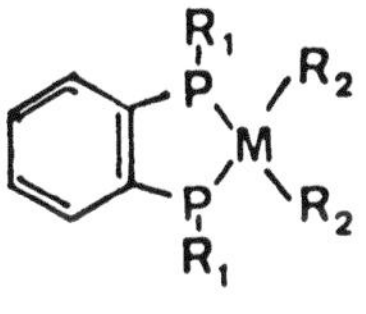 **100** 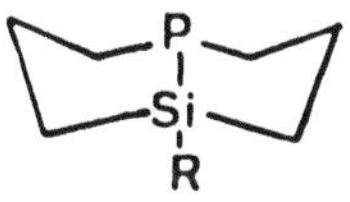 **101**

Rings/ Connectivities	Formula	Structure	^{31}P NMR data (δP [solv, temp] J_{PH} Hz)	Ref.
CC'Ge	C_3,C'_2;C_3	tBuPhPGeMe_3	−13.8(b)	277
CC'Ge/5	H_3;C'_2;C_2P	**100**; R^1 = Me, R^2 = Et, M = Ge	−70.7; −74.2^{I}(b)	48
	CH_2;C'_2;C_2N	PhP*CH_2CH_2NMe-Ge*Me_2	−87.5(b)	263
	CH_2;C'_2;C_2O	PhP*CH_2CH_2OGe*Me_2	−101.8(b)	6
	CH_2;C'_2;C_2P	PhP*CH_2CH_2PPhGe*Me_2	−63.7; −71.3^{I}(b)	266
	CH_2;C'_2;C_2S	PhP*CH_2CH_2SGe*Me_2	−73.2^{I}(b)	6
	C_2H;C'_2;C_2P	**100**; R^1 = iPr, R^2 = Et, M = Ge	−39.7; −41.3^{I}(b)	48
CC'Ge/6	CH_2;C'_2;C_2P	PhP*CH_2CH_2GeMe_2-PPhGe*Me_2	−74.3; −69.2^{I}(b)	6
C'_2Ge	$(C'_2)_2$;C_3	Ph_2PGeMe_3	−48.5(b)	277
C'_2Ge/5	$(C'_2)_2$;C_2P	**100**; R^1 = Ph, R^2 = Me, M = Ge	−33.5; −38.5^{I}(b)	266
C_2Si	$(H_3)_2$;H_3	Me_2PSiH_3	−143(n) J_{PSi}31.0	259
	$(H_3)_2$;CH_2	Me_2PSiH_2Me	−138.0(n) J_{PSi}29.4	259
	$(H_3)_2$;C_2H	Me_2PSiHMe_2	−135.5(n) J_{PSi}23.2	259
	$(H_3)_2$;C_3	Me_2PTms	−132.0(n) J_{PSi}20.3	259
	$(CH_2)_2$;C_3	Et_2PTms	−89.3(b) J_{PSi}23.2	98
	$(C_3)_2$;C_3	tBu_2PTms	−3.2(b)	277
C_2Si/5	CH_2,C_2H;C_3	TmsP*$CH_2CH_2CH_2$-C*HMe	−80.5^{E}; −77.2^{Z}(b)	41
	$C'H_2$,CC'H;C_3	TmsP*CH_2CH=CH-C*HMe	−87^{E}; −75.8^{Z}(b)	41
C_2Si/6	$(CH_2)_2$;C_3	TmsP*CH_2CH_2(NEt_2)-CH_2C*H_2	−100.2	42
C_2Si/5,5	$(CH_2)_2$;C_3	**101**; R = Me	−126.0(b)	43
	$(CH_2)_2$;C_2C'	**101**; R = Ph	−123.4(b)	43
CC'Si	H_3;CO';C_3	MePTmsCO-tBu	−71(pe)	46, 261
	H_3;C'_2;C_3	MePhPTms	−98.2(m)	278
	CH_2;C'_2;C_3	EtPhPTms	−76.3(m)	278
		PrPhPTms	−83.6(m)	278
		BuPhPTms	−82.9(m)	278
	C_2H;CO';C_3	cHexPTmsCO-tBu	−45	261
	C_2H;C'_2;C_3	cPePhPTms	−26.4(b)	277
		cHexPhPTms	−62.2(m)	278
		$(iPrPTms)_2$Dop	−69.5(t/b)	48
	C_3;CO';C_3	tBuPTmsCO-tBu	−107.0(b)	262
CC'Si/5	H3;C'_2;C_2P	**100**; R^1 = R^2 = Me, M = Si	−82.5; −86.5^{I}(b)	48
	CH_2;C'_2;C_2N	PhP*CH_2CH_2NMe-Si*Me_2	−93.7(b)	263
	CH_2;C'_2;C_2S	PhP*CH_2CH_2SSi*Me_2	−79.1(b)	263

TABLE E (continued)
^{31}P NMR Data of Three Coordinate (λ3σ3) Compounds Containing Phosphorus Bonds to Hydrogen and/or Group IV Atoms but Not to Halogen, Chalcogenide or Group V Atoms

Rings/ Connectivities	Formula	Structure	^{31}P NMR data (δP [solv, temp] J_{PH} Hz)	Ref.
	$C_2H;C'_2;C_2P$	**100**; R[1] = iPr, R[2] = H, M = Si	−42.0; −40.2[l](b)	48
C'_2Si	$NP',C'_2;C_3$	4-ClPh'NTmsC(=PPh)-PTmsPh	−40.9(c)	264
		4-TfPh'NTmsC(=PPh)-PTmsPh	−38.2(c)	264
		PhCONTmsC(=PPh)-PTmsPh	−30.4[E]; −41.2[l]	265
	$CO',C'_2;C_3$	tBuCOPTmsPh	−33.0	261
	$(C'_2)_2;C_3$	Ph_2PTms	−56.8(b), −53.7(m)	277, 278
$C'_2Si/5$	$(C'_2)_2;C_2P$	**100**; R[1] = Ph, R[2] = Me, M = Si	−48.1; −51.0[l](b)	266
C_2Sn	$(CH_2)_2;C_3$	Et_2PSnMe_3	−77.4(b) J_{PSn}665	98
	$(C_3)_2;C_3$	tBu_2PSnMe_3	20.7(b) J_{PSn}795	277
CC'Sn	$C_3;C'_2;C_3$	$tBuPhPSnMe_3$	−13.4(b) J_{PSn}690.5	277
CC'Sn/5	$CH_2;C'_2;C_2O$	$PhP^*CH_2CH_2OSn^*Me_2$	−102.0(k)	263
	$CH_2;C'_2;C_2P$	$PhP^*CH_2CH_2PPhSn^*Me_2$	−77.3, −90.5(b)	266
	$CH_2;C'_2;C_2S$	$PhP^*CH_2CH_2SSn^*Me_2$	−86.3	6
	$C_2H;C'_2;C_2P$	**100**; R[1] = iPr, R[2] = Me, M = Sn	−25.5; −28.4[l](b)	48
	$C_2H;C'_2;C_2P$	**100**; R[1] = iPr, R[2] = Ph, M = Sn	−26.7; −29.2[l](b)	48
CC'Sn/6	$CH_2;C'_2;C_2P$	$PhP^*CH_2CH_2Ge$ $Me_2PPhSn^*Me_2$	−69.8, −81.5	6
C'_2Sn	$(C'_2)_2;C_3$	Ph_2PSnMe_3	−57.8(b) J_{PSn}586	277
		Ph_2PSnBu_3	−56	260
	$(C'_2)_2;C'_3$	Ph_2PSnPh_3	−56	260, 276
$C'_2Sn/5$	$(C'_2)_2;C_2P$	**100**; R[1] = Ph, R[2] = Me, M = Sn	−44.2; −50.8[l](b)	266

REFERENCES

1. **Vincent E., Verdonck, L., and Van der Kelen, G. P.,** *Spectrochim. Acta, Part A*, 36A, 699, 1980.
2. **Junkes, P., Baudler, M., Dobbers, J., and Rackwitz, D.,** *Z. Naturforsch.*, 27b, 1451, 1972.
3. **Maier, L.,** Primary, Secondary and Tertiary Phosphines, in *Organic Phosphorus Compounds*, Vol. 1, Kosolapoff, G. M. and Maier, L., Eds., John Wiley & Sons, New York, 106, 1972.
4. **Issleib, K., Schmidt, H., and Wirkner, C.,** *Synth. React. Inorg. Met.-Org. Chem.*, 11, 279, 1981.
5. **Zschunke, A., Bauer, E., Schmidt, H., and Issleib, K.,** *Z. Anorg. Allg. Chem.*, 495, 115, 1982.
6. **Satgé, J., Escudié, J., Couret, C., Ranaivonjatow, H., and Andrianarison, M.,** *Phosphorus Sulfur*, 27, 65, 1986; *J. Am. Chem. Soc.*, 109, 386, 1987.
7. **Issleib, K., Kühne, U., and Krech, F.,** *Phosphorus Sulfur*, 17, 73, 1983.
8. **Hietkamp, S., Sommer, H., and Stelzer, O.,** *Chem. Ber.*, 117, 3400, 1984.
9. **Kühne, U., Krech, F., and Issleib, K.,** *Phosphorus Sulfur*, 13, 153, 1982.
10. **McFarlane, W.,** *Proc. R. Soc. A*, 306, 185, 1968.
11. **Diel, B. N. and Norman, A. D.,** *Phosphorus Sulfur*, 12, 227, 1982.

12. **King, R. B., Cloyd, J. C., and Kappor, P. N.,** *J. Chem. Soc. Perkin Trans. I,* 2226, 1973.
13. **Lasne, M. C., Ripoll, J. L., and Thuillier, A.,** *J. Chem. Soc. Chem. Commun.,* 1428, 1986.
14. **Thamm, R. and Fluck, E.,** *Z. Naturforsch.,* 36b, 910, 1981.
15. **Gordon, M. D. and Quin, L. D.,** *J. Chem. Soc. Chem. Commun.,* 35, 1975.
16. **Gordon, M. D. and Quin, L. D.,** *J. Am. Chem. Soc.,* 98, 15, 1976.
17. **Zschunke, A., Riemer, M., Schmidt, H., and Issleib, K.,** *Phosphorus Sulfur,* 17, 237, 1983.
18. **Goetz, H., Hadamik, H., and Juds, H.,** *Justus Liebigs Ann. Chem.,* 737, 132, 1970.
18a. **Seymour, S. J. and Jonas, J.,** *J. Magn. Resonance,* 89, 376, 1972.
19. **Maier, L.,** *Phosphorus,* 4, 41, 1974.
20. **Issleib, K. and Vollmer, R.,** *Z. Chem.,* 18, 451, 1978.
21. **Batchelor, R. and Birchall, Th.,** *J. Am. Chem. Soc.,* 104, 674, 1982.
22. **Issleib, K. and Vollmer, R.,** *Z. Anorg. Allg. Chem.,* 481, 22, 1981.
23. **Kyba, E. P., Liu, S. T., and Harris, R. L.,** *Organometallics,* 2, 1877, 1983.
24. **Kyba, E. P., Kerby, M. C., and Rines, St. P.,** *Organometallics,* 5, 1189, 1986.
25. **Gerdau, T. and Kramolowsky, R.,** *Z. Naturforsch.,* 37b, 332, 1982.
26. **Baacke, M., Stelzer, O., and Wray, V.,** *Chem. Ber.,* 113, 1356, 1980.
27. **Baacke, M., Hietkamp, S., Morton, S., and Stelzer, O.,** *Chem. Ber.,* 114, 2568, 1981.
28. **Grim, S. O. and Molenda, R. P.,** *Phosphorus,* 4, 189, 1974.
29. **Rithner, C. D. and Bushweller, C. H.,** *J. Am. Chem. Soc.,* 107, 7823, 1985.
30. **Dubois, R. A. and Garrou, P. E.,** *Organometallics,* 5, 466, 1986.
31. **Kapoor, P. N., Pathak, D. D., Gaur, G., and Kutty, M.,** *J. Organomet. Chem.,* 276, 167, 1984.
32. **Grim, S. O. and Yankowsky, A. W.,** *J. Org. Chem.,* 42, 1236, 1977.
33. **Uriarte, R., Mazanec, T. J., Tau, K. D., and Meek, D. W.,** *Inorg. Chem.,* 19, 79, 1980.
34. **Glaser, R., Kountz, D. J., Waid, R. D., Gallucci, J. C., and Meek, D. W.,** *J. Am. Chem. Soc.,* 106, 6324, 1984.
35. **Bookham, J. L., McFarlane, W., and Colquhoun, I. J.,** *J. Chem. Soc. Chem. Commun.,* 1041, 1986.
36. **Cowley, A. H. and Pakulski, M. K.,** *J. Am. Chem. Soc.,* 106, 1491, 1984.
37. **Issleib, K., Schmidt, H., and Wirkner, Ch.,** *Z. Anorg. Allg. Chem.,* 488, 75, 1982.
38. **Cowley, A. H., Kilduff, J. E., Norman, N. C., and Pakulski, M.,** *J. Chem. Soc. Dalton Trans.,* 1801, 1986.
39. **Issleib, K. and Schmidt, H.,** *Z. Anorg. Allg. Chem.,* 459, 131, 1979.
40. **Zschunke, A., Meyer, H., and Issleib, K.,** *Org. Magn. Reson.,* 7, 470, 1975.
41. **Couret, C., Escudie, J., and Thaoubane, S. A.,** *Phosphorus Sulfur,* 20, 81, 1984.
42. **Hackney, M. L. J. and Norman, A. D.,** *J. Chem. Soc., Chem. Commun.,* 850, 1986.
43. **Issleib, K., Kühne, U., and Krech, F.,** *Phosphorus Sulfur* 21, 367, 1985.
44. **Charrier, C., Maigrot, N., Mathey, F., Robert, F., and Jeannin, Y.,** *Organometallics,* 5, 623, 1986.
45. **Issleib, K. and Gans, W.,** *Z. Anorg. Allg. Chem.,* 491, 163, 1982.
46. **Schmidbaur, H. and Schnatterer, S.,** *Chem. Ber.,* 119, 2832, 1986.
47. **Schmidbaur, H. and Schnatterer, S.,** *Chem. Ber.,* 119, 2841, 1986.
48. **Issleib, K., Leissring, E., and Riemer, M.** *Z. Anorg. Allg. Chem.,* 519, 75, 1984.
49. **Burg, A. B.,** *Inorg. Nucl. Lett.,* 13, 199, 1977.
50. **Volcko, E. J. and Verkade, J. G.,** *Phosphorus Sulfur,* 21, 111, 1984.
51. **Colson, J. G. and Marr, D. H.,** *Anal. Chem.,* 45, 370, 1973.
52. **Rothwell, W. P., Shen, W. X., and Lunsford, J. H.,** *J. Am. Chem. Soc.,* 106, 2452, 1984.
53. **Karsch, H. H.,** *Z. Naturforsch.,* 34b, 1178, 1979.
54. **Ignateva, S. N., Nikonov, G. N., Erastov, O. A., and Arbuzov, B. A.,** *Bull. Acad. Sci.,* (USSR), 1005, 1985.
55. **Holt, M. S. and Nelson, J. H.,** *Inorg. Chem.,* 25, 1316, 1986.
56. **Yashina, N. S., Gefel, E. I., Petrosyan, V. S., and Reutov, O. A.,** *Dokl. Akad. Nauk SSSR,* 283, 654, 1985.
57. **Clark, P. W., Curtis, J. L. S., Garrou, P. E., and Hartwell, G. E.,** *Can. J. Chem.,* 52, 1714, 1974.
58. **Rehder, D. and Kececi, A.,** *Inorg. Chim. Acta,* 103, 173, 1985.
59. **Schmidbaur, H. and Inoguchi, Y.,** *Z. Naturforsch.,* 35b, 1329, 1980.
60. **Lindner, E., Scheytt, C., and Wegner, P.,** *J. Organomet. Chem.,* 308, 311, 1986.
61. **Venanzi, L. M.,** *Pure Appl. Chem.,* 52, 1117, 1980.
62. **Schmidbaur, H. and Schier, A.,** *Chem. Ber.,* 114, 3385, 1981.
63. **Allen, D. W. and Taylor, B. F.,** *J. Chem. Soc. Dalton Trans.,* 51, 1982.
64. **Chervin, I. I., Isobaev, M. D., El'natanov, Yu, I., Shikhaliev, Sh. M., Bystrov, L. V., and Kostyanovskii, R. G.,** *Izv. Akad. Nauk SSSR, Ser. Khim.,* 1769, 1981.
65. **Nichols, D. I.,** *J. Chem. Soc. A,* 1471, 1969.
66. **Grim, S. O. and Yankowsky, A. W.,** *Phosphorus Sulfur* 3, 191, 1977.
67. **Pinell, R. P., Megerle, C. A., Manatt, S. L., and Kroon, P. A.,** *J. Am. Chem. Soc.,* 95, 977, 1973.

68. **Stegmann, H. B., Kühne, H. M., Wax, G., and Scheffler, K.,** *Phosphorus Sulfur,* 13, 331, 1982.
69. **Muylle, E. and Van Der Kelen, G. P.,** *Spectrochim. Acta, Part A,* 32A, 599, 1976.
70. **Bemi, L., Clark, H. C., Davies, J. A., Fyfe, C. A., and Wasylishen, R. E.,** *J. Am. Chem. Soc.,* 104, 438, 1982.
71. **Schiemenz, G. P. and Nielsen, P.,** *Phosphorus Sulfur,* 21, 259, 1985.
72. **Dickert, F. L. and Hellmann, S. W.,** *Anal. Chem.,* 52, 996, 1980.
73. **Stumbreviciute, Z., Fedorov, L. A., Dyatkin, B. L., and Martynov, B. I.,** *Zh. Org. Khim.,* 11, 2002, 1975.
74. **Socol, S. M. and Verkade, J. G.,** *Inorg. Chem.,* 23, 3487, 1984.
75. **Miller, G. R., Yankowsky, A. W., and Grim, S. O.,** *J. Chem. Phys.,* 51, 3185, 1969.
76. **Cogne, A., Wiesenfeld, L., Robert, J. B., and Tyka, R.,** *Org. Magn. Reson.,* 13, 72, 1980.
77. **Horner, L. and Simons, G.,** *Phosphorus Sulfur,* 14, 189, 1983.
78. **Colton, R. and Dakternieks, D.,** *Aust. J. Chem.,* 34, 323, 1981.
79. **Verstuyft, A. W., Redfield, D. A., Cary, L. W., and Nelson, J. H.,** *Inorg. Chem.,* 16, 2776, 1977.
80. **Furin, G. G., Rezvukhin, A. I., Fedotov, M. A., and Yakobson, G. G.,** *J. Fluorine Chem.,* 22, 231, 1983.
81. **Wille, E. E., Stephenson, D. S., Capriel, P., and Binsch, G.,** *J. Am. Chem. Soc.,* 104, 405, 1982.
82. **Hengefeld, A. and Nast, R.,** *Chem. Ber.,* 116, 2035, 1983.
83. **Grimley, E. and Meek, D. W.,** *Inorg. Chem.,* 25, 2049, 1986.
84. **Stepanov, B. I., Bokanov, A. I., and Svergun, V. I.,** *Zh. Obshch. Khim.,* 41, 533, 1971.
85. **Grim, S. O., Shah, D. P., Haas, C. K., Ressner, J. M., and Smith, P. H.,** *Inorg. Chim. Acta,* 36, 139, 1979.
86. **Grim, S. O., Satek, L. C., and Mitchell, J. D.,** *Z. Naturforsch.,* 35b, 832, 1980.
87. **Rosenberg, D. and Drenth, W.,** *Tetrahedron,* 27, 3893, 1971.
88. **Burg, A. B.,** *Inorg. Chem.,* 20, 3734, 1981.
89. **Karsch, H. H.,** *Chem. Ber.,* 115, 823, 1982.
90. **Karsch, H. H.,** *Chem. Ber.,* 117, 783, 1984.
91. **Schmidbaur, H. and Zybill, Ch. E.,** *Chem. Ber.,* 114, 3589, 1981.
92. **Holmes-Smith, R. D., Osei, R. D., and Stobart, St. R.,** *J. Chem. Soc. Perkin Trans. I,* 861, 1983.
93. **Abicht, H. P. and Issleib, K.,** *Z. Anorg. Allg. Chem.,* 447, 53, 1978.
94. **Payne, N. C. and Stephan, D. W.,** *Can. J. Chem.,* 58, 15, 1980.
95. **Shaw, B. L., Mann, B. E., and Masters, C.,** *J. Chem. Soc. A,* 1104, 1971.
96. **Tzschach, A. and Uhlig, W.,** *Z. Anorg. Allg. Chem.,* 475, 251, 1981.
97. **Bychkov, N. N., Bokanov, A. I., and Stepanov, B. I.,** *Zh. Obshch. Khim.,* 49, 1460, 1979.
98. **Van Linthoudt, J. P., Van den Berghe, E. V., and Van der Kelen, G. P.,** *Spectrochim. Acta, Part A,* 36A, 17, 1980.
99. **Sommer, H., Hietkamp, S., and Stelzer, O.,** *Chem. Ber.,* 117, 3414, 1984.
100. **Schier, A. and Schmidbaur, H.,** *Chem. Ber.,* 117, 2314, 1984.
101. **Neilson, R. H.,** *Phosphorus Sulfur,* 18, 43, 1983.
102. **Moulton, C. J. and Shaw, B. L.,** *J. Chem. Soc. Dalton Trans.,* 299, 1980.
103. **Karsch, H. H.,** *Chem. Ber.,* 115, 1956, 1982.
104. **Inoguchi, Y., Milewsky-Mahrla, B., and Schmidbaur, H.,** *Chem. Ber.,* 115, 3085, 1982.
105. **Cook, R. L. and Morse, J. G.,** *Inorg. Chem.,* 21, 4103, 1982.
106. **Levason, W., Smith, K. G., and McAuliffe, C. A.,** *J. Chem. Soc. Dalton Trans.,* 1718, 1979.
107. **Verstuyft, A. W., Nelson, J. H., and Cary, L. W.,** *Inorg. Nucl. Chem. Lett.,* 12, 53, 1976.
108. **Heinicke, J., Nietzschmann, E., and Tzschach, A.,** *J. Organomet. Chem.,* 243, 1, 1983.
109. **Heinicke, J., Nietzschmann, E., and Tzschach, A.,** *J. Organomet. Chem.,* 310, C17, 1986.
110. **Costa, T. and Schmidbaur, H.,** *Chem. Ber.,* 115, 1374, 1982.
111. **Goetz, H., Hadamik, H., and Juds, H.,** *Justus Liebigs Ann. Chem.,* 737, 132, 1970.
112. **Al-Sa'ady, A. K. H., McAuliffe, C. A., Moss, K., Parish, R. V., and Fields, R.,** *J. Chem. Soc. Dalton Trans.,* 491, 1984.
113. **Kozlov, E. S., Tolmachev, A. A., Chernega, A. N., and Boldeskul, I. E.,** *Zh. Obshch. Khim.,* 55, 1262, 1985.
114. **Heinicke, J. and Tzschach, A.,** *J. Prakt. Chem.,* 325, 232, 1983.
115. **Lindner, E., Rauleder, H., Scheytt, Ch., Mayer, H. A., Hiller, W., Fawzi, R., and Wegner, P.,** *Z. Naturforsch.,* 39b, 632, 1984.
116. **Singh, G. and Reddy, G. S.,** *J. Org. Chem.,* 44, 1057, 1979.
117. **King, R. B. and Heckley, P. R.,** *Phosphorus,* 3, 209, 1974.
118. **Grim, S. O., Molenda, R. P., and Mitchell, J. D.,** *J. Org. Chem.,* 45, 250, 1980.
119. **Lindner, E., Rauleder, H., and Hiller, W.,** *Z. Naturforsch.,* 38b, 417, 1983.
120. **Carr, S. W., Colton, R., and Dakternieks, D.,** *Inorg. Chem.,* 23, 720, 1984.
121. **Lindner, E. and Merkle, R. D.,** *Z. Naturforsch.,* 40b, 1580, 1985.

122. **Schmidbaur, H., Graf, W., and Müller, G.,** *Helv. Chim. Acta,* 69, 1748, 1986.
123. **Schmidbaur, H., Strunk, S., and Zybill, C. E.,** *Chem. Ber.,* 116, 3559, 1983.
124. **Colquhoun, I. J., McFarlane, H. C. E., and McFarlane, W.,** *Phosphorus Sulfur,* 18, 61, 1983.
125. **Schmidbaur, H. and Deschler, U.,** *Chem. Ber.,* 116, 1386, 1983.
126. **Schmidbaur, H., Deschler, U., and Mahrla, B. M.,** *Chem. Ber.,* 115, 3290, 1982.
127. **Walton, E. D.,** Dissertation, University of Maryland, 1979.
128. **Grim, S. O. and Walton, E. D.,** *Phosphorus Sulfur,* 9, 123, 1980.
129. **Kellner, K., Hanke, W., and Tzschach, A.,** *Z. Chem.,* 24, 193, 1984.
130. **Grim, S. O. and Matienzo, L. J.,** *Tetrahedron Lett.,* 2951, 1973.
131. **Lindner, E. and Küster, E. U.,** *Z. Naturforsch.,* 39b, 115, 1984.
132. **Grim, S. O., Briggs, W. L., Barth, R. C., Tolman, C. A., and Jesson, J. P.,** *Inorg. Chem.,* 13, 1095, 1974.
133. **Grim, S. O. and Walton, E. D.,** *Inorg. Chem.,* 19, 1982, 1980.
134. **Schmidbaur, H. and Deschler, U.,** *Chem. Ber.,* 114, 2491, 1981.
135. **Schmidbaur, H., Deschler, U., Zimmer-Gasser, B., Neugebauer, D., and Schubert, U.,** *Chem. Ber.,* 113, 902, 1980.
136. **Weber, L., Wewers, D., and Lücke, E.,** *Z. Naturforsch.,* 40b, 968, 1985.
137. **Capka, M., Schraml, J., and Jancke, H.,** *Collect. Czech. Chem. Commun.,* 43, 3347, 1978.
138. **Colquhoun, I. J. and McFarlane, W.,** *J. Chem. Soc. Chem. Commun.,* 484, 1982.
139. **Habib, M., Trujillo, H., Alexander, C. A., and Storhoff, B. N.,** *Inorg. Chem.,* 24, 2344, 1985.
140. **Lindner, E., Funk, G., and Hoehne, S.,** *Chem. Ber.,* 114, 2465, 1981.
141. **Al-Baker, S., Hill, W. E., and McAuliffe, C. A.,** *J. Chem. Soc. Dalton Trans.,* 2655, 1985.
142. **Smith, R. T., Ungar, R. K., and Baird, M. C.,** *Transition Met. Chem.,* 7, 288, 1982.
143. **Shinoda, S., Nakamura, K., and Saito, Y.,** *Chem. Lett.,* 1449, 1983.
144. **Edwards, D. A. and Marshalsea, J.,** *J. Organomet. Chem.,* 96, C50, 1975.
145. **Bechthold, H. C. and Rehder, D.,** *J. Organomet. Chem.,* 233, 215, 1982.
146. **Colquhoun, I. J. and McFarlane, W.,** *J. Chem. Soc. Dalton Trans.,* 1915, 1982.
147. **Dean, P. A. W.,** *Can. J. Chem.,* 57, 754, 1979.
148. **Holy, N., Deschler, U., and Schmidbaur, H.,** *Chem. Ber.,* 115, 1379, 1982.
149. **Quayle, W. H. and Pinnavaia, T. J.,** *Inorg. Chem.,* 18, 2840, 1979.
150. **Mazanec, T. J., Tau, K. D., and Meek, D. W.,** *Inorg. Chem.,* 19, 85, 1980.
151. **Colquhoun, I. J., McFarlane, W., and Keiter, R. L.,** *J. Chem. Soc. Dalton Trans.,* 455, 1984.
152. **Evangelidou-Tsolis, E., Ramirez, F., Pilot, J. F., and Smith, C. P.,** *Phosphorus,* 4, 109, 1974.
153. **Clark, H. C., Davies, J. A., Fyfe, C. A., Hayes, P. J., and Wasylishen, R. E.,** *Organometallics,* 2, 177, 1983.
154. **Schmidbaur, H. and Wohleben-Haumer, A.,** *Chem. Ber.,* 112, 510, 1979.
155. **Salvadori, P., Lazzaroni, R., Raffaelli, A., Pucci, S., Bertozzi, S., Pini, D., and Fatti, G.,** *Chim. Ind. (Milan),* 63, 492, 1981.
156. **Allen, D. W. and Taylor, B. F.,** *J. Chem. Res. Synop.,* 220, 1981.
157. **Wohlleben, A. and Schmidbaur, H.,** *Angew. Chem.,* 89, 428, 1977.
158. **Lindner, E. and Tamoutsidis, E.,** *Z. Naturforsch.,* 38b, 726, 1983.
159. **Mann, B. E., Masters, C., Shaw, B. L., Slade, R. M., and Stainbank, R. E.,** *Inorg. Nucl. Chem. Lett.,* 7, 881, 1971.
160. **Antoniadis, A., Bruns, A., and Kunze, U.,** *Phosphorus Sulfur,* 15, 317, 1983.
161. **Lindner, E., Steinwand, M., and Hoehne, S.,** *Chem. Ber.,* 115, 2478, 1982.
162. **Lindner, E., Steinwand, M., and Hoehne, S.,** *Chem. Ber.,* 115, 2181, 1982.
163. **Lindner, E. and Lesiecki, H.,** *Z. Naturforsch.,* 33b, 849, 1978.
164. **Lindner, E., Merkle, R. D., Hiller, W., and Fawzi, R.,** *Chem. Ber.,* 119, 659, 1986.
165. **Lindner, E. and Frey, G.,** *Chem. Ber.,* 113, 2769, 1980.
166. **Lindner, E. and Frey, G.,** *Chem. Ber.,* 113, 3268, 1980.
167. **Lindner, E. and Hübner, D.,** *Chem. Ber.,* 116, 2574, 1983.
168. **Dankowsky, M., Praefcke, K., Nyburg, S. C., and Wong-Ng, W.,** *Phosphorus Sulfur,* 7, 275, 1979.
169. **Martens, J., Praefcke, K., Schwarz, H., and Simon, H.,** *Phosphorus,* 6, 247, 1976.
170. **Dankowski, M. and Praefcke, K.,** *Phosphorus Sulfur,* 8, 105, 1980.
171. **Schmidbaur, H., Herr, R., and Riede, J.,** *Angew. Chem.,* 96, 237, 1984.
172. **Duncan, M. and Gallagher, M. J.,** *Org. Magn. Reson.,* 15, 37, 1981.
173. **Carty, A. J., Johnson, D. K., and Jacobson, S. E.,** *J. Am. Chem. Soc.,* 101, 5612, 1979.
174. **Schmutzler, R., Schomberg, D., Bartsch, R., and Stelzer, O.,** *Z. Naturforsch.,* 39b, 1177, 1984.
175. **McEwen, W. E., Shiau, W. I., Yeh, Y. I., Schulz, D. N., Pagilagan, R. U., Levy, J. B., Symmes, C., Jr., Nelson, G. O., and Granoth, I.,** *J. Am. Chem. Soc.,* 97, 1787, 1975.
176. **Fyfe, C. A., Davies, J. A., Clark, H. C., Hayes, P. J., and Wasylishen, R. E.,** *J. Am. Chem. Soc.,* 105, 6577, 1983.

177. **Fyfe, C. A., Davies, J. A., Clark, H. C., Hayes, P. J., and Wasylishen, R. E.,** *J. Am. Chem. Soc.*, 105, 6581, 1983.
178. **Horner, L. and Simons, G.,** *Phosphorus Sulfur*, 15, 165, 1983.
179. **Grim, S. O., Shah, D. P., Haas, C. K., Ressner, J. M., and Smith, P. H.,** *Inorg. Chim. Acta*, 36, 139, 1979.
180. **Payne, N. C. and Stephan, D. W.,** *Inorg. Chem.*, 21, 182, 1982.
181. **Ivanov, P. Yu., Negrebetskii, V. V., Bychkow, N. N., and Stepanov, B. I.,** *Zh. Obshch. Khim.*, 51, 1533, 1981.
182. **Granoth, I.,** *J. Chem. Soc. Perkin Trans. I*, 735, 1982.
183. **Schiemenz, G. P. and Nielsen, P.,** *Phosphorus Sulfur*, 21, 267, 1985.
184. **Bowmaker, G. A., Herr, R., and Schmidbaur, H.,** *Chem. Ber.*, 116, 3567, 1983.
185. **Bokanov, A. I., Bychkov, N. N., Negrebetskii, V. V., and Stepanov, B. I.,** *Zh. Obshch. Khim.*, 49, 993, 1979.
186. **Packer, K. J.,** *J. Chem. Soc.*, 1555, 1963.
187. **Maier, L.,** *Helv. Chim. Acta*, 46, 2026, 1963.
188. **Mynott, R., Richter, W. J., and Wilke, G.,** *Z. Naturforsch.*, 41b, 85, 1986.
189. **Lindner, E., Merkle, R. D., and Mayer, H. A.,** *Chem. Ber.*, 119, 645, 1986.
190. **Abicht, H. P., Baumeister, U., Hartung, H., Issleib, K., Jacobson, R. A., Richardson, J., Socol, S. M., and Verkade, J. G.,** *Z. Anorg. Allg. Chem.*, 494, 55, 1982.
191. **Shortt, A. B., Durham, L. J., and Mosher, H. S.,** *J. Org. Chem.*, 48, 3125, 1983.
192. **Horner, L. and Simons, G.,** *Phosphorus Sulfur*, 19, 65, 1984.
193. **Schmidbaur, H. and Schnatterer, S.,** *Chem. Ber.*, 116, 1947, 1983.
194. **Richter, W. J.,** *Angew. Chem.*, 94, 298, 1982.
195. **Quin, L. D., Rao, N. S., Topping, R. J., and McPhail, A. T.,** *J. Am. Chem. Soc.*, 108, 4519, 1986.
196. **Schmidt, S., Mathey, F., and Schmidpeter, A.,** *Tetrahedron Lett.*, 27, 2635, 1986.
197. **Marinetti, A. and Mathey, F.,** *J. Am. Chem. Soc.*, 107, 4700, 1985.
198. **Marinetti, A., Mathey, F., Fischer, J., and Mitchler, A.,** *J. Chem. Soc. Chem. Commun.*, 45, 1984.
199. **Heinicke, J. and Tzschach, A.,** *Phosphorus Sulfur*, 25, 345, 1985.
200. **Clennan, E. L. and Heah, P. C.,** *J. Org. Chem.*, 48, 2621, 1983.
201. **Symmes, C., Jr. and Quin, L. D.,** *Tetrahedron Lett.*, 1853, 1976.
202. **Breen, J. J., Engel, J. F., Myers, D. K., and Quin, L. D.,** *Phosphorus*, 2, 55, 1972.
203. **Quin, L. D. and Caster, K. C.,** *Phosphorus Sulfur* 25, 117, 1985.
204. **Beer, P. D., Hammond, P. J., and Hall, C. D.,** *Phosphorus Sulfur*, 10, 185, 1981.
205. **Quin, L. D., Mesch, K. A., and Orton, W. L.,** *Phosphorus Sulfur*, 12, 161, 1982.
206. **Muller, G., Bonnard, H., and Mathey, F.,** *Phosphorus Sulfur*, 10, 175, 1981.
207. **Alcaraz, J. M., Deschamps, E., and Mathey, F.,** *Phosphorus Sulfur*, 19, 45, 1984.
208. **Mathey, F.,** *Tetrahedron*, 28, 4171, 1972.
209. **Mercier, F., Holand, S., and Mathey, F.,** *J. Organomet. Chem.*, 316, 271, 1986.
210. **Nief, F., Charrier, C., Mathey, F., and Simalty, M.,** *Phosphorus Sulfur*, 13, 259, 1982.
211. **Lindner, E. and Frey, G.,** *Z. Naturforsch.*, 35b, 1150, 1980.
212. **Parrinello, G., Deschenaux, R., and Stille, J. K.,** *J. Org. Chem.*, 51, 4189, 1986.
213. **Fritschel, S. J., Ackerman, J. J. H., Keyser, T., and Stille, J. K.,** *J. Org. Chem.*, 44, 3152, 1979.
214. **Zschunke, A., Meyer, H., Leissring, E., Oehme, H., and Issleib, K.,** *Phosphorus Sulfur*, 5, 81, 1978.
215. **Rampal, J. B., Macdonell, G. D., Edasery, J. P., Berlin, K. D., Rahmann, A., Van der Helm, D., and Pietrusiewicz, K. M.,** *J. Org. Chem.*, 46, 1156, 1981.
216. **Navech, J., Majoral, J. P., Meriem, A., and Kraemer, R.,** *Phosphorus Sulfur*, 18, 27, 1983.
217. **Dutasta, J. P., Jurkschat, K., and Robert, J. P.,** *Tetrahedron Lett.*, 22, 2549, 1981.
218. **Horner, L., Walach, P., and Kunz, H.,** *Phosphorus Sulfur*, 5, 171, 1978.
219. **Van Zon, A., Torny, G. J., and Frijns, J. H. G., Recl.:** *J. R. Neth. Chem. Soc.*, 102, 326, 1983.
220. **Quin, L. D., Caster, K. C., Kisalus, J. C., and Mesch, K. A.,** *J. Am. Chem. Soc.*, 106, 7021, 1984.
221. **Quin, L. D. and Mesch, K. A.,** *J. Chem. Soc. Chem. Commun.*, 959, 1980.
222. **Quin, L. D. and Bernhardt, F. Ch.,** *Magn. Reson. Chem.*, 23, 929, 1985.
223. **Benhammou, M., Kraemer, R., Germa, H., Majoral, J. P., and Navech, J.,** *Phosphorus Sulfur*, 14, 105, 1982.
224. **Fluck, E. and Förster, J. E.,** *Chem.-Ztg.*, 99, 246, 1975.
225. **Van der Ploeg, A. F. M. J. and Van Kloten, G.,** *Inorg. Chim. Acta*, 51, 225, 1981.
226. **Girolami, G. S., Mainz, V. V., and Andersen, R. A.,** *Inorg. Chem.*, 19, 805, 1980.
227. **Karsch, H. H. and Schmidbaur, H.,** *Z. Naturforsch.*, 32b, 762, 1977.
228. **Clark, P. W. and Mulraney, B. J.,** *J. Organomet. Chem.*, 217, 51, 1981.
229. **Schmidbaur, H., Schnatterer, S., Dash, K. C., and Aly, A. A. M.,** *Z. Naturforsch.*, 38b, 62, 1983.
230. **Bartsch, R., Hietkamp, S., Peters, H., and Stelzer, O.,** *Inorg. Chem.*, 23, 3304, 1984.
231. **Grim, S. O., Del Gaudio, J., Molenda, R. P., Tolman, C. A., and Jesson, J. P.,** *J. Am. Chem. Soc.*, 96, 3416, 1974.

232. **Lindner, E. and Vordermaier, G.,** *Z. Naturforsch.*, 33b, 1457, 1978.
233. **Langford, G. R., Akhtar, M., Ellis, P. D., MacDiarmid, A. G., and Odom, J. D.,** *Inorg. Chem.*, 14, 2937, 1975.
234. **Koike, Y., Takayama, T., and Watabe, M.,** *Bull. Chem. Soc. Jpn.*, 57, 3595, 1984.
235. **Slack, D. A., Greveling, I., and Baird, M. C.,** *Inorg. Chem.*, 18, 3125, 1979.
236. **Payne, N. C. and Stephan, D. W.,** *J. Organomet. Chem.*, 221, 203, 1981.
237. **Brown, J. M. and Murrer, B. A.,** *J. Chem. Soc. Perkin Trans. II*, 489, 1982.
238. **Nagel, U., Kinzel, E., Andrade, J., and Prescher, G.,** *Chem. Ber.*, 119, 3326, 1986.
239. **Lindner, E. and Kern, H.,** *Chem. Ber.*, 117, 355, 1984.
240. **Briggs, J. C., McAuliffe, C. A., Hill, W. E., Minahan, D. M., and Dyer, G.,** *J. Chem. Soc. Perkin Trans. II*, 321, 1982.
241. **Horn, H. G. and Sommer, K.,** *Spectrochim. Acta, Part A*, 27, 1049, 1971.
242. **Kraihanzel, C. S., Ressner, J. M., and Gray, G. M.,** *Inorg. Chem.*, 21, 879, 1982.
243. **Creaser, C. S. and Kaska, W. C.,** *Inorg. Chim. Acta*, 30, L325, 1978.
244. **Hill, W. E., Minahan, D. M. A., Taylor, J. G., and McAuliffe, C. A.,** *J. Chem. Soc. Perkin Trans. II*, 327, 1982.
245. **Costa, T. and Schmidbaur, H.,** *Chem. Ber.*, 115, 1367, 1982.
246. **Miyashita, A., Takaya, H., Souchi, T., and Noyori, R.,** *Tetrahedron*, 40, 1245, 1984.
247. **Karsch, H. H.,** *Z. Naturforsch.*, 34b, 1171, 1979.
248. **Hietkamp, S., Sommer, H., and Stelzer, O.,** *Angew. Chem.*, 94, 368, 1982.
249. **King, R. B. and Cloyd, J. C., Jr.,** *Inorg. Chem.*, 14, 1550, 1975.
250. **Askham, F. R., Stanley, G. G., and Marques, E. C.,** *J. Am. Chem. Soc.*, 107, 7423, 1985.
251. **Mynott, R. J., Pregosin, P. S., and Venanzi, L. M.,** *J. Coord. Chem.*, 3, 145, 1973.
252. **Antberg, M., Dahlenburg, L., Höck, N., and Prengl, C.,** *Phosphorus Sulfur*, 26, 143, 1986.
253. **Dean, P. A. W., Phillips, D. D., and Polensek, L.,** *Can. J. Chem.*, 59, 50, 1981.
254. **Hietkamp, S. and Stelzer, O.,** *Chem. Ber.*, 119, 2921, 1986.
255. **King, R. B. and Cloyd, J. C., Jr.,** *Phosphorus*, 3, 213, 1974.
256. **Balch, A. L., Fossett, L. A., Olmstead, M. M., Oram, D. E., and Reedy, P. E., Jr.,** *J. Am. Chem. Soc.*, 107, 5272, 1985.
257. **Schmidbaur, H. and Costa, T.,** *Z. Naturforsch.*, 37b, 677, 1982.
258. **Sinou, D., Lafont, D., Descotes, G., and Dayrit, T.,** *Nouv. J. Chim.*, 7, 291, 1983.
259. **Fritz, G. and Schaefer, H.,** *Z. Anorg. Allg. Chem.*, 409, 137, 1974.
260. **Engelhardt, G.,** *Z. Anorg. Allg. Chem.*, 387, 52, 1972.
261. **Becker, G.,** *Z. Anorg. Allg. Chem.*, 423, 242, 1976.
262. **Weber, L., Reizig, K., and Boese, R.,** *Chem. Ber.*, 118, 1193, 1985.
263. **Andriamizaka, J. D., Escudié, J., Couret, C., and Satgé, J.,** *Phosphorus Sulfur*, 12, 279, 1982.
264. **Appel, R., Knoch, F., Laubach, B., and Sievers, R.,** *Chem. Ber.*, 116, 1873, 1983.
265. **Appel, R., Barth, V., Kunze, H., Laubach, B., Paulen, V., and Knoll, F.,** *ACS Symp. Ser.*, 171 (Phosphorus Chem.), 395, 1981.
266. **Andriamizaka, J. D., Couret, C., Escudié, J., and Satgé, J.,** *Phosphorus Sulfur*, 12, 265, 1982.
267. **Romanenko, V. D., Ruban, A. V., Drapailo, A. B., and Markovski, L. N.,** *J. Gen. Chem. USSR*, 55, 2486, 1985.
268. **Hassler, K.,** *Monatsh. Chem.*, 113, 421, 1982.
269. **Issleib, K., Leissring, E., and Schmidt, H.,** *Z. Chemie*, 26, 446, 1986.
270. **Lequan, R. M., Pouet, M. J., and Simonnin, M. P.,** *Org. Magn. Resonance*, 7, 392, 1975.
271. **Baltusis, L., Frye, J. S., and Maciel, G. E.,** *J. Am. Chem. Soc.*, 109, 40, 1987.
272. **Quin, L. D. and Rao, N. S.,** *J. Org. Chem.*, 48, 3754, 1983.
273. **King, R. B., Cloyd, J. C., and Reimann, R. H.,** *J. Org. Chem.*, 41, 972, 1976.
274. **King, R. B. and Cloyd, J. C.,** *Inorg. Chem.*, 14, 1551, 1975.
275. **Schumann, H., Kroth, H. J., and Rösch, L.,** *Z. Naturforsch.*, 29b, 608, 1974.
276. **Norman, A. D.,** *Inorg. Chem.*, 9, 870, 1970; *J. Am. Chem. Soc.*, 90, 6556, 1970; *J. Organomet. Chem.*, 28, 81, 1971.
277. **Schumann, H. and Kroth, H. J.,** *Z. Naturforsch.*, 32b, 513, 1977.
278. **Appel, R. and Geisler, K.,** *J. Organomet. Chem.*, 112, 61, 1976.
279. **Schumann, H. and Benda, H.,** *Angew. Chem.*, 81, 1049, 1969.
280. **Oakley, R. T., Stanislawski, D. A., and West, R.,** *J. Organomet. Chem.*, 157, 389, 1978.
281. **Maier, L.,** *Helv. Chim. Acta*, 49, 1718, 1966.
282. **Grim, S. O. and McFarlane, W.,** *Nature*, 208, 995, 1965.

Chapter 7

THREE COORDINATE FIVE VALENT ($\lambda 5 \sigma 3$) COMPOUNDS

Compiled and presented by
Helene Germa and Jacques Navech

NMR data also donated by
E. Nieke, Universität Bielefeld

The NMR data for the above compounds include information on meta phosphonyl compounds and diylides, and are given in Table F which follows. The relatively small numbers of compounds in this class make it difficult to find a general relationship between ^{31}P chemical shift and structure. There is wide variation of substituents of phosphorus and a corresponding large range of ^{31}P chemical shifts. The ^{31}P chemical shifts vary from 13 to 297 ppm and are generally downfield of those of the corresponding λ^5,σ^4 compounds.

In the same family of compounds, the chemical shifts can vary in a regular way according to the nature of the substituents linked to the P(=X)=Y fragment. Thus, the influence of the substituents, which have a single bond to phosphorus, is particularly clear in the case of some of the bis(methylene)-phosphoranes. Thus shielding increases across the sequence of P-bound atom $C'_2Cl < C'_2Br < C'_2I$ and also along the sequence of $CC'_2 < C'_3 < C'_2C''$. However, other compounds in this series, i.e., those with P-bound atoms C'_2N, C'_2O and C'_2S, have quite similar chemical shifts.

Within a series possessing similar phosphorus double-bond substituents, the bis(methylene)phosphorane (P-bound atoms = C'_2X, listed above) are more deshielded than the amino(methylene)phosphoranes (P-bound atoms = C'NN') and the bis(amino) phosphoranes (P-bound atoms = NN'_2, CN'_2). In general, the deshielding influence of the Group IV elements on the phosphorus atom increases according to the sequence O $<<$ Se $\leqslant$ S (see compounds with P-bound atoms C'N'O', C'N'S', C'N'Se' and $C'O'_2$, $C'S'_2$, $C'Se'_2$ and C'_2O, C'_2S', C'_2Se'). Indeed the λ^5,σ^3-thiophosphoranes have some of the most deshielded phosphorus atoms in Table F.

There is an ambiguity in the behavior of compounds with P-bound atoms = C'P'S' and C'O'P'. Thus the chemical shifts of the two phosphorus atoms are very different for the phosphorane with P-bound atoms = C'O'P', the spectrum of $[\eta^6\text{-}Cr(CO)_3]$SmsP(=PSms)=S showing an AB system. Whereas the spectrum of the phosphorane SmsP(=PSms)=S in chloroform solution exhibits an AB spin system only at −70°C and at room temperature, a singlet is observed. The phosphorus atoms in the last two compounds may not have different coordination states.

TABLE F
Compilation of ^{31}P NMR Data of Three Coordinate ($\lambda 5\ \sigma 3$) Compounds

P-bound atoms/ rings	Connectivities	Structure	NMR data (δp[solv]J_{PZ} Hz)	Ref.
NP'S'	Si_2;N	$(tBuMe_2Si)_2NP$ $[=PN(SiMe_2\text{-}tBu)_2]=S$	297 $^1J_{PP}$ 863	1
NN'P'	Si_2:Si;N	$Tms_2NP(=NTms)=PNTms_2$	62 $^1J_{PP}$ 881	2
NN'S'	Si_2;C	$Tms_2NP(=N\text{-}tBu)=S$	135	3
	CSi;C	$tBuTmsNP(=N\text{-}tBu)=S$	136(o)	4
NN'Se'	Si_2;C	$Tms_2NP(=N\text{-}tBu)=Se$	120(o) $^1J_{PSe}$ 975	3
	CSi;C	$tBuTmsNP(=N\text{-}tBu)=Se$	118.7(o)	5
NN'_2	$Si_2;(Si)_2$	$Tms_2NP(=NTms)_2$	55(r)[a], 55.5(c)[b]	6
	Si_2;Si,C	$Tms_2NP(=NTms)=NMe$	46.7(r)	7
		$Tms_2NP(=NTms)=N\text{-}tBu$	51.5(b)	8
	$Si_2;(C)_2$	$Tms_2NP(=N\text{-}tBu)_2$	52.6(4)	6b
	CSi;Si,C	$tBuTmsNP(=NTms)=N\text{-}tBu$	53.1(r)	6b
	$CSi;(C)_2$	$tBuTmsNP(=N\text{-}tBu)_2$	35.2(b)	8
	$C_2;(Si)_2$	$TmpNP(=NTms)_2$	47.2(b)	8
	C_2;Si,C	$TmpNP(=NTms)=N\text{-}tBu$	51.5(b)	8
	$C_2;(C)_2$	$iPr_2NP(=N\text{-}tBu)_2$	50.5(c)	6b
S'_3		PS_3^-	52.3	9
HN'_2	$(C')_2$	$HP(=NSms)_2[NiCl_2]$	11.2 $^1J_{PH}$ 603	10
C'N'O'	C'_2;C	$SmsP(=N\text{-}tBu)=O$	93.5(b)	11
C'NS'	$HSi;Si_2$	$Tms_2NP(=CHTms)=S$	185.4(c) $^1J_{PC}$92[a],98[b]	12
	$CH;Si_2$	$Tms_2NP(=CH\text{-}tBu)=S$	153.1(c) $^1J_{PC}$143	12b
	$C_2;Si_2$	$Tms_2NP(=CMe\text{-}tBu)=S$	142.2(c) $^1J_{PC}$146	12b
C'N'S'	C'_2;C	$SmsP(=N\text{-}tBu)=S$	162.4(b)	11
	C'_2;C'	$SmsP[=N(2\text{-}Me\ 4,6\text{-}tBu_2C_6H_2)]=S$	109(c)	13
C'NSe'	$HSi;Si_2$	$Tms_2NP(=CHTms)=Se$	172.4(c) $^1J_{PC}$70	12a
C'N'Se'	C'_2;C	$SmsP(=N\text{-}tBu)=Se$	154.8(b) $^1J_{PSe}$903	11
CN'_2	$HSi_2;(Si)_2$	$Tms_2CHP(=NTms)_2$	83.4(c)	6b
C'NN'	$HSi;Si_2$;Si	$Tms_2NP(=CHTms)=NTms$	102.6(c) $^1J_{PC}$149	14
	$HSi;C_2$;Si	$TmpP(=CHTms)=NTms$	100.6	17
	$CH;Si_2$;Si	$Tms_2NP(=CHMe)=NTms$	98.1(r) $^1J_{PC}$201; 105.5(4) $^1J_{PC}$192	15
	$CH;C_2$;C	$iPr_2NP(=CH\text{-}tBu)=N\text{-}tBu$	77.2(b) $^1J_{PC}$199	16
	$C_2;Si_2$;Si	$Tms_2NP(=CMe_2)=NTms$	86.4(r) $^1J_{PC}$216	15
		$Tms_2NP(=CMeEt)=NTms$	83.6(r) $^1J_{PC}$210; 83.9(r) $^1J_{PC}$211	15
	C_2;CSi;Si	$tBuTmsNP(=CMe_2)=NTms$	86.8(r) $^1J_{PC}$218	15
$C'N'_2$	$C'_2;(C')_2$	$SmsP(=NCOOEt)_2$	104	18
C'O'P'	C'_2;C'	$SmsP(=O)=PSms$	69.8(c) $^1J_{PP}$684	19
$C'O'_2$	C'_2	$SmsP(=O)_2$	13(r)	20
C'P'S'	C'_2;C'	$SmsP(=PMes)=S$	226.7; 241.9(b) $^1J_{PP}$604.3	21
		$SmsP(=PSms)=S$	247.8; 255.8(b) $^1J_{PP}$630	21
		$[\eta^6\text{-}Cr(CO_3]SmsP(=PSms)=S$	247.7; 254.1(c) $^1J_{PP}$666	21
$C'S'_2$	C'_2	$SmsP(=S)_2$	295.3(b)	22
$C'Se'_2$	C'_2	$SmsP(=Se)_2$	273(c) $^1J_{PSe}$ 854	23
C'_2Br	$(Si_2)_2$	$BrP(=CTms_2)_2$	122.5(b) $^1J_{PC}$ 31	24
C'_2Cl	$(Si_2)_2$	$ClP(=CTms_2)_2$	136.8(b) $^1J_{PC}$ 39	24b
C'_2I	$(Si_2)_2$	$IP(=CTms_2)_2$	84.7(b) $^1J_{PC}$ 23.5	24b
CC'N'	HSi_2;HSi;Si	$Tms_2CHP(=CHTms)=NTms$	145.3 $^1J_{PC}$22 $^1J_{PC}$'107; 129 $^1J_{PC}$51 $^1J_{PC}$'72.5	25
C'_2N	$(HSi)_2;Si_2$	$Tms_2NP(=CHTms)_2$	161.3	25
	$(Si_2)_2;C_2$	$Me_2NP(=CTms_2)_2$	167(c) $^1J_{PC}$71	26
		$Et_2NP(=CTms_2)_2$	165	27
		$iPr_2NP(=CTms_2)_2$	167	27
	$H_2;Si_2;Si_2$	$Tms_2NP(=CH_2)=CTms_2$	161.3(c) $^1J_{PC}$88.11	25
C'_2N'	$C'_2;Si_2$;Si	$MesP(=CTms_2)=NTms$	120(c)	28

1

P-bound atoms/ rings	Connectivities	Structure	NMR data (δp[solv]J_{PZ} Hz)	Ref.
	C'$_2$;C'$_2$;C'	MesP(=CPh$_2$)=NPh	18.8(b) $^1J_{PC}$166.5	29
		2.6-Me$_2$C$_6$H$_3$P(=fluorenyl)=NPh	17.6(c) $^1J_{PC}$152	30
C'$_2$O	(Si$_2$)$_2$;C	MeOP(=CTms$_2$)$_2$	174(b) $^1J_{PC}$65	24b
	(Si$_2$)$_2$;C'	PhOP(=CTms$_2$)$_2$	162.3(b) $^1J_{PC}$63	24b
C'$_2$O'	C'$_2$;Si$_2$	SmsP(=CTms$_2$)=O	161.1(b)	31
	C'$_2$;C'Si	SmsP(=CPhTms)=O	153.7(b) $^1J_{PC}$113	32
C'$_2$S	(Si$_2$)$_2$;C	nPentSP(=CTms$_2$)$_2$	168.8(b)	24b
	(Si$_2$)$_2$;C'	PhSP(=CTms$_2$)$_2$	161(b)	24b
C'$_2$S'	C'$_2$;Si$_2$	MesP(=CTms$_2$)=S	191.9(c)	28
		SmsP(=CTms$_2$)=S	204.6(b)	31
C'$_2$Se'	C'$_2$;Si$_2$	SmsP(=CTms$_2$)=Se	195.2(b)	31
	C'$_2$;C'$_2$	2,6-Me$_2$ C$_6$H$_3$P(=CPh$_2$)=Se	125.7 $^1J_{PSe}$890 $^1J_{PC}$114	33
CC'$_2$	H$_2$Si;(Si$_2$)$_2$	[Tms$_2$HCP$_{(b)}$=C(Tms)Si(Me$_2$)- - CH$_2$]P$_{(a)}$(=CTms$_2$)$_2$	(a)177.7. (b)413.7 $^4J_{PP}$2.3	34
	C$_2$H;(Si$_2$)$_2$	2-C$_4$H$_9$P(=CTm$_2$)$_2$	198(b)	24b
		C$_6$H$_{11}$P(=CTms$_2$)$_2$	198(c) $^1J_{PC}$'38	35
	C'$_2$H;(Si$_2$)$_2$	Ph$_2$HCP(=CTms$_2$)$_2$	179.1(b) $^1J_{PC}$'41(c)	34
		fluorenylP(=CTms$_2$)$_2$	171.5(b) $^1J_{PC}$43 $^1J_{PC}$'38	36
	CC'$_2$;(Si$_2$)$_2$	Me$_5$C$_5$P(=CTms$_2$)$_2$	176	1
	C$_3$;(Si$_2$)$_2$	tBuP(=CTms$_2$)$_2$	204.5(b) $^1J_{PC}$'37	24b
	C$_3$;C'Si,Si$_2$	tBuP(=CPhTms)=CTms$_2$	188(c) $^1J_{PC}$33 $^1J_{PC}$'72	37
	C$_3$;C'$_2$,Si$_2$	tBuP(=CPh$_2$)=CTms$_2$	167(b)	37
CC'$_2$/4	PSi$_2$;PSi,Si$_2$	1	P(6)138.8 $^2J_{PP}$89(c)(note P(a)15.7)	34
C'$_3$	(Si$_2$)$_2$,C'H	PhHC=CHP(=CTms$_2$)$_2$	166(b) $^1J_{PC}$49	24b
	(Si$_2$)$_2$,C'$_2$	PhP(=CTms$_2$)$_2$	174 $^1J_{PC}$38	35
		2-MeOC$_6$H$_4$P(=CTms$_2$)$_2$	161(b) $^1J_{PC}$34	24b
	Si$_2$,(C'$_2$)$_2$	PhP(=C$_{(a)}$Tms$_2$)=$_{C(b)}$Ph$_2$	133(b) $^1J_{PC(a)}$40 $^1J_{PC(b)}$126	37
		Mes$_{(a)}$P(=C$_{(b)}$Tms$_2$)=C$_{(c)}$Ph$_2$	120(b) $^1J_{PC(a)}$48 $^1J_{PC(b)}$41 $^1J_{PC(c)}$125	37
C'$_2$C"	(Si$_2$)$_2$;C"	PhC≡CP(=CTms$_2$)$_2$	109.8(b)	24b
	Si$_2$,C'$_2$;C"	TmsC≡CP(=CTms$_2$)=CPh$_2$	79.8(c)	38

REFERENCES

1. **Niecke, E., Leuer, M., and Lysek, M.,** unpublished results.
2. **Niecke, E., Rüger, R., Lysek, M., and Schoeller, W. W.,** *Phosphorus Sulfur,* 18, 35, 1983.
3. **Scherer, O. J. and Kuhn, N.,** *J. Organomet. Chem.,* 82, C3, 1974.
4. **Scherer, O. J. and Kuhn, N.,** *Angew. Chem. Int. Ed. Engl.,* 13, 811, 1974.

5. **Scherer, O. J. and Kuhn, N.,** *J. Organomet. Chem.*, 78, C17, 1974.
6. **a) Niecke, E. and Flick, W.,** *Angew. Chem. Int. Ed. Engl.*, 13, 134, 1974. **b) Niecke, E. and Schäfer, H. G.,** *Chem. Ber.*, 115, 185, 1982.
7. **Appel, R. and Halstenberg, M.,** *Angew. Chem. Int. Ed. Engl.*, 14, 768, 1975.
8. **Markowskii, L. N., Romanenko, V. D., and Ruban, A. V.,** *Synthesis*, 811, 1979.
9. **Roesky, H. W., Ahlrichs, R., and Brode, S.,** *Angew. Chem. Int. Ed. Engl.*, 25, 82, 1986.
10. **Hitchcock, P. B., Jasim, H. A., Lappert, M. F., and Williams, H. D.,** *J. Chem. Soc. Chem. Commun.*, 1634, 1986.
11. **Markowskii, L. N., Romanenko, V. D., Ruban, A. V., and Drapailo, A. B.,** *J. Chem. Soc. Chem. Commun.*, 1962, 1984.
12. **(a) Niecke, E. and Wildbredt, D. A.,** *J. Chem. Soc. Chem. Commun.*, 72, 1981; **(b) Niecke, E., Böske, J., Krebs, B., and Dartmann, M.,** *Chem. Ber.*, 118, 3227, 1985.
13. **Yoshifuji, M., Shibayama, K., Toyota, K., Inamoto, N., and Nagase, S.,** *Chem. Lett.*, 237, 1985.
14. **Neilson, R. H.,** *Inorg. Chem.*, 20, 1679, 1981.
15. **Niecke, E. and Wildbredt, D. A.,** *Chem. Ber.*, 113, 1549, 1980.
16. **Niecke, E., Sayer, A., and Wildbredt, D. A.,** *Angew. Chem. Int. Ed. Engl.*, 20, 675, 1981.
17. **Böske, J., Niecke, E., Ocando-Mavarez, E., Majoral, J. P., and Bertrand, G.,** *Inorg. Chem.*, 25, 2695, 1986.
18. **Navech, J. and Revel, M.,** *Tetrahedron Lett.*, 27, 2863, 1986.
19. **Yoshifuji, M., Ando, K., Toyota, K., Shima, I., and Inamoto, N.,** *J. Chem. Soc. Chem. Commun.*, 419, 1983.
20. **Caminade, A. M., El Khatib, F., and Koenig, M.,** *Phosphorus Sulfur*, 18, 97, 1983.
21. **Yoshifuji, M., Shibayama, K., and Inamoto, N.,** *Heterocycles*, 22, 68, 1984.
22. **(a) Navech, J., Majoral, J. P., Meriem, A., and Kraemer, R.,** *Phosphorus Sulfur*, 18, 27, 1983; **(b) Appel, R., Knoch, F., and Kunze, H.,** *Angew. Chem. Int. Ed. Engl.*, 22, 1004, 1983; **(c) Navech, J., Majoral, J. P., and Kraemer, R.,** *Tetrahedron Lett.*, 24, 5885, 1983.
23. **Yoshifuji, M., Shibayama, K., and Inamoto, N.,** *Chem. Lett.*, 603, 1984.
24. **(a) Appel, R. and Westerhaus, A.,** *Tetrahedron Lett.*, 23, 2017, 1982; **(b) Appel, R., Dunker, K. H., Gaitzsch, E., and Gaitzsch, T.,** *Z. Chem.*, 24, 384, 1984.
25. **Niecke, E., Leuer, M., Wildbredt, D. A., and Schoeller, W. W.,** *J. Chem. Soc. Chem. Commun.*, 1171, 1983.
26. **Appel, R., Peters, J., and Westerhaus, A.,** *Angew. Chem. Int. Ed. Engl.*, 21, 80, 1982.
27. **Leuer, M.,** Diplomarbeit, Universität Bielefeld, 1984.
28. **Neilson, R. H.,** *Phosphorus Sulfur*, 18, 43, 1983.
29. **Van der Knapp, T. A., Klebach, T. C., Visser, P., Lourens, R., and Bickelhaupt, F.,** *Tetrahedron*, 40, 991, 1984.
30. **Van der Knapp, T. A. and Bickelhaupt, F.,** *Chem. Ber.*, 117, 915, 1984.
31. **Appel, R. and Casser, C.,** *Tetrahedron Lett.*, 25, 4109, 1984.
32. **Appel, R., Knoch, F., and Kunze, H.,** *Angew. Chem. Int. Ed. Engl.*, 23, 157, 1984.
33. **Van der Knapp, T. A., Vos, M., and Bickelhaupt, F.,** *J. Organomet. Chem.*, 244, 363, 1983.
34. **Appel, R., Gaitzsch, E., Dunker, K. H., and Knoch, F.,** *Chem. Ber.*, 119, 535, 1986.
35. **Appel, R., Peters, J., and Westerhaus, A.,** *Angew. Chem. Int. Ed. Engl.*, 21, 80, 1982.
36. **Appel, R., Gaitzsch, T., and Knoch, F.,** *Angew. Chem. Int. Ed. Engl.*, 24, 589, 1985.
37. **Appel, R., Gaitzsch, T., Knoch, F., and Lenz, G.,** *Chem. Ber.*, 119, 1977, 1986.
38. **Appel, R. and Casser, C.,** *Chem. Ber.*, 118, 3419, 1985.

Chapter 8

^{31}P NMR DATA OF FOUR COORDINATE ($\lambda 5\ \sigma 4$) PHOSPHONIUM SALTS AND BETAINES

Compiled and presented by
Harry R. Hudson, Keith B. Dillon, and Brian J. Walker

NMR data were also donated by
Louis D. Quin, University of Massachusetts, Amherst, MA
John G. Verkade, Iowa State University, Ames, Iowa
Ekkehard Fluck, Gmelin-Institut and Stephan Kleemann, Fachhochschule München, FRG
H. W. Roesky, Universität Göttingen, FRG
C. Dennis Hall, King's College London, U.K.

The NMR data are presented in Table G. It is divided into five sections

G1) Compounds with no P-C bonds
G2) Compounds with one P-C bond
G3) Compounds with two P-C bonds
G4) Compounds with three P-C bonds
G5) Compounds with four P-C bonds

The first four sections were compiled by H. R. Hudson and K. B. Dillon and contain data on quasiphosphonium salts and betaines, with the latter part of section four being concerned with protonated phosphines. The fifth and final section was compiled by B. J. Walker and it concerns quaternary phosphonium salts and betaines.

Results have been included separately for solid state and solution measurements and for compounds with different counter-anions where the chemical shifts differ significantly. Variation of chemical shift with counter-ion is now well established in these types of compound and is not necessarily caused by the presence of an equilibrium concentration of the corresponding phosphorane. Unless stated otherwise the compounds listed in these tables are believed to exist virtually entirely in the tetracoordinate form.

For weakly basic species containing P=O bonds such as $POCl_3$ where the shifts have been measured in highly acidic solvents, the observed shift value generally represents an equilibrium between the protonated and nonprotonated forms, e.g.,

$$POCl_3 + H^+ \rightleftharpoons Cl_3P(OH)^+ \quad (1)$$

These are represented as the protonated forms in the table for clarity. In such systems protonation causes a higher frequency (downfield) shift of the resonance relative to that of the unprotonated form, and the maximum shift observed for a particular solvent is given in this table. These shifts will naturally vary also with the relative concentrations of the solute and solvent, so that experiments under different conditions may lead to widely differing results.

Among trends apparent in the data, the replacement of Cl by Br causes a marked lower frequency (higher field) shift, which is even more accentuated when I is substituted. The shifts for the limited number of species with several P–I bonds such as $I_3P(OH)^+$ lie well outside the range which is usually considered as the "four-coordinate region". The shift differences between chloro- and bromo-phosphonium ions are reduced successively as more organo-groups are attached to phosphorus.

Oxygen-containing substituents such as RO or HO tend to act as good charge delocalizers, particularly as the number of P–O bonds increases, and the shift range for species with four P–O bonds is quite small.

TABLE G
Compilation of ^{31}P NMR Data of Four Coordinate Phosphonium Salts and Betaines

Section G1: Four Coordinate Phosphonium Salts and Betaines with No P–C Bond

P-bound atoms/rings	Connectivities	Structure	NMR Data (δ_P[solv, temp]^aJ_{PZ} Hz)	Ref.
$BrClN_2$	$(C_2)_2$	$BrClP^+(NMe_2)_2\ Br^-$	50.0(r)	1
$BrClN'_2$	$(C'')_2$	$BrClP^+(NCS)_2\ (Br_3^-)$	5.8(Br_2)	2
$BrCl_2N'$	C''	$BrCl_2P^+NCS\ (Br_3^-)$	20.2(Br_2)	2
$BrCl_2O$	H	$BrCl_2P^+OH\ (ClSO_3^-)$	−7($ClSO_3H$)	3
		$(HS_2O_7^-)$	−10(25% oleum)	4
$BrCl_3$		$BrP^+Cl_3\ BCl_4^-$	49(solid)	5
		(BX_4^-)(X=Br and/or Cl)	50 to 38 (solid)	5
		(BCl_4^-)	50.1(liq HCl)	6
		(Br_2I^-/Cl_2I^-)	50 ± 5(solid)	7
		(Br_3^-)	46.6(Br_2)	2
		(FSO_3^-)	50(FSO_3H)	4
		$(HS_2O_7^-)$	50.5(25% oleum)	4, 8
		(CN^-)	46.7(r)	9
$BrIN_2$	$(C_2)_2$	$BrIP^+(NMe_2)_2\ BrI_2^-$	21.1(liq)	1
		I^-	22.7(liq)	1
BrI_2N	C_2	$BrI_2P^+NMe_2\ Br^-$	−21.1(liq)	1
		I^-	−21.0(liq)	1
BrI_2O	H	$BrI_2P^+OH\ (FSO_3^-)$	−222(FSO_3H)	4
		(HSO_4^-)	−222(H_2SO_4)	4
BrN_3	$(C_2)_3$	$BrP^+(NMe_2)_3\ Br^-$	48.3(r or solid)	1
		BrI_2^-	48.3(liq)	1
		Br_2I^-	48.3(solid)	1
BrO_3	$(P^4)_2$,H	$-OP^+(Br)(OH)O-\ (HS_3O_{10}^-)$	−32(65% oleum)	10
	$(H)_3$	$BrP^+(OH)_3\ (HS_2O_7^-)$	3(25% oleum)	10
		$BrP^+(OCH_2CCl_3)_3\ Br^-$	6.5(ce, −100°)	11
	$(C)_3$	$BrP^+(OEt)_3\ Br^-$	11.5(ce, −100°)	11
		$SbBr_6^-$	15.1($PrNO_2$)	11
		$BrP^+(ONeo)_3\ Br^-$	13(ce, −100°)	11
	$(C')_3$	$BrP^+(OPh)_3\ Br^-$	4.5(c); 3.6($PrNO_2$ or r, −90°)	12, 13
		$SbBr_6^-$	4.3($PrNO_2$ or r)	13
$BrO_3/5$	C,$(C')_2$	$BrP^+(OEt)Cat\ SbBr_6^-$	36.5($PrNO_2$)	11
		$BrP^+(ONeo)Cat\ Br_3^-$	34($PrNO_2$)	11
		$SbBr_6^-$	37($PrNO_2$)	11
Br_2ClN	C_2	$Br_2ClP^+NMe_2\ BBr_4^-$	54.2(solid)	1
		Br^-	54.6(solid)	1
Br_2ClO	H	$Br_2ClP^+OH\ (ClSO_3^-)$	−36($ClSO_3H$)	14
		(FSO_3^-)	−36(FSO_3H)	14
		$(HS_2O_7^-)$	−43(25% oleum)	4
Br_2Cl_2		$Br_2P^+Cl_2\ (BCl_4^-)$	5.8 ± 2.3(solid), 9.7(liq HCl)	5, 6
		(Br_2I^-/Cl_2I^-)	12 ± 5(solid)	7
		(Br_3^-)	9.5(Br_2)	2
		$(ClSO_3^-)$	9($ClSO_3H$)	3
		(FSO_3^-)	9.5(FSO_3H)	4
		$(HS_2O_7^-)$	9.5(25% oleum)	4, 8
		(CN^-)	4.9(r)	9
Br_2IO	H	$Br_2IP^+OH\ (HSO_4^-)$	−150(H_2SO_4)	4
Br_2N_2	$(C_2)_2$	$Br_2P^+(NMe_2)_2\ Br^-$	45.8(solid)	1

P-bound atoms/rings	Connectivities	Structure	NMR Data (δ_P[solv, temp][a]J_{PZ} Hz)	Ref.
$Br_2N'_2$	$(C'')_2$	$Br_2P^+(NCS)_2$ (Br_3^-)	−15.8(Br_2)	2
Br_2O_2	P^4,H	$-OP^+Br_2(OH)$ $(HS_3O_{10}^-)$	−14(65% oleum)	10
	$(H)_2$	$Br_2P^+(OH)_2$ (HSO_4^-)	−20(H_2SO_4)	10
		$(HS_2O_7^-)$	−10(25% oleum)	10
		$(HS_3O_{10}^-)$	−1(65% oleum)	10
Br_3Cl		Br_3P^+Cl (BCl_4^-)	−29.5(solid), −34.6(liq HCl)	5, 6
		(Br_2I^-/Cl_2I^-)	−28 ± 5(solid)	7
		(Br_3^-)	−32.9(Br_2)	2
		$(ClSO_3^-)$	−35.5($ClSO_3H$)	3, 10
		(FSO_3^-)	−35(FSO_3H)	4
		$(HS_2O_7^-)$	−36(25% oleum)	4, 8
		(CN^-)	−27.4(r)	9
Br_3N	C_2	$Br_3P^+NMe_2$ Br^-	13.0(solid)	1
Br_3O	P^4	Br_3P^+O- $(HS_3O_{10}^-)$	−25(65% oleum)	3, 10
	H	Br_3P^+OH $(ClSO_3^-)$	−76($ClSO_3H$)	3
		(FSO_3^-)	−71(FSO_3H)	3
		(HSO_4^-)	−78(H_2SO_4)	10
		$(HS_2O_7^-)$	−60(25% oleum)	14
		$(HS_3O_{10}^-)$	−46(65% oleum)	10
Br_4		Br_4P^+ (BCl_4^-)	−66(solid) −81 (liq HCl)	5, 6
		Br^-	−72(solid)	15
		$Br^-.2CCl_4$	−68(solid)	15
		Br_2I^-	−72(solid)	16
		(Br_2I^-/Cl_2I^-)	−71 ± 5(solid)	7
		(Br_3^-)	−76.3(Br_2)	2
		$(ClSO_3^-)$	−84.5($ClSO_3H$)	3, 10
		(FSO_3^-)	−84.5(FSO_3H)	3, 4, 10
		PF_6^-	−80(solid)	15
		(HSO_4^-)	−84.5(H_2SO_4)	3, 4, 10
		$(HS_2O_7^-)$	−84.5(25% oleum)	3, 8, 10, 14
		$(HS_3O_{10}^-)$	−85.5(65% oleum)	3, 4, 14
$ClIN_2$	$(C_2)_2$	$ClIP^+(NMe_2)_2$ ClI_2^-	27.4(liq)	1
ClI_2N	C_2	$ClI_2P^+NMe_2$ I^-	0.0(liq)	1
ClI_2O	H	ClI_2P^+OH $(ClSO_3^-)$	−198($ClSO_3H$)	4
ClN_3	$(C_2)_3$	$ClP^+(NMe_2)_3$ Br^-	53.3(r)	1
		Cl^-	53.3(solid), 53.3 (r or CS_2), 50.3(c)	1, 17, 18
		Cl_2I^-	53.3(solid)	1
		(HCl_2^-)	53.3(liq HCl)	1
		$ClP^+(NEt_2)_2NMeCH_2CH_2Cl$ I^-	52.8(MeI)	19
		$ClP^+(Pip)_3$ $AlCl_4^-$	46(r)	20
ClN'_3	$(N'_3)_3$	$ClP^+(N_3)_3$ $SbCl_6^-$	27.5(fe)	21
	$(P^4)_3$	$[ClP(NPCl_3)_3]^+Cl^-$ or PCl_6^-	−26.8 ($CHCl_2CHCl_2$)	22
ClO_3	$(H)_3$	$ClP^+(OH)_3$ $(ClSO_3^-)$	4.0($ClSO_3H$)	10
		$(HS_3O_{10}^-)$	10.0(65% oleum)	10
	H,$(C')_2$	$ClP^+(OPh)_2(OH)$ (HCl_2^-)	−3.6(liq HCl)	24
	$(C)_3$	$ClP^+(OEt)_3$ Cl^-	16.2(ce, −80°)	23
		$ClP^+(O-nBu)_3$ Cl^-	16.7(ce, −80°)	23
	$(C')_3$	$ClP^+(OPh)_3$ Cl^-	7.7(c)	12
$ClO_3/5$	C,$(C')_2$	$ClP^+(OEt)Cat$ $SbCl_6^-$	45($PrNO_2$)	11
		$ClP^+(ONeo)Cat$ $SbCl_6^-$	45.5($PrNO_2$)	11
Cl_2FO	H	Cl_2FP^+OH (FSO_3^-)	12.0(FSO_3H) J_{PF}1200	10

TABLE G (continued)
Compilation of ^{31}P NMR Data of Four Coordinate Phosphonium Salts and Betaines

Section G1: Four Coordinate Phosphonium Salts and Betaines with No P–C Bond

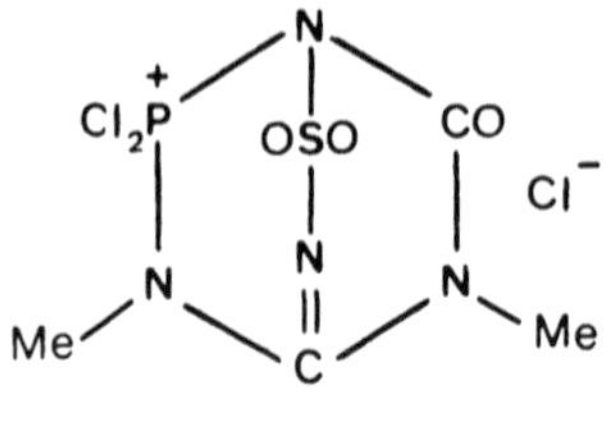

1

P-bound atoms/rings	Connectivities	Structure	NMR Data (δ_P[solv, temp]$^nJ_{PZ}$ Hz)	Ref.
Cl_2IO	H	Cl_2IP^+OH ($ClSO_3^-$)	−89($ClSO_3H$)	10
		(HSO_4^-)	−90(H_2SO_4)	4
Cl_2I_2		$Cl_2P^+I_2$ ($ClSO_3^-$)	−83($ClSO_3H$)	10
$Cl_2NO/4$	CC′;C′	$[Cl_2\dot{P}OC(:NMe)\dot{N}Me]^+(HCl_2^-)$	50(fe)	25
Cl_2N_2	$(C_2)_2$	$Cl_2P^+(NMe_2)_2$ Br^-	51.6(r)	1
		Cl^-	51.6(solid or c)	1
		(HCl_2^-)	51.6(liq HCl)	1
		$Cl_2P^+(NEt_2)_2$ Cl^-	52.6(m), 56.4(fe)	26
		PCl_6^-	57.6(m), 58.6(fe)	26
$Cl_2N_2/6$	$(CC')_2$	$[Cl_2\dot{P}N(Me)CON(Me)CO\dot{N}Me]^+Cl^-$	50.1(fe or $CHCl_2CHCl_2$)	27
	CC′,C′$_2$	$[Cl_2\dot{P}N(Me)CON(Me)CO\dot{N}Ph]^+Cl^-$	49.5(r)	28
$Cl_2N_2/6,6$	C′S_4,CC′	**1**	44.5(fe)	29
$Cl_2N'_2$	$(N')_2$	$Cl_2P^+(N_3)_2$ $SbCl_6^-$	46.7(fe)	21
	$(P^4)_2$	$[Cl_3PN\mathbf{P}Cl_2NPCl_3]^+PCl_6^-$	−13.8, −14(liq)	30, 31
		$[Cl_3PN(\mathbf{P}Cl_2N)_2PCl_3]^+$ BCl_4^-	−10.3 to −11.9(liq)	32
		PCl_6^-	−13.5(liq), −12.3	32, 33
		$[Cl_3PN(\mathbf{P}Cl_2N)_xPCl_3]^+$ PCl_6^-	−13(melt)	34
	$(C'')_2$	$Cl_2P^+(NCS)_2$ (Br_3^-)	24.2(Br_2)	2
Cl_2O_2	$(H)_2$	$Cl_2P^+(OH)_2$ ($ClSO_3^-$)	15.0($ClSO_3H$)	10
		(HSO_4^-)	15.0(H_2SO_4)	10
		($HS_2O_7^-$)	14.0(25% oleum)	10
		($HS_3O_{10}^-$)	19.0(65% oleum)	10
	H,C′	$Cl_2P^+(OH)(OPh)$ (HCl_2^-)	5.9(liq HCl)	24
$Cl_2O_2/5$	$(C')_2$	Cl_2P^+Cat $SbCl_6^-$	71.7(solid), 77.1(ne), 77.8($POCl_3$)	35
Cl_3I		Cl_3P^+I ($ClSO_3^-$)	−6($ClSO_3H$)	10
Cl_3N	C_2	$Cl_3P^+NMe_2$ PCl_6^-	58.9 ± 2.5(solid)	1, 17
		(HCl_2^-)	61.3(liq HCl)	1
		$[Cl_3P^+NMe_2]_2$ PCl_6^-,Cl^-	61.3(solid)	17
		PCl_6^-,Br^-	58.6(solid)	1
		$Cl_3P^+NEt_2$ Cl^-	56.4(fe), 57.9(ne)	36, 37
		PCl_6^-	58.6(fe or ne)	36, 37
Cl_3N'	N′	$Cl_3P^+N_3$ $SbCl_6^-$	66.9(fe)	21
	P^4	$[Cl_3PNPCl_3]^+PCl_6^-$	21.4(fe or ne)	30, 31, 38
		$[Cl_3\mathbf{P}NPCl_2N\mathbf{P}Cl_3]^+PCl_6^-$	12.5(liq), 12.7	30, 31
		$[Cl_3\mathbf{P}N(PCl_2N)_2\mathbf{P}Cl_3]^+BCl_4^-$	12.2 to 13.7(liq)	32
		PCl_6^-	11.3(liq), 12.2	32, 33

P-bound atoms/rings	Connectivities	Structure	NMR Data (δ_P[solv, temp]$^nJ_{PZ}$ Hz)	Ref.
		$[ClP(NPCl_3)_3]^+Cl^-$	6.5($CHCl_2CHCl_2$)	22
		PCl_6^-	6.5($CHCl_2CHCl_2$)	22
		$[Cl_3PN(PCl_2N)xPCl_3]^+PCl_6^-$	13(melt)	34
		$[Cl_3PNPPh_2Cl]^+PCl_6^-$	14.3($POCl_3$)	39
	C″	Cl_3P^+NCS (Br_3^-)	52.0(Br_2)	2
		(NCS^-)	53.3(r)	9
Cl_3O	H	Cl_3P^+OH ($ClSO_3^-$)	27($ClSO_3H$)	14
		(FSO_3^-)	29(FSO_3H)	10
		(HCl_2^-)	9.1(liq HCl)	24
		(HSO_4^-)	21(H_2SO_4)	4
		($HS_2O_7^-$)	35.3(25% oleum)	3
		($HS_3O_{10}^-$)	55.9(65% oleum)	3
Cl_4		Cl_4P^+ $AlCl_4^-$	86.5	40
		$AsCl_6^-$	76(solid)	41
		BCl_4^-	85.5(solid)	17
		(BCl_4^-, X^-)	96 to 73(solid)	5
		$BrClI^-$	81(solid)	16
		(Br_2I^-/Cl_2I^-)	86 ± 5(solid)	7
		(Br_3^-)	83.2(Br_2)	2
		ClO_4^-	87.1(fe)	42
		Cl_2I^-	81(solid)	16
		$NbCl_6^-$	81(solid)	16
		$PCl_2F_4^-$	83(solid)	43
		$PCl_3(CN)_3^-$	87.1(r)	9
		PCl_6^-	86 ± 5, 96(solid), 92(m, −40°)	16, 44-47
		PF_6^-	95(solid)	15
		$PTiCl_{10}^-$	81(solid)	41
		$SbCl_6^-$	88.3(solid), 82.2(r), 86.7(ne), 87.5(fe)	9, 21, 42, 44, 48
		$SnCl_5^-$	73 ± 5(solid)	41
		$TaCl_6^-$	80(solid)	16
		($ClSO_3^-$)	88.5($ClSO_3H$)	3, 10
		(FSO_3^-)	87.5(FSO_3H)	3
		(HCl_2^-)	87.3(liq HCl)	6
		($HS_2O_7^-$)	87.5(25% oleum)	3, 4, 8, 14
		($HS_3O_{10}^-$)	87(65% oleum)	3, 14
		$[Cl_4P^+]_2$ $SnCl_6^{2-}$	81(solid)	41
		$Ti_2Cl_{10}^{2-}$	80	41
FI_2O	H	FI_2P^+OH (FSO_3^-)	−167(FSO_3H) J_{PF}1290	14
FN_3	$(C_2)_3$	$FP^+(NMe_2)_3$	41.8 J_{PF}1020	49
		$PF_3Cl_3^-$	42.5(r) J_{PF}960	1
		$PF_5(NMe_2)^-$	41.2(r) J_{PF}976	1
		$FP^+(NEt_2)_3$	41.4(m) J_{PF}980	50
FO_3	$(H)_3$	$FP^+(OH)_3$ (FSO_3^-)	−7(FSO_3H) J_{PF}990 ± 10	4, 10, 14, 51
F_2N_2	$(C_2)_2$	$F_2P^+(NMe_2)_2$ PF_6^-	21.1(r) J_{PF}1069	1
		$PF_3Cl_3^-$	22.7(r) J_{PF}1060	1
$F_2N_2/5$	$(C_2)_2$	$[F_2\dot{P}NMeCH_2CH_2\dot{N}Me]^+$	−17.3	52
F_2O_2	$(H)_2$	$F_2P^+(OH)_2$ (FSO_3^-)	−21.5(FSO_3H) J_{PF}1004 ± 7	3, 10, 14, 51
IN_3	$(C_2)_3$	$IP^+(NMe_2)_3$ I^-	29.2(solid)	1
		$IP^+(NEt_2)_3$ I^-	−12.7	53
		$I^-.2H_2NC(S)NH_2$	−26.9	53

TABLE G (continued)
Compilation of ^{31}P NMR Data of Four Coordinate Phosphonium Salts and Betaines

Section G1: Four Coordinate Phosphonium Salts and Betaines with No P–C Bond

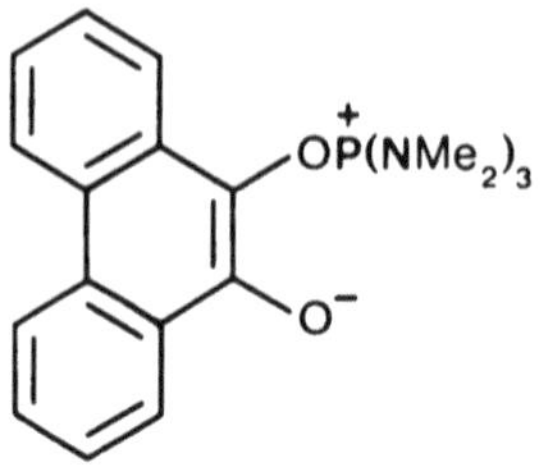

$\overset{+}{O}P(NMe_2)_3$

O^-

2

(open betaine structure only)

P-bound atoms/rings	Connectivities	Structure	NMR Data (δ_P[solv, temp]$^nJ_{PZ}$ Hz)	Ref.
IO_3	$(H)_3$	$IP^+(OH)_3\ (ClSO_3^-)$	$-48(ClSO_3H)$	10
		(HSO_4^-)	$-41(H_2SO_4)$	10
		$(HS_3O_{10}^-)$	-40(65% oleum)	10
	$(C)_3$	$IP^+(OEt)_3\ I^-$	-15.1(ce, $-100°$)	11, 54
		$IP^+(ONeo)_3\ I^-$	-14(ce, $-100°$)	11
I_2O_2	$(H)_2$	$I_2P^+(OH)_2\ (FSO_3^-)$	$-144(FSO_3H)$	14
		(HSO_4^-)	$-149(H_2SO_4)$	10
I_3O	H	$I_3P^+OH\ (FSO_3^-)$	$-308(FSO_3H)$	14
		(HSO_4^-)	$-297(H_2SO_4)$	4
NO_3	$C_2;(C')_3$	$Me_2NP^+(OPh)_3\ Br^-$	-18.1(solid)	1
		Cl^-	-18.5(solid)	17
$NO_3/5$	$C_2;C,(C')_2$	$Et_2NP^+(OEt)Cat\ Cl^-$	18.4(t,h $-50°$)	55
$N'O_3/5$	$C'';P^4,(C')_2$	$SCNP^+(Cat)OP(O)(OEt)_2\ NCS^-$	3.3	56
		$SCNP^+(Cat)OP(S)(OEt)_2\ NCS^-$	-1.9	56
		$SCNP^+(Cat)OP(S)PhtBu\ NCS^-$	1.4	56
	$C'';(C')_3$	$NCSP^+(Cat)OCOMe\ NCS^-$	29	56
N_2O_2	$(C_2)_2;(C')_2$	$(Me_2N)_2P^+(OPh)_2\ PF_6^-$	19.8(m)	57, 58
N_3O	$(C_2)_3;P^4$	$(Me_2N)_3P^+OP(S)Br_2\ Br^-$	47.6(c)	59
		$(Me_2N)_3P^+OP(O)Cl_2\ Cl^-$	55.8(r)	59
		$(Me_2N)_3P^+OP(S)Cl_2\ Cl^-$	53.7(r)	59
	$(C_2)_3;C$	$(Me_2N)_3P^+OMeSbCl_6^-$	38(r)	60
		$(Me_2N)_3P^+OCH_2CCl_3\ ClO_4^-$	36(c)	61
		$(Me_2N)_3P^+OCH_2C{\equiv}CH\ ClO_4^-$	37(c)	61
		$(Me_2N)_3P^+OCH_2CH{:}CHMe\ ClO_4^-$	36(c)	61
		$(Me_2N)_3P^+OPe\ ClO_4^-$	36(c)	61
		$(Me_2N)_3P^+OCH_2CHMe(CH_2)_2Me\ ClO_4^-$	36(c)	61
		$(Me_2N)_3P^+OCH_2Ph\ ClO_4^-$	36(c)	61
		$(Me_2N)_3P^+OCH_2CMe_2(CH_2)_3Me\ ClO_4^-$	35(c)	61
	$(C_2)_3;C'$	$(Me_2N)_3P^+OCH{:}CCl_2\ Cl^-$	34.8	62
		$(Me_2N)_3P^+OC(CO_2Et){:}C(OEt)O^-$	38.2(r)	63
		$(Me_2N)_3P^+OC(Ph){:}C(Ph)O^-$	13.0(r)	64
		$(Me_2N)_3P^+OC(COPh){:}C(Ph)O^-$	35.9(r)	63, 65
		$(Me_2N)_3P^+OC(Ph){:}C(Ph)OP(O)(OMe)_2\ Br^-$	34.2	66
		2	38.5(r)	63, 64

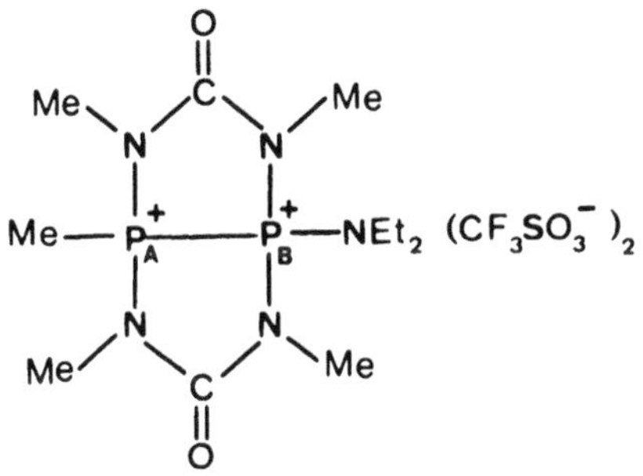

3

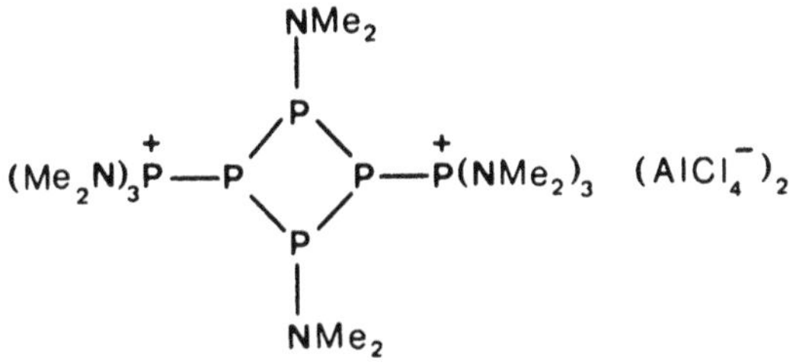

4

P-bound atoms/rings	Connectivities	Structure	NMR Data (δ_P[solv, temp][a] J_{PZ} Hz)	Ref.
$N_3P^4/5,5$	$C_2,(CC')_2;CN_2$	**3** (P_B)	42.7(m) J_{PP}219	67
N_3P	$(C_2)_3;Al,P^4$	$[(Me_2N)_3\mathbf{P}P(AlCl_3)\mathbf{P}(NMe_2)_3]^+$ $AlCl_4^-$	70(r) J_{PP}290	20, 68
		$[(Me_2N)_3\mathbf{P}P(AlCl_3)PPh_3]^+$ $AlCl_4^-$	69(r) J_{PP}297	20
	$(C_2)_3;ClN$	$[(Me_2N)_3\mathbf{P}P(NMe_2)Cl]^+$ $AlCl_4^-$	50.2(r) J_{PP}352	69
	$(C_2)_3;N_2$	$[(Me_2N)_3\mathbf{P}P(NMe_2)_2]^+$ $AlCl_4^-$	38.4(r, −50°)	69
	$(C_2)_3;P^2$	**4**	69	68
	$(C_2)_3;HP^4$	$[(Me_2N)_3\mathbf{P}PH\mathbf{P}(NMe_2)_3]^{2+}$ $(AlCl_4^-)_2$	63(r) J_{PP}270	20, 68
		$[(Me_2N)_3\mathbf{P}PHPPh_3]^{2+}$ $(AlCl_4^-)_2$	61(r) J_{PP}242	20
	$(C_2)_3;CP^4$	$[(Me_2N)_3\mathbf{P}PEt\mathbf{P}(NMe_2)_3]^{2+}$ $(AlCl_4^-)_2$	65(r) J_{PP}318	20
		$[(Me_2N)_3\mathbf{P}P(CH_2Cl)PPh_3]^{2+}$ $(AlCl_4^-)_2$	62(r) J_{PP}279	20
		$[(Me_2N)_3\mathbf{P}PEtPPh_3]^{2+}$ $(AlCl_4^-)_2$	64(r) J_{PP}286	20
	$(C_2)_3;CP$	$[(Me_2N)_3\mathbf{P}\dot{P}CH_2\dot{P}\mathbf{P}(NMe_2)_3]^{2+}$ $(AlCl_4^-)_2$	67.4(r) J_{PP}312	20, 68
		$[(Me_2N)_3\mathbf{P}\dot{P}CH_2\dot{P}PPh_3]^{2+}$ $(AlCl_4^-)_2$	66.4(r) J_{PP}304	20
		$[Pip_3\mathbf{P}\dot{P}CH_2\dot{P}PPh_3]^{2+}$ $(AlCl_4^-)_2$	58.7(r) J_{PP}296	20
$N_3P/5,5$	$C_2,(CC')_2;CN_2$	**5**	42.7(r) J_{PP}171	67
N_3P^2	$(C_2)_3;P^4$	$[(Me_2N)_3\mathbf{P}P\mathbf{P}(NMe_2)_3]^+$ $AlCl_4^-$	85(r) J_{PP}518	20, 70
		BPh_4^-	85 J_{PP}515 ± 3	68, 71
		$[(Me_2N)_3\mathbf{P}PPPh_3]^+$ $AlCl_4^-$	84(r) J_{PP}493	20, 70, 71
		$[Morph_3\mathbf{P}P\mathbf{P}Morph_3]^+$ $AlCl_4^-$	78 J_{PP}566	70
		$[Morph_3\mathbf{P}PPPh_3]^+$ $AlCl_4^-$	76 J_{PP}528	70
		$[Pip_3\mathbf{P}P\mathbf{P}Pip_3]^+$ $AlCl_4^-$	79 J_{PP}560	70
		$[Pip_3\mathbf{P}PPPh_3]^+$ $AlCl_4^-$	77 J_{PP}497	70
N_3S	$(H_2)_3;C$	$(H_2N)_3P^+SMe$ I^-	50.8(v)	72
		$(H_2N)_3P^+SEt$ I^-	49.5(v)	72
		$(H_2N)_3P^+SCH_2CH:CH_2$ Br^-	51.3(v)	72
		$(H_2N)_3P^+SPr$ Br^-	51.8(v)	72
		$(H_2N)_3P^+SCH_2Ph$ Cl^-	50.9(v)	72

TABLE G (continued)
Compilation of ^{31}P NMR Data of Four Coordinate Phosphonium Salts and Betaines

Section G1: Four Coordinate Phosphonium Salts and Betaines with No P–C Bond

5

6

P-bound atoms/rings	Connectivities	Structure	NMR Data (δ_P[solv, temp]$^nJ_{PZ}$ Hz)	Ref.
	$(CH)_3$;C	$(MeNH)_3P^+SPr\ Br^-$	56.9(v)	72
		$(MeNH)_3P^+SCH_2Ph\ Cl^-$	55.2(v)	72
		$(cHexNH)_3P^+SEt\ I^-$	44.4(v)	72
	$(C_2)_3$;C	$(Me_2N)_3P^+SMe\ I^-$	66.8(v)	72
		$SbCl_6^-$	66.7(d)	60
N_3Se	$(C_2)_3$;C	$(Me_2N)_3P^+SeMe\ SbCl_6^-$	63.5(d)	60
N_4	$(H_2)_4$	$(H_2N)_4P^+\ Cl^-$	28.4(w)	73
		I^-	31.6(w)	73
	$(H_2)_2,(C_2)_2$	$(H_2N)_2P^+(Morph)_2\ Cl^-$	33.7	73a
	$H_2,(C_2)_3$	$H_2NP^+(NMe_2)_3\ Cl^-$	42.6(c), 39.0(d)	74
		$H_2NP^+(NEt_2)_3\ Cl^-$	63(c)	75
		$H_2NP^+(NEt_2)_2NMeCH_2CH_2Cl\ I^-$	40.7(r)	19
	$(CH)_2,(C_2)_2$	$(EtNH)(Et_2NCH_2CH_2NH)P^+(NEt_2)_2\ Br^-$	35(r)	19
	$CH,(C_2)_3$	$(Et_2N)_2P^+(NCH_2CH_2)NHEt\ OH^-$	36.3(liq)	19
	$(C_2)_4$	$(Me_2N)_3P^+NEt_2\ I^-$	43.1(r)	76
		$(Me_2N)_3P^+NPr_2\ I^-$	42.3(r)	76
		$(Me_2N)_3P^+NBu_2\ I^-$	42.3(r)	76
		$(Me_2N)_3P^+N(CH_2)_4CH_2I^-$	41.3(r)	76
		$(Me_2N)_2P^+(N(CH_2)_4CH_2)_2\ Br^-$	41.3(r)	76
		$(Me_2N)_3P^+NMeCH_2Ph\ Cl^-$	42.6(r)	76
		$(Et_2N)_3P^+NPr_2\ I^-$	42.1(r)	76
$N_4/5$	$CH,(C_2)_3$	$[(Et_2N)_2\dot{P}NEt(CH_2)_2\dot{N}H]^+Br^-$	45.1(c)	19
		$[(Pr_2N)_2\dot{P}NPr(CH_2)_2\dot{N}H]^+Br^-$	45.8(c)	19
		$[(Bu_2N)_2\dot{P}NBu(CH_2)_2\dot{N}H]^+Br^-$	45.8(c)	19
$N_4/4,5$	$(NP^5)_2,(C_2)_2$	**6**	33.4(r)	79
N_3N'	$(H_2)_3;P^4$	$[(H_2N)_3PNP(NH_2)3]^+Cl^-$	15.6(y or liq NH_3), 15.5(w)	73, 77
		$[(H_2N)_3PNP(NH_2)(NHMe)_2]^+Cl^-$ or I^-	16.5	77
N'_4	$(P^4)_4$	$(Cl_3PN)_4\mathbf{P}^+Cl_2I^-$	$-38.5(C_2Cl_4)$	73
		HgI_3^-	-39.3(ne)	73
	$(N')_4$	$(N_3)_4P^+SbCl_6^-$	11.4(fe), 11.8($MeCHCl_2$)	21, 78
	$(C'')_4$	$(SCN)_4P^+\ (Br_3^-)$	$-19.3(Br_2)$	2

P-bound atoms/rings	Connectivities	Structure	NMR Data (δ_P[solv, temp]$^nJ_{PZ}$ Hz)	Ref.
O_3P^2	$(C)_3;P^4$	$[(EtO)_3$**PPP**$(OEt)_3]^+$ $AlCl_4^-$	82 J_{PP}508	70
		$[(EtO)_3$**PPPPh$_3$**$]^+$ $AlCl_4^-$	80 J_{PP}562	70
O_3S	$(C)_3;Cl$	$(EtO)_3P^+SCl$ Cl^-	45(r or ce, −70°)	23
		$(BuO)_3P^+SCl$ Cl^-	43.6(r or ce, −70°)	23
	$(C)_3;C$	$(MeO)_3P^+SMe$ $SbCl_6^-$	53.2(r)	60
$O_3S/6$	$(C)_3;Cl$	$[(MeO)(\dot{O}CH_2CH_2CHMeO)$-$\dot{P}SCl]^+Cl^-$	46.2^C(ce, −70°), 47.5^T(ce, −70°)	23
O_4	$(H)_4$	$(HO)_4P^+$ (FSO_3^-)	2.3(FSO_3H)	51, 80
		(NO_3^-)	2.0(HNO_3 or HDA)	81
		(HSO_4^-)	2.1(H_2SO_4)	82
	$(H)_3,C'$	$PhOP^+(OH)_3$ (HCl_2^-)	−5.6(liq HCl)	24
	$(H)_2,(C')_2$	$(PhO)_2P^+(OH)_2$ (HCl_2^-)	−12.6(liq HCl)	24
	$H,(C)_3$	$(MeO)_3P^+OH$ (FSO_3^-)	2.0(FSO_3H)	80
		(HCl_2^-)	0.7(liq HCl)	24
		$(EtO)_3P^+OH$ (FSO_3^-)	−1.6(FSO_3H)	80
		$(iPrO)_3P^+OH$ (FSO_3^-)	−4.5(FSO_3H)	80
		$(BuO)_3P^+OH$ (FSO_3^-)	−1.2(FSO_3H)	80
	$H,(C')_3$	$(PhO)_3P^+OH$ (FSO_3^-)	−16.0(FSO_3H)	80
		(HCl_2^-)	−18.2(liq HCl)	24
	$(C)_4$	$(MeO)_4P^+$ BF_4^-	1.9(r)	83
		PF_6^-	0(m), 4.6(m)	84, 85
		$SbCl_6^-$	1.4(m), 1.6(r), 5 ± 3	60, 86, 87
		$(MeO)_3P^+OEt$ $SbCl_6^-$	1.2(m)	86
		$(EtO)_3P^+OMe$ $SbCl_6^-$	−2.6(m)	86
		$(EtO)_4P^+$ BF_4^-	−2.4(r)	83
		PF_6^-	−8(solid), −5(m)	84
		$SbCl_6^-$	−2.7(m), 3 ± 3	86, 87
		$(NeoO)_4P^+$ Cl^-	−2.1(r, −78°)	88
	$(C)_3,C'$	$(1\text{-}NbO)_3P^+OC(:CH_2)$ $C_6H_4NO_2-pBr^-$ or Cl^-	−17(c)	89
	$(C')_4$	$(PhO)_4P^+$ BBr_4^-	−24(solid or ne), −23(r)	90
		BF_4^-	−21(r)	91
		Br^-	−26(r), −22.5(c), −22.6($PrNO_2$ or r)	12, 13, 91
		Br^- or Br_3^-	−23(CH_2BrCH_2Br)	90
		Cl^-	−22/−24(r), −22.5(c)	12, 90-93
		I^-	−23(r)	91
		PCl_6^-	−22(r), −22.6(r)	90, 92, 93
		PF_6^-	−24(m), −28(m), −26(halothane)	58, 84
		$[(PhO)_4P^+]_2$ Br^-,Br_3^-	−20(solid), −23(r), −24(ne)	90
		$[(Ph_2O)_4P^+]_3$ $2Cl_3^-,PCl_6^-$	−23(r)	90
$O_4/5,5$	$(C')_4$	$Cat2P^+$ $SbCl_6^-$	44.0(solid), 42.4(ne)	35
$HBrCl_2$		HP^+BrCl_2 (FSO_3^-)	61.7($FSO_3H/SbF_5/SO_2$, −70°), J_{PH}876	94
HBr_2Cl		HP^+Br_2Cl (FSO_3^-)	31.7($FSO_3H/SbF_5/SO_2$, −70°) J_{PH}841	94
HBr_3		HP^+Br_3 (FSO_3^-)	−3.0($FSO_3H/SbF_5/SO_2$, −70°) J_{PH}810	94

TABLE G (continued)
Compilation of ^{31}P NMR Data of Four Coordinate Phosphonium Salts and Betaines

Section G1: Four Coordinate Phosphonium Salts and Betaines with No P–C Bond

P-bound atoms/rings	Connectivities	Structure	NMR Data (δ_P[solv, temp][a] J_{PZ} Hz)	Ref.
$HClF_2$		HP^+ClF_2 (FSO_3^-)	66.4($FSO_3H/SbF_5/SO_2$, −70°) J_{PF}1273, J_{PH}1069	94
HCl_2F		HP^+Cl_2F (FSO_3^-)	92.0($FSO_3H/SbF_5/SO_2$, −70°) J_{PH}980, J_{PF}1255	94
HCl_3		HP^+Cl_3 (FSO_3^-)	86.9($FSO_3H/SbF_5/SO_2$, −70°) J_{PH}911	94
HFO_2	$(H)_2$	$HP^+F(OH)_2$ (FSO_3^-)	15.7(FSO_3H) J_{PH}899, J_{PF}1099	51
HF_3		HP^+F_3 (FSO_3^-)	16.2($FSO_3H/SbF_5/SO_2$, −70°) J_{PH}1191, J_{PF}1279	94
HO_3	$(H)_3$	$HP^+(OH)_3$ (FSO_3^-)	18.9 ± 1.0 (FSO_3H, −60°) J_{PH}823 ± 2	80
	$(H)_2$,C	$HP^+(OiPr)(OH)_2$ (FSO_3^-)	17.3(FSO_3H, −60°) J_{PH}821	80
		$HP^+(ONeo)(OH)_2$ $CF_3SO_3^-$	15.4(c, excess CF_3SO_3H) J_{PH}786	96
	H,$(C)_2$	$HP^+(OMe)(OCH_2CH_2OH)(OH)$	54.0 J_{PH}833	97
		$HP^+(OPr)_2(OH)$ (FSO_3^-)	19.4(FSO_3H, −60°) J_{PH}820	80
		$HP^+(OiPr)_2(OH)$ (FSO_3^-)	16.3(FSO_3H, −60°) J_{PH}812	80
		$CF_3SO_3^-$	17.6(c, excess CF_3SO_3H) J_{PH}804	96
		$HP^+(OnBu)_2(OH)$ (FSO_3^-)	19.3(FSO_3H, −60°) J_{PH}819	80
		$HP^+(ONeo)_2(OH)$ $CF_3SO_3^-$	17/18(c, excess CF_3SO_3H) J_{PH}792/810	96
	$(C)_3$	$HP^+(OMe)_3$ (FSO_3^-)	24.7, 24.4(FSO_3H, −60°, −50°) J_{PH}827, 826	80, 98
		(HSO_4^-)	26(H_2SO_4), J_{PH}830 ± 2	95
		$HP^+(OEt)_3$ (FSO_3^-)	19.7(FSO_3H, −60°) J_{PH}811	80
		(HSO_4^-)	18(H_2SO_4) J_{PH}806 ± 2	95
		$HP^+(O\text{-}iPr)_3$ (FSO_3^-)	15.7(FSO_3H, −60°) J_{PH}796	80
		(HSO_4^-)	16(H_2SO_4) J_{PH}795 ± 2	95
		$HP^+(OBu)_3$ (FSO_3^-)	20.4(FSO_3H, −60°) J_{PH}812	80
		$HP^+(ONeo)_3$ (HCl_2^-)	20.3(ae, 25°), 24.9(j, −70°) J_{PH}834	96

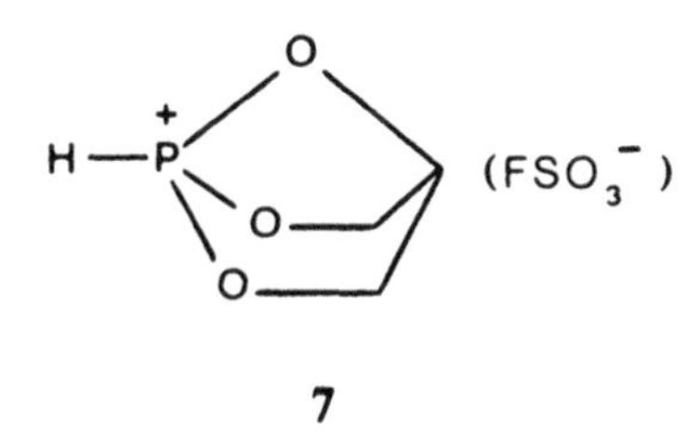

7

P-bound atoms/rings	Connectivities	Structure	NMR Data (δ_P[solv, temp][a] J_{PZ} Hz)	Ref.
		$CF_3SO_3^-$	23.6(c) J_{PH}834 20.0/21.1 (c, CF_3SO_3H) J_{PH}804/810	96
	$(C)_2$,C′	$HP^+(ONeo)_2(OPh)$ (HCl_2^-)	28.2(j/r, 27°) 24.0 (j/r, −80°) J_{PH}837	99
	$(C')_3$	$HP^+(OPh)_3$ (FSO_3^-)	11.7(FSO_3H, −60°) J_{PH}875	80
		(HSO_4^-)	10(H_2SO_4) J_{PH}870 ± 2	95
HO_3/5	$(C)_3$	$[H\dot{P}(OMe)OCH_2CHMe\dot{O}]^+(FSO_3^-)$	44.7(FSO_3H, −50°) J_{PH}890	98
		$[H\dot{P}(OMe)OCH_2CMe_2\dot{O}]^+(FSO_3^-)$	44.8 J_{PH}882 (FSO_3H, −60°)	97
		$[H\dot{P}(OMe)OCMe_2CMe_2\dot{O}]^+(FSO_3^-)$	40.2 J_{PH}856 (FSO_3H, −50°)	97
	$(C)_2$,C′	$[H\dot{P}(OPh)OCH_2CH_2\dot{O}]^+(FSO_3^-)$	43.9 J_{PH}913 (FSO_3H, −50°)	97
		(HSO_4^-)	16(H_2SO_4) J_{PH}822 ± 2	95
HO_3/6	$(C)_3$	$[H\dot{P}(OMe)OCH_2CMe_2CH_2\dot{O}]^+$ (FSO_3^-)	17.1 J_{PH}861 (FSO_3H, −50°)	97
		$[H\dot{P}(OMe)O(CH_2)_3\dot{O}]^+(FSO_3^-)$	18.2(FSO_3H, −50°) J_{PH}870	97
		$[H\dot{P}(OMe)OCHMe$- $CH_2CHMe\dot{O}]^+(FSO_3^-)$	17.3, 18.2[+](FSO_3H, −50°)J_{PH}865	98
		$[H\dot{P}(OMe)OCMe_2CH_2CMe_2\dot{O}]^+$ (FSO_3^-)	15.7 J_{PH}851 (FSO_3H, −50°)	97
HO_3/7	$(C)_3$	$[H\dot{P}(OMe)O(CH_2)_4\dot{O}]^+(FSO_3^-)$	28.7 J_{PH}844 (FSO_3H, −50°)	97
HO_3/5,5,6	$(C)_3$	7	46.1(FSO_3H, −50°) J_{PH}929	98
HO_3/6,6,6	$(C)_3$	$HP^+(OCH_2)_3CMe$ (FSO_3^-)	32.2(FSO_3H, −50°) J_{PH}899	98
H_2FO	H	$H_2P^+F(OH)$ (FSO_3^-)	39.2(FSO_3H), J_{PH}756, J_{PF}1079	51
H_2O_2	$(H)_2$	$H_2P^+(OH)_2$ (FSO_3^-)	36.1(FSO_3H, −60°) J_{PH}687	80
H_4		$H_4P^+(BF_4^-)$	−104.5(BF_3/w or BF_3/y) J_{PH}547(excess BF_3)	100
		(FSO_3^-)	−101.0(FSO_3H) J_{PH}548	101
		I^-	−77.5(solid)	44

TABLE G (continued)
Compilation of ^{31}P NMR Data of Four Coordinate Phosphonium Salts and Betaines

Section G2: Four Coordinate Phosphonium Salts and Betaines with One P–C Bond

P-bound atoms/rings	Connectivities	Structure	NMR Data (δ_P[solv, temp]$^a J_{PZ}$ Hz)	Ref.
CBrClO	H_3;H	$MeP^+BrCl(OH)$ (HSO_4^-)	59.6(H_2SO_4)	102
		$(HS_2O_7^-)$	74.2 ± 1.6 (25% oleum)	102
$CBrCl_2$	H_3	MeP^+BrCl_2 (HSO_4^-)	95.1(H_2SO_4)	102
		$(HS_2O_7^-)$	95.1(25% oleum)	102
$C'BrCl_2$	C'_2	$PhPBrCl_2$ $(HS_2O_7^-)$	80(25% oleum)	8
$C''BrCl_2$	N''	NCP^+BrCl_2 (Br_3^-)	15.4(Br_2)	2
$C'BrN_2$	C'_2;$(C_2)_2$	$PhP^+Br(NMe_2)_2$ (Br^-)	70.9(solid)	1
$CBrO_2$	H_3;$(H)_2$	$MeP^+Br(OH)_2$ (HSO_4^-)	25.8(H_2SO_4)	102
$C'BrO_2$	C'_2;$(H)_2$	$C_6F_5P^+Br(OH)_2$ $(HS_2O_7^-)$	37.1(25% oleum)	103
CBr_2Cl	H_3	MeP^+Br_2Cl (HSO_4^-)	64.5(H_2SO_4)	102
		$(HS_2O_7^-)$	63.7 ± 0.8 (25% oleum)	102
$C'Br_2Cl$	C'_2	PhP^+Br_2Cl $(HS_2O_7^-)$	54(25% oleum)	8
$C''Br_2Cl$	N''	NCP^+Br_2Cl (Br_3^-)	−17.4(Br_2)	2
CBr_2O	H_3;H	$MeP^+Br_2(OH)$ (HSO_4^-)	43.6(H_2SO_4)	102
		$(HS_2O_7^-)$	54.5 ± 1.2 (25% oleum)	102
CBr_3	H_3	MeP^+Br_3 BBr_4^-	27.5(solid), 30.7(ne)	102
		(HSO_4^-)	30.7(H_2SO_4)	102
		$(HS_2O_7^-)$	30.3 ± 1.2 (25% oleum)	102
$C'Br_3$	C'_2	PhP^+Br_3 BBr_4^-	21.1(solid), 24.2(ne)	104
		Br^-	14.6(solid)	104
		$(HS_2O_7^-)$	23(25% oleum)	8
		$C_6F_5P^+Br_3$ BBr_4^-	−38.0(solid), −25.8(Br_2)	103
		Br^-	−42.0(solid), −30.7(Br_2)	103
		Br_3^-	−29.1(Br_2)	103
		$(HS_2O_7^-)$	−25.8(25% oleum)	103
$C''Br_3$	N''	NCP^+Br_3 (Br_3^-)	−50.9 ± 1.8(Br_2)	2
$CClN_2$	C_3;$(C_2)_2$	$tBuP^+Cl(Pip)_2$ $AlCl_4^-$	82(r)	20
$CClN'_2$	H_3;$(N')_2$	$MeP^+Cl(N_3)_2$ $SbCl_6^-$	67.8(fe)	105
	H_3;$(C'')_2$	$MeP^+Cl(NCS)_2$ $SbCl_6^-$	66.2(ne)	103
	CH_2;$(N')_2$	$EtP^+Cl(N_3)_2$ $SbCl_6^-$	62.9(fe)	106
$C'ClN_2$	C'_2;$(C_2)_2$	$PhP^+Cl(NMe_2)_2$ Cl^-	66.2(solid), 67.8(r)	1, 17
		(HCl_2^-)	72.1(liq HCl)	1
$C'ClN'_2$	C'_2;$(N')_2$	$PhP^+Cl(N_3)_2$ BCl_4^-	51.6(fe)	105
		$SbCl_6^-$	51.6(fe)	106
		$C_6F_5P^+Cl(N_3)_2$ BCl_4^- or $SbCl_6^-$	45.3(ne)	103
	C'_2;$(P^4)_2$	$[PhPCl(NPPhCl_2)_2]^+Cl^-$	15.9	107
		$[PhP^+Cl(NPPhCl_2)]_2N$ Cl^-	9.2	107

P-bound atoms/rings	Connectivities	Structure	NMR Data (δ_P[solv, temp]$^nJ_{PZ}$ Hz)	Ref.
	$C'_2;(C'')_2$	$PhP^+Cl(NCS)_2\ SbCl_6^-$	50.0(ne)	103
		$C_6F_5P^+Cl(NCS)_2\ BCl_4^-$	27.5(fe)	103
$CClO_2$	$H_3;(H)_2$	$MeP^+Cl(OH)_2\ (ClSO_3^-)$	65($ClSO_3H$)	108
		(HSO_4^-)	63(H_2SO_4)	108
		$(HS_2O_7^-)$	67(25% oleum)	108
	$CH_2;(C)_2$	$EtP^+Cl(OEt)_2\ Cl^-$	70, 71, 70.3(ce, −40°)	23
$CClO_2/5$	$Cl_3;(C')_2$	$Cl_3CP^+ClCat\ SbCl_6^-$	74.0(ne), 74.4(fe)	109
$C'ClO_2$	$C'_2;(H)_2$	$PhP^+Cl(OH)_2\ (HSO_4^-)$	52(H_2SO_4)	110
		$(ClSO_3^-)$	53($ClSO_3H$)	110
		$HO_3SC_6H_4P^+Cl(OH)_2\ (ClSO_3^-)$	46($ClSO_3H$)	110
		$(HS_2O_7^-)$	43(25% oleum)	110
	$C'_2;(C)_2$	$PhP^+Cl(OBu)_2\ Cl^-$	54/55(ce, −40°)	23
CCl_2N	$H_3;C_2$	$MeP^+Cl_2(NMe_2)\ Cl^-$	87.1(r)	1
CCl_2N'	$H_3;N'$	$MeP^+Cl_2(N_3)\ SbCl_6^-$	90.4(fe)	105
	$H_3;C''$	$MeP^+Cl_2(NCS)\ SbCl_6^-$	96.7(ne)	103
	$CH_2;N'$	$EtP^+Cl_2(N_3)\ SbCl_6^-$	92.0(fe)	106
$C'Cl_2N$	$C'_2;C_2$	$PhP^+Cl_2(NMe_2)\ BCl_4^-$ or Cl^-	72.6(solid)	17
		$SbCl_6^-$	74.0(solid)	17
$C'Cl_2N'$	$C'_2;N'$	$PhP^+Cl_2(N_3)\ BCl_4^-$	72.6(fe)	105
		$SbCl_6^-$	72.6(fe)	106
		$C_6F_5P^+Cl_2(N_3)\ BCl_4^-$ or $SbCl_6^-$	56.5(ne)	103
	$C'_2;P^4$	$[PhPCl_2NPPhCl_2]^+\ Cl^-$	41.7	107, 111
		$[Ph\mathbf{P}Cl_2NPCl(Ph)N\mathbf{P}PhCl_2]^+\ Cl^-$	29.7	107
		$[Ph\mathbf{P}Cl_2(NPPhCl)_2N\mathbf{P}PhCl_2]^+\ Cl^-$	24.8	107
	$C'_2;C''$	$PhP^+Cl_2(NCS)\ SbCl_6^-$	79.1(ne)	103
		$C_6F_5Cl_2(NCS)\ BCl_4^-$	56.5(fe)	103
CCl_2O	$H_3;H$	$MeP^+Cl_2(OH)\ (ClSO_3^-)$	80($ClSO_3H$)	108
		(HSO_4^-)	74.2(H_2SO_4)	102
		$(HS_2O_7^-)$	87(25% oleum)	108
$C'Cl_2O$	$C'_2;H$	$PhP^+Cl_2(OH)\ (ClSO_3^-)$	68($ClSO_3H$)	110
		(HCl_2^-)	41.1(liq HCl)	24
		(HSO_4^-)	57(H_2SO_4)	110
		$(HS_2O_7^-)$	66(25% oleum)	110
		$HO_3SC_6H_4P^+Cl_2(OH)\ (ClSO_3^-)$	52($ClSO_3H$)	110
		(HSO_4^-)	63(H_2SO_4)	110
		$(HS_2O_7^-)$	51(25% oleum)	110
		$(HS_3O_{10}^-)$	70(65% oleum)	110
$C'Cl_2O/5$	$CC';C$	$[Cl_2\dot{P}C(tBu){:}CCl\text{-}C(CH_2Cl)Me\dot{O}]^+Cl^-$	93	112
CCl_3	Cl_3	$Cl_3CP^+Cl_3\ BCl_4^-$	102.9(solid), 105.2(ne)	103
		ICl_4^-	86.9(solid), 96.5(ne)	103
		$SbCl_6^-$	99.7(solid), 104.2(ne), 103.3(fe)	103, 109
	H_3	$MeP^+Cl_3\ AlCl_4^-$	117 ± 1(solid)	113
		$AuCl_4^-$	119.6(solid)	106
		Cl^-	119 ± 2(solid)	113, 114
		ICl_2^-	116.5 ± 1(solid)	113
		ICl_4^-	116.0(solid)	106
		$SbCl_6^-$	120.9(solid, fe, ne), 120.4(r)	105, 106, 114

TABLE G (continued)
Compilation of ^{31}P NMR Data of Four Coordinate Phosphonium Salts and Betaines

Section G2: Four Coordinate Phosphonium Salts and Betaines with One P–C Bond

P-bound atoms/rings	Connectivities	Structure	NMR Data (δ_P[solv, temp]$^nJ_{PZ}$ Hz)	Ref.
		(HSO_4^-)	120.9(H_2SO_4)	102
		($HS_2O_7^-$)	120.5 ± 0.5 (25% ± oleum)	8, 102
	CCl_2	$C_2Cl_5P^+Cl_3$ BCl_4^-	112.1(ne)	103
		ICl_4^-	112.6(solid or ne)	103
		$SbCl_6^-$	109.4(solid), 112.4(ne)	103
	CH_2	EtP^+Cl_3 $AlCl_4^-$	124.4 ± 1.4(solid), 128.6(r)	106, 113, 115
		$Al_2Cl_7^-$	128.6(r)	115
		Cl^-	127.5 ± 2(solid)	114
		ICl_4^-	128.4(solid)	106
		$SbCl_6^-$	127.5(solid), 128.9(fe)	106
		($HS_2O_7^-$)	129(25% oleum)	8
	C_2H	$iPrP^+Cl_3$ $AlCl_4^-$	133.0(ne)	37
		$cHexP^+Cl_3$ Cl^-	127.9(fe), 124.3(m)	26, 36
		PCl_6^-	127.4(fe)	36
	C_3	$tBuP^+Cl_3$ $AlCl_4^-$	139.2(ne)	37
C'Cl_3	C'N'	$Cl_3P^+C(:CCl_2)N:PCl_3$ PCl_6^-	85(ne)	116
		$Cl_3P^+C(:CHCl)N:PCl_3$ PCl_6^-	83.1, 80.7[t](ne)	116
		$Cl_3P^+C(:CMeCl)N:PCl_3$ PCl_6^-	86, 84[t](ne)	116
	C'$_2$	PhP^+Cl_3 $AlCl_4^-$	101(ne)	150
		BCl_4^-	101 ± 2(solid)	113
			101.6(fe)	105
		ClO_4^-	103.0(fe or $EtNO_2$)	42
		PCl_6^-	97 ± 2(solid)	113
			101.0(fe), 95.5(m)	26, 36
		$SbCl_6^-$	88 ± 2(solid)	113
			102.9(fe or $EtNO_2$)	42
			99.7(ne), 103.3(fe)	106, 173
		(HCl_2^-)	101.0(liq HCl)	1
		($HS_2O_7^-$)	103(25% oleum)	8
		$C_6F_5P^+Cl_3$ $AuCl_4^-$	88.7(solid), 82.2(r or ne)	103
		BCl_4^-	82.2(solid, ne, fe)	103
		ICl_4^-	77.5(ne)	103
		$SbCl_6^-$	82.2(solid, ne, fe)	103
		p-$ClC_6H_4P^+Cl_3$ PCl_6^-	99.8, 99.6(fe), 73.1(m)	26, 36
		p-$FC_6H_4P^+Cl_3$ PCl_6^-	100.0(fe)	36
		p-$MeC_6H_4P^+Cl_3$ PCl_6^-	101.1(fe), 94.3(m)	26, 36
		p-$MeOC_6H_4P^+Cl_3$ PCl_6^-	97.1(fe), 89.8(m)	26, 36
C''Cl_3	N''	NCP^+Cl_3 (BCl_4^-)	42.0(BrCN or fe)	9
		(Br^-)	40.4(r or PCl_3)	9
		(Br_3^-)	44.1(Br_2)	2
		(Cl^-)	46.7(cyanogen)	9
		(I^-)	42.0 (r)	9
		PCl_6^-	46.7(solid)	9
		$SbCl_6^-$	44(solid),	9

P-bound atoms/rings	Connectivities	Structure	NMR Data (δ_P[solv, temp][a] J_{PZ} Hz)	Ref.
			45.3(fe),42.0(r)	
		(CN^-)	50.0(cyanogen)	9
		$PCl_5(CN)^-$	48.4(fe)	9
CFN_2	$H_3;(C_2)_2$	$MeP^+F(NMe_2)_2\ MePF_5^-$	71.4(n), J_{PF}1030, 70.9(m), J_{PF}1032	117, 118
	$CH_2;(C_2)_2$	$EtP^+F(NMe_2)_2\ EtPF_5^-$	72.0 J_{PF}1051	52
$CFN_2/4$	$C_3;(CP^5)_2$	$[tBuPFN(tBu)PF_4N(tBu)]^+$	68.5(r), J_{PF}1228	79
$CFN_2/5$	$CH_2;(C_2)_2$	$[EtPFNMeCH_2CH_2NMe]^+EtPF_5^-$	76.7(m) J_{PF}1173	119
CFN'_2	$H_3;(P^4)_2$	$MeP^+F(N{:}P\text{-}iPr_3)_2\ MePF_5^-$	14.3 J_{PF}945 ± 3	120
$C'FN_2$	$C'_2;(C_2)_2$	$PhP^+F(NMe_2)_2\ PhPF_5^-$	56.0(m) J_{PF}1037, 1042	118, 121
		$PhP^+F(Pip)_2\ PhPF_5^-$	49.8, J_{PF}1042	52
$C'FN_2/5$	$C'_2;(CC')_2$	$[PhPFNPhCH_2CH_2NPh]^+PhPF_5^-$	56.1(m), J_{PF}1145	119
CNO_2	$H_3;C'_2;(C')_2$	$MeP^+(NPh_2)(OPh)_2\ BF_4^-$	54	122
		I^-	54	122
CN_2O	$H_3;(C'H)_2;C'$	$MeP^+(NHPh)_2OPh\ BF_4^-$	44	122
		I^-	42	122
	$H_3;(C_2)_2;C$	$MeP^+(NMe_2)_2ONeo\ Cl^-$	60.3(c)	123
		Br^-	60.2(c), 61.5(d)	123
		I^-	59.6(c)	123
	$H_3;(C')_2;C'$	$MeP^+(NPh_2)_2OPh\ BF_4^-$	56	122
		I^-	58	122
$C'N_2O$	$C'_2;(C_2)_2;C$	$PhP^+(NMe_2)_2OMe\ SbCl_6^-$	48.5(r)	60
CN_2P	$C_3;(C_2)_2;CP$	$[tBu\mathbf{P}(Pip)_2PCH_2PPPh_3]^{2+}\ (AlCl_4^-)_2$	89.2(r), J_{PP}340	20
$CN_2P/5$	$H_3;C_2,CC';P$	$[MeP(NMe_2)\text{-}NMeCONMePMe]^+Cl^-$	63.0 J_{PP}232	124
$CN_2P^4/5,5$	$H_3;(CC')_2;N_3$	**3**(P_A)	21.3(m) J_{PP}219	67
$C'N_2P^2$	$C'_2;(C_2)_2;P^4$	$[Ph(Et_2N)_2\mathbf{PPP}(NEt_2)_2Ph]^+\ AlCl_4^-$	64 J_{PP}510	70
		$[Ph(Et_2N)_2\mathbf{PPP}Ph_3]^+\ AlCl_4^-$	77 J_{PP}479	70
$C'N_2S$	$C'_2;(C_2)_2;C$	$Ph(Me_2N)_2P^+SMe\ SbCl_6^-$	74.0(r)	60
$C'NN'S/6$	$C'_2;HP^4;P^4;C$	$[p\text{-}MeOC_6H_4\mathbf{P}(SMe)\text{-}N{:}PPh_2N{:}PPh_2NH]^+I^-$	38.1	125
$C'N_2Se$	$C'_2;(C_2)_2;C$	$Ph(Me_2N)_2P^+SeMe\ MeSO_3^-$	68.9(r)	60
CN_3	$Cl_2F;(C_2)_3$	$Cl_2FCP^+(NMe_2)_3\ Cl^-$	44 ± 1(e)	126
	$N'O';(C_2)_3$	$FSO_2N^-C(O)P^+(NMe_2)_3$	34.2(m)	126a
	$H_3;(C_2)_2;H_2$	$MeP^+(NMe_2)_2NH_2\ Cl^-$	53.0(c)	74
	$H_3;(C_2)_3$	$MeP^+(NMe_2)_3\ Cl^-$	54.4(c)	18
		I^-	58.4(fe), 58.6	60, 127
	$CH_2;(C_2)_3$	$BuP^+(NMe_2)_3\ Br^-$	62 ± 1(n)	126
	$C'H_2;(C_2)_3$	$p\text{-}ClC_6H_4CH_2P^+(NMe_2)_3\ Cl^-$	55 ± 1(e)	126
	$CC'H;(C_2)_3$	$(PhCOCH_2)(PhCO)CHP^+(NMe_2)_3\ Cl^-$	54.9	128
	$C'_3;(C_2)_3$	$Ph_3CP^+(NMe_2)_3\ BF_4^-$	52.4(m, 0°)	129
CN'_3	$H_3;(N')_3$	$MeP^+(N_3)_3\ SbCl_6^-$	50.0($MeCHCl_2$), 51.6(fe)	78, 105
	$H_3;(P^4)_3$	$Me\mathbf{P}^+(N{:}PMe_2Ph)_3\ MePF_5^-$	13.1	120
	$H_3;(C'')_3$	$MeP^+(NCS)_3\ SbCl_6^-$	45.3(ne)	103
	$CH_2;(N')_3$	$EtP^+(N_3)_3\ SbCl_6^-$	51.6(fe)	106
$C'N_3$	$CC';(C_2)_3$	$[PhCOCH_2][Ph(PhCH_2O)\text{-}C{:}]CP^+(NMe_2)_3\ Br^-$	50.4	128
	$C'_2;(H_2)_3$	$PhP^+(NH_2)_3\ Cl^-$	29.4(d)	74
	$C'_2;H_2(CN)_2$	$PhP^+(NH_2)(NMeNMe_2)_2\ Cl^-$	39.9(d)	130
	$C'_2;H_2;(C_2)_2$	$PhP^+(NMe_2)_2NH_2\ Cl^-$	43.0(c),41.0(d)	74
	$C'_2;(C_2)_3$	$PhP^+(NMe_2)_3\ Br^-$	51.4(r)	1
		Cl^-	51.6(r)	1
$C'N_3/5$	$C'_2;(C_2)_3$	$[PhP(NMeCH_2CH_2NMe)\text{-}NMe]^{2+}(PhPF_5^-)_2$	49.5(m)	119

TABLE G (continued)
Compilation of ^{31}P NMR Data of Four Coordinate Phosphonium Salts and Betaines

Section G2: Four Coordinate Phosphonium Salts and Betaines with One P–C Bond

P-bound atoms/rings	Connectivities	Structure	NMR Data (δ_P[solv, temp]$^nJ_{PZ}$ Hz)	Ref.
$C'N'_3$	$C'_2;(N')_3$	$PhP^+(N_3)_3\ BCl_4^-$	37.1(fe)	105
		$SbCl_6^-$	35.5(fe), 36.1($MeCHCl_2$)	78, 106
		$C_6F_5P^+(N_3)_3\ BCl_4^-$ or $SbCl_6^-$	29.1(ne)	103
	$C'_2;(P^4)_3$	$PhP^+(N{:}PMe_2Ph)_3\ PhPF_5^-$	5.8	120
	$C'_2;(C'')_3$	$PhP^+(NCS)_3\ SbCl_6^-$	37.1(ne)	103
		$C_6F_5P^+(NCS)_3\ BCl_4^-$	8.2(fe)	103
CO_2S	$CH_2;(C)_2;Cl$	$EtP^+(OEt)_2SCl\ Cl^-$	88.5(ce, −40°), 89	23
		$SbCl_6^-$	89.5	23
	$C_3;(C)_2;Cl$	$tBuP^+(OEt)_2SCl\ Cl^-$	97	23
$C'O_2S$	$C'_2;(C)_2;Cl$	$PhP^+(OBu)_2SCl\ Cl^-$	72.5(ce, −40°), 73	23
		$SbCl_6^-$	73.5	23
	$C'_2;(C)_2;C$	$PhP^+(OMe)_2SMe\ SbCl_6^-$	83.9(r)	60
$C'O_2Se$	$C'_2;(C)_2;C$	$PhP^+(OMe)_2SeMe\ SbCl_6^-$	84.6(r)	60
CO_3	$H_3;(H)_3$	$MeP^+(OH)_3\ (HSO_4^-)$	48.4(H_2SO_4)	102
		$(HS_2O_7^-)$	45.8(25% oleum)	102
	$H_3;(C)_3$	$MeP^+(OMe)_3\ BF_4^-$	48, 54.0	122, 132
		$SbCl_6^-$	54.3(m), 56(r)	86, 133
		$CF_3SO_3^-$	53.1(c)	134
		$MeP^+(OMe)_2OEt\ SbCl_6^-$	52.6(m)	86
		$MeP^+(OMe)(OEt)_2\ SbCl_6^-$	49.9(m)	86
		$MeP^+(OEt)_3\ SbCl_6^-$	49.5(m)	86
		$MeP^+(OiPr)_3\ I^-$	42.6(m)	131
		$MeP^+(ONeo)_3\ BF_4^-$	46	122
		Br^-, Cl^-, I^-	54(c)	135
		$MeP^+(O\text{-}1Nb)_3\ I^-$	37(c)	89
	$H_3;C,(C')_2$	$MeP^+(OMe)(OPh)_2\ CF_3SO_3^-$	41.5(c)	134
		$MeP^+(OEt)(OPh)_2\ CF_3SO_3^-$	40.8(c)	136
	$H_3;(C')_3$	$MeP^+(OPh)_3\ BF_4^-$	38	122
		Br^-	41(m), 40	122, 135
		Cl^-	41(m)	135
		I^-	39.2(c), 41(m), 40	122, 135, 137
		$CF_3SO_3^-$	43.0(m), 41.4(c)	136, 138
		$MeP^+(OC_6H_4Me\text{-}o)_3\ CF_3SO_3^-$	42.2(c)	138
		$MeP^+(OC_6H_4Me\text{-}p)_3\ CF_3SO_3^-$	41.9(c)	138
	$CH_2;(C)_3$	$EtP^+(OMe)_3\ SbCl_6^-$	54.1(m)	86
		$EtP^+(OMe)_2OEt\ SbCl_6^-$	51.9(m)	86
		$EtP^+(OMe)(OEt)_2\ SbCl_6^-$	50.0(m)	86
		$EtP^+(OEt)_3\ SbCl_6^-$	48.0(m)	86
	$CH_2;(C')_3$	$EtP^+(OPh)_3\ Cl^-$, Br^-, I^-	40(c)	135
		$PrP^+(OPh)_3\ Cl^-$, Br^-, I^-	38(c)	135
		$BuP^+(OPh)_3\ Cl^-$, Br^-, I^-	38(c)	135
		$iBuP^+(OPh)_3\ Cl^-$, Br^-, I^-	38(c)	135
	$C'H_2;(C)_3$	$PhCOCH_2P^+(ONeo)_3\ Br^-$	41.0(c)	139
		$PhCOCH_2P^+(O\text{-}1Nb)_3\ Br^-$, Cl^-	27(c)	89
		$p\text{-}BrC_6H_4COCH_2P^+(O\text{-}1Nb)_3\ Br^-$	25(c)	89
		$p\text{-}O_2NC_6H_4COCH_2P^+(O\text{-}1Nb)_3\ Br^-$	26(c)	89
	$C'H_2;(C')_3$	$PhCH_2P^+(OPh)_3\ Br^-$, I^-	30(c)	135
	$C_2H;(C')_3$	$iPrP^+(OPh)_3\ Cl^-$, Br^-, I^-	38(c)	135

P-bound atoms/rings	Connectivities	Structure	NMR Data (δ_P[solv, temp]$^nJ_{PZ}$ Hz)	Ref.
	$C'_3;(C)_3$	$Ph_3CP^+(OMe)_3$	25.7	140
$CO_3/5$	$H_3;C,(C')_2$	$MeP^+(OMe)Cat\ CF_3SO_3^-$	49.6(c)	134
$CO_3/6,6$	$H_3;(C)_3$	$MeP^+(OCH_2)_3P\ BF_4^-$	51.0(m)	141
		$MeP^+(OCH_2)_3CMe\ BF_4^-$	60.2(m)	141
$C'O_3$	$C'_2;(H)_2;P^4$	$PhP^+(OH)_2OP(O)(OH)Ph\ (ClSO_3^-)$	$23(ClSO_3H)$	110
		(HSO_4^-)	$22(H_2SO_4)$	110
	$C'_2;(H)_3$	$PhP^+(OH)_3\ (ClSO_3^-)$	$33(ClSO_3H)$	110
		(HCl_2^-)	26.3(liqHCl)	24
		(HSO_4^-)	$33(H_2SO_4)$	110
		$(HS_2O_7^-)$	27(25% oleum)	110
		$C_6F_5P^+(OH)_3\ (HS_2O_7^-)$	−8.2(25% oleum)	103
		$HO_3SC_6H_4P^+(OH)_3\ (ClSO_3^-)$	$27(ClSO_3H)$	110
		(HSO_4^-)	$27(H_2SO_4)$	110
		$(HS_2O_7^-)$	23(25% oleum)	110
	$C'_2;(C)_3$	$PhP^+(OMe)_3\ SbCl_6^-$	35.1(r)	60
	$C'_2;(C')_3$	$PhP^+(OPh)_3\ BF_4^-$	24(r)	91
		Br^-	16(r)	91
		Cl^-	19(r)	91
		I^-	23(r)	91
$C'HO_2$	$C'_2;(C)_2$	$PhP^+H(OMe)_2\ (HSO_4^-)$	$54(H_2SO_4)$, $J_{PH}666 \pm 2$	95
		$PhP^+H(ONeo)_2\ (HCl_2^-)$	52.6(jr, excess HCl, 25°)	99
			51.3(jr, excess HCl, −80°) $J_{PH}684$	99
CH_3	Si_3	$(Me_3Si)_3CP^+H_3\ BF_4^-$	−56.3(r, −78°) $J_{PH}502$	142

Section G3: Four Coordinate Phosphonium Salts and Betaines with Two P–C Bonds

P-bound atoms/rings	Connectivities	Structure	NMR Data	Ref.
C'_2B^4N	$(C_{12})_2;H_3;C'H$	$Ph_2P^+(B^-H_3)NHPh$	50.5	209b
C_2BrCl	$(H_3)_2$	$Me_2P^+BrCl\ (HSO_4^-)$	$106.4(H_2SO_4)$	102
		$(HS_2O_7^-)$	109.1(25% oleum)	102
C'_2BrCl	$(C'_2)_2$	$Ph_2P^+BrCl\ (HS_2O_7^-)$	74(25% oleum)	8
C'_2BrN	$(C'_2)_2;P^4$	$Ph_2P^+BrNPPh_2Br\ Br^-$	33.2(r)	143
	$(C'_2)_2;C_2$	$Ph_2P^+Br(NMe_2)\ Br^-$	70.9(r)	1
C_2BrO	$(H_3)_2;H$	$Me_2P^+Br(OH)\ (HSO_4^-)$	$92.7(H_2SO_4)$	102
		$(HS_2O_7^-)$	96.0(25% oleum)	102
C_2Br_2	$(H_3)_2$	$Me_2P^+Br_2\ BBr_4^-$	70.9(solid)	102
		Br_3^-	72.6(solid), 75.8(ne or m)	102
		(HSO_4^-)	$74.2(H_2SO_4)$	102
		$(HS_2O_7^-)$	74.2 ± 1.6 (25% oleum)	102
C'_2Br_2	$(C'_2)_2$	$Ph_2P^+Br_2\ BBr_4^-$	55(solid)	8
		Br^-	56(solid)	8
		Br_3^-	55.6(fe)	144
		$(HS_2O_7^-)$	57.2 ± 0.8 (25% oleum)	8
C''_2Br_2	$(N'')_2$	$(NC)_2P^+Br_2\ Br^-$	$-43.6(r\ or\ PBr_3)$	9
		(Br_3^-)	$-40.3(Br_2)$	2
		(CN^-)	−42.0(BrCN)	9
C_2ClN	$(H_3)_2;C_2$	$Me_2P^+Cl(NMe_2)\ Cl^-$	103.3(r)	1
	$(C_2H)_2;C_2$	$cHex_2P^+Cl(NMe_2)\ Br^-$	104.3(r)	1
		Cl^-	104.7(r)	1
	$(C'H_2)_2;C_2H$	$cHex_2CMe{:}CMeCH_2P^+ClNiPr\ AlCl_4^-$	93.4(r)	145

TABLE G (continued)
Compilation of ^{31}P NMR Data of Four Coordinate Phosphonium Salts and Betaines

Section G3: Four Coordinate Phosphonium Salts and Betaines with Two P–C Bonds

P-bound atoms/rings	Connectivities	Structure	NMR Data (δ_P[solv, temp]$^a J_{PZ}$ Hz)	Ref.
C_2ClN'	$(H_3)_2;N'$	$Me_2P^+Cl(N_3)\ SbCl_6^-$	100.0(fe)	106
	$(CH_2)_2;N'$	$Et_2P^+Cl(N_3)\ SbCl_6^-$	108.0(fe)	106
C'_2ClN	$(C'_2)_2;H_2$	$Ph_2P^+Cl(NH_2)\ Cl^-$	51(y)	146
	$(C'_2)_2;C_2$	$Ph_2P^+Cl(NMe_2)\ BCl_4^-$	71.8(solid)	17
		Cl^-	70.9(solid), 71.3 ± 3.6(r)	1, 17
		$SbCl_6^-$	72.5(fe)	144
		(HCl_2^-)	71.5(liq HCl)	1
$C'_2ClN'_2$	$(C'_2)_2;P^4$	$Ph_2P^+ClNPPh_2Cl\ Cl^-$	43.3(n), 43.6($EtNO_2$)	147, 148
		PCl_6^-	43.3($EtNO_2$)	148
		$SbCl_6^-$	43.5($EtNO_2$)	148
C_2ClO	$(H_3)_2;H$	$Me_2P^+Cl(OH)\ (HSO_4^-)$	106.4(H_2SO_4)	102
		$(HS_2O_7^-)$	108.0 ± 1.6 (25% oleum)	102
	$(H_3)_2;C$	$Me_2P^+Cl(OMe)\ Cl^-$	77	23
C'_2ClO	$(C'_2)_2;H$	$Ph_2P^+Cl(OH)\ (ClSO_3^-)$	76($ClSO_3H$)	108
		(HCl_2^-)	61.0(liq HCl)	24
		(HSO_4^-)	76(H_2SO_4)	108
		$(HS_2O_7^-)$	76(25% oleum)	108
		$Ph(HO_3SC_6H_4)P^+Cl(OH)\ (HS_2O_7^-)$	73(25% oleum)	108
		$(HO_3SC_6H_4)_2P^+Cl(OH)\ (HS_2O_7^-)$	70(25% oleum)	108
		$(HS_3O_{10}^-)$	70(65% oleum)	108
C_2Cl_2	$(Cl_3)_2$	$(Cl_3C)_2P^+Cl_2\ AlCl_4^-$	149.6(r)	103
		$SbCl_6^-$	148.9(ne), 148.4(fe)	103, 109
	$(H_3)_2$	$Me_2P^+Cl_2\ AuCl_4^-$	117.7(solid)	106
		BCl_4^-	119.5 ± 2(solid)	113
		Cl^-	125 ± 6(solid)	113, 114
		Cl_2I^-	121 ± 1(solid)	113
		Cl_4I^-	120.9 (solid)	106
		PCl_6^-	122.6(solid), 124.2(ne)	106
		$SbCl_6^-$	123 ± 4(solid), 124.2(fe or ne)	106, 113
		(HCl_2^-)	131.6(liq HCl)	1
		(HSO_4^-)	123.8 ± 0.5(H_2SO_4)	102
		$(HS_2O_7^-)$	123.4 ± 0.8 (25% oleum)	8, 102
	H_3,C_3	$Me(Me_2CHCMe_2)P^+Cl_2\ Cl^-$	124.7(r)	149
	$(CH_2)_2$	$Et_2P^+Cl_2\ BCl_4^-$	137.4 ± 2(solid)	113
		Cl^-	140.2 ± 3.5(solid)	113, 114
			137.3(m), 138.3, 137.9(fe)	26, 36, 106
			127.3(r), 125.8(ne)	114
		Cl_4I^-	138.7(r)	106
		PCl_6^-	137.4(fe)	36
		$SbCl_6^-$	137.1(r or ne), 138.7(fe)	106
		$(HS_2O_7^-)$	138(25% oleum)	8
	$(C_2H)_2$	$iPr_2P^+Cl_2\ Cl^-$	149.3(fe)	36

P-bound atoms/rings	Connectivities	Structure	NMR Data (δ_P[solv, temp]$^nJ_{PZ}$ Hz)	Ref.
		$cHex_2P^+Cl_2$ Cl^-	138.4(m), 137.7(fe)	26, 36
	$(C_3)_2$	$tBu_2P^+Cl_2$ Cl^-	158.0(m), 157.7(fe)	26, 36
		PCl_6^-	158.0(m), 158.3(fe)	26, 36
C'_2Cl_2	$(C'_2)_2$	$Ph_2P^+Cl_2$ $AlCl_4^-$	92(ne)	150
		BCl_4^-	93.6 ± 5(solid)	113
		ClO_4^-	93.2(fe or $EtNO_2$)	42
		PCl_6^-	89.7 ± 2(solid), 92.5(m), 92.0(fe)	26, 36, 113
		$SbCl_6^-$	92 ± 2(solid), 93.2(fe or $EtNO_2$)	42, 113
		(HCl_2^-)	93.5(liq HCl)	1
		$(HS_2O_7^-)$	95(25% oleum)	8
		$(C_6F_5)_2P^+Cl_2$ BCl_4^-	62.7(ne)	103
		Cl_4I^-	62.9(fe)	103
		$SbCl_6^-$	66.2(solid), 64.6(ne), 64.2(fe)	103
C''_2Cl_2	$(N'')_2$	$(NC)_2P^+Cl_2$ BCl_4^-	9.0(BrCN)	9
		(Br_3^-)	11.0(Br_2)	2
		Cl^-	8.2(cyanogen)	9
		PCl_6^-	6.5(solid)	9
		$SbCl_6^-$	11.0(solid)	9
		CN^-	3.2(cyanogen)	9
		$NCPCl_5^-$	12.9(fe)	9
C_2F_2	$(H_3)_2$	$Me_2P^+F_2$ PF_6^-	155.0(m), J_{PF}1145	151
C'_2NO	$(C'_2)_2$;C_2;H	$Ph_2P^+(NMe_2)OH$ (HSO_4^-)	50.0(H_2SO_4)	144
	$(C'_2)_2$;C_2;C	$Ph_2P^+(NMe_2)OMe$ $SbCl_6^-$	57.5(r)	60, 144
$C'_2N'O$	$(C'_2)_2$;P^4;H	$[Ph_2P^+(OH)NP(OH)Ph_2]$ OAc^-	25.1(u)	143
	$(C'_2)_2$;P^4;C	$[Ph_2P^+(OMe)NP(OMe)Ph_2]$	35.0(r)	143
		$[Ph_2P(OMe)NP(SMe)Ph_2]^+$ $SbCl_6^-$	35.3(r)	152
$C'_2N'O/8$	$(C'_2)_2$;P^4;P^4	$[Ph_2\dot{P}NPPh_2NP\dot{P}h_2N\mathbf{P}Ph_2\dot{O}]^+$ BPh_4^-	25.4(r)	153
		Cl^-	25.7(r)	153
		ClO_4^-	25.7(r)	153
		$SbCl_6^-$	27.6($CHCl_2CHCl_2$)	153
		SbF_6^-	25.0(fe)	153
C'_2NP^2	$(C'_2)_2$;C'H;P^4	$[Ph_2(PhNH)\mathbf{P}\mathbf{P}\mathbf{P}(NHPh)Ph_2]^+$ $AlCl_4^-$	47 J_{PP}498	70
		$[Ph_2(PhNH)\mathbf{P}PPPh_3]^+$ $AlCl_4^-$	49 J_{PP}524	70
	$(C'_2)_2$;C_2;P^4	$[Ph_2(Et_2N)\mathbf{P}P\mathbf{P}(NEt_2)Ph_2]^+$ $AlCl_4^-$	79 J_{PP}508	70
		$[Ph_2(Et_2N)\mathbf{P}PPPh_3]^+$ $AlCl_4^-$	64 J_{PP}501	70
$C'_2N'P$	$(C'_2)_2$;P^4;C'_2	$[Ph_2\mathbf{P}PPh_2N\mathbf{P}Ph_2PPh_2]^+Cl^-$	24.8 J_{PP}260	154
			28.3, 21.8, J_{PP}270	155
$C_2N'S$	$(H_3)_2$;P^4;C	$[Me_2(MeS)PNP(SMe)Me_2]^+$ I^-	43.5(r)	157
		$[Me_2(MeS)\mathbf{P}NP(SMe)Ph_2]^+$ I^-	46.7(r)	157
C'_2NS	$(C'_2)_2$;C_2;C	$Ph_2P^+(NMe_2)SMe$ $SbCl_6^-$	67.5(d)	60, 144
$C'_2N'S$	$(C'_2)_2$;P^4;C	$[Ph_2(MeS)\mathbf{P}NP(SMe)Me_2]^+$ I^-	34.7(r)	157
		$[Ph_2(MeS)\mathbf{P}NP(OMe)Ph_2]^+$ $SbCl_6^-$	38.8(r)	152
		$[Ph_2(MeS)PNP(SMe)Ph_2]^+$ $SbCl_6^-$	38.6(fe)	152, 156
C'_2NSe	$(C'_2)_2$;C_2;C	$Ph_2P^+(NMe_2)SeMe$ $SbCl_6^-$	65.7(r)	60
C_2N_2	$(H_3)_2$;$(Si_2)_2$	$Me_2P^+[N(SiMe_3)_2]_2$ I^-	57.9(c)	158
	$(H_3)_2$;$(H_2)_2$	$Me_2P^+(NH_2)_2$ Cl^-	42.3(d), 71.3	73a, 74
	$(H_3)_2$;H_2,C_2	$Me_2P^+(NH_2)NMe_2$ Cl^-	54.0(c)	74, 159
	$(H_3)_2$;$(C_2)_2$	$Me_2P^+(NMe_2)_2$ Cl^-	68.4 ± 2.5(c)	18
	$(CH_2)_2$;H_2,CH	$Bu_2P^+(NH_2)NHEt$ Cl^-	54	159
$C_2N_2/5$	$(C'H_2)_2$;$(C_2)_2$	$[(Me_2N)_2\dot{P}CH_2CMe{:}CH\dot{C}H_2]^+$ $AlCl_4^-$	84.2(r), 89.6(r)	145, 160
		$[(Me_2N)_2\dot{P}CH_2CMe{:}CMe\dot{C}H_2]^+$ $AlCl_4^-$	82.7(r)	145

TABLE G (continued)
Compilation of ^{31}P NMR Data of Four Coordinate Phosphonium Salts and Betaines

Section G3: Four Coordinate Phosphonium Salts and Betaines with Two P–C Bonds

P-bound atoms/rings	Connectivities	Structure	NMR Data (δ_P[solv, temp]$^n J_{PZ}$ Hz)	Ref.
		$[(Et_2N)_2\dot{P}CH_2CMe{:}CH\dot{C}H_2]^+AlCl_4^-$	78.1(r)	160
		$[(Et_2N)_2\dot{P}CH_2CMe{:}CMe\dot{C}H_2]^+$ $AlCl_4^-$	71.5(r)	160
		$[(iPr_2N)_2\dot{P}CH_2CH{:}CH\dot{C}H_2]^+AlCl_4^-$	76.0(r)	145
		$[(iPr_2N)_2\dot{P}CH_2CMe{:}CH\dot{C}H_2]^+AlCl_4^-$	76.5(r)	145
		$[(iPr_2N)_2\dot{P}CH_2CMe{:}CMe\dot{C}H_2]^+$ $AlCl_4^-$	69.6(r)	145
	$C'H_2,CC'H;(C_2)_2$	$[(iPr_2N)_2\dot{P}CHMeCH{:}CH\dot{C}H_2]^+$ $AlCl_4^-$	79.7(r)	145
	$(CC'H)_2;(C_2)_2$	$[(iPr_2N)_2\dot{P}CHMeCH{:}CH\dot{C}HMe]^+$ $AlCl_4^-$	88.7(r)	145
C_2NN'	$(CH_2)_2;H_2;P^4$	$[Et_2(H_2N)PNP(NH_2)Et_2]^+\ Cl^-$	41(c)	161
$C_2NN'/6$	$(H_3)_2;HP^4;P^4$	$[Me_2\dot{P}NHPPh_2{:}NPPh_2{:}\dot{N}]^+Cl^-$	39.6(c)	162
	$(H_3)_2;C'H;P^4$	$[Me_2\dot{P}NHCMe{:}NPMe_2{:}\dot{N}]^+Cl^-$	36.4(r)	163
		$[Me_2\dot{P}NHCMe{:}NPPh_2{:}\dot{N}]^+Cl^-$	40.3 (r)	163
$C_2N'_2$	$(H_3)_2;(N')_2$	$Me_2P^+(N_3)_2\ SbCl_6^-$	77.0($MeCHCl_2$ or fe)	78, 106
	$(H_3)_2;(P^4)_2$	$[Me_2\mathbf{P}^+(N{:}PMe_3)_2]Me_2PF_4^-$	19.8	120
		$[Me_2\mathbf{P}^+(N{:}PMe_2Ph)_2]\ Me_2PF_4^-$	20.7	120
		$Me_2\mathbf{P}^+[N{:}P(iPr)_3]_2\ Me_2PF_4^-$	5.8	120
	$(CH_2)_2;(N')_2$	$Et_2P^+(N_3)_2\ SbCl_6^-$	82.2(fe)	106
$CC'N_2$	$H_3;C'_2;(C_2)_2$	$MePhP^+(NMe_2)_2\ I^-$	59.4(fe)	60
C'_2N_2	$(C'_2)_2;(H_2)_2$	$Ph_2P^+(NH_2)_2\ Cl^-$	31.9 ± 2(d)	74
	$(C'_2)_2;NP^4,H_2$	$[Ph_2\mathbf{P}^+(NH_2)N(NHMe_2)$-$P({:}NH)Ph_2]^{2+}\ 2Cl^-$	1.5 ± 1.3(d)	130
		$[(Ph_2\mathbf{P}^+NH_2)_2NNMePPh_2]^{2+}\ 2Cl^-$	36.7 ±2.6(d)	130
		$[(Ph_2\mathbf{P}^+NH_2)_2NNMeP({:}NH)Ph_2]^{2+}2Cl^-.CHCl_3$	12.6 ± 2.6(d)	130
	$(C'_2)_2;H_2,HP^4$	$Ph_2\mathbf{P}^+(NH_2)NHP(S)Ph_2\ Cl^-$	40.7(r)	148
	$(C'_2)_2;H_2,CP^4$	$Ph_2\mathbf{P}^+(NH_2)NMeP({:}NH)Ph_2\ Cl^-$	24	159
		$Ph_2\mathbf{P}^+(NH_2)NEtP({:}NH)Ph_2\ Cl^-$	21	159
	$(C'_2)_2;H_2,C_2$	$Ph_2P^+(NH_2)NMe_2\ Cl^-$	42.2(c)	74
	$(C'_2)_2;(C_2)_2$	$Ph_2P^+(NMe_2)_2\ Br^-$	51.6(r)	1
		Cl^-	50.0(r)	1
$C'_2N_2/8$	$(C'_2)_2{:}(C'B^4)_2$	$[Ph_2P^+NPhB^-H_2NPh]_2$	46.6	209b
C'_2NN'	$(C'_2)_2;H_2;P^4$	$[Ph_2P(NH_2)NP(NH_2)Ph_2]^+Cl^-$	20, 20.3	159, 164
	$(C'_2)_2;CH;P^4$	$[Ph_2P(NHMe)NP(NHMe)Ph_2]^+$	20.9(r)	143
	$(C'_2)_2;C_2;P^4$	$[Ph_2P(NMe_2)NP(NMe_2)Ph_2]^+$	25.4(ne)	143
$C'_2NN'/6$	$(C'_2)_2;C'H;P^4$	$[Ph_2\dot{P}NHCMe{:}NPPh_2{:}\dot{N}]Cl^-$	19.0(r)	163, 165
		$[Ph_2\dot{P}NHCMe{:}NPPh_2{:}\dot{N}]$-$BPh_4^-.2MeCN$	20.4(r)	165
		$[Ph_2\dot{P}NHC(CH_2Ph){:}NPPh_2{:}\dot{N}]^+Cl^-$	19.9(r)	163, 165
		$[Ph_2\dot{P}NHCPh{:}NPPh_2{:}\dot{N}]^+Cl^-$	21.9(r)	163, 165
$C'_2N'_2$	$(C'_2)_2;(N')_2$	$Ph_2P^+(N_3)_2\ SbCl_6^-$	51.7($MeCHCl_2$)	78
C_2OS	$(H_3)_2;C;Cl$	$Me_2P^+(OMe)SCl\ Cl^-$	113	23
C'_2OS	$(C'_2)_2;C;C$	$Ph_2P^+(OMe)SMe\ SbCl_6^-$	86.6(r)	60
C'_2OSe	$(C'_2)_2;C;C$	$Ph_2P^+(OMe)SeMe\ SbCl_6^-$	87.8(r)	60
C_2O_2	$(H_3)_2;(H)_2$	$Me_2P^+(OH)_2\ (HSO_4^-)$	85.1 ± 1.3(H_2SO_4)	102
		$(HS_2O_7^-)$	86.3 ± 0.8 (25% oleum)	102

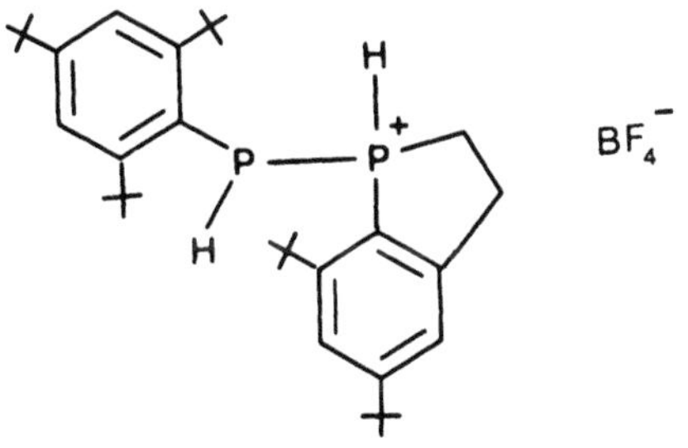

BF_4^-

8

P-bound atoms/rings	Connectivities	Structure	NMR Data (δ_P[solv, temp]^a J_{PZ} Hz)	Ref.
	$(H_3)_2;(C)_2$	$Me_2P^+(OMe)_2\ SbCl_6^-$	98.2(m)	86
		$Me_2P^+(OMe)OEt\ SbCl_6^-$	95.6(m)	86
		$Me_2P^+(OEt)_2\ SbCl_6^-$	91.8(m)	86
	$H_3,CH_2;(C)_2$	$MeEtP^+(OMe)_2\ CF_3SO_3^-$	99.0(c)	134
		$MeEtP^+(OiBu)_2\ I^-$	94(r or c)	166
	$H_3;C_2H;(C)_2$	$MeiPrP^+(OiBu)_2\ I^-$	96(r or c)	166
	$(CH_2)_2;(C)_2$	$Et_2P^+(OEt)_2\ SbCl_6^-$	93.8(m)	86
		$Et_2P^+(OMe)_2\ SbCl_6^-$	101.0(m)	86
		$Et_2P^+(OMe)OEt\ SbCl_6^-$	98.4(m)	86
$C_2O_2/6$	$(CH_2)_2;(C)_2$	$[Et_2\dot{P}OCH_2CMe_2CH_2\dot{O}]^+BF_4^-$	−4.3(n or r)	83
$CC'O_2$	$H_3;C'_2;(C)_2$	$MePhP^+(OMe)_2\ BF_4^-$	76	122
		$CF_3SO_3^-$	78.5(c)	134
		$MePhP^+(O\text{-}iPr)_2\ I^-$	67.6(m)	131
		$MePhP^+(ONeo)_2\ Br^-$ or I^-	74(c)	167
	$H_3;C'_2;(C')_2$	$MePhP^+(OPh)_2\ BF_4^-$	73	122
		Br^-	73	122
		I^-	73, 73.0 ± 1.0(c)	122, 137
	$C'H_2;C'_2;(C)_2$	$PhCOCH_2P^+(Ph)(ONeo)_2$	67.1(c)	139
C'_2O_2	$(C'_2)_2;(H)_2$	$Ph_2P^+(OH)_2\ (ClSO_3^-)$	55($ClSO_3H$)	108
		(HCl_2^-)	48(liq HCl)	24
		(HSO_4^-)	53, 54(H_2SO_4)	108, 144
		$Ph(HO_3SC_6H_4)P^+(OH)_2\ (ClSO_3^-)$	51($ClSO_3H$)	108
		$(HS_2O_7^-)$	51(25% oleum)	108
		$(HO_3SC_6H_4)_2P^+(OH)_2\ (HS_2O_7^-)$	48(25% oleum)	108
		$(HS_3O_{10}^-)$	48(65% oleum)	108
	$(C'_2)_2;(C)_2$	$Ph_2P^+(OMe)_2\ SbCl_6^-$	61.2(r), 61.3(r)	60, 144
	$(C'_2)_2;C,C'$	$Ph_2P^+(ONeo)OC(:CH_2)Ph\ Cl^-$	55.9(c)	139
	$(C'_2)_2;(C'_2)_2$	$Ph_2P^+(OPh)_2\ BF_4^-$	53(r)	91
		Br^-	51(r)	91
		Cl^-	55(r)	91
C_2HF	$(H_3)_2$	$Me_2P^+HF\ BF_4^-$	115.8(fe) J_{PF}925 J_{PH}621	168
		AsF_6^-	115.8(fe) J_{PF}962 J_{PH}608	168
C'_2HO	$(C'_2)_2;C$	$Ph_2P^+H(OMe)(HSO_4^-)$	56(H_2SO_4) J_{PH}553 ± 2	95
		$Ph_2P^+H(ONeo)\ (HCl_2^-)$	51.5(j/r, 25°), 50.4 (j/r, −80°) J_{PH}584	99
$CC'HP$	$CH_2;C'_2;C'H$	**8**	10.6(r, −80°) J_{PH}481, J_{PP}290	142
C'_2H_2	$(C'_2)_2$	$Ph_2P^+H_2\ (FSO_3^-)$	−21.2(FSO_3H) J_{PH}519	101

TABLE G (continued)
Compilation of ^{31}P NMR Data of Four Coordinate Phosphonium Salts and Betaines

Section G4: Four Coordinate Phosphonium Salts and Betaines with Three P–C Bonds

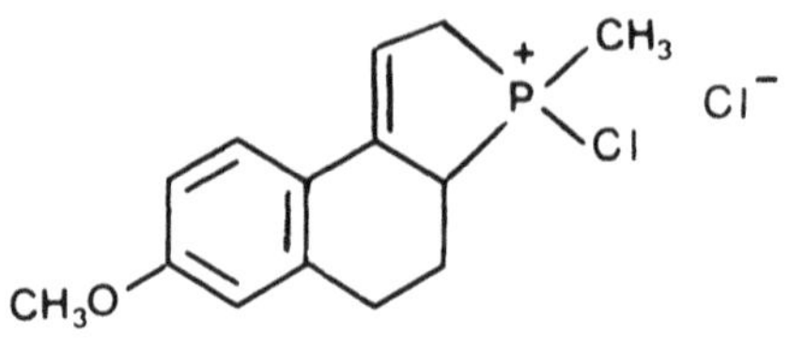

9

P-bound atoms/rings	Connectivities	Structure	NMR Data (δ_P[solv, temp]$^nJ_{PZ}$ Hz)	Ref.
C_3Br	$(H_3)_3$	$Me_3P^+Br\ BBr_4^-$	64.5(solid), 66.2(ne)	102
		Br^-	67.8 ± 1(solid), 66.2(ne)	102
		(HSO_4^-)	66.2(H_2SO_4)	102
		$(HS_2O_7^-)$	67.0 ± 0.8 (25% oleum)	102
	$(CH_2)_3$	$Bu_3P^+Br\ Br^-$	105(m or ne)	169
		$HgBr_3^-$	102(m or ne)	169
$C_2C'Br/5$	$(C'H_2)_2;C'_2$	$[Ph\dot{P}(CH_2CMe:CH\dot{C}H_2)Br]^+Br^-$	89(c)	171
	$CC'H,C'H_2;C'_2$	$[Ph\dot{P}(CH_2CH:CH\dot{C}HMe)Br]^+\ Br^-$	81(c)	170
CC'_2Br	$H_3;(C'_2)_2$	$MePh_2P^+Br\ Br^-$	51.6	73a
C'_3Br	$(C'_2)_3$	$Ph_3P^+Br\ Br^-$	48.6 ± 0.5(solid), 48.3(ne)	172
		$(HS_2O_7^-)$	49(25% oleum)	8
C''_3Br	$(N'')_3$	$(NC)_3P^+Br\ (Br_3^-)$	−33.2(Br_2)	2
		(CN^-)	−30.3(BrCN)	9
C_3Cl	$(H_3)_3$	$Me_3P^+Cl\ BCl_4^-$	87 ± 1(solid)	113
		Cl^-	87 ± 4(solid)	113
		Cl_2I^-	86.7 ± 1(solid)	113
		$(HS_2O_7^-)$	90(25% oleum)	8
	$(H_3)_2,CH_2$	$Me_2P^+(C_{18}H_{37})Cl\ PCl_6^-$	94.7(r)	173
		$SbCl_6^-$	92.5(r)	173
	$(CH_2)_3$	$Et_3P^+Cl\ BCl_4^-$	105 ± 2(solid)	113
		Cl^-	105.8 ± 2.5(solid)	113, 114
			110(c), 112.1(m), 111.4(fe)	26, 36, 174
		Cl_4I^-	111.3(solid)	106
		$SbCl_6^-$	111.3(solid)	114
		$(HS_2O_7^-)$	108(25% oleum)	8
		$Bu_3P^+Cl\ Cl^-$	102.7(r), 103.3(fe)	26, 173
			104(ne), 105 ± 1(m)	26, 175
		PCl_6^-	100.7(r)	173
		$SbCl_6^-$	103.9(r), 102(m)	173, 175
	$(C_2H)_3$	$iPr_3P^+Cl\ Cl^-$	117.4(fe), 118.4(m)	26, 36
		$cHex_3P^+Cl\ Cl^-$	103.2(fe), 104.2(m)	26, 36
$C_3Cl/5$	$H_3,(C'H_2)_3$	$[Me\dot{P}(CH_2CH:CH\dot{C}H_2)Cl]^+Cl^-$	112(c)	171
	$H_3,C'H_2,CC'H$	$[Me\dot{P}(CH_2CH:CH\dot{C}HMe)Cl]^+Cl^-$	121(c)	170
		9	121(r, −70°)	176

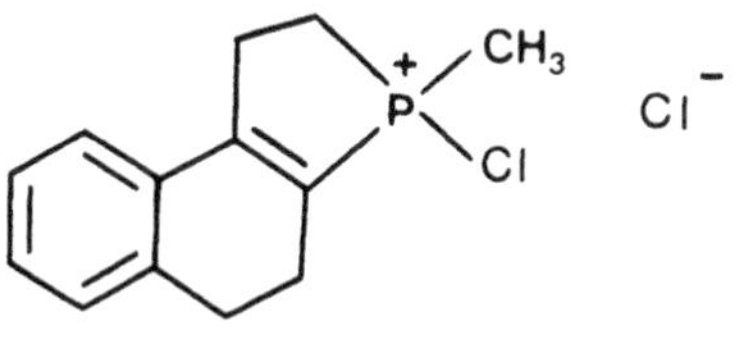

10

P-bound atoms/rings	Connectivities	Structure	NMR Data (δ_P[solv, temp]$^nJ_{PZ}$ Hz)	Ref.
	$H_3,(CC'H)_2$	$[Me\dot{P}(CHMeCH{:}CH\dot{C}HMe)Cl]^+Cl^-$	125(c)	170
$C_2C'Cl$	$(CH_2)_2;C'_2$	$Bu_2PhP^+Cl\ Cl^-$	90.3(fe), 91.0(m)	26
		$Bu_2(p\text{-}MeOC_6H_4)P^+Cl\ Cl^-$	89.7(fe), 90.2(m)	26
$C_2C'Cl/5$	$H_3,CH_2;CC'$	**10**	102(r)	176
	$(C'H_2)_2;C'_2$	$[Ph\dot{P}(CH_2CMe{:}CH\dot{C}H_2)Cl]^+\ Cl^-$	98.8(r, −70° or 0°)	176
CC'_2Cl	$H_3;(C'_2)_2$	$MePh_2P^+Cl\ Cl^-$	28.5	73a
	$CH_2;(C'_2)_2$	$EtPh_2P^+Cl\ Cl^-$	78.1(fe)	36
		$BuPh_2P^+Cl\ Cl^-$	75.2(m), 75.4(fe)	26
	$C_3;(C'_2)_2$	$tBuPh_2P^+Cl\ Cl^-$	85.3(fe), 85.7(m)	26
$CC'_2Cl/3$	$H_3;(C'H)_2$	$[Me\dot{P}(CH{:}\dot{C}H)Cl]^+\ AlCl_4^-$	−68.4(r)	177
	$H_3;C'H,CC'$	$[Me\dot{P}(CMe{:}\dot{C}H)Cl]^+\ AlCl_4^-$	−62.2(r, < − 30°)	177
	$H_3;(CC')_2$	$[Me\dot{P}(CMe{:}\dot{C}Me)Cl]^+\ AlCl_4^-$	−57.3(r)	177
$CC'_2Cl/5$	$H_3;(CC')_2$	$[Me\dot{P}(CMe{:}CMeCMe{:}\dot{C}Me)Cl]^+\ AlCl_4^-$	−78.2(c/r)	178
	$CH_2;C'H,C'_2$	$[Ph\dot{P}(CH{:}CMeCH_2\dot{C}H_2)Cl]^+\ Cl^-$	94.2(r), 99(c)	171, 176.
	$C_2H;CC';C'_2$	$[Ph\dot{P}(CMe{:}CHCH_2\dot{C}HMe)Cl]^+\ Cl^-$	100.5(c)	179
CC''_2Cl	$H_3;(N'')_2$	$MeP^+(CN)_2Cl\ SbCl_6^-$	62.9(fe)	103
C'_3Cl	$(C'_2)_3$	$Ph_3P^+Cl\ AlCl_4^-$	67 ± 7(solid), 65(ne), 64.3(b)	113, 175, 180
		BCl_4^-	81 ± 3(solid)	113
		Cl^-	62 ± 8(solid), 65.2, 65.5(ne)	37, 113, 150
			65 ± 1 ($CHCl_2CHCl_2$)	181
			56.3(fe), 62.9(m)	26,175
		ClO_4^-	65.7(r)	42
		PCl_6^-	64.3 ± 2(solid), 64.7(ne), 65.4(m)	26, 182
			65.0 ± 1.4(r), 65.1(fe or ne)	173, 182, 26, 36, 37
		$SbCl_6^-$	65 ± 2(solid), 63.3(r), 66(m)	113, 173, 175
			65.0(fe or $EtNO_2$)	42
		(HCl_2^-)	66.2(liq HCl)	1
		$(HS_2O_7^-)$	65(25% oleum)	8
		$(p\text{-}MeOC_6H_4)_3P^+Cl\ Cl^-$	63.4(ne)	37
		PCl_6^-	63.3(ne)	37
		$Ph(p\text{-}MeOC_6H_4)_2P^+Cl\ Cl^-$	63.6(m or fe)	26
		$(C_6F_5)_3P^+Cl\ BCl_4^-$	14.5(ne)	103
		Cl_4I^-	25.7(fe)	103
		$SbCl_6^-$	27.3(ne)	103
		$(p\text{-}MeC_6H_4)_3P^+Cl\ Cl^-$	65.0(ne)	37
		PCl_6^-	64.9(ne)	37
$C'_3Cl/3$	$(C'H)_2,C'_2$	$[Ph\dot{P}(CH{:}\dot{C}H)Cl]^+AlCl_4^-$	−69.2(r, <−30°)	177
	$C'H,CC',C'_2$	$[Ph\dot{P}(CMe{:}\dot{C}H)Cl]^+AlCl_4^-$	−57.9(r)	177
	$(CC')_2,C'_2$	$[Ph\dot{P}(CMe{:}\dot{C}Me)Cl]^+AlCl_4^-$	−58.0(r, <−30°)	177
$C'_3Cl/5$	$(CC')_2,C'_2$	$[Ph\dot{P}(CMe{:}CMeCMe{:}\dot{C}Me)Cl]^+\ AlCl_4^-$	−69.3(c/r)	178

TABLE G (continued)
Compilation of ^{31}P NMR Data of Four Coordinate Phosphonium Salts and Betaines

Section G4: Four Coordinate Phosphonium Salts and Betaines with Three P–C Bonds

P-bound atoms/rings	Connectivities	Structure	NMR Data (δ_P[solv, temp]$^nJ_{PZ}$ Hz)	Ref.
C''_3Cl	$(N'')_3$	$(NC)_3P^+Cl$ (Br_3^-)	−19.6(Br_2)	2
		PCl_6^-	−12.9(solid)	9
		$SbCl_6^-$	−26.0(ne)	9
		(CN^-)	−32.5(CNCl)	9
		$NCPCl_5^-$	−24.2(fe)	9
C_3F	$(H_3)_3$	Me_3P^+F AsF_6^-	142.7(SO_2) J_{PF}959	168
		BF_4^-	143.9(SO_2) J_{PF}955	168
C'_3F	$(C'_2)_3$	Ph_3P^+F BF_4^-	93.7(SO_2) J_{PF}998	168
C'_3I	$(C'_2)_3$	Ph_3P^+I I^-	42.5 ± 1.0(solid)	172
C_3N	$(H_3)_3;P^4$	$[Me_3PNPBu3]^+$ I^-	24.6(r)	183
	$(H_3)_3;C_2$	$Me_3P^+NMe_2$ Cl^-	75.0 ± 0.8(solid)	1
	$H_3,(CH_2)_2;H_2$	$MeEt_2P^+NH_2$ Cl^-	38.0(d)	74
	$(CH_2)_3;P^4$	$[Bu_3PNPMe_3]^+$ I^-	35.0(r)	183
		$[Bu_3PNPMe(NMe_2)_2]^+I^-$	34.2(r)	183
		$[Bu_3PNP(NMe_2)_3]^+$ I^-	28.8(r)	183
		PF_6^-	29.1(r)	183
		$[Bu_3PNPBu_3]^+$ I^-	32.9(r)	183
		$[Bu_3PNP(NEt_2)_3]^+$ I^-	28.0(r)	183
		PF_6^-	28.2(r)	183
	$(CH_2)_3;C'H$	$Bu_3P^+NHC_6H_4OSiMe_3$-o Br^-	57.8(r)	184
		$[Bu_3PNHC_6H_4OPBu_3\text{-o}]^{2+}$ $(Br^-)_2$	57.7(r)	184
	$(CH_2)_3;C_2$	$Bu_3P^+NMe_2$ Br^-	66.2(r)	1
		Cl^-	67.8(r)	1
	$(CH_2)_3;C_2$	$Bu_3P^+NEt_2$ Br^-	64.3	184, 185
	$(C_2H)_3;C_2$	$cHex_3P^+NMe_2$ Cl^-	94.6(r)	1
C_3N'	$(H_3)_2;N'$	$Me_3P^+N_3$ $SbCl_6^-$	73.0($MeCHCl_2$)	78
	$(H_3)_2,CH_2;P^4$	$[Me_2(C_{12}H_{25})PNP(C_{12}H_{25})Me_2]^+$ OMe^-	32	186
$C_2C'N$	$(H_3)_2;C'_2;C_2$	$Me_2PhP^+NMe_2$ Br^-	54.9(r)	1
		Cl^-	56.5(r)	1
	$(C_2H)_2;C'_2;C_2$	$cHex_2PhP^+NMe_2$ Br^-	62.9(r)	1
		Cl^-	63.0(r)	1
CC'_2N	$H_3;(C'_2)_2;C_2$	$MePh_2P^+NMe_2$ Br^- or Cl^-	51.6(r)	1
		I^-	50.0(fe)	60
	$C_2H;(C'_2)_2;C_2$	$cHexPh_2P^+NMe_2$ Br^-	56.3(r)	1
		Cl^-	56.5(r)	1
C'_3N	$(C'_2)_3;H_2$	$Ph_3P^+NH_2$ Cl^-	32.0(c), 30.0(d)	74
	$(C'_2)_3;C'H$	Ph_3P^+HNPh Cl^-	34	187
	$(C'_2)_3;C_2$	$Ph_3P^+NMe_2$ BBr_4^-	46.8(r)	1
		BCl_4^-	48.8(solid)	1
		Br^-	46.8(r or w)	1
		Cl^-	46.8(solid), 46.4(w)	1
C'_3N'	$(C'_2)_3;N'$	$Ph_3P^+N_3$ $SbCl_6^-$	47.9($MeCHCl_2$)	78
	$(C'_2)_3;C''$	Ph_3P^+NCS SCN^-	39	56
C_3O	$(H_3)_3;H$	Me_3P^+OH (HSO_4^-)	89.6 ± 0.9(H_2SO_4)	102
		$(HS_2O_7^-)$	110.5 ± 4.1 (25% oleum)	102
	$(H_3)_3;C$	$Me_3P^+OCH(CF_3)_2$ I^-	114.3(m)	189
		Me_3P^+OMe $SbCl_6^-$	101.2(m), 95.8(m)	86, 188
		Me_3P^+OEt $SbCl_6^-$	95.8(m)	86

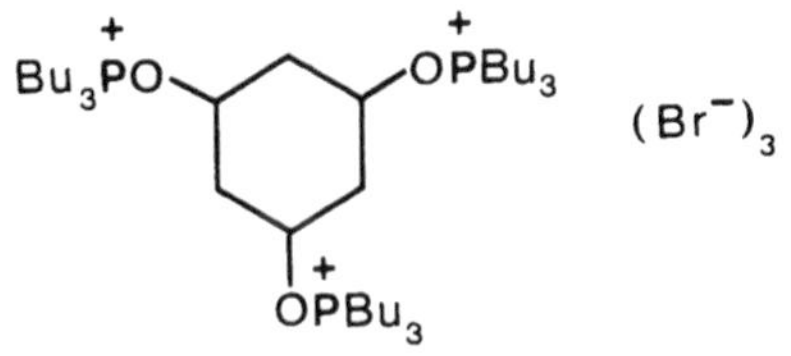

11

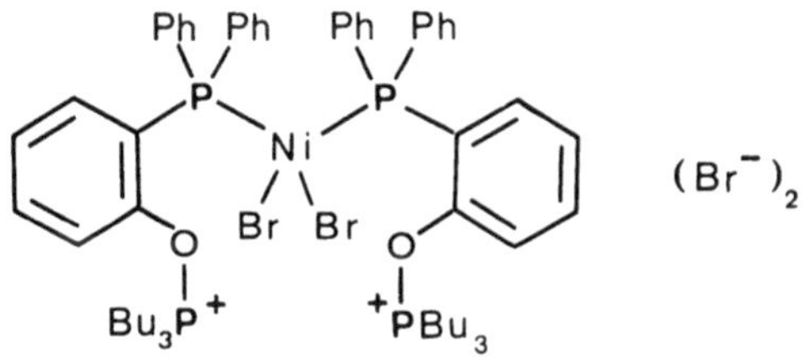

12

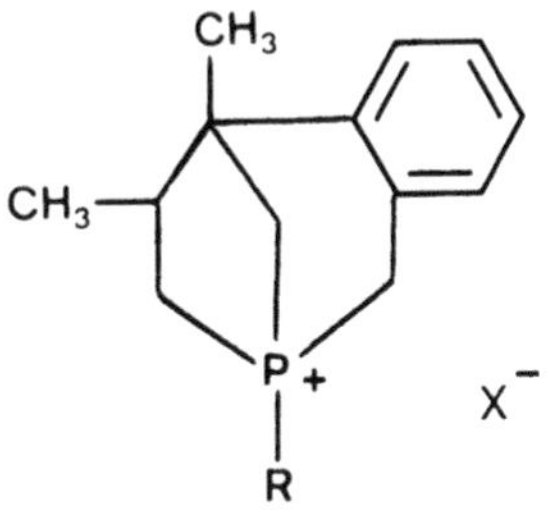

13

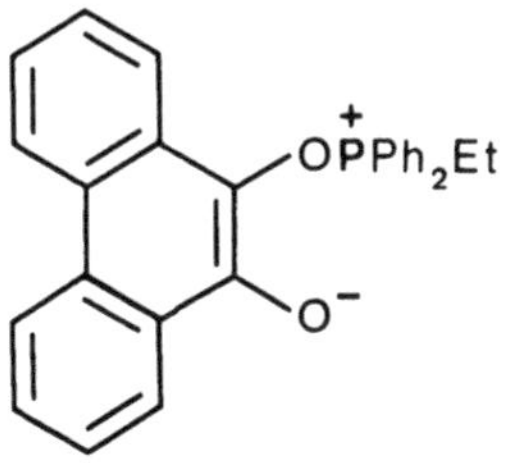

(in equilibrium with phosphorane)

14

P-bound atoms/rings	Connectivities	Structure	NMR Data (δ_P[solv, temp]$^nJ_{PZ}$ Hz)	Ref.
	$(H_3)_2,C_3;C$	$Me_2(tBu)P^+OCH(CF_3)_2\ I^-$	122.2(m)	189
	$H_3,(CH_2)_2;C$	$MeEt_2P^+OMe\ CF_3SO_3^-$	103.5(c)	134
	$H_3,(C_3)_2;C$	$Me(tBu)_2P^+OCH(CF_3)_2\ I^-$	124.5(m)	189
	$(CH_2)_3;P^4$	$Bu_3P^+OP(O)Cl_2\ Cl^-$	104.5	59
	$(CH_2)_3;C$	$Et_3P^+OMe\ SbCl_6^-$	108.4(m)	86, 188
		$Et_3P^+OEt\ SbCl_6^-$	104.0(m)	86, 188
		$Bu_3P^+OEt\ BF_4^-$	98.0(r)	190
		$Bu_3P^+OcHex\ Br^-$	94.7(r)	188
		11	99.5(r)	188
	$(CH_2)_3;C'$	$Bu_3P^+OPh\ Br^-$	105.1(r)	188
		$Bu_3P^+OC_6H_4PPh_2\text{-o}\ Br^-$	103.8(r)	191
		12	103.8(r)	191
		$[Bu_3POC_6H_4NHPBu_3\text{-o}]^{2+}\ (Br^-)_2$	104.5(r)	184
$C_3O/5,6$	$(CH_2)_2,C'H_2$	**13** (R = OEt, X = I)	44.1(c)	192
		13 (R = X = OEt)	45.1(c)	192
$C_2C'O$	$(CH_2)_2;C'_2;C'$	**15**	0.9(b), 10.8(r)	64
$C_2C'O/4$	$CH_2,C_3;C'_2;C$	$[Ph\dot{P}(OEt)CH_2CMe_2\dot{C}Me_2]^+BF_4^-$	91.5(r)	192a
CC'_2O	$H_3;(C'_2)_2;C$	$MePh_2P^+OMe\ CF_3SO_3^-$	74.5(c)	134
		$MePh_2P^+OiPr\ I^-$	68.7(m)	131
		$MePh_2P^+ONeo\ Cl^-,\ Br^-,\ or\ I^-$	72(c)	167
	$CH_2;(C'_2)_2;C'$	**14**	−9.3(b), −7.3(r)	64
	$C'H_2;(C'_2)_2;CH_2$	$Ph_2(PhCOCH_2)P^+ONeo\ Br^-$	68.3(c)	139
C'_3O	$(C'_2)_3;P^4$	$Ph_3P^+OP(O)Cl_2\ Cl^-$	65.0(r)	59
		$Ph_3P^+OP(S)Cl_2\ Cl^-$	65.2(r)	59
	$(C'_2)_3;Si$	$Ph_3P^+OSiMe_3\ Cl^-$	42.2(r)	59
	$(C'_2)_3;H$	$Ph_3P^+OH\ (ClSO_3^-)$	60($ClSO_3H$)	108
		(HCl_2^-)	55.5(liq HCl)	24
		(HSO_4^-)	58(H_2SO_4)	108

TABLE G (continued)
Compilation of ^{31}P NMR Data of Four Coordinate Phosphonium Salts and Betaines

Section G4: Four Coordinate Phosphonium Salts and Betaines with Three P–C Bonds

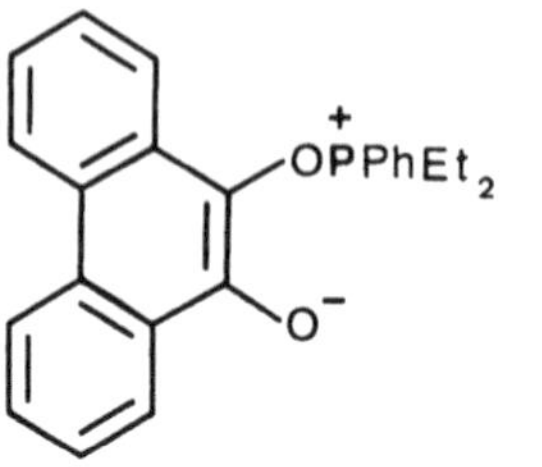

15

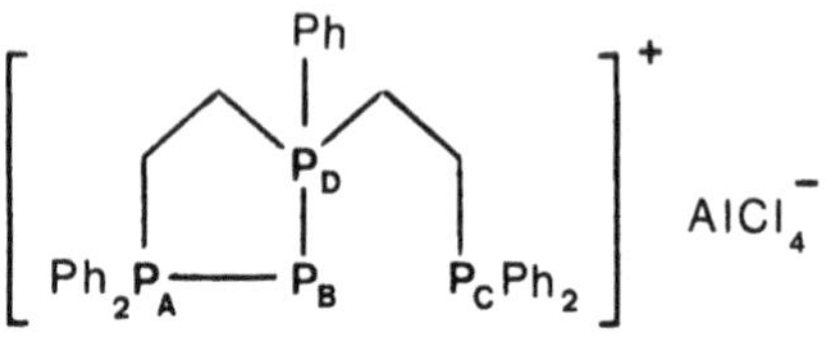

16

P-bound atoms/rings	Connectivities	Structure	NMR Data (δ_P[solv, temp]^aJ_{PZ} Hz)	Ref.
		$(HO_3SC_6H_4)_3P^+OH$ $(HS_2O_7^-)$	56(25% oleum)	108
		$(HS_3O_{10}^-)$	61(65% oleum)	108
	$(C'_2)_3;C$	Ph_3P^+OMe $SbCl_6^-$	65.0(r)	60, 148
		Ph_3P^+OEt BF_4^-	62(r)	190
		SCN^-	61.7	56
		Ph_3P^+OBu SCN^-	61.5	56
	$(C'_2)_3;C'$	$Ph_3P^+OCPh{=}CPh_2$ Cl^-	63.0	193
		$Ph_3P^+OCPh{=}NN{=}PPh_3$ Cl^-	25.4(m)	194
		$Ph_3P^+OCPh{=}CPhOP(O)(OMe)_2$ Br^-	65.9	66
C_3P^4	$(CH_2)_3;C'S_2$	$Bu_3\mathbf{P}^+P(S)(Ph)S^-$	7.0 J_{PP}118	197, 198
		$Bu_3\mathbf{P}^+P(S)(C_6H_4OMe\text{-}p)S^-$	7.4 J_{PP}109	197, 198
		$Bu_3\mathbf{P}^+P(S)(C_6H_4OEt\text{-}p)S^-$	7.0 J_{PP}108	197, 198
		$Bu_3\mathbf{P}^+P(S)(1\text{-}C_{10}H_7)S^-$	9.5 J_{PP}96	198
C_3P	$(CH_2)_3;HP^4$	$[Bu_3PPHPBu_3]^{2+}$ $(AlCl_4^-)_2$	38 J_{PP}277	195
	$(CH_2)_3;CCl$	$Bu_3\mathbf{P}^+PMeCl$ Cl^-	16.5	196
	$(CH_2)_3;CP^4$	$[Bu_3PP(CH_2Cl)\mathbf{P}Bu_3]^{2+}$ $(AlCl_4^-)_2$	39 J_{PP}319	195
	$(CH_2)_3;C_2$	$Et_3\mathbf{P}^+PMe_2$ Cl^-	33.7	196
		$Pr_3\mathbf{P}^+PMe_2$ Cl^-	22.5	196
		$Bu_3\mathbf{P}^+PMe_2$ Cl^-	21.2	196
		$(C_8H_{17})_3\mathbf{P}^+PMe_2$ Cl^-	14.5	196
C_3P^2	$(CH_2)_3;P^4$	$[Bu_3\mathbf{P}PPBu_3]^+$ $AlCl_4^-$	33 J_{PP}473	70
		$[Bu_3\mathbf{P}PPPh_3]^+$ $AlCl_4^-$	32 J_{PP}458	70
$C_2C'P$	$(H_3)_2;C'_2;HP^4$	$[Me_2Ph\mathbf{P}PH\mathbf{P}PMe_2]^{2+}$ $(AlCl_4^-)_2$	18 J_{PP}255	195
	$(H_3)_2;C'_2;CP^4$	$[Me_2PhPP(CH_2Cl)\mathbf{P}PhMe_2]^{2+}$ $(AlCl_4^-)_2$	21 J_{PP}300	195
$C_2C'P^2$	$(H_3)_2;C'2;P^4$	$[Me_2Ph\mathbf{P}P\mathbf{P}PhMe_2]^+$ $AlCl_4^-$	12 J_{PP}451	70
		$[Me_2Ph\mathbf{P}PPPh_3]^+$ $AlCl_4^-$	16 J_{PP}463	70
$C_2C'P^2/5$	$(CH_2)_2;C'_2;P^4$	**16**(P_D)	70.0(r,b −75°) J_{PP}457 70.2(r,b 30°), 70.5(PhCl,b 200°)	199
CC'_2P	$H_3;(C'_2)_2;HP^4$	$[MePh_2\mathbf{P}PH\mathbf{P}Ph_2Me]^{2+}$ $(AlCl_4^-)_2$	18 J_{PP}261	195
	$H_3;(C'_2)_2;CP^4$	$[MePh_2\mathbf{P}P(CH_2Cl)\mathbf{P}Ph_2Me]^{2+}$ $(AlCl_4^-)_2$	20 J_{PP}305	195
$CC'_2P/5$	$CH_2;(C'_2)_2;HP^4$	$[Ph_2\dot{P}PHPPh_2CH_2\dot{C}H_2]^{2+}$ $(AlCl_4^-)_2$	53.1 J_{PP}239	195, 199

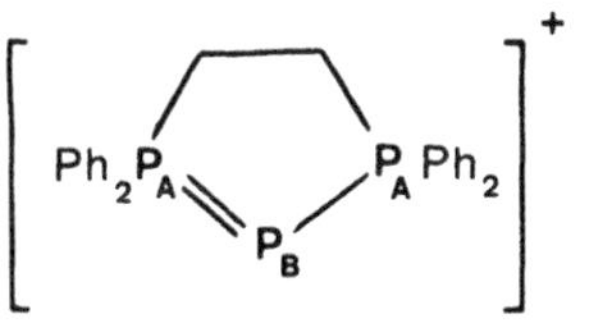

17

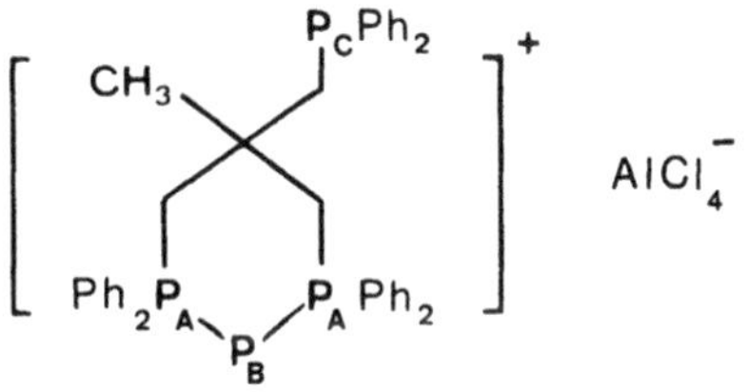

18

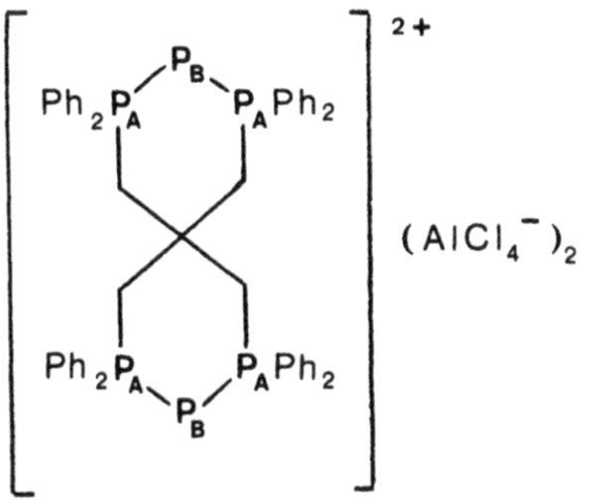

19

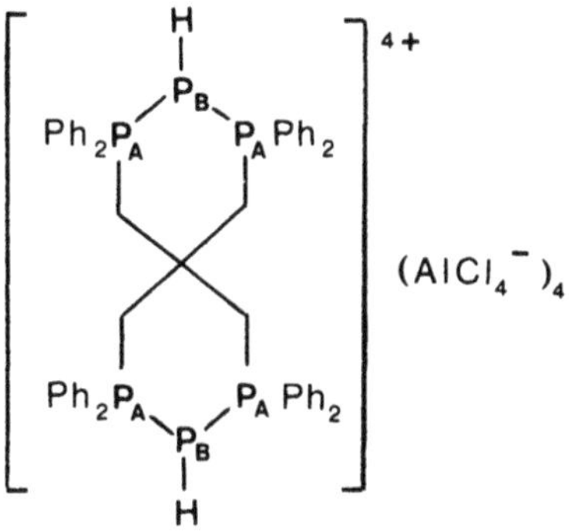

20

P-bound atoms/rings	Connectivities	Structure	NMR Data (δ_P[solv, temp]$^nJ_{PZ}$ Hz)	Ref.
	$CH_2;(C'_2)_2;CP^4$	$[Ph_2\dot{\mathbf{P}}P(CH_2Cl)\mathbf{P}Ph_2CH_2\dot{C}H_2]^{2+}$ $(AlCl_4^-)_2$	52 J_{PP}282	195
		$[Ph_2\dot{\mathbf{P}}P(tBu)\mathbf{P}Ph_2CH_2\dot{C}H_2]^{2+}$ $(AlCl_4^-)_2$	52 J_{PP}283	195
$CC'_2P/6$	$CH_2;(C'_2)_2;HP^4$	**20**(P_A)	17.7 J_{PP}232	199
	$CH_2;(C'_2)_2;CP^4$	**21**(P_A)	8.7 J_{PP}282	199
CC'_2P^2	$H_3;(C'_2)_2;P^4$	$[MePh_2\mathbf{P}P\mathbf{P}Ph_2Me]^+$ $AlCl_4^-$	23 J_{PP}464	70, 71
		$[MePh_2\mathbf{P}PPPh_3]^+$ $AlCl_4^-$	24 J_{PP}480	70, 71
$CC'_2P^2/5$	$CH_2;(C'_2)_2;P^4$	**17**(X = $AlCl_4^-$)(P_A)	64.4 J_{PP}452	199
		17(X = 0.5 $SnCl_6^{2-}$)(P_A)	63.8 J_{PP}449	200
		16(P_A)	65.1(r,b room temp) J_{PP}456	199
$CC'_2P^2/6$	$CH_2;(C'_2)_2;P^4$	**18**(P_A)	19.5 J_{PP}444	199
		19(P_A)	17.6 J_{PP}441	199
C'_3P	$(C'_2)_3;AlP^4$	$[Ph_3\mathbf{P}P(AlCl_3)P(NMe_2)_3]^+$ $AlCl_4^-$	27(r) J_{PP}378	20
	$(C'_2)_3;HP^4$	$[Ph_3\mathbf{P}PHP(NMe_2)_3]^{2+}$ $(AlCl_4^-)_2$	23(r) J_{PP}304	20
		$[Ph_3\mathbf{P}PH\mathbf{P}Ph_3]^{2+}$ $(AlCl_4^-)_2$	23(r) J_{PP}286	20, 70, 195
	$(C'_2)_3;CP^4$	$[Ph_3\mathbf{P}P(CH_2Cl)P(NMe_2)_3]^{2+}$ $(AlCl_4^-)_2$	24(r) J_{PP}350	20
		$[Ph_3\mathbf{P}PEtP(NMe_2)_3]^{2+}$ $(AlCl_4^-)_2$	21(r) J_{PP}356	20
		$[Ph_3\mathbf{P}P(CH_2Cl)\mathbf{P}Ph_3]^{2+}$ $(AlCl_4^-)_2$	25(r) J_{PP}330	20, 195
		$[Ph_3\mathbf{P}PMe\mathbf{P}Ph_3]^{2+}$ $(AlCl_4^-)_2$	23 J_{PP}330	195
		$[Ph_3\mathbf{P}PEt\mathbf{P}Ph_3]^{2+}$ $(AlCl_4^-)_2$	22(r) J_{PP}334	20, 195
		$[Ph_3\mathbf{P}P(iPr)\mathbf{P}Ph_3]^{2+}$ $(AlCl_4^-)_2$	21 J_{PP}354	195
		$[Ph_3\mathbf{P}P(Ph)\mathbf{P}Ph_3]^{2+}$ $(AlCl_4^-)_2$	24 J_{PP}358	195

TABLE G (continued)
Compilation of ^{31}P NMR Data of Four Coordinate Phosphonium Salts and Betaines

Section G4: Four Coordinate Phosphonium Salts and Betaines with Three P–C Bonds

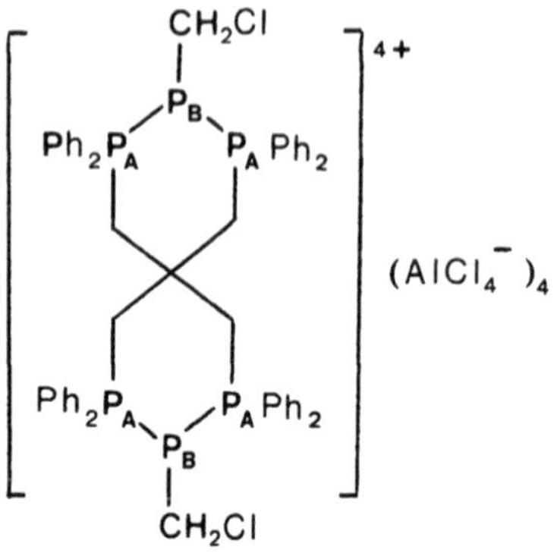

21

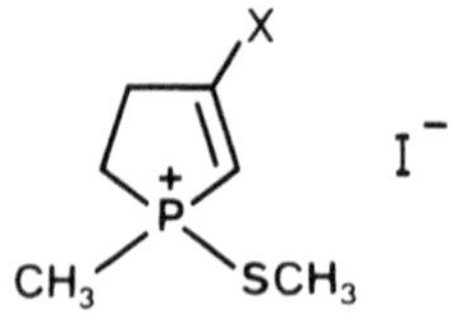

22

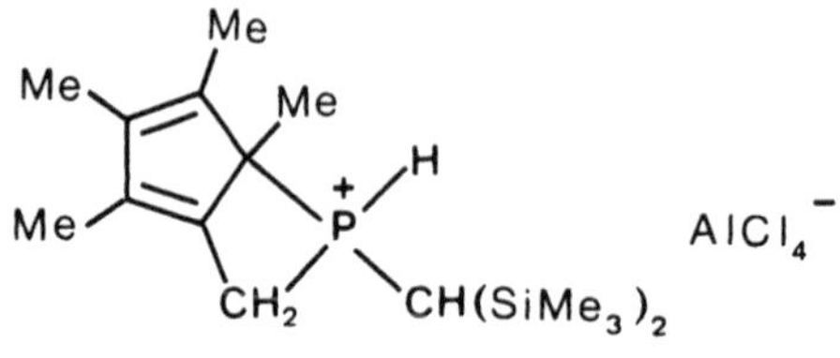

23

P-bound atoms/rings	Connectivities	Structure	NMR Data (δ_P[solv, temp]$^n J_{PZ}$ Hz)	Ref.
	$(C'_2)_3$;CP	$[Ph_3\dot{\mathbf{P}}CH_2\dot{P}P(NMe_2)_3]^{2+}$ $(AlCl_4^-)_2$	28.3(r) J_{PP}335	20
		$[Ph_3\dot{\mathbf{P}}CH_2\dot{P}P(Pip)_2tBu]^{2+}$ $(AlCl_4^-)_2$	28.2(r) J_{PP}329	20
		$[Ph_3\dot{\mathbf{P}}CH_2\dot{P}P(Pip)_3]^{2+}$ $(AlCl_4^-)_2$	27.5(r) J_{PP}333	20
		$[Ph_3\dot{\mathbf{P}}CH_2\dot{\mathbf{P}}PPh_3]^{2+}$ $(AlCl_4^-)_2$	28.6(r) J_{PP}293	20
	$(C'_2)_3$;CH	$[Ph_3\mathbf{P}PhCH_2\mathbf{P}Ph_3]^{2+}$ $(AlCl_4^-)_2$	24 J_{PP}260	20
C'_3P^2	$(C'_2)_3;P^4$	$[Ph_3\mathbf{P}P(NMe_2)_3]^+$ $AlCl_4^-$	29 J_{PP}523	20, 70, 71
		$[Ph_3\mathbf{P}P(OEt)_3]^+$ $AlCl_4^-$	32 J_{PP}437	70
		$[Ph_3\mathbf{P}PMe_2Ph]^+$ $AlCl_4^-$	31 J_{PP}481	70
		$[Ph_3\mathbf{P}PMorph_3]^+$ $AlCl_4^-$	28 J_{PP}526	70
		$[Ph_3\mathbf{P}PBu_3]^+$ $AlCl_4^-$	32 J_{PP}503	70
		$[Ph_3\mathbf{P}PMePh_2]^+$ $AlCl_4^-$	30 J_{PP}482	70, 71
		$[Ph_3\mathbf{P}P(NEt_2)_2Ph]^+$ $AlCl_4^-$	30 J_{PP}524	70
		$[Ph_3\mathbf{P}P(Pip)_3]^+$ $AlCl_4^-$	29 J_{PP}542	70
		$[Ph_3\mathbf{P}P(NEt_2)Ph_2]^+$ $AlCl_4^-$	30 J_{PP}510	70
		$[Ph_3\mathbf{PP}Ph_3]^+$ $AlCl_4^-$	29.6(r) J_{PP}501	71,199
		$[Ph_3\mathbf{P}P(NHPh)Ph_2]^+$ $AlCl_4^-$	29 J_{PP}479	70
$C_2C'S/5$	$H_3,CH_2;C'H$	**22** (X = 1-pyrrolidinyl)	69.8(c)	149
		22 (X = morpholino)	72.2(c)	149
		22 (X = piperidino)	71.0(c)	149
		22 (X = cyclohexylamino)	69.6(c)	149
C'_3S	$(C'_2)_3;C$	Ph_3P^+SMe $SbCl_6^-$	46.6(r)	60
C'_3Se	$(C'_2)_3;C$	Ph_3P^+SeMe $SbCl_6^-$	35.8(r)	60
C_3H	HSi_2,H_2,C'_2	**23**	72.2(r) J_{PH}568	201
	$H_3,(HSi)_2$	$Me[(Me_3Si)_2CH]_2P^+H$ I^-	1.6 J_{PH}482	202
	$(H_3)_3$	Me_3P^+H Cl^-	−2.8(liq) J_{PH}495	203
		(FSO_3^-)	−3.2(FSO_3H)	101

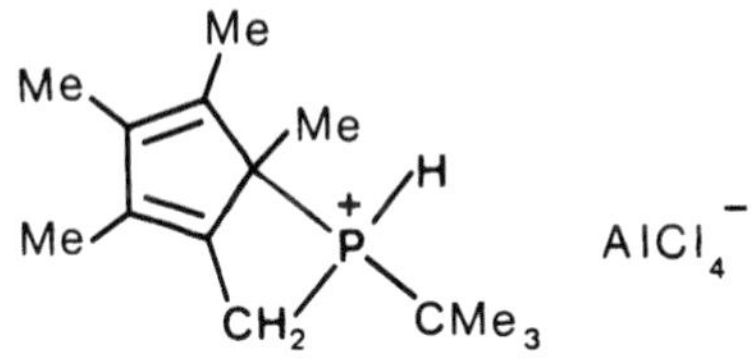

24

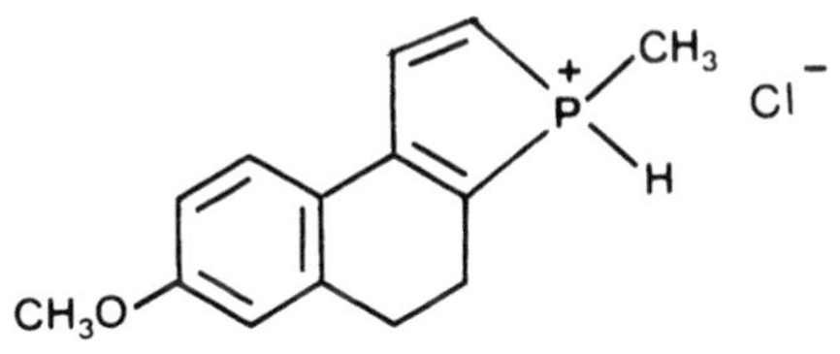

25

P-bound atoms/rings	Connectivities	Structure	NMR Data (δ_P[solv, temp]$^nJ_{PZ}$ Hz)	Ref.
			J_{PH}497	
		(HBr_2^-)	−2.9(w) J_{PH}505	204
	H_2,C_3,CC'_2	**24**	87.1(r) J_{PH}504	201
	$(CH_2)_3$	Et_3P^+H (FSO_3^-)	22.5(FSO_3H) J_{PH}471	101
		(HBr_2^-)	19.7(r) J_{PH}500	204
		Bu_3P^+H (FSO_3^-)	13.7(FSO_3H) J_{PH}470	101
		(HBr_2^-)	11.9(r) J_{PH}475	204
		(HSO_4^-)	12(H_2SO_4) J_{PH}457 ± 10	95
		$(C_8H_{17})_3P^+H$ (FSO_3^-)	13.0(FSO_3H) J_{PH}465	101
	$(C_2H)_3$	iPr_3P^+H (FSO_3^-)	44.4(FSO_3H) J_{PH}448	101
		(HBr_2^-)	39.6(r) J_{PH}455	204
		$cHex_3P^+H$ (FSO_3^-)	32.7(FSO_3H) J_{PH}445	101
		(HBr_2^-)	27.9(r) J_{PH}455	204
	$(C_3)_3$	tBu_3P^+H (FSO_3^-)	58.3(FSO_3H) J_{PH}436	101
$C_2C'H$	$(H_3)_2;C'_2$	$Me_2PhP^+H\ Br^-$	−4.0 J_{PH}525	205
		(HBr_2^-)	−2.1(r) J_{PH}525	204, 206
		(HSO_4^-)	−1(H_2SO_4) J_{PH}500 ± 10	95
	$(CH_2)_2;C'_2$	Et_2PhP^+H (HBr_2^-)	18.3(r) J_{PH}490	204
		Bu_2PhP^+H (HBr_2^-)	12.2(r) J_{PH}490	204
	$(C_2H)_2;C'_2$	iPr_2PhP^+H (HBr_2^-)	32.0(r) J_{PH}485	204
		$cHex_2PhP^+H$ (HBr_2^-)	20.8(r) J_{PH}485	204
CC'_2H	$Si_3;(C'_2)_2$	$Ph_2[Tms_3C]P^+H\ I^-$	−0.5(c) J_{PH}503	207
	$HSi_2;(C'_2)_2$	$Ph_2[(Me_3Si)_2CH]P^+H\ I^-$	−0.3(c) J_{PH}508	207
	$H_2Si;(C'_2)_2$	$Ph_2(Me_3SiCH_2)P^+H\ I^-$	−3.9(c) J_{PH}508	207
	$H_3;(C'_2)_2$	$MePh_2P^+H\ I^-$	−5.6(c)J_{PH}520	207
		(HBr_2^-)	−2.2(r) J_{PH}525	204
	$CH_2;(C'_2)$	$EtPh_2P^+H$ (HBr_2^-)	8.3(r) J_{PH}515	204
		$BuPh_2P^+H$ (HBr_2^-)	2.1(r) J_{PH}520	204
	$C_2H;(C'_2)_2$	$iPrPh_2P^+H$ (HBr_2^-)	14.0(r) J_{PH}510	204
		$cHexPh_2P^+H$ (HBr_2^-)	10.5(r) J_{PH}510	204
$CC'_2H/5$	$H_3;C'H,CC'$	**25**	11.6(r, −90°) J_{PH}518	176
	$CH_2;C'H,C'_2$	**26**	21.8(m) J_{PH}545 (r)(excess CF_3SO_3H)	176
C'_3H	$(C'_2)_3$	$Ph_3P^+H\ AlCl_4^-$	5(r)	20
		Br^-	1.9(liq HBr) J_{PH}510 ± 5	208
		0.5 $Pd_2Cl_6^{2-}$	2.7(liq HCl) J_{PH}504 ± 5	208
		(HCl_2^-)	4.8 ± 1.3(liq HCl) J_{PH}511 ± 6	208

TABLE G (continued)
Compilation of ^{31}P NMR Data of Four Coordinate Phosphonium Salts and Betaines

Section G4: Four Coordinate Phosphonium Salts and Betaines with Three P–C Bonds

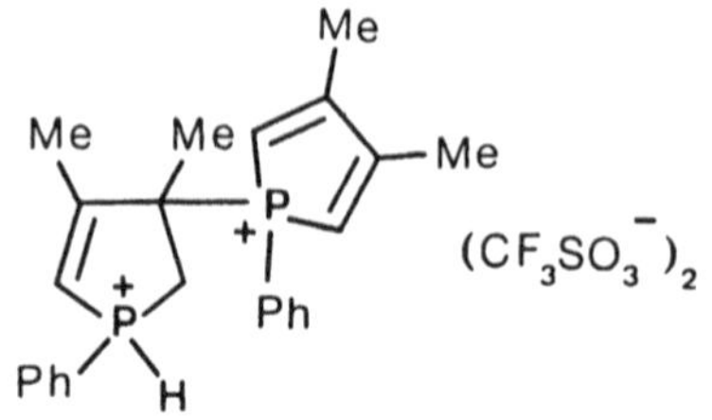

26

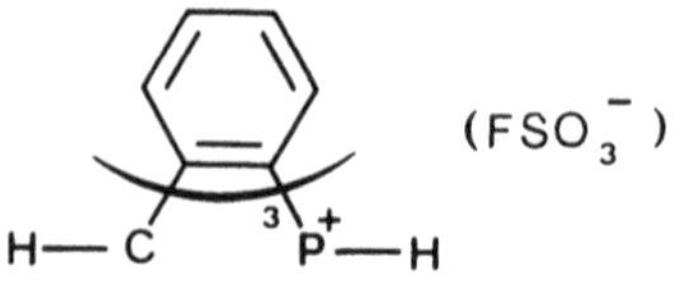

27

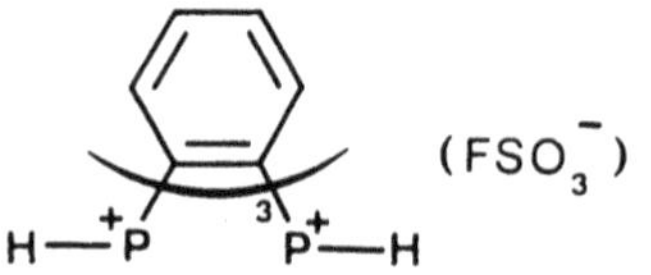

28

P-bound atoms/rings	Connectivities	Structure	NMR Data (δ_P[solv, temp]$^n J_{PZ}$ Hz)	Ref.
		($CF_3CO_2^-$)	5.0(CF_3CO_2H) J_{PH}514 ± 5	208
		(FSO_3^-)	6.8(FSO_3H) J_{PH}510	101
		(HSO_4^-)	5.25(H_2SO_4) J_{PH}510 ± 6	95, 208
		Ph_3P^+D (DCl_2^-)	5.6(liq DCl) J_{PD}77.5	208
		$Ph_2(o\text{-}HOC_6H_4)P^+H\ Br^-$	−5.8(r)	191
		27	−23.3 J_{PH}571	209
		28	−20.1 J_{PH}642	209
C'$_3$H/5	C'$_2$,(C'H$_2$)$_2$	[PhṖ(CH:CMeCH:ĊH)H]$^+$ Cl$^-$	16.3(r, −90°) J_{PH}535	176
		$CF_3SO_3^-$	18.9(r, −70°) (excess CF_3SO_3H) J_{PH}530	176
		[PhṖ(CH:CMeCMe:ĊH)H]$^+$ $CF_3SO_3^-$	9.5(r, −70°) (excess CF_3SO_3H) J_{PH}547	176

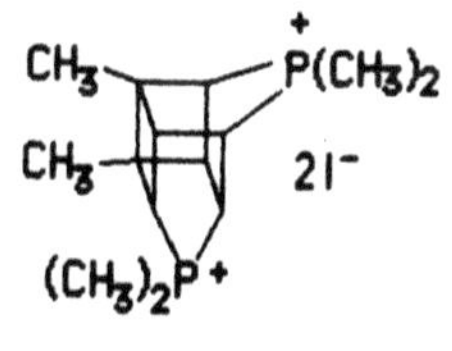

29

Section G5: Four Coordinate Phosphonium Salts and Betaines with Four P–C Bonds

Connectivity of P-bound atoms	Formula	Structure	NMR data (δ_P[solv, temp] $^nJ_{PZ}$ Hz)	Ref.
	P-bound atoms = C4			
$(H_2N)_4$	$C_8H_{20}ClN_4O_4P$	$(H_2NCONHCH_2)_4$ P^+Cl^-	29.5(w)	210
	$C_{12}H_{24}ClN_4O_8P$	(MeO_2CNHCH_2) P^+Cl^-	30.7(d)	211
$(H_2N)_2,(H_2O)_2$	$C_6H_{16}N_4O_4PX$	$(HOCH_2)_2P^+(CH_2NHCONH_2)_2$	29.8(w,65°)	212
$(H_2(O)_4$	$C_4H_{12}O_4PX$	$(HOCH_2)_4P^+X^-$	25.2(X = Cl); 26.5(w, X = I)	210, 213
$(H_2O)_3,H_3$	$C_4H_{12}IOP$	$MeP^+(CH_2OH)_3I^-$	27	213
$(H_2O)_3,CH_2$	$C_7H_{18}ClOP$	$(HOCH_2)_3P^+BuCl^-$	30	213
$(H_3)_4$	$C_4H_{12}PX$	$Me_4P^+X^-$	25.3, 24.4 $^2J_{PH}$15.1	214, 215
$(H_3)_3,C_2F$	$C_9H_9F_{13}IP$	$C_6F_{13}P^+Me_3I^-$	52.4(w) $^2J_{PH}$14.0	216
$(H_3)_3,C_2H$	$C_9H_{20}IP$	2-NbP$^+Me_3I^-$	29.5(w,1)endo; 30.1exo	218
	$C_9H_{18}IP$	7-NbeP$^+Me_3I$	19.8, 24.2	217
$(H_3)_3,C'_2H$	$C_{15}H_{21}O_5P$	$Me_3P^+CHPhC(COMe){=}C(O^-)Me$	10.9	219
$(H_3)_2,(CH_2)_2$	$C_{26}H_{56}IP$	$Me_2P^+(C_{12}H_{25})_2I^-$	29.5	220
	$C_6H_{16}IP$	$Me_2P^+Et_2I^-$	32	213
	$C_{14}H_{24}I_2P_2$	**29**	64.7(w)	221
$H_3,(CH_2)_3$	$C_7H_{18}IP$	$MeP^+Et_3I^-$	37	213
	$C_{37}H_{78}BrP$	(dodecyl)$_3P^+MeBr^-$	32 $^2J_{PH}$13.5	220
$H_3,(CH_2)_2,C_3$	$C_{13}H_{30}BrP$	MeP^+(tBu)(Bu)$_2Br^-$	41.5	222
$H_3,(C'H_2)_3$	$C_{10}H_{18}IP$	$MeP^+(CH_2CH{=}CH_2)_3I^-$	25.2(c)	223
	$C_{22}H_{24}IP$	$MeP^+Bz_3I^-$	27.0(c) $^2J_{PH}$15	224
$H_3,(C_2H)_3$	$C_{10}H_{18}IP$	MeP^+(cPr)$_3I^-$	41.3(1) $^2J_{PH}$15	225
	$C_{10}H_{24}BrP$	MeP^+(iPr)$_3Br^-$	45.5(d)	226
$CHO,(CH_2)_3$	$C_9H_{22}IP$	$Et_3P^+CH(OMe)Me\ I^-$	40	227
$(CH_2)_4$	$C_8H_{20}IP$	$Et_4P^+I^-$	40.1(c)	223, 228
	$C_{14}H_{34}F_{12}P_4$	$(Et_3P^+CH_2)_2(PF^-{}_6)_2$	42.2(m) $^3J_{PH}$19.0	229
	$C_{16}H_{36}IP$	$Bu_4P^+I^-$	31.1(n),26.1(a:w)	230, 231
	$C_{18}H_{42}IPSi$	$Bu_3P^+(CH_2)_3TmsI^-$	33.0(c)	232
$(CH_2)_3,C'H_2$	$C_{19}H_{34}PX$	$Bu_3P^+CH_2PhX^-$	31	233
$(CH_2)_3,C_2H$	$C_9H_{22}PX$	$iPrP^+Et_3X^-$	42.8	228
	$C_{19}H_{38}BrP$	2-NbP$^+Bu_3Br^-$	10.4	234
	$C_{19}H_{38}BrP$	7-NbP$^+Bu_3Br^-$	33.1	234
$(CH_2)_3,CC'H$	$C_{28}H_{40}ClP$	$Bu_3P^+CH(COPh)CH_2COPhCl^-$	35.7(r)	235
$(CH_2)_3,C'_2H$	$C_{16}H_{20}N_3O_2P$	$Et_3P^+CH(4\text{-}NO_2Ph')C^-(CN)_2$	37.2(d)	236
$CH_2,(C'H_2)_3$	$C_{23}H_{26}BrP$	$EtP^+Bz_3Br^-$	28.7	222
$CH_2,(C_3)_3$	$C_{14}H_{32}BrP$	EtP^+(tBu)$_3Br^-$	49.1(1) $^3J_{PH}$14.6	237
$(C'H_2)_2\ (C'H_2)_2$	$C_{18}H_{18}BrN_2P$	$Bz_2P^+(CH_2CN)_2Br^-$	32.2(m)	238
$(C'H_2)_2,\ (C_2H)_2$	$C_{20}H_{28}BrP$	$cPr_2P^+Bz_2Br^-$	37(y)	239
$(C_2H)_4$	$C_{12}H_{20}ClP$	$cPr_4P^+Cl^-$	44.1(1) $^1J_{PC}$89.8	225
$(C_3)_4$	$C_{16}H_{36}IP$	$tBu_4P^+I^-$	56.1(w) $^3J_{PH}$14 $^1J_{PC}$21.5	237

TABLE G (continued)
Compilation of ^{31}P NMR Data of Four Coordinate Phosphonium Salts and Betaines

Section G5: Four Coordinate Phosphonium Salts and Betaines with Four P–C Bonds

$(CH_3)_2P^+$ $2I^-$ $(CH_3)_2P^+$ — **30**

$(CH_3)_2P^+$ $(CH_2)_n$ X^- — **31**

$(CH_3)_2P^+$ $2X^-$ $(CH_3)_2P^+$ — **32**

CH_3 Br^- $PhCH_2$ P CH_3 — **33**

$P^+(CH_2Ph)_2$ $2Br^-$ $P^+(CH_2Ph)_2$ — **34**

Me_2P^+ P^+Me_2 $2Br^-$ — **35**

P-bound atoms/ (rings)	Connectivities	Structure	NMR data (δ_P[solv,temp] $^nJ_{PZ}$ Hz)	Ref.
C_4/5	$(H_3)_2,(C_2H)_2$	**30**	58.3, 52.5(c)	221
	$(H_3)_2,(CC'H)_2$	**31**	50.6(c)n = 1; 60(w)n = 2	221
		32	98.8(c) 58.8(c)	240
	$H_3,(CH_2)_2,C'H_2$	**33**	51.8(c)	241
	$(CHO)_2,(CH_2)_2$	$Bu_2P^+[CH(OH)CH_2CH_2CH(OH)]$ Br^-	46	242
	$(CH_2)_4$	$Et_2P^+[(CH_2)_4]ClO_4^-$	58.7(c) $^2J_{PH}$14	243
	$(C'H_2)_2$, $CC'H,C_2C'$	**34**	89.7,61.5(c) $^3J_{PP}$36	244
C_4/6	$(CHO)_2,(CH_2)_2$	$(iBu)_2P^+$ $[CH(OH)(CH_2)_3CH(OH)]Cl^-$	23	242
C_4/7	$H_2P^4,(H_3)_2,C'H_2$	**35**	23.3(1,25°) $^2J_{PH}$15.8	245
C_4/9	$(H_3)_2,(CH_2)_2$	$Me_2P^+(CH_2)_8I^-$	31.9(c)	246
C_4/6,6	$(H_2N)_4$	$(CH_2NMeCONMeCH_2)_2P^+Cl^-$	29.8(w)	247
	$H_3,(CH_2)_2,C_2H$	**36**	15.3^C(c);21.7^T(c)	222

Connectivity of P-bound atoms	Formula	Structure	NMR data (δ_P[solv,temp] $^nJ_{PZ}$ Hz)	Ref.
	P-bound atoms = C_3C'			
$(H_3)_3;C'_2$	$C_9H_9BrF_5P$	$C_6F_5P^+Me_3Br^-$	25.2(l) $^2J_{PH}$14.6	216
	$C_9H_{14}BrP$	$PhP^+Me_3Br^-$	23.3	222
$(H_3)_2,CH_2;C'_2$	$C_{10}H_{16}BrP$	$EtPhP^+Me_2Br^-$	28.4	222
$H_3,CH_2,C_3;C'_2$	$C_{15}H_{27}ClIPSn$	$tBuMePhP^+(CH_2)_2SnClMeI^-$	41.8(c) $^3J_{SnP}$316.5	248
$H_3,(C'H_2)_2;C'_2$	$C_{21}H_{22}BrP$	$MePhP^+Bz_2Br^-$	27.4	222
$H_3,(C_2H)_2;C'_2$	$C_{13}H_{22}BrP$	$MePhP^+$-iPr_2Br^-	42	222
	$C_{17}H_{26}BrP$	$MePhP^+$-cPe_2Br^-	36.4(d)	226

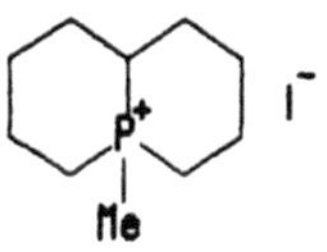

36 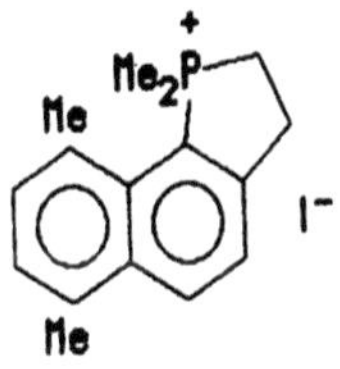 37 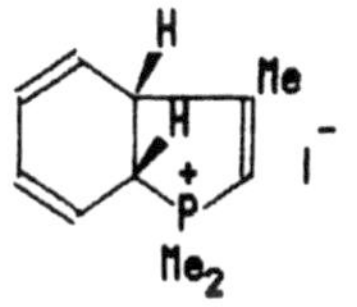 38

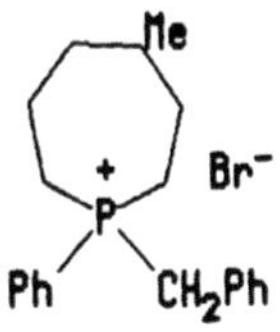

39 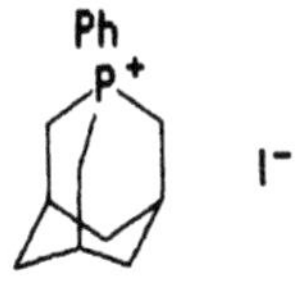 40

P-bound atoms/rings	Connectivities	Structure	NMR Data (δ_P[solv, temp]$^nJ_{PZ}$ Hz)	Ref.
$(CH_2)_3$;SS′	$C_{25}H_{45}F_5S_2P_3Pd$	$Et_3P^+CS^-{}_2Pd(Et_3P)_2C_6F_5$	41.3(c) $^3J_{PP}$3.9	249
$(CH_2)_3$;HP^4	$C_{31}H_{43}ClP_2$	$Bu_3P^+CHPPh_3Cl^-$	25.1(c) $^2J_{PP}$20.2	250
$(CH_2)_3$;CO′	$C_{14}H_{30}ClOP$	$Bu_3P^+COMeCl^-$	26.6(b)	251
$(CH_2)_3$;C′O	$C_{16}H_{30}ClO_5P$	Bu_3P^+(2-fur)$ClO^-{}_4$	34	252
	$C_{19}H_{31}ClFOP$	Bu_3P^+CO(4-F Ph′)Cl^-	39.5(b)	251
$(CH_2)_3$;C′H	$C_{17}H_{32}BrP$	$Bu_3P^+CH{=}CHMeBr^-$ trans	25.4(c)	253
	$C_{28}H_{58}Br_2P_2$	$[Bu_3P^+CH{=}CH]_2Br^-{}_2$	27.8(c)	254
$(CH_2)_3$;CC′	$C_{35}H_{47}BrO_2P$	$Bu_3P^+C(CH_2COPh){=}CHPhOBzBr^-$	28.8 $^3J_{PH}$17	235
$(CH_2)_3$;C′$_2$	$C_{12}H_{20}IP$	$Et_3P^+PhI^-$	36 $^2J_{PH}$13	222, 255, 256
	$C_{15}H_{26}IP$	$Pr_3P^+Phi^-$	29.5(c)	257
	$C_{17}H_{28}IOP$	Pr_3P^+(4-MeCOPh′)I^-	30.2(c)	257
	$C_{17}H_{31}INP$	Pr_3P^+(4-Me_2NPh′)I^-	26.6(c)	257
$(CH_2)_2$,C′H_2;C′$_2$	$C_{17}H_{22}BrP$	$Et_2P^+BzPhBr^-$	32.7(c) $^2J_{PH}$15.2	235
CH_2,(C′H_2)$_2$;C′$_2$	$C_{22}H_{24}BrP$	$EtP^+Bz_2PhBr^-$	30.1	222
CH_2,$(C_2H)_2$;C′$_2$	$C_{11}H_{24}BrP$	$iPr_2P^+EtPhBr^-$	42.2(d)	226
	$C_{12}H_{26}BrP$	$iPr_2P^+PrPhBr^-$	39.9(d)	226

P-bounds atoms (rings)	Connectivities	Structure	NMR data (δ_P[solv,temp] $^nJ_{PZ}$ Hz)	Ref.
C_3C′/5	$(H_3)_2$,CH_2;C′$_2$	**37**	50.1(c)	244
	$(H_3)_2$,CC′H;C′H	**38**	59.5(d) $^1J_{PC}$50.5	221
	CH_2,(C′H_2)$_2$;C′H	$Bz_2P^+[(CH_2)_2CH{=}CH_2]Br^-$	61.8(c)	258
C_3C′/6	H_3,$(C_3)_2$;C′$_2$	$MePhP^+(CMe_2CH_2COCH_2CMe_2)I^-$	35.3(d) $^1J_{PC}$40.4	259
C_3C′/7	$(CH_2)_2$,C′H_2;C′$_2$	**39**	29.9	260
C_3C′/6,6	$(CH_2)_3$;C′$_2$	**40**	4.8(cl)9.1(y) $^2J_{PH}$18	261

TABLE G (continued)
Compilation of ^{31}P NMR Data of Four Coordinate Phosphonium Salts and Betaines

Section G5: Four Coordinate Phosphonium Salts and Betaines with Four P–C Bonds

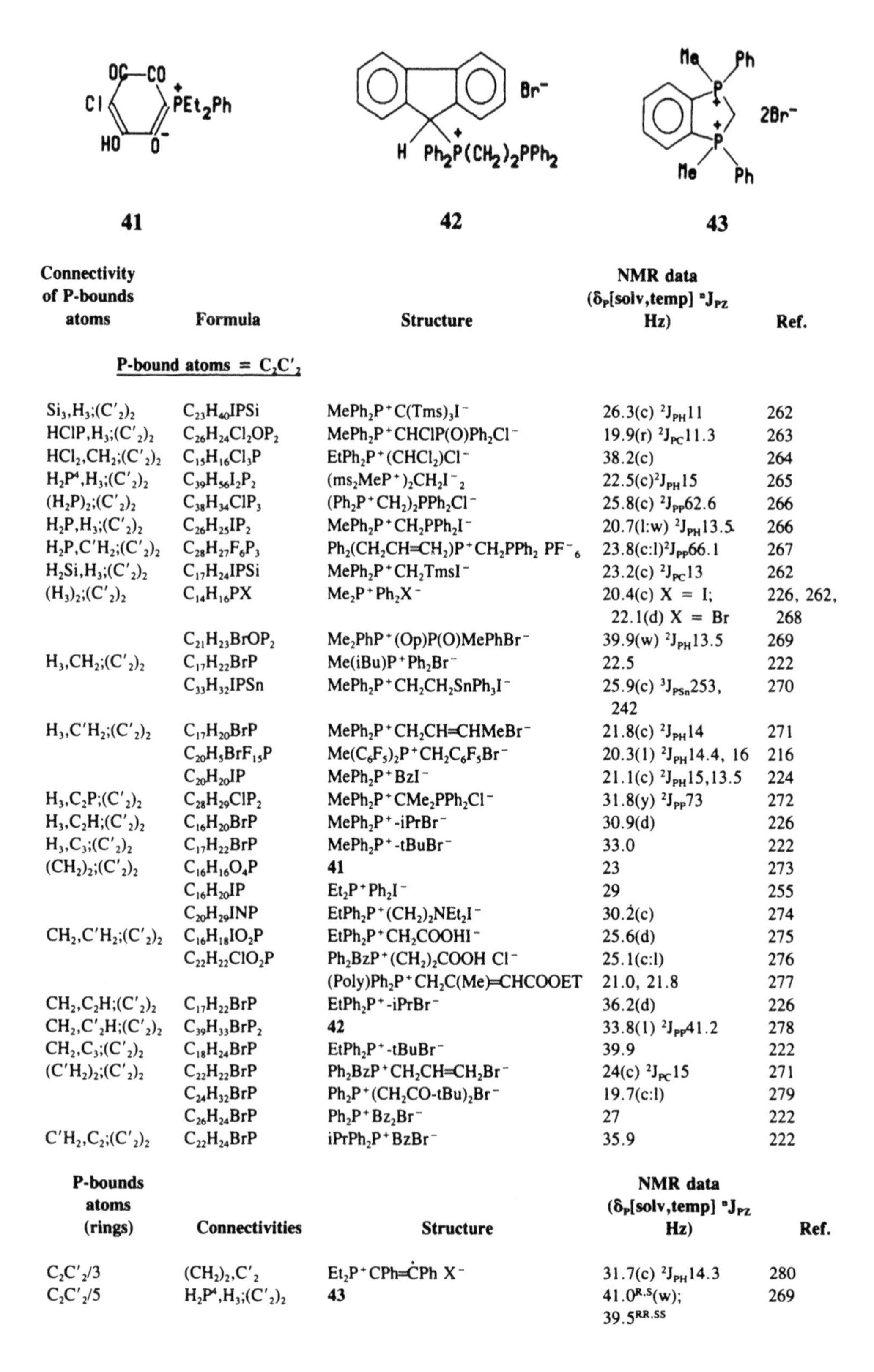

41 42 43

Connectivity of P-bounds atoms	Formula	Structure	NMR data (δ_P[solv,temp] $^nJ_{PZ}$ Hz)	Ref.
	P-bound atoms = C,C'_3			
$Si_3,H_3;(C'_2)_2$	$C_{23}H_{40}IPSi$	$MePh_2P^+C(Tms)_3I^-$	26.3(c) $^2J_{PH}$11	262
$HClP,H_3;(C'_2)_2$	$C_{26}H_{24}Cl_2OP_2$	$MePh_2P^+CHClP(O)Ph_2Cl^-$	19.9(r) $^2J_{PC}$11.3	263
$HCl_2,CH_2;(C'_2)_2$	$C_{15}H_{16}Cl_3P$	$EtPh_2P^+(CHCl_2)Cl^-$	38.2(c)	264
$H_2P^4,H_3;(C'_2)_2$	$C_{39}H_{56}I_2P_2$	$(ms_2MeP^+)_2CH_2I^-{}_2$	22.5(c)$^2J_{PH}$15	265
$(H_2P)_2;(C'_2)_2$	$C_{38}H_{34}ClP_3$	$(Ph_2P^+CH_2)_2PPh_2Cl^-$	25.8(c) $^2J_{PP}$62.6	266
$H_2P,H_3;(C'_2)_2$	$C_{26}H_{25}IP_2$	$MePh_2P^+CH_2PPh_2I^-$	20.7(l:w) $^2J_{PH}$13.5	266
$H_2P,C'H_2;(C'_2)_2$	$C_{28}H_{27}F_6P_3$	$Ph_2(CH_2CH{=}CH_2)P^+CH_2PPh_2\ PF^-{}_6$	23.8(c:l)$^2J_{PP}$66.1	267
$H_2Si,H_3;(C'_2)_2$	$C_{17}H_{24}IPSi$	$MePh_2P^+CH_2TmsI^-$	23.2(c) $^2J_{PC}$13	262
$(H_3)_2;(C'_2)_2$	$C_{14}H_{16}PX$	$Me_2P^+Ph_2X^-$	20.4(c) X = I; 22.1(d) X = Br	226, 262, 268
	$C_{21}H_{23}BrOP_2$	$Me_2PhP^+(Op)P(O)MePhBr^-$	39.9(w) $^2J_{PH}$13.5	269
$H_3,CH_2;(C'_2)_2$	$C_{17}H_{22}BrP$	$Me(iBu)P^+Ph_2Br^-$	22.5	222
	$C_{33}H_{32}IPSn$	$MePh_2P^+CH_2CH_2SnPh_3I^-$	25.9(c) $^3J_{PSn}$253, 242	270
$H_3,C'H_2;(C'_2)_2$	$C_{17}H_{20}BrP$	$MePh_2P^+CH_2CH{=}CHMeBr^-$	21.8(c) $^2J_{PH}$14	271
	$C_{20}H_5BrF_{15}P$	$Me(C_6F_5)_2P^+CH_2C_6F_5Br^-$	20.3(1) $^2J_{PH}$14.4, 16	216
	$C_{20}H_{20}IP$	$MePh_2P^+BzI^-$	21.1(c) $^2J_{PH}$15,13.5	224
$H_3,C_2P;(C'_2)_2$	$C_{28}H_{29}ClP_2$	$MePh_2P^+CMe_2PPh_2Cl^-$	31.8(y) $^2J_{PP}$73	272
$H_3,C_2H;(C'_2)_2$	$C_{16}H_{20}BrP$	$MePh_2P^+$-iPrBr$^-$	30.9(d)	226
$H_3,C_3;(C'_2)_2$	$C_{17}H_{22}BrP$	$MePh_2P^+$-tBuBr$^-$	33.0	222
$(CH_2)_2;(C'_2)_2$	$C_{16}H_{16}O_4P$	**41**	23	273
	$C_{16}H_{20}IP$	$Et_2P^+Ph_2I^-$	29	255
	$C_{20}H_{29}INP$	$EtPh_2P^+(CH_2)_2NEt_2I^-$	30.2(c)	274
$CH_2,C'H_2;(C'_2)_2$	$C_{16}H_{18}IO_2P$	$EtPh_2P^+CH_2COOHI^-$	25.6(d)	275
	$C_{22}H_{22}ClO_2P$	$Ph_2BzP^+(CH_2)_2COOH\ Cl^-$	25.1(c:l)	276
		(Poly)$Ph_2P^+CH_2C(Me){=}CHCOOET$	21.0, 21.8	277
$CH_2,C_2H;(C'_2)_2$	$C_{17}H_{22}BrP$	$EtPh_2P^+$-iPrBr$^-$	36.2(d)	226
$CH_2,C'_2H;(C'_2)_2$	$C_{39}H_{33}BrP_2$	**42**	33.8(1) $^2J_{PP}$41.2	278
$CH_2,C_3;(C'_2)_2$	$C_{18}H_{24}BrP$	$EtPh_2P^+$-tBuBr$^-$	39.9	222
$(C'H_2)_2;(C'_2)_2$	$C_{22}H_{22}BrP$	$Ph_2BzP^+CH_2CH{=}CH_2Br^-$	24(c) $^2J_{PC}$15	271
	$C_{24}H_{32}BrP$	$Ph_2P^+(CH_2CO\text{-}tBu)_2Br^-$	19.7(c:l)	279
	$C_{26}H_{24}BrP$	$Ph_2P^+Bz_2Br^-$	27	222
$C'H_2,C_2;(C'_2)_2$	$C_{22}H_{24}BrP$	$iPrPh_2P^+BzBr^-$	35.9	222

P-bounds atoms (rings)	Connectivities	Structure	NMR data (δ_P[solv,temp] $^nJ_{PZ}$ Hz)	Ref.
$C_2C'_2/3$	$(CH_2)_2,C'_2$	$Et_2P^+CPh{=}\dot{C}Ph\ X^-$	31.7(c) $^2J_{PH}$14.3	280
$C_2C'_2/5$	$H_2P^4,H_3;(C'_2)_2$	**43**	41.0^{R,S}(w); 39.5^{RR,SS}	269

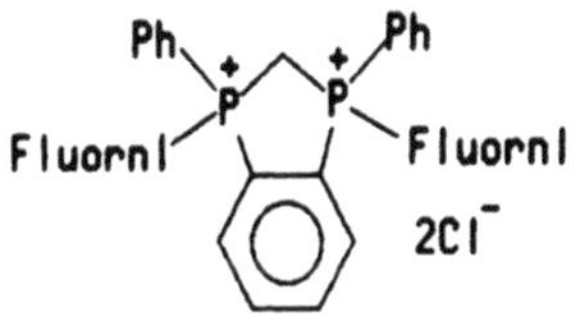

44 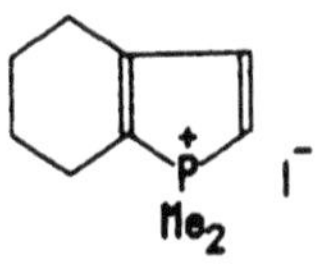 **45** 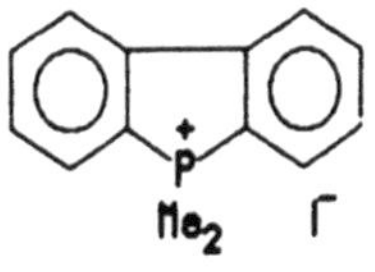 **46**

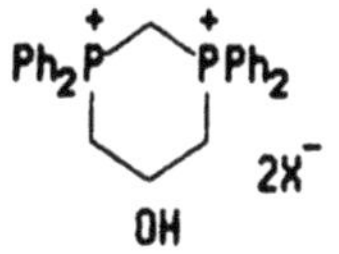

47 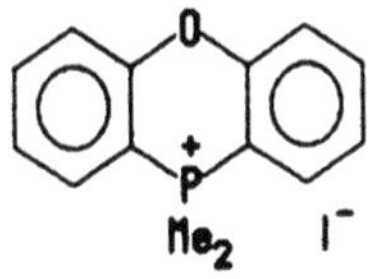 **48** 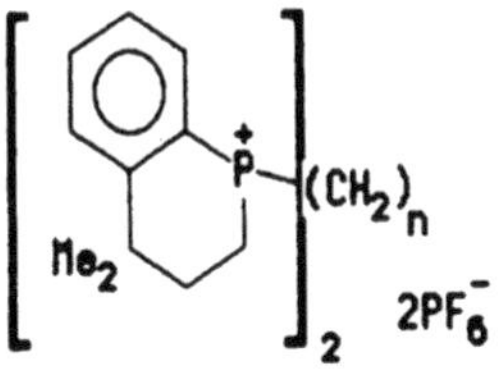 **49**

P-bound atoms/rings	Connectivities	Structure	NMR Data (δ_P[solv, temp] $^nJ_{PZ}$ Hz)	Ref.
	$H_2P^4,C'_2H;(C'_2)_2$	**44**	47.2(l:w)	281
	$(H_3)_2;(C'H)_2$	$Me_2\dot{P}^+$ $CH{=}CMeCMe{=}\dot{C}H$ I^-	34.4(w)	282
	$(H_3)_2;C'H,CC'$	**45**	14.1(c)	244
	$(H_3)_2;(C'_2)_2$	**46**	27(l)	268, 283
$C_2C'_2/6$	$H_2P^4,CH_2;(C'_2)_2$	**47**	16.3(l) 14.4(d)	284
	$H_2P^4,C'H_2;(C'_2)_2$	$Ph_2P^+[CH_2COCH{=}P(Ph_2)CH_2]$	17(d)	285
	$(H_3)_2;(C'_2)_2$	**48**	−7.5(l)	283
	$H_3,CH_2;(C'_2)_2$	$MePhP^+[DopCHMe(CH_2)_2]PF_6^-$	9.8(m:l)	271
	$(CH_2)_2;(C'_2)_2$	$Ph_2P^+(CH_2)_5Cl^-$	16.5(y)	286
		$Ph_2P^+[CH_2)_2PPh_2(CH_2)_2]X^-$	14.4	286
		49; n = 1	10^M, 11.3^{dl} (c:l)	267
		49; n = 2	15.1^M, 16.4^{dl} (c:l)	267
	$CH_2,C'H_2;CC',C'_2$	**50**	23.2(y)	287
	$CH_2,C'H_2;(C'_2)_2$	$Ph_2P^+(CH_2DopCHMeCH_2)PF_6^-$	17.2(c:l) $^4J_{PH}6$	271
$C_2C'_2/7$	$H_2P^4,C'H_2;(C'_2)_2$	$CH_2(Ph_2P^+CH_2DopCH_2P^+Ph_2)Br_2^-$	13(1,25°) $^2J_{PH}15$	245
	$CH_2,C'H_2;(C'_2)_2$	$Ph_2P^+(CH_2DopCHMeCH_2CH_2)PF_6^-$	14.2(c:l) $^2J_{PH}15$	278
		$Ph_2P^+(CH_2DopCOCH_2CH_2)Cl^-$	21.5(c:l)	276
$C_2C'_2/16$	$(CH_2)_2;(C'_2)_2$	**51**; R = CH_2CH_2	25.8(l)	288
		51; R = $CH{=}CH$	28(y)	288
$C_2C'_2/20$	$CH_2,C'H_2;(C'_2)_2$	**52**	27.4(l)	289
$C_2C'_2/26$	$(CH_2)_2;(C'_2)_2$	**53**	27.8	290

Connectivity of P-bound atoms	Formula	Structure	NMR data (δ_P[solv,temp] $^nJ_{PZ}$ Hz)	Ref.
P-bound atoms = CC'_3				
$Cl_3;(C'_2)_3$	$C_{19}H_{15}Cl_3P$	$Ph_3P^+CCl_3X^-$	47.5(r)	291
$H_2Br;(C'_2)_3$	$C_{19}H_{17}Br_2P$	$Ph_3P^+CH_2BrBr^-$	22.6(c) $^2J_{PH}6.2$	226, 292, 293

TABLE G (continued)
Compilation of ^{31}P NMR Data of Four Coordinate Phosphonium Salts and Betaines

Section G5: Four Coordinate Phosphonium Salts and Betaines with Four P–C Bonds

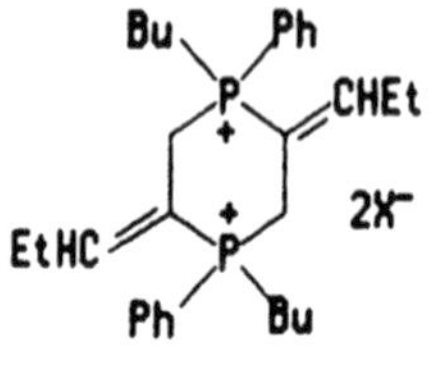

50

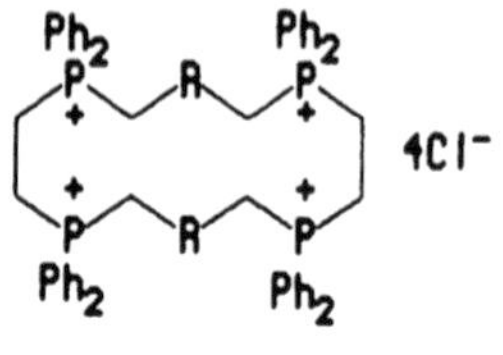

51

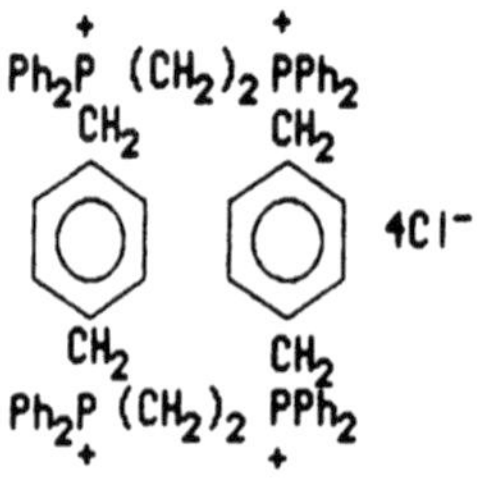

52

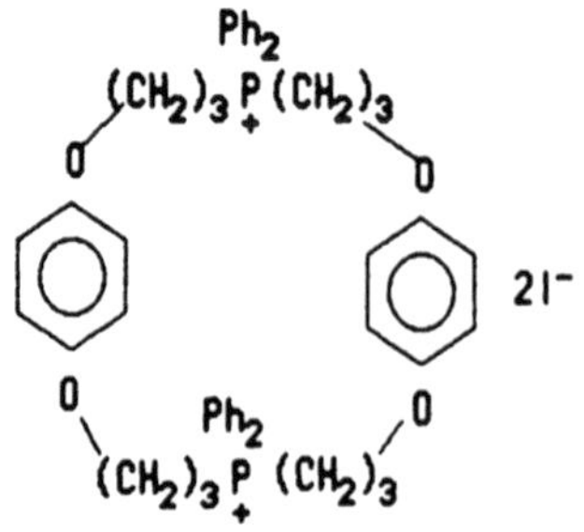

53

P-bound atoms/rings	Connectivities	Structure	NMR Data (δ_P[solv, temp][a] J_{PZ} Hz)	Ref.
$H_2Cl;(C'_2)_3$	$C_{19}H_{17}Cl_2P$	$Ph_3P^+CH_2ClCl^-$	23.8(d) $^2J_{PH}$6.1	226, 292, 293
$H_2Li;(C'_2)_3$	$C_{19}H_{17}ILiP$	$Ph_3P^+CH_2LiCl^-$	22.1(b:j) $^1J_{PC}$51.9	294
$H_2N;(C'_2)_3$	$C_{22}H_{24}ClN_2OP$	$Ph_3P^+CH_2NHCONMeCl^-$	19.9(e)	210
$H_2O;(C'_2)_3$	$C_{19}H_{18}ClOP$	$Ph_3P^+CH_2OHCl^-$	17.7(c) and (d) $^2J_{PH}$O	213, 226, 292
	$C_{20}H_{20}BrOP$	$Ph_3P^+CH_2OMeBr^-$	11.8(a:w)	231
$H_2P_4;(C'_2)_3$	$C_{27}H_{28}F_6P_4$	$Ph_3P^+CH_2PPhMe_2(PF_6^-)_2$	18.9, 25.2(l)	295
$H_2S;(C'_2)_3$	$C_{25}H_{22}ClPS$	$Ph_3P^+CH_2SPhCl^-$	21.5(c) $^2J_{PH}$7.2	296
$H_2Si;(C'_2)_3$	$C_{22}H_{26}IPSi$	$Ph_3P^+CH_2TmsI^-$	23.9(c)	232
$H_3;P^4{}_2,(C'_2)_2$	$C_{45}H_{41}I_2P_3$	$Ph_3P^+C(P^+Ph_2Me)_2I_2^-$	20.1(1,30°)	297
$H_3(C'O)_3$	$C_{13}H_{12}IO_3P$	$(2\text{-Fur})_3P^+MeI^-$	−15.4(l) $^2J_{PH}$14.6	298
$H_3;C'P,(C'_2)_2$	$C_{27}H_{27}IP_2$	$MePh_2P^+C(CH_2)PPh_2I^-$	11.1(c) $^2J_{PH}$15	299
$H_3;(C'S)_3$	$C_{13}H_{12}IPS_3$	$(2\text{-Thiophene})_3P^+MeI^-$	2.3(l)	298
$H_3;C'H,(C'_2)_2$	$C_{16}H_{18}BrP$	$MePh_2P^+CH{=}CHMeBr^-$	9.9^C, 16.2^T	300
$H_3;(C'_2)_3$	$C_{19}H_{18}PX$	$Ph_3P^+MeX^-$	20.8(c), 18.8(l), 22.7(d)	223, 226, 228, 298, 301, 302, 303, 304
	$C_{32}H_{30}F_2O_6P_2S_2$	$(MePh_2P^+)_2Dop[SO_3F^-]_2$	25.7(y) $^1J_{PC}$76.2	295
$CBr_2;(C'_2)_3$	$C_{21}H_{20}Br_3P$	$Ph_3P^+CBr_2EtBr^-$	36.9(c)	231
$CH_2;(C'_2)_3$	$C_{20}H_{20}BrP$	$Ph_3P^+EtBr^-$	26.2	226, 228
	$C_{21}H_{22}BrP$	$Ph_3P^+PrBr^-$	24.1(d)	226
	$C_{21}H_{20}ClO_2P$	$Ph_3P^+(CH_2)_2CO_2HCl^-$	24(y) $^1J_{PC}$86.1	305, 306

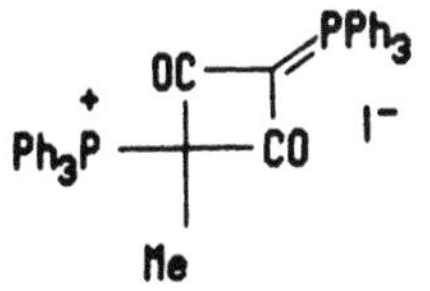

Structures 54, 55, 56:

54: Ph_3P^+ ring with OC, $=PPh_3$, CO, Me; I^-

55: Ph_2P, $^+PPh_2$, O; X^-

56: Me, Ph, P^+, tBu, O, tBu; ClO_4^-

54 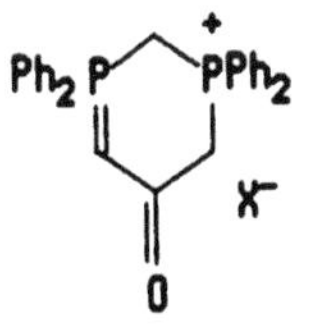 **55** 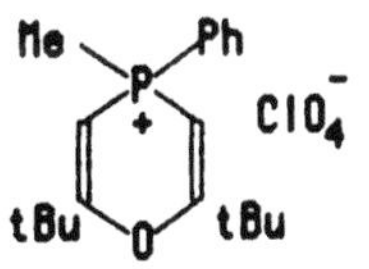 **56**

P-bound atoms/rings	Connectivities	Structure	NMR Data (δ_P[solv, temp] $^nJ_{PZ}$ Hz)	Ref.
	$C_{22}H_{22}ClP$	$Ph_3P^+CH_2C(Me){=}CH_2Cl^-$	21(d)	226
	$C_{22}H_{22}IO_2P$	$Ph_3P^+CH_2CH(OCH_2CH_2O)I^-$	20.7(c)	224, 307
	$C_{25}H_{17}F_6N_2P$	$Ph_3P^+CH_2C(CF_3)_2C^-(CN)_2$	24.8(c) $^2J_{PH}$13.5	308
	$C_{38}H_{34}F_{12}P_4$	$(Ph_3P^+CH_2)_2(Pf_6^-)_2$	24.7(m)	229, 286
$C'H_2;C'H,(C'_2)_2$	$C_{40}H_{36}Br_2P_2$	$(Ph_2BzP^+CH)_2Br_2^-$	20.6(c:l), 3.8(y)?	224, 286
$C'H_2;(C'_2)_3$	$C_{20}H_{17}O_2P$	$Ph_3P^+CH_2COO^-$	23(y)	305
	$C_{20}H_{18}BF_4OP$	$Ph_3P^+CH_2CHOBF_4^-$	19.5(c,20°)	309
	$C_{20}H_{18}ClO_2P$	$Ph_3P^+CH_2COOHCl^-$	20(c)	305
	$C_{21}H_{20}BrOP$	$Ph_3P^+CH_2COMeBr^-$	20(m),13.6(c) enol,21.lketo	310, 311
	$C_{21}H_{20}PX$	$Ph_3P^+CH_2CH{=}CH_2X^-$	21.4	226, 228
	$C_{22}H_{22}BrO_2P$	$Ph_3P^+CH_2COOEtBr^-$	19.2(a:w)	231
	$C_{23}H_{20}BrPS$	$Ph_3P^+CH_2$(2-Thiophene)Br^-	18.3(l) $^2J_{PH}$13	312
	$C_{25}H_{22}BrP$	$Ph_3P^+BzBr^-$	23.5(d), 22, 17.8(a:w)	226, 231, 268
	$C_{27}H_{26}BrN_2P$	$Ph_3P^+CH_2C(Me)NNHPhBr^-$	21.8(28°)	313
$C_2Cl;(C'_2)_3$	$C_{21}H_{21}ClF_6P$	$Ph_3P^+CClMe_2PF_6^-$	35.4(c) $^3J_{PH}$15.6	229
$C_2H;(C'_2)_3$	$C_{21}H_{22}BrP$	Ph_3P^+-iPrBr^-	30.9	228
	$C_{23}H_{24}BrP$	Ph_3P^+-cPeBr^-	30.7(d)	226
	$C_{24}H_{26}BrP$	Ph_3P^+-cHexBr^-	26.6(d)	226, 314
$CC'H;(C'_2)_3$	$C_{36}H_{32}IN_2OP$	$Ph_3P^+CMeCMe{=}NN{=}CPhCOPhI^-$	27.9(c) $^3J_{PH}$18.4	315
$C'_2H;(C'_2)_3$	$C_{25}H_{22}BrP$	Ph_3P^+-c$C_7H_7Br^-$	22.6(d)	226
	$C_{31}H_{26}BrP$	$Ph_3P^+CHPh_2Br^-$	21.4(d)	226
	$C_{48}H_{44}Cl_2O_{16}P_2$	$[Ph_3P^+CH(COOMe)C(COOMe)]_2(ClO_4^-)_2$	28.9(l)	316
$C_3;(C'_2)_3$	$C_{22}H_{24}IP$	Ph_3P^+-tBuI^-	34.7(d)	226
	$C_{25}H_{28}IP$	Ph_3P^+(l-Me-cHex)I^-	35.1(d)	226
$C_2C';(C'_2)_3$	$C_{28}H_{24}ClOP$	Ph_3P^+(l-PhCO)cPrCl^-	21.8(c)	317
$CC'_2;(C'_2)_3$	$C_{41}H_{36}IO_2P_2$	**54**	25.8, 0(c)	318

P-bound atoms/ (rings)	Connectivities	Structure	NMR data (δ_P[solv,temp] $^nJ_{PZ}$ Hz)	Ref.
$CC'_3/6$	$H_2P^4;C'H,C'_2$	**55**	6, 17, 19.9(d)	285
	$H_3;(C'H)_2,C'_2$	**56**	− 13(c)	279
	$CH_2;P_2,(C'_2)_2$	**57**	18, 23.9(l) $^2J_{PP}$12.2	297
	$CH_2;C'H,(C'_2)_2$	**58**	2.8(y) $^2J_{PH}$8	319
		59	3.4?(y)	287
	$CH_2;(C'_2)_2$	Ph_2P^+[DopCHMeCH_2CH_2]PF_6^-	10.7(m:l)	271
	$C'H_2;CP^4,(C'_2)_2$	**60**	20.2, 22.7(c)	299
$CC'_3/9$	$(H_3);(C'H)_2,C'_2$	**61**	1.2(c)	246

TABLE G (continued)
Compilation of ^{31}P NMR Data of Four Coordinate Phosphonium Salts and Betaines

Section G5: Four Coordinate Phosphonium Salts and Betaines with Four P–C Bonds

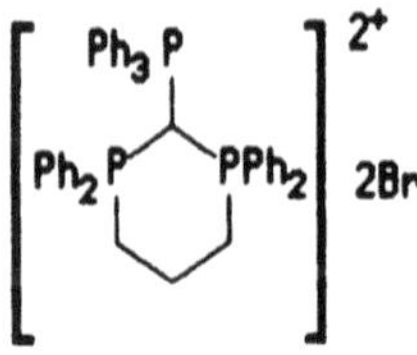

57 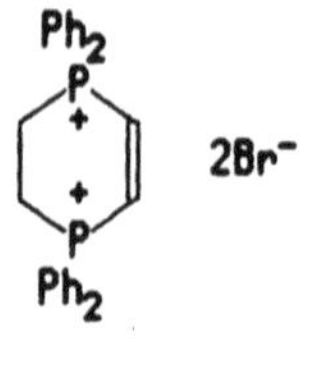 58 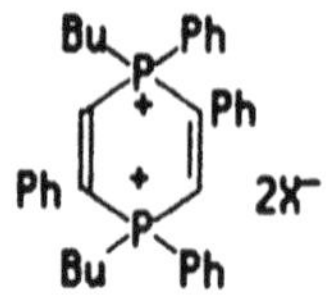 59

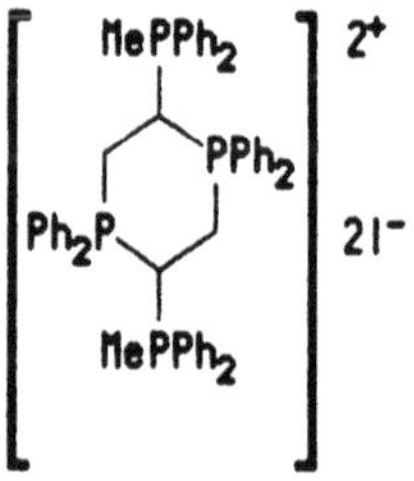

60 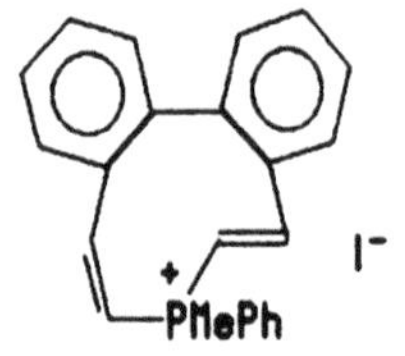 61

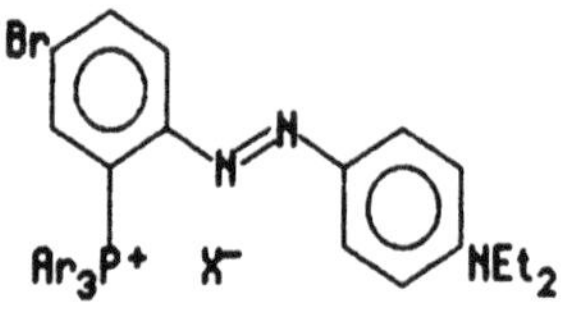

62 Ar= 2-Fur

63 Ar=2-Thiophene

P-bound atoms	Connectivities	Structure	NMR data (δ_P[solv,temp] $^nJ_{PZ}$ Hz)	Ref.
P-bound atoms = CC'C''$_2$				
CC'C''$_2$	H_3;C'_2;$(C'')_2$	$MePhP^+(C{\equiv}C\text{-}tBu)_2I^-$	−22.3(c)	279, 320

Connectivity of P-bound atoms	Formula	Structure	NMR data (δ_P[solv,temp] $^nJ_{PZ}$ Hz)	Ref.
P-bound atoms = C'$_4$				
$ClP^4,(C'_2)_3$	$C_{31}H_{25}Cl_3P_2$	$Ph_3P^+CClPClPh_2Cl^-$	−23.6, 62.6? $^2J_{PP}$42.5	321
$P^4_2,(C'_2)_3$	$C_{49}H_{40}ClOP_3$	$Ph_3P^+C(PPh_3)P(O)Ph_2Cl^-$	23.2, 26.9(y)	322
$PP^4,(C'_2)_3$	$C_{49}H_{40}P_3X$	$Ph_3P^+C(PPh_3)PPh_2X^-$	27.8(r,20°) $^2J_{PP}$80.1	297
$HP^4,(C'_2)_3$	$C_{31}H_{26}Cl_2P_2$	$Ph_3P^+CHPClPh_2Cl^-$	17.6, 56.7(r)	323
$C'N,(C'_2)_3$	$C_{27}H_{21}Cl_3NOP$	$Ph_3P^+C({=}CCl_2)NHCOPhCl^-$	23(y)	324
	$C_{43}H_{42}I_2N_2P_2$	$[Ph_3P^+C(NMe_2){=}]_2Cl_2^-$	−17.7(c)	325
$C'N',(C'_2)_3$	$C_{21}H_{18}BF_4N_2OP$	$Ph_3P^+C({=}N_2)COR\ BF_4^-$	22.6 to 26	326
	$C_{23}H_{19}BrNP$	$Ph_3P^+(2\text{-}Pyr)Br^-$	−15(e)	327
$(C'O)_3,C'_2$	$C_{28}H_{26}BrN_3O_3PX$	**62**	−18.8(c) $^1J_{PC}$107	328
$CP^4,(C'_2)_3$	$C_{38}H_{33}BrP_2$	$Ph_3P^+CMePPh_3Br^-$	27(c) $^3J_{PH}$14	329
	$C_{40}H_{31}BF_{10}OP_2$	$(Ph_3P^+)_2CCTf_2OHBF_4^-$	22(m)	308
$C'P^4,(C'_2)_3$	$C_{39}H_{33}IO_2P_2$	$(Ph_3P^+)_2CCOOMeI^-$	21.4	330
	$C_{44}H_{35}NP_2S$	$(Ph_3P^+)_2CC(S^-){=}NPh$	12.1(r)	331
$(C'S)_3,C'_2$	$C_{28}H_{26}BrN_3PS_3X$	**63**	2(c) $^1J_{PC}$102	328

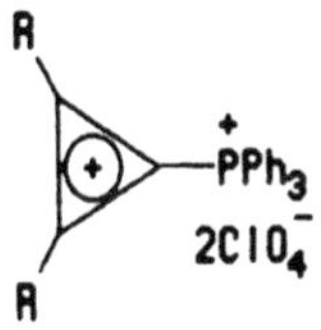

64 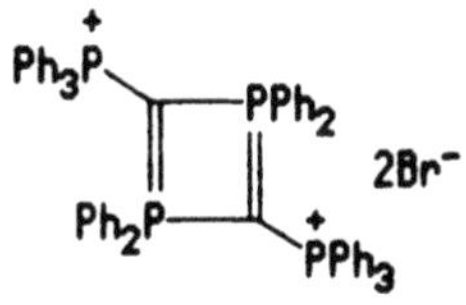 65 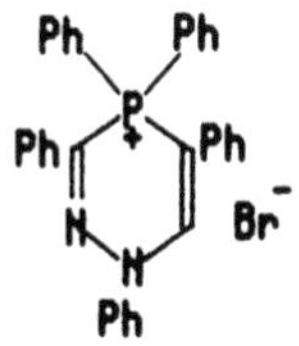 66

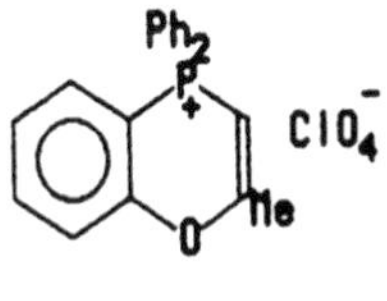

67 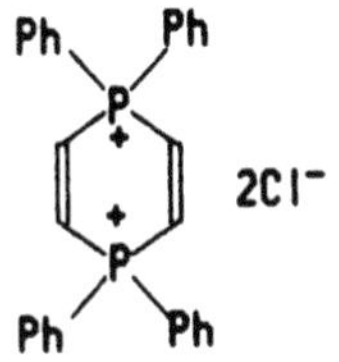 68 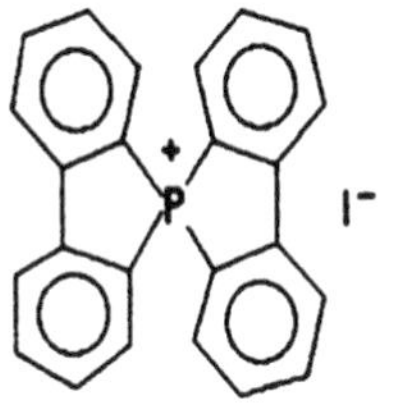 69

P-bound atoms/rings	Connectivities	Structure	NMR Data (δ_P[solv, temp][a]J_{PZ} Hz)	Ref.
$C'S,(C'_2)_3$	$C_{44}H_{38}Cl_2P_2S$	$Ph_3P^+CH_2CH{=}C(SPh)P^+Ph_3Cl_2^-$	21.4,22.3	332
$C'H,(C'_2)_3$	$C_{20}H_{18}BrP$	$Ph_3P^+CH{=}CH_2Br^-$	19.3(c)	253
	$C_{20}H_{18}ClOP$	$Ph_3P^+CH{=}CH(OH)Cl^-$	13.9^Z; 18.9^E(c,20°)	309
	$C_{21}H_{18}BrP$	$Ph_3P^+CH{=}C{=}CH_2Br^-$	18.7(c) $^1J_{PC}$93.3	253
	$C_{21}H_{20}BrOP$	$Ph_3P^+CH{=}CHOMeBr^-$	20.4(c)	307
	$C_{22}H_{16}ClF_6P$	$Ph_3P^+CH{=}CTf_2Cl^-$	17.3(c) $^2J_{PH}$8.5	308
	$C_{23}H_{24}IPS_2$	$Ph_3P^+CH{=}C(SMe)SEtI^-$	11.7(c)	333
	$C_{26}H_{22}BrPS$	$Ph_3P^+CH{=}CHSPhBr^-$	18.7(c)	307
	$C_{26}H_{31}OPSi$	$Ph_3P^+CH{=}CHCH{=}CHMeTmsO^-$	19.8	334
	$C_{39}H_{34}P_2S_2$	$Ph_3P^+CH{=}C(S^-)S^-Ph_3P^+Me$	−6.5, 21.2(c)	333
$CC',(C'_2)_3$	$C_{22}H_{23}NPX$	$Ph_3P^+CH{=}CHNMe_2X^-$	20.7 $^3J_{PH}$16	335
	$C_{23}H_{22}ClO_4P$	Ph_3P^+(1-cPentenyl)ClO_4^-	24(d)	336
$(C'_2)_4$	$C_{22}H_{15}O_3P$	$Ph_3P^+[C{=}C(O^-)COCO]$	7.8(c)	337
	$C_{24}H_{20}BrP$	$Ph_4P^+Br^-$	20.8, 23.1(m)	226, 338
	$C_{25}H_{27}Cl_2N_2O_8P$	**64**; R = NMe_2	6.7(m)	339
	$C_{27}H_{29}Cl_2O_8P$	**64**; R = iPr	53.6(m)	339
	$C_{28}H_{26}FO_5PS$	$Ph_3P^+C(Ph){=}C(OMe)_2FSO_3^-$	24.5 $^3J_{PH}$2	340
	$C_{32}H_{40}IN_4P$	$(4\text{-}Me_2NC_6H_4)_4P^+I^-$	18.7(c)	257

P-bound atoms (rings)	Connectivities	Structure	NMR data (δ_P[solv,temp] [a]J_{PZ} Hz)	Ref.
$C'_4/4$	$(P^4{}_2)_2,(C'_2)_2$	**65**	14.4, 33.6	321
$C'_4/6$	$C'N',(C'_2)_3$	**66**	25.7 $^3J_{PH}$23	341
	$C'H,(C'_2)_3$	**67**	124.8(c)	342
		68	−3.5(y)	287
$C'_4/5,5$	$(C'_2)_4$	**69**	24(v)	343

TABLE G (continued)
Compilation of ^{31}P NMR Data of Four Coordinate Phosphonium Salts and Betaines

Section G5: Four Coordinate Phosphonium Salts and Betaines with Four P–C Bonds

Connectivity of P-bound atoms	Formula	Structure	NMR data (δ_P[solv,temp] $^nJ_{PZ}$ Hz)	Ref.
P-bound atoms = C′$_3$C″				
(C′$_2$)$_3$;C″	$C_{21}H_{18}BrP$	$Ph_3P^+C{\equiv}CMeBr^-$	5.3(c) $^1J_{PC}$191.7	253
	$C_{24}H_{25}BrNP$	$Ph_3P^+C{\equiv}CNEt_2Br^-$	6.3 $^2J_{PC}$40	344
	$C_{26}H_{20}BrP$	$Ph_3P^+C{\equiv}CPhBr^-$	6.6 $^2J_{PC}$31.2	344
	$C_{27}H_{22}BrOP$	$Ph_3P^+C{\equiv}C(4\text{-}MeOC_6H_4)Br^-$	6.1 $^1J_{PC}$190.7	344
	$C_{28}H_{22}IO_2P$	$Ph_3P^+C{\equiv}C(2\text{-}MeOOCC_6H_4)I^-$	−6.6(c)?	345
	$C_{46}H_{34}Cl_2P_2$	$[1,4\text{-di}(Ph_3P^+C{\equiv}C)C_6H_4]Cl_2^-$	−6.5?	346

REFERENCES

1. **Dillon, K. B., Khabbass, N. D., and Ludman, C. J.,** unpublished work; **Khabbass, N. D.,** Ph.D. thesis, Durham, 1981.
2. **Dillon, K. B. and Platt, A. W. G.,** *Polyhedron,* 2, 641, 1983.
3. **Dillon, K. B., Nisbet, M. P., and Waddington, T. C.,** *J. Chem. Soc. Dalton Trans.,* 1455, 1978.
4. **Dillon, K. B., Nisbet, M. P., and Waddington, T. C.,** *J. Chem. Soc. Dalton Trans.,* 1591, 1979.
5. **Dillon, K. B. and Gates, P. N.,** *J. Chem. Soc. Chem. Commun.,* 348, 1972.
6. **Dillon, K. B., Waddington, T. C., and Younger, D.,** *Inorg. Nucl. Chem. Lett.,* 9, 63, 1973.
7. **Grimmer, A-R.,** *Z. Anorg. Allg. Chem.,* 400, 105, 1973.
8. **Dillon, K. B., Nisbet, M. P., and Waddington, T. C.,** *Polyhedron,* 1, 123, 1982.
9. **Dillon, K. B. and Platt, A. W. G.,** unpublished work; **Platt, A. W. G.,** Ph.D. thesis, Durham, 1980.
10. **Dillon, K. B., Nisbet, M. P., and Waddington, T. C.,** *J. Chem. Soc. Dalton Trans.,* 883, 1979.
11. **Michalski, J., Pakulski, M., and Skowronska, A.,** *J. Chem. Soc. Perkin Trans. I,* 833, 1980.
12. **Tseng, C. K.,** *J. Org. Chem.,* 44, 2793, 1979.
13. **Michalski, J., Pakulski, M., Skowronska, A., Gloede, J., and Gross, H.,** *J. Org. Chem.,* 45, 3122, 1980.
14. **Dillon, K. B., Nisbet, M. P., and Waddington, T. C.,** *J. Inorg. Nucl. Chem.,* 41, 1273, 1979.
15. **Wieker, W. and Grimmer, A-R.,** *Z. Naturforsch.,* 21b, 1103, 1966.
16. **Wieker, W. and Grimmer, A-R.,** *Z. Naturforsch.,* 22b, 257, 1967.
17. **Khabbass, N. D. and Ludman, C. J.,** unpublished work; **Khabbass, N. D.,** M.Sc. thesis, Durham, 1978.
18. **Jain, S. R., Krannich, L. K., Highsmith, R. E., and Sisler, H. H.,** *Inorg. Chem.,* 6, 1058, 1967.
19. **Marchenko, A. P., Miroshnichenko, V. V., Povolotskii, M. I., and Pinchuk, A. M.,** *J. Gen. Chem. USSR,* 53, 283, 1983.
20. **Lochschmidt, S., Müller, G., Huber, B., and Schmidpeter, A.,** *Z. Naturforsch.,* 41b, 444, 1986.
21. **Dillon, K. B., Platt, A. W. G., and Waddington, T. C.,** *J. Chem. Soc. Dalton Trans.,* 1036, 1980.
22. **Latscha, H.-P., Haubold, W., and Becke-Goehring, M.,** *Z. Anorg. Allg. Chem.,* 339, 82, 1965.
23. **Michalski, J., Mikolajczak, J., and Skowronska, A.,** *J. Am. Chem. Soc.,* 100, 5386, 1978.
24. **Dillon, K. B., Waddington, T. C., and Younger, D.,** *J. Inorg. Nucl. Chem.,* 43, 2665, 1981.
25. **Latscha, H.-P.,** *Z. Anorg. Allg. Chem.,* 355, 73, 1967.
26. **Timokhin, B. V., Dmitriev, V. I., Glukhikh, V. I., and Korchevin, N. A.,** *Bull. Acad. Sci. USSR Div. Chem. Sci.,* 27, 1160, 1978.
27. **Latscha, H-P.,** *Z. Anorg. Allg. Chem.,* 346, 166, 1966.
28. **Hormuth, P. B. and Latscha, H-P.,** *Z. Anorg. Allg. Chem.,* 365, 26, 1969.
29. **Latscha, H.-P. and Klein, W.,** *Angew. Chem. Int. Ed. Engl.,* 8, 278, 1969.
30. **Becke-Goehring, M. and Fluck, E.,** *Angew. Chem. Int. Ed. Engl.,* 1, 281, 1962.
31. **Fluck, E.,** *Z. Anorg. Allg. Chem.,* 315, 181, 1962.
32. **Moran, E. F.,** *J. Inorg. Nucl. Chem.,* 30, 1405, 1968.
33. **Fluck, E.,** *Z. Anorg. Allg. Chem.,* 320, 64, 1963.

34. **Fluck, E.**, *Z. Anorg. Allg. Chem.*, 315, 191, 1962.
35. **Dillon, K. B., Reeve, R. N., and Waddington, T. C.**, *J. Chem. Soc. Dalton Trans.*, 1465, 1978.
36. **Timokhin, B. V., Feshin, V. P., Dmitriev, V. I., Glukhikh, V. I., Dolgushin, G. V., and Voronkov, M. G.**, *Dokl. Akad. Nauk SSSR (Engl. trans.)*, 236, 966, 1977.
37. **Timokhin, B. V., Vengel'nikova, V. N., Kalabina, A. V., and Donskikh, V. I.**, *J. Gen. Chem. USSR*, 51, 1307, 1981.
38. **Becke-Goehring, M. and Lehr, W.**, *Chem. Ber.*, 94, 1591, 1961.
39. **Becke-Goehring, M., Haubold, W., and Latscha, H-P.**, *Z. Anorg. Allg. Chem.*, 333, 120, 1964.
40. **Van Wazer, J. R.**, unpublished work quoted in **Mark, V., Dungan, C. H., Crutchfield, M. M., and Van Wazer, J. R.**, *Topics Phosphorus Chem.*, 5, 227, 1967.
41. **Wieker, W. and Grimmer, A-R.**, *Z. Naturforsch.*, 22b, 983, 1967.
42. **Schmidpeter, A. and Brecht, H.**, *Angew. Chem.*, 79, 535, 1967.
43. **Kolditz, L., Beierlein, I., Wieker, W., and Grimmer, A-R.**, *Z. Chem.*, 8, 266, 1968.
44. **Dillon, K. B. and Waddington, T. C.**, *Spectrochim. Acta.*, 27A, 1381, 1971.
45. **Andrew, E. R. and Wynn, V. T.**, *Proc. R. Soc. Ser. A*, 291, 257, 1966.
46. **Andrew, E. R., Bradbury, A., Eades, R. G., and Jenks, G. J.**, *Nature*, 188, 1096, 1960.
47. **Schmulbach, C. D. and Ahmed, I. Y.**, *J. Chem. Soc. (A)*, 3008, 1968.
48. **Dillon, K. B., Reeve, R. N., and Waddington, T. C.**, *J. Chem. Soc. Dalton Trans.*, 2382, 1977.
49. **Riesel, L. and Kant, M.**, *Z. Anorg. Allg. Chem.*, 531, 73, 1985.
50. **Riesel, L. and Kant, M.**, *Z. Anorg. Allg. Chem.*, 530, 207, 1985.
51. **Olah, G. A. and McFarland, C. W.**, *Inorg. Chem.*, 11, 845, 1972.
52. **Schmutzler, R.**, unpublished work, quoted in **Murray, M. and Schmutzler, R.**, *Z. Chem.*, 7, 241, 1968.
53. **Mizrakh, L. I., Polonskaya, L. Yu., and Babushkina, T. A.**, *J. Gen. Chem. USSR*, 53, 424, 1983.
54. **Skowronska, A., Pakulski, M., Michalski, J., Cooper, D., and Trippett, S.**, *Tetrahedron Lett.*, 321, 1980.
55. **Denney, D. B., Denney, D. Z., and DiMiele, G.**, *Phosphorus Sulfur*, 4, 125, 1978.
56. **Burski, J., Kieszowski, J., Michalski, J., Pakulski, M., and Skowronska, A.**, *Tetrahedron*, 39, 4175, 1983.
57. **Peake, S. C. and Schmutzler, R.**, *Chem. Ind.*, 1482, 1968.
58. **Peake, S. C., Fild, M., Hewson, M. J. C., and Schmutzler, R.**, *Inorg. Chem.*, 10, 2723, 1971.
59. **Binder, H. and Fluck, E.**, *Z. Anorg. Allg. Chem.*, 365, 170, 1969.
60. **Schmidpeter, A. and Brecht, H.**, *Z. Naturforsch.*, 24b, 179, 1969.
61. **Castro, B. and Selve, C.**, *Bull. Soc. Chim. Fr.*, 4368, 1971.
62. **Mark, V.**, *J. Am. Chem. Soc.*, 85, 1884, 1963.
63. **Ramirez, F., Patwardhan, A. T., and Smith, C. P.**, *J. Am. Chem. Soc.*, 87, 4973, 1965.
64. **Ramirez, F.**, *Acc. Chem. Res.*, 1, 168, 1968.
65. **Ramirez, F., Patwardhan, A. T., Kugler, H. J., and Smith, C. P.**, *Tetrahedron*, 24, 2275, 1968.
66. **Ramirez, F., Tasaka, K., Desai, N. B., and Smith, C. P.**, *J. Org. Chem.*, 33, 25, 1968.
67. **Schomburg, D., Betterman, G., Ernst, L., and Schmutzler, R.**, *Angew. Chem. Int. Ed. Engl.*, 11, 975, 1985.
68. **Schmidpeter, A. and Lochschmidt, S.**, *Angew. Chem. Int. Ed. Engl.*, 25, 252, 1986.
69. **Schultz, C. W. and Parry, R. W.**, *Inorg. Chem.*, 15, 3046, 1976.
70. **Schmidpeter, A., Lochschmidt, S., and Sheldrick, W. S.**, *Angew. Chem. Int. Ed. Engl.*, 24, 226, 1985.
71. **Schmidpeter, A., Lochschmidt, S., Burget, G., and Sheldrick, W. S.**, *Phosphorus Sulfur*, 18, 23, 1983.
72. **Tolkmith, H.**, *J. Am. Chem. Soc.*, 85, 3246, 1963.
73. **Schmidpeter, A. and Weingand, C.**, *Angew. Chem. Int. Ed. Engl.*, 8, 615, 1969.
73a. **Witt, D.**, unpublished work.
74. **Sisler, H. H., Jain, S. R., and Brey, W. S.**, *Inorg. Chem.*, 6, 515, 1967.
75. **Hart, W. A. and Sisler, H. H.**, *Inorg. Chem.*, 3, 617, 1964.
76. **Marchenko, A. P., Koidan, G. N., Povolotskii, M. I., and Pinchuk, A. M.**, *J. Gen. Chem. USSR*, 53, 1364, 1983.
77. **Becke-Goehring, M. and Scharf, B.**, *Z. Anorg. Allg. Chem.*, 353, 360, 1967.
78. **Buder, W. and Schmidt, A.**, *Chem. Ber.*, 106, 3812, 1973.
79. **Schlak, O., Schmutzler, R., Schiebel, H-M., Wazeer, M. I. M., and Harris, R. K.**, *J. Chem. Soc. Dalton Trans.*, 2153, 1974.
80. **Olah, G. A. and McFarland, C. W.**, *J. Org. Chem.*, 36, 1374, 1971.
81. **Addison, C. C., Bailey, J. W., Bruce, S. H., Dove, M. F. A., Hibbert, R. C., and Logan, N.**, *Polyhedron*, 2, 651, 1983.
82. **Dillon, K. B. and Waddington, T. C.**, *J. Chem. Soc. (A)*, 1146, 1970.
83. **Finley, J. H., Denney, D. Z., and Denney, D. B.**, *J. Am. Chem. Soc.*, 91, 5826, 1969.
84. **Kolditz, L., Lehmann, K., Wieker, W., and Grimmer, A-R.**, *Z. Anorg. Allg. Chem.*, 360, 259, 1968.
85. **Ruppert, I.**, *Z. Anorg. Allg. Chem.*, 477, 59, 1981.

86. **Murray, M., Schmutzler, R., Gründemann, E., and Teichmann, H.,** *J. Chem. Soc. (B)*, 1714, 1971.
87. **Teichmann, H., Jatkowski, M., and Hilgetag, G.,** *Angew. Chem. Int. Ed. Engl.*, 6, 372, 1967.
88. **Finley, J. H. and Denney, D. B.,** *J. Am. Chem. Soc.*, 92, 362, 1970.
89. **Henrick, K., Hudson, H. R., Matthews, R. W., McPartlin, E. M., Powroznyk, L., and Shode, O. O.,** *Proc. X Int. Conf. Phosphorus Chem.*, Appel, R., Knoll, F., and Ruppert, I., Eds., *Phosphorus Sulfur*, 30, 157, 1987.
90. **Dillon, K. B. and Nisbet, M. P.,** unpublished work; **Nisbet, M. P.,** Ph.D. thesis, Durham, 1976.
91. **Nesterov, L. V., Mutalapova, R. I., Salikhov, S. G., and Loginiva, E. I.,** *Izv. Akad. Nauk SSSR Ser. Khim.*, 414, 1971.
92. **Gloede, J.,** *Z. Anorg. Allg. Chem.*, 484, 231, 1982.
93. **Gloede, J. and Waschke, R.,** *Phosphorus Sulfur*, 27, 341, 1986.
94. **Vande Griend, L. J. and Verkade, J. G.,** *J. Am. Chem. Soc.*, 97, 5958, 1975.
95. **McFarlane, W. and White, R. F. M.,** *Chem. Commun.*, 744, 1969.
96. **Hudson, H. R. and Roberts, J. C.,** *J. Chem. Soc. Perkin Trans. II*, 1575, 1974.
97. **Weiss, R., Vande Griend, L. J., and Verkade, J. G.,** *J. Org. Chem.*, 44, 1860, 1979.
98. **Vande Griend, L. J., Verkade, J. G., Pennings, J. F. M., and Buck, H. M.,** *J. Am. Chem. Soc.*, 99, 2459, 1977.
99. **Hudson, H. R., Kow, A., and Roberts, J. C.,** *Phosphorus Sulfur*, 19, 375, 1984.
100. **Sheldrick, G. M.,** *Trans. Faraday Soc.*, 63, 1077, 1967.
101. **Olah, G. A. and McFarland, C. W.,** *J. Org. Chem.*, 34, 1832, 1969.
102. **Deng, R. M. K. and Dillon, K. B.,** *J. Chem. Soc. Dalton Trans.*, 1443, 1986.
103. **Ali, R. and Dillon, K. B.,** unpublished work; **Ali, R.,** Ph.D. thesis, Durham, 1987.
104. **Deng, R. M. K. and Dillon, K. B.,** *J. Chem. Soc. Dalton Trans.*, 1917, 1984.
105. **Dillon, K. B., Platt, A. W. G., and Waddington, T. C.,** *J. Chem. Soc. Dalton Trans.*, 2292, 1981.
106. **Deng, R. M. K. and Dillon, K. B.,** unpublished work; **Deng, R. M. K.,** Ph.D. thesis, Durham, 1981.
107. **Fluck, E. and Reinisch, R. M.,** *Z. Anorg. Allg. Chem.*, 328, 165, 1964.
108. **Dillon, K. B., Nisbet, M. P., and Waddington, T. C.,** *J. Chem. Soc. Dalton Trans.*, 465, 1982.
109. **Dmitriev, V. I., Kozlov, É. S., Timokhin, V. B., Dubenko, L. G., and Kalabina, A. V.,** *J. Gen. Chem. USSR*, 50, 1799, 1980.
110. **Dillon, K. B., Nisbet, M. P., and Waddington, T. C.,** *J. Chem. Soc. Dalton Trans.*, 212, 1981.
111. **Fluck, E.,** private communication quoted in **Nöth, H. and Meinel, L.,** *Z. Anorg. Allg. Chem.*, 349, 225, 1967.
112. **Brel', V. K., Ionin, B. I., and Petrov, A. A.,** *J. Gen. Chem. USSR*, 53, 206, 1983.
113. **Dillon, K. B., Lynch, R. J., Reeve, R. N., and Waddington, T. C.,** *J. Chem. Soc. Dalton Trans.*, 1243, 1976.
114. **Deng, R. M. K. and Dillon, K. B.,** *J. Chem. Soc. Dalton Trans.*, 1911, 1984.
115. **Hoffmann, F. W., Simmons, T. C., and Glunz, L. J., III,** *J. Am. Chem. Soc.*, 79, 3570, 1957.
116. **Latscha, H-P., Weber, W., and Becke-Goehring, M.,** *Z. Anorg. Allg. Chem.*, 367, 40, 1969.
117. **Schmutzler, R.,** *J. Chem. Soc.*, 5630, 1965.
118. **Reddy, G. S. and Schmutzler, R.,** *Inorg. Chem.*, 5, 164, 1966.
119. **Schmutzler, R.,** *Inorg. Chem.*, 7, 1327, 1968.
120. **Stadelmann, W., Stelzer, O., and Schmutzler, R.,** *Z. Anorg. Allg. Chem.*, 385, 142, 1971.
121. **Schmutzler, R.,** *J. Am. Chem. Soc.*, 86, 4500, 1964.
122. **Nesterov, L. V., Kessel, A. Ya., Samitov, Yu. Yu. and Musina, A. A.,** *Dokl. Akad. Nauk SSSR*, 180, 116, 1968.
123. **Hudson, H. R., Powroznyk, L., and Qureshi, A. R.,** *Phosphorus Sulfur*, 25, 289, 1985.
124. **Weferling, N., Schmutzler, R., and Sheldrick, W. S.,** *Liebigs. Ann. Chem.*, 167, 1982.
125. **Schmidpeter, A. and Weingand, C.,** *Z. Naturforsch.*, 26b, 187, 1969.
126. **Van Wazer, J. R., Callis, C. F., Shoolery, J. N., and Jones, R. C.,** *J. Am. Chem. Soc.*, 78, 5715, 1956.

126a. **Roesky, H. W. and Sidiropoulos, G.,** *Angew. Chem. Int. Ed. Engl.*, 15, 693, 1976.

127. **Mark, V.,** unpublished work, quoted in **Mark, V., Dungan, C. H., Crutchfield, M. M., and Van Wazer, J. R.,** *Top. Phosphorus Chem.*, 5, 227, 1967.
128. **Ramirez, F., Madan, O. P., and Smith, C. P.,** *Tetrahedron*, 22, 567, 1966.
129. **Kroshefsky, R. D. and Verkade, J. G.,** *Phosphorus Sulfur*, 6, 397, 1979.
130. **Kanamueller, J. M. and Sisler, H. H.,** *Inorg. Chem.*, 6, 1765, 1967.
131. **Edwards, R. C.,** Ph.D. thesis, University of London, 1982.
132. **Bertrand, R. D., Ogilvie, F. B., and Verkade, J. G.,** *J. Am. Chem. Soc.*, 92, 1908, 1970.
133. **Cohen, J. S.,** *J. Am. Chem. Soc.*, 89, 2543, 1967.
134. **Colle, K. S. and Lewis, E. S.,** *J. Org. Chem.*, 43, 571, 1978.
135. **Hudson, H. R., Rees, R. G., and Weekes, J. E.,** *J. Chem. Soc. Perkin Trans. I*, 982, 1974.
136. **Lewis, E. S., Walker, B. J., and Ziurys, L. M.,** *J. Chem. Soc. Chem. Commun.*, 424, 1978.

137. **Fluck, E. and Lorenz, J.,** *Z. Naturforsch.*, 22b, 1095, 1967.
138. **Phillips, D. I., Szele, I., and Westheimer, F. H.,** *J. Am. Chem. Soc.*, 98, 184, 1976.
139. **Petneházy, I., Szakál, G., Töke, L., Hudson, H. R., Powroznyk, L., and Cooksey, C. J.,** *Tetrahedron*, 39, 4229, 1983.
140. **Milbrath, D. S. and Verkade, J. G.,** *J. Am. Chem. Soc.*, 99, 6607, 1977.
141. **Bertrand, R. D., Allison, D. A., and Verkade, J. G.,** *J. Am. Chem. Soc.*, 92, 71, 1970.
142. **Cowley, A. H., Kilduff, J. E., Norman, N. C., and Pakulski, M.,** *J. Am. Chem. Soc.*, 105, 4845, 1983.
143. **Schmidpeter, A. and Brecht, H.,** *Angew. Chem. Int. Ed. Engl.*, 6, 945, 1967.
144. **Schmidpeter, A. and Brecht, H.,** *Z. Naturforsch.*, 23b, 1529, 1968.
145. **Cowley, A. H., Kemp, R. A., Larch, T. G., Norman, N. C., and Stewart, C. A.,** *J. Am. Chem. Soc.*, 105, 7444, 1983.
146. **Becke-Goehring, M. and Haubold, W.,** *Z. Anorg. Allg. Chem.*, 338, 305, 1965.
147. **Fluck, E.,** *Z. Naturforsch.*, 20b, 505, 1965.
148. **Schmidpeter, A. and Groeger, H.,** *Chem. Ber.*, 100, 3979, 1967.
149. **Jain, R. S., Lawson, H. F., and Quin, L. D.,** *J. Org. Chem.*, 43, 108, 1978.
150. **Denney, D. B., Denney, D. Z., and Chang, B. C.,** *J. Am. Chem. Soc.*, 90, 6332, 1968.
151. **Brownstein, M. and Schmutzler, R.,** *J. Chem. Soc. Chem. Commun.*, 278, 1975.
152. **Schmidpeter, A., Brecht, H., and Groeger, H.,** *Chem. Ber.*, 100, 3063, 1967.
153. **Schmidpeter, A. and Stoll, K.,** *Angew. Chem. Int. Ed. Engl.*, 10, 131, 1971.
154. **Fluck, E. and Issleib, K.,** *Chem. Ber.*, 98, 2674, 1965.
155. **Nöth, H. and Meinel, L.,** *Z. Anorg. Allg. Chem.*, 349, 225, 1967.
156. **Schmidpeter, A. and Groeger, H.,** *Z. Anorg. Allg. Chem.*, 345, 106, 1966.
157. **Schmidpeter, A., Brecht, H., and Ebeling, J.,** *Chem. Ber.*, 101, 3902, 1968.
158. **Li, B-L., Engenito, J. S., Jr., Neilson, R. H., and Wisian-Neilson, P.,** *Inorg. Chem.*, 22, 575, 1983.
159. **Clemens, D. F. and Sisler, H. H.,** *Inorg. Chem.*, 4, 1222, 1965.
160. **Soottoo, C. K. and Baxter, S. G.,** *J. Am. Chem. Soc.*, 105, 7443, 1983.
161. **Sisler, H. H. and Frazier, S. E.,** *Inorg. Chem.*, 4, 1204, 1965.
162. **Bermann, M. and Utvary, K.,** *J. Inorg. Nucl. Chem.*, 31, 271, 1969.
163. **Schmidpeter, A. and Ebeling, J.,** *Chem. Ber.*, 101, 3883, 1968.
164. **Sisler, H. H., Ahuja, H. S., and Smith, N. L.,** *Inorg. Chem.*, 1, 84, 1962.
165. **Schmidpeter, A. and Ebeling, J.,** *Angew. Chem. Int. Ed. Engl.*, 6, 87, 1967.
166. **Razumov, A. I., Liorber, B. G., Zykova, T. V., and Bambushek, I. Ya.,** *Zh. Obshch. Khim.*, 40, 2009, 1970.
167. **Hudson, H. R., Kow, A., and Roberts, J. C.,** *J. Chem. Soc. Perkin Trans. II*, 1363, 1983.
168. **Seel, F. and Bassler, H-J.,** *Z. Anorg. Allg., Chem.*, 418, 263, 1975.
169. **Stine, W. R.,** Ph.D. thesis, Syracuse, 1967.
170. **Quin, L. D. and Barket, T. P.,** *J. Am. Chem. Soc.*, 92, 4303, 1970.
171. **Quin, L. D., Gratz, J. P., and Barket, T. P.,** *J. Org. Chem.*, 33, 1034, 1968.
172. **Dillon, K. B. and Waddington, T. C.,** *Nature (Phys. Sci.)*, 230, 158, 1971.
173. **Dillon, K. B., Reeve, R. N., and Waddington, T. C.,** *J. Inorg. Nucl. Chem.*, 38, 1439, 1976.
174. **Axtell, D. D. and Yoke, J. T.,** *Inorg. Chem.*, 12, 1265, 1973.
175. **Wiley, G. A. and Stine, W. R.,** *Tetrahedron Lett.*, 2321, 1967.
176. **Quin, L. D., Belmont, S. E., Mathey, F., and Charrier, C.,** *J. Chem. Soc. Perkin Trans. II*, 629, 1986.
177. **Fongers, K. S., Hogeveen, H., and Kingma, R. F.,** *Tetrahedron Lett.*, 24, 643, 1983.
178. **Fongers, K. S., Hogeveen, H., and Kingma, R. F.,** *Tetrahedron Lett.*, 24, 1423, 1983.
179. **Quin, L. D. and Stocks, R. C.,** *Phosphorus Sulfur*, 3, 151, 1977.
180. **Timokhin, B. V., Dudnikova, V. N., Kron, V. A., and Glukhikh, V. I.,** *Zh. Org. Khim.*, (Engl. transl.), 15, 337, 1979.
181. **Latscha, H.-P.,** *Z. Naturforsch.*, 23b, 139, 1968.
182. **Dillon, K. B., Lynch, R. J., Reeve, R. N., and Waddington, T. C.,** *J. Inorg. Nucl. Chem.*, 36, 815, 1974.
183. **Bartsch, R., Stelzer, O., and Schmutzler, R.,** *Z. Naturforsch.*, 37b, 267, 1982.
184. **Bartsch, R., Weiss, J-V., and Schmutzler, R.,** *Z. Anorg. Allg. Chem.*, 537, 53, 1986.
185. **Bartsch, R.,** Dissertation, Braunschweig, 1980 (quoted in Ref. 184).
186. **Rave, T. W. and Hays, H. R.,** *J. Org. Chem.*, 31, 2894, 1968.
187. **Lancaster, J. E.,** unpublished work, quoted in **Mark, V., Dungan, C. H., Crutchfield, M. M., and Van Wazer, J. R.,** *Top. Phosphorus Chem.*, 5, 227, 1967.
188. **Bartsch, R., Stelzer, O., and Schmutzler, R.,** *Synthesis*, 326, 1982.
189. **Dakternieks, D. and Röschenthaler, G-V.,** *Z. Anorg. Allg. Chem.*, 504, 135, 1983.
190. **Denney, D. B., Denney, D. Z., and Wilson, L. A.,** *Tetrahedron Lett.*, 85, 1968.
191. **Schmutzler, R., Schomburg, D., Bartsch, R., and Stelzer, O.,** *Z. Naturforsch.*, 39b, 1177, 1984.

192. **Quin, L. D. and Spence, S. C.,** *Tetrahedron Lett.*, 23, 2529, 1982.
192a. **Denney, D. B., Denney, D. Z., Hall, C. D., and Marsi, K. L.,** *J. Am. Chem. Soc.*, 94, 245, 1972.
193. **Partos, R. D. and Speziale, A. J.,** *J. Am. Chem. Soc.*, 87, 5068, 1965.
194. **Zhmurova, I. N., Yurchenko, V. G., and Pinchuk, A. M.,** *J. Gen. Chem. USSR*, 53, 1360, 1983.
195. **Schmidpeter, A., Lochschmidt, S., Karaghiosoff, K., and Sheldrick, W. S.,** *J. Chem. Soc. Chem. Commun.*, 1447, 1985.
196. **Spangenberg, S. F. and Sisler, H. H.,** *Inorg. Chem.*, 8, 1006, 1969.
197. **Fluck, E. and Binder, H.,** *Angew. Chem. Int. Ed. Engl.*, 5, 666, 1966.
198. **Fluck, E. and Binder, H.,** *Z. Anorg. Allg. Chem.*, 354, 113, 1967.
199. **Lochschmidt, S. and Schmidpeter, A.,** *Z. Naturforsch.*, 40b, 765, 1985.
200. **Schmidpeter, A., Lochschmidt, S., and Sheldrick, W. S.,** *Angew. Chem. Int. Ed. Engl.*, 21, 63, 1982.
201. **Cowley, A. H. and Mehrotra, S. K.,** *J. Am. Chem. Soc.*, 105, 2074, 1983.
202. **Cowley, A. H. and Kemp, R. A.,** *Inorg. Chem.*, 22, 547, 1983.
203. **Moedritzer, K., Maier, L., and Groenweghe, L. C. D.,** *J. Chem. Eng. Data*, 7, 307, 1962.
204. **Grim, S. O. and McFarlane, W.,** *Can. J. Chem.*, 46, 2071, 1968.
205. **McFarlane, W.,** *Proc. R. Soc. A*, 306, 185, 1968.
206. **McFarlane, W.,** *Chem. Commun.*, 58, 1967.
207. **Eaborn, C., Retta, N., and Smith, J. D.,** *J. Chem. Soc. Dalton Trans.*, 905, 1983.
208. **Dillon, K. B., Waddington, T. C., and Younger, D.,** *J. Chem. Soc. Dalton Trans.*, 790, 1975.
209. **Vande Griend, L. J., Verkade, J. G., Jongsma, C., and Bickelhaupt, F.,** *Phosphorus*, 6, 131, 1976.
209b. **Suess-Fink, G.,** *Chem. Ber.*, 119, 2393, 1986.
210. **Frank, A. W.,** *Phosphorus Sulfur*, 5, 19, 1978.
211. **Frank, A. W. and Drake, G. L., Jr.,** *J. Org. Chem.*, 42, 4040, 1977.
212. **Granzow, A.,** *J. Am. Chem. Soc.*, 99, 2648, 1977.
213. **Crutchfield, M. M., Dungan, C. H., and van Wazer, J. R.,** in *Topics in Phosphorus Chemistry*, Vol. 5, Grayson, M. and Griffiths, E. J., Eds., Interscience, New York, 1968.
214. **Maeir, L.,** *Helv. Chim. Acta*, 49, 2458, 1966.
215. **Quin, L. D. and Breen, J. J.,** *Org. Magn. Reson.*, 5, 17, 1973.
216. **Schmidbaur, H. and Zybill, C. E.,** *Chem. Ber.*, 114, 3589, 1981.
217. **Littlefield, L. B. and Quin, L. D.,** *Org. Magn. Reson.*, 12, 199, 1979.
218. **Quin, L. D., Gallagher, M. J., Cunkle, G. T., and Chestnut, D. B.,** *J. Am. Chem. Soc.*, 102, 3136, 1980.
219. **Ramirez, F.,** *Acc. Chem. Res.*, 1, 168, 1968.
220. **Hays, H. R.,** *J. Org. Chem.*, 31, 3817, 1966.
221. **Quin, L. D., Caster, K. C., Kisalus, J. C., and Mesch, K.,** *J. Am. Chem. Soc.*, 106, 7021, 1984.
222. **Mavel, G.,** in *Annual Reports on N.M.R. Spectroscopy*, Vol. 5B, Mooney, E. F., Ed., Academic Press, New York, 1973.
223. **Fluck, E. and Lorenz, J.,** *Z. Naturforsch. B*, 22, 1095, 1967.
224. **Cristan, H. J., Labaudiniere, L., and Cristol, H.,** *Phosphorus Sulfur*, 15, 359, 1983.
225. **Schmidbaur, H. and Schier, A.,** *Chem. Ber.*, 114, 3385, 1981.
226. **Grim, S. O., McFarlane, W., Davidoff, E. F., and Marks, T. J.,** *J. Phys. Chem.*, 70, 581, 1966.
227. **Hansen, P-E.,** *J. Chem. Soc. Perkin Trans. I*, 1627, 1980.
228. **Riess, J. G., Van Wazer, J. R., and Letcher, J. H.,** *J. Phys. Chem.*, 71, 1925, 1967.
229. **Bowmaker, G. A., Dorzbach, C., and Schmidaur, H.,** *Z. Naturforsch. B*, 39, 618, 1984.
230. **Dillon, K. B. and Waddington, T. C.,** *Spectrochim. Acta*, 27A, 1381, 1971.
231. **Swartz, W. E., Jr. and Hercules, D. M.,** *Anal. Chem.*, 43, 1066, 1971; **Bestmann, H. J. and Arenz, T.,** *Tetrahedron Lett.*, 27, 1995, 1986.
232. **Brovko, V. S., Skvortsov, N. K., and Ivanov, A. Yi.,** *J. Gen. Chem. (USSR)*, 53, 1648, 1983.
233. **Ramirez, F., Madan, O. P., and Smith, C. P.,** *Tetrahedron*, 29, 3721, 1966.
234. **Hanstock, C. C. and Tebby, J. C.,** *Phosphorus Sulfur*, 15, 239, 1983.
235. **Ramirez, F., Madan, O. P., and Smith, C. P.,** *Tetrahedron*, 22, 567, 1966.
236. **Fyfe, C. A. and Zbozny, M.,** *Can. J. Chem.*, 50, 1713, 1972.
237. **Schmidbaur, H., Blaschke, G., Zimmer-Gasser, B., and Schubert, U.,** *Chem. Ber.*, 113, 1612, 1980.
238. **Dahl, O., Henriksen, U., and Trebbien, C.,** *Acta Chem. Scand. B*, 37, 639, 1983.
239. **Foss, V. L., Veits, Yu. A., and Lutsenko, I. F.,** *J. Gen. Chem. (USSR)*, 48, 1558, 1978.
240. **Quin, L. D., Mesch, K. A., Bodalski, R., and Pietrusiewicz, K. M.,** *Org. Magn. Reson.*, 20, 83, 1982.
241. **El-Deek, M., Macdonell, G. D., Venkataramu, S. D., and Berlin, K. D.,** *J. Org. Chem.*, 41, 1403, 1976.
242. **Buckler, S. A. and Epstein, M.,** *J. Org. Chem.*, 27, 1090, 1962.
243. **Purdam, W. R. and Berlin, K. D.,** *J. Org. Chem.*, 40, 2801, 1975.
244. **Quin, L. D., Mesch, K. A., and Orton, W. L.,** *Phosphorus Sulfur*, 12, 161, 1982.
245. **Schmidbaur, H., Costa, T., and Milewski, B.,** *Chem. Ber.*, 114, 1428, 1981.

246. **Quin, L. D., Middlemas, E. D., and Rao, N. S.,** *J. Org. Chem.*, 47, 905, 1982.
247. **Frank, A. W.,** *Phosphorus Sulfur*, 10, 147, 1981.
248. **Weichmann, H.,** *J. Organomet. Chem.*, 262, 279, 1984.
249. **Uson, R., Fonies, J. R., Navarro, M. A. Uson, Garcia, M. P., and Welch, A. J.,** *J. Chem. Soc. Dalton Trans.*, 345, 1984.
250. **Appel, R. and Erbelding, G.,** *Tetrahedron Lett.*, 2689, 1978.
251. **Szpala, A.,** Ph.D. Thesis, North Staffordshire Polytechnic, 1980.
252. **Kargin, M., Nikitin, E. V., Parakin, O. V., Romanov, G. V., and Pudovik, A. N.,** *Phosphorus Sulfur*, 8, 55, 1980.
253. **Albright, T. A., Freeman, W. J., and Schweizer, E. E.,** *J. Am. Chem. Soc.*, 97, 2946, 1975.
254. **Cristan, H-J., Duc, G., and Cristol, H.,** *Synthesis*, 374, 1983.
255. **Al'fonsov, V. A., Nizamor, I. S., Batyeva, E. S. and Pudovik, A. N.,** *J. Gen. Chem. (USSR)*, 53, 1541, 1983.
256. **Tsou, T. T. and Kochi, J. K.,** *J. Am. Chem. Soc.*, 101, 7547, 1979.
257. **Goetz, H., Juds, H., and Marschner, F.,** *Phosphorus Sulfur*, 1, 217, 1972.
258. **Quin, L. D., Borleske, S. G., and Engel, J. F.,** *J. Org. Chem.*, 38, 1858, 1973.
259. **Rampal, J. B., Macdonell, G. D., Edasery, J. P., Berlin, K. D., Rahman, A., Van der Helm, D., and Pietrusiewicz, K. M.,** *J. Org. Chem.*, 46, 1156, 1981.
260. **Marsi, K. L.,** *J. Am. Chem. Soc.*, 93, 6341, 1971.
261. **Meeuwissen, M. J., Van der Knaap, Th. A., and Bickelhaupt, F.,** *Tetrahedron*, 39, 4225, 1983.
262. **Eaborn, C., Retta, N., and Smith, J. D.,** *J. Chem. Soc. Dalton Trans.*, 905, 1983.
263. **Appel, R., Scholer, H-F., and Wihler, H-D.,** *Chem. Ber.*, 112, 462, 1979.
264. **Appel, R. and Huppertz, M.,** *Z. Anorg. Allg. Chem.*, 459, 7, 1979.
265. **Schmidbaur, H. and Schnatterer, S.,** *Chem. Ber.*, 116, 1947, 1983.
266. **Schmidbaur, H. and Deschler, U.,** *Chem. Ber.*, 116, 1386, 1983.
267. **Gurusamy, N., Berlin, K. D., Van der Helm, D., and Hossain, M. B.,** *J. Am. Chem. Soc.*, 104, 3107, 1982.
268. **Wilson, I. F. and Teddy, J. C.,** *J. Chem. Soc. Perkin Trans. I*, 2713, 1972.
269. **Bowmaker, G. A., Herr, R., and Schmidbaur, H.,** *Chem. Ber.*, 116, 3567, 1983.
270. **Weichmann, H., Quell, G., and Tzchach, A.,** *Z. Anorg. Allg. Chem.*, 462, 7, 1980.
271. **Dilbeck, G. A., Morris, D. L., and Berlin, K. D.,** *J. Org. Chem.*, 40, 1150, 1975.
272. **Wohlleben, A. and Schmidbaur, H.,** *Angew. Chem. Int. Ed. Engl.*, 16, 417, 1977.
273. **Ramirez, F., Rhum, D., and Smith, C. P.,** *Tetrahedron*, 21, 1941, 1965.
274. **Issleib, K. and Rieschel, R.,** *Chem. Ber.*, 98, 2086, 1965.
275. **Dahl, O.,** *Acta Chem. Scand. (B)*. 31, 427, 1977.
276. **Macdonell, G. D., Berlin, K. D., Ealick, S. E., and Van der Helm, D.,** *Phosphorus Sulfur*, 4, 187, 1978.
277. **Bernard, M., Ford, W. T., and Nelson, E. C.,** *J. Org. Chem.*, 48, 3164, 1983.
278. **Holy, N., Deschler, U., and Schmidbaur, H.,** *Chem. Ber.*, 115, 1379, 1982.
279. **Skolimowski, J., Skowronski, R., and Simulty, M.,** *Phosphorus Sulfur*, 19, 159, 1984.
280. **Breslow, R. and Deuring, L. A.,** *Tetrahedron Lett.*, 25, 1345, 1984.
281. **Schmidbaur, H., Deschler, U., and Seyferth, D.,** *Z. Naturforsch. B*, 37, 950, 1982.
282. **Wuin, L. D., Borleske, S. G., and Engel, J. F.,** *J. Org. Chem.*, 38, 1954, 1973.
283. **Allen, D. W. and Tebby, J. C.,** *J. Chem. Soc. (B)*, 1527, 1970.
284. **Mastryukova, T. A., Genkina, G. K., Kalyanova, R. M., Shcherkina, T. M., Petrovskii, P. V. and Kabachnik, M. I.,** *J. Gen. Chem. (USSR)*, 48, 233, 1978.
285. **Mastryukova, T. A., Suerbaev, Kh. A., Mastrosov, E. I., Petrovskii, P. V., and Kabachnik, M. I.,** *J. Gen. Chem. (USSR)*, 43, 2593, 1973.
286. **Aguiar, A. M. and Aguiar, H.,** *J. Am. Chem. Soc.*, 88, 4090, 1966.
287. **Chatta, M. S. and Aguiar, A. M.,** *J. Org. Chem.*, 38, 161, 1973.
288. **Vinceus, M., Moron, J. T. G., Pasqualini, R., and Vidal, M.,** *Tetrahedron Lett.*, 28, 1259, 1987.
289. **Venkataramu, S. D., El-Deek, M., and Berlin, K. D.,** *Tetrahedron Lett.*, 3365, 1976.
290. **Cristol, H., Cristau, H-J., Fallouh, F., and Hullot, P.,** *Tetrahedron Lett.*, 2591, 1979.
291. **Appel, R. and Morbach, W.,** *Synthesis*, 699, 1977.
292. **Allen, D. W., Miller, I. T., and Tebby, J. C.,** *Tetrahedron Lett.*, 745, 1968.
293. **Gallagher, M. J.,** *Aust. J. Chem.*, 21, 1197, 1968.
294. **Albright, T. A. and Schweizer, E. E.,** *J. Org. Chem.*, 41, 1168, 1976.
295. **Schmidbaur, H., Herr, R., and Zybill, C. E.,** *Chem. Ber.*, 117, 3374, 1984.
296. **Schmidbaur, H., Zybill, C. E., Kruger, C. C., and Kraus, H-J.,** *Chem. Ber.*, 116, 1955, 1983.
297. **Schmidbaur, H., Strunk, S., and Zybill, C. E.,** *Chem. Ber.*, 116, 3559, 1983.
298. **Allen, D. W., Hutley, B. G., and Mellor, M. T. J.,** *J.Chem. Soc., Perkin Trans. II*, 63, 1972.
299. **Schmidbaur, H., Herr, R., and Riede, J.,** *Angew. Chem. Int. Ed. Engl.*, 23, 247, 1984.

300. **Duncan, M. and Gallagher, M. J.,** *Org. Magn. Reson.*, 15, 37, 1981.
301. **Grim, S. O. and Yankowsky, A. W.,** *J. Org. Chem.*, 42, 1236, 1977.
302. **Graves, G. E. and Van Wazer, J. R.,** *J. Inorg. Nuc. Chem.*, 39, 1101, 1977.
303. **Crews, P.,** *J. Am. Chem. Soc.*, 90, 2961, 1968; **Bestmann, H. J., Liberda, H. G., and Snyder, J. P.,** *J. Am. Chem. Soc.*, 90, 2963, 1968.
304. **Hellwinkel, D. and Schenk, W.,** *Angew. Chem. Int. Ed. Engl.*, 8, 987, 1969.
305. **Denney, D. B. and Smith, L. C.,** *J. Org. Chem.*, 27, 3404, 1962.
306. **Narayanan, K. S. and Berlin, K. D.,** *J. Org. Chem.*, 45, 2240, 1980.
307. **Cristau, H-J., Bottaro, D. F., Plenat, F., Pietrasanta, F., and Cristol, H.,** *Phosphorus Sulfur*, 14, 63, 1983.
308. **Birum, G. H. and Matthews, C. N.,** *J. Org. Chem.*, 32, 3554, 1967.
309. **Nesmeyanov, N. A., Berman, S. T., Petrovsky, P. V., Lutsenko, A. I., and Reutov, O. A.,** *J. Organometallic Chem.*, 129, 41, 1977.
310. **Nesterov, L. V., Aleksandrova, N. A., Temyachev, I. D., Musina, A. A., and Gainullina, R. G.,** *J. Gen. Chem. (USSR).*, 47, 1161, 1977.
311. **Brittain, J. M. and Jones, R. A.,** *Tetrahedron*, 35, 1139, 1979.
312. **Allen, D. W. and Hutley, B. G.,** *J. Chem. Soc. Perkin Trans. I*, 67, 1972.
313. **Schweizer, E. E., De Voe Goff, S., and Murray, W. P.,** *J. Org. Chem.*, 42, 200, 1977.
314. **Cristau, H. J., Voss, J-P., and Cristol, H.,** *Synthesis*, 538, 1979.
315. **Schweizer, E. E. and Lee, K-J.,** *J. Org. Chem.*, 47, 2768, 1982.
316. **Shaw, M. A., Tebby, J. C., Ronayne, J., and Williams, D. H.,** *J. Chem. Soc. (C).*, 944, 1967.
317. **Albright, T. A., Freemann, W. J., and Schweizer, E. E.,** *J. Am. Chem. Soc.*, 97, 2942, 1975.
318. **Birum, G. H. and Matthews, C. N.,** *J. Am. Chem. Soc.*, 90, 3842, 1968.
319. **Aguiar, A. M., Hansen, K. C., and Reddy, G. S.,** *J. Am. Chem. Soc.*, 89, 3067, 1967.
320. **Skolimowski, J. and Simalty, M.,** *Tetrahedron Lett.*, 21, 3037, 1980.
321. **Appel, R., Knoll, F., and Wihler, H-D.,** *Angew. Chem. Int. Ed. Engl.*, 16, 402, 1977.
322. **Birum, G. H. and Matthews, C. N.,** *J. Am. Chem. Soc.*, 88, 4198, 1966.
323. **Appel, R. and Wihler, H-D.,** *Chem. Ber.*, 111, 2054, 1978.
324. **Lobanov, O. P., Martynynuk, A. P., and Drach, B. S.,** *J. Gen. Chem. (USSR)*, 50, 1816, 1980.
325. **Weiss, R., Wolf, H., Schubert, U., and Clark, T.,** *J. Am. Chem. Soc.*, 103, 6142, 1981.
326. **Regitz, M., Tawfik, A. E-R. M., and Heydt, H.,** *Annalen*, 1865, 1981.
327. **Zhmurova, I. N., Kosinskaya, I. M., and Pinchuk, A. M.,** *J. Gen. Chem. (USSR)*, 51, 1304, 1981.
328. **Allen, D. W., Nowell, I. W., March, L. A., and Taylor, B. F.,** *J. Chem. Soc., Perkin Trans. I*, 2523, 1984.
329. **Bestmann, H. J. and Oechsner, H.,** *Z. Naturforsch. B*, 38, 861, 1983.
330. **Matthews, C. N., Driscoll, J. S., and Birum, G. H.,** *J. Chem. Soc. Chem. Commun.*, 736, 1966.
331. **Birum, G. H. and Matthews, C. N.,** *Chem. Ind. (London).*, 653, 1968.
332. **Pariza, R. J. and Fuchs, P. L.,** *J. Org. Chem.*, 48, 2304, 1983.
333. **Bestmann, H. J., Engler, R. H., Hartung, H., and Roth, K.,** *Chem. Ber.*, 112, 28, 1979.
334. **Plenat, F.,** *Tetrahedron Lett.*, 22, 4705, 1981.
335. **Bestmann, H. J., Schmid, G., Oechsner, H., and Ermann, P.,** *Chem. Ber.*, 112, 1561, 1984.
336. **Saheh, G., Minami, T., Ohshiro, Y., and Agawa, T.,** *Chem. Ber.*, 112, 355, 1979.
337. **Schmidt, A. H., Aimene, A., and Hoch, M.,** *Synthesis*, 754, 1984.
338. **Brown, S. J. and Clark, J. H.,** *J. Chem. Soc. Chem. Commun.*, 1256, 1983.
339. **Weiss, R., Priesner, C., and Wolf, H.,** *Angew. Chem. Int. Ed. Engl.*, 18, 472, 1979.
340. **Zaslona, A. T. and Hall, C. D.,** *J. Chem. Soc. Perkin Trans. I.*, 3059, 1981.
341. **Platonov, A. Yu., Maiorova, E. D., Akimova, G. S., and Chistokletov, V. N.,** *J. Gen. Chem. (USSR)*, 52, 395, 1982.
342. **Marszak, M. B. and Simalty, M.,** *Tetrahedron*, 35, 775, 1972.
343. **Hellwinkel, D.,** *Chem. Ber.*, 98, 576, 1965.
344. **Bestmann, H. J. and Kisielowski, L.,** *Chem. Ber.*, 116, 1320, 1983.
345. **Bestmann, H. J. and Kloeters, W.,** *Tetrahedron Lett.*, 3343, 1978.
346. **Bestmann, H. J. and Kloeters, W.,** *Angew. Chem. Int. Ed. Engl.*, 16, 45, 1977.

Chapter 9

^{31}PNMR DATA OF FOUR COORDINATE PHOSPHORUS COMPOUNDS CONTAINING A P=Ch BOND BUT NO BONDS TO H OR GROUP IV ATOMS

The NMR data are presented in Table H. It is divided into four parts: Section H1 contains data of derivatives of phosphoric acid, Section H2 contains data for derivatives of thio-, seleno-, and telluro-phosphoric acids, Section H3 contains data for biologically important phosphates and related compounds, and Section H4 contains data of naturally occurring phosphonates. Supplementary data for Sections H1 and H2 are available at the end of these sections.

SECTIONS H1 AND H2: DERIVATIVES OF PHOSPHORIC ACID AND ITS THIO, SELENO, AND TELLURO COMPOUNDS

Compiled and presented by
Ronald S. Edmundson

NMR data were also donated by
H. W. Roesky, Universität Göttingen, FRG
C. D. Hall, King's College London, U.K.
J. P. Dutasta, Université Scientifique et Médicale de Grenoble, France

The title compounds are represented largely by the esters, halides, anhydrides, and amides of phosphoric acid, together with their thio, seleno, and telluro analogues.

Derivatives of phosphoric acid have chemical shifts within the range ca. 40 to −35 ppm. The most important cause of the variation in the chemical shift is that of change in the atoms bonded directly to phosphorus. For example, progressive deshielding (move to more positive shift values) is observed when ester oxygens are successively replaced by amide nitrogens. More extensive deshielding results when oxygen is replaced by sulfur or selenium but the extent of the deshielding depends on the environment of the sulfur (or selenium) atom. Thus, for dicoordinate sulfur, the deshielding relative to the corresponding oxygen compound is 35 to 40 ppm, but the downfield perturbation induced by a thiophosphoryl group is often 70 to 75 ppm.[1-3]

The extreme upfield portion of the spectral range characterizes the compounds $(RO)_2P(X)Y$, X = 0, and Y = Br or F (δ_P −5 to −23 ppm) and $ROP(X)Y_2$, X = 0, and Y = F (δ_P −20 to −30 ppm), with the silyl esters being even more highly shielded. For many simple esters of the general types indicated, the ethyl esters are shielded relative to the methyl compounds by 1 to 6 ppm for X = 0, Y = halogen, N or O functions, and for X = S, Y = halogen, SH, SMe, STms, S^-, O, or N functions. There are exceptions to this general observation, for example, the esters $(EtO)_2P(S)XR$ where X = S or Se. Extension of a normal alkyl carbon chain beyond three or four carbon atoms in the esters $(RO)_3P(X)$ otherwise produces a negligible effect on the chemical shift, and extensive shielding tends to occur only when the group R is branched at the carbon nearest to phosphorus. Although the chemical shifts of the triethyl and triisopropyl esters (X = S or Se) are essentially the same, the trimethyl ester (X = S) is shielded relative to the corresponding selenide, whereas for the tBu esters the order is reversed. Such a reversal might make it difficult to distinguish between isomeric selenothioic esters on the basis of chemical shift alone (the use of spin-spin coupling constants in this respect is discussed later). On the other hand, isomeric forms of monothio[3] and monoseleno[4] esters are readily differentiated.

Early attempts to explain the order of phosphorus chemical shifts for simple trialkyl phosphates and thiophosphates were based on changes in size of alkyl group or in their electronegativity. It is now widely accepted that the chemical shift of a phosphorus compound is linked to the nature of the atoms directly bonded to the phosphorus, and to the electronic effects transmitted through these atoms. In addition there may be extraneous effects, for example, those of solvent and temperature, but increasingly under investigation are the consequences of variations in bond angles and the interrelated torsional angles, considered together within the term "stereoelectronic effects".[5]

Inductive and resonance effects of substituent organic groups have been correlated with the chemical shift for phosphoric acid derivatives in a series which also includes derivatives of phosphonic and phosphinic acids.[6,7] Electronic effects transmitted through P-O bonds in phosphate esters appear to be relatively slight, but slight shielding is the result of increased electron withdrawal. Thus, for a series of 2-aryloxy-2-thioxo-1,3,2-dioxaphosphorinanes the chemical shifts of the *p*-substituted compounds extend from 54.7 (Me substituent) to only 52.6 (nitro substituent) ppm; the range for the *o*-substituted compounds is 54.7 to 51.1 ppm.[8] A similar variation was noted for acyclic dialkyl aryl phosphates and thiophosphates.[9] The difference in chemical shifts of 4-methylphenyl and 4-nitrophenyl phosphorodifluoridates is only 0.6 ppm.[10] For a series of seven compounds of the general structure $(RO)_nP(O)(OC_6H_4NO_2)_{3-n}$, where R = Et, n = 0, 1, or 2, and for R = Ph, n = 0, 1, 2, or 3, the chemical shifts, extending over 23 ppm, have been correlated with log (hydrolysis rate), increasing electron withdrawal leading to increased shielding.[11]

Theories on the relation between chemical shift and bond angles stem, to a large extent, from considerations of simple cyclic esters. One of the earliest investigations showed that 5- and 6-membered ring phosphorochloridates were quite different in their NMR behavior. Saturated ring compounds give chemical shifts within the ranges 14.5 to 23 and ca. −3 to −6 ppm, respectively; when the phosphorus-containing ring is fused to an aromatic ring as in the 1,2-phenylenedioxy system (5-membered phosphorus ring) and the 1,8-naphthylenedioxy system (6-membered phosphorus ring) the shifts are ca. 18 and −14 ppm[12] The order of increasing shielding for cyclic phenyl phosphates is 5-membered ring < 7-membered ring < 8-membered ring < 6-membered ring < triphenyl phosphate,[13] and for cyclic methyl phosphates the order is 5-membered ring < trimethyl phosphate < 6-membered ring. Replacement of ring hydrogen atoms by methyl groups in 1,3,2-dioxaphospholanes[14-16] and in 1,3,2-dioxaphosphorinanes[17-19] produces perturbations of 0 to 5 ppm.

An examination of a wide range of alkyl phosphates, including acyclic and cyclic esters, neutral as well as ionic esters, revealed an empirical relation between the chemical shift and bond angle, suggesting an explanation for the observations outlined in the previous paragraph. As the bond angle increases from 96 to 107° shielding decreases, but increases again at wider angles.[5,20] Later studies using the fused bicyclic 1,3,2-dioxaphosphorinanes of the type (**9**) showed that the bond angle distortions at phosphorus in 6-membered rings are significant being greater for the ester with the equatorial ester bond and chemical shift downfield from its phosphorus epimer.[21,22] Bond angle changes evidently also account for the unusual downfield shifts on the further ionization of phosphate monoanions. It seems highly probable that the order of shifts for trialkyl phosphates (methyl to *tert*-butyl) is also explicable in terms of bond angle distortions.

Differences of 3 to 6 ppm between the chemical shifts of 6-membered ring derivatives of phosphoric acid epimeric at phosphorus have been observed on many occasions[23-29] and the assignment of the higher field chemical shifts to the epimers having the equatorial P=X bond has been independently confirmed in some cases[30-32] by crystallographic examination. The presence of larger (S, Se) or less electronegative (N) atoms influences the picture of steric and electronic influences and may render such a generalization questionable. Indeed, exceptions to this empirical finding have been recorded, mainly compounds with exocyclic amide groupings, and warnings have been voiced about the dangers in the assignment of conformations at phosphorus in cyclic systems on the basis of chemical shifts alone.[33,34]

Spin-spin coupling[35] often has a role to play in the assignment of chemical structure and molecular geometry. One-bond spin-spin coupling has been studied for phosphorus directly bonded to fluorine, selenium (and tellurium), nitrogen, oxygen, and, of course, other phosphorus atoms, in both cyclic and acyclic compounds. Several of the coupling constants are thought to be negative in sign, but we consider here only the absolute values of the constants.

$^1J_{PF}$ is 940 to 1200 Hz, with higher values for sulfur-containing compounds and lower values for those containing oxygen and/or nitrogen. However, the range is essentially continuous and is likely to have only limited significance in the determination of structure. Based upon the very little data currently available the difference between the coupling constants for axially and equatorially bonded fluorine is believed to be slight (ca. 8 Hz).

On the other hand, coupling between phosphorus and ^{77}Se is of considerable value in distinguishing between isomeric structures and in the assignment of conformation. Compounds possessing the P=Se grouping have spin couplings within the range 870 to 1130 Hz, depending on the nature of the other atoms attached, either directly or indirectly, to phosphorus. Alkyl esters show smaller values than do aryl esters. Values decrease with successive replacement of ester groups by the less electronegative amide functions, mono-, di-, and triamides having constants of 900 to 1000, 825 to 875, and 785 to 790 Hz, respectively. Spin couplings for compounds with dicoordinate selenium, e.g., the *O,O,Se*-triesters, are smaller (410 to 510 Hz). Compounds with both mono- and dicoordinate selenium exhibit both couplings. Intermediate values have been recorded for ionic structures; the arrangements $P(Se)O^-$ and $P(Se)S^-$ show couplings of 770 to 780 and 740 to 755 Hz.[36] A linear relation between $^1J_{P=Se}$ and the phosphorus ionization energy of the corresponding tervalent phosphorus compound has been demonstrated.[37] $^1J_{P=^{123}Te}$ and $^1J_{P=^{125}Te}$ couplings are ca. 1700 to 2000 Hz.

One-bond couplings between phosphorus and ^{15}N provide an indication of electron delocalization from N into P=O and so of nitrogen hybridization. For a series of the *N*-(dimethoxyphosphinyl) derivatives of cyclic secondary amines, $^1J_{PN}$ increases rapidly with progression from aziridine to piperidine, when the constant becomes essentially numerically constant.[38] Assignments of conformation of P-N bonds (and so of configuration at phosphorus) in nucleoside cyclic phosphoramidates have relied[39-41] on earlier chemical shift and coupling constant data derived from monocyclic 2-amino-1,3,2-dioxaphosph(v)orinanes as well as on crystallographic evidence.[31,32] It is very tempting to generalize from the data so obtained and associate the lower field shift and numerically larger $^1J_{PN}$ value with an equatorial P-N bond in such fused bicyclic systems. However, the data from monocyclic 2-amino-1,3,2-dioxaphosphorinanes (**1**; X, Y = O, NHPh or NMe_2; X, Y = S and NHPh) show that the combination of a lower field signal and larger coupling constant is not always associated with equatorial single bonds, nor indeed is the lower field shift necessarily associated with the larger coupling constant.

Few $^1J_{P^{17}O}$ data have yet been recorded for organophosphorus compounds. $^1J_{P=O}$ is 133 ± 7 Hz at 95°C (temperature chosen for experimental reasons) for trimethyl phosphate, and slightly lower values are known for dimethyl phosphate monoanion and methyl phosphate dianion.[42] For diastereoisomeric 6-membered ring phosphates $^1J_{P=O}$ is 138 to 164 Hz; $^1J_{PO(C)}$ is somewhat smaller than for the corresponding thiophosphoryl compounds (78 vs. 92 to 112 Hz).[28] The combination of ^{18}O chemical shifts with line broadening caused by ^{17}O bonded directly to phosphorus has allowed the assignment of absolute configurations in chiral [^{16}O,^{17}O,^{18}O] phosphate esters.[43] Nevertheless, the paucity of data once again suggests caution in the use of phosphorus-^{17}O couplings as a tool in stereochemical assignments. The difference between the coupling constants for axially and equatorially ^{17}O-labeled 2-deoxyadenosine-3′,5′-[^{17}O,^{18}O] monophosphate is ca. 30 Hz at 95°C.

Few data for direct phosphorus-phosphorus couplings have been recorded. Between tervalent and quinquevalent phosphorus splittings are within the range 110 to 250 Hz, but the high value of 451 Hz has also been reported for the one-bond coupling between phosphorus atoms carrying bulky cyclic ester groups.

Longer-range couplings involving phosphorus, particularly in conjunction with one-bond couplings and chemical shifts, are of some value in structure determination, and they are of considerable value in the prediction of molecular conformation or configuration. The values reported for $^2J_{POC}$ couplings for 2-substituted 1,3,2-dioxaphosphacycloalkane 2-sulfides include 0.0 to 2.6 Hz for 5-membered rings and 5.5 to 8.0 Hz for 7-membered rings, and they thus hold some promise in the determination of ring size. The couplings $^2J_{PNC}$ and $^3J_{PNCC}$ lie, respectively, between 1 to 4.5 and 3 to 20 Hz for the amines already referred to.[44] $^2J_{PXSe}$ couplings are 54 to 58 Hz for compounds with the P(O)XSe grouping (X = S or Se). Whatever the nature of the intervening atom X in the systems $(RO)_2P(A)XP(B)(OR)_2$ the value of the $^2J_{PXP}$ coupling depends on the valency of the second phosphorus atom. Thus, for A = O, and B = l.p., O, S, or Se, $^2J_{PNP}$ is 65 to 90, 10 to 15, 2 to 30, and 8 to 14 Hz; when A = S, the corresponding values are 75 to 120, 2 to 30, 5 to 20, and 2.15 to 5.5 Hz. For a diselenide, a spin coupling of 12 Hz is known. When X is also O, S, or Se, $^2J_{PXP}$ is in the range 6 to 25 Hz when both phosphorus atoms are tetracoordinate, but rises to ca. 50 Hz when one is tricoordinate.

Couplings to hydrogen through three bonds have been studied in connection with conformations in 5- and 6-membered ring esters of phosphoric acid;[45] for the 6-membered 1,3,2-dioxaphosph(v)orinane system there exists a clear distinction between axial and equatorial protons as indicated by the $^3J_{POCH(ax)}$ and $^3J_{POCH(eq)}$ values of ca. 1 to 10 and 20 to 30 Hz; the sum of the two couplings varies from ca. 20 Hz (for aralkyl substituents on phosphorus) to 30 to 37 Hz (for hydrogen or a halogen substituent). More recently, the same and similar ring systems have been examined with the aid of $^3J_{PXC^{13}C}$ (X = O or C) couplings.[46] $^4J_{POOCH}$ couplings have been assigned to *cis* and *trans* geometry in vinyl phosphates.[47]

Mention must also be made of the use of the differences in chemical shifts of diastereoisomers in the determination of the stereoisomeric purity of chiral alcohols, thiols, and amines using derivatives which fall within the scope of the present chapter. Epimers at phosphorus based on the structure (4*R*,5*R*)-(**5**) have chemical shifts which differ by 0 to 12 Hz;[48] those of the compounds of the type (**12**) with X and Y = 0, S, OR, or NHR show differences of 0.175 to 0.843 ppm for chiral amines, and 0.111 to 0.307 ppm for chiral alcohols.[49] The determination of enantiomeric purity of chiral compounds using these derivatives produces results comparable with those obtained using other methods.

TABLE H
Four Coordinate Compounds Containing a P=Ch Bond but No Bonds to Group IV Atoms

Section H1: Derivatives of Phosphoric Acid

Rings/ Connectivities	Formula	Structure	NMR data (δ_p [solv] $^nJ_{PZ}$ Hz)	Ref.
	P-bound atoms = BrFNO′			
C_2	$C_4H_{10}BrFNOP$	$Et_2NP(O)BrF$	3 $^1J_{PF}$1058	50
	P-bound atoms = BrO_2O'			
$(C)_2$	$C_2H_6BrO_3P$	$(MeO)_2P(O)Br$	−4.7(m)	51
	$C_4H_{10}BrO_3P$	$(EtO)_2P(O)Br$	−9.0(g)	52
	$C_8H_{18}BrO_3P$	$(tBuO)_2P(O)Br$	−22.9(g)	53
$5/(C')_2$	$C_6H_4BrO_3P$	CatP(O)Br	4.7	54
$6/(C)_2$	$C_4H_8BrO_3P$	**1**; X = O′, Y = Br	−19.0	55
		X = Br, Y = O′	−14.0	
	$C_5H_{10}BrO_3P$	$Br(O)P(OCH_2CMe_2CH_2O)$	−14.7(b)	56
	P-bound atoms = ClFNO′			
C′H	$C_6H_6ClFNOP$	PhNHP(O)ClF	2.4 $^1J_{PF}$1064	57
C_2	$C_4H_{10}ClFNOP$	$Et_2NP(O)ClF$	9 $^1J_{PF}$1088	50
	P-bound atoms = ClFOO′			
C	$C_2H_5ClFO_2P$	EtOP(O)ClF	−4.2 $^1J_{PF}$1082	57
	P-bound atoms = ClNOO′			
CP;C	$C_2H_6Cl_3NO_2P$	Cl(MeO)**P**(O)NMePCl$_2$	11.0 $^2J_{PP}$81	58
CSi;C	$C_5H_{15}ClNO_2PSi$	Cl(MeO)P(O)NMeTms	18.6	58
C′H;C′	$C_{12}H_{11}ClNO_2P$	PhO(PhNH)P(O)Cl	1.5(a)	59
5/CC′;C	$C_9H_{11}ClNO_2P$	**2**; R = Ph, X = O′, Y = Cl	14.6(c)	60
5/CC′;C′	$C_{10}H_{19}ClNO_2P$	**3**; R = tBu, X = O, Y = Cl	21.1	61
6/CH;C	$C_3H_7ClNO_2P$	**4**; R^1 = R^2 = R^3 = H, X = O′, Y = Cl	10.8(c)	62
$6/C_2;C$	$C_{11}H_{15}ClNO_2P$	**4**; R^1 = (S)-CHMePh, R^2 = R^3 = H, X = O′, Y= Cl	8.2^{Sp}; 8.9^{Rp}	63
	P-bound atoms = ClN_2O'			
CP,C_2	$C_3H_9Cl_3N_2OP_2$	Cl(Me$_2$N)**P**(O)NMePCl$_2$	20.3 $^2J_{PP}$87	58
	$C_5H_{15}Cl_2N_3OP_2$	Cl(Me$_2$N)**P**(O)NMeP(NMe$_2$)Cl	24.0 $^2J_{PP}$76	58
CP^4,C_2	$C_3H_9Cl_3N_2O_2P_2$	Cl(Me$_2$N)**P**(O)NMeP(O)Cl$_2$	12.3 or 15.2(r)	64
	$C_5H_{15}Cl_2N_3O_2P_2$	Cl(Me$_2$N)**P**(O)NMeP(O)(NMe$_2$)Cl	20.9 or 21.8	64(b)
	$C_7H_{21}ClN_4O_2P_2$	Cl(Me$_2$N)**P**(O)NMeP(O)(NMe$_2$)$_2$	23.9; 18.9 $^2J_{PP}$14.5	64
CSi,C_2	$C_6H_{18}ClN_2OPSi$	MeTmsNP(O)(NMe$_2$)Cl	29.4	58
$(CH)_2$	$C_4H_{12}ClN_2OP$	$(EtNH)_2P(O)Cl$	24	65
	$C_6H_{16}ClN_2OP$	$(iPrNH)_2P(O)Cl$	19	65
CH,C_2	$C_3H_{10}ClN_2OP$	Me$_2$N(MeNH)P(O)Cl	28.2	64(a)
$(C_2)_2$	$C_4H_{12}ClN_2OP$	$(Me_2N)_2P(O)Cl$	29.6(n)	66
	$C_8H_{16}ClN_2O_3P$	Morph$_2$P(O)Cl	23.4	67
	$C_8H_{20}ClN_2OP$	$(Et_2N)_2P(O)Cl$	34.5	68
$5/(C_2)_2$	$C_4H_{10}ClN_2OP$	Cl(O)P(MeNCH$_2$CH$_2$NMe)	28.8(b)	69

TABLE H (continued)
Four Coordinate Compounds Containing a P=Ch Bond but No Bonds to Group IV Atoms

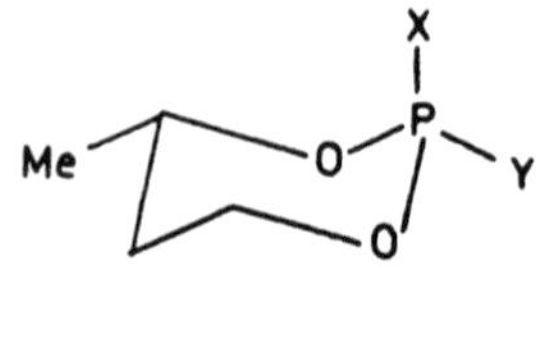

1 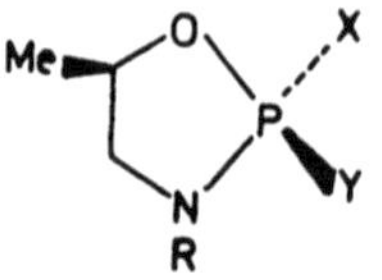 2 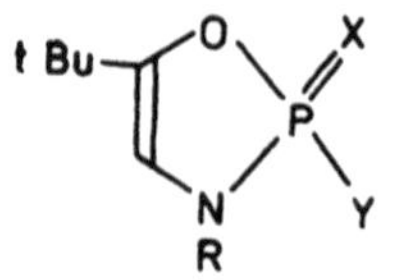 3 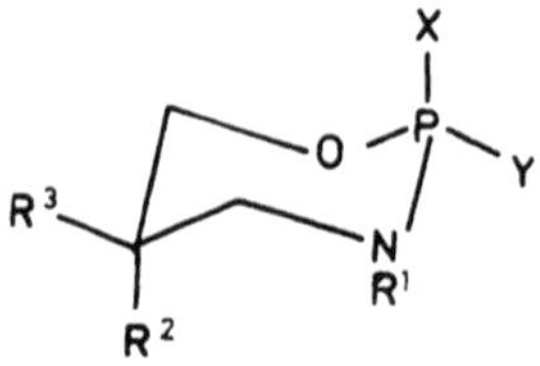 4

Rings/ Connectivities	Formula	Structure	NMR data (δ_p [solv] $^nJ_{PZ}$ Hz)	Ref.
	P-bound atoms = ClO_2O'			
$(C)_2$	$C_2H_6ClO_3P$	$(MeO)_2P(O)Cl$	6.5(g)	70
	$C_4H_4Br_6ClO_3P$	$(Br_3CCH_2O)_2P(O)Cl$	−4.6(ae)	71
	$C_4H_4ClF_6O_3P$	$(TfCH_2O)_2P(O)Cl$	5.1	72
	$C_4H_4Cl_7O_3P$	$(Cl_3CCH_2O)_2P(O)Cl$	−(?)2.5	73
	$C_4H_{10}ClO_3P$	$(EtO)_2P(O)Cl$	2.8(r)[a] 3.3(n)[b]	74
	$C_6H_{14}ClO_3P$	$(iPrO)_2P(O)Cl$	0.7(n)	75
	$C_8H_{18}ClO_3P$	$(BuO)_2P(O)Cl$	3.5(n)	75
	$C_8H_{18}ClO_3P$	$(tBuO)_2P(O)Cl$	−6.6(g)	53
	$C_{10}H_{22}ClO_3P$	$(NeoO)_2P(O)Cl$	3.7(b)	76
	$C_{14}H_{14}ClO_3P$	$(PhCH_2O)_2P(O)Cl$	4.7(n)	66
$(C')_2$	$C_{12}H_{10}ClO_3P$	$(PhO)_2P(O)Cl$	−6.1[a](n) −5.3[b](c)	77
	$C_{14}H_{14}ClO_3P$	$(3\text{-TolO})_2P(O)Cl$	−5.3	79
	$C_{14}H_{14}ClO_3P$	$(4\text{-TolO})_2P(O)Cl$	−4.8	79
	$C_{14}H_{14}ClO_5P$	$(4\text{-MeOC}_6H_4O)_2P(O)Cl$	−3.8	79
$5/(C)_2$	$C_2H_4ClO_3P$	GlcP(O)Cl	22.3 24	16
	$C_4H_8ClO_3P$	5; R = Me, X = O′, Y = Cl	19	16
	$C_6ClF_{12}O_3P$	PfpP(O)Cl	10.5	80
	$C_6H_{12}ClO_3P$	PncP(O)Cl	19	16
	$C_{14}H_{12}ClO_3P$	5; R = Ph; X = O′, Y = Cl X = Cl, Y = O′	17.7 17.5	81
$5/(C')_2$	$C_4H_6ClO_3P$	6; R = Me, X = O, Y = Cl	19(c)	82
	$C_6H_4ClO_3P$	CatP(O)Cl	18.8	83
$6/(C)_2$	$C_3H_4BrClNO_5P$	$Cl(O)P(OCH_2CBrNO_2CH_2O)$	−0.5, −2.7[D](c)	84
	$C_4H_8ClO_3P$	1; X = O′, Y = Cl X = Cl, Y = O′	−5.8 −3.5	55
	$C_5H_{10}ClO_3P$	$Cl(O)P(OCH_2CMe_2CH_2O)$	−4.8[a](b) −1.8[nb](m)	85
	$C_8H_{16}ClO_3P$	7; X = O′, Y = Cl X = Cl, Y = O′	2.4(c) −1.1(c)	29
$7/(C)_2$	$C_4H_4ClF_4O_2P$	$Cl(O)P(OCH_2CF_2CF_2CH_2O)$	7.7	86

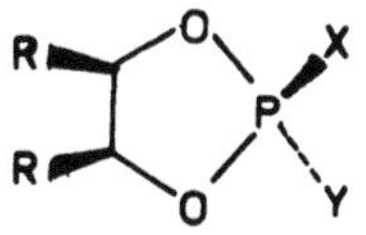

5 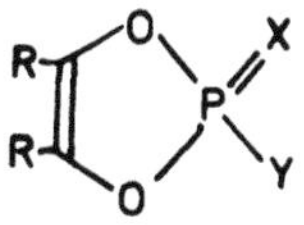 6 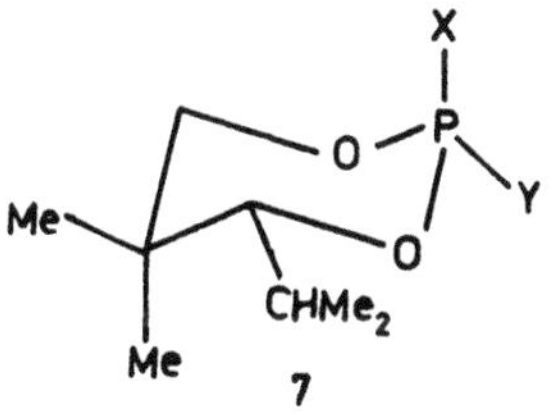 7

Rings/ Connectivities	Formula	Structure	NMR data (δ_p [solv] $^nJ_{PZ}$ Hz)	Ref.
	P-bound atoms = Cl_2NO'			
CP	$CH_3Cl_4NOP_2$	Cl_2**P**$(O)NMePCl_2$	12.9(r) $^2J_{PP}80$	58
	$C_2H_6Cl_3NO_2P_2$	Cl_2**P**$(O)NMeP(OMe)Cl$	14.0 $^2J_{PP}66$	58
CP^4	$CH_3Cl_2F_2NOPS$	Cl_2**P**$(O)NMeP(S)F_2$	10.6 $^2J_{PP}2.3$	87
	$CH_3Cl_4NOP_2S$	Cl_2**P**$(O)NMeP(S)Cl_2$	10.4; [49.4(r)] $^2J_{PP}3$	88
	$CH_3Cl_4NO_2P_2$	$MeN[P(O)Cl_2]_2$	10.3(r)	64
	$C_5H_{15}Cl_2N_3O_2P_2$	Cl_2**P**$(O)NMeP(O)(NMe_2)_2$	13.0 or 14.6 $^2J_{PP}11$	64
$C'P^4$	$C_6H_5Cl_4NOP_2S$	Cl_2**P**$(ONPhP(S)Cl_2$	8; 45.3(r) $^2J_{PP}30$	88
$C'P^4$	$C_6H_5Cl_4NO_2P_2$	$PhN[P(O)Cl_2]_2$	7.4	64(b)
CS^4	$C_2H_6Cl_2NO_3PS$	$MeSO_2NMeP(O)Cl_2$	9.5	89
CH	CH_4Cl_2NOP	$MeNHP(O)Cl_2$	18.4	90
	$C_2H_6Cl_2NOP$	$EtNHP(O)Cl_2$	16	65
	$C_3H_8Cl_2NOP$	$iPrNHP(O)Cl_2$	13	65
	$C_4H_{10}Cl_2NOP$	$tBuNHP(O)Cl_2$	10	65
C_2	$C_2H_6Cl_2NOP$	$Me_2NP(O)Cl_2$	17.4	90
	$C_4H_6Cl_2NOP$	$Cl_2P(O)NMeCH_2C{\equiv}CH$	16.0	91
	$C_4H_8Cl_2NO_2P$	$MorphP(O)Cl_2$	17.5	67
	$C_4H_{10}Cl_2NOP$	$Et_2NP(O)Cl_2$	22.9	68
	P-bound atoms = Cl_2OO'			
C	$CH_2Cl_3O_2P$	$ClCH_2OP(O)Cl_2$	8.7(g)	92
	$CH_3Cl_2O_2P$	$MeOP(O)Cl_2$	4.5[a](g) 5.6[b](n)	93
	$C_2H_2Cl_2F_3O_2P$	$TfCH_2OP(O)Cl_2$	10.1	72
	$C_2H_3Cl_4O_2P$	$ClCH_2CHClOP(O)Cl_2$	6.9(g)	92
	$C_2H_3Cl_6O_4P_3$	$[Cl_2$**P**$(O)O]CMe[P(O)Cl_2]_2$	3.2	94
	$C_2H_4Cl_3O_2P$	$CH_3CHClOP(O)Cl_2$	5.5(c)	94
	$C_2H_4Cl_3O_2P$	$ClCH_2CH_2OP(O)Cl_2$	5.9	95
	$C_2H_4Cl_4O_3P_2$	Cl_2**P**$(O)OCHMeP(O)Cl_2$	7.5	94
	$C_2H_5Cl_2O_2P$	$EtOP(O)Cl_2$	3.4(n)	96
	$C_3HCl_2F_6O_2P$	$Tf_2CHOP(O)Cl_2$	13.9	72
	$C_3H_5Cl_4OP$	$MeCHClCHClOP(O)Cl_2$	7.6(g)	92
	$C_3H_5Cl_4OP$	$ClCH_2CH(CH_2Cl)OP(O)Cl_2$	7.2(g)	92
	$C_3H_7Cl_3O_3P_2$	Cl_2**P**$(O)OCHMeP(O)(OMe)Cl$	5.0	97
	$C_7H_7Cl_2O_2P$	$PhCH_2OP(O)Cl_2$	7.7(b)	98
C'	$C_2H_3Cl_2O_3P$	$MeCOOP(O)Cl_2$	0.9	99
	$C_4H_7Cl_2O_3P$	$iPrCOOP(O)Cl_2$	−1.4	99
	$C_5H_9Cl_2O_3P$	$tBuCOOP(O)Cl_2$	−0.7	99
	$C_6H_5Cl_2O_2P$	$PhOP(O)Cl_2$	1.8(n)	66
	$C_7H_5Cl_2O_3P$	$PhCOOP(O)Cl_2$	−1.6	99
	P-bound atoms = FNOO'			
CSi;C	$C_6H_{17}FNO_2PSi$	$EtTmsNP(O)(OMe)F$	4.5 $^1J_{PF}975$	100

TABLE H (continued)
Four Coordinate Compounds Containing a P=Ch Bond but No Bonds to Group IV Atoms

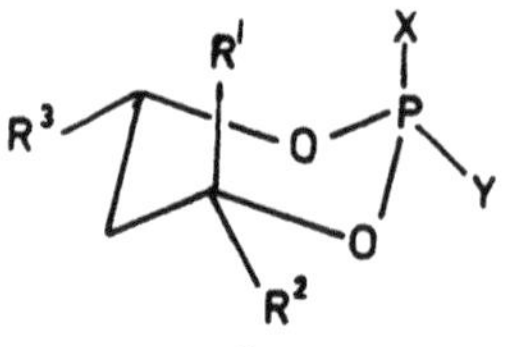

8 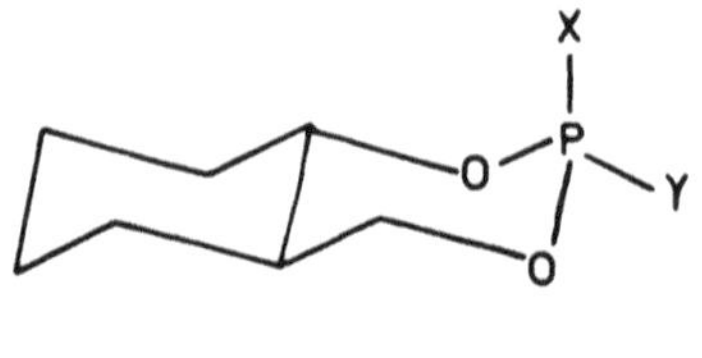 9

Rings/ Connectivities	Formula	Structure	NMR data (δ_p [solv] $^nJ_{PZ}$ Hz)	Ref.
	P-bound atoms = FN_2O'			
$(C)_2$	$C_4H_{12}FN_2OP$	$(Me_2N)_2P(O)F$	17.1(n) $^1J_{PF}$941	101
$5/(C)_2$	$C_4H_{10}FN_2OP$	$F(O)P(MeNCH_2CH_2NMe)$	23.5 $^1J_{PF}$993	101
	P-bound atoms = FO_2O'			
$(C)_2$	$C_2H_6FO_3P$	$(MeO)_2P(O)F$	−5.2 $^1J_{PF}$943	100
	$C_4H_{10}FO_3P$	$(EtO)_2P(O)F$	−8.7 $^1J_{PF}$971	102
	$C_6H_{14}FO_3P$	$(iPrO)_2P(O)F$	−11.2 $^1J_{PF}$968	103
$(C')_2$	$C_{12}H_{10}FO_3P$	$(PhO)_2P(O)F$	−21.5(c) −21.8(n) $^1J_{PF}$1000	78
$6/(C)_2$	$C_4H_8FO_3P$	1; X = O′, Y = F	−15.4[a] $^1J_{PF}$996; −17.4[b] $^1J_{PF}$1024	106
		X = F, Y = O′	−15.6[a] $^1J_{PF}$1003 −17.5[b] $^1J_{PF}$1030	
	$C_5H_{10}FO_3P$	$F(O)P(OCH_2CMe_2CH_2O)$	−17.0 $^1J_{PF}$1008	107
	$C_5H_{10}FO_3P$	8; R[1] = H, R[2] = R[3] = Me, X = F, Y = O′	−16.5 $^1J_{PF}$1000	108
		X = O′, Y = F	−15.2 $^1J_{PF}$987	
	$C_7H_{12}FO_3P$	9; X = F, Y = O′	−17.2(k)	109
	P-bound atoms = FOO'_2			
H_3	CH_3FO_3P	$MeOP(O)F(O^-)$	−5.6(d) $^1J_{PF}$896	103
	P-bound atoms = F_2NO'			
HSi	$C_3H_{10}F_2NOPSi$	$TmsNHP(O)F_2$	−3 $^1J_{PF}$998	110
CN	$C_3H_9F_2N_2OP$	$Me_2NNMeP(O)F_2$	−9.2 $^1J_{PF}$1005	111
CP^4	$CH_3F_4NO_2P_2$	$MeN[P(O)F_2]_2$	−14.4 $^1J_{PF}$ −1041	112
CSi	$C_5H_{14}F_2NOPSi$	$EtTmsNP(O)F_2$	−2.9 $^1J_{PF}$1002	100
CH	CH_4F_2NOP	$MeNHP(O)F_2$	−0.2 $^1J_{PF}$978	90
C_2	$C_2H_6F_2NOP$	$Me_2NP(O)F_2$	−3.7(n) $^1J_{PF}$1004	103
	$C_4H_8F_2NOP$	$F_2(O)P\dot{N}CH_2CH_2CH_2\dot{C}H_2$	6.6 $^1J_{PF}$1012	113
	$C_4H_{10}F_2NOP$	$Et_2NP(O)F_2$	−3.7(n) $^1J_{PF}$1001	103
C'_2	$C_3H_3F_2N_2OP$	$F_2(O)P\dot{N}CH{=}CHN{=}\dot{C}H$	22 $^1J_{PF}$1062	113
	$C_4H_4F_2NOP$	$F_2(O)P\dot{N}CH{=}CHCH{=}\dot{C}H$	17.4 $^1J_{PF}$1052	113
	$C_{12}H_{10}F_2NOP$	$Ph_2NP(O)F_2$	13.3 $^1J_{PF}$1032	113

Rings/ Connectivities	Formula	Structure	NMR data (δ_P [solv] $^nJ_{PZ}$ Hz)	Ref.
	P-bound atoms = F_2OO'			
Si	$C_3H_9F_2O_2PSi$	$TmsOP(O)F_2$	-28.7 $^1J_{PF}1000$	114
Sn	$C_{12}H_{27}F_2O_2PSn$	$Bu_3SnOP(O)F_2$	-25.5 $^1J_{PF}966$	114
C	$CH_3F_2O_2P$	$MeOP(O)F_2$	-19.6 $^1J_{PF}1008$	103
	$C_2H_5F_2O_2P$	$EtOP(O)F_2$	-21.2(n)	96
C′	$C_2H_3F_2O_3P$	$MeCOOP(O)F_2$	-26.3 $^1J_{PF}1030$	99
	$C_4H_7F_2O_3P$	$iPrCOOP(O)F_2$	-28.6 $^1J_{PF}1080$	99
	$C_5H_9F_2O_3P$	$tBuCOOP(O)F_2$	-25.1 $^1J_{PF}1035$	99
	$C_6H_4F_4O_4P_2$	$1,4\text{-}C_6H_4[OP(O)F_2]_2$	-28.8	10
	$C_6H_5F_2O_2P$	$PhOP(O)F_2$	-27.1(n) $^1J_{PF}1027$	103
	$C_7H_5F_2O_3P$	$PhCOOP(O)F_2$	-29.2 $^1J_{PF}1060$	99
	P-bound atoms = NO_2O'			
$Br_2;(C)_2$	$C_4H_{10}Br_2NO_3P$	$(EtO)_2P(O)NBr_2$	$-$(?)10.7(n)	115
$ClO;(C)_2$	$C_{11}H_{17}ClNO_4P$	$(EtO)_2P(O)NCl(OCH_2Ph)$	4.1(n,g)	116
$Cl_2;(C)_2$	$C_4H_{10}Cl_2NO_3P$	$(EtO)_2P(O)NCl_2$	7.0	117
$HN;(C)_2$	$C_6H_{17}N_2O_3P$	$(EtO)_2P(O)NHNMe_2$	5.9	118
	$C_8H_{21}N_2O_3P$	$(tBuO)_2P(O)NHNH_2$	0.0(c)	119
$HO;(C)_2$	$C_{11}H_{18}NO_4P$	$(EtO)_2P(O)NHOCH_2Ph$	8.4(n,g)	116
$HP^4;(C)_2$	$C_8H_{21}NO_6P_2$	$(EtO)_2P(O)NHP(O)(OEt)_2$	-6.1	120
	$C_{16}H_{21}NO_4P_2$	$(EtO)_2P(O)NHP(O)Ph_2$	0.7(b) $^2J_{PP}2.8$	121
$HP^4;(C')_2$	$C_{24}H_{21}NO_6P_2$	$(PhO)_2P(O)NHP(O)(OPh)_2$	-10.7	122
$H_2;(C)_2$	$C_2H_8NO_3P$	$(MeO)_2P(O)NH_2$	15.2(y)	66
	$C_4H_{12}NO_3P$	$(EtO)_2P(O)NH_2$	11.1(ee)	66
$H_2;(C')_2$	$C_{12}H_{12}NO_3P$	$(PhO)_2P(O)NH_2$	2.8(d)	66
$CO;(C)_2$	$C_{14}H_{22}NO_5P$	$(iPrO)_2P(O)N(OMe)COPh$	-2	123
$CP^4;(C)_2$	$C_8H_{22}N_2O_4P_2Se$	$(MeO)_2\mathbf{P}(O)NMeP(Se)(NEt_2)OMe$	6.0; [79.3(m)] $^1J_{PSe}878$, $^2J_{PP}13.4$	124
$CP;(C)_2$	$C_3H_9Cl_2NO_3P_2$	$(MeO)_2P(O)NMePCl_2$	4.9 $^2J_{PP}74$	58
	$C_4H_{12}ClNO_4P_2$	$(MeO)_2\mathbf{P}(O)NMeP(OMe)Cl$	6.3 $^2J_{PP}66$	58
	$C_9H_{23}NO_5P_2S$	$(EtO)_2\mathbf{P}(O)NMeP(S)(OEt)_2$	2.6; [72.0]	125
	$C_{11}H_{27}NO_5P_2Se$	$(MeO)_2\mathbf{P}(O)NMeP(Se)(O\text{-}iBu)_2$	5.2(m) $^2J_{PP}8.5$	124
	$C_{12}H_{29}NO_5P_2S$	$(EtO)_2\mathbf{P}(O)N\text{-}tBuP(S)(OEt)_2$	-1.5; [67.0]	125
$C'P;(C)_2$	$C_{14}H_{25}NO_5P_2$	$(EtO)_2\mathbf{P}(O)NPhP(OEt)_2$	3	125
$C'P^4;(C)_2$	$C_{14}H_{25}NO_5P_2S$	$(EtO)_2\mathbf{P}(O)NPhP(S)OEt)_2$	-1.5	126
$CSi;(C)_2$	$C_6H_{18}NO_3PSi$	$(MeO)_2P(O)NMeTms$	13.2(r)	127
$C'Si;(Si)_2$	$C_{15}H_{32}NO_3PSi_3$	$(TmsO)_2P(O)NPhTms$	-13	128
$C'Si;(C)_2$	$C_{11}H_{20}NO_3PSi$	$(MeO)_2(O)NPhTms$	7	128
$CH;(C)_2$	$C_6H_{16}NO_3P$	$(EtO)_2P(O)NHEt$	9.4(g)	129
	$C_7H_{14}NO_3P$	$(EtO)_2P(O)NHCH_2C{\equiv}CH$	9.1(g)	129
	$C_7H_{16}NO_3P$	$(EtO)_2P(O)NHCH_2CH{=}CH_2$	9.1(g)	129
	$C_9H_{14}NO_3P$	$(MeO)_2P(O)NHCH_2Ph$	11.2(r)	130
	$C_{11}H_{18}NO_3P$	$(EtO)_2P(O)NHCH_2Ph$	8.3(r)	130
	$C_{13}H_{21}BrNO_3P$	$(EtO)_2P(O)NHCHPhCHBrMe$	7.8	131
	$C_{13}H_{22}NO_3P$	$(iPrO)_2P(O)NHCH_2Ph$	22	132
	$C_{15}H_{26}NO_3P$	$(tBuO)_2P(O)NHCH_2Ph$	-0.3(g)	133
	$C_{21}H_{22}NO_3P$	$(PhCH_2O)_2P(O)NHCH_2Ph$	3.5(g)	133
$C'H;(H)_2$	$C_6H_8NO_3P$	$PhNHP(O)(OH)_2$	-12.0(v)	66
$C'H;C',H$	$C_{12}H_{12}NO_3P$	$PhO(PhNH)P(O)OH$	-4.1(v)	66
$C'H;(C)_2$	$C_{10}H_{10}Cl_6NO_3P$	$(Cl_3CCH_2O)_2P(O)NHPh$	$-$(?)1.0	73
	$C_{10}H_{16}NO_3P$	$(EtO)_2P(O)NHPh$	2.0[b](c) 3.0[a](g)	134
$C'H;C,C'$	$C_{14}H_{16}NO_4P$	$PhO(PhNH)P(O)OCH_2CH_2OH$	19.8	135
$C'H;(C')_2$	$C_{18}H_{16}NO_3P$	$(PhO)_2P(O)NHPh$	-7.1(v)	66

TABLE H (continued)
Four Coordinate Compounds Containing a P=Ch Bond but No Bonds to Group IV Atoms

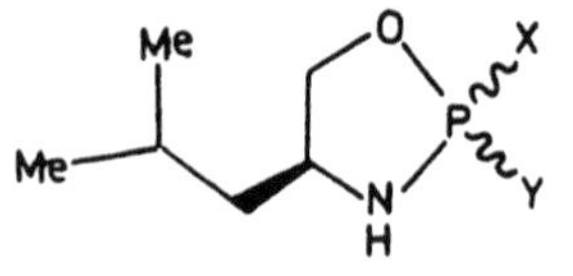

Rings/ Connectivities	Formula	Structure	NMR data (δ_p [solv] $^nJ_{PZ}$ Hz)	Ref.
$C_2;(C)_2$	$C_6H_{16}NO_3P$	$(EtO)_2P(O)NMe_2$	10.0	117
	$C_8H_{18}NO_3P$	$(EtO)_2P(O)NCH_2CH_2CH_2CH_2$	8.5(g)	116
	$C_8H_{18}NO_4P$	$(EtO)_2P(O)Morph$	10.5(g)	134(a)
	$C_8H_{20}NO_3P$	$(EtO)_2P(O)NEt_2$	9.8(g)	133
$C_2;C,C'$	$C_{14}H_{24}NO_4P$	$PhO(Me_2N)P(O)OCMe_2CMe_2OH$	21.1(n)	137
$CC';(C)_2$	$C_{11}H_{24}NO_5P$	$(EtO)_2P(O)NEtCOO\text{-}tBu$	0.5(g)	138
	$C_{12}H_{22}NO_5P$	$(EtO)_2P(O)N(COO\text{-}tBu)CH_2C{\equiv}CH$	−0.3(g)	138
	$C_{12}H_{24}NO_5P$	$(EtO)_2P(O)N(COO\text{-}tBu)CH_2CH{=}CH_2$	0.8	138
	$C_{19}H_{37}N_2O_3PS$	$(iPrO)_2P(O)N(CSNHCh)Ch$	0.0(c)	139
$CC';C,C'$	$C_{14}H_{10}NO_3P$	$MeO(PhO)P(O)NMePh$	8.3	140
$C_2';(C)_2$	$C_7H_{13}N_2O_3P$	$(EtO)_2P(O)NCH{=}CHN{=}CH$	−6.4(c)	141
	$C_{17}H_{28}NO_5P$	$(BuO)_2P(O)NPhCOOEt$	−1	142
$C_2';(C')_2$	$C_{15}H_{13}N_2O_3P$	$(PhO)_2P(O)NCH{=}CHN{=}CH$	−15.7	66
$5/CH;(C)_2$	$C_7H_{16}NO_3P$	**10**; X = O′, Y = OMe	28.0[RP](c); 27.3[SP](c)	143
$5/CH;(C')_2$	$C_{11}H_{14}NO_3P$	**6**; R = Me, X = O, Y = $NHCH_2Ph$	20.8	144
$5/C_2;(C)_2$	$C_4H_{10}NO_3P$	$GlcP(O)NMe_2$	26.9(b)	69
	$C_4H_{10}NO_3P$	**11**; R = Me, X = O, Y = OMe	22.8(c)	145
	$C_7H_{16}NO_3P$	**11**; R = iPr, X = O, Y = OEt	17.8(g)	146
	$C_8H_{18}NO_3P$	**5**; R = Me; X = O′, Y = NEt_2	22.1	147
		X = NEt_2, Y = O′	21.6	
	$C_{12}H_{18}NO_3P$	**12**; X = O′, Y = OEt; (2*R*,4*S*,5*R*)	20.0	148
$5/C_2;CC'$	$C_9H_{12}NO_3P$	**11**; R = Me, X = O, Y = OPh	15.5(c)	140
$5/C_2;(C')_2$	$C_{12}H_{18}NO_3P$	$CatP(O)N(\text{-}iPr)_2$	14.9	149
	$C_{14}H_{16}NO_3P$	**13**; X = O, Y = NEt_2	20.8	150
	$C_{14}H_{16}NO_3P$	**14**; X = O, Y = NEt_2	23.1	150
$5/C'N';(C')_2$	$C_5H_7N_4O_3P$	**6**; R = Me, X = O, Y = 1,2,3,4-tetrazolyl	1.4(c)	151
	$C_6H_8N_3O_3P$	**6**; R = Me, X = O, Y = 1,2,4-triazolyl	4.4(c)	151
$5/CC';C,H$	$C_8H_{10}NO_3P$	**11**; R = Ph, X = O, Y = OH	13.0	140
$5/CC';C,C'$	$C_{12}H_{24}NO_3P$	**3**; R = tBu, X = O, Y = OEt	16.9(c)	152
	$C_{14}H_{14}NO_3P$	**11**; R = Ph, X = O, Y = OPh	9.0	140
	$C_{15}H_{16}NO_3P$	**2**; R = Ph, X = O′ Y = OPh	5.4(c) 3.7(b)	60
$5/C_2';(C')_2$	$C_7H_9N_2O_3P$	**6**; R = Me, X = O, Y = 1-imidazolyl	6.0(c)	82
	$C_8H_{10}NO_3P$	**6**; R = Me, X = O, Y = 1-pyrrolyl	8.8(c)	151
	$C_{15}H_{16}NO_3P$	**15**; $R\text{-}C_6H_3Me_2\text{-}2,6$, X = O, Y = OMe	13.5	153

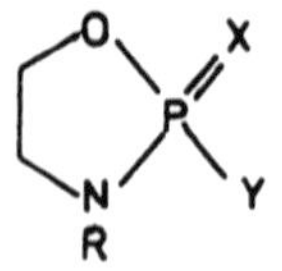

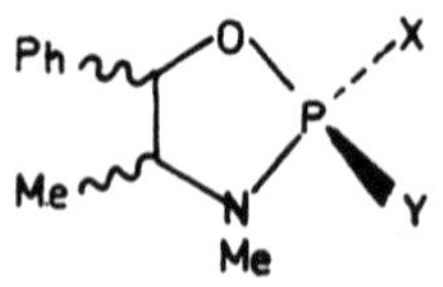

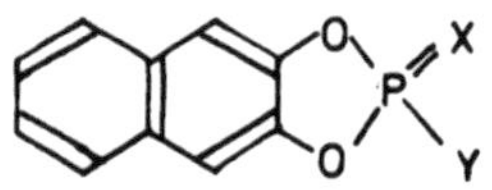

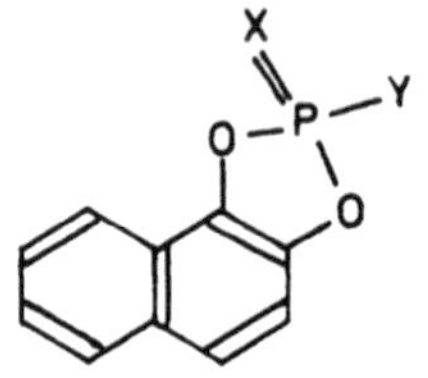

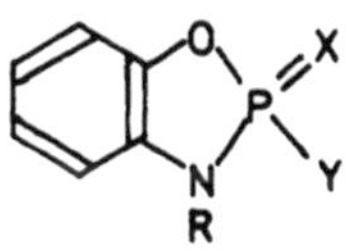

Rings/ Connectivities	Formula	Structure	NMR data (δ_p [solv] $^nJ_{PZ}$ Hz)	Ref.
$6/H_2;(C)_2$	$C_4H_{10}NO_3P$	**1**; X = O′, Y = NH_2	9.1(e)	76
		X = NH_2, Y = O′	6.0(e)	
6/CN′;CC′	$C_{16}H_{13}N_2O_3P$	**16**; X = O, Y = OPh	−14(j)	154
$6/CH;(C)_2$	$C_6H_{14}NO_3P$	**4**; R^1 = H, R^2 = R^3 = Me, X = O′, Y = OMe	3.8(b)	155
	$C_8H_{18}NO_3P$	**1**; X = O′, Y = NH-tBu	4.6	156
		X = NH-tBu, Y = O′	0.4	
	$C_8H_{18}NO_3P$	**4**; R^1 = R^2 = H, R^3 = tBu, X = O′, Y = OMe	7.8(b)	155
		X = OMe, Y = O′	6.3(b)	
$6/C'H;(C)_2$	$C_{10}H_{14}NO_3P$	**1**; X = O′, Y = NHPh	−1.0(d) $^1J_{PN}$49.6	157
		X = NHPh, Y = O′	−4.5(d) $^1J_{PN}$35.4	
	$C_{11}H_{16}NO_3P$	$PhNH(O)P(OCH_2CMe_2CH_2O)$	−2.6	158
$6/C_2;(C)_2$	$C_5H_{12}NO_3P$	$Me_2N(O)P(OCH_2CH_2CH_2O)$	6.2	159
	$C_6H_{14}NO_3P$	**1**; X = O′, Y = NMe_2	7.5 $^1J_{PN}$54.7(b)	55
		X = NMe_2, Y = O′	4.5 $^1J_{PN}$37.6(b)	
	$C_7H_{16}NO_3P$	$Me_2N(O)P(OCH_2CMe_2CH_2O)$	7.1(b)	17
	$C_{10}H_{22}NO_3P$	**4**; R^1 = iPr, R^2 = R^3 = Me, X = O′, Y = OEt	2.5	160
$6/C_2;CC'$	$C_{12}H_{18}NO_3P$	**4**; R^1 = R^2 = R^3 = Me, X = O′, Y = OPh	−2(b)	161
$6/C_2;(C')_2$	$C_{14}H_{16}NO_3P$	**17**; X = O, Y = NEt_2	−(?)5.2	150
$6/CC';(C)_2$	$C_{12}H_{18}NO_3P$	**4**; R^1 = Ph, R^2 = R^3 = Me, X = O′, Y = OMe	0.0(b)	155
	$C_{14}H_{22}NO_3P$	**4**; R^1 = Ph, R^2 = H, R^2 = tBu; X = O′, Y = OMe	2.5(a)	155
		X = OMe, Y = O′	1.2(b)	
6/CC′;C,C′	$C_{17}H_{20}NO_3P$	**4**; R^1 = Ph, R^2 = R^3 = Me, X = O′, Y = OPh	−8.5	160

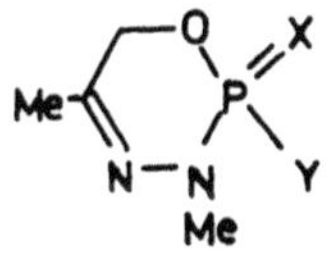

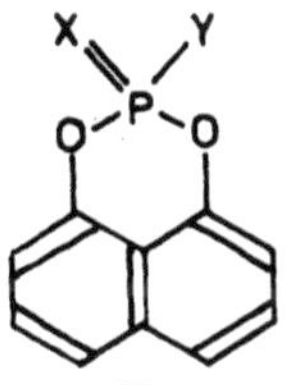

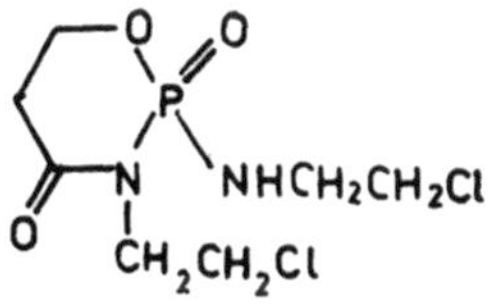

P-bound atoms = NOO′$_2$

Rings/ Connectivities	Formula	Structure	NMR data	Ref.
C′H;H	$C_6H_7NO_3P$	$PhNHP(O)(OH)O^-$	−0.85(m)	162
C′H;C	$C_7H_9NO_3P$	$PhNHP(O)(OMe)O^-$	−1.5(w-y)	51

P-bound atoms = N′O$_2$O′

Rings/ Connectivities	Formula	Structure	NMR data	Ref.
$N';(C)_2$	$C_2H_6N_3O_3P$	$(MeO)_2P(O)N_3$	1.4(m) $^1J_{PN}$14.9	163
$N';(C')_2$	$C_{12}H_{10}N_3O_3P$	$(PhO)_2P(O)N_3$	−10.4(c)	78
$P^4;(C)_2$	$C_5H_{15}NO_6P_2$	$(MeO)_2$**P**$(O)N{=}P(OMe)_3$	1.7 or −0.3 $^2J_{PP}$73	164
	$C_{10}H_{25}NO_6P_2$	$(EtO)_2$**P**$;(O)N{=}P(OEt)_3$	−2.9	120
	$C_{15}H_{35}NO_6P_2$	$(iPrO)_2$**P**$;(O)N{=}P(O\text{-}iPr)_3$	−4.7 or −6.3 $^2J_{PP}$70	164
	$C_{30}H_{25}NO_6P_2$	$(PhO)_2$**P**$;(O)N{=}P(OPh)_3$	−15.2 $^2J_{PP}$75	164

TABLE H (continued)
Four Coordinate Compounds Containing a P=Ch Bond but No Bonds to Group IV Atoms

Rings/ Connectivities	Formula	Structure	NMR data (δ_P [solv] $^nJ_{PZ}$ Hz)	Ref.
C′;(C)$_2$	$C_5H_{10}NO_3PS$	(EtO)$_2$P(O)NCS	−20.0	165
	$C_7H_{14}NO_3PS$	(iPrO)$_2$P(O)NCS	4.2	166
	$C_{11}H_{22}NO_3PS$	(NeoO)$_2$P(O)NCS	−(?)18.7	165
C′;(C′)$_2$	$C_{13}H_{10}NO_3PS$	(PhO)$_2$P(O)NCS	−29.3(n)	66
5/N′;(C′)$_2$	$C_6H_4N_3O_3P$	CatP(O)N$_3$	11.0(b)	167
5/P^4;(C′)$_2$	$C_{24}H_{19}NO_3P_2$	Cat**P**;(O)N=PPh$_3$	15.0(b)	167
5/C′;(C′)$_2$	$C_9H_{18}N_3O_3P$	**6**; X = O, Y = N=C(NMe$_2$)$_2$	13.5(c)	151
6/C′;(C)$_2$	$C_5H_8NO_3PS$	**1**; X = O′, Y = NCS	−27.0(b)	76
		X = NCS, Y = O′	−25.5(b)	
	$C_6H_{10}NO_3P$	CN(O)P(OCH$_2$CMe$_2$CH$_2$O)	−34.5(b) $^1J_{PN}$13.9	56
	$C_6H_{10}NO_3PSe$	SeCN(O)P(OCH$_2$CMe$_2$CH$_2$O)	−29.0(b)	56

P-bound atoms = N$_2$OO′

Rings/ Connectivities	Formula	Structure	NMR data (δ_P [solv] $^nJ_{PZ}$ Hz)	Ref.
NH,C$_2$;C′	$C_{16}H_{20}N_3O_3P$	PhNHNHP(O)(OPh)Morph	6.0(c)	168
(H$_2$)$_2$;C′	$C_6H_9N_2O_2P$	PhOP(O)(NH$_2$)$_2$	15.2(d)	66
CP^4,CH;C	$C_8H_{22}N_2O_4P_2S$	EtO(MeNH)**P**(O)NMeP(S)(OEt)$_2$	8.5 [71.3]	169
(CH)$_2$;C	$C_{16}H_{21}N_2O_2P$	(PhCH$_2$NH)$_2$P(O)OEt	18.4	118
(C′H) $_2$;C	$C_{14}H_{17}N_2O_2P$	(PhNH)$_2$P(O)OEt	2	118
(C′H)$_2$;C′	$C_{18}H_{17}N_2O_2P$	(PhNH)$_2$P(O)OPh	−2.3(v)	66
C′H,C$_2$;P^4	$C_{20}H_{28}N_4O_3P_2$	[(CH$_2$)$_4$NP(O)(NHPh)]$_2$O	−2.3 −2.7^D(c)	168
C′H, C$_2$;C	$C_9H_{15}N_2O_2P$	(Me$_2$N)(PhNH)P(O)OMe	13.1(y)	170
(C$_2$)$_2$;P^4	$C_{10}H_{30}N_5O_4P_3$	[(Me$_2$N)$_2$**P**(O)O]$_2$P NMe$_2$	12.3 $^1J_{PP}$10	171
(C$_2$)$_2$;C′	$C_{10}H_{16}ClN_2O_2P$	(Me$_2$N)$_2$P(O)OC$_6$H$_4$Cl-2	16.3	9
5/CN′, CC′;C	$C_{10}H_{14}N_3O_2P$	MeO(O)ṖNMeN=CPhṄMe	23.0	172
5/C$_2$,C′H;C	$C_{11}H_{17}N_2O_2P$	**2**; R = Ph, X = O, Y = NMe$_2$	19.8, 20.0^D(b)	173
5/(C$_2$)$_2$;C	$C_6H_{15}N_2O_2P$	EtO(O)P(MeNCH$_2$CH$_2$NMe)	23	174
	$C_7H_{17}NO_2P$	**11**; R = iPr, X = O, Y = NMe$_2$	23.7(g)	146
5/(C$_2$)$_2$;C′	$C_{10}H_{15}N_2O_2P$	PhO(O)P(MeNCH$_2$CH$_2$NMe)	20.6(b)	69
5/C$_2$,CC′;C	$C_{12}H_{19}N_2O_2P$	**11**; R = Ph, X = O, Y = NEt$_2$	17.7(b)	173
	$C_{13}H_{21}N_2O_2P$	**2**; R = Ph, X = O, Y = NEt$_2$	15.3(b)	173
6/(CH)$_2$;C′	$C_9H_{13}N_2O_2P$	PhO(O)P(HNCH$_2$CH$_2$CH$_2$NH)	8.6(p)	175
6/CH,C$_2$;C	$C_7H_{15}Cl_2N_2O_2P$	**4**; R^1 = R^2 = R^3 = H, X = O; Y = N(C$_2$H$_4$Cl)$_2$		176
		(*R S*)	12.5(c)	
		(*R*)	12.5(c) 11.7(b)	
	$C_7H_{15}Cl_2N_2O_2P$	**4**; R^1 = C$_2$H$_4$Cl, R^2 = R^3 = H, X= O′,		177
		Y = NHC$_2$H$_4$Cl (*R*)	12.2(c)[b]	
		(*S*)	13.8(y)[a]	
	$C_7H_{17}N_2O_2P$	**4**; R^1 = H, R^2 = R^3 = Me, X = O′, Y = NMe$_2$	13.8(c)	178
	$C_7H_{17}N_2O_2P$	**4**; R^1 = Et, R^2 = R^3 = H, X = O′, Y = NHEt	13.5	160
	$C_9H_{21}N_2O_2P$	**4**; R^1 = R^2 = H, R^3 = tBu, X = O′, Y = NMe$_2$	14.6 14.7^D(b)	178
6/CH,CC′;C	$C_7H_{13}Cl_2N_2O_3P$	**18**	6.5(c)	98
6/(C$_2$)$_2$;C	$C_7H_{17}N_2O_2P$	MeO(O)P(MeNCH$_2$CH$_2$CHMeNMe)	14.8,16 .3^D	179
	$C_8H_{19}N_2O_2P$	**4**; R^1 = iPr, R^2 = R^3 = H; X,Y = O′, NMe$_2$	12.6(g)	146
	$C_{11}H_{20}N_2O_2P$	**4**; R^1 = Et, R^2 = R^3 = Me; X,Y =O′, NEt$_2$	15.5	160

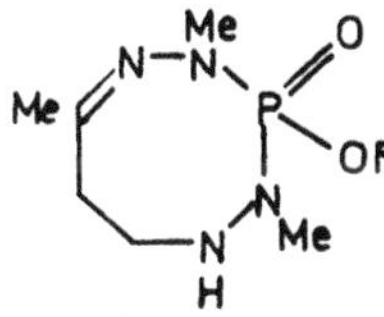

Rings/ Connectivities	Formula	Structure	NMR data (δ_p [solv] $^nJ_{PZ}$ Hz)	Ref.
$6/C_2,CC';C$	$C_{11}H_{17}N_2O_2P$	**4**; R^1 = Ph, R^2 = R^3 = H; X,Y = O′, NMe_2	8.2(b)	173
	$C_{15}H_{25}N_2O_2P$	**4**; R^1 = Ph, R^2 = H, R^3 = tBu, X, Y = O′, NMe_2	8.5, 10.9[D](c)	180
8/ CN,CN′;C′	$C_{12}H_{19}N_4O_2P$	**19**	77.2	181

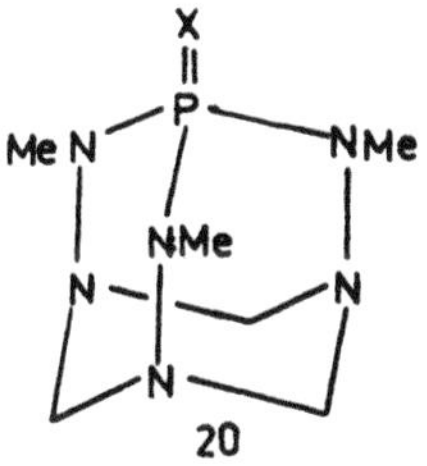

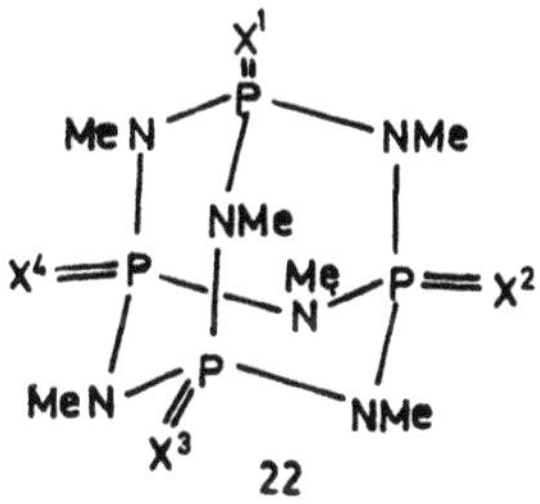

P-bound atoms = N_3O'

Rings/ Connectivities	Formula	Structure	NMR data (δ_p [solv] $^nJ_{PZ}$ Hz)	Ref.
$(NH)_2,C_2$	$C_{14}H_{20}N_5OP$	$(PhNHNH)_2P(O)NMe_2$	15	182
$HO,(C'H)_2$	$C_{12}H_{14}N_3O_2P$	$(PhNH)_2P(O)NHOH$	8.3(y)	170
$(CP)_3$	$C_3H_9F_6N_3OP_4$	$(O)\mathbf{P}(NMePF_2)_3$	4.3	171
$CP,(C_2)_2$	$C_5H_{15}Cl_2N_3OP_2$	$(Me_2N)_2\mathbf{P}(O)NMePCl_2$	21.0 $^2J_{PP}$73	58
$CP^4,(C_2)_2$	$C_9H_{27}N_5OP_2S$	$(Me_2N)_2\mathbf{P}(O)NMeP(S)(NMe_2)_2$	20; [77(r)] $^2J_{PP}$9.7	88
	$C_9H_{27}N_5O_2P_2$	$(Me_2N)_2P(O)NMeP(O)(NMe_2)_2$	20.3(r)	64
$(CH)_3$	$C_3H_{12}N_3OP$	$(MeNH)_3P(O)$	22.8	171
$(C'H)_3$	$C_{18}H_{18}N_3OP$	$(PhNH)_3P(O)$	−4.8(ze)	66
$CH,(C_2)_2$	$C_{10}H_{21}N_3OP$	$(Et_2N)_2P(O)NHEt$	20.1	183
$(C_2)_3$	$C_6H_{12}N_3OP$	$(O)P(\dot{N}CH_2\dot{C}H_2)_3$	41	184
	$C_6H_{18}N_3OP$	$(Me_2N)_3P(O)$	24.8(r) 23.8(n) $^1J_{PN}$ −27	185
	$C_{12}H_{24}N_3O_4P$	$Morph_3P(O)$	26.2	67
	$C_{12}H_{30}N_3OP$	$(Et_2N)_3P(O)$	23.3	68
$5/(C_2)_3$	$C_6H_{16}N_3OP$	$Me_2N(O)P(MeNCH_2CH_2NMe)$	25.7(b)	69
$6/(C_2)_3$	$C_7H_{18}N_3OP$	$Me_2N(O)P(MeNCH_2CH_2CH_2NMe)$	21.1	179
$6,6/(CN)_3$	$C_6H_{15}N_6OP$	**20**; X = O	7.5(c)	186
	$C_6H_{18}N_6OP_2Se$	$(O)\mathbf{P}(NMeNMe)_3P(Se)$	8.0 [67.3](c) $^1J_{PSe}$926	187
	$C_6H_{18}N_6O_2P_2$	$(O)P(NMeNMe)_3P(O)$	8.6(r)	185
$6,6/(C_2)_3$	$C_8H_{18}N_3OP$	$(O)P(MeNCH_2)_3CMe$	19.8	188
$5,5,5,/(C_2)_3$	$C_6H_{12}N_3OP$	**21**; X = O	99	184
$6,6,6/(CP^4)_3$	$C_6H_{18}N_6OP_4S_3$	**22**; X^1 = O′, X^2 = X^3 = X^4 = S′	7.5 [66.1](c) $^2J_{PP}$4.5	189
	$C_6H_{18}N_6O_2Pl_4S_2$	**22**; X^1 = X^2 = O′, X^3 = X^4 = S′	7.0 [63.5](c) $^2J_{PP}$2	189
	$C_6H_{18}N_6O_3P_4S$	**22**; X^1 = X^2 = X^3 = O′, X^4 = S′	5.3 [53.6](c) $^2J_{PP}$<1	189
	$C_6H_{18}N_6O_4P_4$	**22**; X^1 = X^2 = X^3 = X^4 = O′	2.7(c)	189

P-bound atoms = $O_2O'P^4$

Rings/ Connectivities	Formula	Structure	NMR data (δ_p [solv] $^nJ_{PZ}$ Hz)	Ref.
$C_2;O_2O'$	$C_8H_{20}O_6P_2$	$(EtO)_2P(O)P(O)(OEt)_2$	16.7	254
$C_2;O_2S'$	$C_8H_{20}O_5P_2S$	$(EtO)_2\mathbf{P}(O)(P)(S)(OEt)_2$	4.4(c), 4.2(n)	102
$6/C_2;O_2O'$	$C_{10}H_{20}O_6P_2$	**24**; X = Z = O, Y absent	−1.7(c)	255
$6/C_2;O_2S'$	$C_{10}H_{20}O_5P_2S$	**24**; X = O, Z = S, Y absent	−5.5; [58.9](c) $^1J_{PP}$491	255

TABLE H (continued)
Four Coordinate Compounds Containing a P=Ch Bond but No Bonds to Group IV Atoms

Rings/ Connectivities	Formula	Structure	NMR data (δ_P [solv] $^nJ_{PZ}$ Hz)	Ref.
	P-bound atoms = $O_2O'P$			
$(C)_2;P^4_2$	$C_{12}H_{30}O_9P_4$	$[(EtO)_2$**P**$(O)]_3P$	22.7	249
	$C_{18}H_{42}O_9P_4$	$[(iPrO)_2$**P**$(O)]_3P$	21 $^1J_{PP}$157	250
$(C)_2;CP^4$	$C_{10}H_{25}O_6P_3$	$[(EtO)_2$**P**$(O)]_2PEt$	31 $^1J_{PP}$180	250
	$C_{12}H_{29}O_6P_3$	$[(EtO)_2$**P**$(O)]_2P$-tBu	29.4 $^1J_{PP}$200	250
$(C)_2;C_2$	$C_4H_6F_6O_3P_2$	$(MeO)_2$**P**$(O)PTf_2$	16.7 $^1J_{PP}$126	251
	$C_8H_{14}F_6O_3P_2$	$(iPrO)_2$**P**$(O)PTf_2$	11.7 $^1J_{PP}$114	251
	$C_8H_{20}O_3P_2$	$(EtO)_2$**P**$(O)PEt_2$	39	252
$(C)_2;C'_2$	$C_{20}H_{28}O_3P_2$	$(BuO)_2$**P**$(O)PPh_2$	32 $^1J_{PP}$181	253
	P-bound atoms = O_3O'			
$P^4,(Si)_2$	$C_{12}H_{36}O_7P_2Si_4$	$(TmsO)_2P(O)OP(O)(OTms)_2$	−32.7	190
$(Si)_3$	$C_9H_{27}O_4PSi_3$	$(TmsO)_3P(O)$	−26.5(n)	190
$C,(Si)_2$	$C_{15}H_{27}F_{12}O_5PSi_3$	$(TmsO)_2P(O)OCTf_2CTf_2OTms$	−10.3	191
C,H,P^4	$C_2H_8O_7P_2$	MeO(HO)P(O)OP(O)(OH)OMe	−9.8(w)	51
$C,(H)_2$	CH_5O_4P	$MeOP(O)(OH)_2$	−1.2(w)	51
	$C_2H_4F_3O_4P$	$TfCH_2OP(O)(OH)_2$	−1.7	72
	$C_3H_3F_6O_4P$	$Tf_2CHOP(O)(OH)_2$	−2.4	71
	$C_6H_{13}F_{12}O_5P$	$HOCTf_2CTf_2OP(O)(OH)_2$	−8.6	191
$C',(Si)_2$	$C_9H_{18}F_5O_4PSi_2$	$F_2C{=}CTfOP(O)(OTms)_2$	−21.3	191
$C',(H)_2$	$C_6H_7O_4P$	$PhOP(O)(OH)_2$	−4.8(w)	66
	$C_6H_7O_5P$	$2\text{-}HOC_6H_4OP(O)(OH)_2$	−5.5	167
$(C)_2,P^4$	$C_8H_{20}O_6P_2S$	$(EtO)_2$**P**$(O)OP(S)(OEt)_2$	−14.2; [52.5](n) $^2J_{PP}$20	75
	$C_8H_{20}O_6P_2Se$	$(EtO)_2$**P**$(O)OP(Se)(OEt)_2$	−15.5; [58.5] $^2J_{PP}$ 24 $^1J_{PSe}$1020	139
	$C_8H_{20}O_7P_2$	$(EtO)_2P(O)OP(O)(OEt)_2$	−13(n,r)	75
	$C_{12}H_{28}O_6P_2S$	$(iPrO)_2$**P**$(O)OP(S)(O\text{-}iPr)_2$	−18; [51] $^2J_{PP}$22	192
	$C_{12}H_{28}O_6P_2Se$	$(iPrO)_2$**P**$(O)OP(Se)(O\text{-}iPr)_2$	−17.5; [54] $^2J_{PP}$24 $^1J_{PSe}$1020	139
	$C_{20}H_{44}O_6P_2S$	$(NeoO)_2$**P**$(O)OP(S)(ONeo)_2$	−14.2; [53.7] $^2J_{PP}$18	76
	$C_{28}H_{28}O_7P_2$	$(PhCH_2O)P(O)OP(O)(OCH_2Ph)_2$	−12.6(b)	66
$(C)_2,S^4$	$C_5H_{13}O_6PS$	$(EtO)_2P(O)OSO_2Me$	−13.6(n)	141
$(C)_2,Si$	$C_7H_{19}O_4PSi$	$(EtO)_2P(O)OTms$	3	193
$(C)_2,H$	$C_4H_5Cl_6O_4P$	$(Cl_3CCH_2O)_2P(O)OH$	−4.5	194
	$C_4H_5F_6O_4P$	$(TfCH_2O)_2P(O)OH$	−3.3	72
	$C_4H_{11}O_4P$	$(EtO)_2P(O)OH$	0.0(w)	66
	$C_6H_{13}F_{12}O_4P$	$(Tf_2CHO)_2P(O)OH$	−4.8	72
	$C_{14}H_{13}N_2O_8P$	$(4\text{-}O_2NC_6H_4CH_2O)_2P(O)OH$	−1.7(v)	66
C,C',P^4	$C_{14}H_{16}O_7P_2$	PhO(MeO)P(O)OP(O)(OMe)OPh	−17.3(c)	195
C,C',H	$C_8H_{11}O_5P$	$HOCH_2CH_2OP(O)(OH)OPh$	−5.6	13
$(C')_2,P^4$	$C_{24}H_{20}O_6P_2S$	$(PhO)_2$**P**$(O)OP(S)(OPh)_2$	−27.3; [42.5] $^2J_{PP}$24.3	139
	$C_{24}H_{20}O_7P_2$	$(PhO)_2P(O)OP(O)(OPh)_2$	−24.8	196

Rings/ Connectivities	Formula	Structure	NMR data (δ_p [solv] $^nJ_{PZ}$ Hz)	Ref.
$(C')_2$,H	$C_{12}H_9N_2O_8P$	$(4\text{-}O_2NC_6H_4O)_2P(O)OH$	−15.4(a)	66
	$C_{12}H_{11}O_4P$	$(PhO)_2P(O)OH$	−12.0 −12.7(y)	196
	$C_{12}H_{11}O_5P$	$PhO(HO)P(O)OC_6H_4OH\text{-}2$	−3.3	13
$(C)_3$	$C_3H_9O_4P$	$(MeO)_3P(O)$	−3.0(c) −2.4(r)	51
	$C_4H_{11}O_4PS$	$(MeO)_2P(O)OCH_2CH_2SH$	−1	197
	$C_5H_{13}O_4P$	$(EtO)_2P(O)OMe$	−0.8	141
	$C_5H_{13}O_4P$	$(MeO)_2P(O)O\text{-}iPr$	0.1(c)	198
	$C_6H_6Cl_9O_4P$	$(Cl_3CCH_2O)_3P(O)$	−1.3	194
	$C_6H_6F_9O_4P$	$(TfCH_2O)_3P(O)$	−28	72
	$C_6H_{12}Cl_3O_4P$	$(ClCH_2CH_2O)P(O)$	0.0	7
	$C_6H_{15}O_4P$	$(EtO)_3P(O)$	−0.7/−1.2, −1.2(r)	200
	$C_7H_{17}O_4P$	$(EtO)_2P(O)O\text{-}iPr$	−1	199
	$C_8H_{17}O_5P$	$MeO(iPrO)P(O)OCHMeCOMe$	−1.8, -3.6^{D}(r)	202
	$C_8H_{19}O_4P$	$(EtO)_2P(O)O\text{-}tBu$	−5.5	203
	$C_9H_3F_{18}O_4P$	$(Tf_2CHO)_3P(O)$	−4.6	72
	$C_9H_{13}O_4P$	$(MeO)_2P(O)OCH_2Ph$	1.3(c)	198
	$C_9H_{21}O_4P$	$(iPrO)_3P(O)$	−3.2	203
	$C_{10}H_{23}O_4P$	$(BuO)_2P(O)OEt$	−0.8	211
	$C_{10}H_{23}O_4P$	$(tBuO)_2P(O)OEt$	-10.2^{a}(c) -9.5^{b}(g)	204
	$C_{11}H_{25}O_4P$	$(tBuO)_2P(O)O\text{-}iPr$	−10.4(c)	198
	$C_{12}H_{27}O_4P$	$(BuO)_3P(O)$	−1.3(r)	199
	$C_{12}H_{27}O_4P$	$(tBuO)_3P(O)$	−13.3	205
	$C_{12}H_{27}O_4P$	$(NeoO)_2P(O)OEt$	−1.0	203
	$C_{15}H_{33}O_4P$	$(NeoO)_3P(O)$	−0.6	203
	$C_{15}H_{33}O_4P$	$(EtCMe_2O)_3P(O)$	−12.9	203
	$C_{21}H_{18}N_3O_{10}P$	$(4\text{-}O_2NC_6H_4CH_2O)_3P(O)$	−0.7(v)	66
$(C)_2$,C′	$C_4H_7Br_2O_4P$	$(MeO)_2P(O)OCH{=}CBr_2$	3.7(g)	47
	$C_4H_8BrO_4P$	$(MeO)_2P(O))CH{=}CHBr$	2.8^{E}; 3.0^{Z}(g)	47
	$C_4H_9O_4P$	$(MeO)_2P(O)OCH{=}CH_2$	3.0(g)	47
	$C_5H_{11}O_4P$	$(MeO)_2P(O)OCH{=}CHMe$	2.8^{E}; 2.3^{Z}(g)	47
	$C_6H_{11}F_2O_4P$	$(EtO)_2P(O)OCH{=}CF_2$	−6	206
	$C_6H_{12}BrO_4P$	$(EtO)_2P(O)OCH{=}CHBr$	5.4^{E}; 5.4^{Z}(g)	47
	$C_8H_{11}O_4P$	$(MeO)_2P(O)OPh$	−4.1	140
	$C_8H_{15}O_5P$	$(MeO)_2P(O)OCMe{=}CHCOEt$	-6.8^{Z}	207
	$C_9H_{13}O_5P$	$(MeO)_2P(O)OC_6H_4OMe\text{-}4$	3.7	208
	$C_9H_{17}O_6P$	$H_2C{=}C(COOEt)OP(O)(OEt)_2$	−7.1	209
	$C_{10}H_{14}BrO_4P$	$(EtO)_2P(O)OC_6H_4Br\text{-}2$	−7	9
	$C_{10}H_{15}O_4P$	$(EtO)_2P(O)OPh$	−6.8	201
	$C_{14}H_{22}NO_5P$	$(EtO)_2P(O)OCPh{=}NOPr$	−6	123
	$C_{18}H_{29}O_4P$	$(NeoO)_2P(O)OCPh{=}CH_2$	−7.0	210
C,$(C')_2$	$C_{14}H_{14}NO_6P$	$EtO(PhO)P(O)(OC_6H_4NO_2\text{-}4)$	−13.1(c)	211
	$C_{14}H_{15}O_4P$	$(PhO)_2P(O)OEt$	−11.9(c)	79
	$C_{15}H_{17}O_4P$	$(PhO)_2P(O)O\text{-}iPr$	−12.9(c)	198
	$C_{16}H_{19}O_4P$	$EtOP(O)(OTol\text{-}3)_2$	−11.9(c)	79
	$C_{16}H_{19}O_4P$	$EtOP(O)(OTol\text{-}4)_2$	−11.5(c)	79
	$C_{16}H_9O_5P$	$BuO(PhO)P(O)OC_6H_4OH\text{-}2$	−9.8(c)	212
	$C_6H_{19}O_6P$	$EtOP(O)(OC_6H_4OMe\text{-}4)_2$	−10.8(c)	79
$(C')_3$	$C_{18}H_{12}N_3O_{10}P$	$(4\text{-}O_2NC_6H_4O)_3P(O)$	−23(ne)	11
	$C_{18}H_{13}N_2O_8P$	$PhOP(O)(OC_6H_4NO_2\text{-}4)_2$	−18(m)	11
	$C_{18}H_{14}NO_6P$	$(PhO)_2P(O)OC_6H_4NO_2\text{-}4$	−19(ne)	11
	$C_{18}H_{15}O_4P$	$(PhO)_3P(O)$	−18(r), −17.3(n)	199

TABLE H (continued)
Four Coordinate Compounds Containing a P=Ch Bond but No Bonds to Group IV Atoms

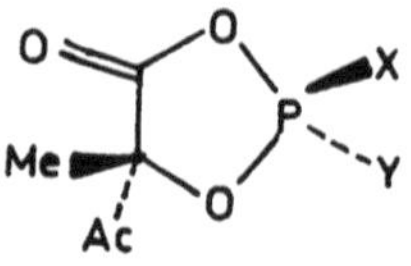

Rings/ Connectivities	Formula	Structure	NMR data (δ_p [solv] $^nJ_{PZ}$ Hz)	Ref.
	$C_{21}H_{21}O_4P$	$(2\text{-TolO})_3P(O)$	−17.1(n)	190
	$C_{21}H_{21}O_4P$	$(3\text{-TolO})_3P(O)$	−17.6(c)	79
	$C_{21}H_{21}O_4P$	$(4\text{-TolO})_3P(O)$	−17.1(c)	79
	$C_{21}H_{21}O_7P$	$(4\text{-MeOC}_6H_4O)_3P(O)$	−16.0	79
$5/(C)_2,P^4$	$C_{12}F_{24}O_7P_2$	PfpP(O)OP(O)Pfp	−34.1	80
$5/(C)_2,H$	$C_6HF_{12}O_4P$	PfpP(O)OH	7.7	191
	$C_6H_{13}O_4P$	PncP(O)OH	14.3	213
$5/CC',H$	$C_5H_7O_6P$	**23**; X, Y = O′, OH	−0.2	214
$5/(C')_2,P^4$	$C_8H_{12}O_7P_2$	**6**; R = Me, X = O, Y = O(O)P(OCMe=CMeO)	−1.5	215
$5/(C')_2,Si$	$C_7H_{15}O_4PSi$	**6**; R = Me, X = O, Y = OTms	3.4	215
$5/(C')_2,H$	$C_4H_7O_4P$	**6**; R = Me, X = O, Y = OH	13.2(r) 15.5(c)	216
	$C_6H_5O_4P$	CatP(O)OH	11.3	217
$5/(C)_3$	$C_3H_7O_4P$	GlcP(O)OMe	18.8[a](c) 17.7[b](m)	219
	$C_4H_9O_4P$	GlcP(O)OEt	17.5(n)	69
	$C_5H_{11}O_4P$	**5**; R = Me; X = O′, Y = OMe	14.6	147
		X = OMe, Y = O′	15.9	
	$C_9HF_{18}O_4P$	$PfpP(O)OCHTf_2$	5.9	220
	$C_{14}H_{28}O_7P_2$	PncP(O)OCHMeP(O)Pnc	12.5	221
$5/(C)_2,C'$	$C_6H_9O_6P$	**23**; X = OMe, Y = O′	2.4(r)	202
		X = O′, Y = OMe	2.7(r)	
	$C_8H_9O_4P$	GlcP(O)OPh	10.6/12.5 11.6(b)	69
	$C_{10}H_{13}O_4P$	**5**; R = Me; X = O′, Y = OPh	7.4	147
		X = OPh, Y = O′	8.7	
	$C_{12}H_{17}O_4P$	PncP(O)OPh	6.5	213
$5/C,(C')_2$	$C_3H_5O_4P$	**6**; R = H, X = O′, Y = OMe	15.7	222
	$C_5H_9O_4P$	**6**; R = Me, X = O′, Y = OMe	11.6 12.4(c,r)	223
	$C_8H_9O_4P$	CatP(O)OEt	8.4[a] 11.8[b]	224
	$C_{11}H_{13}O_4P$	**6**; R = Me, X = O′, Y = OCH_2Ph	11.2(c,r)	215
	$C_{12}H_{11}O_4P$	**13**; X = O, Y = OEt	7.2	150
	$C_{12}H_{11}O_4P$	**14**; X = O, Y = OEt	6.35	150
$5/(C')_3$	$C_{10}H_{11}O_4P$	**6**; R = Me, X = O, Y = OPh	6.1(c)	212
	$C_{12}H_9O_4P$	CatP(O)OPh	5.6/6.7 6.4(c)	212
	$C_{15}H_{17}O_5PSi$	$CatP(O)OC_6H_4OTms\text{-}2$	7.3	83
	$C_{16}H_{11}O_4P$	**13**; X = O, Y = OPh	21.4	150
$6/(C)_2,P^4$	$C_{10}H_{20}O_6P_2S$	**24**; X = S, Y = Z = O	−11[b](c,g) −21[a]: [44[a], 54.5[b]]	226
	$C_{10}H_{20}O_6P_2Se$	**24**; X = Se, Y = Z = O	−22.9; [43.6]	226
	$C_{10}H_{20}O_7P_2$	**24**; X = Y = Z = O	−21.2(p)	227
$6/(C)_2,H$	$C_3H_7O_4P$	$HO(O)P(OCH_2CH_2CH_2O)$	−5.1[a](p) −3.7[b](w)	229

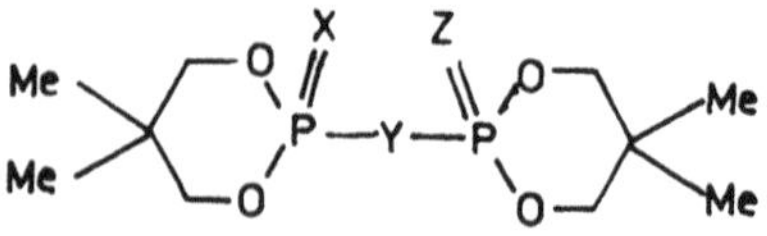

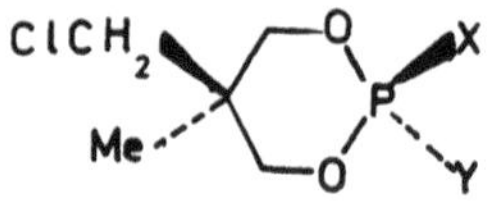

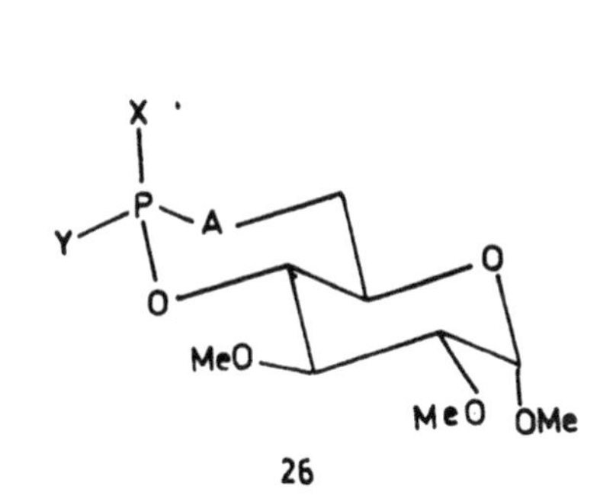

26

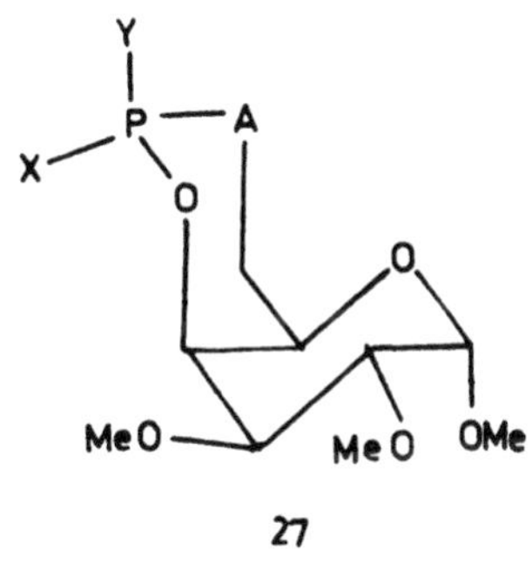

27

28

Rings/ Connectivities	Formula	Structure	NMR data (δ_p [solv] $^nJ_{PZ}$ Hz)	Ref.
	$C_5H_{11}O_4P$	$HO(O)P(OCH_2CMe_2CH_2O)$	−5.2^a(p) −4.0^b(w)	229
	$C_7H_{13}O_4P$	**9**; X, Y = O′, OH	−2.5	230
6/(C)$_3$	$C_4H_9O_4P$	$MeO(O)P(OCH_2CH_2CH_2O)$	−6.7 $^1J_{POC}$78 $^1J_{POMe}$78 $^1J_{P=O}$156	28
	$C_5H_{11}O_4P$	**1**; X = O′, Y = OMe	−4.6(c) $^1J_{P=O}$161	231
		X = OMe, Y = O′	−6.4(c) $^1J_{P=O}$164	
	$C_6H_{12}ClO_4P$	**25**; X = O′; Y = OMe	−8.2	109
		X = OMe, Y = O′	−6.1	
	$C_6H_{13}O_4P$	**8**; R^1 = H, R^2 = R^3 = Me;		28
		X = O′, Y = OMe	−4.9(c)	
		X = OMe, Y = O′	−7.1 $^1J_{P=O}$162	
	$C_6H_{13}O_4P$	**8**; R^1 = R^2 = Me, R^2 = H, X = O′, Y = OMe	−6.1 $^1J_{P=O}$156	28
	$C_7H_{15}O_4P$	$EtO(O)P(OCH_2CMe_2CH_2O)$	−8.5	232
	$C_7H_{15}O_4P$	$tBuO(O)P(OCH_2CH_2CH_2O)$	−12.5	12
	$C_7H_{15}O_4P$	**8**; R^1 = R^2 = R^3 = Me, X = O′, Y = OMe	−6.1(r) $^1J_{P=O}$161	28
		X = OMe, Y = O′	−7.6(r) $^1J_{P=O}$138	
	$C_9H_{17}O_4P$	**9**; X = O′, Y = OEt	−5.3	109
		X = OEt, Y = O′	−7.2	
	$C_{11}H_{21}O_8P$	**26**; A = O, X = O′, Y = OEt	−4.4	23
		X = OEt, Y = O′	−7.5	
	$C_{11}H_{21}O_8P$	**27**; A = O; X = O′, Y = OEt	−6	23
		X = OEt, Y = O′	−8.5	
6/(C)$_2$,C′	$C_9H_{11}O_4P$	$PhO(O)P(OCH_2CH_2CH_2O)$	−14.6(b)	17
	$C_{11}H_{15}O_4P$	$PhO(O)P(OCH_2CMe_2CH_2O)$	−14.8^a −16^b(b)	233
	$C_{12}H_{17}O_4P$	**8**; R^1= R^2 = R^3 = Me, X, Y = O′, OPh	−14, 5, −16.0^D	234
	$C_{13}H_{17}O_4P$	**9**; X = O′, Y = OPh	−10.6(c)	21
		X = OPh, Y = O′	−13.0(c)	
	$C_{14}H_{21}O_4P$	**7**; X = O′, Y = OPh	−9.0(c)	233
		X = OPh, Y = O′	−12.4(c)	
6/C,(C′)$_2$	$C_7H_{11}O_4P$	**28**; X = O, Y = OEt	−18(n)	236
	$C_{10}H_{11}O_4P$	**29**; X = O, Y = OEt	−14.5(n)	237
	$C_{12}H_{11}O_4P$	**17**; X = O, Y = OEt	−17.3	150
6/(C′)$_3$	$C_{16}H_{11}O_4P$	**17**; X = O, Y = OPh	−24.1	150
7/(C)$_2$,H	$C_4H_9O_4P$	$HO(O)P(OCH_2CH_2CH_2CH_2O)$	4.8(w)	228
7/(C)$_2$,C′	$C_{10}H_{13}O_4P$	$PhO(O)P(OCH_2CH_2CH_2CH_2O)$	−3.7(g)	12
	$C_{14}H_{13}O_4P$	**30**; R = H, A = X = O, Y = OPh	−7	238
7/(C′)$_2$H	$C_{12}H_9O_4P$	**31**; X = O, Y = OH	5.7(w)	239

TABLE H (continued)
Four Coordinate Compounds Containing a P=Ch Bond but No Bonds to Group IV Atoms

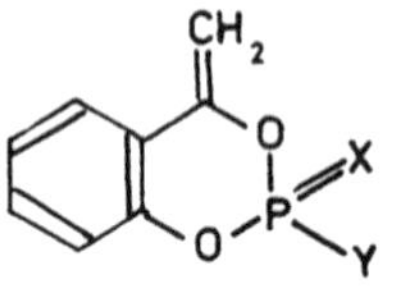

29 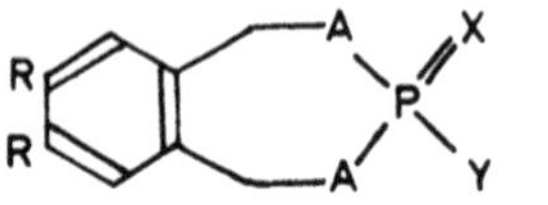 30 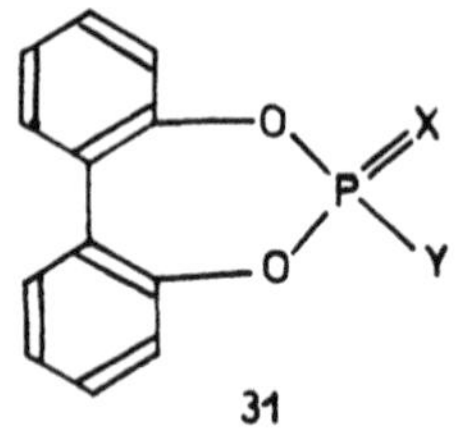 31 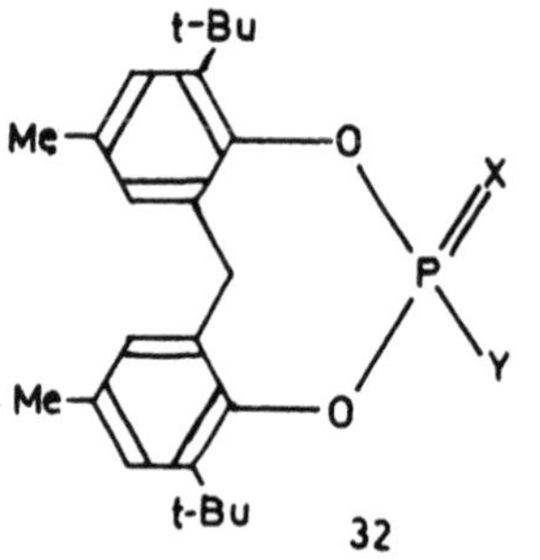 32 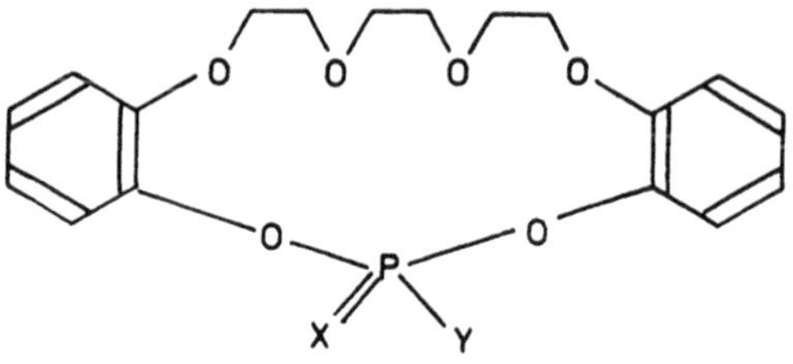 33 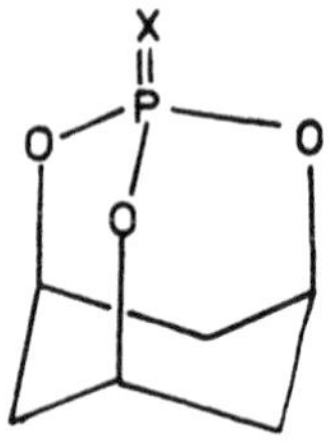 34

Rings/ Connectivities	Formula	Structure	NMR data (δ_p [solv] $^nJ_{PZ}$ Hz)	Ref.
$7/C_3(C')_2$	$C_{13}H_{11}O_4P$	**31**; X = O, Y = OMe	2.6(c)	239
$8/(C)_2$,H	$C_5H_{11}O_4P$	$HO(O)P(OCH_2CH_2CH_2CH_2CH_2O)$	0.0(w)	228
$8/(C')_3$	$C_{29}H_{35}O_4P$	**32**; X = O, Y = OPh	−17.2	13
$17/(C')_3$	$C_{24}H_{25}O_8P$	**33**; X = O, Y = OPh	−16.7(c)	240
$6,6/(C)_3$	$C_3H_6O_4P_2$	$(O)\mathbf{P}(OCH_2)_3P$	−14.25	241
	$C_3H_6O_5P_2$	$(O)\mathbf{P}(OCH_2)_3P(O)$	−18.1(d)	241
	$C_5H_9O_4P$	$(O)P(OCH_2)_3CMe$	−8.0(d)	241
	$C_6H_9O_4P$	**34**; X = O	−10.4(b,d)	186
$8,8/(C)_3$	$C_6H_{12}NO_4P$	$N(CH_2CH_2O)_3P(O)$	−6.6	242
	P-bound atoms = $O_2O'_2$			
C,H	$C_4H_{10}O_4P$	$tBuOP(O)(OH)O^-$	−2.8(m)	243
C′,P[4]	$C_6H_5NO_9P_2$	$4\text{-}O_2NC_6H_4OP(O)(O^-)OP(O)(OH)O^-$	−10.4 or −10.6 $^2J_{PP}$16	244
C′H	$C_6H_4N_2O_8P$	$2,4\text{-}(O_2N)_2C_6H_3OP(O)(OH)O^-$	−3.8(m)	243
	$C_6H_6O_4P$	$PhOP(O)(OH)O^-$	−5.4[a](m) −4.6[b](p)	245
$(C)_2$	$C_2H_6O_4P$	$(MeO)_2P(O)O^-$	−2.4(w)	51
	$C_4H_{10}O_4P$	$(EtO)_2P(O)O^-$	−1.8	246
	$C_8H_{18}O_4P$	$(BuO)_2P(O)O^-$	0.1(y)	247
C.C′	$C_{11}H_{18}NO_4P$	$PhO(O)P(OCH_2CH_2!^+!Me_3)O^-$	−5.7(w)	137
$5/(C)_2$	$C_2H_4O_4P$	$GlyP(O)O^-$	15.9	218
	$C_6H_{12}O_4P$	$PncP(O)O^-$	12.6	248
$5/(C')_2$	$C_4H_6O_4P$	**6**; R = Me, X = O, Y = O^-	−11.2 −12.2	215
$6/(C)_2$	$C_5H_{10}O_4P$	$^-OP(O)(OCH_2CMe_2CH_2O)$	−3.9	246

Rings/ Connectivities	Formula	Structure	NMR data (δ_p [solv] $^nJ_{PZ}$ Hz)	Ref.
	P-bound atoms = OO'$_3$			
C	CH_3O_4P	$MeOP(O)O^-{}_2$	4.4	69
	$C_3H_7O_4P$	$iPrOP(O)(O^-)_2$	0.5	69
	$C_4H_9O_4P$	$tBuOP(O)(O^-)_2$	−3.9	69
	$C_7H_7O_4P$	$PhCH_2OP(O)(O^-)_2$	3.6	69
	$C_3H_5O_4P$	$H_2C{=}CMeOP(O)(O^-)_2$	−1.0(w)	162
	$C_6H_5O_4P$	$PhOP(O)(O^-)_2$	−6.2(v)	66

Section H2: Derivatives of Thio-, Seleno-, and Telluro-Phosphoric Acids

Rings/ Connectivities	Formula	Structure	NMR data	Ref.
	P-bound Atoms = BrO$_2$S'			
$5/(C)_2$	$C_4H_8BrO_2PS$	**5**; R = Me; X = Br, Y = S'	53.3 $^2J_{PC}$1.7	256
		X = S', Y = Br	55.3 $^2J_{PC}$2.4	
	$C_{14}H_{12}BrO_2PS$	**5**; R = Ph; X = Br, Y = S'	56.2(c)	256
		X = S', Y = Br	56.2(c)	
$6/(C)_2$	$C_4H_8BrO_2PS$	**1**; X = Br, Y = S'	43.8(n)	257
	P-bound atoms = Br$_2$SS'			
C	$C_3H_7Br_2PS_2$	$PrSP(S)Br_2$	−0.5	258
	$C_4H_9Br_2PS_2$	$BuSP(S)Br_2$	−4.8	258
	$C_7H_7Br_2PS_2$	$PhCH_2SP(S)Br_2$	−6.1	258
C'	$C_6H_5Br_2PS_2$	$PhSP(S)Br_2$	4.0	258
	P-bound atoms = ClFNS'			
C,Sn	$C_4H_{12}ClFPSSn$	$Me_3SnNMeP(S)FCl$	74.4 $^1J_{PF}$1127 $^2J_{PSn}$18.5	259
	P-bound atoms = ClFOS'			
C	$C_2H_5ClFOPS$	EtOP(S)ClF	59.3 $^1J_{PF}$1150	57
	P-bound atoms = ClFSS'			
C	$C_2H_5ClFPS_2$	EtSP(S)ClF	95.0 $^1J_{PF}$1239	260
C'	$C_6H_5ClFPS_2$	PhSP(S)ClF	84.8 $^1J_{PF}$1226	261
	P-bound atoms = ClNOS'			
C'H;C'	$C_{12}H_{10}ClN_2O_3PS$	$PhNH(4\text{-}O_2NC_6H_4O)P(S)Cl$	57.4	262
$5/C_2$;C	$C_{10}H_{13}ClNOPS$	**12**; X = S, Y = Cl; (2*R*,4*S*,5*R*)	78.5(b)	263
		(2*S*,4*S*,5*R*)	72.9(b)	
5/CC';C	$C_9H_{11}ClNOPS$	**2**; R = Ph, X = Cl, Y = S'	67(c)	60
5/CC';C'	$C_{10}H_{19}ClNOPS$	**3**; R = tBu, X = S, Y = Cl	70.1	61
6/CH;C	$C_3H_7ClNOPS$	**4**; R[1] = R[2] = R[3] = H, X = S', Y = Cl	74[a](c) 70.7[b](y)	264
	P-bound atoms = ClNSS'			
5/CC';C'	$C_7H_7ClNPS_2$	**35**; A = NMe, R = H, X = S', Y = Cl	86	265

TABLE H (continued)
Four Coordinate Compounds Containing a P=Ch Bond but No Bonds to Group IV Atoms

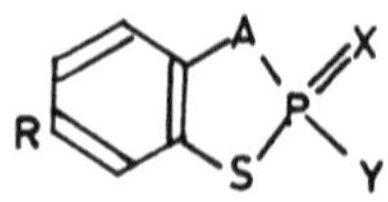

35

Rings/ Connectivities	Formula	Structure	NMR data (δ_p [solv] $^{n}J_{PZ}$ Hz)	Ref.
	P-bound atoms = ClN_2S'			
CP^4,C_2	$C_3H_9Cl_3N_2P_2S_2$	$Cl(Me_2N)P(S)NMeP(S)Cl_2$	70.8; [48.8] $^{2}J_{PP}$19.8	64(b)
	$C_5H_{15}Cl_2N_3P_2S_2$	$Cl(Me_2N)P(S)NMeP(S)(NMe_2)Cl$	73.5 74.8[D]	64(b)
$(C_2)_2$	$C_4H_{12}ClN_2PS$	$(Me_2N)_2P(S)Cl$	91.7	101
$5/(C_2)_2$	$C_4H_{10}ClN_2PS$	$Cl(S)P(MeNCH_2CH_2NMe)$	77.8(b)	69
$6/(CN)_2$	$C_5H_{13}Cl_2N_4PS$	$Cl(S)P(NMeNH)_2CMe(CH_2Cl)$	76.5 76.5[D](b)	154
	P-bound atoms = ClOSS'			
5/C';C'	$C_7H_6ClO_2PS_2$	**35**; R = MeO, A = O, X = S, Y = Cl	91.4	266
	P-bound atoms = $ClO'S_2$			
$(C)_2$	$C_6H_{14}ClOPS_2$	$(PrS)_2P(O)Cl$	55.5	267
$5/(C)_2$	$C_2H_4ClOPS_2$	$Cl(O)P(SCH_2CH_2S)$	109.5(n)	268
	P-bound atoms = ClO_2S'			
$(C)_2$	$C_2H_6ClO_2PS$	$(MeO)_2P(S)Cl$	72.9[b](n) 71.7[a]	269
	$C_4H_{10}ClO_2PS$	$(EtO)_2P(S)Cl$	66.0[a](c) 67.7[b](n)	270
	$C_6H_{14}ClO_2PS$	$(iPrO)_2P(S)Cl$	65[a] 63.0(c)	271
	$C_{10}H_2ClO_2PS$	$(NeoO)_2P(S)Cl$	71.8(b)	76
$5/(C)_2$	$C_2H_4ClO_2PS$	$Cl(S)P(OCH_2CH_2O)$	80	16
	$C_4H_8ClO_2PS$	**5**; R = Me; X = Cl, Y = S'	77.0(c) $^{2}J_{PC}$1.46	256
		X = S', Y = Cl	76.0(c) $^{2}J_{PC}$2.38	
	$C_6H_{12}ClO_2PS$	PncP(S)Cl	80	16
$5/(C')_2$	$C_6H_4ClO_2PS$	CatP(S)Cl	76.5	272
$6/(C)_2$	$C_3H_6ClO_2PS$	$Cl(S)P(OCH_2CH_2CH_2O)$	59.6(c)	273
	$C_4H_8ClO_2PS$	**1**; X = S', Y = Cl	59.0(b)	274
		X = Cl, Y = S'	60.0(b)	
	$C_5H_{10}ClO_2PS$	$Cl(S)P(OCH_2CMe_2CH_2O)$	55.2[b](g) 59.4[a](b)	275
	$C_5H_{10}ClO_2PS$	**8**; $R^1 = R^3$ = Me, R^2 = H, X = S, Y = Cl	56.5	276
	$C_8H_{16}ClO_2PS$	**7**; X = S', Y = Cl	70.4(c)	235
		X = Cl, Y = S'	60.4(c)	
$7/(C)_2$	$C_4H_6ClO_2PS$	$Cl(S)P(OCH_2CH{=}CHCH_2O)$	75.3(b) $^{2}J_{PC}$9.1	277
	$C_4H_8ClO_2PS$	$Cl(S)P(OCH_2CH_2Ch_2Ch_2O)$	72.6(b) $^{2}J_{PC}$9.2	277
	$C_8H_8ClO_2PS$	**30**; A = O, X = S, Y = Cl, R = H	66.6(b) $^{2}J_{PC}$11.3	277
$7/(C')_2$	$C_{12}H_8ClO_2PS$	**31**; X = S, Y = Cl	73.9(c)	239
	P-bound atoms = ClOO'S			
C;C	$C_3H_8ClO_2PS$	EtO(MeS)P(O)Cl	61.5	158
5/C';C'	$C_7H_6ClO_3PS$	**35**; R = MeO, A = X = O, Y = Cl	45	266

Rings/ Connectivities	Formula	Structure	NMR data (δ_p [solv] $^nJ_{PZ}$ Hz)	Ref.
	P-bound atoms = ClO_2Se'			
$(C)_2$	$C_4H_{10}ClO_2PSe$	$(EtO)_2P(Se)Cl$	63.0(n)	278
$(C')_2$	$C_{12}H_{10}ClO_2PSe$	$(PhO)_2P(Se)Cl$	57 $^1J_{PSe}$ − 1042	279
$5/(C')_2$	$C_6H_4ClO_2PSe$	CatP(Se)Cl	65.6(n) $^1J_{PSe}$1130	280
$6/(C)_2$	$C_3H_6ClO_2PSe$	$Cl(Se)P(OCH_2CH_2CH_2O)$	52	281
	$C_5H_{10}ClO_2PSe$	$Cl(Se)P(OCH_2CMe_2CH_2O)$	50	281
	$C_8H_{16}ClO_2PSe$	**7**; X = Cl, Y = Se′	54.6(c) $^1J_{PSe}$1055	235
	P-bound atoms = ClS_2S'			
$5/(C)_2$	$C_2H_4ClPS_3$	$Cl(S)P(SCH_2CH_2S)$	109.2(b)	268
$6/(C)_2$	$C_3H_6ClPS_3$	$Cl(S)P(SCH_2CH_2CH_2S)$	54.7(g)	282
	P-bound atoms = Cl_2NS'			
CP^4	$CH_3Cl_2F_2NP_2S_2$	$Cl_2\mathbf{P}(S)NMeP(S)F_2$	45.5; [56.9] $^1J_{PF}$1139 $^2J_{PP}$5.0	87
	$CH_3Cl_4NP_2S_2$	$MeN[P(S)Cl_2]_2$	47.2	284
CP	$CH_3Cl_4NOP_2S$	$Cl_2\mathbf{P}(S)NMePCl_2$	51.4(r) $^2J_{PP}$122	283
$C'P^4$	$C_6H_5Cl_4NP_2S_2$	$PhN[P(S)Cl_2]_2$	43.5	284
CS	$C_2H_3Cl_5NPS_2$	$Cl_3CSNMeP(S)Cl_2$	65.5	285
CH	CH_4Cl_2NPS	$MeNHP(S)Cl_2$	60.9	284
C′H	$C_2H_4Cl_2NOPS$	$MeCONHP(S)Cl_2$	40.8	286
	$C_6H_6Cl_2NPS$	$PhNHP(S)Cl_2$	48.0	284
C_2	$C_4H_4Cl_2NPS$	$Cl_2P(S)NMeCH_2C{\equiv}CH$	62.0	91
	P-bound atoms = Cl_2NSe'			
C_2	$C_4H_{10}Cl_2NPSe$	$Et_2NP(Se)Cl_2$	39.7 $^1J_{PSe}$996	287
	P-bound atoms = Cl_2OS'			
C	$C_3H_7Cl_2OPS$	$PrOP(S)Cl_2$	57.0	57
C′	$C_6H_5Cl_2OPS$	$PhOP(S)Cl_2$	26	188
	P-bound atoms = $Cl_2O'S$			
C	CCl_2F_3OPS	$TfSP(O)Cl_2$	14.3(c)	289
	$C_3H_7Cl_2OPS$	$PrSP(O)Cl_2$	35.9	290
C′	$C_6H_5Cl_2OPS$	$PhSP(O)Cl_2$	27	288
	P-bound atoms = $Cl_2O'Se$			
C	CCl_2F_3OPSe	$TfSeP(O)Cl_2$	−12.5	291
	P-bound atoms = Cl_2SS'			
C	$CCl_2F_3PS_2$	$TfSP(S)Cl_2$	33.4(c)	289
	P-bound atoms = $FNOS'$			
C′H;C	$C_8H_{11}FNOPS$	$EtO(PhNH)P(S)F$	64.0 $^1J_{PF}$1048	57

TABLE H (continued)
Four Coordinate Compounds Containing a P=Ch Bond but No Bonds to Group IV Atoms

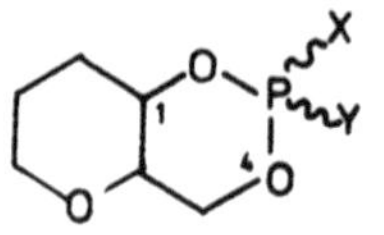

36

Rings/ Connectivities	Formula	Structure	NMR data (δ_p [solv] $^nJ_{PZ}$ Hz)	Ref.
	P-bound atoms = FN′OS′			
C′;C	$C_3H_5FNO_2PS$	EtOP(S) (NCO)F	40.0 $^1J_{PF}$1065	57
	P-bound atoms = FNSS′			
C_2;C	$C_3H_9FNPS_2$	$Me_2N(MeS)P(S)F$	106.0 $^1J_{PF}$1123	260
	$C_6H_{15}FNPS_2$	$Et_2N(EtS)P(S)F$	101.3 $^1J_{PF}$1109	260
C_2;C′	$C_8H_{11}FNPS_2$	$Me_2N(PhS)P(S)F$	94.0 $^1J_{PF}$1080	261
	P-bound atoms = FN_2S'			
$(C_2)_2$	$C_4H_{12}FN_2PS$	$(Me_2N)_2P(S)F$	87.0(n) $^1J_{PF}$1014	103
$5/(C_2)_2$	$C_4H_{10}FN_2PS$	$F(S)P(MeNCH_2CH_2NMe)$	84.2 $^1J_{PF}$1092	101
	P-bound atoms = FOSS′			
C;C	$C_3H_8FOPS_2$	MeO(EtS)P(S)F	102.1 $^1J_{PF}$1158	260
	$C_3H_8FOPS_2$	EtO(MeS)P(S)F	99.1 $^1J_{PF}$1159	260
C;C′	$C_8H_{10}FOPS_2$	ETO(PhS)P(S)F	89.1$^1J_{PF}$1168	261
	P-bound atoms = $FO'S_2$			
$(C)_2$	$C_2H_6FOPS_2$	$(MeS)_2P(O)F$	61.9$^1J_{PF}$1150	111
	P-bound atoms = FO_2S'			
$(C)_2$	$C_2H_6FO_2PS$	$(MeO)_2P(S)F$	65.5(n) $^1J_{PF}$1080	103
	$C_4H_{10}FO_2PS$	$(EtO)_2P(S)F$	62.0 $^1J_{PF}$1072	57
$(C')_2$	$C_{12}H_{10}FO_2PS$	$(PhO)_2P(S)F$	49.5(n) $^1J_{PF}$1105	103
$5/(C)_2$	$C_4H_8FO_2PS$	**5**; R = Me; X = S′, Y = F	74.0(c) $^2J_{PC}$1.8	256
		S = F, Y = S′	76.5(c) $^2J_{PC}$1.65	
$6/(C)_2$	$C_4H_8FO_2PS$	**1**; X = F, Y = S′	57.2(b) $^1J_{PF}$1130	27
		X = S′, Y = F	57.6(b) $^1J_{PF}$1140	
	$C_6H_{10}FO_3PS$	**36**; X = S′, Y = F;		292
		(1*RS*,3*RS*,6*SR*)	56.5	
		(1*RS*,3*SR*,6*SR*)	53.1	
	P-bound atoms = FO_2Se'			
$6/(C)_2$	$C_4H_8FO_2PSe$	**1**; X = Se′, Y = F	60.8(b)	293
		X = F, Y = Se′	61.0(b)	
	P-bound atoms = FS_2S'			
$(C)_2$	$C_2H_6FPS_3$	$(MeS)_2P(S)F$	125.1 $^1J_{PF}$1164	260
	$C_4H_{10}FPS_3$	$(EtS)_2P(S)F$	123.1 $^1J_{PF}$1165	260
$(C')_2$	$C_{12}H_{10}FPS_3$	$(PhS)_2P(S)F$	112 $^1J_{PF}$1175	261

Rings/ Connectivities	Formula	Structure	NMR data (δ_p [solv] $^nJ_{PZ}$ Hz)	Ref.
	P-bound atoms = F_2NS'			
HN	$C_6H_7F_2N_2PS$	PhNHNHP(S)F_2	65.3 $^1J_{PF}$1112	295
HSi	$C_3H_{10}F_2NPSSi$	TmsNHP(S)F_2	66 $^1J_{PF}$1087	294
CP^4	$CH_3Cl_2F_2NOPS$	F_2**P**(S)NMeP(O)Cl_2	57.1 $^1J_{PF}$1136 $^2J_{PP}$2.3	87
CS	$C_2H_3Cl_3F_2NPS_2$	Cl_3CSNMeP(S)F_2	66.3 $^1J_{PF}$1127	285
CSn	$C_4H_{12}F_2NPSSn$	Me_3SnNMeP(S)F_2	71.3 $^1J_{PF}$1099	259
C'H	$C_2H_3F_2NOPS$	MeCONHP(S)F_2	66.3 $^1J_{PF}$1086	286
C_2	$C_2H_6F_2NPS$	Me_2NP(S)F_2	67.9[a](n) 75.7[b] $^1J_{PF}$1081	296
	$C_4H_{10}F_2NPS$	Et_2NP(S)F_2	67.5[a](n) 71.3[b] $^1J_{PF}$1082	296
	P-bound atoms = F_2NSe'			
HSi	$C_3H_{10}F_2NPSeSi$	TmsNHP(Se)F_2	68.3 $^1J_{PF}$1144 $^1J_{PSe}$1075	297
C_2	$C_2H_6F_2NPSe$	Me_2NP(Se)F_2	67.4 $^1J_{PF}$1115	297
	$C_4H_{10}F_2NPSe$	Et_2NP(Se)F_2	73.3 $^1J_{PF}$1122	297
	P-bound atoms = F_2OS'			
C	CH_3F_2OPS	MeOP(S)F_2	52.5(n) $^1J_{PF}$1121	103
	$C_2H_5F_2OPS$	EtOP(S)F_2	50.0 $^1J_{PF}$1120	57
C'	$C_6H_5F_2OPS$	PhOP(S)F_2	43.3(n) $^1J_{PF}$1140	103
	P-bound atoms = $F_2O'S$			
C	CH_3F_2OPS	MeSP(O)F_2	16.9 $^1J_{PF}$1180	111
	$C_2H_5F_2OPS$	EtSP(O)F_2	21.0 $^1J_{PF}$1170	111
C'	$C_6H_5F_2OPS$	PhSP(O)F_2	11.9 $^1J_{PF}$1199	261
	P-bound atoms = F_2SS'			
Sn	$C_3H_9F_2PS_2Sn$	Me_3SnSP(S)F_2	100.4 $^1J_{PF}$1187	259
	P-bound atoms = NOSS'			
5/CC';C;C'	$C_9H_{12}NOPS_2$	**35**; R = H, A = NMe, X = S, Y = OEt	92	298
	P-bound atoms = $NO'S_2$			
5/CC';C,C;	$C_8H_{10}NOPS_2$	**35**; R = H, A = NMe, X = O, Y = SMe	60	298
	P-bound atoms = NO_2S'			
C'O;$(C)_2$	$C_{14}H_{22}NO_3PS_2$	$(iPrO)_2$P(S)N(OMe)CSPh	62	123
CP^4;$(C)_2$	$C_9H_{23}NO_4P_2S_2$	$(EtO)_2$P(S)NMeP(S)$(OEt)_2$	67	169
	$C_{10}H_{26}N_2O_3P_2SSe$	$(MeO)_2$**P**(S)NMeP(Se)(NEt_2)O-iPr	76.3; 73.3(m) $^1J_{PSe}$874 $^2J_{PP}$2.4	124
	$C_{13}H_{31}NO_4P_2SSe$	$(EtO)_2$**P**(S)NMeP(Se)$(O\text{-}iBu)_2$	70.1; 75.3(m) $^1J_{PSe}$926 $^2J_{PP}$5.5	124
C'P;$(C)_2$	$C_{14}H_{25}NO_4P_2S$	$(EtO)_2$**P**(S)NPhP$(OEt)_2$	67.4 $^2J_{PP}$75	299
$C'P^4$;$(C)_2$	$C_{14}H_{25}NO_4P_2S_2$	$(EtO)_2$P(S)NPhP(S)$(OEt)_2$	62.9	169
	$C_{20}H_{30}N_2O_4P_2S$	$(EtO)_2$**P**(S)NPhP$(OEt)_2$=NPh	68.5	169

TABLE H (continued)
Four Coordinate Compounds Containing a P=Ch Bond but No Bonds to Group IV Atoms

Rings/ Connectivities	Formula	Structure	NMR data (δ_P [solv] $^nJ_{PZ}$ Hz)	Ref.
C'H;$(C)_2$	$C_4H_{10}NO_2PS_3$	$(MeO)_2P(S)NHCSSMe$	63	300
	$C_8H_{18}NO_2PS_3$	$(iPrO)_2P(S)NHCSSMe$	54	166
	$C_{10}H_{16}NO_2PS$	$(EtO)_2P(S)NHPh$	65.2	158
	$C_{11}H_{16}NO_2PS_2$	$(EtO)_2P(S)NHCSPh$	61.0	301
	$C_{11}H_{16}NO_3PS_2$	$(EtO)_2P(S)NHCSOPh$	59	192
	$C_{11}H_{17}N_2O_2PS_2$	$(EtO)_2P(S)NHCSNHPh$	56	300
	$C_{16}H_{26}NO_4PS$	$(PrO)_2P(S)NHC_6H_4COOPr$-3	68	302
C_2;$(C)_2$	$C_8H_{18}Cl_2NO_2PS$	$(EtO)_2P(S)N(C_2H_4Cl)_2$	71.9(c)	270(a)
CC';$(C)_2$	$C_7H_{18}NO_3PS_2Si$	$(MeO)_2P(S)NMeCSOTms$	73	303
C'_2;$(C)_2$	$C_{13}H_{20}NO_2PS_2$	$(iPrO)_2P(S)NPhCSH$	64	166
5/HS[4];$(C)_2$	$C_9H_{12}NO_4PS_2$	$PhSO_2NH(S)P(OCHMeCH_2O)$	72.6, 73.0[D]	304
5/CH;$(C)_2$	$C_7H_{16}NO_2PS$	**10**; X = S, Y = OMe; $(R)_P$	88.5[Rp] (c); 87.4[Rp] (c)	143
5/C_2;$(C)_2$	$C_4H_{10}NO_2PS$	$Me_2N(S)PGlc$	91.4(c)	69
	$C_5H_9F_3NO_2PS$	**11**; R = Me, X = S, Y = OCH_2Tf	85.0(c)	305
	$C_{11}H_{16}NO_2PS$	**12**; X, Y = S', OMe; (2*S*,4*S*,5*R*)	83.9(b)	263
5/C_2;$(C')_2$	$C_{14}H_{16}NO_2PS$	**13**; X = S, Y = NEt_2	86.9	150
	$C_{14}H_{16}NO_2PS$	**14**; X = S, Y = NEt_2	89.3	150
5/CC';C,C'	$C_{11}H_{22}NO_2PS$	**3**; R = tBu, X = S, Y = OMe	76.4(c)	152
6/CN';C,C'	$C_{10}H_{13}N_2O_2PS$	**16**; X = S, Y = OPh	52(j)	154
6/CH;C,H	$C_3H_8NO_2PS$	**4**; R[1] = R[2] = R[3] = H, X = S', Y = OH	57.6 (w, pH 7.7)	264(b)
6/CH;$(C)_2$	$C_{14}H_{10}NO_2PS$	**4**; R[1] = R[2] = R[3] = H, X = S', Y = OMe	67.7	264(b)
	$C_8H_{18}NO_2PS$	**4**; R[1] = R[2] = H, R[3] = tBu		155
		X= S', Y = OMe	71.6(b)	
		X = OMe, Y = S'	68.6(b)	
6/C'H;$(C)_2$	$C_{10}H_{14}NO_2PS$	**1**; X = S', Y = NHPh	63.0[a](b); 63.8[b] $^1J_{PN}$24.5(b-p)	307
		X = NHPh, Y = S'	59.5[a](b);61.0[b] $^1J_{PN}$10.8(b-p)	
6/C_2;$(C)_2$	$C_5H_{12}NO_2PS$	$Me_2N(S)P(OCH_2CH_2CH_2O)$	74	308
	$C_6H_{14}NO_2PS$	**1**; X = S', Y = Me_2N	73.5(c) 70.7	310
		X = Me_2N, Y = S'	73.0, 70.2	
	$C_7H_{16}NO_2PS$	$Et_2N(S)P(OCH_2CH_2CH_2O)$	71.3(c)	273
	$C_8H_{16}NO_3PS$	**36**; X = S', Y = NMe_2		292
		(1*RS*,3*RS*,6*SR*)	74.4	
		(1*RS*,3*SR*,6*SR*)	75.1	
6/C_2;C,C'	$C_{12}H_{10}NO_2PS$	**4**; R[1] = R[2] = R[3] = Me, X = S', Y = OPh	66(b)	161
6/C_2;$(C')_2$	$C_{14}H_{16}NO_2PS$	**17**; X = S, Y = NEt_2	60.7	311
6/CC';$(C)_2$	$C_{14}H_{22}NO_2PS$	**4**; R[1] = Ph, R[2] = H, R[3] = tBu;		155
		R = S'; Y = OMe	71.5(b)	
		X = OMe, Y = S'	68.0(b)	
7/C_2;$(C)_2$	$C_6H_{12}NO_2PS$	$Me_2N(S)P(OCH_2CH{=}CHCH_2O)$	81.3(b) $^2J_{PC}$6.7(c)	277
	$C_6H_{14}NO_2PS$	$Me_2N(S)P(OCH_2CH_2CH_2CH_2O)$	79.6(b) $^2J_{PC}$7.0	277
	$C_{10}H_{14}NO_2PS$	**30**; R = H A = O X = S, Y = NMe_2	79.8(b) $^2J_{PC}$5.6	277

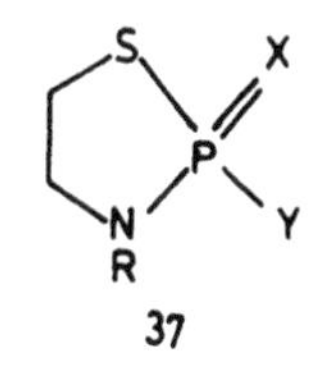

37

38

P-bound atoms = NOO′S

Rings/ Connectivities	Formula	Structure	NMR data (δ_p [solv] $^nJ_{PZ}$ Hz)	Ref.
C_2;C;C	$C_6H_{13}F_3NO_2PS$	MeO(Et$_2$N)P(O)STf	21.6	312
5/C_2;C;C	$C_5H_{11}ClNO_2PS$	**37**; R = CH_2CH_2Cl, X = O, Y = OMe	49.9(r)	270(a)
	$C_{11}H_{16}NO_2PS$	**12**; R = Me, X, Y = O′, SMe		263
		(2*R*,4*S*,5*R*)	40.7(c)	
		(2*S*,4*S*,5*R*)	40.0(c)	
	$C_{12}H_{18}NO_2PS$	**38**; X, Y = O′, OEt		313
		(2*R*,4*S*,5*S*)	45.7	
		(2*R*,4*S*,5*R*)	45.9	
		(2*S*,4*S*,5*R*)	44.3	
5/CC′;C;C	$C_{10}H_{14}NO_2PS$	**2**; R = Ph, X = O, Y = MeS	34.6(c) 32.9(b)	60
		X = MeS, Y = O′	34.1(c) 32.2(b)	
6/CH;C;C	$C_4H_{10}NO_2PS$	**4**; R^1 = R^2 = R^3 = H, X = O′, Y = MeS	27.5(c)	264(b)

P-bound atoms = N′O$_2$S′

Rings/ Connectivities	Formula	Structure	NMR data (δ_p [solv] $^nJ_{PZ}$ Hz)	Ref.
N′;(C)$_2$	$C_{10}H_{15}N_2O_2PS$	(EtO)$_2$P(S)N=NPh	74	314
C′;(C)$_2$	$C_5H_{10}NO_2PS_2$	(EtO)$_2$P(S)NCS	46.0	57
	$C_7H_{14}NO_2PS_2$	(iPrO)$_2$P(S)NCS	42	166
	$C_{11}H_{22}NO_2PS_2$	(NeoO)$_2$P(S)NCS	69.0(n)	76
	$C_{15}H_{21}N_2O_2PS_2$	(iPrO)$_2$P(S)N=CPh(SCH$_2$CN)	57	315

P-bound atoms = NO$_2$Se′

Rings/ Connectivities	Formula	Structure	NMR data (δ_p [solv] $^nJ_{PZ}$ Hz)	Ref.
CP;(C)$_2$	$C_{13}H_{31}NO_4P_2Se$	(iPrO)$_2$**P**(Se)NMeP(O-iPr)$_2$	68.4(m) $^1J_{PSe}$901 $^2J_{PP}$90	124
CP^4;(C)$_2$	$C_{17}H_{39}NO_4P_2Se_2$	(BuO)$_2$P(Se)NMeP(Se)(OBu)$_2$	74.7(m) $^1J_{PSe}$937 $^2J_{PP}$122	124
C′H;(C)$_2$	$C_{10}H_{16}NO_2PSe$	(EtO)$_2$P(Se)NHPh	66.5	158
C_2;(C)$_2$	$C_4H_{12}NO_2PSe$	(MeO)$_2$P(Se)NMe$_2$	87.1 $^1J_{PSe}$903	37
C_2;C,C′	$C_{11}H_{18}NO_2PSe$	MeO(PhO)P(Se)NEt$_2$	78	316
6/CH;(C)$_2$	$C_8H_{18}NO_2PSe$	**1**; X = Se′, Y = NH-tBu	66.0(b) $^1J_{PSe}$870	156
		X = NH-tBu, Y = Se′	59.5(b) $^1J_{PSe}$896	
	$C_{10}H_{22}NO_2PSe$	**8**; R^1 = R^2 = R^3 = Me,		317
		X = NH-tBu, Y = Se′	52.6 $^1J_{PSe}$896	
		X = Se′, Y = NH-tBu	56.6 $^1J_{PSe}$896	
6/C′H;(C)$_2$	$C_{10}H_{14}NO_2PSe$	**1**; X = Se′, Y = NHPh	62.3(b) $^1J_{Se}$903	156
		X = NHPh, Y = Se′	59.0(b) $^1J_{PSe}$994	
	$C_{11}H_{16}NO_2PSe$	PhNH(Se)P(OCH$_2$CMe$_2$CH$_2$O)	62.0	158
6/C_2;(C)$_2$	$C_6H_{14}NO_2PSe$	**1**; X = Se′, Y = NMe$_2$	79.0(b) $^1J_{PSe}$930	309
		X = NMe$_2$, Y = Se′	79.8(b) $^1J_{PSe}$960	
	$C_7H_{16}NO_2PSe$	Me$_2$N(Se)P(OCH$_2$CMe$_2$CH$_2$O)	79.2 $^1J_{PSe}$914	37
	$C_7H_{16}NO_2PSe$	**8**; R^1 = H, R^2 = R^3 = Me,		37
		X = Se′, Y = NMe$_2$	75.1 $^1J_{PSe}$924	
		X = NMe$_2$, Y = Se′	77.6 $^2J_{PSe}$924	

TABLE H (continued)
Four Coordinate Compounds Containing a P=Ch Bond but No Bonds to Group IV Atoms

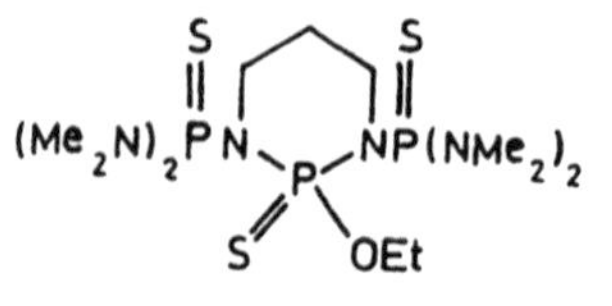

39

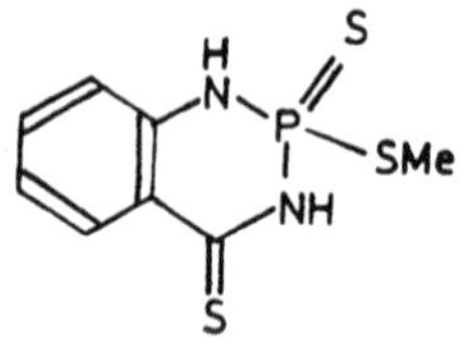

40

Rings/ Connectivities	Formula	Structure	NMR data (δ_p [solv] $^nJ_{PZ}$ Hz)	Ref.
	P-bound atoms = NOO'Se			
C_2;C';C	$C_{11}H_{18}NO_2PSe$	PhO(MeSe)P(O)NEt_2	22	316
	P-bound atoms = NS_2S'			
5/C_2;$(C)_2$	$C_6H_{14}NPS_3$	$Et_2N(S)P(SCH_2CH_2S)$	109.6(b)	268
6/C_2;$(C)_2$	$C_5H_{12}NPS_3$	$Me_2N(S)P(SCH_2CH_2CH_2S)$	82.9(g)	282
	P-bound atoms = N_2OS'			
$(HN)_2$;C'	$C_6H_{11}N_4OPS$	PhOP(S)$(NHNH_2)_2$	76	182
H_2,CN;C'	$C_7H_{12}N_3OPS$	PhOP(S)$(NMeNH_2)NH_2$	69(j)	154
$(CH)_2$;Si	$C_{11}H_{29}N_2OPSSi$	$(tBuNH)_2P(S)OTms$	53.2(b)	318
$(C_2)_2$;C'	$C_{10}H_{16}ClN_2OPS$	$(Me_2N)_2P(S)OC_6H_4Cl$-2	81.0	9
5/C'H,C_2;C	$C_{16}H_{19}N_2OPS$	**12**; X = S', Y = NHPh		263
		(2*R*,4*S*,5*R*)	75.2	
		(2*S*,4*S*,5*R*)	72.2	
	$C_8H_{11}N_2OPS$	**15**; R = H, X = S, Y = NMe_2	80	319
5/$(C_2)_2$;C	$C_6H_{12}F_3N_2OPS$	$TfCH_2O(S)P(MeNCH_2CH_2NMe)$	84.1(c)	305
5/$(C_2)_2$;C'	$C_{10}H_{15}N_2OPS$	$PhO(S)P(MeNCH_2CH_2NMe)$	83.1(b)	69
5/CC',C_2;C	$C_{11}H_{17}N_2OPS$	**2**; R = Ph, X = S', Y = NMe_2	74.3(c)	60
	$C_{12}H_{19}N_2OPS$	**11**; R = Ph, X = S, Y = NEt_2	76	319
6/$(CN)_2$;C'	$C_9H_{15}N_4OPS$	$PhO(S)P(NMeNH)_2CH_2$	68	182
6/CN',C_2;C'	$C_{11}H_{16}N_3OPS$	$PhO(S)P(MeNN{=}CMeCH_2NMe)$	58.5(c)	154
6/$(CP)^4)_2$;C	$C_{13}H_{35}N_6OP_3S_3$	**39**	62.2 or 76.2	320
6/$(CH)_2$;C	$C_5H_{13}N_2OPS$	$EtO(S)P(NHCH_2CH_2CH_2NH)$	65.1	320
6/$(CH)_2$;C'	$C_9H_{13}N_2OPS$	$PhO(S)P(HNCH_2CH_2CH_2NH)$	64(p)	175
6/CH,C_2;C	$C_9H_{21}N_2OPS$	**4**; R^1 = R^2 = H, R^3 = tBu;		180
		X = S', Y = NMe_2	75.2(b)	
		X = NMe_2, Y = S'	73.2(b)	
6/(C_2);C	$C_8H_{19}N_2OPS$	$EtO(S)P(MeNHCH_2CH_2CHMeNMe)$	78.2 77.7[D]	179
6/C_2,CC';C	$C_{15}H_{25}N_2OPS$	**4**; R^1 = Ph, R^2 = H, R^3 = tBu;		180
		X = S', Y = NMe_2	73.7(c)	
		X = NMe_2, Y = S'	68.2(c)	
	P-bound atoms = $N_2O'S$			
5/H_2,C_2;C	$C_{10}H_{15}N_2OPS$	**38**; X, Y = O', NH_2		313
		(2*R*,4*S*,5*S*)	41.7	
		(2*R*,4*S*,5*R*)	43.7	
5/CH,C_2;C	$C_4H_{11}N_2OPS$	**37**; R = H, X = O, Y = NMe_2	54.8(y)	321
5/CC',C_2;C'	$C_9H_{13}N_2OPS$	**35**; R = H, A = NMe, X = O, Y = NMe_2	44	298

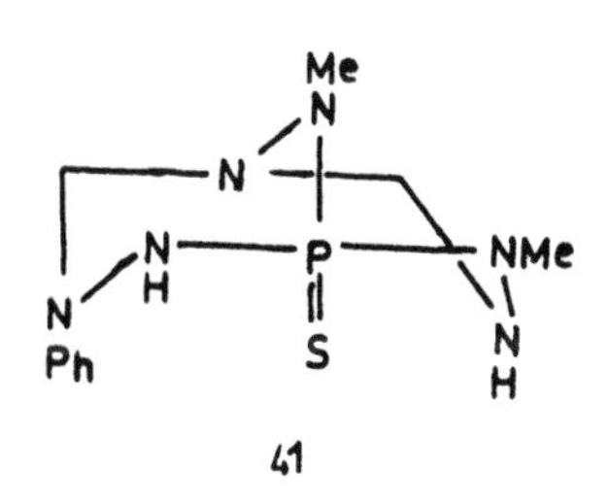

41

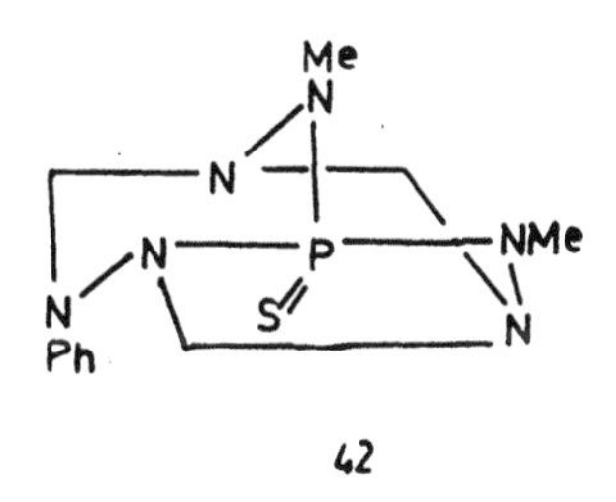

42

Rings/ Connectivities	Formula	Structure	NMR data (δ_p [solv] $^nJ_{PZ}$ Hz)	Ref.
	P-bound atoms = N_2OSe'			
$(CH)_2$;Si	$C_{11}H_{29}N_2OPSeSi$	$(tBuNH)_2P(Se)OTms$	42.6(c) $^1J_{PSe}$824	318
	P-bound atoms = N_2OTe'			
$(C_2)_2$;C	$C_{10}H_{25}N_2OPTe$	$(Et_2N)_2P(Te)OEt$	6.0	322
	P-bound atoms = N_2SS'			
$(C_2)_2$;Si	$C_7H_{21}N_2PS_2Si$	$(Me_2N)_2P(S)STms$	89.9(r)	323
	$C_{11}H_{29}N_2PS_2Si$	$(Et_2N)_2P(S)STms$	95.7	323
5/CH,C_2;C	$C_4H_{11}N_2PS_2$	**37**; R = H, X = S, Y = NMe_2	88.2	321
5/C_2,CC';C'	$C_9H_{13}N_2PS_2$	**35**; R = H, A = NMe, X = S, Y = NMe_2	99	265
6/$(C'H)_2$;C	$C_8H_9N_2PS_3$	**40**	67.5	324
	P-bound atoms = $N_2S'_2$			
$(CH)_2$;1.p.	$C_8H_{20}N_2PS_2$	$(BuNH)_2P(S)S^-$	82.5	325
$(C_2)_2$;1.p.	$C_{16}H_{36}N_2PS_2$	$(Bu_2N)_2P(S)S^-$	101.5	325
	P-bound atoms = N_3S'			
$(NH)_3$	$C_{18}H_{21}N_6PS$	$(PhNHNH)_3P(S)$	78(e) 70	327
$(NH)_2,C_2$	$C_{14}H_{20}N_5PS$	$(PhNHNH)_2P(S)NMe_2$	73	182
	$C_{16}H_{22}N_5OPS$	$(PhNHNH)_2P(S)NHMorph$	71.0(c)	168
$(CN)_3$	$C_3H_{15}N_6PS$	$(H_2NNMe)_3P(S)$	85	326
$CP^4,(C_2)_2$	$C_9H_{27}N_5P_2S_2$	$(Me_2N)_2P(S)NMeP(S)(NMe_2)_2$	77.7	64(b)
CH,$(C_2)_2$	$C_5H_{16}N_3PS$	$MeNHP(S)(NMe_2)_2$	79.5(r)	88
$(C_2)_3$	$C_6H_{12}N_3PS$	$(S)P(NCH_2CH_2)_3$	117	184
	$C_6H_{18}N_3PS$	$(Me_2N)_3P(S)$	82.2[a](r) 82.0[b](n)	328
6/NH,$(CN)_2$	$C_9H_{17}N_6PS$	$PhNHNHP(S)(NMeNH)_2CH_2$	76(c)	326
6/$(CN)_3$	$C_6H_{18}ClN_6PS$	$H_2NNMe(S)P(NMeNH)_2C(CH_2Cl)Me$	82.5 85[D](c)	154
6/$(CN)_2,C_2$	$C_5H_{16}N_5PS$	$Me_2N(S)P(NMeNH)_2CH_2$	62	182
6,6/HN, $(CN)_2$	$C_{10}H_{17}N_6PS$	**41**	66(c)	326
6,6$(CN)_3$	$C_6H_{15}N_6PS$	**20**; X = S	69	182
	$C_6H_{18}N_6P_2S_2$	$(S)P(NMeNMe)_3P(S)$	66.5(r)	185
6,6$(C_2)_3$	$C_8H_{18}N_3PS$	$(S)P(NMeCH_2)_3CMe$	73.2	188
5,5,5,/$(C_2)_3$	$C_6H_{12}N_3PS$	**21**; X = S	148	184
5,6,6/$(CN)_3$	$C_{11}H_{17}N_6PS$	**42**	87(c)	326
6,6,6/$(CP^4)_3$	$C_6H_{18}N_6P_4S_4$	**22**; X^1 = X^2 = X^3 = X^4 = S	67.0	329
6,6,6/(CP), $(CP^4)_2$	$C_6H_{18}N_6P_4S_3$	**22**; X^1 = X^2 = X^3 = S, X^4 = 1.p.	73.2	329
6,6,6/$(CP)_3$	$C_6H_{18}N_6P_4S$	**22**; X^1 = S, X^2 = X^3 = X^4 = 1.p.	52	329

TABLE H (continued)
Four Coordinate Compounds Containing a P=Ch Bond but No Bonds to Group IV Atoms

Rings/ Connectivities	Formula	Structure	NMR data (δ_p [solv] $^nJ_{PZ}$ Hz)	Ref.
		P-bound atoms = N_3Se'		
$(C_2)_3$	$C_6H_{12}N_3PSe$	$(Se)P(NCH_2CH_2)_3$	133 $^1J_{PSe}851$	184
	$C_6H_{18}N_3PSe$	$(Me_2N)_3P(Se)$	82.5(c) $^1J_{PSe}784$	187
	$C_{12}H_{30}N_3PSe$	$(Et_2N)_3P(Se)$	77.3 $^1J_{PSe}790$	330
6,6/$(CN)_3$	$C_6H_{15}N_6PSe$	**20**; X = Se	75.9(b)	186
	$C_6H_{18}N_6P_2Se_2$	$(Se)P(NMeNMe)_3P(Se)$	67.1(b,c) $^1J_{PSe}-922$	331
6,6/$(C_2)_3$	$C_8H_{18}N_3PSe$	$(Se)P(NMeCH_2)_3CMe$	77.7(c) $^1J_{PSe}854$	187
5,5,5/$(C_2)_3$	$C_6H_{12}N_3PSe$	**21**; X = Se	146 $^1J_{PSe}890$	184
6,6,6/$(CP)_3$	$C_6H_{18}N_6P_4Se$	**22**; X^1 = Se, X^2 = X^3 = X^4 = l.p.	59.0	332
6,6,6/ $(CP)_2(CP^4)$	$C_6H_{18}N_6P_4Se_2$	**22**; X^1 = X^2 = Se, X^3 = X^4 = l.p.	70.0	332
6,6,6/$(CP^4)_3$	$C_6H_{18}N_6P_4Se_4$	**22**; X^1 = X^2 = X^3 = X^4 = Se	64.0	332
		P-bound atoms = N_3Te'		
$(C_2)_3$	$C_6H_{18}N_3PTe$	$(Me_2N)_3P(Te)$	52.0 $^1J_{P123Te}1709$ $^1J_{P125Te}2045$	333
	$C_{12}H_{30}N_3PTe$	$(Et_2N)_3P(Te)$	48	322
6,6/$(CN)_3$	$C_6H_{18}N_6P_2Te_2$	$(Te)P(NMeNMe)_3P(Te)$	101.5(b)	185
		P-bound atoms = OS_2S'		
5/C;$(C)_2$	$C_4H_9OPS_3$	$MeO(S)P(SCHMeCH_2S)$	121.7(c)	334
5/C';$(C)_2$	$C_8H_9OPS_3$	$PhO(S)P(SCH_2CH_2S)$	120.7(c)	69
6/C;$(C)_2$	$C_4H_9OPS_3$	$MeO(S)P(SCH_2CH_2CH_2S)$	84.6(g)	282
	$C_6H_{13}OPS_3$	$ETO(S)P(SCHMeCH_2CH_2S)$	86.6, 90.0^D	335
		P-bound atoms = $O'S_3$		
$(P^4)_3$	$C_{18}H_{42}O_7P_4S_6$	$(O)P[S(S)P(O\text{-}iPr)_2]_3$	73.9(c)	336
$(C)_3$	$C_3H_9OPS_3$	$(MeS)_3P(O)$	66.6	337
	$C_6H_{15}OPS_3$	$(EtS)_3P(O)$	61	338
	$C_{21}H_{21}OPS_3$	$(PhCH_2S)_3P(O)$	50.8	339
$(C')_3$	$C_{18}H_{15}OPS_3$	$(PhS)_3P(O)$	20.5	339
		P-bound atoms = $O'Se_3$		
$(C)_3$	$C_3H_9OPSe_3$	$(MeSe)_3P(O)$	16	340
		P-bound atoms = O_2P^4S'		
$(C)_2$;O_2S'	$C_8H_{20}O_4P_2S_2$	$(EtO)_2P(S)P(S)(OEt)_2$	85.3	254
$(C')_2$;O_2S'	$C_{24}H_{20}O_4P_2S_2$	$(PhO)_2P(S)P(S)(OPh)_2$	40	341
		P-bound atoms = O_2SS'		
C,P^4;C	$C_8H_{20}O_3P_2S_4$	$[(EtO)(EtS)P(S)]_2O$	94.5	301
$(C)_2$;As	$C_{14}H_{25}AsO_4P_2S_4$	$PhAs[SP(S)(OEt)_2]_2$	89.0	342
$(C)_2$;Cl	$C_{10}H_{22}ClO_2PS_2$	$(NeoO)_2P(S)SCl$	83.3(r)	76
$(C)_2$;N	$C_{15}H_{32}NO_2PS_2$	$(NeoO)_2P(S)SPip$	96.0(b)	76

Rings/ Connectivities	Formula	Structure	NMR data (δ_p [solv] $^nJ_{PZ}$ Hz)	Ref.
(C)$_2$;P^4	$C_4H_{12}O_4P_2S_3$	(MeO)$_2$P(S)SP(S)(OMe)$_2$	83.5	345
	$C_8H_{20}O_4P_2S_2Se$	(EtO)$_2$**P**(S)SP(Se)(OEt)$_2$	73.6(b-c) $^2J_{PP}$6	225(b)
	$C_8H_{20}O_4P_2S_3$	(EtO)$_2$P(S)SP(S)(OEt)$_2$	78.0	345
	$C_{12}H_{28}O_4P_2S_3$	(iPrO)$_2$P(S)SP(S)(O-iPr)$_2$	76	192
	$C_{18}H_{42}O_6P_4S_7$	[(iPrO)$_2$**P**(S)S]$_3$P(S)	81.5, 82.1(c)	336
(C)$_2$;P	$C_8H_{20}O_2P_2S_4$	(EtO)$_2$P(S)SP(SEt)$_2$	85.5 $^2J_{PP}$51	343
	$C_{12}H_{20}O_2P_2S_2$	(EtO)$_2$P(S)SPEtPh	89	344
	$C_{18}H_{42}O_6P_4S_6$	[(iPrO)$_2$**P**(S)S]$_3$P	81.8(c)	336
(C)$_2$;S	$C_4H_{12}O_4P_2S_4$	(MeO)$_2$P(S)SSP(S)(OMe)$_2$	89.0	345
	$C_8H_{20}O_4P_2S_4$	(EtO)$_2$P(S)SSP(S)(OEt)$_2$	84.0	345
	$C_8H_{20}O_4P_2S_5$	(EtO)$_2$P(S)SSSP(S)(OEt)$_2$	79	139
	$C_{12}H_{28}O_4P_2S_4$	(iPrO)$_2$P(S)SSP(S)(O-iPr)$_2$	81.8(c)	336
	$C_{12}H_{28}O_4P_2S_6$	(iPrO)$_2$P(S)S$_4$P(S)(O-iPr)$_2$	74.4(c)	336
(C)$_2$;Sb	$C_{14}H_{25}O_4P_2S_4Sb$	PhSb[SP(S)(OEt)$_2$]$_2$	92.1	342
(C)$_2$;Se	$C_{12}H_{28}O_4P_2S_4Se$	Se[SP(S)(O-iPr)$_2$]$_2$	80.8	336
(C)$_2$;Si	$C_5H_{15}O_2PS_2Si$	(MeO)$_2$P(S)STms	93	193
	$C_7H_{19}O_2PS_2Si$	(EtO)$_2$P(S)STms	91	193
(C)$_2$;H	$C_2H_7O_2PS_2$	(MeO)$_2$P(S)SH	89.9/90.1(a,c,n)	346
	$C_4H_{11}O_2PS_2$	(EtO)$_2$P(S)SH	85.5	347
	$C_6H_7F_8O_2PS_2$	(CHF$_2$CF$_2$CH$_2$O)$_2$P(S)SH	90.8	348
	$C_6H_{15}O_2PS_2$	(iPrO)$_2$P(S)SH	81.6(c)	336
(C)$_2$;C	$C_3H_9O_2PS_2$	(MeO)$_2$P(S)SMe	99.7(c) 98.5(a)	346
	$C_6H_{15}O_2PS_2$	(EtO)$_2$P(S)SEt	94.1	345
	$C_7H_{19}O_3PS_2Si$	(MeO)$_2$P(S)SCHMeOTms	94	349
	$C_9H_{21}O_2PS_2$	(iPrO)$_2$P(S)S-iPr	84	350
(C)$_2$;C′	$C_{15}H_{30}N_3O_6P_3S_6$	[(EtO)$_2$P(S)SĊ=Ṅ]$_3$	78	350
	$C_{16}H_{26}NO_3PS_2$	(iPrO)$_2$P(S)SCPh=NOPr	76^E 82^Z	123
(C)$_2$;C″	$C_{11}H_{22}NO_2PS_2$	(NeoO)$_2$P(S)SCN	74.9(r)	76
(C′)$_2$;P^4	$C_{24}H_{20}O_4P_2S_3$	(PhO)$_2$P(S)SP(S)(OPh)$_2$	70	351
(C′)$_2$;H	$C_{12}H_{11}O_2PS_2$	(PhO)$_2$P(S)SH	80	351
5/(C)$_2$;H	$C_6H_{13}O_2PS_2$	PncP(S)SH	93.1(c)	350
5/(C)$_2$;C	$C_5H_{11}O_2PS_2$	**5**; R = Me; X = S′, Y = SMe	69.5(c) $^2J_{PC}$0.0	256
		X = SMe, Y = S′	67.8(c) $^2J_{PC}$0.0	
5/(C)$_2$;C′	$C_{10}H_{13}O_2PS_2$	**5**; R = Me; X = S′, Y = SPh	99.0$^2J_{PC}$3.05	256
		X = SPh, Y = S′	101.0 $^2J_{PC}$0.85	
6/(C)$_2$;P^4	$C_{10}H_{20}O_5P_2S_2$	**24**; X = Y = S, Z = O	64.5; 6.0	255
6/(C)$_2$;H	$C_5H_{11}O_2PS_2$	HS(S)P(OCH$_2$CMe$_2$CH$_2$O)	78.6(c)	352
	$C_7H_{15}O_2PS_2$	HS(S)P(OCMe$_2$CH$_2$CMe$_2$O)	75.0	352
	$C_8H_{17}O_2PS_2$	**7**; X, Y = S′, SH	76.0	353
6/(C)$_2$;C	$C_5H_{11}O_2PS_2$	**1**; X = S′, Y = SMe	95.5(b)	26
		S = SMe, Y = S′	88.5(b)	
	$C_6H_{13}O_2PS_2$	MeS(S)P(OCH$_2$CMe$_2$CH$_2$O)	90.0	158
	$C_{11}H_{21}O_6PS_2$	**26**; A = S; X = S′, Y = OEt	66.6	24
		X = OEt, Y = S′	60.5	
7/(C′)$_2$;C	$C_{13}H_{11}O_2PS_2$	**31**; X = S, Y = SMe	111.9(c)	239

P-bound atoms = O$_2$S′$_2$

Rings/ Connectivities	Formula	Structure	NMR data	Ref.
(C)$_2$	$C_2H_6O_2PS_2$	(MeO)$_2$P(S)S$^-$	118.9(a)	354
	$C_4H_{10}O_2PS_2$	(EtO)$_2$P(S)S$^-$	114 (a,m)	354
(C′)$_2$	$C_{14}H_{14}O_2PS_2$	(4-TolO)$_2$P(S)S$^-$	108.8(a)	354
5/(C)$_2$	$C_4H_8O_2PS_2$	**5**; X, Y = S$_2^-$ R = Me (4*R*,5*R*)	126.5(m)	355
6/(C)$_2$	$C_4H_8O_2PS_2$	**1**; X, Y = S$_2^-$	110.4(c)	356

P-bound atoms = OO′S$_2$

Rings/ Connectivities	Formula	Structure	NMR data	Ref.
C;(C)$_2$	$C_4H_{11}O_2PS_2$	MeO(MeS)P(O)SEt (*R*)	56.0(b)	357
5/C;(C)$_2$	$C_3H_7O_2PS_2$	MeO(O)P(SCH$_2$CH$_2$S)	113.4(b)	268
7/C′;(C)$_2$	$C_{16}H_{17}O_2PS_2$	**30**; R = Me, A = S, X = O, Y = OPh	50.4(c)	358

TABLE H (continued)
Four Coordinate Compounds Containing a P=Ch Bond but No Bonds to Group IV Atoms

Rings/ Connectivities	Formula	Structure	NMR data (δ_p [solv] $^nJ_{PZ}$ Hz)	Ref.
	P-bound atoms = OO′SS′			
C,1.p;C	$C_2H_6O_2PS_2$	MeO(MeS)P(S)O$^-$	73.5(a)	357
	P-bound atoms = O$_2$SSe′			
(C)$_2$;H	$C_4H_{11}O_2PSSe$	(EtO)$_2$P(Se)SH	84.7 $^1J_{PSe}$777	330
(C)$_2$;C	$C_5H_{13}O_2PSSe$	(EtO)$_2$P(Se)SMe	94.0 $^1J_{PSe}$887	330
	$C_7H_{15}O_3PSSe$	(EtO)$_2$P(Se)SCH$_2$COMe	71.0	359
6/(C)$_2$;S	$C_{10}H_{20}O_4P_2S_2Se_2$	**24**; X = Z = Se, Y = SS	65.9(b-c)	225(b)
6/(C)$_2$;C	$C_5H_{11}O_2PSSe$	**1**; X = Se′, Y = MeS	94.0(n) $^1J_{PSe}$895	26
		X = MeS, Y = Se′	88.5(n) $^1J_{PSe}$	960
	P-bound atoms = O$_2$S′Se			
(C)$_2$;Se	$C_8H_{20}O_4P_2S_2Se_2$	(EtO)$_2$P(S)SeSeP(S)(OEt)$_2$	72.7(c) $^1J_{PSe}$506	360
(C)$_2$;C	$C_6H_{15}O_2PSSe$	(EtO)$_2$P(S)SeEt	85.5 $^1J_{PSe}$477	330
	$C_7H_{15}O_3PSSe$	(EtO)$_2$(P(S)SeCH$_2$COMe	72.0	359
6/(C)$_2$;Se	$C_{10}H_{20}O_4P_2S_2Se_2$	**24**; X = Z = S, Y = SeSe	63.9(b) $^1J_{PSe}$474	360
6/(C)$_2$;C	$C_5H_{11}O_2PSSe$	**1**; X = SeMe, Y = S′	78.5 $^1J_{PSe}$435	27
		X = S′, Y = SeMe	87.5(b) $^1J_{PSe}$510	
	P-bound atoms = O$_2$S′Se′			
(C)$_2$	$C_4H_{10}O_2PSSe$	(EtO)$_2$P(Se)S$^-$	100.3/100.7(a) $^1J_{PSe}$752	360
	P-bound atoms = O$_2$SeSe′			
(C)$_2$;H	$C_4H_{11}O_2PSe_2$	(EtO)$_2$P(Se)SeH	61.5 $^1J_{PSe}$822	330
(C)$_2$;C	$C_6H_{15}O_2PSe_2$	(EtO)$_2$P(Se)SeEt	73.5 $^1J_{P=Se}$882 $^1J_{P\text{-}Se}$453	330
6/(C)$_2$;C	$C_6H_{13}O_2PSe_2$	MeSe(Se)P(OCH$_2$CMe$_2$CH$_2$O)	79.4 $^1J_{P=Se}$945 $^1J_{P\text{-}Se}$469	262
	P-bound atoms = O$_2$Se′$_2$			
6/(C)$_2$	$C_5H_{10}O_2PSe_2$	$^-$Se(Se)P(OCH$_2$CMe$_2$CH$_2$O)	81.5 $^1J_{PSe}$747	262
	P-bound atoms = O$_3$S′			
(Si)$_3$	$C_9H_{27}O_3PS$	(TmsO)$_3$P(S)	2.8	361
P, (C)$_2$	$C_{16}H_{37}O_6P_3S_2$	[(iPrO)$_2$**P**(S)O]$_2$PO-iBu	52 $^2J_{PP}$14	362
P^4, (C)$_2$	$C_8H_{20}O_5P_2S_2$	(EtO)$_2$P(S)OP(S)(OEt)$_2$	52	347
P^4,C,C′	$C_{14}H_{14}N_2O_9P_2S_2$	[MeO(4-O$_2$NC$_6$H$_4$O)P(S)]$_2$O	67.8(c)	143
Si,(C)$_2$	$C_5H_{15}O_3PSSi$	(MeO)$_2$P(S)OTms	45	193
	$C_7H_{19}O_3PSSi$	(EtO)$_2$P(S)OTms	57	193
H,(C)$_2$	$C_4H_{11}O_3PS$	(EtO)$_2$P(S)OH	63.5[b](n) 57.7[a]	363
	$C_6H_{15}O_3PS$	(iPrO)$_2$P(S)OH	61.5(n)	75
	$C_{10}H_{23}O_3PS$	(NeoO)$_2$P(S)OH	71.2[a](b) 85[b](c)	364
H,C,C′	$C_{11}H_{11}O_3PS$	1-C$_{10}$H$_7$O(MeO)P(S)OH	59.0	365
H,(C′)$_2$	$C_{12}H_{11}O_3PS$	(PhO)$_2$P(S)OH	54.4(ne)	366
(C)$_3$	$C_3H_9O_3PS$	(MeO)$_3$P(S)	73.0(r)	367
	$C_6H_{15}O_3PS$	(EtO)$_3$P(S)	68.1(n)	96

43

Rings/ Connectivities	Formula	Structure	NMR data (δ_p [solv] $^nJ_{PZ}$ Hz)	Ref.
	$C_8H_{17}O_5PS$	$(EtO)_2P(S)OCH_2COOEt$	69.0	363
	$C_8H_{19}O_3PS$	$(EtO)_2P(S)O$-tBu	59.0	203
	$C_9H_{21}O_3PS$	$(iPrO)_3P(S)$	65.1	369
	$C_{10}H_{23}O_3PS$	$(tBuO)_2P(S)OEt$	50.2	203
	$C_{12}H_{27}O_3PS$	$(tBuO)_3P(S)$	41.2	205
	$C_{15}H_{33}O_3PS$	$(NeoO)_3P(S)$	68.5	203
	$C_{21}H_{21}O_3PS$	$(PhCH_2O)_3P(S)$	68.1(c)	370
$(C)_2$,C′	$C_9H_{17}O_4PS$	$(EtO)_2P(S)OCMe{=}CHCOMe$	63	371
	$C_{10}H_{14}NO_5PS$	$(EtO)_2P(S)OC_6H_4NO_2$-4	61	372
C,$(C')_2$	$C_{13}H_{13}O_3PS$	$(PhO)_2P(S)OMe$	59.2(ne)	366
	$C_{14}H_{13}N_2O_7PS$	$(4\text{-}O_2NC_6H_4O)_2P(S)OEt$	55.2(ne)	366
$(C')_3$	$C_{18}H_{15}O_3PS$	$(PhO)_3P(S)$	52.2(ne)	366
5/Si,$(C)_2$	$C_6H_{15}O_3PSSi$	**43;** X = S′, Y = OTms	68.0	373
		X = OTms, Y = S′	68.5	
	$C_7H_{17}O_3PSSi$	**5;** R = Me, X = S′, Y = OTms	67.8(c)	256
		X = OTms, Y = S′	69.5(c)	
		(4*RS*, 5*RS*) form	66.5	
	$C_{17}H_{21}O_3PSSi$	**5;** R = Ph, X = S′, Y = OTms	68.1(c) $^2J_{PC}$2.6	256
		X = OTms, Y = S′	68.7(c) $^2J_{PC}$0.0	
5/$(C)_3$	$C_3H_7O_3PS$	MeO(S)PGlc	85.5	374
	$C_4H_9O_3PS$	**43;** X = S′ Y = OMe	84.0	373
		X = OMe, Y = S′	84.5	
	$C_5H_{11}O_3PS$	**5;** R = Me; X = S′, Y = OMe	80.5(c) $^2J_{PC}$1.7	256
		X = OMe, Y = S′	83.0(c) $^2J_{PC}$1.7	
		(4*RS*,5*RS*) form	80.0(n)	
	$C_{15}H_{15}O_3PS$	**5;** R = Ph; X = S′, Y = OMe	83.2(c) $^2J_{PC}$0.0	256
		X = OMe, Y = S′	84.0(c) $^2J_{PC}$0.0	
5/$(C)_2$,C′	$C_8H_9O_3PS$	PhO(S)PGlc	79.1(c)	69
5/C,$(C')_2$	$C_8H_9O_3PS$	CatP(S)OEt	80.5	375
	$C_8H_{15}O_3PS$	**6;** R = Me, X = S, Y = OBu	76.6	376
	$C_{12}H_{11}O_3PS$	**13;** X = S, Y = OEt	67.9	150
	$C_{12}H_{11}O_3PS$	**14;** X = S, Y = OEt	99.3	150
6/P^4,$(C)_2$	$C_8H_{16}O_5P_2S_2$	**1;** X, Y = S′, $O(S)P(OCHMeCH_2CH_2O)$	45.7, 46.5, 46.6, 48.1, 48.3, 50.0^D(b)	377
6/H,$(C)_2$	$C_4H_9O_3PS$	**1;** X = S, Y = OH	60, 64^D	378
6/$(C)_3$	$C_4H_9O_3PS$	$MeO(S)P(OCH_2CH_2CH_2O)$	64.4(c) $^1J_{POMe}$102 $^1J_{POC}$115	28
	$C_5H_{11}O_3PS$	**1;** X = S′, Y = OMe	66.1(c) $^1J_{POMe}$112 $^1J_{POC}$102, 104	28
		X = OMe, Y = S′	64.1(c) $^1J_{POMe}$95 $^1J_{POC}$123	
	$C_6H_{12}ClO_3PS$	**25;** X = S′, Y = OMe	62.3	109
		X = OMe, Y = S′	64.4	
	$C_6H_{13}O_3PS$	**8;** R^1 = R^3 = Me, R^2 = H, X = OMe, X = S′	63.9(c)	28
	$C_6H_{13}O_3PS$	$MeO(S)P(OCH_2CMe_2CH_2O)$	62.2(c)	2

TABLE H (continued)
Four Coordinate Compounds Containing a P=Ch Bond but No Bonds to Group IV Atoms

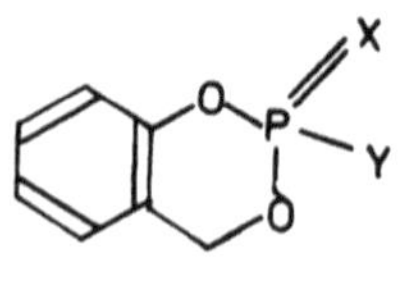

44

Rings/ Connectivities	Formula	Structure	NMR data (δ_p [solv] $^nJ_{PZ}$ Hz)	Ref.
	$C_7H_{15}O_3PS$	**8**; $R^1 = R^2 = R^3$ = Me; X = S′, Y = OMe	62.6(c)	28
		X = OMe, Y = S′	62.5(c)	
	$C_7H_{13}O_3PS$	**36**; X, Y = S′, OMe		292
		(1*RS*,3*RS*,6*SR*)	67.8	
		(1*RS*,3*RS*,6*SR*)	63.6	
	$C_9H_{17}O_3PS$	H_2C=CHCHMeO(S)P($OCH_2CMe_2CH_2O$)	60.4(b)	1
	$C_9H_{17}O_3PS$	MeCH=CHCH_2O(S)P($OCH_2CM_2CH_2O$)	61.7(b)	1
6/(C)$_2$,C′	$C_8H_9O_3PS$	**44**;X = S, Y = OMe	60.0(c)	143
	$C_{11}H_{14}NO_5PS$	4-$O_2NC_6H_4$O(S)P($OCH_2CMe_2CH_2O$)	52.7	8
	$C_{11}H_{15}O_3PS$	PhO(S)P($OCH_2CMe_2CH_2O$)	55.8(b) 53.7(g)	379
	$C_{11}H_{16}NO_3PS$	4-$H_2NC_6H_4$O(S)P($OCH_2CMe_2CH_2O$)	55.5	8
	$C_{12}H_{17}O_3PS$	**8**; $R^1 = R^2 = R^3$ = Me; X, Y = S′, OPh	54.9	234
	$C_{14}H_{21}O_3PS$	**7**; X = S′, Y = OPh	61.8(c)	235
		X = OPh, Y = S′	55.7(c)	
6/C,(C′)$_2$	$C_{12}H_{11}O_3PS$	**17**; X = S, Y = OEt	50.3	150
7/(C)$_2$,C′	$C_{10}H_{11}O_3PS$	PhO(S)P(OCH_2CH=CHCH_2O)	67.7(b) $^2J_{PC}$8.0(c)	277
	$C_{10}H_{13}O_3PS$	PhO(S)P($OCH_2CH_2CH_2CH_2O$)	65.7(b) $^2J_{PC}$7.2(c)	277
	$C_{14}H_{13}O_3PS$	**30**; R = H, A = O, X = S, Y = OPh	65.8(b) $^2J_{PC}$6.6	277
7/(C′)$_2$,H	$C_{12}H_9O_3PS$	**31**; X = S, Y = OH	67.2(w)	239
6,6/(C)$_3$	$C_3H_6O_3P_2S$	(S)**P**(OCH_2)$_3$P	52.1 $^3J_{PP}$48	241
	$C_3H_6O_3P_2S_2$	(S)**P**(OCH_2)$_3$P(S)	49.3 $^3J_{PP}$150	241
	$C_3H_6O_4P_2S$	(S)**P**(OCH_2)$_3$P(O)	49.3 $^3J_{PP}$151	241
	$C_5H_9O_3PS$	(S)P(OCH_2)$_3$CMe	57.4(d)	241
	$C_6H_9O_3PS$	**34**; X = S	64.0(d)	380
8,8/(C)$_3$	$C_6H_{12}NO_3PS$	N(CH_2CH_2O)$_3$P(S)	60.9	242
	P-bound atoms = O,O′S			
(Si)$_2$;Cl	$C_6H_{18}ClO_3PSSi_2$	(TmsO)$_2$P(O)SCl	−4.4	381
(Si)$_2$;S	$C_{12}H_{36}O_6P_2S_2Si_4$	(TmsO)$_2$P(O)SS(O)P(OTms)$_2$	−1.4	381
(Si)$_2$;C	$C_7H_{21}O_3PSSi_2$	(TmsO)$_2$P(O)SMe	5.2	381
(Si)$_2$;C′	$C_{12}H_{23}O_3PSSi_2$	(TmsO)$_2$P(O)SPh	0.8	382
C,H;C	$C_4H_{11}O_3PS$	EtO(EtS)P(O)OH	27.3(g)	383
(C)$_2$;Cl	$C_4H_{10}ClO_3PS$	(EtO)$_2$P(O)SCl	17.8(n) 19	75
	$C_6H_{14}ClO_3PS$	(iPrO)$_2$P(O)SCl	15.7(n)	75
	$C_{10}H_{22}ClO_3PS$	(NeoO)$_2$P(O)SCl	18.5(b)	76
(C)$_2$;P^4	$C_4H_{12}O_6P_2S$	(MeO)$_2$P(O)SP(O)(OMe)$_2$	19.5(c)	370
	$C_8H_{20}O_6P_2S$	(EtO)$_2$P(O)SP(O)(OEt)$_2$	14.3(n)	75
(C)$_2$;S	$C_4H_{12}O_6P_2S_2$	(MeO)$_2$P(O)SS(O)P(OMe)$_2$	23.5	384
	$C_8H_{20}O_6P_2S_2$	(EtO)$_2$P(O)SS(O)P(OEt)$_2$	20(n)	75
	$C_{12}H_{28}O_6P_2S_2$	(iPrO)$_2$P(O)SS(O)P(O-iPr)$_2$	17.2[a] 18.4[b]	385
	$C_{20}H_{44}O_6P_2S_2$	(NeoO)$_2$P(O)SS(O)P(ONeo)$_2$	20.5(r)	76

Rings/ Connectivities	Formula	Structure	NMR data (δ_p [solv] ${}^nJ_{PZ}$ Hz)	Ref.
(C)$_2$;Se	$C_{12}H_{28}O_6P_2S_2Se$	Se[SP(O)(O-iPrO)$_2$]$_2$	18.3(c) ${}^1J_{PSe}$57.5	386
(C)$_2$;C	$C_3H_6F_3O_3PS$	(MeO)$_2$P(O)STf	16.4	312
	$C_3H_9O_3PS$	(MeO)$_2$P(O)SMe	30.5	337
	$C_6H_{15}O_3PS$	(EtO)$_2$P(O)SEt	26.5(n)	75
	$C_7H_{17}O_3PS$	(iPrO)$_2$P(O)SMe	22.7(g)	387
	$C_8H_{17}O_5PS$	(EtO)$_2$P(O)SCH$_2$COOEt	24.7	368
	$C_8H_{19}O_3PS$	(iPrO)$_2$P(O)SEt	24	388
	$C_{19}H_{34}O_6P_2S_2$	[(iPrO)$_2$P(O)S]$_2$CHPh	18	388
(C)$_2$;C′	$C_{10}H_{15}O_3PS$	(EtO)$_2$P(O)SPh	22.6	389
(C)$_2$;C″	$C_5H_{10}NO_3PS$	(EtO)$_2$P(O)SCN	9.3	165
(C′)$_2$;C	$C_{13}H_{11}N_2O_7PS$	(4-O$_2$NC$_6$H$_4$O)$_2$P(O)SMe	20.7(ne)	366
	$C_{13}H_{13}O_3PS$	(PhO)$_2$P(O)SMe	21.1(ne)	366
	$C_{14}H_{13}N_2O_7PS$	(4-O$_2$NC$_6$H$_4$O)$_2$P(O)SEt	20.6(ne)	366
	$C_{14}H_{15}O_3PS$	(PhO)$_2$P(O)SEt	21.0(ne)	366
5/(C)$_2$;C	$C_4H_9O_3PS$	**43**; X = O′, Y = SMe	46.5(n)	373
		X = SMe, Y = O′	46.7(n)	
5/C,C′;C	$C_8H_9O_3PS$	PhO(O)P(OCH$_2$CH$_2$S)	42	14
5/C,C′;C′	$C_5H_9O_3PS$	EtO(O)P(OCMe=CHS)	39.2	390
5/(C′)$_2$;C	$C_{11}H_{13}O_3PS$	**6**; R = Me, X = O, Y = SCH$_2$Ph	42.1	144
6/(C)$_2$;Cl	$C_4H_8ClO_3PS$	**1**; X = SCl, Y = O′	9.2	389
		X = O′, Y = SCl	13.0	
6/(C)$_2$;P^4	$C_{10}H_{20}O_5P_2S_2$	**24**; X = O, Y = Z = S	6.0	255
	$C_{10}H_{20}O_6P_2S$	**24**; X = Z = O, Y = S	6.4(b,c)	225(b)
6/(C)$_2$;S	$C_{10}H_{20}O_6P_2S_2$	**24**; X = Z = O, Y = SS	12.3(c)	391
6/(C)$_2$;C	$C_5H_8F_3O_3PS$	**1**; X = O′, Y = STf	5.1(r)	392
		X = STf, Y = O′	3.3(r)	
	$C_5H_{11}O_3PS$	**1**; X = O′, Y = SMe	23.5(c) 25.8	393
		X = SMe, Y = O′	19.5[a](c) 21.0	
	$C_6H_{13}O_3PS$	MeS(O)P(OCH$_2$CMe$_2$CH$_2$O)	21.8(c)	2
	$C_9H_{17}O_3PS$	H$_2$C=CHCHMeS(O)P(OCH$_2$CMe$_2$CH$_2$O)	19.3(c)	1
	$C_9H_{17}O_3PS$	MeCH=CHCH$_2$S(O)P(OCH$_2$CMe$_2$CH$_2$O)	20.9(c)	1
	$C_{11}H_{21}O_7PS$	**26**; A = S, X = O′, Y = OEt	19	24
		X = OEt, Y = O′	15	
6/(C)$_2$;C′	$C_{10}H_{13}O_3PS$	**1**; X = O′, Y = SPh	15.5	25
		X = SPh, Y = O′	11.9	
6/C,C′;C	$C_9H_{11}O_3PS$	PhO(O)P(OCH$_2$CH$_2$CH$_2$S)	12	394
7/(C′)$_2$;C	$C_{13}H_{11}O_3PS$	**31**; X = O, Y = SMe	37.7(c)	239

P-bound atoms = O$_2$O′S′

Rings/ Connectivities	Formula	Structure	NMR data	Ref.
(Si)$_2$	$C_6H_{18}O_3PSSi_2$	(TmsO)$_2$P(S)O$^-$	35.9	381
(C)$_2$	$C_2H_6O_3PS$	(MeO)$_2$P(S)O$^-$	58.6(r)	1
	$C_4H_{10}O_3PS$	(EtO)$_2$P(S)O$^-$	48.2(c)	246
5/(C)$_2$	$C_4H_8O_3PS$	**5**; R = Me, X = S′, Y = O$^-$	68.5(w) ${}^2J_{PC}$1.34	256
		5; R = Me, X = O$^-$, Y = S′	68.2(w) ${}^2J_{PC}$0.0	
		(*4RS,5RS*) form	67.3 68.1(w)	
	$C_{14}H_{12}O_3PS$	**5**; R = Ph X = S′, Y = O$^-$	71.0(w) ${}^2J_{PC}$0.0	256
		X = O$^-$, Y = S′	68.8(w) ${}^2J_{PC}$0.0	
6/(C)$_2$	$C_4H_8O_3PS$	**1**; X = S′, Y = O$^-$	48.6/50.5	395
		X = O$^-$, Y = S′	52.6/53.5	
	$C_5H_{10}O_3PS$	$^-$O(S)P(OCH$_2$CMe$_2$CH$_2$O)	49.9(r)	1

P-bound atoms = OO′$_2$S

Rings/ Connectivities	Formula	Structure	NMR data	Ref.
H;C	CH_4O_3PS	MeSP(O)(OH)O$^-$	18.9	381

TABLE H (continued)
Four Coordinate Compounds Containing a P=Ch Bond but No Bonds to Group IV Atoms

Rings/ Connectivities	Formula	Structure	NMR data (δ_p [solv] $^nJ_{PZ}$ Hz)	Ref.
	P-bound atoms = OO'$_2$S'			
C	$C_2H_5O_3PS$	$C_nH_{2n+1}OP(S)(O^-)_2$ n = 2 to 9	42.0/42.1(w)	396
	P-bound atoms = O$_3$Se'			
P^4,(C)$_2$	$C_8H_{20}O_6P_2Se$	$(EtO)_2P(Se)O(Se)P(OEt)_2$	58.5, 56.1(n) $^1J_{PSe}$1020	139
P,(C)$_2$	$C_{18}H_{42}O_9P_4Se_3$	$[(iPrO)_2P(Se)O]_3P$	58.3(c) $^1J_{PSe}$874	36
(Si)$_3$	$C_9H_{27}O_3PSe$	$(TmsO)_3P(Se)$	21.4 $^1J_{PSe}$942	382
Si,(C)$_2$	$C_7H_{19}O_3PSeSi$	$(EtO)_2P(Se)OTms$	55 $^1J_{PSe}$940	382
Sn,(C)$_2$	$C_9H_{23}O_3PSeSn$	$(iPrO)_2P(Se)OSnMe_3$	30.4(c) $^1J_{PSe}$644	36
	$C_{24}H_{38}O_6P_2Se_2Sn$	$[(iPrO)_2P(Se)O]_2SnPh_2$	24.4(c) $^1J_{PSe}$546	36
	$C_{24}H_{47}O_9P_3Se_3Sn$	$[(iPrO)_2(Se)O]_3SnPh$	33.1(c)	36
H,(C)$_2$	$C_4H_{11}O_3PSe$	$(EtO)_2P(Se)OH$	62.3(e) $^1J_{PSe}$887	330
	$C_{10}H_{23}O_3PSe$	$(NeoO)_2P(Se)OH$	61(g) $^1J_{PSe}$910	139
(C)$_3$	$C_3H_9O_3PSe$	$(MeO)_3P(Se)$	78.0(c) $^1J_{PSe}$956	36
	$C_4H_{11}O_3PSe$	$(MeO)_2P(Se)OEt$	76.4 $^1J_{PSe}$940	330
	$C_5H_{13}O_3PSe$	$(EtO)_2P(Se)OMe$	73.8 $^1J_{PSe}$935	330
	$C_6H_6Cl_9O_3PSe$	$(Cl_3CCH_2O)_3P(Se)$	71.7(a)	397
	$C_6H_{15}O_3PSe$	$(EtO)_3P(Se)$	68.1(c) $^1J_{PSe}$904	36
	$C_9H_{21}O_3PSe$	$(iPrO)_3P(Se)$	66.6(c) $^1J_{PSe}$924	36
	$C_{10}H_{23}O_3PSe$	$(tBuO)_2P(Se)OEt$	45.2	203
	$C_{12}H_{27}O_3PSe$	$(BuO)_3PSe$	73(n) $^1J_{PSe}$952	280
	$C_{12}H_{27}O_3PSe$	$(tBuO)_3P(Se)$	31.8(a)	397
	$C_{15}H_{33}O_3PSe$	$(NeoO)_3P(Se)$	72.6	203
(C)$_2$,C'	$C_{13}H_{19}O_4PSe$	$(iPrO)_2P(Se)OCOPh$	59.0(c) $^1J_{PSe}$985	36
C,(C')$_2$	$C_{14}H_{15}O_3PSe$	$(PhO)_2P(Se)OEt$	62	316
(C')$_3$	$C_{18}H_{15}O_3PSe$	$(PhO)_3P(Se)$	58.6(c)	158
	$C_{21}H_{21}O_3PSe$	$(2\text{-}TolO)_3P(Se)$	55.0(a)	397
	$C_{24}H_{27}O_3PSe$	$(2,6\text{-}Me_2C_6H_3O)_3P(Se)$	51.0(m)	397
5/(C)$_3$	$C_3H_7O_3PSe$	MeO(Se)PGlc	88.0(c) $^1J_{PSe}$1011	187
	$C_6H_{13}O_3PSe$	tBuO(Se)PGlc	84.6 $^1J_{PSe}$1001	398
5/(C)$_2$,C'	$C_8H_9O_3PSe$	PhO(Se)PGlc	82.0 $^1J_{PSe}$1035	398
6/(C)$_3$	$C_4H_9O_3PSe$	$MeO(Se)P(OCH_2CH_2CH_2O)$	68.6(c) $^1J_{PSe}$985	187
	$C_5H_{11}O_3PSe$	**1**; X = Se', Y = OMe	68.5 $^1J_{PSe}$941	309
		X = OMe, Y = Se'	67.2 $^1J_{PSe}$978	
	$C_9H_{19}O_3PSe$	**7**; X = Se', Y = OMe	72.1(c) $^1J_{PSe}$958	29
		X = OMe, Y = Se'	67.6(c) $^1J_{PSe}$991	
6/(C)$_2$,C'	$C_{14}H_{21}O_3PSe$	**7**; X = Se', Y = OPh	64.8(c) $^1J_{PSe}$979	235
		X = OPh, Y = Se'	58.8(c) $^1J_{PSe}$1020	
7/(C)$_2$,C'	$C_{14}H_{13}O_3PSe$	**30**; R = H, A = O, X = Se, Y = OPh	70	238
5,6/(C)$_3$	$C_4H_7O_3PSe$	**45**; X = Se	81.4(c) $^1J_{PSe}$1099	187
	$C_4H_7O_3PSe$	**46**; X = Se	80.9 $^1J_{PSe}$1047	187
6,6/(C)$_3$	$C_5H_9O_3PSe$	$(Se)P(OCH_2)_3CMe$	60.1(c) $^1J_{PSe}$1053	187
	$C_6H_9O_3PSe$	**34**; X = Se	88.7(b)	186
8,8/(C)$_3$	$C_6H_{12}NO_3PSe$	$N(CH_2CH_2O)_3P(Se)$	58.0 $^1J_{PSe}$973	242

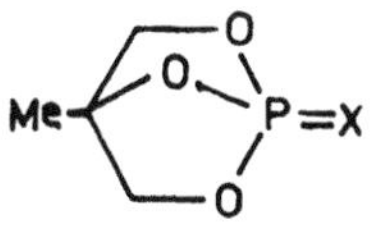

45

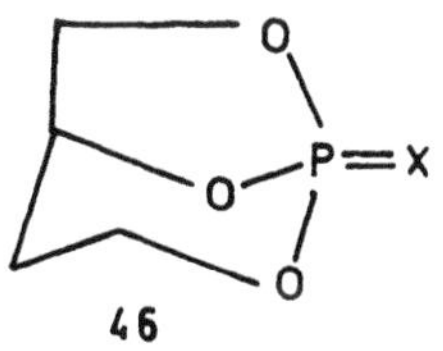

46

Rings/ Connectivities	Formula	Structure	NMR data (δ_p [solv] $^nJ_{PZ}$ Hz)	Ref.
	P-bound atoms = $O_2O'Se$			
$(Si)_2$;C	$C_7H_{21}O_3PSeSi_2$	$(TmsO)_2P(O)Se_3Me$	−7 $^1J_{PSe}$472	382
$(C)_2$;As	$C_{18}H_{42}AsO_9P_3Se_3$	$[(iPrO)_2P(O)Se]^3As$	12.4(c) $^1J_{PSe}$473	36
$(C)_2$;Ge	$C_{24}H_{29}GeO_3PSe$	$(iPrO)_2P(O)SeGePh_3$	19.7(c) $^1J_{PSe}$508	36
$(C)_2$;S^4	$C_{12}H_{19}O_5PSSe$	$(iPrO)_2P(O)SeSO_2Ph$	8.9(c) $^1J_{PSe}$495	36
$(C)_2$;Se	$C_8H_{20}O_6P_2Se_2$	$(EtO)_2P(O)SeSe(O)P(OEt)_2$	11.1 (c) $^1J_{PSe}$498	360
	$C_{12}H_{28}O_6P_2Se_2$	$(iPrO)_2P(O)SeSe(O)(O\text{-}iPr)_2$	11.7(c) $^1J_{PSe}$516	36
	$C_{12}H_{28}O_6P_2Se_3$	$(iPrO)_2P(O)Se_3(O)(O\text{-}iPr)_2$	7.9 $^2J_{PSe}$54.6	386
$(C)_2$;Si	$C_9H_{23}O_3PSeSi$	$(iPrO)_2P(O)SeTms$	18.1(c) $^1J_{PSe}$467	36
$(C)_2$;Sn	$C_{24}H_{29}O_3PSeSn$	$(iPrO)_2P(O)SeSnPh_3$	16.5(c) $^1J_{PSe}$480	36
$(C)_2$;C	$C_3H_6F_3O_3PSe$	$(MeO)_2P(O)SeTf$	15.5 $^1J_{PSe}$412	312
	$C_3H_9O_3PSe$	$(MeO)_2P(O)SeMe$	23.4(c) $^1J_{PSe}$473	36
	$C_6H_{15}O_3PSe$	$(EtO)_2P(O)SeEt$	19.6(c) $^1J_{PSe}$468	36
	$C_8H_{19}O_3PSe$	$(iPrO)_2P(O)SeEt$	17.5(c) $^1J_{PSe}$480	36
$(C')_2$;C	$C_{13}H_{13}O_3PSe$	$(PhO)_2P(O)SeMe$	13.6 $^1J_{PSe}$533	280
5/C,C';C	$C_9H_{11}O_3PSe$	$PhO(O)P(OCHMeCH_2Se)$	32	14
6/$(C)_2$;P^4	$C_{10}H_{20}O_6P_2Se$	**24;** X = Z = O, Y = Se	0.6	226
6/$(C)_2$;Se	$C_{10}H_{20}O_6P_2Se_2$	**24;** X = Z = O, Y = SeSe	−8.8	246
6/$(C)_2$;C	$C_5H_{11}O_3PSe$	**1;** X = O', Y = SeMe	9.1[a], 11.7(b) $^1J_{PSe}$449, 445	399
		X = SeMe, Y = O'	14.3[a] 14(b) $^1J_{PSe}$483, 476	
6/C,C';C	$C_9H_{11}O_3PSe$	$PhO(O)P(OCH_2CH_2CH_2Se)$	3	394
	P-bound atoms = $O_2O'Se'$			
$(C)_2$	$C_4H_{10}O_3PSe$	$(EtO)_2P(Se)O^-$	50.2/51.5(a-c) $^1J_{PSe}$772/787	360
5/$(C)_2$	$C_6H_{12}O_3PSe$	$PncP(Se)O^-$	53.3(w)	373
6/$(C)_2$	$C_4H_8O_3PSe$	**1;** X = Se', Y = O^-	45.3 $^1J_{PSe}$784	262
		X = O^-, Y = Se'	45.4 $^1J_{PSe}$805	
	P-bound atoms = S_3S'			
$(Si)_3$	$C_9H_{27}PS_4Si_3$	$(TmsS)_3P(S)$	65.8(b)	400
$(C)_3$	$C_3H_9PS_4$	$(MeS)_3P(S)$	98.5	337
	$C_6H_{15}PS_4$	$(EtS)_3P(S)$	92	401
	$C_{12}H_{27}PS_4$	$(BuS)_3P(S)$	92.6(n)	96
$(C')_3$	$C_{18}H_{15}PS_4$	$(PhS)_3P(S)$	91.1(t)	96

SECTIONS H3 AND H4: BIOLOGICALLY IMPORTANT COMPOUNDS

Compiled and presented by
John C. Tebby and Thomas Glonek[1]

The majority of compounds in this section fall within the P-bound atom definition of this chapter, i.e., compounds derived from ortho phosphoric acid, its salts, esters, and anhydrides. For completeness, however, it also includes some relevant biologically important phosphonates.

Most naturally occurring phosphates and amides fall within the quite narrow chemical shift range of 5 to −25 ppm. By including the strained cyclic phosphate diesters and the naturally occurring phosphonates the high frequency (low field) end is extended to 22 ppm. The synthetic thiophosphates which have played an important role in the assignment of signals appear at even higher frequency in the region 29 to 45 ppm. At the other extreme polyphosphate branch groups, which have yet to be observed naturally, resonate at −36 to −37 ppm.

The ^{31}P NMR signals of the naturally occurring phosphates, in general, fall into distinct regions depending on the connectivities of the P-bound oxygen atoms. Groups bound to the ester oxygen, which accept electrons, have a shielding effect relative to alkyl groups, moving the phosphorus signal to lower frequency. This effect has been attributed to a higher bond order for the phosphoryl (P=O) group. Phosphorylation of oxygen has the largest influence — moving the signal to low frequency. The effect is greatest when two oxygens are phosphorylated as in polyphosphate middle groups (as typified by the β-phosphorus of ATP), and the signals appear in the unique low frequency region of −21 to −25 ppm. When only one oxygen is phosphorylated the effect is weaker. Most pyrophosphate esters, i.e., one oxygen phosphorylated and one oxygen alkylated — as typified by the α-phosphorus group in ATP or ADP — appear in the range of −10 to −14 ppm. There are exceptions such as glucose pyrophosphate (α-P −7.2 ppm) and guanosine diphosphate (α-P −5.8 ppm). Within the normal −10 to −14 range, the uridine pyrophosphates appear at lowest frequency (−11.4/−13.79), next appears NAD and NADP (−11.3/−12.6), with the nucleotide phosphates at the highest frequency (−10.5/−10.9) for this range. The chemical shifts of the polyphosphate end groups (as typified by the γ-phosphorus of ATP), like all dibasic phosphate groups, are strongly pH dependent. They are well removed from monobasic middle groups and are contained with the well-defined low frequency range −5.5 to −6 ppm at pH 10/7, but move to −10 to −11 ppm at low pH.

The amides resonate over a relatively wide range (−10.9 to 12.08 ppm). However, in the absence of an oxygen connectivity the signals are located within the narrow and well-defined region of −3.1 to −3.6 ppm.

The ortho-phosphate esters are also contained within a quite separate region, i.e., 5 to −3 ppm. Most diesters, i.e., P-bound atoms = $O_2O'_2$ and connectivities = $(C)_2$, appear within the range 1.1 to −3, while the range of the monoesters, i.e., P-bound atoms = $00'_3$ and connectivity = C, extend to the high frequency end, 4.8 to −2.2 ppm. Inorganic phosphate also appears in this region at 2.6 to 0 ppm — its exact position depending on pH. As with the pyrophosphates, overlap of the signals of the mono- and diesters can be reduced by using solutions in the higher pH range. At pH 8 to 12 most monoesters are located between 5 and 2 ppm. Inorganic ortho phosphate appears at 2.6 ppm, and most diesters appear between 1 and 0 ppm, notable exceptions are in the diester group, namely, the phospholipids, e.g., 3 ppm for phosphatidylcholine at pH 10.

The monoesters region for solutions at high pH can be further subdivided. Aldehydes and ketone phosphates, triose phosphates, and pentose 6-phosphates appear in the range 4.8 to 3.9 ppm. While several of the anomeric hexose 1-phosphates appear at lower frequency of 2.3 to 2.5 ppm, others such as fructose 1-phosphate appear are in the higher frequency region. Lowering the pH of solutions of the monoester usually causes the signals to broaden and shift to lower frequency by about 1 ppm.

The signals of the acyclic diesters appear in the region 4 to −4 ppm. Most of the lower molecular weight diesters $<C_{40}$ such as the α-glycerylphosphocholine 3-P appear between 1.0 and −0.3 ppm. They are only weakly affected by change in pH in the range 4 to 10. In contrast, the higher molecular weight diesters are markedly dependent on the medium and resonate across the whole of the diester range. The concentrations of ions such as Ca^{++} and Mg^{++} and other substrates as well as pH changes have a marked influence on conformation and thus on chemical shifts. Oligonucleotides generally appear in the low frequency region −0 to −3 ppm. The absence of the 2′-hydroxyl group in the deoxyribonucleotides causes the phosphorus signal to move to lower frequency by about 0.5 ppm relative to the ribonucleotides. Cyclic esters resonate at higher frequencies. The change is small for 6-membered rings, e.g., cUMP −2.6 ppm, but very marked for the strained 5-membered cyclic phosphates, signals appearing between 10 and 21 ppm.

However, this high frequency region is normally occupied by the primary amides (6 to 12 ppm) and the phosphonates. The phosphonic acid anhydrides, $RP(O)_2OCOR$, appear around 10 ppm and the phosphonic acid anions usually resonate between 16 and 22 ppm.

Presentation and Sequencing of the Data

It is impractical to attempt to relate the chemical shifts of most naturally occurring phosphates to a specific acidic or anionic form, since in solutions between pH 4 and 10 they often occur as an equilibrium mixture of undissociated, partially dissociated, and completely dissociated species; and to attempt to assign specific structures would produce a multitude of entries. As a consequence all the structures in Section H3 of the Table *are assigned the completely dissociated phosphoric acid structures and placed in the appropriate sequence accordingly. Thus the P-bound oxygen atom for the POH group is given as O′ and not O, and the connectivities column does not contain acidic protons nor metal/ammonium cations, nor do these atoms appear in the molecular formulas.* Some authors have assigned anionic or protonated structures, in which case the assignment is indicated as a superscript against the chemical shift. Isomeric compounds appear in alphabetical order of name, e.g., glucose before mannose, and where the name is the same the sequence follows the increasing numerical order of the position of phosphorylation, e.g., glucose 3-P > glucose 4-P > glucose 5-P > glucose 6-P.

The fatty alkyl groups of the *lipids* are present mainly as a mixture of C_{18} to C_{22} saturated and C_{22} unsaturated structures. The chemical shifts of these various fatty esters are not normally resolved and only one chemical shift is observed. Thus, the molecular formulas of the *phosphatidyl* compounds, which consist of a mixture of esters, have been calculated on the basis of the aliphatic chain being $C_{20}H_{41}$. Note, however, that the separate signals for different phosphatidyl side chains are resolved when the side chain is very short, e.g., methyl or ethyl.

ABBREVIATIONS

ba = $KClO_4$-saturated 20% D_2O
bb = aqueous 0.1 *M* Na ethylenediaminetetraacetate (EDTA)
bc = hydrated chloroform-methanol 2:1 EDTA reagent[448] with counterion specified
bd = 50 m*M* NaCl/35 m*M* EDTA
be = 0.1 aq ammonium formate/1 m*M* EDTA
bf = 7 n*M* $Na_2H_2PO_4$/19 m*M* Na_2PO_4/2 m*M* Na_2 EDTA/3 m*M* NaN_3 in D_2O
bg = 8 *M* aqueous urea
bl = living tissue
bp = perchloric acid extracts saturated with $KClO_4$, pH = 10, T = 24°
TMA = tetramethyl ammonium (Me_4N^+)

TABLE H
Four Coordinate (λ5σ4) Phosphorus Compounds Containing No Phosphorus Bonds to Hydrogen nor Group IV Atoms

Section H3: Naturally Occurring Phosphates and Related Compounds[a]

Rings/ Connectivities	Formula	Structure	^{31}P NMR data (δ_P [solv. temp, ph] J_{PH} Hz)	Ref.
	P-bound atoms = NO_2O'			
H_2;$(C)_2$		$H_2NP(O)(dT)_2$ oligonucleotide	12.08(w)	453
	P-bound atoms = NOO'_2			
H_2;C		$H_2NP(O)(OH)dT$-oligonucleotide	9.32(w, pH 13)	453
H_2;C		$H_2NP(O)(OH)dT$-oligonucleotide	8.85(p)	453
CH;C		$B_zNHP(O)(OH)dT(Ac)$ oligonucleotide	6.5(p)	453
C_2;C	$C_{14}N_{22}N_4O_7P$	$PipPO_2Cytosine$	5.5(p)	443
C'_2;C	$C_{13}H_{16}N_4O_8P$	$ImidazoPO_2Guanosine$	−10.9(p)	443
	P-bound atoms = NO'_3			
C'H	$CH_4N_3O_3P$	Guanidine P	−3.1(ba24° pH 10)	427
	$C_4H_8N_3O_5P$	Creatine P	−3.12(ba24° pH 10), −3.10(bp, b1), −3.00(b1))	427, 485
	$C_6H_{13}N_4O_5P$	Arginine P	−3.58(ba24° pH 10)	427
	P-bound atoms = N_2OO'			
6/CH,C_2;O	$C_7H_{15}Cl_2N_2O_2$	Iso-phosphamide	13.38	428
	$C_8H_{17}Cl_2N_2O_2$	Cyclophosphamide 4-methyl	13.5^{SR}; 11.0^{RR}(c)	429
	P-bound atoms = $O_2O'S'$			
$(P^4)_2$	$C_{10}H_{12}N_5O_{13}P_3S$	Guanosine 5'-thio-triP $(GTP_{\alpha s})\beta$-P	29.50^{A-}, 30.42^{HA}	430
P^4,C	$C_{10}H_{12}N_5O_{10}P_2S$	Guanosine 5'-thio-diP $(GDP_{\alpha s})$ α-P	44.85^{A-}, 43.73^{HA}	430
	$C_{10}H_{12}N_5O_{13}P_3S$	Guanosine 5'-thio-triP $(GTP_{\alpha s})$ α-P	43.66^{A-}, 44.25^{HA}	430
5/$(C)_2$	$C_{19}H_{11}N_2O_9PS$	Uridine 2',3'-cyclo-thioP	76.1 & 75.1(w)	453
	P-bound atoms = OO'_2S'			
P^4	$C_{10}H_{12}N_5O_{10}P_2S$	Guanosine 5'-thio-diP $(GDP_{\beta s})$ β-P	33.53^{A-}, 39.55^{HA}	430
	$C_{10}H_{12}N_5O_{13}P_3S$	Guanosine 5'-thio-triP $(GDP_{\tau s})$ τ-P	34.00^{A-}, 39.90^{HA}	430
	P-bound atoms = O_3O'			
$(C)_3$	polymer	R'OP(O)(OR")(OR≈); R' = Me; R" = dT(Ac); R≈ = (Tr)dT	−0.75 and −1.03(p)	453

[a]See text for structural presentation procedure.

Rings/ Connectivities	Formula	Structure	^{31}P NMR data (δ_P [solv. temp, ph] J_{PH} Hz)	Ref.
		R′OP(O)(OR″)(OR≈); R′ = ET; R″ = dT(Ac); R≈ = (Tr)dT	−2.67 and −2.69(c+e)	453
$6/(P^4)_3$	$O_{23}P_8$	μ-Oxo-bis(tetrametaphosphate)	$-37.0J_{PP}27.3$(n)	431
$6/(P^4)_2$,C	$C_9H_{13}N_5O_{13}P_3$	Adenosine-5′ trimetaphosphate	−22.9 (pyridine)	432
$6,6/(P^4)_3$	$O_{11}P_4$	1,5-μ-Oxo-tetrapolyphosphate	$-36.5J_{PP}30.4$(n)	431
	P-bound atoms = $O_2O'_2$			
$(P^4)_2$	O_3P_n	Long chain polyphosphate (middles)	−21.5(bb, pH 7)	433
	$C_{10}H_{12}N_5O_{13}P_3$	Adenosine triP (ATP)[a] βP	−21.45(ba24° pH 10 and bp), −21.33(bp), −19.21(bl)	427, 434
			δP counterion and pH dependence[a]	435, 436
	$C_{10}H_{12}N_5O_{14}P_3$	Guanosine 5′-triP (GTP) β-P	$-21{,}28^{A-}$, $-21{,}58^{HA}$	430
P^4,C	$C_6H_{11}O_{12}P_2$	α-Glucose 1-diP α-P	−14(free), −11.5(bound)	437
	$C_{10}H_{12}N_5O_{11}P_2$	Adenosine diP (ADP)[a] αP	−10.61(ba24° pH 10 and bp), −10.48(bp), −10.65(bl)	427
			δP counter ion and pH dependent[a]	435
	$C_{10}H_{12}N_5O_{13}P_3$	Adenosine triP (ATP)[a] αP	−10.92(ba24° pH 10 and bp), −10.88(bp), −10.63(bl)	427
			δP counter ion and pH dependence[a]	435, 436
	$C_{10}H_{12}N_5O_{11}P_2$	Guanosine 5′-diP (GDP) α-P	-5.81^{A-} -10.20^{HA}	430
	$C_{10}H_{12}N_5O_{14}P_3$	Guanosine 5′-triP (GTP) α-P	-10.53^{A-} -10.83^{HA}	430
	$C_{11}H_{20}N_4O_{11}P_2$	Cytosinediphospho-ethanolamine	−11.15‡(ba24° pH 10)	427
			‡ center of resonance band	
	$C_{12}H_{22}O_{17}P_2$	α-Glucose pyro-P	−7.2	437
	$C_{14}H_{20}N_2O_{16}P_2$	Uridinediphosphoribose (UDP-ribose)	−11.4(ba24° pH 10)	427
	$C_{15}H_{22}N_2O_{17}P_2$	Uridinediphosphogalactose (UDP-galactose)	−12.83(ba24° pH 10), −11.5(bp), −12,8(bl)	427, 434
	$C_{15}H_{22}N_2O_{17}P_2$	Uridinediphosphoglucose (UDP-glucose)	−12.99(ba24° pH 10), −11.5(bp), −12,8(bl)	427, 434
	$C_{15}H_{22}N_2O_{17}P_2$	Uridinediphosphomannose (UDP-mannose)	−13.70(ba24° pH 10), −11.5(bp), −12.8(bl)	427, 434
	$C_{21}H_{26/25}N_7O_{14/17}P_{2/3}$	NAD/NADP multiplet center	−11.37(ba24° pH 10), −11.37 and −12.63(bp), −11.31(bl)	427, 434, 444
		δP counterion and pH dependence[a]		435
	$C_{27}H_{29}N_9O_{15}P_2$	Flavin adenine dinucleotide (FAD)	−11.2	
$(C)_2$	$C_5H_{13}N_2O_6P$	Serine ethanolamine phosphoryldiester (SEP)	0.83 (pH 12), 0.46 (pH 4)	445
	$C_5H_{14}NO_6P$	Glycero 3-phosphorylethanolamine (GPE)	0.81(ba24° pH 10), 0.77(bp), 0.42(bl), 0.85(b1)	427, 434, 445, 490
			0.93 (pH 12), 0.42 (pH 4), 0.92(bp)	
	$C_6H_{12}NO_6P$	Glycerol 3-phosphorylserine (GPS)	0.69 (pH 12), 0.11(pH 7), 0.14 (pH 4), 0.66(b1)	445, 488, 490

TABLE H (continued)
Four Coordinate (λ5σ4) Phosphorus Compounds Containing No Phosphorus Bonds to Hydrogen nor Group IV Atoms

Rings/ Connectivities	Formula	Structure	^{31}P NMR data (δ_P [solv. temp, ph] J_{PH} Hz)	Ref.
	$C_6H_{14}O_8P$	Glycerol 3-phosphorylglycerol (GPG)	0.92 (pH 4/12), 0.98(b1)	445
	$C_6H_{14}N_2O_6P$	Threonine ethanolamine phosphodiester (TEP)	−0.3(pH10), −1.6(pH4/7)	485
	$C_6H_{15}NO_6P$	Glycerol 3-phosphorylmonomethylethanolamine	0.86 (pH 12), 0.29 (pH 4/7)	445
	$C_7H_{17}NO_6P$	Glycerol 3-phosphoryldimethylethanolamine	0.80 (pH 12), 0.16 (pH 4/7)	445
	$C_8H_{20}NO_6P$	Glycerylphosphorylcholine 3-P (GPC)	−0.31(ba24° pH 10), −0.13(bp, pH 4/12), 0.08^{Sn}	427, 434, 445, 446
			0.49(pH4/12, −0.1(b1)	484, 485
	$C_9H_9NO_5P$	Tyrosine-O-phosphate	−0.8(pH7/10), −3(pH4)	487
	$C_9H_{18}O_{11}P$	Glycerol 3-phosphoryl-myo-inositol (GPI)	−0.07 (pH 4/12), 0.10(b1)	445, 490
	$C_9H_{16}O_{17}P_3$	Glycerol 3-phosphoryl-inositol 4,5-di-P	−0.23	484
	$C_9H_{20}O_{11}P_2$	Bis (glyceryl 3-phosphoryl)glycerol (GPGPG)	0.79 (pH 4/12)	445
	$C_{15}H_{27}N_2O_9PS$	Glycerol 3-phosphoryl (N-biotin)ethanolamine	0.71 (pH 11)	445
	$C_{16}H_{32}NO_6P$	$PeCOCH_2CHAcCH_2O$-phosphorylcholine (PAF-acether)	−0.68 (pH 11)	482
	$C_{18}H_{24}N_6O_{12}P$	CpC dinucleoside phosphate	−1.85(bd)	443
	$C_{18}H_{24}N_{10}O_{10}P_2$	ApA dinucleoside phosphate	−0.54(pH 9.21), −2.25(bd, pH 8)	442, 443
	$C_{19}H_{23}N_8O_{13}P_2$	d(pG-3′,5′-pC (Dinucleotide)	−0.5 (pH 5/8)	481
	$C_{20}H_{24}N_{10}O_{12}P$	GpG dinucleoside phosphate	−2.20(bd)	443
	$C_{20}H_{26}N_4O_{14}P$	TpT dinucleoside phosphate	−2.09(bd)	443
	$C_{23}H_{43}O_8P$	Lyso-phosphatidic acid	0.91 (K + bc), 0.95(TMA + bc)	448
	$C_{24}H_{49}DNO_7P$	1-Palmitoyl lyso-phosphatidyl choline (LPC)	0.34	446
	$C_{24}H_{49}DNO_7P$	2-Palmitoyl lyso-phosphatidyl choline (LPC)	0.52	446
	$C_{24}H_{49}DNO_7P$	3-Palmitoyl lyso-phosphatidyl choline (LPC)	1.13	446
	$C_{25}H_{50}NO_8P$	Lyso-phosphatidyl ethanolamine	0.43(K^+/Cs^+ TMA + bc)	448
	$C_{26}H_{51}O_{11}P$	Lyso-phosphatidyl glycerol (PG)	1.14(K^+ + bc), 1.06 (TMA + bc)	448
	$C_{38}H_{46}N_{15}O_{22}P_3$	d(TCGA) - tetranucleoside tri-P 2 conf	6 signals 3.3/5.8 (pH 7)	447
	$C_{40}H_{79}DNO_8P$	1,2-Dipalmitoyl phosphatidyl choline (LPC)	0.86	446
	$C_{40}H_{79}DNO_8P$	1,3-Dipalmitoyl phosphatidyl choline (LPC)	1.45	446
	$C_{40}H_{81}N_2O_6P$	Sphingomyelin (SPH)	−0.09(K^+/Cs^+/TMA + bc)	448
	$C_{43}H_{83}O_8P$	Phosphatidic acid	0.34(K^+ + bc), 0.22 (TMA + bc)	448

Rings/ Connectivities	Formula	Structure	^{31}P NMR data (δ_P [solv. temp, ph] J_{PH} Hz)	Ref.
	$C_{45}H_{90}NO_7P$	Plasmalogen-phosphatidyl ethanolamine	0.07(K^+/Cs^+/TMA + bc)	448
	$C_{45}H_{90}NO_8P$	Phosphatidyl ethanolamine (PE)	3.84(ba24° pH 10), 0.0 ($K+$/TMA + bc)	427, 448
			T. 0.95(Na^+), 3.1(K^+); T_2 0.03(Na^+), 0.23(K^+)	489
	$C_{46}H_{90}NO_{10}P$	Phosphatidyl serine (PS)	3.88(ba24° pH 10), −0.21(K^+ + bc), −0.01(TMA + bc)	427, 448
			T. 1.4(Na^+), 0.9(K^+); T_2 0.15(Na^+), 0.11(K^+)	489
	$C_{46}H_{91}O_{11}P$	Phosphatidyl glycerol (PG)	0.57(K^+ + bc), 0.52 (Cs^+/TMA + bc), −0.7(bl)	448, 490
			T. 2.8(Na^+), 2.3(K^+); T_2 0.22(Na^+), 0.20(K^+)	489
	$C_{47}H_{94}NO_8P$	Phosphatidyl dimethylethanolamine	3.84(ba24° pH 10), −0.18(K^+/Cs^+/ TMA + bc)	427, 448
	$C_{48}H_{96}NO_7P$	Plasmalogen-phosphatidyl choline (plasPC)	−0.78(K^+/Cs^+/ TMA + bc)	448
	$C_{48}H_{96}NO_8P$	Phosphatidyl choline (PC)	3.31(ba24° pH 10), −0.84(K^+/Cs^+/ TMA + bc)	427, 448
			T. 3.1(Na^+), 2.9(K^+); T_2 0.21(Na^+), 0.22(K^+)	489
	$C_{48}H_{96}NO_8P$	Lyso-phosphatidyl choline (LPC)	−0.28(K^+/Cs^+/ TMA + bc)	448
				489
	$C_{49}H_{90}O_{13}P$	Phosphatidyl inositol (PI)	−0.28(K^+/TMA + bc), −0.37(Cs^+ + bc)	448
			T. 2.1(Na^+), 1.8(K^+); T_2 0.26(Na^+), 0.23(K^+)	489
	$C_{49}H_{90}O_{13}P$	Lyso-phosphatidyl inositol (LPI)	0.19(K^+ + bc), 0.06 (TMA + bc)	448
		Phospholipids	Shift reagents; general studies; NOE	449, 450, 451, 493
	$C_{72}H_{98}N_{30}O_{40}P_7$	octamer GGAATTCC c(CpC)	−1.89(bd, pH 8)	457
	$C_{72}H_{98}N_{30}O_{40}P_7$	octamer GGAATTCC d(TpT)	−2.43(bd, pH 8)	457
	$C_{72}H_{98}N_{30}O_{40}P_7$	octamer GGAATTCC d(ApA)	−2.37(bd, pH 8)	457
	$C_{72}H_{98}N_{30}O_{40}P_7$	octamer GGAATTCC d(GpA)	−2.13(bd, pH 8)	457
	$C_{72}H_{98}N_{30}O_{40}P_7$	octamer GGAATTCC d(GpG)	−2.04(bd, pH 8)	457
	$C_{72}H_{98}N_{30}O_{40}P_7$	octamer GGAATTCC d(ApT)	−2.49(bd, pH 8)	457
	$C_{72}H_{98}N_{30}O_{40}P_7$	octamer GGAATTCC d(TpC)	−2.22(bd, pH 8)	457
	$C_{89}H_{174}O_{17}P_2$	Cardiolipin (CL)	0.18(K + bc), 0.10 (TMA + bc)	448, 452
			T. 1.7(Na^+), 1.8(K^+); T_2 0.16(Na^+), 0.16(K^+)	489
		Teichoic acid (polymer glycerol/ribitol-P)	0.6	486
		Nucleotides	pH effects	440
		Oligonucleotides		453—455
	MW >10,000	poly d(A-T)	−2.23 and −2.41(w), −2.43 and −2.84 (+2MCsCl)	456
		poly U	−0.55(w, pH 7)	453
		poly A	−0.77(w, pH 7)	453

TABLE H (continued)
Four Coordinate (λ5σ4) Phosphorus Compounds Containing No Phosphorus Bonds to Hydrogen nor Group IV Atoms

Rings/ Connectivities	Formula	Structure	^{31}P NMR data (δ_P [solv. temp, ph] J_{PH} Hz)	Ref.
		poly G	−0.96(w, pH 7)	453
		poly C	−1.11(w, pH 7)	453
		poly d(ApT)	−2.05(w), −2.23(+2MCsCl)	456
		poly d(TpA)	−2.06(w), −2.28(+2MCsCl)	456
		poly d(ApA)	−2.58(bd, pH 8)	457
		poly d(CpC)	−2.16(bd, pH 8)	457
		poly d(GpG)	−2.16(bd, pH 8)	457
		poly d(TpT)	−2.58(bd, pH 8)	457
		R and S dCTp(S)T poly d(A-T) etc.		454, 455
	MW = 25,000	Transfer ribonucleic acids (tRNA)	24 signals 2 to −2; main 5 −0.2/−1.0	458
		Deoxyribonucleic acids (DNA)	−2.2/−3.2 broad band (bf)	459
		Phosphoglycans	−0.6/−0.9	449, 450, 451
5/(C)$_2$	$C_3H_5O_5P$	Glycerol 1,2-cyclic-P	18.48(w)	484
	$C_5H_8O_7P$	Xylose 1,2-cyclic-P	10.78(w)	484
	$C_6H_7O_9P$	Glucuronic acid 1,2-cyclic-P	10.68(w)	484
		Galactouronic acid 1,2-cyclic-P	10.08(w)	484
	$C_6H_9O_8P$	Glucose 1,2-cyclic-P	10.68(w)	484
		Galactose 1,2-cyclic-P	9.98(w)	484
	$C_6H_{10}O_8P$	Myo-inositol 1,2-cyclic-P	15.88(w)	484
	$C_9H_9N_5O_6P$	Adenosine 2,3-cyclic-P	19.98(w)	484
	$C_9H_9N_5O_7P$	Guanosine 2,3-cyclic-P	19.98(w)	484
	$C_9H_{10}N_2O_8P$	Uridine 2,3-cyclic-P	17.6(v), 20.08(w)	453, 484
	$C_9H_{11}N_3O_7P$	Cytosine 2,3-cyclic-P	20.3,18.1(m), 20.08(w)	453, 460, 484
	$C_{10}H_{14}O_{13}P_2$	Ribose 1,2-cyclic-P 5-bis	18.58(w)	484
	$C_{12}H_{18}O_7P_2$	Fructose 1,2-cyclic-P 6-bis	16.48(w)	484
6/(P^4)$_2$	$O_{11}P_4$	1,5-μ-Oxo-tetrametaphosphate	−28.9J_{PP} 30.4(n)	431
	$O_{23}P_8$	μ-Oxo-bis(tetrametaphosphate)	−26.6 and −26.5J_{PP} 27.3 and 31.5(n)	431
	O_9P_3	Trimetaphosphate	−21.4(bb, pH 7)	433
	$C_{10}H_{12}N_5O_{13}P_3$	Adenosine-5′trimetaphosphate	−24.2 and −24.6J_{PP} 22.9 and 23.7 (pyridine)	432
6/(C)$_2$	$G_3H_6O_5P$	Glycerol-1,3(cyclic)-phosphoryldiester	1.90^D;1.30^D	482
	$C_9H_{10}N_2O_8P$	Uridine 3′,5′-cyclic-P (cUMP)	−2.6	461
	$C_9H_{12}N_2O_8P$	Thymidine 3′,5′-cyclic-P (cTMP)	−2.05	453
	$C_{10}H_{11}N_5O_6P$	Adenosine 3′,5′-cyclic-P (cAMP)	−1.58(w, pH 7)	453
	$C_{10}H_{12}N_2O_7P$	deoxy-Thymidine 3′,5′-cyclic-P 5′-P	11.8(w, pH 6.4)	439
		3′-P	12.5(w, pH 6.4)	439
8/(P^4)$_2$	$O_{12}P_4$	Tetrametaphosphate	−23.6(bb, pH 7)	433
10/(P^4)$_2$	$O_{15}P_5$	Pentametaphosphate	−23.7(bb, pH 7)	433

Rings/ Connectivities	Formula	Structure	^{31}P NMR data (δ_P [solv. temp, ph] J_{PH} Hz)	Ref.
$12/(P^4)_2$	$P_{18}P_6$	Hexametaphosphate	−22.8(bb, pH 7)	433
$14/(P^4)_2$	$O_{21}P_7$	Heptametaphosphate	−22.2(bb, pH 7)	433
$16/(P)_2$	$O_{24}P_8$	Octametaphosphate	−21.5(bb, pH 7)	433
$n/(C)_2$		Cyclic nucleoside diphosphates	Ring size effects	441
	P-bound atoms = OO'_3			
P	$C_6H_{11}O_{12}P_2$	α-Glucose 1-diP β-P	−8(free), −6.2(bound)	437
	$C_{10}H_{12}N_5O_{10}P_2$	Adenosine diP (ADP)[a] β	−6.11(ba24° pH 10 and bp), −6.02(bp), −6.65(bl)	427
		δP counter ion and pH dependence[a]	−10.54 (pH 3.9), −5.94 (pH 8.8) J_{PP} 20/22	435
	$C_{10}H_{12}N_5O_{13}P_3$	Adenosine triP (ATP)[a] τP	−5.80(ba24° pH 10 and bp) −5.70(bp), −5.67(bl)	
			δP counter ion and pH dependence[a]	435, 436
	$C_{10}H_{12}N_5O_{11}P_2$	Guanosine 5′-diP (GDP) β-P	-5.81^{A-} -10.2^{HA}	430
	$C_{10}H_{12}N_5O_{10}P_2S$	Guanosine 5′-thio-diP ($GDP_{\alpha S}$) β-P	-6.00^{A-}, -10.90^{HA}	430
	$C_{10}H_{12}N_5O_{14}P_3$	Guanosine 5′-triP (GTP) τ-P	-5.57^{A-}, -10.24^{HA}	430
C	CH_2NO_5P	Carbamyl P	−1.74(ba24° pH 10)	427
	$C_2H_3O_5P$	Acetyl P	−2.13(ba24° pH 10)	427
	$C_2H_3O_6P$	Glycollic acid 2-P	3.45(ba24° pH 10)	427
	$C_2H_6NO_4P$	Ethanolamine P (PE)	4(ba24° pH 10), 3.85(bl)	427, 490
	$C_3H_3O_6P$	Enolpyruvate (PEP) P	−0.68(ba24° pH 10)	427
	$C_3H_4O_{10}P_2$	Glyceric acid 2,3-diP	3.76 and 3.43(ba24° pH 10)	427
	$C_3H_6NO_6P$	Serine-P	4.00 (pH 8.6), 0.1 (pH 3.9)	463
	$C_3H_5O_6P$	Glyceraldehyde 3-P	4.30(ba24° pH 10)	427
	$C_3H_5O_6P$	Dihydroxyacetone-P	4.16(ba24° pH 10), 4.71(bp) 4.85(bl)	427, 434
	$C_3H_5O_7P$	Glyceric acid 2-P	3.61(ba24° pH 10)	427
	$C_3H_5O_7P$	Glyceric acid 3-P	4.07(ba24° pH 10)	427
	$C_3H_7O_6P$	Glycerol 2-P	4.32(bp), 3.29 (pH 12), 0.15 (pH 4), 3.93(bl)	445, 490
	$C_3H_7O_6P$	Glycerol 3-P	4.29(bp and pH 12), 0.60 (pH 4)	434
	$C_4H_8NO_6P$	Threonine-P	3.2 (pH 9) −0.9 (pH 3.9)	463
$C_5H_8O_{14}P_3$	Ribose-5-P	1-diP	0.27(pH4.94), 3.66(pH9.14)	464
	$C_5H_9O_8P$	Ribose 5-P	3.85(ba24° pH 10), 3.7 (pH 7), 2.6 (pH 6), 0.5 (pH 5), 0.2 (pH 4)	427
	$C_5H_{13}NO_4P$	Choline P (CP)	3.58(ba24° pH 10), 3.32(bp), 3.34(bp), 2.99(bl)	427, 434
	$C_6H_6O_{24}P_6$	Myo-inositol hexa-P Bu_4N+ P-1 and P-3	−2.80 (pH 12), −0.40 (pH 3.8), 0.09 (pH 1)	465
		Myo-inositol hexa-P Bu_4N+ P-2	−2.20 (pH 9), −0.96 (pH 3.8), 0.27 (pH 0)	465
		Myo-inositol hexa-P Bu_4P+ P-4 and P-6	−3.62 (pH 12), −0.12 (pH 3.8), 0.56 (pH 1)	465
		Myo-inositol hexa-P Bu_4N+ P-5	−1.36 (pH 10), −0.58 (pH 3.8), 1.58 (pH 0)	465
	$C_6H_8O_{18}P_4$	Inositol 1,3,4,5-tetraphosphate	4.27 and 4.20 and 4.54 and 5.30(w)	466
	$C_6H_9O_{15}P_3$	Inositol 1,3,4-triphosphate	4.75 and 3.85 and 5.35(w)	466
	$C_6H_{10}O_{12}P_2$	Fructose 1,6-diP and anomers	4.05 and 3.91(ba24° pH 10), 4.10 and 4.05(bl)	427, 467, 490

TABLE H (continued)
Four Coordinate (λ5σ4) Phosphorus Compounds Containing No Phosphorus Bonds to Hydrogen nor Group IV Atoms

Rings/ Connectivities	Formula	Structure	^{31}P NMR data (δ_P [solv. temp, ph] J_{PH} Hz)	Ref.
	$C_6H_{10}O_{12}P_2$	Fructose 2,6-diP	3.99 and −0.43(ba24° pH 10)	427, 483
	$C_6H_{11}O_9P$	Fructose 1-P	4.29(ba24° pH 10)	427
		Fructose 6-P	3.83(ba24° pH 10), 4.5 (pH 10/8), 4.0 (pH 7), 2.5 (pH 6), 0.9 (pH 5), 0.7 (pH 4/3)	427
		Galactose 1-P	2.45(ba24° pH 10)	427
		Galactose 6-P	4.14(ba24° pH 10)	427
		Glucose 1-P	2.32(ba24° pH 10), 2.30(bp), 2.1, 2.05(bl)	427, 434, 437, 490
		Glucose 6-P	4.44(ba24° pH 10), T1/Mn(II)	427, 468
		Mannose 6-P	4.32(ba24° pH 10)	427
	$C_6H_{11}O_9P$	Fructose 3-P	5.2(ba24° pH10)	488
	$C_6H_{13}O_9P$	Sorbitol 3-P	5.9	488
	$C_7H_{13}O_9P$	Heptulose 2-P	−1.1(free), 0.2 (bound AMP)	437
	$C_8H_8NO_6P$	Pyridoxal phosphate	4.5	437, 469
	$C_8H_{14}NO_9P$	N-Acetylglucosamine 1-P	2.04(ba24° pH 10)	427
	$C_9H_{11}N_2O_9P$	Uridine 3′-P (3′-UMP)	3.99(be, pH 8), −0.07 (be, pH 4)	470
	$C_9H_{12}N_3O_8P$	Cytosine 2′-P (2′-CMP)	3.70(be, pH 8), 0.64(be, pH 6), −0.13(be, pH 4)	470
	$C_9H_{12}N_3O_8P$	Cytosine 3′-P (3′-CMP)	3.99(be, pH 8), 1.27(be, pH 6), −0.09(be, pH 4)	470
	$C_9H_{12}N_3O_8P$	Cytosine 5′-P (5′-CMP)	3.69(ba24° pH 10), 3.81 (be, pH 8), 2.76 (pH 6), 0.05(be, pH 4)	427, 472
	$C_9H_{16}O_{17}P_3$	Glycerol 3-phosphoryl-inositol *4,5-diP*	2.22 and 3.18	484
	$C_{10}H_{11}N_4O_8P$	Inosine 5′-P (5′-IMP)	3.96(ba24° pH 10); 3.78(bp, bl)	427, 434, 490
	$C_{10}H_{11}N_4O_9P$	Xanthosine 5′-P (5′-XMP)	3.78(ba24° pH 10)	427
	$C_{10}H_{12}N_5O_7P$	Adenosine (2′-P (2′-AMP)	3.40(ba24° pH 10)	427
	$C_{10}H_{12}N_5O_7P$	Adenosine 3′-P (3′-AMP)	3.96(ba24° pH 10)	427
	$C_{10}H_{12}N_5O_7P$	Adenosine 5′-P (5′-AMP)	3.77(ba24° pH 10), 3.74 (bl)	427, 473, 490
	$C_{10}H_{12}N_5O_8P$	Guanosine 5′-P (5′-GMP)	3.85(pH10), 3.30(pH7.2), 0.39(pH4)	430, 489
	$C_{10}H_{13}N_2O_9P$	Thymidine 5′-P (5′-TMP)	3.4	438, 460
	$C_{17}H_{19}N_4O_9P$	Flavin mono nucleotide 2′-P (2′-FMN)	−3.90 (pH 9), −3.17 (pH 7), 0.20 (pH 4)	474
	$C_{17}H_{19}N_4O_9P$	Flavin mono nucleotide 3′-P (3′-FMN)	−4.92 (pH 9), −4.48 (pH 7), −0.15 (pH 4)	474
	$C_{17}H_{19}N_4O_9P$	Flavin mono nucleotide 4′-P (4′-FMN)	−4.33 (pH 9), −3.65 (pH 7), −0.07 (pH 4)	474
	$C_{17}H_{19}N_4O_9P$	Flavin mono nucleotide 5′-P (FMN)	−4.70 (pH 9), −4.01 (pH 7), −0.90 (pH 4)	474

Rings/ Connectivities	Formula	Structure	^{31}P NMR data (δ_P [solv. temp, ph] J_{PH} Hz)	Ref.
	$C_{21}H_{25}N_7O_{17}P_3$	Nicotinamide adenine dinucleotide 2′-P (NADP 2′-P)	3.52(ba24° pH 10), 3.55 (bp), 3.63(bl)	427, 434, 490
	$C_{26}H_{41}O_8P$	Ecdysteroid-P (3-hydroxy)	1.55(m)	475
	$C_{26}H_{41}O_9P$	Ecdysteroid-P (2,3-dihydroxy)	2.69(m)	475
		Deoxycytidyldeoxyribonucleoside-protected derivatives (dCMP)		471
		Phosphoproteins	Review	476
		α-Casine B	4.3s(bg, pH 10.2), 0.2 (bg, pH 4)	463
		Phosvitn	4.2(bg, pH 9.3), 0.3(bg, pH 3.6)	463
	P-bound atoms = O′$_4$			
None	O′$_4$P	Inorganic phosphate (Pi)	2.61(ba24° pH 10), 2.63(bp, bl), 2.5 (pH 8)	
			1.7 (pH 7), 0.5 (pH 6), 0.15 (pH 5), 0.1 (pH 4/3)	427, 434, 477, 490

Section H4: Naturally Occurring Phosphonates

Rings/ Connectivities	Formula	Structure	^{31}P NMR data	Ref.
	P-bound atoms = COO′$_2$			
H_2P^4;$(C)_2$	$C_{19}H_{26}N_{10}O_{12}P_2$	R′O(HO)POCH$_2$PO(OH)OR″; R′ = R″ = Ad	17.4 (pH 6)	453
		R′O(HO)POCH$_2$**P**O(OH)OR″; R′ = H, R″ = Ad	17.4 (pH 4)	453
CH_2;C′		ROP(O)$_2$OCOR (Phosphonoanhydrides)	10.0/10.3	427, 478
	P-bound atoms = CO′$_3$			
H_2P^4;C	$C_{19}H_{26}N_{10}O_{12}P_2$	R′O(HO)**P**OCH$_2$PO(OH)OR″; R′ = H, R″ = Ad	15.0 (pH 4)	453
CH_2	$C_2H_6NO_3P$	$NH_2CH_2CH_2PO3^{2-}$	16.2	427, 480
	$C_{45}H_{89}NO_7P$	Diacylglyceryl-(2-aminoethyl)phosphonate (DAG-AEP)	21.19(K^+/Cs^+/TMA + bc)	448

CHAPTER H
Supplementary Table

Section H1: Derivatives of Phosphoric Acids

Rings/ Connectivity	Formula	Structure	NMR data (δ_P [solv] $^nJ_{PZ}$ Hz)	Ref.
P-bound atoms = ClO$_2$O′				
C,N′	$C_4H_8Cl_2NO_3P$	Cl(EtO)P(O)ON=CMeCl	7.05(a)	402
P-bound atoms = Cl$_2$OO′				
N′	$C_2H_2Cl_3NO_2P$	$Cl_2P(O)ON{=}C(CH_2Cl)Cl$	12.5(a)	402
P-bound atoms = FO$_2$O′				
(C)$_2$	$C_4H_4F_7O_3P$	$(TfCH_2O)P(O)F$	−10.2 $^1J_{PF}$995	403
P-bound atoms = NO$_2$O′				
C′H;(C)$_2$	$C_{11}H_{16}NO_3PS$	$(EtO)_2P(O)NHCSPh$	−3.5	404
C′$_2$;(C)$_2$	$C_{16}H_{21}F_3NO_5P$	$(EtO)_2P(O)NPhC(Tf){=}CHCOOEt$	−0.4(n)	405
	$C_{16}H_{26}NO_3PS_2$	$(iPrO)_2P(O)NPhCSS(iPr)$	−5	404
P-bound atoms = N$_3$O′				
5/(C$_2$)$_2$,CC′	$C_{13}H_{20}N_3OP$	**47** (3*R*)	18.9(d)	406
		(3*S*)	25.3(d)	
P-bound atoms = N$_2$N′O′				
(C$_2$)$_2$,C′	$C_{15}H_{25}N_4O_3P$	$(Et_2N)_2P(O)N{=}CH(C_6H_4NO_2\text{-}2)$	15.2	407
	$C_{16}H_{25}F_3N_3OP$	$(Et_2N)_2P(O)N{=}C(Tf)Ph$	11.3	407
P-bound atoms = O$_3$O′				
(C)$_2$,N′	$C_3H_7FNO_3P$	$(MeO)_2P(O)ON{=}CHF$	1.8(c)	408
	$C_8H_{17}ClNO_4P$	$(PrO)_2P(O)ON{=}CMeCl$	−0.9(a)	402
C,(C′)$_2$	$C_{16}H_{15}O_6P$	$(PhO)_2P(O)O\dot{C}HCH_2CH_2CO\dot{O}$	−12.0(c)	409
	$C_{20}H_{17}O_5P$	$(PhO)_2P(O)OCH_2COPh$	−11.6	409

Section H2: Derivatives of Thio-, Seleno-, and Telluro-Phosphoric Acids

Rings/ Connectivity	Formula	Structure	NMR data	Ref.
P-bound atoms = NOSS′				
5/C′H;C′;C	$C_7H_8NOPS_2$	**15**; R = H, X = S, Y = SMe	98	410
6/CH;C;C	$C_5H_{12}NOPS_2$	**4**; R^1 = R^2 = R^3 = H, X = S′, Y = SEt	84	410
6/C′H;C;C	$C_8H_{10}NOPS_2$	**48**; R = H, X = S, Y = SMe	82	410
6/C′H;C′;C	$C_9H_{10}NOPS_3$	**49**; R = H, X = Z = S, Y = SEt	72	410
P-bound atoms = NOS′$_2$				
5/CH;C	$C_2H_5NOPS_2$	**11**; R = H, X = S, Y = S$^-$	121	410
6/C′H;C′	$C_7H_5NOPS_3$	**49**; R = H, X = Z = S, Y = S$^-$	88.6	410

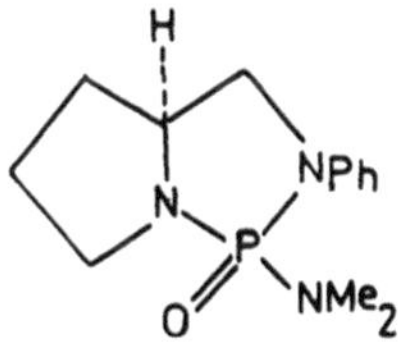

47 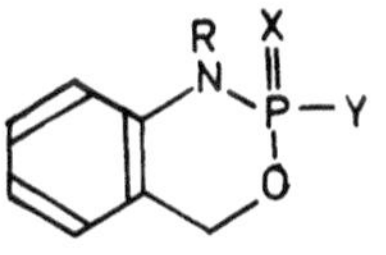 48 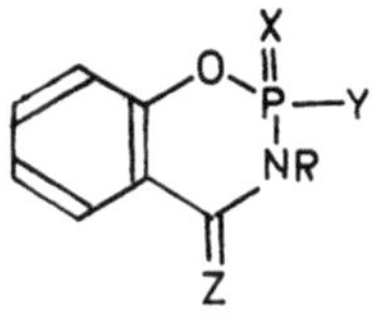 49 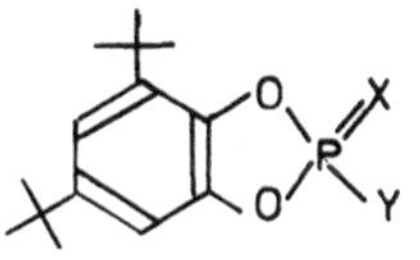 50

P-bound atoms = NO_2S'

$CP^4;(C)_2$	$C_{24}H_{29}NO_2P_2S_2$	$(iPrO)_2P(S)NPhP(S)Ph_2$	65, 66	404
$C'H;(C)_2$	$C_{11}H_{15}N_2O_2PS$	$(EtO)_2P(S)NH\dot{C}{=}NC_6H_4\dot{S}$-2	64	411
$(C')_2;(C)_2$	$C_{17}H_{29}N_2O_2PS_2$	$(iPrO)_2P(S)NPhCSNEt_2$	62	404
$5/CH;(C)_2$	$C_3H_8NO_2PS$	**11;** R = H, X = S, Y = SMe	88.4	412
$5/C_2;(C')_2$	$C_{18}H_{30}NO_2PS$	**50;** X = S, Y = NEt_2	86.8	413

P-bound atoms = NOO'S

$5/C_2;C;C$	$C_5H_{12}NO_2PS$	**51;** n = 0, X = O, Y = NMe_2, R = Me	47	414

P-bound atoms = NOO'S'

$6/C'H;C'$	$C_7H_5NO_2PS_2$	**49;** R = H, X = Z = S, Y = O^-	39.6	410

P-bound atoms = NS_2S'

$5/C'H;C,C'$	$C_{13}H_{12}NPS_3$	**35;** R = H, A = NH, X = S, Y = SCH_2Ph	89	410

P-bound atoms = N_2SS'

$5/(C'H)_2;C$	$C_{13}H_{13}N_2PS_2$	**52;** R^1 = R^2 = H, X = S, Y = SCH_2Ph	72.6	410

P-bound atoms = N_3S'

$(CH)_3$	$C_3H_{12}N_3PS$	$(MeNH)_3P(S)$	68.9(c)	415
	$C_{12}H_{30}N_3PS$	$(tBuNH)_3P(S)$	60.4(c)	415

P-bound atoms = OS_2S'

$5/C;(C)_2$	$C_3H_7OPS_3$	**51;** n = 0, R = H, X = S, Y = SMe	116	410

P-bound atoms = OSS'_2

$5/C;C$	$C_2H_4OPS_2$	**51;** n = 0, R = H, X = S, Y = S^-	125	410

P-bound atoms = O_2SS'

$(C)_2;Br$	$C_2H_6BrO_2PS_2$	$(MeO)_2P(S)SBr$	79.9(c)	416
	$C_6H_{14}BrO_2PS_2$	$(iPrO)_2P(S)SBr$	70.0(r)	416
$(C')_2;Br$	$C_{12}H_{10}BrO_2PS_2$	$(PhO)_2P(S)SBr$	69.8(g) 71.9(r)	418
$(C)_2;Cl$	$C_2H_6ClO_2PS_2$	$(MeO)_2P(S)SCl$	84.3(r)	416
	$C_6H_{14}ClO_2PS_2$	$(iPro)_2P(S)SCl$	75.0(r)	416
$(C)_2;N$	$C_{15}H_{32}NO_2PS_2$	$(NeoO)_2P(S)SN(CH_2)_5$	96.0(b)	416
$(C)_2;P^4$	$C_{20}H_{44}O_4P_2S_3$	$[(NeoO)_2P(S)]_2S$	79(n)	419
$(C)_2;C$	$C_{14}H_{21}O_2PS_2$	$(EtO)_2P(S)SCHMeCH{=}CHPh$	92^E	420

CHAPTER H (continued)
Supplementary Table

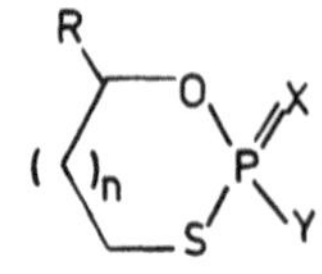

51 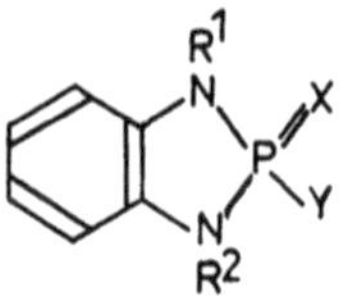 52 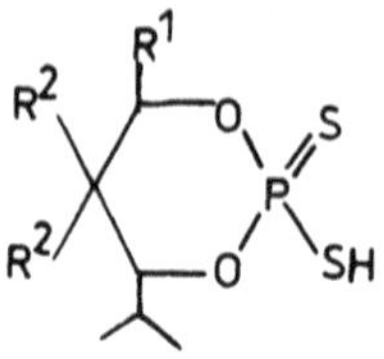 53 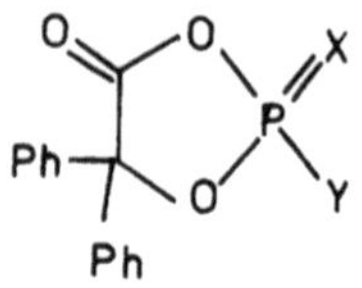 54

Rings/ Connectivity of P-bound atoms	Formula	Structure	NMR data (δ_P [solv] $^nJ_{PZ}$ Hz)	Ref.
(C)$_2$;C′	$C_{15}H_{19}N_2O_3PS_2$	4-TolCOC(CN)=C(NH$_2$)SP(S)(OEt)$_2$	75.2[E]	421
5/(C′)$_2$;C	$C_7H_7O_2PS_2$	CatP(S)SMe	113	410
6/(C)$_2$;P[4]	$C_{10}H_{20}O_4P_2S_3$	**24**; X = Y = Z = S	65.5(m)	419
6/(C)$_2$;S	$C_{10}H_{20}O_4P_2S_4$	**24**; X = Z = S, Y = SS	79.3(r)	419
6/(C)$_2$;H	$C_7H_{15}O_2PS_2$	**53**; R^1 = Me, R^2 = H	113.1	422
	$C_{11}H_{23}O_2PS_2$	**53**; R^1 = iPr, R^2 = Me	120.3	422
	$C_{14}H_{21}O_2PS_2$	**53**; R^1 = Ph, R^2 = Me	84.9	422
6/(C)$_2$;C	$C_9H_{17}O_4PS_2$	(OCH$_2$CH$_2$CH$_2$O)P(S)SCH$_2$ ĊHOCMe$_2$OĊH$_2$	91.0(b), 88.0(c)	423
P-bound atoms = O$_2$S′$_2$				
C,C′	$C_{14}H_{10}O_3PS_2$	**54**; X = S, Y = S$^-$	120.2(m,c)	424
P-bound atoms = OO′S$_2$				
N′;(C)$_2$	$C_7H_{16}NO_2PS_2$	(EtS)$_2$P(O)ON=CMe$_2$	57.2	426
6/C;(C)$_2$	$C_9H_{17}O_4PS_2$	**51**; n = 1, R = H, X = O, Y = SCH$_2$ĊHOCMe$_2$OĊH$_2$	39.7(c)	423
P-bound atoms = O$_3$S′				
Si,(C)$_2$	$C_{22}H_{49}O_3PSSi$	(NeoO)$_2$P(S)OSiBu$_3$	55.3	419
6/(C)$_2$,C′	$C_7H_9O_3PS$	**44**; X = S, Y = OMe *(S)*	59.9(c)	412
		(R)	59.9(c)	
	$C_8H_{10}ClO_3PS$	**44**; X = S, Y = OMe, 6-Cl-4-Me		412
		(2*S*,4*R*)	60.6(c)	
		(2*R*,4*S*)	60.6(c)	
		(2*S*,4*S*)	58.9(c)	
		(2*R*,4*R*)	58.9(c)	
P-bound atoms = O$_2$O′S				
(C)$_2$;C″	$C_{11}H_{22}NO_3PS$	(NeoO)$_2$P(O)SCN	10.4	419
	$C_{12}H_{15}O_3PS$	(EtO)$_2$P(O)SC≡CPh	16.6	426
6/(C)$_2$;C	$C_9H_{17}O_5PS$	**51**; n = 1, R = H, X = O, Y = OCH$_2$ĊHOCMe$_2$OĊH$_2$	13.8(b), 16.8(c)	423
6/CC′;C	$C_{10}H_{13}O_3PS$	**51**; n = 1, R = Me, X = O, Y = OPh	12	414
P-bound atoms = S$_2$S′$_2$				
6/(C)$_2$	$C_3H_6PS_4$	(SCH$_2$CH$_2$CH$_2$S)P(S)S$^-$	77	410

REFERENCES

1. **Bruzik, K. and Stec, W. J.,** *J. Org. Chem.*, 46, 1625, 1981.
2. **Bruzik, K. and Stec, W. J.,** *J. Org. Chem.*, 46, 1618, 1981.
3. **Michalska, M., Michalski, J., and Orlich, I.,** *Tetrahedron*, 34, 617, 1978.
4. **Michalska, M., Michalski, J., and Orlich-Kresel, I.,** *Pol. J. Chem.*, 53, 253, 1979.
5. **Gorenstein, D. G.,** Phosphorus-31 chemical shifts; principles and empirical observations, in *Phosphorus 31-P NMR. Principles and Applications*, Gorenstein, D. G., Ed., Academic Press, London, 1984, chap. 1.
6. **Rezvukhin, A. I., Dolenko, G. N., and Krupoder, S. A.,** *Magn. Reson. Chem.*, 23, 221, 1985.
7. **Bel'skii, V. E., Kudryavtseva, L. A., Gol'dfarb, É. I., and Ivanov, B. E.,** *J. Gen. Chem. U.S.S.R.*, 44, 2612, 1974.
8. **Vlassa, M. and Baraban, A.,** *J. Prakt. Chem.*, 326, 1011, 1984.
9. **Heinicke, J., Böhle, I., and Tzschach, A.,** *J. Organomet. Chem.*, 317, 11, 1986.
10. **Effenberger, F., König, G., and Klenk, H.,** *Synthesis*, p. 70, 1981.
11. **Bel'skii, V. E., Kudryavtseva, L. A., Kurguzova, A. M., Il'ina, O. M., and Ivanov, B. E.,** *Bull. Acad. Sci. U.S.S.R. Div. Chem. Sci.*, p. 1510, 1979.
12. **Blackburn, G. M., Cohen, J. S., and Weatherall, I.,** *Tetrahedron*, 27, 2903, 1971.
13. **Rüger, C., Konig, T., and Schwetlick, K.,** *J. Prakt. Chem.*, 326, 622, 1984.
14. **Nuretdinova, O. N. and Arbuzov, B. A.,** *Bull. Acad. Sci. U.S.S.R. Div. Chem. Sci.*, p. 890, 1981.
15. **Zaslavskaya, N. N., Gilyarov, V. A., and Kabachnik, M. I.,** *Bull. Acad. Sci. U.S.S.R. Div. Chem. Sci.*, p. 662, 1976.
16. **Arshinova, R. P., Mukmenev, É. T., Gurarii, L. I., and Arbuzov, B. A.,** *Bull. Acad. Sci. U.S.S.R. Div. Chem. Sci.*, p. 524, 1978.
17. **Majoral, J.-P. and Navech, J.,** *Bull. Soc. Chim. Fr.*, p. 95, 1971.
18. **Majoral, J.-P. and Navech, J.,** *Bull. Soc. Chim. Fr.*, p. 1331, 1971.
19. **Majoral, J.-P. and Navech, J.,** *Bull. Soc. Chim. Fr.*, p. 2609, 1971.
20. **Gorenstein, D. G.,** *J. Am. Chem. Soc.*, 97, 898, 1975.
21. **Gorenstein, D. G., Rowell, R., and Findlay, J.,** *J. Am. Chem. Soc.*, 102, 5077, 1980.
22. **Gorenstein, D. G. and Rowell, R.,** *J. Am. Chem. Soc.*, 101, 4925, 1979.
23. **Cooper, D. B., Inch, T. D., and Lewis, G. J.,** *J. Chem. Soc. Perkin Trans. 1*, p. 1043, 1974.
24. **Cooper, D. B., Harrison, J. M., Inch, T. D., and Lewis, G. J.,** *J. Chem. Soc. Perkin Trans. 1*, p. 1049, 1974.
25. **Mikołajczyk, M., Krzywanski, J., and Ziemnicka, B.,** *J. Org. Chem.*, 42, 190, 1977.
26. **Okruszek, A. and Stec, W. J.,** *Z. Naturforsch.*, 30B, 430, 1975.
27. **Stec, W. J., Kinas, R., and Okruszek, A.,** *Z. Naturforsch.*, 31B, 393, 1976.
28. **Eliel, E. L., Chandrasekaran, S., Carpenter, L. E., and Verkade, J. G.,** *J. Am. Chem. Soc.*, 108, 6651, 1986.
29. **Edmundson, R. S. and King, T. J.,** *J. Chem. Soc. Perkin Trans. 2*, p. 69, 1985.
30. **Bères, J., Sándor, P., Kálmán, A., Koritsánszky, T., and Ötvös, L.,** *Tetrahedron*, 40, 2405, 1984.
31. **Bères, J., Bentrude, W. G., Parkanji, L., Kálmán, A., and Sopchik, A. E.,** *J. Org. Chem.*, 50, 1271, 1985.
32. **Leśnikowski, Z. J., Stec, W. J., and Zielinski, W. S.,** *J. Am. Chem. Soc.*, 103, 2862, 1981.
33. **Stec, W. J.,** *Z. Naturforsch.*, 29B, 109, 1974.
34. **Verkade, J. G.,** *Phosphorus Sulfur*, 2, 251, 1976.
35. **Gorenstein, D. G.,** Phosphorus-31 spin-spin coupling constants; principles and applications, in *Phosphorus 31-P NMR. Principles and Applications*, Gorenstein, D. G., Ed., Academic Press, London, 1984, chap. 2.
36. **Glidewell, C. and Leslie, E. J.,** *J. Chem. Soc. Dalton Trans.*, p. 527, 1977.
37. **Schiff, D. E., Richardson, J. W., Jacobson, R. A., Cowley, A. H., Lasch, J., and Verkade, J. G.,** *Inorg. Chem.*, 23, 3373, 1984.
38. **Gray, G. A., Buchanan, G. W., and Morin, F. G.,** *J. Org. Chem.*, 44, 1768, 1979.
39. **Stec, W. J. and Zielinski, W. S.,** *Tetrahedron Lett.*, 21, 1361, 1980.
40. **Bottka, S. and Tomasz, J.,** *Tetrahedron Lett.*, 26, 2909, 1985.
41. **Bentrude, W. G. and Tomasz, J.,** *Synthesis*, p. 27, 1984.
42. **Gerlt, J. A., Demou, P. C., and Mehdi, S.,** *J. Am. Chem. Soc.*, 104, 2848, 1982.
43. **Gerlt, J. A.,** Use of chiral [^{16}O,^{17}O,^{18}O] phosphate esters to determine the stereochemical course of enzymatic phosphoryl transfer reactions, in *Phosphorus 31-P NMR. Principles and Applications*, Gorenstein, D. G., Ed., Academic Press, London, 1984, chap. 7.
44. **Buchanan, G. W. and Morin, F. G.,** *Can. J. Chem.*, 57, 21, 1979.
45. **Maryanoff, B., Hutchins, R. O., and Maryanoff, C. A.,** *Top. Stereochem.*, 11, 187, 1979.

46. **Samitov, Yu. Yu. and Karataeva, F. Kh.**, *J. Gen. Chem. U.S.S.R.*, 54, 714, 1984.
47. **Gaydou, E. M. and Llinas, J. R.**, *Org. Magn. Reson.*, 6, 23, 1974.
48. **Anderson, R. C. and Shapiro, M. J.**, *J. Org. Chem.*, 49, 1304, 1984.
49. **Johnson, C. R., Elliott, R. C., and Renning, T. D.**, *J. Am. Chem. Soc.*, 106, 5019, 1984.
50. **Drodz, G. I., Sokal'skii, M. A., Ivin, S. Z., Sosova, E. P., and Strukov, O. G.**, *J. Gen. Chem. U.S.S.R.*, 39, 905, 1969.
51. **Satterthwait, A. C. and Westheimer, F. H.**, *J. Am. Chem. Soc.*, 102, 4464, 1980.
52. **Górecka, A., Leplawy, M., Zabrocki, J., and Zwierzak, A.**, *Synthesis*, p. 474, 1978.
53. **Gajda, T. and Zwierzak, A.**, *Synthesis*, p. 243, 1976.
54. **Gloede, J. and Costisella, B.**, *Z. Anorg. Allg. Chem.*, 471, 147, 1980.
55. (a) **Stec, W. J. and Mikołajczyk, M.**, *Tetrahedron*, 29, 539, 1973; (b) **Stec, W. J. and Łopusinski, A.**, *Tetrahedron*, 29, 547, 1973.
56. **Stec, W. J., Konopka, A., and Uznański, B.**, *J. Chem. Soc. Chem. Commun.*, p. 923, 1974.
57. **Haubold, W. and Fluck, E.**, *Z. Anorg. Allg. Chem.*, 392, 59, 1972.
58. **Keat, R.**, *J. Chem. Soc. Dalton Trans.*, p. 876, 1974.
59. **Leśnikowski, Z. J., Niewiarowski, W., Zielinski, W. S., and Stec, W. J.**, *Tetrahedron*, 40, 15, 1984.
60. **Samitov, Yu. Yu., Musina, A. A., Aminova, R. M., Pudovik, M. A., Khayarev, A. I., and Medvedeva, M. D.**, *Org. Magn. Reson.*, 13, 167, 1980.
61. **Balitskii, Yu. V., Negrebetskii, V. V., and Gololobov, Yu. G.**, *J. Gen. Chem. U.S.S.R.*, 50, 1767, 1980.
62. **Hunston, R. N., Jehangir, M., Jones, A. S., and Walker, R. T.**, *Tetrahedron*, 36, 2337, 1980.
63. **Pankiewicz, K., Kinas, R., Stec, W. J., Foster, A. B., Jarman, M., and Van Maanen, J. M. S.**, *J. Am. Chem. Soc.*, 101, 7712, 1979.
64. (a) **Irvine, I. and Keat, R.**, *J. Chem. Soc. Dalton Trans.*, p. 17, 1972; (b) **Haegele, G., Harris, R. K., Wazeer, M. I. M., and Keat, R.**, *J. Chem. Soc. Dalton Trans*, 1985, 1974.
65. **Bulloch, G. and Keat, R.**, *J. Chem. Soc. Dalton Trans.*, p. 2010, 1974.
66. **Nielsen, M. L., Pustinger, J. V., and Strobel, J.**, *J. Chem. Eng. Data*, 9, 167, 1964.
67. **Shevchenko, M. V. and Kukhar', V. P.**, *J. Gen. Chem. U.S.S.R.*, 54, 1336, 1984.
68. **Coustures, Y., Labarre, M.-C., and Bruniquel, M.-F.**, *Bull. Soc. Chim. Fr.*, p. 926, 1973.
69. **Revel, M. and Navech, J.**, *Bull. Soc. Chim. Fr.*, p. 1195, 1973.
70. **Maier, L.**, *Helv. Chim. Acta*, 56, 492, 1973.
71. **Engels, J. and Krahmer, U.**, *Synthesis*, p. 485, 1981.
72. **Mahmood, T. and Shreeve, J. M.**, *Inorg. Chem.*, 25, 3830, 1986.
73. **Markovskii, L. N., Solov'ev, A. V., Pashinnik, V. E., and Shermolovich, Yu.G.**, *J. Gen. Chem. U.S.S.R.*, 50, 644, 1980.
74. (a) Ref. 76; (b) Ref. 96.
75. **Michalski, J., Mlotkowska, M., and Skowrońska, A.**, *J. Chem. Soc. Perkin Trans. 1*, p. 319, 1974.
76. **Łopusinski, A., Michalski, J., and Stec, W. J.**, *Liebigs Ann. Chem.*, p. 924, 1977.
77. (a) Ref. 66; (b) Ref. 78.
78. **Horner, L. and Gehring, R.**, *Phosphorus Sulfur*, 11, 157, 1981.
79. **Dhawan, B. and Redmore, D.**, *J. Org. Chem.*, 51, 179, 1986.
80. **Röschenthaler, G.-V., Sauerbrey, K., Gibson, J. A., and Schmutzler, R.**, *Z. Anorg. Allg. Chem.*, 450, 79, 1979.
81. **Cullis, P. M. and Lowe, G.**, *J. Chem. Soc. Perkin Trans. 1*, p. 2317, 1981.
82. **Ramirez, F., Okazaki, H., Marecek, J. F., and Tsuboi, H.**, *Synthesis*, p. 819, 1976.
83. **Ramirez, F., Nowakowski, M., and Marecek, J. F.**, *J. Am. Chem. Soc.*, 98, 4330, 1976.
84. **Vilceanu, R. and Neda, I.**, *Phosphorus Sulfur*, 8, 131, 1980.
85. (a) Ref. 17; (b) Ref. 56.
86. **Mahmood, T. and Shreeve, J. M.**, *Inorg. Chem.*, 25, 4081, 1986.
87. **Roesky, H. W. and Schaper, W.**, *Z. Naturforsch.*, 27B, 1137, 1971.
88. **Bulloch, G., Keat, R., and Tennant, N. H.**, *J. Chem. Soc. Dalton Trans.*, p. 2329, 1974.
89. **Roesky, H. W. and Böwing, W. G.**, *Z. Anorg. Allg. Chem.*, 386, 191, 1971.
90. **Robinson, E. A. and Lavery, D. S.**, *Spectrochim. Acta*, 28, 1099, 1972.
91. **Bayandina, E. V., Sadkova, D. N., and Nuretdinov, I. A.**, *J. Gen. Chem. U.S.S.R.*, 51, 798, 1981.
92. **Dogadina, A. V., Efremov, V. A., Belykh, O. A., Komarov, V. Ya., Ionin, B. I., and Petrov, A. A.**, *J. Gen. Chem. U.S.S.R.*, 54, 706, 1984.
93. (a) Ref. 70; (b) Ref. 96.
94. **Maier, L.**, *Helv. Chim. Acta*, 56, 1257, 1973.
95. **Maier, L.**, *Helv. Chim. Acta*, 52, 1337, 1969.
96. **Moedritzer, L., Maier, L., and Groenwegh, L. C. D.**, *J. Chem. Eng. Data*, 7, 307, 1962.
97. **Gazizov, M. B., Khairullin, R. A., Zakharov, V. M., Moskva, V. V., and Savel'eva, E. I.**, *J. Gen. Chem. U.S.S.R.*, 53, 2162, 1983.

98. **Misiura, K., Okruszek, A., Pankiewicz, K., Stec, W. J., Czawnicki, Z., and Ultracka, B.,** *J. Med. Chem.*, 26, 674, 1983.
99. **Effenberger, F. and Konig, G.,** *Chem. Ber.*, 114, 916, 1981.
100. **Riesel, L. and Sturm, D.,** *Z. Anorg. Allg. Chem.*, 539, 187, 1986.
101. **Light, L. and Paine, R. T.,** *Phosphorus Sulfur*, 8, 255, 1980.
102. **Stec, W. J., Van Wazer, J. R., and Goddard, N.,** *J. Chem. Soc. Perkin Trans. 2*, p. 463, 1972.
103. **Reddy, G. S. and Schmutzler, R.,** *Z. Naturforsch.*, 25B, 1199, 1970.
104. **Okruszek, A. and Stec, W. J.,** *Z. Naturforsch.*, 31B, 354, 1975.
105. **Łopusinski, A. and Michalski, J.,** *J. Am. Chem. Soc.*, 104, 290, 1982.
106. (a) Ref. 105; (b) Ref. 104.
107. **Hart, G. J., O'Brien, R. D., Milbrath, D. S., and Verkade, J. G.,** *Pestic. Biochem. Physiol.*, 6, 464, 1976.
108. **Milbrath, D. S., Springer, J. P., Clardy, J. C., and Verkade, J. G.,** *Phosphorus Sulfur*, 11, 19, 1981.
109. **Corriu, R. J. P., Dutheil, J.-P., and Lanneau, G. F.,** *J. Am. Chem. Soc.*, 106, 1060, 1984.
110. **Glemser, O., Biermann, U., and von Halasz, S. P.,** *Inorg. Nucl. Chem. Lett.*, 5, 501, 1969.
111. **Roesky, H. W.,** *Z. Naturforsch.*, 24B, 818, 1969.
112. **Riesel, L. and Sturm, D.,** *Z. Anorg. Allg. Chem.*, 539, 183, 1986.
113. **Fluck, E. and Beuerle, E.,** *Z. Anorg. Allg. Chem.*, 411, 125, 1975.
114. **Roesky, H. W.,** *Chem. Ber.*, 100, 2147, 1967.
115. **Zawadzki, S. and Zwierzak, A.,** *Tetrahedron*, 19, 315, 1973.
116. **Zwierzak, A. and Brylikowska, J.,** *Synthesis*, p. 712, 1975.
117. **Egorov, Yu.P., Borovikov, Yu.Ya., Kreshchenko, E. P., Pinchuk, A. M., and Kovalevskaya, T. V.,** *J. Gen. Chem. U.S.S.R.*, 45, 1683, 1975.
118. **Steinbach, J., Herrmann, E., and Riesel, L.,** *Z. Anorg. Allg. Chem.*, 523, 180, 1985.
119. **Zwierzak, A. and Sulewska, A.,** *Synthesis*, p. 835, 1976.
120. **Steinbach, J., Herrmann, E., and Riesel, L.,** *Z. Anorg. Allg. Chem.*, 511, 51, 1984.
121. **Fluck, E., Richter, H., and Schwarz, W.,** *Z. Anorg. Allg. Chem.*, 498, 161, 1983.
122. **Richter, H., Fluck, E., Riffel, H., and Hess, H.,** *Z. Anorg. Allg. Chem.*, 496, 109, 1983.
123. **Zimin, M. G., Fomakhin, E. V., Sadykov, A. R., Smirnov, V. N., and Pudovik, A. N.,** *J. Gen. Chem. U.S.S.R.*, 55, 1526, 1985.
124. **Nikonorova, L. K., Enikeev, K. M., Grechkin, N. P., Ismaev, I.É., Il'yasov, A. V., and Nuretdinov, I. A.,** *J. Gen. Chem. U.S.S.R.*, 53, 2212, 1983.
125. **Tupchienko, S. K., Dubchenko, T. N., and Gololobov, Yu. G.,** *J. Gen. Chem. U.S.S.R.*, 51, 847, 1981.
126. **Gololobov, Yu. G., Dubchenko, T. N., and Tupchienko, S. K.,** *J. Gen. Chem. U.S.S.R.*, 49, 2328, 1979.
127. **Schlak, O., Stadelmann, W., Stelzer, O., and Schmutzler, R.,** *Z. Anorg. Allg. Chem.*, 419, 275, 1976.
128. **Nesterov, L. V., Krepysheva, N. E., Sabirova, R. A., and Lipkina, G. N.,** *J. Gen. Chem. U.S.S.R.*, 48, 722, 1978.
129. **Zwierzak, A.,** *Synthesis*, p. 921, 1982.
130. **Willeit, A., Müller, E. P., and Peringer, P.,** *Helv. Chim. Acta*, 66, 2467, 1983.
131. **Osowska-Pacewicka, K. and Zwierzak, A.,** *Tetrahedron*, 41, 4717, 1985.
132. **Antokhina, L. A., Mertzalova, F. F., Latypov, Z.Ya., and Nikonorov, K. V.,** *J. Gen Chem. U.S.S.R.*, 53, 2390, 1983.
133. **Zwierzak, A.,** *Synthesis*, p. 507, 1975.
134. (a) **Zwierzak, A. and Osowska, K.,** *Synthesis*, p. 223, 1984; (b) Ref. 141.
135. **Bertrand, G., Majoral, J.-P., and Baceiredo, A.,** *Tetrahedron Lett.*, 21, 5015, 1980.
136. **Zwierzak, A. and Osowska-Pacewicka, K.,** *Monatsh. Chem.*, 115, 117, 1984.
137. **Revel, M., Navech, J., and Mathis, F.,** *Bull. Soc. Chim. Fr.*, p. 105, 1971.
138. **Zwierzak, A. and Pilichowska, S.,** *Synthesis*, p. 922, 1982.
139. **Mikolajczyk, M., Kielbasinski, P., and Basiński, W.,** *J. Org. Chem.*, 49, 899, 1984.
140. **Brown, C., Boudreau, J. A., Hewitson, B., and Hudson, R. F.,** *J. Chem. Soc. Perkin Trans. 2*, p. 888, 1976.
141. **Dabkowski, W., Michalski, J., Radziejewski, C., and Skrzypazynski, Z.,** *Chem. Ber.*, 115, 1636, 1982.
142. **Kolesova, V. A., Aleshnikova, T. V., Strepikheev, Yu. A., and Arbuzova, M. V.,** *J. Gen. Chem. U.S.S.R.*, 55, 1808, 1985.
143. **Hirashima, A. and Eto, M.,** *Agric. Biol. Chem.*, 47, 2831, 1983.
144. **Ramirez, F., Okazaki, H., and Marecek, J. F.,** *Synthesis*, p. 637, 1975.
145. **Devillers, J. and Navech, J.,** *Bull. Soc. Chim. Fr.*, p. 4341, 1970.
146. **Matrosov, E. I., Kryuchkov, A. A., Nifant'ev, É. E., and Kabachnik, M. I.,** *Bull. Acad. Sci. U.S.S.R. Div. Chem. Sci.*, p. 719, 1977.

147. **Corriu, R. J. P., Lanneau, G. F., and Leclerq, D.,** *Tetrahedron Lett.*, 24, 4323, 1984.
148. **Hall, C. R. and Inch, T. D.,** *J. Chem. Soc. Perkin Trans. 1*, p. 2368, 1981.
149. **Gololobov, Yu. G., Gusar', N. I., and Chaus, M. P.,** *Tetrahedron*, 41, 793, 1985.
150. **Voropai, L. M., Ruchkina, N. G., Milliaresi, E. E., and Nifant'ev, É. E.,** *J. Gen. Chem. U.S.S.R.*, 55, 55, 1985.
151. **Ramirez, F., Ricci, J. S., Okasaki, H., Marecek, J. F., and Levy, M.,** *Phosphorus Sulfur*, 20, 279, 1984.
152. **Balitskii, Yu. V., Gololobov, Yu. G., Yurchenko, V. M., Antipin, M.Yu., Struchkov, Yu.T., and Boldeskul, I. E.,** *J. Gen. Chem. U.S.S.R.*, 50, 231, 1980.
153. **Cadogan, J. I. G., Grace, D. S. B., Lim, P. K. K., and Tait, B. S.,** *J. Chem. Soc. Perkin Trans. 1*, p. 2376, 1975.
154. **Majoral, J.-P., Revel, M., and Navech, J.,** *Phosphorus Sulfur*, 4, 317, 1978.
155. **Bentrude, W. G., Setzer, W. N., Sopchik, A. E., Bajwa, G. S., Burright, D. D., and Hutchinson, J. P.,** *J. Am. Chem. Soc.*, 108, 6669, 1986.
156. **Edmundson, R. S. and King, T. J.,** *J. Chem. Soc. Perkin Trans. 1*, p. 1943, 1984.
157. **Stec, W. J. and Okruszek, A.,** *J. Chem. Soc. Perkin Trans. 1*, p. 1828, 1975.
158. **Stec, W. J., Okruszek, Z., Lesiak, K., Uznański, B., and Michalski, J.,** *J. Org. Chem.*, 41, 227, 1976.
159. **Mosbo, J. A. and Verkade, J. G.,** *J. Org. Chem.*, 42, 1549, 1977.
160. **Brault, J. F. and Savignac, P.,** *J. Organomet. Chem.*, 66, 71, 1974.
161. **Durrieu, J., Kraemer, R., and Navech, J.,** *Org. Magn. Reson.*, 4, 709, 1972.
162. **Calvo, K. C. and Westheimer, F. H.,** *J. Am. Chem. Soc.*, 105, 2827, 1983.
163. **Mülle, J. and Schröder, H. Fr.,** *Z. Anorg. Allg. Chem.*, 450, 149, 1979.
164. **Riesel, L., Steinbach, J., and Herrmann, E.,** *Z. Anorg. Allg. Chem.*, 502, 21, 1983.
165. **Łopusinski, A., Łuczak, L., and Michalski, J.,** *Tetrahedron*, 38, 679, 1982.
166. **Pudovik, A. N., Cherkasov, R. A., Zimin, M. G., and Kamalov, R. M.,** *J. Gen. Chem. U.S.S.R.*, 48, 2399, 1978.
167. **Budilova, I. Yu., Gusar', N. I., and Gololobov, Yu. G.,** *J. Gen. Chem. U.S.S.R.*, 53, 247, 1983.
168. **Cremlyn, R. and Akhtar, N.,** *Phosphorus Sulfur*, 7, 247, 1979.
169. **Tupchienko, S. K., Dudchenko, T. N., and Gololobov, Yu. G.,** *J. Gen. Chem. U.S.S.R.*, 53, 570, 1983.
170. **Harger, M. J. P. and Smith, A.,** *J. Chem. Soc. Perkin Trans. 1*, p. 2169, 1986.
171. **Fluck, E. and Kleeman, S.,** *Z. Anorg. Allg. Chem.*, 461, 187, 1980.
172. **Schmidpeter, A., Tautz, H., and Schrebier, F.,** *Z. Anorg. Allg. Chem.*, 475, 211, 1981.
173. **Gilyarev, V. A., Tikhonina, N. A., Shcherbina, T. M., and Kabachnik, M. I.,** *J. Gen. Chem. U.S.S.R.*, 50, 1157, 1980.
174. **Savignac, P., Chenault, J., and Dreux, M.,** *J. Organomet. Chem.*, 66, 63, 1974.
175. **Kraemer, R. and Navech, J.,** *Bull. Soc. Chim. Fr.*, p. 3580, 1971.
176. (a) Ref. 178; (b) **Kawashima, T., Kroshefsky, R. D., Kok, R. A., and Verkade, J. G.,** *J. Org. Chem.*, 43, 1111, 1978.
177. (a) Ref. 98; (b) Ref. 63.
178. **Setzer, W. N., Sopchik, A. E., and Bentrude, W. G.,** *J. Am. Chem. Soc.*, 107, 2083, 1985.
179. **Nifant'ev, É. E., Zavalishkina, A. I., Sorokina, S. F., Borisenko, A. A., Smirnova, E. I., Kurechkin, V. V., and Moiseeva, L. I.,** *J. Gen. Chem. U.S.S.R.*, 49, 53, 1979.
180. **Bajwa, G. S., Chandrasekaran, S., Hargis, J. H., Sopchik, A. E., Blatter, D., and Bentrude, W. G.,** *J. Am. Chem. Soc.*, 104, 6385, 1982.
181. **Merrem, H. J., Majoral, J.-P., and Navech, J.,** *Phosphorus Sulfur*, 11, 241, 1981.
182. **Majoral, J.-P., Kraemer, R., Navech, J., and Mathis, F.,** *Tetrahedron*, 32, 2633, 1976.
183. **Miroshnichenko, V. V., Marchenko, A. P., Kudryavtsev, A. A., and Pinchuk, A. M.,** *J. Gen. Chem. U.S.S.R.*, 52, 2222, 1982.
184. **White, D. W., Karcher, B. A., Jacobson, R. A., and Verkade, J. G.,** *J. Am. Chem. Soc.*, 101, 4921, 1979.
185. **Goetze, R., Nöth, H., and Payne, D. S.,** *Chem. Ber.*, 105, 2637, 1972.
186. **Benhammou, M., Kraemer, R., Germa, H., Majoral, J.-P., and Navech, J.,** *Phosphorus Sulfur*, 14, 105, 1982.
187. **Kroshefsky, R. D., Weiss, R., and Verkade, J. G.,** *Inorg. Chem.*, 18, 469, 1979.
188. **Kroshefsky, R. D., Verkade, J. G., and Pipal, J. R.,** *Phosphorus Sulfur*, 6, 377, 1979.
189. **Casabianca, F., Pinkerton, A. A., and Reiss, J. G.,** *Inorg. Chem.*, 16, 864, 1977.
190. **Muller, N., Lauterbur, P. C., and Goldenson, J.,** *J. Am. Chem. Soc.*, 78, 3557, 1956.
191. **Bohlen, R., Francke, R., Heine, J., Schmutzler, R., and Röschenthaler, G.-V.,** *Z. Anorg. Allg. Chem.*, 533, 18, 1986.

192. **Zimin, M. G., Kamalov, R. M., Cherkasov, R. A., and Pudovik, A. N.,** *J. Gen. Chem. U.S.S.R.*, 52, 423, 1982.
193. **Mizhiritskii, M. D., Lebedev, E. P., and Fufaeva, A. N.,** *J. Gen. Chem. U.S.S.R.*, (Engl. transl.), 52, 1859, 1982.
194. **Maskowska, A., Olejnik, J., and Michalski, J.,** *Bull. Acad. Pol. Sci. Ser. Sci. Chim.*, 27, 115, 1979.
195. **Ramirez, F., Marecek, J. F., Fen Chaw, Yu., and McCaffrey, T.,** *Synthesis,* p. 519, 1978.
196. **Ramirez, F. and Marecek, J. F.,** *J. Org. Chem.*, 48, 847, 1983.
197. **Nuretdinova, O. N. and Guseeva, F. F.,** *Bull. Acad. Sci. U.S.S.R., Div. Chem. Sci.*, p. 1892, 1978.
198. **Badet, B., Julia, M., and Rolando, C.,** *Synthesis,* p. 291, 1982.
199. **Skowrońska, A. and Krawczyk, E.,** *Synthesis,* p. 509, 1983.
200. (a) Ref. 120; (b) Ref. 140; (c) Ref. 199; (d)**Murray, M., Schmutzler, R., Grundemann, E., and Teichmann, H.,** *J. Chem. Soc. Sect. B,* p. 1714, 1971.
201. **Zwierzak, A.,** *Synthesis,* p. 305, 1976.
202. **Ramirez, F., Glaser, S. L., Stern, P., Ugi, I., and Lemmon, P.,** *Tetrahedron,* 29, 3741, 1973.
203. **Mark, V. and Van Wazer, J. R.,** *J. Org. Chem.*, 32, 1187, 1967.
204. (a) Ref. 198; (b)**Mlotkowska, V. and Zwierzak, A.,** *Pol. J. Chem.*, 53, 359, 1979.
205. **Mark, V. and Van Wazer, J. R.,** *J. Org. Chem.*, 29, 1006, 1964.
206. **Kibardin, A. M., Gazizov, T. Kh., Gryaznov, P. I., and Pudovik, A. N.,** *Bull. Acad. Sci. U.S.S.R. Div. Chem. Sci.*, p. 855, 1981.
207. **Arbuzov, B. A., Sakhibullina, V. G., and Polozhaeva, N. A.,** *Bull. Acad. Sci. U.S.S.R., Div. Chem. Sci.*, p. 2650, 1975.
208. **Ramirez, F., Nagabhushanam, M., and Smith, C. P.,** *Tetrahedron,* 24, 1785, 1968.
209. **Zawedzki, M., Cieślak, I., and Śledzinski, B.,** *Organika,* p. 9, 1979.
210. **Petnehazy, I., Szakal, G., Töke, L., Hudson, H. R., Pawroznyk, L., and Cooksey, C. J.,** *Tetrahedron,* 39, 4229, 1983.
211. **Lazarus, R. A. and Benkovic, S. J.,** *J. Am. Chem. Soc.*, 101, 4300, 1979.
212. **Ramirez, F., Marecek, J. F., Tsuboi, H., Okazaki, H., and Nowakowski, M.,** *Phosphorus,* 6, 215, 1976.
213. **Kubayashi, S., Yakoyama, T., Kawabe, K., and Saegusa, T.,** *Polym. Bull. (Berlin),* 3, 585, 1980.
214. **Ramirez, F., Stern, P., Glaser, S., Ugi, I., and Lemmon, P.,** *Phosphorus,* 3, 165, 1973.
215. **Ramirez, F., Marecek, J. F., and Ugi, I.,** *J. Am. Chem. Soc.*, 97, 3809, 1975.
216. (a) **Ramirez, F., Marecek, J. F., Tsuboi, H., and Chan, Y.,** *Phosphorus Sulfur,* 4, 327, 1978; (b) Ref. 215.
217. **Kemp, G. and Trippett, S.,** *J. Chem. Soc. Perkin Trans. 1,* p. 879, 1979.
218. **Kluger, R. and Thatcher, G. R. J.,** *J. Am. Chem. Soc.*, 107, 6006, 1985.
219. (a) Ref. 69; (b) Ref. 218.
220. **Bohlen, R. and Röschenthaler, G.-V.,** *Z. Anorg. Allg. Chem.*, 513, 199, 1984.
221. **Tromelin, A., Manouni, D. el, and Burgada, B.,** *Phosphorus Sulfur,* 27, 301, 1986.
222. **Gaydou, M., Freze, R., and Buono, G.,** *Bull. Soc. Chim. Fr.*, p. 2279, 1973.
223. (a) Ref. 222; (b) Ref. 215.
224. (a) Ref. 167; (b) Ref. 54.
225. (a) Ref. 139; (b)**Katritzky, A. R., Nesbit, M. R., Michalski, J., Tulimowski, Z., and Zwierzak, A.,** *J. Chem. Soc. Sect. B,* p. 140, 1970.
226. **Rycroft, D. S. and White, R. F. M.,** *J. Chem. Soc. Chem. Commun.*, p. 444, 1974.
227. **Brault, J.-F., Majoral, J.-P., Savignac, P., and Navech, J.,** *Bull. Soc. Chim. Fr.*, p. 3149, 1973.
228. **Ramirez, F., Tsuboi, H., Okazaki, H., and Marecek, J. F.,** *Tetrahedron Lett.*, 23, 5375, 1982.
229. (a) Ref. 227; (b) Ref. 228.
230. **Gorenstein, D. G. and Rowell, R.,** *J. Am. Chem. Soc.*, 102, 6165, 1980.
231. (a) Ref. 28; (b) Ref. 55.
232. **Majoral, J.-P., Pujol, R., and Navech, J.,** *Bull. Soc. Chim. Fr.*, p. 606, 1972.
233. (a) Ref. 13; (b) Ref. 232.
234. **Maria, P. C., Elegant, L., Azzaro, M., Majoral, J.-P., and Navech, J.,** *Tetrahedron Lett.*, p. 1485, 1972.
235. **Edmundson, R. S. and King, T. J.,** *J. Chem. Res.(S),* p. 120, 1989.
236. **Mukhametov, F. S., Stepashkina, L. V., Korshin, É. E., Shagidullin, R. R., and Rizpolozhenskii, N. I.,** *J. Gen. Chem. U.S.S.R.*, 53, 1137, 1983.
237. **Mukhametov, F. S. and Korshin, É. E.,** *J. Gen. Chem. .U.S.S.R.*, 55, 1788, 1985.
238. **Arbuzov, B. A., Kadyrov, R. A., Arshinova, R. P., Klochkov, V. V., and Aganov, A. V.,** *Bull. Acad. Sci. U.S.S.R., Div. Chem. Sci.*, p. 1612, 1985.
239. **Kuchen, W. and Mahler, H. F.,** *Phosphorus Sulfur,* 8, 139, 1980.
240. **Kudryavtseva, T. N., Chaikovskaya, A. A., Rozkova, Z. Z., and Pinchuk, A. M.,** *J. Gen. Chem. U.S.S.R.*, 52, 952, 1982.

241. **Allison, D. A. and Verkade, J. G.,** *Phosphorus*, 2, 257, 1973.
242. **Milbrath, D. S. and Verkade, J. G.,** *J. Am. Chem. Soc.*, 99, 6607, 1977.
243. **Ramirez, F., Marecek, J. F., and Yemul, S. S.,** *J. Am. Chem. Soc.*, 104, 1345, 1982.
244. **Ramirez, F. and Marecek, J. F.,** *Tetrahedron*, 36, 3151, 1980.
245. (a) Ref. 243; (b) Ref. 135.
246. **Kudelska, W. and Michalska, M.,** *Tetrahedron*, 37, 2989, 1981.
247. **Jacob, L., Julia, M., Pfeiffer, B., and Rolando, S.,** *Synthesis*, p. 451, 1983.
248. **Leroux, Y., Manouni, D. el, and Burgada, R.,** *Tetrahedron Lett.*, 22, 3393, 1981.
249. **Aaron, H. and Szafraniec, L. L.,** *Phosphorus*, 6, 77, 1976.
250. **Kabachnik, M. I., Novikova, Z. S., and Lutsenko, I. F.,** *J. Gen. Chem. U.S.S.R.*, 52, 662, 1982.
251. **Masslennikov, I. G., Lyubimova, M. W., Krutikov, V. I., and Lavren'tev, A. N.,** *J. Gen. Chem. U.S.S.R.*, 53, 2386, 1983.
252. **Kabachnik, M. I., Prishchenko, A. A., Novikova, Z. S., and Lutsenko, I. F.,** *J. Gen. Chem. U.S.S.R.*, 48, 1088, 1978.
253. **Foss, V. L., Veits, Yu. A., and Lutsenko, I. F.,** *Phosphorus Sulfur*, 3, 299, 1977.
254. **Harris, R. K., Katritzky, A. R., Musierowicz, S., and Ternai, B.,** *J. Chem. Soc. Sect. A*, p. 37, 1967.
255. **Stec, W., Moddeman, W., Albridge, R., and Van Wazer, J. R.,** *J. Phys. Chem.*, 75, 3975, 1971.
256. **Mikołajczyk, M. and Witczak, M.,** *J. Chem. Soc. Perkin Trans. 1*, p. 2213, 1977.
257. **Mikołajczyk, M., Krzywanski, J., and Ziemnicka, B.,** *Phosphorus*, 5, 67, 1974.
258. **Khokhlov, P. S., Berseneva, L. S., and Savenkov, N. F.,** *J. Gen. Chem. U.S.S.R.*, 54, 2274, 1984.
259. **Roesky, H. W. and Wiezer, H.,** *Chem. Ber.*, 104, 2258, 1971.
260. **Roesky, H. W.,** *Chem. Ber.*, 101, 2977, 1968.
261. **Roesky, H. W.,** *Chem. Ber.*, 101, 636, 1968.
262. **Lesiak, K., Lesnikowska, Z. J., Stec, W. J., and Zielinska, B.,** *Pol. J. Chem.*, 53, 2041, 1979.
263. **Lesiak, K. and Stec, W. J.,** *Z. Naturforsch.*, 33B, 782, 1978.
264. (a) **Durrieu, J., Kraemer, R., and Navech, J.,** *Org. Magn. Reson.*, 5, 407, 1973; (b) **Mikołajczyk, M., Omelańczuk, J., and Abdukakharov, W. S.,** *Tetrahedron*, 38, 2183, 1982.
265. **Pudovik, M. A., Morozova, N. P., and Pudovik, A. N.,** *Bull. Acad. Sci. U.S.S.R. Div. Chem. Sci.*, p. 2648, 1976.
266. **Stollberg, D., Gloede, J., and Andra, K.,** *Z. Anorg. Allg. Chem.*, 527, 180, 1985.
267. **Sorokina, S. F., Zavalishina, A. I., and Nifant'ev, É. E.,** *J. Gen. Chem. U.S.S.R.*, 43, 748, 1973.
268. **Peake, S. C., Fild, M., Schmutzler, R., Harris, R. K., Nichols, J. M., and Rees, R. G.,** *J. Chem. Soc. Perkin Trans. 2*, p. 380, 1972.
269. **Omelańczuk, J., Kielbasiński, P., Michalski, J., Mikołajczyk, M.,** and **Skowrońska, A.,** *Tetrahedron*, 31, 2809, 1975; (b) Ref. 96.
270. (a) **Teulade, M.-P. and Savignac, P.,** *Phosphorus Sulfur*, 21, 23, 1984; (b) Ref. 66.
271. (a) Ref. 269; (b) Ref. 270(a).
272. **Munoz, A., Gallagher, M., Klaebe, A., and Wolf, R.,** *Tetrahedron Lett.*, p. 673, 1976.
273. **Predoditelev, D. A., Afanas'eva, D. N., and Nifant'ev, É. E.,** *J. Gen. Chem. U.S.S.R.*, 44, 720, 1974.
274. (a) Ref. 26; (b) Ref. 269(a).
275. (a) Ref. 17; (b) **Dutasta, J. P., Grand, A., Robert, J. B., and Taieb, M.,** *Tetrahedron Lett.*, p. 2659, 1974.
276. **Mikołajczyk, M., Ziemnicka, B., Wieczorek, M. W., and Karolina-Wojcziechowska, J.,** *Phosphorus Sulfur*, 21, 205, 1984.
277. **Guimares, A. C., Robert, J. P., Taieb, C., and Tabony, J.,** *Org. Magn. Reson.*, 11, 411, 1978.
278. **Nuretdinov, I. A., Nikonarova, L. K., Grechkin, N. P., and Gaibullina, R. G.,** *J. Gen. Chem. U.S.S.R.*, 45, 526, 1975.
279. **Enikeev, K. M., Bayandina, E. V., Ismaev, I. É., Buina, N. A., Il'yasov, A. V., and Nuretdinov, I. A.,** *J. Gen. Chem. U.S.S.R.*, 53, 1933, 1983.
280. **Loginova, E. I., Nuretdinov, I. A., and Petrov, Yu. A.,** *Theor. Exp. Chem.*, 10, 47, 1974.
281. **Shagidullin, R. R., Shakirov, I. Kh., Musyakaeva, R. Kh., Vandyukova, I. I., and Nuretdinov, I. A.,** *Bull. Acad. Sci. U.S.S.R. Div. Chem. Sci.*, p. 1234, 1981.
282. **Martin, J. and Robert, J. B.,** *Org. Magn. Reson.*, 9, 637, 1977.
283. **Bulloch, G., Keat, R., Rycroft, D. S., and Thomson, D. G.,** *Org. Magn. Reson.*, 12, 709, 1979.
284. **Keat, R.,** *J. Chem. Soc. Dalton Trans.*, p. 2189, 1972.
285. **Roesky, H. W., Schaper, W., and Tutkunkardes, S.,** *Z. Naturforsch.*, 27B, 620, 1972.
286. **Roesky, H. W., Kuhte, B. H., and Grimm, L. F.,** *Z. Anorg. Allg. Chem.*, 389, 167, 1972.
287. **Dakternicks, D. and Giacomo, R. Di.,** *Phosphorus Sulfur*, 24, 217, 1985.
288. **Aksenov, V. I., Chernyshev, E. A., Bugarenko, E. F., and Borisenko, A. A.,** *J. Gen. Chem. U.S.S.R.*, 41, 478, 1971.
289. **Haas, A. and Kortmann, W.,** *Z. Anorg. Allg. Chem.*, 501, 79, 1983.

290. **Große, J., Pieper, W., and Neumaier, H.,** *Angew. Chem. Int. Ed. Engl.*, 21(Suppl.), 1289, 1982.
291. **Haas, A. and Mikołajczyk, M.,** *Chem. Ber.*, 114, 829, 1981.
292. **Bouchu, D. and Droux, J.,** *Phosphorus Sulfur*, 13, 25, 1982.
293. (a) Ref. 27; (b) Ref. 104.
294. **Glemser, O., Bierman, U., and von Halasz, S. P.,** *Inorg. Nucl. Chem. Lett.*, 5, 643, 1969.
295. **Horn, H.-G.,** *Z. Naturforsch.*, 21B, 617, 1966.
296. (a) Ref. 103; (b) Ref. 295.
297. **Roesky, H. W. and Kloker, W.,** *Z. Naturforsch.*, 28B, 697, 1973.
298. **Pudovik, M. A., Mikhailov, Yu. B., and Pudovik, A. N.,** *Bull. Acad. Sci. U.S.S.R. Div. Chem. Sci.*, p. 868, 1981.
299. **Gololobov, Yu. G., Suvalova, E. A., and Chudakeva, T. I.,** *J. Gen. Chem. U.S.S.R.*, 51, 1215, 1981.
300. **Zimin, M. G., Kamalov, R. M., Cherkasov, R. A., and Pudovik, A. N.,** *Phosphorus Sulfur*, 13, 371, 1982.
301. **Shabana, R., Meyer, H. J., and Lawesson, L.-O.,** *Phosphorus Sulfur*, 25, 297, 1985.
302. **Grechkin, N. P., Nikonorova, L. K., and Zhalonkina, L. A.,** *J. Gen. Chem. U.S.S.R.*, 53, 932, 1983.
303. **Kutyrev, G. A., Lygin, A. V., Cherkasov, R. A., and Pudovik, A. N.,** *J. Gen. Chem. U.S.S.R.*, 52, 439, 1982.
304. **Barabás, A., Mureşalan, V., and Almaśi, L.,** *Org. Magn. Reson.*, 12, 313, 1979.
305. **Denney, D. B., Denney, D. Z., and Liu, L.-T.,** *Phosphorus Sulfur*, 22, 71, 1985.
306. **Gombler, W., Kinas, R. W., and Stec, W. J.,** *Z. Naturforsch., Teil B*, 38B, 815, 1983.
307. (a) Ref. 157; (b) Ref. 306.
308. **Arbuzov, B. A., Arshinova, R. P., Nabidullin, V. N., Il'yasov, A. V., Gubaidullin, R. N., and Sorokina, T. D.,** *Bull. Acad. Sci. U.S.S.R. Div. Chem. Sci.*, p. 725, 1983.
309. **Stec, W. J., Okruszek, A., and Michalski, J.,** *J. Org. Chem.*, 41, 233, 1976.
310. (a) Ref. 55(b); (b) Ref. 309.
311. **Nifant'ev, É. E., Milliaresi, E. E., Ruchkina, N. G., Druyan-Poleshchuk, L. M., Vasyanina, L. K., and Tyan, E. A.,** *J. Gen. Chem. U.S.S.R.*, 51, 1295, 1981.
312. **Haas, A. and Łopusinski, A.,** *Chem. Ber.*, 114, 3176, 1981.
313. **Hall, C. R. and Williams, N. E.,** *J. Chem. Soc. Perkin Trans. 1*, p. 2746, 1981.
314. **Levin, Yu. A., Valeeva, T. G., Gozman, I. P., and Gol'dfarb, E. I.,** *J. Gen. Chem. U.S.S.R.*, 55, 1152, 1985.
315. **Zabirov, N. G., Cherkasov, R. A., and Pudovik, A. N.,** *J. Gen. Chem. U.S.S.R.*, 56, 1089, 1986.
316. **Nuretdinov, I. A., Buina, N. A., Grechkin, N. P., and Loginova, É. I.,** *Bull. Acad. Sci. U.S.S.R. Div. Chem. Sci.*, p. 111, 1971.
317. **Kinas, R., Stec, W. J., and Kruger, C.,** *Phosphorus Sulfur*, 4, 295, 1978.
318. **Scherer, O. J., Forstinger, K., Kaub, J., and Sheldrick, W. S.,** *Chem. Ber.*, 119, 2731, 1986.
319. **Pudovik, M. A., Kibardina, L. K., Alexsandrova, I. A., Khairullin, V. K., and Pudovik, A. N.,** *J. Gen. Chem. U.S.S.R.*, 51, 23, 1981.
320. **Nifant'ev, É. E., Zavalishina, A. I., Kurochkin, V. V., Smirnova, E. I., and Borisenko, A. A.,** *J. Gen. Chem. U.S.S.R.*, 55, 1803, 1985.
321. **Karimova, N. M., Kil'disheva, O. V., and Knunyants, I. L.,** *Bull. Acad. Sci. U.S.S.R. Div. Chem. Sci.*, p. 863, 1983.
322. **Nuretdinov, I. A. and Loginova, É. I.,** *Bull. Acad. Sci. U.S.S.R. Div. Chem. Sci.*, p. 2765, 1973.
323. **Roesky, H. W. and Remmers, G.,** *Z. Anorg. Allg. Chem.*, 431, 221, 1977.
324. **Bryce, M. R. and Mathews, R. S.,** *J. Organomet. Chem.*, 325, 153, 1987.
325. **Becke-Goehring, M. and Hoffman, H.,** *Z. Anorg. Allg. Chem.*, 369, 73, 1969.
326. **Majoral, J.-P., Kraemer, R., Navech, J., and Mathis, F.,** *J. Chem. Soc. Perkin Trans. 1*, p. 2093, 1976.
327. (a) Ref. 326; (b) Ref. 182.
328. (a) Ref. 185; (b) Ref. 367.
329. **Wolff, A. and Riess, J. G.,** *Bull. Soc. Chim. Fr.*, p. 1587, 1973.
330. **Stec, W. J., Okruszek, A., Uznanski, B., and Michalski, J.,** *Phosphorus*, 2, 97, 1972.
331. **Colquhoun, I. J., McFarlane, H. C. E., McFarlane, W., Nash, J. A., Keat, R., Rycroft, D. S., and Thompson, D. G.,** *Org. Magn. Reson.*, 12, 473, 1979.
332. **Elkaim, J.-C., Wolff, A., and Riess, J. G.,** *Phosphorus*, 2, 249, 1973.
333. **Dean, P. A. W.,** *Can. J. Chem.*, 57, 754, 1979.
334. **Revel, M., Roussel, J., Navech, J., and Mathis, F.,** *Org. Magn. Reson.*, 8, 399, 1976.
335. **Borisenko, A. A., Sorokina, S. F., Zavalishina, A. I., Sergeev, N. M., and Nifant'ev, É. E.,** *J. Gen. Chem. U.S.S.R.*, 48, 1144, 1978.
336. **Glidewell, C.,** *Inorg. Chim. Acta*, 25, 159, 1977.
337. **Elegant, L., Azzaro, M., Mankowski-Favelier, R., and Mavel, G.,** *Org. Magn. Reson.*, 1, 471, 1969.

338. **Sinyashin, O. G., Karimullin, Sh. A., Batyeva, É. S., and Pudovik, A. N.,** *J. Gen. Chem. U.S.S.R.*, 54, 475, 1984.
339. **Dwek, R. A., Richards, R. E., Taylor, D., and Shaw, R. A.,** *J. Chem. Soc. Sect. A*, p. 1173, 1970.
340. **Maier, L.,** *Helv. Chim. Acta*, 59, 252, 1976.
341. **Mazitova, F. N. and Khairullin, V. K.,** *J. Gen. Chem. U.S.S.R.*, 50, 652, 1980.
342. **Gupta, R. K., Rai, A. K., Mehrotra, R. C., Jain, V. K., Hoskins, B. F., and Tiekink, E. R. T.,** *Inorg. Chem.*, 24, 3280, 1985.
343. **Simyashin, O. G., Kostin, V. P., Batyeva, É. S., and Pudovik, A. N.,** *J. Gen. Chem. U.S.S.R.*, 53, 415, 1983.
344. **Ofitserov, E. N., Sinyashin, O. G., Batyeva, É. S., and Pudovik, A. N.,** *J. Gen. Chem. U.S.S.R., (Engl. transl.)*, 51, 602, 1981.
345. **Lipman, A. E.,** *J. Org. Chem.*, 31, 471, 1966.
346. **Paasivirta, J., Simainen, J., Vesterinen, R., and Virkki, L.,** *Org. Magn. Reson.*, 9, 708, 1977.
347. **Mazitova, F. N. and Khairullin, V. K.,** *J. Gen. Chem. U.S.S.R.*, 50, 1393, 1980.
348. **Krolevets, A. A.,** *J. Gen. Chem. U.S.S.R.*, 53, 1509, 1983.
349. **Kutyrev, G. A., Kutyrev, A. A., Cherkasov, R. A., and Pudovik, A. N.,** *Dokl. Chem. (Engl. transl.)*, 247, 361, 1979.
350. **Zimin, M. G., Fomakhin, E. V., Islamov, R. G., and Pudovik, A. N.,** *J. Gen. Chem. U.S.S.R.*, 55, 2384, 1985.
351. **Mazitova, F. N. and Khairullin, V. K.,** *J. Gen. Chem. U.S.S.R.*, 50, 652, 1980.
352. **Chauhan, H. P. S., Srivastava, G., and Mehrotra, R. C.,** *Polyhedron*, 3, 1337, 1984.
353. **Samitov, Yu. Yu., Karataeva, F. Kh., and Ovchinnikov, V. V.,** *J. Gen. Chem. U.S.S.R.*, p. 577, 1981.
354. **McLeverty, J. A., Kowalski, R. S. Z., Bailey, N. A., Mulvaney, R., and O'Cleirigh, D. O.,** *J. Chem. Soc. Dalton Trans.*, p. 627, 1983.
355. **Biscarini, P.,** *Inorg. Chim. Acta*, 74, 65, 1983.
356. **Predvoditelev, D. A., Afamas'eva, D. N., and Nifant'ev, É. E.,** *J. Gen. Chem. U.S.S.R.*, 44, 2586, 1974.
357. **Kotynski, A., Lesiak, K., and Stec, W. J.,** *Pol. J. Chem.*, 53, 2403, 1979.
358. **Reddy, C. D., Rao, C. V. N., Reddy, D. B., Thompson, M. D., Jasinska, J., Holt, E. M., and Berlin, K. D.,** *Indian J. Chem. Sect. B*, B24, 481, 1985.
359. **Mastryukova, T. A., Mikhal'skii, Ya., Uryupin, A. B., Skshipshinskii, Z., and Kabachnik, M. I.,** *J. Gen. Chem. U.S.S.R.*, 48, 412, 1978.
360. **Bruzik, K., Katritzky, A. R., Michalski, J., and Stec, W. J.,** *Pol. J. Chem.*, 54, 141, 1980.
361. **Mizhiritskii, M. D. and Reikhsfel'd, V. O.,** *J. Gen. Chem. U.S.S.R.*, 55, 1674, 1985.
362. **Eliseenkov, V. N. and Shumilina, T. N.,** *J. Gen. Chem. U.S.S.R.*, 52, 809, 1982.
363. (a) Ref. 75; (b) **Kudelska, W. and Michalska, M.,** *Tetrahedron*, 42, 629, 1986.
364. (a) Ref. 76, (b) Ref. 139.
365. **Mikolajczyk, M., Omelańczuk, J., Leitloff, M., Drabowicz, J., Ejchart, A., and Jurczak, J.,** *J. Am. Chem. Soc.*, 100, 7003, 1978.
366. **Markowska, A. and Nowicki, T.,** *Nouv. J. Chim.*, 3, 409, 1979.
367. **Schmidpeter, A. and Brecht, H.,** *Z. Naturforsch.* B24, 179, 1969.
368. **Mastryukova, T. A., Uryupin, A. B., Orlov, M., Eremich, D., and Kabachnik, M. I.,** *J. Gen. Chem. U.S.S.R.*, 51, 1251, 1981.
369. **Mark, V., Dungan, C., Crutchfield, M., and Van Wazer, J. R.,** *Top. Phosphorus Chem.*, 5, 227, 1967.
370. **Loewus, D. I. and Eckstein, F.,** *J. Am. Chem. Soc.*, 105, 3287, 1983.
371. **Mukhametov, F. S., Korshin, É. E., Shagidullin, R. R., and Rizpolozhenskii, N. I.,** *J. Gen. Chem. U.S.S.R.*, 53, 2436, 1983.
372. **Gurley, T. W. and Ritchey, W. M.,** *Anal. Chem.*, 48, 1137, 1976.
373. **Mikołajczyk, M., Witczak, M., Wieczorek, M., Bokij, N. G., and Struchkov, Y. T.,** *J. Chem. Soc. Perkin Trans. 1*, p. 371, 1976.
374. **Burgada, R., Lafaille, L., and Mathis, F.,** *Bull. Soc. Chim. Fr.*, p. 341, 1974.
375. **Skowrońska, A., Mikołajczyk, J., and Michalski, J.,** *J. Chem. Soc. Chem. Commun.*, p. 986, 1975.
376. **Kudryavtseva, T. N., Karlstédt, N. B., Proskurnina, M. V., Boganova, N. V., Shostakova, T. G., and Lutsenko, I. F.,** *J. Gen. Chem. U.S.S.R.*, 52, 912, 1982.
377. **Mikołajczyk, M., Ziemnicka, B., Karolak-Wojciechowska, J., and Wieczorek, M.,** *J. Chem. Soc. Perkin Trans. 2*, p. 501, 1983.
378. **Ovchinnikov, V. V., Galkin, V. I., Cherkasov, R. A., and Pudovik, A. N.,** *J. Gen. Chem. U.S.S.R.*, 47, 267, 1977.
379. (a) Ref. 17; (b) Ref. 275(b).
380. **Verkade, J. G. and King, R. W.,** *Inorg. Chem.*, 1, 948, 1962.
381. **Chojnowski, J., Cypryk, M., Fortuniak, W., and Michalski, J.,** *Synthesis* p. 683, 1977.

382. **Borecka, B., Chojnowski, J., Cypryk, M., Michalski, J., and Zielinski, J.**, *J. Organomet. Chem.*, 171, 17, 1979.
383. **Zwierzak, A.**, *Synthesis*, p. 270, 1975.
384. **Krawczyk, E. and Skowrońska, A.**, *Phosphorus Sulfur*, 9, 189, 1980.
385. (a) Ref. 384; (b) Ref. 386.
386. **Glidewell, C.**, *Inorg. Chim. Acta*, 24, 255, 1977.
387. **Kuramshin, I. Ya., Muratova, A. A., Yaskova, É. G., Musina, A. A., Izmailova, F. Kh., and Pudovik, A. N.**, *J. Gen. Chem. U.S.S.R.*, 43, 1446, 1973.
388. **Zimin, M. G., Zabirov, N. G., and Pudovik, A. N.**, *J. Gen. Chem. U.S.S.R.*, 49, 1922. 1979.
389. **Michalski, J., Mikołajczyk, J., and Skowronska, A.**, *J. Am. Chem. Soc.*, 100, 5386, 1978.
390. **Ivanova, Zh. M., Kim, T. V., and Gololobov, Yu. G.**, *J. Gen. Chem. U.S.S.R.*, 48, 2156, 1978.
391. **Drabowitz, J. and Mikołajczyk, M.**, *Synthesis*, p. 32, 1980.
392. **Lopusinski, A. and Haas, A.**, *Chem. Ber.*, 118, 4623, 1985.
393. (a) Ref. 26; (b) Ref. 25.
394. **Arbuzov, B. A. and Nuretdinov, O. N.**, *Bull. Acad. Sci. U.S.S.R., Div. Chem. Sci.*, p. 613, 1983.
395. **Mikołajczyk, M. and Łuczak, J.**, *Tetrahedron*, 28, 5411, 1972.
396. **Bäuerlein, E. and Gaugler, H.**, *Phosphorus Sulfur*, 5, 53, 1978.
397. **Socol, S. M. and Verkade, J. G.**, *Inorg. Chem.*, 23, 3487, 1984.
398. **Socol, S. M. and Verkade, J. G.**, *Inorg. Chem.*, 25, 2658, 1986.
399. (a) Ref. 262; (b) Ref. 395.
400. **Fritz, G. and Hanke, D.**, *Z. Anorg. Allg. Chem.*, 537, 17, 1986.
401. **Sinyashin, O. G., Karimullin, Sh. A., Pudovik, D. A., Batyeva, É. S., and Pudovik, N. A.**, *J. Gen. Chem. U.S.S.R.*, 54, 2195, 1984.
402. **Martynov, I. V., Ivanov, A. N., Episheva, T. A., and Sokolov, V. B.**, *Bull. Acad. Sci. U.S.S.R. Div. Chem. Sci.*, p. 1001, 1987.
403. **Dabkowski, W. and Michalski, J.**, *J. Chem. Soc. Chem. Commun.*, p. 755, 1987.
404. **Zimin, M. G. and Kamalov, R. M.**, *J. Gen. Chem. U.S.S.R.*, 56, 2359, 1986.
405. **Kim, T. V., Kiseleva, E. I., and Sinitsa, A. D.**, *J. Gen. Chem. U.S.S.R.*, 57, 713, 1987.
406. **Peyronel, J.-F., Samuel, O., and Fiaud, J.-C.**, *J. Org. Chem.*, 52, 5320, 1987.
407. **Kozlov, É. S. and Dubenko, L. G.**, *J. Gen. Chem. U.S.S.R.*, 56, 2396, 1986.
408. **Martynov, I. V., Brel', V. K., Uvarov, V. I., Yarkov, A. V., Novikov, V. P., Chepakova, L. A., and Raevskii, O. A.**, *Bull. Acad. Sci. U.S.S.R. Div. Chem. Sci.*, p. 784, 1987.
409. **Kozer, G. F., Lodaya, J. S., Ray, D. G., and Kokil, P. B.**, *J. Am. Chem. Soc.*, 110, 2987, 1988.
410. **Donath, Ch. and Meisel, M.**, *Z. Anorg. Allg. Chem.*, 549, 46, 1987.
411. **Grishina, L. N. and Grechkin, N. P.**, *J. Gen. Chem. U.S.S.R.*, 57, 720, 1987.
412. **Wu, S.-Y., Hirashima, A., Kuwano, E., and Eto, M.**, *Agric. Biol. Chem.*, 51, 537, 1987.
413. **Nifant'ev, É. E., Kukhareva, T. S., Soldatova, I. A., and Chukbar, T. G.**, *J. Gen. Chem. U.S.S.R.*, 56, 2199, 1986.
414. **Nuretdinova, O. N. and Novikova, V. G.**, *Bull. Acad. Sci. U.S.S.R. Div. Chem. Sci.*, p. 2556, 1986.
415. **Hursthouse, M. B., Ibrahim, E. H., Parkes, H. G., Shaw, L. S., Shaw, R. A., and Watkins, D. A.**, *Phosphorus Sulfur*, 28, 261, 1986.
416. **Łopusiński, A. and Potrzebowski, M.**, *Phosphorus Sulfur*, 32, 55, 1987.
417. **Kato, S., Min, S., Beniya, Y., and Ishida, M.**, *Tetrahedron Lett.*, 28, 6473, 1987.
418. (a) Ref. 416; (b) Ref. 417.
419. **Łopusinski, A., Luczak, L., and Brzezińska, E.**, *Phosphorus Sulfur*, 31, 101, 1987.
420. **Obushek, N. D., Vork, M. V., Vengrzhanovskii, V. A., Mel'nik, Ya. I., and Ganushchak, N. I.**, *J. Gen. Chem. U.S.S.R.*, 57, 961, 1987.
421. **Kozlov, V. A., Dol'nikova, T. Yu., Grapov, A. F., and Mel'nikov, N. N.**, *J. Gen. Chem. U.S.S.R.*, 57, 685, 1987.
422. **Samitov, Yu. Yu., Karataeva, F. Kh., Ovchinnikov, V. V., and Cherkasov, R. A.**, *J. Gen. Chem. U.S.S.R.*, 56, 1979, 1986.
423. **Nifant'ev, É. E., Predvoditelev, D. A., Rasadkina, E. N., and Bekker, A. R.**, *J. Gen. Chem. U.S.S.R.*, 57, 186, 1987.
424. **Lamande, L. and Munoz, A.**, *Phosphorus Sulfur*, 32, 1, 1987.
425. **Martynov, I. V., Chepakova, L. A., Brel', V. K., and Sokolov, V. B.**, *J. Gen. Chem. U.S.S.R.*, 56, 2140, 1986.
426. **Vikreva, L. A., Darisheva, A. M., Bubnov, N. N., Godovikov, N. N., and Kabachnik, M. I.**, *Bull. Acad. Sci. U.S.S.R. Div. Chem. Sci.*, p. 626, 1987.
427. **Glonek, T. and Kopp, S. J.**, *Mag. Reson. Imaging*, 3, 359, 1985.
428. **Ludeman, S. M., Bartlett, D. L., and Zon, G.**, *J. Org. Chem.*, 44, 1163, 1979.
429. **Kinas, R. W., Pankiewski, K., Stec, W. J., Farmer, P. B., Foster, A. B., and Jarman, M.**, *J. Org. Chem.*, 42, 1650, 1977.

430. **Rosch, P., Kalbitzer, H. R., and Goody, R. S.,** *FEBS Lett.*, 121, 211, 1980.
431. **Glonek, T., Van Wazer, J. R., Kleps, R. A., and Myers, T. C.,** *J. Am. Chem. Soc.*, 13, 2337, 1974.
432. **Glonek, T., Kleps, R. A., and Myers, T. C.,** *Am. Assoc. Adv. Sci.*, 185, 352, 1974.
433. **Glonek, T., Van Wazer, J. R., Mudgett, M., and Myers, T. C.,** *J. Am. Chem. Soc.*, 11, 567, 1972; **Glonek, T., Lunde, M., Mudgett, M., and Myers, T. C.,** *Arch. Biochem. Biophys.*, 142, 508, 1971.
434. **Greiner, J. V., Kapp, S. J., and Glonek, T.,** *Invest. Ophthal. Vis. Sci.*, 26, 537, 1985.
435. **Prigodich, R. V. and Haake, P.,** *Inorg. Chem.*, 24, 89, 1985.
436. **Karlik, S. J., Elgavish, G. A., and Ichhorn, G. L.,** *J. Am. Chem. Soc.*, 105, 602, 1983: **Sakurai, S. J., Goda, T., and Shimomura, S.,** *Biochem. Biophys. Res. Commun.*, 108, 474, 1982.
437. **Klein, H. W., Im, M. J., Palm, D., and Hemmerich, E. J. H.,** *Biochemistry*, 23, 5853, 1984.
438. **Knorre, D. G., Lebedev, K. K., and Zarytova, V. F.,** *Nucleic Acid Res.*, 3, 1401, 1976.
439. **Vorob'ev, Yu. N., Badashkeeva, A. G., and Lebedev, A. V.,** *Zh. Strukt. Khim.*, 23, 29, 1982.
440. **Haar, W., Thompson, J. C., Maurer, W., and Ruterjans, H.,** *Eur. J. Biochem.*, 40, 259, 1973; **Lebedev, A. V. and Rezvukin, A. I.,** *Biol. Khim.*, 9, 149, 1983.
441. **Gorenstein, D. G. and Kar, D.,** *Biochem. Biophys. Res. Commun.*, 65, 1073, 1975.
442. **Tsuboi, M., Takahashi, S., Kyogoku, Y., Ukita, T., and Kainosho, M. N.,** *Science*, 166, 1505, 1969.
443. **Lebedev, A. V., Rezvukin, A. I., Furin, G. G., and Poleshchuk, O. Kh.,** *Zh. Strukt. Khim.*, 24, 39, 1983.
444. **Sarma, R. H.,** *Fed. Proc.*, 30, 1087, 1971.
445. **Merchant, T. E. and Glonek, T.,** *J. Lipid Res.*, submitted for publication.
446. **Pluckthun, A. and Dennis, E. A.,** *Biochemistry*, 21, 1743, 1982.
447. **Reid, D. G., Salisbury, S. A., Brown, T., and Williams, D. H.,** *Biochemistry*, 24, 4325, 1985.
448. **Meneses, P. and Glonek, T.,** *J. Lipid Res.*, 19, 679, 1988; **Henderson, T. O., Glonek, T., and Myers, T. C.,** *Biochemistry*, 13, 623, 1974; **Glonek, T., Henderson, T. O., Kruski, A. W., and Scanu, A. M.,** *Biochem. Biophys. Acta*, 348, 155, 1974; **Assmann, G., Sokoloski, E. A., and Brewer, H. B.,** *Proc. Natl. Acad. Sci. U.S.A.*, 71, 549, 1974.
449. **Bystrof, V. F., Shapiro, Y. E., Viktorov, A. V., Barsukov, L. I., and Bergelson, L. D.,** *FEBS Lett.*, 25, 337, 1972.
450. **Griffin, R. G.,** *J. Am. Chem. Soc.*, 98, 851, 1976; **Kohler, S. J. and Klein, M. P.,** *Biochemistry*, 15, 976, 1976; **Gally, H. U., Niederberger, W., and Seelig, J.,** *Biochemistry*, 14, 3647, 1975; **de Kruijff, B., Cullis, P. R., and Radda, G. K.,** *Biochim. Biophys. Acta*, 406, 6, 1975; **de Cullis, P. R. and de Kruijff, B.,** *Biochim. Biophys. Acta*, 436, 523, 1976; **Niederberger, W. and Seelig,** *J. Am. Chem. Soc.*, 98, 3704, 1976.
451. **Yeagle, P. L., Hutton, W. C., Huang, C.-H., and Martin, R. B.,** *Proc. Natl. Acad. Sci. U.S.A.*, 72, 3477, 1975; **Yeagle, P. L., Hutton, W. C., Huang, C.-H., and Martin, R. B.,** *Biochemistry*, 15, 2121, 1976.
452. **Henderson, T. O., Glonek, T., and Myers, T. C.,** *Biochemistry*, 13, 623, 1974; **White, D. C. and Tucker, A. N.,** *J. Lipid Res.*, 10, 220, 1969; **Short, S. A. and White, D. C.,** *J. Bacteriol.*, 104, 126, 1970; **Tucker, A. N. and White, D. C.,** *J. Bacteriol.*,108, 1058, 1971.
453. **Connolly, B. A. and Eckstein, F.,** *Biochemistry*, 23, 5523, 1984; **Lebedev, A. V. and Rezvukhin, A. I.,** *Nucleic Acids Res.*, 12, 5514, 1984.
454. **Uznanski, B., Niewiarowski, W., and Stec, W. J.,** *Tetrahedron Lett.*, 23, 4289, 1982.
455. **Chen, C.-W. and Cohen, J. S.,** *Biopolymers*, 22, 879, 1983; **Chen, C.-W., Cohen, J. S., and Behe, M.,** *Biochemistry*, 22, 2136, 1983.
456. **Lerner, D. B., Becktel, W. J., Everett, R., Goodman, M., and Kearns, D. R.,** *Biopolymers*, 23, 2157, 1984; **Joseph, A. P. and Bolton, P. H.,** *J. Am. Chem. Soc.*, 106, 437, 1984: **Eckstein, F. and Jovin, T. M.,** *Biochemistry*, 22, 4546, 1983.
457. **Ott, J., Eckstein, F., and Connolly, B. A.,** *Nucleic Acids Res.*, 13, 6317, 1985; **Ott, J. and Eckstein, F.,** *Biochemistry*, 24, 2530, 1985: **Connolly, B. A. and Eckstein, F.,** *Biochemistry*, 23, 5523, 1984.
458. **Gueron, M.,** *FEBS Lett.*, 19, 264, 1972: **Gorenstein, D. G. and Goldfield, E. M.,** *Biochemistry*, 21, 5839, 1982: **Gorenstein, D. G., Goldfeld, E. M., Chen, R., Kovar, K., and Luxon, B. A.,** *Biochemistry*, 20, 2141, 1981, **Gorenstein, D. G., Luxon, B. A., Goldfeld, E. M., Lai, K., and Vegeais, D.,** *Biochemistry*, 21, 580, 1982.
459. **Chandrasekaran, S., Jones, R. L., and Wilson, D.,** *Biopolymers*, 24, 1963, 1985: **Rill, R. L., Hilliard, P. R., and Levy, G. C.,** *J. Biol. Chem.*, 258, 250, 1983; **Granot, J., Feigon, J., and Kearns, D. R.,** *Biopolymers*, 21, 181, 1982.
460. **Mandel, M. and Westley, J. W.,** *Nature (London)*, 203, 302, 1964.
461. **Gorenstein, D. G.,** *J. Am. Chem. Soc.*, 97, 898, 1975.
462. **Jarvest, R. L., Lowe, G., Baraniak, J., and Stec, W. J.,** *Biochem. J.*, 203, 461, 1982.
463. **Ho, C., Magnuson, J. A., Wilson, J. B., Magnuson, N. S., and Kurland, R. J.,** *Biochemistry*, 8,

464. **Smithers, G. W. and O'Sullivan, W. J.,** *J. Biol. Chem.*, 257, 6164, 1982.
465. **Costello, A. J. R., Glonek, T., and Myers, T. C.,** *Carbohydr. Res.*, 46, 159, 1976.
466. **Lindon, J. C., Baker, D. J., Williams, J. M., and Irvine, R. F.,** *Biochem. J.*, 244, 591, 1987.
467. **Gray, G. R.,** *Biochemistry*, 10, 4705, 1971; **Van den Berg, G. B. and Heerschap, A.,** *Arch. Biochem. Biophys.*, 219, 268, 1982.
468. **Gadian, D. G., Radda, G. K., and Richards, R. E.,** *Biochim. Biophys. Acta*, 358, 57, 1974.
469. **Martinez-Carrion, M.,** *Eur. J. Biochem.*, 54, 39, 1974; **Khomutov, R. M., Severin, E. S., Khurs, E. N., and Galayaev, N. N.,** *Biochem. Biophys. Acta*, 171, 201, 1969.
470. **Gorenstein, D. G., Wyrwicz, A. M., and Bode, J.,** *J. Am. Chem. Soc.*, 98, 2308, 1976.
471. **Ernst, L.,** *Nucleic Acids Res.*, 15, 361, 1987.
472. **Meadows, D. H., Jardetsky, O., Epaud, R. M., Ruterjans, H. H., and Sheraga, H. A.,** *Proc. Natl. Acad. Sci. U.S.A.*, 60, 766, 1968.
473. **Clayden, N. J. and Waugh, J. S.,** *J. Chem. Soc. Chem. Commun.*, p. 292, 1982.
474. **Moonen, C. T. W. and Muller, F.,** *Biochemistry*, 21, 408, 1982.
475. **Isaac, R. E., Rose, M. E., Rees, M. E., and Goodwin, T. W.,** *Biochem. J.*, 213, 533, 1983.
476. **Sykes, B. D.,** *Can. J. Biochem.*, 61, 155, 1983.
477. **Moon, R. B. and Richards, J. H.,** *J. Biol. Chem.*, 248, 7276, 1973.
478. **Glonek, T., Kopp, S. J., Kot, E., Pettegrew, J. W., Harrison, W. H., and Cohen, M. M.,** *J. Neurochem.*, 39, 1210, 1982.
479. **Henderson, T. O., Glonek, T., Hilderbrand, R. L., and Myers, T. C.,** *Arch. Biochem. Biophys.*, 149, 484, 1972.
480. **Hilderbrand, R. L., Henderson, T. O., Glonek, T., and Myers, T. C.,** *Biochemistry*, 12, 4756, 1973: **Kirkpatrick, D. S. and Bishop, S. H.,** *Biochemistry*, 12, 2829, 1973.
481. **Patel, D. J.,** *Biochemistry*, 13, 2388 and 2396, 1974.
482. **Meneses, P. and Navarro, J. N.,** *Phytochemsitry*, submitted for publication.
483. **Ganson, N. J. and Fromm, H. J.,** *J. Biol. Chem.*, 260, 2837, 1984.
484. **Brown, T. R., Graham, R. A., Szwergold, B. S., Thoma, W. J., and Meyer, R. A.,** *Ann. N. Y. Acad. Sci.*, 508, 229, 1987.
485. **Burt, C. T. and Levine, J. F.,** *Biochim. Biophys. Acta*, 1033, 189, 1990.
486. **Ezra, F. S., Lucus, D. S., Mustacich, R. V., and Russel, A. F.,** *Biochemistry*, 22, 3841, 1983.
487. **Moore, R. R., Burt, C. T., and Roberts, M. F.,** *Biochim. Biophys. Acta*, 846, 394, 1985.
488. **Szwergold, B. S., Kappler, F., and Brown, T. R.,** *Science*, 247, 451, 1990.
489. **Meneses, P., Para, P. F., and Glonek, T.,** private communication.
490. **Kasimos, J. N., Merchant, T. E., Gierke, L. W., and Glonek, T.,** *Cancer Res.*, 50, 527, 1990.

Chapter 10

^{31}P NMR DATA OF FOUR COORDINATE PHOSPHORUS COMPOUNDS CONTAINING A P=Ch BOND AND ONE OR TWO P–H BONDS

Compiled and presented by
Michael J. Gallagher

The NMR data are presented in Table I. It is divided into four parts: Section I1 contains data of phosphonic acids and derivatives $HP(Ch)X_2$, Section I2 contains data of phosphinic acids and derivatives $H_2P(Ch)X$, Section I3 contains data of monosubstituted phosphinic acids and derivatives, RHP(Ch)X, and Section I4 contains data of secondary phosphine chalcogenides R_2HP=Ch.

This class constitutes a very large and fairly homogeneous group of substances in which a number of useful general trends can be seen. Substituent effects on chemical shifts are obvious, fairly consistent, and though not giving precise results, allow a reasonable estimate of the δ value for unknown compounds. Solvent and concentration effects on chemical shift are also observable, but they seem to be fairly small other than for anions of acids, at least by comparison with the accuracy of the measurements (±0.5 ppm in most instances). An exception is seen in aqueous solutions when a marked downfield shift (5 to 7 ppm) is observed when compared with the shift in chloroform. This effect is very much less pronounced in other hydroxylic solvents.

The dominant feature of the spectra of these compounds is the large value of $^1J_{PH}$ (300 to 800 Hz). This value is sensitive to the number and nature of substituents on phosphorus and probably provides more information on structure than the chemical shift itself.

SUBSTITUENT EFFECTS ON δ VALUES

These can be estimated roughly by making use of α, β, and γ substituent effects in the same way as with ^{13}C spectral analysis. However, it is important not to extrapolate from one class of compound to another. Provided this is not done, then, in general, α is large (10 to 20 ppm) and positive, β is smaller (ca. 4 ppm) and also positive, while γ effects are smaller still (1 to 2 ppm) and negative. Electronegative substituents, e.g., halogens, produce small (ca. 1 ppm) upfield shifts with respect to the corresponding acid with F > Cl > Br. Anomalies occur with very bulky groups, e.g., t-butyl, and with reactive substituents, e.g., amino-substituted acids. Anions usually absorb upfield of acids, often substantially, but the difference is solvent sensitive. Anions of amino-substituted acids usually absorb downfield of the parent acid.

In general, replacing oxygen on phosphorus with sulfur (and selenium, though fewer data are available) results in a large downfield shift when there is at least one oxygen atom still on phosphorus. The shift is less for selenium than for sulfur and with telleurides is negative in one instance. When both substituents on phosphorus are carbon the effect is negative for all chalcogens and the shift moves further upfield with increasing atomic number. Replacement of hydrogen on phosphorus by deuterium results in a small upfield shift (ca. 0.5 ppm), enough to make the two species clearly distinguishable in admixture.

Occasionally signs of δ values from the older literature seem reversed, but these have not been altered here unless there was supporting evidence.

SUBSTITUENT EFFECTS ON $^1J_{PH}$ VALUES

In general these are sensitive to substituent effects, both α and β, and provide a good indication of the chemical nature of the phosphorus atom in question. In general, $^1J_{PH}$ is markedly dependent on the electronegativity of α-substituents, thus $R_2P(O)H$(R = halogen) > $(RO)_2P(O)H$ > RPH(O)OH > $R_2P(O)H$ (R is alkyl or aryl) and P(O)H > P(S)H > P(Se)H. Within any group increasing electronegativity of substituents also increases $^1J_{PH}$; e.g., $(ArO)_2P(O)H$ > ArO(AlkO)P(O)H > $(AlkO)_2P(O)H$. These are not merely trends since the ranges are well defined and show little overlap, e.g., $(AlkO)_2P(O)H$ have $^1J_{PH}$ 695 ± 15 Hz, whereas $(ArO)_2P(O)H$ have $^1J_{PH}$ ~ 740 Hz and ArO(AlkO)P(O)H are 720 ± 10. Electronegative γ substituents, such as cyano or chlorine, sharply increase $^1J_{PH}$. The difference between types of compounds is much greater; i.e., $(RO)_2PH(O)$ (680 to 760), R(RO)PH(O) (550 ± 20), $R_2P(O)H$ (480 ± 40), $(RO)_2PH(S)$ (ca. 640), R(RO)PH(S) (520), and $R_2P(S)H$ (440 ± 20). Similarly, acids have smaller $^1J_{PH}$ (by ca. 40 Hz) than their esters and anions are smaller again than acids (by ca. 100 Hz). The magnitude of the latter increases as the size of the cation diminishes,[49] but the effect is small (maximum ca. 20 Hz) when compared with the coupling as a whole. $^1J_{PD}$ has been measured in many instances and falls within the expected range (67 to 115 Hz).

In 6-membered cyclic compounds $^1J_{PH}$ imparts useful stereochemical information since $J_{ax} < J_{eq}$. $^1J_{PH}$ eq is usually less than the ranges quoted above for acyclic compounds.

TABLE I

Section I1: Phosphonic Acids and Derivatives $HP(Ch)X_2$

P-bound atoms/rings	Connectivities	Structure	NMR data δP [$^1J_{PH}$ Hz]	Ref.
HBrOO′	H	HP(O)(OH)Br	3.1 J747	43
HBr_2O'		$HP(O)Br_2$	−2.5 J772	43
HClOO′		HP(O)(OH)Cl	5.0 J745	43
HClOO′	C	EtOPH(O)Cl	−2.7 J752	43
HCl_2O'		$HP(O)Cl_2$	−1.5 J772	43
HFOO′	H	HP(O)(OH)F	2.7 J783	23, 49
HF_2O		$HP(O)F_2$	−1 J880	49
HNOO′	$C;C_2$	$t\text{-}AmOPH(O)NEt_2$	6.4	9
HNOO′		$t\text{-}BuOPH(O)NEt_2$	6.5	9
HF_2Se'		$F_2P(Se)H$	74.2 J691	44
HN_2O'	$C_2;C_2$	$(Me_2N)_2P(O)H$	20.5 J570	2, 9
HN_2S'	$H,Si;Si_2$	$HP(S)(NHTms)NTms_2$	35.8 J540	45
HN_2Se'		$HP(Se)(NHTms)NTms_2$	9.5 J526	45
HOSS′	H;H	H(S)P(SH)OH	72.3 J560	40
HO_2S'	Si_2	$(TmsO)_2P(S)H$	44.6 J674	52
HO_2S'	H_2	$(HO)_2P(S)H$	54.7 J538	40
HO_2S'	C_2	$(MeO)_2P(S)H$	74.9	16
HO_2S'		$(EtO)_2P(S)H$	68.4	16, 17
$HO_2S'/5$		$(\cdot SCH_2CH_2O)\dot{P}(O)H$	1.5, 4 J620/680	22
$HO_2S'/6$		**1** (x = S)	$H_{ax.}$ 55.1/61.2 J ca. 590 H eq. 54.76/55.1 J ca. 630	15, 41
HO_2Se'	Si_2	$(TmsO)_2P(Se)H$	36.2 J632	52
$HO_2Se/6$	C_2	**2**	$H_{ax.}$ 66: Heq. 67.7	15, 41
HO_2Te'	Si_2	$(TmsO)_2P(Te)H$	−9.2 J582,	52

P-bound atoms = HO_2O'

P-bound atoms/rings	Connectivities	Structure	NMR data δP [$^1J_{PH}$ Hz]	Ref.
Ge,Si	$C_9H_{25}GeO_3PSi$	$TmsO(Et_3GeO)P(O)H$	~O J664	13
Si_2	$C_6H_{19}O_3PSi_2$	$(TmsO)_2P(O)H$	−14.3, −15.1 J694, 678	13, 52
H,P^4	$H_4O_5P_2$	$[H(OH)P(O)]_2$	3.1/4.4 J668/690	3
	$H_{n+2}P_nO_{2n+1}$	$HO[HP(O)O]_nH$	End 7.4 J710 Middle 14 J760	3
H_2	H_3O_3P	$HP(O)(OH)_2$	1.7/8 J694/700	2
C,P	$C_4H_{12}P_2O_5$	$(EtOP(O)H)_2O$	2.3 J780	51
C,Si	$C_5H_{15}O_3PSi$	TmsO(EtO)P(O)H	−3 J689	13
	$C_6H_{17}O_3PSi$	TmsO(PrO)P(O)H	−5.2 J687	13
	$C_7H_{19}O_3PSi$	TmsO(BuO)P(O)H	−3.5 J686	13
C,H	CH_5O_3P	MeOP(O)H(OH)	5.3(p)/8.6(w) J674, 692	2
	$C_2H_7O_3P$	EtOP(O)H(OH)	5.23 J674	2
	$C_3H_9O_3P$	i-PrOP(O)H(OH)	3.41 J670	2
	$C_4H_{11}O_3P$	BuOP(O)H(OH)	5.83 J678	2
	$C_4H_{11}O_3P$	i-BuOP(O)H(OH)	5.72 J680	2
	$C_4H_{11}O_3P$	t-BuOP(O)H(OH)	−1.1 J674	2
	$C_6H_{15}O_3P$	$C_6H_{13}OP(O)H(OH)$	5.51 (676)	2
C′,H	$C_6H_7O_3P$	PhOP(O)H(OH)	4.0 J719	2
C_2	$C_2H_6DO_3P$	$(MeO)_2P(O)D$	11.2 J107	28
	$C_2H_7O_3P$	$(MeO)_2P(O)H$	9.3(j)/14.3(w) J696/723	1, 2
	$C_4H_{10}DO_3P$	$(EtO)_2P(O)D$	6.8 J106	28
	$C_4H_{11}O_3P$	$(EtO)_2P(O)H$	7.1/8.1 J682/691	1-3
	$C_4H_{11}O_3P$	MeO(i-PrO)P(O)H	7.9 J695	2

TABLE I (continued)

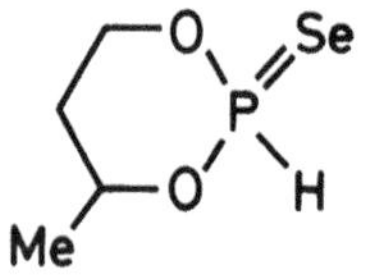

1 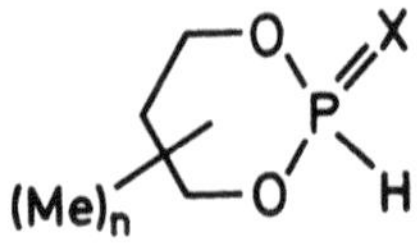 2 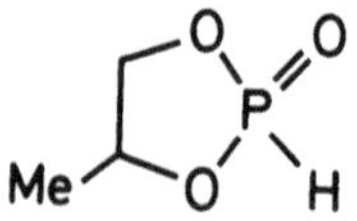 3

P-bound atoms/rings	Connectivities	Structure	NMR data δP [$^1J_{PH}$ Hz]	Ref.
	$C_5H_{11}O_3P$	EtO(CH$_2$=CHCH$_2$O)P(O)H	6.7 J692	25
	$C_5H_{13}O_3P$	MeO(s-BuO)P(O)H	7.6, 7.8 J700	2
	$C_5H_{13}O_3P$	EtO(PrO)P(O)H	5.4 J682	25
	$C_5H_{13}O_3P$	EtO(i-PrO)P(O)H	6.9 J678	25
	$C_6H_{11}O_2P$	(CH$_2$=CHCH$_2$O)$_2$P(O)H	4.9/7.8 J700/708	3
	$C_6H_{12}NO_3P$	EtO(NCMe$_2$CO)P(O)H	3.9 J720	1
	$C_6H_{13}O_3P$	PrO(CH$_2$=CHCH$_2$O)P(O)H	5.4 J686	25
	$C_6H_{14}DO_3P$	(PrO)$_2$P(O)D	6.9 J106	28
	$C_6H_{14}DO_3P$	(i-PrO)P(O)D	3.4 J105	28
	$C_6H_{15}O_3P$	EtO(BuO)P(O)H	6.6 J684	25
	$C_6H_{15}O_3P$	EtO(i-BuO)P(O)H	6.3 J688	25
	$C_6H_{15}O_3P$	EtO(s-BuO)P(O)H	5.2 J684	25
	$C_6H_{15}O_3P$	EtO(t-BuO)P(O)H	1.1/1.6 J680/690	1, 25
	$C_6H_{15}O_3P$	(PrO)$_2$P(O)H	6.5/7.4 J685	3
	$C_6H_5O_3P$	(i-PrO)$_2$P(O)H	3.2/4.5 J685/694	1-3
	$C_6H_{15}O_3P$	PrO(i-PrO)P(O)H	5.7 J678	25
	$C_7H_{14}O_3P$	EtO(CH$_2$=CHCMe$_2$O)P(O)H	1.7 J685	1
	$C_7H_{15}O_3P$	BuO(CH$_2$=CHCH$_2$O)P(O)H	7.7 J694	25
	$C_7H_{17}O_3P$	PrO(BuO)P(O)H	7.4 J686	25
	$C_8H_{11}O_3P$	MeO(PhCH$_2$O)P(O)H	10.2 J713	2
	$C_8H_{13}N_2O_3P$	(NCCMe$_2$O)$_2$P(O)H	−0.1 J730	1
	$C_8H_{15}Cl_4O_3P$	(Me$_2$CHCCl$_2$O)$_2$P(O)H	−3.1 J735	3
	$C_8H_{19}O_3P$	(BuO)$_2$P(O)H	5/8 J670	3
	$C_8H_{19}O_3P$	(i-BuO)$_2$P(O)H	8.6 J700	2
	$C_8H_{19}O_3P$	(s-BuO)$_2$P(O)H	4.7/9.9 J660/690	1
	$C_8H_{19}O_3P$	(t-BuO)$_2$P(O)H	−3.2, −3.4, J678/690	2, 4
	$C_8H_{19}O_3P$	BuO(i-BuO)P(O)H	6.6 J686	25
	$C_8H_{19}O_3P$	(MeEtCHO)$_2$P(O)H	4.7, 9.9 J691, 660	4
	$C_8H_{19}O_3S_2P$	(EtSCH$_2$CH$_2$O)$_2$P(O)H	7.7 J712	3
	$C_{10}H_{19}O_3P$	(H$_2$C=CHCMe$_2$O)P(O)H	−3.2 J690	1
	$C_{10}H_{10}O_3P$	(Me$_3$CCH$_2$O)$_2$P(O)H	7.7 J695	4
	$C_{10}H_{23}O_3P$	(AmO)$_2$P(O)H	6.1	20
	$C_{10}H_{23}O_3P$	(i-AmO)$_2$P(O)H	3.5/4.3 J687	4, 20
	$C_{10}H_{23}O_3P$	(EtCMe$_2$O)$_2$P(O)H	−3.5 J680	1, 27
	$C_{10}H_{23}O_3P$	(Et$_2$CHO)$_2$P(O)H	5.7	20
	$C_{10}H_{23}O_3P$	(Me$_2$CHCHMeO)$_2$P(O)H	4.8	20
	$C_{14}H_7O_3P$	(PhCH$_2$O)$_2$P(O)H	7.9 J713	3
	$C_{16}H_{35}O_3P$	(C$_8$H$_{17}$O)$_2$P(O)H	7	20
5/C$_2$	$C_3H_{13}O_3P$	**3**	21.6, 22.5 J705, 701	28
6/C$_2$	$C_{4+n}H_{7+n}O_3P$	**4**	H eq. −1/−2.5 J ca. 715 H$_{ax}$ −2.5/4.5 J ca. 665	24, 28, 51
	$C_8H_{13}O_6P$	**5**	−0.6, −10.2 J666	29
C,C′	$C_7H_9O_3P$	(PhO)(MeO)P(O)H	7.2 J728	2
	$C_8H_{11}O_3P$	(PhO)(EtO)P(O)H	4.9	2
	$C_9H_{13}O_3P$	(PhO)(i-PrO)P(O)H	2.85, 2.9 J718, 712	2, 3

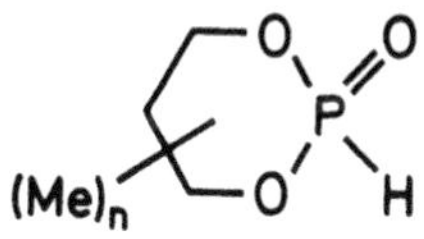

4 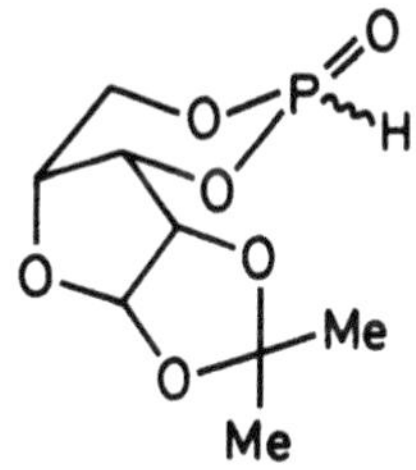 **5**

P-bound atoms/rings	Connectivities	Structure	NMR data δP [$^1J_{PH}$ Hz]	Ref.
	$C_{10}H_{15}O_3P$	(PhO)(t-BuO)P(O)H	−1.5 J712	2
C_2'	$C_{12}H_{11}O_3P$	$(PhO)_2P(O)H$	0.6(n) 1.9 2.2(m) J747, 756	2, 3
	$C_{13}H_{13}O_3P$	$(PhO)(3\text{-}MeC_6H_4O)P(O)H$	1.0 J742	2
	$C_{14}H_{15}O_3P$	$(3\text{-}MeC_6H_4O)_2P(O)H$	1.15 J740	2
	$C_{14}H_{15}O_3P$	$(4\text{-}MeC_6H_4O)_2P(O)H$	−0.65, −1.3 J740	2
	$C_{14}H_{15}O_5P$	$(2\text{-}MeOC_6H_4O)_2P(O)H$	1.87 J762	2

P-bound atoms = HOO'_2

P-bound atoms/rings	Connectivities	Structure	NMR data δP [$^1J_{PH}$ Hz]	Ref.
P^4	HO_6P_2	$[O_3POP(H)O_2]^{3-}$	−1 J620	3
	$H_2O_5P_2$	$[O_2(H)POP(H)O_2]^{=}$	−5 J657	3
H	H_2O_3P	$HP(OH)O_2^-$	−1.7 J571	2, 3
C	CH_4O_3P	$MeOP(H)O_2^-$	2.7, 3.95 J600, 591	2
	$C_2H_6O_3P$	$EtOP(H)O_2^-$	0.9 J627	2
C'	$C_6H_6O_3P$	$PhOP(H)O_2^-$	4.0 J629	
S_3	DPS'_2	$(S_3PPDS_2)^{3-}$	62 J67	48
OS_2	HPS'_2	$(OS_2PPHS_2)^{3-}$	47.1 J434	48
	HS'^-_3	HPS_3^-	41.7 (J514)	40

Section I2: Phosphinic Acids and Derivatives $H_2P(Ch)X$

P-bound atoms/rings	Connectivities	Structure	NMR data δP [$^1J_{PH}$ Hz]	Ref.
H_2Cl	—	$H_2P(O)Cl$	19.2 J601	43
H_2OS'	C	$BuOP(S)H_2$	41.8 J530	29
H_2OS'		$i\text{-}BuOP(S)H_2$	42 J530	29
H_2OO'	Si	$TmsOP(O)H_2$	−4.4	52
H_2OO'	H	$HOP(O)H_2$	3.8/12.7w J565 ± 10	2, 3, 7
H_2OO'	C	$MeOP(O)H_2$	18.7/18.9 J580	3, 7, 30
H_2OO'		$EtOP(O)H_2$	15.0/15.8 J570	3, 7, 30
H_2OO'		$PrOP(O)H_2$	15.6 J568	29
H_2OO'		$i\text{-}PrOP(O)H_2$	12.4 J570	30
H_2OO'		$BuOP(O)H_2$	16 J568	29
H_2OO'		$i\text{-}BuOP(O)H_2$	17.8 J568	29
H_2OO'		$t\text{-}BuOP(O)H_2$	6.6/8.2 J552/560	3
H_2OO'		$EtCMe_2OP(O)H_2$	3.8/6.8 J552	3
H_2OO'		$PhCH_2OP(O)H_2$	15.4 J578	30
H_2O_2'		$H_2PO_2^-$	−0.5, −1.6 J481, 494	2

TABLE I (continued)

Section I3: Monosubstituted Phosphinic Acids and Derivatives RHP(Ch)X

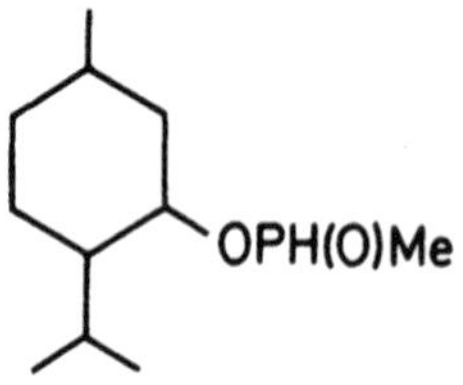

OPH(O)Me

6

PH(O)OH

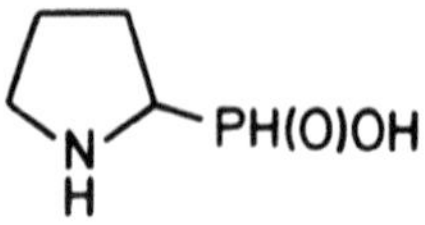

7

P-bound atoms/rings	Connectivities	Structure	NMR data δP [$^1J_{PH}$ Hz]	Ref.
CHBrO′	H_3	MePH(O)Br	35.8 J615	43
C′HBrO′	C'_2	PhPH(O)Br	19.8 J596	43
CHClO′	H_3	MePH(O)Cl	32.2 J598	43
C′HClO′	C'_2	PhPH(O)Cl	19.3 J606	43

P-bound atoms = CHOO′

P-bound atoms/rings	Connectivities	Structure	NMR data δP [$^1J_{PH}$ Hz]	Ref.
HO_2;H	$C_3H_9O_4P$	$(MeO)_2CHPH(O)OH$	21.6 J555	2
	$C_5H_{13}O_4P$	$(EtO)_2CHPH(O)OMe$	23.1 J553	2
HO_2;C	$C_4H_{11}O_4P$	$(MeO)_2CHPH(O)OMe$	30.0 J568	2
	$C_6H_{15}O_4P$	$(EtO)_2CHPH(O)OMe$	31.0 J559	2
	$C_7H_{17}O_4P$	$(EtO)_2CHPH(O)OEt$	26.7(m)/29.0(g) J550	2
	$C_8H_{19}O_4P$	$(EtO)_2CHPH(O)O$-i-Pr	25.3 J555	2
H_2N;H	CH_3NO_2P	$H_2NCH_2PH(O)OH$	14.2 J546	31
	$C_3H_{10}NO_2P$	$Me_2NCH_2PH(O)OH$	9.6	8
	$C_5H_{14}NO_2P$	$Et_2NCH_2PH(O)OH$	6.6	8
4/H_2N	$C_5H_{12}NO_3P$	$O(CH_2CH_2)_2NCH_2PH(O)OH$	8.6	8
H_2O;H	CH_5O_3P	$HOCH_2PH(O)OH$	29.3(m) J540	2
H_2O;C	$C_2H_7O_3P$	$HOCH_2PH(O)OMe$	39.9(m) J553	2
H_3;H	CH_5O_2P	MePH(O)OH	35 J557	6
H_3;C	$C_2H_7O_2P$	MePH(O)OMe	32	9
	$C_{11}H_{23}O_2P$	**6**	26.9, 31.2 J536, 535	34
	$C_3H_9O_2P$	MePH(O)OEt	32.6	9
	$C_4H_{11}O_2P$	MePH(O)OPr	23.2 (33.2?)	9
CHN;H	$C_2H_8NO_3P$	$H_2NCH(R)PH(O)OH$; R = CH_2OH	15.9 J540	31
	$C_2H_8NO_2PS$	$H_2NCH(R)PH(O)OH$; R = CH_2SH	18.1 J541	31
	$C_3H_9BrNO_2P$	$H_2NCH(R)PH(O)OH$; R = CH_2CH_2Br	19.2 J536	31
	$C_3H_{10}NO_3P$	$H_2NCH(R)PH(O)OH$; R = $CH(OH)CH_3$	14.8, 16.5 J543	31
	$C_4H_{10}NO_2P$	**7**	19.2 J538	31
	$C_4H_{19}NO_4P$	$H_2NCH(R)P(H)OH$; R = $CH_2CH_2CO_2H$	18.8 J537	31
	$C_4H_{11}N_2O_2P$	$H_2NCH(R)P(H)OH$; R = $(CH_2)_3NH_2$	31.7 J500	31

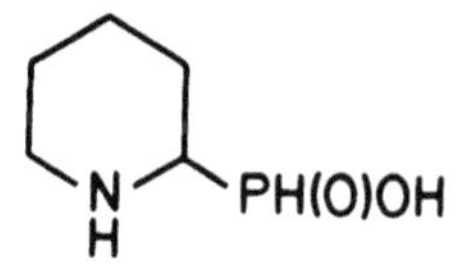

8

P-bound atoms/rings	Connectivities	Structure	NMR data δP [$^1J_{PH}$ Hz]	Ref.
	$C_4H_{12}NO_2P$	$H_2NCH(R)P(H)OH$; R = i-Pr	18.6 J540	31
	$C_4H_{12}NO_2PS$	$H_2NCH(R)P(H)OH$; R = $MeSCH_2CH_2$	19.2 J536	31
	$C_4H_{14}NO_2O_4P_2S_2$	$H_2NCH(R)P(H)OH$; R = CH_2SSCH_2 $(PO_2H_2)CHNH_2$	29.3, 29.4 J506	31
	$C_5H_{12}NO_2P$	**8**	19.0 J533	31
	$C_5H_{14}NO_2P$	$H_2NCH(R)PH(O)OH$; R = i-Bu	21.6 J532	31
	$C_5H_{14}NO_2P$	$H_2NCH(R)PH(O)OH$; R = s-Bu	18.5, 18.8 J534	31
	$C_5H_{13}BrNO_2P$	$H_2NCH(R)PH(O)OH$; R = $(CH_2)_4Br$	19.6 J530	31
	$C_5H_{14}NO_3P$	$H_2NCH(R)PH(O)OH$; R = $(CH_2)_4OH$	20.5 J532	31
	$C_5H_{15}N_4O_2P$	$H_2NCH(R)PH(O)OH$; $NH_2C(=NH)NH$	35.1 J504	31
	$C_8H_{12}NO_2P$	$H_2NCH(R)PH(O)OH$; R = $4\text{-}HOC_6H_4CH_2$	31.8 J507	31
	$C_{10}H_{13}N_4O_2P$	$H_2NCH(R)PH(O)OH$; R = Indol-3-CH_2	32.0 J507	31
CHO;H	$C_2H_7O_3P$	MeCH(OH)PH(O)PH	34.3 J532	7
	$C_4H_{11}O_3P$	$CH_3CH(OH)PH(O)OEt$	40.8 J537	7
CH_2;H	$C_3H_6NO_2P$	$NCCH_2CH_2PH(O)OH$	44	3
	$C_5H_{10}NO_4P$	$HO_2CCH(NH_2)(CH_2)_2PH(O)OH$	32.7 J528	33
CH_2;C	$C_5H_{11}O_4P$	$MeO_2CCH_2CH_2PH(O)OMe$	49.9 J548	32
C_2O;C	$C_4H_{11}O_3P$	$HOCMe_2PH(O)OMe$	45 J552	7
C_2H;C	$C_8H_{17}O_2P$	*c*-$C_6H_{11}PH(O)OEt$	40.3	9
	$C_9H_{19}O_2P$	*c*-$C_6H_{11}PH(O)O$-i-Pr	37.2 J500	3

P-bound atoms = C′HOO′

P-bound atoms/rings	Connectivities	Structure	NMR data δP [$^1J_{PH}$ Hz]	Ref.
OO′;Si	$C_6H_{15}O_4PSi$	EtOCOPH(O)OTms	−3.8 J618	52
OO′;H	CH_3O_4P	HOCOPH(O)OH	10.4 J576	52
OO′;C	$C_2H_5O_4P$	HOCOPH(O)OMe	11.2 J664	52
C′H;H	$C_3H_6O_2P$	$CH_2{=}C{=}CHPH(O)OH$	15.4 J574	36
C'_2;Si	$C_9H_{16}O_2PSi$	PhPH(O)OTms	10.1 J565	13
C'_2;H	$C_6H_7O_2P$	PhPH(O)OH	20(a) 23(w), J560	2, 4
	$C_9H_{14}NO_2P$	$4\text{-}Me_2N\text{-}2\text{-}MeC_6H_3PH(O)OH$	20 J560	14
	$C_{18}H_{32}O_2P$	2,4,6-t-$Bu_3C_6H_2PH(O)OH$	25.7 J576	35
C'_2;C	$C_7H_9O_2P$	PhPH(O)OMe	27.3/29.3 J576	2, 9
	$C_8H_{11}O_2P$	PhPH(O)OEt	26.7 J571	2
	$C_9H_{13}O_2P$	PhPH(O)OPr	26.2	14
	$C_9H_{13}O_2P$	PhPH(O)O-i-Pr	20.8 J549	3
	$C_9H_{13}O_2P$	2,6-$Me_2C_6H_3PH(O)OMe$	27.5 J561	14
	$C_9H_{13}O_2P$	2,4-$Me_2C_6H_3PH(O)OMe$	27.5 J561	14
	$C_9H_{13}O_2P$	2,5-$Me_2C_6H_3PH(O)OMe$	25.1 J556	3
	$C_{10}H_{15}O_2P$	PhPH(O)OBu	23.8 J570	3

TABLE I (continued)

Section I3: Monosubstituted Phosphinic Acids and Derivatives RHP(Ch)X

Me Me
P
O H(D)

9

P-bound atoms/rings	Connectivities	Structure	NMR data δP [$^1J_{PH}$ Hz]	Ref.
	$C_{11}H_{17}O_2P$	2,5-$Me_2C_6H_3PH(O)O$-i-Pr	1.6 (16?) J553	3
C'_2;C'	$C_{12}H_{11}O_2P$	PhPH(O)OPh	24.8 J579	2

P-bound atoms/rings	Connectivities	Structure	NMR data δP [$^1J_{PH}$ Hz]	Ref.
C'HOS'	C_2';C	2,6-$Me_2C_6H_3PH(S)OEt$	58.6 J519	39
C'DOO'	C'_2;D	PhPD(O)OD	19.8 J95	2
$C'HO_2'$	C'H	CH_2=C=CHP(H)O_2^-	12.9, 13.1 J524, 520	36
$C'HO_2'$		Me_2C=C=CHP(H)O_2^-	12.7, 13.4 J540, 542	36
$C'HO'_2$	C_2'	PhP(H)O_2^-	17.4(w)	2
C'HSS'	C'2;P^4	[2,4,6-t-$Bu_3C_6H_2P(S)]_2S$	50.8 J572	38
CH_2O'	C	$NCCH_2CH_2P(O)H_2$	9 J510	3, 5
CH_2O'		i-BuP(O)H_2	6 J494	3, 5
CH_2O'		$C_8H_{17}P(O)H_2$	10 J470	3, 5
CH_2O'	C_2	c-$C_6H_{11}P(O)H_2$	22 J477	3, 5
$C'H_2O'$	C'_2	PhP(O)H_2	7 J493	3, 5
$C'H_2O'$	C_2'	2,4,6-t-$Bu_3C_6H_2P(O)H_2$	−10 J491	37
$C'H_2S$		2,4,6-t-$Bu_3C_6H_2P(S)H_2$	−24.7, −26 J470	37, 38

Section I4: Secondary Phosphine Chalcogenides $R_2P(Ch)H$

P-bound atoms/rings	Connectivities	Structure	NMR data δP [$^1J_{PH}$ Hz]	Ref.
C_2HO'	$(H_3)_2$	$Me_2P(O)H$	25.8 J456	14
C_2HO'	H_3;CH_2	MePH(O)$CH_2CH_2COCH_3$	27.5(c), 38.2(w) J474	41
C_2HO'		MePH(O)$(CH_2)_3CO_2H$	31.8(c), 38.2(w) J476	41
C_2HO'		MePH(O)$C_{12}H_{25}$	33.9	18
C_2HO'	CHO, CH_2	[OctPH(O)CH(OH)]$_2$	44	3
C_2HO'		[OctPH(O)CH(OH)]$_2(CH_2)_3$	46 J502	3
C_2HO'	$(CH_2)_2$	$Et_2P(O)H$	47(w), 41(c) J463	14
C_2HO'		$(NCCH_2CH_2)_2P(O)H$	2.5 J356	3
C_2HO'		$Bu_2P(O)H$	41.8 J462	15
C_2DO'		$Bu_2P(O)D$	40.7 J72	15
C_2HO'		OctPH(O)$CH_2CH_2CONH_2$	39 J454	3
C_2HO'		$Oct_2P(O)H$	28 J308	3
C_2HO'		Oct($CH_2CH_2CO_2Ph$)P(O)H	39 J470	3
C_2HO'	CH_2;C_2O	Oct(1-OH-c-C_6H_{11})P(O)H	52 J454	3
C_2HO'	$(C'H_2)_2$	**9**	39.8(c), 44.4(w) J490	41
C_2DO'		**9**	44.1(w)	41
C_2HO'		$(PhCH_2)_2P(O)H$	41 J341	3
C_2HO'	C_2O, C_2H	i-PrPH(O)CMe_2OH	59 J422	14
C_2HO'		c-C_5H_9(1-OH-c-C_6H_{11})P(O)H	51 J464	3

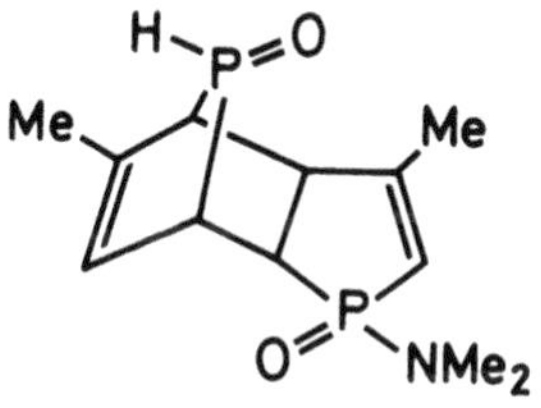

10

P-bound atoms/rings	Connectivities	Structure	NMR data δP [$^{1}J_{PH}$ Hz]	Ref.
C_2HO'	$(C_2H)_2$	$(c\text{-}C_6H_{11})_2P(O)H$	46.4 J427	3
C_2HO'	$(CC'H)_2$	**10**	57.4 J510	47
$CC'HO'$	$C_2H;C'_2$	$c\text{-}C_6H_{11}PH(O)Ph$	34 J423	3
$CC'HO'$	CH_2	$OctPH(O)CONHC_6H_4\text{-}4\text{-}NO_2$	11 J340	3
C'_2HO	$(C'_2)_2$	$Ph_2P(O)H$	25.9 J513	15
C_2HS'	$(H_3)_2$	$Me_2P(S)H$	5 J456	12
C_2HS'	$H_3;C_3$	MePH(S)t-Bu	36.2	19
C_2HS'	$(CH_2)_2$	$Et_2P(S)H$	31 J437	12
C_2HS'		$i\text{-}Bu_2P(S)H$	13.4 J427	12
C_2HS'	$CH_2;C_3$	$Ph_2PCH_2CH_2PH(S)t\text{-}Bu$	52.9 J433	42
C_2HS'	$(C_2H)_2$	$(c\text{-}C_6H_{11})_2P(S)H$	47 J454	14
$CC'HS'$	$CH_2;C_2'$	$Ph_2PCH_2CH_2PH(S)Ph$	43.5 J458	42
C'_2HS'	$(C'_2)_2$	$Ph_2P(S)H$	19.6 J444	12
C_2HSe'	$(CH_2)_2$	$i\text{-}Bu_2P(Se)H$	−5.3 J420	14
$CC'HSe'$	$CH_2;C'_2$	$Ph_2P(Se)CH_2CH_2PH(Se)Ph$	39.6 J427	42
C'_2HSe	$(C'_2)_2$	$Ph_2(Se)H$	5.8 J450	14, 44

REFERENCES

1. **Mark, V. and van Wazer, J. R.,** *J. Org. Chem.*, 32, 1187, 1967.
2. **Gallagher, M. J., Fookes, C. J., Honegger, H., Lee, G.-H., Garbutt, R., and Liu, Y. H.,** unpublished results.
3. **Crutchfield, M. M., Dungan, C. H., Letcher, J. H., Mark, V., and van Wazer, J. R.,** *Topics in Phosphorus Chemistry Vol. 5*, Wiley-Interscience, New York, 1967.
4. **Gerrard, W. and Hudson, H. R.,** *Organic Phosphorus Compounds*, Vol. 5, Kosolapoff, G. M. and Maier, L., Eds., Wiley-Interscience, New York, 1973, 21.
5. **Buckler, S. A. and Epstein, M.,** *Tetrahedron*, 18, 1221, 1962.
6. **Fiat, D., Halmann, M., Kugel, L., and Reuben, J.,** *J. Chem. Soc.*, p. 3837, 1962.
7. **Fitch, S. J.,** *J. Am. Chem. Soc.*, 86, 61, 1964.
8. **Maier, L.,** *Helv. Chim. Acta*, 50, 1742, 1967.
9. **Wolf, R., Houalla, D., and Mathis, F.,** *Spectrochim. Acta (A)*, 23, 1641, 1967.
10. **Finegold, H.,** *Ann. N.Y. Acad. Sci.*, 70, 815, 1958.
11. **Maier, L.,** *Helv. Chim. Acta*, 49, 1000, 1966.
12. **Maier, L.,** *Helv. Chim. Acta*, 49, 1249, 1966.
13. **Brazier, J.-F., Houalla, D., and Wolf, R.,** *Bull. Soc. Chim. Fr.*, p. 1089, 1970.
14. **Frank, W. A., Hamilton, L. A., and Landes, P. S.,** *Organic Phosphorus Compounds Vol. 4*, Kosolapoff, G. M. and Maier, L., Eds., Wiley-Interscience, New York, 1972, chap. 10 and 11.
15. **Stec, W. J., Uznanski, B., and Michalski, J.,** *Phosphorus*, 2, 235, 1973.
16. **Lippmann, H. E.,** *J. Org. Chem.*, 31, 1603, 1966.
17. **Williamson, M. P. and Griffin, C. E.,** *J. Phys. Chem.*, 72, 4043, 1968.

18. **Hays, H. R.,** *J. Org. Chem.*, 31, 3817, 1966.
19. **Kosfled, R., Hagele, G., and Kuchen, W.,** *Angew. Chem. Int. Ed.*, 7, 814, 1968.
20. **Fluck, E. and Binder, H.,** *Z. Anorg. Allg. Chem.*, 354, 139, 1967.
21. **Thorstenson, P. C. and Aaron, H. S.,** *Phosphorus Sulfur*, 22, 145, 1986.
22. **Nifantev, E. E., Zavalishina, A. I., Sorokina, S. F., and Borisenko, A. A.,** *Zh. Obshch. Khim.*, 46, 471, 1976.
23. **Olah, G. A. and MacFarlane, C. W.,** *Inorg. Chem.*, 11, 845, 1972.
24. **Mosbo, J. A. and Verkade, J. G.,** *J. Am. Chem. Soc.*, 95, 204, 1973.
25. **Kluba, M. and Zwierzak, A.,** *Synthesis*, p. 134, 1978.
26. **Gallagher, M. J. and Sussmann, J.,** *Phosphorus*, 5, 91, 1975.
27. **Burgada, R. and Roussel, J.,** *Bull. Soc. Chim. Fr.*, p. 192, 1970.
28. **Stec, W. J.,** *Z. Naturforsch.*, 298, 109, 1974.
29. **Nifantev, E. E., Koroteev, M. P., Pugashova, N. M., Kukhareva, T. S., and Borisenko, A. A.,** *Zh. Obshch. Khim.*, 51, 1900, 1981.
30. **Fookes, C. J. and Gallagher, M. J.,** *J. Chem. Soc. Chem. Commun.*, p. 324, 1978.
31. **Bayles, E. K., Campbell, C. D., and Dingwall, J. G.,** *J. Chem. Soc. Perkin Trans. 1*, p. 2845, 1984.
32. **Karlstedt, N. B., Proskurnina, M. V., and Lutsenko, I. F.,** *Zh. Obshch. Khim.*, 46, 2018, 1976.
33. **Meppelder, F. H., Benschop, H. P., and Kraay, G. W.,** *J. Chem. Soc. Chem. Commun.*, p. 431, 1970.
34. **Maier, L. and Rist, G.,** *Phosphorus Sulfur*, 17, 21, 1983.
35. **Yoshifuji, M., Shibayama, K., Toyota, K., and Inamoto, N.,** *Tetrahedron Lett.*, 24, 4227, 1983.
36. **Belakhov, V. V., Yudelevich, V. I., Komarov, E. V., Ionin, B. I., and Komarov, V. Y., Zakharov, V. I., Lebedev, V. B., and Petrov, A. A.,** *Zh. Obshch. Khim.*, 53, 1503, 1983.
37. **Yoshifuji, M., Shibayama, K., and Toyota, K.,** *Tetrahedron Lett.*, 24, 4727, 1983.
38. **Navech, J., Revel, M., and Kraemer, R.,** *Phosphorus Sulfur*, 21, 105, 1984.
39. **van der Knaap, T. A., Klebach, T. C., Lourens, R., Vos, M., and Bickelhaupt, F.,** *J. Am. Chem. Soc.*, 105, 4026, 1983.
40. **Seel, F. and Zindler, G.,** *Z. Anorg. Allg. Chem.*, 470, 167, 1980.
41. **Quin, L. D. and Roser, C. E.,** *J. Org. Chem.*, 39, 3423, 1974.
42. **Weisheit, R., Stendel, R., Messbauer, B., Langer, C., and Walther, B.,** *Z. Anorg. Allg. Chem.*, 504, 147, 1983.
43. **Kardanov, N. A., Timofeev, A. M., Kvashnina, G. A., Ermanson, L. V., Godovikov, N. N., and Kabachnik, M. I.,** *Phosphorus Sulfur*, 30, 690, 1987; personal communication from N. A. Kardanov.
44. **Macfarlane, W. M. and Rycroft, D. S.,** *J. Chem. Soc. Dalton Trans.*, p. 2162, 1973.
45. **Niecke, E. and Ringel, G.,** *Angew. Chem. Int. Ed.*, 16, 486, 1977.
46. **Mosbo, J. A. and Verkade, J. G.,** *J. Org. Chem.*, 42, 1549, 1977.
47. **Quin, L. D. and Keglevich, G.,** *J. Chem. Soc. Perkin Trans. 2*, p. 1029, 1986.
48. **Krause, W. and Falins, H.,** *Z. Anorg. Allg. Chem.*, 496, 94, 1983.
49. **Spitz, F. R., Cabral, J., and Haake, P.,** *J. Am. Chem. Soc.*, 108, 2802, 1986.
50. **Chun, Z. and Wauzhen, L.,** *Bopuxue Zazhi*, 1, 487, 1984; *Chem. Abstr.*, 105, 172594, 1987.
51. **Stec, W. J. and Mikolaczyk, M.,** *Tetrahedron*, 29, 539, 1973.
52. **Issleib, K., Mogelin, W., and Balszuweit, H.,** *Z. Anorg. Allg. Chem.*, 530, 16, 1986.

Chapter 11

^{31}P NMR DATA OF FOUR COORDINATE PHOSPHORUS COMPOUNDS CONTAINING A P=Ch BOND AND ONE BOND TO A GROUP IV ATOM

Compiled and presented by
Ronald S. Edmundson

NMR data were also donated by
J. Verkade, Iowa State University, Ames, Iowa
H. W. Roesky, Universität Göttingen, FRG
J. P. Dutasta, Université Scientifique et Médicale de Grenoble, France
C. D. Hall, King's College London, U.K.

The NMR data are presented in Table J and it is divided into two parts. Section J1 contains data for derivatives of phosphonic acid and Section J2 contains data for derivatives of thio-, seleno-, and telluro-phosphonic acids. Supplementary data are available at the end of the chapters.

This group of compounds is largely made up of derivatives of the phosphonic acids together with their sulfur-, selenium-, and tellurium-containing analogues. The complete chemical shift range for the group is generally *circa* −25 to +130 ppm. Compounds which possess bonds directly linking phosphorus to sulfur, selenium, or tellurium show signals in the range +20 to +130 ppm and are deshielded relative to those compounds without these bonds for which the chemical shifts are in the general region −25 to +70 ppm.

A consideration of compounds having the structure $R\text{-}(O)_n\text{-}P(O)X_2$ enables a comparison to be made of compounds in the phosphate ($n = 1$) and phosphonate ($n = 0$) series. The deshielding of the phosphorus nucleus (movement of signal to more positive shift values and lower field) for the acid dichlorides ($X = Cl$) amounts to 30 to 50 ppm for phosphonic relative to phosphoric derivatives. The same structural change for acid dialkyl esters results in a deshielding by 30 to 35 ppm for lower primary alkyl groups, rising to 50 ppm for those groups substituted on the α-carbon atom; similar differences in chemical shifts are observed for the parent acids. The phosphorus nucleus in diphenyl phenylphosphonate is deshielded by ca. 40 ppm (estimate) relative to that in triphenyl phosphate.

A reasonable definition of the chemical shift range for a given type of phosphonic acid derivative is sometimes difficult to establish because of the paucity of data. Such is the case, for example, for the alkylphosphonic dibromides; three recorded values include two at ca. 5 ppm and one at ca. 50 ppm. The chemical shifts of alkylphosphonic dichlorides and difluorides are 10 to 66 and 0 to 32 ppm, respectively. Such wide variations for a given structural type clearly indicate a dependence of chemical shift on the alkyl group skeleton; the controlling feature could be steric interactions or changes in electron-donor power, including the possible changes in hyperconjugation. On the other hand, for arylphosphonic compounds with only para substituents (steric effects thus having been eliminated) the chemical shifts of the acid dichlorides and difluorides are in the much narrower ranges 32 to 38 and 7 to 12 ppm, respectively.

Replacement of the phosphonic phosphoryl oxygen by sulfur or selenium results in marked deshielding of the phosphorus nucleus, whereas replacement of a singly bonded oxygen is less marked. The downfield perturbations in chemical shifts extend from quite small differences, e.g., 14 ppm for $RP(S)Br_2$, to fairly large differences, e.g., 78 to 80 ppm for the corresponding difluorides, with intermediate differences for the dichlorides, e.g., 35

(Me), 40 (Ph), and 53 (tBu) ppm. For the methyl-, ethyl-, and t-butylphosphonodichloridoselenoates, the perturbations are 12, 22, and only 8 ppm, respectively, whereas for the difluorides all three compounds are relatively deshielded by 92 to 96 ppm. Thus, the relative deshielding induced on replacement of phosphoryl oxygen by sulfur or selenium depends on the other groups directly bonded to the phosphorus. Moreover, a regular variation in the structure of the phosphonic alkyl group does not always produce the same effects irrespective of the phosphoryl chalcogen atom.

As for phosphate esters, dimethyl esters of phosphonic acids tend to display signals 2 to 4 ppm downfield from the corresponding diethyl esters, and further lengthening of the ester group carbon chain has little influence on the chemical shift.[1] Trimethylsilyl esters have highly shielded phosphorus nuclei, as have silyl phosphate esters. Aryl esters are less shielded than silyl esters, but more so than alkyl esters, in keeping with the greater electron-withdrawing ability of the PhO group.

A similar lack of dependence of chemical shift on variation in R′ is to be seen for the esters RP(X)(OH)OR′, where R = Me or Et, and X = S or Se. Nevertheless, the slight changes in chemical shift when the ester groups are derived from chiral alcohols or thiols have allowed the use of methylphosphonic dichloride as a reagent for the determination of the enantiomeric excess in optically active alcohols and thiols,[2,3] and similarly that of methylphosphonothioic dichloride in reactions with chiral amines, e.g., amino acids.[4] The observed differences in chemical shifts are similar to those found for the P-epimers of (4*S*,5*R*)- and (4*S*,5*S*)-4-methyl-5-phenyl-1,3,2-oxazaphospholidines and 1,3,2-thiazaphospholidines.[5-7] In general, for the diastereoisomeric compounds ABP(Z)Y, the differences in chemical shifts are the largest when A,B = Z = Y and the differences between the groups in either size or electronegativity are maximal.

In considering the effects of structural changes in the phosphonic alkyl group in series such as $RP(X)Cl_2$, $RP(O)(OR')_2$, and RP(O)(XR′)F with R = Me, Et, iPr, or tBu, the feature to be immediately noted is the simple correlation of deshielding resulting from increased α-substitution in the group R with its steric bulk, which in turn could result in distortions in bond angles at phosphorus (see Chapter 9), or, alternatively, the series Me to tBu might be considered as representing a decrease in hyperconjugative activity. The simple relation between the number of carbon atoms on the alkyl α-carbon and chemical shift is quite general: a decrease in the number of free hydrogen atoms, regardless of chain length or convolution, leading to shielding of the phosphorus nucleus.[8]

Certainly, a change in the hybridization of the directly bonded carbon has a pronounced effect on the phosphorus chemical shift. Thus, in series such as $EtP(O)Y_2$, $H_2C{=}CHP(O)Y_2$, and $HC{\equiv}CP(O)Y_2$ (Y = Cl, F, or NMe_2), the phosphorus chemical shifts move rapidly to more negative values as carbon hybridization changes from sp^3 to sp^2 to sp^9, and, in general, alkyne-phosphonic derivatives are highly shielded.[10-12] The presence of electronegative substituents on the α-carbon atom also produces shielding at phosphorus, the extent of which depends on the proximity of the substituent to the phosphorus; thus for the compounds $RO(O)(OR')_2$ shielding increases in the order R = $ClCH_2CH_2$ < $ClCH_2$ < Cl_3C < F_3C.[13]

Phosphonic acid derivatives have formed part of the more extensive examinations[14,15] of the effects of inductive and resonance transmission of negative charge on phosphorus chemical shifts, which were briefly discussed in Chapter 9. Attempts to show the relative influences of inductive and resonance effects on the chemical shift for phosphonic acids derivatives have concerned themselves largely with arylphosphonic derivatives. The phosphorus nucleus is more shielded in dialkyl 2-furylphosphonates than in dialkyl phenylphosphonates and dialkyl 2-thienylphosphonates, but a comparison of the data for these compounds with those for dialkyl (2-*N*-methylpyrrolyl)-phosphonates suggests that factors other than simple inductive effects are important in determining the chemical shift. The chemical shifts for dialkyl arylphosphonates move in a direction opposite to that expected from the electronegativity of the substituents: the more electron-donating the substituent the greater the shielding at phosphorus,[16] an observation confirmed by other authors.[17] Compatible with the much smaller consequences of a change in an ester group and already mentioned, the change in the substituent X for a series of alkyl Y-aryl (X-aryl)phosphonates has five to six times the effect produced by the same change in Y;[18,19] however, the pronounced shielding by the silyl groups in bis(trimethylsilyl) arylphosphonates seems to outweigh any electron effects that the aryl substituents might have had.[20,21] Reasonably systematic data have been provided for arylphosphonic dichlorides[22] and difluorides[23] and for arylphosphonothioic dichlorides,[24] from which it has been concluded that inductive effects have a greater influence on the phosphorus shift than do resonance effects (also concluded in Reference 16) and the correlation of the chemical shift with Hammett constants was proved once again to be in a direction opposite to that expected from the electron-withdrawing ability of the substituent. A detailed study of the compounds $ArCH{=}C(X)P(O)(OEt)_2$ showed that for the *(E)* forms with X = COOEt, the chemical shifts are 12.0 to 16.6 ppm for para-substituents and 12.2 to 13.6 ppm for meta-substituents, and for X = CN, the corresponding ranges are 9.5 to 14.7 and 9.8 to 11.7 ppm.[25] When X = NMe_2 the range for para-substituents is 14.9 to 17.8 ppm.[26] A wider variation is thus observed when resonance effects can operate; electron withdrawal produces upfield shifts, and better correlations can be observed when a dual parameter approach is adopted.

Spin-spin coupling has been observed in phosphonic acid derivatives containing other magnetic nuclei attached to phosphorus through one to four bonds. In phosphonic difluorides and selenophosphonyl compounds, the coupling constants $^1J_{PF}$ and $^1J_{P=Se}$ appear to have values similar to those found in phosphates; $^2J_{PCF}$ is 60 to 120 Hz.* $^1J_{p^4p^3}$, vary from ca. 55 to ca. 330 Hz, increasing with increasing bulk of an alkyl group attached to the tervalent phosphorus atom; $^2J_{p^4cp^4}$ values are 9 to 33 Hz. Included in Table J, but not strictly phosphonic acid derivatives, are some compounds with tetracoordinate phosphorus bonded directly to mercury and which, although thinly exemplified, are noteworthy in having very large coupling constants, viz., $^1J_{p^4(^{199}HgX)}$, being ca. 13,100 Hz when X = halogen and ca. 7600 Hz when X = P^4.

Spin-spin coupling between phosphorus and directly bonded carbon ^{13}C is, not surprisingly, receiving considerable attention. In relation to sp^3 carbon the coupling constant $^1J_{p^{13}c}$ for dimethyl (1-hydroxycycloalkyl)phosphonates has been related to the degree of steric crowding at the α-carbon and hence, in effect, to that at phosphorus itself, becoming less positive as the steric congestion increases.[27] The same interpretation has been placed on the observed differences in coupling constant for a dimethoxyphosphinyl group attached equatorially or axially to cyclohexane, the equatorial coupling being the greater by 5 to 7 Hz. Coupling between phosphorus and carbon in bicyclic systems is larger in the endo position than when oriented exo.[28] In such environments, the coupling constants $^1J_{PC}$ lie within the limits 140 and 190 Hz and are particularly enhanced for three-membered rings. For acyclic linear phosphonates, the coupling constants are normally 100 to 150 Hz, but values as high as 220 Hz have been observed for the compounds $(MeO)_2P(O)CRN_2$ (R = Et_2NSO_2 or Me_2NCO).[29]

The coupling constant $^1J_{PC}$ also depends on the carbon hybridization. For cyclohexyl-, cyclohexenyl-, and phenyl-phosphonic acids, the constants are 136, 177, and 184 Hz;[30] the constants for ethenyl- and ethynylphosphonic dichlorides are 142 and *circa* 300 Hz. In general, for ethynylphosphonic acid derivatives $^1J_{PC}$ is +220 to 300 Hz depending on the other groups attached to phosphorus; $^2J_{PC}$ is +35 to 60 Hz. Couplings to carbon-carbon double bonds depend on molecular geometry; thus for the compounds $ArCH{=}C(NMe_2)P(O)(OEt)_2$ $^1J_{PC}$ is 181 to 186 for the *(E)* compounds and 202 to 207 for

* The magnitude of direct PP couplings involving one three-coordinate phosphorus atom.

the *(Z)* isomers, and the corresponding $^2J_{PCC}$ values are 27 to 32 and 19 to 25 Hz.[26] Coupling between phosphorus and directly bonded carbon appears to decrease as oxygen is replaced by sulfur.

In considering compounds in which the phosphorus is part of a ring system, a true comparison of data from compounds with rings of differing size but essentially the same type, is possible in only a few cases from the available data. One such case comprises the 1,3,2-dioxaphosphacycloalkanes, their 1,3-dithia analogues, as their 2-oxides and 2-sulfides, with Me, tBu, or Ph substituents. Of these, compounds with 6-membered rings (sometimes those with 7-membered rings) resonate downfield of those with 5-membered rings, and those with more than 7 (8), and up to 16-ring atoms, having identical phosphorus-bonded atoms. For the dihydro-1,2-oxaphosphorines (**41**) the chemical shifts are upfield of those of the similar dihydro-1,2-oxaphospholes (**5,6**),[31-35] and the same is true for the perhydro-1,3,2-oxazaphosphorines (**13**) and the 1,3,2-oxazaphospholidines (**8,9**).[36] Chemical shifts of acyclic compounds tend to fall between those of 5-membered ring compounds, on the one hand, and those of 6-membered or larger rings, on the other, assuming identical directly bonded atoms. The chemical shifts of moderate to large ring compounds have been discussed in terms of packing factors, including bond and torsion angles at phosphorus.

A dependence of chemical shift and coupling constant on the stereochemistry at phosphorus in rings has been demonstrated, but it is not possible to reach general conclusions which might be of use in structural assignments. For the 2-benzyl-4-methyl-1,3,2-dioxaphosphorinane 2-oxides and 2-selenides, and their 2,4-dimethyl analogues, epimeric at phosphorus, those compounds with equatorial P-methyl or P-benzyl groups have relatively deshielded phosphorus nuclei, and also have numerically larger coupling constants.[38,39] However, this arrangement of larger coupling constant in association with relative deshielding which is so often applicable in phosphates (but not always; see Chapter 9) does not apply for the bicyclic oxazaphosphorines (**13**).[40] *Cis* and *trans* isomers of compounds with ten or more ring atoms can be differentiated on the basis of chemical shifts, but most pairs of diastereoisomers have identical (or at most only slightly differing) coupling constants for phosphorus to exocyclic groups.

TABLE J

Compilation of ^{31}P NMR Data of Four Coordinate Compounds Containing a P=Ch Bond and One Bond from Phosphorus to a Group IV Atom

Section J1: Derivatives of Phosphonic Acids

Rings/Connec-tivities	Formula	Structure	NMR data (δ_P[solv] $^nJ_{PZ}$ Hz)	Ref.
	P-bound atoms = HgO_2O'			
6/Cl;$(C)_2$	$C_3H_6ClHgO_3P$	$ClHg(O)P(OCH_2CH_2CH_2O)$	69.1(p) $^1J_{PHg}$13,151	41
6/P^4;$(C)_2$	$C_6H_{12}HgO_6P_2$	$Hg[(O)P(OCH_2CH_2CH_2O)]_2$	106(w) $^1J_{PHg}$7625	41
6/C′;$(C)_2$	$C_5H_9HgO_5P$	$MeCOOHg(O)P(OCH_2CH_2CH_2O)$	59.5(p) $^1J_{PHg}$13,757	41
	P-bound atoms = CBrClO′			
H_3	$CH_3BrClOP$	MeP(O)BrCl	25.8	42
	P-bound atoms = CBrOO′			
5/CC′;C′	$C_{11}H_{20}BrO_2P$	**1**; R^1 = R^3 = tBu, R^2 = R^4 = H, X = O, Y = Br	53.0	43
	P-bound atoms = CBr_2O'			
HBr_2	$CHBr_4OP$	$Br_2CHP(O)Br_2$	4.2	44
H_3	CH_3Br_2OP	$MeP(O)Br_2$	6.5	42
C_3	$C_{10}H_9Br_2OP$	$1\text{-}AdP(O)Br_2$	50.7(c) $^1J_{PC}$64.0	45
	P-bound atoms = C′ClFO′			
C'_2	C_6H_5ClFOP	PhP(O)ClF	28.7(n) $^1J_{PF}$1135	46
	P-bound atoms = CClNO′			
H_2P^4;C_2	$C_3H_8Cl_3NO_2P_2$	$Cl(Me_2N)$**P**$(O)CH_2P(O)Cl_2$	29.2 28.4 $^2J_{PP}$11.6	47
	$C_5H_{14}Cl_2N_2O_2P_2$	$H_2C[P(O)(NMe_2)Cl]_2$		47
		(*RS,RS*)	32.0	
		(*RS,SR*)	31.8	
CH_2;C_2	$C_7H_{13}ClNOP$	$EtP(O)(NEtCH_2C{\equiv}CH)Cl$	29	48
4/H_2P^4;CP^4	$C_5H_{11}Cl_2NO_2P_2$	**2**; R = tBu, X = O, Y = Cl	6.1^C 6.9^T(n)	47
6/C′HN;CC′	$C_{26}H_{22}Cl_2N_2O_2P_2$	**3**; X = O, Y = Cl	10,22^D	49
	P-bound atoms = C′ClNO′			
C'_2;CH	$C_{13}H_{21}ClNOP$	$2,4,6\text{-}Me_3C_6H_2P(O)(NH\text{-}tBu)Cl$	25.8	50
C'_2;C_2	$C_8H_{11}ClNOP$	$PhP(O)(NMe_2)Cl$	41.7(r)	50
	$C_{11}H_{17}ClNOP$	$2,4,6\text{-}Me_3C_6H_2P(O)(NMe_2)Cl$	38.7(r)	50
C′H;CC′	$C_5H_7Cl_2N_2O_2P$	**4**	17.1	51
	P-bound atoms = CClOO′			
H_2Cl;C	$C_3H_7Cl_2O_2P$	$EtO(ClCH_2)P(O)Cl$	29.5	52
H_2P^4;C	$C_3H_7Cl_3O_3P_2$	$Cl(EtO)$**P**$(O)CH_2P(O)Cl_2$	24.6 or 19.4 $^2J_{PP}$14.6	53
	$C_5H_{12}Cl_2O_4P_2$	$CH_2[P(O)(OEt)Cl]_2$	23.0, 23.2^D	53
	$C_7H_{17}ClO_5P_2$	$Cl(EtO)$**P**$(O)CH_2P(O)(OEt)_2$	15.1 or 27.2 $^2J_{PP}$8.7	53
H_2P^4;C′	$C_7H_7Cl_3O_2P_2$	$Cl(PhO)$**P**$(O)CH_2P(O)Cl_2$	15.8(b) $^2J_{PP}$15.8	53
H_3;C	$C_2H_6ClO_2P$	MeP(O)(OMe)Cl	31(g)	54
	$C_3H_8ClO_2P$	MeP(O)(OEt)Cl	39.2	52

TABLE J (continued)
Compilation of ^{31}P NMR Data of Four Coordinate Compounds Containing a P=Ch Bond and One Bond from Phosphorus to a Group IV Atom

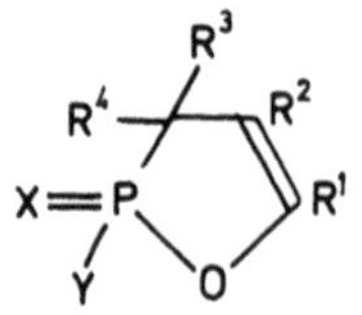

1 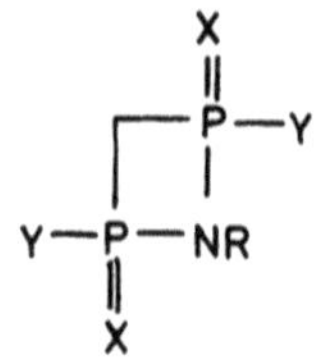 2 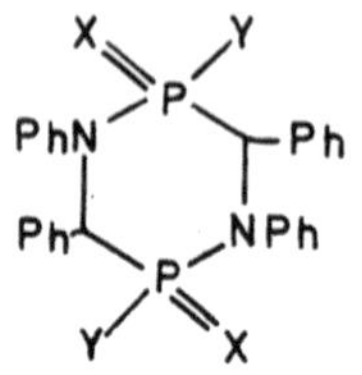 3

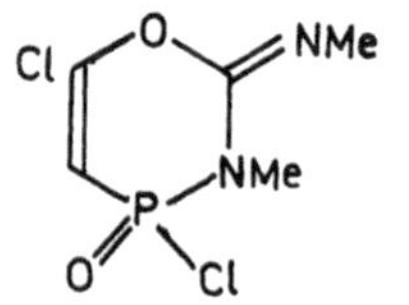

4 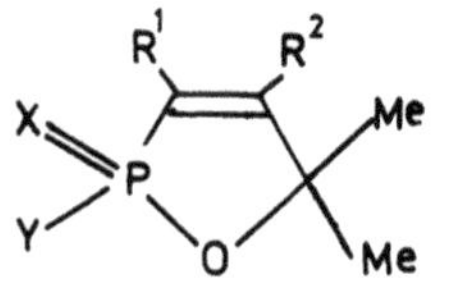 5 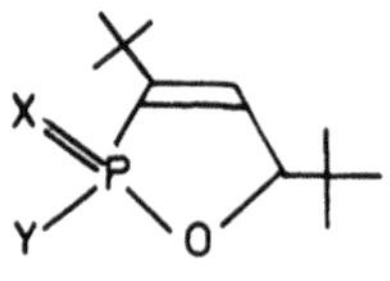 6

Rings/Connectivities	Formula	Structure	NMR data (δ_P[solv] $^nJ_{PZ}$ Hz)	Ref.
H_3;C′	$C_7H_8ClO_2P$	MeP(O)(OPh)Cl	36.4	55
CHO;C	$C_3H_7Cl_3O_3P_2$	Cl(MeO)**P**(O)CHMeOP(O)Cl_2	30.6 or 31.4 $^2J_{PP}$33	56
CH_2;C	$C_4H_9Cl_2O_2P$	$ClCH_2CH_2P(O)(OEt)Cl$	35.6	52
	$C_4H_{10}ClO_2P$	EtP(O)(OEt)Cl	45.2	55
	$C_5H_9ClNO_2P$	$NCCH_2CH_2P(O)(OEt)Cl$	38.4	52
	$C_8H_{18}ClO_2P$	BuP(O)(OBu)Cl	42.8	57
C′H_2;C	$C_9H_{12}ClO_2P$	$PhCH_2P(O)(OEt)Cl$	26.8	52
P-bound atoms = C′ClOO′				
C′H;C	$C_{18}H_{36}Cl_2O_{14}P_6$	[Cl(EtO)**P**(O)CH=CHP(O)(OEt) OP(O)(OEt)CH=]=$_2$	3(g)	58
C′$_2$;C	$C_{10}H_{14}ClO_2P$	PhP(O)(O-iBu)Cl	28.1	55
	$C_{16}H_{24}ClO_2P$	PhP(O)($OC_{10}H_{19}$)Cl[a]*(R)*$_P$	25.6	59
		(S)$_P$	26	
5/C′H;C	$C_5H_7Cl_2O_2P$	**5**; R^1 = H, R^2 = Cl, X = O, Y = Cl	33	60
5/CC′;C	$C_{11}H_{20}ClO_2P$	**6**; X = O, Y = Cl	46.2(c)	61
P-bound atoms = CCl_2O'				
Cl_2S	CCl_5O_3PS	$Cl_2P(O)CCl_2SO_2Cl$	26.5	62
F_3	CCl_2F_3OP	$Cl_2P(O)Tf$	12	63
HClS	$CHCl_4O_3PS$	$Cl_2P(O)CHClSO_2Cl$	23.2	62
HCl_2	$CHCl_4OP$	$Cl_2P(O)CHCl_2$	10.3[a](n), 34.8[b](?)	64
H_2Cl	CH_2Cl_3OP	$Cl_2P(O)CH_2Cl$	36.1[b](g) 39.8[a]	65
H_2I	CH_2Cl_2IOP	$Cl_2P(O)CH_2I$	33.9(c)	66
H_2P^4	$CH_2Cl_4O_2P_2$	$Cl_2P(O)CH_2P(O)Cl_2$	20.9 25.0[b]	69
	$C_5H_{12}Cl_2O_4P_2$	Cl_2**P**(O)$CH_2P(O)(OEt)_2$	29.4, (12.2) $^2J_{PP}$11.7	53
	$C_{13}H_{12}Cl_2O_4P_2$	Cl_2**P**(O)$CH_2P(O)(OPh)_2$	27.3, (5.0) $^2J_{PP}$14.3	53

Footnote explanations may be found following the table (p. 333).

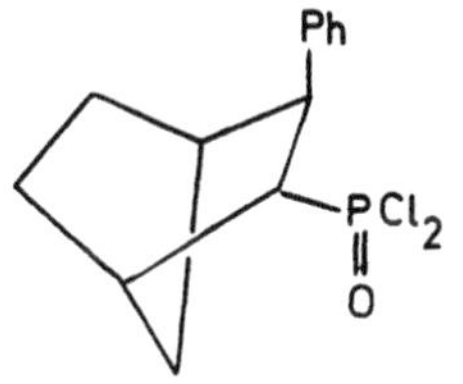

7

Rings/Connec-tivities	Formula	Structure	NMR data (δ_P[solv] $^nJ_{PZ}$ Hz)	Ref.
H_2P	$CH_2Cl_4OP_2$	$Cl_2\mathbf{P}(O)CH_2PCl_2$	23.6 $^2J_{PP}$40	67
H_2S	$CH_2Cl_3O_3PS$	$Cl_2P(O)CH_2SO_2Cl$	20.3	62
H_3	CH_3Cl_2OP	$Cl_2P(O)Me$	42.6[b](c) 43.5[c](k) 44.5[a](g)	70
CCl_2	$C_5H_6Cl_5OP$	$Cl_2P(O)CCl_2CH_2CH{=}CHCH_2Cl$	38.9[E+Z]	71
COP^4	$C_2H_3Cl_6O_4P_2$	$[Cl_2\mathbf{P}(O)]_2CMeOP(O)Cl_2$	33.5	72
CHCl	$C_2H_2Cl_5OP$	$Cl_2P(O)CHClCHCl_2$	8.4(g)	65(b)
	$C_2H_4Cl_3OP$	$Cl_2P(O)CHClMe$	46	72
CHO	$C_2H_4Cl_4O_3P_2$	$Cl_2\mathbf{P}(O)CHMeOP(O)Cl_2$	41	72
	$C_4H_8Cl_3O_2P$	$Cl_2P(O)CH(OCHClMe)Me$	48	73
CH_2	$C_2H_4Cl_3OP$	$Cl_2P(O)CH_2CH_2Cl$	42.9[a] 44.6[b]	74
	$C_2H_4Cl_4O_2P_2$	$Cl_2P(O)CH_2CH_2P(O)Cl_2$	42.5[a] 48.9[b]	75
	$C_2H_5Cl_2OP$	$Cl_2P(O)Et$	52.8[a](n) 54[b](n)	76
	$C_3H_6Cl_4O_2P_2$	$Cl_2P(O)CH_2CH_2CH_2P(O)Cl_2$	46.2	68
	$C_4H_9Cl_2OP$	$Cl_2P(O)Bu$	49.5	57
$C'H_2$	$C_3H_5Cl_2OP$	$Cl_2P(O)CH_2CH{=}CH_2$	45.7	52
	$C_4H_7Cl_2OP$	$Cl_2P(O)CH_2CMe{=}CH_2$	43(n) 42.5(g)	77
	$C_5H_8Cl_3OP$	$Cl_2P(O)CH_2CMe{=}CHCH_2Cl$	40.3[E]	78
C_2H	$C_3H_5Cl_4OP$	$Cl_2P(O)CH(CH_2Cl)_2$	34.4(g)	65(b)
	$C_3H_7Cl_2OP$	$Cl_2P(O)$-iPr	59.2	79
	$C_6H_{11}Cl_2OP$	$Cl_2P(O)$cHex	57	80
	$C_{13}H_{13}Cl_2OP$	**7**	21.5	81
CC'H	$C_4H_6Cl_3OP$	$Cl_2P(O)CHMeCH{=}CHCl$	49.0(g)	65(b)
C_3	$C_4H_9Cl_2OP$	$Cl_2P(O)$-tBu	65.6(g)	82
	$C_{10}H_9Cl_2OP$	$Cl_2P(O)$1-Ad	63.4	45

P-bound atoms = C'Cl_2O'

Rings/Connec-tivities	Formula	Structure	NMR data (δ_P[solv] $^nJ_{PZ}$ Hz)	Ref.
OO'	$C_2H_3Cl_2O_3P$	$Cl_2P(O)COOMe$	11(b)	83
C'S	$C_4H_3Cl_2OPS$	$Cl_2P(O)(\dot{C}{=}CHCH{=}CH\dot{S})$-2	22	84
C'H	$C_2H_2Cl_4O_2P_2$	$Cl_2P(O)CH{=}CHP(O)Cl_2$	23(c)	58
	$C_2H_3Cl_2OP$	$Cl_2P(O)CH{=}CH_2$	30.9 $^1J_{PC}$142.3	10
	$C_3H_3Cl_2OP$	$Cl_2P(O)CH{=}C{=}CH_2$	27	85
	$C_4H_6Cl_3OPS$	$Cl_2P(O)CH{=}C(CH_2Cl)SMe$	19[E] 19[Z]	33
	$C_4H_7Cl_2OP$	$Cl_2P(O)CH{=}CMe_2$	28.5(g, n)	77
	$C_4H_7Cl_2O_2P$	$Cl_2P(O)CH{=}CHOEt$	35	86
	$C_5H_7Cl_2OP$	$Cl_2P(O)CH_A{=}CHCMe{=}CH_2$	27.0[Z] $^2J_{PH_A}$36.2	34
	$C_5H_8Cl_3OP$	$Cl_2P(O)CH{=}CMe(CH_2CH_2Cl)$	28.6[E]	78
	$C_6H_9Cl_2OP$	$Cl_2P(O)CH_A{=}CHCMe{=}CHMe$	39.4 $^2J_{PH_A}$37[Z,Z] 41.5 $^2J_{PH_A}$35[Z,E]	34
	$C_6H_{11}Cl_2OP$	$Cl_2P(O)CH{=}CHBu$	30.7	87
	$C_8H_7Cl_2OP$	$Cl_2P(O)CH{=}CHOPh$	30.6	88
	$C_9H_8Cl_3OPS$	$Cl_2P(O)CH{=}C(CH_2Cl)SPh$	25[E]	60
	$C_{14}H_{15}Cl_2OPS$	$Cl_2P(O)CH{=}C(SPh)C_6H_9$-1[b]	40	60

TABLE J (continued)
Compilation of ^{31}P NMR Data of Four Coordinate Compounds Containing a P=Ch Bond and One Bond from Phosphorus to a Group IV Atom

Rings/Connec-tivities	Formula	Structure	NMR data (δ_P[solv] $^nJ_{PZ}$ Hz)	Ref.
CC′	$C_3H_4Cl_3OP$	$Cl_2P(O)C(CH_2Cl)=CH_2$	24.8(g)	89
	$C_4H_6Cl_3OP$	$Cl_2P(O)C(CH_2Cl)=CHOMe$	36	90
	$C_5H_7Cl_4OP$	$Cl_2P(O)C(CMe_2Cl)=CHCl$	26.4[E] 25.2[Z]	89
	$C_6H_9Cl_2OP$	$Cl_2P(O)C_6H_9$-1[b]	36	91
	$C_9H_{13}Cl_4OP$	$Cl_2P(O)C(\text{-tBu})=CClCMe=CHCl$	34.7	92
	$C_9H_{14}Cl_3OP$	$Cl_2P(O)C(\text{-tBu})=C=C(CH_2Cl)Me$	33.0	92
C′$_2$	$C_4H_3Cl_4OP$	$Cl_2P(O)C(CCl=CHCl)=CH_2$	24.7[E](n)	93
	$C_6H_4Cl_3OP$	$Cl_2P(O)C_6H_4Cl$-4	32.0	1
	$C_6H_5Cl_2OP$	$Cl_2P(O)Ph$	34[a](g) 36[b]	94
	$C_7H_4Cl_2F_3OP$	$Cl_2P(O)C_6H_4Tf$-3	30.7	22
	$C_7H_4Cl_2F_3OP$	$Cl_2P(O)C_6H_4Tf$-4	32.7	22
	$C_7H_7Cl_2OP$	$Cl_2P(O)Tol$-2	35.5(r)	50
	$C_7H_7Cl_2OP$	$Cl_2P(O)Tol$-4	34.0	1
	$C_7H_7Cl_2O_2P$	$Cl_2P(O)C_6H_4OMe$-4	32.0	1
	$C_8H_{10}Cl_2OP$	$Cl_2P(O)C_6H_4NMe_2$-4	37.6	22
	$C_{15}H_{23}Cl_2OP$	$Cl_2P(O)C_6H_2(\text{-tPr})_3$-2,4,6	33.7	50
	$C_{18}H_{29}Cl_2OP$	$Cl_2P(O)C_6H_2(\text{-tBu})_3$-2,4,6	33.9	50
	P-bound atoms = C″Cl$_2$O′			
C″	$C_3H_3Cl_2OP$	$Cl_2P(O)C\equiv CMe$	−12.3[a] −12.8[b]	95
	$C_8H_5Cl_2OP$	$Cl_2P(O)C\equiv CPh$	−12.0	11
	P-bound atoms = CFNO′			
H$_3$;C$_2$	C_3H_9FNOP	$MeP(O)(NMe_2)F$	30 $^1J_{PF}$1000	96
C$_3$;C$_2$	$C_6H_{15}FNOP$	$tBuP(O)(NMe_2)F$	42.5(g)	82
	P-bound atoms = CFOO′			
Cl$_3$;H	$CHCl_3FO_2P$	$Cl_3CP(O)(OH)F$	0.5 $^1J_{PF}$1070	97
H$_3$;H	CH_4FO_2P	$MeP(O)(OH)F$	30.2 $^1J_{PF}$1000	97
C′HN′;C	$C_{10}H_{11}FNO_3P$	PhCH(NCO)P(O)(OEt)F *(S)*	15.3 $^1J_{PF}$1115	98
CH$_2$;C	$C_4H_{10}FO_2P$	$EtP(O)(OEt)F$	31 $^1J_{PF}$1055	99
C$_3$;C	$C_5H_{12}FO_2P$	$tBuP(O)(OMe)F$	36.6(a)	82
	P-bound atoms = C′FOO′			
C′$_2$;H	$C_6H_6FO_2P$	$PhP(O)(OH)F$	17.6 $^1J_{PF}$1010	97
C′$_2$;C	$C_7H_5F_5NO_2P$	$C_5F_4NP(O)(OEt)F$[C]	−2.8 $^1J_{PF}$−1050	100
	P-bound atoms = CF$_2$O′			
HBr$_2$	$CHBr_2F_2OP$	$F_2P(O)CHBr_2$	0.7	44
H$_2$Cl	CH_2ClF_2OP	$F_2P(O)CH_2Cl$	12 $^1J_{PF}$1164	101
H$_2$P^4	$CH_2F_4O_2P_2$	$F_2P(O)CH_2P(O)F_2$	7.7(m) $^2J_{PP}$6.2 $^1J_{PF}$−1117, +9.7	102
H$_2$S	$CH_2ClF_2O_3PS$	$F_2P(O)CH_2SO_2Cl$	0.0 $^1J_{PF}$1142	62
H$_3$	CH_3F_2OP	$F_2P(O)Me$	26.0[a] 26.8[b](n) $^1J_{PF}$1104, 1163	103

Rings/Connec-tivities	Formula	Structure	NMR data (δ_P[solv] $^nJ_{PZ}$ Hz)	Ref.
CH_2	$C_2H_4F_4O_2P_2$	$F_2P(O)CH_2CH_2P(O)F_2$	20.2 $^1J_{PF}-1130$	102
	$C_2H_5F_2OP$	$F_2P(O)Et$	29.2 $^1J_{PF}1130$	11
C_2H	$C_3H_7F_2OP$	$F_2P(O)$-iPr	29.5	8
	$C_6H_{11}F_2OP$	$F_2P(O)$cHex	26.2 $^1J_{PF}1153$	46
C_3	$C_4H_9F_2OP$	$F_2P(O)$-tBu	31.3/31.8 $^1J_{PF}1169(1195)$	104

P-bound atoms = $C'F_2O'$

Rings/Connec-tivities	Formula	Structure	NMR data (δ_P[solv] $^nJ_{PZ}$ Hz)	Ref.
C'Cl	$C_8H_5Cl_2F_2OP$	$F_2P(O)CCl{=}CClPh$	−5.4 $^1J_{PF}1110$	11
C'H	$C_2H_2F_4O_2P_2$	$F_2P(O)CH{=}CHP(O)F_2$	9.0(m) $^1J_{PF}-1107.5$	102
	$C_2H_3F_2OP$	$F_2P(O)CH{=}CH_2$	10.6 $^1J_{PC}196$ $^2J_{Pc}2.8$, $^1J_{PF}1079$	10
	$C_3H_3F_2OP$	$F_2P(O)CH{=}C{=}CH_2$	8.1 $^1J_{PF}-1090$	105
	$C_3H_4ClF_2OP$	$F_2P(O)CH{=}CMeCl$	4.5^E $^1J_{PF}-1082$ $^2J_{PH}11.9$; 3.6^Z $^1J_{PF}-1082$ $^2J_{PH}16.1$	105
	$C_8H_6ClF_2OP$	$F_2P(O)CH{=}CClPh$	5.0^Z $^1J_{PF}-1087$	105
	$C_8H_7F_2OP$	$F_2P(O)CH{=}CHPh$	12^E $^1J_{PF}1082$	101
C'_2	$C_6H_5F_2OP$	$F_2P(O)Ph$	10.8^a(b) $^1J_{PF}1103$ 11.4^b $^1J_{PF}1115$	106
	$C_7H_4F_5OP$	$F_2P(O)C_6H_4Tf$-4	7.7(b) $^1J_{PF}1107$	23
	$C_7H_7F_2OP$	$F_2P(O)C_6H_4OMe$-4	12.3(b) $^1J_{PF}1091$	23

P-bound atoms = $C''F_2O'$

Rings/Connec-tivities	Formula	Structure	NMR data (δ_P[solv] $^nJ_{PZ}$ Hz)	Ref.
C''	C_2HF_2OP	$F_2P(O)C{\equiv}CH$	−21.9 $^1J_{PF}982$	10
	$C_3H_3F_2OP$	$F_2P(O)C{\equiv}CMe$	−20.0 $^1J_{PF}1010$	11
	$C_8H_5F_2OP$	$F_2P(O)C{\equiv}CPh$	−18.7 $^1J_{PF}1010$	11

P-bound atoms = C'INO'

Rings/Connec-tivities	Formula	Structure	NMR data (δ_P[solv] $^nJ_{PZ}$ Hz)	Ref.
C'_2;CC'	$C_{13}H_{13}INOP$	PhP(O)(NMePh)I	31	107

P-bound atoms = CNNO'

Rings/Connec-tivities	Formula	Structure	NMR data (δ_P[solv] $^nJ_{PZ}$ Hz)	Ref.
H_2Cl;C'H;C	$C_8H_{11}ClNO_2P$	$ClCH_2P(O)(NHPh)OMe$	19.6(j)	66
H_3:CH;C	$C_4H_{12}NO_2P$	MeP(O)(NHMe)(OC_nH_{2n+1}) n = 2—4	32	108
	$C_9H_{14}NO_2P$	$MeP(O)(NHCH_2Ph)OMe$	34.8(c) $^1J_{PC}132$	109
	$C_9H_{22}NO_2P$	MeP(O)(NHBu)OBu	28.3	110
H_3;C'Si;C	$C_{12}H_{22}NO_2PSi$	MeP(O)(OEt)NPhTms	30	111
H_3;C'H;C'	$C_{13}H_{13}ClNO_3P$	$MeP(O)(OPh)NHC_6H_4Cl$-2	24.8	110
	$C_{13}H_{13}ClNO_3P$	$MeP(O)(OPh)NHC_6H_4Cl$-3	29.6	110
	$C_{13}H_{13}ClNO_3P$	$MeP(O)(OPh)NHC_6H_4Cl$-4	24.9	110
	$C_{13}H_{14}NO_3P$	MeP(O)(OPh)NHPh	25.3	110
	$C_{14}H_{16}NO_2P$	MeP(O)(OPh)NHTol-2	24.5	110
	$C_{14}H_{16}NO_2P$	MeP(O)(OPh)NHTol-3	25.8	110
	$C_{14}H_{16}NO_2P$	MeP(O)(OPh)NHTol-4	25.4	110
CH_2;C'H;C	$C_{10}H_{16}NO_2P$	EtP(O)(OEt)NHPh *(S)*	34(b)	112
C_3;C'H;C	$C_{10}H_{22}NO_4P$	tBuP(O)(OMe)NHCOO-tBu	36.1(b)	113
	$C_{12}H_{19}N_2O_3P$	tBuP(O)(OMe)NHCONHPh	37.5(c)	113
	$C_{19}H_{34}NO_2P$	tBuP(O)(OMe)NHC_6H_3(-tBu)$_2$-3,5	37	114
5/H_3;C_2;C	$C_{11}H_{22}NO_6P$	**8**; X = O', Y = Me	40.8(c)	36
	$C_{11}H_{22}NO_6P$	**9**; X = O', Y = Me	40.8(c)	36
5/H_3;CC';C'	$C_{11}H_{22}NO_2P$	**H3**, R = tBu, X = O, Y = Me	43.4	115

TABLE J (continued)
Compilation of ^{31}P NMR Data of Four Coordinate Compounds Containing a P=Ch Bond and One Bond from Phosphorus to a Group IV Atom

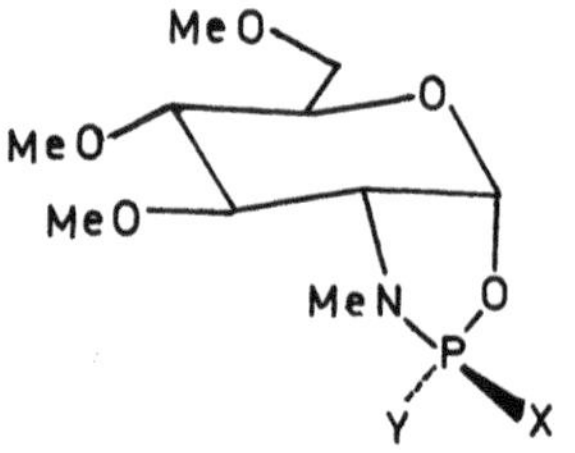

8 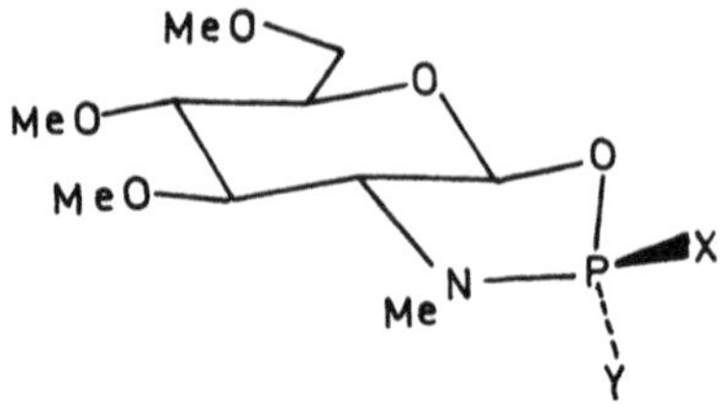 **9** 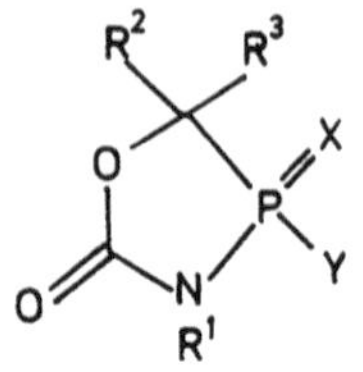 **10**

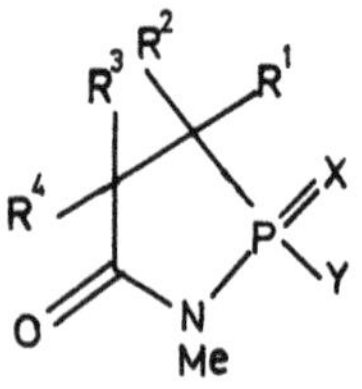

11 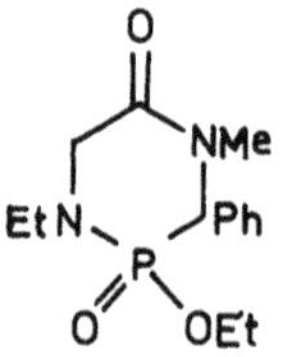 **12** 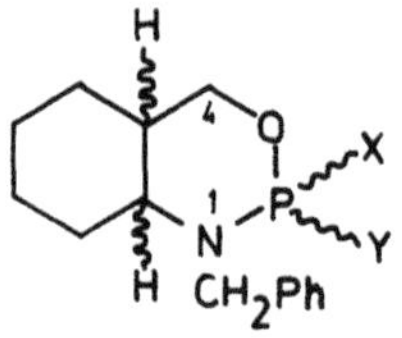 **13**

Rings/Connectivities	Formula	Structure	NMR data (δ_P[solv] $^nJ_{PZ}$ Hz)	Ref.
5/CC'O;CC';C	$C_7H_{12}NO_5P$	**10;** $R^1 = R^2 =$ Me, $R^3 =$ Ac, X = O, Y = OMe	28	116
5/C₂H;CC';C	$C_{12}H_{20}NO_7P$	**11;** $R^1 =$ H, $R^2 =$ Me, $R^3 = R^4 =$ COOEt, X = O, Y = OMe	36 39[D]	117
6/C'HN;C₂;C	$C_{14}H_{21}N_2O_3P$	**12**	16.2	118
6/C'₂;C₂;C	$C_{20}H_{24}NO_2P$	**13;** (2*RS*,4a*RS*,8a*SR*)	13.5	40
		(2*SR*,4a*RS*,8a*SR*)	16.0	
		(2*RS*,4a*RS*,8a*RS*)	17.0	
		(2*SR*,4a*RS*,8a*RS*)	16.4	
	P-bound atoms = CNO'₂			
H₃;CH	$C_8H_{11}NO_2P$	MeP(O)(NHCH₂Ph)O⁻	26.6(a)	109
	P-bound atoms = CN'OO'			
H₃;C';C	$C_6H_{14}NO_3P$	MeP(O)(OMe)N=C(OEt)Me	30.1	119
C₃;C';C	$C_6H_{12}NO_2PS$	tBuP(O)(OMe)NCS *(R)*	28.1(n)	113
	$C_6H_{12}NO_3P$	tBuP(O)(OMe)NCO *(S)*	30.7(b)	113
5/C'O₂;C'₂;C	$C_8H_{14}NO_7P$	**14;** $R^1 =$ Me, $R^2 =$ MeO, Y = OEt	38.2	120
5/C'O₂;C'₂;C	$C_{15}H_{21}N_2O_6P$	**14;** $R^1 =$ Et, $R^2 =$ NHPh, Y = OEt	43.2(g)	120
5/C₂H;C';C	$C_{12}H_{20}NO_6PS$	**15;** $R^1 =$ H, $R^2 =$ Me, $R^3 = R^4 =$ COOEt, $R^5 =$ MeS, X = O, Y = MeO	54 57[D]	117

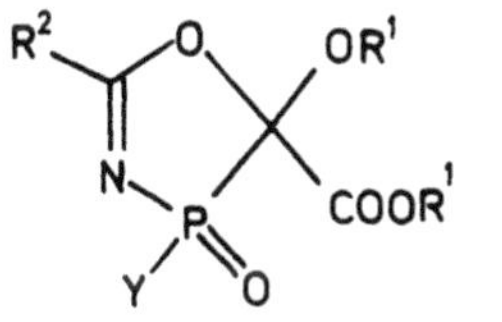

14

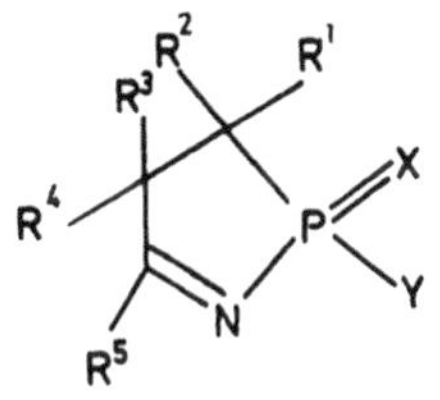

15

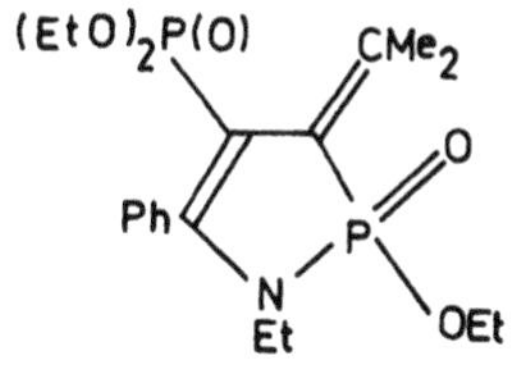

16

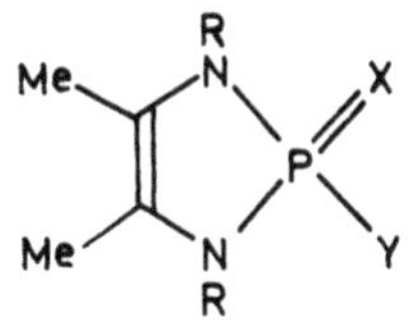

17

Rings/Connectivities	Formula	Structure	NMR data (δ_P[solv] $^nJ_{PZ}$ Hz)	Ref.
5/C_3;C′;C	$C_{11}H_{17}N_2O_5P$	**15**; R^1 = R^2 = Me, R^3 = CN, R^4 = COOEt, R^5 = MeO, X = O, Y = MeO	54.5	121

P-bound atoms = C′NOO′

Rings/Connectivities	Formula	Structure	NMR data (δ_P[solv] $^nJ_{PZ}$ Hz)	Ref.
C'_2;CH;C	$C_{14}H_{24}NO_2P$	PhP(O)(NHt-Bu)O-tBu	15.3(r)	50
	$C_{19}H_{34}NO_2P$	3,5-$(tBu)_2C_6H_3P(O)$(NH-tBu)OMe	22.6	114
5/C'_2;C_2;C	$C_{16}H_{18}NO_2P$	**H12**; X, Y = O′, OPh		122
		(2*R*,4*S*,5*S*)	31.0(c)	
		(2*S*,4*S*,5*S*)	35.8(c)	
5/C'_2;CC′;C	$C_{20}H_{31}NO_5P_2$	**16**	20 34[D]	123
5/C'_2;C'_2;C	$C_{18}H_{14}NO_2P$	**H15**; X = O, R = Y = Ph	29.4	124

P-bound atoms = CN_2O'

Rings/Connectivities	Formula	Structure	NMR data (δ_P[solv] $^nJ_{PZ}$ Hz)	Ref.
Cl_3;$(C_2)_2$	$C_9H_{20}Cl_3N_2OP$	$Cl_3CP(O)(NEt_2)_2$	23.2(n)	125
HBr_2;$(C_2)_2$	$C_5H_{13}Br_2N_2OP$	$Br_2CHP(O)(NMe_2)_2$	24.2	44
HO_2;$(C_2)_2$	$C_{13}H_{31}N_2O_3P$	$(EtO)_2CHP(O)(NEt_2)_2$	23.3	126
H_2P^4;C_2,CH	$C_{13}H_{34}N_4O_2P_2$	$[(tBuNH)(Me_2N)P(O)]_2CH_2$	22.9	47
H_2Si;$(C_2)_2$	$C_{12}H_{31}N_2OPSi$	$TmsCH_2P(O)(NEt_2)_2$	35.4(c) $^1J_{PC}$106.4 $^2J_{PNC}$4.4 $^2J_{PCH}$17.2	127
H_3;H_2,CH	$C_8H_{13}N_2OP$	$MeP(O)(NHCH_2Ph)NH_2$	36.2(w) $^1J_{PC}$112(w) 30.6(c)	109
H_3;$(CH)_2$	$C_9H_{23}N_2OP$	$MeP(O)(NH\text{-}tBu)_2$	21.6(r)	128
CH_2;$(CH)_2$	$C_{10}H_{25}N_2OP$	$EtP(O)(NH\text{-}tBu)_2$	27.8(c)	128
CH_2;$(C_2)_2$	$C_{12}H_{29}N_2OP$	$BuP(O)(NEt_2)_2$	43.5	129
4/H_2P^4;CP^4,C_2	$C_9H_{23}N_3O_2P_2$	**2**; R = tBu, X = O, Y = NMe_2	10.6(n)	47
5/CH_2;$(CC')_2$	$C_{14}H_{29}N_2OP$	**17**; R = Bu, X = O, Y = Et	39.5	130

TABLE J (continued)
Compilation of ^{31}P NMR Data of Four Coordinate Compounds Containing a P=Ch Bond and One Bond from Phosphorus to a Group IV Atom

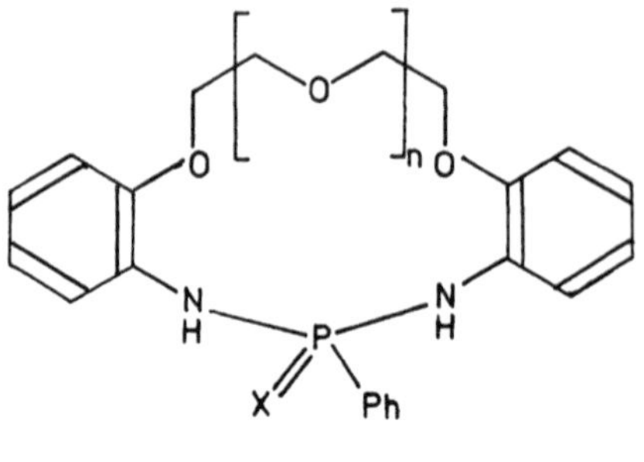

18

Rings/Connec-tivities	Formula	Structure	NMR data (δ_P[solv] $^nJ_{PZ}$ Hz)	Ref.
	P-bound atoms = C′N$_2$O′			
OO′;(C′H)$_2$	$C_{15}H_{17}N_2O_3P$	EtOOCP(O)(NHPh)$_2$	−7.6(k)	83
OO′;(C$_2$)$_2$	$C_7H_{17}N_2O_3P$	EtOOCP(O)(NMe$_2$)$_2$	13.8(b)	83
C′H;(C$_2$)$_2$	$C_6H_{15}N_2OP$	H$_2$C=CHP(O)(NMe$_2$)$_2$	25(a)	131
	$C_7H_{15}N_2OP$	H$_2$C=C=CHP(O)(NMe$_2$)$_2$	23.4	85
C′$_2$;HO,C′H	$C_{12}H_{13}N_2O_2P$	PhP(O)(NHPh)NHOH	21.2(y)	132
C′$_2$;(CH)$_2$	$C_8H_{13}N_2OP$	PhP(O)(NHMe)$_2$	23.6(r)	133
	$C_{20}H_{37}N_2OP$	2,4,6-(tBu)$_3$C$_6$H$_2$P(O)(NHMe)$_2$	24.4(c)	50
C′$_2$;(C′H)$_2$	$C_{18}H_{17}N_2OP$	PhP(O)(NHPh)$_2$	8.4(v)	134
C′$_2$;(C$_2$)$_2$	$C_{10}H_{17}N_2OP$	PhP(O)(NMe$_2$)$_2$	28.1[a](b) 28.9[b](r)	135
	$C_{13}H_{25}N_2O_2PSi$	2-TmsOC$_6$H$_4$P(O)(NMe$_2$)$_2$	26.4	136
	$C_{14}H_{25}N_2OP$	PhP(O)(NEt)$_2$	38.3(n)	134
	$C_{16}H_{30}N_3OP$	4-EtNHC$_6$H$_4$P(O)(NEt$_2$)$_2$	29.8 $^1J_{PC}$162.8	137
C′$_2$;(C′$_2$)$_2$	$C_{30}H_{25}N_2OP$	PhP(O)(NPh$_2$)$_2$	14.8(ze)	134
17/C′$_2$;(C′H)$_2$	$C_{24}H_{27}N_2O_5P$	**18**; X = O, n = 2	10.7	138
	P-bound atoms = C″N$_2$O′			
C″;(C$_2$)$_2$	$C_6H_{13}N_2OP$	HC≡CP(O)(NMe$_2$)$_2$	7.4	12
	$C_7H_{15}N_2OP$	MeC≡CP(O)(NMe$_2$)$_2$	7.6	12
	P-bound atoms = COO′P			
H$_3$;C;CP^4	$C_{12}H_{29}O_4P_3$	tBuP[**P**(O)(O-iPr)Me]$_2$	56.3 56.75[D] $^1J_{PP}$223	139
C$_2$H;C;CP^4	$C_{17}H_{39}O_4P_3$	iPrP[**P**(O)(OBu)(iPr)]$_2$	68.5 $^1J_{PP}$275	140
C$_3$;C;CP^4	$C_{18}H_{33}O_4P_3$	tBuP[**P**(O)(OMe)tBu]$_2$	71.0 71.2[D], $^1J_{PP}$329	139
	P-bound atoms = C′OO′P			
C′$_2$;C;C′P^4	$C_{20}H_{21}O_4P_3$	PhP[**P**(O)(OMe)Ph]$_2$	45.7 $^1J_{PP}$210	141
	$C_{26}H_{33}O_4P_3$	PhP[**P**(O)(OBu)Ph]$_2$	40.3 $^1J_{PP}$216	140
C′$_2$;C;C′$_2$	$C_{20}H_{20}O_2P_2$	Ph(EtO)**P**(O)PPh$_2$	47.0 $^1J_{PP}$205	141
	P-bound atoms = CO$_2$O′			
BrF$_2$;(C)$_2$	$C_5H_{10}BrF_2O_3P$	BrF$_2$CP(O)(OEt)$_2$	−1.1 $^2J_{PF}$92	142
Br$_2$P^4;(C)$_2$	$C_9H_{20}Br_2O_6P_2$	(EtO)$_2$P(O)CBr$_2$P(O)(OEt)$_2$	8.5	143
	$C_{12}H_{25}Br_4O_8P_3$	[(EtO)$_2$**P**(O)CBr$_2$]$_2$P(O)OEt	9.7	64(a)

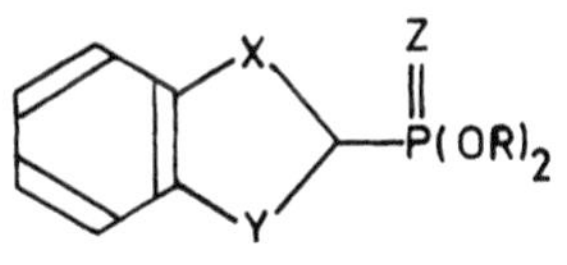

19

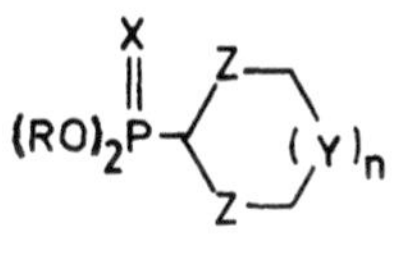

20

Rings/Connec-tivities	Formula	Structure	NMR data (δ_P[solv] $^nJ_{PZ}$ Hz)	Ref.
	$C_{13}H_{28}Br_2O_6P_2$	$(iPrO)_2P(O)CBr_2P(O)(O\text{-}iPr)_2$	6.5	143
$Cl_2P^4;(H)_2$	$CH_4Cl_2O_6P_2$	$(HO)_2P(O)CCl_2P(O)(OH)_2$	7.9	64(a)
$Cl_2P^4;(C)_2$	$C_5H_{12}Cl_2O_6P_2$	$(MeO)_2P(O)CCl_2P(O)(OMe)_2$	10.0	143
	$C_9H_{20}Cl_2O_6P_2$	$(EtO)_2P(O)CCl_2P(O)(OEt)_2$	8.5	143
	$C_{13}H_{28}Cl_2O_6P_2$	$(iPrO)_2P(O)CCl_2P(O)(O\text{-}iPr)_2$	6.5(c)	143
$Cl_3;(C)_2$	$C_3H_6Cl_3O_3P$	$Cl_3CP(O)(OMe)_2$	7.0	144
	$C_5H_4Cl_3F_6O_3P$	$Cl_3CP(O)(OCH_2Tf)_2$	5.7	145
	$C_5H_{10}Cl_3O_3P$	$Cl_3CP(O)(OEt)_2$	5.0	144
$FSi_2;(C)_2$	$C_{13}H_{32}FO_3PSi_2$	$(Tms)_2CFP(O)(O\text{-}iPr)_2$	22.8	146
$F_2P^4;(Si)_2$	$C_{13}H_{36}F_2O_6P_2Si_4$	$(TmsO)_2P(O)CF_2P(O)(OTms)_2$	−15.0 $^2J_{PF}$89.0	147
$F_2P^4;(H)_2$	$CH_4F_2O_6P_2$	$(HO)_2P(O)CF_2P(O)(OH)_2$	3.7 $^2J_{PF}$86	147
$F_2P^4;(C)_2$	$C_9H_{20}F_2O_6P_2$	$(EtO)_2P(O)CF_2P(O)(OEt_2$	3.6[a](c) 4.3[b](c) $^2J_{PF}$86	148
$F_3;(H)_2$	$CH_2F_3O_3P$	$TfP(O)(OH)_2$	6.6	149
$F_3;(C)_2$	$C_3H_6F_3O_3P$	$TfP(O)(OMe)_2$	−1.7	150
	$C_5H_{10}F_3O_3P$	$TfP(O)(OEt)_2$	−2.5(c)	151
$F_3;C,C'$	$C_4H_6F_3O_3P$	$TfP(O)(OMe)OCH{=}CH_2$	−6.1	152
$I_2P^4;(C)_2$	$C_{13}H_{28}I_2O_6P_2$	$(iPrO)_2P(O)CI_2P(O)(O\text{-}iPr)_2$	10.5	143
$NP^4_2;(C)_2$	$C_{14}H_{36}NO_9P_3$	$Me_2NC[P(O)(OEt)_2]_3$	15.8(c)	153
$HBrP^4;(C)_2$	$C_{13}H_{29}BrO_6P_2$	$(iPrO)_2P(O)CHBrP(O)(O\text{-}iPr)_2$	12.0	143
	$C_{15}H_{33}Br_3O_{10}P_4$	$[(EtO)_2\mathbf{P}(O)CHBr]_3P(O)$	17.9	64(a)
$HBr_2;(C)_2$	$C_3H_7Br_2O_3P$	$Br_2CHP(O)(OMe)_2$	12.4	44
	$C_5H_5Br_2Cl_6O_3P$	$Br_2CHP(O)(OCH_2CCl_3)_2$	9.8	44
	$C_5H_5Br_2F_6O_3P$	$Br_2CHP(O)(OCH_2Tf)_2$	12.4	44
	$C_5H_5Br_8O_3P$	$Br_2CHP(O)(OCH_2CBr_3)_2$	8.3	44
	$C_5H_{11}Br_2O_3P$	$Br_2CHP(O)(OEt)_2$	9.7	44
$HClF;(H)_2$	CH_3ClFO_3P	$FClCHP(O)(OH)_2$	6.2(y) $^2J_{PF}$77.8	154
$HClP^4;(C)_2$	$C_5H_{13}ClO_6P_2$	$(MeO)_2P(O)CHClP(O)(OMe)_2$	36.8(b) $^1J_{PC}$141.9	155
	$C_{13}H_{29}ClO_6P_2$	$(iPrO)_2P(O)CHClP(O)(O\text{-}iPr)_2$	11.5	143
	$C_{15}H_{33}Cl_3O_{10}P_4$	$[(EtO)_2P(O)CHCl]_3P(O)$	18.9	64(a)
$HCl_2;(C)_2$	$C_5H_{11}Cl_2O_3P$	$Cl_2CHP(O)(OEt)_2$	10.5	156
$HFP^4;(Si)_2$	$C_{13}H_{37}FO_6P_2Si_4$	$(TmsO)_2P(O)CHFP(O)(OTms)_2$	−7.3 $^2J_{PF}$67	147
$HFP^4;(H)_2$	$CH_5FO_6P_2$	$(HO)_2P(O)CHFP(O)(OH)_2$	10.5 $^2J_{PF}$64	147
$HFP^4;(C)_2$	$C_9H_{21}FO_6P_2$	$(EtO)_2P(O)CHFP(O)(OEt)_2$	12.3(n) $^2J_{PF}$62	147
$HF_2;(C)_2$	$C_5H_{11}F_2O_3P$	$F_2CHP(O)(OEt)_2$	4.1 $^2J_{PF}$91	147
$HIP^4;(C)_2$	$C_{13}H_{29}IO_6P_2$	$(iPrO)_2P(O)CHIP(O)(O\text{-}iPr)_2$	13.5 $^2J_{CH}$17	143
$HNP^4;(C)_2$	$C_{11}H_{27}NO_6P_2$	$Me_2NCH[P(O)(OEt)_2]_2$	6.8(c)	153
$HOS;(C)_2$	$C_{11}H_{15}O_4PS$	**19;** X = Z = O, Y = S, R = Et	12.6	157
$HO_2;(Si)_2$	$C_{11}H_{29}O_5PSi_2$	$(EtO)_2CHP(O)(OTms)_2$	−4.4	126
$HO_2;(C)_2$	$C_9H_{21}O_5P$	$(EtO)_2CHP(O)(OEt)_2$	11.3	126
$HP^4S;(C)_2$	$C_{15}H_{26}O_6P_2S$	$[(EtO)_2P(O)]_2CHSPh$	16.9(c)	158
$HS_2;(C)_2$	$C_5H_{11}O_3PS_3$	**20;** R = Me, X = O, Z = Y = S, n = 1	17.9(b) $^2J_{PH}$20.5	159
	$C_5H_{13}O_3PS_2$	$(MeS)_2CHP(O)(OMe)_2$	22.7(c) $^1J_{PC}$155	160
	$C_6H_{13}O_3PS_2$	**20;** R = Me, X = O, Z = S, Y = CH_2, n = 1	23.2(c) $^1J_{PC}$158.1 21.4(b) $^1J_{PC}$158.4	161
	$C_7H_{15}O_3PS_2$	**20;** R = Et, X = O, Z = S, n = O	22.0(c) $^1J_{PC}$163.3	160
	$C_7H_{17}O_3PS_2$	$(MeS)_2CHP(O)(OEt)_2$	20.0(c)	158

TABLE J (continued)
Compilation of ^{31}P NMR Data of Four Coordinate Compounds Containing a P=Ch Bond and One Bond from Phosphorus to a Group IV Atom

Rings/Connectivities	Formula	Structure	NMR data (δ_P[solv] $^nJ_{PZ}$ Hz)	Ref.
	$C_8H_{17}O_3PS_2$	**20**; R = Et, X = O, Z = S, Y = CH_2, n = 1	18.8(c) $^1J_{PC}$158.2	160
	$C_{11}H_{15}O_3PS_2$	**19**; R = Et, Z = O, X = Y = S	16.2	157
	$C_{17}H_{21}O_3PS_2$	$(PhS)_2CHP(O)(OEt)_2$	18.2(k)	158
$HSi_2;(C)_2$	$C_{11}H_{29}O_3PSi_2$	$(Tms)_2CHP(O)(OEt)_2$	34.3(c)	162
$H_2Cl;(H)_2$	CH_4ClO_3P	$ClCH_2P(O)(OH)_2$	17.8(w)	64(b)
$H_2Cl;(C)_2$	$C_5H_{12}ClO_3P$	$ClCH_2P(O)(OEt)_2$	18.1(n)	64(b)
$H_2Cl;(C')_2$	$C_{13}H_{12}ClO_3P$	$ClCH_2P(O)(OPh)_2$	12.3(n)	64(b)
$H_2F;(C)_2$	$C_5H_{12}FO_3P$	$FCH_2P(O)(OEt)_2$	16.3[a](c) $^2J_{PF}$63.3 17.2[b](c) $^2J_{PF}$62.3	163
$H_2I;(C)_2$	$C_5H_{12}IO_3P$	$ICH_2P(O)(OEt)_2$	20.4(c)	66
$H_2N;(H)_2$	CH_6NO_3P	$H_2NCH_2P(O)(OH)_2$	10.1(w pH 5.5)	164
	$C_2H_9NO_6P_2$	$HN[CH_2P(O)(OH)_2]_2$	9.8(w)	64(b)
	$C_3H_8NO_5P$	$HOOCCH_2NHCH_2P(O)(OH)_2$	8.4 (w pH 3.5/9.3)	164
	$C_3H_{12}NO_9P_3$	$N[CH_2P(O)(OH)_2]_3$	9.5(w)	64(b)
	$C_4H_{14}N_2O_6P_2$	$H_2NC_2H_4N[CH_2P(O)(OH)_2]_2$	8.1 (w pH 1.6) 18.75 (w pH 4.75)	165
	$C_8H_{16}NO_5P$	$(HO)_2P(O)$ $CH_2NH\dot{C}(CH_2CH_2CH_2CH_2\dot{C}H_2COOH$	16.5(w)	166
$H_2N;C,H$	$C_5H_{14}NO_4P$	$Me_2NCH_2P(O)(OH)OCH_2CH_2OH$	9(c)	167
$H_2N;(C)_2$	$C_3H_8NO_5P$	$O_2NCH_2P(O)(OMe)_2$	10.2	168
	$C_4H_{12}NO_3P$	$MeNHCH_2P(O)(OMe)_2$	27.3(c)	66
	$C_4H_{12}NO_5PS$	$(MeO)_2P(O)CH_2NHSO_2Me$	24.7(c)	169
	$C_5H_{12}NO_4P$	$(MeO)_2P(O)CH_2NHAc$	25.6(c)	169
	$C_7H_{16}NO_5P$	$(MeO)_2P(O)CH_2NHCH_2COOEt$	26(n)	170
	$C_9H_{22}NO_3P$	$Et_2NCH_2P(O)(OEt)_2$	25	171
	$C_{10}H_{25}NO_6P_2$	$HN[CH_2P(O)(OEt)_2]_2$	23.3(n)	64(b)
	$C_{15}H_{36}NO_9P_3$	$N[CH_2P(O)(OEt)_2]_3$	22.6(n)	64(b)
$H_2O;(H)_2$	CH_5O_4P	$HOCH_2P(O)(OH)_2$	20.3 (w pH 1.9) 17.5 (w pH 6.8)	164
	$C_2H_8O_7P_2$	$(HO)_2P(O)CH_2OCH_2P(O)(OH)_2$	19(w)	172
	$C_3H_7O_6P$	$(HO)_2P(O)CH_2OCH_2COOH$	19.0(w)	172
$H_2O;(C)_2$	$C_5H_{13}O_4P$	$HOCH_2P(O)(OEt)_2$	24	15
	$C_9H_{19}O_6P$	$(EtO)_2P(O)CH_2OCH_2COOEt$	20.2(g)	172
	$C_9H_{22}O_6P_2$	$(EtO)_2P(O)CH_2OP((OEt)_2$	24.4	173
	$C_{10}H_{24}O_7P_2$	$(EtO)_2P(O)CH_2OCH_2P(O)(OEt)_2$	19.5(g)	172
$H_2P^4;(Si)_2$	$C_{13}H_{38}O_6P_2Si_4$	$(TmsO)_2P(O)CH_2P(O)(OTms)_2$	0.1	53
	$C_{16}H_{34}O_6P_2Si_3$	$(TmsO)_2$**P**$(O)CH_2P(O)(OPh)(OTms)$	−1.5 3.4, $^2J_{PP}$10.2	53
	$C_{19}H_{30}O_6P_2Si_2$	$(TmsO)_2$**P**$(O)CH_2P(O)(OPh)_2$	−3.2(b)12.5(b), $^2J_{PP}$11.7	53
$H_2P^4;(H)_2$	$CH_6O_6P_2$	$(HO)_2P(O)CH_2P(O)(OH)_2$	16.7(w)	174
	$C_2H_8O_6P_2$	$(HO)_2$**P**$(O)CH_2P(O)(OH)(OMe)$	16.2	175
	$C_2H_9O_8P_3$	$[(HO)_2$**P**$(O)CH_2]_2P(O)OH$	17.5(w)	64(a)
	$C_3H_{12}O_{10}P_4$	$[(HO)_2$**P**$(O)CH_2]_3P(O)$	16.0	64(a)
$H_2P^4;(C)_2$	$C_5H_{14}O_6P_2$	$(MeO)_2P(O)CH_2P(O)(OMe)_2$	23.0 $^1J_{PC}$135.3	176
	$C_9H_{22}O_6P_2$	$(EtO)_2P(O)CH_2P(O)(OEt)_2$	19.0(d,n,p) 19.4(c) $^1J_{PC}$135.3	178
	$C_9H_{24}O_{10}P_4$	$[(MeO)_2P(O)CH_2]_3P(O)$	22.8(c)	179
	$C_{12}H_{29}O_7P_3$	$[(EtO)_2$**P**$(O)CH_2]_2P(O)Et$	21.1(c)	180
	$C_{12}H_{29}O_8P_3$	$[(EtO)_2$**P**$(O)CH_2]_2P(O)OEt$	20.3(c)	174
	$C_{13}H_{30}O_6P_2$	$(iPrO)_2P(O)CH_2P(O)(O\text{-}iPr)_2$	15.4[a] 17.5[b]	181

Rings/Connec-tivities	Formula	Structure	NMR data (δ_P[solv] $^nJ_{PZ}$ Hz)	Ref.
	$C_{15}H_{36}O_{10}P_4$	$[(EtO)_2\mathbf{P}(O)CH_2]_3P(O)$	21(c)	179
	$C_{21}H_{48}O_{10}P_4$	$[(iPrO)_2\mathbf{P}(O)CH_2]_3P(O)$	18.1(c)	179
H_2P^4;C',Si	$C_{19}H_{30}O_6P_2Si_2$	(TmsO)(PhO)P(O) $CH_2P(O)(OPh)(OTms)$	4.55 4.8^D(b)	53
	$C_{22}H_{26}O_6P_2Si$	$(TmsO)(PhO)\mathbf{P}(O)CH_2P(O)(OPh)_2$	3.8 12.0(n) $^2J_{PP}$11.7	53
H_2P^4;(C')$_2$	$C_{19}H_{30}O_6P_2Si_2$	$(PhO)_2\mathbf{P}(O)CH_2P(O)(OTms)_2$	12.5(b) $^2J_{PP}$11.7	53
	$C_{25}H_{22}O_6P_2$	$(PhO)_2P(O)CH_2P(O)(OPh)_2$	9.8(b) $^2J_{PH}$21.2	53
H_2S^4;(C)$_2$	CH_5O_6PS	$(HO)_2P(O)CH_2SO_3H$	14.3	62
	$C_4H_{11}O_5PS$	$(MeO)_2P(O)CH_2SO_2Me$	15.8(d)	182
	$C_4H_{11}O_6PS$	$(MeO)_2P(O)CH_2SO_2OMe$	14.4	62
	$C_{10}H_{24}O_8P_2S$	$[(EtO)_2P(O)CH_2]_2SO_2$	11.8(c)	183
	$C_{11}H_{17}O_5PS$	$(EtO)_2P(O)CH_2SO_2Ph$	11.6(d)	182
H_2S^4;(C')$_2$	$C_{19}H_{17}O_6PS$	$(PhO)_2P(O)CH_2SO_2OPh$	3.1	62
	$C_{19}H_{18}NO_5PS$	$(PhO)_2P(O)CH_2SO_2NHPh$	6.0(y)	62
H_2S^3;(C)$_2$	$C_4H_{11}O_4PS$	$(MeO)_2P(O)CH_2SOMe$	21.3(c)	182
	$C_{11}H_{17}O_4PS$	$(EtO)_2P(O)CH_2SOPh$	16,5(n) 19.0(c)	185
H_2S;(H)$_2$	$C_2H_8O_6P_2S$	$(HO)_2P(O)CH_2SCH_2P(O)(OH)_2$	23.5(w)	172
H_2S;(C)$_2$	$C_4H_{11}O_3PS$	$(MeO)_2P(O)CH_2SMe$	26.7(n)	182
	$C_{10}H_{24}O_6P_2S$	$(EtO)_2P(O)CH_2SCH_2P(O)(OEt)_2$	23.7(c)	183
	$C_{11}H_{17}O_3PS$	$(EtO)_2P(O)CH_2SPh$	22.0(n)	182
H_2Si;(C)$_2$	$C_8H_{21}O_3PSi$	$(EtO)_2P(O)CH_2Tms$	30.4(c) 33.0(c)	188
H_2Sn;(C)$_2$	$C_8H_{21}O_3PSn$	$(EtO)_2P(O)CH_2SnMe_3$	36.8(c)	186
H_3;(H)$_2$	CH_5O_3P	$MeP(O)(OH)_2$	23.4/30.5 30.5(c) 27.8(w pH 2.3)	189
H_3;C,H	$C_7H_{17}O_4P$	$MeP(O)(OH)OCMe_2CMe_2OH$	27	190
H_3;(C)$_2$	$C_3H_9O_3P$	$MeP(O)(OMe)_2$	31.5(g) 32.3(n)	191
	$C_5H_7Cl_6O_3P$	$MeP(O)(OCH_2CCl_3)_2$	32.0	192
	$C_5H_{13}O_3P$	$MeP(O)(OEt)_2$	27.7(c) 29.7(k)	194
	$C_8H_{19}O_3P$	MeP(O)(OMe)[OCHMe(-tBu)] $(R)_P$	31.4(c)	195
		$(S)_P$	30.7	
	$C_{11}H_{25}O_3P$	$MeP(O)(ONeo)_2$	29.5 $^2J_{PCH}$17.65	196
	$C_{12}H_{25}O_3P$	$MeOP(O)(OMe)OC_{10}H_{19}$[a]	31.7(c)	195
	$C_{13}H_{29}O_5P$	$MeP(O)(OCMe_2CMe_2OH)_2$	25	190
H_3;(C')$_2$	$C_{13}H_{13}O_3P$	$MeP(O)(OPh)_2$	24(c,n) $^2J_{PCH}$17.9	197
C'ClF;(C)$_2$	$C_8H_{15}ClFO_5P$	$(EtO)_2P(O)CClFCOOEt$	5.5 $^1J_{PC}$184	198
C'Cl_2;(C)$_2$	$C_8H_{15}Cl_2O_5P$	$(EtO)_2P(O)CCl_2COOEt$	8.0 $^1J_{PC}$166	198
CF_2;(H)$_2$	$C_2H_2F_5O_3P$	$(HO)_2P(O)CF_2Tf$	−12.3	149
C'F_2;(C)$_2$	$C_7H_{13}F_2O_3PS_2$	$(EtO)_2P(O)CF_2CSSMe$	2.5	199
	$C_8H_{15}F_2O_5P$	$(EtO)_2P(O)CF_2COOEt$	2.8(c) $^2J_{PF}$97	154
	$C_{11}H_{20}F_4O_7P_2$	$[(EtO)_2P(O)CF_2]_2CO$	6.3(c) $^2J_{PF}$95	154
COP^4;(C)$_2$	$C_6H_{16}O_7P_2$	$[(MeO)_2P(O)]_2C(OH)Me$	22.0	200
C'OP^4;(C)$_2$	$C_{11}H_{18}O_7P_2$	$[(MeO)_2P(O)]_2C(OH)Ph$	18.0	201
CHBr;(H)$_2$	$C_8H_9Br_2O_3P$	$(HO)_2P(O)CHBrCHBrPh$	16.1(y)	202
CHCl;(C)$_2$	$C_6H_{14}ClO_3P$	$(EtO)_2P(O)CHClMe$	20.2(n)	64(b)
C'HCl;(C)$_2$	$C_7H_{14}ClO_3PS$	$(MeO)_2P(O)CHClC(SMe){=}CHMe$	17.8	32
	$C_8H_{16}ClO_5P$	$(EtO)_2P(O)CHClCOOEt$	13.0 $^1J_{PC}$146	198
CHF;(C)$_2$	$C_8H_{18}FO_3P$	$(iPrO)_2P(O)CHFMe$	16.6 $^2J_{PF}$79.3	147
	$C_{14}H_{22}FO_4P$	$(iPrO)_2P(O)CHFCH(OH)Ph$	14.0 14.9^D	147
C'HF;(C)$_2$	$C_8H_{16}FO_4PS$	$(iPrO)_2P(O)CHFCOSH$	6.4	199
	$C_8H_{16}FO_5P$	$(iPrO)_2P(O)CHFCOOH$	9.3	199
	$C_{11}H_{16}FO_3P$	$(EtO)_2P(O)CHFPh$	14.9(c) $^2J_{PF}$85.5	203
CHN;(H)$_2$	$C_3H_{10}NO_3P$	$(HO)_2P(O)CH(NH_2)Et$	14.5(w)	204
	$C_3H_{10}NO_3PS$	$(HO)_2P(O)CH(NH_2)CH_2SMe$	14.0(l)	205
	$C_6H_{12}NO_5P$	$(HO)_2P(O)$ $[\dot{C}HCH_2CH_2CH_2CH(COOH)\dot{N}H]$	10.1 10.6^D(w pH 7)	166

TABLE J (continued)
Compilation of ^{31}P NMR Data of Four Coordinate Compounds Containing a P=Ch Bond and One Bond from Phosphorus to a Group IV Atom

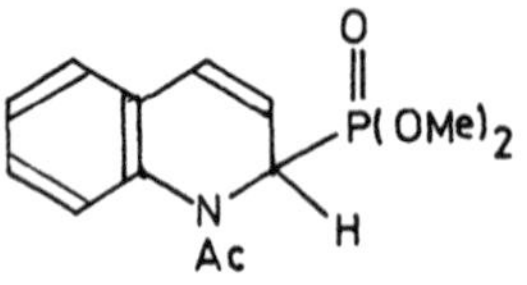

21

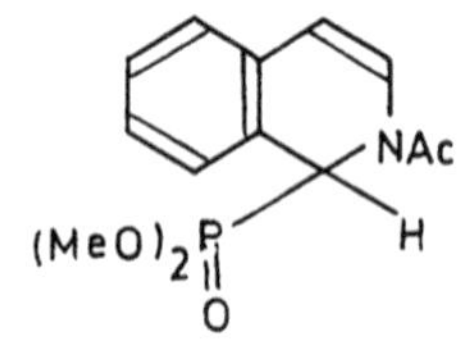

22

Rings/Connec-tivities	Formula	Structure	NMR data (δ_P[solv] $^nJ_{PZ}$ Hz)	Ref.
CHN;(C)$_2$	$C_6H_{16}NO_3P$	(EtO)$_2$P(O)CH(NH$_2$)Me	20.1(u) 28.6(r)	206
CHN;(C')$_2$	$C_{21}H_{21}N_2O_3PS$	(PhO)$_2$P(O)CHMeNHCSNHPh	17.6(u)	207
C'HN;(C)$_2$	$C_{11}H_{18}NO_3P$	(EtO)$_2$P(O)CH(NH$_2$)Ph	15.6(u) 24.7(c)	209(a,b)
	$C_{13}H_{16}NO_4P$	**21**	20.4	210
	$C_{13}H_{16}NO_4P$	**22**	21.3(c)	210
	$C_{23}H_{27}NO_4P_2$	(EtO)$_2$**P**(O)CHPhNHP(O)Ph$_2$	21.7 or 23.9	208
C'HN;(C')$_2$	$C_{26}H_{23}N_2O_3PS$	(PhO)$_2$P(O)CHPhNHCSNHPh	13.5(u)	207
C"HN;(C)$_2$	$C_8H_{17}N_2O_3P$	(EtO)$_2$P(O)CH(CN)NMe$_2$	12.9(k)	211
CHO;(H)$_2$	$C_4H_{11}O_4P$	(HO)$_2$P(O)CH(OH)Pr	24	73
CHO;(C)$_2$	$C_4H_{11}O_4P$	(MeO)$_2$P(O)CH(OH)Me	26 27.8	213
	$C_6H_{15}O_5P$	(MeO)$_2$P(O)CH(OH)CH$_2$CH(OH)Me		214
		(1*RS*,3*SR*)	28.4(c)	
		(1*SR*,3*RS*)	26.5(c)	
	$C_6H_{15}O_7P$	(MeO)$_2$P(O) CH(OH)CH(OH)CH(OH)CH$_2$OH	26.4 27.0 28.4^D	215
	$C_8H_{17}O_4P$	(EtO)$_2$P(O)(ĊHOCH$_2$CH$_2$ĊH$_2$)	24.8	216
	$C_8H_{18}NO_6P$	(PrO)$_2$P(O)CH(OH)CH$_2$NO$_2$	17.1	168
	$C_{10}H_{24}O_8P_2$	(EtO)$_2$P(O)CH(OH)CH(OH)P(O)(OEt)$_2$	22.6(c)	217
	$C_{12}H_{28}O_7P_2$	(EtO)$_2$**P**(O)CH(OEt)CH$_2$P(O)(OEt)$_2$	21.1 27.9	58
C'HO;(Si)$_2$	$C_{19}H_{41}O_5PSi_4$	(TmsO)$_2$P(O)CH(OTms)C$_6$H$_4$OTms-2	6.5	218
C'HO;(H)$_2$	$C_7H_9O_5P$	(HO)$_2$P(O)CH(OH)C$_6$H$_4$OH-2	24.0	218
C'HO;(C)$_2$	$C_{11}H_{17}O_5P$	(EtO)$_2$P(O)CH(OH)C$_6$H$_4$OH-2	23.7	218
	$C_{12}H_{18}NO_5P$	(EtO)$_2$P(O)CHPh(OCONH$_2$)	18.0(c)	19
	$C_{12}H_{19}O_4P$	(EtO)$_2$P(O)CH(OMe)Ph	17.7(c)	219
	$C_{14}H_{25}O_4PSi$	(EtO)$_2$P(O)CH(OTms)Ph	19	220
	$C_{19}H_{34}O_6P_2$	(PrO)$_2$P(O)CHPh[OP(OPr)$_2$]	21	221
CHP^4;(H)$_2$	$C_2H_8O_6P_2$	MeCH[P(O)(OH)$_2$]$_2$	23.0	143
	$C_5H_{12}O_8P_2$	EtOOCCH$_2$CH[P(O)(OH)$_2$]$_2$	20.0	143
	$C_8H_{12}O_6P_2$	PhCH$_2$CH[P(O)(OH)$_2$]$_2$	21.0	143
	$C_8H_{13}O_9P_3$	[(HO)$_2$P$_{A,B}$(O)]$_2$CHCHPhP$_c$(O)(OH)$_2$	18.6(A) 18.1(B) 24.7(C)(v)	222
CHP^4;(C)$_2$	$C_{14}H_{32}O_6P_2$	(iPrO)$_2$P(O)CHMeP(O)(Oi-Pr)$_2$	22.0	222
	$C_{17}H_{36}O_8P_2$	EtOOCCH$_2$CH[P(O)(O-iPr)$_2$]$_2$	20.0	222
	$C_{20}H_{36}O_6P_2$	[(iPrO)$_2$P(O)]$_2$CHCH$_2$Ph	20.5	222
	$C_{20}H_{37}O_9P_3$	[(EtO)$_2$P$_{A,B}$(O)]$_2$ CHCHPhP$_C$(O)(OEt)$_2$	22.1, 20.0 (A,B) 24.45(C)	222
C'HP^4;(C)$_2$	$C_{11}H_{24}O_7P_2$	(EtO)$_2$P(O)CHAcP(O)(OEt)$_2$	13.3	223
CHS;(C)$_2$	$C_6H_{15}O_3PS$	(EtO)$_2$P(O)CH(SH)Me	28.2(c)	224
	$C_{12}H_{19}O_3PS$	(EtO)$_2$P(O)CH(SCH$_2$Ph)Me	25.6(c)	158

Rings/Connectivities	Formula	Structure	NMR data (δ_P[solv] $^nJ_{PZ}$ Hz)	Ref.
	$C_{12}H_{28}O_6P_2S_2$	[(EtO)$_2$P(O)CHMe]$_2$S$_2$	27.3 27.7[D](c)	224
C′HS;(C)$_2$	$C_{11}H_{17}O_3PS$	(EtO)$_2$P(O)CH(SH)Ph	24.2(c)	224
	$C_{17}H_{21}O_3PS$	(EtO)$_2$P(O)CH(SPh)Ph	20.6(c)	158
C′HSi;(C)$_2$	$C_{14}H_{25}O_3PSi$	(EtO)$_2$P(O)CHPhTms	26.4(c)	187
CH$_2$;(Si)$_2$	$C_8H_{23}O_4PSi_2$	(TmsO)$_2$P(O)CH$_2$CH$_2$OH	11.5(c)	225
	$C_{12}H_{30}NO_3PSi_3$	(TmsO)$_2$P(O)CH$_2$CH(Tms)CN	10	226
CH$_2$;(H)$_2$	$C_2H_7O_3P$	EtP(O)(OH)$_2$	31(w)	227
	$C_2H_7O_4P$	(HO)$_2$P(O)CH$_2$CH$_2$OH	26.4(w)	228
	$C_2H_8O_6P_2$	(HO)$_2$P(O)CH$_2$CH$_2$P(O)(OH)$_2$	27.4(w) 26.7	174
	$C_3H_{10}O_6P_2$	(HO)$_2$P(O)CH$_2$CH$_2$CH$_2$P(O)(OH)$_2$	28.2(w)	64(b)
	$C_3H_{11}O_9P_3$	[(HO)$_2$**P**(O)CH$_2$]$_2$CHP(O)(OH)$_2$	27.5	229
	$C_4H_8F_4O_6P_2$	(HO)$_2$P(O)CH$_2$CF$_2$CF$_2$CH$_2$P(O)(OH)$_2$	−1.9	230
	$C_4H_{10}ClO_4P$	(HO)$_2$P(O)CH$_2$CH$_2$OCH$_2$CH$_2$Cl	26.5(w)	172
	$C_4H_{12}O_6P_2$	(HO)$_2$P(O)(CH$_2$)$_4$P(O)(OH)$_2$	31.6(w)	64(b)
	$C_4H_{12}O_7P_2$	(HO)$_2$**P**(O) CH$_2$CH$_2$OP(O)(OH)OCH$_2$CH$_2$OH	22.9 or 27.7	225
	$C_4H_{13}O_8P_3$	[(HO)$_2$P(O)CH$_2$CH$_2$]$_2$P(O)(OH)	27.3	231
	$C_4H_{14}O_{12}P_4$	(HO)$_2$P(O)CH$_2$ | (HO)$_2$P(O)CH$_2$ [CHP(O)(OH)$_2$]	27.2	229
	$C_6H_{18}NO_9P_3$	[(HO)$_2$P(O)CH$_2$CH$_2$]$_3$N	20.8(w)	232
	$C_7H_{17}O_3P$	(HO)$_2$P(O)C$_7$H$_{15\text{-}n}$	36.2(c)	233
	$C_8H_{13}NO_6P_2$	(HO)$_2$P(O)CH$_2$C(NH$_2$)PhP(O)(OH)$_2$	13.1 26.9	234
	$C_{20}H_{21}IO_3P_2$	(HO)$_2$P(O)CH$_2$CH$_2$P$^+$Ph$_3$I$^-$	20.4(d)	235
CH$_2$;(C)$_2$	$C_4H_{11}O_3P$	EtP(O)(OMe)$_2$	34.8	236
	$C_4H_{11}O_4P$	(MeO)$_2$P(O)CH$_2$CH$_2$OH	32.4(c)	228
	$C_6H_{12}Cl_3O_3P$	(ClCH$_2$CH$_2$O)$_2$PCH$_2$CH$_2$Cl	26.4 27.0(n)	237
	$C_6H_{13}O_4P$	(MeO)$_2$P(O)CH$_2$CH$_2$COMe	18	238
	$C_6H_{14}BrO_3P$	(EtO)$_2$P(O)CH$_2$CH$_2$Br	28.5(n)	64(b)
	$C_6H_{14}ClO_3P$	(EtO)$_2$P(O)CH$_2$CH$_2$Cl	24.4(n)	64(b)
	$C_6H_{15}O_3P$	(EtO)$_2$P(O)Et	30.8[a](c) 32.2(k)	193
	$C_6H_{15}O_3PS$	(EtO)(HSCH$_2$CH$_2$O)P(O)Et	33	239
	$C_8H_{18}ClO_4P$	(EtO)$_2$P(O)CH$_2$CH$_2$OCH$_2$CH$_2$Cl	28	172
	$C_9H_{18}ClO_3P$	(EtO)$_2$P(O)CH$_2$CHClCH=CHMe	9.1(g)	240
	$C_{10}H_{24}O_6P_2$	(EtO)$_2$P(O)CH$_2$CH$_2$P(O)(OEt)$_2$	28.6(c) 26.8(n)	241
	$C_{11}H_{26}O_6P_2$	(EtO)$_2$P(O)(CH$_2$)$_3$P(O)(OEt)$_2$	29.3(n)	241
	$C_{12}H_{27}O_3P$	(BuO)$_2$P(O)Bu	31	242
	$C_{12}H_{28}O_6P_2$	(EtO)$_2$P(O)(CH$_2$)$_4$P(O)(OEt)$_2$	31.5(n,c)	241
	$C_{12}H_{28}O_7P_2$	[(EtO)$_2$P(O)CH$_2$CH$_2$]$_2$O	28(n)	172
	$C_{13}H_{28}O_8P_2$	(EtO)$_2$P(O)CH$_2$CH(COOEt)P(O)(OEt)$_2$	21.1 or 27.6(c)	243
	$C_{14}H_{33}O_8P_3$	[(EtO)$_2$**P**(O)CH$_2$CH$_2$]$_2$P(O)OEt	28.6 or 31.1(c)	231
	$C_{15}H_{35}O_9P_3$	[(EtO)$_2$**P**(O)CH$_2$]$_2$CHP(O)(OEt)$_2$[d]	28.5	244
	$C_{18}H_{42}NO_9P_3$	[(EtO)$_2$P(O)CH$_2$CH$_2$]$_3$N	29.8(w)	232
	$C_{19}H_{27}O_4PSi$	(MeO)$_2$P(O)CH$_2$CPh$_2$(OTms)	29.0(c) $^2J_{PCH}$19.6	127
	$C_{20}H_{46}O_{12}P_4$	(EtO)$_2$P(O)CH$_2$[d] | (EtO)$_2$P(O)CH$_2$[CHP(O)(OEt)$_2$]$_2$	30.3	244
CH$_2$;(C′)$_2$	$C_{14}H_{15}O_3P$	(PhO)$_2$P(O)Et	24 26.5 $^2J_{PCH}$21	66
C′H$_2$;(H)$_2$	$C_2H_5O_5P$	(HO)$_2$P(O)CH$_2$COOH	17.6(w)	172
	$C_4H_7O_5P$	(HO)$_2$P(O)CH$_2$C(COOH)=CH$_2$	24.3(w)	245
C′H$_2$;(C)$_2$	$C_4H_9O_4P$	(MeO)$_2$P(O)CH$_2$CHO	18.9(c)(Keto)	246
	$C_5H_{10}ClO_3P$	(MeO)$_2$P(O)CH$_2$CCl=CH$_2$	26.4	34
	$C_5H_{11}O_3P$	(MeO)$_2$P(O)CH$_2$CH=CH$_2$	27.5(g)	247
	$C_6H_{13}O_4P$	(EtO)$_2$P(O)CH$_2$CHO	19.5(n)	58
	$C_7H_{12}F_3O_4P$	(EtO)$_2$P(O)CH$_2$COTf	28.9	248
	$C_7H_{14}ClO_4P$	(EtO)$_2$P(O)CH$_2$COCH$_2$Cl	18.1	223

TABLE J (continued)
Compilation of ^{31}P NMR Data of Four Coordinate Compounds Containing a P=Ch Bond and One Bond from Phosphorus to a Group IV Atom

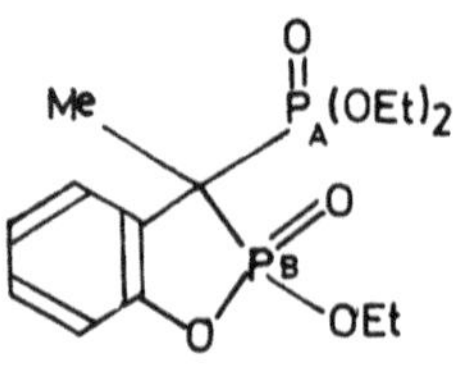

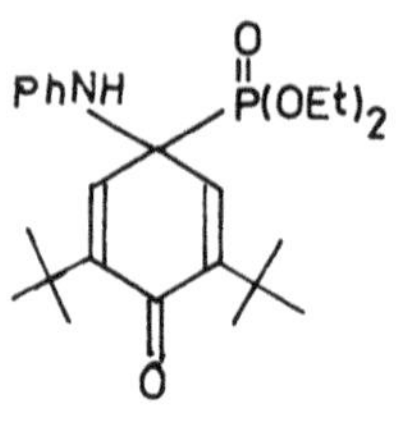

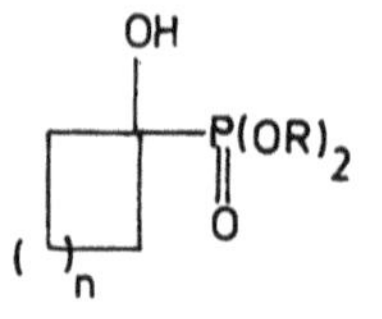

23 **24** **25**

Rings/Connectivities	Formula	Structure	NMR data (δ_P[solv] $^nJ_{PZ}$ Hz)	Ref.
	$C_7H_{15}O_3P$	$(EtO)_2P(O)CH_2CH{=}CH_2$	23.5(n)	247
	$C_7H_{15}O_4P$	$(EtO)_2P(O)CH_2COMe$	17.1(c) 19.5(d,p)	250
	$C_7H_{16}O_7P_2$	$[(MeO)_2P(O)CH_2]_2CO$	21.5	223
	$C_8H_{15}O_6P$	$(MeO)_2P(O)CH_2COCH_2COOEt$	21.5	223
	$C_8H_{17}O_3P$	$(EtO)_2P(O)CH_2CH{=}CHMe$	26.9^E 27.1^Z	251
	$C_8H_{17}O_4P$	$(EtO)_2P(O)CH_2COEt$	20.1(c)	252
	$C_8H_{17}O_5P$	$(EtO)_2P(O)CH_2COOEt$	17.0[a](c) 18.6[b](k) 19.2[b](p,d)	254
	$C_8H_{18}NO_4P$	$(iPrO)_2P(O)CH_2CH{=}NOH$	23.2^E 22.8^Z	255
	$C_8H_{18}NO_4P$	$(EtO)_2P(O)CH_2CONMe_2$	18.7(c)	249
	$C_9H_{15}O_3PS$	$(EtO)_2P(O)CH_2(\dot{C}{=}CHCH{=}CH\dot{S})$-2	21.0(c)	16
	$C_9H_{15}O_4P$	$(EtO)_2P(O)CH_2(\dot{C}{=}CHCH{=}CH\dot{O})$-2	19.8(c)	16
	$C_9H_{18}ClO_3P$	$(EtO)_2P(O)CH_2CCl{=}CMe_2$	23.6	34
	$C_{10}H_{13}O_4P$	$(MeO)_2P(O)CH_2COPh$	22.8	223
	$C_{10}H_{19}O_5P$	$(EtO)_2P(O)CH_2C(COOEt){=}CH_2$	25.0(g)	245
	$C_{10}H_{19}O_6P$	$(EtO)_2P(O)CH_2COCH_2COOEt$	18.6(c)	256
	$C_{11}H_{17}O_3P$	$(EtO)_2P(O)CH_2Ph$	27.1(c) 25.0 $^1J_{PC}$137	258
	$C_{11}H_{17}O_4P$	$(EtO)_2P(O)CH_2C_6H_4OH$-2	29.2(c) $^1J_{PC}$140	259
	$C_{11}H_{24}O_5P_2S_2$	$H_2C{=}C[CH_2P(O)(OEt)_2]$ $[SP(S)(OEt)_2]$	24	35
	$C_{16}H_{17}O_5P$	$(PhCH_2O)_2P(O)CH_2COOH$	22.0(c)	260
	$C_{16}H_{28}O_6P_2$	1,2-$[(EtO)_2P(O)CH_2]_2C_6H_4$	26.5	172
	$C_{18}H_{29}O_4P$	$(NeoO)_2P(O)CH_2COPh$	19.0	261
	$C_{19}H_{33}O_4P$	$(EtO)_2P(O)CH_2C_6H_2(OH)(\text{-}tBu)_2-4,2,5$	23	262
$C'H_2;(C')_2$	$C_{15}H_{15}O_4P$	$(PhO)_2P(O)CH_2COMe$	14.0	263
	$C_{19}H_{17}O_3P$	$(PhO)_2P(O)CH_2Ph$	19.5(c)	21
$C''H_2;(C)_2$	$C_6H_{12}NO_3P$	$(EtO)_2P(O)CH_2CN$	15.0 $^1J_{PC}$141.7(k) 15.9 $^1J_{PC}$138.8(d)	257
	$C_8H_{15}O_3P$	$(EtO)_2P(O)CH_2C{\equiv}CMe$	20.7	251
$CC'Br;(H)_2$	$C_9H_{11}Br_2O_3P$	$(HO)_2P(O)CPhBrCHBrMe$	11.8 13.7^D(y)	264
$CC'Br;C,H$	$C_{10}H_{13}Br_2O_3P$	$(HO)(MeO)P(O)CPhBrCHBrMe$	13.6 15.3^D(y)	264
$CC'Cl;(H)_2$	$C_9H_{12}ClO_3P$	$(HO)_2P(O)CClPhEt$	17.2(y)	264
$CC'Cl;(C)_2$	$C_9H_{18}ClO_5P$	$(EtO)_2P(O)CClMeCOOEt$	16.7 $^1J_{PC}$152	198
$CC'F;(C)_2$	$C_{12}H_{18}FO_3P$	$(EtO)_2P(O)CFPhMe$	1.74 $^2J_{PF}$87.4	203
$CC'P^4;(C)_2$	$C_{14}H_{22}O_6P_2$	**23**	22(A,n) 42(B,n)	265
$CC'S;(C)_2$	$C_{14}H_{23}O_3PS$	$(iPrO)_2P(O)C(SH)PhMe$	24	266
$C_2N';(C)_2$	$C_7H_{16}NO_4P$	$(EtO)_2P(O)CMe_2NO$	17	267
$C'_2N;(C)_2$	$C_{24}H_{36}NO_4P$	**24**	18	268
$C_2O;(C)_2$	$C_5H_{13}O_4P$	$(MeO)_2P(O)CMe_2OH$	30	269
	$C_8H_{17}O_4P$	**25**; n = 1, R = Et	27.8(c) $^1J_{PC}$164.1	270

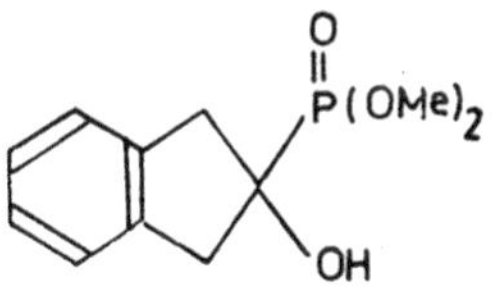

26 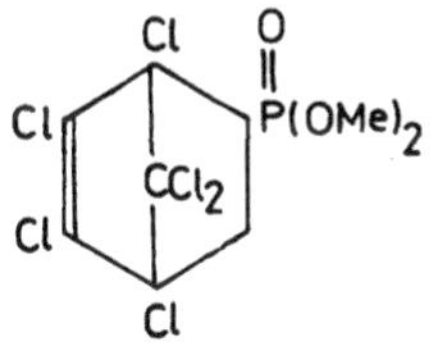 **27** 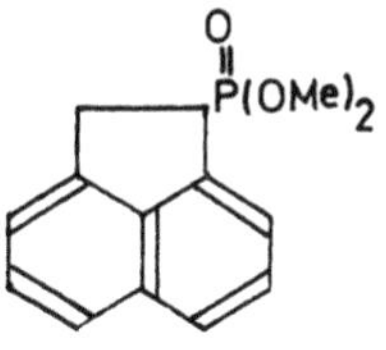 **28**

Rings/Connec-tivities	Formula	Structure	NMR data (δ_P[solv] $^nJ_{PZ}$ Hz)	Ref.
	$C_9H_{19}O_4P$	**25**; n = 2, R = Et	27.6(c) $^1J_{PC}$169.8	270
	$C_{10}H_{21}O_4P$	**25**; n = 3, R = Et	27.3(c) $^1J_{PC}$165.1	270
	$C_{11}H_{15}O_4P$	**26**	28.3	212
	$C_{11}H_{23}O_4P$	**25**; n = 4/9 R = Et	28.1/28.6(c) $^1J_{PC}$157/ 161	270
$C_2P^4;(H)_2$	$C_3H_{10}O_6P_2$	$(HO)_2P(O)CMe_2P(O)(OH)_2$	27.5	143
$C_2P^4;(C)_2$	$C_{15}H_{34}O_6P_2$	$(iPrO)_2P(O)CMe_2P(O)(O\text{-}iPr)_2$	25.0	143
$C_2S;(C)_2$	$C_{10}H_{21}O_3PS$	$(EtO)_2P(O)C(SH)(CH_2)_5$	27	271
$C_2H;(H)_2$	$C_6H_{11}O_3P$	$(HO)_2P(O)C_6H_9$[e]	32.8 $^1J_{PC}$139.9	30
	$C_6H_{13}O_3P$	$cHexP(O)(OH)_2$	33(w)	272
$C_2H;H,C$	$C_8H_4O_3P$	cHexP(O)(OEt)OH	31	273
$C_2H;(C)_2$	$C_7H_{17}O_3P$	$iPrP(O)(OEt)_2$	32.3(c)	193
	$C_9H_9Cl_6O_3P$	**27**	22.7	212
	$C_9H_{21}O_3P$	$iPrP(O)(O\text{-}iPr)_2$	34	173
$C_2H;(C')_2$	$C_{15}H_{17}O_3P$	$iPrP(O)(OPh)_2$	26	197
$CC'H;(H)_2$	$C_6H_{11}O_3P$	$(HO)_2P(O)C_6H_9$[f]	30.3 $^1J_{PC}$137.3	30
$CC'H;(C)_2$	$C_7H_{13}N_2O_5P$	$(MeO)_2P(O)\dot{C}HCH_2NHN{=}\dot{C}COOEt$	27	274
	$C_7H_{15}O_4P$	$(EtO)_2P(O)CHMeCHO$	20.2(c,keto) 24.1(c,enol)	246
$CC'H;(C)_2$	$C_8H_{17}O_4P$	$(EtO)_2P(O)CHMeCOMe$	20.8(c) 23.2	275
	$C_{11}H_{15}O_3P$	$(MeO)_2P(O)C_9H_9$[g]	31.3	276
	$C_{14}H_{15}O_3P$	**28**	23.3	212
	$C_{16}H_{33}N_2O_5P$	$(EtO)_2P(O)CH(CH_2CONEt_2)CONEt_2$	24.4(n)	277
$CC''H;(C')_2$	$C_{15}H_{14}NO_3P$	$(PhO)_2P(O)CH(CN)Me$	13	278
$C'C''H;(C)_2$	$C_{12}H_{15}ClNO_3P$	$(EtO)_2P(O)CH(CN)C_6H_4Cl\text{-}2$	10.15	279
	$C_{12}H_{16}NO_3P$	$(EtO)_2P(O)CH(CN)Ph$	14.5(c)	279
$C'_2H;(C)_2$	$C_{12}H_{15}O_6P$	$(MeO)_2P(O)CH(COPh)COOMe$	18	280
$C_3;(H)_2$	$C_4H_{11}O_3P$	$tBuP(O)(OH)_2$	37.0(a)	82
	$C_9H_{17}O_4P$	$(HO)_2P(O)\dot{C}MeCH_2COCH_2CMe_2\dot{C}H_2$	29.1(d)	281
	$C_9H_{17}O_7P$	$(HO)_2P(O)$ $CMe(CH_2COOH)CH_2CMe_2COOH$	31.3(w)	281
$C_3;(C)_2$	$C_6H_{15}O_3P$	$tBuP(O)(OMe)_2$	38.1(g)	65(b)
	$C_{12}H_{21}O_3P$	$1\text{-}AdP(O)(OMe)_2$	33.3(c) $^1J_{PC}$145.9	45
$C_2C';(H)$	$C_4H_6F_3O_5P$	$(HO)_2P(O)CMeTfCOOH$	21.2	282
$C_2C';(H)_2$	$C_8H_{15}O_7P$	$(HO)_2P(O)C(COOH)Me(CH_2CMe_2$ $COOH)$	23.0(w)	281
$C_2C';(C)_2$	$C_7H_{12}F_3O_5P$	$(MeO)_2P(O)CMeTfCOOMe$	22.2	282
	$C_8H_{17}O_4P$	$(EtO)_2P(O)CMe_2CHO$	23.8	283
	$C_{10}H_{18}F_3O_5P$	$(EtO)_2P(O)CMeTfCOOEt$	20.4	282
$C'_3;(C)_2$	$C_{19}H_{19}O_7P$	$(MeO)_2P(O)C(COPh)_2COOMe$	12	280
$5/H_2N;(C)_2$	$C_9H_{20}NO_3P$	$PncP(O)CH_2NMe_2$	36(m)	167
$5/H_3;(C)_2$	$C_3H_7O_3P$	GlcP(O)Me	42.6(c,v) 52	285
	$C_7H_{15}O_3P$	PncP(O)Me	40(k)	190
	$C_9H_{15}O_7P$	**H5**; R = EtOOC; X, Y = O′, Me	48.8	190
$5/COP^4;(C)_2$	$C_{14}H_{28}O_7P_2$	PncP(O)CMe(OH)P(O)Pnc	36	286

TABLE J (continued)
Compilation of ^{31}P NMR Data of Four Coordinate Compounds Containing a P=Ch Bond and One Bond from Phosphorus to a Group IV Atom

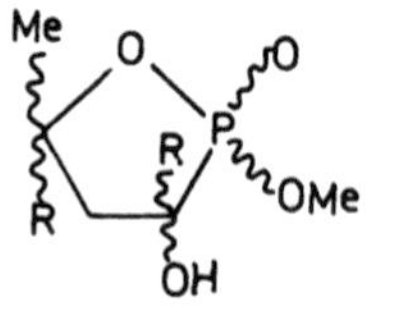

29 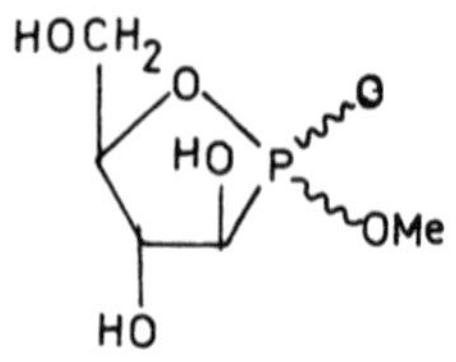 30 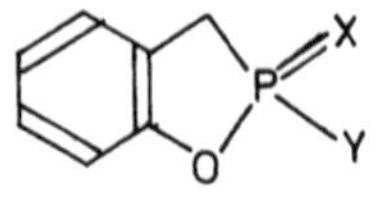 31

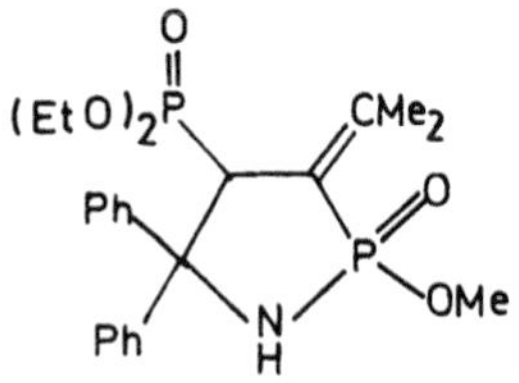

32 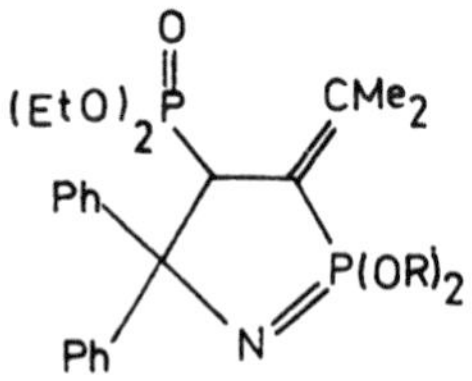 33 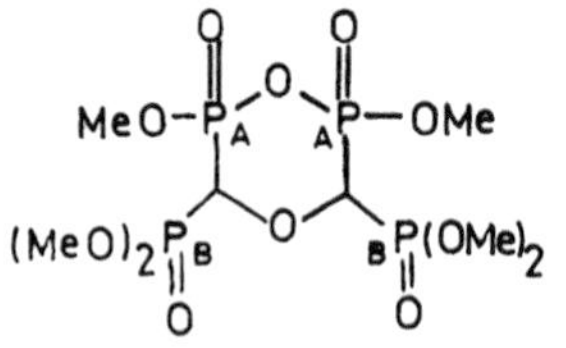 34

Rings/Connectivities	Formula	Structure	NMR data (δ_P[solv] $^nJ_{PZ}$ Hz)	Ref.
5/CHO;(C)$_2$	$C_5H_{11}O_4P$	**29**; R = H (2*SR*,3*SR*,5*RS*)	45.2(c)	214
		(2*RS*,3*SR*,5*RS*)	41.5(c)	
		(2*RS*,3*RS*,5*RS*)	42.7(c)	
		(2*SR*,3*RS*,5*RS*)	38.6(c)	
	$C_5H_{11}O_6P$	**30**; (2*R*,3*R*,4*R*,5*R*)	35.8	287
		(2*S*,3*R*,4*R*,5*R*)	32.5	
	$C_{14}H_{28}O_7P_2$	PncP(O)CHMeOP(O)Pnc	33.6	286
5/CH$_2$;(C)$_2$	$C_6H_{13}O_3P$	**H5**; R = Me, X = O, Y = Et, (4*R*,5*R*)	48.4	288
5/CH$_2$;(C')$_2$	$C_8H_{15}O_3P$	**H6**; R = Me, X = O, Y = Bu	45.0	289
	$C_{10}H_{17}O_3P$	**H6**; RR = (CH$_2$)$_4$, X = O, Y = Bu	48.8	289
	$C_{12}H_{11}O_3P$	**H13**; X = O, Y = Et	7.0	290
5/C'H$_2$;(C)$_2$	$C_8H_{15}O_5P$	**H5**; R = Me, X = O, Y = CH$_2$COOEt		291
		(4*RS*,5*SR*)	32.2(k)	
		(4*RS*,5*RS*)	31.8(k)	
	$C_{21}H_{35}O_4P$	PncP(O)CH$_2$C$_6$H$_2$(OH)(-tBu)$_2$-4,2,5	37	15
5/C'H$_2$;C,C'	$C_9H_{11}O_3P$	**31**; X = O, Y = OEt	45.6(c) $^1J_{PC}$123	259
5/C'H$_2$;(C')$_2$	$C_{10}H_{11}O_5P$	CatP(O)CH$_2$COOEt	36.9	292
5/C$_2$O;(C)$_2$	$C_7H_{15}O_4P$	**29**; R = Me (2*RS*,3*SR*)	40.8(y) 39.8(b)	293
		(2*SR*,3*SR*)	42.3(y) 41.9(b)	
5/CC'H;(C)$_2$	$C_{18}H_{32}O_{10}P_2$	PncP(O)CH (COOMe)CH(COOMe)P(O)Pnc	31	294
	$C_{23}H_{31}NO_5P_2$	**32**	24 34^D	123
	$C_{26}H_{37}NO_5P_2$	**33**; R = Me	25 58	123
5/C$_3$;(C)$_2$	$C_8H_{17}O_3P$	**H5**; R = Me; X, Y = O, tBu (4*R*,5*R*)	30.7	288
6/HOP^4;P^4,C	$C_8H_{20}O_{12}P_4$	**34**	6.5(a) 15.8(B)	200
6/H$_2$O;(C)$_2$	$C_6H_{14}O_6P_2$	**35**	18.8	173
6/H$_2$S^4;(C)$_2$	$C_7H_{15}O_5PS$	MeSO$_2$CH$_2$(O)P(OCH$_2$CMe$_2$CH$_2$O)	5.6(d)	282

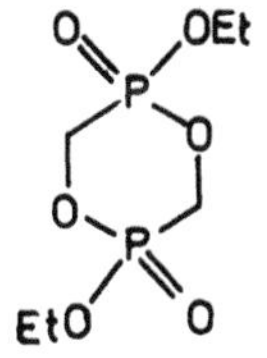

35 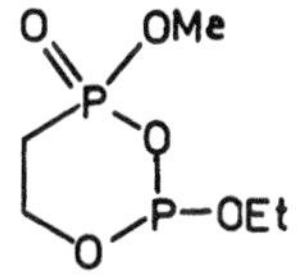 **36** 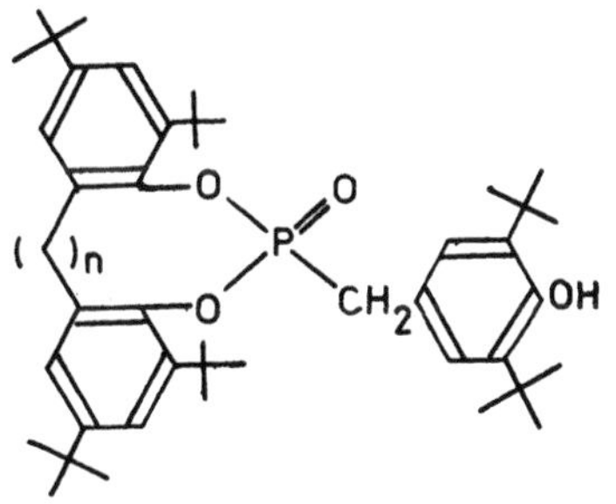 **37**

Rings/Connectivities	Formula	Structure	NMR data (δ_P[solv] $^nJ_{PZ}$ Hz)	Ref.
6/H_2S^3;$(C)_2$	$C_7H_{15}O_4PS$	$MeSOCH_2(O)P(OCH_2CM_2CH_2O)$	10.3(c)	282
6/H_2S;$(C)_2$	$C_7H_{15}O_3PS$	$MeSCH_2(O)P(OCH_2CMe_2CH_2O)$	17.3(n)	282
6/H_3;$(C)_2$	$C_4H_9O_3P$	$Me(O)P(OCH_2CH_2CH_2O)$	23.2(c,v)	284
	$C_5H_{11}O_3P$	**H1**; X = Me, Y = O′	23.0(c) $^1J_{PC}$8.3	297
		X = O′, Y = Me	28.0(c) $^1J_{PC}$11.8	
	$C_6H_{13}O_3P$	$Me(O)P(OCH_2CMe_2CH_2O)$	24.7(c)	193
	$C_7H_{14}O_6P_2$	$Me(O)P(OCH_2)_2C(CH_2O)_2P(O)Me$	27.2(c)	21
	$C_{10}H_{19}O_7P$	**H26**; A = O; X = Me, Y = O′	25.5	298
		X = O′ Y = Me	31.5	
	$C_{10}H_{19}O_7P$	**H27**; X = Me, Y = O′	22.7	298
		X = O′ Y = Me	30.5	
6/H_3;$(C')_2$	$C_6H_9O_3P$	**H28**; X = O, Y = Me	22(n)	299
	$C_9H_9O_3P$	**H29**; X = O, Y = Me	21(g)	300
6/CH_2;C,P	$C_5H_{12}O_5P_2$	**36**	8.0(b) $^2J_{PP}$40.8	301
6/CH_2;$(C)_2$	$C_6H_{13}O_3P$	$EtOP(O)(OCH_2CH_2CH_2CH_2)$	46.2(g)	302
6/CH_2;$(C')_2$	$C_{12}H_{11}O_3P$	**H17**; X = O, Y = Et	−4.1	290
6/$C'H_2$;$(C)_2$	$C_9H_{17}O_5P$	$EtOOCCH_2(O)P(OCH_2CMe_2CH_2O)$	12.2(ze)	291
	$C_{11}H_{15}O_3P$	**H1**; X = O′ Y = CH_2Ph	24.3(c)	303
		X = CH_2Ph, X = O′	19.0(c)	
6/$C''H_2$;$(C)_2$	$C_7H_{12}NO_3P$	$NCCH_2(O)P(OCH_2CMe_2CH_2O)$	7.8(ze)	291
7/$C'H_2$;$(C')_2$	$C_{43}H_{63}O_4P$	**37**; n = O	21.2(b)	304
8/$C'H_2$;$(C')_2$	$C_{44}H_{65}O_4P$	**37**; n = 1	11.2(b)	304

P-bound atoms = CO'_3

Rings/Connectivities	Formula	Structure	NMR data	Ref.
H_2N	CH_4NO_3P	$H_2NCH_2P(O)(O^-)_2$	19.5(w, pH 11.6)	164
	$C_3H_3NO_5P$	$^-OOCCH_2NHCH_2P(O)(O^-)_2$	16,5(w, pH 11.7)	164
	$C_4H_{10}N_2O_6P_2$	$H_2NC_2H_4N[CH_2P(O)(O^-)_2]_2$	15.5 (w, pH 12)	165
CHN	$C_2H_6NO_3P$	$MeCH(NH_2)P(O)(O^-)_2$	22.1(w, 2*M* KOH)	207
	$C_3H_8NO_3PS$	$MeSCH_2CH(NH_2)P(O)(O^-)_2$	18.1(w, HO^-)	205
	$C_8H_{10}NO_3P$	$PhCH_2CH(NH_2)P(O)(O^-)_2$	21.2(w, HO^-)	204
C′HN	$C_7H_8NO_3P$	$PhCH(NH_2)P(O)(O^-)_2$	17.9(w, HO^-)	209(a,c)
CH_2	$C_4H_8O_7P_2$	$[(^-O)_2P(O)CH_2CH_2]_2O$	15(w)	172
	$C_8H_8O_6P_2$	1,2-$[(^-O)_2P(O)CH_2]_2C_6H_4$	18(w)	172
C_2N	$C_3H_8NO_3P$	$H_2NCMe_2P(O)(O^-)_2$	16.8(w, HO^-)	204
C_2H	$C_6H_{11}O_3P$	$cHexP(O)(O^-)_2$	27.9(w, pH 7.3)	305
			24.9(w, pH 10.9)	

P-bound atoms = $C'O_2O'$

Rings/Connectivities	Formula	Structure	NMR data	Ref.
NN′;$(C)_2$	$C_{17}H_{21}N_2O_3P$	$(EtO)_2P(O)C(NHPh)=NPh$	2	306

TABLE J (continued)
Compilation of ^{31}P NMR Data of Four Coordinate Compounds Containing a P=Ch Bond and One Bond from Phosphorus to a Group IV Atom

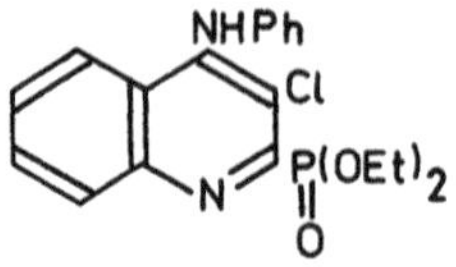

38

Rings/Connec-tivities	Formula	Structure	NMR data (δ_P[solv] $^nJ_{PZ}$ Hz)	Ref.
N'O;(C)$_2$	$C_{23}H_{41}NO_6P_2$	$[(BuO)_2P(O)]_2C=NPh$	−1,7	307
N'S;(C)$_2$	$C_6H_{12}NO_4PS$	$(MeO)_2P(O)C(SMe)=NAc$	2.5(n)	308
OO',(Si)$_2$	$C_9H_{23}O_5PSi_2$	$(TmsO)_2P(O)COOEt$	−23.1	309
OO';(C)$_2$	$C_4H_9O_5P$	$(MeO)_2P(O)COOMe$	−2.9	126
H$_2$N';(C)$_2$	$C_5H_{11}N_2O_3P$	$(EtO)_2P(O)CHN_2$	19.5(c)	58
C'Cl;(H)$_2$	$C_2H_5ClO_6P_2$	$(HO)_2P(O)CCl=CHP(O)(OH)_2$	3.0^E(w), 8.1^E(w)	310
C'Cl;(C)$_2$	$C_{10}H_{21}ClO_6P_2$	$(EtO)_2P(O)CCl=CHP(O)(OEt)_2$	5.6^E 8.8^E(n)	310
C'F;(C)$_2$	$C_8H_{16}FO_3P$	$(iPrO)_2P(O)CF=CH_2$	1.3 $^2J_{PF}$102.2	311
	$C_{14}H_{20}FO_3P$	$(iPrO)_2P(O)CF=CHPh$	3.3^E 4.7^Z $^2J_{PF}96.1^E$ 109.9^Z	311
CN';(C)$_2$	$C_7H_{13}N_2O_5P$	$(MeO)_2P(O)C=NNHCH_2CHCOOMe$	13	274
C'N;(C)$_2$	$C_9H_{16}NO_3P$	$(EtO)_2P(O)C=CHCH=CHNMe$	9.5(c)	16
	$C_{10}H_{22}NO_3P$	$(EtO)_2P(O)C(NMe_2)=CHEt$	16.8^E(c) 15.3^Z(c)	26
	$C_{14}H_{22}NO_3P$	$(EtO)_2P(O)C(NMe_2)=CHPh$	16.9^E(c) 16.8^Z(c)	312
	$C_{18}H_{21}NO_3P$	$(EtO)_2P(O)C(NHPh)=CHPh$	16.1(c)	26
C'N;(C')$_2$	$C_{17}H_{16}NO_3P$	$(PhO)_2P(O)C=CHCH=CHNMe$	2.5(c)	16
C'N';(H)$_2$	$C_5H_6NO_3P$	$(HO)_2P(O)C_5H_4N\text{-}2^h$	−2.3(w)	313
C'N';(C)$_2$	$C_9H_{14}NO_3P$	$(EtO)_2P(O)C_5H_4N\text{-}2^h$	8.2(c) $^1J_{PC}$220	314
	$C_{19}H_{20}ClN_2O_3P$	**38**	7.8	315
CO';(C)$_2$	$C_4H_6F_3O_4P$	$(MeO)_2P(O)COTf$	−5.5	238
	$C_4H_9O_4P$	$(MeO)_2P(O)COMe$	1.5	238
	$C_7H_{15}O_4P$	$(MeO)_2P(O)CO(\text{-}tBu)$	−4.8	238
	$C_{12}H_{17}O_4P$	$(EtO)_2P(O)COCH_2Ph$	−3.0 (keto)	26
C'O;(C)$_2$	$C_5H_{11}O_4P$	$(MeO)_2P(O)C(OMe)=CH_2$	18.6	216
	$C_7H_{17}O_4PSi$	$(MeO)_2P(O)C(OTms)=CH_2$	12	271
	$C_8H_{13}O_4P$	$(EtO)_2P(O)(C=CHCH=CHO)\text{-}2$	3.9(e)	16
	$C_8H_{15}O_5P$	$(EtO)_2P(O)C(OAc)=CH_2$	8	271
	$C_{12}H_{17}O_4P$	$(EtO)_2P(O)C(OH)=CHPh$	12.9	26
	$C_{30}H_{46}O_6P_2$	$(BuO)_2P(O)C[OP(OBu)_2]=CPh_2$	10	307
C'O;(C')$_2$	$C_{16}H_{13}O_4P$	$(PhO)_2P(O)(C=CHCH=CHO)\text{-}2$	−3 0(c)	16
C'O';(C)$_2$	$C_9H_{11}O_4P$	$(MeO)_2P(O)COPh$	1.0	238
C'P^4;(H)$_2$	$C_2H_6O_6P_2$	$H_2C=C[P(O)(OH)_2]_2$	11.2(w) $^1J_{PC}$157	316
C'P^4;(C)$_2$	$C_{10}H_{22}O_6P_2$	$H_2C=C[P(O)(OEt)_2]_2$	12.8(c)	317
	$C_{16}H_{26}O_6P_2$	$PhHC=C[P(O)(OEt)_2]_2$	12.1 16.1^{EZ} $^2J_{PP}$50	318
	$C_{16}H_{28}O_7P_2$	$(MeO)_2P(O)C[=P(OEt)_3]C_6H_4OMe\text{-}4$	31.6 (or 49.4)	276
C'S;(C)$_2$	$C_8H_{13}O_3PS$	$(EtO)_2P(O)(C=CHCH=CHS)\text{-}2$	10.9(c)	16
	$C_{13}H_{18}ClO_3PS$	$(EtO)_2P(O)C((SPh)=CClMe$	13^E 9^Z	319
C'S;(C')$_2$	$C_{16}H_{13}O_3PS$	$(PhO)_2P(O)(C=CHCH=CHS)\text{-}2$	4.1(c)	16
C'H;(H)$_2$	$C_2H_6O_6P_2$	$(HO)_2P(O)CH=CHP(O)(OH)_2$	11.5^E(w)	58
C'H;C,H	$C_6H_{14}O_6P_2$	$HO(EtO)P(O)CH=CHP(O)(OEt)OH$	12.5^E(w)	58
C'H;(C)$_2$	$C_6H_{11}O_3P$	$(MeO)_2P(O)CH=CHCH=CH_2$	28.0	320

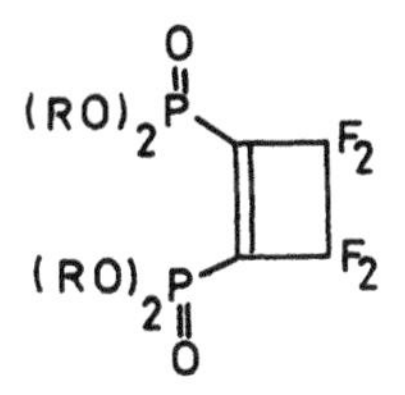

39

Rings/Connectivities	Formula	Structure	NMR data (δ_P[solv] $^nJ_{PZ}$ Hz)	Ref.
	$C_6H_{11}O_5P$	$(MeO)_2P(O)CH{=}\dot{C}OCH_2CH_2\dot{O}$	19.0	321
	$C_6H_{13}O_3P$	$(EtO)_2P(O)CH{=}CH_2$	17.3 $^1J_{PC}$182.1 $^2J_{PC}$1.9	10
	$C_6H_{13}O_5PS$	$(MeO)_2P(O)CH{=}CHSO_2Et$	12.1	322
	$C_6H_{14}NO_3P$	$(MeO)_2P(O)CH{=}CHNMe_2$	28.6(c)	246
	$C_6H_{14}O_6P_2$	$(MeO)_2P(O)CH{=}CH(O)(OMe)_2$	15.5(g)	58
	$C_7H_{11}F_4O_3P$	$(EtO)_2P(O)CH{=}CFTf$	7.5^Z(c)	323
	$C_7H_{13}O_3P$	$(EtO)_2P(O)CH{=}C{=}CH_2$	14.4	85
	$C_7H_{13}O_6P$	$(MeO)_2P(O)CH{=}C(OCOOMe)Me$	15^Z(g)	324
	$C_7H_{14}FO_3P$	$(EtO)_2P(O)CH{=}CHCH_2F$	16.9^E(c)	203
	$C_7H_{15}O_3P$	$(EtO)_2P(O)CH{=}CHMe$	14.7(k)	193
	$C_8H_{16}NO_3P$	$(MeO)_2P(O)CH{=}C{=}N(\text{-tBu})$	25.5	325
	$C_8H_{17}O_3P$	$(EtO)_2P(O)CH{=}CMe_2$	17.5(c)	77
	$C_9H_{17}O_3P$	$(EtO)_2P(O)CH{=}C{=}CMe_2$	15.0	326
	$C_9H_{20}NO_3P$	$(iPrO)_2P(O)CH{=}C(NH_2)Me$	24.5^E 24.9^Z	327
	$C_9H_{21}O_4PSi$	$(EtO)_2P(O)CH{=}CH(OTms)$	20.6^E(p) 18.0^Z(p)	12
	$C_{10}H_{21}O_3PS_2$	$(EtO)_2P(O)CH{=}C(SEt)_2$	11.3	321
	$C_{10}H_{21}O_5P$	$(EtO)_2P(O)CH{=}C(OEt)_2$	20.3	321
	$C_{12}H_{15}O_5P$	$(MeO)_2P(O)CH{=}C(OCOPh)Me$	12(g)	324
	$C_{13}H_{18}FO_3P$	$(EtO)_2P(O)CH{=}CHCHFPh$	16.6^E(c)	203
	$C_{14}H_{21}O_3P$	$(iPrO)_2P(O)CH{=}CHPh$	17.0^E	301
	$C_{18}H_{21}O_5P$	$(EtO)_2P(O)CH{=}C(OPh)_2$	19.0	321
C'H;(C')$_2$	$C_{14}H_{13}O_3P$	$(PhO)_2P(O)CH{=}CH_2$	10(a)	131
CC';(H)$_2$	$C_6H_{11}O_3P$	$(HO)_2P(O)C_6H_9$[b]	19.1 $^1J_{PC}$176.9	30
	$C_{11}H_{21}O_3P$	$(HO)_2P(O)C(\text{-tBu}){=}C{=}CH(\text{-tBu})$	16.1(a)	61
CC';(C)$_2$	$C_8H_{12}F_4O_6P_2$	**39;** R = Me	+(?)0.8	328
	$C_8H_{14}Cl_3O_3P$	$(EtO)_2P(O)C(CH_2Cl){=}C(CH_2Cl)Cl$	14.4^Z	31
	$C_8H_{15}O_3P$	$(EtO)_2P(O)CMe{=}C{=}CH_2$	15.0	326
	$C_{11}H_{20}ClO_3P$	$(MeO)_2P(O)C(\text{-tBu}){=}C{=}C(CH_2Cl)Me$	19.7	92
	$C_{14}H_{27}O_3P$	$(BuO)_2P(O)C_6H_9$[b]	18.2	91
CC';(C')$_2$	$C_{15}H_{15}O_3P$	$(PhO)_2P(O)CMe{=}CH_2$	12	131
C'$_2$;(Si)$_2$	$C_{12}H_{22}FO_3PSi_2$	$(TmsO)_2P(O)C_6H_4F\text{-}2$	−6.3(b)	329
	$C_{12}H_{22}FO_3PSi_2$	$(TmsO)_2P(O)C_6H_4F\text{-}4$	−0.9(b)	329
	$C_{12}H_{23}O_3PSi_2$	$(TmsO)_2P(O)Ph$	0.0(b)	329
	$C_{13}H_{25}O_3PSi_2$	$(TmsO)_2P(O)Tol\text{-}3$	0.6(b)	329
	$C_{13}H_{25}O_3PSi_2$	$(TmsO)_2P(O)Tol\text{-}4$	−0.5(b)	329
	$C_{13}H_{25}O_4PSi_2$	$(TmsO)_2P(O)C_6H_4OMe\text{-}2$	−2.8(b)	329
C'$_2$;P^4,H	$C_{12}H_{12}O_5P_2$	$(HO)PhP(O)OP(O)Ph(OH)$	9.8	331
C'$_2$;(P)$_2$	$C_6H_5F_4O_3P_3$	$PhP(O)(OPF_2)_2$	1.4(c, r) $^2J_{PP}$17.8	330
C'$_2$;(H)$_2$	$C_5H_6NO_3P$	$(HO)_2P(O)C_5H_4N\text{-}3$[h]	3.1(w)	313
	$C_6H_7O_3P$	$(HO)_2P(O)Ph$	17(c) 18.5(w)	333
	$C_6H_8NO_3P$	$(HO)_2P(O)C_6H_4NH_2\text{-}3$	3.6(w)	313
	$C_6H_8NO_3P$	$(HO)_2P(O)C_6H_4NH_2\text{-}4$	12.2	313
	$C_9H_8NO_3P$	$(HO)_2P(O)C_9H_6N\text{-}3$[i]	4.3(w)	313
	$C_9H_{13}O_3P$	$(HO)_2P(O)Mes$	27.5(c)	334

TABLE J (continued)
Compilation of ^{31}P NMR Data of Four Coordinate Compounds Containing a P=Ch Bond and One Bond from Phosphorus to a Group IV Atom

Rings/Connectivities	Formula	Structure	NMR data (δ_P[solv] $^nJ_{PZ}$ Hz)	Ref.
C′$_2$;C,P^4	$C_{16}H_{20}O_5P_2$	(EtO)PhP(O)OP(O)Ph(OEt)	8.6	335
C′$_2$;H,C	$C_8H_{11}O_3P$	PhP(O)(OEt)OH	16.8	59
	$C_{10}H_{15}O_3P$	MesP(O)(OMe)OH	25.4(c)	334
	$C_{16}H_{25}O_3P$	PhP(O)(OC$_{10}$H$_{19}$)OH[a]	16.8	59
C′$_2$;(C)$_2$	$C_6H_{10}ClO_3P$	HClC=C[P(O)(OMe)$_2$]CH=CH$_2$	13.6^E	89
	$C_9H_{11}O_4P$	(MeO)$_2$P(O)C(COMe)=CH$_2$	13.4	336
	$C_8H_{11}O_3P$	PhP(O)(OMe)$_2$	19.3(n) 20.5(r)	337
	$C_8H_{13}Cl_2O_3P$	H$_2$C=C[P(O)(OEt)$_2$]CCl=CHCl	12.4(n)	93
	$C_8H_{15}O_3PS$	H$_2$C=C[P(O)(OMe)$_2$]C(SEt)=CH$_2$	14.3	338
	$C_9H_{10}Cl_4NO_3P$	(EtO)$_2$P(O)C$_5$Cl$_4$N-4[j]	6.1(c)	100
	$C_9H_{10}F_4NO_3P$	(EtO)$_2$P(O)C$_5$F$_4$N-4[c]	0.6(c)	100
	$C_9H_{14}NO_3P$	(EtO)$_2$P(O)C$_5$H$_4$N-3[h]	12.7 25.2	339
	$C_9H_{14}NO_3P$	(EtO)$_2$P(O)C$_5$H$_4$N-4[h]	15.0	340
	$C_{10}H_9F_6O_3P$	PhP(O)(OCH$_2$Tf)$_2$	20.4(t)	341
	$C_{10}H_{10}Br_2N_3O_9P$	(EtO)$_2$P(O)C$_6$Br$_2$(NO$_2$)$_3$-3,5,2,4,6	−0.2(c)	342
	$C_{10}H_{10}F_5O_3P$	(EtO)$_2$P(O)C$_6$F$_5$	4.4(c)	151
	$C_{10}H_{13}O_5P$	(MeO)$_2$P(O)C$_6$H$_4$COOMe-4	19.0	1
	$C_{10}H_{14}BrO_3P$	(EtO)$_2$P(O)C$_6$H$_4$Br-4	15.5(c)	16
	$C_{10}H_{14}ClO_3P$	(EtO)$_2$P(O)C$_6$H$_4$Cl-3	14.1(c)	16
	$C_{10}H_{14}ClO_3P$	(EtO)$_2$P(O)C$_6$H$_4$Cl-4	17.0	1
	$C_{10}H_{14}NO_5P$	(EtO)$_2$P(O)C$_6$H$_4$NO$_2$-2	10.5(e)	16
	$C_{10}H_{15}O_3P$	(EtO)$_2$P(O)Ph	16.9(c,v,n)	343
	$C_{10}H_{15}O_4P$	(EtO)$_2$P(O)C$_6$H$_4$OH-2	22.4	136
	$C_{10}H_{16}NO_3P$	(EtO)$_2$P(O)C$_6$H$_4$NH$_2$-2	18.6	313
	$C_{10}H_{16}NO_3P$	(EtO)$_2$P(O)C$_6$H$_4$NH$_2$-3	18.9	313
	$C_{10}H_{16}NO_3P$	(EtO)$_2$P(O)C$_6$H$_4$NH$_2$-4	18.1	313
	$C_{10}H_{20}NO_3P$	H$_2$C=C[P(O)(OMe)$_2$]C(NEt$_2$)=CH$_2$	16.8	344
	$C_{11}H_{15}O_4P$	(EtO)$_2$P(O)C$_6$H$_4$CHO-3	14.2	16
	$C_{11}H_{17}O_3P$	(EtO)$_2$P(O)Tol-3	17.0	16
	$C_{11}H_{17}O_3P$	(EtO)$_2$P(O)Tol-4	19.0	1
	$C_{11}H_{17}O_4P$	(EtO)$_2$P(O)C$_6$H$_4$OMe-4	17.3(c)	16
	$C_{12}H_{20}NO_3P$	(EtO)$_2$P(O)C$_6$H$_4$NHEt-4	21.9(c) $^1J_{PC}$118.4	141
	$C_{13}H_{15}O_5P$	(EtO)$_2$P(O)C$_9$H$_5$O$_2$[k]	10.8	345
	$C_{13}H_{16}NO_3P$	(EtO)$_2$P(O)C$_9$H$_6$N-3[i]	13.8	313
	$C_{13}H_{23}O_4PSi$	(EtO)$_2$P(O)C$_6$H$_4$OTms-2	15.9	136
	$C_{14}H_{20}Br_2N_2O_{10}P_2$	1,4-[(EtO)$_2$P(O)]$_2$C$_6$Br$_2$(NO$_2$)$_2$-2,6,3,5	0.6	342
	$C_{14}H_{20}ClO_3PS$	(MeO)$_2$P(O)CPh=C(SMe)CH$_2$Cl	14.7^Z	32
	$C_{14}H_{24}O_6P_2$	1,2-[(EtO)$_2$P(O)]$_2$C$_6$H$_4$	15.4	346
	$C_{15}H_{21}O_5P$	(EtO)$_2$P(O)C(COOEt)=CHPh	13.8(c)	25
	$C_{16}H_{24}NO_5PS$	(EtO)$_2$P(O)C(COOEt)=C(SMe)NHPh	13.5 18.6	347
	$C_{16}H_{29}O_7P$	(EtO)$_2$P(O)C(COOEt)=CHCMe$_2$CH$_2$COOEt	13.8^E 10.5^Z	281
	$C_{18}H_{24}O_5P_2$	(EtO)$_2$P(O)C$_6$H$_4$[P(O)Ph(OEt)]-2	15.6	346
C′$_2$;CC′	$C_{16}H_{20}NO_3P$	PhP(O)(OPh)OCH$_2$CH$_2$NMe$_2$	15.6(d)	18
	$C_{19}H_{23}O_3P$	PhP(O)(ONeo)OCPh=CH$_2$	15.0	261
C′$_2$;(C′)$_2$	$C_{10}H_{11}O_3P$	PhP(O)(OCH=CH$_2$)$_2$	11.4	284
	$C_{16}H_{13}Cl_2O_3P$	H$_2$C=C[P(O)(OPh)$_2$]CCl=CHCl	5.6(n)	93
	$C_{18}H_{15}O_3P$	PhP(O)(OPh)$_2$	10.3(c) 11.8(e)	348
	$C_{20}H_{15}O_5P$	PhP(O)(OCOPh)$_2$	8.8	349
C′C″;(C)$_2$	$C_{13}H_{16}NO_3P$	(EtO)$_2$P(O)C(CN)=CHPh	11.9(c) $^1J_{PC}$197.1	25
5/CO′;(C)$_2$	$C_8H_{15}O_4P$	MeCOP(O)Pnc	10.5	286

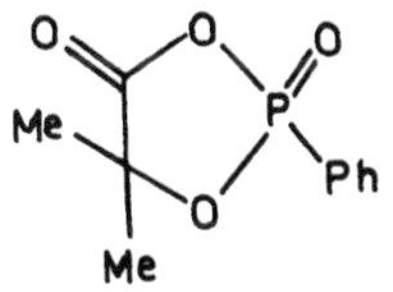

40 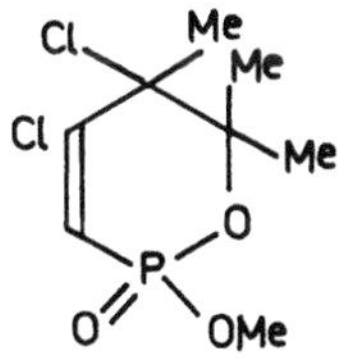 41 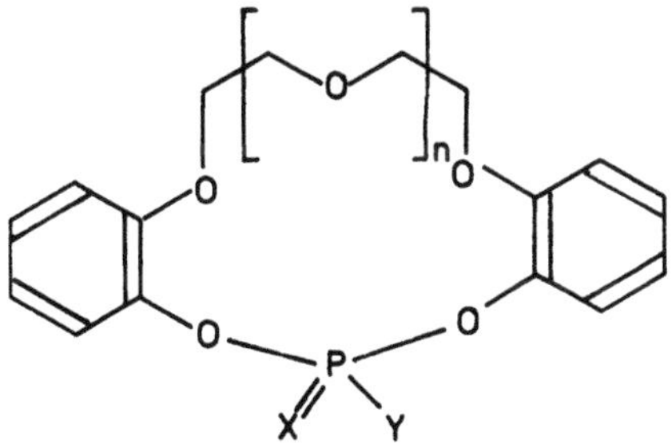 42

Rings/Connectivities	Formula	Structure	NMR data (δ_P[solv] $^nJ_{PZ}$ Hz)	Ref.
5/C'O';(C)$_2$	$C_{13}H_{17}O_4P$	PhCOP(O)Pnc	12.9	286
5/C'H;H,C	$C_5H_9O_3P$	**5**; R[1] = R[2] = H, X = O, Y = OH	41.9(c)	350
5/C'H;(C)$_2$	$C_6H_{11}O_3P$	**5**; R[1] = R[2] = H, X = O, Y = OMe	38 40.4	351
	$C_8H_9O_3PS$	**5**; R[1] = H, R[2] = SMe, X = O, Y = OEt	30	33
5/CC';P[4],C	$C_{22}H_{40}O_5P_2$	**6**; X = O, Y = OP(O)-OtBuCH=C(-tBu)	28.4(c)	61
5/CC';H,C	$C_6H_9Cl_2O_3P$	**5**; R[1] = CH_2Cl, R[2] = Cl, X = O, Y = OH	33.0	31
	$C_{11}H_{21}O_3P$	**6**; X = O, Y = OH	42.0(c)	61
5/CC';(C)$_2$	$C_8H_{13}Cl_2O_3P$	**5**; R[1] = CH_2Cl, R[2] = Cl, X = O, Y = OEt	25.0	31
5/C'$_2$;(C)$_2$	$C_8H_9O_3P$	PhP(O)Glc	31(c, v) 36.0(c) 38.4	352
	$C_8H_{12}ClO_3P$	**5**; R[1] = CH=CH$_2$, R[2] = Cl, X = O, Y = OMe	22.2 24.4(c)	354
	$C_9H_{15}O_3PS$	**5**; R[1] = CH=CH$_2$, R[2] = SMe, X = O, Y = OMe	31.7(c)	253
	$C_{10}H_{13}O_3P$	**H5**; R = Me, X = O, Y = Ph (4*R*, 5*R*)	30.7	288
	$C_{12}H_{17}O_3P$	PhP(O)Pnc	28.7	107
	$C_{13}H_{17}O_3PS$	**5**; R[1] = Ph, R[2] = SMe, X = O, Y = OMe	31.5	32.
5/C'$_2$;CC'	$C_{10}H_{11}O_4P$	**40**	24	355
6/C'H;(C)$_2$	$C_8H_{13}Cl_2O_3P$	**41**	6.8 $^2J_{PH}$ 8.5	356
6/C'$_2$;(C)$_2$	$C_9H_{11}O_3P$	Ph(O)P(OCH$_2$CH$_2$CH$_2$O)	11.2(c,v)	284
	$C_{11}H_{15}O_3P$	Ph(O)P(OCH$_2$CMe$_2$CH$_2$O)	13.7(b)	357
7/C'H;(C)$_2$	$C_5H_9O_4P$	MeO(O)P(CH=CHOCH$_2$CH$_2$O)	24	358
7/C'$_2$;(C)$_2$	$C_{10}H_{13}O_3P$	Ph(O)P(OCH$_2$CH$_2$CH$_2$CH$_2$O)	21.2(c)	359
17/C'$_2$;(C')$_2$	$C_{24}H_{25}O_7P$	**42**; X = O, Y = Ph, n = 2	14.5(c)	360

P-bound atoms = C'O'$_3$

Rings/Connectivities	Formula	Structure	NMR data	Ref.
C'$_2$	$C_6H_5O_3P$	PhP(O)(O$^-$)$_2$	14.75(w pH 7) 10.8, 13.8(w)	64(b)

P-bound atoms = C"O$_2$O'

Rings/Connectivities	Formula	Structure	NMR data	Ref.
N";(C)$_2$	$C_5H_{10}NO_3P$	(EtO)$_2$P(O)CN	−22	361
N";(C')$_2$	$C_{13}H_{10}NO_3P$	(PhO)$_2$P(O)CN	−29.9(c)	362
C";(H)$_2$	$C_2H_4O_6P_2$	(HO)$_2$P(O)C≡CP(O)(OH)$_2$	−12(w)	310
	$C_3H_5O_3P$	(HO)$_2$P(O)C≡CMe	−7.6	11

TABLE J (continued)
Compilation of ^{31}P NMR Data of Four Coordinate Compounds Containing a P=Ch Bond and One Bond from Phosphorus to a Group IV Atom

Rings/Connectivities	Formula	Structure	NMR data (δ_P[solv] $^nJ_{PZ}$ Hz)	Ref.
	$C_6H_{10}ClO_3P$	$(EtO)_2P(O)C{\equiv}CCl$	−3	310
	$C_6H_{11}O_3P$	$(EtO)_2P(O)C{\equiv}CH$	−9.2	12
	$C_7H_{13}O_3P$	$(EtO)_2P(O)C{\equiv}CMe$	−7.9	12
	$C_9H_{13}O_3P$	$(EtO)_2P(O)C{\equiv}CC{\equiv}CMe$	−9	363
	$C_{10}H_{20}NO_3P$	$(iPrO)_2P(O)C{\equiv}CNMe_2$	−2.2	325
	$C_{10}H_{20}O_6P_2$	$(EtO)_2P(O)C{\equiv}CP(O)(OEt)_2$	−11.0(n,c)	310
	$C_{12}H_{15}O_3P$	$(EtO)_2P(O)C{\equiv}CPh$	−7.5	11
6/N″;(C)$_2$	$C_6H_{10}NO_3P$	$NC(O)P(OCH_2CMe_2CH_2O)$	−28.5	364

Section J2: Derivatives of Thio-, Seleno-, and Telluro-Phosphonic Acids

Connectivity of P-bound atoms	Formula	Structure	NMR data (δ_P [solv]$^nJ_{PZ}$ HZ)	Ref.
	P-bound atoms = CBr$_2$S′			
H$_2$Br	CH_2Br_3PS	$BrCH_2P(S)Br_2$	20.0(n)	64(b)
H$_3$	CH_3Br_2PS	$MeP(S)Br_2$	20.5(n)	64(b)
C′H$_2$	$C_7H_7Br_2PS$	$PhCH_2P(S)Br_2$	36.5	365
C$_2$H	$C_3H_7Br_2PS$	$iPrP(S)Br_2$	74.0(r)	366
C$_3$	$C_4H_9Br_2PS$	$tBuP(S)Br_2$	85.3(r) 83.5(c)	366
	P-bound atoms = C′Br$_2$S′			
C′$_2$	$C_6Br_2F_5PS$	$C_6F_5P(S)Br_2$	45.5	367
	$C_6H_5Br_2PS$	$PhP(S)Br_2$	19.8	368
	P-bound atoms = CClFSe′			
H$_3$	$CH_3ClFPSe$	$MeP(Se)ClF$	98.1 $^1J_{PF}$1175 $^1J_{PSe}$1020	39
CH$_2$	$C_2H_5ClFPSe$	$EtP(Se)ClF$	111.0 $^1J_{PF}$1175	39
	P-bound atoms = CClNS′			
H$_3$;C$_2$	$C_5H_{13}ClNPS$	$MeP(S)(NEt_2)Cl$	88	369
CH$_2$;C$_2$	$C_7H_{13}ClNPS$	$EtP(S)(NEtCH_2C{\equiv}CH)Cl$	100	48
C$_2$H;C$_2$	$C_{10}H_{21}ClNPS$	$cHexP(S)(NEt_2)Cl$	108	370
C$_3$;C$_2$	$C_8H_{19}ClNPS$	$tBuP(S)(NEt_2)Cl$	117.1	371
	P-bound atoms = C′ClNS′			
C′$_2$;C$_2$	$C_{10}H_{15}ClNPS$	$PhP(S)(NEt_2)Cl$	85	370
	P-bound atoms = CClNSe′			
C$_3$;C$_2$	$C_8H_{19}ClNPSe$	$tBuP(Se)(NEt_2)Cl$	118.0 $^1J_{PSe}$874	371

Rings/Connec-tivities	Formula	Structure	NMR data (δ_P[solv] $^nJ_{PZ}$ Hz)	Ref.
	P-bound atoms = C′ClNSe′			
C'_2;C_2	$C_{11}H_{13}ClNPSe$	PhP(Se)(NEtCH$_2$C≡CH)Cl	74	48
	P-bound atoms = CClOS′			
H_3;C	C_3H_8ClOPS	MeP(S)(OEt)Cl	96	369
H_3;C′	C_7H_8ClOPS	MeP(S)(OPh)Cl	92	369
CH_2;C	$C_4H_{10}ClOPS$	EtP(S)(OEt)Cl *(S)*	106	112
C_2H;C	$C_4H_{10}ClOPS$	iPrP(S)(OMe)Cl	117.5	79
	P-bound atoms = CClO′S			
H_3;C	C_3H_8ClOPS	MeP(O)(SEt)Cl	60	369
H_3;C′	C_7H_8ClOPS	MeP(O)(SPh)Cl	52	369
	P-bound atoms = C′ClOS′			
C'_2;C	$C_8H_{10}ClOPS$	PhP(S)(OEt)Cl	86.4 87.5	372
	P-bound atoms = CClSS′			
H_3;C′	$C_7H_8ClPS_2$	MeP(S)(SPh)Cl	90	369
	P-bound atoms = CCl$_2$S′			
F_3	CCl_2F_3PS	TfP(S)Cl$_2$	43.3	150
H_2Cl	CH_2Cl_3PS	ClCH$_2$P(S)Cl$_2$	73.0(n) 82.9	373
H_2P^4	$CH_2Cl_4P_2S_2$	Cl$_2$P(S)CH$_2$P(S)Cl$_2$	52.9	68
H_3	CH_3Cl_2PS	MeP(S)Cl$_2$	77.5 79.5	374
CHCl	$C_2H_4Cl_3PS$	MeCHClP(S)Cl$_2$	86.9	74(a)
CH_2	$C_2H_4Cl_3PS$	ClCH$_2$CH$_2$P(S)Cl$_2$	78.8	74(a)
	$C_2H_4Cl_4P_2S_2$	Cl$_2$P(S)CH$_2$CH$_2$P(S)Cl$_2$	78.8	68
	$C_3H_6Cl_4P_2S_2$	Cl$_2$P(S)(CH$_2$)$_3$P(S)Cl$_2$	56.5	68
$C'H_2$	$C_5H_8Cl_3PS$	ClCH$_2$CH=CMeCH$_2$P(S)Cl$_2$	80.3^E	78
	$C_7H_7Cl_2PS$	PhCH$_2$P(S)Cl$_2$	85.3	64(b)
C_2H	$C_3H_7Cl_2PS$	iPrP(S)Cl$_2$	104.1	79
	$C_{10}H_{19}Cl_2PS$	Cl$_2$P(S)C$_{10}$H$_{19}$^{a}	101.6	375
C_3	$C_4H_9Cl_2PS$	tBuP(S)Cl$_2$	118.1(r) 116.4(c)	366
	P-bound atoms = C′Cl$_2$S			
C′H	$C_2H_2Cl_4P_2S_2$	Cl$_2$P(S)CH=CHP(S)Cl$_2$	57(g)	58
	$C_2H_3Cl_2PS$	Cl$_2$P(S)CH=CH$_2$	69.5 $^1J_{PC}$109.9 $^2J_{PC}$3.2	10
	$C_5H_8Cl_3PS$	Cl$_2$P(S)CH=CMeCH$_2$CH$_2$Cl	60.1^E	78
C'_2	$C_6Cl_2F_5PS$	C$_6$F$_5$P(S)Cl$_2$	30.0	367
	$C_6H_4BrCl_2PS$	Cl$_2$P(S)C$_6$H$_4$Br-2	44.0(n)	376
	$C_6H_4BrCl_2PS$	Cl$_2$P(S)C$_6$H$_4$Br-4	71.3(n) 70.2(g)	376
	$C_6H_4Cl_3PS$	Cl$_2$P(S)C$_6$H$_4$Cl-2	65.6(n)	376
	$C_6H_4Cl_3PS$	Cl$_2$P(S)C$_6$H$_4$Cl-4	71.6(n) 69.9(g)	376
	$C_6H_4Cl_2FPS$	Cl$_2$P(S)C$_6$H$_4$F-2	59.7(n)	376
	$C_6H_4Cl_2FPS$	Cl$_2$P(S)C$_6$H$_4$F-4	70.0(g) 71.6(n)	376
	$C_6H_5Cl_2PS$	Cl$_2$P(S)Ph	74.5(n) 73.1(g)	376
	$C_8H_{10}Cl_2NPS$	Cl$_2$P(S)C$_6$H$_4$NMe$_2$-4	77.7(n) 77.6(g)	376
	P-bound atoms = CCl$_2$Se′			
H_3	CH_3Cl_2PSe	MeP(Se)Cl$_2$	55.1 $^1J_{PSe}$955(923)	377

TABLE J (continued)
Compilation of ^{31}P NMR Data of Four Coordinate Compounds Containing a P=Ch Bond and One Bond from Phosphorus to a Group IV Atom

Rings/Connec-tivities	Formula	Structure	NMR data (δ_P[solv] $^nJ_{PZ}$ Hz)	Ref.
CH_2	$C_2H_5Cl_2PSe$	EtP(Se)Cl_2	75.5(n) $^1J_{PSe}$921	377
	$C_4H_9Cl_2PSe$	BuP(Se)Cl_2	71.5(n) $^1J_{PSe}$915	377
	P-bound atoms = C′Cl₂Se′			
C′$_2$	$C_6H_5Cl_2PSe$	PhP(Se)Cl_2	57(n) $^1J_{PSe}$940	377
	P-bound atoms = CFNS′			
H_3;CH	C_2H_7FNPS	MeP(S)(NHMe)F	98.4(n) $^1J_{PF}$1043	46
C_3;C_2	$C_8H_{19}FNPS$	tBuP(S)(NEt_2)F	118.7 $^1J_{PF}$1084	371
	P-bound atoms = C′FNS′			
C′$_2$;C_2	$C_{10}H_{15}FNPS$	PhP(S)(NEt_2)F	90.0(n) $^1J_{PF}$1059	46
	P-bound atoms = CFNSe′			
C_3;C_2	$C_8H_{19}FNPSe$	tBuP(Se)(NEt_2)F	130.0 $^1J_{PF}$1099 $^1J_{PSe}$905	371
	P-bound atoms = CFOS′			
CH_2;C	$C_4H_{10}FOPS$	EtP(S)(OEt)F	108.7(n) $^1J_{PF}$1100	46
	P-bound atoms = C′FOS′			
C′$_2$;C	C_7H_8FOPS	PhP(S)(OMe)F	93.5(n) $^1J_{PF}$1084	46
	$C_8H_{10}FOPS$	PhP(S)(OEt)F	90.0 $^1J_{PF}$1075	378
	P-bound atoms = CFSS′			
H_3;H	CH_4FPS_2	MeP(S)(SH)F	115.1 $^1J_{PF}$1151	379
H_3;C	$C_3H_8FPS_2$	MeP(S)(SEt)F	121.0(n) $^1J_{PF}$1093	46
CH_2;H	$C_2H_6FPS_2$	EtP(S)(SH)F	124.6 $^1J_{PF}$1124	379
	P-bound atoms = CF₂S′			
H_3	CH_3F_2PS	MeP(S)F_2	104.9 $^1J_{PF}$1147	46
C_3	$C_4H_9F_2PS$	tBuP(S)F_2	118.3 $^1J_{PF}$1209	104(a)
	P-bound atoms = C′F₂S′			
C′H	$C_2H_3F_2PS$	F_2P(S)CH=CH_2	90.9 $^1J_{PC}$151.3 $^2J_{PC}$7.9	10
	$C_8H_6ClF_2PS$	F_2P(S)CH=CClPh	78.8^Z $^1J_{PF}$1145	105
	$C_8H_7F_2PS$	F_2P(S)CH=CHPh	95.0(n) $^1J_{PF}$1118	46
C′$_2$	C_6F_7PS	F_2P(S)C_6F_5	61.8 $^1J_{PF}$1158	367
	$C_6H_5F_2PS$	F_2P(S)Ph	91.3(n) $^1J_{PF}$1140	46

Rings/Connec-tivities	Formula	Structure	NMR data (δ_P[solv] $^nJ_{PZ}$ Hz)	Ref.
	P-bound atoms = CF_2Se'			
H_3	CH_3F_2PSe	$MeP(Se)F_2$	112.5 $^1J_{PF}$1190 $^1J_{PSe}$1040	39
CH_2	$C_2H_5F_2PSe$	$EtP(Se)F_2$	121.0 $^1J_{PSe}$1192	39
	P-bound atoms = $C'F_2Se'$			
C'_2	$C_6H_5F_2PSe$	$PhP(Se)F_2$	124.1 $^1J_{PF}$1181	39
	P-bound atoms = CNOS'			
$HF_2;C_2;C$	$C_4H_{10}F_2NOPS$	$F_2CHP(S)(NMe_2)OMe$ *(R)*	72.9(c) $^2J_{PF}$80.6, 89.1	6
$H_3;C_2;C'$	$C_{11}H_{18}NOPS$	$MeP(S)(NEt_2)OPh$	87	369
$CH_2;HS^4;C$	$C_{10}H_{16}NO_3PS_2$	$EtP(S)(NHSO_2Ph)OEt$	86.5(c)	380
$C_2H;C_2;H$	$C_{10}H_{22}NOPS$	$cHexP(S)(NEt_2)OH$	77(g)	381
$C_2H;C_2;C$	$C_9H_{20}NOPS$	$cHexP(S)(NMe_2)OMe$	103.1	382
$5/HCl_2;C_2;C$	$C_{11}H_{14}Cl_2NOPS$	**H12**; X, Y = S′, $CHCl_2$		7
		(2*R*,4*S*,5*R*)	87.2(c)	
		(2*S*,4*S*,5*R*)	86.9(c)	
$5/HF_2;C_2;C$	$C_{11}H_{14}F_2NOPS$	**H12**; X, Y = S′, CHF_2		6
		(2*R*,4*S*,5*R*)	78.9(e) $^2J_{PF}$83.0	
		(2*S*,4*S*,5*R*)	78.8(e) $^2J_{PF}$85.5, 89.1	
$5/H_2F;C_2;C$	$C_{11}H_{15}FNOPS$	**H12**; X, Y = S′, CH_2F		6
		(2*R*,4*S*,5*R*)	91.3(c) $^2J_{PF}$56.0	
		(2*S*,4*S*,5*R*)	90.8(c) $^2J_{PF}$57.4	
$5/H_3;C_2;C$	$C_{11}H_{22}NO_5PS$	**8**; X = S′, Y = Me	102.0(c)	36
		X = Me, Y = S′	99.0(c)	
	$C_{11}H_{22}NO_5PS$	**9**; X = S′, Y = Me	109.6(c)	36
		X = Me, Y = S′	103.8(c)	
$5/H_3;CC';C$	$C_9H_{12}NOPS$	**H11**; R = Ph, X = S, Y = Me	94	383
$5/H_3LCC';C'$	$C_{12}H_{12}NOPS$	**H3**; R = tBu, X = S, Y = Me	99.0	115
	P-bound atoms = CNO′S			
$H_3;C_2;C'$	$C_{11}H_{18}NOPS$	$MeP(O)NEt_2)SPh$	44	369
$C_2H;C_2;C$	$C_9H_{20}NOPS$	$cHexP(O)(NMe_2)SMe$	62.5	382
$5/H_3;C_2;C$	$C_{11}H_{16}NOPS$	**H38**; X, Y = O′, Me		5
		(2*R*,4*S*,5*S*)	58.2	
		(2*S*,4*S*,5*S*)	60.8	
		(2*R*,4*S*,5*R*)	58.2	
		(2*S*,4*S*,5*R*)	54.1	
	P-bound atoms = C′NOS′			
$C'_2;C_2;H$	$C_{10}H_{16}NOPS$	$PhP(S)(NEt_2)OH$	62(c)	332
$C'_2;C_2;C$	$C_{11}H_{18}NOPS$	$PhP(S)(NEt_2)OMe$	82.5	382
$5/C'_2;C'H;C$	$C_{13}H_{12}NOPS$	**43**; X = S, Y = NHPh	83.6(c)	384
$5/C'_2;C_2;C$	$C_9H_{12}NOPS$	**43**; X = S, Y = NMe_2	94.8(c)	384
	$C_9H_{12}NOPS$	**H11**; R = Me, X = S, Y = Ph	94.3	385
	$C_{16}H_{18}NOPS$	**H12**; X, Y = S′, Ph		386
		(2*R*,4*S*,5*R*)	99.4(c)	
		(2*S*,4*S*,5*R*)	92.6	
$5/C'_2;C'_2;C$	$C_{27}H_{22}NO_2PS$	**44**; X = NPh	86.4 91.3[D]	387
$6/C'_2;CH;C'$	$C_{17}H_{18}NO_2PS$	**45**	65.3(c)	388
$10/C'_2;C_2;C$	$C_{18}H_{24}N_2O_2P_2S_2$	**46**	80.9 81.1*	385

TABLE J (continued)
Compilation of ^{31}P NMR Data of Four Coordinate Compounds Containing a P=Ch Bond and One Bond from Phosphorus to a Group IV Atom

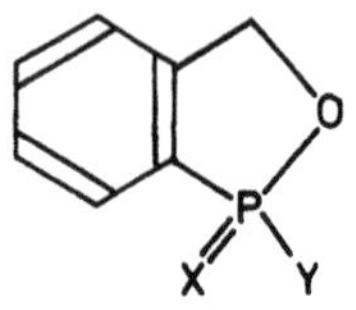

43

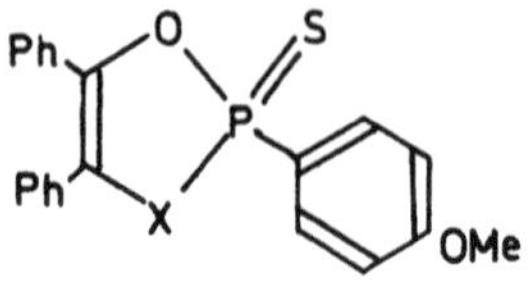

44

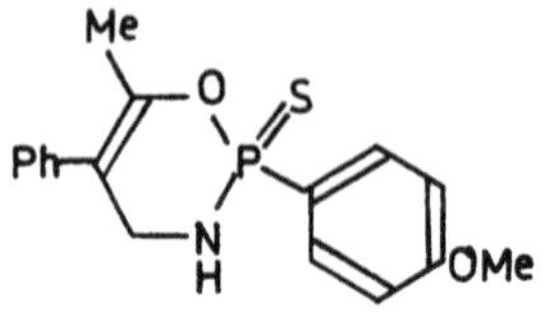

45

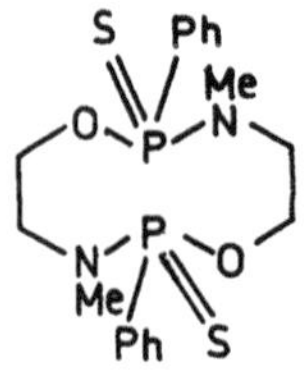

46

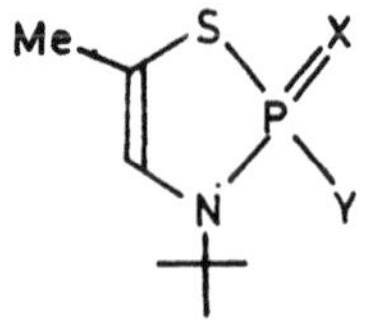

47

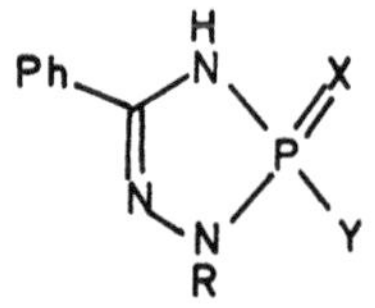

48

Rings/Connec-tivities	Formula	Structure	NMR data (δ_P[solv] $^nJ_{PZ}$ Hz)	Ref.
	P-bound atoms = C'NO'S			
C'$_2$;C$_2$;C	$C_{11}H_{18}NOPS$	PhP(O)(NEt$_2$)SMe	46.8	382
5/C'$_2$;C$_2$;C	$C_{16}H_{18}NOPS$	**H38**; X, Y = O', Ph		5
		(2*S*,4*S*,5*S*)	51.2(c)	
		(2*R*,4*S*,5*S*)	46.3	
	P-bound atoms = C'N'OS'			
C'$_2$;C';C	$C_9H_{10}NOPS_2$	PhP(S)(OEt)NCS	62.3	378
	P-bound atoms = CNSS'			
5/H$_3$;CC';C'	$C_8H_{16}NPS_2$	**47**; X = S, Y = Me	96.4	115
	P-bound atoms = CNS'$_2$			
H$_3$;C$_2$	$C_3H_9NPS_2$	MeP(S)(NMe$_2$)S$^-$	93.4	389
	P-bound atoms = C'NSS'			
C'$_2$;C$_2$;Si	$C_{11}H_{20}NPS_2Si$	PhP(S)(NMe$_2$)STms	80.2	390
	P-bound atoms = CN$_2$S'			
H$_3$;(C$_2$)$_2$	$C_9H_{23}N_2PS$	MeP(S)(NEt$_2$)$_2$	77.3	391
CH$_2$;(C$_2$)$_2$	$C_{10}H_{25}N_2PS$	EtP(S)(NEt$_2$)$_2$	84.7	391
5/H$_3$;C'N',C'H	$C_{14}H_{14}N_3PS$	**48**; R = Ph, X = S, Y = Me	74.5	392
5/CH$_2$;CN',C'H	$C_{10}H_{14}N_3PS$	**48**; R = Me, X = S, Y = Me	91	392

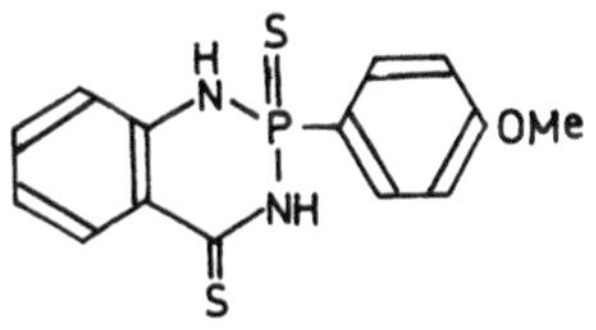

49

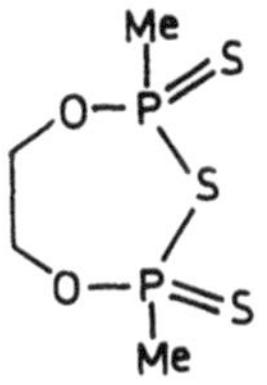

50

Rings/Connectivities	Formula	Structure	NMR data (δ_P[solv] $^nJ_{PZ}$ Hz)	Ref.
P-bound atoms = $C'N_2S'$				
$C'N';(C_2)_2$	$C_{16}H_{28}N_3PS$	$(Et_2N)_2P(S)CPh{=}NMe$	68.3	393
$C'_2;(H_2)_2$	$C_6H_9N_2PS$	$PhP(S)(NH_2)_2$	64.5(e)	134
$C'_2;(C_2)_2$	$C_{10}H_{17}N_2PS$	$PhP(S)(NMe_2)_2$	81.7(r)	135(b)
	$C_{12}H_{22}N_3PS$	$(Me_2N)_2P(S)C_6H_4NHEt$-4	82.6(c) $^1J_{PC}$131.8	137
	$C_{13}H_{25}N_2OPSSi$	$(Me_2N)_2P(S)C_6H_4OTms$-2	78.2	136
5/$C'_2;C'N',C'H$	$C_{19}H_{16}N_3PS$	**48**; R = Y = Ph, X = S	67	392
6/$C'_2;(C'H)_2$	$C_{14}H_{13}N_2OPS$	**49**	48.3	394
17/$C'_2;(C'H)_2$	$C_{24}H_{27}N_2O_4PS$	**18**; n = 2, X = S	54.8	138
20/$C'_2;(C'H)_2$	$C_{26}H_{31}N_2O_5PS$	**18**; n = 3, X = S	55.6	138
P-bound atoms = $C'N'_2S'$				
$C'_2;(C')_2$	$C_8F_5N_2PS_3$	$C_6F_5P(S)(NCS)_2$	6.8	11
P-bound atoms = CN_2Se'				
$H_3;(C_2)_2$	$C_5H_{15}N_2PSe$	$MeP(Se)(NMe_2)_2$	82.5 80.5 $^1J_{PSe}$789, −767	396
$CH_2;(C_2)_2$	$C_{10}H_{25}N_2PSe$	$EtP(Se)(NEt_2)_2$	85.7 $^1J_{PSe}$748	397
	$C_{12}H_{29}N_2PSe$	$BuP(Se)(NEt_2)_2$	74	398
P-bound atoms = $C'N_2Se'$				
$C'_2;(C_2)_2$	$C_{10}H_{17}N_2PSe$	$PhP(Se)(NMe_2)_2$	84.0 84.6(r) $^1J_{PSe}$ −790	399
	$C_{14}H_{25}N_2PSe$	$PhP(Se)(NEt_2)_2$	77.2 $^1J_{PSe}$766	397
P-bound atoms = CN_2Te'				
$H_3;(C_2)_2$	$C_5H_{15}N_2PTe$	$MeP(Te)(NMe_2)_2$	51.8 $^1J_{P^{125}Te}$1950	396(a)
P-bound atoms = COSS'				
$H_3;C;H$	$C_4H_{12}O_2P_2S_4$	$Me(HS)P(S)OCH_2CH_2OP(S)(SH)Me$	94	400
$CH_2;C;C$	$C_8H_{19}OPS_2$	EtP(S)(OEt)(S-iBu)	114	401
	$C_8H_{19}OPS_2$	BuP(S)(OMe)(S-iPr)	114	401
$CH_2;C;C'$	$C_{15}H_{30}N_3O_3P_3S_6$	$[Et(EtO)P(S)\dot{S}C{=}\dot{N}]_3$	106	402
6/$H_3;C;C$	$C_{10}H_{19}O_5PS_2$	**H26**; A = S, X = S′, Y = Me	100	403
		X = Me, Y = S′	90.5	
7/$H_3;C;P^4$	$C_4H_{10}O_2P_2S_3$	**50**	93	404
P-bound atoms = $CO'S_2$				
$F_3;(C)_2$	$C_5H_{10}F_3OPS_2$	$TfP(O)(SEt)_2$	45.9	150

TABLE J (continued)
Compilation of ^{31}P NMR Data of Four Coordinate Compounds Containing a P=Ch Bond and One Bond from Phosphorus to a Group IV Atom

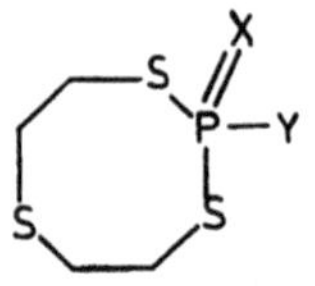

51 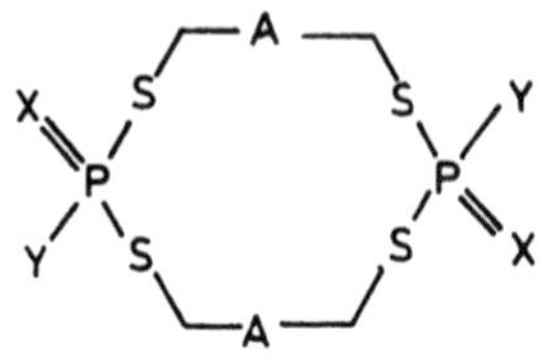 **52** 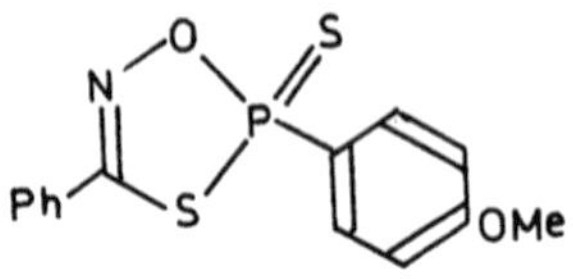 **53**

Rings/Connectivities	Formula	Structure	NMR data (δ_P[solv] $^nJ_{PZ}$ Hz)	Ref.
H_3;(C)$_2$	$C_3H_9OPS_2$	MeP(O)(SMe)$_2$	61.0	405
	$C_7H_{17}OPS_2$	MeP(O)(SPr)$_2$	56.0 $^1J_{PC}$76.9	37
H_3;(C')$_2$	$C_{13}H_{13}OPS_2$	MeP(O)(SPh)$_2$	56	369
5/H_3;(C)$_2$	$C_3H_7OPS_2$	Me(O)P(SCH$_2$CH$_2$S)	96.4(b)	37
5/C_3;(C)$_2$	$C_6H_{13}OPS_2$	tBu(O)P(SCH$_2$CH$_2$S)	107.4 $^1J_{PC}$72.5	37
6/H_3;(C)$_2$	$C_4H_9OPS_2$	Me(O)P(SCH$_2$CH$_2$CH$_2$S)	46.6(b) $^1J_{PC}$76.3	37
8/H_3;(C)$_2$	$C_5H_{11}OPS_3$	**51**; X = O, Y = Me	57.1(b) $^1J_{PC}$78.5	407
8/C_3;(C)$_2$	$C_8H_{17}OPS_3$	**51**; X = O, Y = tBu	83.0(b) $^1J_{PC}$72.0	407
16/H_3;(C)$_2$	$C_{10}H_{22}O_2PS_6$	**52**; A = CH$_2$SCH$_2$, X = O, Y = Me	61.7 62.0*(b) $^1J_{PC}$76.4, 76.5	407
16/C_3;(C)$_2$	$C_{16}H_{34}O_2P_2S_6$	**52**; A = CH$_2$SCH$_2$, X = O, Y = tBu	83.8 83.5*(b) $^1J_{PC}$70.9, 69.9	407
P-bound atoms = C'OSS'				
C'$_2$;N';C	$C_{11}H_{16}NO_2PS_2$	4-MeOC$_6$H$_4$P(S)(SMe)ON=CMe$_2$	101.4(c)	408
C'$_2$;C;S	$C_{14}H_{15}O_2PS_3$	4-MeOC$_6$H$_4$P(S)(OMe)SSPh	90.3	409
C'$_2$;C;H	$C_{19}H_{33}OPS_2$	2,4,6-(tBu)$_3$C$_6$H$_2$P(S)(OMe)SH	95.4(c)	410
5/C'$_2$;N';C'	$C_{14}H_{12}NO_2PS_2$	**53**	117(c)	411
5/C$_2$;C';C'	$C_{21}H_{17}O_3PS$	**44**; X = S	103.4(c)	387
6/C'$_2$;C;C	$C_{15}H_{21}O_2PS_2$	**H26**; A = S; X = S', Y = Ph	89	403
		X = Ph, Y = S'	79	
P-bound atoms = C'O'S$_2$				
C'$_2$;(H)$_2$	$C_6H_7OPS_2$	PhP(O)(SH)$_2$	40.2	331
C'$_2$;(C)$_2$	$C_{10}H_{15}OPS_2$	PhP(O)(SEt)$_2$	52.5(n)	412
5/C'$_2$;(C')$_2$	$C_{13}H_{11}OPS_2$	**H35**; R = Me, A = S, X = O, Y = Ph	58.4(c)	413
P-bound atoms = COSSe'				
CH$_2$;C;C	$C_5H_{13}OPSSe$	EtP(Se)(OEt)SMe	102.7(g)	414
	$C_{17}H_{21}OPSSe$	EtP(Se)(OEt)SCHPh$_2$	102.3(g)	414
P-bound atoms = COS'Se				
CH$_2$;C;C	$C_5H_{13}OPSSe$	EtP(S)(OEt)SeMe	111.1(g)	414
	$C_{17}H_{21}OPSSe$	EtP(S)(OEt)SeCHPh$_2$	109.7(g)	414

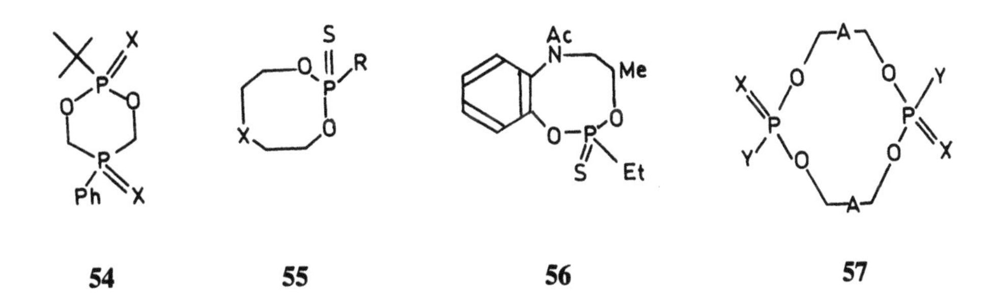

54 55 56 57

Rings/Connectivities	Formula	Structure	NMR data (δ_P[solv] $^nJ_{PZ}$ Hz)	Ref.
P-bound atoms = CO_2S'				
F_3;$(C)_2$	$C_5H_{10}F_3O_2PS$	TfP(S)$(OEt)_2$	62.6	150
HO_2;Si,C	$C_{10}H_{25}O_4PSSi$	$(EtO)_2$CHP(S)(OTms)OEt	70.3	415
HO_2;$(C)_2$	$C_9H_{21}O_4PS$	$(EtO)_2CHP(S)OEt)_2$	82.5	415
H_2Si;$(C)_2$	$C_6H_{17}O_2PSSi$	$TmsCH_2P(S)(OMe)_2$	97.8(c) $^1J_{PC}$100.1, $^2J_{PCH}$20.0 $^2J_{POC}$6.8	127
	$C_8H_{21}O_2PSSi$	$TmsCH_2P(S)(OEt)_2$	80.2	127
H_3;H,C	$C_2H_7O_2PS$	MeP(S)(OMe)OH	89.6 $^1J_{PC}$100.5 $^2J_{PC}$3.9	416
	$C_3H_9O_2PS$	MeP(S)(OH)OC_nH_{2n+1} n = 2/10	88.5/89.5	417
H_3;$(C)_2$	$C_5H_{13}O_2PS$	MeP(S)$(OEt)_2$	96	369
	$C_{12}H_{25}O_2PS$	MeP(S)(OMe)$OC_{10}H_{19}$[a] $(R)_P$	91.8(c)	196
H_3;$(C')_2$	$C_{13}H_{13}O_2PS$	MeP(S)$(OPh)_2$	90	369
CH_2;$(Si)_2$	$C_{12}H_{31}O_2PSSi$	HexP(S)$(OTms)_2$	72.6	418
CH_2;$(H)_2$	$C_6H_{15}O_2PS$	HexP(S)$(OH)_2$	79.5	418
CH_2;C,H	$C_3H_9O_2PS$	EtP(S)(OMe)OH	95.9	416
CH_2;$(C)_2$	$C_6H_{15}O_2PS$	EtP(S)$(OEt)_2$	101.2	419
$C'H_2$;$(C)_2$	$C_8H_{18}NO_3PS$	$(iPrO)_2P(S)CH_2CH$=NOH	89[Z]	255
	$C_{19}H_{33}O_3PS$	$(EtO)_2P(S)CH_2C_6H_2(tBu)_2(OH)$-3,5,4	93	262
C_2H;$(Si)_2$	$C_{12}H_{29}O_2PSSi_2$	cHexP(S)$(OTms)_2$	74.9	418
C_2H;$(H)_2$	$C_6H_{13}O_2PS$	cHexP(S)$(OH)_2$	81.6	418
C_2H;C,H	$C_4H_{11}O_2PS$	iPrP(S)(OMe)OH	98.3	79
	$C_8H_{17}O_2PS$	cHexP(S)(OEt)OH	96.5(y)	381
5/H_3;$(C)_2$	$C_3H_7O_2PS$	MeP(S)Glc	114.2 $^1J_{PC}$96.2	420
5/C_2H;$(C')_2$	$C_{20}H_{21}O_2PS$	cHex(S)P(OCPh=CPhO)	126(c)	412
5/C_3;$(C)_2$	$C_6H_{13}O_2PS$	tBuP(S)Glc	131.0 $^1J_{PC}$90.8	420
6/H_3;$(P^4)_2$	$C_3H_9O_3P_3S_3$	$[MeP(S)O]_3$	76 80.5(c)*	381
6/H_3;$(C)_2$	$C_4H_9O_2PS$	Me(S)P$(OCH_2CH_2CH_2O)$	97 109.4	422
	$C_6H_{13}O_2PS$	Me(S)P$(OCH_2CMe_2CH_2O)$	94.8(m) 95.8(b)	424
6/C_2H;$(P^4)_2$	$C_{18}H_{33}O_3P_3S_3$	$[cHexP(S)O]_3$	91(c)	381
6/C_3;$(C)_2$	$C_{12}H_{18}O_2P_2S_2$	**54**; X = S	113.9(c) $^1J_{PC}$100.1(b)	425
7/H_3;$(C)_2$	$C_5H_9O_2PS$	Me(S)P$(OCH_2CH$=$CHCH_2O)$	101.2(b) $^2J_{PC}$8.7(c)	426
	$C_5H_{11}O_2PS$	Me(S)P$(OCH_2CH_2CH_2CH_2O)$	97.6(b) 109.4 $^2J_{PC}$8.3	427
	$C_9H_{11}O_2PS$	(**H30**); R = H, A = O, X = S, Y = Me	103.0(b) $^2J_{PC}$6.6	426
7/C_3;$(C)_2$	$C_8H_{17}O_2PS$	tBu(S)P$(OCH_2CH_2CH_2CH_2O)$	110.1(b) $^1J_{PC}$107.5	428
8/H_3;$(C)_2$	$C_5H_{11}O_3PS$	**55**; X = O, R = Me	92.7 $^1J_{PC}$125.1	421
	$C_6H_{13}O_2PS$	**55**; X = CH_2, R = Me	112.2	421
	$C_6H_{14}NO_2PS$	**55**; X = NMe, R = Me	89(b) $^1J_{PC}$126.9	429
	$C_6H_{14}O_2P_2S_2$	**55**; X = P(S)Me, R = Me	93.6[C] 93.1[T](b) 98	430
8/CH_2;C,C'	$C_{13}H_{18}NO_4PS$	**56**	98	431
10/H_3;$(C)_2$	$C_6H_{14}O_4P_2S_2$	**57**; A absent; X = S, Y = Me	96.7[C] 95.7[T](b)	420
10/C_3;$(C)_2$	$C_{12}H_{26}O_4P_2S_2$	**57**; A absent; X = S, Y = tBu	109.1[T](c)	420
12/H_3;$(C)_2$	$C_{12}H_{26}O_4P_2S_2$	**57**; A = CMe_2, X = S, Y = Me	93.6(b) 91.8(c)*	424
14/C_3;$(C)_2$	$C_{16}H_{34}O_4P_2S_2$	**57**; A = CH_2CH_2, X = S, Y = tBu	107.2 107.7* $^1J_{PC}$110	428

TABLE J (continued)
Compilation of ^{31}P NMR Data of Four Coordinate Compounds Containing a P=Ch Bond and One Bond from Phosphorus to a Group IV Atom

Rings/Connectivities	Formula	Structure	NMR data (δ_P[solv] $^nJ_{PZ}$ Hz)	Ref.
16/H$_3$;(C)$_2$	$C_{10}H_{22}O_6P_2S_2$	**57**; A = CH$_2$OCH$_2$, X = S, Y = Me	96.1 96.3* $^1J_{PC}$117.2, 116.3	422
	$C_{12}H_{28}N_2O_4P_2S_2$	**57**; A = CH$_2$NMeCH$_2$, X = S, Y = Me	94.6(b)	429
17/H$_3$;(C')$_2$	$C_{19}H_{23}O_6PS$	**42**; X = S, Y = Me, n = 2	101.0(a)	360
	P-bound atoms = COO'S			
F$_3$;C;C	$C_5H_{10}F_3O_2PS$	TfP(O)(OEt)(SEt)	28.3	150
H$_2$F;C;C	$C_4H_{10}FO_2PS$	FCH$_2$P(O)(SMe)(OEt) *(R)*	45.5(c) $^2J_{PF}$66	6
H$_3$;C;C	$C_7H_{15}O_4PS$	MeP(O)(OEt)SCH$_2$COOEt	51.6	432
	$C_8H_{19}O_2PS$	MeP(O)(SMe)OCHMe(tBu) *(R)*$_P$	53.8(c)	195
		(S)$_P$	53.2(c)	
	$C_{12}H_{25}O_2PS$	MeP(O)(SMe)OC$_{10}$H$_{19}$[a]	50.8	195
H$_3$;C';C'	$C_{13}H_{13}O_2PS$	MeP(O)(OPh)(SPh)	46	369
CH$_2$;Si;C	$C_6H_{17}O_2PSSi$	EtP(O)(SMe)OTms	50.3	433
CH$_2$;C;Cl	$C_4H_{10}ClO_2PS$	EtP(O)(OEt)SCl	103	434
CH$_2$;C;C	$C_5H_{13}O_2PS$	EtP(O)(OEt)SMe (−)	61.5(n)	435
C$_3$;C;Cl	$C_5H_{12}ClO_2PS$	tBuP(O)(OMe)SCl *(S)*	63.5(r)	113
C$_3$;C;C	$C_{11}H_{22}ClO_2PS$	tBuP(O)(OMe)SC$_6$H$_{10}$Cl-2	66.4 66.6^D(r)	436
C$_3$;C;C"	$C_6H_{12}NO_2PS$	tBuP(O)COMe)SCN	62.6	113
6/H$_3$;C;C	$C_{10}H_{19}O_6PS$	**H26**; A = S; X = O, Y = Me	46	403
		X = Me, Y = O	43.5	
	P-bound atoms = COO'S'			
H$_3$;C	$C_7H_{16}O_2PS$	MeP(S)[OCHMe(tBu)]O$^-$ *(R)*$_P$	74.2(c)	195
		(S)$_P$	72.4(c)	
	P-bound atoms = C'O$_2$S'			
NN';(C)$_2$	$C_{17}H_{21}N_2O_2PS$	(EtO)$_2$P(S)C(NHPh)=NPh	60.7	393
OO';(Si)$_2$	$C_9H_{23}O_4PSSi_2$	(TmsO)$_2$P(S)COOEt	31.7	309
C'N';(C)$_2$	$C_8H_{15}N_2O_4PS$	(EtO)$_2$P(S)CN$_2$COOEt	70	437
C'H;(C)$_2$	$C_6H_{13}O_2PS$	(EtO)$_2$P(S)CH=CH$_2$	83.7 $^1J_{PC}$148.3 $^2J_{PC}$7.3	10
	$C_{10}H_{22}O_4P_2S_2$	(EtO)$_2$P(S)CH=CHP(S)(OEt)$_2$	78.5^E	193
C'H;(C')$_2$	$C_{14}H_{13}O_2PS$	(PhO)$_2$P(S)CH=CH$_2$	86	438
C'$_2$;(H)$_2$	$C_6H_7O_2PS$	PhP(S)(OH)$_2$	14.6	331
C'$_2$;C,N'	$C_{11}H_{16}NO_3PS$	4-MeOC$_6$H$_4$P(S)(OMe)ON=CMe$_2$	101.4	411
C'$_2$;C,P^4	$C_{16}H_{20}O_3P_2S_2$	Ph(EtO)P(S)OP(S)(OEt)Ph	74.5 75.0^D	335
C'$_2$;C,H	$C_7H_9O_2PS$	PhP(S)(OMe)OH	81.3	335
	$C_8H_{11}O_2PS$	PhP(S)(OEt)OH	67(y) 78.3(n,c)	439
C'$_2$;(C)$_2$	$C_8H_6F_5O_2PS$	C$_6$F$_5$P(S)(OMe)$_2$	72.6	440
	$C_8H_{11}O_2PS$	PhP(S)(OMe)$_2$	89.8(n)	337(a)
	$C_{14}H_{21}NO_4PS$	(PrO)$_2$P(S)C$_6$H$_4$COOMe-2	62	441
C'$_2$;(C')$_2$	$C_{18}H_{10}F_5O_2PS$	C$_6$F$_5$P(S)(OPh)$_2$	58.2	440
5/C'H;(C')$_2$	$C_8H_7O_2PS$	CatP(S)CH=CH$_2$	111	438
5/C'$_2$;(C)$_2$	$C_8H_9O_2PS$	PhP(S)Glc	104.3 $^1J_{PC}$137.1	442
5/C'$_2$;C,C'	$C_{13}H_{11}O_2PS$	**43**; X = S, Y = OPh	94.8	384
5/C$_2$;(C')$_2$	$C_{20}H_{15}O_2PS$	PhP(S)(OCPh=CPhO)	104(c)	412

Rings/Connec-tivities	Formula	Structure	NMR data (δ_P[solv] $^nJ_{PZ}$ Hz)	Ref.
6/C'$_2$;(P^4)$_2$	$C_{18}H_{15}O_3P_3S_3$	[PhP(S)O]$_3$	71(c)	443
	$C_{21}H_{21}O_6P_3S_3$	[4-MeOC$_6$H$_4$P(S)O]$_3$	71	444
6/C'$_2$;(C)$_2$	$C_{11}H_{15}O_2PS$	Ph(S)P(OCH$_2$CMe$_2$CH$_2$O)	86.0(g) 83.8(m)	424(a)
7/C'$_2$;(C)$_2$	$C_{10}H_{11}O_2PS$	Ph(S)P(OCH$_2$CH=CHCH$_2$O)	90.6(b) $^2J_{PC}$8.1	426
	$C_{10}H_{13}O_2PS$	Ph(S)P(OCH$_2$CH$_2$CH$_2$CH$_2$O)	87.1(b) $^2J_{PC}$8.0	426
	$C_{14}H_{13}O_2PS$	**H30**; A = O, R = H, X = S, Y = Ph	94.4(b) $^2J_{PC}$6.8	426
8/C'$_2$;(C)$_2$	$C_{10}H_{13}O_3PS$	**55**; X = O, R = Ph	83.3(b) $^1J_{PC}$168	429
10/C'$_2$;(C)$_2$	$C_{16}H_{18}O_4P_2S_2$	**57**; A absent, X = S, Y = Ph	86.8^C(b) 86.3^T(c)	420
17/C'$_2$;(C')$_2$	$C_{24}H_{25}O_6PS$	**42**; X = S, Y = Ph, n = 2	85.5(a)	360

P-bound atoms = C'OO'S

(C')$_2$;C;C	$C_{10}H_{10}Cl_3O_2PS$	(MeO)(MeS)P(O)C(C$_6$H$_3$Cl$_2$-2,6)=CHCl	27.5	445
6/C'$_2$;C;C	$C_9H_{11}O_2PS$	Ph(O)P(OCH$_2$CH$_2$CH$_2$S)	38	446

P-bound atoms = C'OO'S'

C'$_2$;C, l.p.	$C_{11}H_{12}O_2PS$	1-C$_{10}$H$_7$P(S)(OMe)O$^-$	53(c)	447

P-bound atoms = CO$_2$Se'

H$_3$;C,H	$C_3H_9O_2PSe$	MeP(Se)(OH)OC$_n$H$_{2n+1}$ n = 2/4	87/88	448
H$_3$;(C)$_2$	$C_3H_9O_2PSe$	MeP(Se)(OMe)$_2$	102.3 $^1J_{PSe}-861$	395
CH$_2$;C,H	$C_4H_{11}O_2PSe$	EtP(Se)(OEt)OH	96 $^1J_{PSe}$796	448
CH$_2$;(C)$_2$	$C_6H_{15}O_2PSe$	EtP(Se)(OEt)$_2$	107.0 $^1J_{PSe}$837	397
6/H$_3$;(C)$_2$	$C_6H_{13}O_2PSe$	Me(Se)P(OCH$_2$CMe$_2$CH$_2$O)	97 100	449
6/C'H$_2$;(C)$_2$	$C_{11}H_{15}O_2PSe$	**H1**; X = Se', Y = CH$_2$Ph X = CH$_2$Ph, Y = Se'	96.9(b), $^1J_{PSe}$883 92.0(b), $^1J_{PSe}$909	303
6/C$_3$;(C)$_2$	$C_{12}H_{18}O_2P_2Se_2$	**54**; X = Se	124.4(c) $^1J_{PSe}$894(b) $^1J_{PC}$83.5(c)	425
8/H$_3$;(C)$_2$	$C_6H_{14}O_2P_2Se_2$	**55**; X = P(Se)Me, R = Me	97.3^C(c) 85.5^T(b) $^1J_{PSe}$890.6^C(c) 927.3^T(b)	430

P-bound atoms = COO'Se

CH$_2$;C;C	$C_6H_{15}O_2PSe$	EtP(O)(OEt)SeEt	52	450

P-bound atoms = COO'Se'

CH$_2$;C	$C_4H_{10}O_2PSe$	Et(EtO)P(Se)O$^-$	79 $^1J_{PSe}$706	451

P-bound atoms = C'O$_2$Se'

OO';(Si)$_2$	$C_9H_{23}O_4PSeSi_2$	(TmsO)$_2$P(Se)COOEt	−9.9	309
C'$_2$;(C)$_2$	$C_8H_{11}O_2PSe$	PhP(Se)(OMe)$_2$	97.5 $^1J_{PSe}-876$	395
C'$_2$;(C)$_2$	$C_{10}H_{15}O_2PSe$	PhP(Se)(OEt)$_2$	92.2 $^1J_{PSe}$850	397

P-bound atoms = C'OO'Se

C'$_2$;C;C	$C_8H_{11}O_2PSe$	PhP(O)(OMe)SeMe	38.7	337(a)

TABLE J (continued)
Compilation of ^{31}P NMR Data of Four Coordinate Compounds Containing a P=Ch Bond and One Bond from Phosphorus to a Group IV Atom

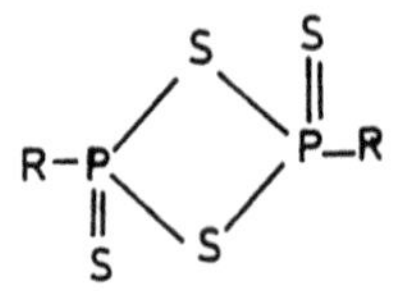

58 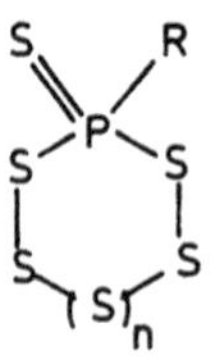 **59**

Rings/Connectivities	Formula	Structure	NMR data (δ_P[solv] $^nJ_{PZ}$ Hz)	Ref.
	P-bound atoms = C″O$_2$Se′			
6/N″;(C)$_2$	$C_5H_8NO_2PSe$	**H1**; X = Se′, Y = CN	23.4	452
		X = CN, Y = Se′	28.2 $^1J_{PSe}$1040	
	P-bound atoms = CS$_2$S′			
H$_3$;(Si)$_2$	$C_7H_{21}PS_3Si_2$	MeP(S)(STms)$_2$	61.8(b) $^1J_{PC}$64.1	453
H$_3$;(Sn)$_2$	$C_7H_{21}PS_3Sn_2$	MeP(S)(SSnMe$_3$)$_2$	70.2(b) $^1J_{PC}$61.6	453
H$_3$;(H)$_2$	CH_5PS_3	MeP(S)(SH)$_2$	57.9(s-b) $^1J_{PC}$59.9	453
H$_3$;(C)$_2$	$C_3H_9PS_3$	MeP(S)(SMe)$_2$	82.5	405
	$C_5H_{13}PS_3$	MeP(S)(SEt)$_2$	77	369
	$C_7H_{17}PS_3$	MeP(S)(SPr)$_2$	77.6 $^1J_{PC}$61.3	37
H$_3$;(C′)$_2$	$C_{13}H_{13}PS_3$	MeP(S)(SPh)$_2$	79	369
C$_2$H;(C)$_2$	$C_{10}H_{21}PS_3$	cHexP(S)(SEt)$_2$	102.0(n)	412
C$_3$;(Si)$_2$	$C_{10}H_{27}PS_3Si_2$	tBuP(S)(STms)$_2$	100.2(b) $^1J_{PC}$53.2	453
C$_3$;(H)$_2$	$C_4H_{11}PS_3$	tBuP(S)(SH)$_2$	96.6(k) 95.8(s-b) $^1J_{PC}$49.7	453
4/H$_3$;(P^4)$_2$	$C_2H_6P_2S_4$	**58**; R = Me	33(de)	454
4/CH$_2$;(P^4)$_2$	$C_4H_{10}P_2S_4$	**58**; R = Et	36	454
4/C$_2$H;(P^4)$_2$	$C_{12}H_{22}P_2S_4$	**58**; R = cHex	45	454
5/H$_3$;(C)$_2$	$C_3H_7PS_3$	Me(S)P(SCH$_2$CH$_2$S)	97.6(b) $^1J_{PC}$56.7	37
5/C$_3$;(C)$_2$	$C_6H_{13}PS_3$	tBu(S)P(SCH$_2$CH$_2$S)	129.0(b) $^1J_{PC}$46.3	37
6/H$_3$;(S)$_2$	CH_3PS_6	**59**; R = Me, n = 1	39.1(s-b, −20°)	455
6/H$_3$;(C)$_2$	$C_4H_9PS_3$	Me(S)P(SCH$_2$CH$_2$CH$_2$S)	49.1(b) $^1J_{PC}$58.5 59.1(g) $^1J_{PC}$51.5	456
6/C$_3$;(C)$_2$	$C_7H_{15}PS_3$	tBu(S)P(SCH$_2$CH$_2$CH$_2$S)	100.8(b,g), $^1J_{PC}$50.1	37
7/C$_3$;(S)$_2$	$C_4H_9PS_7$	**59**; R = tBu, n = 2	166.7(b, 10°C)	455
8/H$_3$;(S)$_2$	CH_3PS_8	**59**; R = Me, n = 3	59.1(s-b, −20°C)	455
8/H$_3$;(C)$_2$	$C_5H_{11}PS_4$	**51**; X = S, Y = Me	77.4(c) $^1J_{PC}$63.6	406
8/C$_3$;(C)$_2$	$C_8H_{17}PS_4$	**51**; X = S, Y = tBu	112.4(b) $^1J_{PC}$51.6	407
12/H$_3$;(C)$_2$	$C_8H_{18}P_2S_6$	**52**; A = CH$_2$, X = S, Y = Me	76.3(b)	37
12/C$_3$;(C)$_2$	$C_{14}H_{30}P_2S_6$	**52**; A = CH$_2$, X = S, Y = tBu	116.1 120.2*(b), $^1J_{PC}$51.4, 50.4	37
16/H$_3$;(C)$_2$	$C_{10}H_{22}P_2S_8$	**52**; A = CH$_2$SCH$_2$, X = S, Y = Me	80.1 80.6*(c,b) $^1J_{PC}$61.7, 61.7	407
16/C$_3$;(C)$_2$	$C_{16}H_{34}P_2S_8$	**52**; A = CH$_2$SCH$_2$, X = S, Y = tBu	117.0 118.5*(b) $^1J_{PC}$50.3, 51.6	407
	P-bound atoms = CSS′$_2$			
H$_3$	$C_2H_6PS_3$	MeP(S)(SMe)S$^-$	81.3	389

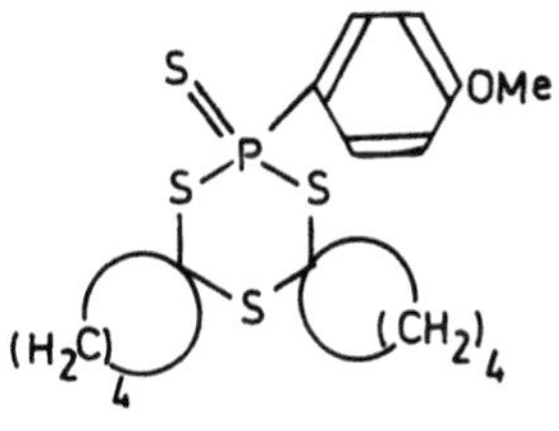

60

Rings/Connec-tivities	Formula	Structure	NMR data (δ_P[solv] $^nJ_{PZ}$ Hz)	Ref.
P-bound atoms = $C'S_2S'$				
$C'_2;(Si)_2$	$C_{12}H_{23}PS_3Si_2$	$PhP(S)(STms)_2$	62.0(b) 63.1	458
$C'_2;(H)_2$	$C_6H_7PS_3$	$PhP(S)(SH)_2$	60.2	331
$C'_2;(C)_2$	$C_8H_{11}PS_3$	$PhP(S)(SMe)_2$	83.5(n)	337(a)
	$C_{10}H_{15}PS_3$	$PhP(S)(SEt)_2$	80.5(n)	412
$4/C'_2;(P^4)_2$	$C_{12}H_{10}P_2S_4$	**58**; R = Ph	15.4(de)	458
	$C_{14}H_{14}P_2S_4$	**58**; R = 4-MeOC_6H_4	14.5	458
$5/C'_2;(C)_2$	$C_8H_9PS_3$	$Ph(S)P(SCH_2CH_2S)$	98.6(b)	459
$6/C'_2;(C)_2$	$C_{17}H_{23}OPS_4$	**60**	59.4(c)	387

[a]L-Menthyl.
[b]1-Cyclohexen-1-yl.
[c]Tetrafluoropyridinyl.
[d]Degenerate.
[e]1-Cyclohexen-4-yl.
[f]1-Cyclohexen-3-yl.
[g]1-Indanyl.
[h]Pyridinyl.
[i]Quinolinyl.
[j]Tetrachloropyridinyl.
[k]3-Coumarinyl.

*Nonidentified stereoisomers.

Chapter J
Supplementary Table

Section J1: Derivatives of Phosphonic Acids

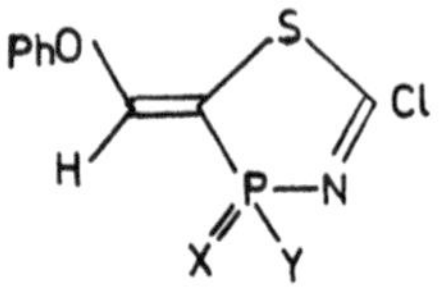

61 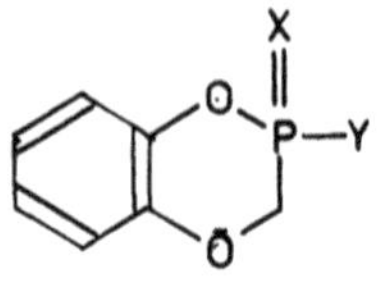 **62** 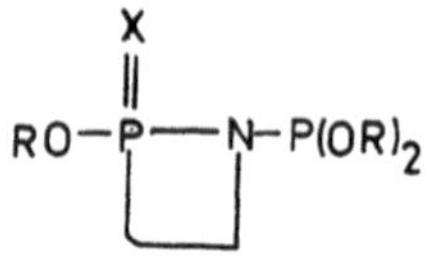 **63**

Rings/ Connectivities	Formula	Structure	NMR data (δ_P [solv] $^nJ_{PZ}$)	Ref.
	P-bound atoms = C′Br$_2$O			
C′$_2$	$C_6H_5Br_2OP$	PhP(O)Br$_2$	1.1(c)	460
	P-bound atoms = C′ClN′O′			
5/C′S;C′	$C_8H_5Cl_2NOPS$	**61**; X = O, Y = Cl	50.9(c)	461
	P-bound atoms = CClOO′			
H$_3$;N′	$C_5H_{10}Cl_2NO_2P$	Me(Cl)P(O)ON=CCl(-iPr)	43.75	462
C′$_2$;C′	$C_{12}H_{10}ClO_2P$	PhP(O)(OPh)Cl	24.4(c)	460
6/H$_2$O;C′	$C_7H_6ClO_3P$	**62**; X = O, Y = Cl	17.7(b)	463
	P-bound atoms = CCl$_2$O′			
H$_2$P^4	$CH_2Cl_4OP_2S$	Cl$_2$**P**(O)CH$_2$P(S)Cl$_2$	25.6 $^2J_{PP}$11.6	464
C$_2$S	$C_5H_{11}Cl_2OPS$	Cl$_2$P(O)CMe$_2$(SEt)	56	465
	P-bound atoms = C′Cl$_2$O′			
C′H	$C_8H_{13}Cl_2OP$	Cl$_2$P(O)CH=C=C(-tBu)Me	28.2	466
CC′	$C_{11}H_{19}Cl_2OP$	Cl$_2$P(O)C(-tBu)=C=CH(-tBu)	35.9	466
	P-bound atoms = CF$_2$O′			
H$_2$P^4	$CH_2Cl_2F_2OP_2S$	F$_2$**P**(O)CH$_2$P(S)Cl$_2$	7.5 $^1J_{PF}$1118	464
	$CH_2F_4OP_2S$	F$_2$**P**(O)CH$_2$P(S)F$_2$	7.0 $^1J_{PF}$1123 $^2J_{PP}$6.3	464
H$_2$P	$CH_2Cl_2F_2OP_2$	F$_2$**P**(O)CH$_2$PCl$_2$	12.2 $^1J_{PF}$1118, $^2J_{PP}$43.0	464
	$CH_2F_4OP_2$	F$_2$**P**(O)CH$_2$PF$_2$	14.0 $^1J_{PF}$1124.7 $^2J_{PP}$27.0	464
	P-bound atoms = C′F$_2$O′			
C′H	$C_4H_7F_2OP$	Me$_2$C=CHP(O)F$_2$	13.7	467
	P-bound atoms = CNOO′			
4/CH$_2$;CP;C	$C_5H_{13}NO_4P_2$	**63**; X = O, R = Me	20.15	468
5/CO;C$_2$;C	$C_8H_{18}NO_3P$	**64**; X = O, Y = NMe$_2$	44, 48^D(c)	469
	P-bound atoms = CN′OO′			
C$_3$;C′;C	$C_6H_{12}NO_2PS$	tBu(MeO)P(O)NCS	97.7(r)	470

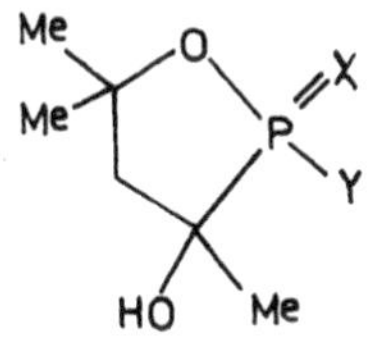

64 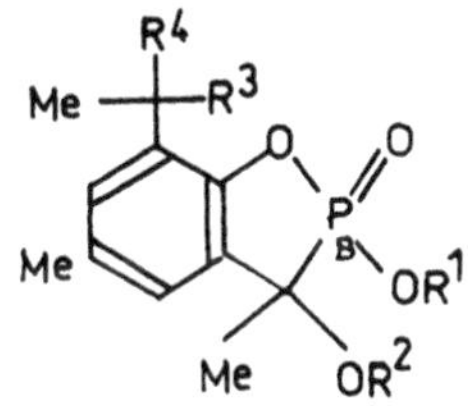 65 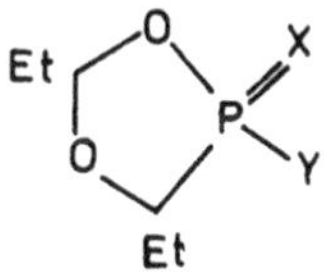 66

Rings/ Connectivities	Formula	Structure	NMR data (δ_P [solv] $^nJ_{PZ}$)	Ref.
P-bound atoms = CO$_2$O′				
$H_2O;(C)_2$	$C_{17}H_{21}O_5P$	$(EtO)_2P(O)CH_2OC_6H_4OH$-2	21.1(b)	463
$H_3;C,N'$	$C_7H_{15}ClNO_3P$	Me(iPrO)P(O)ON= CEtCl	33.9	471
$C'HN;(C)_2$	$C_{11}H_{13}F_3NO_3P$	$(MeO)_2P(O)CHPhN{=}CHTf$	19.8	472
	$C_{21}H_{38}NO_4P$	$(EtO)_2P(O)CH(NMe_2)C_6H_2$ (-tBu)$_2$OH-3,5,4,	24	473
$CHN';(C)_2$	$C_{13}H_{17}F_3NO_3P$	$(EtO)_2P(O)CHTfN{=}CHPh$	12.1	472
$C'HN';(C)_2$	$C_{18}H_{22}NO_3P$	$(EtO)_2P(O)CHPhN{=}CHPh$	19.0(g)	474
$C'HO;C,H$	$C_{15}H_{17}O_5P$	$HO(EtO)P(O)CHPhOC_6H_4OH$-2	19.0(d)	463
$CHP^4;(C)_2$	$C_{16}H_{28}O_6P_2S$	$[(EtO)_2P(O)]_2CHCH_2SPh$	18.0	475
$CH_2;(C)_2$	$C_8H_9O_5P$	$(EtO)_2P(O)CH_2CHOHCHOHCH_3$		476
		(2*RS*,3*RS*)	31.7	
		(2*RS*,3*SR*)	30.8	
$C'H_2;(C)_2$	$C_7H_{15}O_5P$	$(EtO)_2P(O)CH_2COCH_2OH$	19.3	476
$C_2N';(C)_2$	$C_{10}H_8F_3N_2O_3P$	$(EtO)_2P(O)CMeEtN{=}C{=}NTf$	23.65 25.75	477
$CC'O;(Si)_2$	$C_{21}H_{45}O_5PSi_4$	(TmsO)P(O)CMe (OTms)C_6H_3(OTms)Me-2,5	6.8(b)	478
	$C_{25}H_{52}O_8P_2Si_4$	**65**; R[1] = Et, R[2] = Tms, R[3] = OTms, R[4] = $(TmsO)_2\mathbf{P}_A(O)$	5.0(b)	479
$CC'O;(H)_2$	$C_9H_{13}O_5P$	$(HO)_2P(O)CMe(OH)C_3H_3(OH)Me$-2,5	26.3(b)	478
$CC'O;(C)_2$	$C_{13}H_{21}O_5P$	$(EtO)_2P(O)CMe(OH)C_6H_3(OH)Me$-2,5	23.4(b)	478
	$C_{19}H_{37}O_5PSi_2$	$(EtO)_2P(O)CMe$ (OTms)C_6H_3(OTms)Me-2,5	21.5(b)	478
	$C_{23}H_{51}O_8P_2Si_2$	**65**; R[1] = Et, R[2] = Tms, R[3] = OTms, R[4] = $(EtO)_2\mathbf{P}(O)$	21.1, 20.6^D(b)	479
$C_2H;(C)_2$	$C_8H_{17}O_4P$	$(EtO)_2P(O)CHMeCH_2CHO$	30.0(c)	480
$CC'H;(C)_2$	$C_{14}H_{21}O_4P$	$(EtO)_2P(O)CHPhCHMeCHO$	23.6 23.5^D(c)	480
5/CHO;$(C)_2$	$C_8H_{17}O_4P$	**66**; X = O, Y = OEt	24.8 24.75^D(c)	481
5/$C_2O;(C)_2$	$C_9H_{19}O_4P$	**64**; X = O, Y = OiPr	34 38^D(c)	469
5/CC′O; C′,Si	$C_{19}H_{35}O_4PSi_2$	**65**; R[1] = R[2] = Tms, R[3] = R[4] = Me	10.7 11.3^D(b)	478
5/CC′O;C,C′	$C_{18}H_{31}O_4PSi$	**65**; R[1] = Et, R[2] = Tms, R[3] = R[4] = Me	39.1 39.7^D(b)	478
	$C_{23}H_{51}O_8P_2Si_2$	**65**; R[1] = Et, R[2] = Tms, R[3] = OTms, R[4] = $(EtO)_2P_A(O)$	(B)38.8 39.9^D(b)	479
	$C_{25}H_{52}O_8P_2Si_4$	**65**; R[1] = Et, R[2] = Tms, R[3] = OTms, R[4] = $(TmsO)_2P_A(O)$	(B)39.2(b)	479
6/H_2O;C′,H	$C_7H_7O_4P$	**62**; X = O, Y = OH	12.1(c)	463
6/H_2O;C,C′	$C_9H_{11}O_4P$	**62**; X = O, Y = OEt	8.9(b)	463
P-bound atoms = C′O$_2$O′				
$CN';(C)_2$	$C_{13}H_{17}F_3NO_3P$	$(EtO)_2P(O)CTf{=}NCH_2Ph$	−2.0(c)	472
$C'N';(C)_2$	$C_9H_{12}NO_4P$	$(MeO)_2P(O)C({=}NOH)Ph$	11.6	482

Chapter J (continued)
Supplementary Table

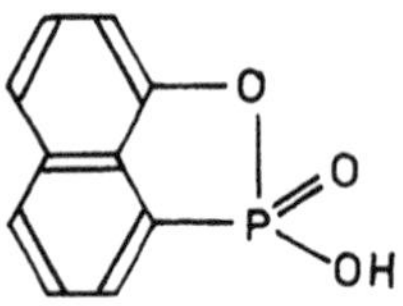

67 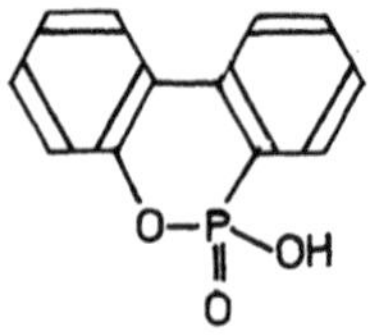 68 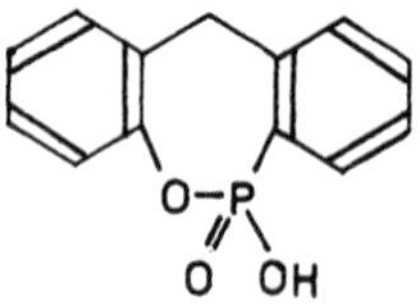 69

Rings/ Connectivities	Formula	Structure	NMR data (δ_P [solv] $^nJ_{PZ}$)	Ref.
	$C_{11}H_{13}F_3NO_3P$	$(MeO)_2P(O)CPh{=}NCH_2Tf$	6.7(g)	472
	$C_{18}H_{22}NO_3P$	$(EtO)_2P(O)CPh{=}NCH_2Ph$	6	474
CO′;(C)₂	$C_8H_{15}O_6P$	$(EtO)_2P(O)COCH_2COOMe$	4.4	483
CC′;(H)₂	$C_3H_7O_3P$	$(HO)_2P(O)CMe{=}CH_2$	20(w) $^1J_{PC}$170.6 $^2J_{PC}$′10.3 $^2J_{PC}$13.2	484
CC′;(C)₂	$C_{12}H_{21}N_2O_3P$	$(EtO)_2P(O)C(CN){=}\dot{C}(CH_2)_4\dot{N}Me$	21.0	485
C′₂;(H)₂	$C_7H_4F_3N_2O_7P$	$(HO)_2P(O)C_6H_2(NO_2)_2$Tf-2,6,4	−2.6	486
C′₂;(C)₂	$C_{11}H_{12}F_3N_2O_7P$	$(EtO)_2P(O)C_6H_2(NO_2)_2$Tf-2,6,4	2.8	486
C′₂;CC′	$C_{13}H_{13}O_3P$	PhP(O)(OPh)OMe	40.1(c)	487
5/C′₂;C′H	$C_{10}H_7O_3P$	**67**	27.1(d)	488
6/C′₂;C′H	$C_{12}H_9O_3P$	**68**	6.0(d)	488
7/C′₂;C′H	$C_{13}H_{11}O_3P$	**69**	8.4(d)	488
	P-bound atoms = C″O₂O′			
C″;C,C′	$C_{14}H_{15}O_5P$	$PhC{\equiv}CP(O)(OMe)OC(COOEt){=}CH_2$	−9	489

Section J2: Derivatives of Thio-, Seleno-, and Telluro-Phosphonic Acids

Rings/ Connectivities	Formula	Structure	NMR Data (δ_P [solv] $^nJ_{PZ}$)	Ref.
	P-bound atoms = C′ClN′S′			
5/C′S;C′	$C_8H_5Cl_2NPS_2$	**61**; X = S, Y = Cl	89.0(c)	461
	P-bound atoms = C′ClOS′			
C′₂;C′	$C_{12}H_{10}ClOPS$	PhP(S)(OPh)Cl	84.3 (1,2-dibromoethane)	460
	P-bound atoms = CCl₂S′			
H₂P⁴	$CH_2Cl_2F_2OP_2S$	$Cl_2\mathbf{P}(S)CH_2P(O)F_2$	54.5	464
	$CH_2Cl_4OP_2S$	$Cl_2\mathbf{P}(S)CH_2P(O)Cl_2$	52.7 $^2J_{PP}$11.6	464
CH₂	$C_4H_8Cl_3PS$	$Me_2CClCH_2P(S)Cl_2$	74.6	467
	P-bound atoms = C′Cl₂S′			
C′H	$C_4H_7Cl_2PS$	$Me_2C{=}CHP(S)Cl_2$	63.7	467

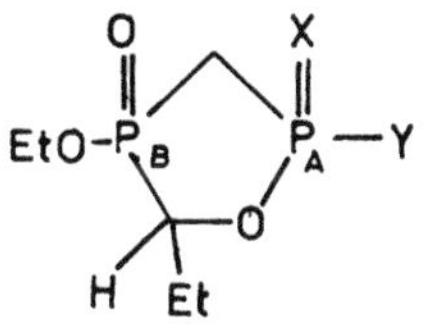

70

Rings/ Connectivities	Formula	Structure	NMR data (δ_P [solv] $^nJ_{PZ}$)	Ref.
P-bound atoms = CF_2S'				
H_2P^4	$CH_2F_4OP_2S$	F_2**P**$(S)CH_2P(O)F_2$	77.9 $^1J_{PF}$1165.6 $^2J_{PP}$6.3	464
	$CH_2F_4P_2S_2$	$F_2P(S)CH_2P(S)F_2$	75.6 $^1J_{PF}$ − 1162.8 $^2J_{PP}$16.1	464
H_2P	$CH_2Cl_2F_2P_2S$	F_2**P**$(S)CH_2PCl_2$	85.4 $^1J_{PF}$1167 $^2J_{PP}$58.9	464
	$CH_2F_4P_2S$	F_2**P**$(S)CH_2PF_2$	86.4 $^1J_{PF}$1169.5 $^2J_{PP}$39.5	464
P-bound atoms = CNOS′				
5/CO;C_2;C	$C_8H_{18}NO_2PS$	**64**; X = S, Y = NMe_2	114	469
P-bound atoms = C-NSS′				
C'_2;C'_2;C′	$C_{27}H_{22}NOPS_2$	**44**; X = NPh	86.1 91.4	490
P-bound atoms = CN_2S'				
H_3;$(CH)_2$	$C_3H_{11}N_2PS$	$MeP(S)(NHMe)_2$	69.3(c)	491
	$C_9H_{23}N_2PS$	$MeP(S)(NH\text{-}tBu)_2$	53.2(c)	491
P-bound atoms = $C'N_2S'$				
C'_2;$(CH)_2$	$C_8H_{13}N_2PS$	$PhP(S)(NHMe)_2$	68.1(c)	491
	$C_{14}H_{25}N_2PS$	$PhP(S)(NH\text{-}tBu)_2$	54.0(c)	491
P-bound atoms = $CO'S_2$				
C′HS;$(C)_2$	$C_{13}H_{21}OPS_3$	$(EtS)_2P(O)CHPh(SEt)$	64	465
P-bound atoms = C′OSS′				
C'_2;C;C	$C_{15}H_{17}O_2PS_2$	$4\text{-}MeOC_6H_4P(S)(SCH_2Ph)OMe$	97.7(c)	492
P-bound atoms = CO_2S'				
5/H_2P^4;$(C)_2$	$C_8H_{18}O_4P_2S$	**70**; X = S, Y = OEt	86.7 or 87.2	493
5/CO;$(C)_2$	$C_8H_{14}O_3PS$	**64**; X = S, Y = OEt	112 (c)	469
P-bound atoms = COO′S				
C_2H;C;C	$C_{12}H_{25}O_2PS$	cHexP(O)(OPr)SPr	57.6	494

REFERENCES

1. **Kharrasova, F. M., Zykova, T. V., Salakhutdinov, R. A., and Rakhimova, G. I.,** *J. Gen. Chem. U.S.S.R.*, 43, 2621, 1973.
2. **Feringa, P. L., Smaardijk, Ab. A., Wynberg, H., Strijtveen, B., and Kellogg, R. M.,** *Tetrahedron Lett.*, 27, 997, 1986.
3. **Strijtveen, B., Feringa, B. L., and Kellogg, R. M.,** *Tetrahedron*, 43, 123, 1987.
4. **Feringa, B. L., Strijtveen, B., and Kellogg, R. M.,** *J. Org. Chem.*, 51, 5484, 1986.
5. **Hall, C. R. and Williams, N. E.,** *J. Chem. Soc. Perkin Trans. 1*, p. 2746, 1981.
6. **Hall, C. R., Inch, T. D., and Williams, N. E.,** *J. Chem. Soc. Perkin Trans. 1*, p. 233, 1985.
7. **Hall, C. R., Inch, T. D., Peacocke, G., Pottage, C., and Williams, N. E.,** *J. Chem. Soc. Perkin Trans. 1*, p. 669, 1984.
8. **Fokin, A. V. and Landau, M. A.,** *Bull. Acad. Sci. U.S.S.R. Div. Chem. Sci.*, p. 2271, 1976.
9. **Zykova, T. V., Moskva, V. V., Sitdikova, T. Sh., and Salakhutdinov, R. A.,** *J. Gen. Chem. U.S.S.R.*, 51, 1843, 1981.
10. **Althoff, W., Fild, M., Rieck, H.-P., and Schmutzler, R.,** *Chem. Ber.*, 111, 1845, 1978.
11. **Fluck, E. and Seng, N.,** *Z. Anorg. Allg. Chem.*, 393, 126, 1972.
12. **Lequan, R.-M., Pouet, M.-J., and Simonnin, M.-P.,** *Org. Magn. Reson.*, 7, 392, 1975.
13. **Belskii, V. E., Kudryavtseva, L. A., and Ivanov, B. E.,** *J. Gen. Chem. U.S.S.R.*, 42, 2421, 1972.
14. Ref. H6.
15. Ref. H7.
16. **Allen, D. W., Hutley, B. G., and Mellor, M. T. J.,** *J. Chem. Soc. Perkin Trans. 2*, p. 789, 1977.
17. **Mitsch, C. C., Freedman, L. D., and Moreland, C. G.,** *J. Magn. Reson.*, p. 446, 1970; p. 140, 1971.
18. **Manninen, P. A.,** *Acta Chem. Scand.*, 35B, 13, 1981.
19. **Tarasova, R. I., Zykova, T. V., Dvoinishnikova, T. A., Salakhutdinov, R. A., and Sinitsyna, N. I.,** *J. Gen. Chem. U.S.S.R.*, 51, 2080, 1981.
20. **Issleib, K., Balszuweit, A., Koetz, J., Richter, S., and Leutloff, S.,** *Z. Anorg. Allg. Chem.*, 529, 151, 1985.
21. **Honig, M. L. and Weil, E. D.,** *J. Org. Chem.*, 42, 379, 1977.
22. **Grabiak, R. C., Miles, J. A., and Schwenzer, G. M.,** *Phosphorus Sulfur*, 9, 197, 1980.
23. **Szafraniec, L. L.,** *Org. Magn. Reson.*, 6, 565, 1974.
24. **Sidky, M. M., Mahran, M. R., El-Kateb, A. A., and Hennawy, I. T.,** *Phosphorus Sulfur*, 10, 409, 1981.
25. **Robinson, C. N. and Slater, C. D.,** *J. Org. Chem.*, 52, 2011, 1987.
26. **Costisella, B., Keitel, I., and Gross, H.,** *Tetrahedron*, 37, 1227, 1981.
27. **Buchanan, G. W. and Morin, F. G.,** *Can. J. Chem.*, 55, 2885, 1977.
28. **Haslinger, E., Öhler, E., and Robien, W.,** *Monatsh. Chem.*, 113, 1321, 1982.
29. **Bartlett, P. A. and Long, H. P.,** *J. Am. Chem. Soc.*, 99, 1267, 1977.
30. **Ohms, G. and Grossmann, G.,** *Z. Anorg. Allg. Chem.*, 544, 232, 1987.
31. **Brel', V. K., Komarov, V. Ya., Ionin, B. I., and Petrov, A. A.,** *J. Gen. Chem. U.S.S.R.*, 53, 52, 1983.
32. **Angelov, Kh. M., Vachkov, K. V., Kirilev, M., and Lebedev, V. B.,** *J. Gen. Chem. U.S.S.R.*, 52, 472, 1982.
33. **Khusainova, N. G., Naumova, L. V., Berdnikov, E. A., and Pudovik, A. N.,** *J. Gen. Chem. U.S.S.R.*, 52, 904, 1982.
34. **Mikhailova, T. S., Zakharov, V. I., Ignat'ev, V. M., Ionin, B. I., and Petrov, A. A.,** *J. Gen. Chem. U.S.S.R.*, 50, 1370, 1980.
35. **Khusainova, N. G., Sippel', I., Berdnikov, E. A., Cherkasov, R. A., and Pudovik, A. N.,** *J. Gen. Chem. U.S.S.R.*, 55, 2377, 1985.
36. **Hall, C. R., Inch, T. D., Pottage, C., Williams, N. E., Campbell, M. M., and Kerr, P. F.,** *J. Chem. Soc. Perkin Trans. 1*, p. 1967, 1983.
37. **Martin, J. and Robert, J. B.,** *Org. Magn. Reson.*, 15, 87, 1981.
38. Ref. H295.
39. Ref. H297.
40. **Goodridge, R. J., Hambley, T. W., and Ridley, D. D.,** *Aust. J. Chem.*, 39, 591, 1986.
41. **Sokolov, V. I., Musaev, A. A., Bashilov, V. V., and Petrovskii, P. V.,** *J. Gen. Chem. U.S.S.R.*, 53, 202, 1983.
42. **Deng, R. M. K. and Dillon, K. B.,** *J. Chem. Soc. Dalton Trans. 1*, p. 1443, 1986.
43. **Kennedy, E. R. and Macomber, R. S.,** *J. Org. Chem.*, 39, 1952, 1974.
44. **Elkain, J.-C., Casabianca, F., and Reiss, J. G.,** *Synth. React. Inorg. Metal-Org. Chem.*, 9, 479, 1979.
45. **Duddeck, H. and Hanna, A. G.,** *Magn. Reson. Chem.*, 23, 41, 1985.

46. Ref. H103.
47. **Bulloch, G. and Keat, R.,** *J. Chem. Soc. Dalton Trans.* p. 1113, 1976.
48. **Shtyrlina, A. A., Enikeev, K. M., Bayandina, E. V., Ismaev, I. É., Il'yasov, A. V., and Nuretdinov, I. A.,** *Bull. Acad. Sci. U.S.S.R. Div. Chem. Sci.*, p. 1718, 1985.
49. **Maidanovich, N. K., Iksanova, S. V., and Gololobov, Yu. G.,** *J. Gen. Chem. U.S.S.R.*, 52, 812, 1982.
50. **Freeman, S. and Harger, M. J. P.,** *J. Chem. Soc. Perkin Trans. 2*, p. 1399, 1987.
51. **Dmitrichenko, M. Yu., Donskikh, V. I., Rozinov, V. G., Rybkina, V. V., and Sergienko, L. M.,** *J. Gen. Chem. U.S.S.R.*, 55, 1672, 1985.
52. **Knunyants, I. L. and Neimysheva, A. A.,** *J. Gen. Chem. U.S.S.R.*, 42, 2415, 1972.
53. **Lang, G. and Herrmann, E.,** *Z. Anorg. Allg. Chem.*, 536, 187, 1986.
54. Ref. H70.
55. Ref. H290.
56. Ref. H97.
57. **Moedritzer, K. and Miller, R. E.,** *Synth. React. Inorg. Metal-Org. Chem.*, 4, 417, 1974.
58. **Maier, L.,** *Phosphorus*, 3, 19, 1973.
59. **Corriu, R. J. P., Lanneau, G. F., and Leclerq, D.,** *Tetrahedron*, 36, 1617, 1980.
60. **Khusainova, N. G., Naumova, L. V., Berdnikov, E. A., Kutyrev, G. A., and Pudovik, A. N.,** *Phosphorus Sulfur*, 13, 147, 1982.
61. **Elder, R. C., Florian, L. R., Kennedy, E. R., and Macomber, R. S.,** *J. Org. Chem.*, 38, 4177, 1973.
62. **Fild, M. and Rieck, H.-P.,** *Chem. Ber.*, 113, 142, 1980.
63. **den Hartog, J. A. J. and van Boom, J. H.,** *Recl. Trav. Chim. Pays Bas*, 100, 285, 1981.
64. (a)**Maier, L.,** *Helv. Chim. Acta*, 53, 1948, 1970; (b) Ref. H96.
65. (a)**Maier, L.,** *Z. Anorg. Allg. Chem.*, 394, 117, 1972; (b) Ref. H92.
66. **Harger, M. J. P. and Williams, A.,** *J. Chem. Soc. Perkin Trans. 1*, p. 1681, 1986.
67. **Mclard, R. W., Fujita, T. S., Stremler, K. E., and Poulter, C. D.,** *J. Am. Chem. Soc.*, 109, 5544, 1987.
68. **Sommer, K.,** *Z. Anorg. Allg. Chem.*, 376, 37, 1970.
69. (a) Ref. 53; (b) Ref. 68.
70. (a) Ref. 54; (b) Ref. 68; (c) Ref. 64(b).
71. **Startsev, V. V., Zubritskii, L. M., Lukin, M. G., and Petrov, A. A.,** *J. Gen. Chem. U.S.S.R.*, 55, 1660, 1985.
72. **Maier, L.,** *Helv. Chim. Acta*, 56, 1257, 1973.
73. **Nurtdinov, S. Kh., Savron, V. I., Zykova, T. V., Salakhutdinov, R. A., and Tsivunin, V. S.,** *J. Gen. Chem. U.S.S.R.*, 49, 2159, 1979.
74. (a) Ref. H95; (b) Ref. 52.
75. (a) Ref. 47; (b) Ref. 68.
76. (a) Ref. 57; (b) Ref. 54.
77. **Maier, L.,** *Phosphorus*, 5, 223, 1975.
78. **Prorubshchikov, A. Yu., Kuzina, N. G., Lykov, A. D., Reshetov, P. N., Mashlyakovskii, L. N., and Ionin, B. I.,** *J. Gen. Chem. U.S.S.R.*, 52, 1587, 1982.
79. **Mikołajczyk, M. and Omelańczuk, J.,** *Phosphorus*, 3, 47, 1973.
80. **Andreev, N. A. and Grishina, O. N.,** *J. Gen. Chem. U.S.S.R.*, 51, 444, 1981.
81. **Krolevets, A. A., Popov, A. G., and Antipova, V. V.,** *J. Gen. Chem. U.S.S.R.*, 55, 1902, 1985.
82. **Schmutzler, R. and Reddy, G. S.,** *Z. Naturforsch.*, 20B, 832, 1965.
83. **Issleib, K. and Steibitz, B.,** *Z. Anorg. Allg. Chem.*, 542, 37, 1986.
84. **Aliev, R. Z. and Khairullin, V. K.,** *J. Gen. Chem. U.S.S.R.*, 46, 263, 1976.
85. **Ayed, N., Baccar, B., Mathis, F., and Mathis, R.,** *Phosphorus Sulfur*, 21, 335, 1985.
86. **Mitrasov, Yu. N. and Kormachev, V. V.,** *J. Gen. Chem. U.S.S.R.*, 53, 1507, 1983.
87. **Rozinov, V. G., Rybkina, V. V., Kolbina, V. E., Glukhikh, V. I., Danskikh, V. I., and Seredkina, S. G.,** *J. Gen. Chem. U.S.S.R.*, 51, 1494, 1981.
88. **Rybkina, V. V., Rozinov, V. G., and Seredkina, S. G.,** *J. Gen. Chem. U.S.S.R.*, 51, 2270, 1981.
89. **Efanov, V. A., Mingaleva, K. S., Dogadina, A. V., Ionin, B. I., and Petrov, A. A.,** *J. Gen. Chem. U.S.S.R.*, 52, 1759, 1982.
90. **Kormachev, V. V., Mitrasov, Yu. N., Kukhtin, V. A., Yakovleva, T. M., and Kurskii, Yu. A.,** *J. Gen. Chem. U.S.S.R.*, 51, 801, 1981.
91. **Fridland, S. V., Dmitrieva, N. V., Vigalo, I. V., Zykova, T. V., and Salakhutdinov, R. A.,** *J. Gen. Chem. U.S.S.R.*, 43, 575, 1973.
92. **Brel', V. K., Ionin, B. I., Petrov, A. A., and Martynov, I. V.,** *J. Gen. Chem. U.S.S.R.*, 55, 2189, 1985.
93. **Lykov, A. D., Mashlyakovskii, L. N., Vafina, G. S., and Ionin, B. I.,** *J. Gen. Chem. U.S.S.R.*, 51, 1454, 1981.
94. (a) Ref. 54; (b) Ref. 22.

95. (a) Ref. 11; (b) Ref. 12.
96. Ref. H50.
97. **Bender, R., Demay, C., Elkain, J.-C., and Reiss, J. G.,** *Phosphorus,* 4, 183, 1974.
98. **Kozhushko, B. N., Lomakina, A. V., Palachuk, Yu.-A., and Shokol, V. A.,** *J. Gen. Chem. U.S.S.R.,* 53, 1768, 1983.
99. **Drozd, G. I., Ivin, S. Z., and Vorshavskii, A. D.,** *J. Gen. Chem. U.S.S.R.,* 39, 1146, 1969.
100. **Boenigk, W. and Haegele, G.,** *Chem. Ber.,* 116, 2418, 1983.
101. **Patsanovskii, I. I., Ishmaeva, E. A., Levin, Yu. A., Gilyarov, M. M., and Pudovik, A. N.,** *J. Gen. Chem. U.S.S.R.,* 51, 922, 1981.
102. **Althoff, W., Fild, M., and Schmutzler, R.,** *Chem. Ber.,* 114, 1082, 1981.
103. (a) Ref. H337; (b) Ref. 46.
104. (a)**Fild, M. and Schmutzler, R.,** *J. Chem. Soc. Sect. A,* p. 2359, 1970; (b) Ref. 82.
105. **Zakharov, V. I., Belov, Yu. V., Ionin, B. I., and Petrov, A. A.,** *Dokl. Chem. (Engl. transl.),* 209, 329, 1973.
106. (a) Ref. 23; (b) Ref. 11.
107. Ref. H135.
108. **Grechkin, N. P., Nikonorova, N. K., Titova, N. S., Nuretdinov, I. A., and Zhilonkina, L. A.,** *J. Gen. Chem. U.S.S.R.,* 48, 1194, 1978.
109. **Jacobsen, N. E. and Bartlett, P. A.,** *J. Am. Chem. Soc.,* 105, 1613, 1983.
110. **Ashkinazi, L. A., Zavlin, P. M., Shek, V. M., Ionin, B. I., and Ignatovich, Ya. A.,** *J. Gen. Chem. U.S.S.R.,* 46, 1699, 1976.
111. Ref. H128.
112. Ref. H158.
113. **Łopusinski, A., Łuczak, L., Michalski, J., Kabachnik, M. M., and Moriyama, M.,** *Tetrahedron,* 37, 2011, 1981.
114. **Baceiredo, A., Bertrand, G., and Majoral, J.-P.,** *Nouv. J. Chim.,* 7, 255, 1983.
115. Ref. H61.
116. **Konovalova, I. V., Buraeva, L. A., Kashtanova, N. M., and Pudovik, A. N.,** *J. Gen. Chem. U.S.S.R.,* 52, 1745, 1982.
117. **Konovalova, I. V., Mikhailova, N. V., Gareev, R. D., Burnaeva, L. A., Anufrieva, T. V., and Pudovik, A. N.,** *J. Gen. Chem. U.S.S.R.,* 51, 826, 1981.
118. **Gololobov, Yu. G. and Nesterova, L. I.,** *J. Gen. Chem. U.S.S.R.,* 52, 2178, 1982.
119. **Muller, A. J. and Aaron, H. S.,** *Phosphorus Sulfur,* 25, 339, 1985.
120. **Poliichuk, Yu. A., Koshushko, B. N., and Shokol, V. A.,** *J. Gen. Chem. U.S.S.R.,* 52, 497, 1982.
121. **Konovalova, I. V., Mikhailova, N. V., Burnaeva, L. A., and Pudovik, A. N.,** *J. Gen. Chem. U.S.S.R.,* 52, 1134, 1982.
122. Ref. H272.
123. **Khusainova, N. G., Bredikhina, Z. A., and Pudovik, A. N.,** *J. Gen. Chem. U.S.S.R.,* 52, 1139, 1982.
124. Ref. H153.
125. **Kozlov, É. S., Dubenko, L. G., and Marchenko, A. P.,** *J. Gen. Chem. U.S.S.R.,* 50, 2156, 1980.
126. **Livantsev, M. V., Proskurnina, M. V., Prishchenko, A. A., and Lutsenko, I. F.,** *J. Gen. Chem. U.S.S.R.,* 54, 2237, 1984.
127. **Kawashima, T., Ishii, T., and Inamoto, N.,** *Bull. Chem. Soc. Jpn.,* 60, 1831, 1987.
128. **Harger, M. J. P.,** *J. Chem. Soc. Perkin Trans. 1,* p. 2127, 1983.
129. Ref. H68.
130. **Kibardin, A. M., Mikhailov, Yu. B., Gryazneva, T. V., and Pudovik, A. N.,** *Bull. Acad. Sci. U.S.S.R. Div. Chem. Sci.,* p. 2461, 1985.
131. **Pudovik, A. N., Stabrovskaya, L. A., Evstaf'ev, G. I., Remizov, A. B., and Gareev, R. D.,** *J. Gen. Chem. U.S.S.R.,* 42, 1847, 1972.
132. Ref. H170.
133. **Harger, M. J. P. and Smith, A.,** *J. Chem. Soc. Perkin Trans. 1,* p. 683, 1987.
134. Ref. H66.
135. (a) Ref. 134; (b) Ref. H367.
136. Ref. H9.
137. **Kozlov, É. S., Tolmachev, A. A., Chernega, A. N., and Boldeskul, I. E.,** *J. Gen. Chem. U.S.S.R.,* 55, 1126, 1955.
138. **Dutasta, J. P. and Simon, P.,** *Tetrahedron Lett.,* 28, 3577, 1987.
139. Ref. H250.
140. **Foss, V. L., Kudinova, V. V., and Lutsenko, I. F.,** *J. Gen. Chem. U.S.S.R.,* 49, 489, 1979.
141. **Fluck, E. and Binder, H.,** *Inorg. Nucl. Chem. Lett.,* 3, 307, 1967.
142. **Davisson, V. J., Woodside, A. B., Neal, T. R., Stremlev, K. E., Muehlbacher, M., and Poulter, C. D.,** *J. Org. Chem.,* 51, 4768, 1986.

143. **Quimby, O. T., Curry, J. D., Nicholson, D. A., Prentice, J. B., and Ray, C. H.,** *J. Organomet. Chem.*, 13, 199, 1968.
144. **Kharrasova, F. M., Zykova, T. V., Salakhutdinov, R. A., Efimova, V. D., and Shagidullina, R. D.,** *J. Gen. Chem. U.S.S.R.*, 44, 2379, 1974.
145. **Krutikov, V. I., Semenova, E. S., Maslennikov, I. G., and Lavrent'ev, A. N.,** *J. Gen. Chem. U.S.S.R.*, 53, 1405, 1983.
146. **Blackburn, G. M. and Parratt, M. J.,** *J. Chem. Soc. Perkin Trans. 1*, p. 1425, 1986.
147. **McKenna, C. E. and Shen, P.,** *J. Org. Chem.*, 46, 4573, 1981.
148. (a) Ref. 128; (b) Ref. 147.
149. **Mahmood, T. and Shreeve, J. M.,** *Inorg. Chem.*, 25, 3128, 1986.
150. **Maslennikov, I. G., Lavrent'ev, A. N., Lyubimova, M. V., Shvedova, Yu. I., and Lebedev, V. B.,** *J. Gen. Chem. U.S.S.R.*, 53, 2417, 1983.
151. **Burton, D. J. and Flynn, R. M.,** *Synthesis*, p. 615, 1979.
152. **Maslennikov, I. G., Shvedova, Yu. I., and Lavrent'ev, A. N.,** *J. Gen. Chem. U.S.S.R.*, 54, 204, 1984.
153. **Gross, H. and Costisella, B.,** *J. Prakt. Chem.*, 328, 231, 1986.
154. **Blackburn, G. M., Brown, D., Martin, S. J., and Parratt, M. J.,** *J. Chem. Soc. Perkin Trans. 1*, p. 181, 1987.
155. **Hutchinson, D. W. and Semple, G.,** *J. Organomet. Chem.*, 309, C7, 1986.
156. **Albrecht, S. and Herrmann, E.,** *Z. Anorg. Allg. Chem.*, 536, 187, 1986.
157. **Gross, H. and Costisella, B.,** *Synthesis*, p. 622, 1977.
158. **Mikołajczyk, M., Bałczewski, P., and Grzejszczak, S.,** *Synthesis*, p. 127, 1980.
159. **Mikołajczyk, M., Bałczewski, P., Wroblewski, K., Karolak-Wojciechowska, J., Miller, A. J., Wieczorek, M. W., Antipin, M. Y., and Struchkov, Y. T.,** *Tetrahedron*, 40, 4885, 1984.
160. **Mlotkowska, B., Gross, H., Costisella, B., Mikołajczyk, M., Grzejszczak, S., and Zatorski, A.,** *J. Prakt. Chem.*, 319, 17, 1977.
161. **Engelhardt, M., Haegele, G., Mikołajczyk, M., Bałczewski, P., and Wendisch, D.,** *Magn. Reson. Chem.*, 23, 18, 1985.
162. **Savignac, P., Teulade, M.-P., and Collignon, N.,** *J. Organomet. Chem.*, 323, 135, 1987.
163. (a) Ref. 154; (b) Ref. 6.
164. **Rueppet, M. L. and Marvel, J. T.,** *Org. Magn. Reson.*, 8, 19, 1976.
165. **Redmore, D. and Dhawan, B.,** *Phosphorus Sulfur*, 16, 233, 1983.
166. **Diel, P. J. and Maier, L.,** *Phosphorus Sulfur*, 20, 313, 1984.
167. **Gonçalves, H. and Majoral, J. P.,** *Phosphorus Sulfur*, 4, 343, 1978.
168. **Berkova, G. A., Baranov, G. M., Zhidkova, L. A., and Perekalin, V. V.,** *J. Gen. Chem. U.S.S.R.*, 51, 619, 1981.
169. **Scharf, D. J.,** *J. Org. Chem.*, 41, 28, 1976.
170. **Grechkin, N. P., Zhelonkina, L. A., and Nikonorova, L. K.,** *J. Gen. Chem. U.S.S.R.*, 52, 2406, 1982.
171. **Ivanov, B. E., Samarina, S. V., Krokhina, S. S., and Anoshina, N. P.,** *Bull. Acad. Sci. U.S.S.R. Div. Chem. Sci.*, p. 1993, 1974.
172. **Maier, L. and Crutchfield, M. M.,** *Phosphorus Sulfur*, 5, 45, 1978.
173. **Ivanov, B. E., Kudryatseva, L. A., Zyablikova, T. A., Bykova, T. G., and Gol'dfarb, É. I.,** *Bull. Acad. Sci. U.S.S.R. Div. Chem. Sci.*, p. 1396, 1971.
174. **Maier, L.,** *Helv. Chim. Acta*, 52, 827, 1969.
175. **Novikova, Z. S., Prishchenko, A. A., and Lutsenko, I. F.,** *J. Gen. Chem. U.S.S.R.*, 47, 2409, 1977.
176. (a) Ref. 143; (b) Ref. 155.
177. **Bottin-Strzalko, T., Corset, J., Froment, J., Pouet, M. J., Seyden-Penne, J., and Simonnin, M. P.,** *Phosphorus Sulfur*, 22, 217, 1985.
178. (a) Ref. 174; (b) Ref. 177.
179. **Maier, L.,** *Helv. Chim. Acta*, 52, 858, 1969.
180. **Maier, L.,** *Helv. Chim. Acta*, 52, 845, 1969.
181. (a) Ref. 68; (b) Ref. 64(a).
182. **Mikołajczyk, M. and Zatorski, A.,** *Synthesis*, p. 669, 1973.
183. **Mikołajczyk, M., Popielarczyk, M., and Grzejszczak, S.,** *Phosphorus Sulfur*, 10, 369, 1981.
184. **Drabowicz, J. and Mikołajczyk, M.,** *Synthesis*, p. 758, 1978.
185. (a) Ref. 182; (b) Ref. 184.
186. **Savignac, P. and Mathey, F.,** *Synthesis*, p. 725, 1982.
187. **Aboujaoude, E. E., Liétjé, S., Collignon, N., Teulade, M. P., and Savignac, P.,** *Synthesis*, p. 934, 1986.
188. (a) Ref. 186; (b) Ref. 187.
189. (a) Ref. 57; (b) Ref. 164.
190. **Gonçalves, H. and Majoral, J. P.,** *Phosphorus Sulfur*, 4, 357, 1978.
191. (a) Ref. 54; (b) Ref. 64(b).

192. **Scott, G., Hammond, P. J., Hall, C. D., and Bramblett, J. D.,** *J. Chem. Soc. Perkin Trans. 2*, p. 882, 1977.
193. **Teulade, M.-P., Savignac, P., Aboujaoude, E. E., and Collignon, N.,** *J. Organomet. Chem.*, 312, 283, 1986.
194. (a) Ref. 193; (b) Ref. 158.
195. **Julia, M., Mestdagh, H., and Rolande, C.,** *Tetrahedron*, 42, 3841, 1986.
196. **Hudson, H. R., Rees, R. G., and Weekes, J. E.,** *J. Chem. Soc. Perkin Trans. 1*, p. 982, 1974.
197. (a) Ref. 21; (b) Ref. 134.
198. **McKenna, C. E. and Khawli, L. A.,** *J. Org. Chem.*, 51, 5467, 1986.
199. **Blackburn, G. M., Brown, D., and Martin, S. J.,** *J. Chem. Res. (S)*, 92, 1985.
200. **Nicholson, D. A., Cilley, W. A., and Quimby, O. T.,** *J. Org. Chem.*, 35, 3149, 1970.
201. **Nicholson, D. A. and Vaughn, H.,** *J. Org. Chem.*, 36, 3843, 1971.
202. **Calvo, K. C.,** *J. Am. Chem. Soc.*, 107, 3690, 1985.
203. **Blackburn, G. M. and Kent, D. E.,** *J. Chem. Soc. Perkin Trans. 1*, p. 913, 1986.
204. **Stec, W. J. and Lesiak, K.,** *J. Org. Chem.*, 41, 3757, 1976.
205. **Kudzin, Z. H.,** *Synthesis*, p. 643, 1981.
206. **Kudzin, Z. H. and Kotynski, A.,** *Synthesis*, p. 1029, 1980.
207. **Kudzin, Z. H. and Stec, W. J.,** *Synthesis*, p. 469, 1978.
208. **Krzyzanowska, B. and Stec, W. J.,** *Synthesis*, p. 521, 1978.
209. (a) Ref. 206; (b) Ref. 208; (c) Ref. 207.
210. **Suezawa, H., Hirota, M., Shibata, Y., Takeuchi, I., and Hamada, Y.,** *Bull. Chem. Soc. Jpn.*, 59, 2362, 1986.
211. **Costisella, B. and Gross, H.,** *Tetrahedron*, 38, 139, 1982.
212. **Evelyn, L., Hall, C. D., Steiner, P. R., and Stokes, D. H.,** *Org. Magn. Reson.*, 5, 141, 1973.
213. (a) Ref. 73; (b) Ref. 212.
214. **Wroblewski, A. E.,** *Liebigs Ann. Chem.*, p. 1448, 1986.
215. **Wroblewski, A. E.,** *Liebigs Ann. Chem.*, p. 1854, 1986.
216. **Thiem, J. and Baulsen, H.,** *Phosphorus*, 6, 51, 1975.
217. **Mikroyannidis, J. A., Tsolis, A. K., and Gourghiotis, D. J.,** *Phosphorus Sulfur*, 13, 279, 1982.
218. **Nifant'ev, É. E., Kirkareva, T. S., Popkova, T. N., and Davydochkina, O. V.,** *J. Gen. Chem. U.S.S.R.*, 56, 164, 1986.
219. **Kim, D. Y. and Oh, D. Y.,** *Synth. Commun.*, 16, 859, 1986.
220. **Gazizov, T. Kh., Karelov, A. A., Sudarev, Yu. I., and Pudovik, A. N.,** *J. Gen. Chem. U.S.S.R.*, 53, 238, 1983.
221. **Konovalova, I. V., Ofitserova, E. Kh., Mironov, V. F., Ofitserov, E. N., Ostanina, I. L., and Pudovik, A. N.,** *J. Gen. Chem. U.S.S.R.*, 52, 2394, 1982.
222. **Fischer, U. and Haegele, G.,** *Z. Naturforsch.*, 40B, 1152, 1985.
223. **Corbel, B., Medinger, L., Haelters, J. P., and Sturtz, G.,** *Synthesis*, p. 1048, 1985.
224. **Mikołajczyk, M., Grzejszczak, S., Chefczynska, A., and Zatorski, A.,** *J. Org. Chem.*, 44, 2967, 1979.
225. **Kleiner, H.-J. and Schumann, C.,** *Phosphorus Sulfur*, 13, 363, 1982.
226. **Lebedev, E. P., Pudovik, A. N., Tsyganov, B. N., Nazmutdinov, R. Ya., and Romanov, G. V.,** *J. Gen. Chem. U.S.S.R.*, 47, 698, 1977.
227. **Andreev, N. A., Grishina, O. N., and Smirnov, V. N.,** *J. Gen. Chem. U.S.S.R.*, 49, 288, 1979.
228. **Gloede, J., Weight, E., and Gross, H.,** *J. Prakt. Chem.*, 322, 327, 1980.
229. **Cilley, W. A., Nicholson, D. A., and Campbell, D.,** *J. Am. Chem. Soc.*, 92, 1685, 1970.
230. Ref. H86.
231. **Maier, L.,** *Phosphorus*, 1, 105, 1971.
232. **Maier, L.,** *Phosphorus*, 1, 67, 1971.
233. Ref. H247.
234. **Issleib, K., Doepfer, K. P., and Balszuweit, A.,** *Phosphorus Sulfur*, 14, 171, 1983.
235. **Anan'eva, L. G., Fedorova, G. K., and Boldeskul, I. E.,** *J. Gen. Chem. U.S.S.R.*, 53, 480, 1983.
236. **Murray, M., Schmutzler, R., Grundemann, E., and Teichmann, H.,** *J. Chem. Soc. Sect. B*, p. 1714, 1971.
237. (a) Ref. 57; (b) Ref. 64(b).
238. **Laskorin, B. N., Yakshin, V. V., and Sokal'skaya, L. I.,** *J. Gen. Chem. U.S.S.R.*, 42, 1256, 1972.
239. Ref. H197.
240. **Gajda, T. and Zwierzak, A.,** *Tetrahedron*, 41, 4953, 1985.
241. (a) Ref. 174; (b) Ref. 64(b).
242. **Mukmeneva, N. A., Cherezova, E. N., Yamalieva, L. N., Kolesov, S. V., Minsker, K. S., and Kirpichnikov, P. A.,** *Bull. Acad. Sci. U.S.S.R. Div. Chem. Sci.*, p. 1009, 1985.
243. **Okamoto, Y., Tone, T., and Sakurai, H.,** *Bull. Chem. Soc. Jpn.*, 54, 303, 1981.
244. (a) Ref. 200; (b) Ref. 229.

245. **Bentley, R. L. and Dingwall, J. G.,** *Synthesis,* p. 552, 1985.
246. **Aboujaoude, E. E., Collignon, N., and Savignac, P.,** *Synthesis,* p. 634, 1983.
247. **Mavel, G., Mankowski-Favelier, R., Lavielle, G., and Sturtz, G.,** *J. Chim. Phys. Phys. Chim. Biol.,* 64, 1698, 1967.
248. **Krawiecka, B., Michalski, J., and Wojna-Tadeusak, E.,** *J. Org. Chem.,* 51, 4201, 1986.
249. **Aboujaoude, E. E., Collignon, N., Teulade, M.-P., and Savignac, P.,** *Phosphorus Sulfur,* 25, 57, 1985.
250. (a) Ref. 249; (b) Ref. 12.
251. **Zakhareva, V. I., Dogadina, A. V., Mashlyakovskii, L. N., Ionin, B. I., and Petrov, A. A.,** *J. Gen. Chem. U.S.S.R.,* 44, 96, 1974.
252. **Mikołajczyk, M. and Bałczewski, P.,** *Synthesis,* p. 691, 1984.
253. **Bottin-Strzalko, T., Corset, J., Fremont, F., Pouet, M.-J., Seyden-Penne, J., and Simonnin, M.-P.,** *J. Org. Chem.,* 45, 1270, 1980.
254. (a) Ref. 249; (b) Ref. 253.
255. **Zyablikova, T. A., Liorber, B. G., Khammatova, Z. T., Il'yasov, A. V., Enikeev, K. M., and Pavlov, V. A.,** *J. Gen. Chem. U.S.S.R.,* 54, 214, 1984.
256. **Bodalski, R., Pietrusiewicz, K. M., Monkiewicz, J., and Koszak, J.,** *Tetrahedron Lett.,* 21, 2287, 1980.
257. **Bottin-Strzalko, T., Seyden-Penne, J., Pouet, M.-J., and Simonnin, M.-P.,** *J. Org. Chem.,* 43, 4346, 1978.
258. (a) Ref. 203; (b) Ref. 257.
259. **Chasar, D. W.,** *J. Org. Chem.,* 48, 4768, 1983.
260. **Alewood, P. F., Johns, R. B., Hoogenraad, N. J., and Sutherland, T.,** *Synthesis,* p. 303, 1984.
261. Ref. H210.
262. **Ovchinnikov, V. V., Cherkasova, O. A., Mukmaneva, N. A., Luakumovich, A. G., and Pudovik, A. N.,** *J. Gen. Chem. U.S.S.R.,* 51, 834, 1981.
263. Ref. H222.
264. **Satterthwait, A. C. and Westheimer, F. H.,** *J. Am. Chem. Soc.,* 100, 3197, 1978.
265. **Mukhametov, F. S., Korshin, É. E., Korshunov, R. L., Efremov, Yu. Ya., and Zyablikova, T. A.,** *J. Gen. Chem. U.S.S.R.,* 56, 1571, 1986.
266. **Zimin, M. G., Burilov, A. R., Islamov, R. G., and Pudovik, A. N.,** *J. Gen. Chem. U.S.S.R.,* 53, 34, 1983.
267. **Levin, Ya. A. and Skorobogatova, M. S.,** *Bull. Acad. Sci. U.S.S.R. Div. Chem. Sci.,* p. 465, 1976.
268. **Pavlichenko, M. G., Ivanov, B. E., Pankukh, B. I., Eliseenkov, V. N., and Gorshanov, F. B.,** *Bull. Acad. Sci. U.S.S.R. Div. Chem. Sci.,* p. 1759, 1985.
269. **Nesterov, L. V., Krepysheva, N. E., and Aleksandrova, A. N.,** *J. Gen. Chem. U.S.S.R.,* 54, 47, 1984.
270. **Buchanan, G. W., Bourque, K., and Seeley, A.,** *Magn. Reson. Chem.,* 24, 360, 1986.
271. **Al'fonsov, V. A., Zamaletdinova, G. U., Nizamov, I. S., Batyeva, É. S., and Pudovik, A. N.,** *J. Gen. Chem. U.S.S.R.,* 54, 1324, 1984.
272. (a) Ref. 80; (b) Ref. 227.
273. **Andreev, N. A. and Grishina, O. N.,** *J. Gen. Chem. U.S.S.R.,* 54, 2410, 1984.
274. **Gareev, R. D. and Pudovik, A. N.,** *J. Gen. Chem. U.S.S.R.,* 52, 2333, 1982.
275. (a) Ref. 249; (b) Ref. 223.
276. **Griffiths, D. V. and Tebby, J. C.,** *J. Chem. Soc. Chem. Commun.,* p. 871, 1986.
277. **McCabe, D. J., Bowen, S. M., and Paine, R. T.,** *Synthesis,* p. 319, 1986.
278. **Gareev, R. D., Loginova, G. M., and Shermergorn, I. M.,** *J. Gen. Chem. U.S.S.R.,* 52, 616, 1982.
279. **Kim, D. Y. and Oh, D. Y.,** *Synth. Commun.,* 17, 953, 1987.
280. **Sakhibullina, V. G., Polezaeva, N. A., and Arbuzov, B. A.,** *J. Gen. Chem. U.S.S.R.,* 52, 1112, 1982.
281. **Cook, B. and Dingwall, J. G.,** *Phosphorus Sulfur,* 22, 211, 1985.
282. **Haegele, G., Worms, K.-H., and Blum, H.,** *Phosphorus,* 5, 277, 1975.
283. **Teulade, M.-P. and Savignac, P.,** *Synth. Commun.,* 17, 125, 1987.
284. **Corfield, G. C. and Monks, H. H.,** *J. Macromol. Sci. Chem.,* A9, 1113, 1975.
285. (a) Ref. 284; (b) **Arbuzov, B. A., Vizel', A. O., Ivanovskaya, K. M., and Gol'dfarb, É. I.,** *J. Gen. Chem. U.S.S.R.,* 43, 2125, 1973.
286. Ref. H221.
287. **Wroblewski, A. E.,** *Z. Naturforsch.,* 41B, 791, 1986.
288. **Richter, W. J.,** *Phosphorus Sulfur,* 10, 395, 1981.
289. Ref. H376.
290. Ref. H150.
291. **Breuer, E. and Bannet, D. M.,** *Tetrahedron,* 34, 997, 1978.
292. **Ismagilov, R. K., Razumov, A. I., Zykova, T. V., and Niyazov, N. A.,** *J. Gen. Chem. U.S.S.R.,* 53, 1805, 1983.
293. **Wroblewski, A. E.,** *Tetrahedron,* 39, 1809, 1983.

294. **Burgada, R. and Mohri, A.,** *Phosphorus Sulfur,* 13, 85, 1982.
295. **Arbuzov, B. A., Arshinova, R. P., Mareev, Yu. M., Mikirov, I. Kh., and Vinogradova, V. S.,** *Bull. Acad. Sci. U.S.S.R. Div. Chem. Sci.,* p. 629, 1974.
296. **Lesiak, K., Uznański, B., and Stec, W. J.,** *Phosphorus,* 6, 65, 1975.
297. (a) Ref. 295; (b) Ref. 296.
298. Ref. H23.
299. Ref. H236.
300. Ref. H237.
301. **Gazizov, M. B., Zakharov, V. M., Khairullin, R. A., Moskva, V. V., Muzin, R. Z., Efremov, Yu. Ya., and Savel'erc, E. I.,** *J. Gen. Chem. U.S.S.R.,* 54, 200, 1984.
302. Ref. H146.
303. **Stec, W. J., Lesiak, K., Mielczarek, D., and Stec, B.,** *Z. Naturforsch.,* 30B, 710, 1975.
304. **Odorisio, P. A., Pastor, S. D., and Spivac, J. D.,** *Phosphorus Sulfur,* 20, 273, 1984.
305. (a) Ref. 272; (b) Ref. 30.
306. **Nikonorov, K. V., Gurylev, É. A., Antokhina, L. A., and Latypov, Z. Ya.,** *Bull. Acad. Sci. U.S.S.R. Div. Chem. Sci.,* p. 2330, 1981.
307. **Foss, V. L., Lokashev, N. V., and Lutsenko, I. F.,** *J. Gen. Chem. U.S.S.R.,* 52, 1946, 1982.
308. **Stec, W. J., Lesiak, K., and Sudol, M.,** *Synthesis,* p. 785, 1975.
309. **Issleib, K., Mögelin, W., and Balszuweit, A.,** *Z. Anorg. Allg. Chem.,* 530, 16, 1985.
310. **Maier, L.,** *Phosphorus,* 2, 229, 1973.
311. **Blackburn, G. M. and Parratt, M. J.,** *J. Chem. Soc. Perkin Trans. 1,* p. 1417, 1986.
312. (a) Ref. 26; (b) Ref. 211.
313. **Bulot, J. J., Aboujaoude, E. E., Collignon, N., and Savignac, P.,** *Phosphorus Sulfur,* 21, 197, 1984.
314. **Redmore, D.,** *Phosphorus Sulfur,* 5, 271, 1979.
315. **Shivanyuk, A. F., Lozinskii, M. O., and Kalinin, V. N.,** *J. Gen. Chem. U.S.S.R.,* 52, 613, 1982.
316. **Berdnikov, E. A., Levin, Ya. A., Gazizova, L. Kh., and Chernov, P. P.,** *J. Gen. Chem. U.S.S.R.,* 55, 512, 1985.
317. **Degenhardt, C. R. and Burdsall, D. C.,** *J. Org. Chem.,* 51, 3488, 1986.
318. **Moskvin, A. V., Korsakov, M. V., Smorygo, N. A., and Ivin, B. A.,** *J. Gen. Chem. U.S.S.R.,* 54, 1987, 1984.
319. **Khusainova, N. G., Naumova, L. V., Berdnikov, E. A., and Pudovik, A. N.,** *J. Gen. Chem. U.S.S.R.,* 54, 1758, 1984.
320. **Mashlyakovskii, L. N., Berezina, G. G., Dogadina, A. V., Ionin, B. I., and Smirnov, S. A.,** *Vysokomol. Soedin. Ser. A,* 18, 308 (Engl. transl. p. 354), 1976.
321. **Garibina, V. A., Leonov, A. A., Dogadina, A. V., Ionin, B. I., and Petrov, A. A.,** *J. Gen. Chem. U.S.S.R.,* 55, 1771, 1985.
322. **Samitov, Yu. Yu., Berdnikov, E. A., Tantasheva, F. R., Margulis, B. Ya., and Kataev, E. G.,** *J. Gen. Chem. U.S.S.R.,* 45, 2097, 1975.
323. **Ishihara, T., Mackawa, T., Yamasaki, Y., and Ando, T.,** *J. Org. Chem.,* 52, 300, 1987.
324. **Sakhibullina, V. G., Polezhaeva, N. A., and Arbuzov, B. A.,** *J. Gen. Chem. U.S.S.R.,* 54, 906, 1984.
325. **Leonov, A. A., Dogadina, A. V., Ionin, B. I., and Petrov, A. A.,** *J. Gen. Chem. U.S.S.R.,* 53, 205, 1983.
326. **Mark, V.,** *Tetrahedron Lett.,* p. 281, 1962.
327. **Alikin, A. Yu., Liorber, B. G., Sokolov, M. P., Razumov, A. I., Zykova, T. V., and Salakhutdinov, R. A.,** *J. Gen. Chem. U.S.S.R.,* 52, 274, 1982.
328. **Bauer, G. and Haegele, G.,** *Z. Naturforsch. Teil B,* 34, 1252, 1979.
329. **Issleib, K., Balszuweit, A., Koetz, J., Richter, S., and Leutloff, R.,** *Z. Anorg. Allg. Chem.,* 529, 151, 1985.
330. **Ebsworth, E. A. V., Hunter, G. M., and Rankin, D. W. H.,** *J. Chem. Soc. Dalton Trans.,* p. 245, 1983.
331. **Fluck, E. and Binder, H.,** *Z. Anorg. Allg. Chem.,* 354, 139, 1969.
332. **Daly, J. J., Maier, L., and Sanz, F.,** *Helv. Chim. Acta,* 55, 1991, 1972.
333. (a) Ref. 332; (b) Ref. 64(b); (c) Ref. 30.
334. **Sigal, I. and Leow, L.,** *J. Am. Chem. Soc.,* 100, 6394, 1978.
335. **Corriu, R. J. P., Lanneau, G. F., and Leclercq, D.,** *Phosphorus Sulfur,* 9, 149, 1980.
336. **Brel', V. K., Ionin, B. I., and Petrov, A. A.,** *J. Gen. Chem. U.S.S.R.,* 53, 1985, 1983.
337. (a) **Mavel, G., Mankowski-Favelier, R., and Tanh, T. N.,** *J. Chim. Phys. Phys. Chim. Biol.,* 64, 1692, 1967; (b) Ref. H367.
338. **Brel', V. K., Dogadina, A. V., Ionin, B. I., and Petrov, A. A.,** *J. Gen. Chem. U.S.S.R.,* 52, 456, 1982.
339. (a) Ref. 313; (b) **Collins, D. J., Hetherington, J. W., and Swan, J. M.,** *Aust. J. Chem.,* 27, 1355, 1974.

340. **Redmore, D.,** *J. Org. Chem.*, 41, 2148, 1976.
341. **Hall, C. D.,** unpublished results.
342. **Onys'ko, P. P., Proklina, N. V., and Gololobov, Yu. G.,** *J. Gen. Chem. U.S.S.R.*, 52, 1270, 1982.
343. (a) Ref. 64(b); (b) Ref. 284.
344. **Brel', V. K., Ionin, B. I., and Petrov, A. A.,** *J. Gen. Chem. U.S.S.R.*, 51, 2263, 1981.
345. **Singh, R. K. and Rogers, M. D.,** *J. Heterocycl. Chem.*, 22, 1713, 1985.
346. **Kyba, E. C., Pines, S. P., Owens, P. W., and Chou, S. S. P.,** *Tetrahedron Lett.*, 22, 1875, 1981.
347. **Kozlov, V. A., Dol'kinova, T. Yu., Ivanchenko, V. I., Negrebitskii, V. V., Grapov, A. F., and Mel'nikov, N. N.,** *J. Gen. Chem. U.S.S.R.*, 53, 2008, 1983.
348. (a) Ref. 16; (b) Ref. 134.
349. **Hennig, H. W. and Sartori, P.,** *Chem. Ztg.*, 107, 257, 1983.
350. **Macomber, R. S. and Kennedy, E. R.,** *J. Org. Chem.*, 41, 3191, 1976.
351. (a) Ref. 35; (b) Ref. 34.
352. (a) Ref. 284; (b) Ref. H69; (c) Ref. H135.
353. **Angelov, C. M. and Christov, C. Z.,** *Synthesis*, p. 664, 1984.
354. (a) **Angelov, Kh. M., Stoyanov, N. M., and Ionin, B. I.,** *J. Gen. Chem. U.S.S.R.*, 52, 162, 1982; (b) Ref. 353.
355. **Munoz, A., Garnigues, B., and Wolf, R.,** *Phosphorus Sulfur*, 4, 47, 1984.
356. **Angelov, C. M., Christov, V. Ch., Petrova, J., and Kirilov, M.,** *Phosphorus Sulfur*, 17, 37, 1983.
357. Ref. H232.
358. **Ismailov, V. M., Moskva, V. V., Zykova, T. V., and Bairomov, R. N.,** *J. Gen. Chem. U.S.S.R.*, 54, 407, 1984.
359. **Kobayashi, S., Tokunoh, M., and Saegusa, T.,** *Macromolecules*, 19, 466, 1986.
360. Ref. H240.
361. **Patsanovskii, I. I., Ishmaeva, É. A., Romanov, G. V., Volkova, V. N., Strelkova, E. N., and Pudovik, A. N.,** *Dokl. Chem. (Engl. transl.)*, 255, 383, 1980.
362. Ref. H78.
363. **Ionin, B. I., Lebedev, V. B., and Petrov, A. A.,** *Dokl. Chem. (Engl. transl.)*, 152, 831, 1963.
364. Ref. H56.
365. **Khokhlov, B. S., Beresenova, L. S., and Savenkov, N. F.,** *J. Gen. Chem. U.S.S.R.*, 54, 867, 1984.
366. **Kuchen, W. and Haegele, G.,** *Chem. Ber.*, 103, 2114, 1970.
367. **Fild, M. and Stankiewicz, T.,** *Z. Naturforsch.*, 29B, 206, 1974.
368. **Maier, L.,** *Helv. Chim. Acta*, 47, 120, 1964.
369. **Nesterov, L. V., Kessel', A. Ya., Samitov, Yu. Yu., and Musina, A. A.,** *J. Gen. Chem. U.S.S.R.*, 40, 1228, 1970.
370. **Andreev, N. A. and Grishina, O. N.,** *J. Gen. Chem. U.S.S.R.*, 49, 623, 1979.
371. Ref. H287.
372. (a) Ref. 59; (b) **Yurchenko, R. I. and Klepa, T. I.,** *J. Gen. Chem. U.S.S.R.*, 54, 632, 1984.
373. (a) Ref. 64(b); (b) Ref. 65(a).
374. (a) Ref. H337; (b) Ref. 54.
375. **Haegele, G., Kuckelhaus, W., Seega, J., and Tossing, G.,** *Z. Naturforsch.*, 40B, 1053, 1985.
376. **Maier, L.,** *Phosphorus*, 4, 41, 1974.
377. **Nuretdinov, I. A., Loginova É. I., and Bayandina, E. V.,** *J. Gen. Chem. U.S.S.R.*, 48, 975, 1978.
378. Ref. H57.
379. **Roesky, H. W.,** *Chem. Ber.*, 101, 3679, 1968.
380. **Barabás, A., Popescu, R., Murezan, V., and Almási, L.,** *Org. Magn. Reson.*, 10, 35, 1977.
381. **Maier, L.,** *Phosphorus*, 5, 253, 1975.
382. **Almási, L., Fenesan, I., and Biró, B.,** *J. Prakt. Chem.*, 321, 913, 1979.
383. **Pudovik, M. A., Terent'eva, S. A., and Pudovik, A. N.,** *J. Gen. Chem. U.S.S.R.*, 51, 402, 1981.
384. **Dahl, B. M., Dahl, O., and Trippett, S.,** *J. Chem. Soc. Perkin Trans. 1*, p. 2239, 1981.
385. **Robert, J. B. and Weichmann, H.,** *J. Org. Chem.*, 43, 3031, 1978.
386. **Hall, C. R. and Williams, N. E.,** *Tetrahedron Lett.*, 21, 4959, 1980.
387. **Scheibye, S., Shabana, R., Lawesson, S.-O., and Rømming, C.,** *Tetrahedron*, 38, 993, 1982.
388. **Pedersen, B. S. and Lawesson, S.-O.,** *Tetrahedron*, 35, 2433, 1979.
389. **Seel, F. and Zindler, G.,** *Chem. Ber.*, 113, 1837, 1980.
390. Ref. H323.
391. Ref. H235.
392. **Haddad, M., N'Gando M'Pando, Th., Malavaud, C., Lopez, L., and Barrans, J.,** *Phosphorus Sulfur*, 20, 333, 1984.
393. **Malenko, D. M. and Sinitsa, A. D.,** *J. Gen. Chem. U.S.S.R.*, 56, 1467, 1986.
394. Ref. H324.
395. **McFarlane, W. and Rycroft, D. S.,** *J. Chem. Soc. Dalton Trans.*, p. 2162, 1973.

396. (a) Ref. 293; (b) Ref. 395.
397. Ref. H330.
398. **Nuretdinov, I. A., Bayandina, E. V., and Vinokurova, G. M.,** *J. Gen. Chem. U.S.S.R.*, 44, 2548, 1977.
399. (a) Ref. 395; (b) Ref. 135(b).
400. **Kutyrev, G. A., Korolev, O. S., Safiullina, N. R., Yarkova, É. G., Lebedeva, O. E., Cherkasov, R. A., and Pudovik, A. N.,** *J. Gen. Chem. U.S.S.R.*, 56, 1081, 1986.
401. **Dimukhametov, M. N. and Nuretdinov, I. A.,** *J. Gen. Chem. U.S.S.R.*, 54, 1704, 1984.
402. Ref. H350.
403. Ref. H24.
404. **Kutyrev, G. A., Korolev, O. S., Yarkova, É. G., Cherkasov, R. A., and Pudovik, A. N.,** *J. Gen. Chem. U.S.S.R.*, 56, 1086, 1986.
405. **Pantzer, R., Schmidt, W., and Goubeau, J.,** *Z. Anorg. Allg. Chem.*, 395, 262, 1973.
406. **Martin, J. and Riber, J. B.,** *Nouv. J. Chim.*, 4, 515, 1980.
407. (a) Ref. 37; (b) Ref. 406.
408. **El-Barbary, A. A., Shabana, R., and Lawesson, S.-O.,** *Phosphorus Sulfur,* 21, 349, 1985.
409. **Shabana, R., Yousif, N. M., and Lawesson, S.-O.,** *Phosphorus Sulfur,* 24, 327, 1985.
410. **Navech, J., Revel, M., and Kraemer, R.,** *Phosphorus Sulfur,* 21, 105, 1984.
411. **El-Barbary, A. A., Shabana, R., and Lawesson, S.-O.,** *Phosphorus Sulfur,* 21, 375, 1985.
412. **Yoshifuji, M., Nakayama, S., Okazaki, R., and Inamoto, N.,** *J. Chem. Soc. Perkin Trans. 1,* p. 2065, 1973.
413. **Acher, F. and Wakselman, M.,** *Bull. Chem. Soc. Jpn.*, 55, 3675, 1982.
414. **Mastryukova, T. A., Michalski, J., Uryupin, A. B., Skrzypcznski, Z., and Kabachnik, M. I.,** *J. Gen. Chem. U.S.S.R.*, 48, 1329, 1978.
415. **Livant'sev, M. V., Prishchenko, A. A., Lutsenko, I. F.,** *J. Gen. Chem. U.S.S.R.*, 55, 2195, 1985.
416. Ref. H365.
417. **Abduvakhabov, A. A., Kamaev, F., Leont'ov, V. B., Inoyatova, K., Sadykov, A. S., Godovikov, N. N., and Kabachnik, M. I.,** *Bull. Acad. Sci. U.S.S.R. Div. Chem. Sci.*, p. 2674, 1971.
418. **Nifant'ev, É. E., Magdeeva, R. K., Shcholpet'va, N. B., and Mastryukova, T. A.,** *J. Gen. Chem. U.S.S.R.*, 50, 2159, 1980.
419. **Radeglia, R., Schulze, J., and Teichmann, H.,** *Z. Chem.*, 15, 357, 1975.
420. **Dutasta, J. -P., Grand, A., Guimares, A. C., and Robert, J. P.,** *Tetrahedron,* 35, 197, 1979.
421. **Dutasta, J. -P. and Robert, J. B.,** *J. Am. Chem. Soc.*, 100, 1925, 1978.
422. (a) **Shagidullin, R. R., Shakirov, I. Kh., Musyakova, R. Kh., Vandryukova, I. I., and Nuretdinov, I. A.,** *Bull. Acad. Sci. U.S.S.R. Div. Chem. Sci.*, p. 916, 1981; (b) Ref. 421.
423. **Albrand, J. P., Dutasta, J. P., and Robert, J. B.,** *J. Am. Chem. Soc.*, 95, 4584, 1974.
424. (a) Ref. 275(b); (b) Ref. 423.
425. **Tzschach, A., Turkschat, K., Zschunke, A., Muegge, C., Altenbrunn, B., Piccinni-Leopardi, C., Germain, G., Declercq, J. P., and van Meersche, M.,** *J. Crystallogr. Spectrosc. Res.*, 15, 423, 1985.
426. Ref. H277.
427. (a) Ref. 426; (b) Ref. 421.
428. **Dutasta, J.-P., Guimares, A. C., and Robert, J. B.,** *Tetrahedron Lett.*, p. 801, 1977.
429. **Dutasta, J.-P.,** *J. Chem. Res. (S),* 22, 1986.
430. **Piccinni-Leopardi, C., Reisse, J., Germain, G., Declercq, J. P., Van Meersche, M., Turkschat, K., Muegge, C., Zschunke, A., Dutasta, J.-P., and Robert, J. B.,** *J. Chem. Soc. Perkin Trans. 2,* p. 85, 1986.
431. **Terent'eva, S. A., Pudovik, M. A., and Pudovik, A. N.,** *J. Gen. Chem. U.S.S.R.*, 53, 1282, 1983.
432. Ref. H368.
433. Ref. H382.
434. Ref. H369.
435. Ref. H309.
436. **Haegele, G., Engelhardt, M., Peters, W., Skowrońska, A., Gwara, J., and Wendisch, D.,** *Phosphorus Sulfur,* 21, 53, 1984.
437. **Polozev, A. M., Polozhaeva, N. A., and Arbuzov, B. A.,** *J. Gen. Chem. U.S.S.R.*, 54, 1922, 1984.
438. **Pudovik, A. N., Cherkasov, R. A., and Kutyrev, G. A.,** *Dokl. Chem. (Engl. transl.),* 208, 98, 1973.
439. (a) Ref. 294; (b) Ref. 332.
440. **Fild, M.,** *Z. Anorg. Allg. Chem.*, 358, 256, 1968.
441. **Grechkin, N. P., Nikonorova, L. K., and Zhelonkina, L. A.,** *J. Gen. Chem. U.S.S.R.*, 52, 503, 1982.
442. (a) Ref. H69; (b) Ref. 420.
443. (a) Ref. 381; (b) Ref. 332.
444. **Zabirov, N. G., Cherkasov, R. A., and Pudovik, A. N.,** *J. Gen. Chem. U.S.S.R.*, 56, 1047, 1986.
445. **Stec, W. J., Uznański, B., Bruzik, K., and Michalski, J.,** *J. Org. Chem.*, 41, 1291, 1976.

446. **Nuretdinova, O. N. and Novikova, V. G.,** *Bull. Acad. Sci. U.S.S.R. Div. Chem. Sci.*, p. 1096, 1985.
447. Ref. H269.
448. **Nuretdinova, I. A., Buina, N. A., Bayandina, E. V., Loginova, E. I., and Giniyatullina, M. A.,** *Bull. Acad. Sci. U.S.S.R. Div. Chem. Sci.*, p. 378, 1978.
449. (a) Ref. 422(a); (b) **Shagidullin, R. R., Shakirov, I. Kh., Musyakaeva, R. Kh., and Nuretdinov, I. A.,** *Bull. Acad. Sci. U.S.S.R. Div. Chem. Sci.*, p. 1557, 1981.
450. **Nuretdinov, I. A., Sadkova, D. N., and Bayandina, E. V.,** *Bull. Acad. Sci. U.S.S.R. Div. Chem. Sci.*, p. 2441, 1977.
451. Ref. H49.
452. **Uznański, B. and Stec, W. J.,** *Synthesis*, p. 735, 1975.
453. **Hahn, J. and Nataniel, T.,** *Z. Anorg. Allg. Chem.*, 543, 7, 1986.
454. **Andreev, N. A. and Grishina, O. N.,** *J. Gen. Chem. U.S.S.R.*, 52, 1581, 1982.
455. **Hahn, J. and Nataniel, T.,** *Z. Anorg. Allg. Chem.*, 547, 180, 1987.
456. (a) Ref. 37; (b) Ref. H282.
457. (a) Ref. 390; (b) Ref. 453.
458. **Fluck, E. and Binder, H.,** *Z. Anorg. Allg. Chem.*, 377, 298, 1970.
459. Ref. H268.
460. **Duddeck, H. and Lecht, R.,** *Phosphorus Sulfur*, 29, 169, 1987.
461. **Dmitrichenko, M. Yu., Rozinov, V. G., Bychkova, T. I., Donskikh, V. I., Dolgushin, G. V., Rabovskii, G. V., and Sergienko, L. M.,** *J. Gen. Chem. U.S.S.R.*, 57, 480, 1987.
462. **Sokolov, V. B., Ivanov, A. N., Epishina, T. A., and Martynov, I. V.,** *J. Gen. Chem. U.S.S.R.*, 57, 1479, 1987.
463. **Tsvetkov, E. N., Degtyrarev, A. N., and Bovin, A. N.,** *J. Gen. Chem. U.S.S.R.*, 56, 2249, 1986.
464. **Fild, M. and Handke, W.,** *Z. Anorg. Allg. Chem.*, 555, 109, 1987.
465. **Al'fonsov, V. A., Nizamov, I. S., Batyeva, É. S., and Pudovik, A. N.,** *Dokl. Chem. (Engl. transl.)*, 292, 9, 1987.
466. **Prudnikova, O. G., Brel', V. K., Ionin, B. I., and Petrov, A. A.,** *J. Gen. Chem. U.S.S.R.*, 57, 1313, 1987.
467. **Krolevets, A. A., Antipova, V. V., and Popov, A. G.,** *J. Gen. Chem. U.S.S.R.*, 56, 1800, 1986.
468. **Gubnitskaya, E. S., Peresypkina, L. P., and Parkhomenko, V. S.,** *J. Gen. Chem. U.S.S.R.*, 56, 1779, 1986.
469. **Nafikova, A. A., Aminova, R. M., Mukhametov, F. S., Enikeev, K. M., Ismaev, I. É., and Gaimullina, R. S.,** *J. Gen. Chem. U.S.S.R.*, 57, 287, 1987.
470. Ref. H419.
471. **Sokolov, V. B., Ivanov, A. N., Epishina, T. A., and Martynov, I. V.,** *J. Gen. Chem. U.S.S.R.*, 57, 845, 1987.
472. **Sinitsa, A. D., Onys'ko, P. P., Kim, T. V., Kiseleva, E. I., and Pirozhenko, V. V.,** *J. Gen. Chem. U.S.S.R.*, 56, 2372, 1986.
473. **Pavlichenko, M. G., Ivanov, B. E., Pantukh, B. I., Eliseenkov, V. N., and Gershamov, F. B.,** *J. Gen. Chem. U.S.S.R.*, 56, 1764, 1986.
474. **Onys'ko, P. P., Kim, T. V., Kiseleva, E. I., and Sinitsa, A. D.,** *J. Gen. Chem. U.S.S.R.*, 57, 1104, 1987.
475. **Hutchinson, D. W. and Thornton, D. M.,** *J. Organomet. Chem.*, 346, 341, 1988.
476. **Pondhaven-Raphalen, A. and Sturtz, G.,** *Phosphorus Sulfur*, 29, 329, 1987.
477. **Aksinenko, A. Yu., Pushin, A. N., Sokolov, V. B., Gontar', A. F., and Martynov, I. V.,** *Bull. Acad. Sci. U.S.S.R. Div. Chem. Sci.*, p. 1090, 1987.
478. **Popkova, T. N., Kukhareva, T. S., Bekker, A. R., and Nifant'ev, É, E.,** *J. Gen. Chem. U.S.S.R.*, 56, 1604, 1986.
479. **Nifant'ev, É. E., Kukhareva, T. S., Popkova, T. N., and Bekker, A. R.,** *J. Gen. Chem. U.S.S.R.*, 57, 1792, 1987.
480. **Teulade, M.-P. and Savignac, P.,** *Synthesis*, p. 1037, 1987.
481. **Novikova, Z. S., Odinets, I. L., and Lutsenko, I. F.,** *J. Gen. Chem. U.S.S.R.*, 57, 623, 1987.
482. **Breuer, E., Karaman, R., Gibson, D., Leader, H., and Goldblum, A.,** *J. Chem. Soc. Chem. Commun.*, p. 504, 1988.
483. **Liorber, B. G., Tarzivolova, T. A., Pavlov, V. A., Zykova, T. V., Kisilev, V. V., Tumasheva, N. A., Slizkii, A. Yu., and Shegvaleev, T. Sh.,** *J. Gen. Chem. U.S.S.R.*, 57, 465, 1987.
484. **Zverev, V. V., Berdnikov, E. A., Levin, Yu. A., Gazizova, L., and Bazhanova, Z. G.,** *J. Gen. Chem. U.S.S.R.*, 57, 454, 1987.
485. **Zozlov, V. A., Churusova, S. G., Ivanchenko, V. I., Negrebetskii, V. V., Grapov, A. F., and Mel'nikov, N. N.,** *J. Gen. Chem. U.S.S.R.*, 56, 1775, 1986.
486. **Maier, L., Duerr, D., and Rempfler, H.,** *Phosphorus Sulfur*, 32, 19, 1987.
487. **Petrov, K. A., Agafonov, S. V., and Pokatum, V. P.,** *J. Gen. Chem. U.S.S.R.*, 57, 83, 1987.

488. **Cadogan, J. I. G., Cowley, A. H., Gosney, I., Pakulinski, M., Wright, P. M., and Yaslak, S.,** *J. Chem. Soc. Chem. Commun.*, p. 1685, 1986.
489. **Konovalova, I. V., Grishin, Yu. G., Burnaeva, L. A., Dokuchaeva, I. S., and Chistokletov, V. N.,** *J. Gen. Chem. U.S.S.R.*, 56, 2470, 1986.
490. **Shabana, R., Mahron, M. R., and Hafez, T. S.,** *Phosphorus Sulfur*, 31, 1, 1987.
491. Ref. H415.
492. **Shabana, R.,** *Phosphorus Sulfur*, 29, 293, 1987.
493. **Novikova, Z. S., Odinets, I. L., and Lutsenko, I. F.,** *J. Gen. Chem. U.S.S.R.*, 57, 459, 1987.
494. **Butorina, L. S., Zalesova, N. N., Nifant'ev, É. E., and Mastryukova, T. A.,** *J. Gen. Chem. U.S.S.R.*, 57, 693, 1987.

Chapter 12

^{31}P NMR DATA OF FOUR COORDINATE PHOSPHORUS COMPOUNDS CONTAINING A P=Ch BOND AND TWO BONDS TO GROUP IV ATOMS

Compiled and presented by
Ronald S. Edmundson

NMR data were also donated by
H. W. Roesky, Universität Göttingen, FRG
J. P. Dutasta, Université Scientifique et Médicale de Grenoble, France
C. D. Hall, King's College London, U.K.

The NMR data are presented in Table K and it is divided into two parts. Section K1 contains data for phosphinic acids and their derivatives and Section K2 contains data for thio-, seleno-, and telluro-phosphinic acids and their derivatives. Supplementary data are available at the end of the chapter.

The most important feature in the nuclear magnetic resonance spectroscopy of the titled compounds, consisting almost entirely of derivatives of the phosphinic acids, and their thio, seleno, and telluro analogues, is the occurrence of very wide variations in the chemical shifts, dependent upon the nature of the directly bonded carbon groups. Nonetheless, such wide variations are simple extensions of the trends already indicated for derivatives of phosphoric acid (Chapter 9) and the phosphonic acids and their derivatives (Chapter 11). Derivatives of phosphoric acid have chemical shifts dependent to a large extent on the directly bonded atoms O, S, Se, Te, N, and halogen, and only to a relatively small extent on the indirectly bonded carbon groups; for the phosphonic acids, the shift is dependent on both the carbon bonded group and the other hetero atoms.

For the compounds $R_2P(O)Hal$, a gradual increase in the bulk of the organic groups, from methyl through to *tert*-butyl, results in deshielding. The extent of the deshielding is dependent on the nature of the halide, being ca. 9, 36, and 30 ppm for the fluoride, chloride, and bromide of di-*tert*-butylphosphinic acid relative to the halides of dimethylphosphinic acid. The corresponding deshieldings for the series $R_2P(S)Hal$ are ca. 27, 60, and 85 ppm, respectively. The phosphorus nucleus in dibutylphosphinothioic iodide is deshielded by ca. 39 ppm relative to that in dimethylphosphinothioic iodide, whereas for the compounds $(R)_2P(S)I$, the difference is only ca. 5 ppm. The limited data suggest that similar variations are to be expected for the selenides $R_2P(Se)Hal$.

In the phosphinic acid ester series $R_2P(O)OR'$, the chemical shift is only slightly dependent on the alkyl group R′, a situation already found for phosphoric and phosphonic esters, but here the dependence appears to increase slightly when O is replaced by S.

With no change in the group Y in the compounds $R_2P(O)Y$, the replacement of an attached sp^3 carbon (R = Et) by an sp^2 carbon (R = vinyl) results in shielding to the extent of 35 to 40 ppm; the shielding in the diphenylphosphinic acid series relative to the dimethylphosphinic acid derivative tends to be slightly less, and is generally in the range 15 to 40 ppm. Data for compounds with P-bonded sp carbons are particularly lacking, but those which are available suggest an even greater shielding effect by such hybridization changes.

In the phosphoric acid and phosphonic acid series a change in the phosphoryl chalcogen atoms from oxygen to sulfur or selenium leads to pronounced deshielding, and it is generally possible to distinguish between each of the isomeric monothio and isomeric monoseleno esters, and also, very often, between isomeric selenothio esters. However, the position with regard to derivatives of the phosphinic acids is not always so clearly defined. Acyclic esters of the type $R_2P(S)OR'$ (R = alkyl or Ph) have chemical shifts in the range 79 to 114 ppm, and are thus well deshielded relative to their *S*-alkyl isomers which have shifts in the range 46 to 70 ppm, a situation comparable to that found, for instance, with the thiophosphoric esters. The differentiation between monothio and dithio compounds, always clear for phosphoric esters, may not be so for the phosphinic esters. Thus, the chemical shifts for the compounds $Me_2P(X)YMe$ are 57, 46, 94, and 47.5 ppm when X, Y are OO, OS, SO, and SS; by comparison the values for the series $(MeO)_2P(X)YMe$ are −3, 30, 73, and 98 ppm. A similar variation in the data for the esters $(EtO)_2P(X)YR$ (R = Me or Et) may be compared with those for the compounds $Et_2P(X)YEt$, for which the values are 60, 58.5, 110, and 80 ppm.[1]

The chemical shifts of dialkylphosphinothioic acids in nonaqueous media are close to those of *O*-alkyl dialkylphosphinothioate esters, suggesting that the dyad tautomeric equilibrium lies well over to the thiophosphoryl side; on the other hand, the shifts of the dialkylphosphinothioate anions lie midway between those of the isomeric monothio esters, unlike the shifts of the dialkyl phosphorothioate anions vis-à-vis isomeric trialkyl phosphorothioate esters. It is interesting to note that, for compounds of the type $Me_2P(X)YP(Z)Me_2$, the phosphorus atom bearing nonidentical chalcogen atoms is more highly shielded than suggested by the average shifts of the phosphorus atoms bearing two oxygen or two sulfur atoms;[1] long-range effects of one phosphorus nucleus on the other are small.[2]

A further example of the unpredictability of the chemical shift can be seen in the comparison of the compounds $Me_2P(X)Y$ and $Ph_2P(X)Y$; for Y = NMe_2 or OMe, the diphenylphosphinic derivatives are the more shielded when X = O or S, the chemical shifts are very close for X = Se, Y = OMe, but the dimethylphosphinoselenoic amide is shielded relative to the diphenylphosphinoselenoic amide.

With careful application, chemical shifts appear to allow a differentiation between isomeric selenothio esters, the phosphorus nucleus in *S*-alkyl dialkylphosphinoselenothioates being shielded relative to that in the isomeric *Se*-alkyl esters. The shielding in trimethylsilyl esters, strongly evident in esters of phosphoric and phosphinic acids, is not so pronounced for the silyl esters of phosphinic acids.

Relatively few attempts have been made to correlate the chemical shifts of phosphinic acid derivatives with structure in a quantitative manner. For the phosphinothioic bromides $R^1R^2P(S)Br$, where $R^1 = CH_{3-k}Me_k$, $R^2 = CH_{3-l}Me_l$, and N = k + 1, there exists a linear correlation between the chemical shift (the range 80 to 135 ppm was covered by the compounds examined) and N (1 to 5), and it was possible to predict the shifts for those compounds for which N = O or 6 (64.6 and 146.6 ppm).[3] The relation between the phosphorus shift and the number of hydrogen atoms attached to an α-carbon atom in the compounds $R^1R^2P(O)Y$ is entirely general[4,5] and may imply the operation of hyperconjugation effects; the increased substitution on the α-carbon leading to increased deshielding as a controlling feature operates regardless of the length or convolution of the alkyl carbon chain.

Other correlations suggest the lesser importance of electronic effects on the determination of the chemical shift in appropriate compounds. Thus, for the series $Ph_2P(O)OAr$, a linear correlation exists between the phosphorus shift (which, for the compounds examined, extends over the range 30.8 to 33.1) and the Hammett-Taft constants for the substituents MeO, Me, F, Br, Ac, CF_3, CN, and NO_2 in the 4-position of the aryl group Ar.[6] A further correlation has been observed for the compounds $4\text{-}O_2NC_6H_4P(O)R^1R^2$, where R^1 and R^2 are alkyl or substituted alkyl groups.[7] The relative phosphorus shifts of ethyl diphenylphosphinate (27.2 ppm), ethyl di-2-thienylphosphinate (16.0 ppm), and ethyl di-2-furanylphosphinate (4.25 ppm) are also thought to indicate the operation of electronic effects.[8]

There are still very few data which allow a satisfactory comparison between acyclic and cyclic compounds, and between cyclic compounds with different phosphorus-containing ring sizes. Five-membered rings containing phosphorus as the only heterocyclic element appear to be deshielded from comparable acyclic phosphinate esters, e.g., 1-ethoxyphospholane 1-oxide vs. ethyl diethylphosphinate, and the isomeric 1-ethoxydihydrophosphole 1-oxides vs. esters of allylic or vinylic phosphinic acids.[9] The very few data on six-membered ring compounds, also with endocyclic P–C bonds, suggest that these, like six-membered ring phosphonates, are shielded relative to both acyclic compounds and also those with five-membered ring structures. Of the series of 1,3-diphosphacycloalkanes (**12**) where X = O, and Y = OH or $OCHMe_2$, the compounds with the five-membered ring are deshielded relative to those compounds with six- or seven-membered rings.

Those coupling constants between phosphorus and nearby atoms, through one or several bonds which have been measured for phosphoric and phosphonic acid derivatives, have also been measured for derivatives of the phosphinic acids. Earlier listings of phosphorus chemical shifts, e.g., that by Mark et al.,[10] contain few coupling constant data in total, but later tabulations by Finer and Harris,[11] Mavel,[12] and Gorenstein[13] contain appreciable data with relevant discussion. For phosphinic acid fluorides, $^1J_{PF}$ has values normally in the range 960 to 1100 Hz, including P–F bonds exocyclic to six-membered phosphorus-containing rings, although higher values (1200 Hz) have been observed for phosphetane derivatives. Any stereochemical significance of $^1J_{PF}$ values appears not to have been ascertained. $^2J_{PF}$ constants have been measured for a few derivatives of bistrifluoromethylphosphinic acid, and are around 70 to 140 Hz. $^1J_{P^{15}N}$ constants have been measured for a few structural types, e.g., ca. 50, 50 to 55, and 55 to 60 Hz for the azides $R_2P(X)N_3$ when X is O, S, and Se, and −0.5, 10 to 25, and ca. 15 Hz for the anilides $R_2P(X)NHPh$ where X has the same significance.

Trends for $^1J_{PC}$ can be discerned. Thus for the compounds $R_2P(X)Y$, $^1J_{PC}$ decreases as X = O → S → Se for a given group R (Me or Et) and group Y (Cl or N_3), and the values for the diethylphosphinic derivatives are ca. 3 to 5 Hz smaller than those for the dimethylphosphinic compounds. A further indication that the coupling constant diminishes with increase in steric bulk of the carbon substituents comes from the tetraalkyldiphosphine disulfides $R_2P(S)P(S)R_2$.

Coupling between two phosphorus atoms is observed when the two nuclei are chemically nonequivalent, and also when they are magnetically nonequivalent. The magnitude and sign of $^1J_{PP}$ appear to depend on several possible factors, including the states of oxidation of the two nuclei, and the position of conformational (rotameric) equilibrium about the P–P, in addition to effects, already described, of the steric bulk of substituents. Many earlier determinations of $^1J_{PP}$ constants are unreliable;[11] more recent determinations indicate values of 200 to 400 Hz for compounds with the $P^4(X)$-P^3 system when X = O, but rather smaller values for compounds with the diphosphine disulfide structure and which range from ca. 18 Hz for the tetramethyl compounds, increasing with the steric bulk of the substituents to ca. 100 Hz for the tetra-butyl compound, the values being probably negative in sign. The steric effects of R are a major factor in the determination of $^2J_{PP}$ as well as the chemical shifts of the two nuclei in the anhydride-type compounds $Me_2P(X)$-Y-$P(Z)Me_2$; when at least one phosphorus atom possesses a lone electron pair, $^2J_{PP}$ is 50 to 55 Hz, but is normally smaller (10 to 50 Hz) for coupling between two quinquevalent thio- or seleno-phosphoryl phosphorus nuclei. The values for $^2J_{POP}$ are about twice as large as those for $^2J_{PSP}$.[2]

The coupling constants most widely reported upon, however, are those involving selenium ^{77}Se; the direct coupling constants have been shown to be negative in sign, at least in many cases. Typical values for $^1J_{P=Se}$ include 820 to 840 Hz for X = Cl, 720 to 730 Hz for X = NMe_2, and 755 to 770 Hz for X = OR in $R_2P(Se)X$. Values for compounds with phosphorus directly bonded to sp^2 carbons, as in the series $Ph_2P(Se)X$, tend to be marginally, but consistently, higher at 850, 750 to 790, and 795 to 810 Hz for X = Cl, NMe_2, and OR. Coupling to dicoordinate selenium is numerically different, and for those compounds which possess both singly and doubly bonded selenium, the two coupling constants are quite distinct. Thus, for the diselenophosphinic esters $R_2P(Se)SeR'$, the values of $^1J_{P=Se}$ and $^1J_{P\text{-}Se}$ are ca. 750 and ca. 360 Hz, respectively. Each of the isolable stereoisomers of the anhydroselenide tBuMeP(Se)SeP(Se)tBuMe has its own distinct values (1) 760 and 350 Hz, and (2) 750 and 390 Hz, for which an explanation has been given.[14] Coupling between phosphorus and selenium thus has important structural implications in the assignment of phosphorus-selenium bond order. $^1J_{P=Se}$ has been extensively examined in connection with the conformation of exocyclic P=Se bonds in cyclic phosphoroselenoic acid derivatives; the same constant may prove to be an aid in the assignment of conformation at phosphorus in phosphacycloalkanes.

In conclusion, the picture presented by compounds of the type ABP(X)Y is one in which, for derivatives of the phosphinic acids, the chemical shifts are governed as much, if not to the greater extent, by the nature (steric bulk, hybridization) of the carbon groups A and B, as by X (O, S, or Se) and Y (Halogen, O, S, Se, or N).

TABLE K
Compilation of ³¹P NMR Data of Four Coordinate Compounds Containing a P=Ch Bond and Two Bonds from Phosphorus to Group IV Atoms

Section K1: Derivatives of Phosphinic Acids

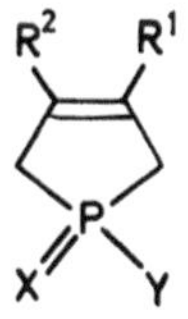

1 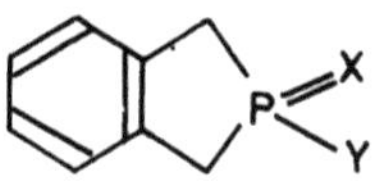 2 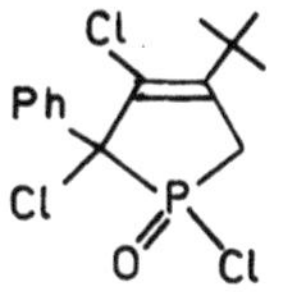 3

Rings/ Connectivities	Formula	Structure	NMR data (δ_P[solv] J_{PZ}Hz)	Ref.
	P-bound atoms = C_2BrO'			
$(H_3)_2$	C_2H_6BrOP	$Me_2P(O)Br$	72.6	15
$(C_3)_2$	$C_8H_{18}BrOP$	$tBu_2P(O)Br$	102.5	16
$5/(C'H_2)_2$	$C_6H_{10}BrOP$	1; R^1 = R^2 = Me, X = O, Y = Br	68.4(C)	17
	P-bound atoms = C'_2BrO'			
$(C'_2)_2$	$C_{12}H_{10}BrOP$	$Ph_2P(O)Br$	38.7	18
	P-bound atoms = C_2ClO'			
$(F_3)_2$	C_2ClF_6OP	$Tf_2P(O)Cl$	10.8 $^2J_{PF}$129.4	19
$(H_2Cl)_2$	$C_2H_4Cl_3OP$	$(ClCH_2)_2P(O)Cl$	49.3(n,c) 51.2(c)	22
H_2Cl, H_3	$C_2H_5Cl_2OP$	$Me(ClCH_2)P(O)Cl$	57.0(n) 58.7	25
$(H_3)_2$	C_2H_6ClOP	$Me_2P(O)Cl$	58.7(b) $^1J_{PC}$80 62.8(g)	27
$(CH_2)_2$	$C_4H_8Cl_3OP$	$(ClCH_2CH_2)_2P(O)Cl$	60.3(c)	28
	$C_4H_{10}ClOP$	$Et_2P(O)Cl$	73.7(b)[a] 76.7(g) $^1J_{PC}$75(b)	27
	$C_8H_{18}ClOP$	$Bu_2P(O)Cl$	71.3	29
$(C_3)_2$	$C_8H_{18}ClOP$	$tBu_2P(O)Cl$	92 94.7	31
$4/(H_2S)_2$	C_2H_4ClOPS	$Cl(O)\dot{P}CH_2S\dot{C}H_2$	44(b-$SOCl_2$)	32
$5/(C'H_2)_2$	C_4H_6ClOP	1; R^1 = R^2 = H, X = O, Y = Cl	83.3(c) 84.0(n)	34
	C_8H_8ClOP	2; X = O, Y = Cl	80.4(c)	35
$5/C'H_2$, C'_2Cl	$C_{14}H_{16}Cl_3OP$	3	47.8	36
$6/(CHO)_2$	$C_{12}H_{24}ClO_3P$	4; R = iPr, X = O, Y = Cl	50(b)	37
	P-bound atoms = $CC'ClO'$			
H_3;OO'	$C_4H_8ClO_3P$	$Me(EtOOC)P(O)Cl$	34.7(c)	38
H_3;C'_2	$C_7H_3ClF_5OP$	$Me(C_6F_5)P(O)Cl$	58.9	39
	C_7H_8ClOP	$MePhP(O)Cl$	52.0(g)	23
CH_2;C'_2	$C_8H_{10}ClOP$	$EtPhP(O)Cl$	59	40
C_3;OO'	$C_7H_{14}ClO_3P$	$tBu(EtOOC)P(O)Cl$	51.6(b)	38
C'_3;C'_2	$C_{10}H_9ClF_5OP$	$tBu(C_6F_5)P(O)Cl$	88.5	39
	$C_{10}H_{14}ClOP$	$tBuPhP(O)Cl$	70.7(b)	41
$5/CH_2$, $C'H$	C_4H_6ClOP	5; X = O, Y = Cl	80.3(c)	34

TABLE K (continued)
Compilation of ^{31}P NMR Data of Four Coordinate Compounds Containing a P═Ch Bond and Two Bonds from Phosphorus to Group IV Atoms

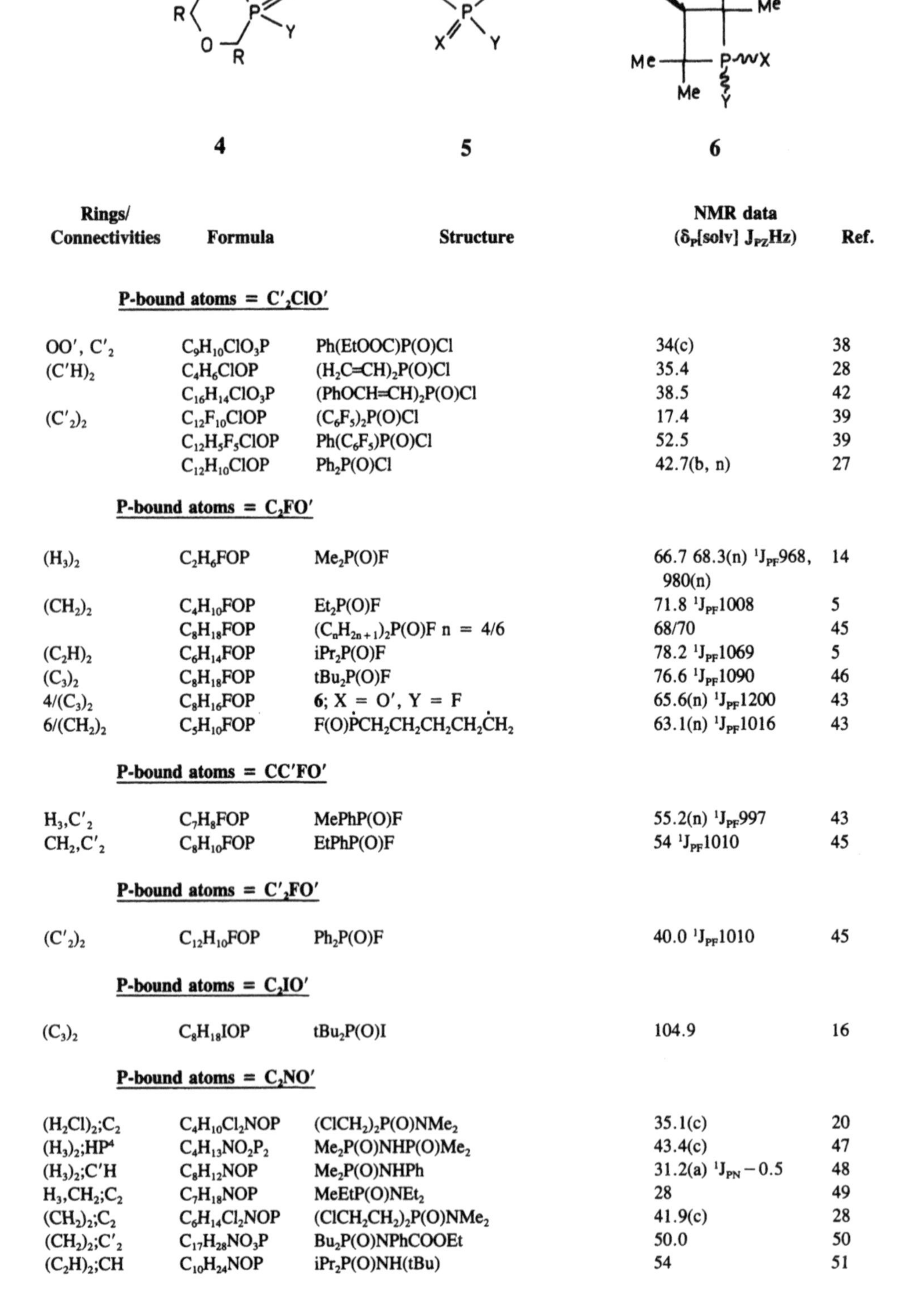

Rings/ Connectivities	Formula	Structure	NMR data (δ_P[solv] J_{PZ}Hz)	Ref.
	P-bound atoms = C′₂ClO′			
OO′, C′₂	$C_9H_{10}ClO_3P$	Ph(EtOOC)P(O)Cl	34(c)	38
(C′H)₂	C_4H_6ClOP	$(H_2C{=}CH)_2P(O)Cl$	35.4	28
	$C_{16}H_{14}ClO_3P$	$(PhOCH{=}CH)_2P(O)Cl$	38.5	42
(C′₂)₂	$C_{12}F_{10}ClOP$	$(C_6F_5)_2P(O)Cl$	17.4	39
	$C_{12}H_5F_5ClOP$	$Ph(C_6F_5)P(O)Cl$	52.5	39
	$C_{12}H_{10}ClOP$	$Ph_2P(O)Cl$	42.7(b, n)	27
	P-bound atoms = C₂FO′			
(H₃)₂	C_2H_6FOP	$Me_2P(O)F$	66.7 68.3(n) $^1J_{PF}$968, 980(n)	14
(CH₂)₂	$C_4H_{10}FOP$	$Et_2P(O)F$	71.8 $^1J_{PF}$1008	5
	$C_8H_{18}FOP$	$(C_nH_{2n+1})_2P(O)F$ n = 4/6	68/70	45
(C₂H)₂	$C_6H_{14}FOP$	$iPr_2P(O)F$	78.2 $^1J_{PF}$1069	5
(C₃)₂	$C_8H_{18}FOP$	$tBu_2P(O)F$	76.6 $^1J_{PF}$1090	46
4/(C₃)₂	$C_8H_{16}FOP$	**6**; X = O′, Y = F	65.6(n) $^1J_{PF}$1200	43
6/(CH₂)₂	$C_5H_{10}FOP$	$F(O)\dot{P}CH_2CH_2CH_2CH_2\dot{C}H_2$	63.1(n) $^1J_{PF}$1016	43
	P-bound atoms = CC′FO′			
H₃,C′₂	C_7H_8FOP	MePhP(O)F	55.2(n) $^1J_{PF}$997	43
CH₂,C′₂	$C_8H_{10}FOP$	EtPhP(O)F	54 $^1J_{PF}$1010	45
	P-bound atoms = C′₂FO′			
(C′₂)₂	$C_{12}H_{10}FOP$	$Ph_2P(O)F$	40.0 $^1J_{PF}$1010	45
	P-bound atoms = C₂IO′			
(C₃)₂	$C_8H_{18}IOP$	$tBu_2P(O)I$	104.9	16
	P-bound atoms = C₂NO′			
(H₂Cl)₂;C₂	$C_4H_{10}Cl_2NOP$	$(ClCH_2)_2P(O)NMe_2$	35.1(c)	20
(H₃)₂;HP⁴	$C_4H_{13}NO_2P_2$	$Me_2P(O)NHP(O)Me_2$	43.4(c)	47
(H₃)₂;C′H	$C_8H_{12}NOP$	$Me_2P(O)NHPh$	31.2(a) $^1J_{PN}-0.5$	48
H₃,CH₂;C₂	$C_7H_{18}NOP$	$MeEtP(O)NEt_2$	28	49
(CH₂)₂;C₂	$C_6H_{14}Cl_2NOP$	$(ClCH_2CH_2)_2P(O)NMe_2$	41.9(c)	28
(CH₂)₂;C′₂	$C_{17}H_{28}NO_3P$	$Bu_2P(O)NPhCOOEt$	50.0	50
(C₂H)₂;CH	$C_{10}H_{24}NOP$	$iPr_2P(O)NH(tBu)$	54	51

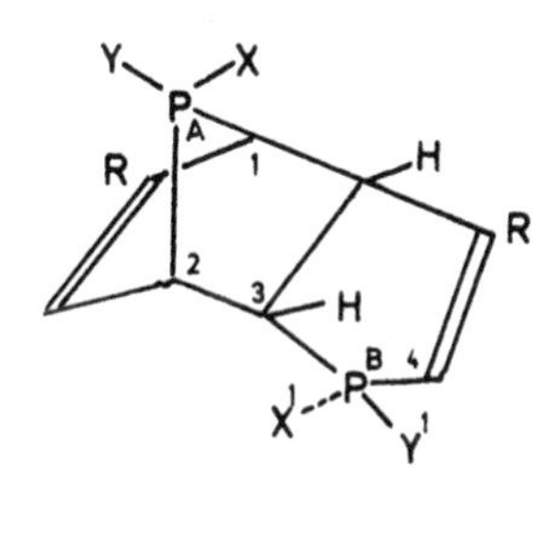

7

8

Rings/ Connectivities	Formula	Structure	NMR data (δ_P[solv] J_{PZ}Hz)	Ref.
$(C_3)_2$;Cl_2	$C_8H_{18}Cl_2NOP$	$tBu_2P(O)NCl_2$	80.6	52
$(C_3)_2$;HCl	$C_8H_{19}ClNOP$	$tBu_2P(O)NHCl$	63.8	52
$(C_3)_2$;H_2	$C_8H_{20}NOP$	$tBu_2P(O)NH_2$	55.0	52
$(C_3)_2$;CH	$C_{12}H_{28}NOP$	$tBu_2P(O)NH(tBu)$	53	51
$(C_3)_2$;C'H	$C_{14}H_{24}NOP$	$tBu_2P(O)NHPh$	58.3(r-me) $^1J_{PN}$11.5	48
4/$(H_2S)_2$;C_2	$C_4H_{10}NOPS$	$Me_2N(O)\dot{P}CH_2S\dot{C}H_2$	38(c)	32
5/$(C'H_2)_2$;C_2	$C_7H_{14}NOP$	**1**; R^1 = H, R^2 = Me, X = O, Y = NMe_2	63.5(b) 68.2(c)	53
	$C'_8H_{16}NOP$	**1**; R^1 = R^2 = Me, X = O, Y = NMe_2	56.9(b) 61.5(c)	53
	$C_9H_{18}NOP$	**1**; R^1 = H, R^2 = Me, X = O, Y = NEt_2	60.9(b) 65.5(c)	53
	$C_{10}H_{16}NOP$	**1**; R^1R^2 = $CH_2CH{=}CHCH_2$, X = O, Y = NMe_2	62.2(c)	53
	$C_{10}H_{20}NOP$	**1**; R^1 = R^2 = Me, X = O, Y = NEt_2	53.5(b) 58.4(c)	53
5,5/$(CC'H)_2$;C_2	$C_{14}H_{24}N_2O_2P_2$	**7**; R = Me, X = X' = O', Y = Y' = NMe_2	83.0(A) $^1J_{PC^1}$80.2 $^1J_{PC^2}$74.7 64.8(B) $^1J_{PC^3}$91.2 $^1J_{PC^4}$118.6	54
	$C_{18}H_{30}N_2O_2P_2$	**7**; R = Me, X = X' = O', Y = Y' = NEt_2	82.5(A)(c) 62.0(B)	55

P-bound atoms = $C_2N'O'$

Rings/ Connectivities	Formula	Structure	NMR data (δ_P[solv] J_{PZ}Hz)	Ref.
$(H_3)_2$;N'	$C_2H_6N_3OP$	$Me_2P(O)N_3$	46.4 48.2(c) $^1J_{PC}$85(b) $^1J_{PN}$47.9 $^2J_{PN}$4.9	56
$(CH_2)_2$;N'	$C_4H_{10}N_3OP$	$Et_2P(O)N_3$	54.8 $^1J_{PN}$51.1 $^2J_{PN}$4.5(b) 55.7(b) $^1J_{PC}$82 $^2J_{PC}$7	56

P-bound atoms = CC'NO'

Rings/ Connectivities	Formula	Structure	NMR data (δ_P[solv] J_{PZ}Hz)	Ref.
H_3;OO';C_2	$C_8H_{18}NO_3P$	$Me(EtOOC)P(O)NEt_2$	24.6(c)	38
CH_2;C'_2;OH	$C_9H_{14}NO_4PS$	$EtPhP(O)NHOSO_2Me$	40.3(d-c)	57
C_3;C'_2;C'_2	$C_{13}H_{17}N_2OP$	$tBuPhP(O)C_3H_3N_2$[a]	42.8(c)	58
5/CC'H;C'H;C_2	$C_{12}H_{18}NOP$	**8**; X = O, Y = NMe_2	68.4(c)	59
5/C_2H;C'H;C_2	$C_{13}H_{21}NO_2P_2$	**7**; R = Me, X = l.p., Y = OMe, X' = O, Y' = NMe_2	60.9(B)(c)	60
	$C_{14}H_{24}N_2O_2P_2S$	**7**; R = Me, X = S, X' = O, Y = Y' = NMe_2	62.5(B) 125.0(A)	61
	$C_{18}H_{23}NO_2P_2S$	**7**; R = Me, X = OPh, Y = S, X = O, Y = NMe_2	62.6(B)	61

TABLE K (continued)
Compilation of ^{31}P NMR Data of Four Coordinate Compounds Containing a P=Ch Bond and Two Bonds from Phosphorus to Group IV Atoms

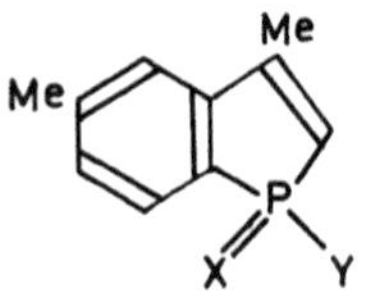

Me, Me, P, X, Y — **9**

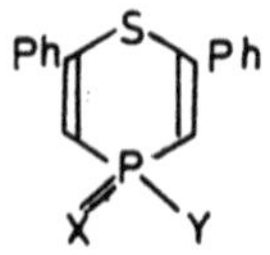

Ph, S, Ph, P, X, Y — **10**

Me, N, Me, P, X, Y — **11**

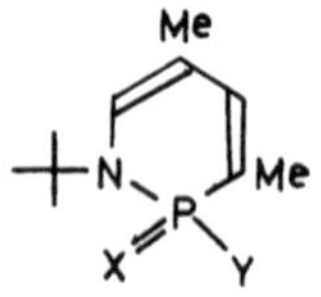

Rings/ Connectivities	Formula	Structure	NMR data (δ_P[solv] J_{PZ}Hz)	Ref.
P-bound atoms = CC'N'O'				
$C_3;C'_2;N'$	$C_{18}H_{30}N_3OP$	$tBu(3,5\text{-}tBu_2C_6H_3)P(O)N_3$	50.2	62
$C_3;C'_2;C'$	$C_{11}H_{14}NOPS$	tBuPhP(O)NCS	39.0(r,b) 42.7(p)	63
P-bound atoms = C'₂NO'				
$OO',C'_2;C_2$	$C_{13}H_{20}NO_3P$	$PhP(O)(COOEt)NEt_2$	16.6	38
$(C'_2)_2;HP^4$	$C_{16}H_{21}NO_4P_2$	$Ph_2\mathbf{P}(O)NHP(O)(OEt)_2$	20.0(b) $^2J_{PP}$2.8	64
$(C'_2)_2;H_2$	$C_{12}H_{12}NOP$	$Ph_2P(O)NH_2$	22.4 25.5(g)	66
$(C'_2)_2;CH$	$C_{18}H_{22}NOP$	$Ph_2P(O)NHcHex$	21.8(c)	65
	$C_{19}H_{18}NOP$	$Ph_2P(O)NHCH_2Ph$	23.5(c)	65
$(C'_2)_2;C'H$	$C_{18}H_{16}NOP$	$Ph_2P(O)NHPh$	15.7(d) 18.8(y-r)	68
$(C'_2)_2;C_2$	$C_{14}H_{16}NOP$	$Ph_2P(O)NMe_2$	29.6(r)	69
	$C_{16}H_{20}NOP$	$Ph_2P(O)NEt_2$	27(d)	67
	$C_{26}H_{24}NOP$	$Ph_2P(O)N(CH_2Ph)_2$	31.1	70
$(C'_2)_2;C'_2$	$C_{24}H_{20}NOP$	$Ph_2P(O)NPh_2$	25.8(c)	20
$5/C'H,C'_2;C_2$	$C_{12}H_{16}NOP$	**9**; X = O, Y = NMe_2	67(c)	61
$6/(C'H)_2;C_2$	$C_{20}H_{22}NOPS$	**10**; X = O, Y = NEt_2	15.5(c)	71
$6/CC',C'_2;CC'$	$C_{16}H_{22}NOP$	**11**; X = O, Y = Ph	18.0	72
P-bound atoms = C'₂N'O'				
$(C'_2)_2;N'$	$C_{12}H_{10}N_3OP$	$Ph_2P(O)N_3$	28.3 29.6	73
$(C'_2)_2;C'$	$C_{13}H_{10}NOPS$	$Ph_2P(O)NCS$	26	74
P-bound atoms = C₂O'P				
$(H_3)_2;C_2$	$C_4H_{12}OP_2$	$Me_2\mathbf{P}(O)PMe_2$	44.9(r) $^1J_{PP}$200	75
$(C_2H)_2;O_2$	$C_{14}H_{32}O_3P$	$iPr_2\mathbf{P}(O)P(OBu)_2$	55.5 $^1J_{PP}$250	76
$(C_2H)_2;CP^4$	$C_{16}H_{37}O_2P_3$	$[iPr_2\mathbf{P}(O)]_2P(tBu)$	73.6 $^1J_{PP}$320	77
$(C_2H)_2;C_2$	$C_{12}H_{28}OP_2$	$iPr_2\mathbf{P}(O)P(iPr)_2$	63.8 $^1J_{PP}$276	76
$(C_3)_2;O_2$	$C_{16}H_{36}O_3P_2$	$tBu_2\mathbf{P}(O)P(OBu)_2$	62.5 $^1J_{PP}$290	76
$(C_3)_2;C_2$	$C_{16}H_{36}OP_2$	$tBu_2\mathbf{P}(O)P(tBu)_2$	86.2 88 $^1J_{PP}$392	76
P-bound atoms = CC'O'P				
$C_3;C'_2;CC'$	$C_{20}H_{28}OP_2$	$tBuPh\mathbf{P}(O)PPh(tBu)$	51.2 $^1J_{PP}$300	78
P-bound atoms = C₂OO'				
$(Br_2P^4)_2;C$	$C_{12}H_{25}Br_3O_8P_3$	$EtO\mathbf{P}(O)[CBr_2P(O)(OEt)_2]_2$	15.2	79
$Cl_3,CH_2;C$	$C_6H_{12}Cli3O_2P$	$Pr(Cl_3C)P(O)OEt$	42.0	80

Rings/ Connectivities	Formula	Structure	NMR data (δ_P[solv] J_{PZ}Hz)	Ref.
$Cl_3,C_3;C$	$C_9H_{18}Cl_3OP$	$tBu(Cl_3C)P(O)OBu$	52	30
$(F_3)_2;P^4$	$C_4F_{12}O_3P_2$	$Tf_2P(O)OP(O)Tf_2$	−5.4 $^2J_{PF}$137.5	19
$(F_3)_2;P$	$C_4F_{12}O_3P_2$	$Tf_2\mathbf{P}(O)OPTf_2$	−0.3$^2J_{PF}$131.8	19
$(F_3)_2;H$	$C_2HF_6O_2P$	$Tf_2P(O)OH$	−7.4 $^2J_{PF}$87.9	81
$F_3,H_3;C$	$C_3H_6F_3O_2P$	$MeTfP(O)OMe$	34	82
$HN,H_3;H$	$C_4H_{10}NO_4P$	$MeP(O)(OH)CH(NH_2)CH_2COOH$	33.0(w pH 2)	83
$(HO_2)_2;C$	$C_{12}H_{27}O_6P$	$[(EtO)_2CH]_2P(O)OEt$	28.0	84
$HO_2,CH_2;C$	$C_9H_{21}O_4P$	$(EtO)_2CHP(O)ET(OEt)$	43.7	85
$(H_2Cl)_2;P^4$	$C_4H_8Cl_4O_3P_2$	$(ClCH_2)_2P(O)OP(O)(CH_2Cl)_2$	37.3(k)	23
$(H_2Cl)_2;H$	$C_2H_5Cl_2O_2P$	$(ClCH_2)_2P(O)OH$	32.0(w) 32.9(w)	86
$(H_2Cl)_2;C$	$C_3H_7Cl_2O_2P$	$(ClCH_2)_2P(O)OMe$	40.9(c)	87
	$C_4H_9Cl_2O_2P$	$(ClCH_2)_2P(O)OEt$	39.7(c,n)	23
	$C_5H_{11}Cl_2O_2P$	$(ClCH_2)_2P(O)O\text{-}iPr$	37.3(c)	87
$(H_2N)_2;H$	$C_2H_9N_2O_2P$	$(H_2NCH_2)_2P(O)OH$	29.2(w)	86
	$C_8H_{25}N_2O_2PSi_2$	$(TmsNHCH_2)_2P(O)OH$	44.0(c)	86
	$C_{10}H_{25}N_2O_2P$	$(tBuNHCH_2)_2P(O)OH$	18.5(w)	86
$H_2N,H_3;H$	$C_4H_{10}NO_4P$	$MeP(O)(OH)CH_2NHCH_2COOH$	32.8(w pH 1); 30.9(w pH 7)	88
$H_2N,C'H_2;H$	$C_5H_{10}NO_6P$	$HOOCCH_2NHCH_2P(O)(OH)CH_2COOH$	20.9(w pH 1); 29.4(w pH 9)	89
$(H_2O)_2;Si$	$C_{11}H_{31}O_4PSi_3$	$(TmsOCH_2)_2P(O)OTms$	32.8	90
$(H_2O)_2;H$	$C_2H_7O_4P$	$(HOCH_2)_2P(O)OH$	44.0(y) 45.8(w)	91
$(H_2O)_2;C$	$C_{16}H_{19}O_4P$	$(PhOCH_2)_2P(O)OEt$	40	49
$H_2O,H_3;H$	$C_2H_7O_3P$	$Me(HOCH_2)P(O)OH$	52.9	24
$H_2O,CH_2;C$	$C_5H_{13}O_3P$	$Et(HOCH_2)P(O)OEt$	62	92
$(H_2P^4)_2;H$	$C_2H_9O_8P_3$	$HO(O)\mathbf{P}[CH_2P(O)(OH)_2]_2$	37.3(w)	79
$(H_2P^4)_2;C$	$C_{11}H_{27}O_8P_3$	$MeO(O)\mathbf{P}[CH_2P(O)(OEt)_2]_2$	34.3	87
	$C_{12}H_{29}O_8P_3$	$EtO(O)\mathbf{P}[CH_2P(O)(OEt)_2]_2$	37.3(c)	79
	$C_{24}H_{37}O_6P_3$	$EtO(O)\mathbf{P}[CH_2P(O)(OBu)Ph]_2$	38.8 33.6	87
$H_2P^4,H_3;C$	$C_2H_8O_5P_2$	$Me(HO)\mathbf{P}(O)CH_2P(O)(OH)_2$	47.3	94
	$C_9H_{22}O_4P$	$H_2C[P(O)(O\text{-}iPr)Me]_2$	41.6 41.95[D]	93
$H_2P^4,CH_2;C$	$C_9H_{22}O_4P_2$	$H_2C[P(O)(OEt)Et]_2$ *(RR,SS)*	48.5(n)	93
		(RS,SR)	48.2(n)	
$H_2P, H_3;C$	$C_8H_{20}O_4P$	$Me(EtO)\mathbf{P}(O)CH_2P(OEt)_2$	45.7(n) $^2J_{PP}$35	93
$H_2P,CH_2;C$	$C_9H_{22}O_4P$	$Et(EtO)\mathbf{P}(O)CH_2P(OEt)_2$	49(n) $^2J_{PP}$38	93
$H_3,Cl_3;C$	$C_4H_5Cl_3F_3O_2P$	$Me(Cl_3C)P(O)OCH_2Tf$	33.4	95
$(H_3)_2;P^4$	$C_4H_{12}O_2P_2S$	$Me_2\mathbf{P}(O)OP(S)Me_2$	54.3(r),[92.1] $^1J_{PC}$93.3, 72.5 $^2J_{PP}$30.7	2
	$C_4H_{12}O_3P_2$	$Me_2P(O)OP(O)Me_2$	52.6(k) 56.5	97
$(H_3)_2;Si$	$C_5H_{15}O_2PSi$	$Me_2P(O)OTms$	39.0(b) 41.8(c)	99
	$C_6H_{18}O_4P_2Si$	$Me_2Si[OP(O)Me_2]_2$	41.6(c)	100
$(H_3)_2;H$	$C_2H_7O_2P$	$Me_2P(O)OH$	48.6(b) 49.4(y) 52.0(c)	101
$(H_3)_2;C$	$C_3H_9O_2P$	$Me_2P(O)OMe$	52.2 57.4 $^2J_{PH}$14.2	102
	$C_8H_7F_{12}O_3P$	$Me_2P(O)OCTf_2CTf_2OH$	68.5(c)	75
$(H_3)_2;C'$	$C_8H_{10}BrO_2P$	$Me_2P(O)OC_6H_4Br\text{-}2$	55.6	103
$H_3,CH_2;H$	$C_4H_{10}NO_4P$	$HOOCCH(NH_2)CH_2P(O)(OH)Me$	42.5(w pH 1)	83
$H_3,CH_2;C$	$C_6H_{13}O_3P$	$Me(MeO)P(O)CH_2CH_2COMe$	56	104
$H_3,C'H_2;C$	$C_9H_{17}O_4P$	$Me(EtO)P(O)CH_2C(COOEt){=}CH_2$	50.1(c)	83
$(CF_2)_2;H$	$C_4HF_{10}O_2P$	$(TfCF_2)_2P(O)OH$	−2.2 $^2J_{PF}$68.3	81
$CF_2,CH_2;C$	$C_7H_{10}F_7O_2P$	$Et(C_3F_7)P(O)OEt$	33.4	105
$(C'HO)_2;H$	$C_{14}H_{15}O_4P$	$[PhCH(OH)]_2P(O)OH$	36.5 38.1[D](y) 41(d)	107
$(CH_2)_2;P^4$	$C_8H_{20}O_2P_2S$	$Et_2\mathbf{P}(O)OP(S)Et_2$	60, [107]	1
$(CH_2)_2;Si$	$C_7H_{19}O_2PSi$	$Et_2P(O)OTms$	59.3(c)	100
	$C_9H_{23}O_2PSi$	$Pr_2P(O)OTms$	46.8(c)	100

TABLE K (continued)
Compilation of ³¹P NMR Data of Four Coordinate Compounds Containing a P=Ch Bond and Two Bonds from Phosphorus to Group IV Atoms

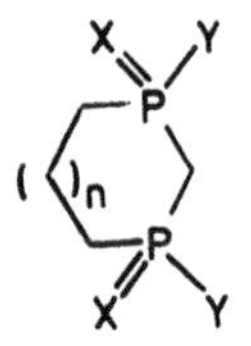

12 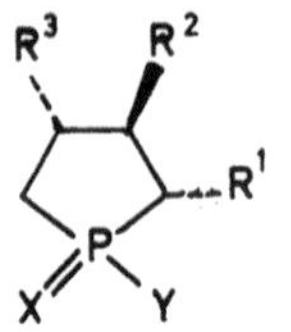 **13** 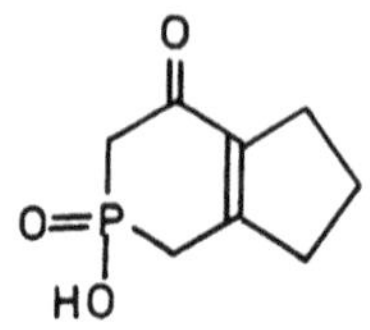 **14**

Rings/ Connectivities	Formula	Structure	NMR data (δ_P[solv] J_{PZ}Hz)	Ref.
$(CH_2)_2$;H	$C_4H_{11}O_2P$	$Et_2P(O)OH$	58.2(a)	1
	$C_4H_{13}O_8P_3$	$HO(O)P[CH_2CH_2P(O)(OH)_2]_2$	53.2	108
	$C_{12}H_{27}O_2P$	$Hex_2P(O)OH$	55	109
$(CH_2)_2$;C	$C_6H_{12}Cl_3O_2P$	$(ClCH_2CH_2)_2P(O)OCH_2CH_2Cl$	49.6(c)	108
	$C_6H_{12}F_3O_2P$	$Et(TfCH_2)P(O)OEt$	43.4	105
	$C_6H_{13}Cl_2O_2P$	$(ClCH_2CH_2)_2P(O)OEt$	47.5(c)	108
	$C_6H_{15}O_2P$	$Et_2P(O)OEt$	60.0	102(a)
	$C_{14}H_{33}O_8P_2$	$EtO(O)P[CH_2CH_2P(O)(OEt)_2]_2$	52.7(c)	108
$(CH_2)_2$;C'	$C_{10}H_{13}Cl_2O_2P$	$(ClCH_2CH_2)_2P(O)OPh$	49.3(c)	28
	$C_{10}H_{15}O_2P$	$Et_2P(O)OPh$	61	110
CH_2,C'H_2;C	$C_{11}H_{16}ClO_2P$	$PhCH_2(ClCH_2CH_2)P(O)OEt$	47.7(c)	111
C_2H,C'H;H	$C_9H_{17}O_2P$	cHex(H_2C=$CHCH_2$)P(O)OH	52.9	112
(C_2H_2);H	$C_{12}H_{23}O_2P$	$cHex_2P(O)OH$	57	109
$(C_3)_2$;Si	$C_{11}H_{27}O_2PSi$	$tBu_2P(O)OTms$	56.6(c)	100
$(C_3)_2$;C	$C_{10}H_{23}O_2P$	$tBu_2P(O)OEt$	64.3	16
4/$(C_3)_2$;C	$C_9H_{19}O_2P$	**6**; X = O', Y = OMe	55.1^T(c)	113
4/$(C_3)_2$;C'	$C_{14}H_{21}O_2P$	**6**; X = O', Y = OPh	57.0	113
5/H_2P^4,CH_2;H	$C_3H_8O_4P_2$	**12**; n = O, X = O, Y = OH	54.5	94
5/H_2P^4,CH_2;C	$C_9H_{20}O_4P_2$	**12**; n = o, X = O, Y = iPrO	53.5	94
5/H_3,C_2C';C	$C_7H_{13}O_2P$	**J1**; R^1 = R^3 = R^4 = Me, R^2 = H, X = O, Y = Me	75.1(c) $^1J_{PC}$78.0	114
5/CHBr,CH_2;H	$C_4H_7Br_2O_2P$	**13**; R^1 = R^2 = Br, R^3 = H, X = O, Y = OH	35.8(c) $^1J_{PC^1}$ 88.2 $^2J_{PC^2}$ 27.9 $^2J_{PC^3}$ 4.4 $^2J_{PC^4}$ 94.1	115
5/$(CH_2)_2$;H	$C_4H_7Br_2O_2P$	**13**; R^1 = H, R^2 = R^3 = Br, X = O, Y = OH	54.4(d) $^1J_{PC^1}$ 89.8 $^2J_{PC^2}$ 12.7	115
	$C_{10}H_{13}O_2$	$HO(O)\dot{P}CH_2CHPhCH_2\dot{C}H_2$	78.1(c)	116
5/$(CH_2)_2$;C	$C_8H_{17}O_2P$	**13**; R^1 = R^2 = R^3 = H, X = O, Y = OBu	75	117
5/$(C'H_2)_2$;P^4	$C_8H_{12}O_3P_2$	**1**; R^1 = R^2 = H, X = O, Y = $O(O)\dot{P}CH_2CH$=$CH\dot{C}H_2$	67.4	118
5/$(C'H_2)_2$;H	$C_4H_7O_2P$	**1**; R^1 = R^2 = H, X = O, Y = OH	75.1(w)	9
	$C_8H_9O_2P$	**2**; X = O, Y = OH	71.2(c)	119
	$C_8H_{11}O_2P$	**1**; R^1R^2 = CH_2CH=$CHCH_2$, X = O, Y = OH	60.1(d)	119
	$C_8H_{13}O_2P$	**1**; R^1R^2 = $(CH_2)_4$, X = O, Y = OH	69.1(c)	120
5/$(C'H_2)_2$;C	$C_6H_{11}O_2P$	**1**; R^1 = R^2 = H, X = O, Y = OEt	68(n)	121
	$C_8H_{15}O_2P$	**1**; R^1 = R^2 = Me, X = O, Y = OEt	68.4(c)	17
5/$(C'H_2)_2$;C'	$C_{10}H_{11}O_2P$	**1**; R^1 = R^2 = H, X = O, Y = OPh	75.1	118
6/H_2P^4,CH_2;H	$C_4H_{10}O_4P_2$	**12**; n = 1, X = O, Y = OH	43.5	94
6/$(C'HO)_2$;H	$C_{21}H_{19}O_4P$	**4**; R = Ph, X = O, Y = OH	16(d)	106
6/$(CH_2)_2$;C	$C_9H_{19}O_2P$	$BuO(O)\dot{P}CH_2CH_2CH_2CH_2\dot{C}H_2$	45	117
6/$(C'H)_2$;C	$C_8H_{11}O_3P$	**14**	46.8(c)	122

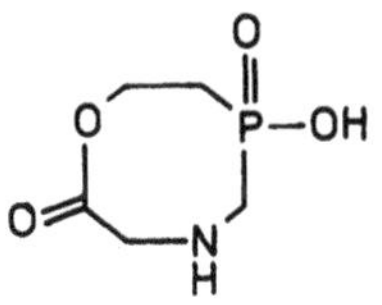

15 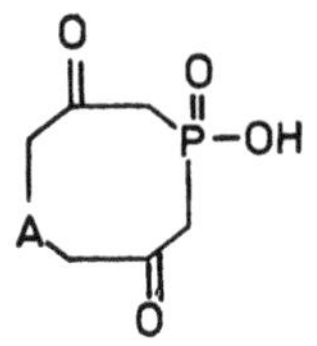 **16** 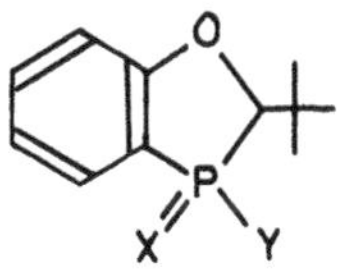 **17**

Rings/ Connectivities	Formula	Structure	NMR data (δ_P[solv] J_{PZ}Hz)	Ref.
$7/H_2P^4,CH_2;H$	$C_5H_{12}O_4P_2$	**12**; n = 2, X = O, Y = OH	45	94
$8/H_2N,CH_2;H$	$C_5H_{10}NO_4P$	**15**	44.8(w pH = 1)	123
$9/(C'H_2)_2;H$	$C_8H_{11}O_4P$	**16**; A = HC=CH	30.0(d)	122
	$C_8H_{13}O_4P$	**16**; A = CH_2CH_2	26.5(d)	120
$5,5/(CC'H)_2;C$	$C_{12}H_{18}O_4P_2$	**7**; R = H, X = X' = O', Y = Y' = OEt	70 or 83(c)	121

P-bound atoms = $C_2O'_2$

Rings/ Connectivities	Formula	Structure	NMR data (δ_P[solv] J_{PZ}Hz)	Ref.
$(H_2N)_2$	$C_2H_8N_2O_2P$	$(H_2NCH_2)_2P(O)O^-$	40.6(w)	21
H_2N,H_3	$C_4H_8NO_4P$	$Me(O)P(O^-)CH_2NHCH_2COO^-$	39.1(w pH = 10)	89
$H_2N,C'H_2$	$C_5H_9NO_6P$	$^-OOCCH_2NHCH_2P(O)(O^-)CH_2COO^-$	33.7(w pH = 11)	89
$(H_3)_2$	$C_2H_6O_2P$	$Me_2P(O)O^-$	36.7 45	126

P-bound atoms = CC'OO'

Rings/ Connectivities	Formula	Structure	NMR data (δ_P[solv] J_{PZ}Hz)	Ref.
$Cl_3;C'_2;C$	$C_9H_{10}Cl_3O_2P$	$Ph(Cl_3C)P(O)OEt$	26.5	80
$OO';C'_2H;C$	$C_{11}H_{19}O_8P$	$(EtOOC)_2CHP(O)(OEt)COOMe$	13.2	127
$HO_2;OO';C$	$C_8H_{20}O_6P$	$EtO(O)P(COOEt)CH(OEt)_2$	15.4	128
$H_2N;C'H_2;H$	$C_9H_{12}NO_4P$	$HOOCCH_2NHCH_2P(O)(OH)Ph$	26.3(w pH = 1)	88
$H_2P^4;C'_2;C$	$C_{24}H_{37}O_6P_2$	$[Ph(BuO)\mathbf{P}(O)CH_2]_2P(O)OEt$	33.6	87
	$C_{33}H_{48}O_7P_4$	$[Ph(BuO)\mathbf{P}(O)CH_2]_3P(O)$	34.7	129
$H_3;CO';C$	$C_4H_9O_3P$	$Me(MeCO)P(O)OMe$	34	104
$H_3;C'H;C$	$C_6H_{13}O_2P$	$Me(EtO)P(O)CH{=}CHMe$	35.1(k)	130
$H_3;C'_2;P^4$	$C_{14}H_{16}O_3P_2$	$MePhP(O)OP(O)MePh$	43.7	131
$H_3;C'_2;Si$	$C_{10}H_{17}O_2PSi$	MePhP(O)OTms	39.5(b)	132
$H_3;C'_2;H$	$C_7H_9O_2P$	MePhP(O)OH	40.1	131
$H_3C'_2;C$	$C_9H_{13}O_2P$	Me(4-Tol)P(O)OMe	45.7(c)	133
$CH_2;C'H;C$	$C_6H_{13}O_2P$	$Et(EtO)P(O)CH{=}CH_2$	40	49
$CH_2;C'_2;Si$	$C_{15}H_{26}NO_2PSi_2$	$TmsCH(CN)CH_2P(O)Ph(OTms)$	31	134
$CH_2;C'_2{}';H$	$C_8H_{11}O_2P$	EtPhP(O)OH	44	40
$CH_2;C'_2;C$	$C_9H_{13}O_2P$	EtPhP(O)OMe	44.3(m)	135
	$C_{10}H_{15}O_2P$	EtPhP(O)OEt *(S)*	42.0(m) 50.0	137
	$C_{11}H_{17}O_2P$	EtPhP(O)O-iPr	41.3(m)	135
	$C_{18}H_{29}O_2P$	$EtPhP(O)OC_{10}H_{19}$[b] (+)	44.2(c)	138
		(−)	43.3(c)	
$CH_2;C'_2;C'$	$C_{14}H_{15}O_2P$	EtPhP(O)OPh	44	110
$C'H_2;C'H;C$	$C_{11}H_{15}O_2P$	$H_2C{=}CHP(O)(OEt)CH_2Ph$	37.7(c)	111
$C'H_2;C'_2;C$	$C_{19}H_{23}O_3P$	$Ph(PhCOCH_2)P(O)ONeo$	32.1	139
$C_2S;C'_2;C$	$C_{12}H_{19}O_2PS$	Ph(EtO)P(O)CMeEt(SH)	41	140
$C'_2H;CO';C$	$C_{11}H_{19}O_7P$	$(EtOOC)_2CHP(O)(COMe)OEt$	17	127
$C_3;C'_2;P^4$	$C_{20}H_{28}Cl_2O_2P$	$tBuPhP(O)OP^+Ph(tBu)\ Cl^-$	67.7 68.3^{D}(r) $^2J_{PP}$41/47	41
$C_3;C'_2;S^4$	$C_{11}H_{17}O_4PS$	$tBuPhP(O)OSO_2Me$	53.7(c)	58
$C_3;C'_2;C$	$C_{19}H_{33}O_2P$	$tBu(3,5\text{-}tBu_2C_6H_3)P(O)OMe$	48	62
$5/H_3;CC';C_2C'$	$C_7H_{13}O_2P$	**J5**; R^1 = Me, R^2 = H, X = O, Y = Me	59.4(c) $^1J_{PC}$103	114
$5/CHO;C'_2;C$	$C_{13}H_{19}O_3P$	**17**; X = O, Y = OEt	51.6 52.4^{D}	141

TABLE K (continued)
Compilation of ^{31}P NMR Data of Four Coordinate Compounds Containing a P=Ch Bond and Two Bonds from Phosphorus to Group IV Atoms

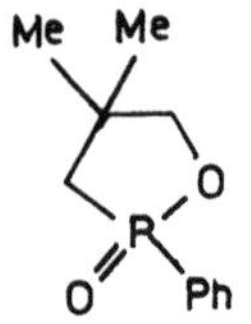

18 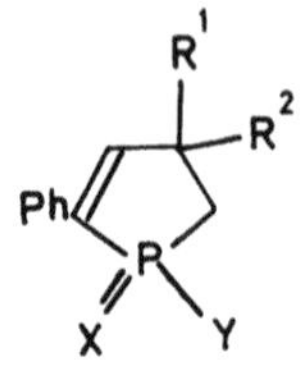 **19** 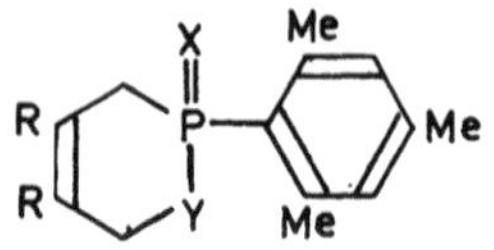 **20**

Rings/ Connectivities	Formula	Structure	NMR data (δ_P[solv] J_{PZ}Hz)	Ref.
5/CH$_2$;C'H;H	$C_4H_7O_2P$	**5**; X = O, Y = OH	76.8(w)	9
5/CH$_2$;C'H, C	$C_6H_{11}O_2P$	**5**; X = O, Y = OEt	69(n)	142
5/CH$_2$;C'$_2$;C	$C_{11}H_{15}O_2P$	**18**	58.6	143
	$C_{13}H_{14}NO_2P$	**19**; R^1 = H, R^2 = CN, X = O, Y = OEt	63.0	144
	$C_{14}H_{17}O_4P$	**19**; R^1 = H, R^2 = COOMe, X = O, Y = OEt	62.7	144
	$C_{15}H_{19}O_4P$	**19**; R^1 = Me, R^2 = COOMe, X = O, Y = OEt	61.1	144
6/CH$_2$;C'$_2$;C	$C_{10}H_{13}O_2P$	Ph(O)ṖOCH$_2$CH$_2$CH$_2$ĊH$_2$	37.6(c)	145
6/C'H$_2$;C'$_2$;C	$C_{13}H_{17}O_2P$	**20**; R = H, X = Y = O	35.6	146
	P-bound atoms = C'$_2$OO'			
(OO')$_2$;Si	$C_9H_{19}O_6PSi$	(EtOOC)$_2$(P(O)OTms	−15.1	147
C'N,C'$_2$;H	$C_{11}H_{12}NO_2P$	Ph(HO)(O)PĊ=CHCH=CHṄMe	9.8(c)	53
(C'O)$_2$;C	$C_{10}H_{11}O_4P$	EtO(O)P(Ċ=CHCH=CHȮ)$_2$	4.25(c)	8
(C'S)$_2$;C	$C_{10}H_{11}O_2PS_2$	EtO(O)P(Ċ=CHCH=CHṠ)$_2$	16.0(c)	8
(C'H)$_2$;H	$C_{16}H_{15}O_4P$	(PhOCH=CH)$_2$P(O)OH	31.2	42
(C'H)$_2$;C	$C_6H_{10}ClO_2P$	(H$_2$C=CH)$_2$P(O)OCH$_2$CH$_2$Cl	30.0(c)	83
	$C_6H_{11}O_2P$	(H$_2$C=CH)$_2$P(O)OEt	26.4	28
(C'H)$_2$;C'	$C_{10}H_{11}O_2P$	(H$_2$C=CH)$_2$P(O)OPh	26.7	28
C'H,C'$_2$;H	$C_{14}H_{13}O_2P$	Ph(PhCH=CH)P(O)OH	29.5^Z(c)	148
C'H,C'$_2$;C	$C_{16}H_{18}O_4P_2$	(MeO)(O)PhPCH=CHP(O)(OMe)Ph	26.9	149
(C'$_2$)$_2$;N	$C_{16}H_{20}NO_2P$	Ph$_2$P(O)ONEt$_2$	32.3(c)	150
(C'$_2$)$_2$;O	$C_{24}H_{20}O_4P_2$	Ph$_2$P(O)OOP(O)Ph$_2$	43(c)	150
(C'$_2$)$_2$;P^4	$C_{24}H_{20}O_3P_2$	Ph$_2$P(O)OP(O)Ph$_2$	32.2 33.1	152
(C'$_2$)$_2$;P	$C_{12}H_{10}F_2O_2P_2$	Ph$_2$P(O)OPF$_2$	30.4(r,c)	151
(C'$_2$)$_2$;H	$C_{12}H_{11}O_2P$	Ph$_2$P(O)OH	15(w) 25.5(e)	153
	$C_{12}H_{11}O_4P$	(2-HOC$_6$H$_4$)$_2$P(O)OH	32.9(y)	154
	$C_{14}H_{15}O_4P$	(4-Tol)$_2$P(O)OH	33.8(y)	154
(C'$_2$)$_2$;C	$C_{13}H_{13}O_2P$	Ph$_2$P(O)OMe	29.8(m) 32.2(r)	155
	$C_{14}H_{12}F_3O_2P$	Ph$_2$P(O)OCH$_2$Tf	33.0(t)	156
	$C_{14}H_{15}O_2P$	Ph$_2$P(O)OEt	27.2(c)	8
	$C_{14}H_{15}O_4P$	(2-HOC$_6$H$_4$)$_2$P(O)OEt	41.9(e)	154
	$C_{15}H_{17}O_2P$	Ph$_2$P(O)O-iPr	26.8(m)	135
	$C_{22}H_{24}O_4P_2$	1,2-C$_6$H$_4$[P(O)OEt)Ph]$_2$	30.4 31.9^D	157
	$C_{25}H_{21}O_2P$	Ph$_2$P(O)OCHPh$_2$	69.5(k)	158
	$C_{23}H_{27}NO_4P_2$	Ph$_2$P(O)NHCHPhP(O)(OEt)$_2$	21.7 or 23.9	65
(C'$_2$)$_2$;C'	$C_{20}H_{17}O_2P$	Ph$_2$P(O)OCPh=CH$_2$	29.7	139
	$C_{18}H_{14}BrO_2P$	Ph$_2$P(O)OC$_6$H$_4$Br-2	30.0	103

Rings/ Connectivities	Formula	Structure	NMR data (δ_P[solv] J_{PZ}Hz)	Ref.
5/C'H,C'$_2$;C	$C_{11}H_{13}O_2P$	**9**; X = O, Y = OMe	74.1(c)	60
6/(C'H)$_2$;H	$C_{16}H_{13}O_2PS$	**10**; X = O, Y = OH	24.8(c)	71
6/(C'H)$_2$;C	$C_{17}H_{15}O_2PS$	**10**; X = O, Y = OMe	25.2(b)	71
	P-bound atoms = C'C"OO'			
C'$_2$,C";C	$C_{14}H_{20}O_5P_2$	Ph(EtO)**P**(O)C≡CP(O)(OEt)$_2$	7.2	157
	$C_{18}H_{20}O_4P_2$	Ph(EtO)P(O)C≡CP(O)(OEt)Ph	7.0 7.1[D]	157

Section K2: Derivatives of Thio-, Seleno-, and Telluro-Phosphinic Acids

Connectivity of P-bound atoms	Formula	Structure	NMR data (δ_P [solv] $^nJ_{PZ}$Hz)	Ref.
	P-bound atoms = C$_2$BrS'			
(H$_3$)$_2$	C_2H_6BrPS	Me$_2$P(S)Br	63.2(c) 64.6	159
H$_3$,C'H$_2$	$C_8H_{10}BrPS$	Me(PhCH$_2$)P(S)Br	73.7(g)	23
H$_3$,C$_3$	$C_5H_{12}BrPS$	tBuMeP(S)Br	103.1(c) 107.5(r)	160
(CH$_2$)$_2$	$C_4H_{10}BrPS$	Et$_2$P(S)Br	96.4 98.3(n)	159
CH$_2$,C$_3$	$C_6H_{14}BrPS$	tBuEtP(S)Br	123.5(r)	160
(C$_2$H)$_2$	$C_6H_{14}BrPS$	iPr$_2$P(S)Br	124.1	3
(C$_3$)$_2$	$C_8H_{18}BrPS$	tBu$_2$P(S)Br	146.2(c) 148.1(r)	160
	P-bound atoms = CC'BrS'			
H$_3$,C'$_2$	$C_7H_3BrF_5PS$	Me(C$_6$F$_5$)P(S)Br	32.5	39
	C_7H_8BrPS	MePhP(S)Br	61.0(g)	23
C$_3$,C'$_2$	$C_{10}H_9BrF_5PS$	tBu(C$_6$F$_5$)P(S)Br	72.1	39
	P-bound atoms = C'$_2$BrS'			
(C'$_2$)$_2$	$C_{12}F_{10}BrPS$	(C$_6$F$_5$)$_2$P(S)Br	13.9	39
	$C_{12}H_5BrF_5PS$	Ph(C$_6$F$_5$)P(S)Br	31.5	39
	$C_{12}H_{10}BrPS$	Ph$_2$P(S)Br	64.4	161
	P-bound atoms = C$_2$ClS'			
(H$_2$Cl)$_2$	$C_2H_4Cl_3PS$	(ClCH$_2$)$_2$P(S)Cl	81.9(n)	20
(H$_3$)$_2$	C_2H_6ClPS	Me$_2$P(S)Cl	84.9(b) 87.3(n) $^1J_{PC}$61(b)	27
H$_3$,C$_2$H	$C_{11}H_{22}ClPS$	Me(L-C$_{10}$H$_{19}$)P(S)Cl[b] $(R)_P$	101.8(b) $^1J_{PC}$51.1	162
		$(S)_P$	97.5(b) $^1J_{PC}$52.0	
H$_3$,C$_3$	$C_5H_{12}ClPS$	tBuMeP(S)Cl	110.8(r)	163
(CH$_2$)$_2$	$C_4H_8Cl_3PS$	(ClCH$_2$CH$_2$)$_2$P(S)Cl	78.6(c)	28
	$C_4H_{10}ClPS$	Et$_2$P(S)Cl	108.6(b) $^1J_{PC}$58	26
(C$_2$H)$_2$	$C_{20}H_{38}ClPS$	(C$_{10}$H$_{19}$)$_2$P(S)Cl[b]		164
		D-Menthyl, L-menthyl $(R)_P$	121.9(b)	
		D-Menthyl, L-menthyl $(S)_P$	118.8(b)	
		L-Menthyl, L-menthyl	111.6	
C$_2$H,C$_3$	$C_{14}H_{28}ClPS$	tBu(L-C$_{10}$H$_{19}$)P(S)Cl[b] $(R)_P$	130.7(t) $^1J_{PC}$15.6	162
		$(S)_P$	134.8(t) $^1J_{PC}$12.1	
(C$_3$)$_2$	$C_8H_{18}ClPS$	tBu$_2$P(S)Cl	144.9(r))143,8(c)	160
4/(H$_2$S)$_2$	$C_2H_4ClPS_2$	Cl(S)ṖCH$_2$SĊH$_2$	90(g) $^2J_{PH}$5.5, 18.0	32

TABLE K (continued)
Compilation of ^{31}P NMR Data of Four Coordinate Compounds Containing a P=Ch Bond and Two Bonds from Phosphorus to Group IV Atoms

Rings/ Connectivities	Formula	Structure	NMR data (δ_P[solv] J_{PZ}Hz)	Ref.
4/$(C_3)_2$	$C_8H_{16}ClPS$	**6**; X = S′, Y = Cl	137^T	165
6/$(CHO)_2$	$C_{12}H_{24}ClO_2PS$	**4**; R = iPr, X = S, Y = Cl	70 80^D	37
	P-bound atoms = CC′ClS′			
H_3,C′$_2$	C_7H_8ClPS	MePhP(S)Cl	81.0	23
C_2H;C′$_2$	$C_{16}H_{24}ClPS$	Ph(L-$C_{10}H_{19}$)P(S)Cl[b] $(R)_P$	97.7(b)	162
		$(S)_P$	101.7(b)	
	P-bound atoms = C′$_2$ClS′			
(C′H)$_2$	C_4H_6ClPS	$(H_2C{=}CH)_2P(S)Cl$	70.9	28
(C′$_2$)$_2$	$C_{12}H_8ClF_2PS$	$(4\text{-}FC_6H_4)_2P(S)Cl$	77.7	161
	$C_{12}H_{10}ClPS$	$Ph_2P(S)Cl$	79.5(n) 80.2(c)	167
	P-bound atoms = C_2ClSe′			
$(H_3)_2$	C_2H_6ClPSe	$Me_2P(Se)Cl$	67.6(b) $^1J_{PC}$51 $^1J_{PSe}$838	26
$(CH_2)_2$	$C_4H_{10}ClPSe$	$Et_2P(Se)Cl$	99.8(b) $^1J_{PC}$46.9 $^1J_{PSe}$834	26
	$C_8H_{18}ClPSe$	$Bu_2P(Se)Cl$	94(n) $^1J_{PSe}$820	168
	P-bound atoms = CC′ClSe′			
CH_2;C′$_2$	$C_8H_{10}ClPSe$	EtPhP(Se)Cl	87(n) $^1J_{PSe}$840 (884)	169
	P-bound atoms = C′$_2$ClSe′			
(C′$_2$)$_2$	$C_{12}H_{10}ClPSe$	$Ph_2P(Se)Cl$	70(n) 71.8(b) $^1J_{PSe}$849(n)	170
	P-bound atoms = C_2FS′			
$(H_3)_2$	C_2H_6FPS	$Me_2P(S)F$	121.3(n) $^1J_{PF}$985	171
$(CH_2)_2$	$C_4H_{10}FPS$	$Et_2P(S)F$	134.6 $^1J_{PF}$1008	5
$(C_3)_2$	$C_8H_{18}FPS$	$tBu_2P(S)F$	157.8 $^1J_{PF}$1093	46
6/$(CH_2)_2$	$C_5H_{10}FPS$	$F(S)\dot{P}CH_2CH_2CH_2CH_2\dot{C}H_2$	117.8(n) $^1J_{PF}$1009	171
	P-bound atoms = CC′FS′			
H_3;C′$_2$	$C_7H_3F_6PS$	$Me(C_6F_5)P(S)F$	91.6 $^1J_{PF}$1034	39
	C_7H_8FPS	MePhP(S)F	110.1(n) $^1J_{PF}$1003	171
C_3;C′$_2$	$C_{10}H_9F_6PS$	$tBu(C_6F_5)P(S)F$	115.1 $^1J_{PF}$1043	39
	P-bound atoms = C′$_2$FS′			
(C′$_2$)$_2$	$C_{12}F_{11}PS$	$(C_6F_5)_2P(S)F$	53.5 $^1J_{PF}$1070	39
	$C_{12}H_5F_6PS$	$Ph(C_6F_5)P(S)F$	82.7 $^1J_{PF}$1043	39
	$C_{12}H_{10}FPS$	$Ph_2P(S)F$	102.3(c) $^1J_{PF}$1008	166

Rings/ Connectivities	Formula	Structure	NMR data (δ_P[solv] J_{PZ}Hz)	Ref.
		P-bound atoms = C_2IS'		
$(H_3)_2$	C_2H_6IPS	$Me_2P(S)I$	17.5	172
$(CH_2)_2$	$C_4H_{10}IPS$	$Et_2P(S)I$	63.0	172
	$C_8H_{18}IPS$	$Bu_2P(S)I$	57.3	172
		P-bound atoms = C'_2IS'		
$(C'_2)_2$	$C_{12}H_{10}IPS$	$Ph_2P(S)I$	27	173
		P-bound atoms = C_2NS'		
$(H_2Cl)_2;C_2$	$C_6H_{14}Cl_2NPS$	$(ClCH_2)_2P(S)NEt_2$	62.4(c)	20
$(H_3)_2;HP^4$	$C_4H_{13}NP_2S_2$	$Me_2P(S)NHP(S)Me_2$	59.6(r)	174
	$C_{14}H_{17}NP_2S_2$	Me_2**P**$(S)NHP(S)Ph_2$	52.5 or 64.1(r)	174
$(H_3)_2;HSi$	$C_5H_{16}NPSSi$	$Me_2P(S)NHTms$	55.6(c)	175
$(H_3)_2;C'H$	$C_8H_{12}NPS$	$Me_2P(S)NHPh$	51.2(C) 54.3(m) $^1J_{PN}$11.3(ae)	176
$(H_3)_2;C_2$	$C_7H_{16}NPS$	$Me_2P(S)Pip$	65.6(m)	67
$(CH_2)_2;HS^4$	$C_{10}H_{16}NO_2PS_2$	$Et_2P(S)NHSO_2Ph$	82.9(c)	177
$(CH_2)_2;HSi$	$C_7H_{20}NPSSi$	$Et_2P(S)NHTms$	55.1(c)	175
$(C_3)_2;C'H$	$C_{14}H_{24}NPS$	$tBu_2P(S)NHPh$	91.0(c + me) $^1J_{PN}$22	48
$5/(C'H_2)_2;C_2$	$C_8H_{16}NPS$	**1**; $R^1 = R^2 = Me$, $X = S$, $Y = NMe_2$	84.1(c)	178
	$C_9H_{18}NPS$	**1**; $R^1 = Me$, $R^2 = H$, $X = S$, $Y = NEt_2$	88.5(c)	178
	$C_{10}H_{20}NPS$	**1**; $R^1 = R^2 = Me$, $X = S$, $Y = NEt_2$	79.6(c)	178
		P-bound atoms = $C_2N'S'$		
$(H_3)_2;N'$	$C_2H_6N_3PS$	$Me_2P(S)N_3$	74.2(b) 75.8(a) $^1J_{PC}$68, $^1J_{P^{15}N}$54.8 $^2J_{P^{15}N}$5.1	179
$(CH_2)_2;N'$	$C_4H_{10}N_3PS$	$Et_2P(S)N_3$	91.4(b) 92.2(b) $^1J_{PC}$65, $^1J_{P^{15}N}$51.1' $^1J_{P^{15}N}$4.5	179
		P-bound atoms = $CC'NS'$		
$H_3;C'_2;C_2$	$C_9H_{14}NPS$	$MePhP(S)NMe_2$	70.7(c) $^1J_{PC}$76.2	180
$CH_2;C_2;C_2$	$C_{10}H_{16}NPS$	$EtPhP(S)NMe_2$ *(R)*$_P$	79.8(c)$^1J_{PC}$73.2 79.6(b)	181
		P-bound atoms = C'_2NS'		
$(C'_2)_2;HP^4$	$C_{24}H_{21}NP_2S_2$	$Ph_2P(S)NHP(S)Ph_2$	55.2(r,k)	183
$(C'_2)_2;CP$	$C_{25}H_{23}NP_2S$	Ph_2**P**$(S)NMePPh_2$	54.0(c)	184
$(C'_2)_2;C_2$	$C_{14}H_{16}NPS$	$Ph_2P(S)NMe_2$	70.9(r)	69
$5/C'H,C'_2;C_2$	$C_{12}H_{16}NPS$	**9**; $X = S$, $Y = NMe_2$	101	61
$6/CC',C'_2;CC'$	$C_{16}H_{22}NPS$	**11**; $X = S$, $Y = Ph$	51.3	72.
		P-bound atoms = $C'_2N'S'$		
$(C'_2)_2;N'$	$C_{12}H_{10}N_3PS$	$Ph_2P(S)N_3$	68.0(b)	26
$(C'_2)_2;P^4$	$C_{27}H_{25}NO_2P_2S_2$	Ph_2**P**$(S)N{=}P(SCH_2COOMe)Ph_2$	43	185
$(C'_2)_2;C'$	$C_{13}F_{10}NPS_2$	$(C_6F_5)_2P(S)NCS$	−4.2	186

TABLE K (continued)
Compilation of ^{31}P NMR Data of Four Coordinate Compounds Containing a P=Ch Bond and Two Bonds from Phosphorus to Group IV Atoms

Rings/ Connectivities	Formula	Structure	NMR data (δ_P[solv] J_{PZ}Hz)	Ref.
P-bound atoms = C_2NSe'				
$(H_3)_2$;C'H	$C_8H_{12}NPSe$	$Me_2P(Se)NHPh$	40.2(r) $^1J_{P^{15}N}$16.5	48
$(H_3)_2$;C_2	$C_4H_{12}NPSe$	$Me_2P(Se)NMe_2$	59.0 $^1J_{PSe}$ −720	187
$(CH_2)_2$;C_2	$C_8H_{20}NPSe$	$Et_2P(Se)NEt_2$	79.0 $^1J_{PSe}$728	188
	$C_{12}H_{18}NPSe$	$Bu_2P(Se)NEt_2$	73	189
$(C_3)_2$;C'H	$C_{14}H_{24}NPSe$	$tBu_2P(Se)NHPh$	93.3(c + me) $^1J_{P^{15}N}$27.2	48
P-bound atoms = $C_2N'Se'$				
$(H_3)_2$;N'	$C_2H_6N_3PSe$	$Me_2P(Se)N_3$	64.0(b) $^1J_{PC}$60.6 $^1J_{PSe}$808 $^1J_{PN}$57.4	26
$(CH_2)_2$;N'	$C_4H_{10}N_3PSe$	$Et_2P(Se)N_3$	89.4(b) $^1J_{PC}$56.7 $^1J_{PSe}$807	26
P-bound atoms = C'_2NSe'				
$(C'_2)_2$;CP^4	$C_{25}H_{23}NP_2Se_2$	$Ph_2P(Se)NMeP(Se)Ph_2$	72.0(c) $^1J_{PSe}$ −791	190
$(C'_2)_2$;C_2	$C_{14}H_{16}NPSe$	$Ph_2P(Se)NMe_2$	71.6 72.0(r) $^1J_{PSe}$ −760	191
	$C_{16}H_{20}NPSe$	$Ph_2P(Se)NEt_2$	66.7(b) $^1J_{PSe}$753	188
P-bound atoms = $C'_2N'Se'$				
$(C'_2)_2$;N'	$C_{12}H_{10}N_3PSe$	$Ph_2P(Se)N_3$	66.5(b)	26
P-bound atoms = C_2NTe'				
$(H_3)_2$;C'H	$C_8H_{12}NPTe$	$Me_2P(Te)NHPh$	−10.3(r-b) $^1J_{PN}$36	48
P-bound atoms = C_2OS'				
$(H_2Cl)_2$;C'	$C_8H_9Cl_2OPS$	$(ClCH_2)_2P(S)OPh$	87.0(c)	20
$(H_3)_2$;P^4	$C_4H_{12}OP_2S_2$	$Me_2P(S)OP(S)Me_2$	93.4 $^1J_{PC}$72.2 $^2J_{PP}$34	2
	$C_{14}H_{16}OP_2S_2$	$Me_2P(S)OP(S)Ph_2$	94.2(r), 78.6(r) $^2J_{PP}$13.4	192
$(H_3)_2$;Si	$C_5H_{15}OPSSi$	$Me_2P(S)OTms$	77.5(b) 79.9(c)	193
	$C_6H_{18}O_2P_2S_2Si$	$Me_2Si[OP(S)Me_2]_2$	79.4(c)	100
$(H_3)_2$;H	C_2H_7OPS	$Me_2P(S)OH$	87.9	1
$(H_3)_2$;C	C_3H_9OPS	$Me_2P(S)OMe$	94.3(c)	1
	$C_6H_{15}OPS$	$Me_2P(S)O$-tBu	81.3(c)	1
H_3,C_3;P^4	$C_{10}H_{24}OP_2S_2$	tBuMeP(S)OP(S)Me(tBu) *(R,R)*	109.7(t)	194
		(R,S)	110.8 $^2J_{PP}$56	
H_3,C_3;C	$C_6H_{15}OPS$	tBuMeP(S)OMe	112.0(r)	195
	$C_9H_{21}OPS$	tBuMeP(S)O-tBu	98.2(r)	195
$(CH_2)_2$;P^4	$C_{12}H_{28}OP_2S_2$	$Pr_2P(S)OP(S)Pr_2$	102.5	196
$(CH_2)_2$;Si	$C_7H_{19}OPSSi$	$Et_2P(S)OTms$	94.7(c)	100
$(CH_2)_2$;H	$C_4H_{11}OPS$	$Et_2P(S)OH$	99.7(a)	1
	$C_8H_{19}OPS$	$Bu_2P(S)OH$	95	197
$(CH_2)_2$;C	$C_6H_{13}Cl_2OPS$	$(ClCH_2CH_2)_2P(S)OEt$	93.2	28
	$C_6H_{15}OPS$	$Et_2P(S)OEt$	106.4(c)	198
$(C_3)_2$;Si	$C_{11}H_{27}OPSSi$	$tBu_2P(S)OTms$	114	100

Rings/ Connectivities	Formula	Structure	NMR data (δ_P[solv] J_{PZ}Hz)	Ref.
4/$(H_2S)_2$;C	$C_3H_7OPS_2$	$MeO(S)\dot{P}CH_2\dot{S}CH_2$	98(g) $^2J_{PH}$12.5, 12.5	32
4/$(C_3)_2$;C	$C_{10}H_{21}OPS$	**6**; X = S′, Y = OEt	120.6^C	165
5/$(CH_2)_2$;H	C_4H_9OPS	$HO(S)\dot{P}CH_2CH_2CH_2\dot{C}H_2$	112	197
5/$(CH_2)_2$;C′	$C_{12}H_{15}OPS$	**1**; R[1] = R[2] = Me, X = S, Y = OPh	108.5(c)	61
6/(CHO)$_2$;H	$C_{12}H_{25}O_3PS$	**4**; R = iPr, X = S, Y = OH	68(b)	37
6/(CHO)$_2$;C	$C_{13}H_{27}O_3PS$	**4**; R = iPr, X = S, Y = OMe	34(b)	37
P-bound atoms = C_2O′S				
$(H_3)_2$;Cl	C_2H_6ClOPS	$Me_2P(O)SCl$	68.7	199
$(H_3)_2$;C	C_3H_9OPS	$Me_2P(O)SMe$	45.9(g)	200
	$C_{11}H_{15}OPS$	$Me_2P(O)SCH_2CH{=}CHPh$	47.6	201
H_3,CH_2;C	$C_6H_{13}O_2PS$	$Me(MeCOCH_2CH_2)P(O)SMe$	58.7(c) $^1J_{PC}$70(Me), 65(CH_2)	202
$(CH_2)_2$;C	$C_5H_{13}OPS$	$Et_2P(O)SMe$	58.6(g)	200
	$C_8H_{19}OPS$	$Pr_2P(O)SEt$	55.0(g)	200
CH_2,C_2S;C	$C_9H_{21}OPS_2$	$Et(EtSCMe_2)P(O)SEt$	69.5	203
$(C_2H)_2$;C	$C_7H_{17}OPS$	$iPr_2P(O)SMe$	70.0(g)	200
4/$(C_3)_2$;C	$C_{10}H_{21}OPS$	**6**; X = SEt, Y = O′	74.6^C	165
P-bound atoms = C_2O′S′				
$(H_3)_2$	C_2H_6OPS	$Me_2P(S)O^-$	62.4(y)	1
H_3,C_3	$C_5H_{12}OPS$	$tBuMeP(S)O^-$	80.6(w)	195
$(CH_2)_2$	$C_4H_{10}OPS$	$Et_2P(S)O^-$	67.7(a)	1
P-bound atoms = CC′OS′				
H_3;C′$_2$;H	C_7H_9OPS	MePhP(S)OH	78.5	204
CH_2;C′$_2$;H	$C_8H_{11}OPS$	EtPhP(S)OH	85.3	204
CH_2;C′$_2$;C	$C_{18}H_{29}OPS$	EtPhP(S)O(L-$C_{10}H_{19}$)^b (−)	90.5 92.0(b)	138
C_3;C′$_2$;H	$C_{10}H_{15}OPS$	tBuPhP(S)OH	96.5(b) 97.0 $^1J_{PC}$73.2	205
5/C_2C′;C′$_2$;C′	$C_{18}H_{23}OPS$	**J1**; R[1]R[2] = $(CH_2)_5$, R[3]R[4] = $(CH_2)_4$, X = S, Y = Ph	116	206
P-bound atoms = CC′O′S				
CH_2;C′$_2$;C	$C_{10}H_{15}OPS$	EtPhP(O)SEt *(R)*	48 53.0	207
C_3;C′$_2$;Cl	$C_{10}H_{14}ClOPS$	tBuPhP(O)SCl *(R)*	69.4(r) 67.0(r)	63
C_3;C′$_2$;S^4	$C_{11}H_{17}O_3PS_2$	$tBuPhP(O)SSO_2Me$	64.0	208
C_3;C′$_2$;C	$C_{11}H_{17}OPS$	tBuPhP(O)SMe	63.3 66.6	41
C_3;C′$_2$;C″	$C_{11}H_{14}NOPS$	tBuPhP(O)SCN *(R)*	73.8	16(a)
P-bound atoms = CC′O′S′				
CH_2;C′$_2$	$C_8H_{10}OPS$	$EtPhP(S)O^-$	66.0(y)	138
P-bound atoms = C′$_2$OS′				
(C′H_2)$_2$;C	$C_6H_{11}OPS$	$(H_2C{=}CH)_2P(S)OEt$	84.4	28
(C′$_2$)$_2$;P^4	$C_{24}H_{20}OP_2S_2$	$Ph_2P(S)OP(S)Ph_2$	79	209
(C′$_2$)$_2$;H	$C_{12}H_{11}OPS$	$Ph_2P(S)OH$	76.0	152(a)
(C′$_2$)$_2$;C	$C_{13}H_3F_{10}OPS$	$(C_6F_5)_2P(S)OMe$	48.8	210
	$C_{13}H_{13}OPS$	$Ph_2P(S)OMe$	83.5(c,r)	69
	$C_{14}H_{15}OPS$	$Ph_2P(S)OEt$	79.1(t)	156
	$C_{19}H_{17}OPS$	$Ph_2P(S)OCH_2Ph$	82.0(c)	211

TABLE K (continued)
Compilation of ^{31}P NMR Data of Four Coordinate Compounds Containing a P=Ch Bond and Two Bonds from Phosphorus to Group IV Atoms

Rings/ Connectivities	Formula	Structure	NMR data (δ_P[solv] J_{PZ}Hz)	Ref.
	$C_{25}H_{21}OPS$	$Ph_2P(S)OCHPh_2$	82.7	211
$(C'_2)_2;C'$	$C_{18}H_5F_{10}OPS$	$(C_6F_5)_2P(S)OPh$	39.8	210
P-bound atoms = $C'_2O'S$				
$(C'_2)_2;C$	$C_{13}H_{13}OPS$	$Ph_2P(O)SMe$	42,8(c)	212
P-bound atoms = C_2OSe'				
$(H_3)_2;C$	$C_3H_9OPS_e$	$Me_2P(Se)OMe$	90.2 $^1J_{PSe}-768$	187
$(CH_2)_2;C$	$C_6H_{15}OPSe$	$Et_2P(Se)OEt$	107.5 $^1J_{PSe}755$	188
P-bound atoms = $CC'OSe'$				
$CH_2;C'_2;C$	$C_{10}H_{15}OPSe$	$EtPhP(Se)OEt$	97	213
	$C_{11}H_{15}OPSe$	$EtPhP(Se)OCH_2CH{=}CH_2$	100	213
P-bound atoms = $CC'O'Se'$				
$C_3;C_2$	$C_{10}H_{14}OPSe$	$tBuPhP(Se)O^-$	74.8(y) $^1J_{PSe}632$	188
P-bound atoms = C'_2OSe'				
$(C'_2)_2;C$	$C_{13}H_{13}OPSe$	$Ph_2P(Se)OMe$	86.8 88.5(c) $^1J_{PSe}-810$	214
	$C_{14}H_{15}OPSe$	$Ph_2P(Se)OEt$	84.1 $^1J_{PSe}795$	188
	$C_{15}H_{15}OPSe$	$Ph_2P(Se)OCH_2CH{=}CH_2$	87	213
	$C_{15}H_{17}OPSe$	$Ph_2P(Se)O\text{-}iPr$	89.4	213
P-bound atoms = C_2OTe'				
$(CH_2)_2;C$	$C_6H_{15}OPTe$	$Et_2P(Te)OEt$	83	215
P-bound atoms = C_2P^4S'				
$(H_3)_2;C_2S'$	$C_4H_{12}P_2S_2$	$Me_2P(S)P(S)Me_2$	34.7(b) 35.0(r) $^1J_{PP}-18.8$	218
$(H_3)_2;C_2Se'$	$C_4H_{12}P_2SSe$	$Me_2P(S)P(Se)Me_2$	33.9 (r) $^1J_{PP}-40$(r)	217
$(H_3)_2;C_2S$	$C_{14}H_{16}P_2S_2$	$Me_2P(S)P(S)Ph_2$	30.9(or 39.0) (b)	217
$H_3,C_3;C_2S'$	$C_{10}H_{24}P_2S_2$	$tBuMeP(S)P(S)Me(tBu)$	53.6 58.9^D $^1J_{PP}-103$, -109	217
$(CH_2)_2;C_2S'$	$C_8H_{20}P_2S_2$	$Et_2P(S)P(S)Et_2$	49.4(b) 51.3(c) $^1J_{PC}43$ $^2J_{PPC}10$(c) $^1J_{PP}53.5$	220
$(C_2H)_2;C_2S'$	$C_{12}H_{28}P_2S_2$	$iPr_2P(S)P(S)iPr_2$	64.6(c) $^1J_{PC}34$ $^2J_{PPC}8$ $^1J_{PP}95$	220
	$C_{24}H_{44}P_2S_2$	$cHex_2P(S)P(S)(\text{-}cHex)_2$	86.1	219
5/$(CH_2)_2;C_2S'$	$C_8H_{16}P_2S_2$	$(CH_2)_4P(S)P(S)(CH_2)_4$	61.6	216
6/$(CH_2)_2;C_2S'$	$C_8H_{18}P_2S_2$	**21**	39.4(b)	219
P-bound atoms = C'_2P^4S'				
$(C'_2)_2;C'_2S'$	$C_{24}H_{20}P_2S_2$	$Ph_2P(S)P(S)Ph_2$	37.9(t)	219

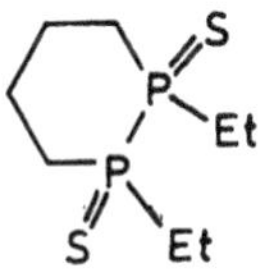

21 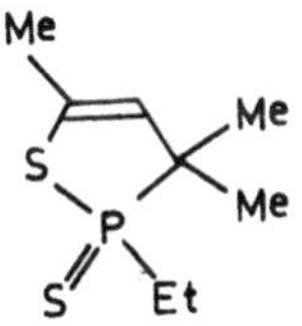 **22** 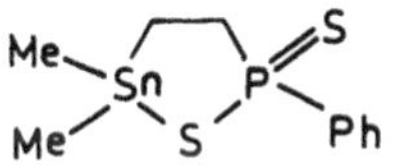 **23**

Rings/ Connectivities	Formula	Structure	NMR data (δ_P[solv] J_{PZ}Hz)	Ref.
P-bound atoms = C_2P^4Se'				
$(H_3)_2;C_2Se'$	$C_4H_{12}P_2S_2$	$Me_2P(Se)P(Se)Me_2$	13.5(r) $^1J_{PP}-67$	217
P-bound atoms = C_2SS'				
$(H_3)_2;P^4$	$C_4H_{12}P_2S_3$	$Me_2P(S)SP(S)Me_2$	61.1 $^2J_{PP}$14.8	2
	$C_{14}H_{16}P_2S_3$	$Me_2\mathbf{P}(S)SP(S)Ph_2$	35.9 32.8 $^2J_{PP}$13.4	192
$(H_3)_2;Si$	$C_5H_{15}PS_2Si$	$Me_2P(S)STms$	51.9(c) 52.1(r)	221
$(H_3)_2;H$	$C_2H_7PS_2$	$Me_2P(S)SH$	52.8(c)	1
$(H_3)_2;C$	$C_3H_9PS_2$	$Me_2P(S)SMe$	47.5	222
$H_3,C_3;P^4$	$C_{10}H_{24}P_2S_3$	tBuMeP(S)SP(S)MetBu *(R), (R)*	90.8(t) $^2J_{PP}-17$	195
		(R), (S)	91.0(t)	
$H_3,C_3;P$	$C_{10}H_{24}P_2S_2$	tBuMe**P**(S)SPMe(tBu) *(RS, SR)*	87.8(b) $^2J_{PP}$54.8	194
		(RR, SS)	85.8(b) $^2J_{PP}$51.3	
$H_3,C_3;H$	$C_5H_{13}PS_2$	tBuMeP(S)SH	83.0(r)	223
$(CH_2)_2;P^4$	$C_8H_{20}P_2S_3$	$Et_2P(S)SP(S)Et_2$	85.4(c)	1
$(CH_2)_2;S$	$C_8H_{20}P_2S_4$	$Et_2P(S)SSP(S)Et_2$	86.7(c)	1
	$C_8H_{20}P_2S_5$	$Et_2P(S)SSSP(S)Et_2$	89.0(c)	1
$(CH_2)_2;Si$	$C_7H_{19}PS_2Si$	$Et_2P(S)STms$	74.5(c)	1
$(CH_2)_2;H$	$C_4H_{11}PS_2$	$Et_2P(S)SH$	76.0(a)	1
$(CH_2)_2;C$	$C_5H_{13}PS_2$	$Et_2P(S)SMe$	80.3(c)	200
	$C_9H_{21}PS_2$	$Bu_2P(S)SMe$	75.9(c)	200
$(C_2H)_2;P^4$	$C_{12}H_{28}P_2S_3$	$iPr_2P(S)SP(S)iPr_2$	106.5 $^2J_{PP}$15.7	2
$(C_2H)_2;C$	$C_7H_{17}PS_2$	$iPr_2P(S)SMe$	97.2(c)	200
$(C_3)_2;P^4$	$C_{16}H_{36}P_2S_3$	$tBu_2P(S)SP(S)tBu_2$	121.3 118.0(b) $^2J_{PP}$14.6	225
$5/CH_2,C_2C';C'$	$C_8H_{15}PS_2$	**22**	119	206
P-bound atoms = $C_2S'_2$				
$(H_3)_2$	$C_2H_6PS_2$	$Me_2P(S)S^-$	52.7(y)	1
H_3,C_3	$C_5H_{12}PS_2$	$tBuMeP(S)S^-$	81.1(a-r)	195
$(CH_2)_2$	$C_4H_{10}PS_2$	$Et_2P(S)S^-$	73.3(c)3(c) 74.4(a)	226
P-bound atoms = $CC'SS'$				
$H_3;C'_2;H$	$C_8H_{12}P_2S_4$	$1,4\text{-}C_6H_4[P(S)Me(SH)]_2$	50.0(c)	227
$H_3;C'_2;C$	$C_{10}H_{16}P_2S_4$	$1,4\text{-}C_6H_4[P(S)Me(SMe)]_2$	58.1(c)	227
$CH_2;C'_2;P^4$	$C_{12}H_{20}O_2P_2S_3$	$EtPhP(S)SP(S)(OEt)_2$	71 or 81 $^2J_{PP}$20.5	218
$CH_2;C'_2;H$	$C_{18}H_{24}O_2P_2S_4$	$1,4\text{-}(CH_2)_4[P(S)(SH)C_6H_4OMe\text{-}4]_2$	61.2(c-y)	227
$CH_2;C'_2;C$	$C_{10}H_{15}PS_2$	$EtPhP(S)SEt$	70	229
$5/CH_2;C'_2;Sn$	$C_{10}H_{15}PS_2Sn$	**23**	82.9(c) J_P119_{Sn}65.4	230
$6/C'H_2;C'_2;C$	$C_{24}H_{35}PS_2$	**20**; X = Y = S, R = Me	42.7(c)	231

TABLE K (continued)
Compilation of ^{31}P NMR Data of Four Coordinate Compounds Containing a P=Ch Bond and Two Bonds from Phosphorus to Group IV Atoms

Rings/ Connectivities	Formula	Structure	NMR data (δ_P[solv] J_{PZ}Hz)	Ref.
	P-bound atoms = CC′S′$_2$			
$H_3;C'_2$	$C_7H_8PS_2$	$MePhP(S)S^-$	54.9	232
	P-bound atoms = C′$_2$SS′			
$(C'_2)_2;P^4$	$C_{24}H_{20}P_2S_3$	$Ph_2P(S)SP(S)Ph_2$	61.9	2
$(C'_2)_2;Si$	$C_{15}H_{19}PS_2Si$	$Ph_2P(S)STms$	54(c)	100
$(C'_2)_2;H$	$C_{12}H_{11}PS_2$	$Ph_2P(S)SH$	56.5	152a
	$C_{20}H_{20}O_2P_2S_4$	$1,4\text{-}C_6H_4[P(S)(SH)C_6H_4OMe\text{-}4]_2$	53.7(c)	227
$(C'_2)_2;C$	$C_{13}H_{13}PS_2$	$Ph_2P(S)SMe$	64.2	158
$(C'_2)_2;C'$	$C_{14}H_{13}OPS_2$	$Ph_2P(S)SCOMe$	60	209
	$C_{18}H_{15}PS_2$	$Ph_2P(S)SPh$	64.8	158
	P-bound atoms = C′$_2$S′$_2$			
$(C'_2)_2$	$C_{12}H_{10}PS_2$	$Ph_2P(S)S^-$	61.3(c)	233
	P-bound atoms = C$_2$S′Se			
$(CH_2)_2;Se$	$C_8H_{20}P_2S_2Se_2$	$Et_2P(S)SeSeP(S)Et_2$	72.0(c)	1
	P-bound atoms = CC′SSe′			
$CH_2;C'_2;C$	$C_{10}H_{15}PSSe$	$EtPhP(Se)SEt$	61	213
$C_3;C'_2;C$	$C_{13}H_{19}OPSSe$	$tBuPhP(Se)SCH_2COMe$	82.5	234
	$C_{23}H_{25}PSSe$	$tBuPhP(Se)SCHPh_2$	80.9	235
	P-bound atoms = CC′S′Se			
$C_3;C'_2;C$	$C_{13}H_{19}OPSSe$	$tBuPhP(S)SeCH_2COMe$	92.7	236
	$C_{23}H_{25}PSSe$	$tBuPhP(S)SeCHPh_2$	90.8(g)	235
	P-bound atoms = C′$_2$SSe′			
$(C'_2)_2;C$	$C_{13}H_{13}PSSe$	$Ph_2P(Se)SMe$	64.2	235
	$C_{14}H_{15}PSSe$	$Ph_2P(Se)SEt$	54	213
	P-bound atoms = C′$_2$S′Se			
$(C'_2)_2;C$	$C_{13}H_{13}PSSe$	$Ph_2P(S)SeMe$	41.2(g)	235
	P-bound atoms = C$_2$SeSe′			
$H_3,C_3;P^4$	$C_{10}H_{24}P_2Se_3$	$tBuMeP(Se)SeP(Se)tBuMe$	75.7 78.0^D(b,c) $^1J_{PSe'}$ −760.4, −751.1; $^1J_{PSe}$ −346.2, −390.6 $^2J_{PP}$12.4, 19.2	14
$(CH_2)_2;P^4$	$C_8H_{20}P_2Se_3$	$Et_2P(Se)SeP(Se)Et_2$	61.5(c)	1

Rings/ Connectivities	Formula	Structure	NMR data (δ_P[solv] J_{PZ}Hz)	Ref.
	P-bound atoms = CC′SeSe′			
C_3;C'_2;C	$C_{11}H_{17}PSe_2$	tBuPhP(Se)SeMe	81.3(y) $^1J_{PSe'}$753 $^1J_{PSe}$358	188
	P-bound atoms = CC′Se′$_2$			
C_3;C'_2	$C_{10}H_{14}PSe_2$	tBuPhP(Se)Se$^-$	58.0(y) $^1J_{PSe}$604	188
	P-bound atoms = C′$_2$SeSe′			
$(C'_2)_2$;P^4	$C_{24}H_{20}P_2Se_3$	$Ph_2P(Se)SeP(Se)Ph_2$	42.2(b,c) $^1J_{PSe'}$ −776.4 $^1J_{PSe}$ −381.8	14

[a]1-Imidazolyl.
[b]Menthyl.

TABLE K
Supplementary Table

Section K1: Derivatives of Phosphinic Acids

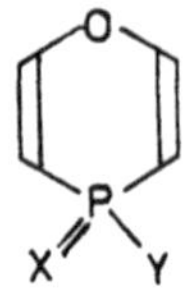

24

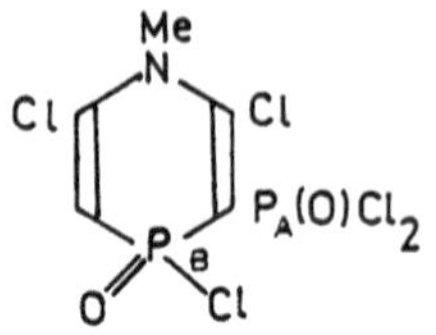

25

Rings/ Connectivities of P-bound atoms	Formula	Structure	NMR data (δ_P [solv] $^nJ_{PZ}$Hz)	Ref.
P-bound atoms = C_2ClO'				
CH_2,C_2S	$C_7H_{16}ClOPS$	EtP(O)(CMe_2SEt)Cl	74	237
P-bound atoms = CC′ClO′				
$C_2S;C'_2$	$C_{11}H_{16}ClOPS$	PhP(O)(CMe_2SEt)Cl	61	237
P-bound atoms = C'_2ClO'				
$6/(C'H)_2$	C_4H_4ClOP	**24**; X = O, Y = Cl	10.7[a] 11.0[b]	240
$6/C'H,C'P^4$	$C_5H_4Cl_5NO_2P_2$	**25**	B 9.0 $^2J_{PP}$21.2	241
P-bound atoms = C'_2FO'				
$6/(C'H)_2$	C_4H_4FOP	**24**; X = O, Y = F	19.0 $^1J_{PF}$979	239
P-bound atoms = C_2OO'				
$Cl_3,H_2O;C$	$C_5H_{10}Cl_3O_3P$	Cl_3CP(O)(OEt)CH_2OMe	32.4	242
$Cl_2H,HO_2;C$	$C_8H_{17}Cl_2O_4P$	Cl_2CHP(O)(OEt)CH(OEt)$_2$	26.3	242
$ClCH_2,HO_2;C$	$C_8H_{18}ClO_4P$	ClCH_2P(O)(OEt)CH(OEt)$_2$	33.1	242
$H_3,CHO;H$	$C_3H_6Cl_3O_3P$	MeP(O)(OH)CH(OH)CCl_3	40.0	243
$H_3,CHO;C$	$C_6H_8Cl_3F_6O_3P$	MeP(O)($OCH_2C_2F_4CF_2$H)CH(OH)CCl_3	34.5	243
$5/H_2P^4,CHO;C$	$C_8H_{18}O_4P_2S$	**J70**; Y = OEt, X = S	B 47.9 47.6[D]	244
$5/H_2P,CHO;C$	$C_8H_{18}O_4P_2$	**J70**; Y = OEt, X = l.p.	B 60.3 60.9[D] J_{PP}1.90, 1.00	244
P-bound atoms = CC′OO′				
$Cl_2H;OO';C$	$C_5H_9Cl_2O_4P$	Cl_2CHP(O)(OEt)COOMe	11.5	242
$H_3;OO';C$	$C_6H_{13}O_4P$	MeP(O)(OEt)COOEt	27.8	245
$C_3;OO';Si$	$C_{10}H_{23}O_4PSi$	tBuP(O)(OTms)COOEt	26.2(g)	245
P-bound atoms = C'_2OO'				
$(OO')_2;C$	$C_6H_{11}O_6P$	EtOP(O)(COOMe)$_2$	−4.3	242
$C'_2,OO';Si$	$C_{12}H_{19}O_4PSi$	PhP(O)(OTms)COOEt	5.0(g)	245
$C'_2,OO';C$	$C_{10}H_{13}O_4P$	PhP(O)(OMe)COOEt	14.3	245
$(C')_2;N'$	$C_{15}H_{15}ClNO_2P$	Ph_2P(O)ON=CClEt	34.7(c)	246
$(C')_2;C'$	$C_{19}H_{15}O_4P$	Ph_2P(O)OCOC_6H_4OH-4	32.4(c)	247

Section K2: Derivatives of Thio-, Seleno-, and Telluro-Phosphinic Acids

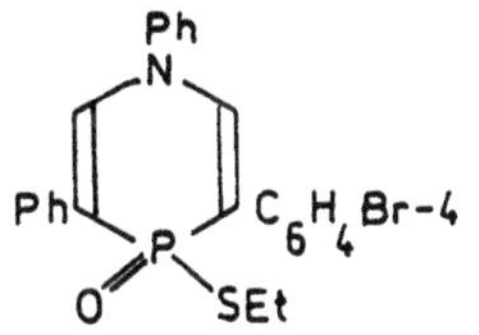

26

Rings/ Connectivities of P-bound atoms	Formula	Structure	NMR data (δ_P [solv] $^nJ_{PZ}$Hz)	Ref.
P-bound atoms = CC′OS′				
C_3;C'_2;Si	$C_{13}H_{23}OPSSi$	tBuPhP(S)OTms	93.0	248
P-bound atoms = C′$_2$O′S				
$(C'_2)_2$;C	$C_{23}H_{20}BrN_2OPS$	**26**	6.2(c)	249

REFERENCES

1. **Haegele, G., Kuchen, W., and Steinberger, H.,** *Z. Naturforsch.*, 29B, 349, 1974.
2. **Harris, R. K., McVicker, E. M., and Haegele, G.,** *J. Chem. Soc. Dalton Trans.*, p. 9, 1978.
3. **Peters, W. and Haegele, G.,** *Z. Naturforsch.*, 38B, 96, 1983.
4. Ref. J8.
5. **Knunyants, I. L., Georgiev, V. I., Galakhov, I. V., Ragulin, L. I., and Neimysheva, A. A.,** *Dokl. Chem., (Engl. transl.)*, 201, 992, 1971.
6. **Hoz, S., Dunn, E. J., Buncel, E., Bannard, R. A. B., and Purdon, J. G.,** *Phosphorus Sulfur*, 24, 321, 1985.
7. Ref. H11.
8. **Allen, D. W., Hutley, B. G., and Mellor, M. T. J.,** *J. Chem. Soc. Perkin Trans. 1*, p. 1705, 1977.
9. **Moedritzer, K.,** *Synth. React. Inorg. Metal-Org. Chem.*, 5, 45, 1975.
10. **Mark, V., Dungan, C., Crutchfield, M., and Van Wazer, J.,** *Top. Phosphorus Chem.*, 5, 227, 1967.
11. **Finer, E. G. and Harris, R. K.,** *Prog. Nucl. Magn. Reson. Spectrosc.*, 6, 61, 1971.
12. **Mavel, G.,** *Annu. Rev. NMR Spectrosc.*, 5b, 1, 1973.
13. **Gorenstein, D. G.,** *Phosphorus 31-P NMR. Principles and Applications*, Gorenstein, D. G., Ed., Academic Press, London, 1984, chap. 2.
14. Ref. H331.
15. Ref. J42.
16. **Dahl, O.,** *J. Chem. Soc. Perkin Trans. 1*, p. 947, 1978.
17. **Hammond, P. J., Scott, G., and Hall, C. D.,** *J. Chem. Soc. Perkin Trans. 2*, p. 205, 1982.
18. Ref. H54.
19. **Burg, A. B.,** *Inorg. Chem.*, 17, 2322, 1978.
20. **Maier, L.,** *Helv. Chim. Acta*, 54, 1651, 1971.
21. **Maier, L.,** *J. Organomet. Chem.*, 178, 157, 1979.
22. (a) Ref. 20; (b) Ref. 21.
23. Ref. H96.
24. Ref. J65(a).
25. (a) Ref. 23; (b) Ref. 24.
26. **Schroeder, H. F. and Mueller, J.,** *Z. Anorg. Allg. Chem.*, 451, 158, 1979.
27. (a) Ref. 26; (b) Ref. 23.
28. **Maier, L.,** *Phosphorus*, 1, 111, 1971.
29. Ref. H142.
30. **Efimova, V. D. and Kharrasova, F. M.,** *J. Gen. Chem. U.S.S.R.*, 54, 1490, 1984.
31. (a) Ref. 30; (b) Ref. 16.
32. **Arshinova, R. P., Zyablikova, T. A., Mukhametzyanova, E. Kh., and Shermergorn, I. M.,** *Dokl. Chem., (Engl. transl.)*, 204, 504, 1972.
33. **Felcht, U. F. H. and Cross, G.,** *Phosphorus Sulfur*, 13, 291, 1982.
34. (a) Ref. 33; (b) Ref. 9.
35. **Quin, L. D. and Berhardt, F. C.,** *Magn. Reson. Chem.*, 23, 929, 1985.
36. **Brel', V. K., Ionin, B. I., and Petrov, A. A.,** *J. Gen. Chem. U.S.S.R.*, 52, 1705, 1982.
37. **Arbuzov, B. A., Erastov, O. A., Ignat'eva, S. N., Romanova, I. P., Arshinova, R. P., and Kadyrov, R. A.,** *Bull. Acad. Sci. U.S.S.R. Div. Chem. Sci.*, p. 381, 1983.
38. Ref. J83.
39. Ref. J367.
40. Ref. H338.
41. Ref. J248.
42. Ref. J88.
43. Ref. H103.
44. (a) Ref. 5; (b) Ref. 43.
45. **Nikitin, E. V., Ignat'ev, Yu. A., Parakin, O. V., Koschev, I. P., Romanov, G. V., Kargin, Yu. M., and Pudovik, A. N.,** *J. Gen. Chem. U.S.S.R.*, 52, 2400, 1982.
46. Ref. J104(a).
47. Ref. H66.
48. **McFarlane, W. and Wrackmeyer, B.,** *J. Chem. Soc. Dalton Trans.*, p. 2351, 1976.
49. Ref. H7.
50. Ref. H142.
51. **Veits, Yu. A., Gurov, M. V., Domnikov, A. P., Foss, V. L., and Lutsenko, I. F.,** *J. Gen. Chem U.S.S.R.*, 54, 419, 1984.
52. **Harger, M. J. P. and Stephen, M. A.,** *J. Chem. Soc. Perkin Trans. 1*, p. 705, 1980.

53. **Szewczyk, J., Lloyd, J. R., and Quin, L. D.,** *Phosphorus Sulfur,* 21, 155, 1984.
54. **Quin, L. D., Szewczyk, J., Szewczyk, K. M., and McPhail, A. T.,** *J. Org. Chem.*, 51, 3341, 1986.
55. **Quin, L. D. and Szewczyk, J.,** *J. Chem. Soc. Chem. Commun.*, p. 1551, 1984.
56. (a) Ref. 26; (b) Ref. H163.
57. Ref. J133.
58. Ref. H141.
59. **Keglevich, G. and Quin, L. D.,** *Phosphorus Sulfur,* 26, 129, 1986.
60. **Quin, L. D. and Keglevich, G.,** *J. Chem. Soc. Perkin Trans. 2,* p. 1029, 1986.
61. **Quin, L. D. and Szewczyk, J.,** *J. Chem. Soc. Chem. Commun.*, p. 844, 1986.
62. Ref. J114.
63. (a) Ref. J113; (b) Ref. H76.
64. Ref. H121.
65. Ref. J208.
66. (a) Ref. 65; (b) Ref. 20.
67. **Appel, R. and Milker, R.,** *Chem. Ber.*, 108, 2349, 1975.
68. (a) Ref. 67; (b) Ref. 20.
69. Ref. H367.
70. **Cristau, H. J., Coste, J., Tuchon, A., and Christol, H.,** *J. Organomet. Chem.*, 241, Cl, 1983.
71. **Proklina, N. V. and Fedorova, G. K.,** *J. Gen. Chem. U.S.S.R.*, 55, 1905, 1985.
72. **Bourdieu, C. and Foucaud, A.,** *Tetrahedron Lett.*, 27, 4725, 1986.
73. (a) Ref. 62; (b) Ref. 26.
74. **Kutyrev, G. A., Lygin, A. V., Cherkasov, R. A., Konovalova, I. V., and Pudovik, A. N.,** *J. Gen. Chem. U.S.S.R.*, 53, 1080, 1983.
75. **Volkholz, M., Stelzer, O., and Schmutzler, R.,** *Chem. Ber.*, 111, 890, 1978.
76. Ref. H253.
77. Ref. H250.
78. **Foss, V. L., Solodenko, V. A., Veits, Yu. A., and Lutsenko, I. F.,** *J. Gen. Chem. U.S.S.R.*, 49, 1510, 1979.
79. Ref. J64a.
80. Ref. J144.
81. Ref. J149.
82. **Maslennikov, I. G., Prokof'eva, G. N., Lavrent'ev, A. N., and Shcherbaeva, M. M.,** *J. Gen. Chem. U.S.S.R.*, 52, 816, 1982.
83. **Maier, L. and Rist, G.,** *Phosphorus Sulfur,* 17, 21, 1983.
84. Ref. J415.
85. Ref. J126.
86. (a) Ref. 23; (b) Ref. 21.
87. **Maier, L.,** *Helv. Chim. Acta,* 52, 827, 1969.
88. **Maier, L.,** *Phosphorus Sulfur,* 11, 139, 1981.
89. **Maier, L.,** *Phosphorus Sulfur,* 11, 149, 1981.
90. **Majewski, P.,** *Synthesis,* p. 455, 1987.
91. (a) Ref. 90; (b) Ref. 86.
92. **Bel'skii, V. E., Zotova, A. M., Kudryavtseva, L. A., and Ivanov, B. E.,** *Bull. Acad. Sci. U.S.S.R. Div. Chem. Sci.*, p. 1978, 1985.
93. **Prishchenko, A. A., Novikova, Z. S., and Lutsenko, I. F.,** *J. Gen. Chem. U.S.S.R.*, 47, 2451, 1977.
94. Ref. J175.
95. Ref. J145.
96. **Seel, F. and Velleman, K. D.,** *Chem. Ber.*, 104, 2972, 1971.
97. (a) Ref. 23; (b) Ref. 96.
98. **Appel, R. and Milker, R.,** *Chem. Ber.*, 110, 3201, 1977.
99. (a) Ref. 98; (b) Ref. 1.
100. **Kuchen, W. and Steinberger, H.,** *Z. Anorg. Allg. Chem.*, 415, 266, 1975.
101. (a) Ref. 23; (b) Ref. 1; (c) Ref. 96.
102. (a) Ref. H201(d); (b) Ref. 96.
103. Ref. H9.
104. Ref. J238.
105. **Garabadzhii, A. V., Rodin, A. A., Lavrent'ev, A. N., and Sochilin, E. G.,** *J. Gen. Chem. U.S.S.R.*, 51, 34, 1981.
106. **Arbuzov, B. A., Erastov, O. A., Romanova, I. P., and Zyablikova, T. A.,** *Bull. Acad. Sci. U.S.S.R. Div. Chem. Sci.*, p. 2336, 1981.
107. (a) Ref. 90; (b) Ref. 106.

108. Ref. J231.
109. **Nifant'ev, E. É., Magdeeva, R. K., and Shchepet'eva, N. P.,** *J. Gen. Chem. U.S.S.R.*, 50, 1416, 1980.
110. **Kharrasova, F. M., Efimova, V. D., and Salakhutdinov, R. A.,** *J. Gen. Chem. U.S.S.R.*, 48, 953, 1978.
111. **Maier, L. and Lea, P. J.,** *Phosphorus Sulfur*, 17, 1, 1983.
112. **Nifant'ev, E. É., Magdeeva, R. K., Maslennikova, V. I., Taber, A. M., and Kolechits, I. V.,** *J. Gen. Chem. U.S.S.R.*, 52, 2173, 1982.
113. **Emsley, J., Middleton, T. B., and Williams, J. K.,** *J. Chem. Soc. Dalton Trans.*, p. 633, 1974.
114. **Moedritzer, K. and Miller, R. E.,** *J. Org. Chem.*, 47, 1530, 1982.
115. **Moedritzer, K. and Miller, R. E.,** *Phosphorus Sulfur*, 10, 279, 1981.
116. **Macdiarmid, J. E. and Quin, L. D.,** *J. Org. Chem.*, 46, 1451, 1981.
117. Ref. H13.
118. **Jansen, W. and Ulses, R.,** *Z. Anorg. Allg. Chem.*, 532, 197, 1986.
119. **Middlemas, E. D. and Quin, L. D.,** *J. Org. Chem.*, 44, 2587, 1979.
120. **Quin, L. D., Middlemas, E. D., Ras, N. S., Miller, R. W., and McPhail, A. I.,** *J. Am. Chem. Soc.*, 104, 1893, 1982.
121. **Kluger, R. and Westheimer, F. H.,** *J. Am. Chem. Soc.*, 91, 4143, 1969.
122. **Quin, L. D., Middlemas, E. D., and Nandakumar, S. R.,** *J. Org. Chem.*, 47, 905, 1982.
123. Ref. H301.
124. **Quin, L. D. and Middlemas, E. D.,** *J. Am. Chem. Soc.*, 99, 8370, 1977.
125. **Appel, R. and Milker, R.,** *Chem. Ber.*, 107, 2658, 1974.
126. (a) Ref. 1; (b) Ref. 125.
127. **Malenko, D. M. and Gololobov, Yu. G.,** *J. Gen. Chem. U.S.S.R.*, 49, 267, 1979.
128. **Livantsev, M. V., Prishchenko, A. A., and Lutsenko, I. F.,** *J. Gen. Chem. U.S.S.R.*, 56, 2146, 1986.
129. Ref. J179.
130. Ref. J193.
131. **White, D. W., Gibbs, D. E., and Verkade, J. G.,** *J. Am. Chem. Soc.*, 101, 1937, 1979.
132. Ref. J329.
133. **Sybert, P. D., Bertelo, C., Bigelow, W. B., Varaprath, S., and Stille, J. K.,** *Macromolecules*, 14, 502, 1981.
134. Ref. J226.
135. **Appel, R. and Warning, U.,** *Chem. Ber.*, 109, 805, 1976.
136. **Mikołajczyk, M. and Luczak, J.,** *J. Org. Chem.*, 43, 2132, 1978.
137. (a) Ref. 135; (b) Ref. 136.
138. **Mikołajczyk, M., Omelańczuk, J., and Perlikowska, W.,** *Tetrahedron*, 35, 1531, 1979.
139. Ref. H210.
140. Ref. J266.
141. **Heinicke, J. and Tzschach, A.,** *Z. Chem.*, 23, 439, 1983.
142. **Fanni, T., Taira, K., Gorenstein, D. G., Vaidyamathaswarmy, R., and Verkade, J. G.,** *J. Am. Chem. Soc.*, 108, 6311, 1986.
143. **Singh, G.,** *J. Org. Chem.*, 44, 1060, 1979.
144. **Maiorova, E. D., Platonov, A. Yu., and Chistokletov, V. N.,** *J. Gen. Chem. U.S.S.R.*, 53, 1506, 1983.
145. Ref. J359.
146. Ref. J334.
147. Ref. J309.
148. **Quin, L. D. and Rao, N. S.,** *J. Org. Chem.*, 48, 3754, 1983.
149. Ref. J322.
150. **Yaouanc, J. J., Masse, G., and Sturz, G.,** *Synthesis*, p. 807, 1985.
151. Ref. J330.
152. (a) Ref. J331; (b) Ref. 131.
153. (a) Ref. J227; (b) Ref. 23.
154. Ref. H79.
155. (a) Ref. 135; (b) Ref. 69.
156. **Hall, C. D.,** unpublished results.
157. Ref. J346.
158. **Goda, K., Okazaki, R., Akiba, K., and Inamoto, N.,** *Bull. Chem. Soc. Jpn.*, 51, 260, 1978.
159. (a) Ref. 23; (b) Ref. 3.
160. Ref. J306.
161. Ref. J368.
162. **Haegele, G., Kueckelhaus, W., Tossing, G., Seega, J., Harris, R. K., Cresswell, C. J., and Jageland, P. T.,** *J. Chem. Soc. Dalton Trans.*, p. 795, 1987.

163. **Haegele, G., Kueckelhaus, W., Tossing, G., Mootz, D., and Wussow, H.-G.,** *Phosphorus Sulfur,* 22, 241, 1985.
164. Ref. J375.
165. **Corfield, J. R., Oram, R. K., Smith, D. J. H., and Trippett, S.,** *J. Chem. Soc. Perkin Trans. 1,* p. 713, 1972.
166. Ref. H78.
167. (a) Ref. 23; (b) Ref. 166.
168. **Nuretdinov, I. A., Loginova, É. I., and Bayandina, E. V.,** *J. Gen. Chem. U.S.S.R.,* 48, 975, 1978.
169. (a) Ref. 168; (b) Ref. H280.
170. (a) Ref. 168; (b) Ref. 26.
171. Ref. H103.
172. **Feshchenko, N. G., Gomelya, N. D., and Matyusha, G.,** *J. Gen. Chem. U.S.S.R.,* 52, 202, 1982.
173. **Al'fonsov, V. A., Zamaletdinova, G. U., Batyeva, É. S., and Pudovik, A. N.,** *J. Gen. Chem. U.S.S.R.,* 52, 1957, 1982.
174. **Schmidpeter, A. and Ebeling, J.,** *Chem. Ber.,* 101, 815, 1968.
175. **Steinberger, H. and Kuchen, W.,** *Z. Naturforsch.,* 29B, 611, 1974.
176. (a) Ref. 48; (b) Ref. 69.
177. Ref. J380.
178. **Quin, L. D. and Szewczyk, J.,** *Phosphorus Sulfur,* 21, 161, 1984.
179. (a) Ref. 26; (b) Ref. 56(b).
180. **Johnson, C. R. and Elliott, R. C.,** *J. Am. Chem. Soc.,* 104, 7041, 1982.
181. (a) Ref. 180; (b) Ref. 138.
182. **Schmidpeter, A. and Groeger, H.,** *Z. Anorg. Allg. Chem.,* 345, 106, 1966.
183. (a) Ref. 174; (b) Ref. 182.
184. Ref. H283.
185. Ref. H315.
186. Ref. J11.
187. Ref. J395.
188. Ref. H330.
189. Ref. J189.
190. Ref. H331.
191. (a) Ref. 187; (b) Ref. 69.
192. **Wheatland, D. A., Clapp, C. H., and Waldron, R. W.,** *Inorg. Chem.,* 11, 2340, 1972.
193. (a) Ref. 75; (b) Ref. 1.
194. **Haegele, G., Tossing, G., Kueckelhaus, W., and Seega, J.,** *J. Chem. Soc. Dalton Trans.,* p. 2803, 1984.
195. **Kuchen, W. and Haegele, G.,** *Chem. Ber.,* 103, 2274, 1970.
196. **Turpin, R., Dagnac, P., and Voigt, D.,** *Bull. Soc. Chim. Fr.,* p. 999, 1977.
197. Ref. H378.
198. **Keck, H., Kuchen, W., and Kuehlborn, S.,** *Phosphorus Sulfur,* 24, 343, 1985.
199. Ref. H389.
200. Ref. H387.
201. Ref. J445.
202. **Jain, R. S., Lawson, H. F., and Quin, L. D.,** *J. Org. Chem.,* 43, 108, 1978.
203. **Al'fonsov, V. A., Nizamov, I. S., Batyeva, É. S., and Pudovik, A. N.,** *J. Gen. Chem. U.S.S.R.,* 55, 1454, 1985.
204. Ref. H365.
205. (a) Ref. 63(b); (b) Ref. 204.
206. **Nurtdinov, S. Kh., Ismagilova, N. M., Fakhrutdinova, R. A., and Zykova, T. V.,** *J. Gen. Chem. U.S.S.R.,* 53, 923, 1983.
207. (a) Ref. 40; (b) Ref. 136.
208. **Dabkowski, W., Lopusinski, A., Michalski, J., and Radzicjewski, C.,** *Phosphorus Sulfur,* 8, 375, 1980.
209. **Al'fonsov, V. A., Pudovik, D. A., Batyeva, É. S., and Pudovik, A. N.,** *J. Gen. Chem. U.S.S.R.,* 55, 1956, 1985.
210. Ref. J440.
211. **Monkiewicz, J., Pietrusiewicz, K. M., and Bodalski, R.,** *Bull. Acad. Pol. Sci. Ser. Sci. Chim.,* 28, 351, 1980.
212. Ref. J337(a).
213. **Bayandina, E. V., Nuretdinov, I. A., Nurmukhamedova, L. V.,** *J. Gen. Chem. U.S.S.R.,* 48, 2424, 1978.

214. (a) Ref. 187; (b) Ref. 212.
215. Ref. H322.
216. **Aime, S., Harris, R. K., McVicker, E. M., and Fild, M.,** *J. Chem. Soc. Dalton Trans.*, p. 2144, 1976.
217. **McFarlane, H. C. E., McFarlane, W., and Nash, J. A.,** *J. Chem. Soc. Dalton Trans.*, p. 440, 1980.
218. (a) Ref. 216; (b) Ref. 217.
219. **Fluck, E. and Issleib, H.,** *Chem. Ber.*, 98, 2674, 1965.
220. (a) Ref. 218; (b) Ref.216.
221. (A) Ref. 100; (b) Ref. 98.
222. Ref. J405.
223. **Haegele, G. and Kuchen, W.,** *Chem. Ber.*, 103, 2885, 1970.
224. **Issleib, K. and Hoffmann, M.,** *Chem. Ber.*, 99, 1320, 1966.
225. (a) Ref. 224; (b) Ref. 2.
226. (a) Ref. H354; (b) Ref. 1.
227. **Diemart, K., Haas, P., and Kuchen, W.,** *Chem. Ber.*, 111, 629, 1978.
228. Ref. H344.
229. **Negrebetskii, V. V., Kal'chenko, V. I., Rudyi, R. B., and Markovskii, L. N.,** *J. Gen. Chem. U.S.S.R.*, 54, 2208, 1984.
230. **Weichmann, H., Meunier-Piret, J., and Van Meersche, M.,** *J. Organomet. Chem.*, 309, 273, 1986.
231. Ref. H410.
232. **Diemert, K. and Kuchen, W.,** *Phosphorus Sulfur*, 3, 131, 1977.
233. Ref. H354.
234. Ref. H359.
235. Ref. J414.
236. H359.
237. Ref. J465.
238. **Rozinov, V. G., Kolbina, V. E., and Donskikh, V. I.,** *J. Gen. Chem. U.S.S.R.*, 57, 847, 1987.
239. **Fridland, S. V., Miftakhov, M. N., and Arkhipov, V. P.,** *J. Gen. Chem. U.S.S.R.*, 57, 1353, 1987.
240. (a) Ref. 238; (b) Ref. 239.
241. **Dmitrichenko, M. Yu., Rozinov, V. G., Donskikh, V. I., Dolgushin, G. V., Feshin, V. P., and Kalabina, A. V.,** *J. Gen. Chem. U.S.S.R.*, 57, 1710, 1987.
242. **Levantsov, M. V., Prishchenko, A. A., and Lutsenko, I. F.,** *J. Gen. Chem. U.S.S.R.*, 57, 928, 1987.
243. **Palagina, T. V., Krutikov, V. I., Naidenov, A. V., and Lavrent'ev, A. N.,** *J. Gen. Chem. U.S.S.R.*, 56, 2385, 1986.
244. Ref. J493.
245. **Issleib, K. and Stiebitz, B.,** *Synth. React. Inorg. Metal-Org. Chem.*, 16, 1253, 1986.
246. **Sokolev, V. B., Epishina, T. A., Ivanov, A. N., Kharitonov, A. V., Brel', V. K., and Martynov, I. V.,** *J. Gen. Chem. U.S.S.R.*, 57, 1478, 1987.
247. **Hennig, H.-W., Neu, P., and Sartori, P.,** *Chem. Ztg.*, 111, 135, 1987.
248. Ref. H419.
249. **Namestnikov, V. I., Tamm, L. A., Trishin, Yu. G., and Chistokletov, V. N.,** *J. Gen. Chem. U.S.S.R.*, 57, 1263, 1987.

Chapter 13

^{31}P NMR DATA OF FOUR COORDINATE PHOSPHORUS COMPOUNDS CONTAINING A P=Ch BOND AND THREE BONDS FROM PHOSPHORUS TO GROUP IV ATOMS

Compiled and presented by
Marie-Paule Simonnin and Tekla Strzalko,

NMR data were also donated by
L. D. Quin, University of Massachusetts, Amherst, MA
C. D. Hall, King's College London, U.K.
J. P. Dutasta, Université Scientifique et Médicale de Grenoble, France

The NMR data are presented in Table L and it is divided into the 44 sections listed below:

L1. Trialkyl phosphine oxides
L2. Trialkyl polyphosphine oxides
L3. Cyclic trialkyl phosphine oxides
L4. Monoalkenyl/aryl phosphine oxides
L5. Monoalkenyl/aryl polyphosphine oxides
L6. Cyclic monoalkenyl/aryl phosphine oxides
L7. Monoalkynyl phosphine oxides
L8. Dialkenyl/aryl phosphine oxides
L9. Dialkenyl/aryl polyphosphine oxides
L10. Cyclic dialkenyl/aryl phosphine oxides
L11. Dialkynyl phosphine oxides
L12. Trialkenyl/aryl phosphine oxides
L13. Trialkenyl/aryl polyphosphine oxides
L14. Cyclic trialkenyl/aryl phosphine oxides
L15. Alkynyl dialkenyl/aryl phosphine oxides
L16. Dialkynyl alkenyl/aryl phosphine oxides
L17. Trialkynyl phosphine oxides
L18. Trialkyl phosphine sulfides
L19. Trialkyl polyphosphine sulfides
L20. Cyclic trialkyl phosphine sulfides
L21. Monoalkenyl/aryl phosphine sulfides
L22. Monoalkenyl/aryl polyphosphine sulfides
L23. Cyclic Monoalkenyl/aryl phosphine sulfides
L24. Monoalkynyl phosphine sulfides
L25. Dialkenyl/aryl phosphine sulfides
L26. Dialkenyl/aryl polyphosphine sulfides
L27. Cyclic dialkenyl/aryl phosphine sulfides
L28. Trialkenyl/aryl phosphine sulfides
L29. Trialkenyl/aryl polyphosphine sulfides
L30. Cyclic trialkenyl/aryl phosphine sulfides
L31. Trialkyl phosphine selenides
L32. Trialkyl polyphosphine selenides
L33. Cyclic trialkyl phosphine selenides
L34. Monoalkenyl/aryl phosphine selenides
L35. Monoalkenyl/aryl polyphosphine selenides
L36. Cyclic monoalkenyl/aryl phosphine selenides
L37. Dialkenyl/aryl phosphine selenides
L38. Dialkenyl/aryl polyphosphine selenides
L39. Cyclic dialkenyl/aryl phosphine selenides
L40. Trialkenyl/aryl phosphine selenides
L41. Trialkenyl/aryl polyphosphine selenides
L42. Cyclic trialkenyl/aryl phosphine selenides
L43. Alkynl dialkenyl/aryl phosphine selenides
L44. Trialkyl phosphine tellurides

The ^{31}P chemical shifts of nearly all tertiary phosphine chalcogenides ($R_3P{=}Ch$) are positive except for a few cases: the most striking examples of compounds having negative ^{31}P chemical shifts are

- Three-membered cyclic phosphine oxides and sulfides (Sections L3 and L30)
- Among acyclic derivatives: $(C_6F_5)_3PO$ (L13), $(C_6F_5)_3PS$ (L28), $(R{-}C{\equiv}C)_3PO$ (L17), and (2-furyl)$_3$PSe and (2-thienyl)$_3$PSe (L40)

The influence of the various substituents on the trends in ^{31}P shifts depends upon the nature of the chalcogenide (Ch) heteroatom (O, S, Se), the case of Te compounds not being discussed as only two examples have been reported (L44). For a given substituent, the back donation of the Ch heteroatom lone pairs depends on its nature, i.e., the double bond P=Ch character decreases along the series P(O) > P(S) > P(Se): in this latter case, it appears, from Se chemical shift, that this bond has little π character, these compounds existing predominantly in the dipolar form $R_3\overset{+}{P}\text{-}\overset{-}{Se}$.[139]

The influence of the nature of the substituents on ^{31}P shifts will be summarized in the case of identically substituted $R_3P{=}Ch$ compounds:

- When R is an alkyl group (L1, L18, L31), ^{31}P moves downfield from R=Me to R=Et but moves upfield when R=Bu is in the three series. However, there are divergent data in the literature for $(tBu)_3PCh$ (Ch=S, Se) which will not be discussed here. In phosphine oxides (L1), electronegative substituents within R (CH_2Cl, CH_2Br, CH_2OH, $CH_2N[alkyl]_2$) induce a lowfield shift, while CHF_2 and CF_3 induce a highfield shift related to Me_3PO.
- When R is an unsaturated group (L12, L28, L40), ^{31}P shielding increases, due to π conjugation: in phosphine oxides (L12), this effect follows the sequence Ph < $H_2C{=}CH$ < C_6F_5 < R–C≡C, while in selenides (L40) it decreases from 2-furyl to 2-thienyl. However, in phosphine sulfides (L28), the aryl-substituted derivatives resonate downfield in relation to Me_3PS except for $(C_6F_5)_3PS$ which suffers a highfield shift.

Interestingly, a good linear Hammett correlation[87] has been observed between ^{31}P shifts and the σ constant of the R substituent in E styrene derivatives, $4\text{-R-}C_6H_4\text{-}CH_A{=}CH_BP(Ch)Ph_2$, Ch=O,S (L12, L28): electron donating (withdrawing) substituents induce a lowfield (highfield) shift while the reverse is observed for H_B.

In the propenyl and styryl PO and PS derivatives (L12, L28), the ^{31}P chemical shifts[86] allow the assignment of the Z or E stereochemistry. The Z isomers resonate to high field of the related E isomers, in accordance with the known effect of steric compression. Moreover, the stereochemical dependence of $^3J_{PH}$ and $^3J_{PC}$ gives an unambiguous assignment of this geometry.

When the P atom is bound to a conformationally rigid six-membered ring (L8, L18), the ^{31}P nucleus of the axially oriented $P(Ch)R_2$ substituent is deshielded related to the equatorial isomer; this observation holds for carbocycles[75,97] or 2-P-substituted 1-3 dithianes.[73]

In 2-norbornyl-Me_2PO or -Me_2PS derivatives,[6] the P nucleus is more shielded when it occupies the endo rather than the exo position (L1, L18).

When the P atom is included within a cycle, the only peculiarities are observed when there are substantial bond angle changes. The ^{31}P chemical shift of the three-membered phosphirane oxide (**1**) resonates at −1.3 ppm[8] (L3) and that of the three-membered phosphirene sulfide (L30), which is highly strained, appears at −79 ppm.[122] This is the most shielded compound described among the P^{IV} compounds of this table.

The bicyclic compounds (**9**), which are dimers of phosphole oxides or sulfides, have two P atoms with different shift values: the bridging P nucleus (L3, L6, L20, L23) resonates at lower field than that of the 2-phospholene moiety (L6, L10, L23, L27). From X-ray determination, the CPC bond angles are 83° and 94°, respectively,[147] showing that in these systems, a larger bond angle corresponds to an upfield shift, although the presence of a double bond α to P in the 2-phospholene moiety might also contribute to the observed upfield shift. However, it is not possible to deduce the stereochemistry around the P atoms from ^{31}P shifts, because the shift difference between many isomeric pairs of compounds is very small, casting doubt on the stereochemical assignments deduced only from ^{31}P NMR for similar compounds in the literature. The stereochemistry appears to be unambiguously established only when X-ray analysis has been performed.[147]

In these bicyclic series and other cyclic systems in which the P atom is included within the cycle, the influence of the substituents follows the previous trends. The conformation of the five- and six-membered rings does not seem to be different from that of the carbocyclic or heterocyclic analogues; a peculiar case is compound (**18**), with an O-B-O fragment, which adopts a sofa conformation[59] (L6, L23).

The possibility of intramolecular coordination of the phosphoryl oxygen to Sn, which only takes place when a Cl atom is substituting the Sn atom, deserves comment: ^{31}P is shifted downfield by 11 ppm relative to a noncoordinated Sn analog (L4); in the same time a drastic reduction of J_{PSn} is observed.[50]

TABLE L
Compilation of ^{31}P NMR Data of Four Coordinate Compounds Containing a P=Ch Bond and Three Bonds from Phosphorus to Group IV Atoms

Section L1: Trialkyl Phosphine Oxides

Connectivities	Formula	Structure	NMR data (δ_P [solv, temp] $^nJ_{PZ}$ Hz)	Ref.
	P-bound atoms = C_3O'			
$(F_3)_3$	C_3F_9OP	$(Tf)_3PO$	3.4 $^2J_{PF}$113.5	1
$(HF_2)_3$	$C_3H_3F_6OP$	$(CHF_2)_3PO$	15.8 $^2J_{PF}$77.6	1
$(H_2Br)_3$	$C_3H_6Br_3OP$	$(CH_2Br)_3PO$	42.5	2(a)
$(H_2Br)_2,H_2O$	$C_5H_9Br_2O_3P$	$(CH_2Br)_2(MeOCOCH_2)PO$	33.6(m) $^1J_{PC}$67	3
$(H_2Cl)_3$	$C_3H_6Cl_3OP$	$(CH_2Cl)_3PO$	38.1	2(a)
$(H_2Cl)_2,H_3$	$C_3H_7Cl_2OP$	$(CH_2Cl)_2MePO$	41.6	2(b)
$(H_2Cl)_2,CH_2$	$C_4H_9Cl_2OP$	$(CH_2Cl)_2EtPO$	45.7	2(b)
	$C_{10}H_{21}Cl_2OP$	$(CH_2Cl)_2C_8H_{17}PO$	45.4	2(b)
$H_2Cl,(H_3)_2$	C_3H_8ClOP	$Me_2(CH_2Cl)PO$	42.0	2
H_2Cl,H_3,CH_2	$C_4H_{10}ClOP$	$MeEtCH_2ClPO$	48.0	2
$(H_2N)_3$	$C_9H_{24}N_3OP$	$(Me_2NCH_2)_3PO$	48.5	2(b)
	$C_{15}H_{36}N_3OP$	$(Et_2NCH_2)_3PO$	43.2	2(a)
	$C_{18}H_{36}N_3OP$	$[Pip\text{-}CH_2]_3PO$	51(e)	2(b)
$(H_2O)_3$	$C_3H_9O_4P$	$(CH_2OH)_3PO$	45.4	2(a)
$H_2O,(CH_2)_2$	$C_{11}H_{25}O_2P$	$Bu_2(EtOCH_2)PO$	45	4
$(H_3)_3$	C_3H_9OP	Me_3PO	36.2 $^1J_{PC}$68.3	2(a), 5
$(H_3)_2,CH_2$	$C_6H_{15}O_2P$	$Me_2(EtCHOHCH_2)PO$	41(b)	2(b)
	$C_{14}H_{31}OP$	$Me_2(C_{12}H_{25})PO$	48	2(b)
	$C_{16}H_{35}OP$	$Me_2(C_{14}H_{29})PO$	37.2	2(b)
$(H_3)_2,C_2H$	$C_9H_{17}OP$	$Me_2(2\text{-}Nb)PO$ *endo*	44.5(c)	6
		exo	45.6(c)	6
$(H_3)_2,C_3$	$C_6H_{15}OP$	$Me_2(tBu)PO$	38.1[b] 47.8[a]	2
$H_3,(CH_2)_2$	$C_{25}H_{53}OP$	$(C_{12}H_{25})_2MePO$	53	2(b)
$H_3,(C'H_2)_2$	$C_7H_{13}O_3P$	$(CH_3COCH_2)_2MePO$	38.4(c)	7
$CHCl,CH_2,C_3$	$C_{14}H_{30}ClOP$	$(tBuCHCl)Neo(tBu)PO$	56.6(c)	8
$(C'HO)_2,CH_2$	$C_{22}H_{31}O_3P$	$(PhCHOH)_2(C_8H_{17})PO$	50	2(b)
$C'HO,(CH_2)_2$	$C_{13}H_{15}N_2O_2P$	$(NCCH_2CH_2)_2(PhCHOH)PO$	46	2
$C'HO,(C'H_2)_2$	$C_{21}H_{21}O_2P$	$(PhCH_2)_2(PhCHOH)PO$	41	2
$(CH_2)_3$	$C_6H_{15}OP$	Et_3PO	48.3	2(a)
	$C_9H_{12}N_3OP$	$(NCCH_2CH_2)_3PO$	37.0	2(a)
	$C_{10}H_{23}OP$	Bu_2EtPO	45.7	2
	$C_{12}H_{27}OP$	Bu_3PO	43.2, 45.8	2(a)
	$C_{13}H_{25}O_5P$	$[HO_2C(CH_2)_2]_2C_7H_{15}PO$	52	2
	$C_{15}H_{33}OP$	$(i\text{-}C_5H_{11})_3PO$	50.4(c)	9(a)
	$C_{16}H_{35}OP$	$Et_2(C_{12}H_{25})PO$	56.1	2
	$C_{36}H_{75}OP$	$(C_{12}H_{25})_3PO$	53.0	2(b)
$(CH_2)_2,C'H_2$	$C_{15}H_{31}O_3P$	$(C_6H_{13})_2(MeO_2CCH_2)PO$	45	2
$(CH_2)_2,C''H_2$	$C_6H_{12}NOP$	$Et_2(NCCH_2)PO$	47.3(c) $^2J_{PH}$14.2	10
$(CH_2)_2,C_2Ge$	$C_{13}H_{25}F_6GeOP$	$Et_2[Et_3GeC(Tf)_2]PO$	16 $^3J_{PF}$ 25.3	11

TABLE L (continued)
Compilation of ^{31}P NMR Data of Four Coordinate Compounds Containing a P=Ch Bond and Three Bonds from Phosphorus to Group IV Atoms

Connectivities	Formula	Structure	NMR data (δ_P [solv, temp] $^nJ_{PZ}$ Hz)	Ref.
$(CH_2)_2,C_2S$	$C_{14}H_{29}OPS$	$Bu_2[(HS)C^*(CH_2)_4C^*H_2]PO$	53	12
$(CH_2)_2,CC'S$	$C_{16}H_{27}OPS$	$Bu_2[(HS)CMePh]PO$	52	12
$(CH_2)_2,C_3$	$C_{14}H_{31}OP$	$Neo_2(tBu)PO$	52.2(c)	8
$(C'H_2)_3$	$C_9H_{15}O_7P$	$(MeO_2CCH_2)_3PO$	34.4	2(a)
	$C_{21}H_{21}OP$	$(PhCH_2)_3PO$	40.9(c)[a], 43.4[b]	9
	$C_{24}H_{21}O_7P$	$(PhO_2CCH_2)_3PO$	36.0	2(a)
$C''H_2,(C_2H)_2$	$C_8H_{16}NOP$	$(iPr)_2(NCCH_2)PO$	53.7(c) $^2J_{PH}$13.5	10
$C''H_2,(C_3)_2$	$C_{10}H_{20}NOP$	$(tBu)_2(NCCH_2)PO$	58.2(c) $^2J_{PH}$12.6	10
$(C_2H)_3$	$C_{18}H_{33}OP$	$(c\text{-}C_6H_{11})_3PO$	50	2
$(C_3)_3$	$C_{12}H_{27}OP$	$(tBu)_3PO$	[−41]	13

Section L2: Trialkyl Polyphosphine Oxides

P-bound atoms = C_3O'

Connectivities	Formula	Structure	NMR data	Ref.
HP^4_2, $(H_3)_2$	$C_{17}H_{23}OP_3S_2$	$[Me_2P(S)][Ph_2P(S)]CH\mathbf{P}(O)Me_2$	46.4(c + r)	14(a)
$(H_2P^4)_3$	$C_9H_{24}O_4P_4$	$[Me_2P(O)CH_2]_3\mathbf{P}O$	36.9	2(b)
$(H_2P^4)_2,H_3$	$C_7H_{19}O_3P_3$	$[Me_2P(O)CH_2]_2\mathbf{P}(O)Me$	39.6	2(b)
$H_2P^4,(H_3)_2$	$C_7H_{19}O_3P_3$	$MeP(O)[CH_2\mathbf{P}(O)Me_2]_2$	42.4	2(b)
	$C_9H_{24}O_4P_4$	$P(O)[CH_2\mathbf{P}(O)Me_2]_3$	43.1	2(b)
$(H_3)_2,CH_2$	$C_{17}H_{22}O_2P_2$	$Ph_2P(O)(CH_2)_3\mathbf{P}(O)Me_2$	39.4(c) $^4J_{PP}$2.0	15
	$C_{18}H_{24}O_2P_2$	$Ph_2P(O)(CH_2)_4\mathbf{P}(O)Me_2$	39.3(c)	15
$(CH_2)_3$	$C_{11}H_{26}O_2P_2$	$Et_2P(O)(CH_2)_3P(O)Et_2$	49.1(c)	15
	$C_{19}H_{26}O_2P_2$	$Ph_2P(O)(CH_2)_3\mathbf{P}(O)Et_2$	45.9(c) $^4J_{PP}$2.0	15
$(CH_2)_3$	$C_{20}H_{28}O_2P_2$	$Ph_2P(O)(CH_2)_4\mathbf{P}(O)Et_2$	47.2(c)	15
$(CH_2)_3$	$C_{20}H_{44}O_2P_2$	$Et_2P(O)(CH_2)_{12}P(O)Et_2$	55.5	2b
$(CH_2)_3$	$C_{21}H_{46}O_2P_2$	$Bu_2P(O)(CH_2)_5P(O)Bu_2$	47.7	2b
$(CH_2)_3$	$C_{42}H_{42}O_4P_4$	$[Ph_2P(O)(CH_2)_2]_3\mathbf{P}(O)$	45.3(r,35°) $^3J_{PP}$46.1	16

Section L3: Cyclic Trialkyl Phosphine Oxides

P-bound atoms/rings = $C_3O'/5$

Connectivities	Formula	Structure	NMR data	Ref.
$(C_2H)_2,C_3$	$C_{14}H_{29}OP$	**1**	−1.3(c) $^1J_{PC}$12.5	8

P-bound atoms/rings = $C_3O'/6$

Connectivities	Formula	Structure	NMR data	Ref.
$H_3,(CH_2)_2$	$C_5H_9O_2P$	**2**; n = O, R = Me	60.5(c)	17
	$C_5H_{11}OP$	$(H_2C^*CH_2CH_2CH_2)P^*(O)Me$	65.8	5
	$C_8H_{13}O_2P$	**2**; n = 3, R = Me	81.6(c)	17
	$C_{12}H_{17}OP$	$[H_2C^*CMe(Ph)CH_2CH_2]P^*(O)Me$	66.2(c)	18
H_3,CH_2,C_2O	$C_8H_{13}O_2P$	**3**; n = 3, R = O	54.4(c)	17
	$C_9H_{15}O_2P$	**3**; n = 4, R = O	60.6(c)	17
$H_3,(C'H_2)_2$	$C_6H_9O_3P$	$[H_2C^*C(CO_2H)=CHCH_2]P^*(O)Me$	70.9(w)	19
	$C_7H_{11}O_3P$	$[H_2C^*C(CO_2H)=CMeCH_2]P^*(O)me$	67.3(w)	19
	$C_8H_{13}O_3P$	$[H_2C^*C(CO_2Me)=CMeCH_2]P^*(O)Me$	68.9(w)	19
	$C_9H_{11}OP$	**4**; R^1 = Me, R^2 = H	73.3(w)	20
$H_3,C'H_2,CC'H$	$C_9H_{13}O_2P$	**5**; R^1 = O, R^2 = Me, R^3 = H, R^4-R^5 = O	72.7(c)	21
		5; R^1 = Me, R^2 = O, R^3 = H, R^4-R^5 = O	64.4(c)	21

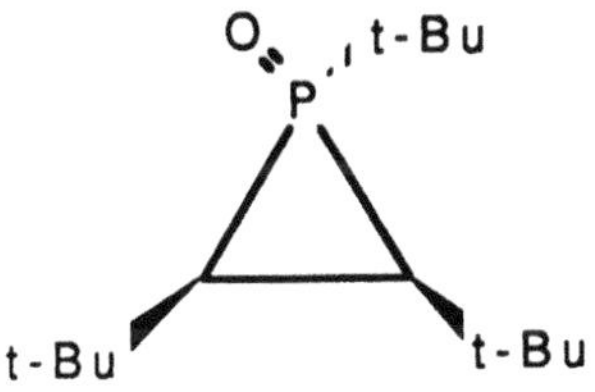

1 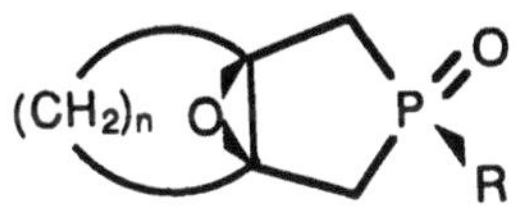 2 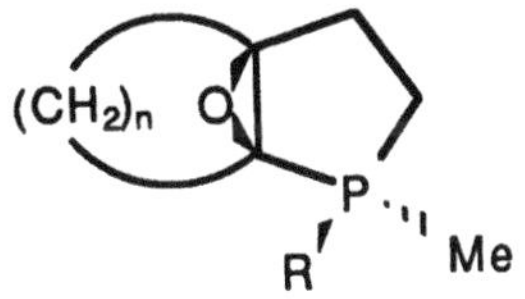 3

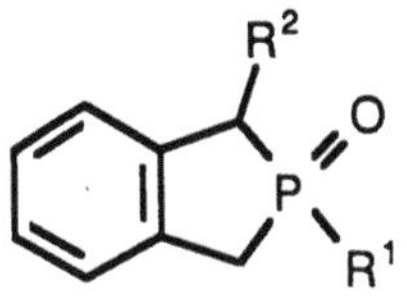

4 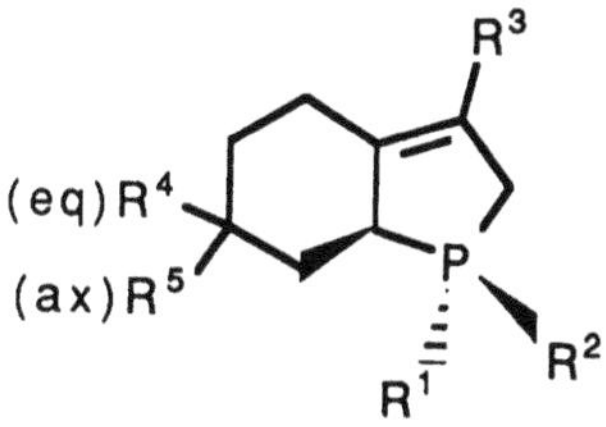 5 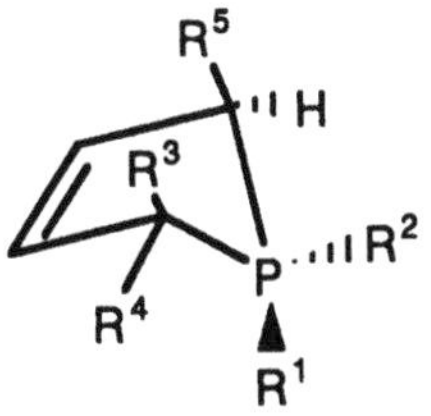 6

Connectivities	Formula	Structure	NMR data (δ_P [solv, temp] $^nJ_{PZ}$ Hz)	Ref.
	$C_9H_{15}OP$	**5**; R^1 = O, R^2 = Me, R^3 = R^4 = R^5 = H	77.1(w)	22
		5; R^1 = Me, R^2 = O, R^3 = R^4 = R^5 = H	70.5(w)	22
	$C_{10}H_{17}OP$	**5**; R^1 = O, R^2 = Me, R^3 = H, R^4 = Me, R^5 = H	68.6(c)	23
		5; R^1 = O, R^2 = Me, R^3 = R^4 = H, R^5 = Me	70.5(c)	23
		5; R^1 = Me, R^2 = O, R^3 = R^4 = H, R^5 = Me	62.8(c)	23
$H_3,(CC'H)_2$	$C_7H_{13}OP$	**6**; R^1 = R^3 = R^5 = Me, R^2 = O, R^4 = H	79(t)	24(a)
		6; R^1 = O, R^2 = R^3 = R^5 = Me, R^4 = H	63(t)	24(a)
$C'HBr,C'H_2,C_3$	$C_{12}H_{16}BrOP$	**4**; R^1 = tBu, R^2 = Br	68.0(c)	25
$(C'H_2)_3$	$C_{11}H_{13}OP$	$(H_2C^{\cdot}CH{=}CHCH_2)P^{\cdot}(O)CH_2Ph$	63.2(c)	26
$(C'H_2)_2,C_3$	$C_{12}H_{17}OP$	**4**; R^1 = tBu, R^2 = H	78.6(c)	25

P-bound atoms/rings = $C_3O'/6$

Connectivities	Formula	Structure	NMR data	Ref.
$H_3,(CH_2)_2$	$C_7H_{15}OP$	$(H_2C^{\cdot}CH_2CHMeCH_2CH_2)P^{\cdot}(O)Me$	40.9^C(c), 38.7^T(c)	27
	$C_7H_{15}OP$	$(H_2C^{\cdot}CHMeCH_2CH_2CH_2)P^{\cdot}(O)Me$	40.6^C(c), 42.2^T(c)	27
$H_3,(C'H_2)_2$	$C_6H_9O_2P$	$(H_2C^{\cdot}CH{=}CHCOCH_2)P^{\cdot}(O)Me$	39.3(c)	7
$(CH_2)_3$	$C_7H_{14}NO_2P$	$[H_2C^{\cdot}CH_2C({=}NOH)CH_2CH_2]P^{\cdot}(O)Et$	54(w)	28

P-bound atoms/rings = $C_3O'/7$

Connectivities	Formula	Structure	NMR data	Ref.
$H_3,(C'H_2)_2$	$C_7H_{11}O_3P$	$(H_2C^{\cdot}COCH_2CH_2COCH_2)P^{\cdot}(O)Me$	29.8(d)	29(a)

P-bound atoms/rings = $C_3O'/8$

Connectivities	Formula	Structure	NMR data	Ref.
$H_3,(C'H_2)_2$	$C_8H_{13}O_3P$	$[H_2C^{\cdot}O(CH_2)_3COCH_2]P^{\cdot}(O)Me$	29.0(d)	29(a)

TABLE L (continued)
Compilation of ^{31}P NMR Data of Four Coordinate Compounds Containing a P=Ch Bond and Three Bonds from Phosphorus to Group IV Atoms

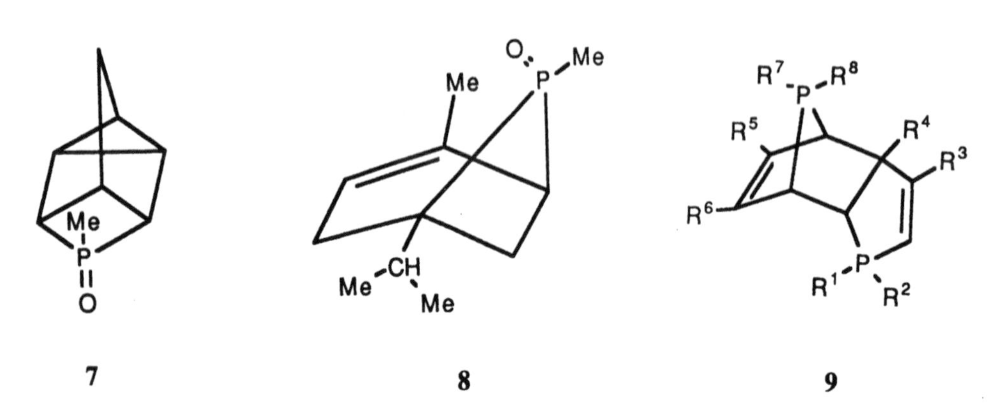

Connectivities	Formula	Structure	NMR data (δ_P [solv, temp] $^nJ_{PZ}$ Hz)	Ref.
		P-bound atoms/rings = C$_3$O′/9		
$H_3,(CH_2)_2$	$C_9H_{19}OP$	$[H_2C^{\cdot}(CH_2)_6CH_2]P^{\cdot}(O)Me$	48.9(c)	30
$H_3,(C'H_2)_2$	$C_9H_{15}O_3P$	$[H_2C^{\cdot}CO(CH_2)_4COCH_2]P^{\cdot}(O)Me$	32.0(c)	29
		P-bound atoms/rings = C$_3$O′/4,5		
$H_3,(C_2H)_2$	$C_8H_{11}OP$	7	71.5	31
		P-bound atoms/rings = C$_3$O′/4,6		
$H_3,CC'H,C_3$	$C_{11}H_{19}OP$	8	35.3	31
		P-bound atoms/rings = C$_3$O′/5,5		
$H_3,(CC'H)_2$	$C_{10}H_{14}O_2P_2$	9; $R^1 = R^8 = O$, $R^2 = R^7 = Me$, $R^3 = R^4 = R^5 = R^6 = H$,	91.4(c) $^3J_{PP}$35.0	32(a)
	$C_{12}H_{18}O_2P_2$	9; $R^1 = R^8 = O$, $R^2 = R^3 = R^5 = R^7 = Me$, $R^4 = R^6 = H$	86.8(c) $^3J_{PP}$34.2	32(a)
		9; $R^1 = R^7 = O$, $R^2 = R^3 = R^5 = R^8 = Me$, $R^4 = R^6 = H$,	87.9(c) $^3J_{PP}$36.9	32(a)
	$C_{13}H_{21}NO_2P_2$	9; $R^1 = R^8 = O$, $R^2 = NMe_2$, $R^3 = R^5 = R^7 = Me$, $R^4 = R^6 = H$	88.4(c) $^3J_{PP}$36.6	33
	$C_{14}H_{22}O_2P_2$	9; $R^1 = R^8 = O$, $R^2 = R^3 = R^4 = R^5 = R^6 = R^7 = Me$	79.0(c) $^3J_{PP}$40.0	32(a)
$CH_2,(CC'H)_2$	$C_{20}H_{34}O_2P_2$	9; $R^1 = R^7 = O$, $R^2 = R^8 = Bu$, $R^3 = R^4 = R^5 = R^6 = Me$	83.0(c)	34
$C'H_2,(CC'H)_2$	$C_{19}H_{25}NO_2P_2$	9; $R^1 = R^8 = O$, $R^2 = NMe_2$, $R^3 = R^5 = Me$, $R^4 = R^6 = H$, $R^7 = CH_2Ph$	90.2(c) $^3J_{PP}$36.6	33
$(CC'H)_2,C_3$	$C_{20}H_{34}O_2P_2$	9; $R^1 = R^7 = O$, $R^2 = R^8 = tBu$, $R^3 = R^4 = R^5 = R^6 = Me$	131.9(c)	35(a)

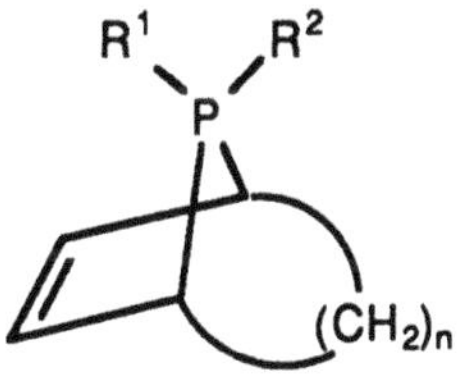

10 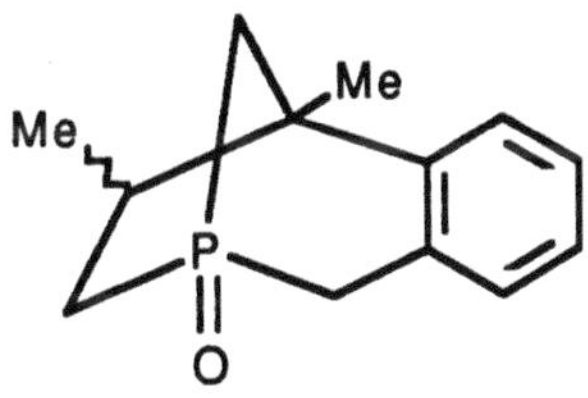 11 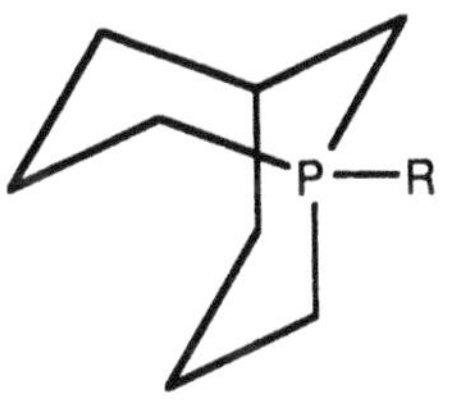 12

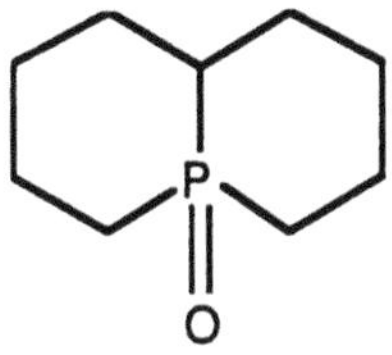

13 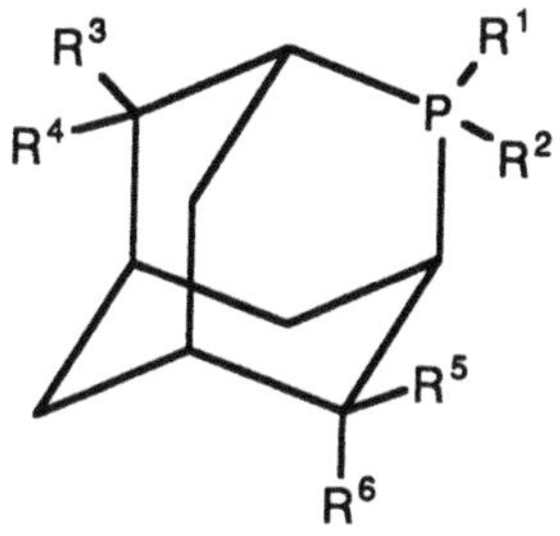 14 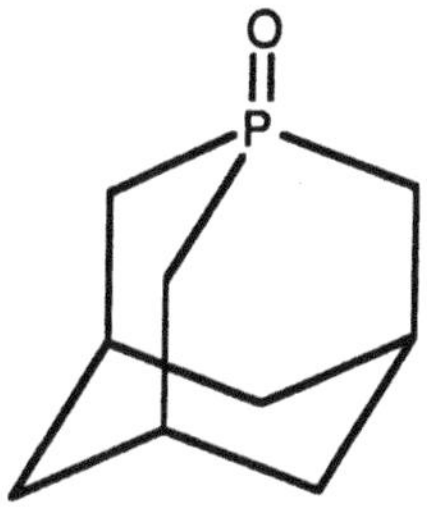 15

Connectivities	Formula	Structure	NMR data (δ_P [solv, temp] $^nJ_{PZ}$ Hz)	Ref.
		P-bound atoms/rings = C_3O'/5,6		
$H_3,(CC'H)_2$	$C_8H_{13}OP$	**10**; n = 3, R^1 = Me, R^2 = O	69.4(c)	36
		10; n = 3, R^1 = O, R^2 = Me	73.7(c)	36
$(CH_2)_2,C'H_2$	$C_{13}H_{17}OP$	**11**	54.5(c)	18
		P-bound atoms/rings = C_3O'/5,7		
$H_3,(CC'H)_2$	$C_9H_{15}OP$	**10**; n = 4, R^1 = Me, R^2 = O	64.4(c)	36
		10; n = 4, R^1 = O, R^2 = Me	67.6(c)	36
		P-bound atoms/rings = C_3O'/6,6		
$(CH_2)_3$	$C_8H_{15}OP$	**12**; R = O	26.0(b)	37
$(CH_2)_2,C'_2H$	$C_9H_{17}OP$	**13**	38.0^C(c), 38.0^T(c)	38
$(C_2H)_3$	$C_{15}H_{25}OP$	(**14**; R^1 = O, R^2 = c-C_6H_{11}, R^3 = R^4 = R^5 = R^6 = H	43.8(c)	39
		P-bound atoms/rings = C_3O'/6,6,6		
$(H_2O)_3$	$C_3H_6O_5P_2$	$OP(OCH_2)_3PO$	6.4	2(a)
$(CH_2)_2$	$C_9H_{15}OP$	**15**	34.8(y)	40

TABLE L (continued)
Compilation of ^{31}P NMR Data of Four Coordinate Compounds Containing a P=Ch Bond and Three Bonds from Phosphorus to Group IV Atoms

Section L4: Monoalkenyl/Aryl Phosphine Oxides

Connectivities	Formula	Structure	NMR data (δ_P [solv, temp] $^nJ_{PZ}$ Hz)	Ref.
P-bound atoms = $C_2C'O'$				
$(H_2Cl)_2;C'_2$	$C_8H_9Cl_2OP$	$(CH_2Cl)_2PhPO$	33.9	2(b)
$H_2Cl,H_3;C'_2$	$C_8H_{10}ClOP$	$MePhCH_2ClPO$	47.2	2
$(H_3)_2;C'H$	$C_8H_{15}OP$	$Me_2(CH_2{=}CMeCMe{=}CH)PO$	28.6(c)	41
$(H_3)_2;C''H$	C_5H_9OP	$Me_2(H_2C{=}C{=}CH)PO$	41.0	2
$(H_3)_2;C'_2$	$C_8H_6F_5OP$	$Me_2C_6F_5PO$	29.7	42(a)
	$C_8H_{11}OP$	Me_2PhPO	34.3(c)	9(a)
$H_3,CH_2;C'_2$	$C_{12}H_{19}OP$	MeNeoPhPO	34.6(c, 28°)	43(a)
$H_3,C'H_2;C'H$	$C_{14}H_{19}OP$	$Me(CH_2{=}CMeCMe{=}CH)PhCH_2)PO$	30.6(c)	41
$H_3,C'H_2;C'_2$	$C_{20}H_{23}O_2P$	$Me(PhCOCH_2CMe{=}CMeCH_2)PhPO$	32.4[Z](c)	44
$H_3,C_2H;C'_2$	$C_{17}H_{27}OP$	MePh(Men)PO *R*	35.9(b)	45
		S	37.6(b)	45
	$C_{17}H_{27}OP$	MePh(NeoMen)PO *R*	38.8(b)	45
		S	36.2(b)	45
$(CF_2)_2;C'_2$	$C_{12}H_4F_{15}OP$	$(C_3F_7)_2(4\text{-}FC_6H_4)PO$	19.0(g) $^2J_{PF}$74	46
$CHCl,CH_2;C'_2$	$C_{16}H_{26}ClOP$	PhNeo(tBuCHCl)PO (*R,R/S,S*)	36.3(g)	43(b)
		(*R,S/S,R*)	36.7(c + b)	43(b)
$(CH_2)_2;C'_2$	$C_{10}H_{15}OP$	Et_2PhPO	42.4[2] 48.8(c)[9a]	2, 9(a)
	$C_{12}H_{20}NOP$	$Et_2(4\text{-}NHEtC_6H_4)PO$	43.8(c)	47
	$C_{13}H_{19}O_3P$	$Et_2(3\text{-}EtO_2CC_6H_4)PO$	48.5(c)	9(a)
	$C_{14}H_{23}OP$	Bu_2PhPO	41.5	2
	$C_{15}H_{25}OP$	$Bu_2(4\text{-}MeC_6H_4)PO$	42.5(c)	9(a)
	$C_{16}H_{27}OP$	Neo_2PhPO	34.8(c, 28°)[a], 35.1(r)[b]	43
$CH_2,C'H_2;C'_2$	$C_{10}H_{13}O_3P$	$Et(HO_2CCH_2)PhPO$	42.3	48
	$C_{12}H_{17}O_3P$	$Bu(HO_2CCH_2)PhPO$	40.4	48
	$C_{13}H_{18}BrO_3P$	$[Br(CH_2)_3](EtO_2CCH_2)PhPO$	32.1(c)	49
$CH_2,C_3;C'_2$	$C_{20}H_{36}ClOPSn$	$[(tBu)_2ClSn(CH_2)_2](tBu)PhPO$	57.6(c) $^nJ_{PSn}$20.0	50
	$C_{26}H_{41}OPSn$	$[(tBu)_2PhSn(CH_2)_2](tBu)PhPO$	46.5(c) $^3J_{PSn}$160.4	50
$(C'H_2)_2;C'H$	$C_{20}H_{23}OP$	$(PhCH_2)_2(CH_2{=}CMeCMe{=}CH)PO$	29.8(c)	41
$(C'H_2)_2;C'_2$	$C_{11}H_{13}O_3P$	$(HOCCH_2)$ $(MeCOCH_2)PhPO$	28.7(c)	7
	$C_{12}H_{15}O_3P$	$(MeCOCH_2)_2PhPO$	28.9(c)	7
	$C_{22}H_{19}O_3P$	$(PhCOCH_2)_2PhPO$	30.4(c)	51
$C'H_2,C_2H;C'_2$	$C_{11}H_{15}O_3P$	$iPr(HO_2CCH_2)PhPO$	46.9	48
	$C_{12}H_{17}O_3P$	$sBu(HO_2CCH_2)PhPO$	45.9	48
$(C_3)_2;C'H$	$C_{11}H_{19}ClF_3OP$	$t(Bu)_2(CF_3CCl{=}CH)PO$	50.8	52
$(C_3)_2;C'_2$	$C_{15}H_{25}OP$	$(tBu)_2(4\text{-}MeC_6H_4)PO$	53.2(c)	9(a)

Section L5: Monoalkenyl/Aryl Polyphosphine Oxides

Connectivities	Formula	Structure	NMR data (δ_P [solv, temp] $^nJ_{PZ}$ Hz)	Ref.
P-bound atoms = $C_2C'O'$				
$(H_2P^4)_2;C'_2$	$C_{32}H_{29}O_3P_3$	$[Ph_2P(O)CH_2]_2\mathbf{P}(O)Ph$	27.6	2(b)
$(CH_2)_2;C'_2$	$C_{42}H_{42}O_4P_4$	$[Ph_2P(O)(CH_2)_2\mathbf{P}(O)PhCH_2\text{-}]_2$	47.9(r, 35°) $^3J_{PP}$55.8	16
$CH_2,C_3;C'_2$	$C_{28}H_{46}O_2P_2Sn$	$Et_2Sn[(CH_2)_2P(O)(tBu)Ph]_2$	51.6(c) $^3J_{PSn}$161.3	53

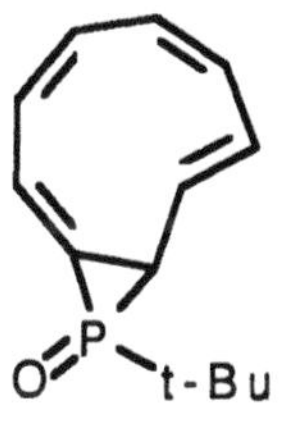

16 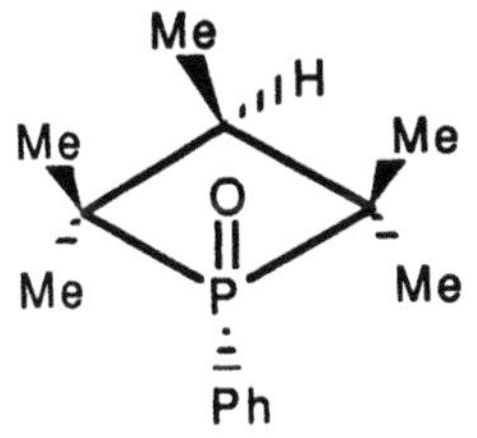 17 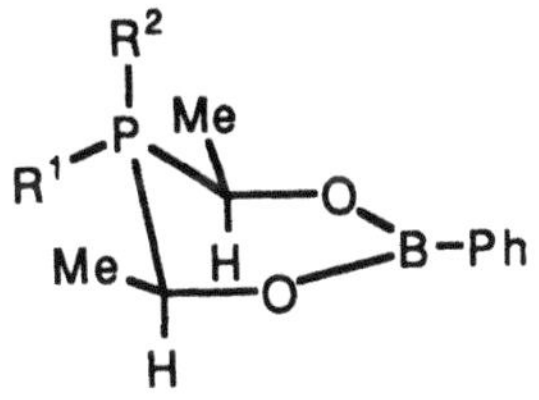 18

Section L6: Cyclic Monoalkenyl/Aryl Phosphine Oxides

Connectivities	Formula	Structure	NMR data (δ_P [solv, temp] $^nJ_{PZ}$ Hz)	Ref.
P-bound atoms/rings = $C_2C'O'/3$				
$C'_2H,C_3;CC'$	$C_{13}H_{17}OP$	**16**	33.2(c, −20°)	54
P-bound atoms/rings = $C_2C'O'/4$				
$CH_2,C_3;C'_2$	$C_{13}H_{19}OP$	$(Me_2C^*CMe_2CH_2)P^*(O)Ph$	50(r)	55
$(C_3)_2;C'_2$	$C_{14}H_{21}OP$	**17**	53.6(c)$^1J_{PC}$59.3	31
P-bound atoms/rings = $C_2C'O'/5$				
$H_3,CH_2;C'H$	C_5H_9OP	$(HC^*{=}CHCH_2CH_2)P^*(O)Me$	66.5	5
	$C_6H_{11}O_2P$	$[HC^*{=}C(OMe)CH_2CH_2]P^*(O)Me$	58.5(c)	56
	$C_7H_{11}O_4P$	$[HC^*{=}C(OMe)CH(CO_2H)CH_2]P^*(O)Me$	69.7(w)	19
$H_3,CH_2;C'_2$	$C_6H_9O_3P$	$[(HO_2C)C^*{=}CHCH_2CH_2]P^*(O)Me$	74.3(w)	19
	$C_7H_{11}O_3P$	$[(MeO_2C)C^*{=}CHCH_2CH_2]P^*(O)Me$	70.9(w)	19
	$C_7H_{11}O_4P$	$[(HO_2C)C^*{=}C(OMe)CH_2CH_2]P^*(O)Me$	66.5(w)	19
$H_3,C_2H;C'H$	$C_{12}H_{18}O_2P_2$	**9**; $R^1 = R^8 = O, R^2 = R^3 = R^5 = R^7 = Me, R^4 = R^6 = H$	60.8(c) $^3J_{PP}$34.2	32(a)
		9; $R^1 = R^7 = O, R^2 = R^3 = R^5 = R^8 = Me, R^4 = R^6 = H$	63.4(c) $^3J_{PP}$36.9	32(a)
	$C_{14}H_{22}O_2P_2$	**9**; $R^1 = R^8 = O, R^2 = R^3 = R^4 = R^5 = R^6 = R^7 = Me$	56.8(c) $^3J_{PP}$40.0	32(a)
$H_3,C_2H;CC'$	$C_7H_{13}OP$	$(MeC^*{=}CHCH_2CHMe)P^*(O)Me$	66.5^C(t),63.5^T(t)	24(a)
$(CH_2)_2;C'_2$	$C_{13}H_{15}O_2P$	**2**; n = 3, R = Ph	76.0	57
$CH_2,C'H_2;C'H$	$C_{11}H_{13}OP$	$(HC^*{=}CHCH_2CH_2)P^*(O)CH_2Ph$	67.7(c)	26
$CH_2,C_2H;C'H$	$C_{20}H_{34}O_2P_2$	**9**; $R^1 = R^7 = O, R^2 = R^8 = Bu, R^3 = R^4 = R^5 = R^6 = Me$	61.5(c)	34
$C'HBr,C'H_2;C'_2$	$C_{14}H_{12}BrOP$	**4**; $R^1 = Ph, R^2 = Br$	46.7(c)	25
$(C'H_2)_2;C'_2$	$C_{11}H_{13}OP$	$(H_2C^*CMe{=}CHCH_2)P^*(O)Ph$	57.2(c)	58
$C_2H,C_3;C'H$	$C_{20}H_{34}O_2P_2$	**9**; $R^1 = R^7 = O, R^2 = R^8 = tBu, R^3 = R^4 = R^5 = R^6 = Me$	74.4(c)	35(a)
$(CC'H)_2;C'_2$	$C_{12}H_{15}OP$	**6**; $R^1 = Ph, R^2 = O, R^3 = R^5 = Me, R^4 = H$	74(t)^a, 72.9(c)^b	24
		6; $R^1 = O, R^2 = Ph, R^3 = R^5 = Me, R^4 = H$	56(t)^a, 55.8(c)^b	24
P-bound atoms/rings = $C_2C'O'/6$				
$(CHO)_2;C'_2$	$C_{16}H_{18}BO_3P$	**18**; $R^1 = O, R^2 = Ph$	23(d)	59
		18; $R^1 = Ph, R^2 = O$	30(d)	59

TABLE L (continued)
Compilation of ^{31}P NMR Data of Four Coordinate Compounds Containing a P=Ch Bond and Three Bonds from Phosphorus to Group IV Atoms

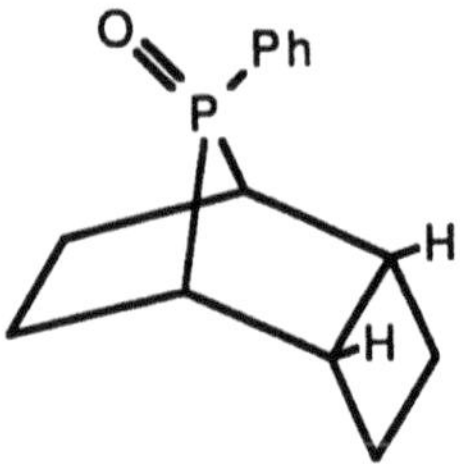

19

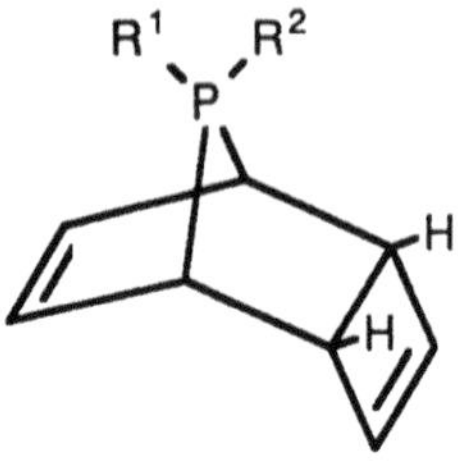

20

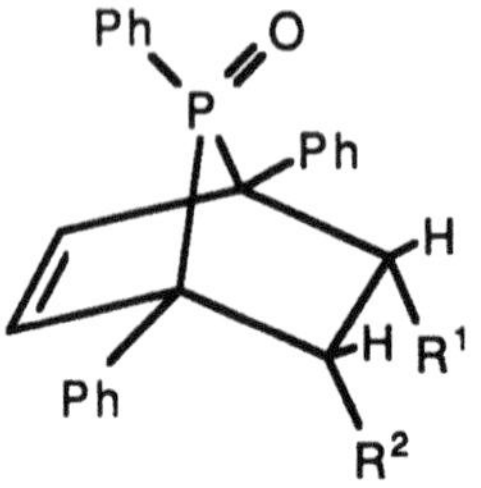

21

Connectivities	Formula	Structure	NMR data (δ_P [solv, temp] $^nJ_{PZ}$ Hz)	Ref.
$(CH_2)_2;C'_2$	$C_{11}H_{13}F_2OP$	$(H_2C^*CH_2CF_2CH_2CH_2)P^*(O)Ph$	24.5(c)	60
	$C_{12}H_{15}OP$	$[H_2C^*CH_2C(=CH_2)CH_2CH_2]P^*(O)Ph$	30.2(c)	61
	$C_{15}H_{23}OP$	$[H_2C^*CH_2CH(tBu)CH_2CH_2]P^*(O)Ph$	30.0^{C}(de), 28.2^{T}(de)	62
$(C'H_2)_2;C'_2$	$C_{11}H_{11}O_2P$	$(H_2C^*CH{=}CHCOCH_2)P^*(O)Ph$	32.2(c)	7
$C'H_2,C'_2O;C'H$	$C_{18}H_{17}O_2P$	$[HC^*{=}CHCH{=}CHC(OH)Ph]$ $P^*(O)CH_2Ph$	28.0(c)	63
	$C_{20}H_{21}O_2P$	$[HC^*{=}CMeCMe{=}CHC$ $(OH)Ph]P^*(O)CH_2Ph$	29(c)	64
P-bound atoms/rings = $C_2C'O'/9$				
$(C'H_2)_2;C'_2$	$C_{14}H_{15}Br_2O_3P$	$(H_2C^*COCH_2CHBr$ $CHBrCH_2COCH_2)P^*(O)Ph$	24.3^{T}(d)	65
	$C_{14}H_{17}O_3P$	$[H_2C^*CO(CH_2)_4COCH_2]P^*(O)Ph$	25.4(c)	29(b)
P-bound atoms/rings = $C_2C'O'/5,5$				
$(C_2H)_2;C'_2$	$C_{14}H_{17}OP$	**19**	67.0(c)	36
$(CC'H)_2;C'_2$	$C_{14}H_{13}OP$	**20**; R^1 = Ph, R^2 = O	95.0(c)	54
		20; R^1 = O, R^2 = Ph	98.8(c)	36
	$C_{18}H_{23}NO_2P_2$	**9**; $R^1 = R^8$ = O, R^2 = NMe_2, $R^3 = R^5$ = Me, $R^4 = R^6$ = H, R^7 = Ph	81.8(c) $^3J_{PP}$36.6	66
	$C_{20}H_{18}O_2P_2$	**9**; $R^1 = R^8$ = O, $R^2 = R^7$ = Ph, $R^3 = R^4 = R^5 = R^6$ = H	84.9(c) $^3J_{PP}$36.6	32(a)
	$C_{22}H_{22}O_2P_2$	**9**; $R^1 = R^8$ = O, $R^2 = R^7$ = Ph, $R^3 = R^5$ = Me, $R^4 = R^6$ = H	81.7(c) $^3J_{PP}$40.0	32(a)
		9; $R^1 = R^7$ = O, $R^2 = R^8$ = Ph, $R^3 = R^5$ = Me, $R^4 = R^6$ = H	79.7(c)	34
$(CC'_2)_2;C'_2)$	$C_{25}H_{20}NOP$	**21**; R^1 = H, R^2 = CN	75.5(c)	36
	$C_{28}H_{25}O_5P$	**21**; $R^1 = R^2 = CO_2Me$	74.5(c)	36
$(C'_3)_2;C'_2$	$C_{40}H_{31}O_5P$	**22**	105.5	67

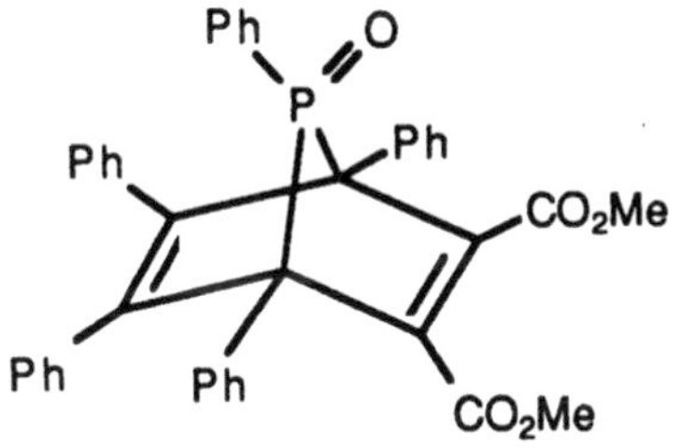

Ph O P Ph Ph CO_2Me Ph Ph CO_2Me

22

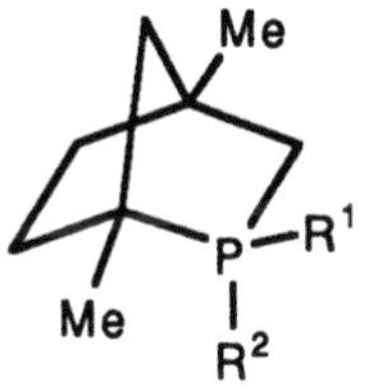

Me P R^1 Me R^2

23

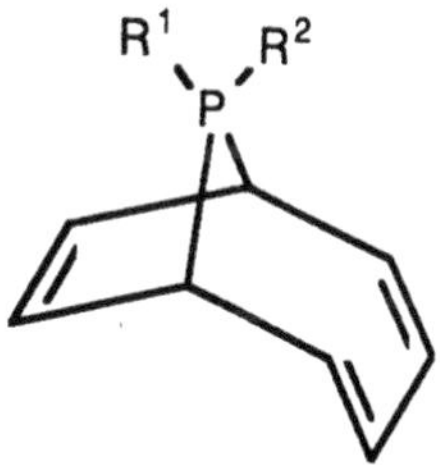

R^1 R^2 P

24

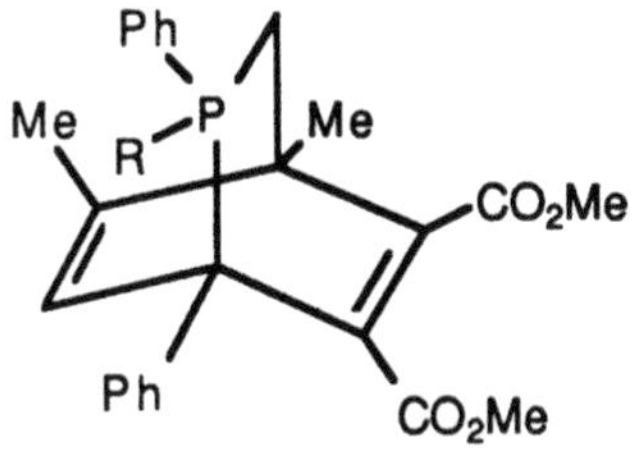

Ph Me R P Me CO_2Me Ph CO_2Me

25

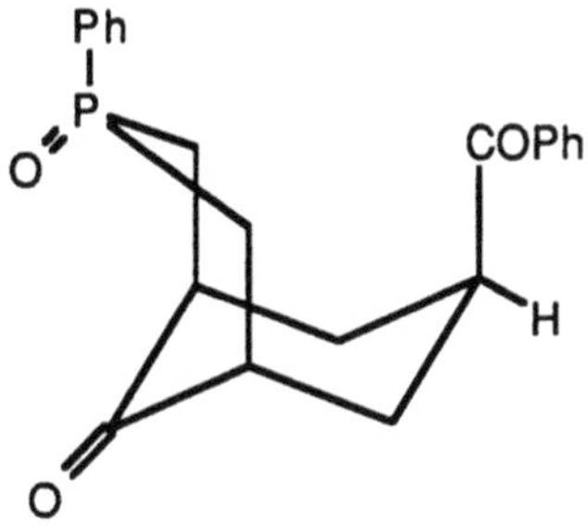

Ph P O COPh H O

26

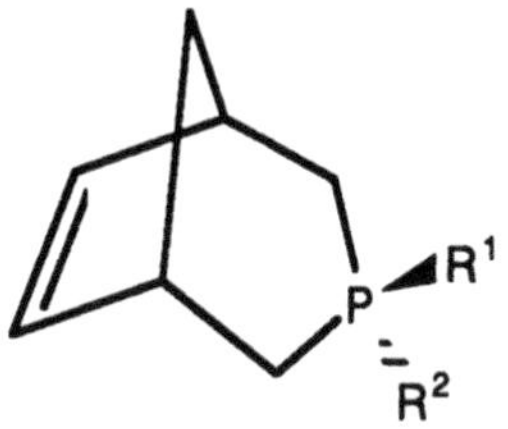

R^1 P R^2

27

Connectivities	Formula	Structure	NMR data (δ_P [solv, temp] $^nJ_{PZ}$ Hz)	Ref.
P-bound atoms/rings = $C_2C'O'/5,6$				
$CH_2,C_3;C'_2$	$C_{14}H_{19}OP$	**23**; R^1 = Ph, R^2 = O	53.1(b)	68
		23; R^1 = O, R^2 = Ph	53.9(b)	68
P-bound atoms/rings = $C_2C'O'/5,7$				
$(C'_2H)_2;C'_2$	$C_{14}H_{13}OP$	**24**; R^1 = Ph, R^2 = O	26(c)	69
		24; R^1 = O, R^2 = Ph	38(c)	69
P-bound atoms/rings = $C_2C'O'/6,6$				
$CH_2,C'_3;C'_2$	$C_{25}H_{25}O_5P$	**25**; R = O	46.2(c)	70
$(C_2H)_2;C'_2$	$C_{15}H_{19}OP$	**14**; R^1 = O, R^2 = Ph, R^3 = R^4 = R^5 = R^6 = H	36.8(c)	39
$(CC'H)_2;C'_2$	$C_{15}H_{15}O_3P$	**14**; R^1 = O, R^2 = Ph, R^3-R^4 = O, R^5-R^6 = O	41.9(c)	71
P-bound atoms/rings = $C_2C'O'/6,7$				
$(CH_2)_2;C'_2$	$C_{13}H_{15}OP$	**26**; R^1 = Ph, R^2 = O	28.5(c)	72
		26; R^1 = O, R^2 = Ph	30.9(c)	72
P-bound atoms/rings = $C_2C'O'/6,8$				
$(CH_2)_2;C'_2$	$C_{21}H_{21}O_3P$	**27**	25.5(c)	40
		28	26.6(c)	40

TABLE L (continued)
Compilation of ^{31}P NMR Data of Four Coordinate Compounds Containing a P=Ch Bond and Three Bonds from Phosphorus to Group IV Atoms

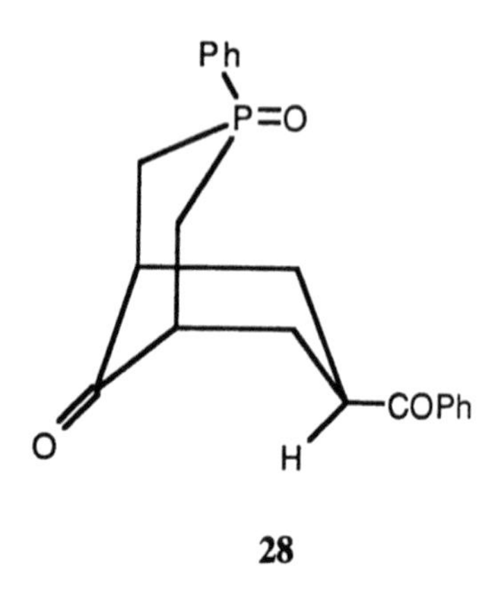

28

Section L7: Monoalkynyl Phosphine Oxides

P-bound atoms = $C_2C''O'$

Connectivities	Formula	Structure	NMR data (δ_P [solv, temp] $^nJ_{PZ}$ Hz)	Ref.
$(C_3)_2;C''$	$C_{10}H_{19}OP$	$(tBu)_2(HC{\equiv}C)PO$	48.2	13(a)

Section L8: Dialkenyl/Aryl Phosphine Oxides

P-bound atoms = CC'_2O'

Connectivities	Formula	Structure	NMR data (δ_P [solv, temp] $^nJ_{PZ}$ Hz)	Ref.
$HO_2(C'_2)_2$	$C_{18}H_{21}O_3P$	$Ph_2(HC^{\cdot}OCH_2CMe_2CH_2O^{\cdot})PO$	23.2(c) $^3J_{PC}$10.4	73
$HS_2;(C'_2)_2$	$C_{15}H_{15}OPS_3$	$Ph_2(HC^{\cdot}SCH_2SCH_2S^{\cdot})PO$	47.9(d) $^3J_{PC}$O	73
	$C_{18}H_{21}OPS_2$	$Ph_2(HC^{\cdot}SCH_2CMe_2CH_2S^{\cdot})PO$	32.8(c) $^3J_{PC}$O	73
	$C_{20}H_{25}OPS_2$	$Ph_2[HC^{\cdot}SCH_2CH(tBu)CH_2S^{\cdot}]PO$	34.1^C(c) $^3J_{PC}$O	73
			28.5^T $^3J_{PC}$7.1	73
$H_2Cl;(C'_2)_2$	$C_{13}H_{12}ClOP$	$Ph_2(CH_2Cl)PO$	30.4	2
$H_3;C'H,C'_2$	$C_{13}H_{17}OP$	$Me(CH_2{=}CMeCMe{=}CH)PhPO$	23.7(c)	41
$H_3;(C'_2)_2$	$C_{13}H_3F_{10}OP$	$(C_6F_5)_2MePO$	18.7	42(a)
	$C_{21}H_{19}OP$	$Me(CHPh{=}CPh)PhPO$	28.9^E(r)	74
$CHO;(C'_2)_2$	$C_{18}H_{25}O_2PSi$	$Ph_2[(Me_3SiO)EtCH]PO$	28	2(b)
$C'HO;(C'_2)_2$	$C_{19}H_{17}O_2P$	$Ph_2(PhCHOH)PO$	30	2
	$C_{20}H_{19}O_2P$	$Ph_2(PhCHOMe)PO$	28	2
$CH_2;(NO')_2$	$C_{22}H_{27}N_4O_7P$	$(4\text{-}NO_2C_6H_4NHCO)_2(C_8H_{17})PO$	15	2(b)
	$C_{22}H_{29}N_2O_3P$	$(PhNHCO)_2(C_8H_{17})PO$	23	2(b)
$CH_2;(C'_2)_2$	$C_{14}H_5F_{10}OP$	$(C_6F_5)_2EtPO$	19.3	42(a)
	$C_{14}H_{15}OP$	Ph_2EtPO	35	2
	$C_{16}H_{19}OP$	$Ph_2(iBu)PO$	21	2
	$C_{17}H_{21}OP$	Ph_2NeoPO	27.9(c, 28°)	43(a)
$C'H_2;C'H,C'_2$	$C_{19}H_{21}OP$	$PhCH_2(CH_2{=}CMeCMe{=}CH)PhPO$	23.6(c)	41
$C'H_2;(C'_2)_2$	$C_{14}H_{13}O_3P$	$Ph_2(HO_2CCH_2)PO$	36.9	48
$C''H_2;(C'_2)_2$	$C_{14}H_{12}NOP$	$Ph_2(NCCH_2)PO$	25.0(c) $^2J_{PH}$15.1	10
$C_2Ge;(C'_2)_2$	$C_{21}H_{25}F_6GeOP$	$Ph_2[Et_3GeC(Tf)_2]PO$	8$^3J_{PF}$34.8	11

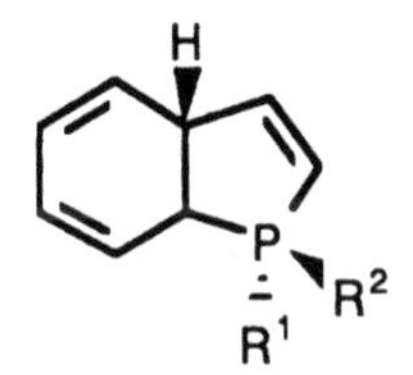

29

Connectivities	Formula	Structure	NMR data (δ_P [solv, temp] $^nJ_{PZ}$ Hz)	Ref.
$C_2H;(C'_2)_2$	$C_{15}H_{17}OP$	$Ph_2(iPr)PO$	37	2
	$C_{17}H_{21}OP$	$Ph_2(Et_2CH)PO$	38	2
	$C_{18}H_{21}OP$	$Ph_2(c\text{-}C_6H_{11})PO$	33	2
	$C_{22}H_{29}OP$	$Ph_2(4\text{-}tBu\ c\text{-}C_6H_{10})PO$	34.5^T, 36.3^C	75
$CC'H;(C'_2)_2$	$C_{20}H_{19}OP$	$Ph_2(PhMeCH)PO$	31	2(b)
$C_3;(C'_2)_2$	$C_{16}H_{19}OP$	$Ph_2(tBu)PO$	39	76

Section L9: Dialkenyl/Aryl Polyphosphine Oxides

P-bound atoms = CC'_2O'

Connectivities	Formula	Structure	NMR data	Ref.
$HP^4_2;(C'_2)_2$	$C_{37}H_{31}OP_3SSe$	$[Ph_2P(S)][Ph_2P(Se)]CH\mathbf{P}(O)Ph_2$	23.1(c)	14(b)
	$C_{37}H_{31}OP_3S_2$	$[Ph_2O(S)]_2CH\mathbf{P}(O)Ph_2$	23.7(c)	14(b)
	$C_{37}H_{31}O_2P_3S$	$[Ph_2P(S)]CH[\mathbf{P}(O)Ph_2]_2$	23.1(c)	14(b)
	$C_{37}H_{31}O_2P_3Se$	$[Ph_2P(Se)]CH[\mathbf{P}(O)Ph_2]_2$	24.4(c)	14(b)
$HP^3P^4;(C'_2)_2$	$C_{37}H_{31}OP_3S$	$[Ph_2P][Ph_2P(S)]CH\mathbf{P}(O)Ph_2$	25.8(c)	14(b)
$H_2P^4;(C'_2)_2$	$C_{25}H_{22}OP_2S$	$Ph_2P(S)CH_2\mathbf{P}(O)Ph_2$	23.1(c + r) $^2J_{PP}$15.6	77
	$C_{25}H_{22}OP_2Se$	$Ph_2P(Se)CH_2\mathbf{P}(O)Ph_2$	23.0(c + r) $^2J_{PP}$18.0	77
	$C_{25}H_{22}O_2P_2$	$Ph_2P(O)CH_2P(O)Ph_2$	24.2(c + r)	77
	$C_{26}H_{25}BrOP_2$	$[MePh_2P^+Br^-]CH_2\mathbf{P}(O)Ph_2$	24.6(c + r) $^2J_{PP}$12.0	77
	$C_{32}H_{29}O_3P_3$	$PhP(O)[CH_2\mathbf{P}Ph_2]_2)$	27.6	2(b)
$H_2P^3;(C'_2)_2$	$C_{25}H_{22}OP_2$	$Ph_2PCH_2\mathbf{P}(O)Ph_2$	27.7(c + r) $^2J_{PP}$50.0	77
$CH_2;(C'_2)_2$	$C_{17}H_{22}O_2P_2$	$Me_2P(O)(CH_2)_3\mathbf{P}(O)Ph_2$	29.0(c) $^4J_{PP}$2.0	15
	$C_{18}H_{24}O_2P_2$	$Me_2P(O)(CH_2)_4\mathbf{P}(O)Ph_2$	29.1(c)	15
	$C_{18}H_{24}O_2P_2$	$H(tBu)P(O)(CH_2)_2\mathbf{P}(O)Ph_2$	30.7(b) $^3J_{PP}$45.2	78
	$C_{19}H_{26}O_2P_2$	$Et_2P(O)(CH_2)_3\mathbf{P}(O)Ph_2$	27.5(c) $^4J_{PP}$2.0	15
	$C_{20}H_{28}O_2P_2$	$Et_2P(O)(CH_2)_4\mathbf{P}(O)Ph_2$	28.2(c)	15
	$C_{26}H_{24}O_2P_2$	$Ph_2P(O)(CH_2)_2P(O)Ph_2$	35.8	2(b)
	$C_{32}H_{38}O_2P_2Sn$	$Et_2Sn[(CH_2)_2P(O)Ph_2]_2$	33.6(c) $^3J_{PSn}$190.4	53
	$C_{42}H_{42}O_4P_4$	$P(O)[(CH_2)_2\mathbf{P}(O)Ph_2]_3$	27.8(r, 35°) $^3J_{PP}$46.1	16
	$C_{42}H_{42}O_4P_4$	$[Ph_2\mathbf{P}(O)(CH_2)_2P(O)PhCH_2-]_2$	38.5(r, 36°) $^3J_{PP}$55.8	16
$C_2P^4;(C'_2)_2$	$C_{27}H_{24}O_2P_2$	$(H_2C^{\cdot}CH_2C^{\cdot})[P(O)Ph_2]_2$	34.3(c, 30°)	79

Section L10: Cyclic Dialkenyl/Aryl Phosphine Oxides

P-bound atoms/rings = $CC'_2O'/5$

Connectivities	Formula	Structure	NMR data	Ref.
$H_3;(CC')_2$	$C_9H_{15}OP$	$(MeC^{\cdot}{=}CMeCMe{=}CMe)P^{\cdot}(O)Me$	51.0(c)	80
$C_2H;C'H,C'_2$	$C_{20}H_{18}O_2P_2$	(**9**;R^1 = R^8 = O, R^2 = R^7 = Ph, R^3 = R^4 = R^5 = R^6 = H)	55.5(c) $^3J_{PP}$36.6	32(a)
	$C_{22}H_{22}O_2P_2$	(**9**; R^1 = R^8 = O, R^2 = R^7 = Ph, R^3 = R^5 = Me, R^4 = R^6 = H)	55.1(c) $^3J_{PP}$40.0	32(a)
		(**9**; R^1 = R^7 = O, R^2 = R^8 = Ph, R^3 = R^5 = Me, R^4 = R^6 = H)	53.7(c)	34

TABLE L (continued)
Compilation of ^{31}P NMR Data of Four Coordinate Compounds Containing a P=Ch Bond and Three Bonds from Phosphorus to Group IV Atoms

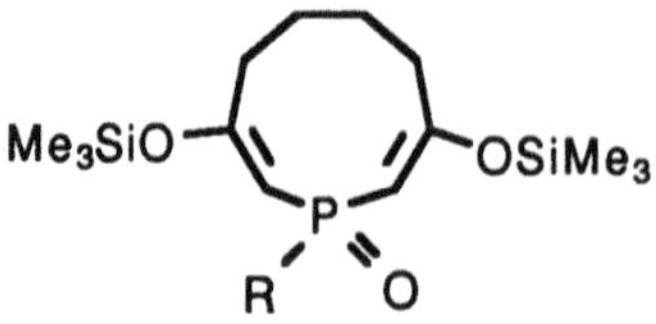

Me$_3$SiO ... OSiMe$_3$, P, R, O

30

Connectivities	Formula	Structure	NMR data (δ_P [solv, temp] $^nJ_{PZ}$ Hz)	Ref.
$C_2H;CC',C'_2$	$C_{12}H_{15}OP$	(MeC$^\cdot$=CHCH$_2$CHMe)P$^\cdot$(O)Ph	64^C(t)^a, 63.1^C(c)^b	24
			58.5^T(t)^a, 57.4^T(c)^b	24
$CC'H;C'H,C'_2$	$C_{14}H_{13}OP$	(**29**; R^1 = Ph, R^2 = O) *cis*-fused	71.1^T(c)	54
		(**29**; R^1 = O, R^2 = Ph)*cis*-fused	62.4^C(c)	54
		(**29**; R^1 = Ph, R^2 = O)*trans*-fused	46.9(c)	54
$C_3;(CC')_2$	$C_{12}H_{21}OP$	(MeC$^\cdot$=CMeCMe=CMe)P$^\cdot$(O)(tBu)	72.1(c)	80
P-bound atoms/rings = $CC'_2O'/6$				
$H_3;(C'H)_2$	$C_6H_{10}NOP$	(HC$^\cdot$=CMeNHCH=CH)P$^\cdot$(O)Me	9.4(c)	51
$C'_2O;C'H,C'_2$	$C_{18}H_{18}NO_2P$	[HC$^\cdot$=CMeCMe=CHC(OH) (2-Pyridyl)]P$^\cdot$(O)Ph	20.4(c)	81
	$C_{17}H_{17}O_2PS$	[HC$^\cdot$=CMeCMe=CHC(OH) (2-Thienyl)]P$^\cdot$(O)Ph	20.1(c)	63
	$C_{17}H_{17}O_3P$	[HC$^\cdot$=CMeCMe=CHC(OH) (2-Furyl)]P$^\cdot$(O)Ph	19.6(c)	63
P-bound atoms/rings = $CC'_2O'/7$				
$CH_2;CC',C'_2$	$C_{19}H_{30}OP_2$	(tBu)C$^\cdot$=C(tBu)PH(CH$_2$)$_3$P$^\cdot$(O)Ph	41.2(c) $^3J_{PP}$8	82
P-bound atoms/rings = $CC'_2O'/9$				
$H_3;(C'H)_2$	$C_{15}H_{31}O_3PSi_2$	(**30**; R = Me)	20.3(c)	30

Section L11: Dialkynyl Phosphine Oxides

Connectivities	Formula	Structure	NMR data	Ref.
P-bound atoms = CC''_2O'				
$C_3;(C'')_2$	$C_8H_{11}OP$	(HC≡C)$_2$(tBu)PO	6.5	13(a)

Section L12: Trialkenyl/Aryl Phosphine Oxides

Connectivities	Formula	Structure	NMR data	Ref.
P-bound atoms = C'_2O'				
$NS',(C'_2)_2$	$C_{15}H_{16}NOPS$	Ph$_2$[Me$_2$NC(S)]PO	27.9(k)	83
$C'O',(C'_2)_2$	$C_{19}H_{14}ClO_2P$	Ph$_2$(4-ClC$_6$H$_4$CO)PO	22.6(c)	84
	$C_{21}H_{17}O_3P$	Ph$_2$(4-MeCOC$_6$H$_4$CO)PO	22.2(c)	84
$(C'H)_3$	C_6H_9OP	(H$_2$C=CH)$_3$PO	16.5(c)	85

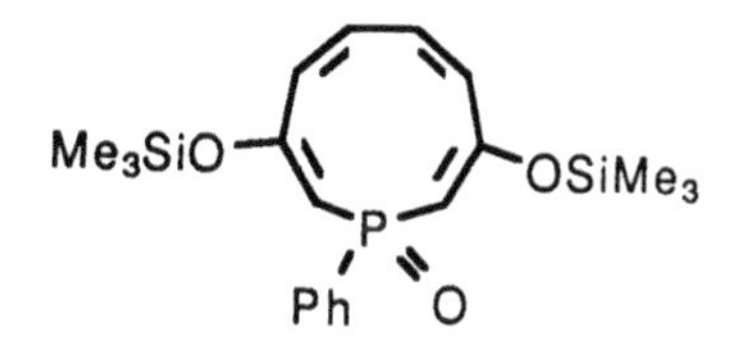

31

Connectivities	Formula	Structure	NMR data (δ_P [solv, temp] $^nJ_{PZ}$ Hz)	Ref.
$C'H,(C'_2)_2$	$C_{14}H_{13}OP$	$Ph_2(H_2C{=}CH)PO$	22.4(c)	76
	$C_{15}H_{15}OP$	$Ph_2(MeCH{=}CH)PO$	21.3^Z(c) $^3J_{PC}$7.3	86
			23.7^E(c) $^3J_{PC}$18.3	86
	$C_{20}H_{17}OP$	$Ph_2(PhCH{=}CH)PO$	19.9^Z(c) $^3J_{PC}$6.9	86
			25.4^E(c) $^3J_{PC}$18.4	86
	$C_{21}H_{16}NOP$	$Ph_2(4\text{-}NCC_6H_4CH{=}CH)PO$	18.3^E(b)	87
	$C_{22}H_{22}NOP$	$Ph_2(4\text{-}Me_2NC_6CH{=}CH)PO$	19.9^E(b)	87
$(C'_2)_3$	$C_{18}F_{15}OP$	$(C_6F_5)_3PO$	−8.2(c)	2(b)
	$C_{18}H_5F_{10}OP$	$(C_6F_5)_2PhPO$	0.7	42(b)
	$C_{18}H_{10}F_5OP$	$Ph_2C_6F_5PO$	18.7	42(b)
	$C_{18}H_{12}Cl_3OP$	$(4\text{-}ClC_6H_4)_3PO$	23	2(a)
	$C_{18}H_{12}F_3OP$	$(2\text{-}FC_6H_4)_3PO$	11.1 $^3J_{PF}$4	88
	$C_{18}H_{13}F_2OP$	$(2\text{-}FC_6H_4)_2PhPO$	15.6 $^3J_{PF}$4	88
	$C_{18}H_{14}FOP$	$Ph_2(2\text{-}FC_6H_4)PO$	20.4 $^3J_{PF}$4	88
	$C_{18}H_{15}OP$	Ph_3PO	23.0,[2a] 29.3(c)[9a]	2(a), 9(a)
	$C_{20}H_{20}NOP$	$Ph_2(4\text{-}NHEtC_6H_4)PO$	28.9(c)	47

Section L13: Trialkenyl/Aryl Polyphosphine Oxides

P-bound atoms = C'_3O'

Connectivities	Formula	Structure	NMR data	Ref.
$C''P^4,(C'_2)_2$	$C_{51}H_{40}O_4P_4$	$[Ph_2P(O)]_2C{=}C{=}C[P(O)Ph_2]_2$	23.5(c)	89
$C'H,(C'_2)_2$	$C_{26}H_{22}O_2P_2$	$Ph_2P(O)CH{=}CHP(O)Ph_2$	20.4^Z(c), 21.7^E(c)	86

Section L14: Cyclic Trialkenyl/Aryl Phosphine Oxides

P-bound atoms/rings = $C'_3O'/4$

Connectivities	Formula	Structure	NMR data	Ref.
$C'O,(C'_2)_2$	$C_{21}H_{15}O_2P$	$(PhC^{\cdot}{=}CPhCO)P^{\cdot}(O)Ph$	63(r)	90

P-bound atoms/rings = $C'_3O'/5$

Connectivities	Formula	Structure	NMR data	Ref.
$(CC')_2,C'_2$	$C_{14}H_{17}OP$	$(MeC^{\cdot}{=}CMeCMe{=}CMe)P^{\cdot}(O)Ph$	46.4(c)	80
$(C'_2)_3$	$C_{22}H_{17}OP$	$(PhC^{\cdot}{=}CHCH{=}CPh)P^{\cdot}(O)Ph$	41.5(c)	30

P-bound atoms/rings = $C'_3O'/6$

Connectivities	Formula	Structure	NMR data	Ref.
$(C'H)_2,C'_2$	$C_{11}H_{12}NOP$	$(HC^{\cdot}{=}CMeNHCH{=}CH)P^{\cdot}(O)Ph$	10.5(c)	51
	$C_{22}H_{18}NOP$	$(HC^{\cdot}{=}CPhNHCPh{=}CH)P^{\cdot}(O)Ph$	6.0(c)	51

P-bound atoms/rings = $C'_3O'/9$

Connectivities	Formula	Structure	NMR data	Ref.
$(C'H)_2,C'_2$	$C_{20}H_{29}O_3PSi_2$	**31**	30.3(c)	91
	$C_{20}H_{33}O_3PSi_2$	**30**; R = Ph	16.1(c)	30

TABLE L (continued)
Compilation of ^{31}P NMR Data of Four Coordinate Compounds Containing a P=Ch Bond and Three Bonds from Phosphorus to Group IV Atoms

Connectivities	Formula	Structure	NMR data (δ_P [solv, temp] $^nJ_{PZ}$ Hz)	Ref.
		Section L15: Alkynyl Dialkenyl/Aryl Phosphine Oxides		
	P-bound atoms = C′$_2$C″O′			
(C′$_2$)$_2$;C″	$C_{14}H_{11}OP$	$Ph_2(HC{\equiv}C)PO$	5.7(c) $^1J_{PC}$164.6	92
	$C_{15}H_{13}OP$	$Ph_2(MeC{\equiv}C)PO$	6.6(c) $^1J_{PC}$174.4	92
		Section L16: Dialkynyl Alkenyl/Aryl Phosphine Oxides		
	P-bound atoms = C′C″$_2$O′			
C′$_2$;(C″)$_2$	$C_{22}H_{10}F_5OP$	$(PhC{\equiv}C)_2C_6F_5PO$	−38.9(c)	9(a)
		Section L17: Trialkynyl Phosphine Oxides		
	P-bound atoms = C″$_3$O′			
(C″)$_3$	C_6H_3OP	$(HC{\equiv}C)_3PO$	−56	13(a)
	C_9H_9OP	$(MeC{\equiv}C)_3PO$	−54.3(c) $^1J_{PC}$239.8	92
	$C_{24}H_{15}OP$	$(PhC{\equiv}C)_3PO$	−51.5(c)	9(a)
		Section L18: Trialkyl Phosphine Sulfides		
	P-bound atoms = C$_3$S′			
$(H_2Cl)_2,CH_2$	$C_4H_9Cl_2PS$	$(CH_2Cl)_2EtPS$	55.2(c)	32(b)
$H_2Cl,(H_3)_2$	C_3H_8ClPS	$Me_2(CH_2Cl)PS$	44.3	32(b)
$H_2N,(CH_2)_2$	$C_{14}H_{30}NPS$	(iBu)$_2$(Pip-CH$_2$)PS	47.4(c)	32(b)
$(H_3)_3$	C_3H_9PS	Me_3PS	30.9(c),[93] 59(solid)[94]	93, 94
$(H_3)_2,CH_2$	$C_4H_{11}PS$	Me_2EtPS	57	32(b)
	$C_6H_{15}PS$	Me_2BuPS	36.5(c)	95
$(H_3)_2,C_2H$	$C_9H_{15}PS$	Me$_2$(7-Nbe)PS anti	30.2(c)	96
		syn	36.5(c)	96
	$C_9H_{17}PS$	Me$_2$(2-Nb)PS endo	40.2(c)	6
		exo	42.9(c)	6
	$C_9H_{19}PS$	Me$_2$(4-Me c-C$_6$H$_{10}$)PS	42.7^C(c, 27°), 42.7^T(c, 27°)	97
	$C_{12}H_{25}PS$	Me$_2$(4-tBu c-C$_6$H$_{10}$)PS	43.3^C(c, 27°), 42.5^T(c, 27°)	97
$H_3,(CH_2)_2$	$C_9H_{21}PS$	Bu_2MePS	43.3(c)	95
C′HN,$(CH_2)_2$	$C_{14}H_{21}N_2O_2PS$	Et$_2$[(4-NO$_2$C$_6$H$_4$)(HNCH$_2$CH=CH$_2$)CH]PS	61.1(e + a)	32(b)
	$C_{18}H_{29}N_2O_2PS$	Bu$_2$[(4-NO$_2$C$_6$H$_4$)(HNCH$_2$CH=CH$_2$)CH]PS	57.4(e + a)	32(b)
$(CH_2)_3$	$C_6H_{15}PS$	Et_3PS	51.9(a)	32(b)
	$C_{12}H_{27}PS$	Bu_3PS	48.1(c)	76
	$C_{12}H_{27}PS$	$Et_2(C_8H_{17})PS$	52.6	32(b)
	$C_{15}H_{33}PS$	Neo_3PS	36.9	98

Connectivities	Formula	Structure	NMR data (δ_P [solv, temp] $^nJ_{PZ}$ Hz)	Ref.
	$C_{20}H_{43}PS$	$(iBu)_2(C_{12}H_{25})PS$	44.2	32(b)
$(C_2H)_3$	$C_{18}H_{33}PS$	$(c\text{-}C_6H_{11})_3PS$	$66.3(SO_2)$,[99] 60(solid)[94]	94, 99
$(C_3)_3$	$C_{12}H_{27}PS$	$(tBu)_3PS$	[41],[13b] $90.2(SO_2)$[99]	13(b), 99

Section L19: Trialkyl Polyphosphine Sulfides

P-bound atoms = C_3S'

Connectivities	Formula	Structure	NMR data	Ref.
$HP^4_2,(H_3)_2$	$C_7H_{19}P_3S_3$	$HC[P(S)Me_2]_3$	41.7(c)	32(b)
	$C_{17}H_{23}OP_3S_2$	$Me_2P(O)[Ph_2P(S)]CH\mathbf{P}(S)Me_2$	46.4(c + r)	14(a)
	$C_{17}H_{23}P_3S_3$	$Ph_2P(S)CH[\mathbf{P}(S)Me_2]_2$	40.1(c + r)	14(a)
$HP^3P^4,(H_3)_2$	$C_{17}H_{23}P_3S_2$	$Me_2P[Ph_2P(S)]CH\mathbf{P}(S)Me_2$	45.4(c + r)	14(a)
	$C_{27}H_{27}P_3S_2$	$Ph_2P[Ph_2P(S)]CH\mathbf{P}(S)Me_2$	43.7(c + r)	14(a)
$H_2P^4,(H_3)_2$	$C_5H_{14}P_2S_2$	$Me_2P(S)CH_2P(S)Me_2$	32.7(c)	100
	$C_{15}H_{18}P_2SSe$	$Ph_2P(Se)CH_2\mathbf{P}(S)Me_2$	36.4(c + r)	77(b)
	$C_{15}H_{18}P_2S_2$	$Ph_2P(S)CH_2\mathbf{P}(S)Me_2$	35.3(r) $^2J_{PP}17$	101
$H_2P^4,(CH_2)_2$	$C_9H_{22}P_2S_2$	$Et_2P(S)CH_2P(S)Et_2$	49.5(e)	32(b)
$H_2P^4,(C_2H)_2$	$C_{19}H_{26}P_2S_2$	$Ph_2P(S)CH_2\mathbf{P}(S)(iPr)_2$	69.5(r) $^2J_{PP}16$	101
$H_2P^4,(C_3)_2$	$C_{21}H_{30}P_2S_2$	$Ph_2P(S)CH_2\mathbf{P}(S)(tBu)_2$	73.4(r) $^2J_{PP}16$	101
$(H_3)_3,CH_2$	$C_{14}H_{32}P_2S$	$(Neo)_2PCH_2CH_2\mathbf{P}(S)Me_2$	38.1 $^3J_{PP}35$	98
$H_3,(CH_2)_2$	$C_9H_{23}P_3S$	$(Me_2PCH_2CH_2)_2\mathbf{P}(S)Me$	46.0(r)	102
$(CH_2)_3$	$C_{11}H_{26}P_2S_2$	$Et_2P(S)(CH_2)_3P(S)Et_2$	53.5(e)	32(b)
	$C_{36}H_{78}P_4S_3$	$P[CH_2CH_2\mathbf{P}(S)(Neo)_2]_3$	43.6 $^3J_{PP}32$	98
	$C_{42}H_{42}P_4S_4$	$[Ph_2P(S)(CH_2)_2]_3\mathbf{P}(S)$	51.6(r, 35°) $^3J_{PP}55.4$	16

Section L20: Cyclic Trialkyl Phosphine Sulfides

P-bound atoms/rings = $C_3S'/5$

Connectivities	Formula	Structure	NMR data	Ref.
$H_3,CH_2,C'H_2$	C_5H_9OPS	$H_2C^*COCH_2CH_2P^*(S)Me$	42.4(c)	103
H_3,CH_2,C_2O	C_9H_5OPS	**3**; n = 4, R = S	59.9(c)	17
$H_3(C'H_2)_2$	C_5H_9PS	$H_2C^*CH{=}CHCH_2P^*(S)Me$	54.8(c)	103
$H_3,C'H_2,CC'H$	$C_6H_{11}PS$	$MeHC^*CH{=}CHCH_2P^*(S)Me$	63.4^C(c), 60.8^T(c)	103
	$C_9H_{14}BrPS$	**5**; R^1 = S, R^2 = Me, R^3 = Br, R^4 = R^5 = H	51.4(c)	103
	$C_9H_{15}PS$	**5**; R^1 = S, R^2 = Me, R^3 = R^4 = R^5 = H	59.9(c)	103
		5; R^1 = Me, R^2 = S, R^3 = R^4 = R^5 = H	58.0(c)	103
$H_3,(CC'H)_2$	$C_7H_{13}PS$	**6**, R^1 = S, R^2 = R^3 = R^5 = Me, R^4 = H	74(t)	104
		6, R^1 = R^3 = R^5 = Me, R^2 = S, R^4 = H	66(c)	104

P-bound atoms/rings = $C_3S'/6$

Connectivities	Formula	Structure	NMR data	Ref.
$(H_2O)_2,H_3$	$C_5H_{13}O_2PSSi$	$H_2C^*OSiMe_2OCH_2P^*(S)Me)$	28(a, 50°)	105
$H_3,(CH_2)_2$	$C_6H_{13}PS$	$H_2C^*(CH_2)_4P^*(S)Me$	31.3(c)	2(b)
	$C_7H_{15}PS$	$H_2C^*CHMeCH_2CH_2CH_2P^*(S)Me$	32.6^C(c), 30.5^T(c)	27
	$C_{10}H_{21}OPS$	$H_2C^*CH_2C(tBu)(OH)CH_2CH_2P^*(S)Me$	29.4^C(c), 31.9^T(c)	106

TABLE L (continued)
Compilation of ^{31}P NMR Data of Four Coordinate Compounds Containing a P=Ch Bond and Three Bonds from Phosphorus to Group IV Atoms

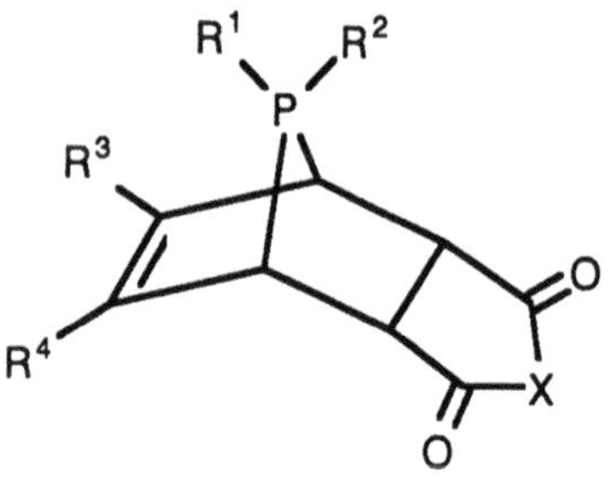

32 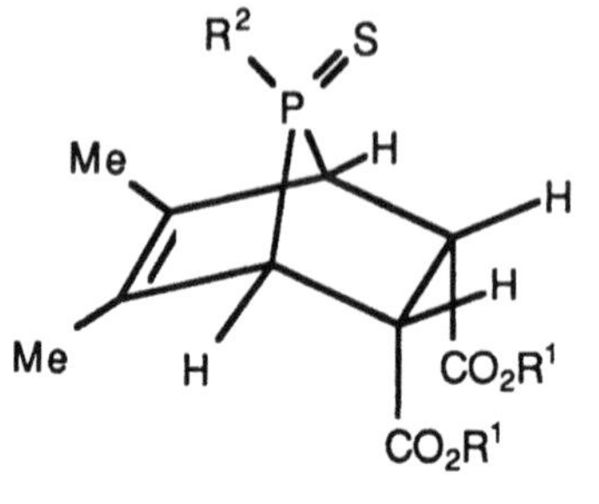 **33** 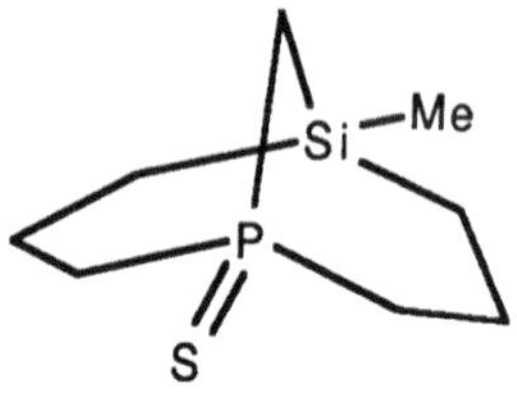 **34**

Connectivities	Formula	Structure	NMR data (δ_P [solv, temp] $^nJ_{PZ}$ Hz)	Ref.
		P-bound atoms/rings = C_3S'/8		
$H_3,(CH_2)_2$	$C_6H_{14}O_2P_2S_2$	$[H_2C^*CH_2OP(S)(Me)OCH_2CH_2]P^*(S)Me$	35.9^C(b)	65
			36.0^T(b) $^4J_{PP}$1.1	65
		P-bound atoms/rings = C_3S'/5,5		
$H_3,(CC'H)_2$	$C_{12}H_{18}P_2S_2$	**9**; $R^1 = R^8 = S$, $R^2 = R^3 = R^5 = R^7 = Me$, $R^4 = R^6 = H$	106.5(c) $^3J_{PP}$41.5	36
		9, $R^1 = R^7 = S$, $R^2 = R^3 = R^5 = R^8 = Me$, $R^4 = R^6 = H$	104.4(c) $^3J_{PP}$36.6	36
	$C_{14}H_{22}P_2S_2$	**9**; $R^1 = R^8 = S$, $R^2 = R^3 = R^4 = R^5 = R^6 = R^7 = Me$	101.0(c) $^3J_{PP}$49.0	36
$CH_2,(CC'H)_2$	$C_{14}H_{19}O_3PS$	**32**; X = O, R^1 = Bu, R^2 = S, $R^3 = R^4$ = Me	114.7(c)	107
	$C_{16}H_{25}O_4PS$	**33**; R^1 = Me, R^2 = Bu	102(a)	107
	$C_{20}H_{24}NO_2PS$	**32**; X = NPh, R^1 = Bu, R^2 = S, $R^3 = R^4$ = Me	112(c)	107
$C''H_2,(CC'H)_2$	$C_{16}H_{20}N_2P_2S_2$	**9**; $R^1 = R^8 = S$, $R^2 = R^7 = CH_2CN$, $R^3 = R^4 = R^5 = R^6$ = Me	98.9(c) $^3J_{PP}$51.9	108
$(CC'H)_2,C_3$	$C_{18}H_{30}P_2S_2$	**9**; $R^1 = R^7 = S$, $R^2 = R^8$ = tBu, $R^3 = R^6$ = Me, $R^4 = R^5$ = H	134.5(c)	35(a)
		P-bound atoms/rings = C_3S'/6,6		
$H_2Si,(CH_2)_2$	$C_8H_{17}PSSi$	**34**	26.0(y)	109
$(CH_2)_3$	$C_8H_{15}PS$	**12**; R = S	22.1(b)	37

Section L21: Monoalkenyl/Aryl Phosphine Sulfides

Connectivities	Formula	Structure	NMR data	Ref.
		P-bound atoms = $C_2C'S'$		
$(H_2Cl)_2;C'_2$	$C_8H_9Cl_2PS$	$(CH_2Cl)_2PhPS$	44.2(c)	32(b)

Connectivities	Formula	Structure	NMR data (δ_P [solv, temp] $^nJ_{PZ}$ Hz)	Ref.
$(H_2O)_2;C'_2$	$C_8H_{11}O_2PS$	$(CH_2OH)_2PhPS$	41.8(y)	32(b)
$(H_2Si)_2;C'_2$	$C_{14}H_{27}PSSi_2$	$(Me_2SiCH_2)_2PhPS$	38.8	32(b)
$H_2Si,H_3;C'_2$	$C_{11}H_{19}PSSi$	$Me(Me_3SiCH_2)PhPS$	36.6	32(b)
$(H_3)_2;C'_2$	$C_8H_6F_5PS$	$Me_2C_6F_5PS$	30.5	42(a)
	$C_8H_{11}PS$	Me_2PhPS	32.5(c)	110
$H_3,CH_2;C'_2$	$C_9H_{13}PS$	MeEtPhPS	45.8	32(b)
$H_3,C'H_2;C'_2$	$C_9H_{11}OPS$	$Me(OHCCH_2)PhPS$	28.7(c)	111
	$C_9H_{11}O_2PS$	$Me(HO_2CCH_2)PhPS$	37.7(a)	32(b)
$H_3,C_2H;C'_2$	$C_{13}H_{19}PS$	$Me(c\text{-}C_6H_{11})PhPS$	43.7(c)	112
$(CH_2)_2;C'H$	$C_{10}H_{21}PS$	$Bu_2(H_2C{=}CH)PS$	44(c)	32(b)
	$C_{12}H_{25}PS$	$(Neo)_2(H_2C{=}CH)PS$	33.1	98
$(CH_2)_2;C'_2$	$C_{10}H_{10}F_5PS$	$Et_2C_6F_5PS$	52.6,[42(a)] [33.6][32(b)]	42(a), 32(b)
	$C_{10}H_{15}PS$	Et_2PhPS	52.0[b], 50.4(b)[a]	32
	$C_{16}H_{27}PS$	$(Neo)_2PhPS$	34.9	98
$C'H_2,C_2H;C'_2$	$C_{16}H_{23}O_2PS$	$(EtO_2CCH_2)(c\text{-}C_6H_{11})PhPS$	46.1(c)	112
$(C_2H)_2;NS'$	$C_{19}H_{28}NPS_2$	$(c\text{-}C_6H_{11})_2[PhHNC(S)]PS$	80.2(r)	113
$(C_2H)_2;SS'$	$C_{14}H_{25}PS_3$	$(c\text{-}C_6H_{11})_2(MeS_2C)PS$	73.6(r)	113

Section L22: Monoalkenyl/Aryl Polyphosphine Sulfides

P-bound atoms = $C_2C'S'$

Connectivities	Formula	Structure	NMR data (δ_P [solv, temp] $^nJ_{PZ}$ Hz)	Ref.
$(CH_2)_2;C'_2$	$C_{34}H_{33}P_3S_3$	$[Ph_2P(S)(CH_2)_2]_2$ **P**(S)Ph	44.6(SO_2)	114
	$C_{42}H_{42}P_4S_4$	$[Ph_2P(S)(CH_2)_2$**P**$(S)PhCH_2\text{-}]_2$	47.5(r, 35°) $^3J_{PP}$58.5	16

Section L23: Cyclic Monoalkenyl/Aryl Phosphine Sulfides

P-bound atoms/rings = $C_2C'S'/5$

Connectivities	Formula	Structure	NMR data (δ_P [solv, temp] $^nJ_{PZ}$ Hz)	Ref.
$H_3,CH_2;C'H$	C_5H_8ClPS	$HC^{\cdot}{=}CClCH_2CH_2P^{\cdot}(S)Me$	58.5(c)	103
	C_5H_9PS	$HC^{\cdot}{=}CHCH_2CH_2P^{\cdot}(S)Me$	65.5(c)	103
	$C_6H_{11}OPS$	$HC^{\cdot}{=}C(OMe)CH_2CH_2P^{\cdot}(S)Me$	58.7(c)	103
	$C_9H_{16}NOPS$	$HC^{\cdot}{=}C(Morph)CH_2CH_2P^{\cdot}(S)Me$	57.8(c)	115
	$C_9H_{16}NPS$	$HC^{\cdot}{=}C[N^{\cdot}(CH_2)_3$ $C^{\cdot}H_2]CH_2CH_2P^{\cdot}(S)Me)$	58.7(c)	115
	$C_{10}H_{18}NPS$	$HC^{\cdot}{=}C(Pip)CH_2CH_2P^{\cdot}(S)Me$	58.1(c)	115
	$C_{11}H_{20}NPS$	$HC^{\cdot}{=}C[NH(c\text{-}C_6H_{11})]$ $CH_2CH_2P^{\cdot}(S)Me$	60.1(c)	115
$H_3,C_2H;C'H$	$C_{10}H_{14}P_2S_2$	**9**; $R^1 = R^8 = S$, $R^2 = R^7 = Me$, $R^3 = R^4 = R^5 = R^6 = H$	62.0(c) $^3J_{PP}$ 41.5	32(a)
	$C_{12}H_{18}P_2S_2$	**9**; $R^1 = R^8 = S$, $R^2 = R^3 = R^5 = R^7 = Me$, $R^4 = R^6 = H$	60.4(c) $^3J_{PP}$41.5	36
		9; $R^1 = R^7 = S$, $R^2 = R^3 = R^5 = R^8 = Me$, $R^4 = R^6 = H$	63.1(c) $^3J_{PP}$36.6	36
	$C_{14}H_{22}P_2S_2$	**9**; $R^1 = R^8 = S$, $R^2 = R^3 = R^4 = R^5 = R^6 = R^7 = Me$	56.1(c) $^3J_{PP}$49.0	36
$H_3,C_2H;CC'$	$C_7H_{13}PS$	$MeC^{\cdot}{=}CHCH_2CHMeP^{\cdot}(S)Me$	66.9^C(b), 63.3^T(b)	104
$H_3,CC'H;C'H$	$C_9H_{11}PS$	**29**; $R^1 = S$, $R^2 = Me$)*cis*-fused	72.1^C(c)	36
$H_3,CC'H;C'_2$	$C_{29}H_{25}PS$	$PhC^{\cdot}{=}CPhCHPhCHPhP^{\cdot}(S)Me$	61.7(c)	116
$C''H_2,C_2H;C'H$	$C_{16}H_{20}N_2P_2S_2$	**9**; $R^1 = R^8 = S$, $R^2 = R^7 = CH_2CN$, $R^3 = R^4 = R^5 = R^6 = Me$	55.7(c) $^3J_{PP}$51.9	108
$C_2H,C_3;C'H$	$C_{18}H_{30}P_2S_2$	**9**; $R^1 = R^7 = S$, $R^2 = R^8 = tBu$, $R^3 = R^6 = Me$, $R^4 = R^5 = H$	86.5(c)	35(a)

TABLE L (continued)
Compilation of ^{31}P NMR Data of Four Coordinate Compounds Containing a P=Ch Bond and Three Bonds from Phosphorus to Group IV Atoms

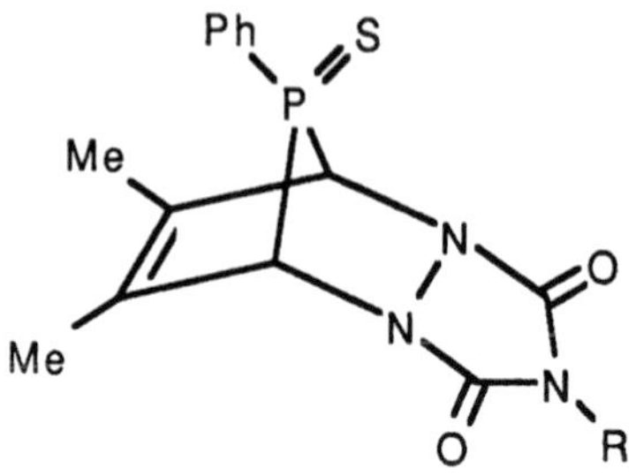

35

Connectivities	Formula	Structure	NMR data (δ_P [solv, temp] $^nJ_{PZ}$ Hz)	Ref.
$(CC'H)_2;C'_2$	$C_{12}H_{15}PS$	**6**; R^1 = Ph, R^2 = S, R^3 = R^5 = Me, R^4 = H	77.3(b)	24(a)
		6; R^1 = S, R^2 = Ph, R^3 = R^5 = Me, R^4 = H	66.7(b)	24(a)
		6; R^1 = Ph, R^2 = S, R^3 = H, R^4 = R^5 = Me	67.4(b)	24(a)
P-bound atoms/rings = $C_2C'S'/6$				
$(H_2O)_2;C'_2$	$C_{10}H_{15}O_2PSSi$	$H_2C^{\cdot}OSiMe_2OCH_2P^{\cdot}(S)Ph$	26(a, 24°)	105
$CHN,C'H_2;C'_2$	$C_{15}H_{22}NPS$	$Me_2NC^{\cdot}HCH_2CMe{=}CMeCH_2P^{\cdot}(S)Ph$	31	117
$(CHO)_2;C'_2$	$C_{16}H_{18}BO_2PS$	**18**; R^1 = S, R^2 = Ph	30.6(b)	59
		18; R^1 = Ph, R^2 = S	36.0(b)	59
$(CH_2)_2;C'_2$	$C_{11}H_{13}F_2PS$	$H_2C^{\cdot}CH_2CF_2CH_2CH_2P^{\cdot}(S)Ph$	27.8(c)	60
	$C_{12}H_{15}PS$	$H_2C^{\cdot}CH_2C({=}CH_2)CH_2CH_2P^{\cdot}(S)Ph$	33.3(c)	61
	$C_{14}H_{16}NOPS$	$H_2C^{\cdot}CH(CH_2CH_2CN)COCH_2CH_2P^{\cdot}(S)Ph$	33.1^C(c)	118
	$C_{15}H_{23}OPS$	$H_2C^{\cdot}CH_2C(tBu)(OH)CH_2CH_2P^{\cdot}(S)Ph$	31^C(c), 34.4^T(c)	119
$CH_2,C'H_2;C'_2$	$C_{13}H_{17}PS$	$H_2C^{\cdot}CMe{=}CMeCH_2CH_2P^{\cdot}(S)Ph$	29.8(r)	70
	$C_{15}H_{21}PS$	$H_2C^{\cdot}CH{=}C(tBu)CH_2CH_2P^{\cdot}(S)Ph$	31.9(c)	119
$(C'H_2)_2;C'_2$	$C_{18}H_{17}PS$	$PhC^{\cdot}{=}CHCH{=}CHCH_2P^{\cdot}(S)CH_2Ph$	24.2(c)	63
$(CC'H)_2;C'_2$	$C_{23}H_{21}OPS$	$PhC^{\cdot}HCH_2COCH_2CHPhP^{\cdot}(S)Ph$	47.9(c)	120
$(C_3)_2;C'_2$	$C_{15}H_{21}OPS$	$Me_2C^{\cdot}CH_2COCH_2CMe_2P^{\cdot}(S)Ph$	64.4(c)	120
P-bound atoms/rings = $C_2C'S'/5,5$				
$(C'HN)_2;C'_2$	$C_{15}H_{16}N_3O_2PS$	**35**; R = Me	79.7(c)	35(b)
	$C_{20}H_{18}N_3O_2PS$	**35**; R = Ph	80.0(c)	35(b)
	$C_{18}H_{21}O_4PS$	**33**; R^1 = Me, R^2 = Ph	93(c)	121
	$C_{20}H_{18}P_2S_2$	**9**; R^1 = R^8 = S, R^2 = R^7 = Ph, R^3 = R^4 = R^5 = R^6 = H	112.7(c) $^3J_{PP}$44.0	36
		9; R^1 = R^7 = S, R^2 = R^8 = Ph, R^3 = R^4 = R^5 = R^6 = H	106.5(c) $^3J_{PP}$34.1	36

Connectivities	Formula	Structure	NMR data (δ_P [solv, temp] $^nJ_{PZ}$ Hz)	Ref.
	$C_{22}H_{20}NO_2PS$	**32**, X = NPh, R^1 = Ph, R^2 = S, R^3 = R^4 = Me	107.1(c)	107
		32; X = NPh, R^1 = S, R^2 = Ph, R^3 = R^4 = Me	97(c)	121

P-bound atoms/rings = $C_2C'S'/6,6$

Connectivities	Formula	Structure	NMR data	Ref.
$CH_2,C'_3;C'_2$	$C_{25}H_{25}O_4PS$	**25**; R = S	58.7(r)	70
$(CC'H)_2;C'_2$	$C_{15}H_{15}O_2PS$	**14**; R^1 = S, R^2 = Ph, R^3-R^4 = 0, R^5-R^6 = 0	56.3(1)	39

Section L24: Monoalkynyl Phosphine Sulfides

P-bound atoms = $C_2C''S'$

Connectivities	Formula	Structure	NMR data	Ref.
$(C_2H)_2;C''$	$C_8H_{15}PS$	$(iPr)_2(HC{\equiv}C)PS$	53	13(a)
$(C_3)_2;C''$	$C_{10}H_{19}PS$	$(tBu)_2(HC{\equiv}C)PS$	67.0	13(a)

Section L25: Dialkenyl/Aryl Phosphine Sulfides

P-bound atoms = CC'_2S'

Connectivities	Formula	Structure	NMR data	Ref.
$HS_2;(C'_2)_2$	$C_{20}H_{25}PS_3$	$Ph_2[HC^{\cdot}SCH_2CH(tBu)CH_2S^{\cdot}]PS$	49.8^C(c) $^3J_{PC}$O	73
			45.7^T(c) $^3J_{PC}$8.8	73
$H_2N;(C'_2)_2$	$C_{17}H_{20}NPS$	$Ph_2[H_2C^{\cdot}(CH_2)_3N^{\cdot}]CH_2PS$	36.1(c)	32(b)
	$C_{17}H_{22}NPS$	$Ph_2(Et_2NCH_2)PS$	34.4(b)	32(b)
$H_3;(C'_2)_2$	$C_{13}H_3F_{10}PS$	$(C_6F_5)_2MePS$	13.1,[42a][52.6][32b]	32(b), 42(a)
	$C_{13}H_{13}PS$	Ph_2MePS	35.2(c)	110
	$C_{21}H_{19}PS$	Me(PhCH=CPh)PhPS	36.3^E(b)	122
	$C_{22}H_{21}PS$	Me(PhCMe=CPh)PhPS	32.8^E(t)	122
$C'HN;(C'_2)_2$	$C_{21}H_{22}NPS$	Ph_2(PhNHEtCH)PS	48.8(e)	32(b)
$C'HO;(C'_2)_2$	$C_{19}H_{15}Cl_2OPS$	$Ph_2[(3,4\text{-}Cl_2C_6H_3)CH(OH)]PS$	50.3(e + a)	32(b)
$CH_2;(C'H)_2$	$C_9H_{17}PS$	$(H_2C{=}CH)_3NeoPS$	30.9	98
$CH_2;(C'_2)_2$	$C_{14}H_5F_{10}PS$	$(C_6F_5)_2EtPS$	26.6,[42a][13.1][32b]	32(b), 42(a)
	$C_{15}H_{14}NPS$	$Ph_2(NCCH_2CH_2)PS$	34.7(a)	32(b)
	$C_{17}H_{21}PS$	Ph_2NeoPS	35.6	98
	$C_{18}H_{13}PS$	$Ph_2(C_6H_{23})PS$	42.0	32(b)
	$C_{24}H_{25}PS$	Bu(PhCH=CPh)PhPS	45.3^E(b), 37.9^Z(b)	122
	$C_{25}H_{27}PS$	Bu(PhCMe=CPh)PhPS	38.2^Z(b)	122
$C'H_2;(C'_2)_2$	$C_{14}H_{13}O_2PS$	$Ph_2(HO_2CCH_2)PS$	37.3(d)	32(b)
$C_2H;(C'_2)_2$	$C_{18}H_{21}PS$	$Ph_2(c\text{-}C_6H_{11})PS$	48.6(b)	32(b)
$CC'H;(C'_2)_2$	$C_{18}H_{21}OPS$	$Ph_2[(CH_2{=}CH)(Me_2(HO)C)CH]PS$	38.8(c)	123
	$C_{18}H_{21}O_2PS$	$Ph_2[(tBu)(HO_2C)CH]PS$	48.3(a)	32(b)

Section L26: Dialkenyl/Aryl Polyphosphine Sulfides

P-bound atoms = CC'_2S'

Connectivities	Formula	Structure	NMR data	Ref.
$HP^4_2;(C'_2)_2$	$C_{17}H_{23}OP_3S_2$	$Me_2P(O)[Me_2P(S)]CH\mathbf{P}(S)Ph_2$	39.8(c + r)	14(a)
	$C_{17}H_{23}P_3S_3$	$[Me_2P(S)]_2CH\mathbf{P}(S)Ph_2$	39.3(c + r)	14(a)
	$C_{37}H_{31}OP_3S_2$	$Ph_2P(O)CH[\mathbf{P}(S)Ph_2]_2$	42.5(c)	14(b)
	$C_{37}H_{31}P_3S_3$	$HC[P(S)Ph_2]_3$	41.9(c + r, 25°)	14

TABLE L (continued)
Compilation of ^{31}P NMR Data of Four Coordinate Compounds Containing a P=Ch Bond and Three Bonds from Phosphorus to Group IV Atoms

Connectivities	Formula	Structure	NMR data (δ_P [solv, temp] $^nJ_{PZ}$ Hz)	Ref.
$HP^3P^4;(C'_2)_2$	$C_{27}H_{27}P_3S_2$	$Ph_2P[Me_2P(S)]CH\mathbf{P}(S)Ph_2$	44.1(c + r)	14(a)
	$C_{37}H_{31}OP_3S$	$Ph_2P[Ph_2P(O)]CH\mathbf{P}(S)Ph_2$	43.8(c)	14(b)
	$C_{37}H_{31}P_3SSe$	$Ph_2P[Ph_2P(Se)]CH\mathbf{P}(S)Ph_2$	43.6(c + r)	14(b,c)
$HP^3_2;(C'_2)_2$	$C_{37}H_{31}P_3S$	$(Ph_2P)_2CH\mathbf{P}(S)Ph_2$	46.3(c + r)[a], 46.6(c)[b]	14(a,b)
$H_2P^4;(C'_2)_2$	$C_{15}H_{18}P_2SSe$	$Me_2P(Se)CH_2\mathbf{P}(S)Ph_2$	32.9(c + r)	77
	$C_{15}H_{18}P_2S_2$	$Me_2P(S)CH_2\mathbf{P}(S)Ph_2$	32.4 $^3J_{PP}$17	101
	$C_{19}H_{26}P_2S_2$	$(iPr)_2P(S)CH_2\mathbf{P}(S)Ph_2$	31.9 $^2J_{PP}$16	101
	$C_{21}H_{30}P_2S_2$	$(tBu)_2P(S)CH_2\mathbf{P}(S)Ph_2$	37.8 $^2J_{PP}$16	101
	$C_{25}H_{22}OP_2S$	$Ph_2P(O)CH_2\mathbf{P}(S)Ph_2$	35.6(c + r)	77(b)
	$C_{25}H_{22}P_2S_2$	$Ph_2P(S)CH_2P(S)Ph_2$	34.5(c)	77(b)
$H_2P^3;(C'_2)_2$	$C_{15}H_{18}P_2S$	$Me_2PCH_2\mathbf{P}(S)Ph_2$	39.0(r) $^2J_{PP}$56	101
	$C_{19}H_{26}P_2S$	$(iPr)_2PCH_2\mathbf{P}(S)Ph_2$	41.6(r) $^2J_{PP}$77	101
	$C_{21}H_{30}P_2S$	$(tBu)_2PCH_2\mathbf{P}(S)Ph_2$	42.9(r) $^2J_{PP}$87	101
	$C_{39}H_{36}P_4S_3$	$P[CH_2\mathbf{P}(S)Ph_2]_3$	38.5(c + r)	14(a)
$CHP^3;(C'_2)_2$	$C_{26}H_{24}P_2S$	$Ph_2PCHMe\mathbf{P}(S)Ph_2$	51.8(c + r)	14(a)
	$C_{30}H_{32}P_2S$	$Ph_2PCH(Neo)\mathbf{P}(S)Ph_2$	56.0(c + r)	14(a)
$CH_2;(C'_2)_2$	$C_{18}H_{24}P_2S_2$	$H(tBu)P(S)(CH_2)_2\mathbf{P}(S)Ph_2$	44.0(b) $^3J_{PP}$53.2	78
	$C_{26}H_{24}P_2S_2$	$Ph_2P(S)(CH_2)_2P(S)Ph_2$	44.1(c)	32(b)
	$C_{41}H_{39}P_3S_3$	$MeC[CH_2P(S)Ph_2]_3$	35.2(a)	114(b)
	$C_{42}H_{42}P_4S_4$	$P(S)[CH_2CH_2\mathbf{P}(S)Ph_2]_3$	41.0(r, 35°) $^3J_{PP}$55.4	16
	$C_{42}H_{42}P_4S_4$	$[Ph_2\mathbf{P}(S)(CH_2)_2P(S)PhCH_2-]_2$	40.8(r, 35°) $^3J_{PP}$58.5	16
$C_2P^4;(C'_2)_2$	$C_{27}H_{24}P_2S_2$	$(H_2C^{\cdot}CH_2C^{\cdot})[P(S)Ph_2]_2$	53.3(c, 30°)	79
$C_2H;(C'_2)_3$	$C_{32}H_{36}P_2S_2$	$Ph_2P(S)(CH_2)_2CH(C_5H_{11})\mathbf{P}(S)Ph_2$	47.7	32(b)

Section L27: Cyclic Dialkenyl/Aryl Phosphine Sulfides

<u>P-bound atoms/rings = CC'$_2$S'/5</u>

Connectivities	Formula	Structure	NMR data	Ref.
$H_3;(C'H)_2$	$C_7H_{11}PS$	$HC^{\cdot}$=CMeCMe=CHP$^{\cdot}$(S)Me	44.0(c)	34
$H_3;(C'_2)_2$	$C_{29}H_{23}PS$	$PhC^{\cdot}$=CPhCPh=CPhP$^{\cdot}$(S)Me	51.9(c)	116
$CH_2;(C'H)_2$	$C_8H_{13}OPS$	$HC^{\cdot}$=CMe CMe=CHP$^{\cdot}$(S)CH_2CH_2OH	45.5(c)	108
	$C_9H_{12}NPS$	$HC^{\cdot}$=CMe CMe=CHP$^{\cdot}$(S)CH_2CH_2CN	46.5(c)	108
	$C_{10}H_{16}BrPS$	$HC^{\cdot}$=CMeCMe=CHP$^{\cdot}$(S)$(CH_2)_4Br$	48.7(c)	107
	$C_{10}H_{17}PS$	$HC^{\cdot}$=CMeCMe=CHP$^{\cdot}$(S)Bu	52.2(c)	124
	$C_{11}H_{18}BrPS$	$HC^{\cdot}$=CMeCMe=CHP$^{\cdot}$(S)$(CH_2)_5Br$	49.3(c)	107
	$C_{12}H_{20}BrPS$	$HC^{\cdot}$=CMeCMe=CHP$^{\cdot}$(S)$(CH_2)_6Br$	49(c)	107
$CH_2;C'H,C'_2$	$C_{12}H_{19}OPS$	(COMe)$C^{\cdot}$=CMeCMe=CHP$^{\cdot}$(S)Bu	55.3(c)	124(a)
$C'H_2;(C'H)_2$	$C_9H_{13}OPS$	$HC^{\cdot}$=CMeCMe=CHP$^{\cdot}$(S)CH_2COMe	38.8(c)	108
	$C_{10}H_{15}O_2PS$	$HC^{\cdot}$=CMeCMe=CHP$^{\cdot}$(S)CH_2CO_2Et	41.2(c)	108
	$C_{14}H_{15}OPS$	$HC^{\cdot}$=CMeCMe=CHP$^{\cdot}$(S)CH_2COPh	40.7(c)	108
$C'H_2;C'H,C'_2$	$C_{12}H_{13}PS$	$HC^{\cdot}$=CMeC(=CH_2)$CH_2P^{\cdot}$(S)Ph	47.5(c)	35(a)
$C''H_2;(C'H)_2$	$C_8H_{10}NPS$	$HC^{\cdot}$=CMeCMe=CHP$^{\cdot}$(S)CH_2CN	40.3(c)	108
$C_2H;C'H,C'_2$	$C_{20}H_{18}P_2S_2$	**9**; $R^1 = R^8 = S$, $R^2 = R^7 = Ph$, $R^3 = R^4 = R^5 = R^6 = H$	64.1(c) $^3J_{PP}$44.0	36
		9; $R^1 = R^7 = S$, $R^2 = R^8 = Ph$, $R^3 = R^4 = R^5 = R^6 = H$	75.7(c) $^3J_{PP}$34.1	36

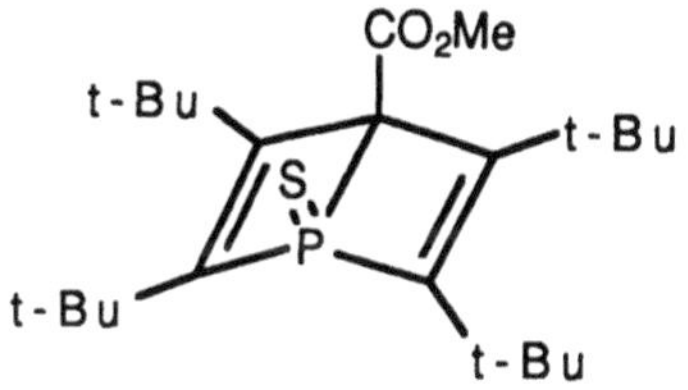

36

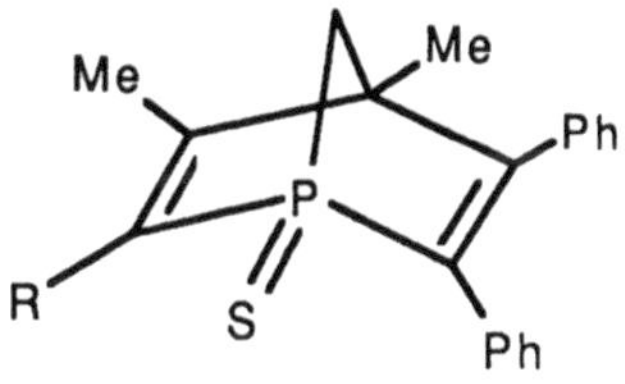

37

Connectivities	Formula	Structure	NMR data (δ_P [solv, temp] $^nJ_{PZ}$ Hz)	Ref.
	$C_{22}H_{22}P_2S_2$	9; $R^1 = R^8 = S$, $R^2 = R^7 = Ph$, $R^3 = R^5 = Me$, $R^4 = R^6 = H$	62.7(c) $^3J_{PP}$43.9	32(a)
	$C_{24}H_{26}P_2S_2$	9; $R^1 = R^8 = S$, $R^2 = R^7 = Ph$, $R^3 = R^4 = R^5 = R^6 = Me$	57.6(c) $^3J_{PP}$45	124(b)
$C_2H;CC',C'_2$	$C_{12}H_{15}PS$	MeC*=CHCH$_2$CHMeP*(S)Ph	68.6^C(b), 69.6^T(b)	104
$C_3;(C'H)_2$	$C_{10}H_{17}PS$	HC*=CMeCMe=CHP*(S)tBu	70.5(c)	35(a)
$C_3;C'H,C'_2$	$C_{12}H_{19}OPS$	(COMe)C*=CMeCMe=CHP*(S)tBu	73.8(c)	124(a)
	$C_{13}H_{21}O_2PS$	(CO$_2$Et)C*=CMeCMe=CHP*(S)tBu	73.7(c)	124(a)
	P-bound atoms/rings = CC'$_2$S'/6			
$C'H_2;(C'_2)_2$	$C_{17}H_{15}PS$	PhC*=CHCH=CHCH$_2$P*(S)Ph	21.6(c)	63
	$C_{17}H_{17}OPS$	(2-Furyl)C*=CH · CMe=CMeCH$_2$P*(S)Ph	21.3(c)	63
	$C_{19}H_{19}PS$	PhC*=CHCMe=CMeCH$_2$P*(S)Ph	26.7(c)	125
	P-bound atoms/rings = CC'$_2$S'/4,4			
$C'_3;(CC')_2$	$C_{23}H_{39}O_2PS$	**36**	39.3(c)	126
	P-bound atoms/rings = CC'$_2$S'/5,5			
$CH_2;C'H,C'_2$	$C_{20}H_{19}PS$	**37**; R = H	44.7(c)	127
$CH_2;(C'_2)_2$	$C_{26}H_{23}PS$	**37**;R = Ph	50.3(c)	128

Section L28: Trialkenyl/Aryl Phosphine Sulfides

Connectivities	Formula	Structure	NMR data	Ref.
	P-bound atoms = C'$_3$S'			
$NS',(C'_2)_2$	$C_{15}H_{16}NPS_2$	Ph$_2$[Me$_2$NC(S)]PS	48.1(k)	83, 129
	$C_{20}H_{18}NPS_2$	Ph$_2$[PhMeNC(S)]PS	48.6^Z(k), 47.5^E(k)	129
$N'S,(C'_2)_2$	$C_{15}H_{16}NPS_2$	Ph$_2$[MeN=C(SMe)]PS	43.0^Z(k), 30.3^E(k)	129
	$C_{20}H_{18}NPS_2$	Ph$_2$[PhN=C(SMe)]PS	42.5^Z(k), 24.2^E(k)	129
$SS',(C'_2)_2$	$C_{14}H_{13}PS_3$	Ph$_2$(MeS$_2$C)PS	49.4(r)	113
$C'H,(C'_2)_2$	$C_{14}H_{13}PS$	Ph$_2$(H$_2$C=CH)PS	37.5(c)	32(b)
	$C_{15}H_{15}PS$	Ph$_2$(MeCH=CH)PS	27.9^Z(c) $^3J_{PC}$11.0	86
			35.3^E(c) $^3J_{PC}$25.7	86
	$C_{16}H_{16}NPS$	Ph$_2$(MeN=CHCH=CH)PS	69.8^Z(b)	130
	$C_{18}H_{19}PS$	Ph$_2$(Me$_2$C=CHCH=CH)PS	34.3^E(c)	123
	$C_{21}H_{16}NPS$	PH$_2$(4-NCC$_6$H$_4$CH=CH)PS	35.1^E(b)	87
	$C_{22}H_{19}PS$	Ph$_2$(PhCH=CHCH=CH)PS	33.9^{E,E}(c)	123
	$C_{22}H_{22}NPS$	Ph$_2$(4-Me$_2$NC$_6$H$_4$CH=CH)PS	36.2^E(b)	87
$CC',(C'_2)_2$	$C_{17}H_{15}PS$	Ph$_2$(H$_2$C*CH=CHCH=C*)PS	34.0(c)	131

TABLE L (continued)
Compilation of ^{31}P NMR Data of Four Coordinate Compounds Containing a P=Ch Bond and Three Bonds from Phosphorus to Group IV Atoms

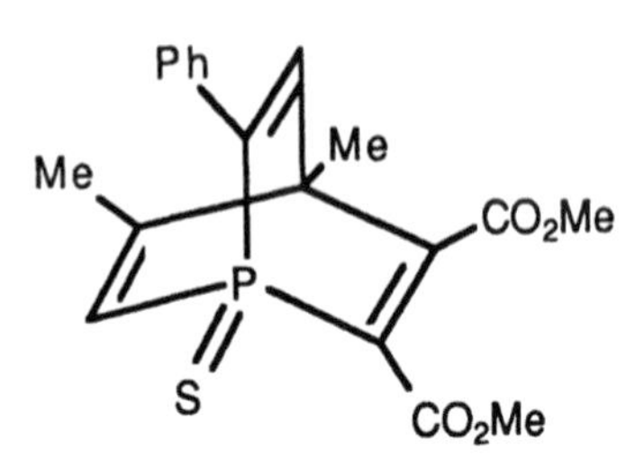

38

Connectivities	Formula	Structure	NMR data (δ_P [solv, temp] $^nJ_{PZ}$ Hz)	Ref.
$(C'_2)_3$	$C_{18}F_{15}PS$	$(C_6F_5)_3PS$	−8.6(c)[32b], −7.8[42]	32(b), 42
	$C_{18}H_5F_{10}PS$	$(C_6F_5)_2PhPS$	15.6	42(a)
	$C_{18}H_{10}F_5PS$	$Ph_2(C_6F_5)PS$	33.6,[42a][26.6][32b]	32(b), 42(a)
	$C_{18}H_{12}F_3PS$	$(4\text{-}FC_6H_4)_3PS$	37.1	32(b)
	$C_{18}H_{15}PS$	Ph_3PS	42.6(c),[32b] 43.2(c)[76]	32(b), 76
			43.0 (solid)	94
	$C_{19}H_{17}OPS$	$Ph_2(4\text{-}MeOC_6H_4)PS$	41.4(b)	32
	$C_{20}H_{19}O_2PS$	$(4\text{-}MeOC_6H_4)_2PhPS$	39.6(b)	32(a)
	$C_{21}H_{21}O_3PS$	$(4\text{-}MeOC_6H_4)_3PS$	35.7	32(b)

Section L29: Trialkenyl/Aryl Polyphosphine Sulfides

P-bound atoms = C'$_3$S'

Connectivities	Formula	Structure	NMR data	Ref.
$C''P^4,(C'_2)_2$	$C_{51}H_{40}P_4S_4$	$[Ph_2P(S)]_2C{=}C{=}C[P(S)Ph_2]_2$	45.6(c)	89
$C'H,(C'_2)_2$	$C_{26}H_{22}P_2S_2$	$Ph_2P(S)CH{=}CHP(S)Ph_2$	29.0^Z(c), 36.3^E(c)	86

Section L30: Cyclic Trialkenyl/Aryl Phosphine Sulfides

P-bound atoms/rings = C'$_3$S'/3

Connectivities	Formula	Structure	NMR data	Ref.
$(C'_2)_3$	$C_{20}H_{15}PS$	$PhC^*{=}CPhP^*(S)Ph$	−79(b)	122

P-bound atoms/rings = C'$_3$S'/5

Connectivities	Formula	Structure	NMR data	Ref.
$(C'H)_2,C'_2$	$C_{12}H_{13}PS$	$HC^*{=}CMeCMe{=}CHP^*(S)Ph$	46.1(c)[a], 45.5(c)[b]	124
$C'H,(C'_2)_2$	$C_{13}H_{13}O_2PS$	$(CO_2H)C^*{=}CMeCMe{=}CHP^*(S)Ph$	53.2(c)	124(a)
	$C_{14}H_{15}OPS$	$(COMe)C^*{=}CMeCMe{=}CHP^*(S)Ph$	48.0(c)	124(a)
	$C_{15}H_{17}O_2PS$	$(CO_2Et)C^*{=}CMeCMe{=}CHP^*(S)Ph$	48.8(c)	124(a)

P-bound atoms/rings = C'$_3$S'/6

Connectivities	Formula	Structure	NMR data	Ref.
$C'H,(C'_2)_2$	$C_{21}H_{21}OPS$	$PhC^*{=}CHCMe(COMe)CMe{=}CHP^*(S)Ph$	10.6(c)	63
	$C_{26}H_{25}PS$	$PhC^*{=}CHCMe(CH_2Ph)CMe{=}CHP^*(S)Ph$	9.7(c)	63

Connectivities	Formula	Structure	NMR data (δ_P [solv, temp] $^nJ_{PZ}$ Hz)	Ref.
	P-bound atoms/rings = C'$_3$S'/6,6			
C'H,(C'$_2$)$_2$	$C_{19}H_{19}O_4PS$	**38**	6.6(c)	132

Section L31: Trialkyl Phosphine Selenides

Connectivities	Formula	Structure	NMR data	Ref.
	P-bound atoms = C$_3$Se'			
(H$_3$)$_3$	C_3H_9PSe	Me_3PSe	8.0(r) $^1J_{PSe}-684$	114(a)
			8.0 (solid)	94
(CH$_2$)$_3$	$C_6H_{15}PSe$	Et_3PSe	45.8 $^1J_{PSe}$691	32(b)
	$C_9H_{21}PSe$	Pr_3PSe	72.6	32(b), 133
	$C_{12}H_{27}PSe$	Bu_3PSe	36.3(r) $^1J_{PSe}$689	134
(C$_2$H)$_3$	$C_{18}H_{33}PSe$	(c-C_6H_{11})$_3$PSe	58(r) $^1J_{PSe}$683	135
			7.0 (solid)	94
(C$_3$)$_3$	$C_{12}H_{27}PSe$	(tBu)$_3$PSe	91.7(r, 40°) $^1J_{PSe}$692	99
			92.5(b + t) $^1J_{PSe}$711.6	136
			[31]	13(b)

Section L32: Trialkyl Polyphosphine Selenides

Connectivities	Formula	Structure	NMR data	Ref.
	P-bound atoms = C$_3$Se'			
H$_2$P^4,(H$_3$)$_2$	$C_{15}H_{18}P_2SSe$	$Ph_2P(S)CH_2$**P**(Se)Me_2	14.6(c + r)	77
(CH$_2$)$_3$	$C_{42}H_{42}P_4Se_4$	[$Ph_2P(Se)(CH_2)_2$]$_3$**P**Se	39.8(r, 35°) $^1J_{PSe}$740	16

Section L33: Cyclic Trialkyl Phosphine Selenides

Connectivities	Formula	Structure	NMR data	Ref.
	P-bound atoms/rings = C$_3$Se'/6			
(H$_2$O)$_2$,H$_3$	$C_5H_{13}O_2PSeSi$	$H_2C^*OSiMe_2OCH_2P^*(Se)Me$	11(a, 24°) $^1J_{PSe}-734.1$	105
	P-bound atoms/rings = C$_3$Se'/8			
H$_3$,(CH$_2$)$_2$	$C_6H_{14}O_2P_2Se_2$	[$H_2C^*CH_2OP$(Se)(Me)OCH_2CH_2]**P***(Se)Me	21.3^C(c) $^1J_{PSe}$737.6	65
			21.6^T(b) $^1J_{PSe}$734.1	65

Section L34: Monoalkenyl/Aryl Phosphine Selenides

Connectivities	Formula	Structure	NMR data	Ref.
	P-bound atoms = C$_2$C'Se'			
(H$_3$)$_2$;C'$_2$	$C_8H_{11}PSe$	Me_2PhPSe	15.1(r) $^1J_{PSe}-710$	114(b)
H$_3$,CH$_2$;C'$_2$	$C_9H_{13}PSe$	MeEtPhPSe	29.8(b) $^1J_{PSe}$693	137
	$C_{10}H_{15}PSe$	MePrPhPSe	26.1(b) $^1J_{PSe}$704	137
H$_3$,C$_3$;C'$_2$	$C_{11}H_{17}PSe$	Me(tBu)PhPSe	47.4(b) $^1J_{PSe}$702	137
(CH$_2$)$_2$;C'$_2$	$C_{10}H_{15}PSe$	Et_2PhPSe	44.6(b) $^1J_{PSe}$705	137
	$C_{14}H_{23}PSe$	Bu_2PhPSe	37.6(r) $^1J_{PSe}$715	138

Section L35: Monoalkenyl/Aryl Polyphosphine Selenides

Connectivities	Formula	Structure	NMR data	Ref.
	P-bound atoms = C$_2$C'Se'			
(CH$_2$)$_2$;C'$_2$	$C_{34}H_{33}P_3Se_3$	[$Ph_2P(Se)CH_2CH_2$]$_2$**P**(Se)Ph	41.5(r) $^1J_{PSe}$734	135
	$C_{42}H_{42}P_4Se_4$	[$Ph_2P(Se)(CH_2)_2$**P**(Se)$PhCH_2-$]$_2$	39.0(r, 35°) $^1J_{PSe}$738	16

TABLE L (continued)
Compilation of ^{31}P NMR Data of Four Coordinate Compounds Containing a P=Ch Bond and Three Bonds from Phosphorus to Group IV Atoms

Connectivities	Formula	Structure	NMR data (δ_P [solv, temp] $^nJ_{PZ}$ Hz)	Ref.
Section L36: Cyclic Monoalkenyl/Aryl Phosphine Selenides				
P-bound atoms/rings = C₂C′Se′/6				
$(H_2O)_2$; C'_2	$C_{10}H_{15}O_2PSeSi$	$H_2C^{*}OSiMe_2OCH_2P^{*}(Se)Ph$	12(b, 50°) $^1J_{PSe}$751.5	105
$(CHO)_2$;C'_2	$C_{16}H_{18}BO_2PSe$	**18**; R¹ = Se, R² = Ph	21(b)	59
		18; R¹ = Ph, R² = Se	36(b)	59
P-bound atoms/rings = C₂C′Se′/5,5				
$(CC'H)_2$;C'_2	$C_{14}H_{11}O_3PSe$	**32**; X = O, R¹ = Ph, R² = Se, R³ = R⁴ = H	102.6(c) $^1J_{PSe}$757	35(b)
	$C_{20}H_{16}NO_2PSe$	**32**; X = NPh, R¹ = Ph, R² = Se, R³ = R⁴ = H	104.3(c) $^1J_{PSe}$754	35(b)
Section L37: Dialkenyl/Aryl Phosphine Selenides				
P-bound atoms = CC′₂Se′				
H_3;$(C'_2)_2$	$C_{13}H_{13}PSe$	Ph_2MePSe	23.8(r) $^1J_{PSe}-725$	114(b), 139
CH_2;$(C'_2)_2$	$C_{14}H_{15}PSe$	Ph_2EtPSe	35.8(r) $^1J_{PSe}$725	139
	$C_{16}H_{19}PSe$	Ph_2BuPSe	34.6(r) $^1J_{PSe}$730	138
	$C_{26}H_{24}AsPSe$	$Ph_2(Ph_2AsCH_2CH_2)PSe$	36.4(r) $^1J_{PSe}$735	134
C_2H;$(C'_2)_2$	$C_{15}H_{17}PSe$	$Ph_2(iPr)PSe$	51.5(r) $^1J_{PSe}$735	139
Section L38: Dialkenyl/Aryl Polyphosphine Selenides				
P-bound atoms = CC′₂Se′				
HP^4_2;$(C'_2)_2$	$C_{37}H_{31}OP_3SSe$	$HC[Ph_2P(O)][Ph_2P(S)]$**P**$(Se)Ph_2$	35.8(c)	14(b)
	$C_{37}H_{31}O_2P_3Se$	$HC[Ph_2P(O)]_2$**P**$(Se)Ph_2$	35.4(c)	14(b)
	$C_{37}H_{31}P_3S_2Se$	$HC[Ph_2P(S)]_2$**P**$(Se)Ph_2$	37.1(c)	14(b)
HP^3P^4;$(C'_2)_2$	$C_{37}H_{31}P_3SSe$	$HC(Ph_2P)[Ph_2P(S)]$**P**$(Se)Ph_2$	37.8(c + r)	14(c), 77
	$C_{37}H_{31}P_3Se_2$	$HC(Ph_2P)[$**P**$(Se)Ph_2]_2$	37.8(c + r)	14(c), 77
HP^3_2;$(C'_2)_2$	$C_{37}H_{31}P_3Se$	$HC(Ph_2P)_2$**P**$(Se)Ph_2$	42.1(c + r)	14(c), 77
H_2P^4;$(C'_2)_2$	$C_{15}H_{18}P_2SSe$	$Me_2P(S)CH_2$**P**$(Se)Ph_2$	21.6(c + r)	77
	$C_{25}H_{22}OP_2Se$	$Ph_2P(O)CH_2$**P**$(Se)Ph_2$	24.6(c + 4)	77
	$C_{25}H_{22}P_2SSe$	$Ph_2P(S)CH_2$**P**$(Se)Ph_2$	24.4(c + r)	77
	$C_{25}H_{22}P_2Se_2$	$Ph_2P(Se)CH_2P(Se)Ph_2$	24.2(r) $^1J_{PSe}$753	134
	$C_{26}H_{25}IP_2Se$	$[MePh_2P^+I^-]CH_2$**P**$(Se)Ph_2$	21.9(c + r)	77
H_2P^3;$(C'_2)_2$	$C_{25}H_{22}P_2Se$	Ph_2PCH_2**P**$(Se)Ph_2$	30.4(r) $^1J_{PSe}$733	77, 134
CHP^4;$(C'_2)_2$	$C_{38}H_{33}P_3Se_3$	$Ph_2P(Se)CH_2CH[$**P**$(Se)Ph_2]_2$	45.8 $^1J_{PSe}-776.6$	140
CHP^3;$(C'_2)_2$	$C_{38}H_{33}P_3Se$	$Ph_2PCH_2CH(Ph_2P)$**P**$(Se)Ph_2$	52.8	140
	$C_{38}H_{33}P_3Se_2$	$Ph_2P(Se)CH_2CH(Ph_2P)$**P**$(Se)Ph_2$	54.7	140
CH_2;$(C'_2)_2$	$C_{18}H_{24}P_2Se_2$	$H(tBu)P(Se)(CH_2)_2$**P**$(Se)Ph_2$	36.1(b) $^3J_{PP}$54.0	78
	$C_{26}H_{24}P_2Se$	$Ph_2P(CH_2)_2$**P**$(Se)Ph_2$	36.4(r) $^1J_{PSe}-749.3$	135, 141
	$C_{26}H_{24}P_2Se_2$	$Ph_2P(Se)(CH_2)_2P(Se)Ph_2$	35.6(r) $^1J_{PSe}$741	134, 142
	$C_{34}H_{33}P_3Se_3$	$PhP(Se)[CH_2CH_2$**P**$(Se)Ph_2]_2$	35.2(r) $^1J_{PSe}$737	135

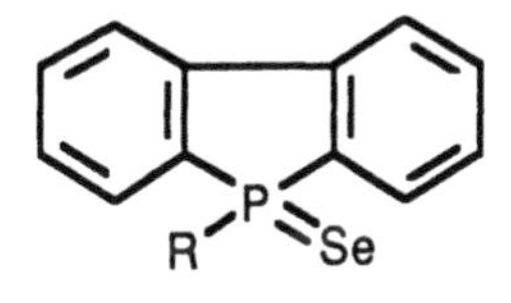

39

Connectivities	Formula	Structure	NMR data (δ_P [solv, temp] $^nJ_{PZ}$ Hz)	Ref.
	$C_{38}H_{33}P_3Se$	$(Ph_2P)_2CHCH_2\mathbf{P}(Se)Ph_2$	39.8	140
	$C_{38}H_{33}P_3Se_2$	$(Ph_2P)Ph_2P(Se)CHCH_2\mathbf{P}(Se)Ph_2$	41.8	140
	$C_{38}H_{33}P_3Se_3$	$[Ph_2P(Se)]_2CHCH_2\mathbf{P}(Se)Ph_2$	42.3 $^1J_{PSe}$ −737.3	140
	$C_{41}H_{39}P_3Se_3$	$MeC[CH_2P(Se)Ph_2]_3$	22.6(r) $^1J_{PSe}$722	135
	$C_{42}H_{42}P_4Se_4$	$[Ph_2\mathbf{P}(Se)(CH_2)_2]_3P(Se)$	33.5(r, 35°) $^1J_{PSe}$739	16

Section L39: Cyclic Dialkenyl/Aryl Phosphine Selenides

P-bound atoms/rings = CC′₂Se′/5

Connectivities	Formula	Structure	NMR data	Ref.
$H_3;(C'_2)_2$	$C_{13}H_{11}PSe$	**39**; R = Me	20.7(r) $^1J_{PSe}$735	139
$CH_2;(C'_2)_2$	$C_{14}H_{13}PSe$	**39**; R = Et	33.1(r)$^1J_{PSe}$762	139
$C_2H;(C'_2)_2$	$C_{15}H_{15}PSe$	**39**; R = iPr	43.2(r) $^1J_{PSe}$739	139

Section L40: Trialkenyl/Aryl Phosphine Selenides

P-bound atoms = C′₃Se′

Connectivities	Formula	Structure	NMR data	Ref.
$C'N,(C'_2)_2$	$C_{17}H_{16}NPSe$	Ph_2(NMe 2-Pyrryl)PSe	17.5(r) $^1J_{PSe}$728	143(a)
$(C'O)_3$	$C_{12}H_9O_3PSe$	(2-Furyl)$_3$PSe	−22.1(c, 21°) $^1J_{PSe}$793	143
$(C'O)_2,C'_2$	$C_{14}H_{11}O_2PSe$	(2-Furyl)$_2$PhPSe	−2.0(c, 21°)^b $^1J_{PSe}$774	143
$C'O(C'_2)_2$	$C_{16}H_{13}OPSe$	Ph_2(2-Furyl)PSe	16.9(C, 21°)^b $^1J_{PSe}$754	143
$(C'S)_3$	$C_{12}H_9S_3PSe$	(2-Thienyl)$_3$PSe	−4.2(c, 21°) $^1J_{PSe}$757	143
$(C'S)_2,C'_2$	$C_{14}H_{11}PS_2Se$	(2-Thienyl)$_2$PhPSe	9.3(c, 21°)^b $^1J_{PSe}$752	143
$C'S,(C'_2)_2$	$C_{16}H_{13}PSSe$	Ph_2(2-Thienyl)PSe	20.6(c, 21°)^b $^1J_{PSe}$743	139, 143
$(C'_2)_3$	$C_{18}H_{15}PSe$	Ph_3PSe	34.1(c)^a $^1J_{PSe}$ −735	114(a), 144
			33 (solid)	94
	$C_{19}H_{17}PSe$	$Ph_2(2\text{-}MeC_6H_4)PSe$	30.5(r, 40°) $^1J_{PSe}$730	99
	$C_{19}H_{17}OPSe$	$Ph_2(2\text{-}MeOC_6H_4)PSe$	32.8(c, 21°) $^1J_{PSe}$718	143(b)
	$C_{19}H_{17}OPSe$	$Ph_2(4\text{-}MeOC_6H_4)PSe$	34.7(c, 21°) $^1J_{PSe}$722	143(b)
	$C_{20}H_{20}NPSe$	$Ph_2(4\text{-}EtHNC_6H_4)PSe$	33.7(c) $^1J_{PSe}$709	47
	$C_{21}H_{12}F_9PSe$	$(3\text{-}TfC_6H_4)_3PSe$	35.5(c, 21°)^b $^1J_{PSe}$766	143
	$C_{21}H_{21}PSe$	$(4\text{-}MeC_6H_4)_3PSe$	34.4(c, 21°)^b $^1J_{PSe}$715	143(b), 145
	$C_{21}H_{21}O_3PSe$	$(2\text{-}MeOC_6H_4)_3PSe$	−20.2(c, 21°)$^1J_{PSe}$720	134(b)
	$C_{21}H_{21}O_3PSe$	$(4\text{-}MeOC_6H_4)_3PSe$	−39.7(c, 21°)^b $^1J_{PSe}$710	143(b), 145
	$C_{24}H_{27}O_6PSe$	$[2,6(MeO)_2C_6H_3]_3PSe$	−15.2(c, 21°) $^1J_{PSe}$717	143(b)

Section L41: Trialkenyl/Aaryl Polyphosphine Selenides

P-bound atoms = C′₃Se′

Connectivities	Formula	Structure	NMR data	Ref.
$C'P^4,(C'_2)_2$	$C_{26}H_{22}P_2Se_2$	$Ph_2P(Se)C({=}CH_2)P(Se)Ph_2$	38.4(c) $^1J_{PSe}$756	142
$C'P^3,(C'_2)_2$	$C_{26}H_{22}P_2Se$	$Ph_2PC({=}CH_2)P(Se)Ph_2$	39.2(r) $^1J_{PSe}$ −752.4	141
$C''P^4,(C'_2)_2$	$C_{51}H_{40}P_4Se_4$	$[Ph_2P(Se)]_2C{=}C{=}C[P(Se)Ph_2]_2$	37.2(c)	89
$C'H,(C'_2)_2$	$C_{26}H_{22}P_2Se$	$Ph_2PCH{=}CH\mathbf{P}(Se)Ph_2$	20.9^Z(r) $^1J_{PSe}$ −751.6	141
			29.6^E(r) $^1J_{PSe}$ −753	141

TABLE L (continued)
Compilation of ^{31}P NMR Data of Four Coordinate Compounds Containing a P=Ch Bond and Three Bonds from Phosphorus to Group IV Atoms

Connectivities	Formula	Structure	NMR data (δ_P [solv, temp] $^nJ_{PZ}$ Hz)	Ref.
	$C_{26}H_{22}P_2Se_2$	$Ph_2P(Se)CH{=}CHP(Se)Ph_2$	22.6^Z(c) $^1J_{PSe}$735	142
			28.6^E(c) $^1J_{PSe}$750	142

Section L42: Cyclic Trialkenyl/Aryl Phosphine Selenides

P-bound atoms/rings = C'$_3$Se'/5

Connectivities	Formula	Structure	NMR data	Ref.
C'S,(C'$_2$)$_2$	$C_{16}H_{11}PSSe$	**39**; R = 2-Thienyl	12.2(r) $^1J_{PSe}$745	139
(C'H)$_2$,C'$_2$	$C_{12}H_{13}PSe$	(HC*=CMeCMe=CH)P*(Se)Ph	28.5(c)[a], 31.8(c)[b]	35
(C'$_2$)$_3$	$C_{22}H_{17}PSe$	(PhC*=CHCH=CPh)P*(Se)Ph	35.7(r) $^1J_{PSe}$745	139

Section L43: Alkynl Dialkenyl/Aryl Phosphine Selenides

P-bound atoms = C'$_2$C"Se'

Connectivities	Formula	Structure	NMR data	Ref.
(C'$_2$)$_2$;C"	$C_{26}H_{20}P_2Se$	$Ph_2PC{\equiv}CP(Se)Ph_2$	5.7(r) $^1J_{PSe}$ −761	141
	$C_{26}H_{20}P_2Se_2$	$Ph_2P(Se)C{\equiv}CP(Se)Ph_2$	7.5(c) $^1J_{PSe}$774	142

Section L44: Trialkyl Phosphine Tellurides

P-bound atoms = C$_3$Te'

Connectivities	Formula	Structure	NMR data	Ref.
(CH$_2$)$_3$	$C_{12}H_{27}PTe$	(Bu)$_3$PTe	11 $^1J_{PTe}$1720	146
(C$_3$)$_3$	$C_{12}H_{27}PTe$	(tBu)$_3$PTe	77.5(b + t) $^1J_{PTe}$1600	136, 146

REFERENCES

1. **Burg, A. B.,** *Inorg. Chem.*, 24, 3342, 1985.
2. (a) **Mark, V., Dungan, C. H., Crutchfield, M. M., and Van Wazer, J. R.,** *Topics in Phosphorus Chemistry,* Vol. 5, Crutchfield, M. M., Dungan, C. H., Letcher, J. H., Mark, V., and Van Wazer, J. R., Eds., Interscience, New York, 1967, chap. 4; (b) **Hays, H. R. and Peterson, D. J.,** *Organic Phosphorus Compounds,* Vol. 3, Kosolapoff, G. M. and Maier, L., Eds., Wiley-Interscience, New York, 1972, chap. 6.
3. **Clardy, J. C., McEwen, G. K., Mosbo, J. A., and Verkade, J. G.,** *J. Am. Chem. Soc.*, 93, 6937, 1971.
4. **Lavielle, G. and Reisdorf, D.,** *C. R. Seances Acad. Sci. Ser. C,* 272, 100, 1971.
5. **Gorenstein, D. G. and Shah, D. O.,** *Phosphorus-31 NMR,* Part 4, Gorenstein, D. G., Ed., Academic Press, Orlando, FL, 1984.
6. **Quin, L. D., Gallagher, M. J., Cunkle, G. T., and Chesnut, D. B.,** *J. Am. Chem. Soc.*, 102, 3136, 1980.
7. **Quin, L. D. and Kisalus, J. C.,** *Phosphorus Sulfur,* 22, 35, 1985.
8. (a) **Quast, H. and Heuschmann, M.,** *Angew. Chem. Int. Ed. Engl.*, 17, 867, 1978; (b) **Quast, H. and Heuschmann, M.,** *Liebigs Ann. Chem.*,P. 977, 1981.
9. (a) **Rezvukhin, A. I., Dolenko, G. N., and Krupoder, S. A.,** *Magn. Reson. Chem., 23, 221, 1985; (b)* **Cristau, H. J., Labaudinière, L., and Christol, H.,** *Phosphorus Sulfur,* 15, 359, 1983.
10. **Dahl, O. and Jensen, F. K.,** *Acta Chem. Scand. Ser. B,* 29, 863, 1975.
11. **Satgé, J., Couret, C., and Escudié, J.,** *J. Organomet. Chem.*, 24, 633, 1970.
12. **Zimin, M. G., Burilov, A. R., Islamov, R. G., and Pudovik, A. N.,** *Zh. Obshch. Khim.*, 53, 46, 1983.
13. (a) **Rosenberg, D. and Drenth, W.,** *Tetrahedron,* 27, 3893, 1971; (b) **Letcher, J. H. and Van Wazer, J. R.,** *J. Chem. Phys.*, 44, 815, 1966.
14. (a) **Grim, S. O., Satek, L. C., and Mitchell, J. D.,** *Z. Naturforsch.*, 35B, 832, 1980; (b) **Grim, S. O., Sangokoya, S. A., Colquhoun, I. J., McFarlane, W., and Khanna, R. K.,** *Inorg. Chem.*, 25, 2699, 1986; (c) **Grim, S. O. and Walton, E. D.,** *Phosphorus Sulfur,* 9, 123, 1980.
15. **Benn, F. R., Briggs, J. C., and McAuliffe, C. A.,** *J. Chem. Soc. Dalton Trans.*, p. 293, 1984.
16. **Dean, P. A., Philipes, D. D., and Polensek, L.,** *Can J. Chem.*, 59, 50, 1981.
17. **Quin, L. D., Symmes, C., Jr., Middlemas, E. D., and Lawson, H. F.,** *J. Org. Chem.*, 45, 4688, 1980.
18. **MacDiarmid, J. E. and Quin, L. D.,** *J. Org. Chem.*, 46, 1451, 1981.
19. **Borleske, S. G. and Quin, L. D.,** *Phosphorus,* 5, 173, 1975.
20. **Middlemas, E. D. and Quin, L. D.,** *J. Org. Chem.*, 44, 2587, 1979.
21. **Quin, L. D. and MacDiarmid, J. E.,** *J. Org. Chem.*, 46, 461, 1981.
22. **Symmes, C. and Quin, L. D.,** *J. Org. Chem.*, 41, 238, 1976.
23. **Quin, L. D. and MacDiarmid, J. E.,** *J. Org. Chem.*, 47, 3248, 1982.
24. (a) **Hammond, P. J., Lloyd, J. R., and Hall, C. D.,** *Phosphorus Sulfur,* 10, 67, 1981; (b) **Quin, L. D. and Stocks, R. C.,** *Phosphorus Sulfur,* 3, 151, 1977.
25. **Quin, L. D. and Bernhardt, F. C.,** *Magn. Reson. Chem.*, 23, 929, 1985.
26. **Quin, L. D., Borleske, S. G., and Engel, J. F.,** *J. Org. Chem.*, 38, 1858, 1973.
27. **Quin, L. D. and Lee, S. O.,** *J. Org. Chem.*, 43, 1424, 1978.
28. **Martz, M. D. and Quin, L. D.,** *J. Org. Chem.*, 34, 3195, 1969.
29. (a) **Quin, L. D. and Middlemas, E.,** *J. Am. Chem. Soc.*, 99, 8370, 1977; (b) **Quin, L. D., Middlemas, E. D., Rao, N. S., Miller, R. W., and McPhail, A. T.,** *J. Am. Chem. Soc.*, 104, 1893, 1982.
30. **Quin, L. D., Middlemas, E. D., and Rao, N. S.,** *J. Org. Chem.*, 47, 905, 1982.
31. **Quin, L. D., Kisalus, J. C., and Mesch, K. A.,** *J. Org. Chem.*, 48, 4466, 1983.
32. **(a) Quin, L. D., Mesch, K. A., Bodalski, R., and Pietrusiewicz, K. M.,** *Org. Magn. Res.*, 20, 83, 1982; (b) **Maier, L.,** *Organic Phosphorus Compounds,* Vol. 4, Kosolapoff, G. M. and Maier, L., Eds., Wiley-Interscience, New York, 1972, chap. 7.
33. **Quin, L. D. and Keglevich, G.,** *J. Chem. Soc. Perkin Trans. 2,* p. 1029, 1986.
34. **Mathey, F. and Mankowski-Favelier, R.,** *Org. Magn. Reson.*, 4, 171, 1972.
35. **(a) Mathey, F.,** *Tetrahedron,* 28, 4171, 1972; (b) **Hussong, R., Heydt, H., and Regitz, M.,** *Phosphorus Sulfur,* 25, 201, 1985.
36. **Quin, L. D., Caster, K. C., Kisalus, J. C., and Mesch, K.,** *J. Am. Chem. Soc.*, 106, 7021, 1984.
37. **Krech, F. and Issleib, K.,** *Z. Anorg. Allg. Chem.*, 425, 209, 1976.
38. **Cowles, J. M., Ph. D. thesis,** Marquette University, Milwaukee, WI, August 1982.
39. **Zemlyanoi, V. N., Aleksandrov, A. M., and Kukhar', V. P.,** *Zh. Obshch. Khim.*, 55, 2667, 1985.
40. **Meeuwissen, H. J., Sirks, G., Bickelhaupt, F., Stam, C. H., and Spele, A. L.,** *Recl. Trav. Chim. Pays Bas J. R. Netherlands Chem. Soc.*, 101, 443, 1982.
41. **Mathey, F. and Muller, G.,** *Can. J. Chem.*, 56, 2486, 1978.

42. (a) **Fild, M.**, *Z. Anorg. Allg. Chem.*, 358, 257, 1968; (b) **Fild, M., Hollenberg, I., and Glemser, O.**, *Z. Naturforsch.*, 22B, 253, 1967.
43. (a) **Singh, G. and Reddy, G. S.**, *J. Org. Chem.*, 44, 1057, 1979; (b) **Heuschmann, M. and Quast, H.**, *Chem. Ber.*, 115, 3384, 1982.
44. **Santini, C. and Mathey, F.**, *Can. J. Chem.*, 61, 21, 1983.
45. **Baran Shortt, A., Durham, L. J., and Mosher, H. S.**, *J. Org. Chem.*, 48, 3125, 1983.
46. **Yagupol'skii, L. M., Pavlenko, N. V., Ignat'ev, N. V., Matyushecheva, G. I., and Semenii, V. Y.**, *Zh. Obshch. Khim.*, 54, 334, 1984.
47. **Kozlov, E. S., Tolmauev, A. A., Tcherneva, A. N., and Boldeskul, I. E.**, *Zh. Obshch. Khim.*, 55, 1262, 1985.
48. **Grim, S. O. and Satek, L. C.**, *J. Inorg. Nucl. Chem.*, 39, 499, 1977.
49. **Mathey, F. and Mercier, F.**, *J. Chem. Soc. Chem. Commun.*, p. 191, 1980.
50. **Weichmann, H., Mügge, C., Grand, A., and Robert, J. B.**, *J. Organomet. Chem.*, 238, 343, 1982.
51. **Quin, L. D., Henderson, C. C., Rao, N. S., and Kisalus, J. C.**, *Synthesis*, 12, 1074, 1984.
52. **Kolodyazhnyi, O. I.**, *Zh. Obshch. Khim.*, 52, 447, 1982.
53. **Von Weichmann, H. and Rensch, B.**, *Z. Anorg. Allg. Chem.*, 503, 106, 1983.
54. **Quin, L. D., Rao, N. S., Topping, R. J., and McPhail, A. T.**, *J. Am. Chem. Soc.*, 108, 4519, 1986.
55. **Denney, D. B., Denney, D. Z., Hall, C. D., and Marsi, K. L.**, *J. Am. Chem. Soc.*, 94, 245, 1972.
56. **Quin, L. D. and Caputo, J. A.**, *Chem. Commun.*, p. 1463, 1968.
57. **Quin, L. D. and Lawson, H. F.**, *Phosphorus Sulfur*, 15, 195, 1983.
58. **Quin, L. D., Belmont, S. E., Mathey, F., and Charrier, C.**, *J. Chem. Soc. Perkin Trans. 2*, p. 629, 1986.
59. **Arbuzov, B. A., Erastov, O. A., Nikonov, G. N., Romanova, I. P., Arshinova, R. P., and Ovodova, O. V.**, *Izv. Akad. Nauk. SSSR Ser. Khim.*, 11, 2535, 1983.
60. **Mathey, F. and Muller, G.**, *C. R. Seances Acad. Sci. C*, 277, 45, 1973.
61. **Mathey, F. and Muller, G.**, *C. R. Seances Acad. Sci. C*, 273, 305, 1971.
62. **MacDonell, G. D., Berlin, K. D., Baker, J. R., Ealick, S. E., Van der Helm, D., and Marsi, K. L.**, *J. Am. Chem. Soc.*, 100, 4535, 1978.
63. **Alcaraz, J. M., Deschamps, E., and Mathey, F.**, *Phosphorus Sulfur*, 19, 45, 1984.
64. **Mathey, F.**, *Tetrahedron Lett.*, 20, 1753, 1979.
65. (a) **Dutasta, J. P., Jurkschat, K., and Robert, J. B.**, *Tetrahedron Lett.*, 22, 2549, 1981; (b) **Piccinni-Leopardi, C., Reisse, J., Germain, G., Declercq, J. P., Van Meerssche, M., Jurkschat, K., Mügge, C., Zschunke, A., Dutasta, J. P., and Robert, J. B.**, *J. Chem. Soc. Perkin Trans. 2*, p. 85, 1986.
66. **Keglevich, G. and Quin, L. D.**, *Phosphorus Sulfur*, 26, 129, 1986.
67. **Matsumoto, K., Hashimoto, S., and Otani, S.**, *Heterocycles*, 22, 2713, 1984.
68. **Gamliel, A., Ph.D. thesis**, Marquette University, Milwaukee, WI, November 1983.
69. **Katz, T. J., Nicholson, C. R., and Reilly, C. A.**, *J. Am. Chem. Soc.*, 88, 3832, 1966.
70. **Deschamps, E. and Mathey, F.**, *J. Chem. Soc. Chem. Commun.*, p. 1214, 1984.
71. **Kukhar', V. P., Zemlyanoi, V. N., and Aleksandrov, A. M.**, *Zh. Obshch. Khim.*, 54, 220, 1984.
72. **Haque, M. U., Horne, W., Cremer, S. E., and Most, J. T.**, *J. Chem. Soc. Perkin Trans. 2*, 7, 1000, 1981.
73. **Mikolajczyk, M., Graczyk, P., and Balczewski, P.**, *Tetrahedron Lett.*, 28, 573, 1987.
74. **Mathey, F. and Marinetti, A.**, *Bull. Soc. Chim. Belg.*, 93, 533, 1984.
75. **Juaristi, E. and Löpez-Nunez, N. A.**, *J. Org. Chem.*, 51, 1357, 1986.
76. **Albright, T. A., Freeman, W. J., and Schweizer, E. E.**, *J. Org. Chem.*, 40, 3437, 1975.
77. (a) **Grim, S. O. and Walton, E. D.**, *Inorg. Chem.*, 19, 1982, 1980; (b) **Walton, E. D.**, Dissertation, University of Maryland, College Park, 1979.
78. **Von Weisheit, R., Stendel, R., Messbauer, B., Langer, C., and Walther, B.**, *Z. Anorg. Allg. Chem.*, 504, 147, 1983.
79. **Schmidbaur, H. and Pollok, T.**, *Helv. Chim. Acta*, 67, 2175, 1984.
80. **Fongers, K. S., Hogeveen, H., and Kingma, R. F.**, *Tetrahedron Lett.*, 24, 1423, 1983.
81. **Alcaraz, J. M., Brecque, A., and Mathey, F.**, *Tetrahedron Lett.*, 23, 1565, 1982.
82. **Charrier, C., Maigrot, N., Mathey, F., Robert, F., and Jeannin, Y.**, *Organometallics*, 5, 623, 1986.
83. **Bruns, A., Hiller, W., and Kunze, U.**, *Z. Naturforsch.*, 39B, 14, 1984.
84. **Lindner, E. and Hübner, D.**, *Chem. Ber.*, 116, 2574, 1983.
85. **Maier, L.**, *Phosphorus, 1*, 245, 1972.
86. **Duncan, M. and Gallagher, M. J.**, *Org. Magn. Reson.*, 15, 37, 1981.
87. **Gloyna, D., Köppel, H., Schleinitz, K. D., Berndt, K. G., and Radeglia, R.**, *J. Prakt. Chem.*, 325, 269, 1983.
88. **Stegmann, H. B., Kühne, H. M., Wax, G., and Scheffler, K.**, *Phosphorus Sulfur*, 13, 331, 1982.
89. **Schmidbaur, H. and Pollok, T.**, *Angew. Chem. Int. Ed. Engl.*, 25, 348, 1986.
90. **Marinetti, A., Fischer, J., and Mathey, F.**, *J. Am. Chem. Soc.*, 107, 5001, 1985.

91. **Rao, N. S. and Quin, L. D.,** *J. Org. Chem.*, 49, 3157, 1984.
92. **Lequan, R. M., Pouet, M. J., and Simonnin, M. P.,** *Org. Magn. Reson.*, 7, 392, 1975.
93. **McFarlane, W.,** *Proc. R. Soc. London Ser. A*, 306, 185, 1968.
94. **Dutasta, J. P., Robert, J. B., and Wiesenfeld, L.,** *A.C.S. Symp. Ser. Phosphorus Chem.*, 171, 581, 1981.
95. **Quin, L. D., Gordon, M. D., and Lee, S. H.,** *Org. Magn. Res.*, 6, 503, 1974.
96. **Littlefield, L. B. and Quin, L. D.,** *Org. Magn. Res.*, 12, 199, 1979.
97. **Gordon, M. D. and Quin, L. D.,** *J. Am. Chem. Soc.*, 98, 15, 1976.
98. **King, R. B., Cloyde, J. C., Jr., and Reimann, R. H.,** *J. Org. Chem.*, 41, 972, 1976.
99. **Dean, P. A. W. and Polensek, L.,** *Can. J. Chem.*, 58, 1627, 1980.
100. **Karsch, H. H.,** *Chem. Ber.*, 115, 818, 1982.
101. **Grim, S. O., Smith, P. H., Colquhoun, I. J., and McFarlane, W.,** *Inorg. Chem.*, 19, 3195, 1980.
102. **King, R. B. and Cloyd, J. C.,** *J. Am. Chem. Soc.*, 97, 53, 1975.
103. **Symes, C., Jr. and Quin, L. D.,** *J. Org. Chem.*, 41, 1548, 1976.
104. **Beer, P. D., Hammond, P. J., and Hell, C. D.,** *Phosphorus Sulfur*, 10, 185, 1981.
105. **Patsanovskii, I. I., Ishmaeva, E. A., Strelkova, E. N., Il'yasov, A. V., Zyablikova, T. A., Ismaev, I. E., Kudyakov, N. M., Voronkov, M. G., and Pudovik, A. N.,** *Zh. Obshch. Khim.*, 54, 1738, 1984.
106. **Featherman, S. I., Lee, S. O., and Quin, L. D.,** *J. Org. Chem.*, 39, 2899, 1974.
107. **Holand, S. and Mathey, F.,** *J. Org. Chem.*, 46, 4386, 1981.
108. **Muller, G., Bonnard, H., and Mathey, F.,** *Phosphorus Sulfur*, 10, 175, 1981.
109. **Kühne, U., Krech, F., and Issleib, K.,** *Phosphorus Sulfur*, 13, 153, 1982.
110. **Peterson, D. J. and Hays, H. R.,** *J. Org. Chem.*, 30, 1939, 1965.
111. **Mathey, F. and Mercier, F.,** *J. Organometal. Chem.*, 177, 255, 1979.
112. **Mathey, F. and Mercier, F.,** *Tetrahedron Lett.*, 33, 3081, 1979.
113. **Carr, St. W. and Colton, R.,** *Inorg. Chem.*, 23, 720, 1984.
114. (a) **Dean, P. A. W. and Hughes, M. K.,** *Can. J. Chem.*, 58, 180, 1980; (b) **McFarlane, W. and Rycroft, D. S.,** *J. Chem. Soc. Dalton Trans.*, P. 2162, 1973; (c) **McFarlane, W. and Rycroft, D. S.,** *J. Chem. Soc. Chem. Commun.*, p. 902, 1972.
115. **Jain, R. S., Lawson, H. F., and Quin, L. D.,** *J. Org. Chem.*, 43, 108, 1978.
116. **Charrier, C., Bonnard, H., de Lauzou, G., and Mathey, F.,** *J. Am. Chem. Soc.*, 105, 6871, 1983.
117. **Meriem, A., Majoral, J. P., Revel, M., and Navech, J.,** *Tetrahedron Lett.*, 24, 1975, 1983.
118. **Rampal, J. B., Berlin, K. D., and Satayamurthy, N.,** *Phosphorus Sulfur*, 13, 179, 1982.
119. **Mathey, F. and Santini, C.,** *Can. J. Chem.*, 57, 723, 1979.
120. **Rampal, J. B., Mcdonell, G. D., Edasery, J. P., Berlin, K. D., Rahman, A., Van der Helm, D., and Pietrusiewicz, K. M.,** *J. Org. Chem.*, 46, 1156, 1981.
121. **Mathey, F. and Mercier, F.,** *Tetrahedron Lett.*, 22, 319, 1981.
122. **Marinetti, A. and Mathey, F.,** *J. Am. Chem. Soc.*, 107, 4700, 1985.
123. **Mathey, F., Mercier, F., and Santini, C.,** *Inorg. Chem.*, 19, 1813, 1980.
124. **(a) Mathey, F.,** *Tetrahedron*, 32, 2395, 1976; (b) **Santini, C., Fisher, J., Mathey, F., and Mitschler, A.,** *J. Am. Chem. Soc.*, 102, 5809, 1980.
125. **Mathey, F.,** *Tetrahedron Lett.*, 2, 133, 1978.
126. **Blatter, K., Rösch, W., Vogelbacher, U., Fink, J., and Regitz, M.,** *Angew. Chem.*, 99, 67, 1987.
127. **De Lauzon, G., Charrier, C., Bonnard, H., Mathey, F., Fisher, J., and Mitschler, A.,** *J. Chem. Soc. Chem. Commun.*, p. 1272, 1982.
128. **Mathey, F., Mercier, F., and Charrier, C.,** *J. Am. Chem. Soc.*, 103, 4595, 1981.
129. **Kunze, U., Bruns, A., Hiller, W., and Mohyla, J.,** *Chem. Ber.*, 118, 227, 1985.
130. **Angelov, C. M. and Dahl, O.,** *Tetrahedron Lett.*, 24, 1643, 1983.
131. **Mathey, F. and Lampin, J. P.,** *J. Organometal. Chem.*, 128, 297, 1977.
132. **Alcaraz, J. F. and Mathey, F.,** *J. Chem. Soc. Chem. Commun.*, p. 508, 1984.
133. **Jones, R. A. Y. and Katritzky, A. R.,** *Angew. Chem.*, 74, 60, 1962.
134. **Carr, S. W. and Colton, R.,** *Aust. J. Chem.*, 34, 35, 1981.
135. **Dean, P. A. W.,** *Can. J. Chem.*, 57, 754, 1979.
136. **Du Mont, W. W.,** *Z. Naturforsch.*, 40B, 1453, 1985.
137. **Stec, W. J., Okruszek, A., Uznanski, B., and Michalski, J.,** *Phosphorus*, 2, 97, 1972.
138. **Grim, S. O., Walton, E. D., and Satek, L. C.,** *Can. J. Chem.*, 58, 1476, 1980.
139. **Allen, D. W. and Taylor, B. F.,** *J. Chem. Res. (S)*, p. 220, 1981.
140. **Colquhoun, I. J. and McFarlane, W.,** *J. Chem. Soc. Chem. Commun.*, p. 484, 1982.
141. **Colquhoun, I. J. and McFarlane, W.,** *J. Chem. Soc. Dalton Trans.*, p. 1915, 1982.
142. **Colquhoun, I. J., Christina, H., McFarlane, E., McFarlane, W., Nash, J. A., Keat, R., Rycroft, D. S., and Thompson, D. G.,** *Org. Magn. Res.*, 12, 473, 1979.

143. (a) **Allen, D. W. and Taylor, B. F.,** *J. Chem. Soc. Dalton Trans.*, p. 51, 1982; (b) **Allen, D. W. and Nowell, I. W.,** *J. Chem. Soc. Dalton Trans.*, p. 2505, 1985.
144. **Schmidpeter, A. and Brecht, H.,** *Z. Naturforsch.*, 24B, 179, 1969.
145. **Pinnell, R. P., Megerle, C. A., Mannatt, S. L., and Kroon, P. A.,** *J. Am. Chem. Soc.*, 95, 977, 1973.
146. (a) **Nuretdinov, I. A. and Loginova, E. I.,** *Izv. Akad. Nauk SSSR Ser. Khim.*, p. 2827, 1973; (b) **Du Mont, W. W. and Kroth, H. J.,** *Z. Naturforsch.*, 36B, 332, 1981.
147. **Quin, L. D., Szewczyk, J., Linehan, K., and Harris, D. L.,** *Magn. Reson. Chem.*, 25, 271, 1987.

Chapter 14

^{31}P NMR DATA OF FOUR COORDINATE PHOSPHORUS COMPOUNDS CONTAINING A FORMAL MULTIPLE PHOSPHORUS BOND TO A GROUP V ATOM

Compiled and presented by
John C. Tebby and S. S. Krishnamurthy

NMR data were also donated by
H.W. Roesky, Universität Göttingen, FRG
S.G. Kleemann, Fachhochschule München, FRG
E. Fluck, Gmelin-Institut, Frankfurt/Main, FRG

The NMR data of phosphazenes and phosphoniumphosphides are presented in Table M. It is divided into three sections. Section M1 contains data of all phosphazenes (phosphinimines) acyclic and cyclic, which are not cyclopolyphosphazenes. The latter are presented in Section M2. Discussions of the trends in the NMR parameters of the phosphazenes with variation of structure precede Sections M1 and M2. Section M3 contains the NMR data of the phosphoniumphosphides (Y_3P=PZ).

TRENDS IN ^{31}P CHEMICAL SHIFTS FOR PHOSPHAZENES (PHOSPHINIMINES)

The chemical shifts of the majority of phosphazenes fall within the range +50 to −50 ppm. This is, in general, slightly upfield of phosphonium ylides and phosphoryl compounds, most of which resonate in the region +100 to −25 ppm. Within the phosphazene class of compounds (Y_3P=NZ) the three singly bound substituents, Y, have a very pronounced influence on the chemical shift. Thus, when the substituent, Z, which resides on the anionic nitrogen, remains constant as phenyl, the range is nevertheless +63 to −50, and when Z is trimethylsilyl, the range is almost the same as that for the whole series, i.e., +94 to −85 ppm. The wider range for the latter is mainly due to a wider range of P-bound atoms, Y, and the chemical shifts are similar when the P-bound atoms are the same. Compounds with three bromine substituents, $Y_3 = Br_3$, δP ~ −100, resonate at lowest frequency (highest field), while oxygen, chlorine, and hydrogen P-substituents have only modest shielding effects, their mean chemical shifts being ca. −10 ppm. Care must be exercised when comparing the extent of ranges for different series, because for any particular series, the more compounds that are known the more likely compounds at the extremities of the range are to be included. Thus examples of the triaza compounds are abundant and the greater variety of subtituents must be at least partially responsible for the wide chemical shift range observed. One of the P-bound atoms which cause the chemical shift to move to a higher frequency is carbon. Within the carbon series (Y = organyl) bulky organic P-substituents, e.g., Y = t-butyl, produce resonances at the highest frequencies, e.g., tBu_3P=NH δP 56 cf. Me_3P=NH δP 13 ppm, with the triphenyl compound having an intermediate value ie δP 21 ppm. In contrast bulky groups on the anionic nitrogen appear to have a shielding influence, i.e., the opposite to their effect when bound to phosphorus. As is usual for four coordinate compounds, phosphazenes which have phosphorus or sulfur bound to phosphorus also resonate at high frequency. However, the largest shifts to high frequency are observed for the iododiorganyl (i.e., P-bound atoms = C_2IN') and silyldiorganyl (P-bound atoms = $C_2N'Si$) compounds. In contrast a diiodoorganylphosphazene is reported to resonate at very low frequency (−228).

The influence of the substituents on the anionic nitrogen atom was evaluated using the trichlorophosphazenes ($Cl_3P{=}NZ$). Bulky alkyl substituents on nitrogen shielded the phosphorus to the greatest extent δP −88.4 and −98 when Z = tBu and Et_3C, respectively. Fluorinated alkyl and fluoroaryl groups also have a marked shielding effect and signals between −20 to −45 are observed. The N-aryl compounds appear in the central part of the trichlorophosphazene region at ca. −10 ppm. Strong delocalization of the anion by carbonyl groups causes marked deshielding and signals between +10 (Z = CO_2Et) and + 25 (Z = $COCCl_3$) ppm are observed. The trichlorophosphazenes appearing at highest frequency (lowest field), δP 40 to 23 ppm, were the congugated diphosphazenes (Z = $CX{=}NPX_3$).

Increasing the electron availability at the anionic nitrogen atom of N-arylphosphazenes has a shielding or deshielding influence depending on the nature of the phosphorus substituents.[134] For most phosphorus substituents (e.g., alkyl, aryl, heteroaryl, chloro) shielding occurs. The magnitude is most pronounced when the phosphorus atom bears chloro groups. However, increasing the electron availability causes marked deshielding for the methyl(bisdimethylamino)phosphazenes and deshielding to a lesser extent for the methyldiphenoxyphosphazenes. The chemical shifts of these compounds as a group are also furthest downfield. The study involved ten series of compounds and the electron availability was controlled through the aryl ring by varying *para* and *meta* substituents (OMe, Me, H, F, CO_2Me, COMe, CN, NO_2). The Hammett substituent constants σ were used to estimate the extent of electron release through the aryl ring.

A summary of the trends in chemical shifts for the phosphazenes (excluding cyclopolyphosphazenes), as they vary according to the nature of the P-bound atoms, is given in the form of a bar chart, in Figure 1.5f of Chapter 1.

TRENDS IN ^{31}P CHEMICAL SHIFTS AND P-P COUPLING CONSTANTS FOR CYCLOPOLYPHOSPHAZENES

Cyclopolyphosphazenes are characterised by the presence of ($N{=}PX_2$) repeating units and contain four coordinated phosphorus nuclei "multiply bonded" to two nitrogen atoms. They occur in various ring sizes and shapes and are some of the best examples of multispin systems studied by NMR techniques.[18,128]

The ^{31}P chemical shifts of most cyclopolyphosphazenes range from −78 δ for $N_5P_5Br_{10}$ to +59 δ for the $P(SEt)_2$ grouping in *gem*-$N_3P_3F_4(SEt)_2$. The exceptions to this general rule are provided by cyclophosphazenes that contain direct covalent bonds between the skeletal phosphorus and a transition metal. These metallocyclophosphazenes are characterized by pronounced downfield shifts (100 to 270δ). Figure 1 depicts the ^{31}P chemical shift ranges for cyclotriphosphazenes containing various substituent groups; the inset shows the downfield shifts for metallocyclophosphazenes.

The ^{31}P chemical shifts of cyclophosphazenes move upfield with increasing ring size (Figure 2). There is a considerable increase in shielding on passing from the trimer to the tetramer. The chemical shift reaches a limiting value for the pentamer and higher oligomers.

A summary of the trends in chemical shifts for the cyclopolyphosphazenes, as they vary according to the nature of the P-bound atoms, is also given in the form of a bar chart in Figure 1.5g of Chapter 1.

The two-bond P-P coupling constants increase with increasing electronegativity of the substituents. Highest values are found for $N_3P_3F_6$ (160 to 180 Hz) and $N_4P_4F_8$ (155 Hz).[128] The values for three series of compounds are shown in Figure 3. Four-bond P-P couplings for cyclotetraphosphazenes are small; they are negative relative to $^2J_{PP}$. The values range from −0.7 to −2.2 Hz.[18]

ACKNOWLEDGMENT

Thanks are due to Mr. A. Chandrasekaran for checking the data and help in the preparation of the manuscript.

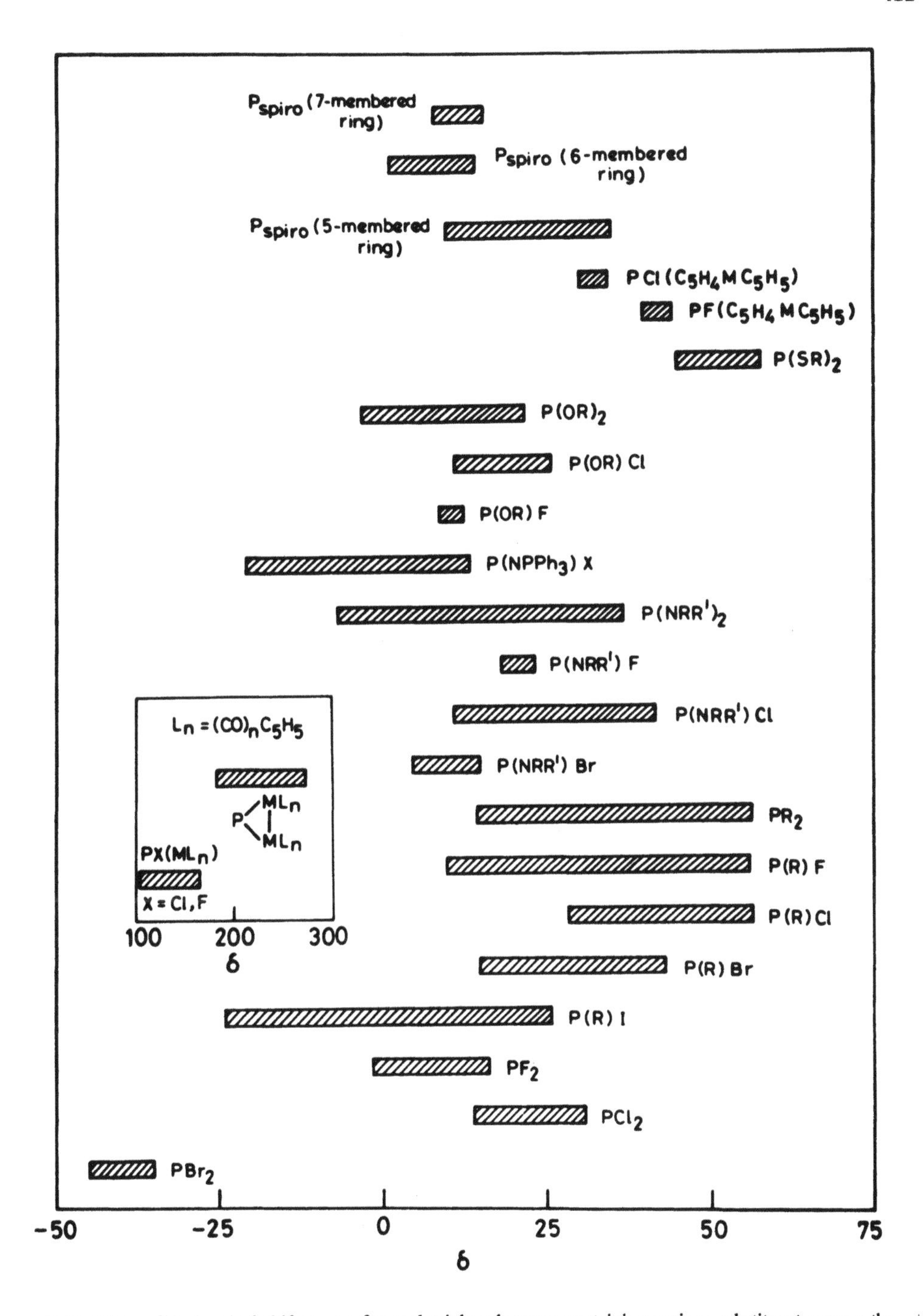

FIGURE 1. [31]P chemical shift ranges for cyclotriphosphazenes containing various substituent groups; the inset shows values for cyclotriphosphazenes containing direct phosphorus to transition metal covalent bonds.

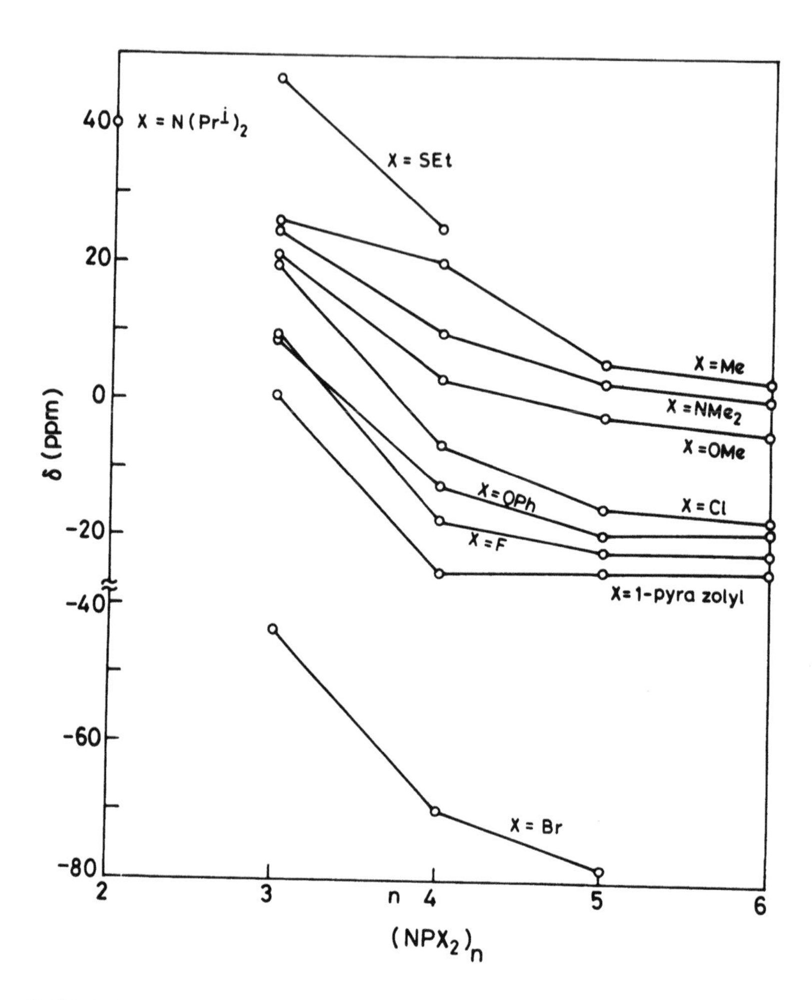

FIGURE 2. The variation of ^{31}P chemical shifts of cyclophosphazenes with ring size; the number (n) of repeating NPR_2 units is plotted on the abcissa. The ^{31}P chemical shift value for only one cyclodiphosphazene ($n = 2$) is available.

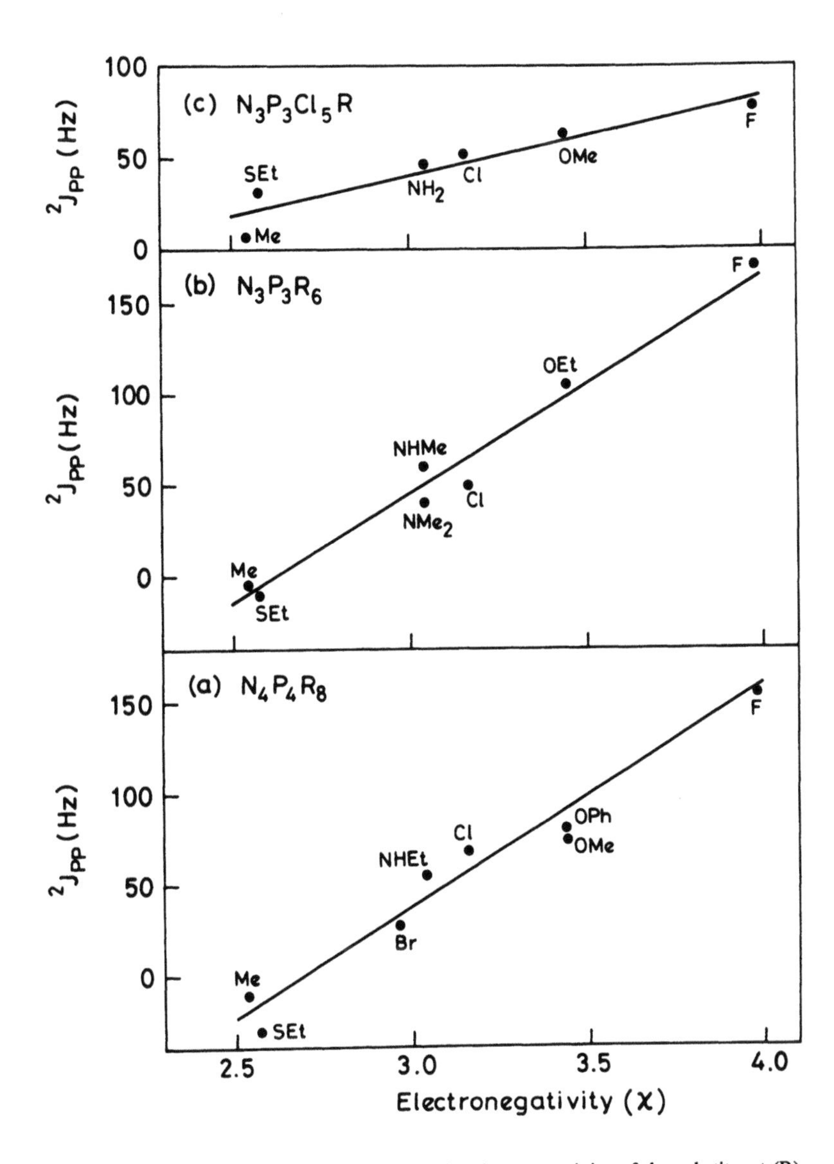

FIGURE 3. The variation of $^2J_{PP}$(Hz) with the electronegativity of the substituent (R) or the atom of the substituent (R) group attached to phosphorus for three classes of cyclophosphazenes; (a) $N_4P_4R_8$, (b) $N_3P_3R_6$, (c) $N_3P_3Cl_5R$. In (a) and (b), the values for the methyl (Me) and thioethyl (SEt) derivatives are assumed to be negative to obtain a smooth linear relationship. The signs of these coupling constants have not been determined experimentally (see References 1 and 2* for sources of data and a detailed discussion).

TABLE M

Compilation of ^{31}P NMR Data of Four Coordinate Phosphorus Compounds Containing a Formal Multiple Bond from Phosphorus to a Group V Atom

Section M1: Phosphazenes (Excluding Cyclopolyphosphazenes)

P-bound atoms/rings	Connectivities	Structure	NMR data (δ_P[solv,temp] J_{pz} Hz	Ref.
BrN'O$_2$	Si;(C)$_2$	(TfCH$_2$O)$_2$BrP=NTms	−34.90(r)	1
BrNN'O	Si$_2$;Si;C	(Tms$_2$N)(tBuO)BrP=NTms	−36.9(n)	2
	Si$_2$;Si;C	(Tms$_2$N)(AdO)BrP=NTms	−38.2(n)	2
	Si$_2$;C;C	(Tms$_2$N)(tBuO)BrP=N-tBu	−37.5(n)	2
	Si$_2$;C;C	(Tms$_2$N)(AdO)BrP=NAd	−39.3(n)	2
	CSi;C;C	(tBuTmsN)(tBuO)BrP=N-tBu	−63.0(n)	2
BrN$_2$N'	(Si$_2$)$_2$;Si	(Tms$_2$N)$_2$BrP=NTms	−23.6(b)	3
	(C$_2$)$_2$;C	(Et$_2$N)$_2$BrP=NCH$_2$CH$_2$Br	13.0(n)	4
	(C$_2$)$_2$;C	(Pr$_2$N)$_2$BrP=NCH$_2$CH$_2$Br	13.9(n)	4
	(C$_2$)$_2$;C	(Bu$_2$N)$_2$BrP=NCH$_2$CH$_2$Br	14.3(n)	4
	(C$_2$)$_2$;C'	(Me$_2$N)$_2$BrP=NPh	5.0(n)	5
BrN$_2$N'/4	(PSi)$_2$;Si	**1**; R = Y = Tms, X = Br	(+?)70.2^c(he), 71.0^T(he)	6
Br$_2$NN'	Si$_2$;Si	(Tms$_2$N)Br$_2$P=NTms	(+?)85.0(he)	6
	HSi;S^4	(TmsNH)P(Br$_2$)=NSO$_2$Tf	−40.6(b)	7
	HSi;S^4	(TmsNH)P(Br$_2$)=NSO$_2$Me	−35.4	7
	HSi;S^4	(TmsNH)Br$_2$P=NSO$_2$Tol	−51.1(m)	7
	C$_2$;Si	[Me$_2$C*(CH$_2$)$_3$CMe$_2$N*]Br$_2$P=NTms	−102.3(b)	8
Br$_3$N'	S^4	Br$_3$P=NSO$_2$Tf	−85.1(g)	7
	S^4	Br$_3$P=NSO$_2$Me	−105.1(b)	7
	S^4	Br$_3$P=NSO$_2$Ph'Cl−4	−99.5(m)	7
	S^4	Br$_3$P=NSO$_2$Ph	−101.4(m)	7
	S^4	Br$_2$P=NTos	−108.9(g)	7
ClFNO	Si;C	ClF(EtO)P=NSiCl(tBu)$_2$	−20.8(c)	8b
	Si;C	ClF(PhO)P=NSiCl(tBu)$_2$	−27.4(c)	8b
ClFNS	P^4;C	(MeS)ClFP=NP(O)F$_2$	32.5 J_{PF}1175 J_{PP}27	9
	P^4;C	(MeS)ClFP=NP(O)Cl$_2$	38.8 J_{PF}1153J_{PP}74	9
	P^4;C	(EtS)ClFP=NP(O)Cl$_2$	33.1 J_{PF}1176 J_{PP}24	9
ClFNN'	HSi;P^4	(TmsNH)ClFP=NP(S)Cl$_2$	5.1 J_{PF}1076 J_{PP}29	10
	CSi;Si	(Et2N)ClFP=NSiCl(tBu)$_2$	−9.7(c)	8b
	CH;Si	(EtHN)ClFP=NSiCl(tBu)$_2$	−10.1(c)	8b
	CH;Si	(tBuHN)ClFP=NSiCl(tBu)$_2$	−13.6(c)	8b
	C$_2$;Si	(EtTmsN)ClFP=NSiCl(tBu)$_2$	−6.4(c)	8b
ClFN'$_2$	(P^4)$_2$	Cl$_3$P=NPClF=NP(S)Cl$_2$	−20.5 J_{PF}1006 J_{PP}38	10
ClF$_2$N'	P^4	ClF$_2$P=NPF$_2$=NP(S)F$_2$	−13.2 J_{PF}1095 J_{PP}148	10
ClN'O$_2$	P$_4$;(C)$_2$	Cl(EtO)$_2$P=NP(O)Cl$_2$	−1 J_{PP}48	10b
ClN'O$_2$/5	S^4;(C')$_2$	ClCatP=NSO$_2$Ph	9.7	11
	C;(C')$_2$	ClCatP=NCHClPh	18.4	11b
	C';(C')$_2$	ClCatP=N(Pyr-4-Cl)	−2.7	12
	C';(C')$_2$	ClCatP=NAr	−3 to 2(ae)	12
	C';(C')$_2$	ClCatP=NPh'NO$_2$−2	14.6	12
ClNN'O	Si$_2$;Si;C	(Tms$_2$N)(tBuO)ClP=NTms	−16.3(n)	2
	Si$_2$;Si;C	(Tms$_2$N)(AdO)ClP=NTms	−18.0(n)	2
	Si$_2$;C;C	(Tms$_2$N)(tBuO)ClP=N-tBu	−17.4(n)	2
	Si$_2$;C;C	(Tms$_2$N)(tBuO)ClP=NAd	−16.8(n)	2
	CSi;C;C	(tBuTmsN)(tBuO)ClP=N-tBu	−39.5(n)	2
	C$_2$;Si;C	[Me$_2$C*(CH$_2$)$_3$ CMe$_2$N*]Cl(tBuO)P=NTms	−24.3(b)	8
ClN'$_2$O/6	(S^4)$_2$;C	**2**; X = OMe, Y = Z = SOCl	17.9, 18.9	12b
	(S^4)$_2$;C	**2**; X = OMe, Y = Z = SOPh	17.4	12b
	(S^4)$_2$;C'	**2**; X = OPh, Y = Z = SOCl	13.0	12b
	(S^4)$_2$;C'	**2**; X = OPh, Y = Z = SOPh	13.1	12b

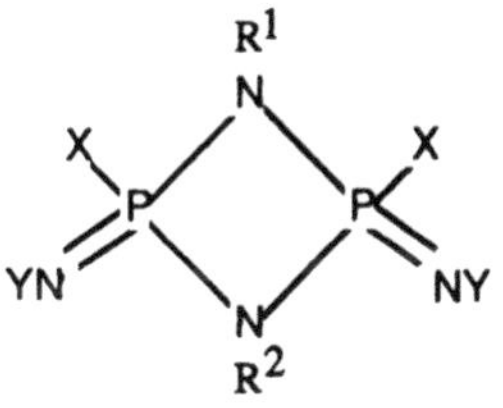

1 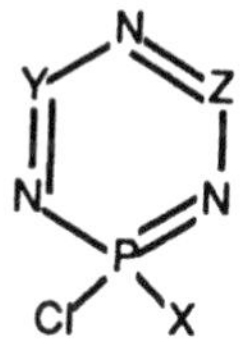 2

P-bound atoms/rings	Connectivities	Structure	NMR data (δ_P[solv,temp] J_{pz} Hz	Ref.
ClN_2N'	$(Si_2)_2;Si$	$(Tms_2N)_2ClP=NTms$	−6.2(b)	3
	$(Si_2)_2;C$	$(Tms_2N)_2ClP=N$-tBu	−3.8(b)	3
	$Si_2,C_2;Si$	$(Me_2N)(Tms_2N)ClP=NTms$	−2.6(n)	13
	$Si_2,C_2;Si$	$(Et_2N)(Tms_2N)ClP=NTms$	−2.0(n)	13
	$Si_2,C_2;Si$	$[Me_2C^*(CH_2)_3$-$CMe_2N^*]Cl(Tms_2N)P=NTms$	−7.7(b)	3
	$Si_2,C_2;C$	$Tms_2NPCl(NMe_2)=N$-tBu	−15.5(n), −12.1 (n, −70°)	13
	$Si_2,C_2;C$	$[Me_2C^*(CH_2)_3$-$CMe_2N^*]Cl(Tms_2N)P=N$-tBu	−4.4(b)	3
	$(CSi)_2;C$	$(tBuTmsN)_2ClP=N$-tBu	−18.1(b)	3
	$CSi,C_2;Si$	$tBuTmsNPCl(NMe_2)=NTms$	2.0(n), 5.5(n, −70°)	13
	$CSi, C_2;Si$	$tBuTmsNPCl(NEt_2)=N$-tBu	−10.9(n)	13
	$(C_2)_2;N'$	$(Et_2N)_2ClP=NN=CPh_2$	39.4(c)	14
	$(C_2)_2;P'$	$(Et_2N)_2ClP=NP(O)ClF$	0.6 J_{PP}45	15
	$(C_2)_2;S^4$	$[(Me_2N)_2ClP=N]_2SO_2$	7.6	16
	$(C_2)_2;S^4$	$[(Et_2N)_2ClP=N]_2SO_2$	3.7	16
	$(C_2)_2;S^4$	$(Me_2N)_2ClP=NSO_2Ph$	24.9	17
	$(C_2)_2;S^4$	$Morph_2ClP=NSO_2Ph$	13.5	17
	$(C_2)_2;Si$	$(iPr_2N)(Me_2N)ClP=NTms$	4.8(m)	18
	$(C_2)_2;Si$	$(iPr_2N)_2ClP=NTms$	−7.9(b)	19
	$(C_2)_2;Si$	$[Me_2C^*(CH_2)_3$ $CMe_2N^*]Cl(Me_2N)P=NTms$	−7.5(b)	8
	$(C_2)_2;C'$	$(Me_2N)_2ClP=NCOCCl_3$	41.8	17
	$(C_2)_2;C'$	$Morph_2ClP=NCOCCl_3$	32.4	17
	$(C_2)_2;C'$	$(Me_2N)_2ClP=NPh$	9.8	17
	$(C_2)_2;C'$	$(Me_2N)(Et_2N)ClP=NPh$	12.1	20
	$(C_2)_2;C'$	$(Morph)_2ClP=NSO_2Ph'NO_2-4$	9.3	17
	$(C_2)_2;C'$	$(Et_2N)_2ClP=NPh$	9.8,11	21, 22
	$(C_2)_2;C'$	$(Et_2N)(Pip)ClP=NPh$	7.2	20
	$(C_2)_2;C'$	$(Me_2N)(Bu_2N)ClP=NPh$	8.0	20
	$C_2,C'_2;C'$	$(Et_2N)(Pyrrolyl)ClP=NPh$	6.7	20
$ClN_2N'/5$	$(C_2)_2;C'$	$(MeN^*CH_2CH_2MeN)P^*Cl=NPh$	6	21
$ClNN'_2/6$	$H_2;P^4,S^4$	**2**; X = NH_2, Y = $PClNH_2$, Z = SOCl	16.6(m)	22b
	$H_2;P^4,S^4$	**2**; X = NH_2, Y = $P(NH_2)_2$, Z = SOCl	21.5(d)	22b
	$H_2;(S^4)_2$	**2**; X = NH_2, Y = Z = SOCl	17.6(m)	22b
	$CH;(S^4)_2$	**2**; X = NHBu, Y = Z = SOCl	15.5	22c
	$CH;(S^4)_2$	**2**; X = NH-tBu, Y = Z = SOCl	7.6	22c
	$C_2;P^4,S^4$	**2**; X = Morph, Y = $P(Morph)_2$, Z = S	25.5	22d
	$C_2;(S^4)_2$	**2**; X = Pip, Y = Z = SOF	19.7	22d

TABLE M (continued)
Compilation of ^{31}P NMR Data of Four Coordinate Phosphorus Compounds Containing a Formal Multiple Bond from Phosphorus to a Group V Atom

Section M1: Phosphazenes (Excluding Cyclopolyphosphazenes)

P-bound atoms/rings	Connectivities	Structure	NMR data (δ_P[solv,temp] J_{pz} Hz	Ref.
	C_2;S,S^4	2; X = Pip, Y = S, Z = SOPip	19.7^T/21.7^I(c)	22e
	C_2;$(S^4)_2$	2; X = Pip, Y = Z = SOPip	23.7^T/24.3^I(c)	22e
	C_2;$(S^4)_2$	2; X = NMe_2, Y = Z = SOR	20.1/28.3(c)	22f
	C_2;$(C')_2$	2; X = NMe_2, Y = Z = C(NMe_2)	−59.0(r)	22g
	C_2;$(C')_2$	2; X = NEt_2, Y = CCl, Z = CPh	−47.6	22h
	C'_2;$(S^4)_2$	2; X = Pyr, Y = Z = SOPh	17.1(c)	22i
ClN'$_3$	$(P^4)_3$	$(Cl_3P{=}N)_2Cl$**P**NP^+Cl_3	−27.0, −26.8	22, 23
	$(P^4)_3$	$(Cl_3P{=}N)_2Cl$**P**$=NCl_2P{=}NSO_2Cl$	−26.4	23
ClN'$_3$/6	P^4,$(S^4)_2$	2; X = N=PPh_2, Y = Z = SOCl	−8.0(c)	23b
	P^4,$(S^4)_2$	2; X = N=PPh_3, Y = Z = SOF	−5.1(c)	23b
	P^4,$(C')_2$	2; X = $NPCl_2NR_2$, Y = Z = CCl	−20.6(r)	22g
	P^4,$(C')_2$	2; X = $NPCl_2NR_2$, Y = CCl, Z = C(NMe_2)	−25.3(r)	22g
	P^4,$(C'')_2$	2; X = NCS, Y = PCl2, Z = SOF	−2.9	23c
Cl_2FN'	P^4	Cl_2F**P**$=NPF_2{=}NP(S)F_2$	4.3 J_{PF}1130 J_{PP}116	10
	Si	$Cl_2FP{=}NPSi(iPr)_3$	−31.7	8b
	Si	$Cl_2FP{=}NPSi(tBu)_3$	−33.9	8b
	P^4	$Cl_2FP{=}NP(S)Cl_2$	−1.8 J_{PP}35	24
$Cl_2N'O$	P^4;C	$Cl_2(EtO)P{=}NP(O)Cl_2$	1 J_{PP}39	10b
	S^4;S^4	$(ClSO_2O)Cl_2P{=}NSO_2Cl$	−4	22
	S^4;C	$(MeO)Cl_2P{=}NSO_2F$	18.1	25
	S^4;C	$(MeO)Cl_2P{=}NSO_2Cl$	17.8	25
	S^4;C	$(EtO)Cl_2P{=}NSO_2F$	15.4	25
	S^4;C	$(EtO)Cl_2P{=}NSO_2Cl$	14.6	25
	S^4;C	$[(MeO)Cl_2P{=}N]_2SO_2$	4.6	26
	S^4;C	$(PrO)Cl_2P{=}NSO_2F$	15.4	25
	S^4;C	$(BuO)Cl_2P{=}NSO_2Cl$	15.2	25
	S^4;C	$[(EtO)Cl_2P{=}N]_2SO_2$	5.5	26
	S^4;C	$[(PrO)Cl_2P{=}N]_2SO_2$	6.5	26
	S^4;C	$[(PhO)Cl_2P{=}N]_2SO_2$	0.2	26
	Si;Si	$(TmsO)Cl_2P{=}NTms$	41.8(n)	27
$Cl_2N'S$	P^4;C	$(MeS)Cl_2P{=}NP(O)F_2$	36.5 J_{PP}42	9
	P^4;C	$(MeS)Cl_2P{=}NP(O)Cl_2$	30.1 J_{PP}6.8	9
	P^4;C	$(EtS)Cl_2P{=}NP(O)Cl_2$	27.4 J_{PP}9.2	9
	P^4;C	$(EtS)Cl_2$**P**$=NPCl_2{=}NP(O)Cl_2$	34.1 J_{PP}28	28
Cl_2NN'	Si_2;Si	$(Tms_2N)Cl_2P{=}NTms$	24.0(he)	6
	HSi;P^4	$(TmsHN)Cl_2P{=}NP(S)F_2$	7.8 J_{PP}65	10
	HSi;P^4	$(TmsHN)Cl_2P{=}NP(S)ClF$	4.9 J_{PP}42	10
	HSi;P^4	$(TmsHN)Cl_2P{=}NP(S)Cl_2$	1.1 J_{PP}9.5	10
	HSi;P^4	$(TmsHN)Cl_2$**P**$=NPCl_2{=}NP(S)Cl_2$	6.9 J_{PP}31	29
	HSi;P^4	$(TmsHN)Cl_2$**P**$NPCl_2{=}NP$-$Cl_2{=}NP(S)Cl_2$	8.4 J_{PP}34.2	29
	HSi;S^4	$(TmsHN)Cl_2P{=}NSO_2F$	−14.3	30
	CSi;P^4	$(MeTmsN)Cl_2$**P**$=NP(O)F_2$	11.8 J_{PP}62	15
	CSi;P^4	$(MeTmsN)Cl_2$**P**$=NPCl_2{=}NP(S)Cl_2$	10.9 J_{PP}30.5	29
	CH;S^4	$[(EtHN)Cl_2P{=}N]_2SO_2$	12.8	31
	CH;S^4	$[(PrHN)Cl_2P{=}N]_2SO_2$	12.9	31
	CH;S^4	$[(BuHN)Cl_2P{=}N]_2SO_2$	13.0	31
	C_2;P^4	$(Me_2N)Cl_2P{=}NP(O)F_2$	14.1 J_{PP}69	15
	C_2;P_4	$(Me_2N)Cl_2P{=}NP(O)ClF$	11.5 J_{PP}54	15

P-bound atoms/rings	Connectivities	Structure	NMR data (δ_P[solv,temp] J_{pz} Hz	Ref.
	$C_2;P_4$	$(Me_2N)Cl_2$**P**$=NPCl_2=NP(O)F_2$	11.7 J_{PP}38	15
	$C_2;P^4$	$(Et_2N)Cl_2P=NP(O)F_2$	10.7 J_{PP}68	15
	$C_2;P^4$	$(Et_2N)Cl_2P=NP(O)ClF$	7.7 J_{PP}50	15
	$C_2;P^4$	$(Et_2N)Cl_2$**P**$=NPCl_2=NP(O)F_2$	7.8 J_{PP}38	15
	$C_2;S^4$	$[(Me_2N)Cl_2P=N]_2SO_2$	7.6	16
	$C_2;S^4$	$[Et_2N)Cl_2P=N]_2SO_2$	3.7	16
	$C_2;S^4$	$(C^*H_2CH_2OCH_2$ $CH_2N^*)Cl_2P=NSO_2Ph$	7.2	17
	$C_2;Si$	$[Me_2C^*(CH_2)_3CMe_2N^*]Cl_2P=NTms$	−36.5(b)	8
	$C_2;C$	$(Me_2N)Cl_2P=NCCl_2CCl_3$	4.5	17
	$C_2;C$	$MorphCl_2P=NCCl_2CCl_3$	0.0	17
	$C_2;C'$	$(Me_2N)Cl_2P=NCOCCl_3$	27.9	17
	$C_2;C'$	$MorphP(Cl_2)=NCOCCl_3$	24.3	17
	$C_2;C'$	$(Me_2N)Cl_2P=NPh$	−19.2, −18.6(n)	17, 32
	$C_2;C'$	$MorphPCl_2=Ph'NO_2-4$	−16.1	17
	$C_2;C'$	$(Et_2N)Cl_2P=NPh$	−21.6(n)	22, 32
	$C_2;C'$	$(Bu_2N)Cl_2P=NPh$	−22.1(n)	32
	$C_2;C'$	$(PhCH_2MeN)Cl_2P=NPh$	−19.8(n)	32
$Cl_2NN'/6$	$CC';C'$	$Cl_2P^*=NCONMeCON^*Me$	0	32b
$Cl_2N'_2$	$(P^4)_2$	$Cl_3P=N$**P**$Cl_2=NP(O)F_2$	−1.3 J_{PP}74,35	15
	$(P^4)_2$	$Cl_3P=N$**P**$Cl_2=NP(S)F_2$	−15.1 J_{PP}47.6	10
	$(P^4)_2$	$Cl_3P=N$**P**$Cl_2=NP(S)ClF$	−25.9 J_{PP}47.5	10
	$(P^4)_2$	$Cl_3P=N$**P**$Cl_2=NP(O)Cl_2$	−20.0	22
	$(P^4)_2$	$Cl_3P=N$**P**$Cl_2=NP(S)Cl_2$	−26.5	22
	$(P^4)_2$	$Cl_3P=NCl_2$**P**$=NCl_2P=NSO_2Cl$	−13.2	23
	$(P^4)_2$	$Cl_3P=NCl_2$**P**$=NCl_2P=NP^+Cl_3$ PCl_6^-	−13.5	22
	$(P^4)_2$	$(EtS)Cl_2P=N$**P**$Cl_2=NP(O)Cl_2$	−11.7 J_{PP}28	28
	$(P^4)_2$	$(Me_2N)Cl_2P=N$**P**$Cl_2=NP(O)F_2$	−12.7 J_{PP}70	15
	$(P^4)_2$	$(TmsHN)Cl_2P=N$**P**$Cl_2=NP(S)Cl_2$	−20.4 J_{PP}4.5 J_{PP}31	29
	$(P^4)_2$	(TmsHN)- $Cl_2P=N$**P**$Cl_2=NPCl_2=NP(S)Cl_2$	−15.0 J_{PP}38.5	29
	$(P^4)_2$	(TmsHN)- $Cl_2P=NPCl_2=N$**P**$Cl_2=NP(S)Cl_2$	−22.0 J_{PP}6	29
	$(P^4)_2$	$(Et_2N)Cl_2P=N$**P**$Cl_2=NP(O)F_2$	−13.3 J_{PP}71	15
	$(P^4)_2$	$(TmsMeN)MeClP=N$**P**$Cl_2=NP(S)Cl_2$	−20.4 J_{PP}12 and 30.5	29
	$(P^4)_2$	$Ph_2ClP=NCl_2$**P**$=NCl_2P=NClP^+$ Ph_2 $Cl-$	−11.5(re)	33
	P^4,S^4	$Cl_3P=NCl_2$**P**$=NSO_2F$	−6.8	23
	P^4,S^4	$Cl_3P=NCl_2$**P**$=NSO_2Cl$	−7.0(n)	33
	P^4,S^4	$Cl_3P=NCl_2P=NCl_2$**P**$=NSO_2Cl$	−9.7	23
	P^4,S^4	$(Cl_3P=N)_2ClP=NCl_2$**P**$=NSO_2Cl$	−12.6	23
	P^4,S^4	$(Cl_3P=NCl_2$**P**$=N)_2SO_2$	−17.5	33b
	P^4,S^4	$(Cl_3P=N)Cl_2P=NSO_2Ph$	−16.4	33b
	P^4,S^4	$Ph_2ClP=NCl_2$**P**$=NSO_2Cl$	−5.2	23
	P^4,S^4	$Ph_2ClP=NCl_2$**P**$=NSO_2Ph$	−11.5(re)	33
	$P^4;C'$	$(SCN)Cl_2P=NP(O)F_2$	−16.8 J_{PP}81	15
	S^4,C'	$(Cl_3P=N)_2C=NCl_2$**P**$=NSO_2Cl$	4.2	23
	S^4,C''	$Cl_3P=N=C=NCl;_2$**P**$=NSO_2Cl$	4.0	23
$Cl_2N'_2/6$	P^4,S^4	**2**; X = Cl, Y = PCl_2, Z = SOCl	26	33b
	P^4,S^4	**2**; X = Cl, Y = PCl_2, Z = SOPh	20.7	22f
	P^4,S^4	**2**; X = Cl, Y = $PClN=PPh_3$, Z = SOPh	21.8	23b
	$(S^4)_2$	**2**; X = Cl, Y = Z = SOR	24.3/25.4[C](c); 20.7/21.9[T]	22e, 33c
	$(S^4)_2$	**2**; X = Cl, Y = Z = SOF	31.9(m), 30.9(c)	22b, 23b
	$(S^4)_2$	**2**; X = Cl, Y = Z = SOPip	28.1/28.6(c)	22e
	$(S^4)_2$	**2**; X = Cl, Y = SOCl, Z = SO-sBu	27.7/28.6	22c

TABLE M (continued)
Compilation of ^{31}P NMR Data of Four Coordinate Phosphorus Compounds Containing a Formal Multiple Bond from Phosphorus to a Group V Atom

Section M1: Phosphazenes (Excluding Cyclopolyphosphazenes)

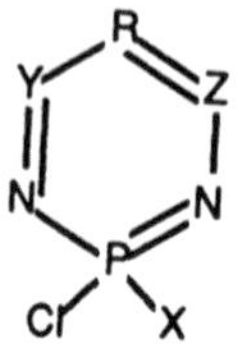

3

P-bound atoms/rings	Connectivities	Structure	NMR data (δ_P[solv,temp] J_{pz} Hz	Ref.
	$(S^4)_2$	2; X = Cl, Y = SOCl, Z = SO-tBu	26.5/28.1	22c
	$(S^4)_2$	2; X = Cl, Y = Z = SOPh	22.6(m), 22.1[C], 25.4[T]	22b, 22f
	$(C')_2$	2; X = Cl, Y = Z = CCl	55.6(m)	33d
	$(C')_2$	3; R = H, X = Cl, Y = Z = CCl	49.2(m)	33d
	$(C')_2$	3; X = Cl, Y = Z = $CCCl_3$	68.5(m)	33d
	$(C')_2$	3; X = Cl, Y = Z = CTf	82.7(m)	33d
	$(C')_2$	2; X = Cl, Y = CCl, Z = $C(NMe_2)$	52.6(r)	22g
	$(C')_2$	2; X = Cl, Y = Z = $C(NMe_2)$	56.2(r)	22g
	$(C')_2$	3; R = Cl, X = Cl, Y = CCl, Z = $C(NEt_2)$	39.4	33e
	$(C')_2$	3; R = H, X = Cl, Y = CCl, Z = $C(NEt_2)$	46.5	33e
	$(C')_2$	3; R = CN, X = Cl, Y = C(Ph), Z = $C(NMe_2)$	50.0	33e
	$(C')_2$	2; X = Cl, Y = CPh, Z = $C(NEt_2)$	57.5	22h
Cl_3N'	N'	Cl_3**P**=NNC(Cl)N*P(O)ClN(P*OCl)ClC(Cl)NNPCl_3	14.2	22
	P^4	Cl_3**P**=NP(O)F_2	6.4, 6.5 J_{PP}70	34, 35
	P^4	Cl_3**P**=NP(S)F_2	−4.4 J_{PP}70	36
	P^4	Cl_3**P**=NPF_2=NP(S)F_2	6.5 J_{PP}86	10
	P^4	Cl_3**P**=NP(O)ClF	3.9 J_{PP}46.5	34, 35
	P^4	Cl_3**P**=NP(SO)ClF	−0.8	36
	P^4	Cl_3**P**=NPF_2=NP(S)ClF	7.7 J_{PP}82	10
	P^4	Cl_3**P**=NCl_2P=NSO_2F	10.1	23
	P^4	Cl_3**P**=NPCl_2=NP(O)F_2	0.8 J_{PP}35	15
	P^4	Cl_3**P**=NPCl_2=NP(S)F_2	3.3 J_{PP}32.4	15
	P^4	Cl_3**P**=NPF_2=NP(S)Cl_2	7.6 J_{PP}78.8	10
	P^4	Cl_3**P**=N$(PF_2=N)_2$P(S)Cl_2	12.7 J_{PP}90	24
	P^4	Cl_3**P**=NP(O)Cl_2	−0.1, −1	10b, 22, 23
	P^4	Cl_3**P**=NP(S)Cl_2	−2.9, −3.4	22
	P^4	Cl_3**P**=NPCl_2=NP(S)ClF	4.3 J_{PP}32.4	10
	P^4	Cl_3**P**=NPClF=NP(S)Cl_2	4.2 J_{PP}54.6	10
	P^4	Cl_3**P**NP^+Cl_3	21.4	22
	P^4	Cl_3**P**NCl_2P=NSO_2Cl	11.0, 9.3	23, 33
	P^4	Cl_3**P**NPCl_2=NP(O)Cl_2	7.1	22
	P^4	Cl_3**P**=NPCl_2=NP(S)Cl_2	0.0, 6.0	22

P-bound atoms/rings	Connectivities	Structure	NMR data (δ_P[solv,temp] J_{pz} Hz	Ref.
	P^4	Cl_3**P**$=NPCl_2=NP^+Cl_3$ PCl_6^-	14.0	22
	P^4	Cl_3**P**$=NCl_2P=NCl_2P=NSO_2Cl$	10.0	23
	P^4	$(Cl_3P=N)_2ClP=NP^+Cl_2$	6.2, 6.5	22, 23
	P^4	Cl_3**P**$=NCl_2P=NCl_2P=NPCl_3$ PCl_6^-	12.5	22
	P^4	$(Cl_3$**P**$=N)_2ClP=NCl_2P=NSO_2Cl$	0.9	23
	P^4	$(Cl_3$**P**$=N)_3P=NP^+Cl_3$	4.0	23
	P^4	$Cl_3P=N(PN_3C_2Cl_3)$	22.7(re)	33
	P^4	Cl_3**P**$=NP(O)PhF$	-10.9 $J_{PP}10$	15
	P^4	Cl_3**P**$=NP(O)(OPh)_2$	-11.9 $J_{PP}47$	10b
	S^4	$Cl_3P=NSO_2F$	20.8	23
	S^4	Cl_3**P**$=NSOFP=NSO_2F$	28.5	38
	S^4	$Cl_3P=NSO_2Cl$	20.5(n), 21.0	23, 33
	S^4	$Cl_3P=NSO_2NHPh'Cl(p)$	4.2(r)	33
	S^4	$Cl_3P=NSO_2NHPh$	3.8(c)	33
	S^4	$Cl_3P=NSO_2Tol$	2.4	39
	S^3	$Cl_3P=NSOBu$	7.9	37
	Si	$Cl_3P=NSiF(tBu)_2$	-53.8(c)	8b
	Te	$Cl_3P=NTeF_3$	17.3	40
	C	$Cl_3P=NCCl_2Tf$	-6.5	41
	C	$Cl_3P=NCCl_2CCl_3$	$-11.1(POCl_3)$, -10.7	33, 42
	C	$Cl_3P=NCCl_2CCl_2N=PCl_3$	-18.0(re)	33
	C	$Cl_3P=NCClTf_2$	-21.2	41
	C	$Cl_3P=NCTf_2NSO_2Cl$	-8.4	34
	C	$Cl_3P=NCTf_3$	-44.8	41
	C	$Cl_3P=NCCl_2CClMe_2$	8.8(n)	33
	C	$Cl_3P=NCHMeEt$	-38.7	41
	C	$Cl_3P=NCMe_3$	-88.4	41
	C	$Cl_3P=NCHEt_2$	-76.5	41
	C	$Cl_3P=NCHPr_2$	-80.3	41
	C	$Cl_3P=NCEt_3$	-98	41
	C′	$Cl_3P=NCCl=NP^+Cl_3$ PCl_6^-	38.5(re)	33
	C′	$(Cl_3$**P**$=N)_2C=NP^+Cl_3$ Cl^-	33.5(re)	33
	C′	$(Cl_3$**P**$=N)_2C=NCl_2P=NSO_2Cl$	23.0	23
	C′	$Cl_3P=NCOCCl_3$	25.1	42
	C′	$Cl_3P=NCCl=CCl_2$	-23.8(n)	33
	C′	$Cl_3P=NC(P^+Cl_3)=CCl_2$	14.2(ne)	33
	C′	$Cl_3P=NCCl=CHCl$	-24.0(n)	33
	C′	$Cl_3P=NC(P^+Cl_3)=CHCl$	15.9(ne)	33
	C′	$Cl_3P=NCO_2Me$	8.9	42
	C′	$Cl_3P=NCO_2Et$	7.3	42
	C′	$Cl_3P=NC_6F_5$	-26.3	44
	C′	$Cl_3P=NPh'Cl-4$	-10.0(r)	33
	C′	$Cl_3P=NPh$	-47.6(re), -52.5(r)	41, 43
	C′	$Cl_3P=NCOPh'F-4$	13.2	42
	C′	$Cl_3P=NPh'Tf(p)$	-8.2	44
	C′	$Cl_3P=NCOPh'NO_2-4$	18.7	42
	C′	$Cl_3P=NCOPh$	13.2	42
	C′	$Cl_3P=NCOPh'Me-4$	12.2	42
	C″	Cl_3**P**$=N=C=NCl_2P=NSO_2Cl$	30.0	23
$FN'O_2$	Si;$(C)_2$	$(EtO)_2FP=NTms$	15.5	45
FN_2N'	$(C_2)_2$;S^4	$(Me_2N)_2FP=NSO_2Ph$	18.6(g) $J_{PF}939$	46
	$(C_2)_2$;H	$(Et_2N)_2FP=NH$	32.6 $J_{PF}969$	47
	$(C_2)_2$;C	$(Me_2N)_2FP=NMe$	17.7 $J_{PF}986$	46
	$(C_2)_2$;C	$(Et_2N)_2FP=NEt$	14.6 $J_{PF}991$	48
	$(C_2)_2$;C	$(Et_2N)_2FP=NPr$	12.5 $J_{PF}993$	48
	$(C_2)_2$C	$(Pip)_2FP=N(CH_2)_3CH=CH_2$	6.2 $J_{PF}992$	48
	$(C_2)_2$;C′	$(Et_2N)_2FP=NPh$	6.2 $J_{PF}992$	48
	$(CC')_2$;C′	$(EtPhN)_2FP=NPh$	13.0 $J_{PF}1000$	48

TABLE M (continued)
Compilation of ^{31}P NMR Data of Four Coordinate Phosphorus Compounds Containing a Formal Multiple Bond from Phosphorus to a Group V Atom

Section M1: Phosphazenes (Excluding Cyclopolyphosphazenes)

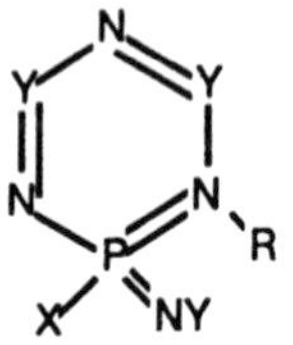

4

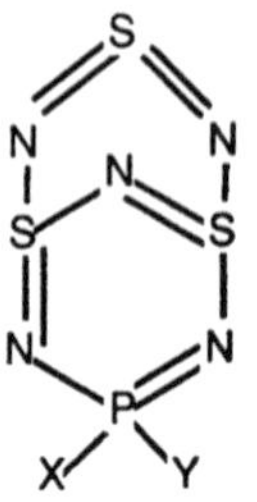

5

P-bound atoms/rings	Connectivities	Structure	NMR data (δ_P[solv,temp] J_{pz} Hz	Ref.
$FNN'_2/6$	CC';Si;C'	**4**; R = Me, X = F, Y = Tf	−30.3 J_{PF}1002	49
	CC';Si;C'	**4**; R = Et, X = F, Y = Tf	−29.2 J_{PF}1012	49
$F_2N'O$	Si;C	$(PrO)F_2P{=}NTms$	28.5	45
	Si;C	$(BuO)F_2P{=}NTms$	28.8	45
$F_2N'S$	P^4;C	$MeSF_2P{=}NP(O)Cl_2$	22.3 J_{PF}1142 J_{PP}54	9
	P^4;C	$MeSF_2P{=}NP(O)F_2$	26.3 J_{PF}1133 J_{PP}107	9
	P^4;C	$EtSF_2P{=}NP(O)Cl_2$	21.8 J_{PF}1150 J_{PP}53	9
F_2NN'	Si_2;Si	$(Tms_2N)F_2P{=}NTms$	(+?)14.4(he)	6
	$H_2;P^4$	$(H_2N)F_2P{=}NP(S)ClF$	3.5 J_{PF}960 J_{PP}89	24, 50
	$H_2;P^4$	$(H_2N)F_2P{=}NP(S)Cl_2$	−5.9 J_{PF}985 J_{PP}57	24, 50
	$H_2;P^4$	$(H_2N)F_2P{=}NP(S)F_2$	2.3 J_{PF}960 J_{PP}120	24, 50
	$CP^4;P^4$	$[(F_2SP)MeN]F_2\mathbf{P}{=}NP(S)ClF$	−17.9 J_{PF}1050 $^2J_{PF}$35	24
	$CP^4;P^4$	$[(F_2SP)MeN]F_2\mathbf{P}{=}NP(S)F_2$	−15.9 J_{PF}1040 $^2J_{PF}$34	24, 50
	$C_2;P^4$	$(Et_2N)F_2P{=}NP(O)Cl_2$	8.7	45
	C_2;P	$(Et_2N)F_2P{=}NPCl_2$	4.9	45
	C_2;Si	$(Et_2N)F_2P{=}NSiCl_3$	9.3	45
	C_2;Si	$(Et_2N)F_2P{=}NTms$	14.6	45
	C_2;Si	$(Pip)F_2P{=}NTms$	15.5	45
	C_2;C'	$[(MeOCOCH_2)t\text{-}BuN]F_2P{=}NPh$	22(g)	51
$F_2N'_2$	$(P^4)_2$	$ClF_2P{=}N\mathbf{PF_2}{=}NP(S)F_2$	−23.2 J_{PF}924 J_{PP}122	10
	$(P^4)_2$	$Cl_2FP{=}N\mathbf{PF_2}{=}NP(S)F_2$	−24.3 J_{PF}936 J_{PP}118	10
	$(P^4)_2$	$F_3P{=}N\mathbf{PF_2}{=}NP(S)Cl_2$	−29 J_{PF}925 J_{PP}77	10
	$(P^4)_2$	$Cl_3P{=}N\mathbf{PF_2}{=}NP(S)F_2$	−25.1 J_{PP}114 J_{PF}947	10
	$(P^4)_2$	$Cl_3P{=}N\mathbf{PF_2}{=}P(S)ClF$	−27.6 J_{PF}948	10
	$(P^4)_2$	$Cl_3P{=}N\mathbf{PF_2}{=}NP(S)Cl_2$	−30.9 J_{PF}959 J_{PP}56	10
	$(P^4)_2$	$Cl_3P{=}N(PF_2{=}N)_2P(S)Cl_2$	−25.9 J_{PP}53 J_{PF}927	24
$F_2N'_2/6$	$(S^3)_2$	**5**; X = Y = F	−11.5(b) J_{PF}965, 890	51b
	$(S^3)_2$	**5**; X = F, Y = Ph	−10.1(b) J_{PF}1050	51b
F_3N'	P^4	$F_3\mathbf{P}{=}NPF_2{=}NP(S)Cl_2$	−40.3 J_{PF}1030 J_{PP}175	10
	Si	$F_3\mathbf{P}{=}NSiF(tBu)_2$	46.4(c)	8b
IN_2N'	$(C_2)_2$;C	$(Et_2N)_2IP{=}NCH_2CH_2I$	−1.6(n)	4
	$(C_2)_2$;C	$(Pr_2N)_2IP{=}NCH_2CH_2I$	0.6(n)	4
	$(C_2)_2$;C'	$(Et_2N)_2IP{=}NPh$	−20.6(?n)	52
	$(C_2)_2$;C'	$(Pr_2N)_2IP{=}NPh$	−22.1(?n)	52
$IN_2N'/4$	$(PSi)_2$;Si	**1**; R = Y = Tms, X = I	(+?)128.4^C(he); 129.4^T(he)	6
I_2NN'	Si_2;Si	$(Tms_2N)I_2P{=}NTms$	(+?)227.8(he)	6

P-bound atoms/rings	Connectivities	Structure	NMR data (δ_P[solv,temp] J_{pz} Hz)	Ref.
N′O$_2$P	S^4;(C)$_2$;C$_2$	Et$_2$PP(OEt)$_2$=NTos	37.9 J_{PP}286	53
	Si;(C)$_2$;C$_2$	Et$_2$PP(OEt)$_2$=NTms	24.0 J_{PP}200	53
	Si;(C)$_2$;C$_2$	iPr$_2$PP(OEt)$_2$=NTms	20.0 J_{PP}215	53
	H;(C)$_2$;C$_2$	iPr$_2$PP(OEt)$_2$=NH	31.0 J_{PP}235	53
	C;(C)$_2$;C$_2$	Et$_2$PP(OEt)$_2$=N-tBu	7.5 J_{PP}180	53
	C;(C)$_2$;C$_2$	iPr$_2$PP(OEt)$_2$=N-tBu	2.2 J_{PP}210	53
	C′;(C)$_2$;C$_2$	Et$_2$PP(OEt)$_2$=NPh	24.0 J_{PP}250	53
N′O$_2$S	C′;(C)$_2$	[Me$_2$C=C(SMe)S]P(OEt)$_2$=NPh	6.0(n)	54
N′O$_3$	N′;(C)$_3$	(MeO)$_3$P=NN=C(COPh)$_2$	22	55
	N′;(C)$_3$	(EtO)$_3$P=NN=CPh$_2$	18.6(c)	14
	P^4;Ba,(C′)$_2$	Ba[OP(OPh)$_2$=NP(OPh)$_2$]$_2$	−8.9	56
	P^4;Fe,(C′)$_2$	Fe[OP(OPh)$_2$=NP(OPh)$_2$]$_3$	−1.0	56
	P^4;Si,(C)$_2$	(TmsO)(EtO)$_2$P=NP(OEt)$_2$=NTms	7.6 J_{PNP}61	57
	P^4;(C)$_3$	(MeO)$_3$P=NP(S)F$_2$	2 J_{PNP}87	24
	P^4;(C)$_3$	(MeO)$_3$P=NP(O)OMe)$_2$	−0.3 J_{PNP}73.3	58
	P^4;(C)$_3$	(EtO)$_3$P=NP(O)Cl$_2$	−4 J_{PNP}51/54	10b
	P^4;(C)$_3$	(MeO)$_3$P=NP(O)(OEt)$_2$	−2.7 J_{PNP}70	58
	P^4;(C)$_3$	(EtO)$_3$P=NP(O)(OMe)$_2$	−2.5 J_{PNP}70	58
	P^4;(C)$_3$	(ClCH$_2$CH$_2$O)$_3$P=NP(O)(OMe)(SMe)	−4.7 J_{PNP}48.6	58
	P^4;(C)$_3$	(ClCH$_2$CH$_2$O)$_3$P=NP(S)(OMe)$_2$	−4.6 J_{PNP}67.4	58
	P^4;(C)$_3$	(EtO)$_3$P=NP(O)Cl(OEt)	−3.7	10b
	P^4;(C)$_3$	(MeO)$_3$P=NP(O)(O-iPr)$_2$	−4.1 J_{PNP}70	58
	P^4;(C)$_3$	(EtO)$_3$P=NP(S)Me(SEt)	−2.1	59
	P^4;(C)$_3$	(EtO)$_3$P=NP(O)(OEt)$_2$	−3.2 J_{PNP}70, −2.9.	10b, 60
	P^4;(C)$_3$	(iPrO)$_3$P=NP(O)OMe)$_2$	−5.7 J_{PNP}70	58
	P^4;(C)$_3$	(MeO)$_3$P=NP(O)(NEt$_2$)$_2$	0.0	61
	P^4;(C)$_3$	(EtO)$_3$P=NP(O)(NMe$_2$)$_2$	−2.3	61
	P^4;(C)$_3$	(iBuO)$_3$P=NP(O)(OMe)$_2$	−1.7 J_{PNP}71.4	58
	P^4;(C)$_3$	(EtO)$_3$P=NP(O)(NEt$_2$)$_2$	−3.7	61
	P^4;(C)$_3$	(BuO)$_3$P=NP(O)(NEt$_2$)$_2$	−3.0	61
	P^4;(C)$_3$	(EtO)$_3$P=NP(O)(OPh)$_2$	−3.2 J_{PNP}73.5	58
	P^4;(C)$_2$,(C′)	(nBuO)$_2$(PhO)P=NP(O)(OMe)$_2$	−8.5 J_{PNP}70	58
	P^4;(C′)$_3$	(PhO)$_3$P=NP(O)(OPh)$_2$	−21.6 J_{PNP}75	58
	Si;Si,(C)$_2$	(TmsO)(EtO)$_2$P=NTms	−14.9	57
	Si;Si,(C)$_2$	(TmsO)(BuO)$_2$P=NTms	−14.7	57
	Si;(C)$_3$	(MeO)$_3$P=NTms	−2.9(r)	62
	Si;(C)$_3$	(TfCH$_2$O)$_3$P=NTms	−15.3	63
	Si;(C)$_3$	(EtO)$_3$P=NTms	−8.3	63
	Si;(C)$_3$	(iPrO)$_3$P=NTms	−11.6	63
	Si;(C)$_3$	(BuO)$_3$P=NTms	−8.7	63
	C;Si,(C)$_2$	(EtO)$_2$(TmsO)P=NCH(OTms)CCl$_3$	−15	64
	C;(C)$_3$	(MeO)$_3$P=NMe	1.1(n)	65
	C;(C)$_3$	(MeO)$_3$P=NEt	0.1(n), −1.3(t)	65, 66
	C;(C)$_3$	(EtO)$_3$P=NPr	−3.7	60
	C;(C)$_3$	(EtO)$_3$P=NBu	−3.9	60
	C;(C)$_3$	(EtO)$_3$P=NCH$_2$Ph	−0.8	60
	C;(C)$_3$	(MeO)$_3$P=NCTf$_2$CH$_2$CR$_2$N=CTf$_2$	−15.1	67
	C;(C)$_3$	(iPrO)$_3$P=NCTf$_2$CH$_2$CR$_2$N=CTf$_2$	−19.4	67
	C′;(C)$_3$	(MeO)$_3$P=NPh	−3(b)	68
	C′;(C)$_3$	(EtO)$_3$P=NPh	−5.5, −2.6(t), −5.7(b)	60, 66, 68
	C′;(C)$_3$	(EtO)$_3$P=NPh′OMe − 4	−8.0	69
	C′;(C)$_3$	(iPrO)$_3$P=NPh	−6.5(c)	68
	C′;(C)$_2$,C′	(CH$_2$=CHO)(PrO)$_2$P=NPh	−14.8(n)	70
	C′;(C)$_2$,C′	(EtO)$_2$(CHCO$_2$Et=CMeO)P=NPh	−20.1(n)	70
	C′;(C)$_2$,C′	H$_2$C=C[P(O)OEt$_2$]OP(O-iPr)$_2$=NPh	−18.0(n)	70
	C′;(C)$_2$,C′	CH$_2$=CPhOP(OEt)$_2$=NPh	−17.3(n)	71
	C′;(C)$_2$,C′	(EtOOC)$_2$C=CMeOP(OEt)$_2$=NPh	−22.0(n)	70

TABLE M (continued)
Compilation of ^{31}P NMR Data of Four Coordinate Phosphorus Compounds Containing a Formal Multiple Bond from Phosphorus to a Group V Atom

Section M1: Phosphazenes (Excluding Cyclopolyphosphazenes)

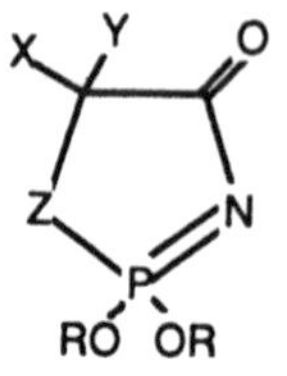

X Y O Z N P RO OR

6

Ph N MeN N P X X

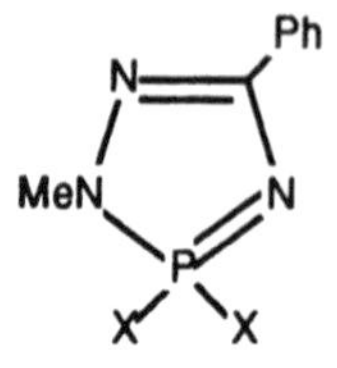

7

P-bound atoms/rings	Connectivities	Structure	NMR data (δ_P[solv,temp] J_{pz} Hz	Ref.
$N'O_3/5$	P^4;$(C)_3$	GlcP(OEt)=NP(O)$(OEt)_2$	19.2	11, 72
	P^4;C,$(C')_2$	(EtO)CatP=NP(O)$(OEt)_2$	14.9	73
	S^4;$(C)_3$	GlcP(OMe)=NTos	17.5(t)	74
	C';$(C)_3$	**6**; R = $CH_2CF_2CHF_2$, X = OEt, Y = COPh, Z = O	13	74b
	C';$(C)_3$	**6**; R = $CH_2CF_2CHF_2$, X = OPh, Y = COPh, Z = O	12	74b
	C';$(C)_3$	(MeO)PncP=NPh	5.3(t)	66, 74
	C';$(C)_3$	(PrO)PncP=NPh	4.0(t)	74
	C';$(C)_3$	$tBuCH_2OP$(Pnc)=NPh	−5.5(b)	68
	C';$(C')_3$	CatPCl=NC_6H_2(4-Tf)(2,6-NO_2)	5.6	13
$N'O_3/6$	C';$(C)_3$	O*$CH_2CMe_2CH_2OP$*(OMe)=NPh	−10.3(b)	68
$N'O_3/6,6$	C;$(C)_3$	EtC$(CH_2O)_3$ P=$NCTf_2CH_2CR_2N$ = CTf_2	−29.7	67
$NN'O_2$	C'Si;Si;$(C)_2$	(PhTmsN)$(EtO)_2$P=NTms	−4.4	57
	C_2;Si;$(C)_2$	$(TfCH_2O)_2(Me_2N)$P=NTms	0.4(c)	1
	C_2;C';$(C)_2$	$tBuCOCH_2N$(iPr)P$(OEt)_2$=NPh	4.5	75
	C_2;C';C,C'	(Et_2N)(PrO)$(CCl_2$=CMeO)P=NPh	−6.9(n)	70
	C_2;C';C,C'	(Et_2N)(EtO)$(CCl_2$=C-tBuO)P=NPh	−10.9(n)	70
	C_2;C';$(C)_2$	$(neo\text{-}PeO)_2P(NMe_2)$=NPh	5.9(c)	68
	C_2;C';C,C'	(Et_2N)(PrO)(CHCl=C-tBuO)P=NPh	−12.1(n)	70
	C_2;C';C,C'	(Et_2N)(EtO)$(CCl_2$=CPhO)P=NPh	−7.5(n)	71
	C_2;C';C,C'	(Et_2N)(PrO)(CHCl=CPhO)P=NPh	−7.8(n)	71
	CC';C';$(C)_2$	(PhCONMe)$(EtO)_2$P=NPh	−8.6;	75b
	CC';C';$(C)_2$	MeN=CPhNMeP$(OEt)_2$=NPh	-3.4^E(b) -9.2^Z(b)	76
$NN'O_2/5$	P^4;C_2;$(C)_2$	(Et_2N)GlcP=NP(O)$(OEt)_2$	28.7	72
	Si_2;Si;$(C)_2$	PncP$(NTms_2)$=NTms	8.6	11
	Si_2;Si;$(C')_2$	(O*CTf=CTfOP*$(NTms_2)$=NTms	−5.3	11
	CN';C';$(C')_2$	**7**; X = O-pTol	35.5(c)	76b
	C_2;P^4;$(C)_2$	GlcP(NMe_2)=NP(O)$(OEt)_2$	28.7	11
	C_2;P^4;$(C')_2$	CatP(NMe_2)=NP(O)$(OEt)_2$	19.2	11
	C_2;Si;$(C')_2$	O*CMe=CMeOP*(NMe_2)=NTms	15.9(c)	11, 77
	C_2;C';$(C)_2$	(N*HCH_2CH_2OP*(OMe)=NPh	19.3(v)	74
	C_2;C';$(C)_2$	GlcP(NMe_2)=NPh	20.6(b)	72
	C_2;C';$(C)_2$	(O*CH_2CH_2NMeP*(OEt)=NPh'F-4	15.9	76
	C_2;C';$(C)_2$	(O*CH_2CH_2NMeP*(OEt)=NPh	15.9	78
	C_2;C';$(C)_2$	(O*$CHMeCH_2OP$*(NMe_2)=NPh	20.9^C(b); 20.6^T(b)	79
	C_2;C';$(C)_2$	(O*CH_2CH_2NMeP*(O-iPr)=NPh'F-4	14.5	78
	C_2;C';$(C)_2$	(O*CH_2CH_2NMeP*(OPr)=NPh	14.8	78
	C_2;C';$(C)_2$	GlcP(NEt_2)=NPh	20.6	11

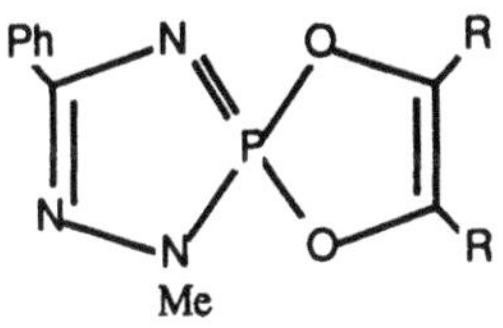

8

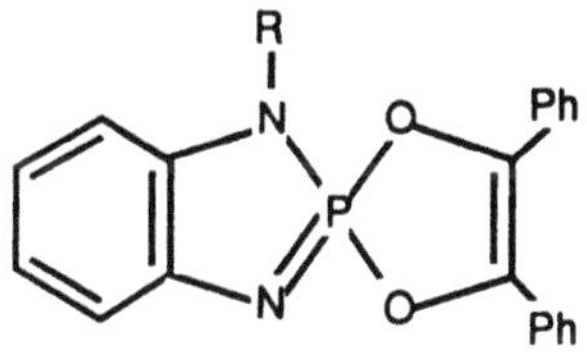

9

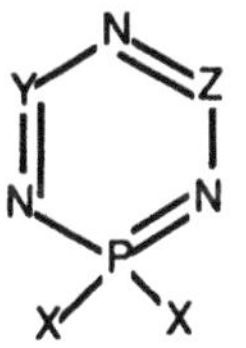

10

P-bound atoms/rings	Connectivities	Structure	NMR data (δ_P[solv,temp] J_{pz} Hz	Ref.
	C_2;C′;$(C)_2$	(O*CHMeCH$_2$OP*(NEt$_2$)=NPh	18.1^{C}(n); 17.7(n)	79
	C_2;C′;$(C)_2$	PncP(NMe$_2$)=NPh	13.8(t)	72
	CC′;C′;$(C)_2$	(O*CH$_2$CH$_2$NPhP*(OEt)=NPh′F-4	4.2	78
	CC′;C′;$(C)_2$	O*CH$_2$CH$_2$NPhP*(OEt)*NPh	3.9	78
	CC′;C′;$(C)_2$	MeN=CPhNMePGlc=NPh	8.4^{E}(b); 4.2^{Z}(b)	76
	CC′;C′;C,C′	O*C(tBu)=CHN(tBu)P*(OMe)=NPh	10.2	75
NN′O_2/5,5	CN′;C′;$(C')_2$	**8**; R = Ph	10	80
	CC′;C′;$(C')_2$	**9**; R = sBu	15.3; 14.9^{D}(xylene)	80
N′$_2O_2$	$(P^4)_2$;$(C)_2$	(Me$_2$N)$_3$P=N(TfCH$_2$O)$_2$P=NP$^+$MePh	−3.7(a)	81
	P^4,P;$(C)_2$	(Me$_2$N)$_3$P=N(TfCH$_2$O)$_2$**P**=NPPh$_2$	3.4(a)	81
	P^4;Si;Si,$(C)_2$	(TmsO)(EtO)$_2$P=N**P**(OEt)$_2$=NTms	12.8	57
N′$_2O_2$/6	$(S^4)_2$;$(C)_2$	**10**; X = OMe, Y = Z = SOCl	4.1^{C}	12b
	$(S^4)_2$;$(C')_2$	**10**; X = OPh, Y = Z = SOCl	5.7^{C}	12b
	$(S^4)_2$;$(C')_2$	**10**; X = OPh, Y = Z = SOPh	10.5^{T}	12b
	$(S^3)_2$;$(C')_2$	**10**; X = OPh, YZ = SNbeS	−23.4	81b
	$(S)_2$;C	**10**; X = OPh, Y = Z = S	−3.4	81b
	$(C')_2$;$(C)_2$	**10**; X = OMe, Y = CH, Z = C(CCl$_3$)	33.7	81c
	$(C')_2$;$(C')_2$	**10**; X = OMe, Y = CH, X = C(OMe)	38.5	81c
	$(C')_2$;$(C)_2$	**10**; X = OEt, Y = CH, Z = C(OEt)	35.0	81c
NN′S_2/5	C_2;C′;$(C)_2$	S*CH$_2$CH$_2$SP*(NMe$_2$)=NPh	58(b)	73
N_2N′O	Si_2,HSi;Si;C	Tms$_2$NP(NHTms)(Nox)=NTms	−1.6	83
	Si_2,HSi;Si;C′	Tms$_2$HP(NHTms)O-CMe=CH$_2$)=NTms	6.9(r)	84
	Si_2,HSi;Si;C′	Tms$_2$NP(NHTms)-(OCPh=CH$_2$)=NTms	7.9(b)	84
	HSi,CSi;Si;C′	MeSCH$_2$N(Tms)P-(NHTms)(OTms)=NTms	4.9(c)	84
	HSi,CSi;Si;C′	PhSCH$_2$N(Tms)P-(NHTms)(OTms)=NTms	5.6(c)	84
	HSi,C_2;Si;C	TmsNHP(NMe$_2$)(OMe)=NTms	1.8	85
	HSi;C_2Si;C	TmsHNP(Pip)(Nox)=NTms	−4.3	83
	CSi;CH;C;C	tBuTmsNP(NHtBu)(Nox)=N-tBu	13.0	83
	$(C_2)_2$;P^4;C	(Me$_2$N)$_2$(EtO)P=NP(O)(OEt)$_2$	19.1	61
	$(C_2)_2$;P^4;C	(Et$_2$N)$_2$(MeO)P=NP(O)(OMe)$_2$	17.5	61
	$(C_2)_2$;P^4;C	(Et$_2$N)$_2$(EtO)P=NP(O)(OEt)$_2$	15.6	61
	$(C_2)_2$;P^4;C	(Et$_2$N)$_2$(BuO)P=NP(O)(OBu)$_2$	15.2	61
	$(C_2)_2$;H;C	(iPr$_2$N)$_2$(MeO)P=NH	34(b)	19
	$(C_2)_2$;H;C′	(Et$_2$N)$_2$P(OPh)=NH	29.2	86
	$(C_2)_2$;H;C′	(Et$_2$N)$_2$P(OPh′Y-4)=NH Y = Me, OMe, F, Cl, NO$_2$,-CH$_2$ CF$_2$CHF$_2$	27.1/33.3	86
	$(C_2)_2$;C;C′	(Et$_2$N)$_2$(PhO)P=NCH$_2$CH$_2$OPh	18.3(phenol)	4
	$(C_2)_2$;C′;C′	(Et$_2$N)$_2$(PhO)P=NPh	5.0	52
	$(C_2)_2$;C′;C′	(Pr$_2$N)$_2$(PhO)P=NPh	4.1	52

TABLE M (continued)
Compilation of ^{31}P NMR Data of Four Coordinate Phosphorus Compounds Containing a Formal Multiple Bond from Phosphorus to a Group V Atom

Section M1: Phosphazenes (Excluding Cyclopolyphosphazenes)

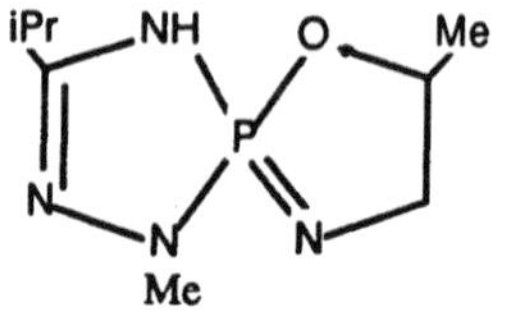

11

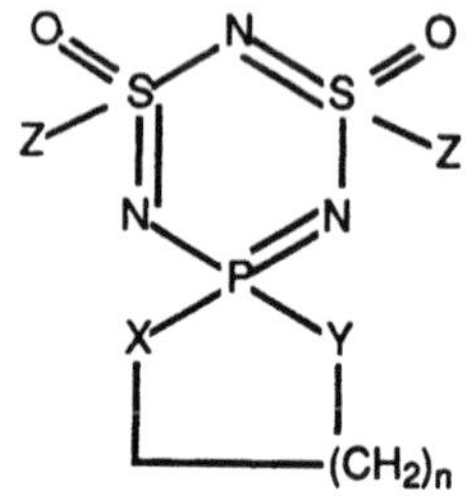

12

P-bound atoms/rings	Connectivities	Structure	NMR data (δ_P[solv,temp] J_{px} Hz	Ref.
$N_2N'O/5$	$C'N,C_2;C';C'$	$O^*CMe{=}NNPhP^*(NEt_2){=}NPh$	3.1(c)	87
	$(C_2)_2;C';C$	$N^*MeCH_2CH_2NMeP^*(OMe){=}NPh$	19.5(t)	74
	$(C_2)_2;C';C$	$O^*CH_2CH_2NMeP^*(NEt_2){=}NPh'F\text{-}4$	20.2	78
	$(C_2)_2;C';C$	$O^*CH_2CH_2NMeP^*$-$(NEt_2){=}NPh$	19.8	78
	$(C_2)_2;C';C$	$N^*MeCHMeCHPhOP^*(NMe_2){=}NPh$	18.7(m)	74
	$(C_2)_2;C';C$	$O^*CH_2CH_2NPhP^*(NEt_2){=}NPh'F\text{-}4$	8.0	78
	$C_2,CC';C';C$	$O^*CH_2CH_2NPhP^*(NEt_2){=}NPh$	8.1	78
$N_2N'O/5,5$	$CN,C'H;C;C$	**11**	21.2(c), 18.4(t)	87b
$NN'_2O/5,6$	$CH;(S^4)_2;C$	**12**; n = 1, X = O, Y = NH, Z = Cl	27.0^C	87c
	$CH;(S^4)_2;C$	**12**; n = 1, X = O, Y = NH, Z = Ph	22.5^T(m)	87c
$NN'_2O/6,6$	$CH;(S^4)_2;C$	**12**; n = 2, X = O, Y = NH, Z = Cl	5.2^C(m)	87c
	$CH;(S^4)_2;C$	**12**; n = 2, X = O, Y = NH, Z = Ph	4.6^T(d)	87c
N_3N'	$H_2,(CH)_2;P^4$	$(MeHN)_2(NH_2)P{=}NP^+\text{-}(NH_2)_3$ Cl^-	−16.5	33
	$Si_2,HSi,C_2;Si$	$(TmsHN)(Tms_2N)(Et_2N)P{=}NTms$	3.1	88
	$Si_2,HSi,C_2;Si$	$(TmsHN)(Tms_2N)(iPr_2N)P{=}NTms$	−4.6	88
	$CP^4,(C_2)_2;C$	$(Me_2N)_2P(O)N(Me)\mathbf{P}(NMe_2)_2{=}NMe$	21.6(b) J_{PP}18.4	90
	$CP^4,(C_2)_2;C$	$(Et_2N)_2P(O)N(Et)\mathbf{P}(NEt_2)_2{=}NEt$	13.6	89
	$CP,(C_2)_2;C$	$(Et_2N)_2PN(Et)\mathbf{P}(NEt_2)_2{=}NEt$	19.3	89
	$CSi,CH,C_2;C$	$tBuHNP(NTms\text{-}tBu)(NMe_2)P{=}N\text{-}tBu$	15.9	88
	$CSi,(C_2)_2;C$	$TmsEtNP(NEt_2)_2{=}NEt$	18.8	91
	$CH,(C_2)_2;P^4$	$EtHNP(NEt_2)_2{=}NP(NEt_2)_2{=}NEt$	18.7(tautomeric mix.)	89
	$CH,(C_2)_2;Si$	$tBuHNP(Mpip)(NEt_2){=}N\text{-}tBu$	−0.6	88
	$CH,(C_2)_2;H$	$EtNH(Et_2N)_2P{=}NH$	36.9	89
	$CH,(C_2)_2;C$	$EtNH(Et_2N)_2P{=}NEt$	19.3(b)	89
	$CH,(C_2)_2;C$	$EtHNP(NEt_2)_2{=}NCH_2CH_2NEt_2$	18.2	92
	$(C_2)_2,C'P^4;C'$	$(Me_2N)_2BrP^+NPh(Me_2N)_2\mathbf{P}{=}NPh$	21.6(c)	4
	$(C_2)_3;Cl$	$(Me_2N)_3P{=}NCl$	47.3(n)	93
	$(C_2)_3;Cl$	$(Bu_2N)_3P{=}NCl$	47.7(n)	93
	$(C_2)_3;Hg$	$(Me_2N)_3P{=}NHgMe$	33.5(b)	94
	$(C_2)_3;I$	$(Et_2N)_3P{=}NI$	46.5(n)	93
	$(C_2)_3;N'$	$(Me_2N)_3P{=}NN{=}C(COMe)_2$	40	55
	$(C_2)_3;N'$	$(Me_2N)_3P{=}NN{=}NPh$	42.8	95

P-bound atoms/rings	Connectivities	Structure	NMR data (δ_P[solv,temp] J_{pz} Hz	Ref.
	$(C_2)_3$;N′	$(Me_2N)_3P{=}NN{=}C^*CODopC^*O$	41	55
	$(C_2)_3$;N′	$(Me_2N)_3P{=}NN{=}C$-($COCH_2CH_2CMe_2CO$)	36	55
	$(C_2)_3$;N′	$Morph_3P{=}NN{=}NPh$	33	96
	$(C_2)_3$;N′	$(Me_2N)_3P{=}NN{=}CPh_2$	37.8(c)	14
	$(C_2)_3$;N′	$(Me_2N)_3P{=}NN{=}NPOPh_2$	25.1	95
	$(C_2)_3$;N′	$(Me_2N)_3P{=}NN{=}NPO(OPh)_2$	23.5	95
	$(C_2)_3$;N′	$(Me_2N)_3P{=}NN{=}C(COPh)_2$	38	55
	$(C_2)_3;P^4$	$(Me_2N)_3\mathbf{P}{=}N(Me_2N)_2P{=}NTms$	18.9(r)	62
	$(C_2)_3;P^4$	$(Me_2N)_3\mathbf{P}{=}NP^+(NEt_2)_3$	14.5(r)	93
	$(C_2)_3;P^4$	$(R_2N)_3P{=}NP^+(NR_2)_3$ Hal^- R = Et, Pr, Bu	13.0/14.5(r)	93
	$(C_2)_3;P^4$	$(Me_2N)_3\mathbf{P}{=}NP^+(NMePh)_3$ Hal^-	15.6(r)	93
	$(C_2)_3;P^4$	$(Me_2N)_3\mathbf{P}{=}NP^+Bu_3$	22.5(r)	99
	$(C_2)_3;P^4$	$[(Me_2N)_3\mathbf{P}{=}N]_3P(O)$	13.2(c) J_{PP}30	100
	$(C_2)_3;P^4$	$[(Me_2N)_3\mathbf{P}{=}N]_3P^+H$	20.1 J_{PP}28	97
	$(C_2)_3;P^4$	$(Me_2N)_3\mathbf{P}{=}NP(O)Ph_2$	25.1 J_{PP}64.5	95
	$(C_2)_3;P^4$	$(Me_2N)_3\mathbf{P}{=}NP(O)(OPh)_2$	23.5 J_{PP}22.2	95
	$(C_2)_3;P^4$	$(Et_2N)_3\mathbf{P}{=}NP^+Bu_3$	16.6(r)	99
	$(C_2)_3;P^4$	$(Me_2N)_3\mathbf{P}{=}NP(OCH_2Tf)_2{=}NPPh_2$	22.3(a)	81
	$(C_2)_3;P^4$	$(Me_2N)_3\mathbf{P}{=}NP(OCH_2Tf)_2{=}NP^+MePh_2$	25.3(a)	81
	$(C^2)_3;P^4$	$(Me_2N)_3\mathbf{P}{=}NPPh_2{=}NTms$	26.2(a)	81
	$(C_2)_3;P^4$	$(Me_2N)_3\mathbf{P}{=}NP(NMe_2)_2{=}NPPh_2$	17.0(a)	81
	$(C_2)_3;P^4$	$(Me_2N)_3\mathbf{P}{=}NPPh_2{=}NP(OCH_2Tf)_2$	24.7(a)	81
	$(C_2)_3;P^4$	$(Me_2N)_3P{=}NPPh_2{=}NP^+Me(OCH_2Tf)_2$	25.9(a/m)	81
	$(C_2)_3;P^4$	$(Me_2N)_3P{=}NP(NMe_2)_2{=}NP^+MePh_2$	18.7(a)	81
	$(C_2)_3$;P	$(Me_2N)_3\mathbf{P}{=}NPCl_2$	24.3 J_{PP}84	97
	$(C_2)_3$;P	$(Me_2N)_3\mathbf{P}{=}NPCl(NMe_2)$	27.6 J_{PP}85	97
	$(C_2)_3$;P	$(Et_2N)_3\mathbf{P}{=}NPCl_2$	19.6 J_{PP}79	97
	$(C_2)_3$;P	$(Me_2N)_3\mathbf{P}{=}NP(OCH_2Tf)_2$	24.3(n)	98
	$(C_2)_3$;P	$(Me_2N)_3\mathbf{P}{=}NP(NMe_2)_2$	23.6 J_{PP}107	97
	$(C_2)_3$;P	$[(Me_2N)_3\mathbf{P}{=}N]_2PNMe_2$	11.7 J_{PP}68	97
	$(C_2)_3$;P	$[(Me_2N)_3\mathbf{P}{=}N]_3P$	16.6 J_{PP}13	97
	$(C_2)_3$;P	$(Me_2N)_3\mathbf{P}{=}NPPh_2$	32.6(a)	81
	$(C_2)_3$;P	$(Me_2N)_3\mathbf{P}{=}NP(OPh)_2$	22.0(a)	98
	$(C_2)_3$;P	$[(Et_2N)_3\mathbf{P}{=}N]_2PCl$	19.0 J_{PP}12	97
	$(CC')_3$;P	$(MePhN)_3\mathbf{P}{=}NPCl_2$	6.6 J_{PP}83	97
	$(CC')_3$;P	$[(EtPh)_3\mathbf{P}{=}N]_2PCl$	4.7 J_{PP}10	97
	$(C_2)_3;S^4$	$(Me_2N)_3P{=}NSO_2Ph$	22.9(b)	101
	$(C_2)_3;S^4$	$Morph_3P{=}NSO_2Ph$	15.5	17
	$(C_2)_3;S^4$	$(Et_2N)_3P{=}NSO_2Ph$	19.9(b)	101
	$(C_2)_3$;S~	$(C^*H_2CH_2N^*)_3P(NEt_2)_2{=}NSO{\cdot}Ph$	25.6(b)	102
	$(C_2)_3;S^4$	$(Pr_2N)_3P{=}NSO_2Ph$	15.6(b)	102
	$(C_2)_3$;Si	$(Me_2N)_3P{=}NTms$	14.8(r)	62
	$(C_2)_3$;Si	$(Et_2N)_3P{=}NTms$	5.5, 8.5(n)	63, 93
	$(C_2)_3$;H	$(Me_2N)_3P{=}NH$	40.0	103
	$(C_2)_3$;H	$(C^*H_2CH_2N^*)_3P{=}NH$	53.8(n)	104
	$(C_2)_3$;H	$(C^*H_2CH_2N^*)_2P(NEt_2){=}NH$	50.3(n)	104
	$(C_2)_3$;H	$(C^*H_2CH_2N^*)P(NEt_2)_2{=}NH$	42.6(n)	104
	$(C_2)_3$;H	$(R_2N)_3P{=}NH$ R = Et, Pr, Bu, Hex, Allyl	37.0/37.5(b)	103
	$(C_2)_3$;H	$(Et_2N)_2P(Pip){=}NH$	33.5(b)	103
	$(C_2)_3$;H	$(iPr_2N)_2(Me_2N)P{=}NH$	32.2(b)	19
	$(C_2)_3$;H	$(iPr_2N)_3P{=}NH$	37.5	102
	$(C_2)_3$;C	$(Me_2N)_3P{=}NMe$	32.6(n), 23.5(n), 26.0	65, 101, 105
	$(C_2)_3$;C	$(Me_2N)_3P{=}NEt$	21.5(t) J_{PN} − 32.7	66, 106
	$(C_2)_3$;C	$(Me_2N)_3P{=}NCH_2CH_2NHTf$	32.0(b)	107

TABLE M (continued)
Compilation of ^{31}P NMR Data of Four Coordinate Phosphorus Compounds Containing a Formal Multiple Bond from Phosphorus to a Group V Atom

Section M1: Phosphazenes (Excluding Cyclopolyphosphazenes)

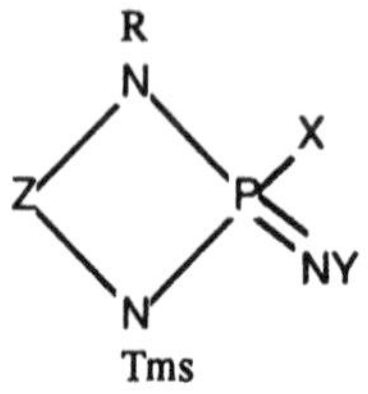

13

P-bound atoms/rings	Connectivities	Structure	NMR data (δ_P[solv,temp] J_{pz} Hz	Ref.
	$(C_2)_3$;C	$(Me_2N)_3P$=NCH_2CH_2NHMe	25.2(b)	107
	$(C_2)_3$;C	$(Me_2N)_3P$=$NCH_2CH_2NHP(O)(OEt)_2$	26(b)	107
	$(C_2)_3$;C	$(C^*H_2CH_2N^*)P(NEt_2)_2$=NEt	23.2	92
	$(C_2)_3$;C	$(R_2N)_3P$=NR R = Et, Pr, Bu	16.8/19.0(n)	101, 108
	$(C_2)_3$;C	$(Me_2N)_3P$=NCH_2CH_2NHPh	24.8(b)	107
	$(C_2)_3$;C	$(Et_2N)_3P$=NEt	17.9	101
	$(C_2)_3$;C	$(Et_2N)_2P(NPr_2)$=NEt	16.8	105
	$(C_2)_3$;C	$(Me_2N)_3P$=N(C*HCONPhC*HPh)	27	109
	$(C_2)_3$;C	$(Pr_2N)_3P$=NPr	16.8	82, 101
	$(C_2)_3$;C	$(C^*H_2CH_2N^*)P(NBu_2)_2$=NBu	22.3	92
	$(C_2)_3$;C	$(Et_2N)(iPr_2N)_2P$=$NCH_2CH_2NEt_2$	18.3(n)	4
	$(C_2)_3$;C	$YN(CH_2CH_2)_2NY$; Y = $P(NPr_2)_2$=NPr	15.7	89
	$(C_2)_3$;C′	$(Me_2N)_3P$=NCHO	35.6(b)	103
	$(C_2)_3$;C′	$(Me_2N)_3P$=NCOMe	33.4(b)	103
	$(C_2)_3$;C′	$(Et_2N)_3P$=NCOMe	30.0(b)	103
	$(C_2)_3$;C′	$(Me_2N)_3P$=NPh	15.9, 16.6(t), 19.5	17, 66, 95
	$(C_2)_3$;C′	$(Et_2N)_3P$=$NCOCCl_3$	37.3(n)	93
	$(C_2)_3$;C′	$(Et_2N)_3P$=NCOMe	31.7(n)	93
	$(C_2)_3$;C′	$(Me_2N)_3P$=NTol	25.3(t)	66
	$(C_2)_3$;C′	$Morph_3P$=$NC_6H_2Br3(2,4,6)$	1.4($PhNO_2$)	96
	$(C_2)_3$;C′	$Morph_3P$=$NPh'NO_2-4$	13.5	17
	$(C_2)_3$;C′	$Morph_3P$=NPh	7.7(b)	96
	$(C_2)_3$;C′	$(Et_2N)_3P$=NPh	14.2(t)	110
	$(C_2)_3$;C′	$Morph_3P$=$NCOCCl_3$	15.7	17
	$(C_2)_2$,CC′;H	$(Me;_2N)_2P(NMePh)$=NH	23.5(b)	19
	$(C_2)_2$;CC′;C	$(CH_2$=$CHNBu)P(NBu)_2$=NBu	10.8(n)	111
	$(CC')_3$;Cl	$(MePhN)_3P$=NCl	37.5(n)	93
$N_3N'/3$	(Si_2),C_2;Si	N*(tBu)N(tBu)P*$(NTms_2)$=NTms	−40.2(c)	112
$N_2N'/4$	$(P^5Si)_2$, Si_2;Si	**13**; R = Y = Tms, Z = R_3P	−8.6/−9.8	115
	P^5Si,Si_2,CP^5;Si	**13**; R = Me, Y = Tms, Z = $(Me_2N)_3P$	5.4(b)	113
	$(P^4Si)_2$,Si_2;Si	**13**; R = Y = Tms, Z = $(Me_2N)PMe$	−5.4	115
	$(P^4Si)_2$,Si_2;Si	**13**; R = Y = Tms, Z = (Me/Ph)$_2$P	−7.9/−9.0	115
	$(P^4Si)_2$,Si_2;Si	**13**; R = Tms, Y = Me, Z = Tms_2P(=NMe)	−23.5	115

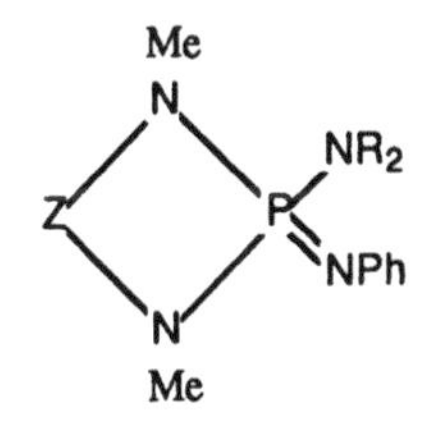

14

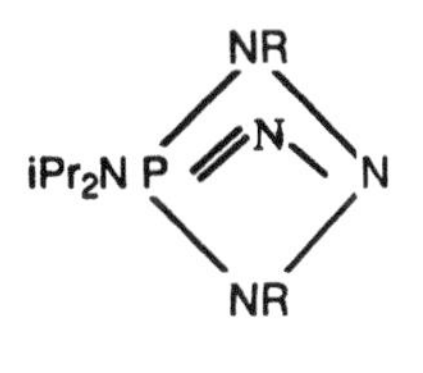

15

P-bound atoms/rings	Connectivities	Structure	NMR data (δ_P[solv,temp] J_{pz} Hz	Ref.
	$P^4Si,Si_2,CP^4;Si$	**13**; R = nAlk/Me, Y = Tms, Z = Ph_2P	−4.9/−3.3	115
	$P^4Si,Si_2CP^4;Si$	**13**; R = iPr, Y = Tms, Z = Ph_2P	−17.3	115
	$(S^4Si)_2,Si_2;Si$	**13**; R = Y = Tms, Z = Me_2S	6.8(r)	114
	$Si_2,(C'Si)_2;Si$	N*TmsC(NTms)-NTmsP*($Ntms_2$)=NTms	−9.8(r)	116
	$Si_2,C'Si,CC';Si$	N*(tBu)C(NTms)N-(Tms)P*($NTms_2$)=NTms	−7.9(r)	116
	$(CP^4)_2,CH;Si$	**1**; X = NH-tBu, Y = Tms, R = tBu	−29.1	117
	$(CC')_2;C'$	**14**; R = Et, Z = CO	11.6(c)	118
$N_3N'/5$	$N'Si,Si_2,CN';Si$	N*TmsN=NNEtP*($NTms_2$)=NTms	−0.3(c)	112
	$N'Si,Si_2,CN';Si$	N*TmsN=NN-(tBu)P*($NTms_2$)=NTms	−3.6(t)	112
	$(CN')_3;Si$	N*(tBu)N=NNEtP*(NTmsEt)=NTms	−9.1(c)	112
	$(CN')_2,CSi;Si$	N*(tBu)N=NN-(tBu)P*(NTms-tBu)=NTms	−5.6(b)	112
	$(CN')_2,C_2;C$	N*tBuN=NNEtP*(N-iPr_2)=NEt	−3.3(c)	112
	$(CN')_2,C_2;C$	N*tBuN=NNEtP*(N-iPr_2)=N-tBu	−28.3(t)	112
	$(CN')_2,C_2;C$	N*tBuN=NN(tBu)P*(N-iPr_2)=N-tBu	−36.8(c)	112
	$CN',(CH)_2;C'$	**7**; R = NHMe	47.5(c)	76b
	$CN',(C_2)_2;C'$	**7**; R = NMe_2	60.5(c)	76b
	$CH,(CC')_2;C'$	PhN*CH_2CH_2PhN-P*(NHtBu)=NPh	2.8(c)	74
	$(C_2)_3;C'$	(Me_2N)(MeN*CH_2CH_2MeN)P*=NPh	19.6(t)	66, 74
	$C_2,(CC')_2;S^4$	PhN*CH_2CH_2NPh-P*-(NMe_2)=NSO_2Tol	11.9(c)	74
	$C_2,(CC')_2;C$	PhN*CH_2CH_2PhNP*(NMe_2)=NEt	15.6(c)	74
	$C_2,(CC')_2;C'$	PhN*CH_2CH_2PhNP*(NMe_2)=NPh	1.5(t)	74
$N_3N'/4,4$	$(CN)_2,C_2;N$	**15**; R = tBu	23.9	118b
$N_3N'/5,5$	$CN',(CH)_2;C'$	**7**; XX = N(Me)CH_2CH_2N(Me)	55.4(c)	76b
$N_3N'/6,6$	$(C_2)_3;N'$	MeC(CH_2NMe)$_3$P=NN=NPh	33.7	95
	$(CN)_3;N'$	P(NMeNMe)$_3$P=NN=NPh	25.4, J_{PP}74.4	119
	$(CN)_3;P^4$	ZN=P(NMeNMe)$_3$P=NZ (Z = P[O]Ph_2)	2.5, 2 J_{PP}19.9, $^3J_{PP}$110.5	95
	$(CN)_3;P^4$	ZN=P(NMeNMe)$_3$P=NZ (Z = PO[OPh]$_2$)	1.0	95
	$(CN)_3;C'$	BrP(NMeNMe)$_3$P=NPh	−16.2	119
	$(CN)_3;C'$	P(NMeNMe)$_3$P=NPh	−1.8, J_{PP}78.2	119
	$(CN)_3;C'$	ChP(NMeNMe)$_3$P=NPh Ch = O, S, Se	−6.7/−7.3, J_{PP}102/106.5	119
	$(CN)_3;C'$	PhN=P(NMeNMe)$_3$P=NPh	−60, −6.6	95, 120
	$(C_2)_3;P^4$	MeC(CH_2NMe)$_3$P=NP(O)Ph_2	14.6	95
	$(C_2)_3;P^4$	MeC(CH_2NMe)$_3$P=NPO(OPh)$_2$	13.8 J_{PP}74.4	95
	$(C_2)_3;C'$	MeC(CH_2NMe)$_3$P=NPh	11.4(c)	95

TABLE M (continued)
Compilation of ^{31}P NMR Data of Four Coordinate Phosphorus Compounds Containing a Formal Multiple Bond from Phosphorus to a Group V Atom

Section M1: Phosphazenes (Excluding Cyclopolyphosphazenes)

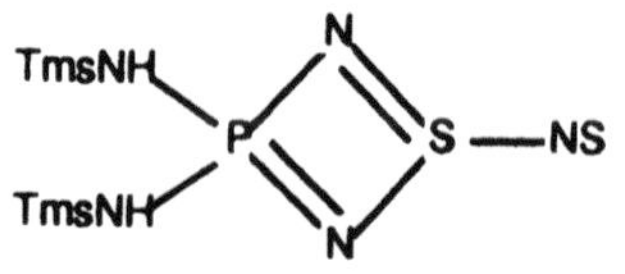

16 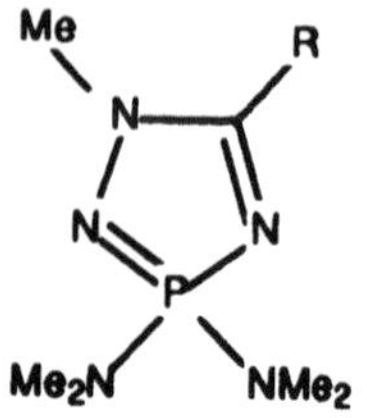 17 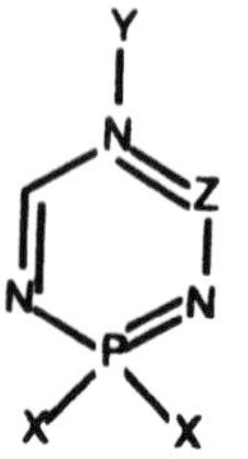 18

P-bound atoms/rings	Connectivities	Structure	NMR data (δ_P[solv,temp] J_{pz} Hz	Ref.
$N_2N'_2$	$(Si_2)_2$;N′,Si	TmsN=C=NP$(NTms_2)_2$=NTms	−11.8(r)	116
	$(C_2)_2$;N′,Si	$(iPr_2N)_2P(N_3)$=NTms	−9.3(b)	121
	$(C_2)_2$;N′,C′	$(iPr_2N)_2P(N_3)$=NPh	−8(b)	121
	$(C_2)_2;(P^4)_2$	$(Me_2N)_3P$=NP$(NMe_2)_2$=NP$^+$MePh$_2$	6.9(a)	81
	$(C_2)_2;P^4$,P	$(Me_2N)_3P$=NP$(NMe_2)_2$=NPPh$_2$	4.4(a)	81
	$(C_2)_2;P^4$,Si	$(Me_2N)_3P$=N$(Me_2N)_2$PNTms	1.3(r) J_{PP}45	62
	$(C_2)_2$;C′,N′	$(iPr_2N)_2$(OCN)P=NPh	6.4(b)	19
	$(C_2)_2$;(C′)$_2$	$(iPr_2N)_2$P(NCO)=NPh	6.4(b)	121
$N_2N'_2$/4	$(HSi)_2;(S^3)_2$	**16**	25.1(b)	121b
$N_2N'_2$/5	$(C_2)_2$;N,C′	**17**; R = Me	74.5	76b
	$(C_2)_2$;N,C′	**17**; R = Ph	73.0	76b
$N_2N'_2$/6	$(HSi)_2;(S)_2$	**10**; X = TmsHN, Y = Z = S	−1.5(b)	121b
	$(H_2)_2;(S^4)_2$	**10**; X = NH_2, Y = Z = SOF	10.3^C(m)	87c
	$(H_2)_2;(S^4)_2$	**10**; X = NH_2, Y = Z = SOPh	7.9^C; 7.1^T(m)	87c
	$(CH)_2;(S^4)_2$	**10**; X = NHMe, Y = Z = SOF	13.0^C(c)	121c
	$(CH)_2;(S^4)_2$	**10**; X = NHBu, Y = Z = SOCl	7.4	22b
	$(CH)_2;(S^4)_2$	**10**; X = NHBu, Y = Z = SOF	9.7^C; 9.7^T	22b
	$(C_2)_2$; P^4,C′	**18**; X = Me_2N, Y = Lp, Z = $P(NMe_2)_2$	27.7(n)	121d
	$(C_2)_2;P^4$,C′	**18**; X = Me_2N, Y = $P^B(NMe_2)$, Z = Me	18.3; P^B25.8	121d
	$(C_2)_2;(S^4)_2$	**10**; X = Pip, Y = Z = SOF	11.0(c)	22c
	$(C_2)_2;(S^4)_2$	**10**; X = NMe_2, Y = Z = SOAr	20.6^C; 14.1/14.7^T	33c
	$(C_2)_2;(S^4)_2$	**10**; X = Morph, Y = P(Morph)$_2$, Z = SOF	16.9(c)	22d
	$(C_2)_2;(S^4)_2$	**10**; X = Pip, Y = Z = SOPip	14.1^C; 14.9^T(c)	22e
	$(C_2)_2;(S^4)_2$	**10**; X = NMe_2, Y = Z = SOCl	19.3/19.5(c)	22f
	$(C_2)_2;(S^4)_2$	**10**; X = NMe_2, Y = Z = SOPh	15.1(c)	22f
	$(C_2)_2$;(C′)$_2$	**18**; X = Et_2N, Y = Lp, Z = CH	27.9(n)	121d
	$(C_2)_2$;(C′)$_2$	**18**; X = Et_2N, Y = Me, Z = CH	17.7(n)	121d
$N_2N'_2$/5,6	$(CH)_2;(S^4)_2$	**12**; n = 1, X = Y = NH, Z = Cl	18.7^C	87c
	$(CH)_2;(S^4)_2$	**12**; n = 1, X = Y = NH, X = F	27.0^C(m)	87c
	$(CH)_2;(S^4)_2$	**12**; n = 1, X = Y = NH, Z = Ph	20.0^T(d)	87c
	$(C_2)_2;(S^4)_2$	**12**; n = 1, X = Y = NMe, Z = Cl	13.1; 14.5^C(c)	121e
$N_2N'_2$/6,6	$(CH)_2;(S^4)_2$	**12**; n = 2, X = Y = NH, Z = Cl	8.0^C	87c
	$(CH)_2;(S^4)_2$	**12**; n = 2, X = Y = NH, Z = F	10.6^C(c)	87c
	$(CH)_2;(S^4)_2$	**12**; n = 2, X = Y = NH, Z = Ph	3.6^T(d)	87c

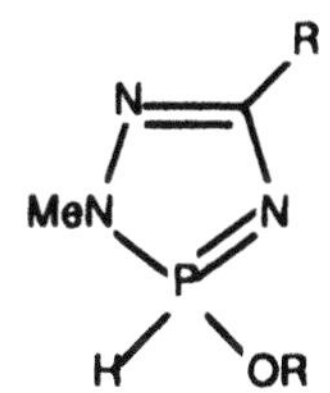

19

20

P-bound atoms/rings	Connectivities	Structure	NMR data (δ_P[solv,temp] J_{pz} Hz	Ref.
$HN'O_2/6$	$C;(C_2H)_2$	$(O^*CH_2CH_2CHMeOP^*H{=}N$- $C^*ClCHNRNO^*$	19.4 J_{PH}569	122
HNN'O	$Si_2;Si;Si$	$(Tms_2N)(TmsO)PH{=}NTms$	−17.1(c) J_{PH}596	123
	$Si_2;Si;C$	$(Tms_2N)tBuOPH{=}NTms$	−15.7(n) J_{PH}579	2
	$Si_2;Si;C$	$(Tms_2N)AdOPH{=}NTms$	−16.9(n) J_{PH}586	2
	$C_2;Si;C$	Mpip(MeO)PH=NTms	−6.1(b) J_{PH}577	8
HNN'O/5	N';C';C	**19**; R = $CHMeCH_2N_3$, R' = iPr	8.9(c) J_{PH}665	87b
$HN'_2O/6$	$P^4,S^4;C$	**20**; R = H, X = iPr, Y = PCl_2, Z = SOPh	1.4(c)	123b
	$(S^4)_2;C$	**20**; R = H, X = iPr, Y = Z = SOPh	−1.0(c) J_{PH}731.6	123b
HN_2N'	$(Si_2)_2;Si$	$(Tms_2N)_2PH{=}NTms$	−13.0(c) J_{PH}543	123
	$Si_2,CH;Si$	$Tms_2NPH(NH\text{-}tBu){=}NTms$	−21.1 J_{PH}519	88
	$Si_2,C_2;Si$	$Tms_2NPH(NEt_2){=}NTms$	−11.7 J_{PH}511	88
	$Si_2,C_2;Si$	$Tms_2NPH(Pip){=}NTms$	−8.9 J_{PH}521	88
	$(C_2)_2;Si$	$Mpip(Me_2N)HP = NTms$	7.3(b)	8
CBrClN'	$C_3;Si$	tBuPBrCl=NTms	3.7	124
CBrFN'	$C_3;Si$	tBuPBrF=NTms	9.8	124
CBrNN'	$Cl_3;C_2;Si$	$Mpip(Cl_3C)BrP{=}NTms$	−23.7(b)	8
	$H_3;Si_2;Si$	$MePBr(NTms_2){=}NTms$	2.6(c)	125
	$C_2H;Si_2;Si$	$(Tms_2N)iPrBrP{=}N\text{-}tBu$	−19.7(he)	6
CBr_2N'	$C_3;Si$	$tBuPBr_2{=}NTms$	−11.4	124
	$C_3;C$	$tBuPBr_2{=}N\text{-}tBu$	−44.4(c)	126
CClFN'	$C_3;Si$	tBuPClF=NTms	23.2	124
CClIN'	$C_3;Si$	tBuPClI=NTms	−29.0	124
CClN'O	$HSi_2;Si;C$	$Tms_2CHPCl(OMe){=}NTms$	15.2(c)	127
	$HSi_2;Si;C$	$Tms_2CHPCl(OCH_2Tf){=}NTms$	12.1(c)	127
CClNN'	$Cl_3;Si_2;Si$	$(Tms_2N)(Cl_3C)ClP{=}NTms$	−3.8(he)	6
	$HClSi;C_2;Si$	$(TmsClCH)PCl(NEt_2){=}NTms$	7.4	128
	$HCl_2;CH;C$	$(Cl_2CH)PCl(NH\text{-}tBu){=}N\text{-}tBu$	−35.1(b/c)	129
	$HCl_2;C_2;C$	$Cl_2CHPCl(NEt_2){=}NAllyl$	6.3(b/c)	129
	$HCl_2;C_2;C$	$(CHCl_2)PCl(NEt_2){=}N\text{-}iPr$	−6.7(b/c)	129
	$HCl_2;C_2;C$	$(CHCl_2)PCl(NEt_2){=}N\text{-}tBu$	−25.5(b/c)	129
	$HSi_2;Si_2;Si$	$Tms_2CHPCl(NTms_2){=}NTms$	15.6(c)	18
	$HSi_2;C_2;Si$	$Tms_2CHPCl(NMe_2){=}NTms$	21.8(c)	127
	$HSi_2;C_2;Si$	$(Tms_2CH)(Mpip)ClP{=}NSiMe_2Cl$	17.8(t)	18
	$HSi_2;C_2;Si$	$(Tms_2CH)(Mpip)ClP{=}NTms$	15.6(c)	18
	$HSi_2;C_2;Si$	$(Tms_2CH)(Mpip)ClP{=}NTms$	12.5(b)	18
	$H_2Cl;CH;C$	$ClCH_2PCl(NH\text{-}tBu){=}N\text{-}tBu$	−23.3(b/c)	129
	$H_2Cl;C_2;C$	$ClCH_2PCl(NEt_2){=}N\text{-}tBu$	−14.7(b/c)	129
	$H_2Cl;C_2;Si$	$ClCH_2PCl(NEt_2){=}NTms$	5.6(b/c)	129
	$H_2Si;Si_2;Si$	$TmsCH_2PCl(NTms_2){=}NTms$	20.1(c)	130
	$H_2Si;C_2;Si$	$TmsCH_2PCl(NEt_2){=}NTms$	17.2(r)	131
	$H_3;Si_2Si$	$MePCl(NTms_2){=}NTms$	20.3(c)	132
	$CHSi;Si_2;Si$	$TmsCHMePCl(NTms_2){=}NTms$	27.2(c)	132
	$C'HSi;Si_2;Si$	$TmsCHPhPCl(NTms_2){=}NTms$	11.4(c); 14.8^D(c)	132

TABLE M (continued)
Compilation of ^{31}P NMR Data of Four Coordinate Phosphorus Compounds Containing a Formal Multiple Bond from Phosphorus to a Group V Atom

Section M1: Phosphazenes (Excluding Cyclopolyphosphazenes)

P-bound atoms/rings	Connectivities	Structure	NMR data (δ_P[solv,temp] J_{pz} Hz	Ref.
	CH_2;Si_2;Si	EtPCl($NTms_2$)=NTms	25.1(c)	132
	$C'H_2$;Si_2;Si	$PhCH_2PCl$($NTms_2$)=NTms	15.1(c)	132
	C'_2;Si_2Si	(Tms_2N)PhP(Cl)=NTms	12.3(c)	123
CClNN'/5	$C'H_2$;HN';C'	N*HN=CMeCH_2P*(Cl)=NPh	36(c)	133
CClNN'/6	CH_2;$(C')_2$	**20**; R = Et, X = Cl, Y = CNH_2, Z = $CNMe_2$	47.0(r)	133b
CCl_2N'	Cl_3;Cl	CCl_3PCl_2=NCl	−35.8	42
	Cl_3;H	CCl_3PCl_2=NH	−28.0	42
	Cl_3;C	CCl_3PCl_2=NMe	−28.1	42
	Cl_3;C	CCl_3PCl_2=NEt	−34.6	42
	Cl_3;C	CCl_3PCl_2=N-iPr	−42.5	42
	Cl_3;C	CCl_3PCl_2=N-tBu	−53.5	42
	Cl_3;C'	CCl_3PCl_2=NPh	−28.4	42
	Cl_3;C'	CCl_3PCl_3=NTol(o)	−33.2	42
	Cl_3;C'	CCl_3PCl_2=NPh'NO_2 − 2	−16.8	42
	Cl_3;C'	CCl_3PCl_2=NAr	δP = −28.8 + 14.4σ	42
	H_3;C'	$MePCl_2$=NPh'Y − 4 Y = OMe/NO_2	−19.0/−3.1(c)	134
	H_3;C'	$MePCl_2$=NPh'CN(o)	−3.0	135
	H_3;C'	$MePCl_2$=NC_6H_3(2-CN,5-Me)	−4.4	135
	C_3;Si	$tBuPCl_2$=NTms	−19.7	124
	C'_2;S^4	$PhPCl_2$=NTos	13.7	136
$C'Cl_2N'$	C'_2;C'	$PhPCl_2$=$NCOCCl_3$	31.6	137
	C'_2;C'	$PhPCl_2$=NCOPh	31.6	124
C'FNN'	C'_2;C_2;C	Ph(Me_2N)FP=NMe	19.8(b) J_{PF}1033	46
$C'F_2N'$	C'_2;P^4	$PhPF_2$=NP(S)ClF	9.2 J_{PF}1105 J_{PP}47	138
	C'_2;P^4	$PhPF_2$=NP(S)Cl_2	5.9 J_{PF}1106 J_{PP}21	138
	C'_2;P^4	$PhPF_2$=NP(S)F_2	12.5 J_{PF}1095 J_{PP}	138
	C'_2;S^4	$PhPF_2$=NSO_2F	6.5 J_{PF}1127	138
	C'_2;C"	$PhPF_2$=NCN	15.3 J_{PF}1140	139
CINN'	H_3;Si_2;Si	MePI($NTms_2$)=NTms	−27.0(c)	125
	CH_2;Si_2;Si	EtPI($NTms_2$)=NTms	8.9(he); 10.7(c)	6, 125
	C_2H;Si_2;Si	iPrPI($NTms_2$)=NTms	−0.8(c)	125
	C_3;Si_2;Si	tBuPI($NTms_2$)=NTms	7.7(c)	125
CN'OS	H_3;P^4;C;C	MeP(OEt)(SEt)=NP(O)(OEt)2	46.6 J_{PP}34	59
	CH_2;S^4;C;C	EtP(OEt)(SMe)=NSO_2Ph'Cl-4	58.2(g)	140
$CN'O_2$	H_3;C';$(C')_2$	$MeP(OPh)_2$=NPh'Y − 4 Y = NO_2/OMe	26.3/27.7(c)	134
	C'HN';C';$(C)_2$	PhCH=NCHPhP$(OEt)_2$=NPh'Tf-3	2(b)	141
	C'HN';C';$(C)_2$	PhCH=NCHPhP$(OEt)_2$=NPh	10(b)	141
	C'HO;C';$(C)_2$	PhCH(OTms)P$(OEt)_2$=NCOMe	35	142
	CH_2;Si;$(C)_2$	$NCCH_2CH_2P(OEt)_2$=NTms	12	143
	CH_2;C';$(C)_2$	$MeCO_2CH_2CH_2P(OEt)_2$=NCOMe	32	144
	CH_2C';$(C)_2$	$Et_2NCOCH_2CH_2P(OEt)_2$=NPh	23	145
	CH_2;C';$(C)_2$	$ArNHCH_2CH_2P(OEt)_2$=NPh	29	145
	$C'H_2$;Si;Si,C'	MeO_2CCH_2P(OTms)-(OPh'OH-2)=NTms	23(c)	146
	C_2H;Si;Si_2	MeCHOCHMeP$(OTms)_2$=NTms	20(c)	146
	C_3;C';Si,C	tBuP(OMe)($OSiMe_2$Ph)=NCO_2tBu	29.1(j)	147

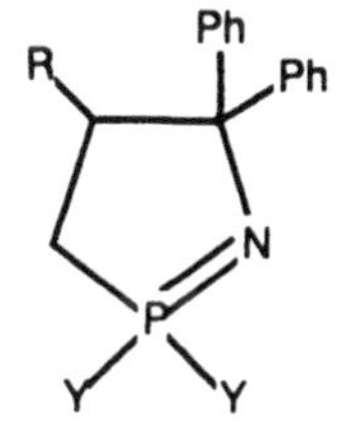

21 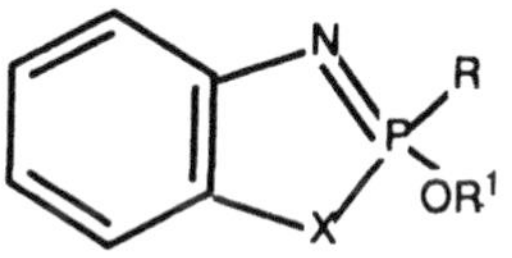 22 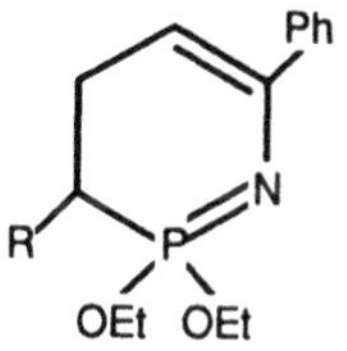 23 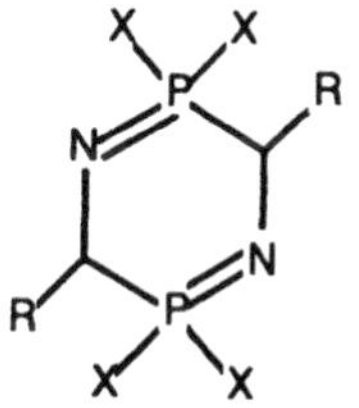 24

P-bound atoms/rings	Connectivities	Structure	NMR data (δ_P[solv,temp] J_{pz} Hz	Ref.
$CN'O_2/5$	$CH_2;C;(C)_2$	**21**; R = CO_2Me, Y = OMe	75.9	147b
	$CH_2;C';(C)_2$	**6**; R = Me, X = Br, Y = CN/CO_2Me, Z = CH_2	24.0	147c
	$CH_2;C';C,C'$	**22**; R = CH_2CH_2CN, R' = iPr, X = O	26	147d
	$C_3;C';(C')_2$	CatP(tBu)=NPh	37.2(PhCl, 20°)	12
$CN'O_2/6$	$C_2H;C';(C)_2$	**23**; R = C(=CHMe)P(O)$OMe)_2$	37	147e
$CN'O_2/5,6$	$H_2N';C;(C')_2$	**24**; XX = Cat, R = H	42.1, 40.7(b)	147f
	$C'HN';C;(C')_2$	**24**; XX = Glc, R = Ph	40.7, 38.3(b)	147f
$CN'O_2/5$	$Cl_3;C';(C')_2$	CatP(CCl_3)=NPh	−11.5(PhCl, 20°)	12
$C'N'O_2$	$N'S^4;C';(C)_2$	PhNHC(S)P$(OEt)_2$=NPh	10	148
	$C'N';C';(C)_2$	MeN=CPhP$(OEt)_2$=NPh	10(b)	141
	$C'N';C';(C)_2$	PhN=CPhP$(OEt)_2$=NPh	3(b)	141
	$CO';Si;(C)_2$	MeCOP$(OEt)_2$=NTms	16	149
	$C'_2;C;(C)_2$	Ph$(MeO)_2$P=NMe	16.7(n)	65
$C'N'O_2/5$	$C'_2;C;(C')_2$	PhP(Cat)=$NCH_2CH_2NHCOTf$	14.9	150
$C'N'O_2/8$	$C'_2;C';(C)_2$	O*CH_2CH_2N(tBu)-CH_2CH_2OP*Ph=NPh	4.3(b)	151
$CN'S_2$	$CH_2;C';(C)_2$	$NCCH_2CH_2P(SEt)_2$=NPh	63(n)	152
CNN'Cl	$H_2Si;C_2;Si$	$TmsCH_2P(NEt_2)$Cl=NTms	17.2(r)	153
CNN'O	$H_2Si;Si_2;Si;C$	$TmsCH_2P$(OMe)$(NTms_2)$=NTms	19.2(c)	130
	$C_3;Si_2;Si;C$	tBuP(OMe)$(NTms_2)$=NTms	25.2(c)	125
	$C_3;Si_2;Si;C$	t-BuP(OCH_2Tf)$(NTms_2)$=NTms	24.7(c)	125
CNN'O/3	$C_2O;Si_2;Si;C$	O*CTf_2P*$(NTms_2)$=NTms	−58.9	154
CNN'O/5	$H_3;C_2;C';C$	O*CH(tBu)CH_2N(tBu)P*Me=NPh	28.5	75
	$H_3;CC';C';C$	O*CH_2CH_2NPhP*Me=NPh'F(p)	19.0	78
	$H_3;CC';C';C'$	O*CtBu=CHN-tBuP*Me=NPh	30.8	75
	$C'HO;C_2;C';C$	PhCH(OTms)P*-(NMeCH_2CH_2O*)=NCOMe	47	142
	$C'HO;CC';Si;C$	PhCH(OTms)-P*(NPhCH_2CH_2O*)=NTms	16	155
	$C'HO;CC';C';C$	PhCH(OTms)-P*(NPhCH_2CH_2O*)=NPh	14	155
	$CH_2;C'H;C';C$	**22**; R = CH_2CH_2CN, R' = Et, X = NH	36	155b
$CN'_2O/5$	$CHO;(C')_2;C$	**25**; X = NCO, Y = $OCH_2(CF_2)_2H$, Z = CCl_3	18	155c
$CN'_2O/6$	$CH_2;(C')_2;C$	**20**; R = Et, X = OMe, Y = C(NH_2), Z = C(NMe_2)	55.9(r)	133b
C'NN'O	$C'_2;C'H;S^4;C$	PhP(NHPh)(OMe)=NTos	14.6	156
CNN'P	$H_2Si;C_2;Si;C'_2$	$TmsCH_2$**P**$(PPh_2)(NEt_2)$=NTms	17.1(r)	131
	$H_2Si;C_2;Si;N_2$	$TmsCH_2$**P**[P$(NTms_2)_2$]NEt_2)=NTms	26.9(r)	131
CNN'P/3	$C_3;CP;C;CN$	tBuN*P(tBu)**P***(tBu)=N-tBu	28.2 J_{PP}226	157
CNN'P/4	$C_3;SiP^4;Si;CP^4$	tBuP***P**-tBu(=NTms)-NTms**P***(tBu)=NTms	−3.7 J_{PP}174	158

TABLE M (continued)
Compilation of ^{31}P NMR Data of Four Coordinate Phosphorus Compounds Containing a Formal Multiple Bond from Phosphorus to a Group V Atom

Section M1: Phosphazenes (Excluding Cyclopolyphosphazenes)

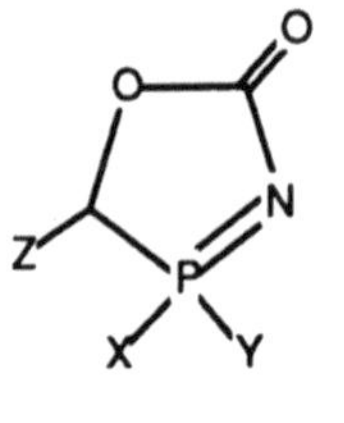

25

P-bound atoms/rings	Connectivities	Structure	NMR data (δ_P[solv,temp] J_{px} Hz	Ref.
$CN'_2P^4/6$	$H_3;(P^4)_2;C_2S'$	**20**; R = Me, X = P(S)Me_2, Y = Z = PPh_2	27.5(r) J_{PP}96	158b
	$H_3;(P^4)_2;C_2S'$	**20**; R = Me, X = P(=NTos)Me_2, Y = Z = PPh_2	24.4(r) J_{PP}144.5	158b
	$H_3;(P^4)_2;C_2S'$	**20**; R = Me, X = P(S)Ph_2, Y = Z = PPh_2	30.4(r) J_{PP}95	158b
CNN'S/3	HSSi;Si_2;Si;C	S*CHTmsP*($NTms_2$)=NTms	−43.6	159
$CN'_2Si/6$	$H_3;(P^4)_2;C_2S'$	**20**; R = Me, X = Tms Y = Z = PPh_2	26.6(r)	158b
$CN'_2Sn/6$	$H_3;(P^4)_2;C_2S'$	**20**; R = Me, X = $SnMe_3$, Y = Z = PPh_2	29.5(r) J_{PSn}1000	158b
CNN'S/3	HSSi;Si_2;Si;C	S*CH-tBuP*($NTms_2$)=NTms	−42.8	159
CN_2N'	HClSi;$(C_2)_2$;Si	(CHClTms)P(NEt_2)$_2$=NTms	7.1	128
	$H_2Cl;(C_2)_2$;C	ClCH_2P(NEt_2)$_2$=N-tBu	−8.2(b/c)	129
	$H_3;Si_2$,SiSn;Si	MeP($NTms_2$)(NTmsSnCl_3)=NTms	36.6	160
	$H_3;(Si_2)_2$;Si	MeP($NTms_2$)$_2$=NTms	8.5	132
	$H_3;Si_2$,HSi;Si	MeP(NHTms)($NTms_2$)=NTms	6.3(b)	161
	$H_3;Si_2$,HSi;C_3	MeP(NHTms)($NTms_2$)=N-tBu	0.7(b)	161
	$H_3;(C_2)_2;P^4$	Me(Me_2N)$_2$P=NPMe(NMe_2)=NTms	30.4(r)	99
	$H_3;(C_2)_2;P^4$	Me(Me_2N)$_2$P=NP^+Bu_3 I^-	34.0(r)	99
	$H_3;(C_2)_2$;C'	MeP(NMe_2)$_2$=NPh'Y−4 Y = NO_2/OMe	35.7/39.7(c)	134
	$H_3;(C_2)_2$;Si	Me(Me_2N)$_2$P=NTms	17.5(r)	99
$CN_2N'/3$	$H_2N;Si_2$,CSi;Si	C*H_2NTmsP*($NTms_2$)=NTms	−60.1	162
$CN_2N'/4$	$CHP^4;P^4Si;Si_2$;Si	TmsN*P($NTms_2$)-CHMeP*($NTms_2$)=NTms	−5.0(r)	163
	CHSi;$(P^4Si)_2$;Si	**13**; X = CHMeTms, Y = R = Tms, Z = P(X)=NTms	−5.6(r)	163
	$C'_2;(P^4Si)_2$;Si	**13**; X = Mes, Y = R = Tms, Z = PX=NTms	−19.7^C; −24.7^T(c)	164
$CN_2N'/5$	H_3;(CSi)$_2$;Si	MeN*$SiMe_2SiMe_2$NMP*Me=NTms	22.2	165
$CN_2N'/6$	H_3;(CSi)$_2$;Si	MeN*$SiMe_2$-O$SiMe_2$NMeP*Me=NTms	19.7(c)	165b
	H_3;(CSi)$_2$;Si	MeN*$SiMe_2$NMe$SiMe_2$-NMeP*Me=NTms	19.5(c)	165b
CNN'_2	$H_3;C_2;P^4$,Si	Me(Me_2N)2P=NPMe(NMe_2)=NTms	6.7(r)	99
$CNN'_2/6$	$H_3;C_2$;(C')$_2$	**20**; R = Me, X = NMe_2, Y = C(NH_2), Z = C(NMe_2)	47.5(r)	133b

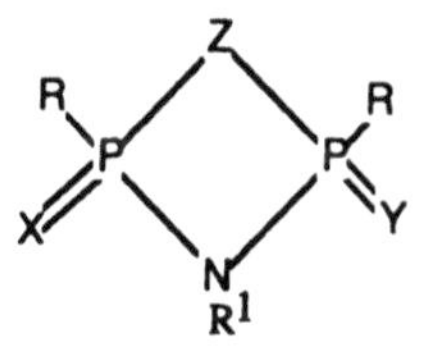

26

P-bound atoms/rings	Connectivities	Structure	NMR data (δ_P[solv,temp] J_{pz} Hz	Ref.
	$H_3;C_2;(C')_2$	**20**; R = Me, X = $N(CH_2)_4$, Y = $C(NH_2)$, Z = $C(NMe_2)$	41.3(r)	133b
	$CH_2;C_2;(C')_2$	**20**; R = Et, X = $N(CH_2)_2$, Y = $C(NH_2)$, Z = $C(NMe_2)$	55.7(r)	133b
$C'N_2N'$	$C'C;Si_2;HN';Si$	CH_2=CMeP-($NTms_2$)(NHN=CH_2)=NTms	3.1(r)	163
	$C'C;Si_2,HN';Si$	CH_2=CMeP($NTms_2$)(NHNHN*CH_2-NTmsP*($NTms_2$)=NTms	20.5(r)	163
	$C'C;Si_2,HN';Si$	CH_2=CMeP($NTms_2$)-(NHN=CMe_2)=NTms	1.6	166
	$C'C;HN',Si_2;Si$	CH_2=C(tBu)P($NTms_2$)-(NHN=CMe-tBu)=NTms	−5.8	166
	$C'_2;CSi,CH;Si$	PhP(NTms-tBu)(NH-tBu)=NTms	2.2(b)	161
	$C'_2;CSi,CH;C_3$	PhP(NTms-tBu)(NH-tBu)=N-tBu	−21.1(b)	161
	$C'_2;(C_2)_2;C$	$Ph(Me_2N)_2P$=NMe	29.3(n)	65
	$C'_2;(C_2)_2;C$	$Ph(Me_2N)_2P$=N(C*HCONPhC*HPh)	30	109
	$C'_2;(C_2)_2;C'$	$Ph(cC_4H_8N)_2P$=NPh'Y−4 Y = OMe/NO_2	11.5/15.7	134
$C'N_2N'/4$	$C'_2;(C'P^4)_2;C'$	**26**; R = Sms, R' = CO_2Et, X = Y = Z = NCO_2Et	42.6; 57.2[I]	166b
CHNN'	$Si_2;C_2;Si$	(Tms_2N)iPrHP=NTms	9.9(cd), J_{PH}444	123
	$Si_2;C_2;Si$	(Tms_2N)tBuHP=NTms	13.5(c) J_{PH}446	123
$CHN'_2/6$	$H_3;(C')_2$	**20**; R = H, X = Me, Y = $C(NH_2)$, Z = $C(NMe_2)$	23.9(w) J_{PH}512	133b
	$H_3;(C')_2$	**20**; R = D, X = Me, Y = $C(NH_2)$, Z = $C(NMe_2)$	23.4(w) J_{PD}77.5	133b
	$H_3;(C')_2$	**20**; R = H, X = Me, Y = $C(NH_2)$, Z = C(NHPh)	21.0(y) J_{PH}501	133b
	$CH_2;(C')_2$	**20**; R = H, X = Et, Y = $C(NH_2)$, Z = $C(NMe_2)$	37.5(w) J_{PH}502	133b
C_2BrN'	$(F_3)_2;Si$	Tf_2PBr=NTms	−50.5 $^2J_{PF}$125	167
	$(H_3)_2;Si$	Me_2BrP=NTms	6.8(b)	1
	$(CF_2)_2;Si$	$(C_2F_5)_2PBr$=NTms	−49.1, $^2J_{PF}$85	167
	$(C_3)_2;P$	$(tBu)_2$**P**Br=NP(iPr)($NTms_2$)	73.9(b), $^2J_{PP}$77.4	168
$CC'BrN'$	$H_2Si;C'_2;Si$	$TmsCH_2PBr$(Mes)=NTms	0.8(b)	1
	$(H_3);C'_2;Si$	MePhBrP=NTms	−0.1(b)	1
C'_2BrN'	$(C'_2);Si$	Ph_2BrP=NTms	−0.1(b)	1
C_2ClN'	$(Cl_3)_2;H$	$(Cl_3C)_2PCl$=NH	28.0	42
	$(Cl_3)_2;Cl$	$(Cl_3C)_2PCl$=NCl	35.8	42
	$(Cl_3)_2;C$	$(Cl_3C)_2PCl$=NMe	18.7	42
	$(Cl_3)_2;C$	$(Cl_3C)_2PCl$=NPr	27.4	42
	$(Cl_3)_2;C$	$(Cl_3C)_2PCl$=N-cHex	36.5	42
	$(F_3)_2;Si$	Tf_2PCl=NTms	−37.9	169
	$(HCl_2)_2;C$	$(ClCH_2)_2PCl$=N-tBu	58(b/c)	129
	$HCl_2,C_2H;Si$	Cl_2CH(iPr)PCl=NTms	22.2(b/c)	129
	$HSi_2,H_2Si;Si$	$Tms_2CHPCl(CH_2Tms)$=NTms	29.1(c)	127

TABLE M (continued)
Compilation of ^{31}P NMR Data of Four Coordinate Phosphorus Compounds Containing a Formal Multiple Bond from Phosphorus to a Group V Atom

Section M1: Phosphazenes (Excluding Cyclopolyphosphazenes)

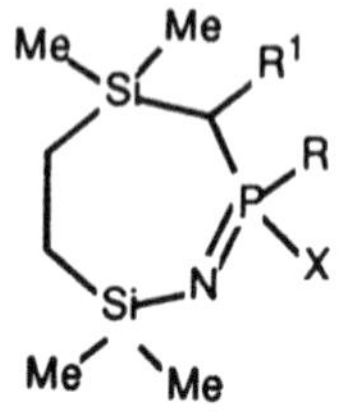

27

P-bound atoms/rings	Connectivities	Structure	NMR data (δ_P[solv,temp] J_{pz} Hz	Ref.
	$HSi_2,H_3;Si$	$Tms_2CHPClMe{=}NTms$	27.0(c)	127
	$HSi_2,C_3;Si$	$Tms_2CHPCl(tBu){=}NTms$	49.7(c)	127
	$H_2Cl,C_2H;Si$	$ClCH_2(iPr)PCl{=}NTms$	26.8(b/c)	129
	$H_2Cl,C_3;Si$	$ClCH_2(tBu)PCl{=}NTms$	29.9(b/c)	129
	$H_2Cl,C_3;P$	$ClCH_2(tBu)PCl{=}NPCl_2$	58.5(b/c) J_{PP}27	129
	$H_2Si,H_3;Si$	$tBuMe_2SiCH_2PClMe{=}NTms$	54.6(c)	127
	$(H_3)_2;S^4$	$Me_2ClP{=}NSO_2Cl$	59.2	23
	$H_3,C_3;Si$	$Me(tBu)PCl{=}NTms$	40(b/c)	129
	$(C'F_2)_2;Si$	$(C_2F_5)_2PCl{=}NTms$	−37.3	169
	$C'HSi,C'H_2;Si$	$CH_2{=}CHCH(Tms)PCl(Allyl){=}NTms$	25.5, 23.4^D(c)	127
	$C'HSi,C'H_2;Si$	$PhCH(Tms)PCl(CH_2Ph){=}NTms$	22.6, 24.0^D(c)	127
	$C_2Si,C_2H;Si$	$TmsCMe_2PCl(iPr){=}NTms$	54.0(c)	127
	C_2Si,C_3Si	$TmsCMe_2PCl(tBu){=}NTms$	57.4(c)	127
	$C_2H,C_2H;Si$	$iPr_2PCl{=}NTms$	50.7(c)	127
	$C_2H,C_3;Si$	$iPr(tBu)PCl{=}NTms$	54.6(c)	127
$C_2ClN'/6$	$H_3;(S^3)_2$	**20**; R = X = Me, YZ = SNbeS	1.8(c)	81b
	$H_3;(S)_2$	**20**; R = X = Me, Y = Z = S	6.2(c)	81b
$C_2ClN'/7$	$CHSi_2,C_3;Si$	**27**; R = tBu, R' = Tms, X = Cl	52.1; 44.2^D(c)	127
$CC'ClN'$	$H_2Si;C'_2;Si$	$TmsCH_2PClPh{=}NTms$	16.6(c)	127
	$HSi_2;C'_2;Si$	$Tms_2CHPClPh{=}NTms$	29.1(c)	127
	$H_3,C'_2;Si$	$iPrPClPh{=}NTms$	30.6(c)	127
	$CHSi;C'_2;Si$	$TmsCHMePClPh{=}NTms$	25.8; 27.0^D(c)	127
	$C'HSi;C'_2;Si$	$CH_2{=}CHCH(Tms)PClPh{=}NTms$	17.7, 16.5^D(c)	127
	$C'HSi;C'_2;Si$	$PhCH(Tms)PClPh{=}NTms$	18.6, 16.9^D(c)	127
	$CH_2,C'_2;Si$	$EtPClPh{=}NTms$	24.3(c)	127
	$C_2Si;C'_2;Si$	$TmsCMe_2PClPh{=}NTms$	35.9(c)	127
C'_2ClN'	$(H_3)_2;S^4$	$Ph_2ClP{=}NSO_2Cl$	36.5, 36.0($C_2H_2Cl_4$)	23, 33
	$(C'_2)_2;P^4$	$Ph_2ClP{=}NCl_2P(O)Cl_2$	30.0(n)	33
	$(C'_2)_2;P^4$	$Ph_2ClP{=}NSO_2Ph$	31.5($C_2H_2Cl_4$)	33
	$(C'_2)_2;P^4$	$Ph_2ClP{=}NCl_2P{=}NSO_2Ph$	34.0(re)	33
	$(C'_2)_2;P^4$	$Ph_2ClP{=}NCl_2P(O)Ph_2$	25.8(re)	33
	$(C')_2;P^4$	$Ph_2ClP{=}NCl_2P{=}NSO_2Cl$	8.3	23
	$(C'_2)_2;P^4$	$Ph_2ClP{=}NCl_2P{=}NCl_2P{=}NClP^+Ph_2$ Cl^-	37.0(re)	33
	$(C'_2)_2;P^4$	$Ph_2PCl{=}NPPh_2P{=}NPClPh_2$	33.9 J_{PP}7.9	39
	$(C'_2)_2;S^4$	$Ph_2PCl{=}NSO_2Cl$	36.5	170
	$(C'_2)_2;S^4$	$Ph_2PCl{=}NSO_2F$	38.5	33b
	$(C'_2)_2;C'$	$Ph_2PCl{=}NCOCCl_3$	40.0	137

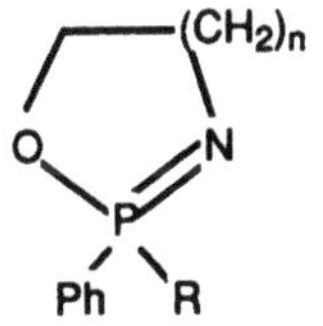

28

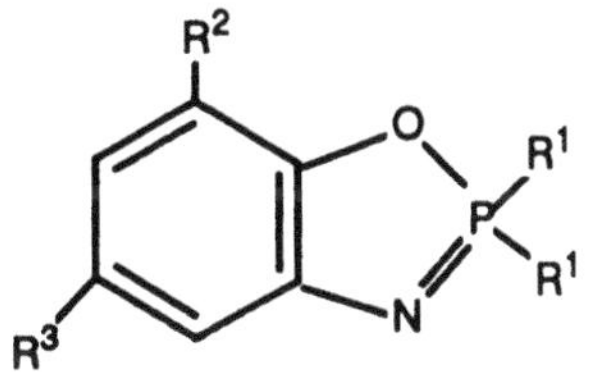

29

P-bound atoms/rings	Connectivities	Structure	NMR data (δ_P[solv,temp] J_{pz} Hz	Ref.
	$(C'_2)_2;C'$	$Ph_2PCl{=}NC(CCl_3){=}C(CN)_2$	21.4	171
	$(C'_2)_2;C'$	$Ph_2PCl{=}NPh$	14(re)	170
	$(C'_2)_2;C'$	$Ph_2PCl{=}NPh'CN$-4	16.9(r)	135
	$(C'_2)_2;C'$	$Ph_2PCl{=}NCOPh$	40.9	137
	$(C'_2)_2;C'$	$Ph_2PCl{=}NC_6H_3$(2-CN,5-Me)	18.5(r)	135
	$(C'_2)_2;C'$	$Ph_2PCl{=}NCPh{=}C(CN)_2$	28.3	171
C'_2FN'	$(C'_2)_2;C''$	$Ph_2PF{=}NCN$	50.3 J_{PF}1074	139
C_2IN'	$(C_2H)_2;S^4$	$iPr_2PI{=}NTos$	44.0	172
	$(C_2H)_2;C'$	$iPr_2PI{=}NPh$	37.5	172
	$(C_3)_2;Si$	$tBu_2PI{=}NP$(1-Ad)$(NTms_2)$	86.5(b), $^2J_{PP}$143	168
$C_2N'O$	$(H_3)_2;Si;Si$	$Me_2P(OTms){=}NTms$	13.5(b), 14.5(n)	173
	$(H_3)_2;Si;C$	$Me_2(TfCH_2O)P{=}NTms$	32.3(c)	1
	$(H_3)_2;Si;C$	$Me_2(TfCH_2O)P{=}NSiMe_2Cl$	38.8(r)	173b
	$(C_2H)_2;C;Si$	$iPr_2P(OSi\text{-}iPr_3){=}N\text{-}cC_5H_9$	25	174
	$(C_2H)_2;C';Si$	$iPr_2P(OSi\text{-}iPr_3){=}NPh$	28	174
	$(C_3)_2;C;Si$	$tBu_2P(OTms){=}N$-cPe	21	174
	$(C_3)_2;C';Si$	$tBu_2P(OTms){=}NPh$	24	174
$C_2N'O/7$	$H_2Si,H_3;Si;C$	**27**; R = Me, R' = H, X = OCH_2Tf	39.6(r)	173b
$CC'N'O$	$H_2Si;C'_2;Si;C$	$TmsCH_2PMes(OMe){=}NTms$	22.9(c)	164
	$H_2Si;C'_2;Si;C$	$TmsCH_2PMes(OCH_2Tf){=}NTms$	23.4(c)	164
	$C_3;C'_2;C';Si$	$tBuPPh(OTms){=}NCO_2$-tBu	41.9(j)	147
$CC'N'O/5$	$CH_2;C'_2;C;C$	**28**; n = 1, R = $CH_2CH_2P(O)(OEt)_2$	30(d)	174b
	$CH_2;C';C;C$	**28**; n = 1, R = $CH_2CH_2P(O)(OPh)_2$	20(d)	174b
$CC'N'O/6$	$CH_2;C'_2;C;C$	**28**; n = 2, R = CH_2CH_2CN	30(d)	174b
$C'_2N'O$	$(C'_2)_2;P^4;Ba$	$Ba[OP(Ph_2){=}NP(O)(OPh)_2]_2$	−8.9	56
	$(C'_2)_2;C;C$	$Ph_2P(OEt){=}NCTf_2CH_2CMe_2N{=}CTf_2$	4.9	67
$C'_2N'O/5$	$(C'_2)_2;C'C'$	**29**;R^1 = Ph, R^2 = R^3 = H	18.4(d)	198
	$(C'_2)_2;C';C'$	**29**; R^1 = Ph, R^2 = tBu, R^3 = CPh_3	23.6(d)	198
C2N'P	$(H_3)_2;P^4;C_2$	$Me_2\mathbf{P}(PMe_2){=}NSEt_2{=}NP(S)(OPh)_2$	27.8	175
	$(H_3)_2;S^4;C_2$	$Me_2\mathbf{P}(PMe_2){=}NTos$	30.9	175
	$(H_3)_2;S^4;C_2$	$Me_2\mathbf{P}(PMe_2){=}NSEt_2{=}NTms$	41.4	175
	$(H_3)_2;C';C_2$	$Me_2\mathbf{P}(PMe_2){=}NAc$	27.4	175
	$(CH_2)_2;P;(C)_2$	$Et_2P(PEt_2){=}NPEt_2$	32.0	172
	$(CH_2)_2;S^4;(C)_2$	$Et_2P(PEt_2){=}NTos$	41.0	172
	$(CH_2)_2;Si;(C)_2$	$Et_2P(PEt_2){=}NTms$	16.0	172
	$(CH_2)_2;C;(C)_2$	$Et_2P(PEt_2){=}NMe$	43.0	172
	$(C_2H)_2;P;(C)_2$	$iPr_2P(P\text{-}iPr_2){=}NP(iPr)_2$	37.0	172
	$(C_2H)_2;S^4;(C)_2$	$iPr_2P(P\text{-}iPr_2){=}NTos$	50.5	172
	$(C_2H)_2;H;(C)_2$	$iPr_2P(PEt_2){=}NH$	59.0	172
	$(C_2H)_2;C;(C)_2$	$iPr_2P(P\text{-}iPr_2){=}NMe$	52.0	172
	$(CH_2)_2;C;(C)_2$	$Et_2P(PEt_2){=}NPr$	40.0	172
	$(C_2H)_2;C;(C)_2$	$iPr_2P(P\text{-}iPr_2){=}NPr$	33.0	172
	$(CH_2)_2;C;(C)_2$	$Et_2P(PEt_2){=}NBu$	40.5	172

TABLE M (continued)
Compilation of ^{31}P NMR Data of Four Coordinate Phosphorus Compounds Containing a Formal Multiple Bond from Phosphorus to a Group V Atom

Section M1: Phosphazenes (Excluding Cyclopolyphosphazenes)

P-bound atoms/rings	Connectivities	Structure	NMR data (δ_P[solv,temp] J_{pz} Hz	Ref.
	$(CH_2)_2$;C;$(C)_2$	$Et_2P(PEt_2)$=N-tBu	3.0	172
	$(CH_2)_2$;C';$(C)_2$	$Et_2P(P\text{-}iPr_2)$=NPh	17.0	172
	$(C_2H)_2$;Ge;$(C)_2$	$iPr_2P(P\text{-}iPr_2)=NGeMe_3$	30.0	172
	$(C_2H)_2$;Si;$(C)_2$	$iPr_2P(P\text{-}iPr_2)$=NTms	26.0	172
	$(C_2H)_2$;C;$(C)_2$	$iPr_2P(P\text{-}iPr_2)$=N-tBu	11.0	172
	$(C_2H)_2$;C';$(C)_2$	$iPr_2P(PEt_2)$=NPh	27.0	172
$C'_2N'P$	$(C'_2)_2$;P^4;C'_2	$Ph_2P(PPh_2)=NP(S)(OPh)_2$	16.3	175
	$(C'_2)_2$;S^4;C'_2	$Ph_2P(PPh_2)$=NTos	21.0	175
	$(C')_2$;C';C'_2	$Ph_2P(PPh_2)$=NPh	17.2	175
$C_2N'S$	$(C'_2)_2$;P^4;C	$Ph_2P(SCH_2CN)=NP(S)Ph_2$	44	176
	$(C'_2)_2$;P^4;C	$Ph_2P(SR)=NP(S)Ph_2$	27/32	177
	$(C_2H)_2$;C;Ge	$iPr_2P(SP\text{-}iPr_2)$=N-tBu	27	178
	$(C_3)_2$;C;Ge	$tBu_2P(SGeMe_3)$=N-tBu	20	178
	$(C_3)_2$;C;P	$tBu_2P(SP\text{-}iPr_2)$=N-tBu	10	178
	$(C_3)_2$;C;P	$tBu_2P(SPPh_2)$=N-tBu	12.7	178
	$(C_3)_2$;C';P	$tBu_2P(SP\text{-}iPr_2)$=NPh	38	178
	$(C_3)_2$;C;Si	$tBu_2P(STms)$=N-tBu	18	178
$C'_2N'S$	$(C'_2)_2$;P^4;C	$Ph_2P(SCH_2Y)=NP(S)Ph_2$	26/27	178b
$C_2N'Si$	$(C_3)_2$;Si;Si	$tBu_2P(STms)$=NTms	93.5(b)	179
C_2NN'	$H_2Cl,C_2H;C_2$;Si	$ClCH_2P(iPr)(NEt_2)$=NTms	18.8(b/c)	129
	$(H_2N)_2$;C'	$(Me_2NCH_2)_2P(NMe_2)$=NPh	7.6	180
	$H_2Si,H_3;(Si_2)_2$;Si	$TmsCH_2PMe(NTms_2)$=NTms	17.3	132
	$(H_3)_2$;C'O;Si	$Me_2P(NPhOTms)$=NTms	27.2(cd)	181
	$(H_3)_2$;CP;C	$Me_2P(NMePMe_2)$=NMe	31.8(b)	182
	$(H_3)_2$;C^4;C	$Me_2P(NMePOMe_2)$=NMe	28.9(b)	182
	$(H_3)_2$;CH;C	$Me_2P(NHMe)$=NMe	27.3(r)	182
	$(H_3)_2$;C_2;Si	$Me_2P(NMe_2)$=NTms	21.2(c)	1
	$H_3,CH_2;Si_2$;Si	$MePEt(NTms_2)$=NTms	21.1(cd)	125
	$H_3,C_2H;Si_2$;Si	$MeP(iPr)(NTms_2)$=NTms	25.1(cd)	125
	$H_3,C_3;Si_2$;Si	$MeP(tBu)(NTms_2)$=NTms	30.2(cd)	125
	H_3,C_3;CP;C	$tBuMeP(NMePMe_2)$=NMe	45.8(b)	182
	H_3,C_3;CP^4;C	$tBuMeP(NMePSMe_2)$=NMe	41.1(b)	182
	H_3,C_3;CH;C	tBuMeP(NHMe)=NMe	42.8(r)	182
	$H_3,C_3;C_2$;C	$tBuMeP(NEt_2)$=NMe	42	183
	$H_3,C_3;C_2$;C'	$tBuMeP(NEt_2)$=NPh	29	183
$C_2NN'/3$	$(CH_2)_2;Si_2$;Si	$C^*H_2CH_2P^*(NTms_2)$=NTms	−66.9(r)	163
	$(CH_2)_2;C_2$;Si	$C^*H_2CMe_2P^*(NMe_2)$=NTms	−66.9(b)	184
	$CH_2,C_3;Si_2$;Si	$C^*H_2CMe_2P^*(NTms_2)$=NTms	−42.7(r)	163
$C_2NN'/4$	$(H_2P^4)_2;Si_2$;Si	$C^*H_2P(NTms_2)$(=NTms)- $CH_2P^*(NTms_2)$=NTms	−23.8(r)	163
	$H_3,H_2P^4;CP^4$;C	**26**; R = Me, R' = tBu, X = NH-tBu, Y = N-tBu, Z = CH_2	35.1(c)	184b
$C_2N'_2$	(H_3);P^4,Si	Me_3P=N**P**Me_2=NTms	6.7	190
	$(C_3)_2$;P^4,Si	Me_3P=N**P**$(tBu)_2$=NTms	27.8	190
$C_2N'_2/5$	$(H_3)_2;(S^3)_2$	**30**	−77(c)	184c
$C_2N'_2/6$	$H_3,C_2O;(C')_2$	**20**; R = Me, X = CMe_2OH, Y = $C(NH_2)$, Z = $C(NMe_2)$	53.1(ke)	133b
	$(C_2)_2;(S)_2$	**5**; X = Y = Me	−4.4(c)	184d
CC'NN'/4	$HP^4Si;C'_2;SiP^4$;Si	C*HTmsPMes(=CHTms)- NTmsP*Mes=NTms	−10.6(c)	164

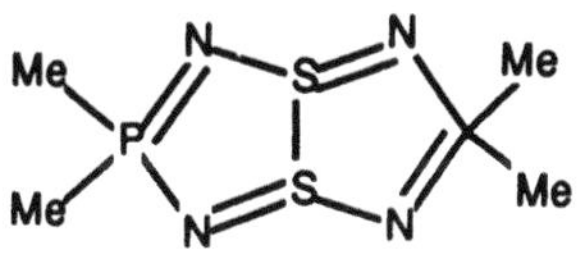

30

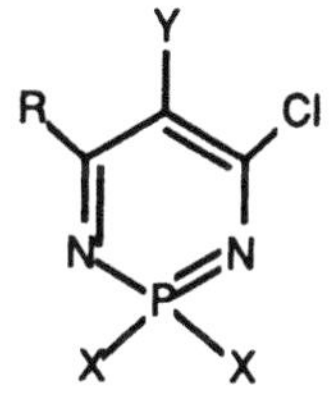

31

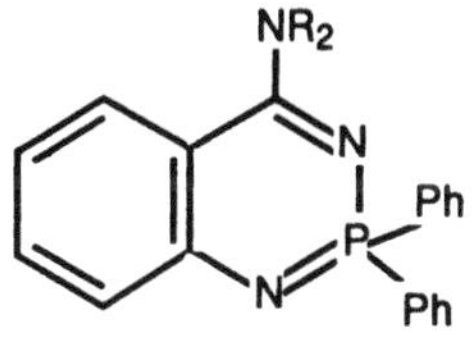

32

P-bound atoms/rings	Connectivities	Structure	NMR data (δ_P[solv,temp] J_{pz} Hz	Ref.
C'_2NN'	$(C'_2)_2;H_2;P^4$	$Ph_2P(NH_2)$=NP$^+$-$Ph_2NH_2Mo\ Cl_6^-$	19.8	39
	$(C'_2)_2;H_2;P^4$	Ph_2**P**(NH_2)=$NPPh_2$=$NPPh_2$=NH	17.7(y)	185
	$(C'_2)_2;H_2;P^4$	$Ph_2P(NH_2)$=$NPPh_2$=NP$^+Ph_2NH_2$	14.0, 14.7, J_{PP}0.6	39
	$(C'_2)_2$;CH;Si	Ph_2P(NH-tBu)=NTms	−6.4	186
	$(C'_2)_2$;CH;W	Ph_2P(NH-tBu) = NWF_4	15.9, J_{PW}115	187
	$(C'_2)_2;C_2$;C	$Ph_2(Me_2N)$P=N(C*HCONPhC*HPh)	2D	109
$C'_2N'_2$	$(C'_2)_2;N'_2$	TmsN(Na)**P**Ph_2=NTms	7.8	188
	$(C'_2)_2;(P^4)_2$	Me_3PN**P**Ph_2=NTms	2.9	188
	$(C'_2)_2;(P^4)_2$	Et_3PN**P**Ph_2=NTms	−4.1	190
	$(C'_2)_2;(P^4)_2$	Me_2PNPPh_2=NP$^+Me_3Cl^-$	5.8, J_{PP}1.5(m)	188
	$(C'_2)_2;(P^4)_2$	$(Me_2N)_3$P=$NPPh_2$=NP$^+$Me$(OCH_2Tf)_2$	0.3(a/m)	81
	$(C'_2)_2;(P^4)_2$	Ph_2PCl=$NPPh_2$=NP$^+$ClPh_2 Cl^-	14.0	39
	$(C'_2)_2;(P^4)_2$	$Ph_2P(NH_2)$=$NPPh_2$=NP$^+(NH_2)$Ph·Cl^-	4.0	39
	$(C'_2)_2;P^4$,P	$(Me_2N)_3$P=$NPPh_2$=NP$(OCH_2Tf)_2$	−1.0(a)	81
	$(C'_2)_2;P^4$,Si	$(Me_2N)_3$P=N**P**Ph_2=NTms	8.0(a)	81
	$(C'_2)_2;P^4$,Si	Me_3P=N**P**Ph_2=NTms	−2.9(r)	188
	$(C'_2)_2;Si_2$;Si	$Ph_2P(NTms_2)$=N**P**Ph_2=NTms	7.6	39
$C'_2N'_2/6$	$(C'_2)_2;(P^4)_2$	**20**; R = Me, X = P(S)Me_2, Y = Z = PPh_2	13.5(r), J_{PP}96	158b
	$(C'_2)_2;(P^4)_2$	**20**; R = Me, X = Tms, Y = Z = **P**Ph_2	9.9(r)	158b
	$(C'_2)_2;(S^3)_2$	**5**; X = Y = Ph	−21.3(c)	184d
	$(C'_2)_2;(S)_2$	**20**; R = X = Ph, Y = Z = S	−21.2(c)	190b
	$(C'_2)_2;(C')_2$	**31**; X = Ph, R = CCl_3	−37.3	171
	$(C'_2)_2;(C')_2$	**31**; X = Ph, R = Ph	−35.4	171
	$(C'_2)_2;(C')_2$	**32**; R = H	27.4(r)	135
	$(C'_2)_2;(C')_2$	**32**; R = Me	25.6(r)	135
C_2HN'	$(C'_2)_2;P^4$	tBu_2PH=NP$^+$MePh_2	47.1 J_{PH}444	189
	$(C'_2)_2;P^4$	t-Bu_2PH=NP$^+(CPh_3)Ph_2$	45.7 J_{PH}456	189
C'_2HN'	$(C'_2)_2;P^4$	Ph_2PH=NP$^+$Me$(tBu)_2$	5.1 J_{PH}498	189
	$(C'_2)_2;P^4$	Ph_2PH=NP$^+$MePh_2	6.5 J_{PH}509	189
C_3N'	$(H_3)_3;P^4$	Me_3**P**=NP$(tBu)_2$N=AsMe_3	8.0(r)	191
	$(H_3)_3;P^4$	Me_3**P**=NP$(tBu)_2$N=PMe_3	13.2(r)	191
	$(H_3)_3;P^4$	Me_3**P**=NP$^+Bu_3$	24.6(r)	99
	$(H_3)_3;P^4$	Me_3**P**=$NPPh_2$=NAsMe_3	22.7(c)	191
	$(H_3)_3;P^4$	Me_3**P**=$NPPh_2$=NTms	18.3(m)	188
	$(H_3)_3;P^4$	Me_3**P**=$NPPh_2$N=PMe_3	24.4(m)	188
	$(H_3)_3$;P	Me_3P=NP$(OCH_2Tf)_2$	15.1(n)	98
	$(H_3)_3$;P	Me_3P=NP-tBu_2	13.3	190
	$(H_3)_3$;P	Me_3P=NPPh_2	22.8(a)	81
	$(H_3)_3$;P	Me_3P=NP$(OPh)_2$	13.2(a)	98
	$(H_3)_3$;Si	Me_3P=NTms	−3.5	192
	$(H_3)_3$;W	$(Me_3P=N)_2WF_4$	37.5 J_{PW}80	187
	$(H_3)_3$;H	Me_3P=NH	12.8	193
	$(H_3)_3$;C″	Me_3**P**=NCN	36.0	194
	$(H_3)_2$,C′H_2;Si	$PhCH_2PMe_2$=NTms	0.3(r)	195

TABLE M (continued)
Compilation of ^{31}P NMR Data of Four Coordinate Phosphorus Compounds Containing a Formal Multiple Bond from Phosphorus to a Group V Atom

Section M1: Phosphazenes (Excluding Cyclopolyphosphazenes)

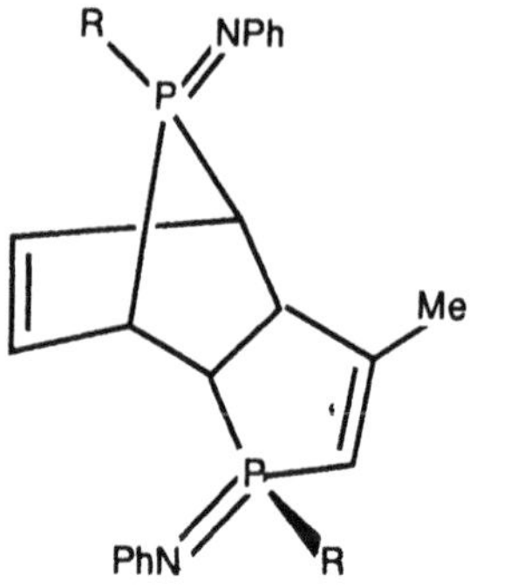

33

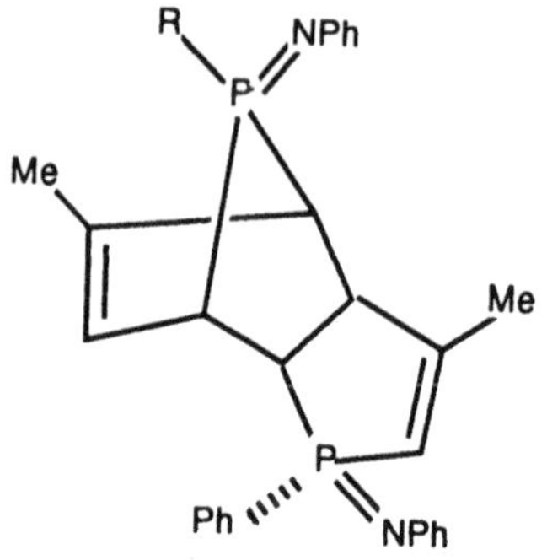

34

P-bound atoms/rings	Connectivities	Structure	NMR data (δ_P[solv,temp] J_{pz} Hz	Ref.
	$(H_3)_2,C'H_2;C$	$PhCH_2PMe_2{=}NMe$	13.2(r)	195
	$H_3,CH_2,C'H_2;Si$	$PhCH_2PMeEt{=}NTms$	9.3(r)	195
	$H_3,CH_2,C'H_2;C$	$PhCH_2PMeEt{=}NMe$	18.2(r)	195
	$H_3,(C_3)_2;C'$	$tBu_2MeP{=}NPh$	20	183
	$(CH_2)_3;P^4$	$Et_3P{=}NP(tBu)_2N{=}PEt_3$	38.1(c)	191
	$(CH_2)_3;P^4$	$Et_3P{=}NPPh_2{=}NTms$	−32.9	190
	$(CH_2)_3;P^4$	$Bu_3P{=}NP^+Bu_3$	32.9(r)	99
	$(CH_2)_3;P^2$	$Bu_3P{=}NP^+N(iPr_2)_2$	46.7	196
	$(CH_2)_3;Si$	$Et_3P{=}NTms$	14.1	192
	$(CH_2)_3;C$	$Bu_3P{=}N\text{-}iPr$	11.0	197
	$(CH_2)_3;C$	$Bu_3P{=}NBu$	15.1	197
	$(CH_2)_3;C$	$Bu_2P{=}N\text{-}sBu$	10.1	197
	$(CH_2)_3;C$	$Bu_3P{=}N\text{-}cHex$	11.6	197
	$(CH_2)_3;C$	$Bu_3P{=}NCH_2tBu$	13.3	197
	$(CH_2)_3;C'$	$Bu_3P{=}NAr$	0.7(b)	198
	$(C_2H)_3;C'$	$iPr_3P{=}NAr$	0.0(b)	198
	$(C_3)_3;Ge$	$tBu_3P{=}NGeMe_3$	32	199
	$(C_3)_3;P^2$	$tBu_3P{=}NP^+N(iPr)_2$	46.7 J_{PP}50	199b
	$(C_3)_3;Si$	$tBu_3P{=}NTms$	31.9	192
	$(C_3)_3;H$	$tBu_3P{=}NH$	55.8	200
$C_3N'/5$	$(H_3)_2;CH_2;C$	**21**; R = CO_2Me, Y = Me	40.1	147b
$C_3N'/5,5$	$H_3,(CC'H)_2;C'$	**33**; R = Me	79.7 J_{PP}41.5	200b
$C_2C'N'$	$H_2Si,H_3;C'_2;Si$	$TmsCH_2PMe(Mes){=}NTms$	3.8(c)	201
	$(H_3)_2;SS';Si$	$Me_2P(CS_2Me){=}NTms$	10.5	201b
	$(H_3)_2;C'_2;Ge$	$Me_2PhP{=}NGeMe_3$	−0.2	199
$C_2C'N'/5$	$H_3,C_2H;C'H;C'$	**33**; R = Me	65.9 J_{PP}41.5	200b
$C_2C'N'/5,5$	$(CC'H)_2;C'_2;C'$	**33**; R = Ph	71.9 J_{PP}41.5	200b
	$(CC'H)_2;C'_2;C'$	**34**; R = Ph	60.5 J_{PP}36.6	200b
	$(CC'H)_2;C'_2;C'$	**35**	79.9 J_{PP}36.6	200b
CC'_2N'	$H_2P^4;(C'_2)_2;Si$	$CH_2(PPh_2{=}NTms)_2 \times SnCl_2$	27.7	186
	$H_3;(C'_2)_2;C'$	$MePh_2P{=}NPh'Y-4$ Y = Me/NO_2	6.6/9.9(c)	134
	$H_3;(C'_2)_2;C'$	$2\text{-}Tol_2MeP{=}NPh'Y-4$ Y = Me/NO_2	5.9/9.8(c)	134

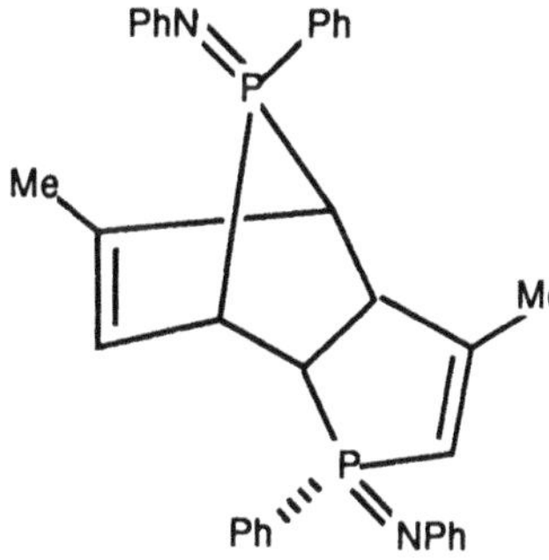

35

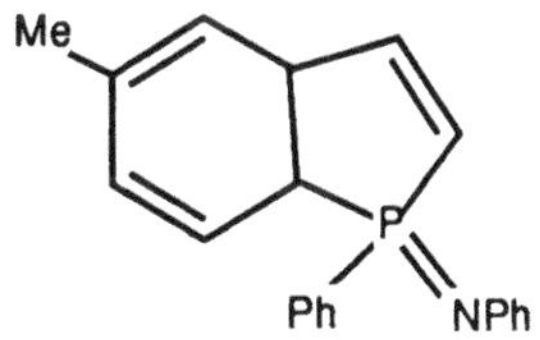

36

P-bound atoms/rings	Connectivities	Structure	NMR data (S_P[solv,temp] J_{pz} Hz	Ref.
	$H_3;(C'_2)_2;C'$	$Mes_2MeP{=}NPh'Y-4$ Y = OMe/ NO_2	3.9/9.9(c)	134
	$CH_2;(C'_2)_2;Si$	Ph_2**P**$({=}NTms)CH_2CH_2P(Ph_2){=}NCN$	1.5 J_{PP}50	194
	$CH_2;(C'_2)_2;C'$	$EtPPh_2{=}NPh$	8.5	202
	$CH_2;(C'_2)_2;C'$	$AllylCH_2PPh_2{=}NPh$	6.7	202
	$CH_2;(C'_2)_2;C''$	YCH_2CH_2Y (Y = $PPh_2{=}NCN$)	27.2	194
	$C_2H;(C'_2)_2;C'$	$iPrPPh_2{=}NPh$	13.8	202
	$C'H;(C'_2)_2;C'$	$PhEtCHPPh_2{=}NPh$	8.7	202
$CC'_2N'/_5$	$CH_2;(C'_2)_2;C$	**21**; R = CN, Y = Ph	48.2	147b
	$C_2H;C'H,C'_2;C'$	**33**; R = Ph	57.3 J_{PP}41.5	200b
	$C_2H;C'H,C'_2;C'$	**34**; R = Ph	33.6 J_{PP}36.6	200b
	$C_2H;C'H,C'_2;C'$	**35**	33.6 J_{PP}36.6	200b
	$CC'H;C'H,C'_2;C'$	**36**	64.3	200b
C'_3N'	$C'H,(C'_2)_2;C'$	$PhC(NH_2){=}CHPPh_2{=}NPh$	5.0	202
	$(C'_2)_3;B$	$Ph_3P{=}NBF_2$	36	203
	$(C'_2)_3;B$	$Ph_3P{=}NBFPh$	7.8	203
	$(C'_2)_3;Fe$	$(Ph_3P{=}NFeCl_2)_2$	37.1	186
	$(C'_2)_3;N$	$Ph_3P{=}NN(CO_2Me)_2$	45.9	204
	$(C'_2)_3;N'$	$Ph_3P{=}NN{=}CH_2$	21.4	205
	$(C'_2)_3;N'$	$Ph_3P{=}NN{=}NC_6H_2(2,4,6\text{-}NO_2)$	25.5(v)	96
	$(C'_2)_3;N'$	$Ph_3P{=}NN{=}CPh_2$	18.3(r)	14
	$(C'_2)_3;N'$	$Ph_3P{=}NN{=}CPhNHAr$	19.3/20.9	206
	$(C'_2)_3;N'$	$Ph_3P{=}NN{=}CPhOP^+Ph_3\ Cl^-$	25.4	206
	$(C'_2)_3;P$	$Ph_3P{=}NP(OCH_2Tf)_2$	10.8(a)	98
	$(C'_2)_3;P^4$	$Ph_3P{=}NP(O)Glc$	13.5 J_{PP}27.5	73
	$(C'_2)_3;P^4$	$Ph_3P{=}NP(O)Cat$	14.6, 15.0	73
	$(C'_2)_3;P$	$Ph_3P{=}NPPh_2$	17.4(a)	81
	$(C'_2)_3;P$	$Ph_3P{=}NP(OPh)_2$	7.6(a)	98
	$(C'_2)_3;S^4$	$Ph_3P{=}NSO_2Ph'Y$ Y = NO_2;NH_2	16.7; 14.3(c)	207
	$(C'_2)_3;Se$	$Ph_3P{=}NSeCl_3$	33.2 J_{PSe}87	208
	$(C'_2)_3;Si$	$Ph_3P{=}NTms$	−1.5, −1.8	192, 205
	$(C'_2)_3;V$	$Ph_3P{=}NV(O)Cl_2$	18.9	186
	$(C'_2)_3;V$	$(Ph_3P{=}N)_4VCl$	6.0 J_{PV}119	186
	$(C'_2)_3;W$	$Ph_3P{=}NWF_3$	35.5	186
	$(C'_2)_3;W$	$Tol_3P{=}NWF_3$	36.2	187
	$(C'_2)_3;W$	$(Tol_3P{=}N)_2WF_4$	23.3	187
	$(C'_2)_3;H$	$Ph_3P{=}NH$	20.8	209
	$(C'_2)_3;C$	$Ph_3P{=}NMe$	11.7	209
	$(C'_2)_3;C$	$Ph_3P{=}N\text{-}iPr$	−2.2	197
	$(C'_2)_3;C$	$Ph_3P{=}NCH_2tBut$	5.6	197
	$(C'_2)_3;C$	Ph_3**P**$=NCPh_3$	−10.3	205
	$(C'_2)_3;C'$	$Ph_3P{=}NCOCCl_3$	23.8	137

TABLE M (continued)
Compilation of ^{31}P NMR Data of Four Coordinate Phosphorus Compounds Containing a Formal Multiple Bond from Phosphorus to a Group V Atom

Section M1: Phosphazenes (Excluding Cyclopolyphosphazenes)

P-bound atoms/rings	Connectivities	Structure	NMR data (δ_P[solv,temp] J_{px} Hz	Ref.
	$(C'_2)_3$;C'	Ph_3P=NCONH$_2$	20.0	209
	$(C'_2)_3$;C'	Ph_3P=NCONMe$_2$	18.2	209
	$(C'_2)_3$;C'	2-ThienylP(Ph)=NPh'Y−4 Y = Me/NO$_2$	11.9/14.2(c)	134
	$(C'_2)_3$;C'	Ph_3P=NC$_6$H$_2$(2,4,6-Br$_3$)	−2.1(ae)	96
	$(C'_2)_3$;C'	Ph_2P=NC$_6$H$_3$(2-OH)(5-Br)	8.0	69
	$(C'_2)_3$;C'	Ph_3P=NC$_6$H$_3$(2-OH)(5-F)	9.6	69
	$(C'_2)_3$;C'	Ph_3P=NPh'Y−4 Y = OMe/NO$_2$	2.4/7.3(c)	134
	$(C'_2)_3$;C'	Ph_3P=NPh	−1.2(b), 3.7(d)	198, 209
	$(C'_2)_3$;C'	Ph_3P=NPh'OH-2	7.1, 7.6(p)	69, 198
	$(C'_2)_3$;C'	Ph_3P=NPh'OMe-2	3.8, 0.8(bd), 1.0(b)	69, 198
	$(C'_2)_3$;C'	Ph_3P=NCOPh	14.9	137
	$(C'_2)_3$;C'	4-MeOPh'$_3$P=NAr	−7.4(b)	198
	$(C'_2)_3$;C'	Ph_3P=NC*=CHC-(CN)=C(CN)CXC*(CO$_2$R)Y	10.3, 11.5	210
	$(C'_2)_3$;C'	PhC(NH$_2$)=CMePPh$_2$=NPh	14.6	202
	$(C'_2)_3$;C"	Ph_3P=NCN	23.1	194
C'$_2$N'/5	$(C'_2)_3$;C'	PhC*=CHCH=CPhP*(Ph)=NY−4 Y = OMe/NO$_2$	12.6/16.4(c)	134
C'C"$_2$N'	C'$_2$;(N")$_2$;S^4	PhP(CN)$_2$=NTos	−19.0	136

Section M2: Cyclopolyphosphazenes, (N=PX$_2$)n, and Their Derivatives

All compounds in this section are cyclic, the ring size being defined by the heading of each subsection. Any additional rings, e.g., Spiro derivatives, have been ignored. Two of the four P-bound atoms (N'$_2$) with connectivities P_2^4 are common to all compounds; hence only the two remaining P-bound atoms and their connectivities are shown here.

Connectivities	Molecular formula	Structure	NMR Data (δ_P[solv, temp] $^nJ_{PP}$Hz)	Ref.
		Section M2.1: Cyclodiphosphazenes		
	Remaining P-bound atoms = N$_2$			
$(C_2)_2$	$C_{24}H_{56}N_6P_2$	1; X = N-iPr$_2$	40(t)	1
		Section M2.2: Cyclotriphosphazenes		
	Remaining P-bound atoms = BrCl			
—	$BrCl_5N_3P_3$	2; R = Br, X = Cl	−7.8(b)	2
—	$Br_2Cl_4N_3P_3$	3; R = Br, X = Cl	−8.7(b)	2
—	$Br_3Cl_3N_3P_3$	4; R = Br, X = Cl	−9.8(b)	3
—	$Br_3Cl_3N_3P_3$	5; R = Br, X = Cl	−10.0(b)	2
—	$Br_4Cl_2N_3P_3$	3; R = Cl, X = Br	−12.1(b)	2
—	$Br_5ClN_3P_3$	2; R = Cl, X = Br	−14.0(b)	2

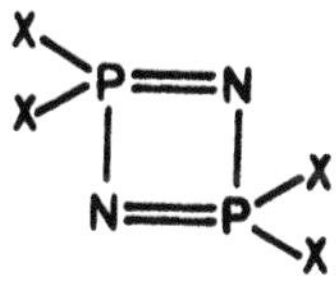

Connectivities	Molecular formula	Structure	NMR Data (δ_P[solv, temp] $^nJ_{PP}$Hz)	Ref.
	Remaining P-bound atoms = BrF			
—	$Br_2F_4N_3P_3$	3; R = Br, X = F	2.64(2) 3.31^T(1)[a]	4
	Remaining P-bound atoms = BrN			
C_2	$C_2H_6Br_5N_4P_3$	2; R = NMe_2, X = Br	4.5(k) $^2J_{PP}$18	5
	$C_4H_{12}Br_4N_5P_3$	3; R = NMe_2, X = Br	10.0^T(k)	5
	$C_6H_{18}Br_3N_6P_3$	4; R = NMe_2, X = Br	15.4^T(k)	5
	Remaining P-bound atoms = Br_2			
—	$Br_3Cl_3N_3P_3$	5; R = Br, X = Cl	−39.8(b)	2
—	$Br_4Cl_2N_3P_3$	3; R = Cl, X = Br	−41.3(b)	2
—	$Br_5ClN_3P_3$	2; R = Cl, X = Br	−42.5(b)	2
—	$Br_6N_3P_3$	2; R = X = Br	−45.4(b)	2
—	$Br_4H_4N_5P_3$	6; R = NH_2, X = Br	−43(k)	6
—	$C_2H_6Br_5N_4P_3$	2; R = NMe_2, X = Br	−39.3(k)	5
—	$C_4H_{12}Br_4N_5P_3$	3; R = NMe_2, X = Br	−36.6^T(k)	5
	Remaining P-bound atoms = ClCr			
C'_8	$C_8H_5Cl_5CrN_3O_3P_3$	7; R = X = Cl, ML_n = $Cr(CO)_3(Cp)$	140.3 $^2J_{PP}$88	7
	Remaining P-bound atoms = ClF			
—	$Cl_4F_2N_3P_3$	3; R = F, X = Cl	18.4^T	4
—	$Cl_5FN_3P_3$	2; R = F, X = Cl	14.5	8
	Remaining P-bound atoms = ClN			
H_2	$Cl_5H_2N_4P_3$	2; R = NH_2; X = Cl	19.0(c) $^2J_{PP}$46.5	9
CH	$CH_4Cl_5N_4P_3$	2; R = NHMe, X = Cl	21.0 $^2J_{PP}$47.3	10, 11
	$C_2H_5Cl_6N_4P_3$	2; R = $NHCH_2CH_2CH_2Cl$, X = Cl	19.2(c) $^2J_{PP}$47	12
	$C_2H_6Cl_5N_4P_3$	2; R = NHEt, X = Cl	18.7(r)	13

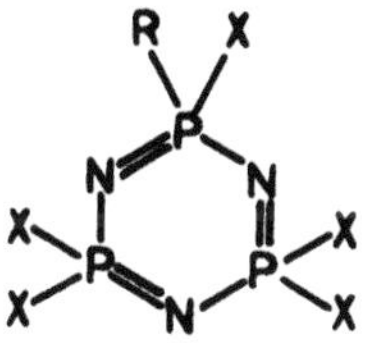

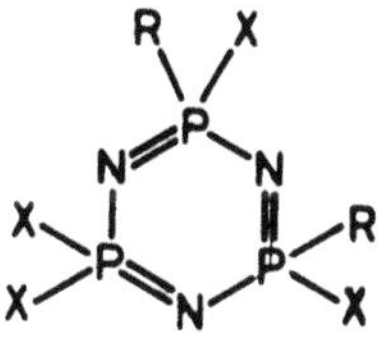

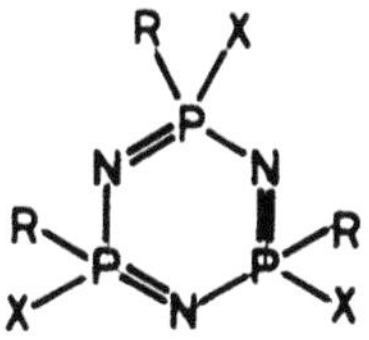

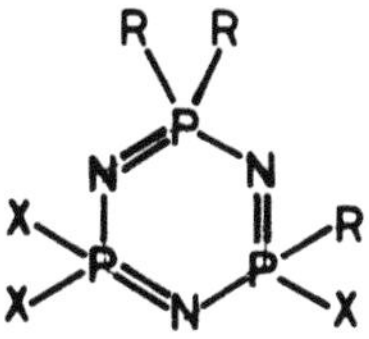

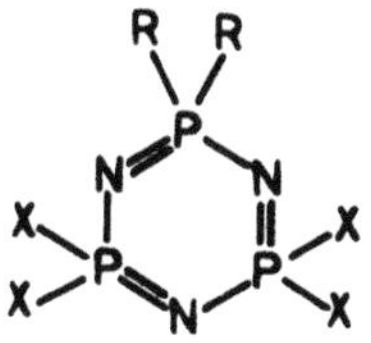

[a]Nonequivalent PFBr; relative intensities in parentheses.

TABLE M (continued)
Compilation of ^{31}P NMR Data of Four Coordinate Phosphorus Compounds Containing a Formal Multiple Bond from Phosphorus to a Group V Atom

Section M2: Cyclopolyphosphazenes, $(N{=}PX_2)n$, and Their Derivatives

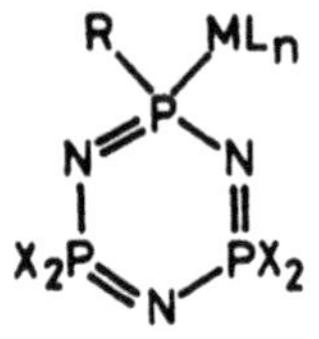

7 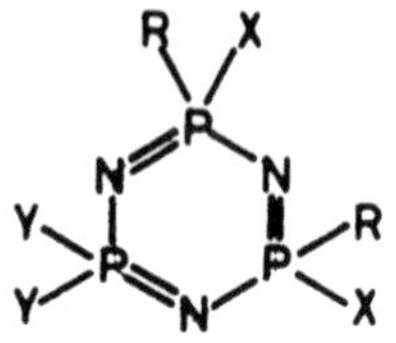 8 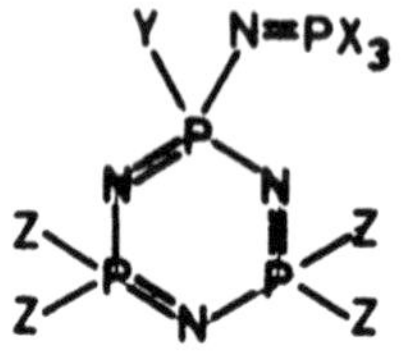 9

Connectivities	Molecular formula	Structure	NMR Data (δ_P[solv, temp] $^nJ_{PP}$Hz)	Ref.
	$C_2H_8Cl_4N_5P_3$	**3**; R = NHMe, X = Cl	22.2^T, 21.6^C	10
	$C_3H_8Cl_5N_4P_3$	**2**; R = NHiPr, X = Cl	17.0(r) $^2J_{PP}$46.5	13
	$C_6H_{16}Cl_5N_5P_3$	**3**; R = NHiPr, X = Cl	19.1^T(r) $^2J_{PP}$48.8	14
	$C_7H_8Cl_5N_4P_3$	**2**; R = NCH_2Ph, X = Cl	18.1 $^2J_{PP}$46.6	13
C'H	$C_6H_6Cl_5N_4P_3$	**2**; R = NHPh, X = Cl	11.7(c) $^2J_{PP}$47.1	15
	$C_7H_8Cl_5N_4P_3$	**2**; R = NHTol-4, X = Cl	12.2(c)	15
	$C_7H_8Cl_5N_4OP_3$	**2**; R = NHC_6H_4 (OMe)-4, X = Cl	13.0(c)	15
	$C_{14}H_{16}Cl_4N_5O_2P_3$	**3**; R = NHC_6H_4(OMe)-4, X = Cl	15.2^C, 15.0^T(c)	15
	$C_{21}H_{24}Cl_3N_6O_3P_3$	**4**; R = NHC_6H_4(OMe)-4, X = Cl	18.0, 17.2^T(c) $^2J_{PP}$43.6	15
	$C_{21}H_{24}Cl_3N_6O_3P_3$	**5**; R = NHC_6H_4(OMe)-4, X = Cl	17.1(c)	15
C_2	$C_2H_4Cl_5N_4P_3$	**2**; R = NC_2H_4, X = Cl	31.2(c) $^2J_{PP}$39	16
	$C_2H_6Cl_5N_4P_3$	**2**; R = NMe_2, X = Cl	21.6(r)	13
	$C_4H_8Cl_4N_5P_3$	**3**; R = NC_2H_4, X = Cl	34.7(c), $^2J_{PP}$38	16
	$C_4H_{12}Cl_4N_5P_3$	**3**; R = NMe_2, X = Cl	25.2^T, 24.9^C(c)	13
	$C_4H_{12}Cl_2F_2N_5P_3$	**8**; R = NMe_2, X = Cl, Y = F	31.3^T, 31.9^C	4
	$C_5H_{10}Cl_5N_4P_3$	**2**; R = NC_5H_{10}, X = Cl	18.7(r)	13
	$C_6H_{12}Cl_3N_6P_3$	**4**; R = NC_2H_4, X = Cl	38.3, 37.9^T(c) $^2J_{PP}$34 37.2^C(c)	16
	$C_6H_{12}Cl_3N_6P_3$	**5**; R = NC_2H_4, X = Cl	35.8(c)	16
	$C_6H_{18}Cl_3N_6P_3$	**4**; R = NMe_2, X = Cl	27.7, 28.3^T(r) $^2J_{PP}$44.4	13
	$C_6H_{18}Cl_3N_6P_3$	**5**; R = NMe_2, X = Cl	27.3(r)	13
	$C_8H_{16}Cl_2N_7P_3$	**3**; R = Cl, X = NC_2H_4	37.6^T $^2J_{PP}$29.4	16
	$C_8H_{24}Cl_2N_7P_3$	**3**; R = Cl, X = NMe_2	32.2^C(r) $^2J_{PP}$38.4	13
	$C_{10}H_{20}ClN_8P_3$	**2**; R = Cl, X = NC_2H_4	42.3(c), $^2J_{PP}$29.3	16
	$C_{12}H_{30}Cl_3N_6P_3$	**4**; R = NEt_2, X = Cl	22.5(r)	13
	$C_{25}H_{50}ClN_8P_3$	**2**; R = Cl, X = NC_5H_{10}	30.4(c), $^2J_{PP}$40.7	17
CC'	$C_7H_8Cl_5N_4P_3$	**2**; R = NMePh, X = Cl	16.9(r), $^2J_{PP}$48.6	18
CP^4	$C_{15}H_{39}Cl_3N_6O_6P_6$	**5**; R = $NMeP(O)(OEt)_2$, X = Cl	16.9(r)	19

Remaining P-bound atoms = ClN'

Connectivities	Molecular formula	Structure	NMR Data	Ref.
P^4	$N_4P_4Cl_8$	**9**; X = Y = Z = Cl	−2.2(ae)	9
	$C_3H_9Cl_5N_4P_4$	**9**; X = Me; Y = Z = Cl	1.1(c)	20
	$C_6H_{18}Cl_5N_7P_4$	**9**; X = NMe_2; Y = Z = Cl	−8.2	21
	$C_9H_{21}Cl_5N_4O_3P_4$	**9**; X = O-iPr; Y = Z = Cl	−5.3	21
	$C_{18}H_{15}Cl_5N_4P_4$	**9**; X = Ph; Y = X = Cl	0.1(c)	20

Remaining P-bound atoms = C1O

Connectivities	Molecular formula	Structure	NMR Data	Ref.
C	$CH_3Cl_5N_3OP_3$	**2**; R = OMe, X = Cl	16.7, $^2J_{PP}$63.3	22
	$C_2H_2Cl_5F_3NOP_3$	**2**; R = OCH_2CF_3, X = Cl	16.1(a), $^2J_{PP}$66	23

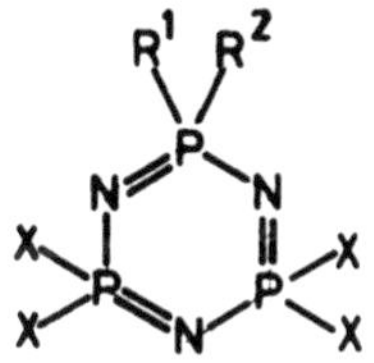

10

Connectivities	Molecular formula	Structure	NMR Data (δ_P[solv, temp] $^nJ_{PP}$Hz)	Ref.
	$C_2H_5Cl_5N_3OP_3$	**2**; R = OEt, X = Cl	13.6, $^2J_{PP}$63.3	22
	$C_3H_7Cl_5N_3OP_3$	**2**; R = O-iPr, X = Cl	12.6, $^2J_{PP}$62.7	22
	$C_4H_4Cl_4F_6N_3O_2P_3$	**3**; R = OCH_2CF_3, X = Cl	19.0^T(a), $^2J_{PP}$69.6	23
	$C_6H_6Cl_3F_9N_3O_3P_3$	**4**; R = OCH_2CF_3, X = Cl	21.9, 21.7^T(a)	23
	$C_6H_6Cl_3F_9N_3O_3P_3$	**5**; R = OCH_2CF_3, X = Cl	21.0(a)	23
	$C_8H_8Cl_2F_{12}N_3O_4P_3$	**3**; R = Cl, X = OCH_2CF_3	24.3^C(a), $^2J_{PP}$85.4	23
	$C_{10}H_{10}ClF_{15}N_3O_5P_3$	**2**; R = Cl, X = OCH_2CF_3	26.3(a), $^2J_{PP}$87	23
C	$C_2H_3Cl_5N_3OP_3$	**2**; R = $OCH{=}CH_2$, X = Cl	13.2(c), $^2J_{PP}$64.7	24
	$C_4H_6Cl_4N_3O_2P_3$	**3**; R = $OCH{=}CH_2$, X = Cl	15.6^T(c), $^2J_{PP}$67.5	25
	$C_6H_9Cl_3N_3O_3P_3$	**4**; R = $OCH{=}CH_2$, X = Cl	19.2^T, 19.0^T 18.9^C(c), $^2J_{PP}$64.5	26
	$C_6H_9Cl_3N_3O_3P_3$	**5**; R = $OCH{=}CH_2$, X = Cl	18.4(c)	26
	$C_7H_7Cl_5N_3OP_3$	**2**; R = OTol-4, X = Cl	12.2(c), $^2J_{PP}$60	27
	$C_8H_{12}Cl_2N_3O_4P_3$	**3**; R = Cl, X = $OCH{=}CH_2$	21.6^C(c), $^2J_{PP}$83.4	25
	$C_{10}H_{15}ClN_3O_5P_3$	**2**; R = Cl, X = $OCH{=}CH_2$	23.9(c), $^2J_{PP}$85	25
	$C_{14}H_{14}Cl_4N_3O_2P_3$	**3**; R = OTol-4, X = Cl	14.4^T, 14.2^C(c)	27
	$C_{21}H_{21}Cl_3N_3O_3P_3$	**4**; R = OTol-4, X = Cl	18.3(c)	27
	$C_{28}H_{28}Cl_2N_3O_4P_3$	**3**; R = Cl, X = OTol-4	19.9^T, 20.1^C	27
	$C_{35}H_{35}Cl_2N_3O_5P_3$	**2**; R = Cl, X = OTol-4	22.4(c), $^2J_{PP}$83	27

Remaining P-bound atoms = ClS

Connectivities	Molecular formula	Structure	NMR Data	Ref.
C	$CH_3Cl_5N_3P_3S$	**2**; R = SMe, X = Cl	41(b)	28
	$C_2H_5Cl_5N_3P_3S$	**2**; R = SEt, X = Cl	33.1(c), $^2J_{PP}$32.9	29
	$C_6H_{15}Cl_3N_3P_3S_3$	**5**; R = SEt, X = Cl	38.5(c)	29
	$C_{10}H_{25}ClN_3P_3S_5$	**2**; R = Cl, X = SEt	37.1(c), $^2J_{PP}$29	29

Remaining P-bound atoms = Cl_2

Connectivities	Molecular formula	Structure	NMR Data	Ref.
	$BrCl_2N_3P_3$	**2**; R = Br, X = Cl	17.7(b)	2
	$Br_2Cl_4N_3P_3$	**2**; R = Br, X = Cl	16.1(b)	2
	$Br_3Cl_3N_3P_3$	**5**; R = Br, X = Cl	14.0	2
	$Cl_4F_2N_3P_3$	**3**; R = F, X = Cl	22.8^T	4
	$Cl_5FN_3P_3$	**2**; R = F, X = Cl	22.8	8
	$Cl_6N_3P_3$	**2**; R = X = Cl	19.3(r)	13
	$Cl_8N_4P_4$	**9**; X = Y = Z = Cl	20.5(ae)	9
	$Cl_{10}N_5P_5$	**9**; X = Z = Cl, Y = $N{=}PCl_3$	17.5(ae)	9
	$H_2Cl_5N_4P_3$	**2**; R=NH_2, X = Cl	19.0(c)	9
	$H_4Cl_4N_5P_3$	**6**; R =NH_2, X = Cl	18.3(ae)	9
	$CH_3BrCl_4N_3P_3$	**10**; R^1 = Me, R^2 = Br, X = Cl	21.0(c)	30
	$CH_3Cl_4IN_3P_3$	**10**; R^1 = Me, R^2 = I, X = Cl	21.3(c)	30
	$CH_3Cl_5N_3OP_3$	**2**; R = OMe, X = Cl	22.5	22
	$CH_3Cl_5N_3P_3$	**2**; R = Me, X = Cl	21.2(c)	30
	$CH_3Cl_5N_3P_3S$	**2**; R = SMe, X = Cl	21(b)	28
	$CH_4Cl_4N_3P_3$	**10**; R^1 = Me, R^2 = H, X = Cl	17.6(k)	31
	$CH_4Cl_5N_4P_3$	**2**; R = NHMe, X = Cl	21.7	11
	$C_2Cl_4N_5P_3S_2$	**6**; R = NCS, X = Cl	22.0(c)	32
	$C_2H_2Cl_5F_3N_3OP_3$	**2**; R = OCH_2CF_3, X = Cl	22.5(a)	23
	$C_2H_3Cl_5N_3OP_3$	**2**; R = OCH CH_2, X = Cl	23.4(c)	24

TABLE M (continued)
Compilation of ^{31}P NMR Data of Four Coordinate Phosphorus Compounds Containing a Formal Multiple Bond from Phosphorus to a Group V Atom

Section M2: Cyclopolyphosphazenes, $(N{=}PX_2)n$, and Their Derivatives

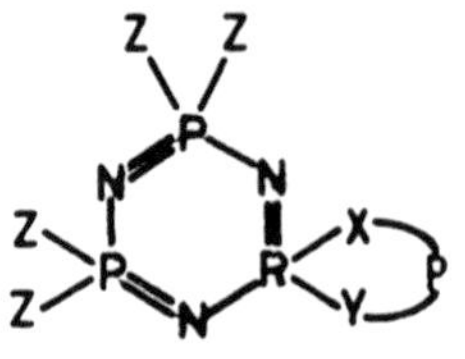

11

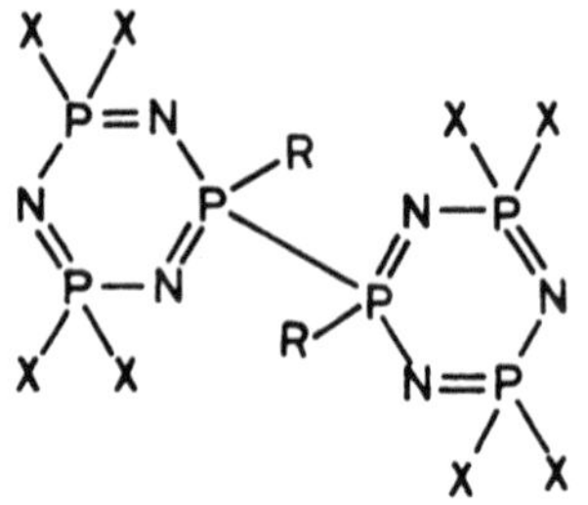

12

Connectivities	Molecular formula	Structure	NMR Data (δ_P[solv, temp] $^nJ_{PP}$Hz)	Ref.
	$C_2H_4Cl_4N_3O_2P_3$	**11**; X = Y = O, R = $(CH_2)_2$, Z = Cl	22.5(c), $^2J_{PP}$67	33
	$C_2H_4Cl_5N_4P_3$	**2**; R = NC_2H_4, X = Cl	22.3	16
	$C_2H_5Cl_4N_4OP_3$	**11**; X = NH, Y = O, Z = Cl, R = $(CH_2)_2$	24.9(c)	34
	$C_2H_5Cl_5N_3OP_3$	**2**; R = OEt, X = Cl	21.3	22
	$C_2H_5Cl_5N_3P_3$	**2**; R = Et, X = Cl	21.7	30
	$C_2H_5Cl_5N_3P_3S$	**2**; R = SEt, X = Cl	21.0(c)	29
	$C_2H_5Cl_6N_4P_3$	**2**; R = $NHCH_2CH_2Cl$, X = Cl	21.9(c), $^2J_{PP}$47.0	12
	$C_2H_6Cl_4N_3P_3$	**6**; R = Me, X = Cl	18.0	35
	$C_2H_6Cl_4N_3P_3$	**10**; R[1] = Et, R[2] = H, X = Cl	18.4(k)	31
	$C_2H_6Cl_4N_3P_3S_2$	**6**; R = SMe, X = Cl	21(b)	28
	$C_2H_6Cl_4N_5P_3$	**11**; X = Y = NH, R = $(CH_2)_2$, Z = Cl	23.5(c), $^2J_{PP}$45.7	34
	$C_2H_6Cl_5N_4P_3$	**2**; R = NHEt, X = Cl	20.3(r)	13
	$C_2H_6Cl_5N_4P_3$	**2**; R = NMe_2, X = Cl	20.5(r)	13
	$C_2H_6Cl_8N_6P_6$	**12**; R = Me, X = Cl	19.8(k)	36
	$C_2H_8Cl_4N_5P_3$	**6**; R = NHMe, X = Cl	21.3	10
	$C_2H_8Cl_4N_5P_3$	**3**; R = NHMe, X = Cl	22.2^T	10
	$C_3H_6Cl_4N_3O_2P_3$	**11**; X = Y = O, R = $(CH_2)_3$, Z = Cl	23.0	37
	$C_3H_7Cl_4N_4OP_3$	**11**; X = NH, Y = O, R = $(CH_2)_3$, Z = Cl	22.4(c)	38
	$C_3H_7Cl_4N_4OP_3$	**11**; X = NMe, Y =O, R = $(CH_2)_2$, Z = Cl	25.1(c), $^2J_{PP}$ 53.9	39
	$C_3H_7Cl_5$ N_3OP_3	**2**; R = OPr^i, X = Cl	21.7	22
	$C_3H_7Cl_5N_3P_3$	**2**; R = O-iPr, X = Cl	21.7	30
	$C_3H_8Cl_4N_3P_3$	**10**; R[1] = Me, R[2] = Et, X = Cl	18.3	35
	$C_3H_8Cl_4N_3P_3$	**10**; R[1] = iPr, R[2] = H, X = Cl	18.4(k), $^2J_{PP}$8	31
	$C_3H_8Cl_4N_3OP_3$	**10**; R[1] = Me, R[2] = CH(OH)Me, X = Cl	19.3 and 18.7[a]	40
	$C_3H_8Cl_4N_3OP_3$	**10**; R[1] = H, R[2] = O-iPr, X = Cl	20.9(c), $^2J_{PP}$26.0	41
	$C_3H_8Cl_4N_5P_3$	**11**; X = Y = NH, R = $(CH_2)_3$, Z = Cl	21.5(c)	42
	$C_3H_8Cl_5N_4P_3$	**2**; R = NH-iPr, X = Cl	20.4(r)	13
	$C_3H_9Cl_3N_3O_3P_3$	**5**; R = $OCH{=}CH_2$, X = Cl	27.1	26

[a] Nonequivalent PCl_2 groups.

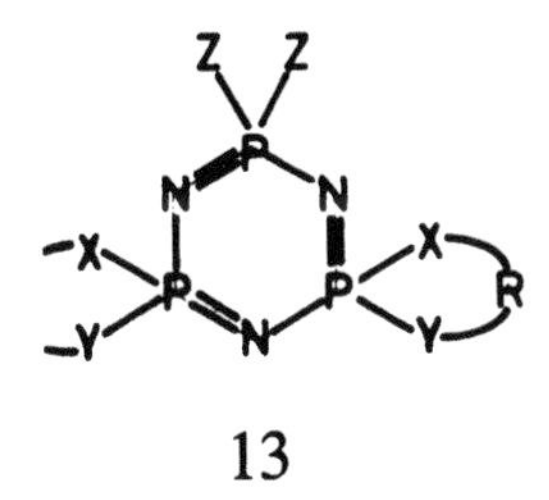

13

Connectivities	Molecular formula	Structure	NMR Data (δ_P[solv, temp] $^nJ_{PP}$Hz)	Ref.
	$C_3H_9Cl_5N_4P_4$	**9**; X = Me, Y = Z = Cl	20.9(c)	20
	$C_4H_4Cl_4F_6N_3O_2P_3$	**3**; R = OCH_2CF_3, X = Cl	24.8(a)	23
	$C_4H_6Cl_4N_3O_2P_3$	**6**; R = OCH=CH_2, X = Cl	24.5(c)	25
	$C_4H_6Cl_4N_3O_2P_2$	**3**; R = OCH=CH_2, X = Cl	24.8^T(c)	25
	$C_4H_6Cl_4N_3P_3$	**10**; R^1 = Me, R^2 = CH=C=CH_2, X = Cl	18.6	43
	$C_4H_6Cl_4N_3P_3$	**10**; R^1 = Me, R^2 = C≡CMe, X = Cl	18.8	43
	$C_4H_8Cl_2N_3O_4P_3$	**13**; X = Y = O, R = $(CH_2)_2$, Z = Cl	30.3	18
	$C_4H_8Cl_4N_3OP_3$	**10**; R^1 = Me, R^2 = $CH_2C(O)Me$, X = Cl	19.2(c), $^2J_{PP}$9.7	44
	$C_4H_8Cl_4N_3O_2P_3$	**11**, X = Y = O, R = $(CH_2)_4$, Z = Cl	23.0(c)	37
	$C_4H_8Cl_4N_3P_3$	**10**; R^1 = Me, R^2 = CH_2CH=CH_2	18.5	35
	$C_4H_8Cl_4N_5P_3$	**6**; R = NC_2H_4, X = Cl	21.9(c)	16
	$C_4H_8Cl_4N_5P_3$	**3**; R = NC_2H_4, X = Cl	25.0^T(c)	16
	$C_4H_8Cl_6N_3O_2P_3$	**6**; R = OCH_2CH_2Cl, X = Cl	23.6(c)	12
	$C_4H_9Cl_5N_3P_3$	**2**; R = tBu, X = Cl	21.8	30
	$C_4H_{10}Cl_2N_5O_2P_3$	**13**; X = NH, Y = O, R = $(CH_2)_2$, Z = Cl	29.0	34
	$C_4H_{10}Cl_4N_3P_3$	**6**; R = Et, X = Cl	19.2	35
	$C_4H_{10}Cl_4N_3P_3$	**10**; R^1 = Me, R^2 = iPr, X = Cl	17.9	35
	$C_4H_{10}Cl_4N_3P_3$	**10**; R^1 = tBu, R^2 = H, X = Cl	18.4(k)	31
	$C_4H_{10}Cl_4N_3P_3S_2$	**6**; R = SEt, X = Cl	18.7(c)	29
	$C_4H_{10}Cl_6N_5P_3$	**6**; R = $NHCH_2CH_2Cl$, X = Cl	21.2(c), $^2J_{PP}$46.6	12
	$C_4H_{10}Cl_4N_5P_3$	**11**; X = Y = N (Me), R = $(CH_2)_2$, Z = Cl	23.8(c), $^2J_{PP}$41.8	45
	$C_4H_{10}Cl_8N_6P_6$	**12**; R = Et, X = Cl	20.4(c)	46
	$C_4H_{11}Cl_5N_3P_3Si$	**2**; R = CH_2Tms, X = Cl	20.6(c)	47
	$C_4H_{12}Cl_2N_3P_3$	**6**; R = Cl, X = Me	16.6(c), $^2J_{PP}$3.6	48
	$C_4H_{12}Cl_2N_3O_4P_3$	**6**; R = Cl, X = OCH=CH_2	28.9(c)	25
	$C_4H_{12}Cl_2N_3P_3S_4$	**6**; R = Cl, X = SMe	17(b)	28
	$C_4H_{12}Cl_4N_5P_3$	**3**; R = NMe_2, X = Cl	21.5^T, 21.6^C(r)	13
	$C_5H_{10}Cl_4N_3OP_3$	**10**; R^1 = Me, R^2 = $CH_2C(OMe)$=CH_2, X = Cl	18.3(c), $^2J_{PP}$4.1	44
	$C_5H_{10}Cl_4N_3OP_3$	**10**; R^1 = Me, R^2 = CH=C(OMe)Me	18.1(c), $^2J_{PP}$12.2	44
	$C_5H_{12}Cl_4N_3P_3$	**10**; R^1 = Me, R^2 = tBu, X = Cl	18.1	35
	$C_6H_5BrCl_4N_3P_3$	**10**; R^1 = Ph, R^2 = Br, X = Cl	21.1, $^2J_{PP}$8.6	49
	$C_6H_5Cl_4FN_3P_3$	**10**, R^1 = Ph, R^2 = F, X = Cl	22.4, $^2J_{PP}$34.1	49
	$C_6H_5Cl_4IN_3P_3$	**10**; R^1 = Ph, R^2 = I, X = Cl	20.6	49
	$C_6H_5Cl_5N_3P_3$	**2**; R = Ph, X = Cl	21.2, $^2J_{PP}$15.5	49
	$C_6H_6Cl_3F_9N_3O_3P_3$	**5**; R = OCH_2CF_3, X = Cl	26.5(a)	23
	$C_6H_6Cl_4N_3P_3$	**10**; R^1 = Ph, R^2 = H, X = Cl	18.2	49
	$C_6H_6Cl_4N_3P_3$	**6**; R = CH_2C≡CH, X = Cl	19.4(k)	43
	$C_6H_6Cl_4N_5P_3$	**11**; X = Y = NH, R = 1,2-C_6H_4, Z = Cl	19.0(k)	50
	$C_6H_6Cl_5N_4P_3$	**2**; R = NHPh, X = Cl	21.2(c)	15

TABLE M (continued)
Compilation of ^{31}P NMR Data of Four Coordinate Phosphorus Compounds Containing a Formal Multiple Bond from Phosphorus to a Group V Atom

Section M2: Cyclopolyphosphazenes, $(N{=}PX_2)n$, and Their Derivatives

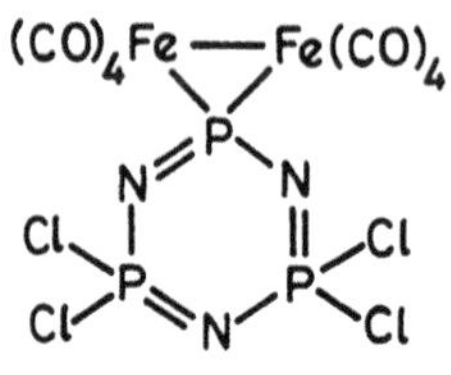

14

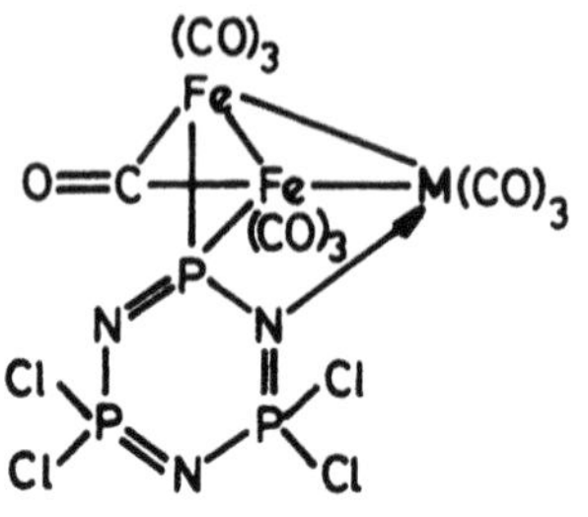

15

Connectivities	Molecular formula	Structure	NMR Data (δ_P[solv, temp] $^nJ_{PP}$Hz)	Ref.
	$C_6H_{12}Cl_2N_3O_4P_3$	**13**; X = Y = O, R = $(CH_2)_3$, Z = Cl	25.9	18
	$C_6H_{12}Cl_3N_6P_3$	**5**; R = NC_2H_4, X = Cl	25.0	16
	$C_6H_{14}Cl_2N_5O_2P_3$	**13**; X = NH, Y = O, R = $(CH_2)_3$, Z = Cl	23.8	38
	$C_6H_{14}Cl_2N_5O_2P_3$	**13**; X = NMe, Y = O, R = $(CH_2)_2$, Z = Cl	30.7(c), $^2J_{PP}$62.3	39
	$C_6H_{14}Cl_4N_3P_3$	**6**; R = iPr, X = Cl	18.8	49
	$C_6H_{15}Cl_3N_3P_3S_2$	**5**; R = SEt, X = Cl	18.1(c)	29
	$C_6H_{16}Cl_2N_7P_3$	**13**; X = Y = NH, R = $(CH_2)_3$, Z = Cl	23.1(c)	42
	$C_6H_{16}Cl_4N_5P_3$	**3**; R = NH-iPr, X = Cl	21.9^C, 21.3^T(r)	14
	$C_6H_{16}Cl_4N_5P_3$	**6**; X = NH-iPr, R = Cl	19.8, $^2J_{PP}$45.5	14
	$C_6H_{18}Cl_3N_6P_3$	**5**; R = NMe_2, X = Cl	21.7(r)	13
	$C_6H_{18}Cl_5N_7P_4$	**9**; X = NMe_2, Y = Z = Cl	19.4	21
	$C_7H_7Cl_5N_3OP_3$	**2**; R = OTol-4, X = Cl	22.5(c)	27
	$C_7H_8Cl_4N_3P_3$	**10**; R^1 = Me, R^2 = Ph	18.6, $^2J_{PP}$11	49
	$C_7H_8Cl_5N_4P_3$	**2**; R = NMePh, X = Cl	19.7(r)	18
	$C_8Cl_4Fe_2N_3O_8P_3$	**14**	13.1(c)	51
	$C_8H_5Cl_5CrN_3O_3P_3$	**7**; R = X = Cl, ML_n = $Cr(CO)_3(Cp)$	12.6	7
	$C_8H_{16}Cl_4N_5O_4P_3$	**6**; R = $NHCH_2COOEt$, X = Cl	21.6(c), $^2J_{PP}$47.4	52
	$C_8H_{20}Cl_2N_3P_3S_4$	**6**; R = Cl, X = SEt	17.4(c)	29
	$C_8H_{20}Cl_2N_7P_3$	**13**; X = Y = NMe, R = $(CH_2)_2$, Z = Cl	29.1	45
	$C_8H_{20}Cl_4N_5P_3$	**6**; R = NH-tBu, X = Cl	17.5(r)	13
	$C_8H_{20}Cl_6N_7P_3$	**6**; R = Cl, X = $NHCH_2CH_2Cl$	24.6(c), $^2J_{PP}$50.5	12
	$C_9H_8Cl_4N_3O_3P_3W$	**7**; R = Me, X = Cl, ML_n = $W(CO)_3(Cp)$	8.8	7
	$C_9H_{21}Cl_5N_4O_3P_4$	**9**; X = O-iPr, Y = Z = Cl	20.5	21
	$C_{10}Cl_4Fe_2N_3O_{10}P_3Ru$	**15**; M = Ru	13.6, 24.6(c)	51
	$C_{10}Cl_4Fe_3N_3O_{10}P_3$	**15**; M = Fe	14.5, 27.1(c)	51
	$C_{10}H_8Cl_8FeN_6P_6$	**16**	18.9	53
	$C_{10}H_8Cl_{10}N_6P_6$	**17**; M = Fe, R = Cl	21.1	53
	$C_{10}H_9Cl_5FeN_3P_3$	**18**; R = X = Cl, M = Fe	20.2	53
	$C_{10}H_9Cl_5N_3P_3Ru$	**18**; R = X = Cl, M = Ru	21.0	53
	$C_{10}H_9Cl_9FeN_6P_6$	**19**; M = Fe	21.1	53

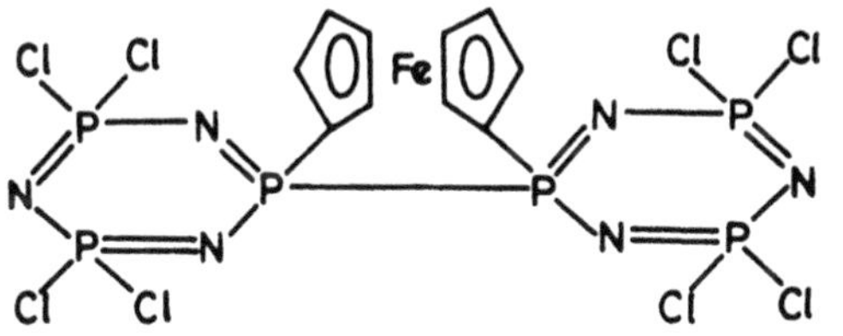

16

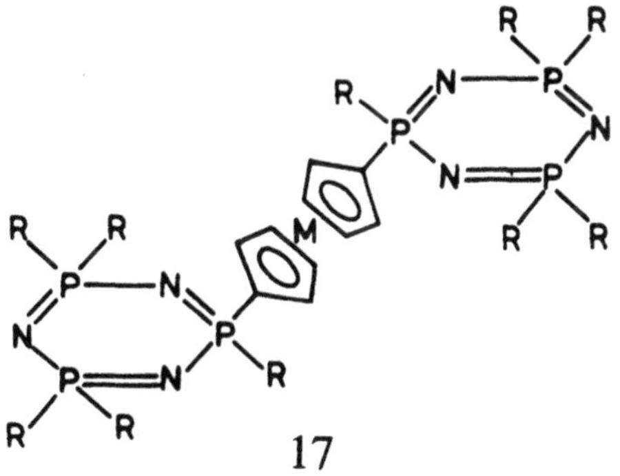

17

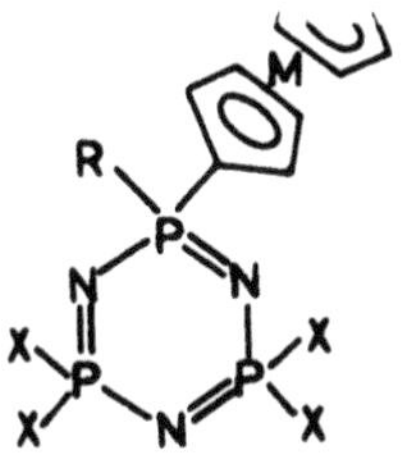

18

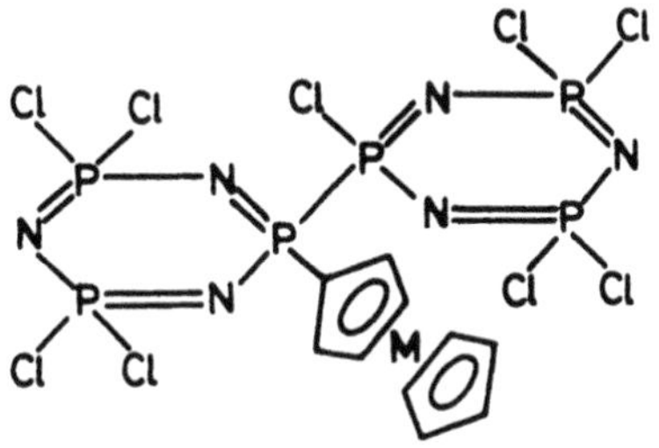

19

Connectivities	Molecular formula	Structure	NMR Data (δ_P[solv, temp] $^nJ_{PP}$Hz)	Ref.
	$C_{10}H_9Cl_9N_6P_6Ru$	**19**; M = Ru	21.1	53
	$C_{10}H_{10}Cl_4N_3P_3Ru$	**18**; M = Ru, R = H, X = Cl	8.0	53
	$C_{10}H_{14}Cl_4N_3OP_3$	**10**; R[1] = iPr, R[2] = CH(OH)Ph, X = Cl	19.3(c)	40
	$C_{11}H_{12}Cl_4FeN_3P_3$	**18**; M = Fe, R = Me, X = Cl	20.7	53
	$C_{12}H_{10}Cl_4N_3$	**6**; R = Ph, X = Cl	17.1(r)	13
	$C_{12}H_{10}Cl_8N_6P_6$	**12**; R = Ph, X = Cl	19.8	36
	$C_{12}H_{12}Cl_4N_5P_3$	**6**; R = NHPh, X = Cl	20.5(c)	15
	$C_{12}H_{12}Cl_4N_5P_3$	**3**; R = NHPh, X = Cl	22.2^T(c), 22.3^C	15
	$C_{12}H_{32}Cl_2N_7P_3$	**6**; R = Cl, X = NH-iPr	22.2(r), $^2J_{PP}$49.4	13
	$C_{13}H_{10}Cl_4FeN_3O_2P_3$	**7**; R = Ph, X = Cl, ML$_n$ = Fe(CO)$_2$(Cp)	7.6(c)	54
	$C_{13}H_{10}Cl_4MoN_3O_3P_3$	**7**; R = Ph, X = Cl, ML$_n$ = Mo(CO)$_3$(Cp)	9.0	7
	$C_{13}H_{10}Cl_4N_3O_2P_3Ru$	**7**; R = Ph, X = Cl, ML$_n$ = Ru(CO)$_2$(Cp)	8.2(c)	54
	$C_{13}H_{10}Cl_4N_3O_3P_3W$	**7**; R = Ph, X = Cl, ML$_n$ = W(CO)$_3$(Cp)	12.4	7
	$C_{14}H_{10}Cl_4CrN_3O_2P_3$	**7**; R = Ph, X = Cl, ML$_n$ = Cr(CO)$_3$(Cp)	8.9(c)	54
	$C_{14}H_{10}Cl_4MoN_3O_3P_3$	**7**; R = Ph, X = Cl, ML$_n$ = Mo(CO)$_3$(Cp)	8.8(c)	54
	$C_{14}H_{10}Cl_4N_3O_3P_3W$	**7**; R = Ph, X = Cl, ML$_n$ = W(CO)$_3$(Cp)	9.0	54
	$C_{14}H_{14}Cl_4N_3O_2P_3$	**3**; R = OTol-*p*, X = Cl	23.9^T(c)	27
	$C_{16}H_{32}Cl_2N_7O_8P_3$	**6**; R = Cl, X = NHCH$_2$COOEt	24.7(c)	52
	$C_{16}H_{40}Cl_2N_7P_3$	**6**; R = Cl, X = NH-tBu	19.7(r)	13
	$C_{18}H_{15}Cl_5N_4P_4$	**9**; X = Ph, Y = Z = Cl	20.3(c)	21
	$C_{18}H_{20}Cl_4N_5P_4$	**9**; X = Ph, Y = NH$_2$, Z = Cl	18.3(c)	55
	$C_{19}H_{22}Cl_4N_5P_4$	**9**; X = Ph, Y = NHMe, Z = Cl	18.2(c)	56
	$C_{20}H_{19}Cl_4N_5P_4$	**9**; X = Ph, Y = NC$_2$H$_4$, Z = Cl	18.1(c)	57

TABLE M (continued)
Compilation of ^{31}P NMR Data of Four Coordinate Phosphorus Compounds Containing a Formal Multiple Bond from Phosphorus to a Group V Atom

Section M2: Cyclopolyphosphazenes, $(N{=}PX_2)n$, and Their Derivatives

Connectivities	Molecular formula	Structure	NMR Data (δ_P[solv, temp] $^nJ_{PP}$Hz)	Ref.
	$C_{20}H_{20}Cl_4N_4OP_4$	**9**; X = Ph, Y = OEt, Z = Cl	18.5	20
	$C_{20}H_{21}Cl_4N_5P_4$	**9**; X = Ph, Y = NMe_2, Z = Cl	16.6(c)	56
	$C_{22}H_{25}Cl_4N_5P_4$	**9**; X = Ph, Y = NH-tBu, Z = Cl	15.9(c)	56
	$C_{24}H_{20}Cl_2N_3P_3$	**6**; R = Cl, X = Ph	14.8, $^2J_{PP}$9.3	13
	$C_{24}H_{20}Cl_4N_4P_4$	**9**; X = Y = Ph, Z = Cl	16.6(c)	20
	$C_{36}H_{30}Cl_4N_5P_5$	**9**; X = Ph, Y = N=PPh_3, Z = Cl	13.4	55
	Remaining P-bound atoms = FFe			
C'$_7$	$C_7H_5F_5FeN_3P_3$	**7**; R = X = F, ML_n = $Fe(CO)_2(Cp)$	133.2(c)	58
	Remaining P-bound atoms = FN			
C_2	$C_2H_6F_5N_4P_3$	**2**; R = NMe_2, X = F	24.2(n)	59
	$C_4H_{12}Cl_2F_2N_5P_3$	**8**; R = NMe_2, X = F, Y = Cl	19.0^C, $^2J_{PP}$69.1, 75.8; 18.9^T, $^2J_{PP}$67.1, 70.5	4
	$C_4H_{12}F_4N_5P_3$	**3**; R = NMe_2, X = F	23.6^T, $^2J_{PP}$102.8	60
	Remaining P-bound atoms = FN'			
C'	$CF_5N_4P_3S$	**2**; R = NCS, X = F	−4, $^1J_{PF}$890	61
P^4	$CH_3F_7N_4OP_4$	**9**; X_3 = F_2 and OMe, Y = Z = F	4.0	62
	$C_{18}H_{15}F_5N_4P_4$	**9**; X = Ph, Y = Z = F	7.4(c)	63
	$C_{22}H_{27}FN_4O_4P_4$	**9**; X = Ph, Y = F, Z = OMe	13.3(c)	63
	Remaining P-bound atoms = FO			
C	$CH_3F_5N_3OP_3$	**2**; R = OMe, X = F	13.5	64
	$C_{10}H_{10}F_{16}N_3O_5P_3$	**2**; R = F, X = OCH_2CF_3	16.7, $^2J_{PP}$103.4	65
C'	$C_{30}H_{25}FN_3O_5P_3$	**2**; R = F, X = OPh	10.4, $^2J_{PP}$100	66
	Remaining P-bound atoms = FRu			
C'$_7$	$C_7H_5F_5N_3P_3Ru$	**7**; R = X = F, ML_n = $Ru(CO)_2(Cp)$	103.9(c)	58
	Remaining P-bound atoms = FS			
C	$CH_3F_5N_3P_3S$	**2**; R = SMe, X = F	47.9	64
	$C_2H_5F_5N_3P_3S$	**2**; R = SEt, X = F	50.0	67
C'	$C_6H_5F_5N_3P_3S$	**2**; R = SPh, X = F	44.1	64
C	$C_6H_{15}F_3N_3P_3S_3$	**5**; R = SEt, X = F	40.0	67
	$C_{10}H_{25}FN_3P_3S_5$	**2**; R = F, X = SEt	33.0	67
	Remaining P-bound atoms = F_2			
	$Cl_3F_7N_5P_5$	**9**; X = F_2 and N=PCl_3, Y = Z = F	8.5	68
	$Cl_5F_5N_5P_5$	**9**; X_3 = Cl_2 and N=PCl_3, Y = Z = F	9	68

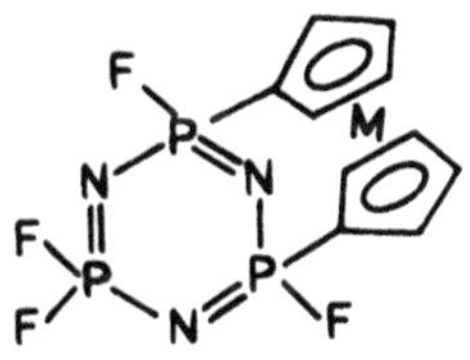

20

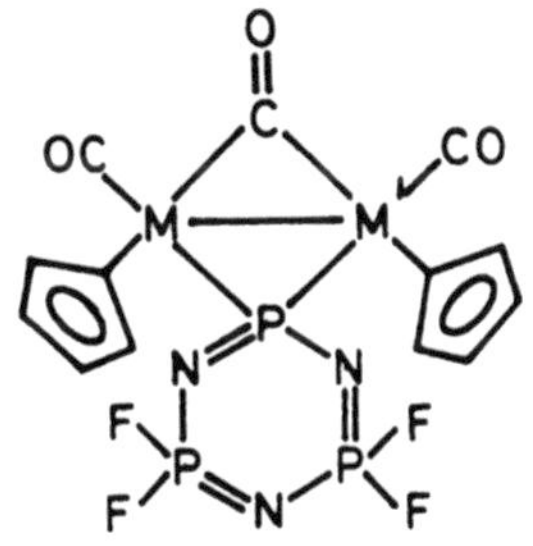

21

Connectivities	Molecular formula	Structure	NMR Data (δ_P[solv, temp] $^nJ_{PP}$Hz)	Ref.
	$F_6N_3P_3$	**2**; R = X = F	13.9(c)	69
	$CF_5N_4P_3S$	**2**; R = NCS, X = F	9.0, $^1J_{PF}$880	61
	$CH_3F_5N_3OP_3$	**2**; R = OMe, X = F	9.7	64
	$CH_3F_5N_3P_3S$	**2**; R = SMe, X = F	6.3	64
	$CH_3F_7N_4OP_4$	**9**; X_3 = F_2 and OMe, Y = Z = F	9.0	62
	$C_2H_5F_5N_3P_3S$	**2**; R = SEt, X = F	6.0	67
	$C_2H_6F_4N_3P_3$	**6**; R = Me, X = F	8.0	70
	$C_2H_6F_4N_5P_3$	**11**; X = Y = NH, R = $(CH_2)_2$, Z = F	10.2(c)	38
	$C_2H_6F_5N_4P_3$	**2**; R = NMe_2, X = F	10.3(n)	59
	$C_2H_7F_4N_4OP_3$	**11**; X = NH, Y = O, R = $(CH_2)_3$, Z = F	10.2(c)	38
	$C_3H_6F_4N_3O_2P_3$	**11**; X = Y = O, R = $(CH_2)_3$, Z = F	9.0(c)	38
	$C_3H_7F_4N_4OP_3$	**11**; X = NMe, Y = O, R = $(CH_2)_2$, Z = F	14.0(c)	38
	$C_3H_8F_4N_5P_3$	**11**; X = Y = NH, R = $(CH_2)_3$, Z = F	10.6(c)	38
	$C_4H_9F_5N_3P_3$	**2**; R = tBu, X = F	9.1	71
	$C_4H_{10}F_4N_3P_3S_2$	**6**; R = SEt, X = F	5.0	67
	$C_4H_{12}F_4N_5P_3$	**6**; R = NMe_2, X = F	14.0(n)	59
	$C_4H_{12}F_4N_5P_3$	**3**; R = NMe_2, X = F	8.0^T	60
	$C_4H_{12}Cl_2F_2N_5P_3$	**8**; R = NMe_2, X = Cl, Y = F	3.6^C, 4.6^T	4
	$C_5H_6F_5N_4P_3$	**2**; R = 1-methyl pyrrol-2-yl, X = F	6.9(c)	72
	$C_6H_5F_5N_3P_3S$	**2**; R = SPh, X = F	7.3	64
	$C_6H_{14}F_2N_5O_2P_3$	**13**; X = NMe, Y = O, R = $(CH_2)_2$, Z = F	14.4(c)	38
	$C_6H_{15}F_3N_3P_3S_3$	**5**; R = SEt, X = F	1.0	67
	$C_6H_{16}F_2N_7P_3$	**13**; X = Y = NH, R = $(CH_2)_3$, Z = F	10.1(c)	38
	$C_7H_5F_5FeN_3O_2P_3$	**7**; R = X = F, ML_n = $Fe(CO)_2(Cp)$	1.4(c)	58
	$C_7H_5F_5N_3O_2P_3Ru$	**7**; R = X = F, ML_n = $Ru(CO)_2(Cp)$	1.6(c)	58
	$C_8H_{18}F_4N_3P_3$	**3**; R = tBu, X = F	9.6, $^2J_{PP}$51.3	71
	$C_8H_{20}F_2N_3P_3S_4$	**6**; R = F, X = SEt	−1.0	67
	$C_9H_9F_5N_3P_3$	**2**; R = $C_6H_4(CMe{=}CH_2)$-4, X = F	9.3(c), $^2J_{PP}$78.1	73
	$C_{10}H_9F_4FeN_3P_3$	**20**; M = Fe	15.4	53
	$C_{10}H_8F_4N_3P_3Ru$	**20**; M = Ru	16.7	53
	$C_{10}H_8F_{10}N_6P_6Ru$	**17**; M = Ru, R = F	8.5	53
	$C_{10}H_9F_5FeN_3P_3$	**18**; M = Fe, R = X = F	8.9	53
	$C_{10}H_9F_5N_3P_3Ru$	**18**; M = Ru, R = X = F	9.1	53
	$C_{12}H_{10}F_4N_3P_3$	**6**; R = Ph, X = F	12.1	69
	$C_{12}H_{10}F_4N_3P_3$	**6**; R = Ph, X = F	12.4^c	74
	$C_{13}H_{10}F_4Fe_2N_3O_3P_3$	**21**; M = M′ = Ru	3.3	58

TABLE M (continued)
Compilation of ^{31}P NMR Data of Four Coordinate Phosphorus Compounds Containing a Formal Multiple Bond from Phosphorus to a Group V Atom

Section M2: Cyclopolyphosphazenes, $(N{=}PX_2)n$, and Their Derivatives

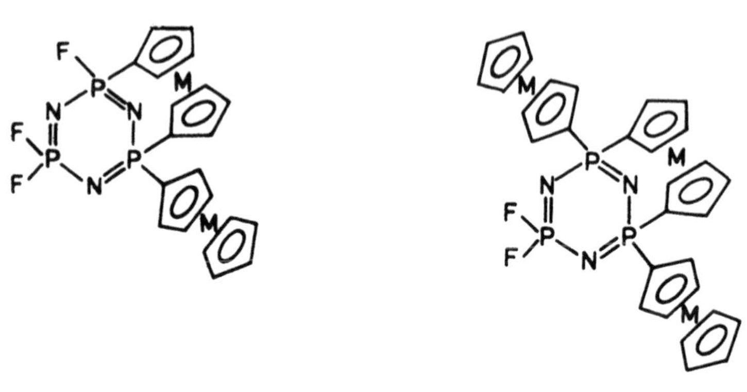

22 23

Connectivities	Molecular formula	Structure	NMR Data (δ_P[solv, temp] $^nJ_{PP}$Hz)	Ref.
	$C_{13}H_{10}F_4N_3O_3P_3Ru_2$	**21** M = M′ = Ru	2.5	58
	$C_{14}H_{10}F_4Fe_2N_3O_4P_3$	**7**; X = F, R = ML_n = $Fe(CO)_2(Cp)$	−0.3	58
	$C_{18}H_{15}F_5N_4P_4$	**9**; X = Ph, Y = Z = F	10.2(c)	63
	$C_{18}H_{15}F_3N_3P_3$	**5**; R = Ph, X = F	8.6	69
	$C_{18}H_{18}F_4N_3P_3$	**3**; R = $C_6H_4(CMe_2{=}CH_2)$-4, X = F	7.2^C(c)	73
	$C_{20}H_{17}F_3Fe_2N_3P_3$	**22**; M = Fe	11.2	75
	$C_{24}H_{20}F_2N_3P_3$	**6**; R = F, X = Ph	6.2, $^2J_{PP}$32.9	69
	$C_{30}H_{26}F_2Fe_3N_3P_3$	**23**; M = Fe	7.7	75
	$C_{30}H_{26}F_2N_3P_3Ru_3$	**23**; M = Ru	7.1	75
Remaining P-bound atoms = Fe_2				
	$C_8Cl_4Fe_4N_3O_4P_3$	**14**	222.5(c)	51
	$C_{10}Cl_4Fe_3N_3O_{10}P_3$	**15**; M = Fe	217.9(c)	51
	$C_{10}Cl_4Fe_2N_3O_{10}$-P_3Ru	**15**; M = Ru	226.3(c)	51
	$C_{13}H_{10}F_4Fe_2N_3O_3P_3$	**21**; M = M′ = Fe	272.3(c)	58
	$C_{14}H_{10}F_4Fe_2N_3O_4P_3$	**7**; X = F, R = ML_n = $Fe(CO)_2(Cp)$	158.4(c)	58
Remaining P-bound atoms = NO				
CH;C	$C_2H_5Cl_4N_4OP_3$	**11**; X = NH, Y = O, R = $(CH_2)_2$, Z = Cl	24.3(c)	18
	$C_3H_7Cl_4N_4OP_3$	**11**; X = NH, Y = O, R = $(CH_2)_3$, Z = Cl	7.2, $^2J_{PP}$53.9	38
	$C_3H_7F_4N_4OP_3$	**11**; X = NH, Y = O, R = $(CH_2)_3$, Z = F	15.1(c)	38
	$C_4H_{10}Cl_2N_5O_2P_3$	**13**; X = NH, Y = O, R = $(CH_2)_2$, Z = Cl	29.0(r)	34
H;C	$C_6H_{14}Cl_2N_5O_2P_3$	**13**; X = NH, Y = O, R = $(CH_2)_3$, Z = Cl	12.7, $^2J_{PP}$50.5	38
CH′;C	$C_{10}H_{29}N_8OP_3$	**11**; X = NH, Y = O, R = $(CH_2)_2$, Z = NMe_2	36.5(r)	34

Connectivities	Molecular formula	Structure	NMR Data (δ_P[solv, temp] $^nJ_{PP}$Hz)	Ref.
CH′;C	$C_7H_7F_3N_5OP_3$	**2**; R = NHC_5H_4N, X = OCH_2CF_3	15.3(k)	76
C_2;C	$C_3H_7Cl_4N_4OP_3$	**11**; X = NMe, Y = O, R = $(CH_2)_2$, Z = Cl	22.4(c)	39
	$C_3H_7F_4N_4OP_3$	**11**; X = NMe, Y = O, R = $(CH_2)_2$, Z = F	30.6(c)	38
	$C_6H_{14}Cl_2N_5O_2P_3$	**13**; X = NMe, Y = O, R = $(CH_2)_2$, Z = Cl	28.5^C(c)	39
	$C_6H_4F_2N_5O_2P_3$	**13**; X = NMe, Y = O, R = $(CH_2)_2$, Z = F	31.6, $^2J_{PP}$98	38

Remaining P-bound atoms = N′O

Connectivities	Molecular formula	Structure	NMR Data	Ref.
C,P^4	$C_{20}H_{20}Cl_4N_4OP_4$	**9**; X = Ph, Y = OEt, Z = Cl	−1.6(c)	20
	$C_{23}H_{30}N_4O_5P_4$	**9**; X = Ph, Y = Z = OMe	13.5(c)	77
C′$_2$	$C_{21}H_{15}N_6O_3P_3S_3$	**4**; R = OPh, X = NCS	−2.8(c)	78

Remaining P-bound atoms = N_2

Connectivities	Molecular formula	Structure	NMR Data	Ref.
$(H_2)_2$	$H_4Br_4N_5P_3$	**6**; R = NH_2, X = Br	5(k)	6
	$H_4Cl_4N_5P_3$	**6**; R = NH_2, X = Cl	9(ae)	9
	$H_{12}N_9P_3$	**2**; R = X = NH_2	18.0(w)	79
	$C_8H_{24}N_9P_3$	**6**; R = NH_2, X = NC_2H_4	18.6(c)	80
	$C_8H_{28}N_9P_3$	**6**; R = NH_2, X = NMe_2	19.2(r), $^2J_{PP}$41.5	13
$(CH)_2$	$C_2H_6Cl_4N_5P_3$	**11**; X = Y = NH, R = $(CH_2)_2$, Z = Cl	22.8(c), $^2J_{PP}$45.7	33, 34
	$C_2H_6F_4N_5P_3$	**11**; X = Y = NH; R = $(CH_2)_2$, Z = F	30.1(c)	38
	$C_2H_8Cl_4N_5P_3$	**6**; R = NHMe, X = Cl	12.3	10
	$C_3H_8Cl_4N_5P_3$	**11**; X = Y = NH, R = $(CH_2)_3$, Z = Cl	7.5(c)	42
	$C_3H_8F_4N_5P_3$	**11**; X = Y = NH, R = $(CH_2)_3$, Z = F	16.3(c)	38
	$C_4H_{10}Cl_6N_5P_3$	**6**; R = $NHCH_2CH_2Cl$, X = Cl	11.3(c), $^2J_{PP}$46.6	12
	$C_6H_{16}Cl_2N_7P_3$	**13**; X = Y = NH, R = $(CH_2)_3$, Z = Cl	12.3(c)	42
	$C_6H_{16}Cl_4N_5P_3$	**6**; R = NH-iPr, X = Cl	6.2(r), $^2J_{PP}$45.5	14
	$C_6H_{16}F_2N_7P_3$	**13**; X = Y = NH, R = $(CH_2)_3$, Z = F	17.1(c)	38
	$C_6H_{24}N_9P_3$	**2**; R = X = NHMe	23.0(w)	79
	$C_8H_{16}Cl_4N_5O_4P_3$	**6**; R = $NHCH_2COOEt$, X = Cl	10.0, $^2J_{PP}$47.4	52
	$C_8H_{20}Cl_4N_5P_3$	**6**; R = NH-tBu, X = Cl	0.7(r), $^2J_{PP}$44.7	13
	$C_8H_{20}Cl_6N_7P_3$	**6**; R = Cl, X = $NHCH_2CH_2Cl$	13.3(c), $^2J_{PP}$50.5	12
	$C_{10}H_{30}N_9P_3$	**11**; X = Y = NH, R = $(CH_2)_2$, Z = NMe_2	35.5, $^2J_{PP}$40	34
	$C_{11}H_{32}N_9P_3$	**11**; X = Y = NH, R = $(CH_2)_3$, Z = NMe_2	17.8(c)	18
	$C_{12}H_{18}F_{18}N_9P_3$	**2**; R = X = $NHCH_2CF_3$	17.4(c)	81
	$C_{12}H_{32}Cl_2N_7P_3$	**6**; R = Cl, X = NH-iPr	9.4(r), $^2J_{PP}$49.4	13
	$C_{12}H_{36}N_9P_3$	**2**; R = X = NHEt	18.0(r)	13
	$C_{12}H_{32}Cl_2N_7O_8P_3$	**6**; R = Cl, X = $NHCH_2COOEt$	14.0, $^2J_{PP}$51.4	52
	$C_{16}H_{40}Cl_2N_7P_3$	**6**; R = Cl, X = NH-tBu	3.9(r)	13
	$C_{18}H_{48}N_9P_3$	**2**; R = X = NH-iPr	12.6(r)	13
	$C_{24}H_{48}N_9O_{12}P_3$	**2**; R = X = $NHCH_2COOEt$	18.9(w)	79
$(CH')_2$	$C_{12}H_{12}Cl_4N_5P_3$	**6**; R = NHPh, X = Cl	−2.6(c)	15
	$C_{14}H_{16}Cl_4N_5O_2P_3$	**6**; R = NHC_6H_4(OMe)-4	2.1(c)	15
	$C_{18}H_{18}N_9P_3$	**24**; X = Y = NH, R = 1,2-C_6H_4	22.0(xylene)	50
	$C_{42}H_{48}N_9O_6P_3$	**2**; R = X = NHC_6H_4(OMe)-4	6.4(c)	15
	$C_{54}H_{60}N_9O_{12}P_3$	**2**; R = X = NHC_6H_4COOEt-4	2.9(ae)	82

TABLE M (continued)
Compilation of ^{31}P NMR Data of Four Coordinate Phosphorus Compounds Containing a Formal Multiple Bond from Phosphorus to a Group V Atom

Section M2: Cyclopolyphosphazenes, $(N{=}PX_2)n$, and Their Derivatives

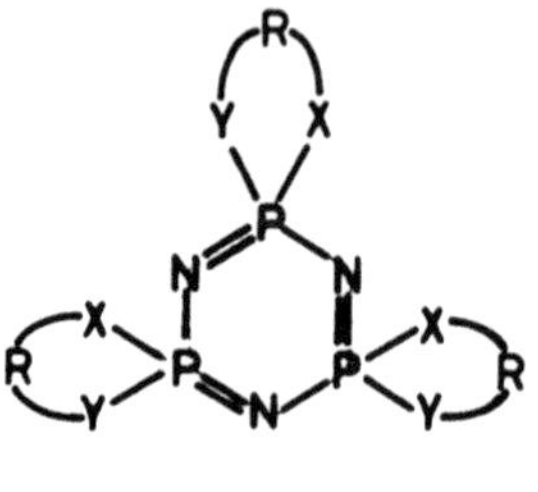

24

Connectivities	Molecular formula	Structure	NMR Data (δ_P[solv, temp] $^nJ_{PP}$Hz)	Ref.
$(C_2)_2$	$C_4H_8Cl_4N_5P_3$	**6**; R = NC_2H_4; K = Cl	34.2	16
	$C_4H_{10}Cl_4N_5P_3$	**11**; X = Y = NMe, R = $(CH_2)_2$, Z = Cl	20.4	45
	$C_5H_{12}Cl_4N_5P_3$	**11**; X = Y = NMe, R = $(CH_2)_3$, Z = Cl	16.5	83
	$C_6H_{12}Cl_3N_6P_3$	**5**; R = NC_2H_4, X = Cl	35.8(c)	16
	$C_6H_{18}Cl_3N_6P_3$	**5**; R = NMe_2; X = Cl	21.7(r)	13
	$C_8H_{16}Cl_2N_7P_3$	**3**; R = Cl, X = NC_2H_4	40.0^T	16
	$C_8H_{16}Cl_2N_7P_3$	**6**; R = Cl, X = NC_2H_4	35.6	16
	$C_8H_{20}Cl_2N_7P_3$	**13**; X = Y = NMe, R = $(CH_2)_2$, Z = Cl	24.9(c)	45
	$C_8H_{24}Cl_2N_7P_3$	**3**; R = Cl, X = NMe_2	24.8^C(r)	13
	$C_{10}H_{20}ClN_8P_3$	**2**; R = Cl, X = NC_2H_4	37.0	16
	$C_{10}H_{28}N_7O_2P_3$	**11**; X = Y = O, R = $(CH_2)_3$, Z = NMe_2	27.9	18
	$C_{10}H_{29}N_8OP_3$	**11**; X = NH, Y = O, R = $(CH_2)_2$, Z = NMe_2	27.3(r)	34
	$C_{10}H_{30}N_9P_3$	**11**; X = Y = NH, R = $(CH_2)_2$, Z = NMe_2	26.7(r)	34
	$C_{11}H_{32}N_9P_3$	**11**; X = Y = NH, R = $(CH_2)_3$, Z = NMe_2	26.7(c)	18
	$C_{12}H_{24}N_9P_3$	**2**; R = X = NC_2H_4	37.0(c)	16, 84
	$C_{12}H_{30}N_9P_3$	**24**; X = Y = NMe, R = $(CH_2)_2$	29.4(c)	45
	$C_{12}H_{36}N_9P_3$	**2**, R = X = NMe_2	24.6(r)	13
	$C_{24}H_{48}N_9P_3$	**2**, R = X = NC_4H_8	17.8(c)	18
	$C_{24}H_{60}N_9P_3$	**2**; R = X = NEt_2	22.5(r)	13
$(CC')_2$	$C_{42}H_{48}N_9P_3$	**2**, R = X = NMePh	13.3(r)	18
$(C_2')_2$	$C_{18}H_{18}N_{15}P_3$	**2**; R = X = imidazolyl	−2.2(k)	85
$(C'N')_2$	$C_{18}H_{18}N_{15}P_3$	**2**; R = X = 1-pyrazoyl	0.4(c)	86
	$C_{30}H_{42}N_{15}P_3$	**2**; R = X = 2,5-dimethyl-1-pyrazolyl	−2.0(c)	86
$(CP^4)_2$	$C_{15}H_{39}Cl_3N_6O_9P_6$	**5**; R = N(Me)P(O)(OEt)$_2$, X = Cl	10.9	19
	Remaining P-bound atoms = NN'			
$H_2;P^4$	$C_{18}H_{17}Cl_4N_5P_4$	**9**; X = Ph, Y = NH_2, Z = Cl	−4.9(c)	55
	$C_{19}H_{19}Cl_4N_5P_4$	**9**; X = Ph, Y = NHMe, Z = Cl	0.8(c)	56
	$C_{22}H_{25}Cl_4N_5P_4$	**9**; X = Ph, Y = NHtBu, Z = Cl	−5.9(c)	56
$C_2;P^4$	$C_{20}H_{19}Cl_4N_5P_4$	**9**p X = Ph, Y = NC_2H_4, Z = Cl	10.3(c)	57
	$C_{20}H_{21}Cl_4N_5P_4$	**9**; X = Ph, Y = NMe_2, Z = Cl	3.9(c)	56

Connectivities	Molecular formula	Structure	NMR Data (δ_P[solv, temp] $^nJ_{PP}$Hz)	Ref.
	$C_{28}H_{35}N_9P_4$	**9**; X = Ph, Y = Z = NC_2H_4	18.7(c)	57
	$C_{28}H_{45}N_9P_4$	**9**; X = Ph, Y = Z = NMe_2	15.3(c)	56
	Remaining P-bound atoms = N'$_2$			
(C')$_2$	$C_2Cl_4N_5P_3S_2$	**6**; R = NCS; X = Cl	−31.5(c)	32
	$C_6N_9P_3S_6$	**2**; R = X = NCS	−30(t)	87
(N')$_2$	$N_{21}P_3$	**2**; R = X = N_3	11.4(m)	88
(P^4)$_2$	$Cl_{10}N_5P_5$	**9**; X = Cl; Y = $NPCl_3$, Z = Cl	−20.4(ae)	9
	$C_{36}H_{30}Cl_4N_5P_5$	**9**; X = Ph, Y = $NPPh_3$, Z = Cl	−10.9(c)	55
	$C_{44}H_{54}N_9P_5$	**9**; X = Ph, Y = $NPPh_3$, Z = NMe_2	1.9(c)	55
	Remaining P-bound atoms = O$_2$			
(C)$_2$	$C_2H_4Cl_4N_3O_2P_3$	**11**; X = Y = O, R = $(CH_2)_2$, Z = Cl	23.8, $^2J_{PP}$67	33
	$C_3H_6Cl_4N_3O_2P_3$	**11**; X = Y = O, R = $(CH_2)_3$, Z = Cl	2.2, $^2J_{PP}$70.8	37
	$C_3H_6F_4N_3O_2P_3$	**11**; X = Y = O, R = $(CH_2)_3$, Z = F	15.2(c)	38
	$C_4H_4Cl_6N_3O_2P_3$	**6**; R = OCH_2CH_2Cl, X = Cl	16.2(c)	12
	$C_4H_8Cl_2N_3O_4P_3$	**13**; X = Y = O, R = $(CH_2)_2$, Z = Cl	30.3	18
	$C_4H_8Cl_4N_3O_2P_3$	**11**; X = Y = O, R = $(CH_2)_4$, Z = Cl	9.3, $^2J_{PP}$74.3	37
	$C_6H_6Cl_3F_9N_3O_3P_3$	**5**; R = OCH_2CF_3, X = Cl	8.4(a)	23
	$C_6H_{12}Cl_2N_3O_4P_3$	**13**; X = Y = O, R = $(CH_2)_3$, Z = Cl	8.3	18
	$C_6H_{12}N_3O_6P_3$	**24**; X = Y = O, R = $(CH_2)_2$	36.5	18
	$C_6H_{16}N_3O_6P_3$	**11**; X = Y = O, R = $(CH_2)_2$, Z = OMe	31.9, 17.0[a] $^2J_{PP}$79.5	37
	$C_6H_{18}N_3O_6P_3$	**2**; R = X = OMe	21.7(c)	89
	$C_8H_8Cl_2F_{12}N_3O_4P_3$	**3**; R = Cl, X = OCH_2CF_3	11.6^C(a)	23
	$C_9H_{18}N_3O_6P_3$	**24**; X = Y = O, R = $(CH_2)_3$	13.2	18
	$C_{10}H_{10}ClF_{15}N_3O_5P_3$	**2**; R = Cl, X = OCH_2CF_3	14.2(a)	23
	$C_{10}H_{10}F_{16}N_3O_5P_3$	**2**; R = F, X = OCH_2CF_3	14.4, $^2J_{PP}$103.4	65
	$C_{10}H_{28}N_7O_2P_3$	**11**; X = Y = O, R = $(CH_2)_2$, Z = NMe_2	35.9, $^2J_{PP}$52.3	18
	$C_{12}H_{12}F_{18}N_3O_6P_3$	**2**; R = X = OCH_2CF_3	16.7(a)	23
	$C_{12}H_{30}N_3O_6P_3$	**2**; R = X = OEt	14.3(c)	89
	$C_{15}H_{15}F_{15}N_5O_5P_3$	**2**; R = NHC_5H_4N, X = OCH_2CF_3	17.1(k)	76
	$C_{22}H_{27}FN_4O_4P_4$	**9**; X = Ph, Y = F, Z = OMe	20.6(c)	63
[a]P(OMe)$_2$				
	$C_{23}H_{30}N_4O_5P_4$	**9**; X = Ph, Y = Z = OMe	21.0(c)	77
C;C'	$C_{12}H_{22}N_3O_6P_3$	**2**; X = OMe, R = OTol-4	15.5, $^2J_{PP}$84.2	18
C;C	$C_{24}H_{30}N_3O_6P_3$	**4**; R = OMe, X = OTol-4	14.3^T, 14.1^C(c)	18
	$C_{36}H_{38}N_3O_6P_3$	**2**; R = OMe, X = OTol-4	13.3(c), $^2J_{PP}$87.2	18
C',Si	$C_{53}H_{50}N_3O_6P_3Si$	**2**; R = $OSiPh_3$, X = OTol-4	2.9	90
(C')$_2$	$C_4H_6Cl_4N_3O_2P_3$	**6**; R = OCH=CH$_2$, X = Cl	−0.6, $^2J_{PP}$68.4	25
	$C_6H_4Cl_4N_3O_2P_3$	**11**; X = Y = O, R = 1,2-C_6H_4	10.8(d)	50
	$C_6H_9Cl_3N_3O_3P_3$	**5**; R = OCH=CH$_2$, X = Cl	2.9(c)	25, 26
	$C_8H_{12}Cl_2N_3O_4P_3$	**6**; R = Cl, X = OCH=CH$_2$	5.7(c), $^2J_{PP}$75.7	25
	$C_{12}H_{18}N_3O_6P_3$	**2**; R = X = OCH=CH$_2$	11.3(c)	25
	$C_{18}H_{12}N_3O_6P_3$	**24**; X = Y = O, R = C_6H_4-1,2	11.8(100°C)	50

TABLE M (continued)
Compilation of ^{31}P NMR Data of Four Coordinate Phosphorus Compounds Containing a Formal Multiple Bond from Phosphorus to a Group V Atom

Section M2: Cyclopolyphosphazenes, (N=PX$_2$)n, and Their Derivatives

Connectivities	Molecular formula	Structure	NMR Data (δ_P[solv, temp] [a]J_{PP}Hz)	Ref.
	$C_{28}H_{28}Cl_2N_3O_4P_3$	3; R = Cl, X = OTol-4	4.4[T](c)	27
	$C_{30}H_{25}FN_3O_6P_3$	2; R = F, X = OPh	8.6	66
	$C_{35}H_{35}ClN_3O_5P_3$	2; R = Cl, X = OTol-4	6.9(c)	27
	$C_{36}F_{30}N_3O_6P_3$	2; R = X = OC_6F_5	10.3	91
	$C_{36}H_{30}N_3O_6P_3$	2; R = X = OPh	8.3(g)	92
	$C_{42}H_{42}N_3O_6P_3$	2; R = X = OTol-4	9.2(c)	27
	$C_{53}H_{50}N_3O_6P_3Si$	2; R = $OSiPh_3$, X = OTol-4	8.4	90
	Remaining P-bound atoms = Ru_2			
$(C'_7Ru)_2$	$C_{13}H_8N_3F_4P_3O_3Ru_2$	21; M = M' = Ru	229.9(c)	
	Remaining P-bound atoms = S_2			
$(C)_2$	$C_2H_6Cl_4N_3P_3S_2$	6; R = SMe, X = Cl	59(b)	28
	$C_4H_{10}Cl_4N_3P_3S_2$	6; R = SEt, X = Cl	51.4, $^2J_{PP}$5.0	29
	$C_4H_{10}F_4N_3P_3S_2$	6; R = SEt, X = F	59.0	67
	$C_4H_{12}Cl_2N_3P_3S_4$	6; R = Cl, X = SMe	50(b)	28
	$C_6H_{15}Cl_3N_3P_3S_3$	5; R = SEt, X = Cl	50.6(c)	29
	$C_6H_{15}F_3N_3P_3S_3$	5; R = SEt, X = F	55.0	67
	$C_6H_{18}N_3P_3S_6$	2; R = X = SMe	46(b)	28
	$C_8H_{20}Cl_2N_3P_3S_4$	6; R = Cl, X = SEt	49.2, $^2J_{PP}$5.2	29
	$C_8H_{20}F_2N_3P_3S_4$	6; R = F, X = SEt	52.0	67
	$C_{10}H_{25}X1N_3P_3S_5$	2, R = Cl, X = SEt	48.0(c)	29
	$C_{10}H_{25}FN_3P_3S_5$	2; R = F, X = SEt	48.0	67
	$C_{12}H_{30}N_3P_3S_6$	2; R = X = SEt	46.5(c)	29
	$C_{12}H_{66}B_{60}N_3P_3S_6$	2; R = X = $SCB_{10}H_{10}CH$*	30.0	93
	Remaining P-bound atoms = HO			
C	$C_3H_8Cl_4N_3OP_3$	10; R[1] = H, R[2] = O-iPr, X = Cl	2.5, $^2J_{PP}$26 $^1J_{PP}$719(c)	41
	Remaining P-bound atoms = CBr			
H_3	$CH_3BrCl_4N_3P_3$	10; R[1] = Me, R[2] = Br, X = Cl	24.4(c)	30
CH_2	$C_2H_5BrCl_4N_3P_3$	10; R[1] = Et, R[2] = Br, X = Cl	34.8(c)	30
CH_2	$C_3H_7BrCl_4N_3P_3$	10; R[1] = iPr, R[2] = Br, X = Cl	43.6(c)	30
C_3	$C_4H_9BrCl_4N_3P_3$	10; R[1] = tBu, R[2] = Br, X = Cl	51.4, $^2J_{PP}$15.0	30
	Remaining P-bound atoms = C'Br			
C'_2	$C_6H_5BrCl_4N_3P_3$	10; R[1] = Ph, R[2] = Br, X = Cl	15.3(c)	49
	$C_{18}H_{15}Br_3N_3P_3$	4; R = Ph, X = Br	16.4[C](b); 20.1, 18.0[Ta]	94
	Remaining P-bound atoms = CCl			
H_2Si	$C_4H_{11}Cl_5N_3P_3Si$	2; R = CH_2Tms, X = Cl	42.0(c)	47
H_3	$CH_3Cl_5N_3P_3$	2; R = Me, X = Cl	39.2(c)	30

*Carborane derivative.
[a]Relative intensities 1:2, respectively.

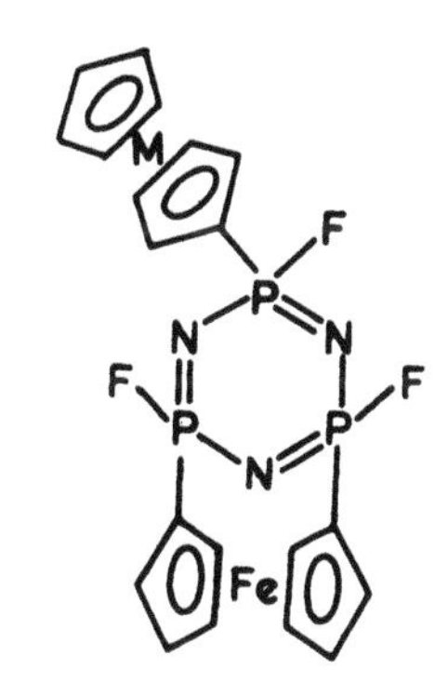

25

Connectivities	Molecular formula	Structure	NMR Data (δ_P[solv, temp] $^nJ_{PP}$Hz)	Ref.
	$C_{25}H_{23}Cl_4N_3P_3$	**10**; R′ = Me, R² = Cl, X = Ph	38.4(r)	95
CH_2	$C_3H_5Cl_5N_3P_3$	**2**; R = Et, X = Cl	46.0(c)	30
C_2H	$C_3H_7Cl_5N_3P_3$	**2**; R = iPr, X = Cl	51.8(c)	30
C_3	$C_4H_9Cl_5N_3P_3$	**2**; R = tBu, X = Cl	57.1(c), $^2J_{PP}$6.9	30
	Remaining P-bound atoms = C′Cl			
C'_2	$C_6H_5Cl_5N_3P_3$	**2**; R = Ph, X = Cl	28.9(c), $^2J_{PP}$15.5	49
	$C_{18}H_{15}Cl_3N_3P_3$	**5**; R = Ph, X = Cl	29.6^C(r); 30.2, 32.8^T	13
	$C_{24}H_{20}Cl_2N_3P_3$	**3**; R = Cl, X = Ph	28.5^T	96
C'_2	$C_{30}H_{25}ClN_3P_3$	**2**; R = Cl, X = Ph	28.6(c), $^2J_{PP}$2.5	97
C'_2Fe	$C_{10}H_8Cl_{10}FeN_6P_6$	**17**; M = Fe, R = Cl	35.0	53
	$C_{10}H_9Cl_5FeN_3P_3$	**18**; M = Fe, R = X = Cl	36.1, $^2J_{PP}$15	53
CRu	$C_{10}H_9Cl_5N_3P_3Ru$	**18**; M = Ru, R = X = Cl	33.7, $^2J_{PP}$17	53
	Remaining P-bound atoms = C′Cr			
C'_2,C'_8	$C_{14}H_{10}Cl_4CrN_3O_3P_3$	**7**; R = Ph, X = Cl, ML_n = $Cr(CO)_3(Cp)$	110.4(c)	54
	Remaining P-bound atoms = CF			
H_3	$CH_3F_5N_3P_3$	**2**; R = Me, X = F	46.2(c)	64
C_3	$C_4H_9F_5N_3P_3$	**2**; R = tBu, X = F	56.5(c)	71
	$C_8H_{18}F_4N_3P_3$	**3**; R = B′, X = F	56.2(c), $^2J_{PP}$51.3	71
	$C_{12}H_{27}F_3N_3P_3$	**4**; R = tBu, X = F	33.2, 31.1(c)	71
	Remaining P-bound atoms = C′F			
C′N	$C_5H_6F_5N_4P_3$	**2**; R = 1-methyl 2-pyrol-2-yl, X = F	23.8(c)	72
C'_2	$C_6H_5Cl_4FN_3P_3$	**10**; R¹ = Ph, R² = F, X = Cl	23.8(c), $^2J_{PP}$34.1	49
	$C_9H_9F_5N_3P_3$	**2**; R = C_6H_4(CMe=CH_2)-4, X = F	35.4(c)	73
	$C_{12}H_{10}F_4N_3P_3$	**3**; R = Ph, X = F	38.4^C	74
	$C_{18}H_{15}F_3N_3P_3$	**5**; R = Ph, X = F	27.3	69
	$C_{18}H_{18}F_4N_3P_3$	**3**; R = C_6H_4(CMe=CH_2)-4, X = F	33.5^C(c)	73
C'_2Fe	$C_{10}H_8F_4FeN_3P_3$	**20**; M = Fe	46.6	53
	$C_{10}H_9F_5FeN_3P_3$	**18**; M = Fe, R = X = F	44.3	53
	$C_{20}H_{17}F_3Fe_2N_3P_3$	**22**; M = Fe	35.5	75
	$C_{20}H_{17}F_3Fe_2N_3P_3$	**25**	48.2, 45.7^C, 48.3, 44.9^T	75

TABLE M (continued)
Compilation of ^{31}P NMR Data of Four Coordinate Phosphorus Compounds Containing a Formal Multiple Bond from Phosphorus to a Group V Atom

Section M2: Cyclopolyphosphazenes, $(N{=}PX_2)n$, and Their Derivatives

Connectivities	Molecular formula	Structure	NMR Data (δ_P[solv, temp] $^nJ_{PP}$Hz)	Ref.
C'_2Ru	$C_{10}H_8F_4N_3P_3Ru$	**20**; M = Ru	43.6	53
	$C_{10}H_8F_{10}N_6P_6$	**17**; M = Ru, R = F	38.3	53
	$C_{10}H_9F_5N_3P_3Ru$	**18**; M = Ru, R = K = F	42.9	53
	Remaining P-bound atoms = C'Fe			
$C'_2;C'_7$	$C_{13}H_{10}Cl_4FeN_3O_2P_3$	**7**; R = Ph, X = Cl, ML_n = $Fe(CO)_2(Cp)$	103.6(c)	54
	Remaining P-bound atoms = CI			
H_3	$CH_3Cl_4IN_3P_3$	**10**; R^1 = Me, R^2 = I, X = Cl	−16.3(c)	30
CH_2	$C_2H_5Cl_4IN_3P_3$	**10**; R^1 = Et, R^2 = I, X = Cl	0.7(c)	30
CH_2	$C_3H_7Cl_4IN_3P_3$	**10**; R^1 = Pr^i, R^2 = I, X = Cl	16.3(c)	30
	Remaining P-bound atoms = C'I			
C'_2	$C_6H_5Cl_4IN_3P_3$	**10**; R^1 = Ph, R^2 = I, X = Cl	−23.9(c) $^2J_{PP}$8.6	49
	Remaining P-bound atoms = C'Mo			
C'_2,C'_8	$C_{13}H_{10}Cl_4MoN_3O_3P_3$	**7**; R = C_5H_5, X = Cl, ML_n = $Mo(CO)_3(Cp)$	86.0, $^2J_{PP}$49	7
	$C_{14}H_{10}Cl_4MoN_3O_3P_3$	**7**; R = Ph, X = Cl, ML_n = $Mo(CO)_3(Cp)$	89.2(c)	54
	Remaining P-bound atoms = C″N			
N'',C_2	$C_9H_{18}N_9P_3$	**4**; R = CN, X = NMe_2	-7.8^T	98
	Remaining P-bound atoms = C″O			
N'',C'	$C_{31}H_{25}N_4O_5P_3$	**2**; R = CN, X = OPh	−9.2, $^2J_{PP}$55	98
$H_3;N'_2$	$C_2H_6Cl_8N_6P_6$	**12**; R = Me, X = Cl	26.4(k)	36
	$C_{50}H_{46}N_6O_8P_6$	**12**; R = Me, X = OPh	29.9(k)	36
	$C_{50}H_{46}N_6P_6$	**12**; R = Me, X = Ph	24.7(r)	99
$CH_2;N'_2$	$C_4H_{10}Cl_8N_6P_6$	**12**; R = Et, X = Cl	31.9(c)	46
	Remaining P-bound atoms = C'P^4			
$C'_2;N'_2$	$C_{12}H_{10}Cl_8N_6P_6$	**12**; X = Ph, $\dot{Y}$ = Cl	17.7(k)	36
$C'_2Fe;CP^4$	$C_{10}H_8Cl_8N_6P_6$	**16**	22.5	53
$C'_2Fe;ClN'_2$	$C_{10}H_9Cl_9FeN_6P_6$	**19**; M = Fe	25.8	53
$C'_2Ru;ClN'_2$	$C_{10}H_9Cl_9N_6P_6Ru$	**19**; M = Ru	24.9	53
	Remaining P-bound atoms = CW			
$H_3;C'_8$	$C_9H_8Cl_4N_3O_3P_3W$	**7**; R = Me, X = Cl, ML_n = $W(CO)_3(Cp)$	58.9, $^2J_{PP}$39	7

Connectivities	Molecular formula	Structure	NMR Data (δ_P[solv, temp] $^nJ_{PP}$Hz)	Ref.
	Remaining P-atoms = C′W			
$C'_2;C'_8$	$C_{13}H_{10}Cl_4N_3O_3P_3W$	**7**; R = C_5H_5, X = Cl, ML_n = $W(CO)_3(Cp)$	58.5, $^2J_{PP}48$	7
	$C_{14}H_{10}Cl_4N_3O_3P_3W$	**7**; R = Ph, X = Cl, ML_n = $W(CO)_3(Cp)$	52.3(c)	54
	Remaining P-bound atoms = CH			
H_3	$CH_4Cl_4N_3P_3$	**10**; R[1] = Me, R[2] = H, X = Cl	13.8(k), 12.1(c)	31, 41
	$C_{25}H_{24}N_3P_3$	**10**; R[1] = Me, R[2] = H, X = Ph	6.9	95
CH_2	$C_2H_6Cl_4N_3P_3$	**10**; R[1] = Et, R[2] = H, X = Cl	20.4(k), $^2J_{PP}8$	31
C_2H	$C_3H_8Cl_4N_3P_3$	**10**; R[1] = iPr, R[2] = H, X = Cl	26.5(k), $^2J_{PP}8$	31
C_3	$C_4H_{10}Cl_4N_3P_3$	**10**; R[1] = Bu′, R[2] = H, X = Cl	32.3(k)	31
	Remaining P-bound atoms = C′H			
C'_2	$C_6H_6Cl_4N_3P_3$	**10**; R[1] = Ph, R[2] = H, X = Cl	6.9(c), $^2J_{PP}10.3$	49
C'_2Ru	$C_{10}H_{10}Cl_4N_3P_3Ru$	**18**, M = Ru, R = H, X = Cl	18.6, $^2J_{PP}11$	53
	Remaining P-bound atoms = C_2			
$(H_3)_2$	$C_2H_6Cl_4N_3P_3$	**6**; R = Me, X = Cl	35.7, $^2J_{PP}7.8$	35
	$C_2H_6F_4N_3P_3$	**6**; R = Me, X = F	41.3(c)	70
	$C_4H_{12}Cl_2N_3P_3$	**6**; R = Cl, X = Me	31.6(c), $^2J_{PP}3.6$	48
	$C_6H_{18}N_3P_3$	**2**, R = X = Me	25.9(c), 31.7(w)	100, 101
	$C_{26}H_{26}N_3P_3$	**6**; R = Me, X = Ph	27.1	102
$H_3;CH_2$	$C_3H_8Cl_4N_3P_3$	**10**; R[1] = Me, R[2] = Et, X = Cl	41.8(c)	35
$H_3;C'H_2$	$C_4H_8Cl_4N_3P_3$	**10**; R[1] = Me, R[2] = $CCH_2H{=}CH_2$, X = Cl	36.9(c)	35
	$C_4H_8Cl_4N_3OP_3$	**10**; R[1] = Me, R[2] = $CCH_2(O)Me$, X = Cl	31.4(c)	44
	$C_5H_{10}Cl_4N_3OP_3$	**10**; R[1] = Me, R[2] = $CCH_2(OMe){=}CH_2$; X = Cl	37.4(c)	44
$H_3;C''H_2$	$C_5H_8Cl_4N_3P_3$	**10**; R[1] = Me, R[2] = $CCH_2{\equiv}CMe$, X = Cl	36.3(k)	43
$H_3;C_2H$	$C_4H_{10}Cl_4N_3P_3$	**10**; R[1] = Me, R[2] = iPr, X = Cl	46.1(c)	35
$H_3;CHO$	$C_8H_{10}Cl_4N_3OP_3$	**10**; R[1] = Me, R[2] = CH(OH)Ph, X = Cl	39.3(c)	40
	$C_{32}H_{30}N_3OP_3$	**10**; R[1] = Me, R[2] = CH(OH)Ph, X = Ph	30.8	95
$H_3;C_3$	$C_5H_{12}Cl_4N_3P_3$	**10**; R[1] = Me, R[2] = tBu, X = Cl	50.4(c)	35
$(CH_2)_2$	$C_4H_{10}Cl_4N_3P_3$	**10**; R = Et, X = Cl	48.1(c)	35
$CH_2;C_2H$	$C_5H_{12}Cl_4N_3P_3$	**10**; R[1] = Et, R[2] = iPr, X = Cl	52.5(c)	35
$CH_2;C_3$	$C_6H_{14}Cl_4N_3P_3$	**10**; R[1] = Et, R[2] = tBu, X = Cl	56.8(c), $^2J_{PP}7$	35
$(C_2H)_2$	$C_6H_{14}Cl_4N_3P_3$	**6**; R = iPr, X = Cl	56.6(c), $^2J_{PP}7.4$	49
$C_2H;HCO$	$C_{10}H_{14}Cl_4N_3OP_3$	**10**; R[1] = iPr, R[2] = CH(OH)Ph, X = Cl	47.3(c), $^2J_{PP}32.6$	40
$(H_2Si)_2$	$C_8H_{22}Cl_4N_3P_3Si_2$	**6**; R = CH_2Tms, X = Cl	40.3(c)	47
	Remaining P-bound atoms = CC′			
$H_3;HC'$	$C_4H_6Cl_4N_3P_3$	**10**; R[1] = Me, R[2] = $CH{=}C{=}CH_2$, X = Cl	25.9(x), $^2J_{PP}9.0$	43
	$C_5H_{10}Cl_4N_3OP_3$	**10**; R[1] = Me, R[2] = CH=C(OMe)Me, X = Cl	27.0(k), $^2J_{PP}12.2$	44
$H_3;C'_2$	$C_7H_8Cl_4N_3P_3$	**10**; R[1] = Me, R[2] = Ph, X = Cl	29.0(c), $^2J_{PP}11$	49

TABLE M (continued)
Compilation of ^{31}P NMR Data of Four Coordinate Phosphorus Compounds Containing a Formal Multiple Bond from Phosphorus to a Group V Atom

Section M2: Cyclopolyphosphazenes, $(N{=}PX_2)n$, and Their Derivatives

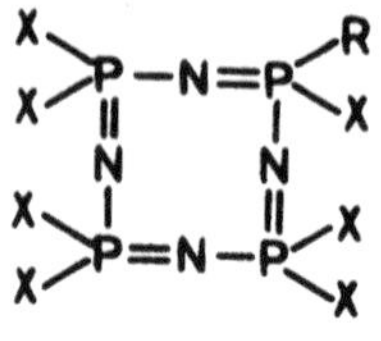

26 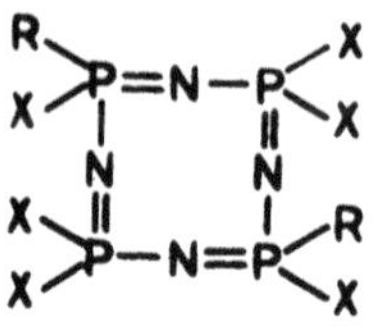 27 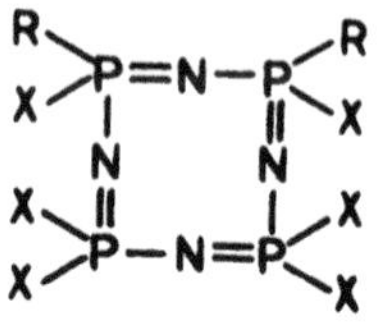 28

Connectivities	Molecular formula	Structure	NMR Data (δ_P[solv, temp] $^nJ_{PP}$Hz)	Ref.
H_3;C'$_2$Fe	$C_{11}H_{12}Cl_4FeN_3P_3$	**18**; M = Fe, R = Me, X = Cl	29.2, $^2J_{PP}$22	53
H_3;C'$_2$Ru	$C_{11}H_{12}Cl_4N_3P_3Ru$	**18**; M = Ru, R = Me, X = Cl	28.8, $^2J_{PP}$22	53
	Remaining P-bound atoms = CC″			
H_3;C″	$C_4H_6Cl_4N_3P_3$	**10**; R[1] = Me, R[2] = C≡CMe, X = Cl	2.5(k), $^2J_{PP}$14.0	44
	Remaining P-bound atoms = C'$_2$			
(C'$_2$)$_2$	$C_{12}H_{10}Cl_4N_3P_3$	**6**; R = Ph, X = Cl	19.5(r)	13
	$C_{12}H_{10}F_4N_3P_3$	**6**; R = Ph, X = F	30.4, $^2J_{PP}$86	69
	$C_{18}H_{15}F_3N_3P_3$	**5**; R = Ph, X = F	27.3	69
	$C_{24}H_{20}Cl_2N_3P_3$	**6**; R = Cl, X = Ph	17.1, $^2J_{PP}$9.3	13
	$C_{24}H_{20}Cl_2N_3P_3$	**3**; R = Cl, X = Ph	19.9^T	96
	$C_{24}H_{20}F_2N_3P_3$	**6**; R = F, X = Ph	27.3, $^2J_{PP}$32.9	69
	$C_{25}H_{23}ClN_3P_3$	**10**; R[1] = Me, R[2] = Cl, X = Ph	15.3(r)	95
	$C_{25}H_{24}N_3P_3$	**10**; R[1] = Me, R[2] = H, X = Ph	12.9(r)	95
	$C_{26}H_{26}N_3P_3$	**6**; R = Me, X = Ph	14.1(c)	102
	$C_{30}H_{25}ClN_3P_3$	**2**; R = Cl, X = Ph	16.8, $^2J_{PP}$2.5	97
	$C_{32}H_{30}N_3OP_3$	**10**; R[1] = Me, R[2] = CH(OH)Ph, X = Ph	14.3(r)	95
	$C_{36}H_{30}N_3P_3$	**2**; R = X = Ph	15.2	96
(C'$_2$Fe)$_2$	$C_{20}H_{17}F_3Fe_2N_3P_3$	**22**; M = Fe	35.5	75
	$C_{30}H_{26}F_2Fe_3N_3P_3$	**23**; M = Fe	29.8	75
(C'$_2$Ru)$_2$	$C_{30}H_{26}F_2N_3P_3Ru_3$	**23**; M = Ru	25.6	75

Section M2.3: Cyclotetraphosphazenes

Connectivities	Molecular formula	Structure	NMR Data (δ_P[solv, temp] $^nJ_{PP}$Hz)	Ref.
	Remaining P-bound atoms = Br$_2$			
	$Br_8N_4P_4$	**26**; R = X = Br	−69.9(s)	103, 104
	Remaining P-bound atoms = ClN			
H_2	$H_4Cl_6N_6P_4$	**27**; R = NH$_2$, X = Cl	−5.8	105
CH	$C_2H_8Cl_6N_6P_4$	**27**; R = NHMe, X = Cl	−2.2(r)	106
	$C_4H_{12}Cl_6N_6P_4$	**27**; R = NHEt, X = Cl	−4.9^T(r), $^2J_{PP}$46	107
	$C_6H_{16}Cl_6N_6P_4$	**27**; R = NH-iPr, X = Cl	−7.4(r), $^2J_{PP}$38.3	106
	$C_8H_{20}Cl_6N_6P_4$	**27**; R = NH-tBu, X = Cl	−10.6(r), $^2J_{PP}$38.1	108
	$C_8H_{20}Cl_6N_6P_4$	**28**; R = NH-tBu, X = Cl	−7.3(c)	109

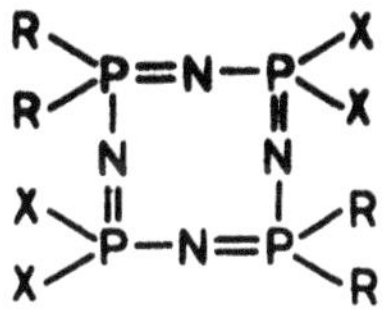

29

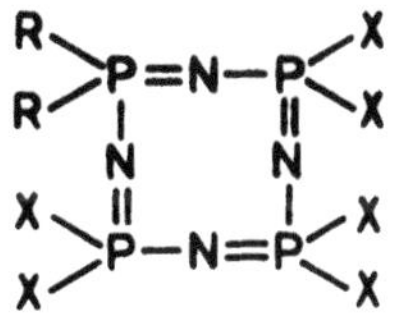

30

Connectivities	Molecular formula	Structure	NMR Data (δ_P[solv, temp] $^nJ_{PP}$Hz)	Ref.
	$C_{10}H_{24}Cl_6N_6P_4$	**27**; R = $NHCH_2CMe_3$, X = Cl	−4.2	18
	$C_{14}H_{16}Cl_6N_6P_4$	**27**; R = $NHCH_2$ Ph, X = Cl	−5.5(c)	106
C′H	$C_6H_6Cl_7N_5P_4$	**26**; R = NHPh, X = Cl	−5.1(c)	110
	$C_{12}H_{12}Cl_6N_6P_4$	**27**; R = NHPh, X = Cl	−12.0(r), $^2J_{PP}$40.3	106
	$C_{12}H_{12}Cl_6N_6P_4$	**28**; R = NHPh, X = Cl	−5.6(c)	110
	$C_{14}H_{16}Cl_6N_6P_4$	**27**; R = NHC_6H_4Me-2, X = Cl	−11.6(j)	111
C_2	$C_2H_4Cl_7N_5P_4$	**26**; R = NC_2H_4, X = Cl	8.6(c)	16
	$C_4H_8Cl_6N_6P_4$	**27**; R = NC_2H_4, X = Cl	8.4^T(c), $^2J_{PP}$27.9	16
	$C_4H_8Cl_6N_6P_4$	**28**; R = NC_2H_4, X = Cl	11.8^T(c)	16
	$C_2H_{12}Cl_6N_6P_4$	**27**; R = NMe_2, X = Cl	−0.2(r)	13
	$C_{12}H_{24}Cl_2N_{10}P_4$	**27**; R = Cl, X = NC_2H_4	13.7^T(c), $^2J_{PP}$24.3	16
	$C_{12}H_{36}Cl_2N_{10}P_4$	**27**; R = Cl, X = NMe_2	4.4(r)	13
	$C_{28}H_{28}Cl_6N_6P_4$	**27**; R = $N(CH_2Ph)_2$, X = Cl	−3.9(c)	112
	$C_{28}H_{28}Cl_6N_6P_4$	**28**; R = $N(CH_2Ph)_2$, X = Cl	−0.2(c)	112
CC′	$C_{14}H_{16}Cl_6N_6P_4$	**27**; R = NMePh, X = Cl	−5.3(r)	113
	$C_{14}H_{16}Cl_6N_6P_4$	**28**; R = NMePh, X = Cl	−3.2(r)	109
	Remaining P-bound atoms = ClN′			
P^4	$C_3H_9Cl_7N_5P_5$	**26**; R = $NPMe_3$, X = Cl	−18.9	18
	$C_{18}H_{15}Cl_7N_5P_5$	**26**; R = $NPPH_3$, X = Cl	−19.0(c)	18
	$Cl_{12}N_6P_6$	**27**; R = $NPCl_3$, X = Cl	−23.5	105
	$C_{36}H_{30}Cl_6N_6P_6$	**27**; R = $NPPh_3$, X = Cl	−16.8(c)	20
	Remaining P-bound atoms = ClO			
C′	$C_2H_3Cl_7N_4OP_4$	**26**; R = OCH=CH$_2$, X = Cl	−10.0(c)	26
	$C_4H_6Cl_6N_4O_2P_4$	**27**; R = OCH=CH$_2$, X = Cl	−10.1(c)	26
	$C_4H_6Cl_6N_4O_2P_4$	**28**; R = OCH=CH$_2$, X = Cl	−8.2(c)	26
	$C_{12}H_{10}Cl_6N_4O_2P_4$	**27**; R = OPh, X = Cl	−11.6(c)	114
	$C_{12}H_{10}Cl_6N_4O_2P_4$	**28**; R = OPh, X = Cl	−9.5(c)	114
	$C_{36}H_{30}Cl_2N_4O_6P_4$	**27**; R = Cl, X = OPh	−6.9(c)	114
	Remaining P-bound atoms = Cl_2			
	$Cl_8N_4P_4$	**26**; R = X = Cl	−6.5(c)	13, 104
	$Cl_{12}N_6P_6$	**27**; R = $NPCl_3$, X = Cl	−8.0	105
	$H_6N_6P_4$	**27**; R = NH_2, X = Cl	−5.8	105
	$H_8N_8P_4$	**29**; R = NH_2, X = Cl	−13.1	105
	$CH_4Cl_6N_4P_4$	**30**; R_2 = H, Me	−6.4	115
	$CH_3Cl_7N_4P_4$	**26**; R = Me, X = Cl	-7.8^a, −4.5	115
	$C_2H_3Cl_7N_4OP_4$	**26**; R = OCH=CH$_2$, X = Cl	−3.7, 5.6(c)[b]	26
	$C_2H_4Cl_7N_5P_4$	**26**; R = NC_2H_4, X = Cl	-7.2^C, 4.7(c)	16
	$C_2H_8Cl_6N_6P_4$	**27**; R = NHMe, X = Cl	−2.2(r)	106
	$C_3H_9Cl_7N_5P_5$	**26**; R = $NPMe_3$, X = Cl	−7.0, −11.0	18
	$C_4H_6Cl_6N_4O_2P_4$	**27**; R = OCH=CH$_2$, X = Cl	−13.1(c)	26
	$C_4H_6Cl_6N_4O_2P_4$	**28**; R = OCH=CH$_2$, X = Cl	−4.5(c)	26
	$C_4H_6Cl_6N_4O_2P_4$	**30**; R = OCH=CH$_2$, X = Cl	−3.9, −5.6[b]	26
	$C_4H_8Cl_6N_6P_4$	**27**; R = NC_2H_4, X = Cl	-1.9^T	16
	$C_4H_8Cl_6N_6P_4$	**28**; R = NC_2H_4, X = Cl	-4.9^T(c)	16

TABLE M (continued)
Compilation of ^{31}P NMR Data of Four Coordinate Phosphorus Compounds Containing a Formal Multiple Bond from Phosphorus to a Group V Atom

Section M2: Cyclopolyphosphazenes, $(N{=}PX_2)n$, and Their Derivatives

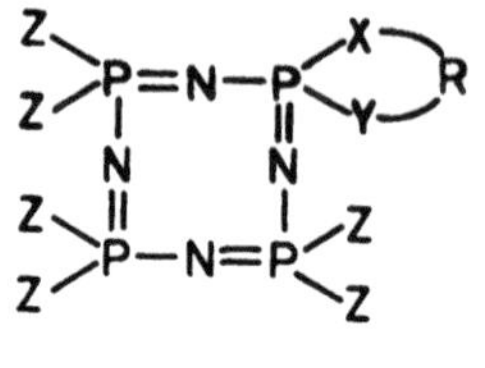

31

Connectivities	Molecular formula	Structure	NMR Data (δ_P[solv, temp] $^nJ_{PP}$Hz)	Ref.
	$C_4H_8Cl_6N_6P_4$	**30**; R = NC_2H_4, X = Cl	−6.5[c], −5.9	16
	$C_4H_{10}Cl_6N_4P_4S_2$	**30**; R = SEt, X = Cl	−7.1, −8.2(c)	29
	$C_4H_{10}Cl_6N_6P_4$	**31**; X = Y = NMe, R = $(CH_2)_2$, Z = Cl	−6.2(c) and −8.3	116
	$C_4H_{12}Cl_6N_6P_4$	**27**; R = NHEt, X = Cl	−4.9(r)	107
	$C_4H_{12}Cl_6N_6P_4$	**27**; R = NMe_2, X = Cl	−3.7(r)	13
	$C_6H_6Cl_7N_5P_4$	**26**; R = NHPh, X = Cl	−11.1(c) and −7.7	110
	$C_6H_{16}Cl_6N_6P_4$	**27**; R = NH-iPr, X = Cl	−4.0(r)	106
	$C_8H_{20}Cl_4N_4P_4S_4$	**29**; R = SEt, X = Cl	−8.8(c), $^2J_{PP}$11.8	29
	$C_8H_{20}Cl_6N_6P_4$	**27**; R = NH-tBu, X = Cl	−5.8(r)	108
	$C_8H_{20}Cl_6N_6P_4$	**28**; R = NH-tBu, X = Cl	−8.7(c)	109
	$C_{12}H_{10}Cl_6N_4O_2P_4$	**27**; R = OPh, X = Cl	−4.7(c)	114
	$C_{12}H_{10}Cl_6N_4O_2P_4$	**28**; R = OPh, X = Cl	−5.1(c)	114
	$C_{12}H_{10}Cl_6N_4P_4S_2$	**30**; R = SPh, X = Cl	−7.5, −8.9	117
	$C_{12}H_{12}Cl_6N_6P_4$	**27**; R = NHPh, X = Cl	−3.0(r)	106
	$C_{12}H_{12}Cl_6N_6P_4$	**28**; R = NHPh, X = Cl	−8.5(c)	110
	$C_{12}H_{30}Cl_2N_4P_4S_6$	**30**; R = Cl, X = SEt	−9.4(c)	29
	$C_{14}H_{16}Cl_6N_6P_4$	**27**; R = NHTol-2, X = Cl	−5.9(j)	111
	$C_{14}H_{16}Cl_6N_6P_4$	**27**; R = NMePh, X = Cl	−5.3(r)	113
	$C_{14}H_{16}Cl_6N_6P_4$	**28**; R = NMePh, X = Cl	−3.2(r)	109
	$C_{18}H_{15}Cl_7N_5P_5$	**26**; R = $NPPh_3$, X = Cl	−11.3, −7.4	18
	$C_{24}H_{20}Cl_4N_4P_4$	**29**; R = Ph, X = Cl	−3.1(r)	13
	$C_{24}H_{20}Cl_4N_4P_4S_4$	**29**; R = SPh, X = Cl	−9.5, $^2J_{PP}$17.6(c)	118
	$C_{28}H_{28}Cl_6N_6P_4$	**27**; R = $N(CH_2Ph)_2$, X = Cl	−3.9(c)	112
	$C_{28}H_{28}Cl_6N_6P_4$	**28**; R = $N(CH_2Ph)_2$, X = Cl	−6.8	112
	$C_{28}H_{32}Cl_4N_8P_4$	**29**; R = NMePh, X = Cl	−11.5	113
	$C_{36}H_{30}Cl_6N_6P_6$	**27**; R = $NPPh_3$, X = Cl	−12.2(c)	20
	Remaining P-bound atoms = FN			
H_2	$H_2F_7N_5P_4$	**26**; R = NH_2, X = F	2.7	119
	Remaining P-bound atoms = FN′			
C′	$CF_7N_5OP_4$	**26**; R = NCO, X = F	−22	61
	$CF_7N_5P_4S$	**26**; R = NCS, X = F	−28	61

[a] PCl_2 antipodal to PCl(Me).
[b] PCl_2 antipodal to PCl(OR).
[c] PCl_2 antipodal to $PCl(NC_2H_4)$.

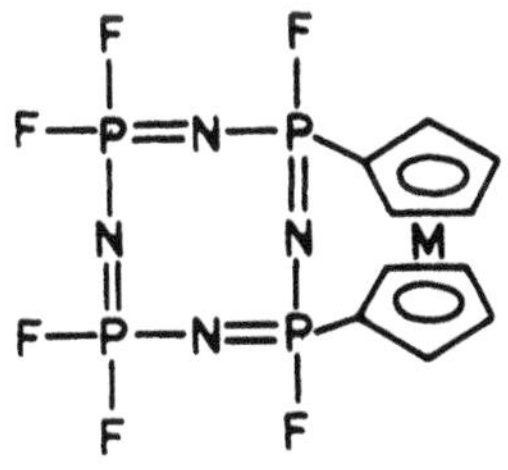

32

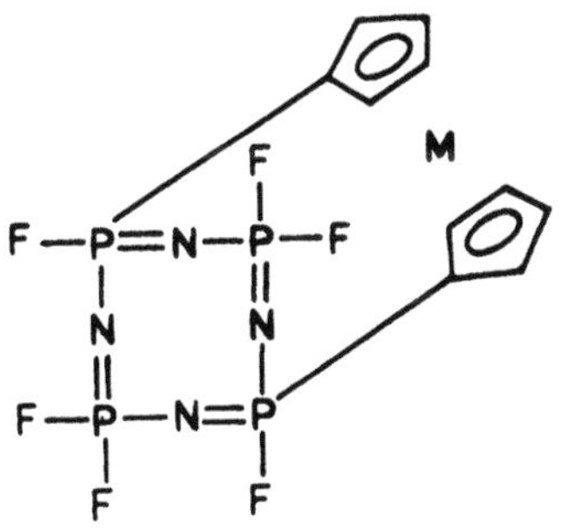

33

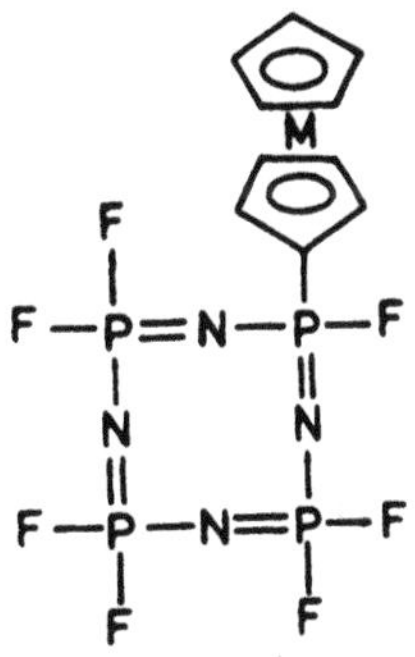

34

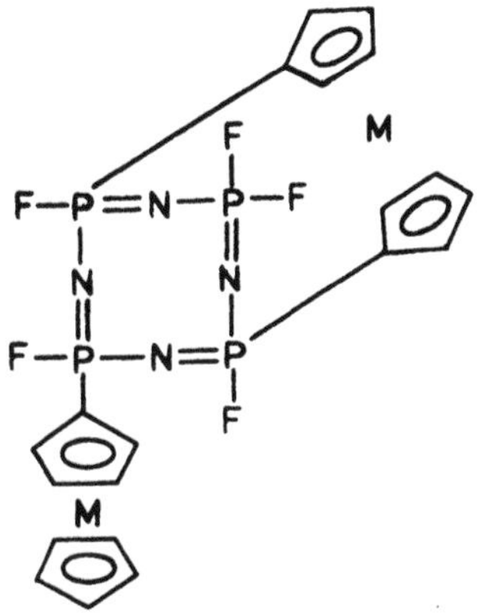

35

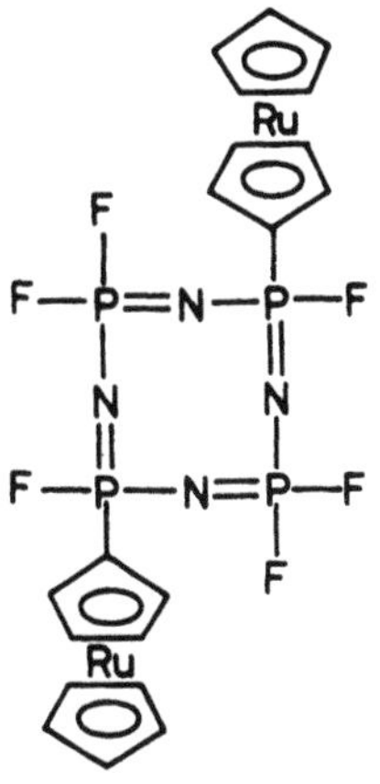

36

Connectivities	Molecular formula	Structure	NMR Data (δ_P[solv, temp] $^nJ_{PP}$Hz)	Ref.
Remaining P-bound atoms = F_2				
	$F_7N_7P_4S_3$	**26**; R = NS_3N_2[a], X = F	−17.0(c)	119
	$F_8N_4P_4$	**26**; R = X = F	−17.7(c)	104
	$H_2F_7N_5P_4$	**26**; R = NH_2, X = F	−15	62
	$CF_7N_5OP_4$	**26**; R = NCO, X = F	−18	61
	$CF_7N_5P_4S$	**26**; R = NCS, X = F	−21	61
	$CH_3F_7N_4P_4$	**26**; R = Me, X = F	−12.9	120
	$C_2H_6F_6N_4P_4$	**30**; R = Me, X = F	−14.2(n)	70, 120
	$C_2H_6F_7N_5P_4$	**26**; R = NMe_2, X = F	−15.7(n)	59
	$C_4H_{12}F_4N_4P_4$	**29**; R = Me, X = F	−8.5	120
	$C_4H_{12}F_6N_6P_4$	**27**; R = NMe_2, X = F	−15.5(n)	59
	$C_5H_6F_7N_5P_4$	**26**; R = 1-methylpyrrol-2-yl, X = F	−15.6	72
	$C_{10}H_8F_6FeN_4P_4$	**32**; M = Fe	−13.3	53
	$C_{10}H_8F_6FeN_4P_4$	**33**; M = Fe	−15.0	53
	$C_{10}H_8F_6N_4P_4Ru$	**33**; M = Ru	−12.7	53
	$C_{10}H_9F_7FeN_4P_4$	**34**; M = Fe	−16	53
	$C_{10}H_9F_7N_4P_4Ru$	**34**; M = Ru	−16.9	53
	$C_{20}H_{17}F_5Fe_2N_4P_4$	**35**; M = Fe	−11.7	75
	$C_{20}H_{17}F_5N_4P_4Ru_2$	**35**; M = Ru	−12.7	75
	$C_{20}H_{18}F_6N_4P_4Ru_2$	**36**	−22.8	53

[a]*N*-(1,2,4,3,5-Trithiadiazol-1-ylidene)amino

TABLE M (continued)
Compilation of ^{31}P NMR Data of Four Coordinate Phosphorus Compounds Containing a Formal Multiple Bond from Phosphorus to a Group V Atom

Section M2: Cyclopolyphosphazenes, $(N{=}PX_2)n$, and Their Derivatives

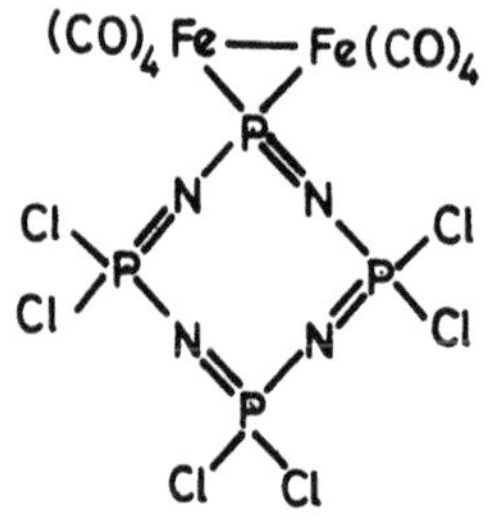

37 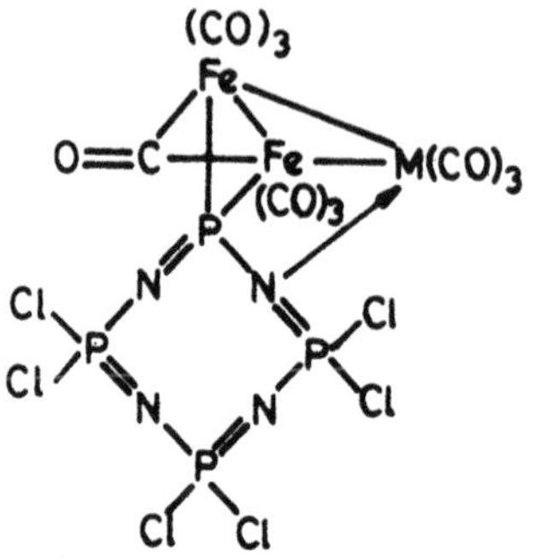 **38**

Connectivities	Molecular formula	Structure	NMR Data (δ_P[solv, temp] $^nJ_{PP}$Hz)	Ref.
	Remaining P-bound atoms = Fe_2			
$(C'_4Fe)_2$	$C_8Cl_6Fe_2N_4O_8P_4$	**37**	185.0(c)	51
$(C'_4Fe_2)_2$	$C_{10}Cl_6Fe_3N_4O_{10}P_4$	**38**; M = Fe	183.8(c)	51
	Remaining P-bound atoms = N_2			
$(H_2)_2$	$H_4N_6P_4$	**27**; R = NH_2, X = Cl	−5.8	105
	$H_8N_8P_4$	**29**; R = NH_2, X = Cl	0.0	105
	$H_{16}N_{16}P_4$	**26**; R = X = NH_2	10.0	121
$(CH)_2$	$C_8H_{32}N_{12}P_4$	**26**; R = X = NHMe	12.2	122
	$C_{15}H_{44}N_{12}P_4$	**31**; X = Y = NH, R = $(CH_2)_3$, Z = NMe_2	1.6(c)	116
	$C_{16}H_{48}N_{12}P_4$	**26**; R = X = NHEt	4.3(r)	107
	$C_{16}H_{48}N_{12}P_4$	**27**; R = NMe_2, X = NHEt	3.6(c), $^2J_{PP}$37.5	123
	$C_{20}H_{56}N_{12}P_4$	**27**; R = NH-tBu, X = NHEt	−0.9, 2.2(c)$^2J_{PP}$32.4	124
	$C_{24}H_{64}N_{12}P_4$	**26**; R = X = NH-iPr	1.1(c)	125
	$C_{26}H_{52}N_{12}P_4$	**27**; R = NMePh, X = NHEt	2.6(c), $^2J_{PP}$37.4	124
	$C_{32}H_{80}N_{12}P_4$	**26**; R = X = NH-tBu	−3.1(r)	108
CH;C_2	$C_{16}H_{48}N_{12}P_4$	**27**; R = NHEt, X = NMe_2	6.8(r), $^2J_{PP}$41.2	107
	$C_{16}H_{48}N_{12}P_4$	**27**; R = NMe_2, X = NHEt	6.2(c)	123
CH;CC′	$C_{26}H_{52}N_{12}P_4$	**27**; R = NMePh, X = NHEt	−0.2(c)	124
CH′;C_2	$C_{24}H_{48}N_{12}P_4$	**27**; R = NHPh, X = NMe_2	0.8(c), $^2J_{PP}$40.8	106
$(C_2)_2$	$C_4H_{10}Cl_6N_6P_4$	**31**; R = $(CH_2)_2$, Z = Cl, X = Y = NMe	6.1(c)	116
	$C_{12}H_{24}Cl_2N_{10}P_4$	**27**; R = Cl, X = NC_2H_4	19.5(c), $^2J_{PP}$24.3	16
	$C_{12}H_{36}Cl_2N_{10}P_4$	**27**; R = Cl, X = NMe_2	9.9(r)	13
	$C_{16}H_{32}N_{12}P_4$	**26**; R = X = NC_2H_4	17.8(c)	16
	$C_{16}H_{48}N_{12}P_4$	**26**; R = X = NMe_2	9.6(r)	13
$(C)_2$	$C_{16}H_{48}N_{12}P_4$	**27**; R = NHEt, X = NMe_2	9.2(c)	107
$(C'N')_2$	$C_{24}H_{24}N_{20}P_4$	**26**; R = X = 1-pyrazolyl	−25.5(c)	86
$(C)_2$	$C_{24}H_{48}N_{12}P_4$	**27**; R = NHPh, X = NMe_2	11.6(c)	106
$(CC')_2$	$C_{28}H_{32}Cl_4N_8P_4$	**29**; R = NMePh, X = Cl	−5.4(r)	113
$(C'N')_2$	$C_{40}H_{56}N_{20}P_4$	**26**; R = X = 2, 5-dimethyl-1-pyrazolyl	−18.8(c)	8

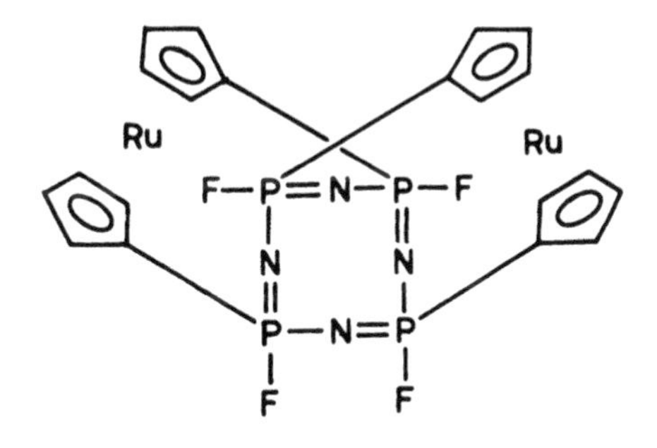

39

Connectivities	Molecular formula	Structure	NMR Data (δ_P[solv, temp] $^nJ_{PP}$Hz)	Ref.
	Remaining P-bound atoms = O_2			
$(C)_2$	$C_8H_{24}N_4O_8P_4$	**26**; R = X = OMe	2.8(c)	89
	$C_{16}H_{16}F_{24}N_4O_8P_4$	**26**; R = X = OCH_2CF_3	−2.0(c)	126
	$C_{16}H_{40}N_4O_8P_4$	**26**; R = X = OEt	−0.6(c)	127
$(C')_2$	$C_4H_6Cl_6N_4O_2P_4$	**30**; R = $OCH{=}CH_2$, X = Cl	−16.7(c)	26
	$C_{48}F_{40}N_4O_8P_4$	**26**; R = X = OC_6F_5	−10.4	91
	$C_{48}H_{40}N_4O_8P_4$	**26**; R = X = OPh	−12.6(c)	114
	Remaining P-bound atoms = S_2			
$(C)_2$	$C_4H_{10}Cl_6N_4P_4S_2$	**30**; R = SEt, X = Cl	30.4(c)	29
	$C_8H_{20}Cl_4N_4P_4S_4$	**29**; R = SEt, X = Cl	29.4(c), $^2J_{PP}$11.8	29
	$C_{12}H_{30}Cl_2N_4P_4S_6$	**30**; R = Cl, X = SEt	26.9(c)	29
	$C_{16}H_{40}N_4P_4S_8$	**26**; R = X = SEt	25.6(c)	29
$(C')_2$	$C_{12}H_{10}Cl_6N_4P_4S_2$	**30**; R = SPh, X = Cl	23.2(c)	29
	$C_{24}H_{20}Cl_4N_4P_4S_4$	**29**; R = SPh, X = Cl	22.9(c)	29
	Remaining P-bound atoms = CCl			
H_3	$CH_3Cl_7N_4P_4$	**26**; R = Me, X = Cl	12.8(c)	115
	Remaining P-bound atoms = CF			
H_3	$CH_3F_7N_4P_4$	**26**; R = Me, X = F	25.6(c)	120
C'N	$C_5H_6F_7N_5P_4$	**26**; R = 1-methylpyrrol-2-yl, X = F	0.7(c)	72
	Remaining P-bound atoms = C'F			
C'_2Fe	$C_{10}H_8F_6FeN_4P_4$	**32**; M = Fe	35.8	53
	$C_{10}H_8F_6FeN_4P_4$	**33**; M = Fe	23.3	53
C'_2Fe	$C_{10}H_9F_7FeN_4P_4$	**34**; M = Fe	21.4	53
C'_2Fe	$C_{20}H_{17}F_5Fe_2N_4P_4$	**35**; M = Fe	19.9 and 17.6	75
C'_2Ru	$C_{10}H_8F_6N_4P_4Ru$	**33**; M = Ru	15.0	53
C'_2Ru	$C_{20}H_{16}F_4N_4P_4Ru$	**39**; M = Ru	15.9	75
C'_2Ru	$C_{20}H_{17}F_5N_4P_4Ru_2$	**35**; M = Ru	18.3 and 13.3	75
	$C_{20}H_{18}F_6N_4P_4Ru_2$	**36**; M = Ru	13.7	53
	Remaining P-bound atoms = CH			
	$CH_4Cl_6N_4P_4$	**30**; R_2 = H, Me; X = Cl	4.3(c)	115

TABLE M (continued)
Compilation of ^{31}P NMR Data of Four Coordinate Phosphorus Compounds Containing a Formal Multiple Bond from Phosphorus to a Group V Atom

Section M2: Cyclopolyphosphazenes, $(N{=}PX_2)n$, and Their Derivatives

Connectivities	Molecular formula	Structure	NMR Data (δ_P[solv, temp] $^nJ_{PP}$Hz)	Ref.
		Remaining P-bound atoms = C_2		
$(H_3)_2$	$C_2H_6F_6N_4P_4$	**30**; R = Me, X = F	19.3(n)	70, 120
	$C_4H_{12}F_4N_4P_4$	**29**; R = Me, X = F	19.5	120
	$C_8H_{24}N_4P_4$	**26**; R = X = Me	19.9(c)	104
	$C_{14}H_{40}N_4P_4Si_2$	**27**; X = Me, R = CH_2Tms	17.2(c), $^2J_{PP}$12	102
	$C_{22}H_{32}N_4O_2P_4$	**28**; R = CH_2COPh, X = Me	16.9(c)	102
$(C'H_2)_2$	$C_{22}H_{32}N_4O_2P_4$	**28**; R = CH_2COPh, X = Me	10.0	102
$(H_2Si)_2$	$C_{14}H_{40}N_4P_4Si_2$	**27**; R = CH_2Tms, X = Me	12.7	102
		Remaining P-bound atoms = C'_2		
$(C'_2)_2$	$C_{24}H_{20}N_4P_4$	**29**; R = Ph, X = Cl	2.9(r)	13

Section M2.4: Higher Homologues of Cyclophosphazenes, $(NPCl_2)_n$ ($n > 4$)[a]

Remaining P-bound atoms	Connectivities	Formula/ structure	NMR data (δ_P[solv, temp]$J_{pp}H_z$)	Ref.
Br_2		$N_5P_5Br_{10}$	−78.1(c)	128
Cl_2		$N_5P_5Cl_{10}$	−15.5(c)	128
		$N_6P_6Cl_{12}$	−17.5(c)	128
F_2		$N_5P_5F_{10}$	−21.9(c)[b]	128, 129
		$N_6P_6F_{12}$	−22.0(b)[b]	129
N_2	$(C_2)_2$	$N_5P_5(NMe_2)_{10}$	−3.1(c)	128
	$(C_2)_2$	$N_6P_6(NMe_2)_{12}$	0.5(c)	128
O_2	$(C_2)_2$	$N_5P_5(OMe)_{10}$	−2.5(c)	89
	$(C_2)_2$	$N_6P_6(OMe)_{12}$	−4.2(c)	89
	$(C'_2)_2$	$N_5P_5(OPh)_{10}$	−19.5(c)	130
	$(C'_2)_2$	$N_6P_6(OPh)_{12}$	−18.5(c)	130
C_2	$(H_3)_2$	$N_5P_5Me_{10}$	5.8(b)[b]	129
	$(H_3)_2$	$N_6P_6Me_{12}$	3.2(b)[b]	129

Section M2.5: Bicyclic Phosphazenes (20, 21)

Connectivity	Formula	Structure	NMR data	Ref.
	Remaining P-bound atoms = N_2			
CP^4;CH	$C_7H_{27}N_{11}P_4$	**40**; R = Me	18.5	122
	$C_{12}H_{37}N_{11}P_4$	**41**; R^1 = R^2 = Me	20.6	125
	$C_{14}H_{41}N_{11}P_4$	**40**; R = Et	18.6	131
	$C_{14}H_{41}N_{11}P_4$	**41**; R^1 = R^2 = Et	18.9(r)	131
	$C_{16}H_{45}N_{11}P_4$	**41**; R^1 = tBu, R^2 = Et	16.1(c)	125
	$C_{24}H_{45}N_{11}P_4$	**41**; R^1 = R^2 = CH_2Ph	19.3(c)	125
	$C_{21}H_{55}N_{11}P_4$	**40**; R = iPr	12.0	125

[a] Phosphorus-31 shielding increases with ring size and reaches limiting value (see Figure 2).
[b] Values for $(NPMe_2)_n$ (n = 3—12) and $(NPF_2)_n$ (n = 3—9) are listed in this paper.

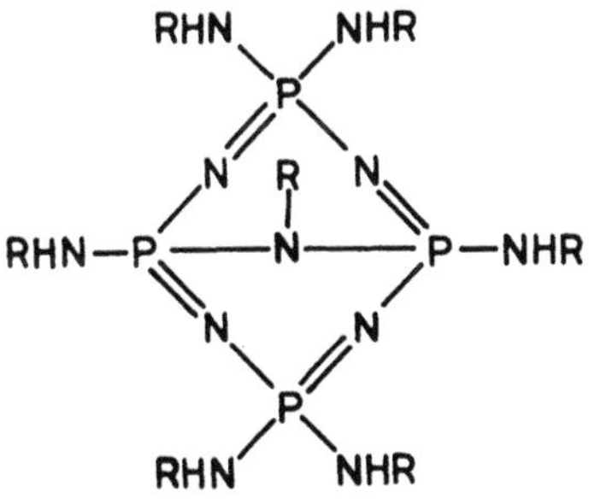

40

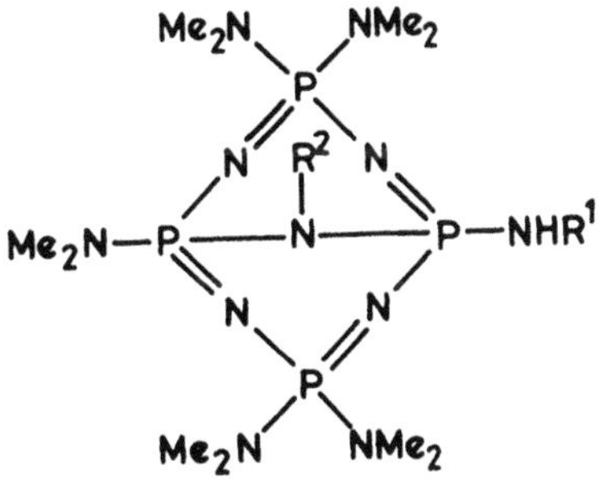

41

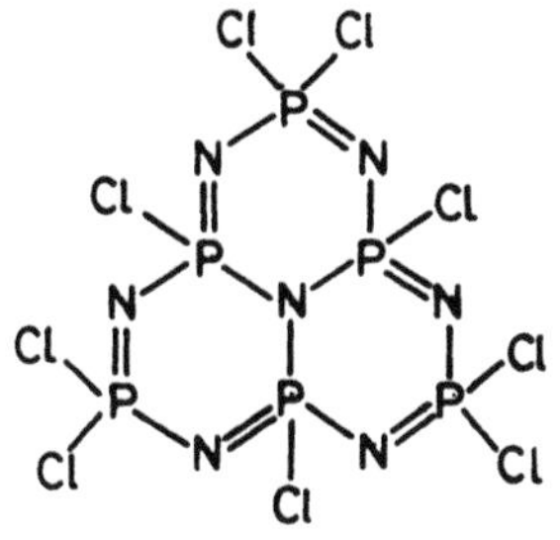

42

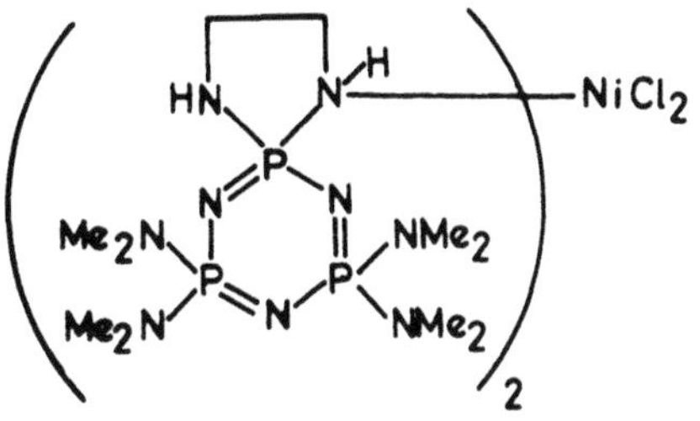

43

Connectivity	Formula	Structure	NMR data	Ref.
$(CH)_2$	$C_7H_{27}N_{11}P_4$	**40**; R = Me	21.5, $^2J_{PP}39$	122
	$C_{14}H_{41}N_{11}P_4$	**40**; R = Et	15.3(r), $^2J_{PP}40.9$	131
	$C_{21}H_{55}N_{11}P_4$	**40**; R = iPr	12.0(c)	125
$(C_2)_2$	$C_{12}H_{37}N_{11}P_4$	**41**; $R^1 = R^2$ = Me	21.7(c)	125
	$C_{14}H_{41}N_{11}P_4$	**41**; $R^1 = R^2$ = Et	22.5(r)	131
	$C_{16}H_{45}N_{11}P_4$	**41**; R^1 = tBu, R^2 = Et	20.6	125
	$C_{24}H_{45}N_{11}P_4$	**41**; $R^1 = R^2 = CH_2Ph$	22.0(c)	125
$C_2;CP^4$	$C_{12}H_{37}N_{11}P_4$	**41**; $R^1 = R^2$ = Me	20.7(c)	125
	$C_{14}H_{41}N_{11}P_4$	**41**; $R^1 = R^2$ = Et	19.7(r)	131
	$C_{16}H_{45}N_{11}P_4$	**41**; R^1 = tBu, R^2 = Et	19.6(c)	125
	$C_{24}H_{45}N_{11}P_4$	**41**; $R^1 = R^2 = CH_2Ph$	19.3(c)	125

Section M2.6: Tricyclic Phosphazene (22)

Remaining P-bound atoms = ClN

Connectivity	Formula	Structure	NMR data	Ref.
$P^4{}_3$	$Cl_9N_7P_6$	**42**	−3.5(c)	132

Remaining P-bound atoms = Cl_2

Connectivity	Formula	Structure	NMR data	Ref.
	$Cl_9N_7P_6$	**42**	20.2(c)	132

Section M2.7: Metal Complexes of Cyclophosphazenes

Remaining P-bound atoms = N_2

Connectivity	Formula	Structure	NMR data	Ref.
CH;CHNi	$C_{20}H_{60}Cl_2N_9P_3Ni$	**43**	25.5(c), $^2J_{PP}31.6$	133
$(C_2)_2$	$C_{20}H_{60}Cl_2N_9P_3Ni$	**43**	21.9	133
$(C'N)_2$	$C_{30}H_{34}Cl_2N_7P_3Pd$	**44**	−10.4(c)	86

TABLE M (continued)
Compilation of ^{31}P NMR Data of Four Coordinate Phosphorus Compounds Containing a Formal Multiple Bond from Phosphorus to a Group V Atom

Section M2: Cyclopolyphosphazenes, $(N{=}PX_2)n$, and Their Derivatives

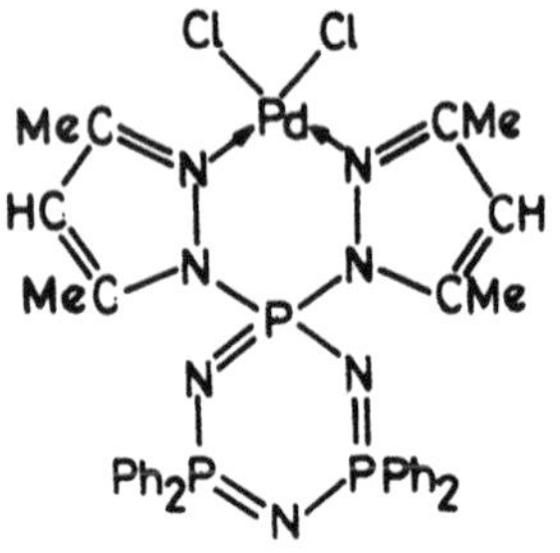

44

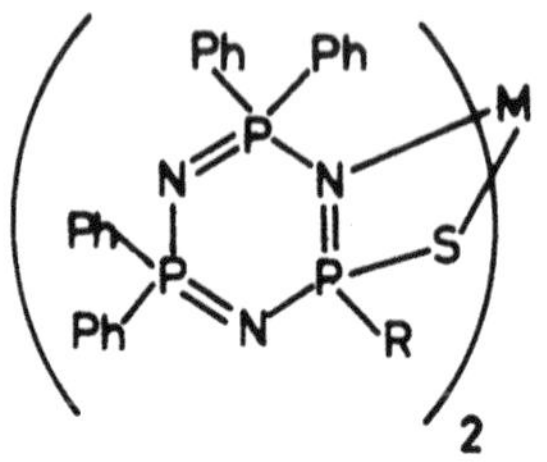

45

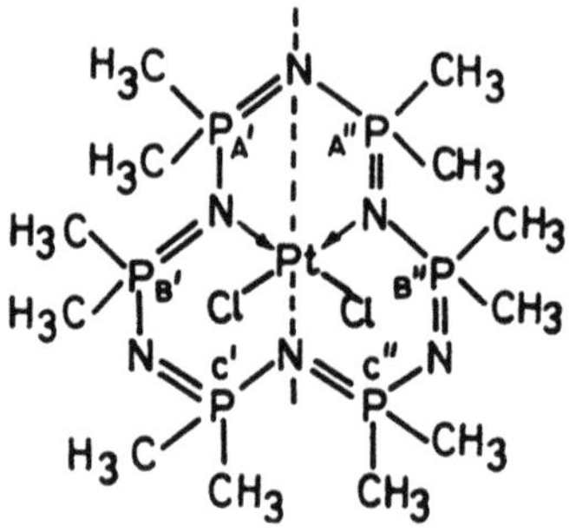

46

Connectivity	Formula	Structure	NMR data	Ref.
	Remaining P-bound atoms = OS			
C';Pd	$C_{30}H_{25}N_3OP_3PdS$	**45**; R = OPh, M = Pd	49.0(ne)	134
	Remaining P-bound atoms = CS			
H_3;Pt	$C_{25}H_{23}N_3P_3PtS$	**45**; R = Me, M = Pt	57.9(ne)	134
	Remaining P-bound atoms = C'S			
C'_2;Ni	$C_{31}H_{27}N_3NiOP_3S$	**45**; R = C_6H_4OMe-4, M = Ni	60.6(ne)	134
	Remaining P-bound atoms = C_2			
$(H_3)_2$	$C_{12}H_{36}Cl_2N_6P_6$	**46**	29.5[a], 29.0[a], 27.1[b], 7.6[c], 7.1(c)[c]	135

[a] AA'.
[b] BB'.
[c] CC' **46**.

Connectivity	Formula	Structure	NMR data	Ref.
	Remaining P-bound atoms = C'$_2$			
$(C'_2)_2$	$C_{25}H_{23}N_3P_3PtS$	**45**; R = Me, M = Pt	9.1 and 27.1(ne)	134
	$C_{30}H_{25}N_3OP_3PdS$	**45**; R = OPh, M = Pd	11.9 and 27.9(ne)	134
	$C_{30}H_{34}Cl_2N_7P_3Pd$	**44**	19.3, 17.3(c)	86
	$C_{31}H_{27}N_3ONiP_3S$	**45**; R = $C_6H_4(OMe)$-4, M = Ni	16.5 and 26.6	134

Section M3: Phosphoniumphosphides (Compounds with Formal P^4=P^2 Multiple Bonds)

P-bound atoms	Connectivities	Structure	NMR data	Ref.
N_3P'	$(C_2)_3;P^4$	$(Me_2N)_3P{=}PP^+(NMe_2)_3\ Ph_43^-$	85 J_{PP}513	1
	$(C_2)_3;\ P^4$	$(Me_2N)_3P{=}PP^+Ph_2Me\ AlCl_4^-$	84 J_{PP}493, $^2J_{PP}$30	1
C_3P'	$(H_3)_3;C$	$Me_3P{=}PTf$	12.7 J_{pp}-436.5	2
CC'_2P'	$(CH_2)_3$	$Et_3P{=}P{-}P(O)(OEt)_2$	44.2 J_{pp}444	3
		$Bu_3P{=}P{-}(O)(OEt)_2$	38.7 J_{pp}456	3
CC'_2P'	$H_3;(C'_2)_2;P^4$	$Ph_2PMe{=}PP^+Ph_2Me\ AlCl_4^-$	28 J_{PP}467	1
	$H_3;(C'_2)_2;P^4$	$Ph_2PMe{=}PP^+Ph_3\ AlCl_4^-$	28 J_{PP}480, J_{PP}25	1
C'_3P'	$(C'_2)_3;P^4$	$Ph_3P{=}PP^+(NMe_2)_3\ AlCl_4^-$	29 J_{PP}523	1
	$(C'_2)_3;P^4$	$Ph_3P{=}PP^+Ph_3\ AlCl_4^-$	30 J_{PP}502	1

REFERENCES

Section M1

1. **Wisian-Neilson, P. and Neilson, R. H.,** *Inorg. Chem.*, 19, 1875, 1980.
2. **Markovskii, L. N., Romanenko, V. D., Dzyuba, V. I, Ruban, A. V., Kalibabchuk, N. N., and Iksanova, S. V.,** *Zh. Obshch. Khim.*, 51, 331, 1981.
3. **Markovskii, L. N., Romanenko, V. D., and Ruban, A. V.,** *Synthesis*, 10, 811, 1979.
4. **Miroshnichenko, V. V., Marchenko, A. P., and Pinchuk, A. M.,** *Zh. Obshch. Khim.*, 55, 2024, 1985.
5. **Marchenko, A. P., Miroshnichenko, V. V., and Pinchuk, A. M.,** *Zh. Obshch. Khim.*, 54, 1213, 1984.
6. **Neicke, E. and Bitter, W.,** *Chem. Ber.*, 109, 415, 1976.
7. **Roesky, H. W. and Remmers, G.,** *Z. Naturforsch.*, 26B, 75, 1971.
8. **Romanenko, V. D., Ruban, A. V., Iksanova, S. V., and Markovskii, L. N.,** *Zh. Obshch. Khim.*, 54, 313, 1984.

8b. **Kliebisch, U. and Klingebiel, U.,** *J. Organomet. Chem.*, 314, 33, 1986.

9. **Roesky, H. W., Kuhtz, B. H., and Grimm, L. F.,** *Z. Anorg. Allg. Chem.*, 389, 167, 1972.
10. **Roesky, H. W., Grimm, L. F., and Niecke, E.,** *Z. Anorg. Allg. Chem.*, 385, 102, 1971.

10b. **Riesel, L., Herrmann, E., Pfutzner, A., Steinbach, J., and Thomas, B.,** *Phosphorus Chemistry Proc. 1981 Int. Conf. ACS Symp. Series 171*, p. 297.

11. **Kukhar, V. P. and Gilyarov, V. A.,** *Pure Appl. Chem.*, 52, 891, 1980.

11b. **Nesterova, L. I. and Sinitsa, A. D.,** *Zh. Obshch. Khim.*, 55, 2624, 1985.

12. **Kukhar, V. P., Grishkun, E. V., and Rudavskii, V. P.,** *Zh. Obshch. Khim.*, 48, 5627, 1978; **Kukhar, V. P., Grishkun, E. V., and Kalibabchuk, N. N.,** *Zh. Obshch. Khim.*, 52, 2227, 1982.
13. **Romanenko, V. D., Ruban, A. V., Iksanova, S. V., and Markovskii, L. N.,** *Zh. Obshch. Khim.*, 52, 581, 1982.
14. **Bellan, J., Marre-Mazieres, M. R., Sanchez, M., and Songstad, J.,** *C. R. Acad. Sci. Ser. 2*, 30, 785, 1985.
15. **Roesky, H. W. and Kloker, W.,** *Z. Naturforsch.*, 27B, 486, 1971.
16. **Klingebiel, U. and Glemser, O.,** *Chem. Ber.*, 104, 3804, 1971.
17. **Shevchenko, M. V. and Kukhar, V. P.,** *Zh. Obshch. Khim.*, 56, 85, 1986.
18. **Boeske, J., Niecke, E., Ocando-Mavarez, E., Majoral, J. P., and Bertrand, G.,** *Inorg. Chem.*, 25, 2695, 1986.

19. **Sicard, G., Baceiredo, A., Bertrand, G., and Majoral, J. P.,** *Angew. Chem.*, 96, 450, 1984.
20. **Bermann, M. and Utvary, K.,** *Monatch. Chem.*, 100, 1280, 1969.
21. **Marre, M. R., Sanchez, M., and Wolf, R.,** *Phosphorus Sulfur*, 13, 327, 1982.
22. **Mark, V., Dungan, M., Crutchfield, M., and Van Wazer, J.,** *Topics in Phosphorus Chemistry*, Vol. 5, Grayson, M. and Griffiths, E. J., Eds., Wiley-Interscience, New York, 1976, chap. 4.
22b. **Winter, H. and Van de Grampel, J. C.,** *Z. Naturforsch.*, 38B, 1652, 1983.
22c. **Van den Berg, J. B., Klei, E., De Ruiter, B., Van de Grampel, J. C., and Kruk, C.,** *Recl. Trav. Chim. Pays Bas*, 95, 206, 1976.
22d. **Baalmann, H. H. and Van de Grampel, J. C.,** *Z. Naturforsch.*, 33B, 964, 1978.
22e. **Baalmann, H. H., Keizer, R., Van de Grampel, J. C., and Kruk, K. C.,** *Z. Naturforsch.*, 33B, 959, 1978.
22f. **Van den Berg, J. B., De Ruiter, B., and Van de Grampel, J. C.,** *Z. Naturforsch.*, 31B, 1216, 1976; **De Ruiter, B. and Van de Grampel, J. C.,** *Inorg. Chim. Acta*, 31, 195, 1978.
22g. **Pinkert, W., Schoening, G., and Glemser, O.,** *Z. Anorg. Allg. Chem.*, 436, 136, 1977.
22h. **Kornuta, P. P., Kolotilo, N. V., and Kalinin, V. N.,** *Zh. Obshch. Khim.*, 49, 1777, 1979.
22i. **Van den Berg, J. B. and Van de Grampel, J. C.,** *Z. Naturforsch.*, 34B, 27, 1979.
23. **Haubold, W. and Fluck, E.,** *Z. Naturforsch. Teil B*, 27, 368, 1972.
23b. **Jekel, A. P. and Van de Grampel, J. C.,** *Z. Naturforsch.*, 34B, 569, 1979.
23c. **Klei, E. and Van de Grampel, J. C.,** *Recl. Trav. Chim. Pays Bas*, 97, 307, 1978.
24. **Grimm, L. F., Ph.D. thesis,** Gottingen, 1971.
25. **Roesky, H. W. and Grosse Bowing, W.,** *Chem. Ber.*, 104, 3204, 1971.
26. **Schoning, G., Klingerbiel, U., and Glemser, O.,** *Chem. Ber.*, 107, 592, 1974.
27. **Czieslic, G., Flaskerud, G., and Rainer, G. O.,** *Chem. Ber.*, 106, 399, 1973.
28. **Roesky, H. W.,** *Z. Naturforsch. Teil B*, 27, 1569, 1972.
29. **Roesky, H. W.,** *Chem. Ber.*, 105, 1439, 1972.
30. **Grosse Bowing, Ph.D. thesis,** Gottingen.
31. **Klingebiel, U. and Glemser, O.,** *Chem. Ber.*, 105, 1510, 1972.
32. **Utvary, K. and Bermann, M.,** *Inorg. Chem.*, 8, 1038, 1969.
32b. **Latscha, H. P. and Klein, W.,** *Z. Anorg. Allg. Chem.*, 377, 225, 1970.
33. **Latscha, H. P., Hormuth, P. B., and Vollmer, H.,** *Z. Naturforsch.*, 24B, 1237, 1969.
33b. **Haubold, W., Fluck, E., and Becke-Goering, M.,** *Z. Anorg. Allg. Chem.*, 397, 269, 1973.
33c. **Viersen, F. J., Bosma, E., De Ruiter, B., Dhathathregan, K. B., Van de Grampel, J. C., and Van Bolhuis, F.,** *Phosphorus Sulfur*, 26, 285, 1986.
33d. **Kukhar, V. P. and Kasheva, T. N.,** *Zh. Obshch. Chim.*, 46, 243, 1976.
33e. **Kornuta, P. P., Kuz'menko, L. S., and Kalinin, V. N.,** *Zh. Obshch. Khim.*, 50, 1313, 1980.
33f. **Winter, H. and Van de Grampel, J. C.,** *Z. Naturforsch.*, 38B, 7, 1983.
34. **Roesky, H. W. and Niecke, E.,** *Z. Naturforsch Teil B*, 24, 1101, 1969.
35. **Glemser, O., Biermann, U., and von Halasz, S. P.,** *Inorg. Nucl. Chem. Lett.*, 5, 501, 1969.
36. **Roesky, H. W.,** *Chem. Ber.*, 101, 3679, 1968.
37. **Tutkunkardes, S., Ph.D. thesis,** Gottingen, 1973.
38. **Roesky, H. W. and Holtschneider, G.,** *Z. Anorg. Allg. Chem.*, 378, 168, 1970.
39. **Witt, M.,** unpublished data.
40. **Hartl, H., Huppmann, P., Lentz, D., and Seppelt, K.,** *Inorg. Chem.*, 22, 2183, 1983.
41. **Kozlov, E. S., Gaidamaka, S. N., Povolotskii, M. I., Kyuntsel, I. A., Mokeeva, V. A., and Soifer, G. B.,** *Zh. Obshch. Khim.*, 48, 1263, 1978.
42. **Kozlov, E. S., Gaidamaka, S. N., and Sadykov, R. Kh.,** *Zh. Obshch. Khim.*, 46, 552, 1976.
43. **Klein, H. A. and Latscha, H. P.,** *Z. Anorg. Allg. Chem.*, 406, 214, 1974.
44. **Ambrosius, K., Ph.D. thesis,** Frankfurt, 1978.
45. **Filonenko, L. P., Kudryavtsev, A. A., and Pinchuk, A. M.,** *Zh. Obshch. Khim.*, 51, 1971, 1981.
46. **Marchenko, A. P., Kudryavtsev, A. A., Tsymbal, I. F., and Pinchuk, A. M.,** *Zh. Obshch. Khim.*, 55, 2627, 1985.
47. **Kozlov, E. S., Dubenko, L. G., and Kudryavtsev, A. A.,** *Zh. Obshch. Khim.*, 51, 962, 1981.
48. **Marchenko, A. P., Kovenya, V. A., and Pinchuk, A. M.,** *Zh. Obshch. Khim.*, 53, 698, 1983.
49. **Schoning, G. and Glemser, O.,** *Chem. Ber.*, 110, 1148, 1977.
50. **Roesky, H. W. and Grimm, L. F.,** *Chem. Ber.*, 103, 3114, 1970.
51. **Nesterova, L. I. and Gololobov, Yu. G.,** *Zh. Obshch. Khim.*, 49, 2625, 1979.
51b. **Appel, R., Ruppert, I., Milker, R., and Bastian, V.,** *Chem. Ber.*, 107, 380, 1974.
52. **Gorbatenko, Zh. K. and Mitel'man, I. E.,** *Zh. Obshch. Khim.*, 51, 717, 1981.
53. **Foss, V. L., Veits, Yu. A., Chernykh, T. E., Staroverova, I. N., and Lutsenko, I. F.,** *Zh. Obshch. Khim.*, 53, 2489, 1983.
54. **Danchenko, M. N. and Gololobov, Yu. G.,** *Zh. Obshch. Khim.*, 52, 2205, 1982.

55. **Arbuzov, B. A., Polozov, A. M., and Polezhaeva, N. A.,** *Zh. Obshch. Khim.*, 54, 1968, 1984.
56. **Richter, H., Fluck, E., Riffel, H., and Hess, H.,** *Z. Anorg. Allg. Chem.*, 496, 109, 1983.
57. **Tikhonina, N. A., Gilyarov, V. A., and Kabachnik, M. I.,** *Zh. Obshch. Khim.*, 52, 760, 1982.
58. **Riesel, L., Steinbach, J., and Herrmann, E.,** *Z. Anorg. Allg. Chem.*, 502, 21, 1983.
59. **Khodak, A. A., Gilyarov, V. A., Shcherbina, T. M., and Kabachnik, M. I.,** *Zh. Obshch. Khim.*, 46, 2482, 1976.
60. **Steinbach, J., Herrmann, E., and Riesel, L.,** *Z. Anorg. Allg. Chem.*, 523, 180, 1985.
61. **Zaslavskaya, N. N., Gilyarov, V. A., and Kabachnik, M. I.,** *Izv. Acad. Nauk. SSSR Ser. Khim.*, p. 662, 1976.
62. **Schlak, O., Stadelmann, W., Stelzer, O., and Schmutzler, R.,** *Z. Anorg. Allg. Chem.*, 419, 275, 1976.
63. **Flindt, E. P., Rose, H., and Marsmann, H. C.,** *Z. Anorg. Allg. Chem.*, 430, 155, 1977.
64. **Nesterov, L. V., Krepysheva, N. E., Sabirova, R. A., and Lipkina, G. N.,** *Zh. Obshch. Khim.*, 48, 790, 1978.
65. **Goldwhite, H., Gysegem, P., Schow, S., and Swyke, C.,** *J. Chem. Soc. Dalton Trans.*, p. 12, 1975.
66. **Martin, G. J., Sanchez, M., and Marre, M. R.,** *Tetrahedron Lett.*, 24, 4989, 1983.
67. **Burger, K., Thenn, W., and Schickaneder, H.,** *Chem. Ber.*, 108, 1468, 1975.
68. **Bellan, J., Sanchez, M., Marre-Mazieres, M. R., and Murillo, B. A.,** *Bull. Soc. Chim. Fr.*, p. 491, 1985.
69. **Tikhonina, N. A., Timofeeva, G. I., Matrosov, E. I., Gilyarov, V. A., and Kabachnik, M. I.,** *Zh. Obshch. Khim.*, 45, 2414, 1975.
70. **Kasukhin, L. F., Ponomarchuk, M. P., Kolodka, T. V., Malenko, D. M., Kim, T. V., Repina, L. A., Kiseleva, E. I., and Gololobov, Yu. G.,** *Zh. Obshch. Khim.*, 53, 1022, 1983.
71. **Kolodka, T. V. and Gololobov, Yu. G.,** *Zh. Obshch. Khim.*, 53, 1013, 1983.
72. **Gilyarov, V. A., Tikhonina, N. A., Andrianov, V. G., Struchkov, Yu. T., and Kabachnik, M. I.,** *Zh. Obshch. Khim.*, 48, 732, 1978.
73. **Budilova, I. Yu., Gusar, N. I., and Gololobov, Yu. G.,** *Zh. Obshch. Khim.*, 53, 285, 1983.
74. **Marre, M. R., Sanchez, M., Brazier, J. F., Wolf, R., and Bellan, J.,** *Can. J. Chem.*, 60, 456, 1982.
74b. **Konovalova, I. V., Burnaeva, L. A., Khusnutdinova, E. K., and Pudovic, A. N.,** *Zh. Obshch. Khim.*, 56, 1245, 1986.
75. **Balitskii, Yu. V., Kasukhin, L. F., Ponomarchuk, M. P., and Gololobov, Yu. G.,** *Zh. Obshch. Khim.*, 49, 42, 1979.
75b. **Sinitsa, A. D., Malenko, D. M., Repina, L. A., Lokionova, R. A., and Shurbura, A. K.,** *Zh. Obshch. Khim.*, 56, 1262, 1986.
76. **Negrebetskii, V. V., Bogel'fer, L. Ya., Sinitsa, A. D., Kal'chenko, V. I., Krishtal, V. S., and Markovskii, L. N.,** *Zh. Obshch. Khim.*, 52, 1496, 1982.
76b. **Schmidpeter, A., Tautz, H., and Schreiber, F.,** *Z. Anorg. Allg. Chem.*, 475, 211, 1981.
77. **Zeiss, W.,** *Angew. Chem.*, 88, 582, 1976.
78. **Kabachnik, M. I., Tikhonina, N. A., Gilyarov, V. A., Korolev, B. A., Pudovik, M. A., Kibardina, L. K., and Pudovik, A. N.,** *Zh. Obshch. Khim.*, 52, 1033, 1982.
79. **Gilyarov, V. A., Tikhonina, N. A., Shcherbina, T. M., and Kabachnik, M. I.,** *Zh. Obshch. Khim.*, 50, 1438, 1980.
80. **Diallo, O., Boisdon, M. T., Malavaud, C., Lopez, L., Haddad, M., and Barrens, J.,** *Tetrahedron Lett.*, 25, 5521, 1984.
81. **Flindt, E. P.,** *Z. Anorg. Allg. Chem.*, 487, 119, 1982.
81b. **Burford, N., Chivers, T., Cordes, A. W., Laidlaw, W. G., Noble, M. C., Oakley, R. T., and Swepston, P. N.,** *J. Am. Chem. Soc.*, 104, 1282, 1982.
81c. **Kornuta, P. P., Kolotilo, N. V., and Markovski, L. N.,** *Zh. Obshch Khim.*, 52, 590, 1982.
82. **Majoral, J. P., Kraemer, R., N'Gando M'Pondo, T., and Na[illegible]h. [illegible].,** *Tetrahedron Lett.*, 21, 1307, 1980.
83. **Romanenko, V. D., Ruban, A. V., Polumbrik, D. M., and Markovskii, L. N.,** *Zh. Obshch. Khim.*, 51, 1923, 1981.
84. **Appel, R. and Halstenberg, J.,** *Chem. Ber.*, 110, 2374, 1977.
85. **Niecke, E. and Flick, W.,** *Angew. Chem.*, 86, 128, 1974.
86. **Kozlov, E. S., Dubenko, L. G., and Tsymbal, I. F.,** *Zh. Obshch. Khim.*, 53, 2348, 1983.
87. **Razhabov, A. and Yusupov, M. M.,** *Zh. Obshch. Khim.*, 55, 748, 1985.
87b. **Marre, M. R., Boisden, M. T., and Sanchez, M.,** *Tetrahedron Lett.*, 23, 853, 1982.
87c. **Hoeve, W., Dhathathregan, K. B., Van de Grampel, J. C., and Van Bolhuis, F.,** *Phosporus Sulfur*, 26, 293, 1986.
88. **Markovskii, L. N., Romanenko, V. D., and Ruban, A. V.,** *Phosphorus Sulfur*, 9, 221, 1980.
89. **Miroshnichenko, V. V., Marchenko, A. P., Kudryavtsev, A. A., and Pinchuk, A. M.,** *Zh. Obshch. Khim.*, 52, 2517, 1982.

90. **Kovenya, V. A., Marchenko, A. P., and Pinchuk, A. M.,** *Zh. Obshch. Khim.*, 55, 700, 1985.
91. **Marchenko, A. P., Bespal'ko, G. K., Koidan, G. N., and Pinchuk, A. M.,** *Zh. Obshch. Khim.*, 53, 1436, 1983.
92. **Marchenko, A. P., Miroshnichenko, V. V., Povolotskii, M. I., and Pinchuk, A. M.,** *Zh. Obshch. Khim.*, 53, 327, 1983.
93. **Marchenko, A. P., Koidan, G. N., and Pinchuk, A. M.,** *Zh. Obshch. Khim.*, 53, 670, 1983.
94. **Lorberth, J.,** *J. Organomet. Chem.*, 71, 159, 1974.
95. **Kroshefsky, R. D. and Verkade, J. D.,** *Inorg. Chem.*, 14, 3090, 1975.
96. **Ponomarchuk, M. P., Kasukhin, L. F., Shevchenko, M. V., Sologub, L. S., and Kukhar, V. P.,** *Zh. Obshch. Khim.*, 54, 2468, 1984.
97. **Marchenko, A. P., Koidan, G. N., Pinchuk, A. M., and Kirsanov, A. V.,** *Zh. Obshch. Khim.*, 54, 1774, 1984.
98. **Flindt, E. P.,** *Z. Anorg. Allg. Chem.*, 447, 97, 1978.
99. **Bartsch, R., Stelzer, O., and Schmutzler, R.,** *Z. Naturforsch. B.Anorg. Chem. Org. Chem.*, 37, 267, 1982.
100. **Koidan, G. N., Marchenko, A. P., and Pinchuk, A. M.,** *Zh. Obshch. Khim.*, 55, 1633, 1985.
101. **Marchenko, A. P., Koidan, G. N., and Pinchuk, A. M.,** *Zh. Obshch. Khim.*, 54, 2691, 1984.
102. **Zal'tsman, I. S., Marchenko, A. P., Koidan, G. N., and Pinchuk, A. M.,** *Zh. Obshch. Khim.*, 55, 1882, 1985.
103. **Koidan, G. N., Marchenko, A. P., Kudryavtsev, A. A., and Pinchuk, A. M.,** *Zh. Obshch. Khim.*, 52, 2001, 1982.
104. **Zal'tsman, I. S., Koidan, G. N., Marchenko, A. P., and Pinchuk, A. M.,** *Zh. Obshch. Khim.*, 54, 2788, 1984.
105. **Marchenko, A. P., Koidan, G. N., Povolotskii, M. I., and Pinchuk, A. M.,** *Zh. Obshch. Khim.*, 53, 1513, 1983.
106. **Wrackmeyer, B.,** *Z. Naturforsch.*, 39B, 533, 1984.
107. **Gusar, N. I., Chaus, M. P., and Gololobov, Yu. G.,** *Zh. Obshch. Khim.*, 49, 1782, 1979.
108. **Kovenya, V. A., Marchenko, A. P., and Pinchuk, A. M.,** *Zh. Obshch. Khim.*, 51, 2678, 1981.
109. **Dimukhametov, M. N., Buina, N. A., and Nuretdinov, I. A.,** *Zh. Obshch. Khim.*, 52, 2797, 1982.
110. **Gutmann, V., Kemenator, Ch., and Utvary, K.,** *Mh. Chem.*, 96, 836, 1965.
111. **Marchenko, A. P., Miroshnichenko, V. V., Kudryavtsev, A. A., and Pinchuk, A. M.,** *Zh. Obshch. Khim.*, 54, 2685, 1984.
112. **Niecke, E. and Schaefer, H. G.,** *Chem. Ber.*, 115, 185, 1982.
113. **Appel, R. and Halstenberg, J.,** *Angew. Chem.*, 89, 268, 1977.
114. **Appel, R. and Halstenberg, J.,** *Angew. Chem.*, 87, 810, 1975.
115. **Appel, R. and Halstenberg, J.,** *Chem. Ber.*, 111, 1815, 1978.
116. **Appel, R. and Halstenberg, J.,** *J. Organomet. Chem.*, 116, C13, 1976.
117. **Reitzel, M., Katti, K. V., and Roesky, H. W.,** unpublished data.
118. **Bermann, M. and Van Wazer, J. R.,** *J. Chem. Soc. Dalton Trans.*, p. 813, 1973.
118b. **Niecke, E. and Schaefer, H. G.,** *Angew. Chem.*, 89, 817, 1977.
119. **Kroshefsky, R. D. and Verkade, J. D.,** *Phosphorus Sulfur*, 6, 397, 1979; **Kroshefsky, R. D., Weiss, R., and Verkade, J. D.,** *Inorg. Chem.*, 18, 469, 1979.
120. **Bermann, M. and Van Wazer, J. R.,** *Inorg. Chem.*, 13, 737, 1974.
121. **Baceiredo, A., Bertrand, G., Majoral, J. P., Anba, F., and Manuel, G.,** *J. Am. Chem. Soc.*, 197, 3945, 1985.
121b. **Appel, R. and Halstenberg, M.,** *Angew. Chem.*, 88, 763, 1976.
121c. **Cnossen-Voswijk, C. and Van de Grampel, J. C.,** *Z. Naturforsch.*, 34B, 850, 1979.
121d. **Schmidpeter, A. and Weingand, C.,** *Angew. Chem. Int. Ed. Engl.*, 10, 397, 1971.
121e. **De Ruiter, B., Kuipers, G., Bijlaart, J. H., and Van de Grampel, J. C.,** *Z. Naturforsch.*, 37B, 1425, 1982.
122. **Borisov, E. V., Aklebinin, A. K., Borisenko, A. A., and Nifant'ev, E. E.,** *Zh. Obshch. Khim.*, 53, 1406, 1983.
123. **O'Neal, H. R. and Neilson, R. H.,** *Inorg. Chem.*, 23, 1372, 1984.
123b. **Winter, H. and Van de Grampel, J. C.,** *J. Chem. Soc. Chem. Commun.*, p. 489, 1984.
124. **Gololobov, Yu. G., Gusar, N. I., and Randina, L. V.,** *Zh. Obshch. Khim.*, 52, 1260, 1982.
125. **Neilson, R. H. and Engenito, J. S.,** *Organometallics*, 1, 1270, 1982.
126. **Niecke, E., Gudat, D., and Symalla, E.,** *Angew. Chem.*, 98, 817, 1986.
127. **Ford, R. R., Goodman, M. A., Neilson, R. H., Roy, A. K., Wettermark, U. G., and Wisian-Neilson, P.,** *Inorg. Chem.*, 23, 2063, 1984.
128. **Prishchenko, A. A., Gromov, A. V., Luzikov, Yu. N., Borisenko, A. A., Lazhko, E. I., Klaus, K., and Lutsenko, I. F.,** *Zh. Obshch. Khim.*, 54, 1520, 1984.

129. **Prishchenko, A. A., Gromov, A. V., Kadyko, M. I., and Lutsenko, I. F.**, *Zh. Obshch. Khim.*, 54, 2517, 1984; 53, 1188, 1983.
130. **Neilson, R. H.**, *Inorg. Chem.*, 20, 1679, 1981.
131. **Li, B. L. and Neilson, R. H.**, *Inorg. Chem.*, 25, 358, 1986.
132. **Li, B. L., Engineto, J. S., Neilson, R. H., and Wisian-Neilson, P.**, *Inorg. Chem.*, 22, 575, 1983.
133. **Arbuzov, B. A., Dianova, E. N., and Sharipova, S. M.**, *Izv. Akad. SSSR Ser. Khim.*, p. 1600, 1981.
133b. **Ebeling, J., Leva, M. A., Stary, H., and Schmidpeter, A.**, *Z. Naturforsch.*, 26B, 650, 1971.
134. **Murphy, P. C. and Tebby, J. C.**, *Phosphorus Chemistry, Proc. Int. Conf.*, Quin, L. D. and Verkade, J. G., Eds., A.C.S. Symp. Ser., 1981, 593; unpublished results.
135. **Schmidpeter, A. and Schindler, N.**, *Chem. Ber.*, 102, 2201, 1969.
136. **Baceiredo, A., Bertrand, G., Majoral, J. P., and Dillon, K. B.**, *J. Chem. Soc. Chem. Commun.*, p. 562, 1985.
137. **Shokol, V. A., Kisilenko, A. A., and Derkach, G. I.**, *Zh. Obshch. Khim.*, 39, 874, 1969.
138. **Roesky, H. W.**, *Z. Naturforsch.*, 25B, 777, 1970.
139. **Glemser, O., Niecke, E., and Stenzel, J.**, *Angew. Chem. Int. Ed. Engl.*, 6, 709, 1967.
140. **Almasi, L., Popescu, R., and Grecu, R.**, *Tetrahedron*, 33, 1327, 1977.
141. **Tupchienko, S. K., Dudchenko, T. N., and Sinitsa, A. D.**, *Zh. Obshch. Khim.*, 55, 776, 1985.
142. **Pudovik, M. A., Kibardina, L. K., Medvedeva, M. D., Pestova, T. A., Kharlampidi, Kh. E., and Pudovik, A. N.**, *Zh. Obshch. Khim.*, 46, 1944, 1976.
143. **Pudovik, A. N., Batyeva, E. S., and Ofitserov, E. N.**, *Zh. Obshch. Khim.*, 45, 2095, 1975.
144. **Pudovik, A. N., Batyeva, E. S., Selivanova, A. S., Nesterenko, V. D., and Finnik, V. P.**, *Zh. Obshch. Khim.*, 45, 1692, 1975.
145. **Pudovik, A. N., Batyeva, E. S., and Yastremskaya, N. V.**, *Zh. Obshch. Khim.*, 43, 2631, 1973.
146. **Ovchinnikov, V. V., Valitova, V. M., Yarkova, E. G., Cherkasov, R. A., and Pudovik, A. N.**, *Zh. Obshch. Khim.*, 52, 1487, 1982.
147. **Lopusinski, A., Luczak, L., Michalski, J., Kabachnik, M. M., and Moriyama, M.**, *Tetrahedron*, 37, 2011, 1981.
147b. **Schmidpeter, A. and Zeiss, W.**, *Angew. Chem. Int. Ed. Engl.*, 10, 396, 1971.
147c. **Konovalova, I. V., Burnaeva, L. A., Temnikova, E. N., and Pudovik, A. N.**, *Zh. Obsch. Khim.*, 53, 931, 1983.
147d. **Pudovic, M. A., Terent'eva, S. A., and Pudovik, A. N.**, *Zh. Obshch. Khim.*, 43, 2619, 1973.
147e. **Bredikhina, Z. A., Khazainova, N. G., Efremov, Yu. Yu., Korshunov, R. L., and Pudovik, A. N.**, *Zh. Obshch. Khim.*, 55, 1710, 1985.
147f. **Nesterova, L. I. and Sinitsa, A. D.**, *Zh. Obshch. Khim.*, 55, 1193, 1985.
148. **Pudovik, A. N., Batyeva, E. S., Ofitserov, E. N., Sinyashin, O. G., and Ivasyuk, N. V.**, *Izv. Akad. Nauk SSSR Ser. Khim.*, p. 1177, 1977.
149. **Pudovik, M. A., Kibardina, L. K., Medvedeva, M. D., and Pudovik, A. N.**, *Zh. Obshch. Khim.*, 49, 988, 1979.
150. **Gusar', N. I., Chaus, M. P., and Gololobov, Yu. G.**, *Zh. Obshch. Khim.*, 53, 2481, 1983.
151. **Osman, F. H., Gawad, M. M. A., and Abbasi, M. M.**, *J. Chem. Soc. Perkin Trans. 1*, p. 1189, 1984.
152. **Kostin, V. P., Sinyashin, O. G., Batyeva, E. S., and Pudovik, A. N.**, *Zh. Obshch. Khim.*, 52, 498, 1982.
153. **Li, B. L. and Neilson, R. H.**, *Inorg. Chem.*, 25, 358, 1986.
154. **Roeschenthaler, G. V., Sauerbrey, K., and Schmutzler, R.**, *Chem. Ber.*, 111, 3105, 1978.
155. **Pudovik, M. A., Kibardina, L. K., Medvedeva, M. D., and Pudovik, A. N.**, *Izv. Akad. Nauk SSSR Ser. Khim.*, p. 1093, 1979.
155b. **Pudovic, M. A., Pestova, T. A., and Pudovik, A. N.**, *Zh. Obshch. Khim.*, 46, 230, 1976.
155c. **Konovalova, I. V. and Burnaeva, L. A.**, *Zh. Obshch. Khim.*, 55, 2189, 1985.
156. **Baceiredo, A., Bertrand, G., and Majoral, J. P.**, *Nouv. J. Chim.*, 7, 255, 1983.
157. **Fritz, G., Haerer, J., and Sheider, K. H.**, *Z. Anorg. Allg. Chem.*, 487, 44, 1982.
158. **Niecke, E., Rueger, R., Krebs, B., and Dartmann, M.**, *Angew. Chem.*, 95, 570, 1983.
158b. **Hoegel, J. and Schmidpeter, A.**, *Z. Anorg. Allg. Chem.*, 458, 168, 1979.
159. **Niecke, E., Boeske, J., Krebs, B., and Dartmann, M.**, *Chem. Ber.*, 118, 3227, 1985.
160. **Romanenko, V. D., Shul'gin, V. F., Brusilovets, A. I., Skopenko, V. V., and Markovskii, L. N.**, *Zh. Obshch. Khim.*, 53, 1428, 1983.
161. **Romanenko, V. D., Shul'gin, V. F., Skopenko, V. V., and Markovskii, L. N.**, *Zh. Obshch. Khim.*, 55, 538, 1985.
162. **Niecke, E. and Flick, W.**, *Angew. Chem.*, 87, 355, 1975.
163. **Niecke, E.and Wildbredt, D. A.**, *Chem. Ber.*, 113, 1549, 1980.
164. **Xie, Z.-M. and Neilson, R. H.**, *Organometallics*, 2, 1406, 1983.
165. **Autzen, H. and Wannagat, U.**, *Z. Anorg. Allg. Chem.*, 420, 139, 1967.

165b. **Wannagat, U., Giesen, K. P., and Falius, H. H.,** *Monatsh. Chem.*, 104, 1444, 1973.
165c. **Falius, H. H., Giesen, K. P., and Wannagat, U.,** *Z. Anorg. Allg. Chem.*, 402, 139, 1973.
166. **Niecke, E. and Wildbredt, D. A.,** *Angew. Chem.*, 90, 209, 1978.
166b. **Navech, J. and Revel, M.,** *Tetrahedron Lett.*, 27, 2863, 1986.
167. **Lucas, J., Ph.D. thesis,** Frankfurt, 1983.
168. **Markovskii, L. N., Romanenko, V. D., Klebanskii, E. O., Chernega, A. N., Antipin, M. Yu., Struchkov, Yu. T., and Boldeskul, I. E.,** *Zh. Obshch. Khim.*, 55, 219, 1985.
169. **Roesky, H. W., Lucas, J., Noltemeyer, M., and Sheldrick, G. M.,** *Chem. Ber.*, 117, 1583, 1984.
170. **Haubold, W. and Becke-Goehring, M.,** *Z. Anorg. Allg. Chem.*, 372, 273, 1970.
171. **Kornuta, P. P., Kuz'menko, L. S., and Markovskii, L. N.,** *Zh. Obshch. Khim.*, 49, 2201, 1979.
172. **Foss, V. L., Veits, Yu. A., Chernykh, T. E., and Lutsenko, I. F.,** *Zh. Obshch. Khim.*, 54, 2670, 1984.
173. **Volkholtz, M., Stelzer, O., and Schmutzler, R.,** *Chem. Ber.*, 111, 890, 1978; **Livantsov, M. V., Prishchenko, A. A., and Lutsenko, I. F.,** *Zh. Obshch. Khim.*, 55, 2226, 1985.
173b. **Wettermark, U. G. and Neilson, R. H.,** *Inorg. Chem.*, 26, 929, 1987.
174. **Foss, V. L., Veits, Yu. A., Leksunkin, V. A., and Gurov, M. V.,** *Zh. Obshch. Khim.*, 55, 1639, 1985.
174b. **Pudovic, M. A., Mironova, T. A., and Pudovik, A. N.,** *Zh. Obshch. Khim.*, 55, 1185, 1985.
175. **Rossknecht, H., Lehmann, W. P., and Schmidpeter, A.,** *Phosphorus*, 5, 195, 1975.
176. **Zabirov, N. G., Gol'dfarb, E. L., Cherkasov, R. A., and Pudovki, A. N.,** *Zh. Obshch. Khim.*, 54, 2792, 1984.
177. **Schmidpeter, A., Brecht, H., and Groeger, H.,** *Chem. Ber.*, 100, 3063, 1967.
178. **Veits, Yu. A., Foss, V. L., Leksunkin, V. A., and Gurov, M. V.,** *Zh. Obshch. Khim.*, 55, 1641, 1985.
178b. **Zabirov, N. G., Cherkasov, R. A., and Pudovik, A. N.,** *Zh. Obshch. Khim.*, 56, 1237, 1986.
179. **Scherer, O. and Gick, W.,** *Chem. Ber.*, 104, 1490, 1971.
180. **Utvary, K., Gutmann, V., and Kemenater, Ch.,** *Inorg. Nucl. Chem. Lett.*, 1, 75, 1965.
181. **Buynac, J. D., Jadhav, K. P., and Lively, D. L.,** *Phosphorus Sulfur*, 20, 145, 1984.
182. **Scherer, O. J., Schnabl, G., and Lenhard, T.,** *Z. Anorg. Allg. Chem.*, 449, 167, 1979.
183. **Kolodyazhni, O. I.,** *Zh. Obshch. Khim.*, 52, 1086, 1982.
184. **Niecke, E. and Flick, W.,** *Angew. Chem.*, 87, 355, 1975.
184b. **Monin, E. A., Kabachnik, M. M., and Novikova, Z. S.,** *Zh. Obshch. Khim.*, 56, 221, 1986.
184c. **Burford, N., Chivers, T., Codding, P. W., and Oakley, R. T.,** *Inorg. Chem.*, 21, 982, 1982.
184d. **Burford, N., Chivers, T., Oakley, R. T., and Oswald, T.,** *Can. J. Chem.*, 62, 712, 1984.
185. **Schmidpeter, A. and Ebeling, J.,** *Angew. Chem. Int. Ed. Engl.*, 6, 565, 1967.
186. **Seseke, U., Ph.D. thesis,** Gottingen, 1986.
187. **Roesky, H. W., Katti, U., Seseke, U., Scholz, R., Herbst, E., Egert, E., and Sheldrick, M.,** *Z. Naturforsch.*, 42B, 1987.
188. **Wolfsberger, W. and Hager, W.,** *Z. Anorg. Allg. Chem.*, 425, 169, 1976.
189. **Caminade, A. M., Ocando, E., Majoral, J. P., Cristante, C., and Bertrand, G.,** *Inorg. Chem.*, 25, 712, 1986.
190. **Wolfsberger, W. and Hager, W.,** *J. Organomet. Chem.*, 118, C65, 1976.
190b. **Burford, N., Chivers, T., Oakley, R. T., Cordes, A. W., and Swepston, P. N.,** *J. Chem. Soc. Chem. Commun.*, p. 1204, 1980.
191. **Wolfsberger, W. and Hager, W.,** *Z. Anorg. Allg. Chem.*, 433, 247, 1977.
192. **Buchner, W. and Wolfsberger, W.,** *Z. Naturforsch.*, 29B, 328, 1974; 32B, 967, 1977.
193. **Schmidbauer, H., Buchner, W., and Scheutzow, D.,** *Chem. Ber.*, 106, 1251, 1973.
194. **Ruppert, I. and Appel, R.,** *Chem. Ber.*, 111, 751, 1978.
195. **Muenstedt, R. and Wannagat, U.,** *Monatsh. Chem.*, 116, 7, 1985.
196. **Mazieres, M. R., Sanchez, M., Bellan, J., and Wolf, R.,** *Phosphorus Sulfur*, 26, 97, 1986.
197. **Ponomarchuk, M. P., Sologub, L. S., Kasukhin, L. F., and Kukhar', V. P.,** *Zh. Obshch. Khim.*, 55, 1725, 1985.
198. **Stegmann, H. B., Bauer, G., Breitmaier, E., Herrmann, E., and Scheffler, K.,** *Phosphorus*, 5, 207, 1975.
199. **Wolfsberger, W.,** *Z. Naturforsch.*, 32B, 152, 1977.
199b. **Mazieres, M. R., Sanchez, M., Bellan, J., and Wolf, R.,** *Phosphorus Sulfur*, 26, 97, 1986.
200. **Wolfsberger, W.,** *Z. Naturforsch.*, 33B, 1452, 1978.
200b. **Quin, L. D., Keglevich, G., and Caster, K. C.,** *Phosphorus Sulfur*, 31, 133, 1987.
201. **Xie, X. M. and Neilson, R. H.,** *Organometallics*, 2, 921, 1983.
201b. **Morton, D. W. and Neilson, R. H.,** *Phosphorus Sulfur*, 25, 315, 1985.
202. **Barluenga, J., Lopez, F., and Palacios, F.,** *J. Chem. Res. Synop.*, 7, 211, 1985.
203. **Maringgele, W., Mellor, A., Noth, H., and Schroen, R.,** *Z. Naturforsch.*, 33B, 673, 1978.
204. **Guthrie, R. D. G. and Jenkins, I. D.,** *Aust. J. Chem.*, 35, 767, 1982.
205. **Albright, T. A., Freeman, W. J., and Schweizer, E. E.,** *J. Org. Chem.*, 41, 2716, 1976.

206. Zhurmova, I. N., Yurchenko, V. G., and Pinchuk, A. M., *Zh. Obshch. Khim.*, 53, 1509, 1983.
207. Pomerantz, M., Chou, W. N., Witczak, M. K., and Smith, C. G., *J. Org. Chem.*, 52, 159, 1987.
208. Roesky, H. W., Weber, K. L., Seseke, U., Pinkert, W., Noltmeyer, M., Clegg, W., and Sheldrick, G. M., *J. Chem. Soc. Dalton Trans.*, p. 565, 1985.
209. Buder, W. and Schmidt, A., *Spectrochim. Acta*, 31A, 1813, 1975.
210. Wamhoff, H., Fassbender, F. J., Hermes, D., Knock, F., and Appel, R., *Chem. Ber.*, 119, 2723, 1986.

Section M2

1. Baceiredo, A., Bertrand, G., Majoral, J. P., Sicard, G., Jaud, J., and Galy, J., *J. Am. Chem. Soc.*, 106, 6088, 1984.
2. Engelhardt, G., Steger, E., and Stahlberg, R., *Z. Naturforsch.*, 21B, 1231, 1966.
3. Coxon, G. E. and Sowerby, D. B., *Inorg. Chim. Acta*, 1, 381, 1967.
4. Clare, P., Sowerby, D. B., Harris, R. K., and Wazeer, M. I. M., *J. Chem. Soc. Dalton Trans.*, p. 625, 1975.
5. Engelhardt, G., Steger, E., and Stahlberg, R., *Z. Naturforsch.*, 21B, 586, 1966.
6. Dieck, R. L. and Moeller, T., *J. Inorg. Nucl. Chem.*, 35, 737, 1973.
7. Allcock, H. R., Riding, G. H., and Whittle, R. R., *J. Am. Chem. Soc.*, 106, 5561, 1984.
8. Paddock, N. L. and Patmore, D. J., *J. Chem. Soc. Dalton Trans.*, p. 1029, 1976.
9. Feistel, G. R. and Moeller, T., *J. Inorg. Nucl. Chem.*, 29, 2731, 1967.
10. Lehr, W., *Z. Anorg. Allg. Chem.*, 352, 27, 1967.
11. Thomas, B. and Scheler, U. H., *Z. Chem.*, 18, 342, 1978.
12. Allen, C. W. and Mackay, J. A., *Inorg. Chem.*, 25, 4628, 1986.
13. Keat, R., Shaw, R. A., and Woods, M., *J. Chem. Soc. Dalton Trans.*, p. 1582, 1976.
14. Linglay, D. J., Shaw, R. A., Woods, M., and Krishnamurthy, S. S., *Phosphorus Sulfur*, 4, 379, 1978.
15. Ganapathiappan, S. and Krishnamurthy, S. S., *J. Chem. Soc. Dalton Trans.*, p. 579, 1987.
16. Van der Huizen, A., Van de Grampel, J. C., Rusch, J. W., and Wilting, T., *J. Chem. Soc. Dalton Trans.* p. 1317, 1986.
17. Van der Huizen, A. A., Jekel, A. P., Bolhuis, J. K., Keestra, D., Ousema, W. H., and Van de Grampel, J. C., *Inorg. Chim. Acta*, 66, 85, 1982.
18. Krishnamurthy, S. S. and Woods, M., *Annual Reports in NMR Spectroscopy*, Vol. 19, Webb, G. R., Ed., Academic Press, London, 1987, 175.
19. Thomas, B., Gehlert, P., Schadow, H., and Scheler, H., *Z. Anorg. Allg. Chem.*, 418, 171, 1975.
20. Biddlestone, M., Keat, R., Parkes, R. G., Rose, H., Pycroft, D. S., and Shaw, R. A., *Phosphorus Sulfur*, 25, 25, 1985.
21. Dahmann, D., Rose, H., and Walz, W., *Z. Naturforsch.*, 35B, 964, 1980.
22. Heatley, M. F. and Todd, S. M., *J. Chem. Soc. A*, p. 1152, 1966.
23. Schmutz, J. L. and Allcock, H. R., *Inorg. Chem.*, 14, 2433, 1975.
24. Allen, C. W., Ramachandran, K., Bright, R. P., and Shaw, J. C., *Inorg. Chim. Acta*, 64, L109, 1982.
25. Ramachandran, K. and Allen, C. W., *Inorg. Chem.*, 22, 1445, 1983.
26. Allen, C. W. and Brown, D. E., *Inorg. Chem.*, 26, 934, 1987.
27. Karthikeyan, S. and Krishnamurthy, S. S., *Z. Anorg. Allg. Chem.*, 513, 231, 1984.
28. Thomas, B., Schadow, H., and Scheler, H., *Z. Chem.*, 15, 26, 1975.
29. Thomas, B. and Grossman, G., *Z. Anorg. Allg. Chem.*, 448, 100, 1979.
30. Allcock, H. R. and Harris, P. J., *Inorg. Chem.*, 20, 2844, 1981.
31. Allcock, H. R. and Harris, P. J., *J. Am. Chem. Soc.*, 101, 6221, 1979.
32. Dieck, R. L. and Moeller, T., *J. Inorg. Nucl. Chem.*, 35, 75, 1973.
33. Contractor, S. R., Hurthouse, M. B., Shaw, L. S., Shaw, R. A., and Yilmaz, H., *Acta Crystallogr.*, 41B, 122, 1985.
34. Krishnamurthy, S. S., Ramachandran, K., Vasudeva Murthy, A. R., Shaw, R. A., and Woods, M., *J. Chem. Soc. Dalton Trans.*, p. 840, 1980.
35. Allcock, H. R., Harris, P. J., and Connolly, M. S., *Inorg. Chem.*, 20, 11, 1981.
36. Allcock, H. R., Connolly, M. S., and Harris, P. J., *J. Am. Chem. Soc.*, 104, 2482, 1982.
37. Manns, H. and Specker, H., *Z. Anal. Chem.*, 103, 275, 1975.
38. Kumara Swamy, K. C. and Krishnamurthy, S. S., *Indian J. Chem.*, A23, 717, 1984.
39. Chandrasekhar, V., Krishnamurthy, S. S., Manohar, H., Vasudeva Murthy, A. R., Shaw, R. A., and Woods, M., *J. Chem. Soc. Dalton Trans.*, p. 621, 1984.
40. Bulwalda, P. L. and Van de Grampel, J. C., *J. Chem. Soc. Chem. Commun.*, p. 1793, 1986.
41. Winter, H. and Van de Grampel, J. C., *J. Chem. Soc. Dalton Trans.*, p. 1269, 1986.
42. Chandrasekhar, V., Krishnamurthy, S. S., Vasudeva Murthy, A. R., Shaw, R. A., and Woods, M., *Inorg. Nucl. Chem. Lett.*, 17, 181, 1981.

43. **Allcock, H. R., Harris, P. J., and Nissan, R. A.,** *J. Am. Chem. Soc.*, 103, 2256, 1981.
44. **Harris, P. J., Schwalker, M. A., Liu, V., and Fischer, B. L.,** *Inorg. Chem.*, 22, 1812, 1983.
45. **de Ruiter, B., Kuipers, G., Bijlaart, J. H., and Van de Grampel, J. C.,** *Z. Naturforsch.*, 37B, 1425, 1982.
46. **Allcock, H. R., Desorcie, J. L., and Harris, P. J.,** *J. Am. Chem. Soc.*, 105, 2814, 1983.
47. **Allcock, H. R., Brennan, D. J., Graaskamp, J. M., and Parvez, M.,** *Organometallics*, 5, 2434, 1986.
48. **Harris, P. J. and Jackson, L. A.,** *Organometallics*, 2, 1477, 1983.
49. **Allcock, H. R., Connolly, M. S., and Whittle, R. R.,** *Organometalics*, 2, 1514, 1983.
50. **Allcock, H. R. and Kugel, R. L.,** *Inorg. Chem.*, 5, 1016, 1966.
51. **Allcock, H. R. Suszko, P. R., Wagner, L. J., Whittle, R. R. and Boso, B.** , *J. Am. Chem. Soc.*, 106, 4966, 1984.
52. **Smaardijk, A. A., de Ruiter, B., Van der Huizen, A. A., and Van de Grampel, J. C.,** *Recl. Trav. Chim.*, 101, 270, 1982.
53. **Allcock, H. R., Lavin, K. D., Riding, G. H., Suzko, P. R., and Whittle, R. R.,** *J. Am. Chem. Soc.*, 106, 2337, 1984.
54. **Allcock, H. R., Mang, M. N., Riding, G. H., and Whittle, R. R.,** *Organometallics*, 5, 2244, 1986.
55. **Lensink, C., de Ruiter, B., and Van de Grampel, J. C.,** *J. Chem. Soc. Dalton Trans.*, p. 1521, 1984.
56. **Krishnamurthy, S. S., Ramabrahman, P., Vasudeva Murthy, A. R., Shaw, R. A., and Woods, M.,** *Z. Anorg. Allg. Chem.*, 522, 226, 1985.
57. **Kumara Swamy, K. C., Poojary, M. D., Krishnamurthy, S. S., and Manohar, H.,** *J. Chem. Soc. Dalton Trans.*, p. 1881, 1985.
58. **Allcock, H. R., Wagner, L. J., and Levin, M. L.,** *J. Am. Chem. Soc.*, 105, 1321, 1983.
59. **Chivers, T., Oakley, R. T., and Paddock, N. L.,** *J. Chem. Soc. A*, p. 2324, 1970.
60. **Niecke, E., Thamm, H., and Bohler, D.,** *Inorg. Nucl. Chem. Lett.*, 8, 261, 1972.
61. **Janssen, E., Ph.D. thesis, Frankfurt, 1975; Roesky, H. W.,** personal communication.
62. **Grosse-Bowing, W., Ph.D. thesis,** University of Gottingen, FRG, 1971.
63. **Kumara Swamy, K. C. and Krishnamurthy, S. S.,** *Inorg. Chem.*, 25, 920, 1986.
64. **Niecke, E., Thamm, H., and Glemser, O.,** *Z. Naturforsch.*, 26B, 366, 1971.
65. **Horn, H. G. and Kolkmann, F.,** *Makromol. Chem.*, 88, 1843, 1982.
66. **Volodin, A. A., Kireev, V. V., Fomin, A. A., Edelev, M. G., and Korshak, V. V.,** *Dokl. Akad. Nauk SSSR*, 209, 98, 1973.
67. **Niecke, E., Glemser, O., and Roesky, H. W.,** *Z. Naturforsch.*, 24B, 1187, 1969.
68. **Roesky, H. W., Grosse Bowing, W., and Niecke, E.,** *Chem. Ber.*, 104, 653, 1971.
69. **Allen, C. W., Tsang, F. Y., and Moeller, T.,** *Inorg. Chem.*, 7, 2183, 1968.
70. **Cordes, A. W., Swepston, P. N., Oakley, R. T., Paddock, N. L., and Ranganathan, T. N.,** *Can. J. Chem.*, 59, 2364, 1981.
71. **Ramachandran, K. and Allen, C. W.,** *J. Am. Chem. Soc.*, 104, 2396, 1982.
72. **Sharma, R. D., Rettig, S. S., Paddock, N. L., and Trotter, J.,** *Can. J. Chem.*, 60, 535, 1982.
73. **Allen, C. W. and Shaw, J. C.,** *Inorg. Chem.*, 25, 4632, 1986.
74. **Allen, C. W. and Moeller, T.,** *Inorg. Chem.*, 7, 2177, 1968.
75. **Allcock, H. R., Lavin, K. D., Riding, G. H., Whittle, R. R., and Parvez, M.,** *Organometalics*, 5, 1626, 1986.
76. **Allcock, H. R., Lavin, M. L., and Austin, P. E.,** *Inorg. Chem.*, 25, 2281, 1986.
77. **Kumara Swamy, K. C. and Krishnamurthy, S. S.,** *J. Chem. Soc. Dalton Trans.*, p. 1431, 1985.
78. **Brandt, K., Kireev, V. V., Korshak, V. V., and Tsokolaev, B. R.,** *Zh. Obshch. Khim.*, 48, 358, 1978.
79. **Allcock, H. R., Fuller, T. J., and Matsumara, K.,** *Inorg. Chem.*, 21, 515, 1982.
80. **Van der Huizen, A. A., Van de Grampel, J. C., Akkerman, W., Lelieveld, P., Van der Meer Kelverkemp, A., and Lamberts, H. B.,** *Inorg. Chim. Acta*, 78, 239, 1983.
81. **Horn, H. G. and Kolkmann, F.,** *Chem. Ztg.*, 105, 213, 1981.
82. **Allcock, H. R., Austin, P. E., and Neenan, T. Y.,** *Macromolecules*, 15, 689, 1982.
83. **Alkubaisi, A. A., Deutsch, W. F., Hursthouse, M. B., Parkes, H. G., Shaw, L. S., and Shaw, R. A.,** *Phosphorus Sulfur*, 28, 229, 1986.
84. **Butour, J. L., Labarre, J. F., and Sournies, F.,** *J. Mol. Struct.*, 65, 51, 1980.
85. **Allcock, H. R. and Fuller, T. J.,** *J. Am. Chem. Soc.*, 103, 2250, 1981.
86. **Gallicano, K. D. and Paddock, N. L.,** *Can. J. Chem.*, 60, 521, 1982.
87. **Latscha, H. P.,** *Z. Anorg. Allg. Chem.*, 362, 7, 1968.
88. **Dillon, K. B., Platt, A. W. G., and Waddington, T. C.,** *J. Chem. Soc. Dalton Trans.*, p. 1036, 1980.
89. **Dathathreyan, K. S., Krishnamurthy, S. S., Vasudeva Murthy, A. R., Shaw, R. A., and Woods, M.,** *J. Chem. Soc. Dalton Trans.*, p. 1928, 1981.
90. **Volodin, A. A., Zelenetskii, S. N., Kireev, V. V., and Korshak, V. V.,** *Zh. Obshch. Khim.*, 45, 37, 1975.

91. **Gitel, P. O., Osipova, L. F., and Solovova, O. P.,** *Zh. Obshch. Khim.*, 45, 1749, 1975.
92. **Ali, S. A., Herrmann, E., and Thomas, B.,** *Z. Chem.*, 24, 133, 1984.
93. **Korshak, V. V., Solomatina, A. I., Bekasova, N. I., Andreyeva, M. A., Bulychova, Ye. G., Vinogradova, S. V., Kalinin, V. N., and Zakharkin, L. I.,** *Polym. Sci. (U.S.S.R.)*, 22, 2180, 1980.
94. **Moeller, T. and Nannelli, P.,** *Inorg. Chem.*, 2, 659, 1963.
95. **Schmidpeter, A., Blanck, K., Eiletz, H., Smetana, H., and Weingand, C.,** *Synth. React. Inorg. Metal Org. Chem.*, 7, 1, 1977.
96. **Grushkin, B., Sanchez, M. G., Ernest, M. V., McClanahan, J. L., Ashby, G. E., and Rice, R. G.,** *Inorg. Chem.*, 4, 1538, 1965.
97. **Keat, R., Rycroft, D. S., Miller, V. R., Schmulbach, C. D., and Shaw, R. A.,** *Phosphorus Sulfur*, 10, 121, 1981.
98. **Rutt, J. S., Parvez, M., and Allcock, H. R.,** *J. Am. Chem. Soc.*, 108, 6089, 1986.
99. **Schmidpeter, A., Hogel, J., and Ahmed, F. R.,** *Chem. Ber.*, 109, 1911, 1976.
100. **Ramamoorthy, V., Ranganathan, T., Rao, G. S., and Manoharan, P. T.,** *J. Chem. Res. (S)*, 316, (*M*)3074, 1982.
101. **Seerle, H. T., Dyson, J., Ranganathan, T. N., and Paddock, N. L.,** *J. Chem. Soc. Dalton Trans.*, p. 203, 1975.
102. **Gallicano, K. D., Oakley, R. T., Paddock, N. L., and Sharma, R. D.,** *Can. J. Chem.*, 59, 2654, 1981.
103. **John, K. and Moeller, T.,** *J. Inorg. Nucl. Chem.*, 22, 199, 1961.
104. **Thomas, B., Grossman, G., and Meyer, H.,** *Phosphorus Sulfur*, 10, 375, 1981.
105. **Lehr, W. and Pietschmann, J.,** *Chem. Ztg.*, 94, 362, 1970.
106. **Krishnamurthy, S. S., Ramachandran, K., and Woods, M.,** *Phosphorus Sulfur*, 9, 323, 1981.
107. **Krishnamurthy, S. S., Sau, A. C., Vasudeva Murthy, A. R., Keat, R., Shaw, R. A., and Woods, M.,** *J. Chem. Soc. Dalton Trans.*, p. 1405, 1976.
108. **Krishnamurthy, S. S., Sau, A. C., Vasudeva Murthy, A. R., Keat, R., Shaw, R. A., and Woods, M.,** *J. Chem. Soc. Dalton Trans.*, p. 1980, 1977.
109. **Krishnamurthy, S. S., Ramachandran, K., Sau, A. C., Sudheendra Rao, M. N., Vasudeva Murthy, A. R., Keat, R., and Shaw, R. A.,** *Phosphorus Sulfur*, 5, 117, 1978.
110. **Thomas, B., Grossmann, G., Bieger, W., and Porzel, A.,** *Z. Anorg. Allg. Chem.*, 504, 138, 1983.
111. **John, K., Moeller, T., and Audrieth, L. F.,** *J. Am. Chem. Soc.*, 82, 5616, 1960.
112. **Krishnamurthy, S. S., Sundram, P. M., and Woods, M.,** *Inorg. Chem.*, 21, 406, 1982.
113. **Krishnamurthy, S. S., Shaw, R. A., Sudheendra Rao, M. N., Vasudeva Murthy, A. R., and Woods, M.,** *Inorg. Chem.*, 17, 1527, 1978.
114. **Dhathathreyan, K. S., Krishnamurthy, S. S., and Woods, M.,** *J. Chem. Soc. Dalton Trans.*, p. 2151, 1982.
115. **Winter, H. and Van de Grampel, J. C.,** *Recl. Trav. Chim.*, 103, 241, 1984.
116. **Chandrasekhar, V., Karthikeyan, S., Krishnamurthy, S. S., and Woods, M.,** *Ind. J. Chem.*, 24A, 379, 1985.
117. **Thomas, B. and Grossmann, G.,** *Z. Anorg. Allg. Chem.*, 523, 112, 1985.
118. **Carrol, A. P., Shaw, R. A., and Woods, M.,** *J. Chem. Soc. Dalton Trans.*, p. 2736, 1973.
119. **Roesky, H. W. and Janssen, E.,** *Chem. Ber.*, 108, 2531, 1975.
120. **Ranganathan, T. N., Todd, S. M., and Paddock, N. L.,** *Inorg. Chem.*, 12, 316, 1973.
121. **Mark, V., Dungan, C. H., Crutchfield, M. M., and Van Wazer, J. R.,** *Top. Phosphorus Chem.*, 5, 391, 1967.
122. **Krishnamurthy, S. S., Ramachandran, K., and Woods, M.,** *J. Chem. Res. (S)*, 92, (*M*)1258, 1979.
123. **Narayanaswamy, P. Y., Dhathathreyan, K. S., and Krishnamurthy, S. S.,** *Inorg. Chem.*, 24, 640, 1985.
124. **Narayanaswamy, P. Y. and Krishnamurthy, S. S.,** unpublished results.
125. **Ramabrahmam, P., Dhathathreyan, K. S., Krishnamurthy, S. S., and Woods, M.,** *Ind. J. Chem.*, 22A, 1, 1983.
126. **Kumara Swamy, K. C., Krishnamurthy, S. S., Vasudeva Murthy, A. R., Shaw, R. A., and Woods, M.,** *Ind. J. Chem.*, 25A, 1004, 1986.
127. **Dhathathreyan, K. S.,** Synthetic and Spectroscopic Studies of Alkoxycyclophosphazenes, Phosphazadienes and Phosphazanes, Ph.D. thesis, Indian Institute of Science, Bangalore, 1981.
128. **Thomas, B. and Grossmann, G.,** *Russ. Chem. Rev.*, (Engl. transl.), 55, 622, 1986.
129. **Oakley, R. T., Rettig, S. J., Paddock, N. L., and Trotter, J.,** *J. Am. Chem. Soc.*, 107, 6923, 1985.
130. **Mitropol'skaya, G. I., Kireev, V. V., Korshak, V. V., and Goryaev, A. A.,** *Zh. Obshch. Khim.*, 52, 2486, 1982.
131. **Krishnamurthy, S. S., Sau, A. C., Vasudeva Murthy, A. R., Shaw, R. A., Woods, M., and Keat, R.,** *J. Chem. Res. (S)*, 70, (*M*)0860, 1977.
132. **Oakley, R. T. and Paddock, N. L.,** *Can. J. Chem.*, 51, 520, 1973.

133. **Chandrasekhar, V., Krishnamurthy, S. S., and Woods, M.,** *ACS Symp. Ser. (Phosphorus Chem.),* 171, 418, 1981.
134. **Schmidpeter, A., Blanck, K., and Ahmed, F. R.,** *Angew. Chem. Int. Ed. Engl.,* 15, 488, 1976.
135. **Paddock, N. L., Ranganathan, T. N., Rettig, S. S., Sharma, R. D., and Trotter, J.,** *Can. J. Chem.,* 59, 2429, 1981.

Section M3

1. **Schmidpeter, A., Lochschmidt, S., Burget, G., and Sheldrick, W. S.,** *Phosphorus Sulfur,* 18, 23, 1983; **Schmidpeter, A. and Lochschmidt, S.,** *Angew. Chem.,* 98, 271, 1986.
2. **Cowley, A. H. and Cushner, M. C.,** *Inorg. Chem.,* 19, 515, 1980; **Burg, A. P.,** *J. Inorg. Nucl. Chem.,* 33, 1575, 1971.
3. **Weber, D. and Fluck, E.,** *Z. Anorg. Allg. Chem.,* 103, 424, 1976.

Chapter 15

^{31}P NMR DATA OF FOUR COORDINATE PHOSPHORUS COMPOUNDS CONTAINING A FORMAL MULTIPLE BOND FROM PHOSPHORUS TO A GROUP IV ATOM

Compiled and presented by
D. Vaughan Griffiths and Penelope A. Griffiths

NMR data was also donated by
H. J. Bestmann, Universität Erlangen-Nurnberg, FRG
E. Fluck, Gmelin-Institut, Frankfurt/Main, FRG
S. G. Kleemann, Fachhochschule München, FRG
C. D. Hall, King's College London, U.K.

The NMR data are presented in Table N in two sections. Section N1 contains the data for phosphonium ylides. Section N2 contains the data for the germanium ylides.

FACTORS INFLUENCING ^{31}P NMR CHEMICAL SHIFTS OF PHOSPHONIUM YLIDES

There are clearly many factors influencing the ^{31}P NMR shifts of phosphorus ylides, and despite the large amount of NMR data now available on these compounds, it is still extremely difficult to develop general quantitative expressions which can be used in a predictive way with any reliability. The shifts of some key ylides have yet to be reported and the shifts of certain others must be viewed with some suspicion until confirmed by further study. Perhaps the most striking observation from this data compilation is the wide range of chemical shifts displayed by phosphorus ylides — from δ_P 126 and δ_P 127 for $tBu_2BrP{=}CHTms$ and $tBu_2ClP{=}GeCl_2$, respectively, to δ_P -29.6 for the carbodiphosphorane $Me_3P{=}C{=}PMe_3$. Moreover, even within a more restricted group of ylides such as the methylene phosphoranes ($XYZP{=}CH_2$) there is still a wide shift range from δ_P 116.7 for $tBu_2ClP{=}CH_2$ to $\delta_P - 8.9$ for $(C_6F_5)Me_2P{=}CH_2$.

Within more restricted groups of ylides, however, some trends become clear. Within the trialkylmethylenephosphoranes, for example, increasing the number of carbons β to the phosphorus causes deshielding of the phosphorus (e.g., for $R_3P{=}CH_2$ we see R = Me [δ_P -2.1], Et [δ_P 23.6], iPr [δ_P 40.5], and tBu [δ_P 57.1]), whereas additional carbons γ to the phosphorus appear to increase shielding (e.g., cf. R = Et [δ_P 23.6] and R = nBu [δ_P 16.4]). Such γ-shielding probably also contributes to the dramatic upfield shift of the phosphorus in the six-membered ring in $(CH_2)_5PMe{=}CH_2$ (δ_P 0.2) relative to its value in the corresponding five-membered ring compound $(CH_2)_4PMe{=}CH_2$ (δ_P 24.9). Other factors, however, clearly contribute to the ^{31}P NMR chemical shifts in the trialkylmethylenephosphoranes, since it is not easy to develop simple β and γ substituent parameters which give accurate predictions of chemical shift in this series. Thus, for example, in those cases where the alkyl groups are bulky such as in $tBu_3P{=}CH_2$, there is likely to be distortion from the normal tetrahedral arrangement around the phosphorus. This would be expected to make π electron feedback from the ylide carbon to the phosphorus more difficult, thereby reducing the shielding at the phosphorus.

The electronegativities of the substituents on the phosphorus also clearly have a marked effect on chemical shifts as shown, for example, in the comparison of $Me_3P{=}C(Tms)_2$ (δ_P -0.5) with $Me_2ClP{=}C(Tms)_2$ (δ_P 49), but electronegativity effects may be markedly altered by other factors. Thus, while in the series $(Me_2N)_nMe_{3-n}P{=}CH_2$ we observe δ_P -2.1

(n = 0), 35.6 (n = 1), 62.5 (n = 2), and 70.8 (n = 3), apparently showing the electronic effects of increasing nitrogen substitution, there would also appear to be a significant contribution to the downfield shift arising from the increase in the number of carbons β to the phosphorus due to the N-methyl groups. The deshielding effects of the N-methyl group would seem to be confirmed by the comparison of $(Me_2N)(iPrO)_2P{=}CHCO_2Et$ (δ_P 60) with $(MeHN)(iPrO)_2P{=}CHCO_2Et$ (δ_P 55). As a consequence of this, alkoxy substituents which have only one carbon β to the phosphorus are therefore less deshielding than an amino group with two carbons β to the phosphorus despite the greater electronegativity of oxygen than nitrogen. This is shown clearly in the case of $R_3P{=}C(CO_2Me)CH(OMe)CO_2Me$ where we observe a chemical shift for R = Me_2N of δ_P 63.5, while for R = MeO the shift is at higher field at δ_P 57.5.

The ^{31}P NMR shift is also sensitive to changes in the substituents on the ylide carbon. Alkyl substitution on the ylide carbon increases shielding at phosphorus (e.g., for $Ph_3P{=}CH_{2-n}Me_n$ we observe δ_P 19.6 [n = 0], 15.0 [n = 1], and 9.8 [n = 2]), although other effects are clearly apparent when the ylide carbon forms part of a cyclic system (e.g., for $Ph_3P{=}C[CH_2]_n$ we observe δ_P 15.6 [n = 2], 16.5 [n = 3], 4.8 [n = 4], and 6.4 [n = 5]).

The presence of a phenyl substituent produces a more marked upfield shift than an alkyl substituent, although the introduction of a second such substituent has a relatively small effect (e.g., for $Ph_3P{=}CH_{2-n}Ph_n$ we observe δ_P 19.6 [n = 0], 7.0 [n = 1], and 6.6 [n = 2]).

The effects of electron withdrawing substituents on the ylide carbon are less easy to summarize since both upfield and downfield shifts are observed. This is due to the varying abilities of such substituents to delocalize the negative charge from the ylide carbon by producing additional resonance structures which can contribute to the bonding in these systems.

At this point it is worth noting that ylides possessing carbonyl groups on the ylide carbon can exhibit a temperature-dependent NMR spectrum due to the partial double-bond character which exists between the ylide carbon and the carbonyl-containing substituents. In some cases this effect only becomes apparent at low temperatures. However, in ylides such as (**5**; R = MeO) line broadening is observed at room temperature and in cases such as $(MeO)_3P{=}C(CO_2Me)CH(CO_2Me)_2$ the rate of rotation around the bond to the α-alkoxycarbonyl group is sufficiently slow to enable signals for both rotamers to be observed at ambient temperatures. Although such effects can cause unexpectedly complex NMR spectra, as in the case of the diylide $(MeO)_3P{=}C(CO_2Me)C(CO_2Me){=}P(OMe)_3$ (δ_P 60.2, 61.3 [J_{PP} 10 Hz], 62.5 [J_{PP} 10 Hz], 64.2),[35] such temperature-dependent spectra can be invaluable for rapidly identifying such ylides.

The shifts in asymmetrical carbodiphosphoranes are also not easy to predict. Thus, for example, while $Ph_3P{=}C{=}PPh_3$ and $(Me_2N)_3P{=}C{=}P(NMe_2)_3$ are reported to have shifts of δ_P −3.5 and 27.7 respectively, those for the mixed system $Ph_3P{=}C{=}P(NMe_2)_3$ are observed at δ_P −20.7 and 39.3. Further work is clearly needed to assess the factors influencing chemical shift in these systems.

λ^5-Phosphorins have also been included in this data compilation, although it is generally considered that the dipolar ylide structure is probably an oversimplification of the true state of the bonding in these systems.

Finally, a section on germanium ylides is also included in this chapter, but no reports of the corresponding silicon analogues were found during a computer search of *Chemical Abstracts*.

TABLE N

Compilation of ^{31}P NMR Data of Four Coordinate Phosphorus Compounds Containing a Formal Multiple Bond from Phosphorus to a Group IV Atom

Section N1: Carbon Ylides

P-bound atoms/ rings	Connectivities	Structure	NMR Data (δ_P[solv, temp] J_{PP} or J_{PF} Hz)	Ref.
$C'ClN_2$	$Cl_2;(C_2)_2$	$(Me_2N)_2ClP{=}CCl_2$	62.3(r)	1
	$Si_2;(C_2)_2$	$(Me_2N)_2ClP{=}CTms_2$	65.5(c)	2
	$Si_2;(C_2)_2$	$(iPr_2N)_2ClP{=}CTms_2$	72.2(b)	3
	$HCl;(C_2)_2$	$(Me_2N)_2ClP{=}CHCl$	37(r)	1
	$HSi;(C_2)_2$	$(Et_2N)_2ClP{=}CHTms$	71.8(c)	4
	$HSi;(C_2)_2$	$(iPr_2N)_2ClP{=}CHTms$	60.6(c)	4
	$H_2;(C_2)_2$	$(Et_2N)_2ClP{=}CH_2$	32.6(b)	1
	$C'H;(C_2)_2$	$(Me_2N)_2ClP{=}CHPh$	62.0	5
	$C'H;(C_2)_2$	$(Et_2N)_2ClP{=}CHPh$	62.2	5
$C''ClN_2$	$P^4;(C_2)_2$	$(Me_2N)_2ClP{=}C{=}P(NMe_2)_3$	39.5/11.3(b) J154	6
	$P^4;(C_2)_2$	$(Me_2N)_2ClP{=}C{=}PPh_3$	31.6(b) J94	6
	$P^4;(C_2)_2$	$(Et_2N)_2ClP{=}C{=}PPh_3$	26.2(b) J99	6
$C'FN_2$	$HSi;(C_2)_2$	$(Me_2N)_2FP{=}CHTms$	70.3 J957	7
	$H_2;(C_2)_2$	$(Me_2N)_2FP{=}CH_2$	70.9 J975	7
	$CH;(C_2)_2$	$(Me_2N)_2FP{=}CHCH(nBu)P(NMe_2)_2$	60.7(b,31°) J1016 + 35	8
	$C'H;(C_2)_2$	$(Me_2N)_2FP{=}CHCO_2Et$	70.5 J948	7
	$C'H;(C_2)_2$	$(Et_2N)_2FP{=}CHPh$	60.4 J950	7
$C'NO_2$	$C'N';C'_2;C_2$	$PhCH_2N{=}CPh$ $(4\text{-}Tol)N(EtO)_2P{=}CPhN{=}CHPh$	20(b)	9
	$C'Si;C_2;C_2$	$(Me_2N)(iPrO)_2P{=}C(CO_2Et)Tms$	59.3	10
	$C'Si;C_2;C_2$	$(Et_2N)(iPrO)_2P{=}C(CO_2Et)Tms$	57	10
	$C'H;C'Si;C_2$	$(TmsPhN)(iPrO)_2P{=}CHCO_2Et$	52.8	10
	$C'H;CH;C_2$	$(MeHN)(iPrO)_2P{=}CHCO_2Et$	55.3	11
	$C'H;C_2;C_2$	$(Me_2N)(iPrO)_2P{=}CHCO_2Et$	60	10
	$C'H;C_2;C_2$	$(Et_2N)(iPrO)_2P{=}CHCO_2Et$	57.8	10
	$C'_2;CH;C_2$	$(MeHN)(iPrO)_2P{=}C(CO_2Me)_2$	51	11
	$C'_2;C'H;C_2$	$(PhNH)(EtO)_2P{=}C(CO_2Et)_2$	47.2	12
	$C'_2;C'H;C_2$	$(PhNH)(iPrO)_2P{=}C(CO_2Me)_2$	42.4	11
	$C'_2;C_2;C_2$	$(Me_2N)(EtO)_2P{=}C(CO_2Et)_2$	57	13
	$C'_2;C_2;C_2$	$(Me_2N)(NeoO)_2P{=}C$ $(CO_2Me)C(CO_2Me){=}NPh$	54.5(c)	14
$C'NO_2/5$	$CC';C_2;C_2$	$(Me_2N)(Glc)P{=}C(CO_2Me)CH(OMe)CO_2Me$	77.8	15
	$CC';C_2;C_2$	$(Me_2N)(Pnc)P{=}C(CO_2Me)CH(OMe)CO_2Me$	67.0(c)	15
	$C'_2;C'N';C'_2$	$(PhO)_2P^{\cdot}{=}CPhCPh{=}NN^{\cdot}Ph$	43.7(c)	16
$C'NO_2/6$	$CH;C_2;C_2$	$(Me_2N)(OCHMeCH_2CH_2O)P{=}CHCH_2Ac$	8.2	17
	$CH;C_2;C_2$	$(Et_2N)(OCHMeCH_2CH_2O)P{=}CHCH_2Ac$	6.9	17
	$CC';C'_2;C_2$	$(MeO)_2P^{\cdot}{=}C(CO_2Me)C(OMe)CO_2MeC$ $(CO_2Me){=}C(CO_2Me)N^{\cdot}Ph$	43(c)	14
	$C'_2;C'H;C_2$	$(iPrO)_2P^{\cdot}{=}C(CO_2Et)(C{=}O)DopN^{\cdot}H$	27(k)	12
$C'N_2O$	$C'H;(C_2)_2;C'$	$(Me_2N)_2[C^{\cdot}H_2(CH_2)_3CH{=}C^{\cdot}{-}O]P{=}CHCO_2Et$	62.3(n)	18
	$C'H;(C_2)_2;C'$	$(Et_2N)_2[C^{\cdot}H_2(CH_2)_3CH{=}C^{\cdot}{-}O]P{=}CHCO_2Et$	61.6(n)	18
$C'N_2O/5$	$C'H;HS^4,C'N';C$	$(TosNH)(MeO)P^{\cdot}{=}CH{-}CMe{=}NN^{\cdot}Ac$	42(r)	19
	$CC';(C_2)_2;C$	$(OCHPhCHMeNMe)(Me_2N)P{=}C$ $(CO_2Me)CH(OMe)CO_2Me$	65.5, 64.9[I](g, −20°)	15
$C'N_3$	$ClP^4;(C_2)_3$	$[(Me_2N)_3P{=}CCl{-}PCl(NMe_2)_2]^+Cl^-$	65.5/47.9(c) J97	6
	$P^4_2;(C_2)_3$	$[(Me_2N)_3P{=}C^{\cdot}{-}P^{\cdot}(NMe_2)_2]_2^{2+}2Cl^-$	42.6/37.9(c) J9	6
	$HP^4;(C_2)_3$	$[(Me_2N)_3P{=}CH{-}PCl(NMe)_2]^+Cl^-$	65.8/52.2(c) J23	6
	$HP^4;(C_2)_3$	$(Me_2N)_3P{=}CH{-}P(O)(NMe_2)_2$	40.5/64.0(k) J50	6
	$HP^4;(C_2)_3$	$[(Me_2N)_3P{\cdots}CH{\cdots}P(NMe_2)_3]^+Cl^-$	54.2(c)	6
	$H_2;(C_2)_3$	$(Me_2N)_3P{=}CH_2$	70.8(n)	20

TABLE N (continued)
Compilation of ^{31}P NMR Data of Four Coordinate Phosphorus Compounds Containing a Formal Multiple Bond from Phosphorus to a Group IV Atom

Section N1: Carbon Ylides

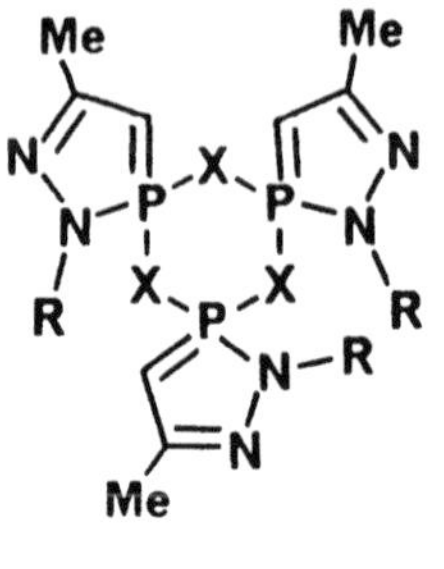

1

P-bound atoms/ rings	Connectivities	Structure	NMR Data (δ_P[solv, temp] J_{PP} or J_{PF} Hz)	Ref.
	$H_2;(C_2)_3$	$(Et_2N)_3P{=}CH_2$	67.2(b, 25°)	21
	$H_2;(C_2)_3$	$(iPr_2N)_2(Me_2N)P{=}CH_2$	59.2(b)	3
	$CP^4;(C_2)_3$	$[(Me_2N)_3P{=}CMe{-}PCl(NMe_2)_2]^+I^-$	71.4/54.1(c) J84	6
	$CP^4;(C_2)_3$	$[(Me_2N)_3P{\cdots}CMe{\cdots}P(NMe_2)_3]^+I^-$	60.3(c)	6
	$C'P^4;(C_2)_3$	$(Me_2N)_3P{=}C(CO_2{}^-)P^+(NMe_2)_3$	59.9(c)	6
	$CH;(C_2)_3$	$(Me_2N)_3P{=}CHMe$	64.5(n)	20
	$CC';(C_2)_3$	$(Me_2N)_3P{=}C(CO_2Me)CH(OMe)CO_2Me$	63.5(g)	15
	$CC';(C_2)_3$	$(Me_2N)_3P{=}C(CO_2Me)CH(CO_2Me)NPhth$	63.6, 64.8[I]	15
	$CC';(C_2)_3$	$(Me_2N)_3P{=}C(COPh)CH_2COPh$	63.2(c/r)	22
	$C'_2;(C_2)_3$	$(Me_2N)_3P{=}C(CO_2Me)C(CO_2Me){=}NEt$	60.7	23
	$C'_2;(C_2)_3$	$(Me_2N)_3P{=}C(CO_2Me)CH{=}NPh$	58.1	23
$C'N_3/4$	$CH;(CP^4)_2, C_2$	$[(tBuN^{\cdot})(iPr_2N)P^{\cdot}{=}CH{-}tBu]_2$	21.5(b/t)	24
$C'N_3/5$	$C'_2;(C_2)_3$	$(Me_2N)(CH_2MeN)_2P{=}C$ $(CO_2Me)C(CO_2Me){=}NPh$	56.3	23
	$C'_2;C_2,(CC')_2$	$(Me_2N)(CH_2PhN)_2P{=}C$ $(CO_2Me)C(CO_2Me){=}NEt$	47	23
	$C'_2;C_2,(CC')_2$	$(Me_2N)(CH_2PhN)_2P{=}C$ $(CO_2Me)C(CO_2Me){=}NPh$	47.5	23
$C'N_3/5,5$	$C'H;C'N',(CC')_2$	$(nPrNCMe{=}CMeN{-}nPr)P^{\cdot}{=}CH{-}$ $CMe{=}NN^{\cdot}Ac$	50	25
	$C'H;C'N',(CC')_2$	$(nPrNCMe{=}CMeN{-}nPr)P^{\cdot}{=}CHCMe{=}NN^{\cdot}Ph$	33	25
$C'N_3/5,6$	$C'H;C'N',(C'P^4)_2$	**1**; R = Ac, X = NPh	17(c)	26
	$C'H;C'N', (C'P^4)_2$	**1**; R = Ph, X = NPh	20(r)	26
$C''N_3$	$P^4;(C_2)_3$	$(Me_2N)_3P{=}C{=}PCl(NMe_2)_2$	39.5/11.3(b) J154	6
	$P^4;(C_2)_3$	$(Me_2N)_3P{=}C{=}P(NMe_2)_3$	27.7(k)	6
	$P^4;(C_2)_3$	$(Me_2N)_3P{=}C{=}PPh_3$	39.3(b) J145	6
	$P^4;(C_2)_3$	$(Et_2N)_3P{=}C{=}PPh_3$	37.4(b) J153	6
$C'O_2S/5$	$C'_2;C_2;H$	$(OCHMeCHMeO)(HS)P{=}CPh_2$	36	27
$C'O_2Si$	$C'_2;C_2;C_3$	$(iPrO)_2TmsP{=}C(CO_2Me)_2$	80.8	28
$C'O_3$	$P^4S;C_3$	$(EtO)_3P{=}C[P(O)(OEt)_2]SP(O)(OEt)_2$	51.8 J118	29
	$P^4S;C_3$	$(iPrO)_3P{=}C[P(O)(O{-}iPr)_2]SP(O)(O{-}iPr)_2)$	49.2 J133	29
	$S_2;C_3$	$(EtO)_3P{=}C(SPh)SP(O)(OEt)_2$	42.7	29
	$S_2;C_3$	$(iPrO)_3P{=}C(SPh)SP(O)(O{-}iPr)_2$	43.3	29
	$C'P^4;C_3$	$(MeO)_3P{=}C(Ph)P(O)(OMe)_2$	54.5(c) J89	30
	$C'P^4;C_3$	$(MeO)_3P{=}C(2{-}EtPh)P(O)(OMe)_2$	50.5(c) J98	31
	$C'P^4;C_3$	$(EtO)_3P{=}C(Ph)P(O)(OMe)_2$	49.5(c) J88	30

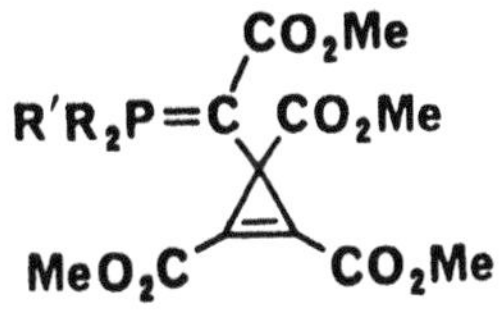

2

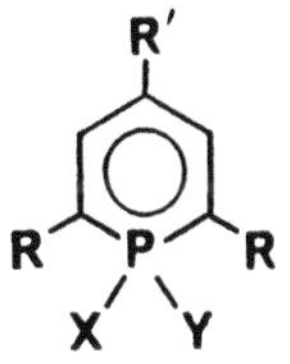

3

P-bound atoms/ rings	Connectivities	Structure	NMR Data (δ_P[solv, temp] J_{PP} or J_{PF} Hz)	Ref.
	$C'P^4;C_3$	$(EtO)_3P{=}C(4\text{-}MeOPh)P(O)(OMe)_2$	49.4(c) J94	31
	$CS;C_3$	$(MeO)_3P{=}C^{\cdot}CTf_2S(C{=}CTf_2)S^{\cdot}$	38.7	32
	$CC';C_3$	$(EtO)_3P{=}C^{\cdot}CH_2CO_2C^{\cdot}O$	41(r)	33
	$CC';C_3$	$(MeO)_3P{=}C(CO_2Me)CH(OMe)CO_2Me$	57.5(c)	15
	$CC';C_3$	$(MeO)_3P{=}C(CO_2Me)CH(CO_2Me)_2$	60.4, 58.5[1](c, 26°)	34
	$CC';C_3$	$(MeO)_3P{=}C(CO_2Me)CMe$ $(CO_2Me)P(O)(OMe)_2$	57.2(b) J6	35
	$CC';C_3$	$(EtO)_3P{=}C(CO_2Me)CH(CO_2Me)_2$	53.1, 51.1[1](t, 26°)	34
	$CC';C_3$	**2**; R′ = R = OMe	61.6, 59.4[1](t, −50°)	36
	$CC';C_3$	**2**; R = R′ = OEt	54.1, 55.8[1](t, −50°)	36
	$CC';C_3$	$(MeO)_3P{=}C(COPh)CH_2COPh$	56.2(c/r)	22
	$CC';C_3$	**2**; R = R′ = O-iPr	49.3, 51.0[1](t, −50°)	36
	$C'_2;C_3$	$(MeO)_3P{=}C(CO_2Me)C(S)CO_2Me$	45.3(b)	34
	$C'_2;C_3$	$(MeO)_3P{=}C^{\cdot}CO_2C(OMe){=}C^{\cdot}(CO_2Me)$	47.4(c)	34
	$C'_2;C_3$	$(EtO)_3P{=}C(CO_2Me)C(S)CO_2Me$	37.5(b)	34
	$C'_2;C_3$	$(EtO)_2(MeO)P{=}C(CO_2Et)_2$	51	13
	$C'_2;C_3$	$(EtO)_3P{=}C^{\cdot}CO_2C(OMe){=}C^{\cdot}(CO_2Me)$	40.2(c)	34
	$C'_2;C_3$	$(iPrO)_3P{=}C(CO_2Me)_2$	57	13
	$C'_2;C_3$	$(MeO)_3P{=}C(CO_2Me)C(CO_2Me){=}NPh$	53.8(b)	14
	$C'_2;C_3$	$(iPrO)_3P{=}C^{\cdot}CO_2C(OMe){=}C^{\cdot}(CO_2Me)$	40.3(t)	34
	$C'_2;C_3$	$(EtO)_3P{=}C(CO_2Me)C(CO_2Me){=}NPh$	47.7(b)	14
	$C'_2;C_3$	$(iPrO)_3P{=}C(CO_2Me)C(CO_2Me){=}NPh$	42.5(r)	14
	$C'C'';C_3$	$(MeO)_3P{=}C(CN)[3,5\text{-di-tBu-4-HOPh}]$	47(ae)	37
	$C'C'';C_3$	$(iPrO)_3P{=}C(CN)[3,5\text{-di-tBu-4-HOPh}]$	50(g)	37
$C'O_3/5$	$CC';C_3$	$(Pnc)(MeO)P{=}C(CO_2Me)CH(CO_2Me)OMe$	69.7	38
	$CC';C_3$	$(Pnc)(tBuO)P{=}C(CO_2Me)CH(CO_2Me)OMe$	57	15
	$CC';C_2,C'$	$(Pnc)(PhO)P{=}C(CO_2Me)CH(CO_2Me)OPh$	62.6	15
	$C'_2;C_3$	$(Pnc)(MeO)P{=}C(CO_2Me)C(CO_2Me){=}NPh$	65.5(r)	14
$C'O_3/6$	$C'_2;C_3$	$[Me_2C(CH_2O)_2](MeO)$- $P{=}C(CO_2Me)C(CO_2Me){=}NPh$	47.7(r)	14
$C'HO_2$	$C''_2;C_2$	$(EtO)_2HP=C(CN)_2$	8.4	39
$C'_2ClF/6$	$(C'_2)_2$	**3**; X = F, Y = Cl, R = R′ = Ph	59(c)	40
$CC'ClN$	$C_2H;Cl_2;C_2$	$iPrCl(Me_2N)P{=}CCl_2$	82.5	1
	$C_3;HCl;C_2$	$tBuCl(Me_2N)P{=}CHCl$	91.4(c)	1
	$C_3;HP;C_2$	$tBuCl(Et_2N)P{=}CH(PCl\text{-}tBu)$	104.4 J130	41
	$C_3;HSi;C_2$	$tBuCl(Et_2N)P{=}CHTms$	103.3	41
	$C_3;H_2;C_2$	$tBuCl(Et_2N)P{=}CH_2$	103.3(r)	1
	$C_3;C'H;C_2$	$tBuCl(Et_2N)P{=}CHCOCCl_3$	95.5	41
C'_2ClN	$Si_2,C'_2;C_2$	$PhCl(Me_2N)P{=}CTms_2$	62.2(c)	4
	$HSi,C'_2;C_2$	$PhCl(iPr_2N)P{=}CHTms$	60.6(c)	4
	$C'Si,C'_2;C_2$	$PhCl(Me_2N)P{=}CPhTms$	58.9(c)	4
$C'_2Cl_2/6$	$(C'_2)_2$	**3**; X = Y = Cl, R = R′ = Ph	17(c)	40
$C'_2FO/6$	$(C'_2)_2;C$	**3**; X = OMe, Y = F, R = R′ = Ph	70(c)	40
$C'_2F_2/6$	$(C'_2)_2$	**3**; X = Y = F, R = R′ = Ph	73(c)	40
$C'_2NO/6$	$(C'_2)_2;C_2;C$	**3**; X = OEt, Y = NEt_2, R = R′ = Ph	45(b)	42
	$(C'_2)_2;C'_2;C$	**3**; X = OMe, Y = $(4\text{-}Tol)_2N$, R = R′ = Ph	42(b)	42

TABLE N (continued)
Compilation of ^{31}P NMR Data of Four Coordinate Phosphorus Compounds Containing a Formal Multiple Bond from Phosphorus to a Group IV Atom

Section N1: Carbon Ylides

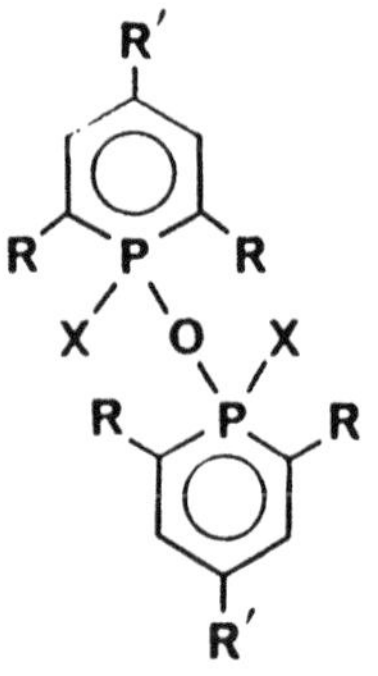

4

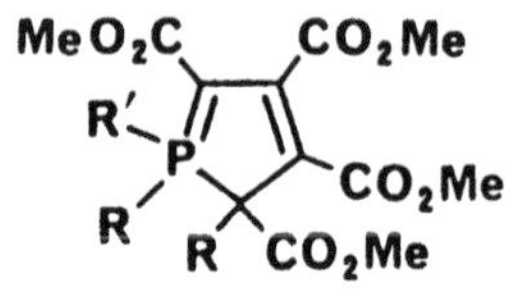

5

P-bound atoms/ rings	Connectivities	Structure	NMR Data (δ_P[solv, temp] J_{PP} or J_{PF} Hz)	Ref.
	$(C'_2)_2;C_2;P^4$	**4**; X = Me_2N, R = R′ = Ph	42.5(b)	43
	$(C'_2)_2;C_2;P^4$	**4**; X = Et_2N, R = R′ = Ph	40.0(b)	43
$CC'N_2$	$H_3;H_2;(C_2)_2$	$(Me_2N)_2MeP{=}CH_2$	62.5(n)	20
	$H_3;H_2;(C_2)_2$	$(Et_2N)_2MeP{=}CH_2$	58.4(b, 25°)	21
$C'_2N_2/6$	$(C'_2)_2;(C_2)_2$	**3**; X = Y = Me_2N, R = R′ = Ph	42.5(b)	43
	$(C'_2)_2;(C_2)_2$	**3**; X = Y = Et_2N, R = R′ = Ph	42(c)	40
	$(C'_2)_2;(C'_2)_2$	**3**; X = Y = Ph_2N, R = R′ = Ph	29.5(p)	44
$C'_2OP/5$	$(C'_2)_2;C;CC'$	$(PhP^*)Ph$-$(MeO)P{=}C(CO_2Me)COC^*(CO_2Me)_2$	86 J198	45
	$(C'_2)_2;C;CC'$	$(PhP^*)Ph(EtO)P{=}C(CO_2Et)COC^*(CO_2Et)_2$	80 J207	45
$CC'O_2$	$HO_2;C'_2;C_2$	$[(EtO)_2CH](EtO)_2P{=}C(CO_2Et)_2$	69.5	46
	$H_2O;C'_2;C_2$	$(MeOCH_2)(EtO)_2P{=}C(CO_2Et)_2$	76.9	47
	$H_3;CC';C_2$	**2**; R′ = Me, R = OMe	85.1, 86.8^{t}(t, 70°)	48
	$H_3;C'_2;C_2$	$(MeO)_2$-$MeP{=}C(CO_2Me)C(CO_2Me){=}PMe(OMe)_2$	90.1, 89.5 J13, 88.9 J13, 88.2^{t}(c, 26°)	34
	$CH_2;C'_2;C_2$	$(MeO_2CCH_2CH_2)(EtO)_2P{=}C(CO_2Et)_2$	82.8	49
$CC'O_2/5$	$C'O;C'_2;C_2$	**5**; R = R′ = OMe	85.9, 85.3, 83.1, 82.4^{t}(c, −50°)	36
	$C'O;C'_2;C_2$	**5**; R = R′ = OEt	80.5	50
	$C'O;C'_2;C_2$	**5**; R = R′ = O-iPr	75.6(c, 26°)	34
C'_2O_2	$OO',C'_2;C_2$	$(MeO_2C)(EtO)_2P{=}C(CO_2Et)_2$	50	51
	$P^4S,C'_2;C_2$	$Ph(EtO)_2P{=}C[P(O)(OEt)Ph]SP(O)(OEt)Ph$	69.5 J112	29
	$S_2,C'_2;C_2$	$Ph(EtO)_2P{=}C(SC_6Cl_5)SP(O)(OEt)Ph$	55.4	29
	$CO',C'_2;C_2$	$Ac(EtO)_2P{=}C(CO_2Et)_2$	49.2	51
	$CO',C'_2;C_2$	$(tBuCO)(EtO)_2P{=}C(CO_2Et)_2$	52.8	51
	$CC',C'_2;C_2$	**2**; R′ = Ph, R = OMe	73.3, 74.5^{t}(t, −70°)	48
	$CC',C'_2;C_2$	**2**; R′ = Ph, R = OEt	69.8, 70.9^{t}(t, −70°)	48
	$(C'_2)_2;C_2$	$Ph(MeO)_2P{=}C^*CO_2C(OMe){=}C^*(CO_2Me)$	65.3(t)	34
	$(C'_2)_2;C_2$	$Ph(MeO)_2P{=}C(CO_2Me)C(CO_2Me){=}PPh(OMe)_2$	75.7, 74.4 J12, 72.9 J12, 71.8^{t}(c, 26°)	34
	$(C'_2)_2;C_2$	$Ph(EtO)_2P{=}C(CO_2Me)C(CO_2Me){=}PPh(OEt)_2$	71.0, 69.9 J10, 68.5 J10, 67.4^{t}(c, 26°)	34
$C'_2O_2/6$	$(C'_2)_2;C_2$	**3**; X = Y = OMe, R = Ph, R′ = NO_2	65(c)	40
	$(C'_2)_2;C_2$	**3**; X = Y = OMe, R = Ph, R′ = Me	65(c)	40

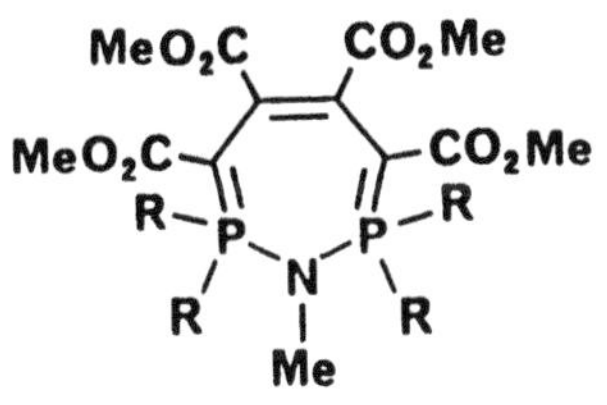

6

P-bound atoms/ rings	Connectivities	Structure	NMR Data (δ_P[solv, temp] J_{PP} or J_{PF} Hz)	Ref.
	(C'$_2$)$_2$;C,C'	**3**; X = iPrO, Y = AcO, R = R' = Ph	53(b)	43
C'C"O$_2$/6	C'N';C';C'$_2$	(4-BrPhO)$_2$P$^\cdot$=C=CPhNPhN=C$^\cdot$(4-FPh)	31.2	52
C$_2$C'Br	(H$_3$)$_2$;Si$_2$	Me$_2$BrP=CTms$_2$	40.0(b)	53
	(C$_2$H)$_2$;C'$_2$	iPr$_2$BrP=C(CO$_2$Me)$_2$	97.6	54
	C$_2$H,C$_3$;C'$_2$	tBu(iPr)BrP=C(CO$_2$Me)$_2$	98	54
	(C$_3$)$_2$;HSi	tBu$_2$BrP=CHTms	126	41
	(C$_3$)$_2$;H$_2$	tBu$_2$BrP=CH$_2$	105	41
	(C$_3$)$_2$;C'$_2$	tBu$_2$BrP=C(CH=CH)$_2$	101.8	41
	(C$_3$)$_2$;C'$_2$	tBu$_2$BrP=Flu	92	41
C$_2$C'Cl	(H$_3$)$_2$;ClSi	Me$_2$ClP=C(Cl)SiCl$_3$	60(b)	55
	(H$_3$)$_2$;P$_2$	Me$_2$ClP=C(PCl$_2$)$_2$	65	56
	(H$_3$)$_2$;Si$_2$	Me$_2$ClP=C(SiCl$_3$)$_2$	64.5(b)	55
	(H$_3$)$_2$;Si$_2$	Me$_2$ClP=C(SiF$_2$CH$_2$)$_2$SiF$_2$	68.0	57
	(H$_3$)$_2$;Si$_2$	Me$_2$ClP=C(SiCl$_2$CH$_2$)$_2$SiCl$_2$	71	57
	(H$_3$)$_2$;Si$_2$	Me$_2$ClP=C(SiClMe$_2$)$_2$	58.3(b)	53
	(H$_3$)$_2$;Si$_2$	Me$_2$ClP=CTmsSiMeCl$_2$	57.0(b)	53
	(H$_3$)$_2$;Si$_2$	Me$_2$ClP=CTmsSiMe$_2$Cl	52.0(b)	53
	(H$_3$)$_2$;Si$_2$	Me$_2$ClP=CTms$_2$	49(b)	53
	(C$_2$H)$_2$;C'H	iPr$_2$ClP=CHPh	74.8	58
	(C$_3$)$_2$;HS	tBu$_2$ClP=CHSMe	109.5	41
	(C$_3$)$_2$;HSi	tBu$_2$ClP=CHTms	114.3	41
	(C$_3$)$_2$;H$_2$	tBu$_2$ClP=CH$_2$	116.7	41
	(C$_3$)$_2$;CH	tBu$_2$ClP=CHPr	101.5	41
	(C$_3$)$_2$;C'H	tBu$_2$ClP=CHCOTf	108.5 J3	41
	(C$_3$)$_2$;C'$_2$	tBu$_2$ClP=CPh$_2$	92.5	41
CC'$_2$Cl	H$_2$Si;HSi,C'$_2$	(TmsCH$_2$)PhClP=CHTms	60.4(c)	4
	H$_3$;Si$_2$,C'$_2$	PhMeClP=CTms$_2$	53.0(c)	4
	CH$_2$;Si$_2$,C'$_2$	PhEtClP=CTms$_2$	65.5(c)	4
	CH$_2$;HSi,C'$_2$	PhEtClP=CHTms	69.1(c)	4
	C$_2$H;Si$_2$,C'$_2$	Ph(iPr)ClP=CTms$_2$	78.0(c)	4
	C$_2$H;HSi,C'$_2$	Ph(iPr)ClP=CHTms	81.8(c)	4
C'$_3$Cl	S$^4{}_2$,(C'$_2$)$_2$	Ph$_2$ClP=C(SO$_2$Ph)$_2$	55	59
	Si$_2$,(C'$_2$)$_2$	Ph$_2$ClP=CTms$_2$	54.5(b)	60
	HP^4,(C'$_2$)$_2$	[Ph$_2$ClP$\cdots$CH$\cdots$PClPh$_2$]$^+$Cl$^-$	57.8(r)	61
	HSi, (C'$_2$)$_2$	Ph$_2$ClP=CHTms	57.3(b)	60
C'$_2$C"Cl	(C'$_2$)$_2$;P^4	Ph$_2$ClP=C=PClPh$_2$	20.7(b)	62
C'$_3$F/6	(C'$_2$)$_3$	**3**; X = F, Y = 4-MePh, R = R' = Ph	49.0 J1113	63
CC'$_2$I	H$_3$;(C'$_2$)$_2$	Me(Mes)IP=CPh$_2$	44.4(k)	64
C$_2$C'N	(H$_3$)$_2$;H$_2$;C$_2$	Me$_2$(Me$_2$N)P=CH$_2$	35.6(n)	20
	(H$_3$)$_2$;H$_2$;C$_2$	Me$_2$(Et$_2$N)P=CH$_2$	61.8(b, 25°)	21
	(CH$_2$)$_2$;C'$_2$;C$_2$	Et$_2$(Et$_2$N)P=C(CO$_2$Et)$_2$	59	65
C$_2$C'N'/6	(CH$_2$)$_2$;H$_2$;P^4	[(CH$_2$)$_5$PMe=N](CH$_2$)$_5$P=CH$_2$	16.7/24.7(t, −40°)	66
C$_2$C'N/7	(H$_3$)$_2$;C'$_2$;CP^4	**6**; R = R' = Me	50.0(c)	67
C$_2$C'N/5,6	(H$_2$P^4)$_2$;C'H;C'N'	**1**; R = Ph, X = CH$_2$	49	68
	(C'$_2$P^4)$_2$;C'H;C'N'	**1**; R = Ac, X = Flu	75	68
C'$_3$N	HP^4,(C'$_2$)$_2$;CH	[Ph$_2$(MeHN)P$\cdots$CH$\cdots$P(NHMe)Ph$_2$]$^+$Cl$^-$	32.1(c)	69

TABLE N (continued)
Compilation of ^{31}P NMR Data of Four Coordinate Phosphorus Compounds Containing a Formal Multiple Bond from Phosphorus to a Group IV Atom

Section N1: Carbon Ylides

P-bound atoms/ rings	Connectivities	Structure	NMR Data (δ_P[solv, temp] J_{PP} or J_{PF} Hz)	Ref.
	$HP^4,(C'_2)_2;CH$	nPrNHPh$_2$P---CH---PPh$_2$N---nPr	27.5(b)	69
	$HP^4,(C'_2)_2;C'H$	[(PhNH)Ph$_2$P---CH---P(NHPh)Ph$_2$]$^+$Cl$^-$	23.2(c)	69
	$HP^4,(C'_2)_2;C_2$	Ph$_2$(Me$_2$N)P=CHP(O)Ph$_2$	40.3(c) J18	70
	$HP^4,(C'_2)_2;C_2$	Ph$_2$(Et$_2$N)P=CHP(O)Ph$_2$	40.1(c) J16	70
$C'_3N/7$	$(C'_2)_3;CP^4$	**6**; R = R′ = Ph	44.0(c)	67
$C'_2C''N$	$(C'_2)_2;P^4;C_2$	Ph$_2$(Me$_2$N)P=C=P(NMe$_2$)Ph$_2$	12.9(b)	70
	$(C'_2)_2;P^4;C_2$	Ph$_2$(Me$_2$N)P=C=PPh$_3$	16.7/−10.5(b) J104	71
	$(C'_2)_2;P^4;C_2$	Ph$_2$(Et$_2$N)P=C=P(NEt$_2$)Ph$_2$	11.9(b)	70
	$(C'_2)_2;P^4;C_2$	Ph$_2$(Et$_2$N)P=C=PPh$_3$	16.1/−10.0(b) J111	71
$C_2C'O$	$(H_3)_2;Si_2;C$	Me$_2$(MeO)P=C(SiCl$_3$)SiCl$_2$OMe	25	55
	$(C_3)_2;C''H;C$	tBu$_2$(EtO)P=CHCN	91.8(c, 30°)	72
$C_2C'O/5$	$H_3,C'O;C'_2;C$	**5**; R′ = Me, R = OMe	96.2, 96.9[I](t, −70°)	48
$CC'_2O/5$	$C'O;(C'_2)_2;C$	**5**; R′ = Ph, R = OMe	86.1(c)	48
	$C'O;(C'_2)_2;C$	**5**; R′ = Ph, R = OEt	83.9(c)	48
$CC'_2O/6$	$H_3;(C'_2)_2;C$	**3**; X = Me, Y = OMe, R = R′ = Ph	30(c)	40
	$H_3;(C'_2)_2;C$	**3**; X = Me, Y = OCH$_2$CH=CH$_2$, R = R′ = Ph	35(c)	40
C'_3O	$S^4_2,(C'_2)_2;C$	Ph$_2$(MeO)P=C(SO$_2$Ph)$_2$	53	59
	$CC',(C'_2)_2;C$	Ph$_2$(EtO)P=C(COPh)CH$_2$COPh	54.2(c/r)	22
	$(C'_2)_3;C$	Ph$_2$(MeO)P=C(CO$_2$Me)C(CO$_2$Me)=P(OMe)Ph$_2$	66.3, 65.6 J10, 64.8 J10, 64.1[I](c, 26°)	34
$C'_3O/6$	$(C'H)_2C'_2;C$	**3**; X = OMe, Y = Ph, R = H, R′ = tBu	40.2(c)	73
$C_2C'P$	$(H_3)_2;P_2;C_2$	Me$_2$PMe$_2$P=C(PMe$_2$)$_2$	14.2(b) J140 + 214	74
	$(H_3)_2;Si_2;HSi$	TmsPHMe$_2$P=C(SiMe$_2$CH$_2$)$_2$SiMe$_2$	1.0(b) J230	53
	$(H_3)_2;Si_2;CP^4$	MeP[Me$_2$P=CTms$_2$]$_2$	8(b) J294	53
	$(H_3)_2;Si_2;CSi$	TmsMePMe$_2$P=CTms$_2$	7.9(b) J258	53
	$(H_3)_2;Si_2;C_2$	Me$_2$PMe$_2$P=C(SiCl$_3$)$_2$	12.9(b) J281	55
	$(H_3)_2;Si_2;C_2$	Me$_2$PMe$_2$P=C(Tms)SiMeCl$_2$	12(b) J227	53
	$(H_3)_2;Si_2;C_2$	Me$_2$PMe$_2$P=C(SiMe$_2$F)$_2$	9.1(b) J213 + 3	75
	$(H_3)_2;Si_2;C_2$	Me$_2$PMe$_2$P=CTms$_2$	7.0(b) J217	53
	$(H_3)_2;HP;C_2$	tBu$_2$PMe$_2$P=CHPPh$_2$	3.0(b/t) J290 + 150	76
	$(CH_2)_2;C'_2;C_2$	Et$_2$PEt$_2$P=C(CO$_2$Me)$_2$	29.8 J271	13
$C_2C'P/5$	$(H_3)_2;Si_2;CP$	Me$_2$P*=C(SiCl$_3$)SiCl$_2$PMeP*Me	29(b) J258 + 9	77
CC'_2P	$H_3;HP,C'_2;C_2$	tBu$_2$PPhMeP=CHPPh$_2$	7.6(c) J285 + 140	78
C'_3P	$Si_2,(C'_2)_2;CP^4$	tBuP(Ph$_2$P=CTms$_2$)$_2$	15.6/17.6(b) J423	79
	$Si_2,(C'_2)_2;C'P^4$	PhP(Ph$_2$P=CTms$_2$)$_2$	35.2(b) J441	80
	$Si_2,(C'_2)_2;C_2$	tBu$_2$PPh$_2$P=CTms$_2$	22.6(b) J366	79
	$Si_2,(C'_2)_2;C_2$	cHex$_2$PPh$_2$P=C(Tms)$_2$	27.0(b) J283	79
	$Si_2,(C'_2)_2;C'_2$	Ph$_2$PPh$_2$P=CTms$_2$	22.5(b) J297	79
	$HP,(C'_2)_2;C_2$	Me$_2$PPh$_2$P=CHPPh$_2$	32.2(b/t, −20°) J211 + 140	76
	$HP,(C'_2)_2;C_2$	tBu$_2$PPh$_2$P=CHPPh$_2$	11.9(b/t) J302 + 110	76
	$HP,(C'_2)_2;C_2$	cHex$_2$PPh$_2$P=CHPPh$_2$	19.4(c) J251 + 138	78
	$HP,(C'_2)_2;C'_2$	Ph$_2$PPh$_2$P=CHPPh$_2$	27.0(b/t, −20°) J238 + 143	76
	$HSi,(C'_2)_2;C_2$	tBu$_2$PPh$_2$P=CHTms	9.1(b) J294	79
$C_2C'P^4$	$(H_3)_2;Si_2;C_2C'$	[Me$_2$P*=C(SiCl$_3$)$_2$]$_2$	19(b)	77
$C_2C'S/5$	$H_3,CC'H;CC';C$	(MeS)MeP=CMeCH=CHCHMe	8.2, 7.4[D](b)	81
$CC'_2S/5$	$CC'H;CC',C'_2;C$	(MeS)PhP=CMeCH=CHCHMe	25.7, 26.6[D](n)	81

P-bound atoms/ rings	Connectivities	Structure	NMR Data (δ_P[solv, temp] J_{PP} or J_{PF} Hz)	Ref.
$C_2C'Si$	$(C_2H)_2;C'_2;C_3$	$iPr_2TmsP{=}C(CO_2Me)_2$	37	54
$C_2C'H$	$(CH_2)_2;C'P^4$	$[nBu_2HP{=}C(CO_2Et)PPh_3]^+Cl^-$	5.7 J32, 2.4 J32[I] (r, −80°)	82
	$C_2H,C_3;C'_2$	$tBu(iPr)HP{=}C(CO_2Me)_2$	40(c)	83
C'_3H	$C'P^4,(C'_2)_2$	$[Ph_2HP{=}C(CO_2Et)PPh_3]^+Cl^-$	0.5 J42, −2.4 J39[I] (r, −90°)	82
C_3C'	$H_2Ge,(H_3)_2;Ge_2$	$(Me_3GeCH_2)Me_2P{=}C(GeMe_3)_2$	5.1(b, 35°)	83a
	$H_2P, (H_3)_2;P_2$	$Me_2PCH_2Me_2P{=}C(PMe_2)_2$	13.1(b) J139 + 47	74
	$(H_3)_3;Ge_2$	$Me_3P{=}C(GeMe_3)_2$	−1.0(b, 35°)	83a
	$(H_3)_3;P^4Si$	$[Me_3P{=}C(Tms)PMe_2PMe_2]^+Cl^-$	15.0(c) J12	74
	$(H_3)_3;PSi$	$Me_3P{=}CTms(PMe_2)$	5.4(b) J156	74
	$(H_3)_3;P_2$	$Me_3P{=}C(PMe_2)_2$	8.2(b) J123	74
	$(H_3)_3;Si_2$	$Me_3P{=}C(SiCl_3)_2$	13.7(r)	55
	$(H_3)_3;Si_2$	$Me_3P{=}C(SiClMe\text{-}CH_2)_2$	6.7, 7.1[D](b, 30°)	84
	$(H_3)_3;Si_2$	$Me_3P{=}C^{\cdot}SiMe_2CH_2Si^{\cdot}Me_2$	−10.3(b, 30°)	84
	$(H_3)_3;Si_2$	$Me_3P{=}C^{\cdot}SiMe_2CH{=}PMe_2Si^{\cdot}Me_2$	1.3/1.5(b, 35°) J6	85
	$(H_3)_3;Si_2$	$Me_3P{=}C[Si^{\cdot}Cl(CH_2)_2C^{\cdot}H_2]_2$	? −30(b)	86
	$(H_3)_3;Si_2$	$Me_3P{=}CTms_2$	−0.5(b, 25°)	74
	$(H_3)_3;Si_2$	$[Me_3P{=}C^{\cdot}(Si^{\cdot}H_2)]_3$	−1.5(b)	87
	$(H_3)_3;Si_2$	$[Me_3P{=}C(SiClMe_2]_2SiMe_2$	−3.2	85
	$(H_3)_3;HGe$	$Me_3P{=}CHGeMe_3$	−0.9(b, 35°)	83a
	$(H_3)_3;HP$	$Me_3P{=}CHPMe_2$	0(b) J142	74
	$(H_3)_3;HP$	$Me_3P{=}CHP(cHex)CH_2CH{=}CHMe$	1.6(b) J147	88
	$(H_3)_3;HP^4$	$[Me_3P{\cdots}CH{\cdots}PMe_3]^+Cl^-$	10.5(r, 30°)	89
	$(H_3)_3;HP^4$	$Me_3P{=}CHPMe_2{=}CHTms$	1.2/2.4 J27	89
	$(H_3)_3;HP^5$	$Me_3P{=}CHPFMe_3$	−32.8(b)	90
	$(H_3)_3;HSi$	$Me_3P{=}CHSiH_3$	4.9(t, −10°)	87
	$(H_3)_3;HSi$	$Me_3P{=}CHTms$	−1.2	91
	$(H_3)_3;H_2$	$Me_3P{=}CH_2$	−2.1	91
	$(H_3)_3;C'Cl$	$Me_3P{=}C(Cl)C(Cl){=}PMe_3$	16.5	92
	$(H_3)_3;C'H$	$Me_3P{=}CHAc$	−0.3(b)	93
	$(H_3)_3;C'H$	$Me_3P{=}CHCMe{=}CH_2$	−10.2(t)	94
	$(H_3)_3;C'H$	$Me_3P{=}CH(C_6F_5)$	−4.2(b)	95
	$(H_3)_3;C'_2$	$Me_3P{=}CAc_2$	7.2(b)	93
	$(H_3)_2,C'HNa;C'H$	$Na^+[PhCH{\cdots}Me_2P{\cdots}CHPh]^-$	−16.6(k)	96
	$(H_3)_2,CH_2;H_2$	$Me_2nPrP{=}CH_2$	4.4	94
	$(H_3)_2,CH_2;C'H$	$EtMe_2P{=}CHAc$	7.4(b)	93
	$(H_3)_2,C'H_2;C'H$	$PhCH_2Me_2P{=}CHPh$	−4.5(b)	96
	$(H_3)_2,C_2H;H_2$	$cPrMe_2P{=}CH_2$	11.1(n, 30°)	66
	$(H_3)_2,C_2H;H_2$	$(cHex)Me_2P{=}CH_2$	10.2(b, 30°)	97
	$(H_3)_2,C_3;H_2$	$Me_2tBuP{=}CH_2$	20.3(b, 30°)	98
	$H_3,(C_2H)_2;H_2$	$cPr_2MeP{=}CH_2$	23.5(b)	99
	$H_3,(C_2H)_2;C'_2$	$Me(iPr)_2P{=}C(CO_2Me)_2$	36.3	54
	$H_3,C_2H,C_3;C'_2$	$Me(iPr)(tBu)P{=}C(CO_2Me)_2$	40.6	83
	$H_3,(C_3)_2;H_2$	$Me(tBu)_2P{=}CH_2$	39.7(b, 30°)	98
	$C'HNa,(C'H_2)_2;C'H$	$Na^+[PhCH\text{-}(PhCH_2)_2P\text{-}CHPh]^-$	−1.1(k)	96
	$(CH_2)_3;H_2$	$Et_3P{=}CH_2$	23.6	94
	$(CH_2)_3;H_2$	$nBu_3P{=}CH_2$	16.4(t)	94
	$(CH_2)_3;CSi$	$Et_3P{=}CMeSi^{\cdot}Me(CH_2)_2C^{\cdot}H_2$	? −10.6(b)	86
	$(CH_2)_3;CSi$	$(Et_3P{=}CMe)_2Si^{\cdot}(CH_2)_2C^{\cdot}H_2$	? −21.3(b)	86
	$(CH_2)_3;CH$	$Et_3P{=}CHMe$	16.9	94
	$(CH_2)_3;CH$	$nBu_2EtP{=}CHPr$	11.1(t)	94
	$(CH_2)_3;C'H$	$Et_3P{=}CHAc$	22.4(b)	93
	$(CH_2)_3;C'H$	$nBu_3P{=}CHAc$	16.3(b)	93
	$(CH_2)_3;CC'$	$Et_3P{=}CMeAc$	23.1(b)	93
	$(CH_2)_3;CC'$	$nBu_3P{=}C(COPh)CH_2COPh$	21.3(c/r)	22
	$(CH_2)_3;C'_2$	$Et_3P{=}C(CO_2Et)_2$	33	65

TABLE N (continued)
Compilation of ^{31}P NMR Data of Four Coordinate Phosphorus Compounds Containing a Formal Multiple Bond from Phosphorus to a Group IV Atom

Section N1: Carbon Ylides

P-bound atoms/ rings	Connectivities	Structure	NMR Data (δ_P[solv, temp] J_{PP} or J_{PF} Hz)	Ref.
	$(CH_2)_3;C'_2$	$nBu_3P{=}CAc_2$	23.4(b)	93
	$(CH_2)_2,CC'H;HP$	$(nPrPhCH)(nBu)_2P{=}CHPPh_2$	25.6(b) J132	100
	$(C'H_2)_3;C'H$	$(PhCH_2)_3P{=}CHPh$	5.5(b)	96
	$(C_2H)_3;HSi$	$iPr_3P{=}CHSi^{\cdot}Me(CH_2)_2C^{\cdot}H_2$	?16.8(b)	86
	$(C_2H)_3;H_2$	$cPr_3P{=}CH_2$	34.4(b)	99
	$(C_2H)_3;H_2$	$iPr_3P{=}CH_2$	40.5	94
	$(C_2H)_3;C'H$	$iPr_3P{=}CHCMe{=}CH_2$	30.0	94
	$(C_2H)_3;C'H$	$iPr_3P{=}CH(C_6F_5)$	32.3(b)	95
	$(C_2H)_3;C'H$	$cPr_3P{=}CHPh$	19.5(b)	99
	$(C_2H)_3;C_2$	$cPr_3P{=}C^{\cdot}CH_2C^{\cdot}H_2$	20.9(b)	101
	$(C_2H)_3;C_2$	$cPr_2iPrP{=}CMe_2$	19.2(b)	101
	$(C_2H)_3;C_2$	$iPr_3P{=}CMe_2$	26.4(b)	101
	$(C_3)_3;H_2$	$tBu_3P{=}CH_2$	57.1(b, 30°)	98
$C_3C'/5$	$H_3,(CH_2)_2;Si_2$	$(CH_2)_4MeP{=}CTms_2$	27.4(b, 25°)	91
	$H_3,(CH_2)_2;H_2$	$(CH_2)_4MeP{=}CH_2$	24.9	91
	$H_3,(CH_2)_2;C'H$	$(CH_2)_4MeP{=}CHPh$	15.3	91
	$(H_3)_2,C'H_2;C'H$	$Me_2P^{\cdot}{=}CHDopC^{\cdot}H_2$	20.1(b)	102
$C_3C'/6$	$H_2Si,(H_3)_2;Si_2$	$Me_2Si^{\cdot}CH_2Me_2P{=}C(Tms)SiMe_2C^{\cdot}{=}PMe_3$	−2.9/2.4(b) J4	85
	$(H_3)_2,CH_2;HSi$	$Me_2P^{\cdot}{=}CHSiMe_2(C^{\cdot}H_2)_3$	1.1(b)	103
	$H_3,(CH_2)_2;H_2$	$(CH_2)_5MeP{=}CH_2$	0.2(b)	66
	$(CH_2)_3;CSi$	$Et_2P^{\cdot}{=}C(Me)SiFEt(C^{\cdot}H_2)_3$	8.9(b)	103
	$(CH_2)_2,C_2H;H_2$	$(CH_2)_5cBuP{=}CH_2$	7.3(b)	104
$C_3C'/6,6$	$(H_2Si)_2,H_3;Si_2$	$MeP^{\cdot}{=}C(SiMe_2CH_2SiMe_2C^{\cdot}H_2)_2$	6.2(b, 30°)	84
	$(CH_2)_3;CH$	$(CH_2)_5P^{\cdot}{=}CH(C^{\cdot}H_2)_4$	−12.5(b)	104
	$(CH_2)_2,C_2H;H_2$	$C^{\cdot}H(CH_2CH_2CH_2CH_2)_2P^{\cdot}{=}CH_2$	11.0(b)	104
$C_3C'/7$	$(H_3)_2,C'H_2;HP^4$	$Me_2\mathbf{P}^{\cdot}{=}CHPMe_2{=}CHDopC^{\cdot}H_2$	4.2/−4.5(b, 25°) J21	105
	$(H_3)_2,C'H_2;HP^4$	$[C^{\cdot}H_2Me_2P{\cdots}CH{\cdots}PMe_2CH_2Dop^{\cdot}]^+Br^-$	23.9(c)	105
C_3C''	$(H_3)_3;P^4$	$Me_3P{=}C{=}PMe_3$	−29.6(b, 35°)	89
	$(H_3)_3;P^4$	$Me_3P{=}C{=}PPh_3$	−21.3(b, 30°) J67	106
	$(CH_2)_3;P^4$	$nBu_3P{=}C{=}PPh_3$	14.2(b) J77	100
$C_2C'_2$	$(H_3)_2;HP^4,HSi$	$Me_3P{=}CHMe_2P{=}CHTms$	1.2/2.4 J27	89
	$(H_3)_2;HSi,C'_2$	$Me_2PhP{=}CHTms$	2.9(b, 30°)	107
	$(H_3)_2;H_2,C'_2$	$(C_6F_5)Me_2P{=}CH_2$	−8.9(t)	95
	$(H_3)_2;H_2,C'_2$	$Me_2PhP{=}CH_2$	2.5(b, 30°)	107
	$(CH_2)_2;CC',C'_2$	$Et_2PhP{=}C(COPh)CH_2COPh$	21.2(c/r)	22
	$(CH_2)_2;(C'_2)_2$	$Et_2PhP{=}C(CO{\cdot}CO{\cdot}CCl{\cdot}COH{\cdot}CO)$	23(r)	108
	$(C_2H)_2;NO',C'_2$	$PhNHCO(iPr)_2P{=}C(CO_2Me)_2$	40.8	54
	$(C_2H)_2;CO',C'_2$	$(Ph_2CHCO)iPr_2P{=}C(CO_2Me)_2$	53.3	54
$C_2C'_2/6$	$(H_3)_2;C'H,C'_2$	$Me_2P{=}CH(2,2'\text{-}biPh{\cdot})$	−8.7(b, 70°)	109
	$(H_3)_2;(C'_2)_2$	3; X = Y = Me, R = R′ = Ph	−8(c)	40
$C_2C'_2/7$	$(H_3)_2;HP^4,C'H$	$Me_2P^{\cdot}{=}CHDopCH_2PMe_2{=}C^{\cdot}H$	4.2/−4.5(b, 25°) J21	105
$C_2C'C''$	$(H_3)_2;C'_2;P^4$	$Me_2PhP{=}C{=}PPh_3$	−18.3(b, 30°) J76	106
$C_2C'C''/6$	$H_3,C'H_2;C'_2;P^4$	$MesMe\mathbf{P}^{\cdot}{=}C{=}PMesMe(2,4\text{-}diMe\text{-}6\text{-}C^{\cdot}H_2Ph)$	5.5/−11.1 J75, 3.4/−13.1 J75[D]	110
CC'_3	$HKP;P_2,(C'_2)_2$	$K^+[Ph_2PCH^-Ph_2P{=}C(PPh_2)_2]$	35.3(b/k)	111
	$HSi_2;HP^4,(C'_2)_2$	$Tms_2CHPh_2\mathbf{P}{=}CHPPh_2{=}CTms_2$	19.2/19.8(t, 10°) J6	79
	$H_2P;P_2,(C'_2)_2$	$Ph_2PCH_2Ph_2P{=}C(PPh_2)_2$	24.9(t/k, 70°) J52 + 70	111
	$H_2P;\ C'H,\ (C'_2)_2$	$Ph_2PCH_2Ph_2P{=}CH(2\text{-}pyr)$	8.5(b) J64	112
	$H_2P;(C'_2)_3$	$Ph_2PCH_2Ph_2P{=}Flu$	3.9 J60	112

P-bound atoms/ rings	Connectivities	Structure	NMR Data (δ_P[solv, temp] J_{PP} or J_{PF} Hz)	Ref.
	$H_2P^4;Si_2,(C'_2)_2$	$Tms_2C{=}PPh_2CH_2Ph_2P{=}CTms_2$	18.7(b)	79
	$H_3;HP,(C'_2)_2$	$MePh_2P{=}CHPPh_2$	14.7(b) J159	113
	$H_2Si;Si_2,(C'_2)_2$	$TmsCH_2Ph_2P{=}CTms_2$	14.7(b)	79
	$H_3;Si_2,(C'_2)_2$	$MePh_2P{=}CTms_2$	9.6(b)	79
	$H_3;HP,(C'_2)_2$	$MePh_2P{=}CHPPh_2$	15.1(b) J154	114
	$H_3;HP^4,(C'_2)_2$	$MePh_2P{=}CHP(O)MePh$	15.3(b) J9	114
	$H_3;HP^4,(C'_2)_2$	$MePh_2\mathbf{P}{=}CHPPh_2{=}CHPh$	12.1(b) J31	112
	$H_3;HSi,(C'_2)_2$	$MePh_2P{=}CHTms$	10.2(b, 30°)	107
	$H_3;H_2,(C'_2)_2$	$MePh_2P{=}CH_2$	10.0(b, 30°)	107
	$H_3;CP,(C'_2)_2$	$MePh_2P{=}CMePPh_2$	18.2(b) J188	115
	$H_3;C'H,(C'_2)_2$	$MePh_2P{=}CH(2\text{-}pyr)$	4.8(b)	112
	$H_3;C'H,(C'_2)_2$	$Me(C_6F_5)_2P{=}CH(C_6F_5)$	20.2(b)	95
	$C'HNa;C'H,(C'_2)_2$	$Na^+[PhCH\cdots Ph_2P\cdots CHPh]^-$	−3.0(k)	96
	$CH_2;Si_2,(C'_2)_2$	$nPrPh_2P{=}CTms_2$	16.3(b)	79
	$CH_2;(C'_2)_3$	$nBuPh_2P{=}C(COCF_2)_2$	11.8[$(EtO)_3PO$]	116
	$CH_2;(C'_2)_3$	$(Ph_2PCH_2CH_2)Ph_2P{=}C(CH{=}CH)_2$	11.7(c) J47	117
	$C'H_2;HP,(C'_2)_2$	$(PhCH_2)Ph_2P{=}CHPPh_2$	23.0(b, 30°) J150	118
	$C'H_2;HP^4,(C'_2)_2$	$(PhCH_2)Ph_2P{=}CHP(O)Ph_2$	21.1/28.8(b, 30°) J21	118
	$C'H_2;HP^4,(C'_2)_2$	$(PhCH_2)Ph_2\mathbf{P}{=}CH\text{-}PPh_2{=}CHPh$	17.7(t, −40°) J38	118
	$C'H_2;C'H,(C'_2)_2$	$(PhCH_2)Ph_2P{=}CHPh$	6(b)	96
	$C_2H;Si_2,(C'_2)_2$	$iPrPh_2P{=}CTms_2$	26.9(b)	79
	$C_2H;HP,(C'_2)_2$	$iPrPh_2P{=}CH\text{-}PPh_2$	20.0/29.5(b) J136	115
	$C'_2H;HP^4,(C'_2)_2$	$FluPh_2\mathbf{P}{=}CHPPh_2{=}Flu$	25.3(b) J35	112
$CC'_3/5$	$H_3;HP^4(C'_2)_2$	$MePhP^*{=}CHPPh({=}CH_2)Dop$	26.7 J57, 27.0 J57[D](b)	119
	$H_3;HP^4,(C'_2)_2$	$[PhMeP^*\cdots CH\cdots PMePhDop^*]^+Br^-$	29.9, 30.7[D](c)	119
	$C'H_2;C'H,(C'_2)_2$	$Ph_2P^*{=}CDopC^*H_2$	27.6(b)	102
	$C'_2;(C'_2)_3$	**5**; R = R′ = Ph	48.5(c)	120
$CC'_3/6$	$H_3;(C'_2)_3$	**3**; X = Me, Y = R = R′ = Ph	−6.5(c)	40
	$C'H_2;HP^4,(C'_2)_2$	$Ph_2P^*{=}CHPPh_2{=}CHCOC^*H_2$	11.2/16.5(c) J39	121
	$C'H_2;C'H,(C'_2)_2$	$C^*H_2Ph_2P{=}CH(1,8\text{-}naph^*)$	−9.5(b)	102
$CC'_3/7$	$C'H_2;HP^4,(C'_2)_2$	$Ph_2P^*{=}CHPPh_2{=}CHDopC^*H_2$	4.7/10.2(b, 25°) J21	105
	$C'H_2;HP^4,(C'_2)_2$	$[C^*H_2Ph_2P\cdots CH\cdots PPh_2CH_2Dop^*]^+Br^-$	14.1(c)	105
	$C'H_2;HP^4,(C'_2)_2$	$Ph_2P^*{=}CHP^+Ph_2CH(BH_3^-)DopC^*H_2$	13.3/25.4(c, 25°) J19	105
CC'_2C''	$H_3;(C'_2)_2;P^4$	$MePh_2P{=}C{=}PMePh_2$	−6.7(b)	114
	$H_3;(C'_2)_2;P^4$	$MePh_2P{=}C{=}PPh_3$	−13.3(b, 30°) J82	106
	$H_3;(C'_2)_2;P^4$	$(Ph_2Me\mathbf{P}{=}C{=}PPh_2CH_2CH_2)_2$	−5.5/−15.6(b) J78	122
	$CH_2;(C'_2)_2;P^4$	$(CH_2CH_2Ph_2\mathbf{P}{=}C{=}PMePh_2)_2$	−5.5/−15.6(b) J78	122
	$C'H_2;(C'_2)_2;P^4$	$MesCH_2Ph_2P{=}C{=}PPh_2Me$	−11.7/−17.3(b) J86	112
	$C'H_2;(C'_2)_2;P^4$	$MesCH_2Ph_2P{=}C{=}PPh_2CH_2Mes$	−14.6(b)	112
	$C_2H;(C'_2)_2;P^4$	$iPrPh_2P{=}C{=}PPh_2Me$	5.0/−17.8(b) J88	115
$CC'_2C''/5$	$CH_2;(C'_2)_2;P^4$	$Ph_2P^*{=}C{=}PPh_2(C^*H_2)_2$	−22.5	123
$CC'_2C''/6$	$H_3;(C'_2)_2;P^4$	$MesMeP^*{=}C{=}PMesMeCH_2(3,5\text{-}diMeDop^*)$	5.5/−11.1 J75, 3.4/−13.1 J75[D]	110
	$CH_2;(C'_2)_2;P^4$	$Ph_2P^*{=}C{=}PPh_2(C^*H_2)_3$	−9.6(b)	124
$CC'_2C''/7$	$CH_2;(C'_2)_2;P^4$	$Ph_2P^*{=}C{=}PPh_2(C^*H_2)_4$	−1.7	123
C'_4	$As_2,(C'_2)_3$	$Ph_3P{=}C(AsPh_2)_2$	25.6(b)	125
	$BrP^4,(C'_2)_3$	$[Ph_3P\cdots CBr\cdots PPh_3]^+Br^-$	24.7	108
	$ClP^4,(C'_2)_3$	$[Ph_3P\cdots CCl\cdots PCl(NMe_2)_2]^+Cl^-$	20.9(c) J77	6
	$P_2,(C'_2)_3$	$Ph_3P{=}C(PMe_2)_2$	26.5(b) J108	126
	$P_2,(C_2')_3$	$Ph_3P{=}C(PPh_2)_2$	27.8(c, 70°) J68	111
	$PP^4,(C'_2)_3$	$Ph_3P{=}C(PPh_2)P(NMe_2)_3{}^+Cl^-$	22.5 J65 + 110	6
	$SSe,(C'_2)_3$	$Ph_3P{=}C(SPh)SePh$	25.4(b)	127
	$Si_2HP^4,(C'_2)_2$	$Tms_2CHPPh_2{=}CHPh_2P{=}CTms_2$	19.2/19.8(t, 10°) J6	79
	$Si_2,CO',(C'_2)_2$	$tBuCOPh_2P{=}CTms_2$	19.4(b)	79
	$Si_2,C'O',(C'_2)_2$	$PhCOPh_2P{=}CTms_2$	12.7(b)	79
	$Si_2,C'O',(C'_2)_2$	$MesCOPh_2P{=}CTms_2$	20.5(b)	79
	$HAs,(C'_2)_3$	$Ph_3P{=}CHAsPh_2$	22.0(b)	125

TABLE N (continued)
Compilation of ^{31}P NMR Data of Four Coordinate Phosphorus Compounds Containing a Formal Multiple Bond from Phosphorus to a Group IV Atom

Section N1: Carbon Ylides

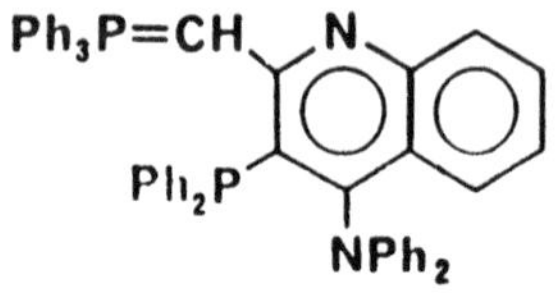

7

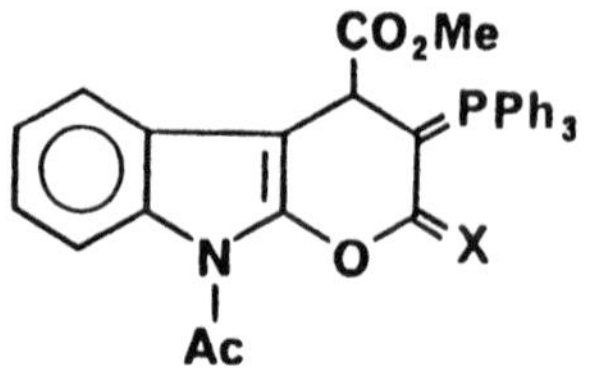

8

P-bound atoms/ rings	Connectivities	Structure	NMR Data (δ_P[solv, temp] J_{PP} or J_{PF} Hz)	Ref.
	$HP,(C'_2)_3$	$Ph_3P=CHPMe_2$	21.0(b) J126	126
	$HP,(C'_2)_3$	$Ph_3P=CH-P(tBu)CH_2CH=CHCH_3$	21.0(b) J133	88
	$HP,(C'_2)_3$	$Ph_3P=CHPPh_2$	21.5(b/t) J150	76
	$HP^4,C'H,(C'_2)_2$	$MePPh_2=CHPh_2P=CHPh$	1.8(b) J30	112
	$HP^4,C'H,(C'_2)_2$	$PhCH_2PPh_2=CHPh_2P=CHPh$	3.2(t, −40°) J38	118
	$HP^4,(C'_2)_3$	$Ph_3P=CHP(O)(NMe_2)_2$	25.0(c) J31	6
	$HP^4,(C'_2)_3$	$[Ph_3P{\cdots}CH{\cdots}P(NMe_2)_3]^+Cl^-$	16.1(c) J37	6
	$HP^4,(C'_2)_3$	$Ph_3P=CHP(O)Ph_2$	18.4(c) J19	71
	$HP^4,(C'_2)_3$	$FluPPh_2=CHPh_2P=Flu$	2.9(b) J35	112
	$H_2,(C'_2)_3$	$Ph_3P=CH_2$	19.6(k)	128
	$CB,(C'_2)_3$	$Ph_3P=CMeB(cPe)_2$	26.2(b)	129
	$CB,(C'_2)_3$	$Ph_3P=CMeB(nPe)_2$	24.2(b)	129
	$C'B,(C'_2)_3$	$Ph_3P=CPhB(nPe)_2$	19.2(b)	129
	$C'Br,(C'_2)_3$	$Ph_3P=C(Br)COPh$	19.8(c)	130
	$C''Br,(C'_2)_3$	$Ph_3P=C(Br)CN$	21.9(c)	130
	$C'Cl,(C'_2)_3$	$Ph_3P=C(Cl)CO_2Et$	22.7(c)	130
	$C'I,(C'_2)_3$	$Ph_3P=C(I)COPh$	19.5(c)	130
	$C'N',(C'_2)_3$	$Ph_3P=C^{\bullet}C(NHPh)=C(COPh)N=N^{\bullet}$	7.1(c)	131
	$C'N',(C'_2)_3$	$Ph_3P=C^{\bullet}C(NHPh)=CP(O)Ph_2N=N^{\bullet}$	18.3/10.5(c)	131
	$CP^5,(C'_2)_3$	$Ph_3P=C^{\bullet}CTf_2OP^{\bullet}Ph_3$	7.3(r)J47	132
	$CH,(C'_2)_3$	$Ph_3P=CHMe$	15.0(b, 28°)	133
	$CH,(C'_2)_3$	$Ph_3P=CHEt$	12.2(d)	134
	$CH,(C'_2)_3$	$Ph_3P=CHCH(OEt)_2$	13.7(b)	135
	$C'H,(C'_2)_3$	$Ph_3P=CHCHO$	15.0, 19.0^I(c)	136
	$C'H,(C'_2)_3$	$Ph_3P=CHAc$	14.8(c) 15.8(b)	136
	$C'H,(C'_2)_3$	$Ph_3P=CHCO_2Me$	17.8(b, 28°)	133
	$C'H,(C'_2)_3$	$Ph_3P=CHCH=CH_2$	10.7(b, 28°)	133
	$C'H,(C'_2)_3$	$Ph_3P=CHCONEt_2$	17.6	137
	$C'H,(C'_2)_3$	$Ph_3P=CHC_6F_5$	14.4(b)	95
	$C'H,(C'_2)_3$	$Ph_3P=CHPh$	7.0(b, 28°)	133
	$C'H,(C'_2)_3$	$Ph_3P=CHCOPh$	17.2(c) 17.9(b)	136
	$C'H,(C'_2)_3$	$Ph_3P=CHC(OEt)=CPhCO_2Me$	13.4(c)	138
	$C'H,(C'_2)_3$	**7**	11.4(p)	139
	$C''H,(C'_2)_3$	$Ph_3P=CHCN$	22.6(c)	130
	$C_2,(C'_2)_3$	$Ph_3P=C(CH_2)_2$	15.6(b, 30°)	140
	$C_2,(C'_2)_3$	$Ph_3P=CMe_2$	9.8(b, 28°)	133
	$C_2,(C'_2)_3$	$Ph_3P=C(CH_2)_3$	16.5(b, 28°)	133
	$C_2,(C'_2)_3$	$Ph_3P=C(CH_2)_4$	4.8(d)	134
	$C_2,(C'_2)_3$	$Ph_3P=C(CH_2)_5$	6.4(d)	134
	$CC',(C'_2)_3$	$Ph_3P=C^{\bullet}(CO{\cdot}O{\cdot}COC^{\bullet}H_2)$	13.7	29
	$CC',(C'_2)_3$	**8**; X = O	23.1(c)	141

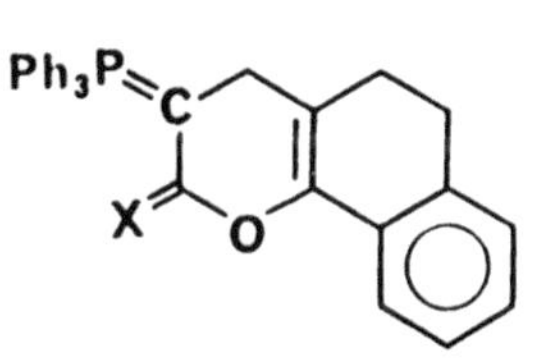

9

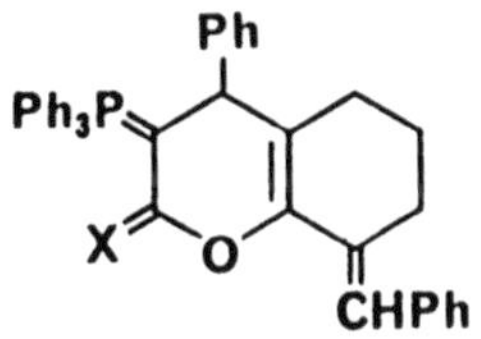

10

P-bound atoms/ rings	Connectivities	Structure	NMR Data (δ_P[solv, temp] J_{PP} or J_{PF} Hz)	Ref.
	$CC',(C'_2)_3$	$Ph_3P{=}C(COPh)CH_2COPh$	16.9(c/r)	22
	$CC',(C'_2)_3$	**9**; X = NPh	16.4(c)	141
	$CC',(C'_2)_3$	$Ph_3P{=}C^{\cdot}C(SEt)_2CPh_2C^{\cdot}(O)$	0.7(c)	142
	$CC',(C'_2)_3$	**8**; X = NPh	17.6(c)	141
	$CC',(C'_2)_3$	$Ph_3P{=}C^{\cdot}CHPhCH{=}CPhOC^{\cdot}{=}NPh$	17.0(p)	141
	$CC',(C'_2)_3$	$Ph_3P{=}C^{\cdot}CH(COPh)CH{=}CPhOC^{\cdot}{=}NPh$	15.7(p)	141
	$CC',(C'_2)_3$	**9**; X = Flu	23.9(c)	141
	$CC',(C'_2)_3$	**8**; X = Flu	8.7(c)	141
	$CC',(C'_2)_3$	**10**; X = NPh	16.1(c)	141
	$CC',(C'_2)_3$	$Ph_3P{=}C^{\cdot}CHPhCH{=}CPhOC^{\cdot}{=}Flu$	23.4(c)	141
	$CC',(C'_2)_3$	**10**; X = Flu	29.1(c)	141
	$(C'_2)_4$	$Ph_3P{=}C^{\cdot}(C{=}O)CH_2(C^{\cdot}{=}O)$	−4.4(c)	143
	$(C'_2)_4$	$Ph_3P{=}C(COCF_2)_2$	9.2[$(EtO)_3PO$]	116
	$(C'_2)_4$	$Ph_3P{=}C(CO)_2C{=}CTf_2$	1.9(b)	143
	$(C'_2)_4$	**11**; X = O, Y = Z = S, R = Me	−15.0(c)	144
	$(C'_2)_4$	$Ph_3P{=}C(CH{=}CH)_2$	12(r)	145
	$(C'_2)_4$	$Ph_3P{=}CAc_2$	16.8(c)	136
	$(C'_2)_4$	$Ph_3P{=}C(CO_2Me)C(O)(CO_2Me)$	16	146
	$(C'_2)_4$	$Ph_3P{=}C(CO_2Me)C(S)CO_2Me$	11.3(c), 9.9(b)	147
	$(C'_2)_4$	$Ph_3P{=}CAcCO_2Et$	18.1(c)	136
	$(C'_2)_4$	$Ph_3P{=}C(CO_2Me)C(CO_2Me){=}C{=}C{=}S$	15.9	148
	$(C'_2)_4$	$Ph_3P{=}C(CO)_2C{=}CH(4\text{-}NO_2Ph)$	1.2(c)	143
	$(C'_2)_4$	$Ph_3P{=}C(Ph)Ac$	15.3(r)	145
	$(C'_2)_4$	$Ph_3P{=}CAcCOPh$	18.8(c)	136
	$(C'_2)_4$	$Ph_3P{=}C(CO_2Me)(4\text{-}MeOPh)$	22.7(c)	149
	$(C'_2)_4$	**11**; X = O, Y = Z = S, R = 4-Tol	−14.9(c)	144
	$(C'_2)_4$	$Ph_3P{=}C(COPh)(CO_2Et)$	19.4(c)	136
	$(C'_2)_4$	$Ph_3P{=}CPh_2$	6.6(k)	145
	$(C'_2)_4$	$Ph_3P{=}C(Ph)COPh$	16.0(c), 15.0(b)	136
	$(C'_2)_4$	$Ph_3P{=}C(CO_2Me)C(CO_2Me){=}C{=}C{=}NPh$	16.9	148
	$(C'_2)_4$	$Ph_3P{=}C(COPh)C(CO_2Me){=}CHCO_2Me$	17.0(ae)	120
	$(C'_2)_4$	$(4\text{-}Tol)_3P{=}C^{\cdot}(CO)C(COMe)(CO_2Me)C(CO_2Me){=}C^{\cdot}CO_2Me$	15	150
	$(C'_2)_4$	**11**; X = Z = O, Y = S, R = Me	−13.1(c)	144
	$(C'_2)_4$	$Ph_3P{=}C(COPh)_2$	19.0(c)	136
	$(C'_2)_4$	**11**; X = PhN, Y = S, Z = O, R = Ph	−12.4(c)	144
	$(C'_2)_4$	**11**; X = Y = O, Z = NPh, R = Ph	15.9(c)	144
	$(C'_2)_4$	$Ph_3P{=}C(CONPh)_2C{=}O$	17.1	143
	$(C'_2)_4$	$Ph_3P{=}C^{\cdot}(C{=}O)CPh_2(C^{\cdot}{=}O)$	−3.4(c)	143
	$(C'_2)_4$	$Ph_3P{=}C(COPh)C(S)COPh$	11.0(c)	147
	$(C'_2)_4$	**11**; X = Y = S, Z = O, R = $N(cHex)_2$	10.9(c)	144
	$(C'_2)_4$	**11**; X = Y = Z = S, R = N $(cHex)_2$	11.3(c)	144
	$(C'_2)_4$	**11**; X = Y = S, Z = NPh, R = NEtPh	11.5(p)	144
	$(C'_2)_4$	**11**; X = Y = O, Z = Flu, R = Ph	17.6(c)	144
	$(C'_2)_4$	**12**; X = Z = NPh, Y = Ph	−4.1(p)	144
	$(C'_2)_4$	**12**; X = PhN, Y = $N(cHex)_2$, Z = O	5.1(p)	144
	$(C'_2)_4$	**12**; X = PhN, Y = Ph, Z = 4-MeOPhN	−5.0(p)	144

TABLE N (continued)
Compilation of ^{31}P NMR Data of Four Coordinate Phosphorus Compounds Containing a Formal Multiple Bond from Phosphorus to a Group IV Atom

Section N1: Carbon Ylides

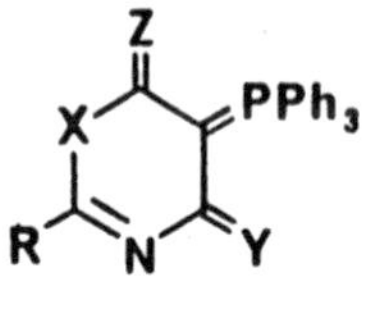

11 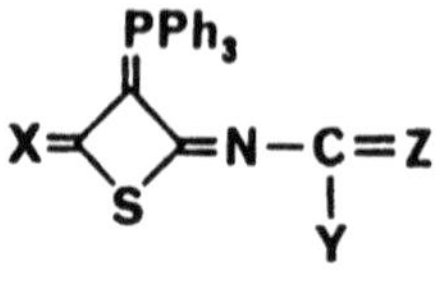 **12**

P-bound atoms/ rings	Connectivities	Structure	NMR Data (δ_P[solv, temp] J_{PP} or J_{PF} Hz)	Ref.
	$(C'_2)_4$	$Ph_3P{=}C(C{=}O)_2CMeP^+Ph_3I^-$	−1.4 J7	151
	$(C'_2)_4$	**11**; X = Y = S, Z = Flu, R = NEtPh	11.3(c)	144
	$(C'_2)_4$	**11**; X = PhN, Y = S, Z = Flu, R = PhO	10.6(p)	144
	$(C'_2)_4$	**12**; X = Flu, Y = Ph, Z = NPh	−0.2(p)	144
	$(C'_2)_4$	**12**; X = Flu, Y = $N(cHex)_2$, Z = O	−5.0(p)	144
	$(C'_2)_4$	**11**; X = Y = S, Z = Flu, R = $N(cHex)_2$	12.0(p)	144
	$(C'_2)_4$	$(Ph_3)P{=}C(CO_2Me)C(CO_2Me){=})_2$	22.9(c)	120
	$(C'_2)_4$	$Ph_3P{=}C(C{=}NPh)_2CMeP^+Ph_3I^-$	−4.4(c)	143
	$(C'_2)_3,C'C''$	$Ph_3P{=}CCN$(3,5-di-tBu-4-HOPh)	25(ae)	37
$C'_4/5$	$HP^4,H_2,(C'_2)_2$	$C^{\cdot}H{=}PPhMe(Dop)PhP^{\cdot}{=}CH_2$	3.9 J57, 5.6 $J57^D$(b)	119
	$C'H,(C'_2)_3$	$DopPh_2P{=}CHC{=}CHCN$	21.5(c)	152
	$C'H,(C'_2)_3$	$DopPh_2P{=}CHC{=}CHAc$	26.1(c)	152
	$(C'H,(C'_2)_3$	(2,2′-biPh)PhP=CH(COPh)	17	153
	$HP^4,(C'_2)_3$	$[Ph_2P^{\cdot}{\cdots}CH{\cdots}PPh_2Dop^{\cdot}]^+Br^-$	31.3(c)	119
$C'_4/6$	$HP^4,C'H,(C'_2)_2$	$Ph_2P^{\cdot}{=}CHCOCH_2PPh_2{=}C^{\cdot}H$	11.2(c)J39	121
$C'_4/7$	$HP^4,C'H,(C'_2)_2$	$Ph_2P^{\cdot}{=}CH(Dop)CH_2PPh_2{=}C^{\cdot}H$	4.7/10.2(b, 25°) J21	105
C'_3C''	$(C'_2)_3;P^4$	$Ph_3P{=}C{=}PMe_3$	−10.1(b, 30°) J67	106
	$(C'_2)_3;P^4$	$Ph_3P{=}C{=}PPhMe_2$	−7.5(b, 30°) J76	106
	$(C'_2)_3;P^4$	$Ph_3P{=}C{=}PCl(NMe_2)_2$	−4.5(b) J94	6
	$(C'_2)_3;P^4$	$Ph_3P{=}C{=}P(NMe_2)_3$	−20.7(b) J145	6
	$(C'_2)_3;P^4$	$Ph_3P{=}C{=}P(nBu)_3$	−2.7 (b) J77	100
	$(C'_2)_3P^4$	$Ph_3P{=}C{=}P(NEt_2)_3$	−23.8(b) J153	6
	$(C'_2)_3;P^4$	$Ph_3P{=}C{=}PPh_2Me$	−5.6(b,30°) J82	106
	$(C'_2)_3;P^4$	$Ph_3P{=}C{=}PPh_2(NMe_2)$	−10.5/16.7(b) J104	71
	$(C'_2)_3;P^4$	$Ph_3P{=}C{=}PPh_3$	−3.5(r)	114
	$(C'_2)_3;C'$	$Ph_3P{=}C{=}CTf_2$	4.1(b)	132
	$(C'_2)_3;C'$	$Ph_3P{=}C{=}CHOEt$	11.7(b)	135
	$(C'_2)_3;C'$	$Ph_3P{=}C{=}C(SMe)_2$	−4.9(b)	142
	$(C'_2)_3;C''$	$Ph_3P{=}C{=}C{=}O$	5.4(c)	138
	$(C'_2)_3;C''$	$Ph_3P{=}C{=}C{=}S$	−8.1(c)	138
	$(C'_2)_3;C''$	$Ph_3P{=}C{=}C{=}NMe$	6.2(p)	154
	$(C'_2)_3;C''$	$Ph_3P{=}C{=}C{=}C(CH{=}CH)_2$	0.2(b)	138
	$(C'_2)_3;C''$	$Ph_3P{=}C{=}C{=}NPh$	2.3(p)	154
	$(C'_2)_3;C''$	$Ph_3P{=}C{=}C{=}CPhCN$	1.9(b)	138
	$(C'_2)_3;C''$	$Ph_3P{=}C{=}C{=}CPhCO_2Me$	1.5(b)	138
	$(C'_2)_3;C''$	$Ph_3P{=}C{=}C{=}Flu$	1.6(p)	155
$C'_3C''/5$	$(C'_2)_3;P^4$	$Ph_2P^{\cdot}{=}C{=}PPh_2Dop^{\cdot}$	29.5(k)	119
	$(C'_2)_3;C'$	$Ph_2P^{\cdot}{=}C{=}C(OEt)Dop^{\cdot}$	38.5(b)	152

TABLE N
Four Coordinate Compounds Containing a Formal Multiple Bond to a Group IV Element

Section N2: Germanium Ylides

P-bound atoms (rings)	Connectivities	Structure	NMR data (δ_P[solv, temp] J_{PP} or J_{PF})	Ref.
C_2BrGe'	$(C_3)_2;Br_2$	$tBu_2P(Br){=}GeBr_2$	118(t/b)	156
C_2ClGe'	$(C_3)_2;Cl_2$	$tBu_2ClP{=}GeCl_2$	127.1(d)	157
C_2GeGe'	$(C_3)_2;Br_3;Br_2$	$tBu_2(Br_3Ge)P{=}GeBr_2$	92.7(t/b)	156
	$(C_3)_2;Cl_3;Cl_2$	$tBu_2(Cl_3Ge)P{=}GeCl_2$	84.4(t/b)	156
C_3Ge'	$(H_3)_3;F_2$	$Me_3P{=}GeF_2$	65	158
	$(H_3)_3;C'Cl$	$Me_3P{=}Ge(Ph)Cl$	55	158
	$(CH_2)_3;Cl_2$	$Et_3P{=}GeCl_2$	0.9	159
	$(CH_2)_3;Cl_2$	$nBu_3P{=}GeCl_2$	10.6(d)	159
	$(C_3)_3;Cl_2$	$tBu_3P{=}GeCl_2$	35.6	159
CC'_2Ge'	$CH_2;(C'_2)_2;Cl_2$	$[-CH_2Ph_2P{=}GeCl_2]_2$	−7(t)	160
C'_3Ge'	$(C'_2)_3;Br_2$	$Ph_3P{=}GeBr_2$	−8.2	159
	$(C'_2)_3;Cl_2$	$Ph_3P{=}GeCl_2$	−3.9	159

REFERENCES

1. **Prishchenko, A. A., Gromov, A. V., Luzikov, Yu. N., Lazhko, E. I., and Lutsenko, I. F.,** *Zh. Obshch. Khim.*, 55, 1194, 1985.
2. **Appel, R., Huppertz, M., and Westerhaus, A.,** *Chem. Ber.*, 116, 114, 1983.
3. **Baceiredo, A., Bertrand, G., and Sicard, G.,** *J. Am. Chem. Soc.*, 107, 4781, 1985.
4. **Appel, R., Peters, J., and Schmitz, R.,** *Z. Anorg. Allg. Chem.*, 475, 18, 1981.
5. **Kolodyazhnyi, O. I.,** *Tetrahedron Lett.*, 26, 439, 1985.
6. **Appel, R., Baumeister, U., and Knoch, F.,** *Chem. Ber.*, 116, 2275, 1983.
7. **Svara, J. and Fluck, E.,** *Phosphorus Sulfur*, 25, 129, 1985.
8. **Heckmann, G., Neumuller, B., and Fluck, E.,** *Z. Naturforsch.*, 42B, 260, 1987.
9. **Tupchienko, S., Dudchenko, T. N., and Sinitsa, A. D.,** *Zh. Obshch. Khim.*, 55, 776, 1985.
10. **Kolodyazhnyi, O. I., Yakovlev, V. N., and Kukhar, V. P.,** *Zh. Obshch. Khim.*, 49, 2458, 1979.
11. **Kolodyazhnyi, O. I. and Kukhar, V. P.,** *Zh. Obshch. Khim.*, 49, 1992, 1979.
12. **Kolodyazhnyi, O. I., Repina, L. A., Kukhar, V. P., and Gololobov, Yu. G.,** *Zh. Obshch. Khim.*, 49, 1004, 1979.
13. **Kolodyazhnyi, O. I.,** *Zh. Obshch. Khim.*, 49, 104, 1979.
14. **Bellan, J., Sanchez, M., Marre-Mazieres, M. R., and Murillo Beltran, A.,** *Bull. Soc. Chim. Fr.*, p. 491, 1985.
15. **Burgada, R., El Khoshnieh, Y. O., and Leroux, Y.,** *Tetrahedron*, 41, 1207, 1985.
16. **Baccolini, G., Todesco, P. E., and Bartoli, G.,** *Phosphorus Sulfur*, 9, 203, 1980.
17. **Voznesenskaya, A. Kh. and Razumova, N. A.,** *Zh. Obshch. Khim.*, 45, 1235, 1975.
18. **Kim, T. V., Kiseleva, E. I., Struchkov, Yu. T., Antipin, M. Yu., Chernega, A. N., and Gololobov, Yu. G.,** *Zh. Obshch. Khim.*, 54, 2486, 1984.
19. **Arbuzov, B. A., Dianova, E. N., and Zabotina, E. Yu.,** *Izv. Akad. Nauk SSSR Ser. Khim.*, p. 2635, 1981.
20. **Issleib, K., Lischewski, M., and Zschunke, A.,** *Org. Magn. Reson.*, 5, 401, 1973.
21. **Schmidbaur, H. and Pichl, R.,** *Z. Naturforsch.*, 40B, 352, 1985.
22. **Ramirez, F., Madan, O. P., and Smith, C. P.,** *Tetrahedron*, 22, 567, 1966.
23. **Bellan, J., Marre, M. R., Sanchez, M., and Wolf, R.,** *Phosphorus Sulfur*, 12, 11, 1981.
24. **Niecke, E., Seyer, A., and Wildbredt, D. A.,** *Angew. Chem. Int. Ed. Engl.*, 20, 675, 1981.
25. **Diallo, O., Boisdon, M. T., Lopez, L., Malavaud, C., and Barrans, J.,** *Tetrahedron Lett.*, 27, 2971, 1986.

26. **Arbuzov, B. A., Dianova, E. N., and Sharipova, S. M.,** *Izv. Akad. Nauk SSSR Ser. Khim.*, p. 1600, 1981.
27. **Ovchinnikov, V. V., Galkin, V. I., Cherasov, R. A., and Pudovik, A. N.,** *Zh. Obshch. Khim.*, 49, 1693, 1979.
28. **Kolodyazhnyi, O. I. and Kukhar, V. P.,** *Zh. Obshch. Khim.*, 49, 949, 1979.
29. **Mark, V., Dungan, C. H., Crutchfield, M. M., and Van Wazer, J. R.,** Compilation of P^{31} NMR data, in *Topics in Phosphorus Chemistry. P^{31} Nuclear Magnetic Resonance*, Vol. 5, Grayson, M. and Griffith, E. J., Eds., Interscience, New York, 1967.
30. **Griffiths, D. V. and Matthews, B. J.,** unpublished results.
31. **Griffiths, D. V. and Tebby, J. C.,** *J. Chem. Soc. Chem. Commun.*, p. 871, 1986.
32. **Raasch, M. G.,** *J. Org. Chem.*, 43, 2500, 1978.
33. **Denney, D. B. and Denney, D. Z.,** *Phosphorus Sulfur*, 13, 315, 1982.
34. **Griffiths, D. V. and Caesar, J. C.,** unpublished results.
35. **Caesar, J. C., Griffiths, D. V., Griffiths, P. A., and Tebby, J. C.,** *Phosphorus Sulfur*, 34, 155, 1987.
36. **Caesar, J. C., Griffiths, D. V., Tebby, J. C., and Willets, S. E.,** *J. Chem. Soc. Perkin Trans. 1* p. 1627, 1984.
37. **Kolesnikov, V. T., Kopel'tsiv, Yu. A., Kudryavtsev, A. A., and Shermolovich, Yu. G.,** *Zh. Obshch. Khim.*, 53, 1265, 1983.
38. **Burgada, R., El Khoshnieh, Y. O., and Leroux, Y.,** *Tetrahedron*, 41, 1223, 1985.
39. **Rymareva, T. G., Khaskin, B. A., Sandakov, V. B., and Plomonenka, V. K.,** *Zh. Obshch. Khim.*, 51, 2633, 1981.
40. **Dimroth, K., Berger, S., and Kaletsch, H.,** *Phosphorus Sulfur*, 10, 305, 1981.
41. **Kolodyazhnyi, O. I.,** *Zh. Obshch. Khim.*, 51, 2466, 1981.
42. **Dimroth, K., Hettche, A., Kanter, H., and Stade, W.,** *Tetrahedron Lett.*, p. 835, 1972.
43. **Hettche, A. and Dimroth, K.,** *Tetrahedron Lett.*, p. 829, 1972.
44. **Dimroth, K., Hettche, A., Stade, W., and Steuber, F. W.,** *Angew. Chem. Int. Ed. Engl.*, 8, 770, 1969.
45. **Bergerhoff, G., Hammes, O., Falbe, J., Tihanyi, B., and Weber, J.,** *Tetrahedron*, 27, 3593, 1971.
46. **Malenko, D. M. and Gololobov, Yu. G.,** *Zh. Obshch. Khim.*, 51, 1432, 1981.
47. **Malenko, D. M., Kasukhin, L. F., Ponomarchuk, M. P., and Gololobov, Yu. G.,** *Zh. Obshch. Khim.*, 50, 1946, 1980.
48. **Caesar, J. C., Griffiths, D. V., and Tebby, J. C.,** *Phosphorus Sulfur*, 29, 123, 1987.
49. **Malenko, D. M. and Gololobov, Yu. G.,** *Zh. Obshch. Khim.*, 48, 2793, 1978.
50. **Tebby, J. C., Willets, S. E., and Griffiths, D. V.,** *J. Chem. Soc. Chem. Commun.*, p. 420, 1981.
51. **Malenko, D. M. and Gololobov, Yu. G.,** *Zh. Obshch. Khim.*, 49, 308, 1979.
52. **Panevin, A. S., Trishin, Yu. G., and Chistokletov, V. N.,** *Zh. Obshch. Khim.*, 54, 1438, 1984.
53. **Fritz, G. and Schick, W.,** *Z. Anorg. Allg. Chem.*, 511, 108, 1984.
54. **Kolodyazhnyi, O. I.,** *Zh. Obshch. Khim.*, 50, 1485, 1980.
55. **Fritz, G., Braun, U., Schick, W., Hoenle, W., and Von Schnering, H. G.,** *Z. Anorg. Allg. Chem.*, 472, 45, 1981.
56. **Fritz, G. and Schick, W.,** *Z. Anorg. Allg. Chem.*, 511, 132, 1984.
57. **Fritz, G., Schick, W., Hoenle, W., and Von Schnering, H. G.,** *Z. Anorg. Allg. Chem.*, 511, 95, 1984.
58. **Kolodyazhnyi, O. I.,** *Tetrahedron Lett.*, 26, 439, 1985.
59. **Kolodyazhnyi, O. I.,** *Zh. Obshch. Khim.*, 52, 1538, 1982.
60. **Appel, R. and Schoeler, H. F.,** *Chem. Ber.*, 112, 1068, 1979.
61. **Appel, R., Geisler, K., and Schoeler, H. F.,** *Chem. Ber.*, 112, 648, 1979.
62. **Appel, R. and Waid, K.,** *Angew. Chem. Int. Ed. Engl.*, 18, 169, 1979.
63. **Schaffer, O. and Dimroth, K.,** *Angew. Chem. Int. Ed. Engl.*, 11, 1091, 1972.
64. **van der Knapp, Th. A. and Bickelhaupt, F.,** *Tetrahedron Lett.*, 23, 2037, 1982.
65. **Repina, L. A., Loktionova, R. A., and Gololobov, Yu. G.,** *Zh. Obshch. Khim.*, 46, 2683, 1976.
66. **Schmidbaur, H. and Scherm, H. P.,** *Chem. Ber.*, 110, 1576, 1977.
67. **Zeiss, W. and Henjes, H.,** *Chem. Ber.*, 111, 1655, 1978.
68. **Arbuzov, B. A., Dianova, E. N., and Sharipova, S. M.,** *Izv. Akad. Nauk SSSR Ser Khim.*, p. 1113, 1981.
69. **Appel, R. and Waid, K.,** *Z. Naturforsch.*, 36B, 127, 1981.
70. **Appel, R. and Waid, K.,** *Z. Naturforsch.*, 36B, 131, 1981.
71. **Appel, R. and Baumeister, U.,** *Z. Naturforsch.*, 35B, 513, 1980.
72. **Dahl, O.,** *J. Chem. Soc. Perkin Trans. I*, p. 947, 1978.
73. **Markl, G., Liebl, R., and Huttner, A.,** *Angew. Chem. Int. Ed. Engl.*, 17, 528, 1978.
74. **Karsch, H. H.,** *Chem. Ber.*, 115, 1956, 1982.
75. **Fritz, G. and Bauer, H.,** *Angew. Chem. Int. Ed. Engl.*, 22, 730, 1983.
76. **Karsch, H. H.,** *Z. Naturforsch.*, 34B, 1171, 1979.
77. **Fritz, G. and Schick, W.,** *Z. Anorg. Allg. Chem.*, 518, 14, 1984.

78. Appel, R., Wander, M., and Knoll, F., *Chem. Ber.*, 112, 1093, 1979.
79. Appel, R., Haubrich, G., and Knoch, F., *Chem. Ber.*, 117, 2063, 1984.
80. Appel, R. and Haubrich, G., *Angew. Chem. Int. Ed. Engl.*, 19, 213, 1980.
81. Hammond, P. J., Lloyd, J. R., and Hall, C. D., *Phosphorus Sulfur*, 10, 67, 1981.
82. Maskyukova, T. A., Aladzheva, I. M., Leont'eva, I. V., Petrovski, P. V., Fedin, E. I., and Kabachnik, M. I., *Tetrahedron Lett.*, 21, 2931, 1980.
83. Kolodiazhnyi, O. I., *Tetrahedron Lett.*, 21, 2269, 1980.
83a. Schmidbaur, H., Eberlein, J., and Richter, W., *Chem. Ber.*, 110, 677, 1977.
84. Schmidbaur, H. and Heimann, M., *Chem. Ber.*, 111, 2696, 1978.
85. Malisch, W. and Schmidbaur, H., *Angew. Chem. Int. Ed. Engl.*, 13, 540, 1974.
86. Schmidbaur, H. and Wolf, W., *Chem. Ber.*, 108, 2851, 1975.
87. Schmidbaur, H. and Zimmer-Gasser, B., *Z. Naturforsch.*, 32B, 603, 1977.
88. Benn, R., Mynott, R., Richter, W. J., and Schroth, G., *Tetrahedron*, 40, 3273, 1984.
89. Schmdibaur, H., Gasser, O., and Hussain, M. S., *Chem. Ber.*, 110, 3501, 1977.
90. Gasser, O. and Schmidbaur, H., *J. Am. Chem. Soc.*, 97, 6281, 1975.
91. Schmidbaur, H. and Scherm, H. P., *Z. Anorg. Allg. Chem.*, 459, 170, 1979.
92. Fluck, E. and Kazenwadel, W., *Phosphorus*, 6, 195, 1976.
93. Malisch, W., Blau, H., and Haaf, F. J., *Chem. Ber.*, 114, 2956, 1981.
94. Koster, R., Simic, D., and Grassberger, M. A., *Liebigs Ann. Chem.*, 739, 211, 1970.
95. Schmidbaur, H. and Zybill, C. E., *Chem. Ber.*, 114, 3589, 1981.
96. Schmidbaur, H., Deschler, U., Milewski-Mahrla, B., and Zimmer-Gasser, B., *Chem. Ber.*, 114, 608, 1981.
97. Schmidbaur, H. and Scherm, H. P., *Z. Naturforsch.*, 34B, 1347, 1979.
98. Schmidbaur, H., Blaschke, G., and Kohler, F. H., *Z. Naturforsch.*, 32B, 757, 1977.
99. Schmidbaur, H. and Schier, A., *Chem. Ber.*, 114, 3385, 1981.
100. Appel, R. and Erbelding, G., *Tetrahedron Lett.*, p. 2689, 1978.
101. Schier, A. and Schmidbaur, H., *Chem. Ber.*, 117, 2314, 1984.
102. Schmidbaur, H. and Mortl, A., *J. Organomet. Chem.*, 250, 171, 1983.
103. Schmidbaur, H. and Wolf, W., *Chem. Ber.*, 108, 2842, 1975.
104. Schmidbaur, H. and Mortl, A., *Z. Naturforsch.*, 35B, 990, 1980.
105. Schmidbaur, H., Costa, T., and Milewski-Mahrla, B., *Chem. Ber.*, 114, 1428, 1981.
106. Schmidbaur, H., Herr, R., and Zybill, C. E., *Chem. Ber.*, 117, 3374, 1984.
107. Schmidbaur, H. and Heimann, M., *Z. Naturforsch.*, 29B, 485, 1974.
108. Ramirez, F., Rhum, D., and Smith, C. D., *Tetrahedron*, 21, 1941, 1965.
109. Costa, T. and Schmidbaur, H., *Chem. Ber.*, 115, 1367, 1982.
110. Schmidbaur, H. and Schnatterer, S., *Chem. Ber.*, 116, 1947, 1983.
111. Schmidbaur, H., Deschler, U., and Milewski-Mahrla, B., *Chem. Ber.*, 116, 1393, 1983.
112. Schmidbaur, H. and Deschler, U., *Chem. Ber.*, 114, 2491, 1981.
113. Schmidbaur, H. and Deschler, U., *Chem. Ber.*, 116, 1386, 1983.
114. Hussain, M. S. and Schmidbaur, H., *Z. Naturforsch.*, 31B, 721, 1976.
115. Schmidbaur, H. and Wohlleben-Hammer, A., *Chem. Ber.*, 112, 510, 1979.
116. Stockel, R. F., Megson, F., and Beachem, M. T., *J. Org. Chem.*, 33, 4395, 1968.
117. Holy, N., Deschler, U., and Schmidbaur, H., *Chem. Ber.*, 115, 1379, 1982.
118. Schmidbaur, H., Deschler, U., Zimmer-Glasser, B., Neugebauer, D., and Schubert, U., *Chem. Ber.*, 113, 902, 1980.
119. Bowmaker, R., Herr, R., and Schmidbaur, H., *Chem. Ber.*, 116, 3567, 1983.
120. Waite, N. E., Tebby, J. C., Ward, R. S., and Williams, D. H., *J. Chem. Soc. (C)*, p. 1100, 1969.
121. Mastryukova, T. A., Suerbaev, Kh. A., Petrovskii, P. V., Fedin, E. I., and Kabachnik, M. I., *Zh. Obshch. Khim.*, 44, 2398, 1974.
122. Schmidbaur, H. and Costa, T., *Z. Naturforsch.*, 37B, 677, 1982.
123. Schmidbaur, H., Costa, T., Milewski-Mahrla, R., and Schubert, U., *Angew. Chem. Int. Ed. Engl.*, 19, 555, 1980.
124. Schmidbaur, H. and Costa, T., *Chem. Ber.*, 114, 3063, 1981.
125. Schmidbaur, H., NuBstein, P., and Muller, G., *Z. Naturforsch.*, 39B, 1456, 1984.
126. Weber, L. and Wewers, D., *Chem. Ber.*, 118, 541, 1985.
127. Schmidbaur, H., Zybill, C., Kruger, C., and Kraus, H. J., *Chem. Ber.*, 116, 1955, 1983.
128. Vedejs, E., Meier, G. P., and Snoble, K. A. J., *J. Am. Chem. Soc.*, 103, 2823, 1981.
129. Bestmann, H. J. and Arenz, T., *Angew. Chem. Int. Ed. Engl.*, 23, 381, 1984.
130. Speziale, A. J. and Ratts, K. W., *J. Am. Chem. Soc.*, 85, 2790, 1963.
131. Bestmann, H. J. and Schmid, G., *Tetrahedron Lett.*, 22, 1679, 1981.
132. Birum, G. H. and Matthews, C. N., *J. Org. Chem.*, 32, 3554, 1967.
133. Albright, T. A., Gordon, M. D., Freeman, W. J., and Schweizer, E. E., *J. Am. Chem. Soc.*, 98, 6249, 1976.

134. **Grim, S. O., Mcfarlane, W., and Marks, T. J.,** *Chem. Commun.*, p. 1191, 1967.
135. **Bestmann, H. J., Roth, K., and Ettlinger, M. E.,** *Chem. Ber.*, 115, 161, 1982.
136. **Brittain, J. M. and Jones, A. R.,** *Tetrahedron*, 35, 1139, 1979.
137. **Bestmann, H. J. and Kisielowski, L.,** *Chem. Ber.*, 116, 1320, 1983.
138. **Bestmann, H. J. and Sandmeier, D.,** *Chem. Ber.*, 113, 274, 1980.
139. **Bestmann, H. J., Schmid, G., Bohme, R., Wilhelm, E., and Burzlaff, H.,** *Chem. Ber.*, 113, 3937, 1980.
140. **Schmidbaur, H., Schier, A., Milewski-Mahrla, B., and Schubert, U.,** *Chem. Ber.*, 115, 722, 1982.
141. **Bestmann, H. J. and Schmid, G.,** *Tetrahedron Lett.*, 25, 1441, 1984.
142. **Bestmann, H. J. and Roth, K.,** *Tetrahedron Lett.*, 22, 1681, 1981.
143. **Birum, G. H. and Matthews, C. N.,** *J. Am. Chem. Soc.*, 90, 3842, 1968.
144. **Bestmann, H. J., Schmid, G., Sandmeier, D., and Geismann, C.,** *Tetrahedron Lett.*, 21, 2401, 1980.
145. **Caminade, A. M., El Khatib, F., and Koenig, M.,** *Phosphorus Sulfur*, 14, 381, 1983.
146. **Ketari, R. and Foucaud, A.,** *Tetrahedron Lett.*, p. 2563, 1978.
147. **Tebby, J. C., Wilson, I. F., and Griffiths, D. V.,** *J. Chem. Soc. Perkin Trans. 1*, p. 2133, 1979.
148. **Bestmann, H. J., Schmid, G., and Sandmeier, D.,** *Ang. Chem. Int. Ed. Engl.*, 14, 53, 1975.
149. **Hall, C. D.,** personal communication.
150. **Waite, N. E., Tebby, J. C., Ward, R. S., Shaw, M. A., and Williams, D. H.,** *J. Chem. Soc. C*, p. 1620, 1971.
151. **Bestmann, H. J., Besold, R., and Sandmeier, D.,** *Tetrahedron Lett.*, p. 2293, 1975.
152. **Bestmann, H. J., Roth, K., and Saalfrank, R. W.,** *Angew. Chem. Int. Ed. Engl.*, 16, 877, 1977.
153. **Wilson, I. F. and Tebby, J. C.,** *J. Chem. Soc. Perkin Trans. 1*, p. 2713, 1972.
154. **Bestmann, H. J. and Schmid, G.,** *Chem. Ber.*, 113, 3369, 1980.
155. **Bestmann, H. J., Schmid, G., Sandmeier, D., Schade, G., and Oeschner, H.,** *Chem. Ber.*, 118, 1709, 1985.
156. **Du Mont, W.-W. and Gero, R.,** *Inorg. Chim. Acta*, 35, L341, 1979.
157. **Du Mont, W.-W. and Schumann, H.,** *J. Organomet. Chem.*, 128, 99, 1977.
158. **Escudie, J., Couret, C., Riviere, P., and Satge, J.,** *J. Organomet. Chem.*, 124, C45, 1977.
159. **Du Mont, W.-W. and Rudolph, G.,** *Chem. Ber.*, 109, 3419, 1976.
160. **Bruncks, N., Du Mont, W.-W., Pickardt, J., and Rudolph, G.,** *Chem. Ber.*, 114, 3572, 1981.

Chapter 16

^{31}P NMR DATA OF FOUR COORDINATE PHOSPHORUS COMPOUNDS CONTAINING A FORMAL NEGATIVELY CHARGED ATOM α TO PHOSPHORUS

Compiled and presented by
Jacqueline Seyden-Penne, Marie-Paule Simonnin, and Tekla Strzalko

The NMR data are presented in Table O with no subdivisions. Carbanions stabilized by phosphoryl groups form the major part of the table.

All the results reported are dealing with anionic species associated to alkali metal cations: Li, Na, K, K complexed by [2,2,2]cryptand, quoted K(2,2,2), and ammonium ions.

As a general trend, when the ^{31}P chemical shift of the neutral precursor is available, deprotonation of a nitrogen α to P (CNN$'_2$, CNN'S', C'NN'S') induces a highfield shift (−0.5 to −26 ppm) so that the most shielded phosphorus (5.9) is observed in $Me[(Tms)_2N]P(=NTms)\overline{N}Tms,Li^+$; in contrast, deprotonation of a carbon α to P induces a lowfield shift, which is usually greater for phosphoryl derivatives (+5 to +30 ppm) than for thiophosphoryl ones, for which a small upfield shift can be observed (+12 to −4 ppm); the shift may depend upon the associated cation and the solvent.

For a given carbon substituent, the ^{31}P nuclei of carbanionic species formed from phosphonates (C'O_2O') and phosphonamides (C'N_2O') resonate at lower field than those formed from phosphine oxides (C$'_3$O'), while the most deshielded ^{31}P are observed for thiophosphonate carbanion (C'O_2S').

The most extensive series examined are in the phosphonate case (C'O_2O'): the ^{31}P chemical shift decreases, according to the substituent on the anionic carbon, in the following order: $\overline{C}H_2$ > $\overline{C}HMe$ > $\overline{C}HSiMe_3$ > $\overline{C}Me_2$ > $\overline{C}(SiMe_3)_2$ > $\overline{C}HCl$ > $\overline{C}HPh$ > $\overline{C}ClSiMe_3$ > $\overline{C}HCN$ ≥ $\overline{C}HP(O)(OEt)_2$ > $\overline{C}HCO_2Me$ > $\overline{C}HCOMe$ > $\overline{C}ClCO_2Et$ > $\overline{C}ClCOEt$ > $\overline{C}ClCHO$, in which the ^{31}P shift is the smallest (at highest field) in this series. Although less examples are available, similar trends are apparent for phosphonamides and phosphine oxides.

Therefore the presence on the anionic carbon either of an electronegative substituent, or of a group able to delocalize the negative charge, induces a highfield shift of the ^{31}P nucleus.

In the case of carbanions formed from phosphonates (C'O_2O'), phosphonamides (C'N_2O'), and phosphine oxides (C$'_3$O'), substituted by groups able to delocalize the negative charge, a more detailed study by ^{13}C and ^{1}H NMR has shown that the anionic carbon is planar.[9,13,14,16,17,19] Similarly, an X-ray diffraction study of $[Ph_2P(S)]_3C^-$, Bu_4N^+ has shown that the anion is trigonal planar at the central carbon with an average PCP bond angle of 119°9.[L14b]

When the anionic carbon α to P bears a carbonyl substituent, the organic framework of the system can adopt either a Z or an E geometry around the partial double bond C_1⋯C_2, according to the nature of the associated cation and of the solvent:

Z

E   Y = Me,OMe,NMe_2

In associating medium, Z isomers can thus chelate the associated cation. Similarly, diphosphorylated derivatives can also form chelates.

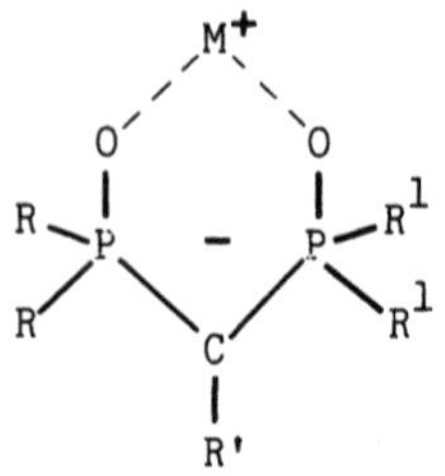

M^+

O O

R P − P R^1

R C R^1

R'

TABLE O

P-bound atoms/rings	Connectivities	Structure	NMR Data (δ_P[solv, temp] $^nJ_{PZ}$Hz)	Ref.
C′N$_2$O′	$ClP^4;(C_2)_2$	$(Me_2N)_2\mathbf{P}(O)\overline{C}ClP(O)(OEt)_2,Li^+$	40.8(k) $^2J_{PP}$100	1
	$HP^4;(C_2)_2$	$(Me_2N)_2\mathbf{P}(O)\overline{C}HP(O)(OEt)_2,Li^+$	47.5(k) $^2J_{PP}$55	1
	$HP^4;(C_2)_2$	$[(Me_2N)_2P(O)]_2\overline{C}H,Li^+$	48.3(k)	1
	$H_2;(C_2)_2$	$(Me_2N)_2P(O)\overline{C}H_2,Li^+$	63.7(k)	2
	$CP^4;(C_2)_2$	$(Me_2N)_2\mathbf{P}(O)\overline{C}MeP(O)(OEt)_2Li^+$	48.6(k) $^2J_{PP}$85	1
	$CH;(C_2)_2$	$(Me_2N)_2P(O)\overline{C}HMe,Li^+$	57.2(k)	2
	$CH;(C_2)_2$	$(Me_2N)_2P(O)\overline{C}HBu,Li^+$	54.1(k)	2
	$C''H;(C_2)_2$	$(Me_2N)_2P(O)\overline{C}HCN,K^+$	45.9(d, 31°) $^1J_{PC}$202.2	L16
CNN′S′	$H_3;Si_3;C$	$Me[(Tms)_2N]P(S)\overline{N}$-tBu,$Li^+$	41.1(j)	3
	$H_3;CSi;C$	$Me[(tBu)(Tms)N]P(S)\overline{N}$-tBu,$Li^+$	47.0(j)	3
C′NN′S′	$C'_2;Si_2;C$	$Ph[(Tms)_2N]P(S)\overline{N}$-tBu,$Li^+$	33.1(j)	3
	$C'_2;CSi;C$	$Ph[(tBu)(Tms)N]P(S)\overline{N}$-tBu,$Li^+$	42.0(j)	3
CNN′$_2$	$H_3;Si_2;(Si)_2$	$Me[(Tms)_2N]P(=NTms)\overline{N}Tms,Li^+$	5.9	3
C′O$_2$S′	$HCl;(C)_2$	$(EtO)_2P(S)\overline{C}HCl,Li^+$	105.5(k)	2
	$HP^4;(C)_2$	$(EtO)_2\mathbf{P}(S)\overline{C}HP(O)Et_2,Li^+$	92.9(k) $^2J_{PP}$83	1
	$HSi;(C)_2$	$(EtO)_2P(S)\overline{C}HSiMe_3,Li^+$	99.4(k)	4
	$H_2;(C)_2$	$(EtO)_2P(S)\overline{C}H_2,Li^+$	112.5(k)[a], 116.6(k + h, −30°)[b]	5
C′O$_2$O′	$ClP^4;(C)_2$	$[(EtO)_2P(O)]_2\overline{C}Cl$, Li^+	32.2(k)	1
	$ClP^4;(C)_2$	$(EtO)_2\mathbf{P}(O)\overline{C}ClP(O)(NMe_2)_2,Li^+$	31.8(k)	1
	$ClSi;(C)_2$	$(EtO)_2P(O)\overline{C}ClSiMe_3,Li^+$	44.2(k)	2
	$OS;(C)_2$	$(MeO)_2P(O)\overline{C}(OMe)(SMe),Li^+$	42.0(k, −70°)	6
	$S_2;(C)_2$	$(MeO)_2P(O)\overline{C}(SMe)_2,Li^+$	46.3(k, −70°)	6
	$Si_2;(C)_2$	$(MeO)_2P(O)\overline{C}(SiMe_3)_2,Li^+$	48.9(k)	7
	$Si_2;(C)_2$	$(EtO)_2P(O)\overline{C}(SiMe_3)_2,Li^+$	45.0(k)	7
	$Si_2;(C)_2$	$(iPrO)_2P(O)\overline{C}(SiMe_3)_2;Li^+$	43.0(k)	7
	$HCl;(C)_2$	$(EtO)_2P(O)\overline{C}HCl,Li^+$	45.2(k)	8
	$HCl;(C)_2$	$(iPrO)_2P(O)\overline{C}HCl,Li^+$	42.7(k)	8
	$HP^4;(C)_2$	$[(EtO)_2P(O)]_2\overline{C}H,M^+$		
		M = K	40.1(d, 31°) $^1J_{PC}$214	9
			41.0(p, 31°)	9
		M = Li	40.3(k)	1
		M = Na	38(k)	10
	$HP^4;(C)_2$	$(EtO)_2\mathbf{P}(O)\overline{C}HP(O)(NMe_2)_2,Li^+$	41.3(k)	1
	$HSi;(C)_2$	$(MeO)_2P(O)\overline{C}HSiMe_3,Li^+$	55.2(k)	4
	$HSi;(C)_2$	$(EtO)_2P(O)\overline{C}HSiMe_3,Li^+$	51.9(k)	4
	$HSi;(C)_2$	$(EtO)_2P(O)\overline{C}HSiMe_2$-tBu,$Li^+$	52.8(k)	7
	$HSi;(C)_2$	$(EtO)_2P(O)\overline{C}HSiPh_2$-tBu,$Li^+$	50.7(k)	7
	$H_2;(C)_2$	$(MeO)_2P(O)\overline{C}H_2,Li^+$	61.9(k)	5(a)
	$H_2;(C)_2$	$(EtO)_2P(O)\overline{C}H_2,Li^+$	59.1(k), 60.6(k, −70°)	5(a), 6
	$H_2;(C)_2$	$(iPrO)_2P(O)\overline{C}H_2,Li^+$	57.4(k)	5(a)
	$C'Cl;(C)_2$	$(MeO)_2P(O)\overline{C}ClCHO,Li^+$	20.1(k)	8
	$C'Cl;(C)_2$	$(EtO)_2P(O)\overline{C}ClCHO,Li^+$	16.5(k), 24.8(w)	8
	$C'Cl;(C)_2$	$(iPrO)_2P(O)\overline{C}ClCHO,Li^+$	13.7(k)	8
	$C'Cl;(C)_2$	$(EtO)_2P(O)\overline{C}ClCOEt,Li^+$	19.2(k)	8
	$C'Cl;(C)_2$	$(EtO)_2P(O)\overline{C}ClCO_2Et;Li^+$	25.8(k)	8
	$C'Cl;(C)_2$	$(EtO)_2P(O)\overline{C}ClCOCO_2Et,Li^+$	18.7(k)	8
	$C'Cl;(C)_2$	$(iPrO)_2P(O)\overline{C}ClCOMe,Li^+$	17.9(k)	8
	$C'Cl;(C)_2$	$(EtO)_2P(O)\overline{C}ClCOCH(OEt)_2,Li^+$	17.1(k)	8
	$C'Cl;(C)_2$	$(EtO)_2P(O)\overline{C}ClCOPh,Li^+$	19.0(k)	8
	$C'N;(C')_2$	$(PhO)_2P(O)\overline{C}(Ph'Y)NHPh'Z$	8/11(d) Y, Z = H, Br, NO_2(meta and para)	21
	$CP^4;(C)_2$	$[(EtO)_2P(O)]_2\overline{C}Me,Li^+$	41.0(k)	1
	$CP^4;(C)_2$	$(EtO)_2\mathbf{P}(O)\overline{C}MeP(O)(NMe_2)_2,Li^+$	40.8(k)	1

TABLE O (continued)

P-bound atoms/rings	Connectivities	Structure	NMR Data (δ_P[solv, temp] $^nJ_{PZ}$Hz)	Ref.
	$CP^4;(C)_2$	$[(EtO)_2P(O)]_2\overline{C}Et,Li^+$	40.8(k)	1
	$CP^4;(C)_2$	$[(EtO)_2P(O)]_2\overline{C}Pr,Li^+$	40.8(k)	1
	$CP^4(C)_2$	$[(EtO)_2P(O)]_2\overline{C}Bu,Li^+$	40.8(k)	1
	$C'P^4;(C)_2$	$[(EtO)_2P(O)]_2\overline{C}Ph,Li^+$	35.9(k)	1
	$CSi;(C)_2$	$(MeO)_2P(O)\overline{C}MeSiMe_3,Li^+$	55.9(k)	4
	$CSi;(C)_2$	$(EtO)_2P(O)\overline{C}MeSiMe_3,Li^+$	50.6(k)	4
	$C'Si;(C)_2$	$(EtO)_2P(O)\overline{C}PhSiMe_3,Li^+$	44.1(k)	4
	$CH;(C)_2$	$(EtO)_2P(O)\overline{C}HMe,Li^+$	53.5(k)	11
	$CH;(C)_2$	$(EtO)_2P(O)\overline{C}HEt,Li^+$	50.9(k)	11
	$CH;(C)_2$	$(iPrO)_2P(O)\overline{C}HMe,Li^+$	51.0(k)	5(a)
	$CH;(C)_2$	$(EtO)_2P(O)\overline{C}H(iPr),Li^+$	48.8(k)	11
	$CH;(C)_2$	$(EtO)_2P(O)\overline{C}HBu,Li^+$	51.2(k)	11
	$CH;(C)_2$	$(BuO)_2P(O)\overline{C}HMe,Li^+$	53.5(k)	11
	$CH;(C)_2$	$(BuO)_2P(O)\overline{C}HPr,Li^+$	50.2(k)	11
	$CH;(C)_2$	$(iBuO)_2P(O)\overline{C}HPr,Li^+$	49.6(k)	11
	$C'H;(C)_2$	$(CF_3CH_2O)_2P(O)\overline{C}HCO_2Me,Li^+$	43.6(k)	12
	$C'H;(C)_2$	$(EtO)_2P(O)\overline{C}HCOMe,M^+$	34.1(p, 31°) $^1J_{PC}$192.0	13
		M = K, chelate	33.3(d, 31°) $^1J_{PC}$194.9	
		M = K, Z	35.7(d, 31°) $^1J_{PC}$200.6	
		M = K, E	33.4(d, 31°) $^1J_{PC}$195.0	
		M = K(2,2,2),Z	36.0(d, 31°) $^1J_{PC}$200.5	
		M = K(2,2,2),E		
	$C'H;(C)_2$	$(EtO)_2P(O)\overline{C}HCO_2Me,M^+$		14
		M = K, chelate	38.7(k, 31°) $^1J_{PC}$218	
			38.9(p, 31°)	
		M =K(2,2,2),Z	37.9(k, 31°) $^1J_{PC}$219.7	
			37.4(d, 31°)	
		M = K(2,2,2),E	36.4(k, 31°) $^1J_{PC}$215.0	
			36.0(d, 31°)	
		M = Li, chelate	38.7(k, 31°)	
			38.4(d, 31°) $^1J_{PC}$219.6	
	$C'H;(C)_2$	$(EtO)_2P(O)\overline{C}HCO_2Et,Li^+$	40.7(m)	15(a)
	$C'H;(C)_2$	$(EtO)_2P(O)\overline{C}HCONMe_2,Li^+$	39.9(k)	12
	$C'H;(C)_2$	$(EtO_2P(O)\overline{C}HCONEt_2,Na^+$	45.3	15(b)
	$C'H;(C)_2$	$(EtO)_2P(O)\overline{C}HPh,Li^+$	44.7(k, 31°) $^1J_{PC}$224.8	16
	$C'H;(C)_2$	$(BuO)_2P(O)\overline{C}HCONEt_2,Na^+$	43.7	15(b)
	$C''H;(C)_2$	$(EtO)_2P(O)\overline{C}HCN,M^+$		16
		M = K	44.1(k, 31°)	
			41.9(d, 31°) $^1J_{PC}$235.5	
		M = K(2,2,2)	41.8(d, 31°) $^1J_{PC}$235.7	
		M = Li	42.8(k, 31°)	
			41.6(d, 31°) $^1J_{PC}$238.7	
	$C_2;(C)_2$	$(EtO)_2P(O)\overline{C}Me_2,Li^+$	45.6(k)	5(a)
$C'O_2O'/5$	$C'H;(C)_2$	$^{\cdot}OCHMeCHMeOP^{\cdot}(O)\overline{C}HCO_2Me,K^+$	49.4(d, 31°)	17
	$C''H;(C)_2$	$^{\cdot}OCHMeCHMeOP^{\cdot}(O)\overline{C}HCN,Na^+$	50(k)	18
$C'O_2O'/6$	$H_2;(C)_2$	$^{\cdot}OCH_2CMe_2CH_2OP^{\cdot}(O)\overline{C}H_2,Li^+$	51.0(k)	5(a)
	$C'H;(C)_2$	$^{\cdot}OCH_2CMe_2CH_2OP^{\cdot}(O)\overline{C}HCO_2Et,K^+$	33.1(d, 31°) $^1J_{PC}$228.3	17
	$C''H;(C)_2$	$^{\cdot}OCH_2CMe_2CH_2OP^{\cdot}(O)\overline{C}HCN,K^+$	37.4(d, 31°) $^1J_{PC}$250.0	17
$C_2C'O'$	$HP^4;(CH_2)_2$	$[Et_2\mathbf{P}(O)]\overline{C}H[P(S)(OEt)_2],Li^+$	39.3(k)	1
C'_3O'	$P^4{}_2,(C'_2)_2$	$[Ph_2P(O)]\overline{C}[P(S)Ph_2]_2,Li^+$	35.1(r) $^2J_{PP}$23.9	L14(b)
			34.6(c + y)	
	$P^4{}_2,(C'_2)_2$	$[Ph_2\mathbf{P}(O)]_2\overline{C}[P(S)Ph_2],M^+$		L14(b)
		M = Li	39.4(c) $^2J_{PP}$23.5	
			36.6(c + y)	
		M = NBu_4	35.6(c + y) $^2J_{PP}$23.0	

P-bound atoms/rings	Connectivities	Structure	NMR Data (δ_P[solv, temp] $^nJ_{PZ}$Hz)	Ref.
	$P^4{}_2,(C'_2)_2$	$[Ph_2P(O)]_3\overline{C},M^+$		L14(b)
		M = Li	34.3(c) $^2J_{PP}$10	
			35.3(c + y)	
		M = NBu_4	34.8(c + y)	
	$H_2,(C'_2)_2$	$Ph_2P(O)\overline{C}H_2,Li^+$	44.5(k)	2
	$CP^4,(C'_2)_2$	$[Ph_2P(O)]_2\overline{C}Me,Li^+$	43.5(k)	1
	$CH,(C'_2)_2$	$Ph_2P(O)\overline{C}HMe,Li^+$	43.3(k)	2
	$CH,(C'_2)_2$	$Ph_2P(O)\overline{C}HEt,Li^+$	39.2(k)	2
	$CH,(C'_2)_2$	$Ph_2P(O)\overline{C}H\text{-}iPr,Li^+$	35.3(k)	2
	$C'H,(C'_2)_2$	$Ph_2P(O)\overline{C}HCO_2Me,M^+$		19
		M = K, Z	25.2(d, 31°) $^1J_{PC}$143.8	
		M = Li, Z	34.8(k, 31°) $^1J_{PC}$142.7	
	$C''H,(C'_2)_2$	$Ph_2P(O)\overline{C}HCN,M^+$		19
		M = K	28.9(d, 31°) $^1J_{PC}$145.6	
			34.6(p, 31°)	
		M = K(2,2,2)	28.1(p, 31°) $^1J_{PC}$147.1	
C'_3S'	$P^4{}_2,(C'_2)_2$	$[Ph_2\mathbf{P}(S)]_2\overline{C}[P(O)Ph_2],Li^+$	43.8(r) $^2J_{PP}$23.9	L14(b)
			44.2(c + y)	
	$P^4{}_2,(C'_2)_2$	$[Ph_2\mathbf{P}(S)]\overline{C}[P(O)Ph_2]_2,M^+$		
		M = Li	39.4(c)$^2J_{PP}$23.5	
			40.7(c + y)	
		M = NBu_4	40.9(c + y) $^2J_{PP}$23.0	
	$P^4{}_2,(C'_2)_2$	$[Ph_2P(S)]_3\overline{C},M^+$		L14(b)
		M = Li	43.3(c) $^2J_{PP}$25.4	
			45.2(c + y)	
		M = NBu_4	44.0(r)	
	$HP^3,(C'_2)_2$	$[Ph_2\mathbf{P}(S)]\overline{C}H[PPh_2],Li^+$	45.6(j, −43°) $^2J_{PP}$166	20

REFERENCES

1. **Teulade, M. P., Savignac, P., Aboujaoude, E. E., Liétgé, S., and Collignon, N.,** *J. Organomet. Chem.,* 304, 283, 1986.
2. **Savignac, P. and Teulade, M. P.,** unpublished data.
3. **Romanenko, V. D., Shulgin, V. F., Skopenko, V. V., and Markovskii, L. W.,** *Zh. Obshch. Khim.,* 55, 538, 1985.
4. **Aboujaoude, E. E., Liétgé, S., Collignon, N., Teulade, M. P., and Savignac, P.,** *Synthesis,* p. 934, 1986.
5. (a) **Aboujaoude, E. E., Collignon, N., and Savignac, P.,** *J. Organomet. Chem.,* 264, 9, 1984; (b) **Mathey, F. and Savignac, P.,** *Tetrahedron,* 34, 649, 1978.
6. **Mikolajczyk, M., Grzejszczak, S., Zatorski, A., Mlotkowska, B., Gross, H., and Costisella, B.,** *Tetrahedron,* 34, 3081, 1978.
7. **Savignac, P., Teulade, M. P., and Collignon, N.,** *J. Organomet. Chem.,* 323, 135, 1987.
8. **Teulade, M. P., Savignac, P., Aboujaoude, E. E., and Collignon, N.,** *J. Organomet. Chem.,* 287, 145, 1985.
9. **Bottin-Strzalko, T., Corset, J., Froment, F., Pouet, M. J., Seyden-Penne, J., and Simonnin, M. P.,** *Phosphorus Sulfur,* 22, 217, 1985.
10. **Cotton, F. A. and Schunn, R. A.,** *J. Am. Chem. Soc.,* 85, 2394, 1963.
11. **Teulade, M. P. and Savignac, P.,** *Tetrahedron Lett.,* 28, 405, 1987.
12. **Strzalko, T., Seyden-Penne, J., Froment, F., Corset, J., and Simonnin, M. P.,** unpublished data.
13. **Bottin-Strzalko, T., Seyden-Penne, J., Pouet, M. J., and Simonnin, M. P.,** *Organ. Magn. Reson.,* 19, 69, 1982.
14. (a) **Bottin-Strzalko, T., Corset, J., Froment, F., Pouet, M. J., Seyden-Penne, J., and Simonnin, M. P.,** *J. Org. Chem.,* 45, 1270, 1980. (b) **Strzalko, T., Seyden-Penne, J., Froment, F., Corset, J., and Simonnin, M. P.,** *J. Chem. Soc. Perkin Trans. 2,* p. 783, 1987.

15. (a) **Blanchette, M. A., Choy, W., Davis, J. T., Essenfeld, A. P., Masamune, S., Roush, W. R., and Sakai, T.,** *Tetrahedron Lett.*, 25, 2183, 1984. (b) **Bowen, S. M., Duesler, E. N., Paine, R. T., and Campana, C. F.,** *Inorg. Chim. Acta*, 59, 53, 1982.
16. **Bottin-Strzalko, T., Seyden-Penne, J., Pouet, M. J., and Simonnin, M. P.,** *J. Org. Chem.*, 43, 4346, 1978.
17. **Bottin-Strzalko, T., Seyden-Penne, J., Breuer, E., Pouet, M. J., and Simonnin, M. P.,** *J. Chem. Soc. Perkin Trans. 2*, p. 1801, 1985.
18. **Zbaida, S. and Breuer, E.,** *Experientia*, 35, 851, 1979.
19. **Bottin-Strzalko, T., Etemad-Moghadam, G., Seyden-Penne, J., Pouet, M. J., and Simonnin, M. P.,** *Nouv. J. Chim.*, 7, 155, 1983.
20. **Colquhoun, I. J., McFarlane, H. C. E., and McFarlane, W.,** *Phosphorus Sulfur*, 18, 61, 1983.
21. **Smith, S., Zimmer, H., Fluck, E., and Fischer, P.,** *Phosphorus Sulfur*, 30, 327, 1987.

Chapter 17

^{31}P NMR DATA OF PHOSPHORANIDES

Compiled and presented by
Jean G. Riess and Alfred Schmidpeter

NMR data were also donated by
Keith Dillon, University of Durham, U.K.

The NMR data are presented in Table P with no subdivisions.

Four-coordinate tervalent phosphorus (with one exception) is restricted to anions, commonly called phosphoranides.[18] Two groups are known: halogeno and pseudohalogeno (primarily cyano) phosphoranides, and five-membered cyclic or spirobicyclic oxyphosphoranides. Phosphoranide anions tend to dissociate ($PX_4^- \rightleftarrows PX_3 + X^-$) and often exist in solution only within a mobile equilibrium,[4] together with the neutral halogeno or pseudohalogeno phosphine, or in the case of a cyclic oxyphosphoranide together with the ring-opened phosphinoalkoxide R_2P O^-. In such cases limiting shift is taken for the phosphoranide. Some shifts are only tentatively assigned.

Tetrahalogeno (pseudohalogeno) phosphoranides PX_4^- (X = Cl, Br, I, NCO, NCS) have a low-field phosphorus chemical shift like the parent halogeno (pseudohalogeno) phosphines PX_3. Substitution of X for C_6F_5 and even more so for CN moves $\delta^{31}P$ to higher field. As shown in Figure 1 characteristic ranges may thus be distinguished for the different phosphoranides. The addition of a halide X^- to a phosphine to give a phosphoranide is, in general, accompanied by a more or less pronounced high-field chemical shift,[4] which in general increases in the order[7] $I^- < NCS^- < Br^- < Cl^-$.

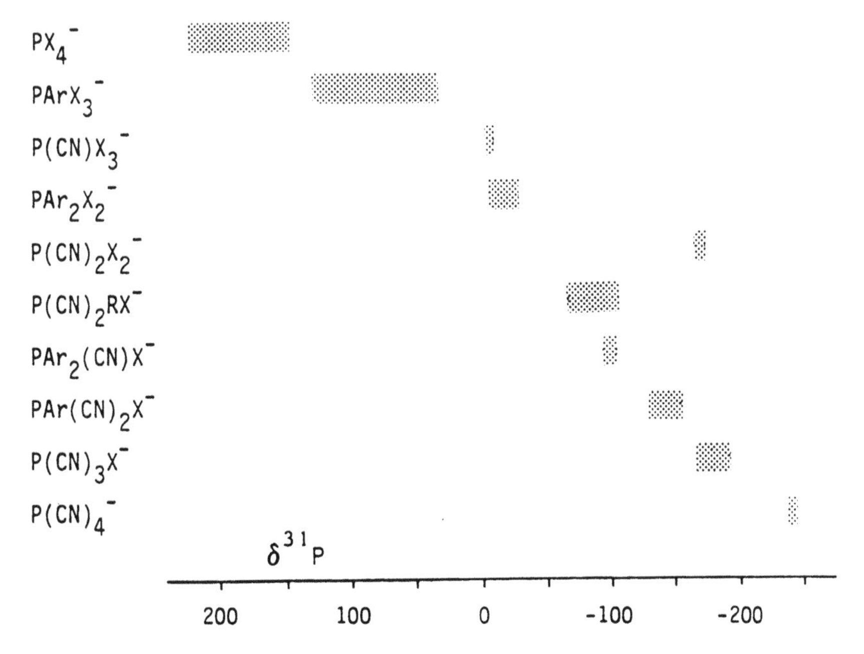

FIGURE 1. ^{31}P NMR shift ranges of tetrahalogeno (pseudohalogeno) phosphoranides (X = Cl, Br, I, NCO, NCS) and their pentafluorophenyl (Ar = C_6F_6) and cyano (R = Me, Et, Ph) derivatives.

TABLE P
Compilation of ^{31}P NMR Data of Anionic Four Coordinate Phosphorus Compounds, Phosphoranides

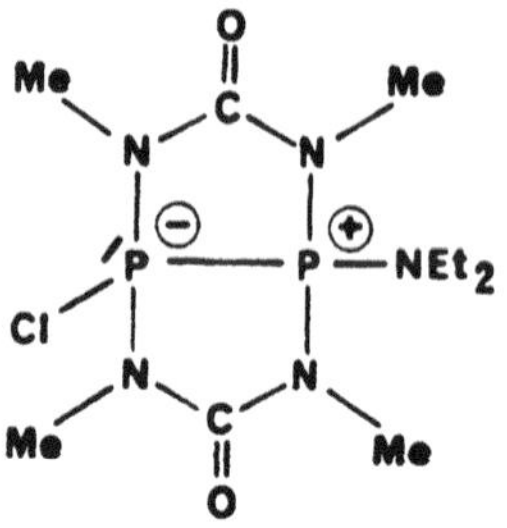

1 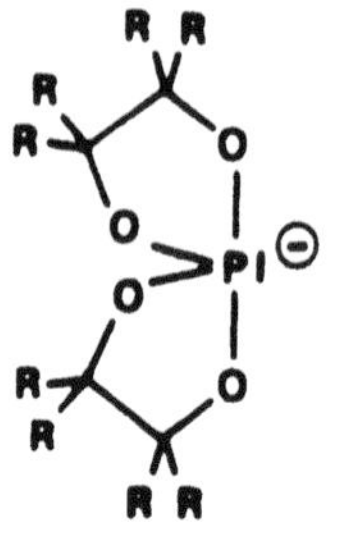 2 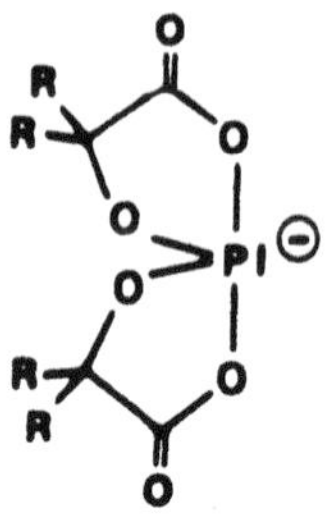 3 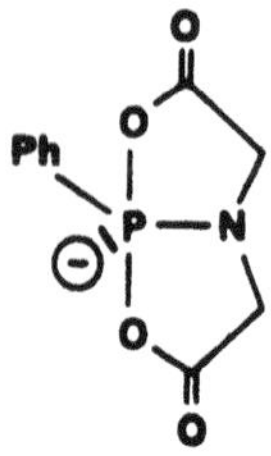 4

P-bound atoms/rings	Connectivities	Structure	NMR data (δ_P[solv, temp])	Ref.
$BrCl_3$		$PBrCl_3^-$ Pr_4N^+	211.1	1
Br_3N'	C″	$PBr_3(NCS)^-$ Et_4N^+	222.2(r)	1
Br_4		PBr_4^- Pr_4N^+	150(r), 217(solid)	2, 3
ClN_2P^4/5,5	$(CC')_2$;N_3	**1**	42.6	9
$Cl_2N'_2$	$(C'')_2$	$PCl_2(NCO)_2^-$ Et_4N^+	143.5	1
Cl_3N'	C″	$PCl_3(NCO)^-$ Et_4N^+	201.5	1
Cl_4		PCl_4 $^-Et_4N^+$	209.5(r), 201.5(solid)	4
O_4/5,5	$(C)_4$	**2**; R = Tf	−5(solid), 81.5(soln)	10
	$(C)_2$,$(C')_2$	**3**; R = Ph, Et_3HN^+	86.4(90°), 79.2(−60°)	11
$C'BrCl_2$	$(C')_2$	$(C_6F_5)PBrCl_2$ $^-Pr_4N^+$	124.2	5
$C'BrN'_2$	C'_2;$(C'')_2$	$(C_6F_5)PBr(NCS)_2$ $^-Pr_4N^+$	37.1	5
$C'Br_2I$	C'_2	$(C_6F_5)PBr_2I$ $^-Pr_4N^+$	90.0	5
$C''Br_3$	N″	$P(CN)Br_3^-Pr_4N^+$	−1.6(r)	4
	N″	$P(CN)Br_3^-Na^+$18Crown	−2.2(k)	4
$C'ClN'_2$	C'_2;$(C'')_2$	$(C_6F_5)PCl(NCS)_2^-$ Et_4N^+	32.5	5
$C'Cl_2I$	C'_2	$(C_6F_5)PCl_2I_5^-$ Pr_4N^+	131.9	5
$C'Cl_3$	C'_2	$(C_6F_5)PCl_3^-$ Pr_4N^+	100.0	5
$C''Cl_3$	N″	$P(CN)Cl_3^-$ Et_4N^+	−9.8(r)	4
$C'IN'_2$	C'_2;$(C'')_2$	$(C_6F_5)PI(NCS)_2^-$ Pr_4N^+	58.0	5
$C'NO_2$	C'_2;C_2;$(C')_2$	**4**	−46	17
$C'N'_3$	C'_2;$(C'')_3$	$(C_6F_5)P(NCS)_3^-$ Et_4N^+	51.6	5
C'_2BrCl	$(C'_2)_2$	$(C_6F_5)_2PBrCl^-$ Pr_4N^+	−6.3	5
C'_2BrI	$(C'_2)_2$	$(C_6F_5)_2PBrI^-$ Pr_4N^+	−16.2	5
C'_2Br_2	$(C'_2)_2$	$(C_6F_5)_2PBr_2^-$ Pr_4N^+	−30.7	5
C''_2Br_2	$(N'')_2$	$P(CN)_2Br_2^-$ R_4N^+	−164.4(r)	4
	$(N'')_2$	$P(CN)_2Br_2^-$ Na^+18Crown	−165.5(k)	6
C'_2ClI	$(C'_2)_2$;$(C'')_3$	$(C_6F_5)_2PClI^-$ Pr_4N^+	29.5	5
C'_2Cl_2	$(C'_2)_2$	$(C_6F_5)_2PCl_2^-$ Pr_4N^+	−25.6	5
C''_2Cl_2	$(N'')_2$	$P(CN)_2Cl_2^-$ Et_4N^+	−156.4(m)	4
C''_2I_2	$(N'')_2$	$P(CN)_2I_2^-$ Na^+18Crown	−171.9(k)	6
C'_2O_2/5,5	$(C'_2)_2$;$(C)_2$	**5** Li^+	−35	12, 13
	$(C'_2)_2$;$(C)_2$	**6**; R = H	−18.5	14
	$(C'_2)_2$;$(C)_2$	**6**; R = Me	−17.9	13, 14
CC''_2Br	H_3;$(N'')_2$	$MeP(CN)_2Br^-$ Pr_4N^+	−98.4	7
	CH_2;$(N'')_2$	$EtP(CN)_2Br^-$ Pr_4N^+	−70.9	7
$C'_2C''Br$	$(C'_2)_2$;N″	$(C_6F_5)_2P(CN)Br^-$ Pr_4N^+	−101.6	5
$C'C''_2Br$	C'_2;$(N'')_2$	$PhP(CN)_2Br^-$ Pr_4N^+	−90.4	7
	C'_2;$(N'')_2$	$(C_6F_5)P(CN)_2Br^-$ Pr_4N^+	−140.2	7
C''_3Br	$(N'')_3$	$P(CN)_3Br^-$ Pr_4N^+	−182(solid), −179.4(r)	3,4
	$(N'')_3$	$P(CN)_3Br^-$ Na^+18Crown	−182.4(k), −177(solid)	6, 8

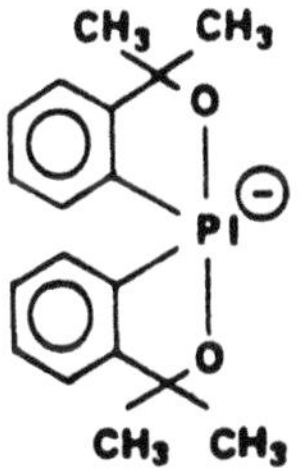

5 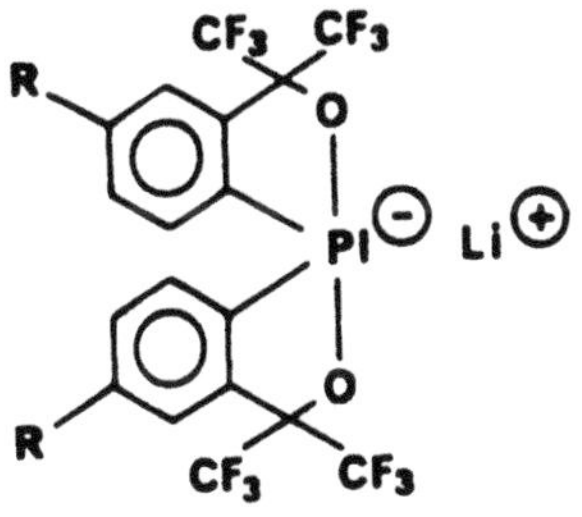 6 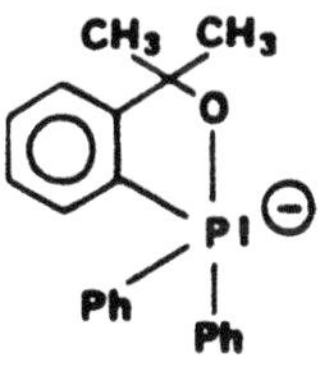 7 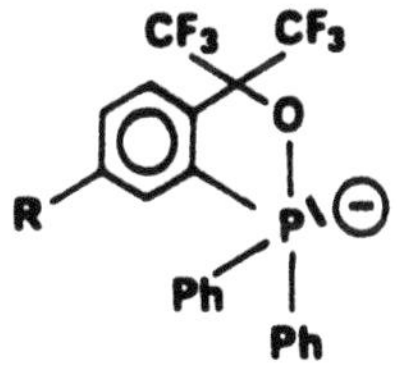 8

P-bound atoms/rings	Connectivities	Structure	NMR data (δ_P[solv, temp])	Ref.
	$(N'')_3$	$P(CN)_3Br^-$ $Ph_3P^+CHPPh_3$	−176.5(m)	4
CC''_2Cl	$H_3;(N'')_2$	$MeP(CN)_2Cl^-$ Pr_4N^+	−101.3	7
	$CH_2;(N'')_2$	$EtP(CN)_2Cl^-$ Et_4N^+	−78.0	7
$C'_2C''Cl$	$(C'_2)_2;N''$	$(C_6F_5)_2P(CN)Cl^-$ Pr_4N^+	−112.9	5
$C'C''_2Cl$	$C'_2;(N'')_2$	$PhP(CN)_2Cl^-$ Pe_4N^+	−103.3	7
	$C'_2;(N'')_2$	$(C_6F_5)P(CN)_2Cl^-$ Pr_4N^+	−154.7	7
C''_3Cl	$(N'')_3$	$P(CN)_3Cl^-$ Na^+18Crown	−177.0(k)	6
	$(N'')_3$	$P(CN)_3Cl^-$ R_4N^+	−191.8(r)	4
	$(N'')_3$	$P(CN)_3Cl^-$ $Ph_3P^+CHPPh_3$	−192.7(m)	4
CC''_2I	$H_3;(N'')_2$	$MeP(CN)_2I^-$ Pr_4N^+	−90.4	7
	$CH_2;(N'')_2$	$EtP(CN)_2I^-$ Pr_4N^+	−66.2	7
$C'_2C''I$	$(C'_2)_2;N''$	$(C_6F_5)_2P(CN)I^-$ Pr_4N^+	−95.1	5
$C'C''_2I$	$C'_2;(N'')_2$	$PhP(CN)_2I^-$ Pr_4N^+	−77.5	7
	$C'_2;(N'')_2$	$(C_6F_5)P(CN)_2I^-$ Pr_4N^+	−128.9	7
C''_3I	$(N'')_3$	$P(CN)_3I^-$ R_4N^+	−163(r)	4
	$(N'')_3$	$P(CN)_3I^-$ Na^+18Crown	−164.3(k), −168(solid)	6, 8
CC''_2N'	$H_3;(N'')_2;C'$	$MeP(CN)_2(NCS)^-$ Pr_4N^+	−92.0	7
	$CH_2;(N'')_2;C''$	$EtP(CN)_2(NCS)^-$ Et_4N^+	−66.2	7
$C'_2C''N'$	$(C'_2)_2;N'';C''$	$(C_6F_5)_2P(CN)(NCS)^-$ Pr_4N^+	−98.4	5
$C'C''_2N'$	$C'_2;(N'')_2;C''$	$PhP(CN)_2(NCS)^-Et_4N^+$	−83.8	7
	$C'_2;(N'')_2;C''$	$(C_6F_5)P(CN)_2(NCS)^-$ Et_4N^+	−135.5	7
$C'_3O/5$	$(C'_2)_3;C$	**7**	−30.3	15
	$(C'_2)_3;C$	**8**; R = H	−11.1	14, 16
	$(C'_2)_3;C$	**8**; R = tBu	−9.9	16
C''_4	$(N'')_3$	$P(CN)_4^-$ R_4N^+	−239(r)	3
	$(N'')_4$	$P(CN)_4^-$ Na^+18Crown	−242(k)	3

REFERENCES

1. **Platt, A. W. G.,** Ph.D. thesis, University of Durham, U.K., 1980.
2. **Dillon, K. B. and Waddington, T. C.,** *J. Chem. Soc. Chem. Commun.*, p. 1317, 1969.
3. **Sheldrick, W. S., Schmidpeter, A., Zwaschka, F., Dillon, K. B., Platt, A. W. G., and Waddington, T. C.,** *J. Chem. Soc. Dalton Trans.*, p. 413, 1981.
4. **Dillon, K. B., Platt, A. W. G., Schmidpeter, A., Zwaschka, F., and Sheldrick, W. S.,** *Z. Anorg. Allg. Chem.*, 7, 488, 1982.
5. **Ali, R.,** Ph.D. thesis, University of Durham, U.K., 1987.

6. **Schmidpeter, A. and Zwaschka, F.,** *Angew. Chem.*, 91, 441, 1979; *Angew. Chem. Int. Ed. Engl.*, 18, 411, 1979.
7. **Deng, R. M. K., Dillon, K. B., and Sheldrick, W. S.,** *J. Chem. Soc. Dalton Trans.*, p. 551, 1990.
8. **Sheldrick, W. S., Zwaschka, F., and Schmidpeter, A.,** *Angew. Chem.*, 91, 1000, 1979; *Angew. Chem. Int. Ed. Engl.*, 18, 935, 1979.
9. **Betterman, G., Buhl, H., Schmutzler, R., Schomburg, D., and Wermuth, U.,** *Phosphorus Sulfur*, 18, 77, 1983.
10. **Schomburg, D., Storzer, W., Bohlen, R., Kuhn, W., and Röschenthaler, G.-V.,** *Chem. Ber.*, 116, 3301, 1983.
11. **Garrigues, B., Koenig, M., and Munoz, A.,** *Tetrahedron Lett.*, p. 4205, 1979.
12. **Granoth, I. and Martin, J. C.,** *J. Am. Chem. Soc.*, 100, 7434, 1978.
13. **Granoth, I. and Martin, J. C.,** *J. Am. Chem. Soc.*, 101, 4623, 1979.
14. **Granoth, I. and Martin, J. C.,** *J. Am. Chem. Soc.*, 103, 1234, 1981.
15. **Granoth, I., Alkabets, R., Shirin, E., Margalit, Y., and Bell, P.,** *A.C.S. Symp. Ser.*, 171, 435, 1981.
16. **Ross, M. R. and Martin, J. C.,** *A.C.S. Symp. Ser.*, 171, 429, 1981.
17. **Garrigues, B. and Munoz, A.,** *C. R. Acad. Sci. Ser. B*, 293, 677, 1981.
18. **Riess, J. G.,** *Phosphorus-31 NMR Spectroscopy in Stereochemical Analysis*, Verkade, J. G. and Quin, L. D., Eds., VCH Publishers, Deerfield Beach, FL, 1987, chap. 20.

Chapter 18

^{31}P NMR DATA OF FIVE COORDINATE ($\lambda5\sigma5$) PHOSPHORUS COMPOUNDS

Compiled and presented by
Jean Francois Brazier, Lydia Lamandé, and Robert Wolf

NMR data was also donated by
E. Nieke, Universität Bielefeld, FRG
R. Schmutzler, Technische Universität Braunschweig, FRG
C. D. Hall, King's College London, U.K.
H. W. Roesky, Universität Göttingen, FRG

The NMR data of ($\lambda5\sigma5$) pentacovalent phosphoranes are presented in Table Q with no subdivisions. Polycyclic compounds are abundant. The definitions of the ring systems which incorporate phosphorus differ from other sections in that all rings involving the phosphorus are identified and those which are fused are identified by placing the letter f immediately following the ring size.

Table Q contains almost 1300 phosphorus chemical shifts which represents less than half the data available (1988). With respect to the variety of atoms directly connected to phosphorus, about 170 different environments are listed and it is hoped that few unique structures have been missed.

A frequency diagram of the chemical shifts of pentacovalent phosphorus compounds shows that the majority (~1200) occurs in the range 0 to −80 ppm. This is within a small part of the overall range for phosphorus compounds which occur mostly between 700 and −600 ppm. There is a useful relationship between chemical shift and structure for this class of compound. Specifically, replacement of a P-bound oxygen atom by sulfur or of a proton by a halogen, or a six-membered ring by a five-membered ring leads often to a systematic change in chemical shift. Used with caution, such rules are useful for chemical structure determination, particularly in solution.

Attention must be given to the fact that pentacovalent phosphorus compounds can exist in equilibrium with other phosphorus structures. The dissociation of phosphoranes to phosphonium salts, which involves rupture of a covalent bond, is particularly important. Usually only one peak is observed for compounds in solution. The existence of such rapid equilibria can be detected and studied by variation of the solvent, concentration, and temperature. A search of Table G for the existence of phosphonium salts with similar structures may also be of assistance.

Another equilibrium which can cause a problem occurs when a nucleophile is present in solution. Hexacoordinate compounds can be formed especially when phosphorus is incorporated in small rings and/or there are several electronegative P-bound atoms. Reference to the following Table R on hexacoordinated phosphorus compounds is recommended.

The least frequent equilibrium to be observed is that involving the formation of tricoordinated compounds by reductive elimination. All the examples studied so far have exhibited two phosphorus signals, one each for the five and three coordinate species. Thus this type of slowly interconverting equilibrium has not caused any problem.

TABLE Q

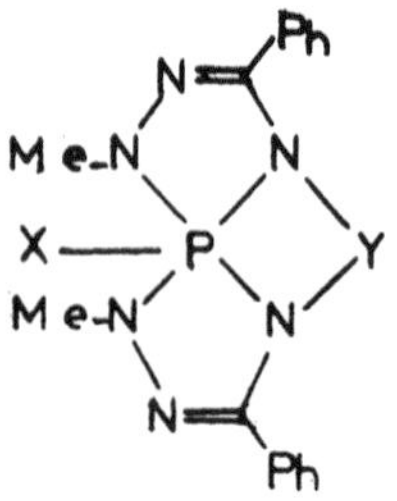

1

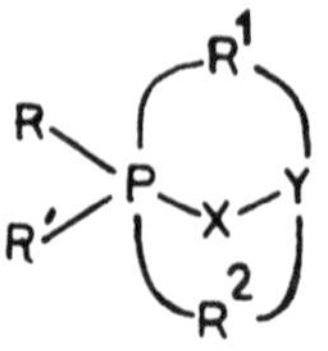

2

P-bound atoms/rings	Connectivities	Structure	NMR data (δP[solv temp] J Hz)	Ref.
$BrCl_2O_2/5$	C'_2	$BrCl_2P(Cat)$	−75.8(c)	1
BrF_4		$BrPF_4$	−72.6(−70°)	2
$BrN_4/5,4f,5f$	$(CN')_2,(C'P)_2$	1; X = Br, Y = PCl_3	−48.8	3
	$(CN')_2,(C'P^4)_2$	1; X = Br, Y = $P(NMe_2)_2{}^+Br^-$	−47.4	3
	$(CN')_2,(C'P^4)_2$	1; X = Br, Y = $PPh_2{}^+Br^-$	−35.9	3
$BrO_4/5,5$	C_4	$(Pfp)_2PBr$	−37.9	4
	C'_4	$BrP(Cat)_2$	−28.2	5
$Br_2ClO_2/5$	C'_2	$(Br_2)ClP(Cat)$	−131(c)	1
Br_2O_3	C_3	$Br_2P(OCHTf_2)_3$	−80	6
$Br_2O_3/5$	C,C'_2	$Br_2(Cat)POCHMe(CH_2)_5Me$	−195	7
	C'_3	$Br_2(PhO)P(Cat)$	−108.5(−100°C)	8
$Br_2O_3/6,6f$	C_3	2; R = R′ = Br, R^1 = R^2 = $OC(CCl_3)H$, X-Y = $OC(CCl_3)HP$	−32	9
$Br_3O_2/5$	C'_2	$Br_3P(Cat)$	−189(c)	1
Br_5		Br_5P	−101	10
$ClF_2O_2/5$	C'_2	$ClF_2P(Cat)$	−30.8 J_{PF}979	11
$ClNO_3$	$C_2;C_3$	$Cl(Cl_3CCH_2O)_3PN^{\cdot}(CH_2)_2OCH_2C^{\cdot}H_2$	−4.5	12
	$C_2;C_3$	$Cl(Cl_3CCH_2O)_3PN^{\cdot}(CH_2)_4C^{\cdot}H_2$	−12	12
$ClNO_3/5,5$	$C'H;C'_3$	ClP(Oam)(Cat)	−20.7	13
$ClN_2O_2/4,5$	$(CP^5)_2;C'_2$	$[(Cat)ClP\text{-}N(CHClPh)]_2$	−47.1	14
$ClN_2S_2/5,5$	$(CC')_2;C'_2$	$ClP(NMeDopS)_2$	−24.8	15
$ClN_4/4,4$	$(CP^5)_2,(CS^4)_2$	$[NMeSO_2NMe)ClP\text{-}NMe]_2$	−81.2(r)	16a
	$(CP^5)_2,(CC')_2$	$[(NMeCONMe)ClP\text{-}NMe]_2$	−67(c)	16b
	$(CS^4)_4$	$ClP(NMeSO_2NMe)_2$	−86.0(r)	16a
	$(CC')_4$	$ClP(NMeCONMe)_2$	−56.6	17
	$(CC')_2,C'_2$	(NMeCONMe)P(NPhCONPh)Cl	−62.4	17
$ClN_4/5,4f,5f$	$(CN')_2,(C'P^4)_2$	1; X = Cl, Y = P(O)Ph	−30.2 −25.7*	18
	$(CN')_2,(C'P^4)_2$	1; X = Cl, Y = $P^{\cdot}$=NCPh=NN$^{\cdot}$Me	−34, −36	19
	$(CN')_2,(C'P)_2$	1; X = Cl, Y = PCl	−13.8	3
	$(CN')_2,(C'P)_2$	1; X = Cl, Y = PCl_3	−35.7	3
	$(CN')_2,(C'P)_2$	1; X = Cl, Y = $PNMe_2$	−23.9	3
ClO_4	C'_4	$ClP(OPh)_4$	−22.8(r) −23.3(n)	20
$ClO_4/5$	C'_4	$Cl(Cat)P(OPh)_2$	−43	21
$ClO_4/5,5$	N'_2,C'_2	$ClP(OCPh{=}NO)_2$	−2.9	22
	C_4	$(Pfp)_2PCl$	−12.4	4

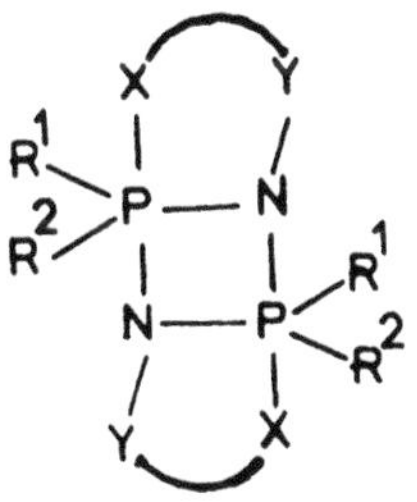

3

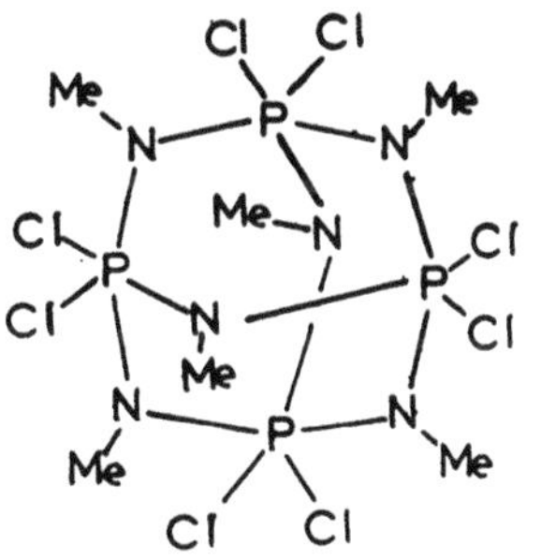

4

P-bound atoms/rings	Connectivities	Structure	NMR data (δP[solv temp] J Hz)	Ref.
	C'_4	$ClP(Cat)_2$	9.7(c) −11(n)	1, 20
$Cl_2FN_2/4$	$(CP^5)_2$	$(Cl_2FPNMe)_2$	−59.3 J_{PF}970	11
$Cl_2FO_2/5$	C'_2	$Cl_2FP(Cat)$	−14.5 J_{PF}974	11
$Cl_2NO_2/5$	$H_2;C_2$	$Cl_2P(NH_2)(Pfp)$	−38.6(c)	23
$Cl_2N'O_2$	$N';C_2$	$(Cl_3CCMe_2O)_2Cl_2PN_3$	−22.2	24
$Cl_2N'O_2/5$	$C';C'_2$	$(Cat)Cl_2PN{=}CHPh$	−30.5	14
$Cl_2N_2O/4,5f$	$(N'P^5)_2;C'$	**3**; $R^1 = R^2 = Cl$, X-Y = OCMe=N	−52.7(g)	25
Cl_2N_3	$(C_2)_3$	$Cl_2P(NMe_2)_3$	−52.8	26
$Cl_2N_3/4,5f$	$(NP^5)_2,C'$	**3**; $R^1 = R^2 = Cl$, X-Y = N=CPhNMe	−65.5(c)	27
	$CN',(C'P^5)_2$	**3**; $R^1 = R^2 = Cl$, X-Y = NMeN=CMe	−67.5(c)	27
	$CN', (C'P^5)_2$	**3**; $R^1 = R^2 = Cl$, X-Y = NMeN=CPh	−63	28
$Cl_2N_3/6,6f$	$(CP^5)_3$	**4**	−74.3	29
Cl_2O_3	C_3	$Cl_2P(OCHTf_2)_3$	−76	6
$Cl_2O_3/5$	CC'_2	$Cl_2(Cat)POCHMe(CH_2)_5Me$	−35	7
	C'_3	$Cl_2(Cat)POC(CCl_3){=}CCl_2$	−32.5	30
	C'_3	$Cl_2(Cat)POC(tBu)C{=}CCl_2$	−35	30
	C'_3	$Cl_2(PhO)P(Cat)$	−34	21
	C'_3	$(Cat)Cl_2P\text{-}Cat\text{-}PCl_2(Cat)$	−33.5(r)	31
$Cl_2O_3/6,6f$	C_3	**2**; $R = R' = Cl$, $R^1 = R^2 = OC(CCl_3)H$, X-Y = $OC(CCl_3)HP$	−38	9
$Cl_3NO/5$	$HN';C'$	$Cl_3P(NHN{=}CMeO)$	−59(g)	32
$Cl_3N_2/4$	$(CP^5)_2$	$[Cl_3P\text{-}NMe]_2$	−78.2	29
	$(C'P^5)_2$	$[Cl_3P\text{-}N(2,6\text{-}Me_2C_6H_3)]_2$	−72.4	33
	$(C'P^5)_2$	$[Cl_3P\text{-}N(3,5\text{-}Me_2C_6H_3)]_2$	−82.5(b)	34
	$(C'P^5)_2$	$[Cl_3P\text{-}N(2\text{-}FC_6H_4)]_2$	−79.9	33
	$(C'P^5)_2$	$[Cl_3P\text{-}N(3,5\text{-}Tf_2C_6H_3)]_2$	−81.5	33
	$(CP^4)_2$	$Cl_3P[NMeP^{\cdot}({=}NCTf{=}NCTf{=}N^{\cdot})NMe]$	−62(r)	35
	$(C_2)_2$	$Cl_3P[NMeC(CCl_3)ClNMe]$	−50.1(m)	36
	$(C_2)_2$	$Cl_3P[NMeCTf(Cl)NtBu]$	−59.8	37
	$(CC')_2$	$Cl_3P[NMeC(O)NMe]$	−60	38
$Cl_3N_2/5$	$CN',C'H$	$Cl_3P(NHCPh{=}NNMe)$	−75(c)	27
	$C'N',C'_2$	$Cl_3P[NPhC(Cl){=}NNPh]$	−77(re)	39
$Cl_3O_2/5$	N',C'	$Cl_3P(OCPh{=}NO)$	−22.7	22
	C_2	$Cl_3P(Glc)$	−29.4(−100°C)	40
	C'_2	$Cl_3P(Cat)$	−26.4(c)	1
Cl_4F		Cl_4PF	−46 J_{PF}1010	41

TABLE Q (continued)

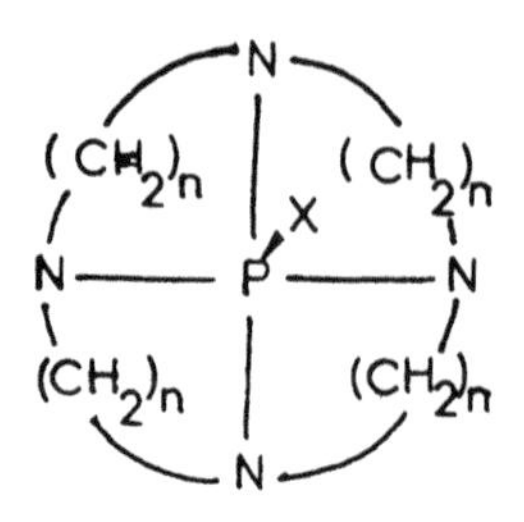

5

P-bound atoms/rings	Connectivities	Structure	NMR data (δP[solv temp] J Hz)	Ref.
Cl_4N	C_2	Cl_4PNEt_2	−71.9	42
Cl_5		PCl_5	−82	43
$FN_2O_2/5$	$(H_2)_2;C_2$	$FP(Pfp)(NH_2)_2$	−44.5	44
$FN_2O_2/4,5$	$(P^5Si)_2;C_2$	$[(Pfp)FP\text{-}NTms]_2$	−47.5(g) J_{PF}994	45
	$(CP^5)_2;C_2$	$[(Pfp)FP\text{-}N\text{-}tBu]_2$	−44.3 J_{PF}1032	46
	$(CP^5)_2;C'_2$	$(Cat)FP(NMePF_3NMe)$	−51.7	47
$FN_2O_2/5,5$	$(C'H)_2;C'_2$	$FP(Oam)_2$	−36.4(t, −25°)	48
$FN_4/4,4$	$(CS)_4$	$FP(NMeSO_2NMe)_2$	−85.0 J_{PF}1030	49
	$(CS)_2;(CC')_2$	$FP[NMeC(O)NMe]$ $(NMeSO_2NMe)$	−67 J_{PF}1035	49
$FN_4/5,5f,5f,5f$	$(C_2)_4$	**5**; X = F, R = lp, n = 2,2,2,2	−14.2(b) J_{PF}793	50
$FN_4/5,5f,5f,6f$	$(C_2)_4$	**5**; X = F, R = lp, n = 2,2,2,3	−32.2 J_{PP}872	50
$FN_4/5,5f,6f,6f$	$(C_2)_4$	**5**; X = F, R = lp, n = 2,2,3,3	−42.8 J_{PF}872	50
$FN_4/5,6f,5f,6f$	$(C_2)_4$	**5**; X = F, R = lp, n = 2,3,2,3	−46.3 J_{PF}927	50
$FO_4/5$	C_4	$(Glc)FP(OMe)OTf$	−50.8 J_{PF}749	51
	C_4	$F(OCHMeCH_2O)P$ $(OMe)(OTf)$	−52 J_{PF}769	51
	C_4	$F(OCHPhCHPhO)P$ $(OMe)(OTf)$	−53 J_{PF}729	51
$FO_4/5,5$	$C'_2N'_2$	$FP(OCPh{=}NO)_2$	−19 J_{PF}1047	22
	C_4	$FP(Pfp)_2$	−33.0 J_{PF}1020	52
	$C_2C'_2$	$FP(Glc)(OCTf{=}CTfO)$	−25 J_{PF}1003	53
	C'_4	$FP[O(2,4\text{-}tBu_2C_6H_3)O]_2$	−25.9 J_{PF}992	54
$F_2NO_2/5$	$P^4;C_2$	$F_2(Pfp)PN{=}PPhF_2$	−55.8 J_{PF}846	55
	$Si_2;C_2$	$F_2(Tms_2N)P(Pfp)$	−40.5(r) J_{PF}876	45
	$H_2;C_2$	$F_2(NH_2)P(Pfp)$	−47.3	44
	$CSi;C_2$	$F_2(NTmstBu)P(Pfp)$	−39.0	56
	$CH;C_2$	$F_2(NtBuH)P(Pfp)$	−48.8	56
	$C_2;C_2$	$F_2(NEt_2)P(Pfp)$	−48.5 J_{PF}871	53
	$C_2;C_2$	$(Pfp)F_2PN^{\cdot}CMe_2$ $(CH_2)_3C^{\cdot}Me_2$	−40.1	56
	$C_2;C_2$	$F_2(Pfp)PN(CH_2CH{=}CH_2)_2$	−50.2 J_{PF}881	53
	$C_2;C'_2$	$F_2(OCTf{=}CTfO)PN$ $(CH_2CH{=}CH_2)_2$	−47.9 J_{PF}876	53
F_2N_3	$(C_2)_3$	$F_2P(NMe_2)_3$	−65.7 J_{PF}700	57
$F_2N_3/4$	$CH,(CP^5)_2$	$[(NHMe)F_2P\text{-}NMe]_2$	−68.1 J_{PF}803	58
	$CH,(CP^5)_2$	$[(NHMe)F_2P\text{-}NEt]_2$	−66.0 J_{PF}808	58
$F_2N_3/5$	$(C_2)_3$	$F_2P(NMeCH_2CH_2NMe)$ NMe_2	−26.4	59

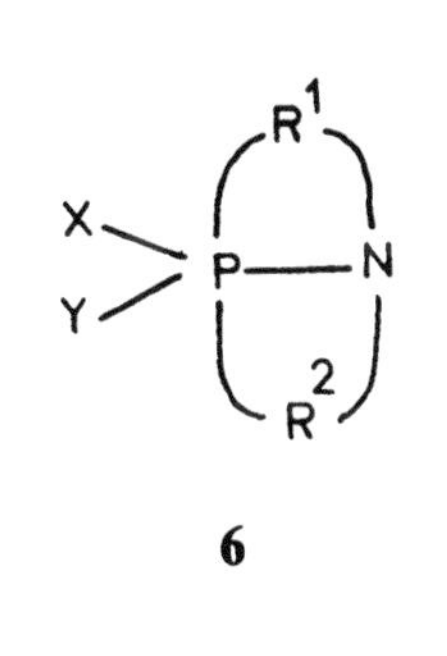

6

P-bound atoms/rings	Connectivities	Structure	NMR data (δP[solv temp] J Hz)	Ref.
	$(C_2)_3$	$F_2P(NMeCH_2CH_2NMe)N(CH_2)_4$	−50.6 $J_{PF}796$	57
$F_2N'_3/6,6f$	S_3	**2**; X-Y = N=S, R[1] = R[2] = N=S=N, R = R′ = F	−11.5(f) $J_{PF}925/1008$	60
$F_2O_2S/5$	$C_2;C$	$F_2P(Pfp)SMe$	−14.6 $J_{PF}1032$	61
F_2O_3	C_3	$F_2P(OMe)_3$	−72 $J_{PF}719$	62
	C_3	$F_2P(OCHTf_2)_3$	−82 $J_{PF}790$	6
	C'_3	$(C_6F_5O)_2PF_2$	−84 (k) $J_{PF}809$	63
	C'_3	$PF_2(PhO)_3$	−88.4(b) $J_{PF}768$	63
$F_2O_3/5$	C_3	$F_2(MeO)P(Pfp)$	−52.8(n) $J_{PF}900$	55
	C_3	$F_2(MeO)P(OCHPhCHPhO)$	−48	64
	C,C'_2	$F_2(nPrO)P(OCTf{=}CTfO)$	−54.8 $J_{PF}907$	53
$F_3NO/5$	$C'H;C'$	$F_3P(Oam)$	−51.5 (t, −88°)	48
F_3N_2	$(H_2)_2$	$F_3P(NH_2)_2$	−58.6 $J_{PF}652/848$	65
	$(C_2)_2$	$F_3P(NMe_2)_2$	−65.0(n) $J_{PF}883$	57
	$(C_2)_2$	$F_3P[NMe(CH_2Ph)]_2$	−66.2(n) $J_{PF}753/868$	66
$F_3N_2/4$	$(CP^5)_2$	$(F_3P\text{-}NMe)_2$	−71.5(n)	67
	$(CS)_2$	$F_3P(NMeSO_2NMe)$	−76.8 $J_{PF}968$	49
	$(CC')_2$	$F_3P[NMeC(O)NMe]$	−55.8	6, 38
F_3O_2	C_2	$F_3P(OCHTf_2)_2$	−79 $J_{PF}845$	11
$F_3O_2/5$	C_2	$F_3P(Pfp)$	−54.5(n) $J_{PF}935$	45
	C'_2	$F_3P(Cat)$	−51.9 $J_{PF}926$	11
	C'_2	$F_3P[O(2,4\text{-}tBu_2C_6H_2)O]$	−51.7 $J_{PF}915$	54
F_4N	HP^3	$F_4PN(H)PF_2$	−71.3(r)	68
	CN	$F_4PN(Me)N(Me)PF_4$	−70.4(30°)	69
	CH	F_4PNHMe	−68.2	70
	C_2	$(Me_2N)PF_4$	−69.7(n, 25°) $J_{PF}845$	71
	C_2	$(iPr_2N)PF_4$	−60.3(c, −20°) $J_{PF}868$	72
	C_2	$F_4PN(CH_2Ph)Me$	−69.2(n) $J_{PF}850$	66
	C'_2	$F_4PN^{\cdot}CH{=}CH\text{-}CH{=}C^{\cdot}H$	−51.7	73
	C'_2	$(Ph_2N)PF_4$	−75.1(b)	71
F_4O	C	$F_4P(OMe)$	−79	74
F_4S	C	$F_4P(STf)$	−40 $J_{PF}923/1058$	75
	C	$F_4P(SMe)$	−34.2(0°) $J_{PF}1032$	76
	C	$F_4P(SC_4H_3)$	−57.5	77
F_5		PF_5	−80.3(−90°) $J_{PF}938$	78
I_2O_3	C_3	$(EtO)_3PI_2$	−42.5(j)	79
$NO_2S_2/5,5$	$C'N';CC';C_2$	$(SCH_2CH_2S)P[OC(OEt){=}NN(CO_2Et)]OMe$	+11.7	80
	$C_2;C_2;C_2$	$(Glc)P(SCH_2CH_2S)NMe_2$	+25	80
	$C_2;C_2;C'_2$	$(NMeCH_2CH_2O)P(SCTf{=}CTfS)OCH_2Tf$	−10.7(r)	81
$NO_2S_2/5,5f$	$C_2;C_2;C'_2$	**6**; X = Y = SPh, R[1] = R[2] = OCH_2CH_2	−19.2	82

TABLE Q (continued)

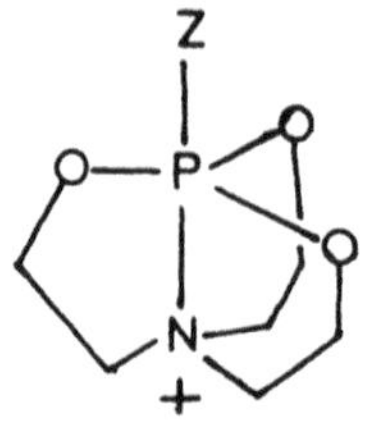

7

P-bound atoms/rings	Connectivities	Structure	NMR data (δP[solv temp] J Hz)	Ref.
NO_2S_2/5,5f,6	C'_2;C'_2;C_2	**6**; XY = $S(CH_2)_3S$, R^1 = ODop, R^2 = OCPh=CPh	−21.2(c)	83
N^4O_3S/5,5f,5f	C_3;C_3;C	**7**; Z = SMe	+5,2(m)	84
NO_3S/5,5f	C_2;C_3;C'	**6**; X = OMe, Y = SPh, R^1 = R^2 = OCH_2CH_2	−28	85
NO_3S/5,5	C_2;C_2C';C'	(Pnc)P(OCPh=CPhS)NMe_2	+10.3(c)	86
	C_2;CC'_2;C'	(OCH_2CH_2NMe)P[O (2-MeC_6H_3)O]SPh	−23.4	87
$N^4O^3O_3$/5,5f,5f	C_3;C_3Si_2	**7**; Z = $O^+(SiEt_3)_2$	−25.6	88
N^4O_4/5,5f,5f	C_3;C_4	**7**; Z = OMe	−20.2(m)	84
	C_3;C_3B^4	**7**; Z = O-BF_3	−18.1	88
	C_3;C_3Si	**7**; Z = $OSiEt_3$	−18.7	88
	C_3;H,C_3	**7**; Z = OH	−11	88
NO_4	C_2;C_4	$(Cl_3CCH_2O)_4PN^{\cdot}(CH_2)_2OCH_2C^{\cdot}H_2$	−74.9	12
	C_2;C_4	$(Cl_3CCH_2O)_4PN^{\cdot}(CH_2)_4C^{\cdot}H_2$	−71.7	12
	C_2;C_4	$(Tf_2CHO)_4PNEt_2$	−75.6	89
NO_4/4	CC';$C_2C'_2$	$(PhO)_2P(NPhCTf_2O)OCHTf_2$	−63.7	90
NO_4/5	H_2;C_4	NH_2(Pfp)$P(OCHTf_2)_2$	−50.3	23
	C'N';C_3C'	$(MeO)_3$P(OCPh=NNPh)	−62	91
	C_2;C_4	($NMeCH_2CH_2O$)$P(OCH_2Tf)_3$	−63.1(c)	81
	C_2C_4	(Pfp)$(Tf_2CHO)_2PNMe_2$	−53.3	89
	C_2;$C_2C'_2$	(Pfp)$P(OPh)_2NiPr_2$	−51	92
	CC';$C_2C'_2$	(Pfp)$P(OPh)_2$NMePh	−57	92
NO_4/4,5	CC';C_4	**6**; X = Y = OEt, R^1 = $OCTf_2$, R^2 = $OCTf_2$C(O)	−36	90
	CC';C_4	**6**; X = Y = OEt, R^1 = $OCTf_2$, R^2 = $OCTf_2$C(S)	−30	90
	CC';$C_2C'_2$	**6**; X = Y = OPh, R^1 = $OCTf_2$, R^2 = $OCTf_2$C(O)	−49	90
NO_4/5,5	HS;C'_4	4-$ClC_6H_4SO_2NH$-P$(Cat)_2$	−36.6(ae)	93
	H_2;C_4	H_2NP$(Pfp)_2$	−30(c)	23
	CN;C'_4	$(Cat)_2$PN$(NEt_2)CH_2Ph$	−25.9	14
	C'N';C'_4	$(Cat)_2PN^{\cdot}$CMe=CHCMe=$N^{\cdot}$	−27(c) −30(b)	94
	C'P;C'_4	$(EtO)_2$PN(Ph)P$(Cat)_2$	−27.9	95
	C'P;C'_4	$(Cat)_2$PN$(POEt_2)$(Me-C_6H_4)	−32	95
	C'H;N';C_3	(NHCPh=NO)P(Glc)OMe	−20	96
	C'H;C_4	PhHNP$(OCHMeCH_2O)_2$	−34(c)	97
	CH;$C_2C'_2$	(NHCHMeCHMeO)P (OCPh=CPhO)OMe	−36.5 −39*	98
	C'H;$C_2C'_2$	PhHNP$[OC(O)CMe_2O]_2$	−57.6(c)	99
	C'H;CC'_3	MeOP(Cat)(Oam)	−32.2	13
	C_2;N'_2,C'_2	Et_2NP$(OCPh{=}NO)_2$	−16.9(c)	100
NO_4/5,5f	C_2;C_4	**6**; X = Y = OMe, R^1 = R^2 = OCH_2CH_2	−42	85
	C_2;C_4	**6**; X = Y = OEt, R^1 = R^2 = OCH_2CH_2	−43.5	82
NO_4/5,5	C_2;C_4	(Pnc)P(Pfp)NMe_2	−24	101

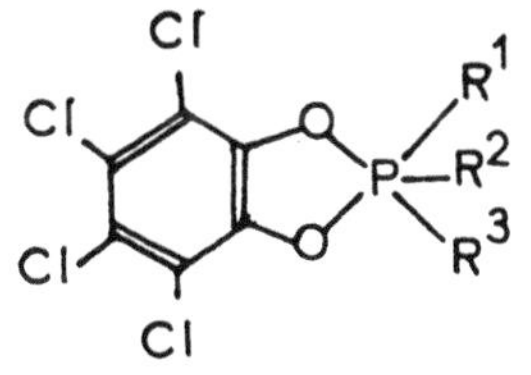

8 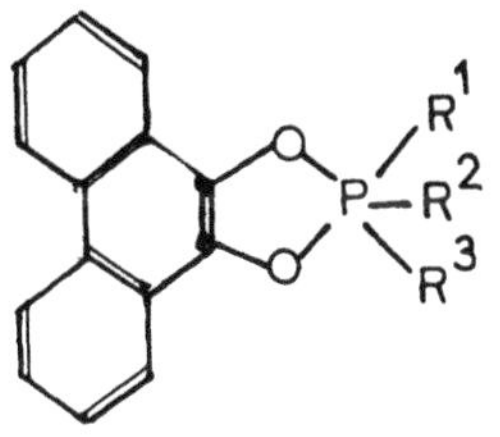 **9**

P-bound atoms/rings	Connectivities	Structure	NMR data (δP[solv temp] J Hz)	Ref.
	$C_2;C_4$	(Glc)P[OCPh(CN)CPh(CN)O]NMe_2	−24, −29(b)	102
	$C_2;C_4$	(Glc)P[OCPh(CO_2Me)CPh (CO_2Me)O]NMe_2	−31.3	103
	$C_2;C_4$	(Pnc)P[OCPh(CN)CPh(CN)O]	−33, −38(t)	102
	$C_2;C_3C'$	(Pfp)P(OCH_2CH_2NMe)OPh	−41.6	104
	$C_2;C_2C'_2$	(OCH_2CH_2NMe)P (OCMe=CMeO)OMe	−40	105
	$C_2;C_2C'_2$	(Pnc)P(OCTf=CTfO)NMe_2	+11	86
	$C_2;C_2C'_2$	(Cat)P(Pnc)NMe_2	−31.6(k)	106
	$C_2;C_2C'_2$	**8**; R^1 = NMe_2, R^2R^3 = Pfp	−29.8	101
	$C_2;C_2C'_2$	**9**; R^1 = NMe_2, R^2R^3 = Glc	−21.3	107
	$C_2;C_2C'_2$	(OCH_2CH_2NMe)P (OCPh=CPhO)$O(CH_2)_2$NHMe	−40	108
	$C_2;C_2C'_2$	(Glc)P(OCPh=CPhO)NMe_2	−27	107
	$C_2;CC'_3$	($NMeCH_2CH_2O$)P[O (2-MeC_6H_3)O]OPh	−41.2	87
	$C_2;CC'_3$	($NMeCH_2CH_2O$)P (OCPh=CPhO)Oquinoleine	−43.7	109
	$C_2;C'_4$	(OCMe=CMeO)P (OCPh=CPhO)NEt_2	−30	110
	$C_2;C'_4$	(Cat)P(OCPh=CPhO)NMe_2	−27	111
	$CC';C'_4$	$(Cat)_2$PNMeC(Tf)=N(i-Bu)	−33.7(b)	112
NO_4/5,5f	$C'_2;C_2C'_2$	**6**; X = Y = OCH_2Ph, R^1 = ODop, R^2 = OCPh=CPh	−41(b)	83
	$C'_2;C'_4$	**6**; X = Y = OPh, R^1 = ODop, R^2 = OCPh=CPh	−48.1(r)	83
NO_4/5,6	$C_2;C_4$	$Et_2N[O(CH_2)_3O]$P[OCPh (CN)CPh(CN)O]	−50	113
	$C_2;C_3C'$	(Pfp)(PhO)P[NMe$(CH_2)_3$O]	−58	86
	$C_2;C_2C'_2$	**9**; R' = NMe_2, R^2R^3 = $O(CH_2)_3O$	−40.7	107
NO_4/5,7	$C_2;C_2C'_2$	(Pnc)P(OCPh=CHCH=CPhO)NMe_2	−26	114
NO_4/6,6f	$C_2;C_4$	**6**; X = Y = OMe, R^1 = R^2 = $O(CH_2)_3$	−66	85
	$C_2;C_4$	**6**; X = Y = OEt, R^1 = R^2 = $O(CH_2)_3$	−69.9	82
NO_4/5,5,5f	$C_2;C_2C'_2$	**6**; XY = OCMe=CMeO, R^1 = R^2 = OCH_2CH_2	−3	115
	$C_2;C_2C'_2$	**6**; XY = OC(4-$NO_2C_6H_4$)=C (4-$NO_2C_6H_4$)O, R^1 = R^2 = OCH_2CH_2	−5.7	116
	$C_2;C_2C'_2$	**6**; XY = OC(4-MeC_6H_4)=C (4-MeC_6H_4)O R^1 = R^2 = OCH_2CH_2	−7.4	116
	$C'_2;C_2C'_2$	**6**; XY = Glc, R^1 = ODop, R^2 = OCPh=CPh	−16.4(r)	117
	$C'_2;C'_4$	**6**; XY = Cat, R^1 = ODop, R^2 = OCPh=CPh	−16.9(k) −29.9(p)	83

TABLE Q (continued)

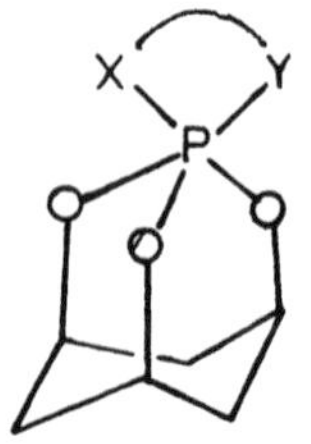

10

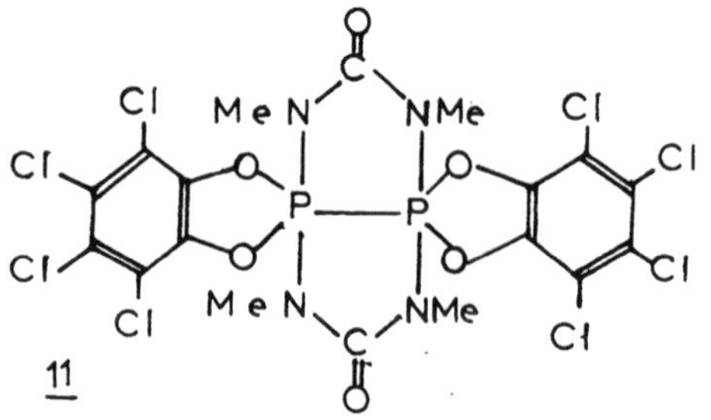

11

P-bound atoms/rings	Connectivities	Structure	NMR data (δP[solv temp] J Hz)	Ref.
NO_4/5,5f,6	C'_2;$C_2C'_2$	**6**; XY = $O(CH_2)_3$, R^1 = ODop, R^2 = OCPh=CPh	−37.6(r)	83
NO_4/5,6,6f	C_2;$C_2C'_2$	**6**; XY = OCMe=CMeO, R^1 = R^2 = $O(CH_2)_3$	−42	85
	C_2;$C_2C'_2$	**6**; XY = OCPh=CPhO, R^1 = R^2 = $O(CH_2)_3$	$-41.2(N_{ax})$	82
	C_2;$C_2C'_2$	**6**; XY = OPhenanthrenylO, R^1 = R^2 = $O(CH_2)_3$	$-36.5(N_{ax})$	82
NO_4/5,6,6f,6f	C'N';C_3C'	**10**; X-Y = OCPh=NNPh	−56.8(r)	118
	C'N';C_3C'	**10**; X-Y = N(COOEt)N=C(OEt)O	−50.7	119a
$N'O_4$/5	C';C_4	$(MeO)_3P$[N=CHCMe(tBu)O]	−32.7	119b
$N'O_4$/5,5	N';C'_4	$(Cat)_2PN_3$	−27(b), −27.4	120, 24
	S^4;$C_2C'_2$	$[OC(O)CMe_2O]_2PNTos^-HNEt_3^+$	−58(r)	121
	C;$C_2C'_2$	**8**; R^1 = N_3, R^2R^3 = Pnc	+7.8	122
	C;$C_2C'_2$	**8**; R^1 = NCO, R^2R^3 = Pnc	−14.5	122
	C;$C_2C'_2$	**8**; R^1 = NCS, R^2R^3 = Pnc	−42.7	122
	C';C'_4	$(Cat)_2P$-N=CHPh	−24.7 −28.6*	14
N_2O_2P/5,5,5f	$(CC')_2$;C'_2;N_2O_2	**11**	−44.0	123
N_2O_2S/5,5	$(C_2)_2$;C_2;C'	(Pfp)(PhS)P($NMeCH_2CH_2NMe$)	−13.6	101
N_2O_3/5	$(C_2)_2$;C_3	($NMeCH_2CH_2NMe$)P$(OCH_2Tf)_3$	−57.8(c)	81
	$(C_2)_2$;C_3	($NMeCH_2CH_2NMe$)P$(OCHTf_2)_3$	−60.8(r)	81
	$(C_2)_2$;C'_3	(PhO)$(NMe_2)_2$P(OCH=CHO)	−34.8(r)	124
N_2P_3/4,5f	$(N'P^5)_2$;C_2C'	**3**; R^1 = R^2 = OMe, X-Y =OCPh=N	−47.8	125
	$(CP^5)_2$;C_3	**3**; R^1 = R^2 = OMe, X-Y = OC(OEt)CNC(tBu)Ph	−59	126
	$(CP^5)_2$,C_3	**3**; R^1 = R^2 = OMe, X-Y = OCH(4-$NO_2C_6H_4$)CPh_2	−57.9	127
	$(CP_5)_2$,C_3	**3**; R^1 = R^2 = OEt, X-Y = OC(OEt)CNCPh (4-MeC_6H_4)	−48	126
N_2O_3/4,5	$(C'P^5)_2$,C_3	[(Glc)EtOP-NPh]$_2$	−58.2 −59	128
N_2O_3/4,5f	$(C'P^5)_2$;C_2C'	**3**; R^1 = R^2 = OMe, X-Y = ODop	−53.4	129
	$(C'P^5)_2$;C_2C'	**3**; R^1 = R^2 = OEt, X-Y = O(4-$NO_2C_6H_3$)	−55.9	130
	$(C'P^5)_2$;C_2C'	**3**; R^1 = R^2 = OEt, X-Y = ODop	−56.7(b)	131
N_2O_3/4,5	$(C'P^5)_2$;CC'_2	[(Cat)EtOP-NPh]$_2$	−60.2(c)	93
	$(C'P^5)_2$;C'_3	[(Cat)PhOP-NPh]$_2$	−62.6(b)	93
	$(C'P^5)_2$;C'_3	[(Cat)(CH_2=CPhO)P-NPh]	−66.4	132
N_2O_3/4,5f	$(C'P^5)_2$;C'_3	**3**; R^1 = R^2 = OPh, X-Y = O(4-BrC_6H_3)	−61.6(b)	133

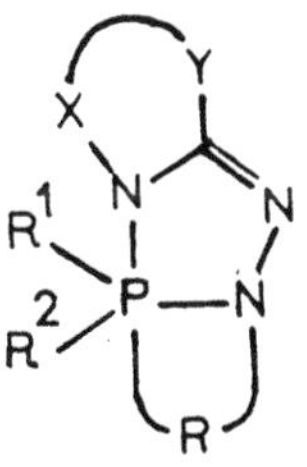

12

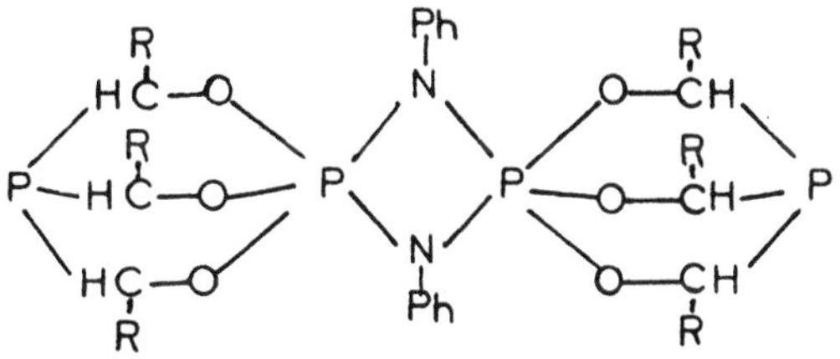

13

P-bound atoms/rings	Connectivities	Structure	NMR data (δP[solv temp] J Hz)	Ref.
	$(C'P^5)_2;C'_3$	**3**; $R^1 = R^2 =$ OPh, X-Y = ODop	−63.4(b)	131
N_2O_3/5,5	$HS^4,C'H;C'_3$	(Cat)(Oam)$PNHSO_2C_6H_4Cl$	−42.6(m)	93
	$C'N',C_2;C_2C'$	(Cat)P[N(CO_2Et) N=C(CO_2Et)O]NMe_2	−36.3	134
	$C'N',C_2;C_2C'$	(Pnc)P[N(CO_2Et)N=C (CO_2Et)O]NMe_2	−43.0	134
	$C'N',C_2;C_2C'$	(Glc)P(NPhN=CPhO)NMe_2	−38(b)	102
	$C'N',C_2;C_2C'$	(Pnc)P(NPhN=CPhO)NMe_2	−48(b)	102
N_2O_3/5,5f	$C'N',C'_2;C_2C'$	**12**; $R^1 = R^2 =$ OMe, R = OCPh=CPh, X-Y = CH=CH-CH=CH	−52.5	135
	$C'N'C'_2;C_2C'$	**12**; $R^1 = R^2 =$ OPh, R = Onaphtyl, X-Y = CH=CHCH=CH	−59.1	135
	$C'N',C'_2;C_2C'$	**12**; $R^1 = R^2 =$ OPh, R = Onaphtyl, X-Y = CH=CHS	−54.8	135
N_2O_3/5,5	$CH,C'H;C_3$	(Glc)P(OCH_2CHMeNH)NHPh	−46.2(t)	97
	$CH,C'H;C_3$	(Pnc)P(OCH_2CH_2NH)NHPh	−52.6(c)	97
	$(C'H)_2;N'_2C$	MeOP(ON=CPhNH)$_2$	−17.7	96
	$CH,C_2;C_3$	(Glc)P(OCH_2CH_2NMe)NHEt	−41.1	97
	$(C'H)_2;C'_3$	(Cat)P(Oam)NHPh	−39.4	13
	$(C'H)_2;C'_3$	$(Oam)_2PO(2\text{-}NH_2C_6H_4)$	−47	136
	$C'H,C_2;C_3$	(Glc)P(OCH_2CH_2NMe)NHPh	−45.9	137
	$C'H,C_2;C_3$	(Pnc)P(OCH_2CH_2NMe)NHPh	−52.6	137
	$(C_2)_2;C_2C'$	(Pfp)P(NMeCH_2CH_2NMe)OPh	−36.4	101
	$(C_2)_2;CC'_2$	(OCH_2CH_2NMe)P[O (2-MeC_6H_3)O]NMe_2	−37.9	87
	$(C_2)_2;CC'_2$	**9**; $R^1 =$ OMe, $R^2R^3 =$ NMeCH_2CH_2NMe	−33.5	138
	$(C_2)_2;CC'_2$	(NMeCHMeCHPhO)P [OCPh=C(OMe)O]NMe_2	−42.7(r)	139
	$(C_2)_2;CC'_2$	(OCHPhCHMeNMe)P- [O(2,4-$tBu_2C_6H_2$)O]NMe_2	−38(b)	140
	$(C_2)_2;CC'_2$	**8**; $R^1 = NMe_2$, $R_2R_3 =$ OCHPhCHMeNMe	−34(b)	140
	$(C_2)_2;CC'_2$	(NMeCHMeCHPhO)P (OCPh=CPhO)NMe_2	−43	141
	$(CC')_2;C_2C'$	PhOP[OCMe_2C(O)NMe]$_2$	−54.5(c)	142
N_2O_3/5,6	$(C_2)_2;C_2C'$	[NMe(CH_2)$_3$NM3]P(Pfp)OPh	−37	86
N_2O_3/4,5f,5	$(CP^5)_2;C_3$	**3**; $R^1R^2 =$ Glc, X-Y = OC(Ph)CNC(OEt)Ph	−10(c)	143
N_2O_3/4,6,6f	$(C'P^5)_2;C_3$	**13**; R = Tf	−73.9	144
	$(C'P^5)_2;C_3$	**13**; R = CF_2CF_2H	−71.7	144
N_2O_3/5,5,5f	$C'N',C'_2;C_2C'$	**12**; $R^1R^2 =$ Glc, R = OCPh=CPh, XY = CH=CHCH=CH	−28.4	135

TABLE Q (continued)

P-bound atoms/rings	Connectivities	Structure	NMR data (δP[solv temp] J Hz)	Ref.
$N'_2O_3/5$	$C'_2;C'_3$	(Cat)(EtO)P(N=CClSCl)$_2$	−39	145
	$C''_2;CC'_2$	(Cat)(EtO)P(NCS)$_2$	−77	145
	$C'_2;C'_3$	(Cat)(PhO)P(N=CCl$_2$)$_2$	−44	145
$N_3O_2/5$	$(H_2)_3;C_2$	(Pfp)P(NH$_2$)$_3$	−58.0	44
	$(C_2)_3;C_2$	(NMeCH$_2$CH$_2$NMe)P(OEt)$_2$NMe$_2$	−49	146
$N_3O_2/4,5$	$(P^5Si)_2,C_2;C'_2$	[(OCMe=CMeO)NMe$_2$P-Ntms]$_2$	−53.1	147
	$(PSi)_2,Si_2;C'_2$	(OCTf=CTfO)P[NTmsP (NTms$_2$)NTms]NTms$_2$	−28.7	148
	$(PSi)_2,C_2;C'_2$	(OCMe=CMeO)P[NTmsP (NMe)$_2$NTms]NMe$_2$	−40.6	147
	$(CP^5)_2,C_2;C'_2$	[(Cat)NMe$_2$P-N(CH$_2$)$_2$NHCOTf]$_2$	−51	149
	$(CP^5)_2,C_2;C'_2$	[(Cat)NEt$_2$P-N(CH$_2$)$_2$NHCOTf]$_2$	−45	149
	$(C'P^5)_2,C_2;C_2$	[(Glc)C˙H$_2$CH$_2$N˙P-NPh]$_2$	−53	150
	$(C'P^5)_2,C_2;C_2$	[(Cat)C˙H$_2$CH$_2$N˙P-NPh]$_2$	−52.4	150
	$(C'P^5)_2,C_2;C_2$	[Bu$_2$N(Cat)P-NPh]$_2$	−53.4(b) −55(c)	93
$N_3O_2/4,5f$	$(CP^5)_2,CC';C'_2$	**3**; R^1 = R^2 = OPh, X-Y = N(CH$_2$Ph)COCH$_2$	−65(c)	151
$N_3O_2/4,5$	$(CC')_2,C_2;C'_2$	[NMeC(O)NMe]P (OCPh=CPhO)NMe$_2$	−44	152
$N_3O_2/5,5$	$HS^4,(C'H)_2;C'_2$	(Cat)[NH(Dop)NH]PNHSO$_2$C$_6$H$_4$Cl	−49.5(monoglyme)	93
	$(C'N')_2,C_2;C'_2$	[N(CO$_2$Et)N=C(OEt)O]$_2$PNMe$_2$	−34.7	80
	$C'N',(C'H)_2;C'_2$	(Oam)P(NPhN=CMeO)NHPh	−55.2(c)	153
	$C'N',C'H,C_2;C'_2$	(Oam)P(NPhN=CMeO)NEt$_2$	−46.2(b)	153
	$C'N',(C_2)_2;CC'$	(NMeCHMeCHPhO)P[N (CO$_2$Et)N=C(CO$_2$Et)O]NMe$_2$	−49.5 −49.7*	134
	$CH,(C_2)_2;C'_2$	(Cat)P(NHCH$_2$CH$_2$NEt)NEt$_2$	−25.3	149
	$(CH)_2,C'H;C_2$	(OCH$_2$CHMeNH)$_2$PNHPh	−49.3(b)	97
	$(CH)_2,C'H;C_2$	(OCH$_2$CMe$_2$NH)$_2$PNHPh	−52.8(t)	97
	$C'H,(C_2)_2;C_2$	(OCH$_2$CH$_2$NMe)$_2$PNHPh	−59.5(c)	137
	$(C_2)_3;C_2$	(NMeCH$_2$CH$_2$NMe)P(Pfp)NMe$_2$	−29.6 −28.3	101, 59
	$(C_2)_3;C_2$	(NMeCH$_2$CH$_2$NMe)P[OC (Ph)TfC(Ph)TfO]NMe$_2$	−34.9	59
	$(C_2)_3;C_2$	(NMeCH$_2$CH$_2$NMe)P[OCPh (CO$_2$Me)CPh(CO$_2$Me)O]NMe$_2$	−36.8	103
	$(C_2)_3;C_2$	(NMeCH$_2$CH$_2$NMe)P[OC (Ph)TfC(Ph)TfO]N(CH$_2$)$_4$	−38.5 −40.5*	59
	$(C_2)_3;C'_2$	**9**; R^1, R^2 = NMeCH$_2$CH$_2$NMe, R^3 = NMe$_2$	−29.8	138
	$(C_2)_3;C'_2$	(NMeCH$_2$CH$_2$NMe)P (OCPh=CPhO)NMe$_2$	−36.9	154
	$(C_2)_3;C'_2$	(NMeCH$_2$CH$_2$NMe)P [OCPh=C(COPh)O]NMe$_2$	−31.9	138
	$(C_2)_3;C'_2$	**9**; R^{1}R^2 = NMeCH$_2$CH$_2$NMe, R^3 = N(CH$_2$)$_4$	−35.1	138
	$(C_2)_3;C'_2$	(NMeCH$_2$CH$_2$NMe)P (OCPh=CPhO)N(CH$_2$)$_4$	−41.8	154
	$(C_2)_3;C'_2$	(NMeCH$_2$CH$_2$NMe)P [OCPh=C(COPh)O]N(CH$_2$)$_4$	−37.5	138
	$(C_2)_2,C'_2;C'_2$	**6**; X = Y = NMe$_2$, R^1 = ODop, R^2 = OCPh=CPh	−36(r)	83
$N_3O_2/4,5f,5$	$(CP^5)_2,C_2;C'_2$	**3**; R^{1}R^2 = Cat, X-Y = N(iPr)CH$_2$CH$_2$	−48.6	149
$N_3O_2/5,5,5f$	$CN',C'H,C'_2;C'_2$	**6**; XY = NHCPh=NNMe, R^1 = ODop, R^2 = OCPh=CPh	−30.2	155
	$(C_2)_3;C'_2$	**6**; XY = OCPh=CPhO, R^1 = R^2 = NMeCH$_2$CH$_2$	−30.6(r)	156

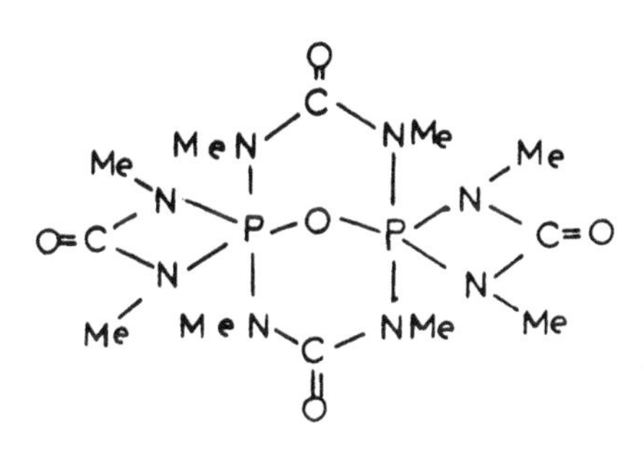

14

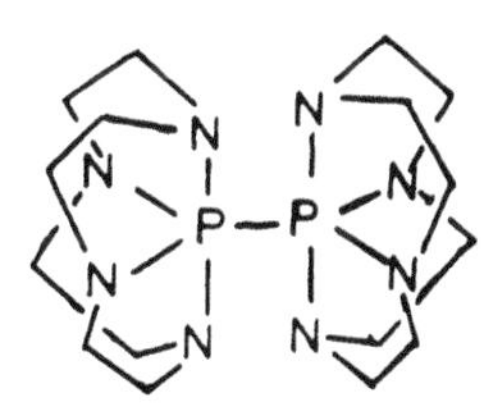

15

P-bound atoms/rings	Connectivities	Structure	NMR data (δP[solv temp] J Hz)	Ref.
	$(C_2)_2,C'_2;C'_2$	6; XY = $NMeCH_2CH_2NMe$, R^1 = ODop, R^2 = OCPh=CPh	−26.2(b)	83
N_3O_2/5,6,6f	$(C_2)_3;C'_2$	6; XY = OCPh=CPhO, $R^1 = R^2 = NMe(CH_2)_3$	−37.1(r)	156
	$(C_2)_3;C'_2$	6; XY = OphenanthrenylO, $R^1 = R^2 = NMe(CH_2)_3$	−28.9	115
	$(C_2)_3;C'_2$	6; XY = $OC(4\text{-}NO_2C_6H_4)$=C$(4\text{-}NO_2C_6H_4)O$, $R^1 = R^2 = NMe(CH_2)_3$	−33.5	157
	$(C_2)_3;C'_2$	6; XY = $OC(4\text{-}OMeC_6H_4)$=C$(4\text{-}OMeC_6H_4)O$, $R^1 = R^2 = NMe(CH_2)_3$	−37.7	157
N_3S_2/5,5	C'H,C_2;CC'	$(SCH_2CH_2S)P[NHDopN(CH_2Ph)]NMe_2$	+10	80
N_4O/4,5f	$(N'P^5)_2,(C_2)_2;C'$	3; $R^1 = R^2 = NMe_2$, X-Y = OCMe=N	−40.5	25
N_4O/5,5f	C'N,$(C_2)_2,C'_2;C'$	12; R = OCPh=CPh, $R^1R^2 = NMeCH_2CH_2NMe$, X-Y = CH=CHCH=CH	−34.2	135
N_4O/4,5f,5	$(CN')_2,(C'P^5)_2;C'$	3; R^1R^2 = OC(OMe)=NN(CO_2Me), X-Y = NMeN=CPh	−63.4(c)	158
N_4O/4,6,6f	$(CC')_4;P^5$	**14**	−70.1	159
N_4P/4,4	$(CC')_4;O_2O'$	$[NMeC(O)NMe]_2P\text{-}P(O)(OEt)_2$	−80.9 J_{PP}708	160
	$(CC')_4;C'_2$	$[NMeC(O)NMe]_2P\text{-}PPh_2$	−65.8 J_{PP}270	161
N_4P/5,5f,5f,5f	$(C_2)_4;N_4$	**15**	−36.8(b)	162
N_4S/4,5f	$(N'P^5)_2,(C_2)_2;C'$	3; $R^1 = R^2 = NMe_2$, X-Y = SCPh=N	−10.8(c)	163
N_5/4	$P^4Si,CP^4,(C_2)_3$	$[NMeP(NTms_2)(=NTms)NTms]P(NMe_2)_3$	−38.3	164
N_5/4,4	$(CC')_4,N'$	$N_3P[NMeC(O)NMe]_2$	−66.8(b)	120
N_5/5,4f,5f	$H_2,(CN')_2,(C'P^4)_2$	1; X = NH_2, Y = P(O)Ph	−45.6(c)	3
	$(CN')_2,(C'P^4)_2,C_2$	1 X = NMe_2, Y = P(O)Ph	−30.6(c)	3
	$(CN')_3,(C'P^4)_2$	1; X = $NMeN=CMe_2$, Y = P(O)Ph	−35.9(c)	3
N_4N'/4,4	$(CC')_4;P^4$	$Ph_3P=NP[NMeC(O)NMe]_2$	−71.3	165
OS_4/5,5	$C;C_2C'_2$	$(SCH_2CH_2S)P(SCTf=CTfS)OCH_2Tf$	+20.3(c)	81
O_3S_2/5	$CC'_2;C'_2$	$(Cat)P(OMe)[S(4\text{-}MeC_6H_4)]SPh$	−47.6(h)	166
	$C'_3;C'_2$	$(PhO)(EtS)_2P(OCH=CHO)$	−3.2(r)	124
O_3S_2/5,5	$C_3;C_2$	$(Glc)P(SCH_2CH_2S)OMe$	+24.2	80
	$C_2C';C'_2$	(Pfp)(PhO)P(SCTf=CTfS)	+13.6(r)	86
	$CC'_2;C'_2$	(MeO)(Cat)P(SCTf=CTfS)	−17.5	167

TABLE Q (continued)

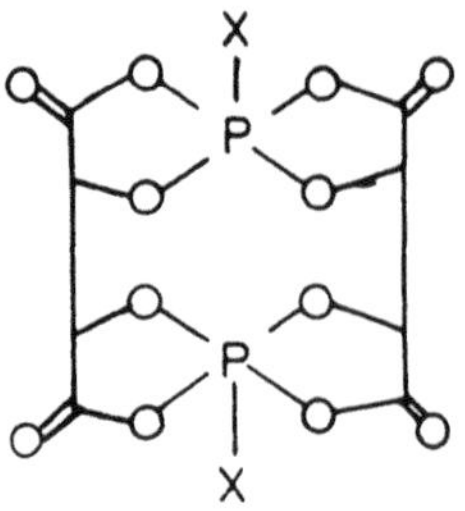

16

P-bound atoms/rings	Connectivities	Structure	NMR data (δP[solv temp] J Hz)	Ref.
O_4P^4/5,5	$C_2C'_2$;C'SS$^-$	[OC(O)CMe$_2$O]$_2$PP(S)(4-MeOC$_6$H$_4$)S$^-$HNEt$_3^+$	−44 J_{PP}221	121
	C'_4;N'O_2	(Cat)$_2$PP(OEt)$_2$=NC$_6$H$_4$NO$_2$	−29.8 J_{PP}470	95
O_4S	C_4;C'	(TfCH$_2$O)$_4$PS(4-ClC$_6$H$_4$)	−51.7(t)	168
	C_4;C'	(TfCH$_2$O)$_4$P(SPh)	−50.6(c, 30°C) −52.5(t)	169, 168
	C_4;C'	(Tf$_2$CHO)$_4$PSPh	−54.0(c)	170
	C_4;C'	(HCF$_2$CF$_2$CH$_2$O)$_4$PSPh	−49	171
O_4S/5,5	C_4;H	(Pfp)$_2$PSH	+14.2	172
	C_4;C	(Pfp)P(Glc)SMe	−2.5	61
	C_4;C	(Pnc)P(Glc)SEt	−10.5	173
	C_4;C	(Pnc)P(Pfp)SEt	−7.2	173
	C_4;C	(Pnc)$_2$PSEt	−23.0	173
	C_4;C'	(Glc)P(Pfp)SPh	−3.4	101
	C_4;C'	(Pnc)P(Pfp)SPh	−14	101
	C_3C';C	(Pfp)P(OCH$_2$CH$_2$S)OPh	+2.1, +2.7(r)	104, 87
	$C_2C'_2$;C	[OC(O)CMe$_2$O]PSMe	−32.5(r)	121
	$C_2C'_2$;C'	(Pfp)P(ODopS)OPh	−2.4	87
O_4S''/5,5	C_4	(Pfp)$_2$PS$^-$HNEt$_3^+$	+17.4	172
	$C_2C'_2$	**16**; X = S$^-$HNEt$_3^+$	+7(r)	121
	$C_2C'_2$	[OC(O)CMe$_2$O]$_2$PS$^-$HNEt$_3^+$	−4.3(r)	174
O_4Se/5,5	C_4;C'	(Pnc)P(Pfp)SePh	−22.0(r) J_{PSe}620	175
	C_4;C'	(Pnc)$_2$PSePh	−30.3(r) J_{PSe}459	175
	$C_2C'_2$;C	[OC(O)CMe$_2$O]$_2$PSeMe	−49.8(r) J_{PSe}618	121
	$C_2C'_2$;C'	(Pnc)P(Cat)SePh	−20.4(r)	175
O_4Se''/5,5	$C_2C'_2$	**16**; X = Se$^-$HNEt$_3^+$	−19.3 J_{PSe}988	121
	$C_2C'_2$	[OC(O)CMe$_2$O]$_2$PSe$^-$HNEt$_3^+$	−32.3(r) J_{PSe}953	121
	$C_2C'_2$	[OC(O)CPh$_2$O]$_2$PSe$^-$HNEt$_3^+$	−27.7(r) J_{PSe}968	121
O_4Si/5,5	C'_4;C_3	(Cat)$_2$PTms	−132.6	176
•O_5	$(P^4)_2$,C'_3	(PhO)$_3$P[OP(S)(OEt)$_2$]$_2$	−94.4	145
	C_5	P(OMe)$_5$	−66	177
	C_5	(MeO)$_3$P(OEt)$_2$	−68	178
	C_5	(MeO)$_3$P[OC•H(CH$_2$)$_3$C•HO]	−41.6(−50°)	179
	C_5	P(OCH$_2$CCl$_3$)$_5$	−79.6	180
	C_5	P(OCH$_2$Tf)$_5$	−76.6(c, 30°)	169
	C_5	(Me$_2$CHO)$_3$P(OEt)$_2$	−73	178
	C_5	P(OCHTf$_2$)$_5$	−84	6
	C_5	P(OCH$_2$CF$_2$CHF$_2$)$_5$	−56	181
	C_5	P[OCH$_2$(CF$_2$)$_3$CHF$_2$]$_5$	−76	181
	C_5	P(OCH$_2$tBu)$_5$	−72	177
	CC'_4	MeOP(OPh)$_4$	−81.5	182
	C'_5	(C$_6$F$_5$O)$_5$P	−86.1(c)	183

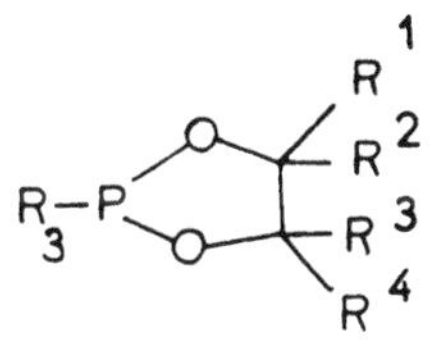

17

P-bound atoms/rings	Connectivities	Structure	NMR data (δP[solv temp] J Hz)	Ref.
	C'_5	$(PhO)_5P$	−85.6, −84.2(n)	184, 20
$O_5/4$	O_2,C'_3	$(OOO)P(OPh)_3$	−63	185
$O_5/5$	$(P^4)_2,C,C'_2$	$(Cat)P[OP(S)(OEt)_2]_2OEt$	−61.3	145
	$P^4,C_2C'_2$	$[OCPh{=}C(COPh)O]$ $P[OP(O)(OMe)_2](OMe)_2$	−57.7	186
	C_5	$(Glc)P(OCH_2Tf)_3$	−52.5(c, 30°)	169
	C_5	$(Glc)P(OEt)_3$	−52	178
	C_5	$(Pfp)P(OMe)_3$	−50.1	59
	C_5	$(Pnc)P(OMe)_3$	−54.3(b)	187
	C_5	$(MeO)_3P[OCHEtCMe(COMe)O]$	−51.3	188
	C_5	$(Glc)P(OCHTf_2)_3$	−62.6(c)	170
	C_5	**17**; R = OMe, R^1 = R^3 = Me, R^2 = R^4 = C(O)Me	−52.6 −54.8	189
	C_5	**17**; R = OMe, R^1 = R^3 = Me, R^2 = R^4 = CO_2Me	−52.2 −50.4	138
	C_5	$(Pnc)P(OEt)_3$	−60	187
	C_5	**17**; R = OMe, R^1R^2 = $(CH_2)_5$, R^3 = Me, R^4 = C(O)Me	−54.3	190
	C_5	**17**; R = OMe, R^1 = Me, R^2 = C(O)Me, R^3R^4 = NPh	−57.8	191
	C_5	**17**; R = OMe, R^1 = H, R^2 = C(O)Ph, R^3 = Me, R^4 = C(O)Me	−50.1	189
	C_5	**17**; R = OMe, R^1 = Ph, R^2 = Tf, R^3 = Me, R^4 = C(O)Me	−49.6	190
	C_5	$(PrO)_3P(Pnc)$	−58.3(b)	192
	C_5	**17**; R = OEt, R^1 = R^3 = H, R^2 = R^4 = 4-$NO_2C_6H_4$	−52.4 −53.4	193
	C_5	**17**; R = OMe, R^1R^2 = R^3R^4 = fluorenyl	−46.7(c)	194
	C_5	**17**; R = OEt, R^1R^2 = R^3R^4 = fluorenyl	−49.3(c)	194
	C_5	**17**; R = OiPr, R^1R^2 = R^3R^4 = fluorenyl	−52.2(c)	194
	$C_3C'_2$	$(MeO)_3P(OCTf{=}CTfO)$	−46.7(c)	195
	$C_3C'_2$	$(MeO)_3P(OCMe{=}CMeO)$	−48.9(n)	196
	$C_3C'_2$	$(MeO)_3P(OCMe{=}CEtO)$	−46(n)	197
	$C_3C'_2$	$(Cat)P(OMe)_3$	−51.6	87
	$C_3C'_2$	$(Cat)P(OMe)(OEt)_2$	−49	198
	$C_3C'_2$	$(MeO)_3P(OCH{=}CPhO)$	−45.4(g)	196
	$C_3C'_2$	$(Cat)P(OEt)_3$	−50	178
	$C_3C'_2$	$(PhCH_2O)P(OCMe{=}CMeO)(OMe)_2$	−49.8	199
	$C_3C'_2$	$(MeO)_3P(OnaphtylO)$	−45.5	200
	$C_3C'_2$	**9**; R^1 = R^2 = R^3 = OMe	−44.7	138
	$C_3C'_2$	$(MeO_3P(OCPh{=}CPhO)$	−53(b)	197
	$C_3C'_2$	$(MeO)_3P[OCPh{=}C(C(O)Ph)O]$	−50.2	201
	$C_3C'_2$	$(PhCH_2O)_2P(OCMe{=}CMeO)OMe$	−50.6	199

TABLE Q (continued)

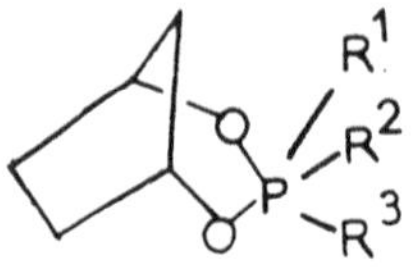

R[1] R[2] R[3] P

18

R' R' (CH$_2$)$_n$ P(OR)$_3$ R' R' 19

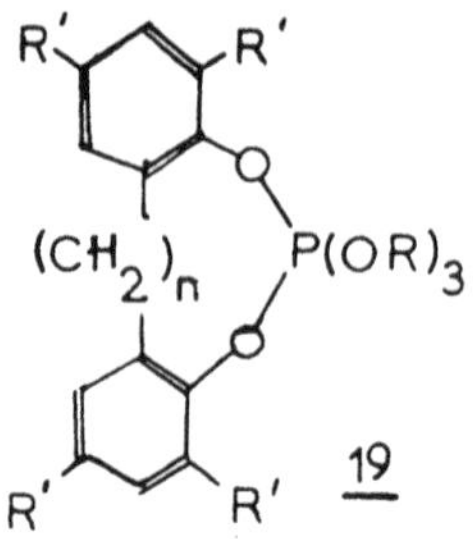

19

P-bound atoms/rings	Connectivities	Structure	NMR data (δP[solv temp] J Hz)	Ref.
	$C_3C'_2$	(iPrO)$_3$P(OCPh=CPhO)	−53.9(r)	196
	$C_3C'_2$	(OCPh=CPhO) (EtO)$_2$POCH$_2$C˙HO(CH$_2$)$_2$C˙H$_2$	−46.6	202
	$C_3C'_2$	**9**; R^1 = R^2 = R^3 = O-iPr	−49.2(b)	203
	$C_3C'_2$	(PhCH$_2$O)$_3$P(OCMe=CMeO)	−51.3	199
	$C_3C'_2$	(TMPO-O)$_3$P(OCMe=CMeO)	−29(b)	204
	$C_2C'_3$	(MeO)$_2$P(OCMe=CMeO)OPh	−53.3(b)	182
	CC'_4	(MeO)P(OCMe=CMeO)(OPh)$_2$	−57.4(b)	182
	C'_5	(PhO)$_3$P(OCTf=CTfO)	−61.8(c)	195
	C'_5	(PhO)$_3$P(OCH=CHO)	−61.3(r)	124
	C'_5	(Cat)P(OPh)$_3$	−60	205
	C'_5	**9**; R^1 = R^2 = R^3 = OPh	−58.6	43
	C'_5	(PhO)$_3$P(OCPh=CPhO)	−62(b)	197
O_5/6	C_5	(EtO)$_3$P[O(CH$_2$)$_3$O]	−72	178
	C_5	**18**; R^1 = R^2 = R^3 = OMe	−72.4(b)	206
O_5/7	C_3,C'_2	**19**; n = 0, R' = H, R = CH$_2$Tf	−63.1	207
	C_3,C'_2	**19**; n = 0, R' = tBu, R = CH$_2$Tf	−67.6	207
O_5/8	C_3,C'_2	**19**; n = 1, R' = R = CH$_2$Tf	−78.5	207
O_5/4,5	O_2,C_2C'	(PhO)P(OCHPhCHPhO)(OOO)	−28.3 −37.3	208
	O_2,C'_3	(Cat)(PhO)P(OOO)	−37	209
O_5/4,6	O_2,C_2C'	(PhO)(OCH$_2$CMe$_2$CH$_2$O)P(OOO)	−53	208
O_5/5,5	N'_2,C'_3	(OCPh=NO)$_2$POPh	−25.3(b)	100
	Si,C_4	(Pfp)$_2$POTms	−42.0	52
	$Si,C_2C'_2$	**8**; R^{1}R^2 = Pnc, R^3 = OTms	−35.9	122
	Si,C'_4	(Cat)$_2$POTms	−30.5	210
	Si,C'_4	(Cat)P(OCPh=CPhO)OTms	−39	211
	H,C_4	(Pfp)$_2$POH	−34.5	4
	$H,C_2C'_2$	**16**; X = OH	−44(v)	212
	H,C'_4	(Cat)$_2$POH	−27(c, 30°)	210
	C_5	(Glc)$_2$POCH$_2$Tf	−28.5(c, 30°)	169
	C_5	(Glc)$_2$POEt	−27.9(c, 30°)	87
	C_5	(Glc)$_2$POCHTf$_2$	−29.7(r)	170
	C_5	[OCH$_2$CH(Me)O]$_2$POEt	−31.5	213
	C_5	(Glc)P(Pfp)OMe	−27.7	61
	C_5	[OCH(Me)CH(Me)O]$_2$POEt	−37.5	213
	C_5	(Glc)$_2$PO(CH$_2$)$_2$OP(Glc)$_2$	−27	214
	C_5	(Pnc)$_2$POEt	−42.4	173
	C_5	(Pfp)$_2$POCHTf$_2$	−32.6	52
	C_5	(Pnc)(Glc)PO(CH$_2$)$_2$OP(Glc)(Pnc)	−35	214
	C_4C'	(Glc)P[OCH$_2$C(O)O]OMe	−38(j)	215
	C_4C'	(Glc)P(Pfp)OPh	−31.3	101
	C_4C'	(Glc)P(Pnc)OPh	−37.2(k)	106

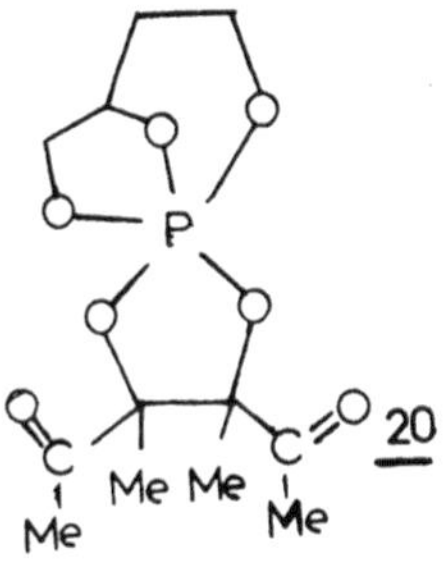

20 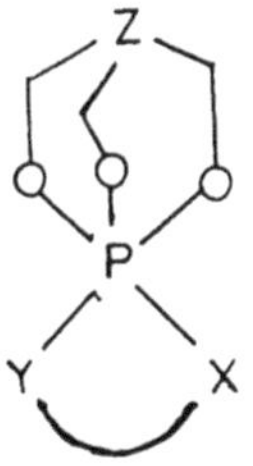 **21**

P-bound atoms/rings	Connectivities	Structure	NMR data (δP[solv temp] J Hz)	Ref.
	C_4C'	(Pnc)P(Pfp)OPh	−39.0	101
	C_4C'	$(Pnc)_2POPh$	−44.2(k)	106
	$C_3C'_2$	(Glc)P(OCMe=CMeO)OMe	−27.3	184
	$C_3C'_2$	(Pnc)P[OCPh=C(OMe)O]OEt	−41(r)	139
	$C_3C'_2$	(Glc)P(OCMe=CMeO)O-TMPO	−24(b)	204
	$C_3C'_2$	(Glc)P(OCPh=CPhO)OMe	−28	108
	$C_3C'_2$	**9**; R^1 = OMe, R^2R^3 = Glc	−23	107
	$C_3C'_2$	(Pnc)P(OCMe=CMeO)O-TMPO	−32(q)	204
	$C_3C'_2$	tBuN-[$(CH_2)_2O$**P**(Glc)(OCMe=CMeO)]$_2$	−28.5(b)	216
	$C_3C'_2$	(Pnc)P(OCPh=CPhO)OMe	−36(n)	217
	$C_3C'_2$	(Pnc)P(OCPh=CPhO)OCH_2CH_2OH	−36	108
	$C_3C'_2$	(Pnc)P(Cat)OCHPhC(O)Ph	−35	111
	$C_2C'_3$	(Glc)P(OCTf=CTfO)OPh	−51	86
	$C_2C'_3$	(Pnc)P(OCTf=CTfO)OPh	−37(j)	86
	$C_2C'_3$	**8**; R^1R^2 = Pfp, R^3 = OPh	−32.9	101
	$C_2C'_3$	[$OCMe_2C(O)O$]P(OCPh=CPhO) OMe	−44	98
	$C_2C'_3$	**9**; R^1R^2 = Glc, R^3 = OPh	−27	107
	$C_2C'_3$	(Pnc)P(OCPh=CPhO)OPh	−39	111
	$C_2C'_3$	(Pnc)P(OCPh=CPhO)OC(O)Ph	−39	217
	$C_2C'_3$	(Glc)P[O(3,5-$tBu_2C_6H_2$)O] $OC(O)CH_2NHPh$	−28	109
	CC'_4	$(Cat)_2POMe$	−24.8	210
	CC'_4	[$OCMe_2C(O)O$]P(OCPh=CPhO)OPh	−47.5	218
	C'_5	(Cat)P(OCTf=CTfO)OPh	−29.5	219
	C'_5	$(Cat)_2POPh$	−29.1(c) −25.9(n)	20
	C'_5	$(Cat)_2PO(4-MeC_6H_4)$	−30(r)	220
	C'_5	$(Cat)_2P(Cat)P(Cat)_2$	−30.5	220
O_5/5,6	C_5	(Pfp)P[$O(CH_2)_3O$]OMe	−47.9	221
	C_5	(Pfp)P[$O(CH_2)_3O$]$OCHTf_2$	−50.8	221
	C_4C'	(Pfp)P[$O(CH_2)_3O$]OPh	−68	86
	$C_2C'_3$	**9**; R^1R^2 = $OCH_2CMe_2CH_2O$, R^3 = OPh	−48.7	107
O_5/5,7	C_3,C'_2	(Pnc)P(OCPh=CHCH=CPhO)OMe	−33.8	114
O_5/5,14	Si,C'_4	(Cat)P[(Cat)$CH_2CH_2OCH_2CH_2$ (Cat)]OTms	−39	222
O_5/5,5,6f	C_5	**20**	−33.5	152
O_5/5,6,6f	C_5	**21**; X-Y = Pfp, Z = P	−39.2	223
	C_5	**22**	−55.5	152
	C_5	**21**; X-Y = Pfp, Z = CMe	−36.1	223
O_5/5,6,6f,6f	C_5	**10**; X-Y = Pfp	−41.7	223
	$C_3C'_2$	**10**; X-Y = OCTf=CTfO	−41.7(c, r)	118, 195

TABLE Q (continued)

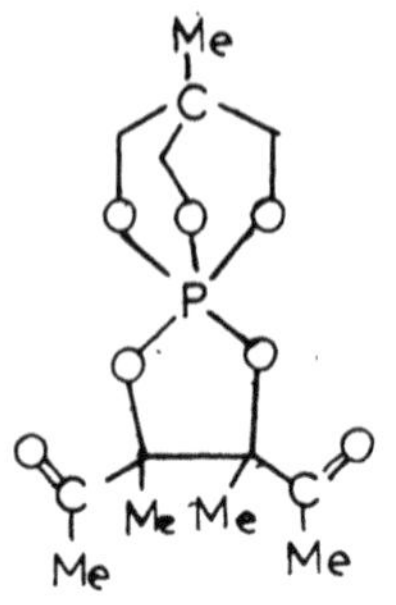

22

P-bound atoms/rings	Connectivities	Structure	NMR data (δP[solv temp] J Hz)	Ref.
	$C_3C'_2$	**10**; X-Y = $O(2,4\text{-}tBu_2C_6H_2)O$	−43.3	119a
$O_4O'/5,5$	C_4	$(Pfp)_2PO^-K^+$	−37.9	4
	C_4	$(Pfp)_2PO^-HNEt_3^+$	−33.6	52
	$C_2C'_2$	$[OC(O)CH_2O]_2PO^-HNEt_3^+$	−36	224
	$C_2C'_2$	**16**; X = $O^-HNEt_3^+$	−42.7(k)	225
	$C_2C'_2$	$[OC(O)CMe_2O]_2PO^-HNEt_3^+$	−49	224
$S_5/5,5$	$C_3C'_2$	$(SCH_2CH_2S)P(SCTf = CTfS)SC_4H_9$	−7.0(c)	81
	CC'_4	(SCTf = CTfS) P[S (4-MeC_6H_3)S]SC_4H_9	−4.4(f)	81

P-bound atoms/rings	Connectivities	Structure	NMR data (δP[solv, temp] J_{PH}Hz)	Ref.
$HFNO_2/5$	$H_2;C_2$	$H(F)P(Pfp)NH_2$	−44.0(c), J915	23
HF_4		HPF_4	−53.6(−98°) J1115	226
			J_{PF}986	227
			−50, J1084	
$HNO_2S/5,5f$	CC';CC';C	**6**; X = H, Y = SEt, R^1 = ODop, R^2 = $OCHMeCH_2$	−35, J800	228
$HN^4O_3/5,5f,5f$	$C_3;C_3$	**7**; Z = H	−20.9, J779	229
$HNO_3/5$	$C'_2;CC'_2$	(Cat)HP(OBu)N˙CH=CHCH=C˙H	−39, J700	230
$HNO_3/5,5$	$CH;C_2N'$	HP(Glc)(ON=CtBuNH)	−28, J788	231
	$HN';C_2C'$	(Glc)HP(OCPh=NNH)	−38.3, J847	232
	$HN';C_2C'$	(Pnc)HP(OCPh=NNH)	−46, J820	232
	$HN';C'_3$	HP[O(2-$MeOC_6H_4$)O](OCPh=NNH)	−36.5, J879	232
	$CH;C_3$	HP(Glc)(OCH_2CH_2NH)	−36.4, J780	233
	$CH;C_3$	HP(Glc)(OCH_2CMe_2NH)	−41.4, J768	234
	$CH;C_3$	HP(Pnc)(OCH_2CH_2NH)	−47.3, J777	233
	$CH;C_3$	HP(Pnc)($PCHPhCH_2NH$)	−49.5(b), J810	235
	$CH;C_2C'$	HP(Glc)[OC(O)CH_2NH]	−38, J825	236
	$CH;C_2C'$	HP(Glc)[OC(O)CHMeNH]	−43, J829	237
	$CH;C_2C'$	HP(Glc)[OC(O)CMe_2NH]	−47, J828	237
	$C'H;C_2C'$	HP(Glc)(Oam)	−37.7, J822	238
	$CH;C'_3$	HP(Cat)[OC(O)CMe_2NH]	−47, J885	236
	$CH;C'_3$	HP[O(30MeC_6H_3)O] [OC(O)CMe_2NH]	−46, J880	236
	$CH;C'_3$	HP(Cat)[OC(O)CHPhNH]	−42, J900	236
	$C'H;C'_3$	HP(Cat)(Oam)	−35, J884	239

P-bound atoms/rings	Connectivities	Structure	NMR data (δP[solv temp] J Hz)	Ref.
	$C_2;C_3$	HP(Pnc)(OCHMeCHMeNMe)	−54.5, J790	240
	$C_2;C_3$	HP(Pnc)(OCHPhCH$_2$NMe)	−52, J792	240
HNO$_3$/5,5f	$C_2;C_3$	**6**; X = H, Y = OtBu, R[1] = OCHPhCHMe, R[2] = OCH$_2$CH$_2$	−42.6, J795	241
HNO$_3$/5,5	$C_2;C_2C'$	HP(Glc)[OC(O)C˙H(CH$_2$)$_3$N˙]	−42, J825	242
	$C_2;C_2C'$	HP(OCHPhCHMeNMe) [OC(O)CMe$_2$O]	−50, J860	243
	$C_2;CC'_2$	HP(Cat)(OCH$_2$CH$_2$NMe)	−49.5, J846	239
	$C_2;CC'_2$	HP(Cat)(OCHMeCHMeNMe)	−42, J868	239
	$C_2;CC'_2$	HP(Cat)(OCHPhCH$_2$NMe)	−40.5, J862	240
	$C_2;CC'_2$	HP(Cat)(OCHPhCHPhNMe)	−39.5, J863	240
	$CC';C_3$	HP(Pnc)[OCHMeC(O)NMe]	−48.5, J768	244
	$CC';C_2C'$	HP(Glc)[OC(O)CH$_2$NPh]	−45, J855	236
HNO$_3$/5,5f	$CC';C_2C'$	**6**; X = H, Y = OMe, R[1] = ODop, R[2] = OCHMeCH$_2$	−40, J850	228
	$CC';C_2C'$	**6**; X = H, Y = OCH$_2$CH$_2$OH, R[1] = ODop, R[2] = OCHMeCH$_2$	−42, J850	228
	$CC';C_2C'$	**6**; X = H, Y = OCH$_2$CH$_2$Cl, R[1] = ODop, R[2] = OCHMeCH$_2$	−38, J840	228
HNO$_3$/5,5	$CC';C_2C'$	HP(Pnc)(ODopNMe)	−46, J824	239
	$CC';C'_3$	HP(Cat)(ODopNMe)	−36, J880	239
DNO$_3$/5,5f	$C'_2;CC'_2$	**6**; X = D, Y = OCD$_3$, R[1] = R[2] = OC(tBu)=CH	−34.4, J_{PD}133.7	245
HNO$_3$/5,6	$CH;C_2C'$	HP[OC(O)CMe$_2$NH][O(CH$_2$)$_3$O]	−55, J855	236
HN$_2$O$_2$/5,5	$HN',C'H;C_2$	HP(Pnc)(NHCPh=NNH)	−62, J766	246
	$HN',C'H;C_2$	HP(NHCPh=NNH)(OCHPhCHPhO)	−56, J743 773	247
	$HN',C'H;C'_2$	HP(Cat)(NHCPh=NNH)	−48, J797	247
	$(CH)_2;C_2$	HP(OCH$_2$CH$_2$NH)(OCHMeCH$_2$NH)	−39.5(n), J785	235
	$(CH)_2;C_2$	HP(OCH$_2$CHMeNH)$_2$	−56(b), J734	235
	$(CH)_2;C_2$	HP[O(cycloC$_6$H$_{10}$)O]$_2$	−58, J729	239
	$(CH)_2;C'_2$	HP[OC(O)CMe$_2$NH]$_2$	−75, J816	248
	$(C'H)_2;N'_2$	HP[ON=C(Ph)NH]$_2$	−31, J782	231
	$(C'H)_2;C_2$	HP(Pnc)(NHDopNH)	−64.4, J750	247
	$(C'H)_2;C_2$	HP(OCHPhCHPhO)(NHDopNH)	−54.5, J743; −52.5, J772*	247
	$(C'H)_2;CC'$	HP(Oam)(OCHPhCONH)	−53.5, J802; −51.5, J826*	242
	$(C'H)_2;C'_2$	HP(Oam)$_2$	−47.5, J835	238
	$C'H,CC';CC'$	HP(Oam)[OCH$_2$C(O)NMe]	−52, J816	249
HN$_2$O$_2$/5,5f	$C'H,CC';CC'$	**6**; X = H, Y = NHPh, R[1] = ODop, R[2] = OCHMeCH$_2$	−54, J820	228
HN$_2$O$_2$/5,5	$(C_2)_2;C_2$	HP(OCH$_2$CH$_2$NMe)$_2$	−62, J756	108
	$(C_2)_2;C_2$	HP(OCHMeCHMeNMe)$_2$	−67, J760	239
HN$_2$O$_2$/5,5f	$(C_2)_2;C_2$	**6**; X = H, Y = NMe$_2$, R[1] = OCH$_2$CH$_2$, R[2] = OCH$_2$CMe$_2$	−41.4, J820	250
HN$_2$O$_2$/5,5	$(C_2)_2;C_2$	HP(OCHPhCHPhNMe)$_2$	−63, J778	239
	$(C_2)_2;C_2$	HP(OCHPhCH$_2$NMe)$_2$	−60.5, J790	240
	$(C_2)_2;C_2$	HP[OCH(CONHMe)CHPhNMe]$_2$	−63.5, J784	240
	$(C_2)_2;CC'$	HP(OCHPhCHMeNMe) [OC(O)C˙H(CH$_2$)$_3$N˙]	−60, J790	242
HN$_2$O$_2$/5,5f	$C_2,CC';CC'$	**6**, X = H, Y = NEt$_2$, R[1] = ODop, R[2] = OCHMeCH$_2$	−42, J820	228
HN$_2$O$_2$/5,5	$(CC')_2;C_2$	HP[OCHMeC(O)NMe]$_2$	−57.5, J765	244
	$(CC')_2;C_2$	HP[OCMe$_2$C(O)NMe]$_2$	−62, J750	142

TABLE Q (continued)

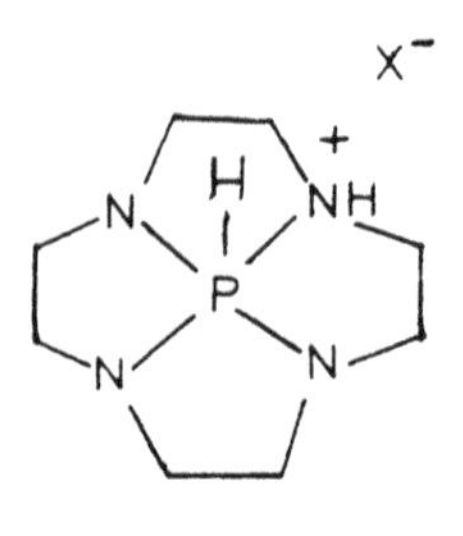

23

P-bound atoms/rings	Connectivities	Structure	NMR data (δP[solv temp] J Hz)	Ref.
	$(CC')_2;C'_2$	$HP(ODopNMe)_2$	−51.5, J884	239
$HN_2O_2/5,6$	$(C'H)_2;C'_2$	HP(Oam)[OC(O)DopNH]	−59, J860	218
$HNN'O_2/5,5$	$HN,C';C_2$	HP(Pnc)(N=CPhNHNH)	−55, J752	246
$HN^4{}_2N_2/5,5f,5f,5f$	$(C_2B)_2,(C_2)_2$	**5**; X = H, R = BH_3, n = 2,2,2,2	−28, J790	251
$HN^4{}_2N_2/5,6f,5f,6f$	$(C_2B)_2,(C_2)_2$	**5**; X = H, R = BH_3, n = 2,3,2,3	−45, J780	252
$HN^4N_3/5,5f,5f,5f$	$(C_2)_3,C_2H$	**23**; X = I^-	−45, J738	253
	$(C_2)_3,C_2H$	**23**; X = $Co(CO)_4$	−40, J729	253
	$(C_2)_3,C_2H$	**23**; X = $Mn(CO)_5$	−50, J653	253
	$(C_2)_3,C_2H$	**23**; X = $MoCp(CO)_3$	−47, J696	253
	$(C_2)_3,C_2H$	**23**; X = $WCp(CO)_3$	−44, J717	253
$HN_4/5,5f,5f,5f$	$(C_2)_4$	**5**; X = H, R = lp, n = 2,2,2,2	−54.5, J621	254
$HO_2S_2/5,5$	$C_2;C_2$	$HP(OCH_2CH_2S)_2$	+8, J711	255
$HO_4/5$	C_4	$HP(Pnc)(OMe)_2$	−26, J694	256
	$C_2C'_2$	$HP(Cat)(OBu)_2$	−28, J740	230
$HO_4/5,5$	$N'_2C'_2$	$HP(OCPh{=}NO)_2$	−18.3, J894	22
	$N'C'_3$	HP(Cat)(OCPh=NO)	−19, J890	224
	C_4	$HP(Glc)_2$	−27, J828	257
	C_4	HP(Glc)(Pfp)	−23.4, J894	61
	C_4	HP(Glc)(Pnc)	−35, J814	257
	C_4	$HP[OCH(CO_2Me)CH(CO_2Me)O]_2$	−22, J870	243
	C_4	$HP(Pnc)_2$	−42.0, J795	257
	C_4	$HP(OCHPhCH_2O)_2$	−25.2, J840	258
	C_4	$HP(OCHPhCHPhO)_2$	−28, J839, 824, 783*	258
	C_3C'	$HP(Glc)[OC(O)CH_2O]$	−28, J887	234
	C_3C'	HP(Glc)[OC(O)CHMeO]	−32, J894	234
	C_3C'	$HP(Glc)[OC(O)CMe_2O]$	−35, J900	243
	C_3C'	HP(Glc)[OC(O)CHPhO]	−31, J893	234
	C_3C'	HP(Pnc)[OC(O)CHPhO]	−37, J877, −36, J881	243
	$C_2C'_2$	$HP[OC(O)CH_2O]_2$	−35, J951	243
	$C_2C'_2$	**16**; X = H	−38, J980	212
	$C_2C'_2$	HP(Glc)(Cat)	−23, J875	259
	$C_2C'_2$	$HP[OC(O)CH(CH_2CO_2H)O]_2$	−40, J951	212
	$C_2C'_2$	$HP[OC(O)CMe_2O]_2$	−51, J920	243
	$C_2C'_2$	HP(Pnc)(Cat)	−29, J850	259
	$C_2C'_2$	$HP[OC(O)C(CH_2CO_2H)_2O]_2$	−45, J951	212
	$C_2C'_2$	$HP[OC(O)CHPhO]_2$	−42, J970	243
	CC'_3	$HP(Cat)[OC(O)CMe_2O]$	−36, J910	224
	CC'_3	HP(Cat)[OC(O)CHPhO]	−32, J930	243
	C'_4	$HP(OcycloC_6H_{10}O)_2$	−32, J821/804/783*	258
$HO_4/5,6$	C_4	$HP(Pfp)[O(CH_2)_3O]$	−35.8, J925	221

P-bound atoms/rings	Connectivities	Structure	NMR data (δP[solv temp] J Hz)	Ref.
H_2F_2N	H_2	$H_2F_2PNH_2$	−57.5	260
H_2F_3		H_2PF_3	−24.1(−40°), J866, J_{PF}865	226
H_2NO_2/5,5f	C_2;C_2	**6**; X = Y = H, R^1 = R^2 = OCH_2CH_2	−61.8, J655	261
	C_2;C_2	**6**; X = Y = H, R^1 = $OCHPhCH_2$, R^2 = OCH_2CMe_2	−62.4	261
H_2O_3/5	C_3	H_2(Pnc)$POCH_2$tBu	−44.2, J691	262
CBr_2F_2	Cl_3	$(Cl_3C)PBr_2F_2$	−27 J_{PF}1109	263
C′ClF_3	C′$_2$	$C_6F_5PClF_3$	−27 J_{PF}981	264
$CClNO_2$/5	C_3;C_2;C_2	(Pfp)tBuP(Cl)NEt_2	+14.9	265
$CClN_2O$/4,5f	H_3;(N′P^5)$_2$;C′	**3**; R^1 = Cl, R^2 = Me, X-Y = ODop	−35.9(ne)	129
	H_3;(N′P^5)$_2$;C′	**3**; R^1 = Cl, R^2 = Me, X-Y = OCPh=N	−32.8 −31	125
$CClN_2S$/4,5f	H_3;(N′P^5)$_2$;C′	**3**; R^1 = Cl, R^2 = Me, X-Y = SCMe=N	−25.1	163
C′ClN_2S/4,5f	C′$_2$;(N′P^5)$_2$;C′	**3**; R^1 = Cl, R^2 = Ph, X-Y = SCPh=N	−33.2 −27.5(ce)	163
CCl_2F_2	Cl_3	$(Cl_3C)PCl_2F_2$	−7.4 J_{PF}1106	263
	F_3	$TfPCl_2F_2$	−23.7 J_{PF}1130	266
	CF_2	$(nC_3F_7)PCl_2F_2$	−14.2 J_{PF}1095	263
C′Cl_2F_2	C′$_2$	$C_6F_5PCl_2F_2$	−14.3 J_{PF}960	264
CCl_2NO/5	H_3;HN′;C′	(Me)Cl_2P(NHN=CMeO)	−26.6(r)	32
C′Cl_2N_2/4	C′$_2$;$(C_2)_2$	(Ph)Cl_2P[NMeC(CCl_3)ClNMe]	−33.2	37
	C′$_2$;$(C_2)_2$	(Ph)Cl_2P(NMeCTfClNMe)	−35.0	37
CCl_2O_2/5	C′H_2;C_2	($PhCH_2$)Cl_2P(Pfp)	+7.3	52
C′Cl_2S_2/5	C′$_2$;C′$_2$	(SCTf=CTfS)P(Cl)$_2$Ph	−18	267
C′Cl_3F	C′$_2$	$C_6F_5P(F)Cl_3$	−37.5 J_{PF}892	264
C′Cl_4	C′$_2$	$C_6F_5PCl_4$	−70.9	264
	C′$_2$	$PhPCl_4$	−45.4(b) −36.8(m)	268
$CFNO_2$	C_3;C_2;C_2	tBu(F)P(Pfp)NEt_2	−13.3	265
C′FNO_2/5	C′$_2$;C_2;C_2	(4-MeC_6H_4)FP(Pfp)NEt_2	−34.4 J_{PF}826	53
	C′$_2$;C_2;C′$_2$	(OCTf=CTfO)PF(Ph)NEt_2	−34.2 J_{PF}821	53
CFN_2O/4	CH_2;$(C_2)_2$;C	($OCHPhCH_2$)P(NEt_2)$_2$F	−44 J_{PF}765	269
C′FO_3	CC′;C_3	(MeO)$_3$P(F)C=CClCF$_2$C′F$_2$	−72.8 J_{PF}845	270
CF_2NO/5	H_3;C′H;C′	(Me)F_2P(Oam)	−21.4(r, −60°) J_{PF}789, 1029	48
	H_3;CC′;C′	(Me)F_2P(N-tBuCH=C-tBuO)	−20.5(n) J_{PF}843, 1042	271
C′F_2NO/5	C′$_2$;C′H;C′	(Ph)F_2P(Oam)	−37.0(t, −50°) J_{PF}809/1025	48
CF_2N_2	H_3;$(C_2)_2$	(Me)F_2P(NMe_2)$_2$	−49.1 J_{PF}712	272
CF_2N_2/4	HCl_2;$(CP^5)_2$	[(Cl_2HC)F_2P-NMe]$_2$	−64.8 gauche −62.1 *trans* −62.3 average	273
	H_2Cl;$(CP^5)_2$	[($ClCH_2$)F_2P-NMe]$_2$	−56.3(b)	67
	H_3;$(CP^5)_2$	[(Me)F_2P-NMe]$_2$	−51.2 gauche −47.4 *trans* −50.4 average	273
	H_3;(CC′)$_2$	(Me)F_2P[NMeC(O)NEt]	−34.1 J_{PF}922	274
	CH_2;$(CP^5)_2$	[(Et)F_2P-NMe]$_2$	−45.6(n)	67
	CH_2;(CC′)$_2$	(Et)F_2P[NMeC(O)NMe]	−32.0 J_{PF}939	274
	CH_2;CC′;C′$_2$	(Et)F_2P[NMeC(O)NPh]	−29.6 J_{PF}885	274

TABLE Q (continued)

P-bound atoms/rings	Connectivities	Structure	NMR data (δP[solv temp] J Hz)	Ref.
$C'F_2N_2/4$	$C'_2;(CP^5)_2$	$[(C_6F_5)F_2P\text{-}NMe]_2$	−65.5	275
	$C'_2;(CP^5)_2$	$[(Ph)F_2P\text{-}NMe]_2$	−56.1 J_{PF}900	266
	$C'_2;(CP^5)_2$	$[(Ph)F_2P\text{-}NEt]_2$	−53.6 J_{PF}900	266
	$C'_2;(CC')_2$	$(Ph)F_2P[NMeC(O)NMe]$	−46.8 J_{PF}928	274
	$C'_2;(CC')_2$	$(Ph)F_2P[NMeC(O)NEt]$	−43.7 J_{PF}925	274
$C''F_2N_2$	$C'';C_2$	$(NEt_2)_2F_2PC{\equiv}CPF_2(NEt_2)_2$	−79.2 J_{PF}669	276
	$C'';C_2$	$Co(CO)_6[(NEt_2)_2F_2PC{\equiv}CPF_2(NEt_2)_2]$	−61.1 J_{PF}732	276
CF_2O_2	$H_3;C_2$	$(Me)F_2P(OCHTf_2)_2$	−41.6	277
	$H_3;C'_2$	$(Me)F_2P(OC_6F_5)_2$	−54.9(k) J_{PF}870	63
$CF_2O_2/5$	$H_3;C'_2$	$[O(2,4\text{-}tBu_2C_6H_2)O]PF_2Me$	−12.1 J_{PF}955	54
$C'F_2O_2$	$C'_2;C_2$	$(Ph)F_2P(OCH_2CCl_3)_2$	−59.1	278
	$C'_2;C_2$	$(Ph)F_2P(OCH_2Tf)_2$	−58.5	278
	$C'_2;C'_2$	$(Ph)F_2P(OC_6F_5)_2$	−60.7(k) J_{PF}817	63
	$C'_2;C'_2$	$(Ph)F_2P(OPh)_2$	−69(b) J_{PF}829	63
$C'F_2O_2/5$	$C'_2;C_2$	$(Ph)F_2P(Glc)$	−31.4	64
	$C'_2;C_2$	$(Ph)F_2P(Pfp)$	−32.4 J_{PF}959	53
	$C'_2;C'_2$	$(Ph)F_2P(OCTf{=}CTfO)$	−30.8 J_{PF}976	53
$C'F_2S_2/5$	$C'_2;C'_2$	$(SCTf{=}CTfS)PF_2Ph$	+15 J_{PF}980	267
CF_3N	$H_3;C_2$	$(Me)F_3PNMe_2$	−37.2	279
	$H_3;C_2$	$F_3(Me)PN^{\cdot}CHMe(CH_2)_3C^{\cdot}H_2$	−38.4	280
	$H_3;C_2$	$(Me)F_3PN(CH_2Ph)Me$	−36.8(n) J_{PF}812, 962	66
	$H_3;C'_2$	$(Me)F_3PN^{\cdot}CH{=}CHCH{=}C^{\cdot}H$	−38.9	73
	$CH_2;C'_2$	$(Et)F_3PN^{\cdot}CH{=}CHCH{=}C^{\cdot}H$	−38.6 J_{PF}868, 991	73
CF_3N'	$H_3;C'$	$(Me)F_3PN{=}CTf_2$	−30.3	281
$C'F_3N$	C'_2,C_2	$(Ph)F_3PNMe_2$	−53.0 J_{PF}819, 959	282
	$C'_2;C_2$	$(C_6F_5)F_3PNEt_2$	−57.3 J_{PF}808/970	264
	$C'_2;C_2$	$(Ph)F_3PN^{\cdot}CHMe(CH_2)_3C^{\cdot}H_2$	−53.5	280
	$C'_2;C_2$	$(Ph)F_3PN(CH_2Ph)Me$	−53.3(n) J_{PF}828, 966	66
	$C'_2;CC'$	$(Ph)F_3PN(Me)C(CCl_3){=}NMe$	−57.6 −59.6 (de, 20°)	283
	$C'_2;CC'$	$(Ph)F_3PN(Me)CTf{=}NMe$	−58.1 −58.9 (de, −20°)	283
	$C'_2;C'_2$	$(Ph)F_3PN^{\cdot}CH{=}CHCH{=}C^{\cdot}H$	−60.0 J_{PF}860, 978	73
	$C'_2;C'_2$	$(Ph)F_3PN^{\cdot}CMe{=}CHCH{=}C^{\cdot}Me$	−46.2	73
$C'F_3O$	$C'_2;C$	$(Ph)F_3POCH_2CCl_3$	−55.7	278
	$C'_2;C$	$(Ph)F_3POCH_2Tf$	−54.2	278
CF_3S	F_3,C	$(Tf)F_3PSMe$	−28(0°) J_{PF}1028	284
	$H_3;C$	$(Me)F_3PSMe$	+2 J_{PF}925, 1062	76
$C'F_3S$	$C'_2;C$	$(Ph)F_3PSMe$	−18.9 J_{PF}942, 1042	76
	$C'_2;C$	$(G_6F_5)F_3PSEt$	−18 J_{PF}909, 1046	264
CF_4	F_3	$TfPF_4$	−66.4 J_{PF}1103	266
	H_3	$MePF_4$	−29.9 J_{PF}965	266
	H_2Cl	$ClCH_2PF_4$	−43.7	263
	H_2P^5	$F_4PCH_2PF_4$	−41.9(r)	285
	CH_2	$F_4PCH_2CH_2PF_4$	−35.3(r)	285
	CH_2	$EtPF_4$	−30	286
	C_3	$1\text{-}AdPF_4$	−35.6 J_{PF}1040	287
$C'F_4$	$C'H$	$F_4PCH{=}CHPF_4$	−53.6(r)	285
	$C'H$	$F_4PCH{=}CH_2$	−53.4	286
	$C'N$	$F_4PC^{\cdot}{=}CHCHCMeN^{\cdot}H$	−61.5(b)	73
	$C'S$	$F_4PC^{\cdot}{=}CHCH{=}CHS^{\cdot}$	−57.5	288

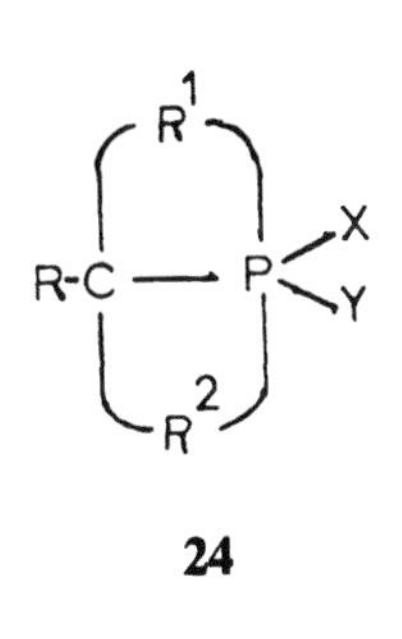

24

P-bound atoms/rings	Connectivities	Structure	NMR data (δP[solv temp] J Hz)	Ref.
	C'_2	$(C_6F_5)PF_4$	−52.2	289
	C'_2	$PhPF_4$	−51.7	263
	C'_2	$(3\text{-}TfC_6H_4)PF_4$	−52.9	279
	C'_2	$(2,5\text{-}Me_2C_6H_3)PF_4$	−42.6 J_{PF}980	266
$C''F_4$	C''	$HC{\equiv}CPF_4$	−67.8	286
CNO_2P^4/5,5	$H_3;CC';C'_2;CS'N$	**8**; R^1 = Me, R^2R^3 = NMeC(O)NMeP(S)Me	−16.5 J_{PP}232	290
$C'NO_2S$/5,5f	$C'_2;C_2;C_2;C$	**6**; X = Ph, Y = SnBu, $R^1 = R^2 = OCH_2CH_2$	−29.6	291
	$C'_2;C_2;C_2;C'$	**6**; X = Ph, Y = SPh, $R^1 = R^2 = OCH_2CH_2$	−30.4	291
$C'NO_2S$/5,5	$C'_2;C'_2;C_2;C'$	(Glc)PhP[SDopN(2,6-$Me_2C_6H_3$)]	−10.6	292
CN^4O_3/5,5f,5f	$C'_3;C_3;C_3$	**7**; Z = CPh_3	+5.9	229
CNO_3/5	$C'H_2,C_2;C_2C'$	$(CH_2CH{=}CMeO)P(OMe)_2NMe_2$	−22.7	293
	$CC'H;C_2;C_3$	(Pnc)(MeO)NMe_2PCH(CO_2Me)CH(CO_2Me)OMe	−40.5	294
CNO_3/5,5	$H_3;C'N';C'_3$	Me(NPhN=CMeO)P(OCMe=CMeO)	−23	295
	$H_3;CH;C'_3$	Me-[NHCMe_2C(O)O]P(OCPh=CPhO)	−28	218
CNO_3/5,5f	$H_3;CC';CC'Si$	**6**; X = Me, Y = OTms, $R^1 = OCH_2CH_2$, R^2 = ODop	−24	296
CNO_3/5,5	$H_3;CC';C'_3$	(Cat)MeP(ODopNMe)	−14.4(b)	297
	$C'HN;C_2;C_3$	(Pnc)P(OCMe_2CH_2NMe)CH(Ph)NMe_2	−36, −40(c)	298
CNO_3/5,5f	$C'HO;C'H;C_2C'$	**24**; X = Y = OCH_2CH_2Cl, R = H, R^1 = Dop, R^2 = NCH(O)	−40.5(c)	299
CNO_3/5,5	$C'H_2;C'H;CC'_2$	(Oam)P(CH_2CH=CMeO)OEt	−12.8	300
CNO_3/5,5f	$CC'O;C'H;C_2C'$	**24**; X = Y = OMe, R = Tf, R^1 = OCTf=CH, R^2 = NHC(O)O	−46	301
CNO_3/5,5	$CC''H;C_2;C_3$	(Pnc)P(OCMe_2CH_2NMe)-CH(CN)CH_2CN	−46.5	302
	$C'_2H;C'H;CC'_2$	(Oam)P[CHPhC-(COMe)=CMeO]OEt	−21.9	300
	$C'_2H;C_2;C_2C'$	(Pnc)P(OCPh=CHCHPh)NMe_2	−16.2 −21.4*	114
	$C'_2H;C_2;C_2C'$	(OCH_2CH_2NMe)P[CHPhC-(COMe)=CMeO]OMe	−26	105
	$C'_2H;C_2;C_2C'$	(Pnc)P[CHPhC(COMe)=CMeO]-NMe_2	−17 −24.5*	105
	$C'_2H;C_2;C_2C'$	(NMeCHMeCHPhO)P-(OCMe=CMeCHPh)OMe	−26 −31.5*	141
	$C'_2H;C_2;C_2C'$	(Pnc)P[OC(Ph)=CHCH(COPh)]-NMe_2	−18.1 −21.9*	114
	$C'_2H;C_2;C_2C'$	(Pnc)P[OCMe=CHCH(CO_2Me)]-NMe_2	−17.2 −22.3*	114
	$C'_2H;C_2;C_2C'$	(Glc)P[OCPh=CHCH(COPh)]NMe_2	−12.6	114

TABLE Q (continued)

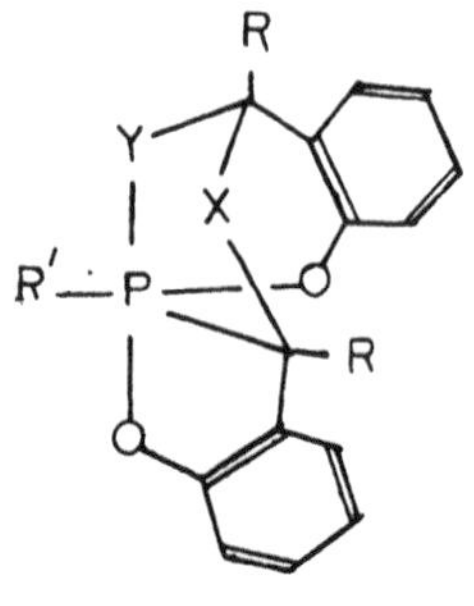

25 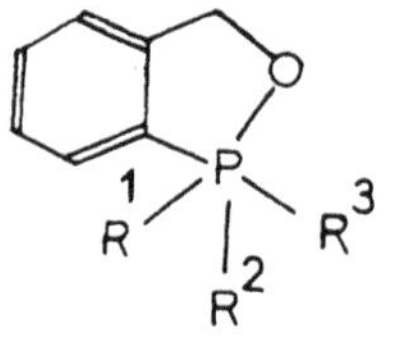 **26**

P-bound atoms/rings	Connectivities	Structure	NMR data (δP[solv temp] J Hz)	Ref.
	$C'_2H;C'_2;C_2C'$	**6**; X = Y = OMe, R^1 = CHPhCMe=CH, R^2 = ODop	−19.4	303
CNO_3/5,5,5f	$C'HO;C'H;C_2C'$	**24**; XY = Glc, R = H, R^1 = ODop, R^2 = NHC(O)O	−17.8	299
	$CC'O;C'H;C_2C'$	**24**; XY = Glc, R = Tf, R^1 = OCMe=CH, R^2 = NHC(O)O	−29	301
	$C'_2H;C'_2;C_2C'$	**6**; XY = Glc, R^1 = CHPhCH=CH, R^2 = ODop	+4.7	303
	$C'_2H;C'_2;C'_3$	**6**; XY = Cat, R^1 = CHPhCH=CH, R^2 = ODop	+7.2	303
CNO_3/5,5f,6	$C'HO;C'H;C_2C'$	**24**; XY = $OCH(Me)CH_2CH_2O$, R = H, R^1 = ODop, R^2 = NHC(O)O	−37.2	299
CNO_3/5,5f,6f	$CC'O;C_2;CC'_2$	**25**; R′ = NEt_2, X = Y = O, R = Me	−22	304
$CN'O_3$/5	$C_2N';C'_2;C_3$	$(MeO)_3P[CTf_2N{=}CPhN(Mes)]$	−36.9(b)	305
	$C_2N';C'_2;C_3$	$(EtO)_3P[CTf_2N{=}CPhN(Mes)]$	−39.2(b)	305
	$C'_2;C'_2;C_2C'$	$(MeO)_2PhP[ODopN(Mes)]$	−46.8, −47.9*	306
$C'NO_3$/5,5	$CO;C_2;C_2C'$	(Glc)P[C(O)-$CMe_2C({=}CMe_2)O]NMe_2$	−53	307
	$C'_2;CH;C_3$	$(Cat)P[NHCH_2C(O)O]Ph$	−32	308
	$C'_2;C'H;C_2N'$	(ON=CPhNH)P(Glc)Ph	−19.3	309
	$C'_2;C'H;C_3$	**26**; R^1R^2 = Pnc, R^3 = NHPh	−39(c, 25°)	310
	$C'_2;C'H;C_2C'$	(Glc)P(OCPh=CHNH)Ph	−25.4	309
	$C'_2;C'H;C_2C'$	(Glc)P(OCPh=CPhNH)Ph	−29.7	309
	$C'_2;C'H;CC'_2$	**26**; R^1R^2 = $O(3,4\text{-}Me_2C_6H_2)O$, R^3 = NHPh	−27.4(c, 25°)	310
	$C'_2;C'H;CC'_2$	**26**; R^1R^2 = $OC(4\text{-}MeC_6H_4){=}C(4\text{-}MeC_6H_4)O$, R^3 = NHPh	−32.6(c, 25°)	310
	$C'_2;C_2;C_3$	$(Glc)P(NMeCH_2CH_2O)Ph$	−35(r)	215
	$C'_2;C_2;CC'_2$	**26**; R^1R^2 = OCMe=CMeO, R^3 = NMe_2	−24.4(c, 25°)	310
	$C'_2;C_2;CC'_2$	**26**; R^1 = NMe_2, R^2R^3 = $O(3,4\text{-}Me_2C_6H_2)O$	−20.1(c, 25°)	310
$C'NO_3$/5,6	$C'_2;C'H;C_2N'$	$[O(CH_2)_3O]PhP(ON{=}CPhNH)$	−35	309
	$C'_2;C'H;C_3$	$(Glc)P(OCH_2DopNH)Ph$	−44.5	309
	$C'_2;C'H;C_2C'$	$[O(CH_2)_3O]P(OCPh{=}CHNH)Ph$	−41.1	309
	$C'_2;C'H;C_2C'$	$[O(CH_2)_3O]P(OCPh{=}CPhNH)Ph$	−45.6	309

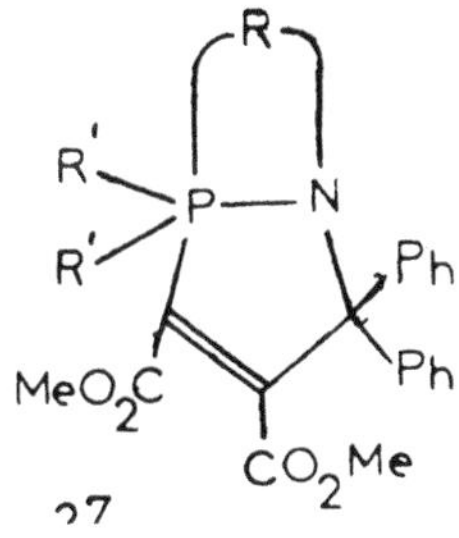

27

P-bound atoms/rings	Connectivities	Structure	NMR data (δP[solv temp] J Hz)	Ref.
$C'N'O_3/5$	$C'_2;C';C_3$	$(MeO)_2PhP[N{=}CHC(Me)tBuO]$	−18.9	119b
	$C'_2;C';C_3$	$(MeO)_2PhP(N{=}CHCPh_2O)$	−15.9	119b
$CN_2O_2/4,5$	$H_3;(CP^5)_2;C_2$	$[(Pfp)MeP\text{-}NMe]_2$	−27.1 −34.8(c)	120
$CN_2O_2/4,5f$	$CHO;(C'P^5)_2;C_2$	**3**; $R^1 = R^2 = OCH_2CF_2CF_2H$, X-Y = $CH(CCl_3)OC(O)$	−40	311
	$C'C''O;(C'P^5)_2;C_2$	**3**; $R^1 = R^2 = OCH_2CF_2CF_2H$, X-Y = $CPh(CN)OC(O)$	−61	311
$CN_2O_2/5,5$	$H_3;(C'H)_2;C'_2$	$(Oam)_2PMe_3$	−27.9(t, −25°)	48
$CN_2O_2/5,5f$	$H_3;C_2,C'_2;C'_2$	**6**; X = Me, Y = NMe_2, R^1 = ODop, R^2 = OCPh=CPh	−22(r)	297
$CN_2O_2/5,5$	$H_3;(CC')_2;C'_2$	(Cat)MeP(NMeDopNMe)	−25.1(b)	297
	$H_3;(CC')_2;C'_2$	$MeP(NMeDopO)_2$	−32(b)	297
	$CH_2;(CH)_2;C_2$	$(OCH_2CH_2NH)_2PCH_2CH_2CN$	−38	302
	$C'_2H;(C_2)_2;CC'$	$(OCH_2CH_2NMe)P$ $[CHPhC(COMe){=}CMeO]NMe_2$	−31.5 −36*	105
	$C'_2H;(C_2)_2;CC'$	(NMeCHMeCHPhO)P $(OCMe{=}CMeCHPh)NMe_2$	−32 −39.5*	141
$CN_2O_2/5,5,5f$	$C'_2H;C'H,CN';C'_2$	**24**; XY = NHCPh=NNMe, R = H, R^1 = ODop, R^2 = OCMe=C(COMe)	−9.7	155
$C'N_2O_2/5,5$	$CO;(C_2)_2;CC'$	$(OCH_2CH_2NMe)P$ $[C(O)CMe_2C({=}CMe_2)O]NMe_2$	−73	307
	$C'_2;(CH)_2;C_2$	$(OCH_2CMe_2NH)_2PPh$	−42	312
	$C'_2;CH,C_2;C'_2$	$(Cat)P(NHCH_2CH_2NBu)Ph$	−27	149
	$C'_2;(C'H)_2;C_2$	(Glc)P(NHDopNH)Ph	−39	309
	$C'_2;(C'H)_2;C'_2$	$(Oam)_2PPh$	−35.1(t, −25°)	48
$C'N_2O_2/5,5f$	$C'_2;C'H,C_2;C_2$	**6**; $R^1 = R^2 = OCH_2CH_2$, X = Ph, Y = NHPh	−49.8	97
	$C'_2;C'H,C_2;C'_2$	**6**; $R^1 = R^2 = OC(O)CH_2$, X = Ph, Y = NHPh	−50.8	99, 313
$C'N_2O_2/5,5$	$C'_2;(CC')_2;C_2$	$PhP[OCMe_2C(O)NMe]_2$	−46.5(c)	142
$CN_2S_2/5,5$	$CH_2;(C'H)_2;C'_2$	$(NHDopS)_2PEt$	−28	314
$CN_3O/4,5$	$C'_2H;C_2,(CC')_2;C'$	[NMeC(O)NMe]P $[CHPhC(COMe){=}CMeO]NMe_2$	−36.5 −40.5*	105
$CN_3O/5,5f$	$H_3;C_2,C'_2,C'N';C'$	**12**; R^1 = Me, R^2 = NMe_2, R = OCPh=CPh, X-Y = CH=CHCH=CH	−37.1	135
$CN_3O/5,5,5f$	$C'_2H;CN'$, $C'H,C'_2;C'$	**6**; XY = NHCPh=NNMe, R^1 = CHPhCH=CH, R^2 = ODop	−26.4	155
	$C'_2H;(C_2)_2,C'_2;C'$	**6**; XY = $NMe(CH_2)_2NMe$, R^1 = CHPhCH=CH, R^2 = ODop	−17.9	303
$C'N_3O/5,5f$	$C'_2;CC',(C_2)_2;C'$	**27**; R = OCPh=CH, R′ = NMe_2	−31.5	315
$C'N_3O/5,5,5f$	$C'_2;(C_2)_2,CC';C'$	**6**; XY = $NMe(CH_2)_2NMe$, R^1 = $CPh{=}CHCH_2$, R^2 = ODop	−21.5	303

TABLE Q (continued)

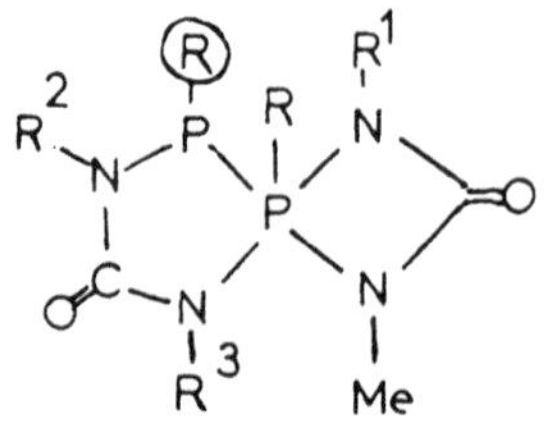

28

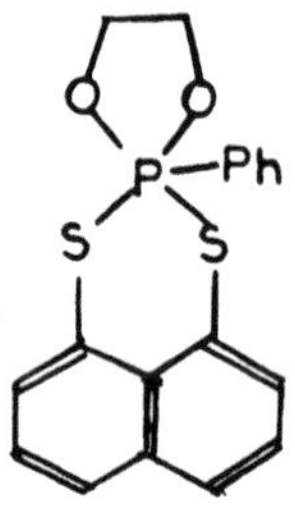

29

P-bound atoms/rings	Connectivities	Structure	NMR data (δP[solv temp] J Hz)	Ref.
	$C'_2;(C_2)_2,CC';C'$	**6**; XY = $NMe(CH_2)_2NMe$, R^1 = $CPh{=}CMeCH_2$, R^2 = ODop	−18.6	303
$C'N_3P/4,5$	$C'_2;C_3;C'N$	**28**; R = Ph, R^1 = R^2 = R^3 = Me	−69 J_{PP}156	316
	$C'_2;C_2C';C'N$	**28**; R = C_6F_5, R^1 = R^2 = 3-TfC_6H_4, R^3 = Me	−72.6 J_{PP}184.5	317
	$C'_2;CC'_2;C'N$	**28**; R = R^1 = R^3 = Ph, R^2 = Me	−64.3 J_{PP}190	316
$CN_3S/4,5f$	$H_3;(N'P^5)_2,C_2;C'$	**3**; R^1 = Me, R^2 = NMe_2, X-Y = SC(Ph)=N	−18.7 −22 (ce)	163
$CN_4/5,5$	$H_3;(CC')_4$	$MeP(NMeDopNMe)_2$	−39.5(c)	297
$C'N_4/4,4$	$C'_2;(CS)_4$	$PhP(NEtSO_2NEt)_2$	−89.5	318
$C'N_4/4,5f$	$C'_2;(C_2)_2(CC')_2$	**27**; R = NMeC(O), R′ = NMe_2	−31(c)	319
$C'N_4/5,5f,5f,5f$	$C'_2;(C_2)_4$	**5**; X = Ph, R = 1p, n = 2,2,2,2	−45.3(b)	320
$C'O_2S_2/5,6$	$C'_2;C'_2;C'_2$	$PhP(ODopS)_2$	+18.2(r)	321
	$C'_2;C'_2;C'_2$	**9**; R^1 = Ph, R^2R^3 = S(3-MeC_6H_4)S	+29.5	321
	$C'_2;C_2;C'_2$	**29**	−17.6(b)	322
$C'O_2S_2/5,7$	$C'_2;C_2;C'_2$	**30**	+17(b)	322
$CO_3S/5,5$	$C'_2H;C_2C';C'$	(Glc)P [OCMe=C(COMe)CHPh]OMe	−9.6 −7.6*	323
	$C'_2H;C'_3;C'$	[CHPhC(COMe)=CMeO]P (ODopS)O(4-FC_6H_4)	+13.5(r)	175
$C'O_3S/5$	$C'_2;C_2C';C$	$(MeO)_2P[C({=}CH_2)CH{=}C(Me)O]SMe$	−26.5	293
$CO_3Se/5,5$	$C_2N';C_3;C'$	$(Glc)P[SeC(NMe_2){=}NCTf_2]OMe$	−26.1	324
CO_4	$Cl_3;C'_4$	$(PhO)_4PCCl_3$	−76.5(b)	325
	$H_3;C_4$	$MeP(OCHTf_2)_4$	−53.0	277
$CO_4/4$	$CH_2;C_4$	$(CH_2CTf_2O)P(OMe)_2OCHTf_2$	−35.7	184
	$C_3;C_4$	$(CTf_2CTf_2O)P(OCHTf_2)_3$	−4	326
$CO_4/5$	$H_2N;C_4$	$(MeO)_3P[CH_2N(Tos)CH_2O]$	−30.3	327
	$H_2N;C_4$	$(EtO)_3P[CH_2N(Tos)CH_2O]$	−33.6	327
	$H_2P^5;C_4$	$CH_2[P(OEt)_2$(OCMe (COMe)CMe(COMe)O$]_2$	−30.0 −30.5 −30.2*	328
	$H_2P^5;C_2C'_2$	$CH_2[P(OEt)_2(OCMe{=}CMeO)]_2$	−29.9	328
	$H_3;C_4$	$(Pfp)P(OCHTf_2)_2Me$	−24	326
	$CHO;C_4$	$(EtO)_3P[OCH(CCl_3)OCH(CCl_3)]$	−42	329
	$CHO;CC'_3$	$(PhO)_3P[OCH(CCl_3)OCH(CCl_3)]$	−55	330
	$C_2N';C_3,C'$	$(MeO)_3P(CTf_2N{=}CtBuO)$	−34.7	331
	$C_2N';C_3,C'$	$(EtO)_3P(CTf_2N{=}CtBuO)$	−38.4	331
	$C_2N';C_3,C'$	$(MeO)_3P[C(CClF_2)_2N{=}CPhO]$	−34.1(b)	332
	$C_2N';C'_4$	$(PhO)_3P[CTf_2N{=}C(OEt)O]$	−50.6(b)	332
	$C_2N';C'_4$	$(PhO)_3P[C(CClF_2)_2N{=}CtBuO]$	−46.8(b)	332
	$C_2H;C_4$	$(MeO)_3P$[OCH $(CF_2CF_2H)CH_2CH(CF_2CF_2H)]$	−37.5	333

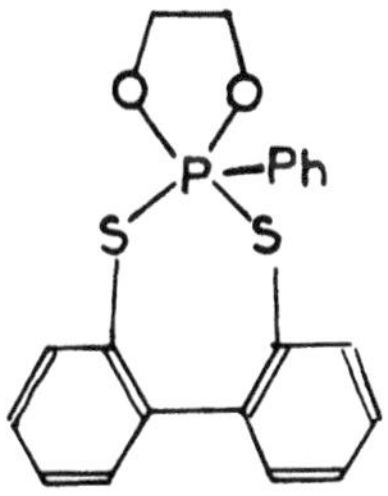

30

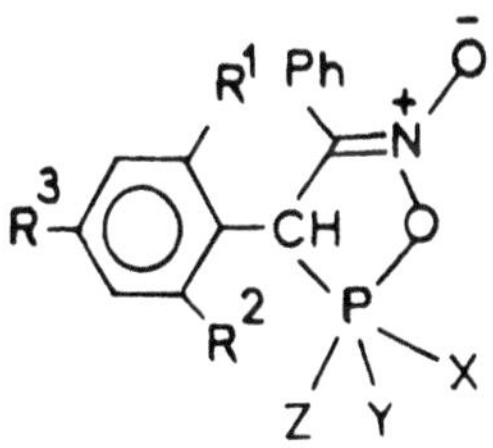

31

P-bound atoms/rings	Connectivities	Structure	NMR data (δP[solv temp] J Hz)	Ref.
	$C_2H;C_4$	$[H(CF_2)_2CH_2]_3P[OCH(CF_2CF_2H)CH_2CH(CF_2CF_2H)]$	−43	333
	$CC'H;C_4$	$(Pnc)P(OMe)_2CH(CO_2Me)CH(CO_2Me)OMe$	−46.5	294
	$CC'H;C_4$	$(Pnc)P(OMe)_2CH(CONMe)_2)CH(CO_2Me)OMe$	−38.8 −41.4*	334
	$CC'H;C_3C'$	$(MeO)_3P[CHMeC(COMe){=}CMeO]$	−27	201
	$C'_2H;C_3;N^4$	**31**; X = Y = Z = OMe, $R^1 = R^2 = R^3 = H$	−36.6(c)	336
	$C'_2H;C_3,N^4$	**31**; X = Y = Z = OMe, $R^1 = R^2 = Cl$, $R^3 = H$	−42.7(c)	336
	$C'_2H;C_3,N^4$	**31**; $R^1 = R^2 = R^3 = X = Y = Z = OMe$	−39.1(c)	336
	$C'_2H;C_3C'$	$(MeO)_3P[CHPhC(COMe){=}CMeO]$	−37	201
	$C'_2H;C_3C'$	$(MeO_3P[CH(NO_2C_6H_4)C(COMe){=}CPhO]$	−27.8	201
	$C'_2H;C_3C'$	$(MeO)_3P[CH(NMe_2C_6H_4)C(COMe){=}CPhO]$	−36.1	201
	$C_3;(P^4)_2,C'_2$	$(Cat)tBuP[OP(S)(OEt)_2]_2$	−25.5	145
$CO_4/5,5$	$H_2N;C_4$	$(Glc)_2PCH_2NMe_2$	−9.5	336
	$H_3;C_4$	$(Pfp)_2PMe$	+1.9	337
	$H_3;C_2C'_2$	$[OC(O)CPh_2O]_2PMe$	−24(k)	99
	$H_3;C'_4$	$(Cat)_2PMe$	+3.2(c) +1.8(b) +2.4(n)	20
	$H_3;C'_4$	**8**; $R^1 = Me$, $R^2R^3 = Cat$	−17.8	338
	$H_3;C'_4$	$[O(2,4\text{-}tBu_2C_6H_2)O]_2PMe$	+3	54
	CHN,C_4	$(Glc)_2PCH(NMe_2)CH_2CO_2Me$	−7.9 −9.3	339
	$C'HN;C_4$	$(Glc)_2PCH(Ph)NHMe$	−14	336
	$C'HN;C_4$	$(Pnc)_2PCH(Ph)NHMe$	−26	336
	$C'HN;C_4$	$(Pnc)_2PCH(Ph)N(CH_2)_4$	−21.5	336
	$C'HN;C_4$	$(Glc)(Pnc)PCH(Ph)NHMe$	−20	336
	$C'HO;C_4$	$(Glc)_2PCH(OH)Ph$	−18(r)	340
	$C'HO;C_4$	$(Pnc)_2PCH(OH)Ph$	−28.3	340
	$CH_2;C_4$	$(Glc)_2PCH_2CH_2CN$	−12	302
	$CH_2;C_4$	$(OCHMeCHMeO)_2PEt$	−12.6 −13.7*	341
	$CH_2;C_4$	$(Pnc)_2PEt$	−22	342
	$C'H_2;C_3C'$	$(Pnc)P[OC(OMe){=}CHCH_2]OMe$	−12.4	334
	C_2O,C_4	$(Pnc)_2PC(OH)Tf_2$	−34.5(r)	340
	C_2O,C_4	$(Glc)P(CTf_2OCTf_2O)OMe$	−16.3	61
	$CC'O;C_3C'$	$(Pnc)P[CTf(OH)CH{=}CTfO]OEt$	−22 −25*	343
	$CC'H;C_4$	$(Glc)_2PCH(CO_2Me)CH(NMe_2)CO_2Me$	−13 −15*	339
	$CC'H;C_4$	$(Pnc)_2PCH(CO_2Me)CH_2CO_2Me$	−29	302
	$CC'H;C_4$	$(Pnc)_2PCH(CO_2Me)CH(NMe_2)CO_2Me$	−28.5 −30.3*	339

TABLE Q (continued)

P-bound atoms/rings	Connectivities	Structure	NMR data (δP[solv temp] J Hz)	Ref.
	CC″H;C_4	$(Glc)_2PCH(CN)CH_2CN$	−27	302
	$C'_2H;C_3C'$	(Glc)P[CHPhC (COMe)=CMeO]OMe	−9, −11.2 −8.7*	105, 323
	$C'_2H;C_3C'$	(Glc)P[OCPh=CHCH(COPh)]OMe	−8	114
	$C_3;C'_4$	$(Cat)_2PtBu$	−5.6(b)	344
CO_4/6,6	$H_3;C_4$	$MeP(OCH_2CMe_2CH_2O)_2$	−40	87
	$C'H_2;C_4$	$PhCH_2P(OCH_2CMe_2CH_2O)_2$	−46.2	104
CO_4/5,5f,6f	CC′O;$C_2C'_2$	**25**; R′ = OEt, R = Me, X = Y = O	−20	304
	CC′O;CC'_3	**25**; R′ = OPh, R = Me, X = Y = O	−22	304
CO_4/5,6f,6f,6f	$C_2N';C_3C'$	**10**; XY = $OCPh{=}NCTf_2$	−17.9	345
$C'O_4$	$C'_2;C_4$	$(MeO)_4PC(CO_2Me){=}CH(CO_2Me)$	−55.7	346
	$C'_2;C_4$	$PhP(OCH_2Tf)_4$	−58.7 −61.7 (c, 30°)	278, 169
	$C'_2;C_4$	$PhP(OEt)_4$	−55	146
	$C'_2;C_4$	$Ph(EtO)_2P(OiPr)_2$	−58.9	347
	$C'_2;C_4$	$PhP(OCHTf_2)_4$	−65 −62(c)	30, 170
	$C'_2;C_4$	$PhP(OC^{\cdot}HCH_2CH_2CH_2C^{\cdot}H_2)_4$	−57.5	348
	$C'_2;C_4$	$PhP(OCH_2tBu)_4$	−60	348
$C'O_4$/5	C′O;C_3C'	$(EtO)_3P[C({=}CPh_2)OC({=}CPh_2O]$	−55(b)	349
	$C'P^2;C_2C'_2$	$(MeO)_2P$ $(OCPh{=}CPhO)C^{\cdot}{=}PNMeN{=}C^{\cdot}Me$	−36	350
	$C'_2;C_4$	(Glc) $(MeO)_2PC(CO_2Me){=}CH(CO_2Me)$	−34.9	346
	$C'_2;C_4$	$(Pfp)PhP(OMe)_2$	−31.2	351
	$C'_2;C_4$	(Pnc) $(MeO)_2PC(CO_2Me){=}CH(CO_2Me)$	−50, −52.6*	346
	$C'_2;C_4$	$(Pfp)PhP(OCHTf_2)_2$	−38.6	351
	$C'_2;C_3C'$	$(EtO)_3P$ $[C(CO_2Me){=}C(CO_2Me)C(O)O]$	−53.8	352
	$C'_2;C_3C'$	(Pnc)(MeO)(OCOPh)PC $(CO_2Me){=}CH(CO_2Me)$	−49	346
	$C'_2;C_2C'_2$	$(MeO)_2P[C({=}CH_2)CH{=}CMeO]OPh$	−49.6	293
	$C'_2;C_2C'_2$	$(Cat)P(OMe)_2Ph$	−29.5	87
	$C'_2;C_2C'_2$	$(Pnc)(PhO)_2PC(CO_2Me){=}CHCO_2Me$	−52.1 −55.5*	346
$C'O_4$/7	$C'_2;C_4$	**18**; R^1 = R^2 = OMe; R^3 = Ph	−55.4(c)	206
$C'O_4$/5,5	$NO';C_4$	$(Glc)_2PC(O)NHPh$	−31	336
	$NO';C_4$	$(Pnc)_2PC(O)NHPh$	−49	336
	$S'_2;C_2C'_2$	$[OCMe_2C(O)O]_2PCS_2{}^-HNEt_3{}^+$	−48.2(c)	224
	$CO';C_4$	$(Pfp)_2PC(O)Me$	−36.9	52
	$C'H;C_4$	$(Glc)_2PCH{=}CH(COOMe)$	−22.9 −19.5*	353
	$C'_2;C_4$	$(Glc)_2PC(COOH){=}CH(COOH)$	−17	353
	$C'_2;C_4$	$(Glc)_2PPh$	−20 −19.3	312
	$C'_2;C_4$	$(Glc)_2P(4\text{-}MeC_6H_4)$	−19.2	354
	$C'_2;C_4$	$(OCHMeCHMeO)_2PPh$	−26.8 −27.5*	341
	$C'_2;C_4$	$(Glc)(Pnc)PC(CO_2Me){=}CH(CO_2Me)$	−33	353
	$C'_2;C_4$	$(Pnc)_2PPh$	−35	312
	$C'_2;C_4$	(Pnc)P[OC $(OMe)_2CH{=}C(CONMe_2)$]OMe	−30.4	334
	$C'_2;C_4$	$(Pnc)_2PC(CO_2Me){=}CH(CO_2Me)$	−41 −39.7	339
	$C'_2;C_3C'$	$(Glc)P[OCH_2C(O)O]Ph$	−22(j)	215
	$C'_2;C_3C'$	$(Glc)P[OCMe_2C(O)O]Ph$	−27(j)	215
	$C'_2;C_3C'$	**26**; R^1R^2 = Pnc, R^3 = OPh	−28.1(c, 25°)	310
	$C'_2;C_2C'_2$	**8**; R^1R^2 = Pfp, R^3 = Ph	−9.2	101

P-bound atoms/rings	Connectivities	Structure	NMR data (δP[solv temp] J Hz)	Ref.
	C'_2;$C_2C'_2$	(Cat)P(Glc)Ph	−14(r)	215
	C'_2;CC'_3	**26**; R[1] = OPh, R[2]R[3] = O(3,4-$Me_2C_6H_2$)O	−18.3(c, 25°)	310
	C'_2;C'_4	$(Cat)_2PPh$	−9.8(c) −9(b) −9.5(n)	20
$C'O_4$/5,6	C'_2;C_4	(Glc)P($OCH_2CMe_2CH_2O$)Ph	−35(r)	215
$C'O_4$/6,6	C'O;C_4	$(OCH_2CMe_2CH_2O)_2PFu$	−65.3(f)	355
	C'_2;C_4	$PhP(OCH_2CH_2CH_2O)_2$	−50.2(c)	356
	C'_2;C_4	$PhP(OCH_2CMe_2CH_2O)_2$	−48.4	87
$C''O_4$/5,5	N'';$C_2C'_2$	**8**, R[1]R[2] = Pnc, R[3] = CN	−38.9(r)	122
$C'S_4$/5,5	C'_2;C'_4	$(SCTf{=}CTfS)_2PPh$	−11.5	267
CHF_3	H_3	Me(H)PF_3	−5.4(−60°) J_{PH}850	357
			−7.7(−50°) J_{PH}860 J_{PF}805, 965	358
	CH_2	nBu(H)PF_3	−7.1(−60°) J_{PH}840	357
	$C'H_2$	$(CH_2{=}CHCH_2)P(H)F_3$	−11(−80°) J_{PH}886 J_{PF}800, 966	359
$C'HF_3$	C'_2	Ph(H)PF_3	−30.2(−60°) J_{PH}863	357
$CHNO_2$/5,5f	H_3;C_2;C_2	**6**; X = H, Y = Me, R[1] = R[2] = OCH_2CH_2	−43.7 J_{PH}680	360
	H_3;C_2;C_2	**6**; X = H, Y = Me, R[1] = R[2] = $OCHMeCH_2$	−43.2 J_{PH}700 −44.6(t) J_{PH}707*	361
	H_3;C_2;C_2	**6**; X = H, Y = Me, R[1] = R[2] = OCH_2CH_2	−44.3 J_{PH}663	362
$C'HNO_2$/5	C'_2;C'H;C'_2	(Oam)P(H)[O(5-$NH_2C_6H_4$)]Ph	−39.5 J_{PH}736	363
$C'HNO_2$/5,5f	C'_2;C_2;C_2	**6**; X = H, Y = Ph, R[1] = R[2] = OC(O)CH_2	−47(u) J_{PH}780	313
	C'_2;C_2;C_2	**6**; X = H, Y = Ph, R[1] = R[2] = OCH_2CH_2	−44.8(c) J_{PH}705	361
	C'_2;C_2;C_2	**6**; X = H, Y = Ph, R[1] = OCH_2CHMe, R[2] = $OCMe_2CH_2$	−42.9 J_{PH}703 −48.1(c) J_{PH}715*	364
	C'_2;C_2;C_2	**6**; X = H, Y = 3,4-$Me_2C_6H_3$, R[1] = R[2] = $OCHMeCH_2$	−42.8 J_{PH}698 −47.4(c) J_{PH}703*	361
	C'_2;C_2;C_2	**6**; X = H, Y = Ph, R[1] = OCHPhCMe, R[2] = OCH_2CH_2	−54.1 J_{PH}687 −49.2 J_{PH}712*	362
	C'_2;C_2;C_2	**6**; X = H, Y = Ph, R[1] = $OCHPhCH_2$, R[2] = OCH_2CMe_2	−50.6 J_{PH}687 −45.3 J_{PH}700*	362
	C'_2;C'_2;C'_2	**6**; X = H, Y = Ph, R[1] = R[2] = ODop	−37.4 J_{PH}769	365
CHN_2O/5	CH_2;CN',C'H;C	(NHCPh=NNMe)HP(OEt)Et	−35.6 J_{PH}468	366
$C'HN_2O$/5	C'_2;CN',C'H;C	(NHCPh=NNMe)HP(OMe)Ph	−41.9 J_{PH}507	366
CHO_3/5	H_3;C_3	(Pnc)HP(Me)$OCMe_2CMe_2OH$	−47 J_{PH}690	214
	H_3;C'_3	(Cat)P(H)[O(5-OHC_6H_4)]Me	−9.5 J_{PH}765	363
$C'HO_3$/5	C'_2;C_3	($OCHMeCH_2O$)P(H)(Ph)$OCHMeCH_2OH$	−24 J_{PH}656	363
$C'HO_3$/5,5	C'_2;C_3	**26**; R[1] = H, R[2]R[3] = Pnc	−33(c, 25°) J_{PH}731	310
	C'_2;CC'_2	**26**; R[1] = H, R[2]R[3] = O(3,4-$Me_2C_6H_2$)O	−15.4(c, 25°) J_{PH}788	310
C_2Br_2F	$(F_3)_2$	Tf_2Br_2F	−41(r)	367
C_2Br_3	$(F_3)_2$	Tf_2Br_3	−51(r)	367
C'_2ClF_2	$(C'_2)_2$	$(C_6F_5)_2PClF_2$	−25 J_{PF}835	264
C_2ClNO/4	C_2H,C_3;C_2;C	(tBu)NEt_2P[$OCTf_2C$(Me)H]	+2.5 +5.4*	368
C_2ClNO/5	H_2,N'H;C'	Me_2ClP(NHN=CMeO)	−14.2(r)	32
C'_2ClN_2/4	$(C'_2)_2$;$(C_2)_2$	Ph_2P[NMeCPh(Cl)NMe]Cl	−28	37

TABLE Q (continued)

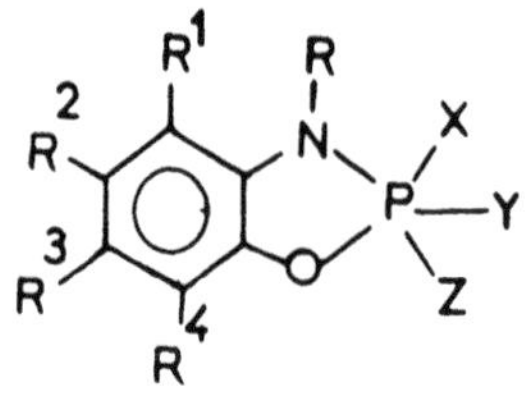

32

P-bound atoms/rings	Connectivities	Structure	NMR data (δP[solv temp] J Hz)	Ref.
$C_2ClO_2/4$	$CH_2,C_2H;C_2$	$Et(OCHTf_2)PCl(OCTf_2CHMe)$	+18.4 +13.5	369
$C'_2ClO_2/5$	$(C'_2)_2;C_2$	$Ph_2(Cl)P(Pfp)$	−5.4	351
	$(C'_2)_2;C'_2$	$Ph_2ClP(OCH{=}CHO)$	−4.9(q)	124
$C'_2ClS_2/5$	$(C'_2)_2;C'_2$	$(SCTf{=}CTfS)P(Cl)Ph_2$	−28	267
C'_2Cl_2F	$(C'_2)_2$	$(C_6F_5)_2PCl_2F$	−49 J_{PF}894	264
C_2Cl_2N	$(F_3)_2;C_2$	$Tf_2P(NMe_2)Cl_2$	−31.5	370
C_2Cl_3	$H_3;F_3$	$Tf(Me)PCl_3$	−36.4(30°)	371
C'_2Cl_3	$(C'_2)_2$	$(C_6F_5)_2PCl_3$	−80	264
$C'_2FNO/5$	$(C'_2)_2;C'H;C'$	**32**; X = F, Y = Z = Ph, R = R^1 = R^4 = H, R^2 = R^3 = NO_2	−36.9(a)	372
	$(C'_2)_2;C'H;C'$	**32**; X = F, Y = Z = Ph, R = R^2 = R^4 = H, R^1 = OMe, R^3 = CPh_3	−42(b)	372
	$(C'_2)_2;CC';C'$	**32**; X = F, Y = Z = Ph, R = Et, R^1 = R^2 = R^3 = R^4 = H	−45.3(b)	372
	$(C'_2)_2;CC';C'$	**32**; X = F, Y = Z = Ph, R = Et, R^1 = tBu, R^2 = R^4 = H, R^3 = CPh_3	−42.8(b)	372
C_2FN_2	$(H_3)_2;(CP^5)_2$	$(Me_2FP\text{-}NMe)_n$	−59.3	373
C'_2FN_2	$(C'_2)_2;(CP^5)_2$	$[(C_6F_5)_2FP\text{-}NMe]_2$	−80(c) J_{PF}772	264
	$(C'_2)_2;(CP^5)_2$	$(Ph_2FPNMe)_n$	−64.3	373
C_2FO_2	$(H_3)_2;C_2$	$Me_2(F)P(OCHTf_2)_2$	−11.5(n, 33°) J_{PF}726	374
$C_2FO_2/4$	$H_3,CH_2;C_2$	$(OCTf_2CH_2)FP(Me)OCHTf_2$	−2.1(b) J_{PF}832	375
$C_2FO_2/5$	$(H_3)_2;C_2$	$(Pfp)P(F)Me_2$	−12(n) J_{PF}789	376
C_2F_2N	$(H_3)_2;C_2$	$Me_2F_2PN(CH_2Ph)Me$	−27.4(n) J_{PF}653	66
C'_2F_2N	$(C'_2)_2;C_2$	$Ph_2F_2PNMe_2$	−54 J_{PF}709	377
	$(C'_2)_2;C_2$	$(C_6F_5)_2PF_2(NEt_2)$	−60.3 J_{PF}720	264
	$(C'_2)_2;C_2$	$Ph_2F_2PN(CH_2Ph)Me$	−55.1(n) J_{PF}711	66
	$(C'_2)_2;C'_2$	$Ph_2F_2PN^{\cdot}CH{=}CHCH{=}C^{\cdot}H$	−57.8(b)	73
C_2F_2O	$(H_3)_2;C'$	$Me_2F_2P(OC_6F_5)$	−3.9(m) J_{PF}747	63
	$(H_3)_2C'$	Me_2F_2POPh	−11.7(n) J_{PF}736	63
C'_2F_2O	$(C'_2)_2;C$	Ph_2F_2POEt	−45.8 J_{PF}756	378
	$(C'_2)_2;C'$	$Ph_2F_2P(OC_6F_5)$	−45.5(m) J_{PF}812	63
	$(C'_2)_2;C'$	Ph_2F_2POPh	−34.4(b) J_{PF}797	63
C_2F_2S	$(F_3)_2,C$	Tf_2F_2PSMe	−30.4(33°)	284
	$F_3,H_3;C$	$Tf(Me)F_2PSMe$	−15.2(30°)	284
C'_2F_2S	$(C'_2)_2;C$	$(C_6F_5)_2F_2PSEt$	−35.8 J_{PF}760	284
C_2F_3	$(F_3)_2$	Tf_2PF_3	−50.9 J_{PF}1260	263
	F_3,H_3	$(Me)TfPF_3$	−12.2(0°) J_{PF}909/1026	371
	$(H_3)_2$	Me_2PF_3	+8 J_{PF}787/975	263

P-bound atoms/rings	Connectivities	Structure	NMR data (δP[solv temp] J Hz)	Ref.
C_2F_3/5	$(CH_2)_2$	$[(CH_2)_4]PF_3$	−29.8(25°) J_{PF}915	263
C_2F_3/6	$(CH_2)_2$	$[(CH_2)_5]PF_3$	−22	86
$CC'F_3$	$H_3;C'_2$	$Me(Ph)PF_3$	−13.4	263
	$CH_2;C'N$	$(Et)F_3PC^{\cdot}$=CHCH=CMeN$^{\cdot}$H	−26.6	73
C'_2F_3	$C'N,C'_2$	$(Ph)F_3PC^{\cdot}$=CHCH=CMeN$^{\cdot}$H	−51.8	73
	$(C'_2)_2$	$(C_6F_5)_2PF_3$	−28.8	289
	$(C'_2)_2$	Ph_2PF_3	−34.8	263
C_2NO_2/5	$(H_3)_2;C_2;C_2$	$Me_2P(Pfp)NMe_2$	−1.5(n)	376
C_2NO_2/5,5f	$(H_3)_2;C'_2;C'_2$	**6**; X = Y = Me, R^1 = ODop, R^2 = OCPh=CPh	−9.5(r)	83
	$(CH_2)_2;C_2;C_2$	**6**; X = Y = CH_2Tf, R^1 = R^2 = OCH_2CH_2	−43.3	82
C_2NO_2/6,6f	$(CH_2)_2;C_2;C_2$	**6**; X = Y = CH_2Tf, R^1 = R^2 = $O(CH_2)_3$	−72.2	82
	$(CH_2)_2;C_2;C_2$	**6**; X = Y = $CHTf_2$, R^1 = R^2 = $O(CH_2)_3$	−77.1	82
$C_2N'O_2$/5	$(H_3)_2;N'';C_2$	$Me_2P(Pfp)N_3$	+3(n)	376
$CC'NO_2$/5	$C_2N';C'_2;C'_2;C_2$	$(MeO)_2PhP[CTf_2N$=CPhN(Mes)]	−11.6(c)	305
$CC'NO_2$/5,5f	$H_3;C'_2;C'_2;C'_2$	**6**; X = Me, Y = Ph, R^1 = ODop, R^2 = OCPh=CPh	−21.6(r)	83
$CC'NO_2$/5,5	$C'H_2;C'_2;C_2;CC'$	$[CH_2(3\text{-}NO_2C_6H_3)O]$ PhP(OCHPhCHMeNMe)	−28.9 −29.5*	379
$CC'NO_2$/5,5f	$C'H_2;C'_2;C_2;C'_2$	**6**; X = Ph, Y = CH_2Ph, R^1 = R^2 = $OC(O)CH_2$	−31(v)	99
$CC'NO_2$/5,5	$C_2O;C'_2;C_2;C_2$	**26**; R^1R^2 = CTf_2OCTf_2O, R^3 = NMe_2	−13.7(25°, c)	310
$CC'NO_2$/5,5f,6f	$C'HN;C'_2;C_2;C'_2$	**25**; X = Y = NMe, R = H, R' = Ph	−1.7(c)	380
C'_2NO_2/5	$(C'_2)_2;C'H;CC'$	(Oam)P(OMe)Ph_2	−36.1	381
	$(C'_2)_2;C'H;CC'$	(Oam)P(OCH_2tBu)Ph_2	−38.7	381
	$(C'_2)_2C'H;CC'$	**32**; X = OMe, Y = Z = Ph, R = R^1 = R^2 = R^4 = H, R^3 = CPh_3	−30(c)	382
	$(C'_2)_2;C'H;CC'$	**32**; X = OMe, Y = Z = Ph, R = R^2 = R^4 = H, R^1 = OMe, R^3 = CPh_3	−34.9(c)	382
	$(C'_2)_2;C'H;CC'$	**32**; X = OEt, Y = Z = Ph, R = R^2 = R^4 = H, R^1 = tBu, R^3 = CPh_3	−39.5(b)	382
C'_2NO_2/4,5f	$(C'_2)_2;CC';C_2$	**6**; X = Y = Ph, R^1 = $OCTf_2$, R^2 = $OCTf_2C(S)$	−17.5(c)	383
C'_2NO_2/5,5f	$C'_2;S'S;C_2;C'_2$	**6**; X = Ph, Y = $CS_2^-HNEt_3$, R^1 = R^2 = $OC(O)CH_2$	−48.6 −46(r)	313, 99
	$(C'_2)_2;C'_2;C'_2$	**6**; X = Y = Ph, R^1 = ODop, R^2 = OCPh=CPh	−33.4(r) −33.9(p)	83
$C'_2N'O_2$/5	$(C'_2)_2$;C:pr;C_2	Ph_2(MeO)P(N=CHCPh_2O)	−9.2	119b
C_2N_2O/4,5f	$(H_3)_2;(N'P^5)_2;C'$	**3**; R^1 = R^2 = Me, X-Y = OCMe=N	−42.5	25
	$(H_3)_2;C'N',CC';C'$	**6**; X = Y = Me, R^1 = OCMe=N, R^2 = NMeC(O)	−30.2(r)	384
	$(H_3)_2;C'N'C'_2;C'$	**6**; X = Y = Me, R^1 = OCMe=N, R^2 = NPhC(O)	−27.2(r)	384
	$(H_3)_2;(C'P^5)_2;C'$	**3**; R^1 = R^2 = Me, X-Y = ODop	−44.1	129
	$(H_3)_2;CC',C'_2;C'$	**6**; X = Y = Me, R^1 = OCMe=CH, R^2 = NMeC(O)	−30.2	125
	$(H_3)_2;CC',C'_2;C'$	**6**; X = Y = Me, R^1 = OCPh=CH, R^2 = NMeC(O)	−29.7	125

TABLE Q (continued)

P-bound atoms/rings	Connectivities	Structure	NMR data (δP[solv temp] J Hz)	Ref.
	$(CH_2)_2;(C'P^5)_2;C'$	3; $R^1 = R^2 =$ Et, X-Y = O(4-BrC_6H_3)	−32.7(b)	133
	$(CH_2)_2;(C'P^5)_2;C'$	3; $R^1 = R^2 =$ Et, X-Y = ODop	−35.7, −35.7(b)	131, 133
	$(C'_2)_2;(N'P^5)_2;C'$	3; $R^1 = R^2 =$ Ph, X-Y = OCPh=N	−53	125
$C_2N_2O/5,5f$	$(H_3)_2;C'N,C'_2;C'$	12; $R^1 = R^2 =$ Me, R = ONapht, X-Y = CH=CHS	−34.9	135
	$(H_3)_2;C'N,C'_2;C'$	12; $R^1 = R^2 =$ Me, R = ONapht, X-Y = CH=CHCH=CH	−35.2	135
$C_2N_2O/5,5$	$H_3,C'_2H;(C_2)_2;C'$	[CHPhC(COMe)=CMeO]-P($NMeCH_2CH_2NMe$)Me	−19	105
$CC'N_2O/5,5$	$C'_2H;C'_2;$ $C'N'C'H;C'$	(Oam)P(NPhN=CPhCHPh)Ph	−33.0 −34.3*	385
$C'_2N_2O/4,5f$	$(C'_2)_2;(C'P^5)_2;C'$	3; $R^1 = R^2 =$ Ph, X-Y = O(2,4-$tBu_2C_6H_2$)	−50.6	386
$C'_2N_2O/5,5f$	$(C'_2)_2;C'_2;C'N';C'$	12; $R^1 = R^2 =$ Ph, R = ONapht, X-Y = CH=CHS	−45.8	135
	$(C'_2)_2;C'_2;C'N';C'$	12; $R^1 = R^2 =$ Ph, R = OCPh=CPhO, X-Y = CH = CHCH=CH	−48	135
$C_2N_2S/4,5f$	$(H_3)_2;(N'P^5)_2;C'$	3; $R^1 = R^2 =$ Me, X-Y = SCMe=N	−42.9(c)	163
$C'_2N_2S/4,5f$	$(C'_2)_2;(N'P^5)_2;C'$	3; $R^1 = R^2 =$ Ph, X-Y = SCPh=N	−38.4	163
C_2O_3	$(F_3)_2;C_3$	$Tf_2P(OCHTf_2)_3$	−72.8	277
	$(H_3)_2;C_3$	$Me_2P(OCHTf_2)_3$	−20.3	277
$C_2O_3/4$	$H_3,CH_2;C_3$	$(CH_2CTf_2O)P(OCHTf_2)_2Me$	−10(b)	375
	$H_3,CH_2;C_3$	$Mo(CO)_2[(Tf_2CHO)(Me_2PCTf_2O)P(OCTf_2CH_2)Me]_2$	−10.5	387
	$H_3,C_3;C_3$	$(Tf_2CHO)_2P(OCTf_2CTf_2)Me$	−10	326
	$CH_2,C_2H;C_3$	$(CHMeCTf_2O)P(OCHTf_2)_2Et$	−5.8	369
$C_2O_3/5$	$(H_3)_2;C_2P^5$	(Pfp)Me_2POPMe_2(Pfp)	−2.3(r)	376
	$(H_3)_2;C_2Si$	(Pfp)Me_2POTms	−2(n)	367
	$(H_3)_2;C_3$	(Pfp)Me_2POMe	−8.5(n)	376
	$(H_3)_2;C'_3$	(OCTf=CTfO)P(OPh)Me_2	+8	388
	$H_3,C_3;C_3$	(Pfp)MeP($OCHTf_2$)tBu	+12.3	387
	$(CH_2)_2;C_3$	(Pfp)P($OCHTf_2$)Et_2	−11.2	369
	$(C_2H)_2;C'_3$	(OCTf=CTfO)P(OPh)iPr_2	+13	388
$C_2O_3/4,5$	$(C_3)_2;C_3$	(Pnc)P($CMe_2CHMeCMe_2$)$OCHTf_2$	+10.2	389
	$(C_3)_2;C_3$	(Pfp)P($CMe_2CHMeCMe_2$)$OCHTf_2$	+23	390
	$(C_3)_2;C_2C'$	(Pfp)P($CMe_2CHMeCMe_2$)OPh	+24	390
$C_2O_3/5,5$	$CH_2,C'_2H;C_2C'$	(Glc)P[OCMe=C(COMe)CHPh]Et	+12.8 +13.5*	323
$C_2O_3/5,6$	$H_3,CH_2;C_2,P^5$	(Pfp)MeP[$(CH_2)_3$P(Pfp)(Me)O]	+4.7 +5.3	391
$CC'O_3/5$	$H_3;C'_2;C_2C'$	[C(=CH_2)CH=CMeO]P$(OMe)_2$Me	−21.6	293
	$H_3;C'_2;C'_3$	(OCTf=CTfO)MeP(OPh)Ph	−5	388
	$C'H_2;C'_2;C_2C'$	(CH_2CH=CMeO)P$(OMe)_2$Ph	−11.2	293
	$C_2N';C'_2;C_2C'$	[CTf_2N=C(4-ClC_6H_4)O]P$(OMe)_2$Ph	−21.5(c, b)	332
	$C_2N':C'_2;C_2C'$	[CTf_2N=C(4-MeC_6H_4)O]P$(OMe)_2$Ph	−26.7	332
	$C'_2H;C'_2;C_2N^+$	31; X = Ph, Y = Z = OMe, $R^1 = R^2 =$ Cl, $R^3 =$ H	−24.5	335
	$C'_2H;C'_2;C_2N^+$	31; X = Ph, Y = Z = OMe, $R^1 = R^2 = R^3 =$ Me	−16	335
	$C'H_2;C'O;C_2C'$	(CH_2CPh=CPhO)P$(OMe)_2$Fu	−34.9	355
	$C'_2H;C'O;C_2C'$	[OCMe=C(COMe)CHPh]P$(OMe)_2$Fu	−28.4 −34.7*	355

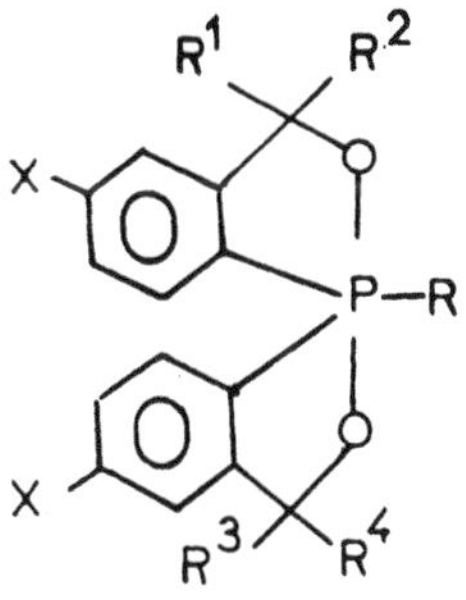

33

P-bound atoms/rings	Connectivities	Structure	NMR data (δP[solv temp] J Hz)	Ref.
	$C'_2H,C'_2;C_2C'$	[CHPhC (COMe)=CMeO]P$(OMe)_2$Ph	−16.7 −13.3* (r, 25°)	392
	$C'_2H,C'_2;C'_3$	[CHPhC(COMe)=CMeO]P$(OPh)_2$Ph	−24.8 −30*(r, 25°)	392
	$C_2C';C'O;C_3$	$(MeO)_3P[CMe_2C(O)OC(=CMe_2)]$	−53.8	393
$CC'O_3/5,5$	$CH_2;C'O;C_2C'$	(Pnc)P$[CH_2CH_2C(O)O]$Fu	−30.7	355
	$CH_2;C'_2;C_3$	(Pfp)P$[O(CH_2)_3]$Ph	−3.8	394
	$C'_2H;C'_2;C_2C'$	(Glc)P[OCMe=C(COMe)CPhH]Ph	−4.5	395
$CC'O_3/5,5f,6f$	$C'HO;C'_2;CC'_2$	**25**; X = Y = O, R = H, R′ = Ph	−7.9(c)	380
	$CC'O;C'_2;CC'_2$	**25**; X = Y = O, R = Tf, R′ = Ph	−12.7(c)	380
	$C'_2O;C'_2;CC'_2$	**25**; X = Y = O, R = R′ = Ph	−14	380
	$C'_2O;C'_2;CC'_2$	**25**; X = Y = O, R = 2,4-$(MeO)_2C_6H_3$, R′ = Ph	−18.6(b)	380
C'_2O_3	$(C'_2)_2;C_3$	$Ph_2P(OCH_2Tf)_3$	−45.2(30°, c)	169
	$(C'_2)_2;C_3$	$Ph_2P(OEt)_3$	−41	146
	$(C'_2)_2;C_3$	$Ph_2(OEt)_2$POiPr	−43.9	347
$C'_2O_3/5$	$(C'O)_2;C'_3$	(OCTf=CTfO)P$[O(4\text{-}FC_6H_4)]Fu_2$	−60.3	355
	$(C'_2)_2;C_3$	(Pfp)P(OMe)Ph_2	−18	87
	$(C'_2)_2;C_3$	(Pnc)P(OMe)Ph_2	−37.4 −45	187, 87
	$(C'_2)_2;C_3$	(Pfp)P$(OCHTf_2)Ph_2$	−18.7, −19.4(r)	351, 378
	$(C'_2)_2;C_2C'$	$[OCMe=CHC(=CH_2)]P(OMe)_2Ph$	−33.7	293
	$(C'_2)_2;CC'_2$	(Cat)P(OMe)Ph_2	−19	87
	$(C'_2)_2;CC'_2$	(OCMe=CMeO) $Ph_2POCH_2C^{\cdot}HOCH_2CH_2C^{\cdot}H_2$	−24.4	202
	$(C'_2)_2;CC'_2$	(OCMe=CMeO) $Ph_2POCH_2C^{\cdot}HOCH_2CH_2O^{\cdot}$	−26(c)	396
	$(C'_2)_2;CC'_2$	(OCMe=CMeO) $Ph_2POCH_2C^{\cdot}H(CH_2)_3C^{\cdot}H_2$	−26.1	202
	$(C'_2)_2;C'_3$	(OCH=CHO)P(OPh)Ph_2	−18.8(r)	124
	$(C'_2)_2;C'_3$	(OCTf=CTfO)P(OPh)Ph_2	−18.8	388
$C'_2O_3/6$	$(C'_2)_2;C_3$	**18**; R^1 = OMe, $R^2 = R^3$ = Ph	−49.2(b)	206
$C'_2O_3/5,5$	$(C'_2)_2;H,C_2$	**33**; $R^1 = R^2 = R^3 = R^4$ = Me, R = OH, X = H	−27	397
	$(C'_2)_2;H,C_2$	**33**; $R^1 = R^2 = R^3 = R^4$ = Me, R = OH, X = Me	−16.0(k)	398
	$(C'_2)_2;H,C_2$	**33**; $R^1 = R^2 = R^3 = R^4$ = Tf, R = OH, X = H	−17.6	399
	$(C'_2)_2;H,C_2$	**33**; $R^1 = R^2 = R^3 = R^4$ = Tf, R = OH, X = Me	−17.8	400
	$(C'_2)_2;H,C'_2$	**33**; $R^1R^2 = R^3R^4$ = O, R = OH, X = H	−38.5(c)	401

TABLE Q (continued)

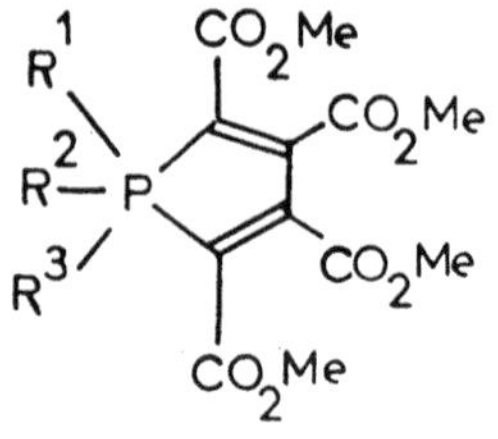

34

P-bound atoms/rings	Connectivities	Structure	NMR data (δP[solv temp] J Hz)	Ref.
	$(C'_2)_2;C_3$	**33**; $R^1 = R^2 = R^3 = R^4 = Me$, $R = OMe$, $X = H$	−25.6	402
	$(C'_2)_2;C_3$	**34**; $R^1R^2 = Pnc$, $R^3 = OMe$	−35.6	403
	$(C'_2)_2;C_3$	**33**; $R^1 = R^2 = R^3 = R^4 = Me$, $R = OMe$, $X = Me$	−25.8	398
	$(C'_2)_2;CC'_2$	**33**; $R^1R^2 = R^3R^4 = O$, $R = OMe$, $X = H$	−46.3	400
	$(C'_2)_2;CC'_2$	**26**; $R^1 = Ph$, $R^2R^3 = O(3,4\text{-}Me_2C_6H_2)O$	−12.4(c, 25°)	310
$C'_2O_2O'/5,5$	$(C'_2)_2;C_2$	**33**; $R^1 = R^2 = R^3 = R^4 = Tf$, $R = O^-K^+$, $X = H$	−16.1	399
	$(C'_2)_2;C_2$	**33**; $R^1 = R^2 = R^3 = R^4 = Tf$, $R = O^-Na^+$, $X = Me$	−14.8	402
$C'C''O_3/5$	$C'_2;N'';C_2C'$	$[C({=}CH_2)CH{=}CMeO]P(OMe)_2(CN)$	−47.3	293
C''_2O_3	$N''_2;C_3$	$(CHF_2CF_2CH_2O)_3P(CN)_2$	−76.8	404
C_2HF_2	$(H_3)_2$	$Me_2P(H)F_2$	−31.7(0°) J_{PH}733 J_{PF}535	358
$C'_2HO_2/5$	$(C'_2)_2;C_2$	$Ph_2(H)P(Pfp)$	−20.7 J_{PH}462	405
	$(C'_2)_2;C'_2$	$Ph_2(H)P[O(2,5\text{-}tBu_2C_6H_2)O]$	−30.8 J_{PH}449	406
$C'_2HO_2/5,5$	$(C'_2)_2;C_2$	**33**; $R^1 = R^2 = R^3 = R^4 = Tf$, $R = H$, $X = H$	−48.2 J_{PH}733	407
	$(C'_2)_2;C_2$	**33**; $R^1 = R^2 = R^3 = R^4 = Me$, $R = H$, $X = H$	−60.1(k) J_{PH}680	408
	$(C'_2)_2;C_2$	**33**; $R^1 = R^2 = R^3 = R^4 = Tf$, $R = H$, $X = Me$	−47.6 J_{PH}730	398, 407
	$(C'_2)_2;C_2$	**33**; $R^1 = R^2 = R^3 = R^4 = Me$, $R = H$, $X = Me$	−58.4 J_{PH}680	398
C_3BrF	$(CH_2)_3$	$(nBu)_3P(F)Br$	−57.8	
$C'_3BrO/5$	$(C'_2)_3;C$	$Ph_2(Br)P(OCMe_2Dop)$	−17.8(c)	409
C_3Br_2	$(F_3)_3$	Tf_3PBr_2	−65(r,f)	367
C_3ClN	$(F_3)_3;C_2$	$Tf_3P(NMe_2)Cl$	−44.5	370
$C_3ClO/4$	$(H_3)_2,CH_2;C$	$Me_2P(CH_2CTf_2O)Cl$	−13.6	410
	$CH_2,(C_3)_2;C$	$(tBu)_2P(OCTf_2CH_2)Cl$	+0.3	368
	$CH_2,(C_3)_2;C$	$(tBu)_2P[OCTf(Ph)CH_2]Cl$	+9.9	411
$C_3ClO/5$	$(CH_2)_2,C_2O;C$	$Et_2P(CTf_2OCTf_2O)Cl$	−6.4	369
$C'_3ClO/5$	$(C'_2)_3;C'$	$[OC(O)Dop]P(Cl)Ph_2$	−30.4(c)	412
C_3Cl_2	$(CH_2)_3$	Et_3PCl_2	−110.1	26
	$(CH_2)_3$	$(nPr)_3PCl_2$	−101	26
$C_2C'Cl_2$	$(H_3)_2;C'_2$	$(Me)_2PhPCl_2$	−53.3	26
	$(CH_2)_2;C'_2$	$(Et)_2PhPCl_2$	−91.3	26
CC'_2Cl_2	$CH_2;(C'_2)_2$	$(Et)Ph_2PCl_2$	−74.3	26
	$C_2H;(C'_2)_2$	$(iPr)Ph_2PCl_2$	−86.7	26

P-bound atoms/rings	Connectivities	Structure	NMR data (δP[solv temp] J Hz)	Ref.
C'_3Cl_2	$(C'_2)_3$	Ph_3PCl_2	−61.5	26
			−46.4(t) +63.7(m) (12 solvents)	413
C_3FN	$(F_3)_3$;CH	$Tf_3PF(NHMe)$	−74(55°)	70
	$(F_3)_3;C_2$	$Tf_3PF(NMe_2)$	−62.7(30°)	414
	$(F_3)_3;C_2$	$Tf_3(F)PN(CH_2Ph)Me$	−68.8	70
	$(F_3)_2;H_3$;CH	$MeTf_2P(F)NHMe$	−65.1	70
	$(F_3)_2,H_3;C_2$	$MeTf_2P(F)NMe_2$	−54.4	70
C_3FO	$(F_3)_3$;C	$Tf_3PF(OMe)$	−57(30°)	414
$C_3FO/4$	$(H_3)_2,CH_2$;C	$Me_2P(CH_2CTf_2O)F$	−20.8 J_{PF}650	410
$C'_3FO/5$	$(C'_2)_3;C'$	[OC(O)Dop]P(F)Ph_2	−50.7 J_{PF}700	412
C_3FS	$(F_3)_3$;C	$Tf_3PF(SMe)$	−49.5(30°)	414
C_3F_2	$(F_3)_3$	Tf_3PF_2	−59.8 J_{PF}988	263
	$(H_3)_3$	Me_3PF_2	−13.9 J_{PF}553	415
	$(H_3)_2,CH_2$	Me_2F_2**P**$(CH_2CH_2PF_2Ph_2)$	−14.1(c) J_{PF}546	416
	$(H_3)_2,CH_2$	$Me_2F_2P(CH_2)_3PF_2Me_2$	−12.8(c) J_{PF}543	416
	$(CH_2)_3$	Et_3PF_2	−13.2	288
	$(CH_2)_3$	$(nBu)_3PF_2$	−14.5 J_{PF}585	62
	$(CH_2)_3$	F_2**P**$(CH_2CH_2PF_2Ph_2)_3$	−18.8(c) J_{PF}630	416
	$(C_2H)_3$	iPr_3PF_2	−17.1(n) J_{PF}652	417
$C_2C'F_2$	$(H_3)_2;C'_2$	Me_2PhPF_2	−27.9 J_{PF}589	415
	$(H_3)_2;C'_2$	$Me_2(2\text{-}MeOC_6H_4)PF_2$	−16.5	389
	$(CH_2)_2;C'_2$	$F_2(Ph)P(CH_2CH_2PF_2Ph_2)_2$	−31.1(c) J_{PF}632	416
	$(C_3)_2;C'_2$	$(tBu)_2P(Ph)F_2$	−30.5 J_{PF}759	62
$C_2C'F_2/4$	$CH_2,C_3;C'_2$	$(CMe_2CMe_2CH_2)P(Ph)F_2$	+7.5(−80°) J_{PF}973 872	418
	$(C_3)_2;C'_2$	$(CMe_2CHMeCMe_2)P(Ph)F_2$	−1 J_{PF}911	418
$C_2C'F_2/5$	$(CH_2)_2;C'_2$	$(Ph)F_2P(CH_2CHMeCH_2CH_2)$	−2.3	288
CC'_2F_2	$H_2P^5;(C'_2)_2$	CH_2(**PF_2Ph_2**$)_2$	−45.2(c) J_{PF}650	416
	$H_3;(C'_2)_2$	$(Me)Ph_2PF_2$	−41.1 J_{PF}638	415
	$CH_2;(C'_2)_2$	$Ph_2F_2P(CH_2)_4PF_2Ph_2$	−41.3(c) J_{PF}642	416
	$CH_2;(C'_2)_2$	$F_2P(CH_2CH_2$**P**$F_2Ph_2)_3$	−46.8(c) J_{PF}650	416
C'_3F_2	$C'N,(C'_2)_2$	$(Ph)_2F_2PC^{\cdot}$=CHCH=C(Me)$N^{\cdot}$H	−63.2(b)	73
	$C'O',(C'_2)_2$	$Ph_2F_2PC(O)Ph$	−61.0 J_{PF}705	419
	$(C'_2)_3$	Ph_3PF_2	−56 J_{PF}666	62
	$(C'_2)_3$	$Ph_2F_2PC^{\cdot}$=C(Me)NHC(Me)=$C^{\cdot}$H	−47.8(b)	73
$C_2C'NO/4,5f$	$(H_3)_2;C'_2;C_2$;C	**27**; R = $OCMe_2$, R′ = Me	−55.4(c)	420
$C_2C'NO/5,5f$	$(H_3)_2;C'_2;C_2$;N	**27**; R = ON(Me)CHPh, R′ = Me	−57.9(c)	421
	$(H_3)_2;C'_2;C_2$;N	**27**; R = ON(Ph)CH(COPh), R′ = Me	−48.6(c) −49.2(d)	421
	$(H_3)_2;C'_2;C_2$;N	**27**; R = $ON^{\cdot}CH_2CH_2DopC^{\cdot}H$, R′ = Me	−54.2	421
	$(H_3)_2;C'_2;CC';N'$	**27**; R = ON=CPh, R′ = Me	−34.5(c) −34.1(d)	421
	$(H_3)_2;C'_2;CC',C'$	**27**; R = OCPh=CH, R′ = Me	−46.0	315
$CC'_2NO/5$	$H_3;(C'_2)_2;C'H;C'$	$Ph_2MeP[O(2\text{-}tBu,4\text{-}CHPh_2C_6H_2)NH]$	−52.5 −58.3*	422
	$C'H_2;(C'_2)_2;C'H;C'$	$Ph_2(CH_2Ph)P[O(2,4\text{-}tBu_2C_6H_2)NH]$	−49.6 −52.4*	422
	$C'H_2;(C'_2)_2;CC',C'$	$Ph_2(CH_2Ph)P[O(2\text{-}tBu, 4\text{-}MeC_6H_4)NEt]$	−42.4(c)	382
	$C_2H;(C'_2)_2;C'H;C'$	$Ph_2(iPr)P[2,4\text{-}tBu_2C_6H_2)NH]$	−41.6	422
	$C_2H;(C'_2)_2;C'H;C'$	$Ph_2(iPr)P[P(2\text{-}tBu,4\text{-}CPh_3C_6H_2)NH]$	−39.8	422
$CC'_2NO/5,5f$	$C'HO;(C'_2)_2;C'H;C'$	**24**; X = Y = Ph, R = H, R^1 = ODop, R^2 = NHC(O)O	−41.9	299
	$C'_2H;(C'_2)_2;C'_2;C'$	**6**; X = Y = Ph, R^1 = CHPhCH=CH, R^2 = ODop	−22.5	303
$C'_3NO/5$	$(C'_2)_3;C'H;C'$	$Ph_3P[NH(3,5\text{-}(NO_2)_2C_6H_2)O]$	−15.4(d)	423
	$(C'_2)_3;C'H;C'$	$Ph_3P(Oam)$	−44.4(p)	423

TABLE Q (continued)

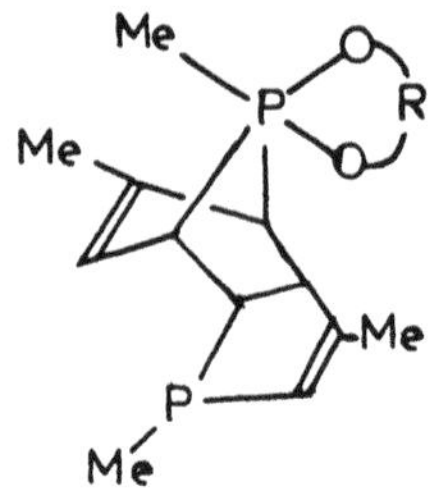

35

P-bound atoms/rings	Connectivities	Structure	NMR data (δP[solv temp] J Hz)	Ref.
	$(C'_2)_3$;C'H;C'	$Ph_3P[NH(3,5\text{-}tBu_2C_6H_2)O]$	−48(b)	423
	$(C'_2)_3$;C'H;C'	$Ph_3P[NH(2,5\text{-}tBu_2,3\text{-}OHC_6H)O]$	−51.9	423
	$(C'_2)_3$;C'H;C'	$Ph_3P[NH(3\text{-}tBu,5\text{-}NPPh_3C_6H_2)O]$	−50.3(t)	424
	$(C'_2)_3$;CC';C'	$Ph_3P[O(4\text{-}ClC_6H_3)NEt]$	−31.6(c)	382
	$(C'_2)_3$;CC';C'	$Ph_3P[O(2,4\text{-}tBu_2C_6H_2)NEt]$	−38.7(c)	382
$C'_3NO/4,5f$	$(C'_2)_3;C_2;N$	**27**; R = $ON^{\cdot}CH_2CH_2DopC^{\cdot}H$, R' = Ph	−50(c)	421
	$(C'_2)_3;C_2;C$	**27**; R = $OCMe_2$; R' = Ph	−49.3(c)	420
	$(C'_2)_3;C_2;C$	**27**; R = OCMeTf, R' = Ph	−41.2(c)	420
	$(C'_2)_3;C_2;C$	**27**; R = $OCTf_2$, R' = Ph	−34.9(c)	420
$C'_3NO/5,5f$	$(C'_2)_3;C_2;C$	**27**; R = ON(Ph)CH(Ph), R' = Ph	−53.2(c)	421
	$(C'_2)_3;C_2;C$	**27**; R = ONPhC(COPh)H, R' = Ph	−49.9(c)	421
	$(C'_2)_3$;CC'N'	**27**; R = ON=CPh, R' = Ph	−30(c)	421
$C'_3N'O/5$	$(C'_2)_3$;N';C'	$Ph_2(N_3)P[OC(O)Dop]$	−60(b)	120
	$(C'_2)_3$;C';C	$Ph_3P(N{=}CHCPh_2O)$	−24.8	119b
C_3N_2	$(F_3)_3;(C_2)_2$	$Tf_3P(NMe_2)_2$	−57.6	70
$C_3N_2/4$	$(F_3)_2,H_3,(CP^5)_2$	$(MeTf_2P\text{-}NMe)_2$	−81.7	425
$C_3N_2/4,5f$	$(H_3)_2,CH_2;(CC')_2$	**6**; X = Y = Me, R^1 = NMeC(O), R^2 = $CH_2CH(CN)CPh_2$	−64.2(c)	319
$C_2C'N_2/4,5f$	$(H_3)_2;C'_2;(CC')_2$	**27**; R = NMeC(O), R' = Me	−67.3(c)	319
$C_2C'N_2/5,5f$	$(H_3)_2;C'_2$;C'N',CC'	**27**; R = N(Ph)N=CPh, R' = Me	−62.4(c)	421
$CC'_2N_2/4,5f$	$CH_2;(C'_2)_2;(CC')_2$	**6**; X = Y = Ph, R^1 = NMeC(O), R^2 = $CH_2CH(CO_2Me)CPh_2$	−53.1(c)	319
$C'_3N_2/4$	$(C'_2)_3;CP^4,P^4Si$	$Ph_3P[NMeP({=}NTms)(NTms_2)NTms]$	−0.6 +3.3	426a
	$(C'_2)_3;(CP^4)_2$	$Ph_3P[NMeP({=}NTms)(NTms_2)NMe]$	−6 −4.9	426a
$C'_3N_2/4,5f$	$(C'_2)_3;(CC')_2$	**27**; R = NMeC(O), R' = Ph	−59.2(c)	319
$C'_3N_2/5,5f$	$(C'_2)_3$;C'N',CC'	**27**; R = N(Ph)N=CPh, R' = Ph	−48.0	421
$C_3OS/5$	$(CH_2)_3$;C';C	$Et_2P[CH_2CHMeC(O)O]SEt$	−14	426b
C_3O_2	$(F_3)_3;Si_2$	$Tf_3P(OTms)_2$	−93	427
	$(F_3)_3;C_2$	$Tf_3P(OMe)_2$	−75(30°)	414
	$(H_3)_3;C_2$	$Me_3P(OEt)_2$	−39	146
$C_3O_2/4$	$(H_3)_2,CH_2;C_2$	$Me_2P(CH_2CTf_2O)OCHTf_2$	−23.7	428
	$H_3,CH_2,C_2P^3;C_2$	$Me(Tf_2HCO)P\text{-}(OCTf_2CH_2)CTf_2PMe_2$	+9.4	375
$C_3O_2/5$	$(CH_2)_2,C_2O;C_2$	$Et_2P(CTf_2OCTf_2O)OCHTf_2$	−0.9	369
	$CH_2,(C_3)_2;C_2$	$(tBu)_2P[OC(Ph)TfCH_2]OMe$	−14.6	411
$C_3O_2/5,5$	$H_3,(C'H_2)_2;C_2$	$(Pfp)MeP(CH_2CMe{=}CMeCH_2)$	+14.0	394
	$H_3,(CC'H)_2;C'_2$	**35**; R = Phenanthrenyl	+0.6	429
	$H_3,(CC'_2)_2;C'_2$	**9**; R^1 = Me, R^2R^3 = CHMeCH=CHCHMe	+30 +12.3*(c) +28.4 +11*(t)	168

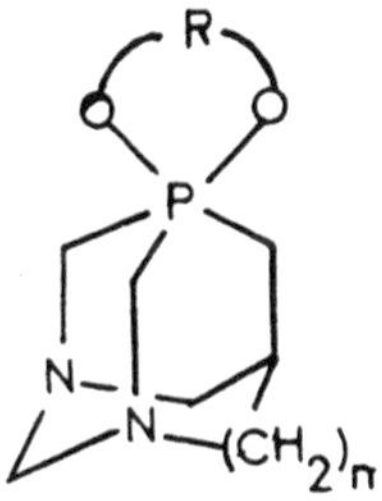

36

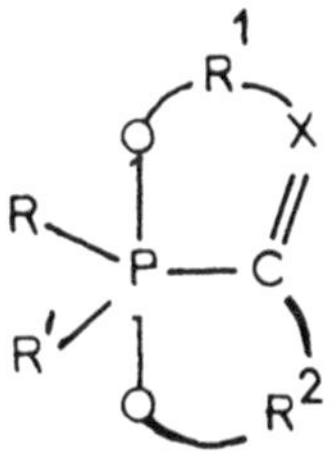

37

P-bound atoms/rings	Connectivities	Structure	NMR data (δP[solv temp] J Hz)	Ref.
C_3O_2/5,5,6f	$(CH_2)_2$,$C'H_2$;C_2	**2**; X-Y = CH_2CMe, RR′ = Pnc, R^1 = CH_2CHMe, R^2 = CH_2Dop	−29.1(b)	430
C_3O_2/5,5,6f,6f	$(H_2N)_3$;C'_2	**36**; R = 2,4-$tBu_2C_6H_2$, n = 0	−52.3	119a
C_3O_2/5,6,6f,6f	$(H_2N)_3$;C'_2	**36**; R = 2,4-$tBu_2C_6H_2$, n = 1	−69.7	119a
C_3O_2/5,6,6f,7f	$(H_2N)_3$;C'_2	**36**; R = Dop, n = 2	−52	119a
	$(H_2N)_3$;C'_2	**36**; R = 2,4-$tBu_2C_6H_2$, n = 2	−54	119a
	$(H_2N)_3$;C'_2	**36**; R = Phénanthrenyl, n = 2	−48	119a
C_3O_2/5,6,6f,11f	$(H_2N)_3$;C'_2	**36**; R = 2,4-$tBu_2C_6H_2$, n = 6	−39	119a
$C_2C'O_2$/4	H_3,CH_2;C'_2;C_2	$(CH_2CTf_2O)MeP(OCHTf_2)Ph$	?+31.4(r)	378
	CH_2,C_3;C'_2;C_2	$(CH_2CMe_2CMe_2)P(OEt)_2Ph$	−36	431
	$(C_3)_2$;C'_2;C_2	$(CMe_2CH_2CMe_2)P(OEt)_2Ph$	−22	431
$C_2C'O_2$/5	$(H_3)_2$;C'_2;C_2	$Me_2PhP(Pfp)$	?+10.9(r)	378
	$(CH_2)_2$;C'_2;C_2	$(EtO)_2P[CH_2)_4]Ph$	−12	432
$C_2C'O_2$/6	$(CH_2)_2$;C'_2;C_2	$(EtO)_2P[CH_2)_5]Ph$	−48	432
$C_2C'O_2$/4,5	$(C_3)_2$;$C'O$;C_2	$(Pfp)P(CMe_2CHMeCMe_2)Fu$	−7.9 −8.9*	355
	$(C_3)_2$;$C'H$;C_2	$(Pfp)P(CMe_2CHMeCMe_2)CH{=}CMe_2$	−0.5 +7*	389, 390
	$(C_3)_2$;C'_2;C_2	$(Glc)P(CMe_2CHMeCMe_2)Ph$	?+14.1(b)	433
	$(C_3)_2$;C'_2;C_2	$(Pfp)P(CMe_2CHMeCMe_2)Ph$	+7.3 +3.4*	389, 390
	$(C_3)_2$;C'_2;C'_2	$(Cat)P(CMe_2CHMeCMe_2)Ph$	−5.7	87
$C_2C'O_2$/4,6	CH_2,C_3;C'_2;C_2	$(CH_2CMe_2CMe_2)P(OCH_2CH_2CH_2O)Ph$	−42	434
	$(C_3)_2$;C'_2;C_2	$(CMe_2CH_2CMe_2)P(OCH_2CMe_2CH_2O)Ph$	−35	434
	$(C_3)_2$;C'_2;C_2	**18**; R^1 = Ph, R^2R^3 = $CMe_2CH_2CMe_2$	−42.5 −44.8(b)*	435
$C_2C'O_2$/4,7	$(C_3)_2$;C'_2;C_2	$(CMe_2CH_2CMe_2)P[O(CH_2)_4O]Ph$	−21(r)	434
$C_2C'O_2$/5,5	$(C'H_2)_2$;C'_2;C_2	$(Pfp)P(CH_2CMe{=}CMeCH_2)Ph$	+7.8	394
	$(CC')_2$;C'_2;C'_2	**9**; R^1 = Ph, R^2R^3 = $CHMeCH{=}CHCHMe$	+26.2 +4.1*(c)	168
$C_2C'O_2$/4,4,5f	$(C_3)_2$;CC';C_2	**37**; RR′ = $CMe_2CHMeCMe_2$, X = CPh, R^1 = CH_2, R^2 = CTf_2	+31.6(c)	436
CC'_2O_2/4	CH_2;$(C'_2)_2$;C_2	$Ph_2P(OCTf_2CH_2)OCHTf_2$	−35.5	437
	C_2H;$(C'_2)_2$;C_2	$Ph_2P(CHTfCTf_2O)OCHTf_2$	−32.1(b)	428
CC'_2O_2/5	$H_3$$(C'_2)_2$;$C_2$	**34**; R^1 = R^2 = OMe, R^3 = Me	−34.5(t)	438
	CH_2;$(C'_2)_2$;C_2	$EtPh_2P(Pfp)$	−6.1(b)	428
	CH_2;$(C'_2)_2$;C'_2	**9** R^1 = R^2 = Ph, R^3 = Et	−9.3(b)	439
	C'_2H;$(C'_2)_2$;CC'	$(MeO)P[CH(Ph)C(COMe){=}CMeO]Ph_2$	−25.7(r, 25°)	392
	C'_2H;$(C'_2)_2$;C'_2	$(PhO)P[CH(Ph)C(COMe){=}CMeO]Ph_2$	−40.1	392
	C_2C';$(C'_2)_2$;CN'	$(MeO)P[CMe_2CH{=}N(O)O]Ph_2$	−20	440
CC'_2O_2/5,5	H_3;$(C'_2)_2$;C_2	**33**; R = R^1 = R^2 = R^3 = R^4 = Me, X = H	−39.2	402

TABLE Q (continued)

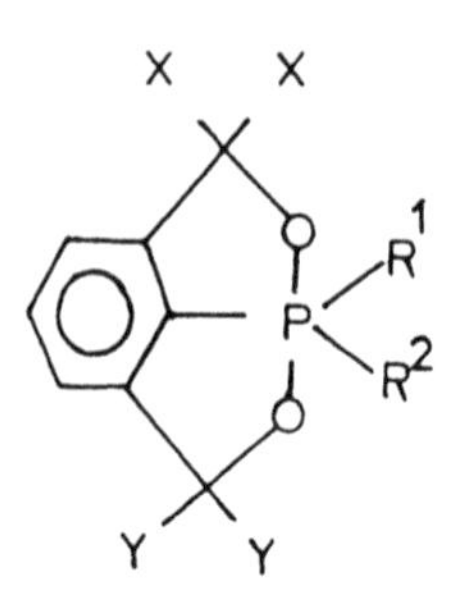

38

P-bound atoms/rings	Connectivities	Structure	NMR data (δP[solv temp] J Hz)	Ref.
	H_3;$(C'_2)_2$;C'_2	**33**; R = Me, R^1R^2 = R^3R^4 = O, X = H	−51(c)	441
	H_3;$(C'_2)_2$;C'_2	**33**; R = R^1 = R^2 = Me, R^3R^4 = O, X = H	−33.5	400
	CH_2;$(C'_2)_2$;C_2	**33**; R = $(CH_2)_4OH$, R^1 = R^2 = R^3 = R^4 = Tf, X = Me	−20.3	398
	CH_2;$(C'_2)_2$;C_2	**33**; R = CH_2Ph, R^1 = R^2 = R^3 = R^4 = X = Me	−35	398
	C_2H;$(C'_2)_2$;C'_2	**33**; R = sBu, R^1R^2 = R^3R^4 = O, X = H	−44.8	400
	C_2H;$(C'_2)_2$;C'_2	**33**; R = sBu, R^1R^2 = O, R^3 = R^4 = Me, X = H	−19.7	400
C'_3O_2	$(C'_2)_3$;O_2	$Ph_3P(OOtBu)_2$	−42.3(c)	442
	$(C'_2)_3$;C_2	$Ph_3P(OEt)_2$	−54(r), −50.9	146, 347
	$(C'_2)_3$;C_2	$Ph_3P(OiPr)_2$	−48(k)	443
	$(C'_2)_3$;C_2	$Ph_3P(OBu)_2$	−55(k)	443
	$(C'_2)_3$;C_2	$Ph_3P(OCHMe_2)_2$	−49.6	444
	$(C'_2)_3$;C_2	$CH_2[CHOP(OMe)Ph_3]_2$	−53.9	444
	$(C'_2)_3$;C'_2	$Ph_3P(OPh)_2$	−65.5(k)	443
	$(C'_2)_3$;C'_2	$Ph_3P[O(4\text{-}NO_2C_6H_4)]_2$	−61.9	445
	$(C'_2)_3$;C'_2	$Ph_3P[O(4\text{-}MeC_6H_4)]_2$	−67.1	445
$C'_3O_2/5$	$(C'_2)_3$;C_2	**34**; R^1 = R^2 = OMe, R^3 = Ph	−31.9(t, −70°)	438
	$(C'_2)_3$;C_2	$(Glc)PPh_3$	−36.2	444
	$(C'_2)_3$;C_2	$(Pnc)PPh_3$	−48.4(b)	187
	$(C'_2)_3$;C_2	$Ph_3P[OC^*H(CH_2)_4C^*HO]$	−37.7	446
	$(C'_2)_3$;C'_2	**9**; R^1 = R^2 = R^3 = Ph	−15.6(b)	439
$C'_3O_2/6$	$(C'_2)_3$;C_2	$Ph_3P[O(CH_2)_3O]$	−54.9	444
	$(C'_2)_3$;C_2	**18**; R^1 = R^2 = R^3 = Ph	−64(b)	206
$C'_3O_2/7$	$(C'_2)_3$;C_2	$Ph_3P[O(CH_2)_4O]$	−46.9 −45.4	444
$C'_3O_2/4,5f$	$(C'_2)_2$,CN';C_2	**37**; R = R' = Ph, R^1 = CTf_2, R^2 = CTf_2, X = N	+1.3(c)	447
$C'_3O_2/5,5$	$(C'_2)_3$;C_2	**33**; X = R^1 = R^2 = R^3 = R^4 = H, R = Ph	−36.6(c)	448
	$(C'_2)_3$;C_2	**33**; X = R^1 = R^3 = R^4 = H, R^2 = $CH_2C(O)Me$, R = Ph	−38.1 −43.9*(c)	449
	$(C'_2)_3$;C_2	**33**; X = H, R^1 = R^2 = R^3 = R^4 = Me, R = Ph	−44.5(c)	450
$C'_3O_2/5,5f$	$(C'_2)_3$;C_2	**38**; XX = YY = H, R^1 = R^2 = Ph	−22.9(c)	448

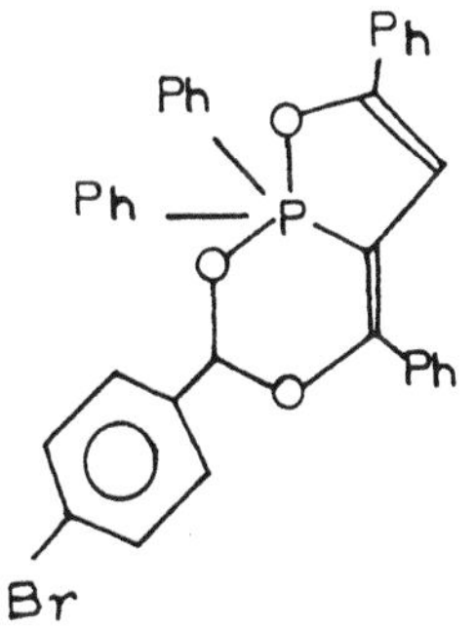

39

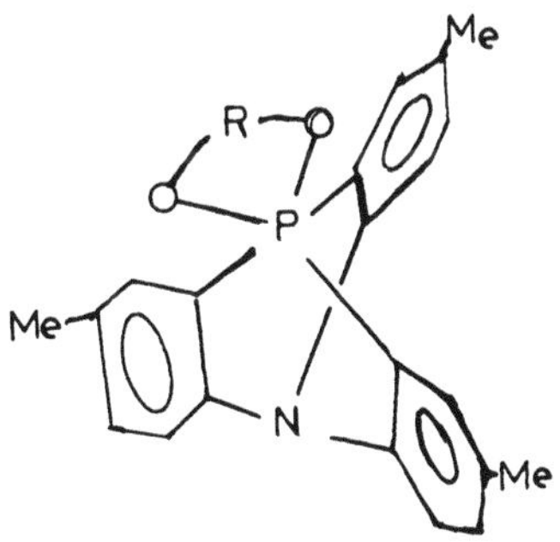

40

P-bound atoms/rings	Connectivities	Structure	NMR data (δP[solv temp] J Hz)	Ref.
	$(C'_2)_3;C_2$	**38**; XX = YY = H, $R^1 = R^2$ = Biphenyle	−14.5(c)	448
	$(C'_2)_3;CC'$	**33**; R^1R^2 = O, $R^3 = R^4$ = X = H, R = Ph	−38.4(c)	448
	$(C'_2)_3;C'_2$	**38**; XX = YY = O, $R^1 = R^2$ = Ph	−46.7(c) −51.5(b)	401, 448
$C'_3O_2/5,5$	$(C'_2)_3;C'_2$	**33**; $R^1R^2 = R^3R^4$ = O, R = Ph, X = H	−60.8(c)	441
	$(C'_2)_3;C'_2$	**33**; $R^1R^2 = R^3R^4$ = O, X = H, R = 2-$CO_2HC_6H_4$	−46.5	441
$C'_3O_2/5,6f$	$(C'_2)_2,CN';C_2$	**37**; R = R′ = Ph, X = N, $R^1 = CTf_2$, $R^2 = CTf_2OCTf_2$	−41.6(c)	383
	$(C'_2)_3;CC'$	**39**	−50.2(b)	439
$C'_3O_2/5,5,5f$	$(C'_2)_3;C_2$	**38**; XX = YY = H, R^1R^2 = Fluorenyl	−4.4(c)	448
$C'_3O_2/5,5f,6$	$(C'_2)_3;C_2$	**38**; XX = YY = H, R^1R^2 = (3-MeDop)O(3-MeDop)	−32.6(c)	448
$C'_3O_2/5,6,6f$	$(C'_2)_3;C'_2$	**40**; R = C_6Cl_4	−43.3	451
	$(C'_2)_3;C'_2$	**40**; R = 2,4-$tBu_2C_6H_2$	−54	451
$C'_3OO'/5$	$(C'_2)_3;C$	$Ph_2(OCMe_2Dop)PO^-Na^+$	−30.3	452
$C'_2C''O_2/5$	$(C'_2)_2;N'';C'_2$	$(OCTf{=}CTfO)P(CN)Ph_2$	−43.6(r)	122
$C_2C'S_2/4,5$	$(C_3)_2;C'_2;C'_2$	$(CMe_2CH_2CMe_2)P(SCTf{=}CTfS)Ph$	−17.5	267
$C'_3HO/5$	$(C'_2)_3;C$	$(OCTf_2Dop)P(H)Ph_2$	−49.3(c) J_{PH}266	407
C_4Cl	$(F_3)_3,H_3$	$Tf_3(Me)PCl$	−53.8(30°)	453
	$(F_3)_2,(H_3)_2$	Me_2Tf_2PCl	−50.1 + (33°)	454
C_4F	$(F_3)_3,H_3$	$Tf_3(Me)PF$	−52.1(30°) J_{PF}843	453
	$(F_3)_2,(H_3)_2$	Me_2Tf_2PF	−54.3(−20°)	454
	C_4	Et_4PF	−60(t, 35°)	455
$C_4F/5$	$(H_3)_2,(CH_2)_2$	$Me_2PF[(CH_2)_4]$	−47.3 (t,25°) −55(−65°)	456
C_4N	$(F_3)_3,H_3;CH$	$MeTf_3PNHMe$	−77.6	70
	$(F_3)_3,H_3;C_2$	$MeTf_3PNMe_2$	−65.3(33°)	457
	$(F_3)_2,(H_3)_2;C_2$	$Me_2Tf_2PNMe_2$	−78.4(−30°)	454
C_4O	$(F_3)_3,H_3;C$	$Tf_3MeP(OMe)$	−59.3(30°)	453
	$(F_3)_2,(H_3)_2;C$	$Tf_2Me_2P(OMe)$	−70.6(33°)	454
	$(H_3)_4;C$	$Me_4P(OMe)$	−88 (b, 30°)	458
	$(H_3)_4;C$	$Me_4P(OEt)$	−92(b, 30°)	458
$C_4O/5$	$(H_3)_3;CH_2;C$	$Me_3P[O(CH_2)_3]$	−78.9(b, 30°)	459
	$(CH_2)_4;C$	$Et_3P[O(CH_2)_3]$	−59.1(b, 30°)	459
$C_4O/6$	$(H_3)_3CH_2;C$	$Me_3P[O(CH_2)_4]$	−98.9(t)	460
$C_4O/5,5$	$H_3,(CH_2)_3;C$	$[(CH_2)_4]P[O(CH_2)_3]Me$	−50.6(t)	460
	$(CH_2)_4;C$	$MeOP[(CH_2)_4]_2$	−37(t)	456

TABLE Q (continued)

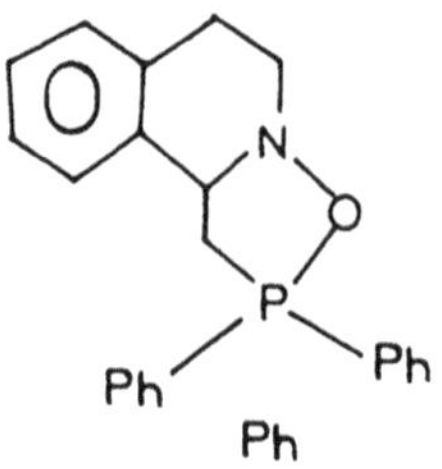

41 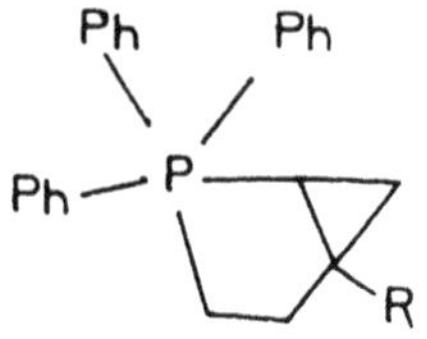 **42** 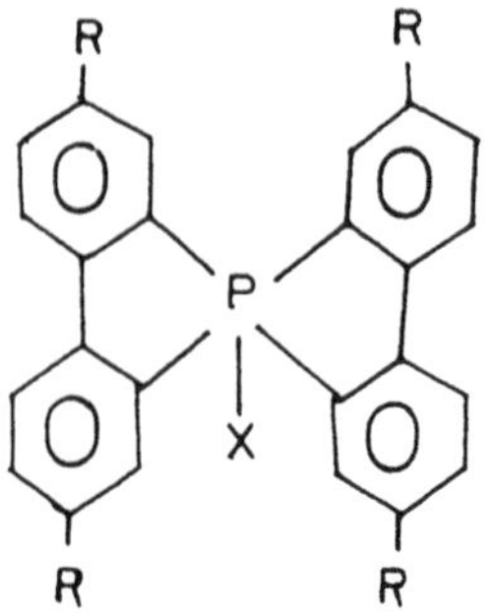 **43**

P-bound atoms/rings	Connectivities	Structure	NMR data (δP[solv temp] J Hz)	Ref.
$C_4O/5,6$	$H_3,(CH_2)_3;C$	$[(CH_2)_5]P[O(CH_2)_3]Me$	−72.9(t, 35°)	460
$C_4O/6,6$	$H_3,(CH_2)_3;C$	$[(CH_2)_5]P[O(CH_2)_4]Me$	−93.1(t, 35°)	460
C_4S	$(F_3)_3,H_3;C$	$Tf_3MePSMe$	−55.1(−10°)	457
	$(F_3)_2,(H_3)_2;C$	Tf_2Me_2SMe	−85.4(33°)	454
$C_2C'_2O/5$	$(H_3)_2;(C'_2)_2;C$	$Me_2PhP(OCMe_2Dop)$	−78.7	409
CC'_3O	$H_3;(C'_2)_3;C$	$Ph_3(Me)POEt$	−67.4	458
$CC'_3O/4$	$CH_2;(C'_2)_3;C$	$Ph_3P[CH_2C^{\cdot}(CH_2CH_2CH_2CH_2C^{\cdot}H_2)O]$	−73 −74(k)	461
	$C_2H;(C'_2)_3;C$	$Ph_3P[CH(tBu)CH(Me)O]$	−64.2	462
	$C_2H;(C'_2)_3;C$	$Ph_3P[CH(Me)CH(Ph)O]$	−63	463
	$C_2H;(C'_2)_3;C$	$Ph_3P[CH(nPr)CH(Ph)O]$	−61.4 −63.8*	464
	$C_2H;(C'_2)_3;C$	$Ph_3P[CH(Me)CH(CH_2CH_2Ph)O]$	−61(k)	461
$CC'_3O/5$	$H_3;(C'_2)_3;C$	$MePh_2P(OCMe_2Dop)$	−65.4	409
	$CH_2;(C'_2)_3;N$	$Ph_3P[CH_2CH(Ph)NMeO]$	−60.3	465
	$CH_2;(C'_2)_3;N$	**41**	−56.4(c) −59(d)	465
	$C_2H;(C'_2)_3;C$	**42**; R = H	−58.9	466
	$C_2H;(C'_2)_3;C$	**42**; R = Me	−56.9	466
$C'_4O/4$	$CP^4,(C'_2)_3;C$	$Ph_3P[C(=PPh_3)CTf_2O]$	−54	467
$C'_4O/5$	$C'H,(C'_2)_3;C$	$Ph_3P(CH{=}CMeCMe_2O)$	−59.7(c)	468
C_4H	$(F_3)_3,H_3$	$MeTf_3PH$	−103.7(f, −50°) J_{PH}608	469
$C'_4H/5,5$	$(C'_2)_4$	**43**; X = R = H	−112 J_{PH}482	470
	$(C'_2)_4$	**43**; X = H, R = Me	−110 J_{PH}483	470
$CC'_4/5,5$	$H_3;(C'_2)_4$	**43**; X = Me, R = H	−98.8(s)	470
	$H_3;(C'_2)_4$	**43**; X = R = Me	−99.6(s)	470
	$CH_2;(C'_2)_4$	**43**; X = Et, R = H	−88.8(s)	470
	$CH_2;(C'_2)_4$	**43**; X = Bu, R = Me	−88.4(s)	470
C'_5	$(C'_2)_5$	$(4\text{-}ClC_6H_4)_5P$	−87(s)	471
	$(C'_2)_5$	PPh_5	−89(j, k)	471
	$(C'_2)_5$	$(4\text{-}MeC_6H_4)_5P$	−88(s)	471

REFERENCES

1. **Flück, E., Gross, H., Binder, H., and Gloede, J.,** *Z. Naturforsch.*, 21B, 1125, 1966.
2. **Rogowski, R. and Cohn, K.,** *Inorg. Chem.*, 7, 2193, 1968.
3. **Schmidpeter, A., Nayibi, M., Mayer, P., and Tautz, H.,** *Chem. Ber.*, 116, 1468, 1983.

4. **Röschenthaler, G. V. and Storzer, W.,** *Angew. Chem. Int. Ed.*, 21, 208, 1982.
5. **Gloede, J. and Costisella, B.,** *Z. Anorg. Allg. Chem.*, 471, 147, 1980.
6. **Dakternieks, D., Röschenthaler, G. V., and Schmutzler, R.,** *J. Fluorine Chem.*, 11, 387, 1978.
7. **Michalski, J., Mikolajczyk, J., Pakulski, M., and Skowronska, A.,** *Phosphorus Sulfur*, 4, 233, 1978.
8. **Gloede, J., Gross, H., Michalski, J., Pakulski, M., and Skowronska, A.,** *Phosphorus Sulfur*, 13, 157, 1982.
9. **Markovskii, L. N., Solov'ev, A. V., Pirozhenko, V. V., and Chermolovitch, Yu G.,** *Zh. Obshch. Khim.*, 51, 1950, 1981.
10. **Flück, E.,** unpublished results.
11. **Binder, H.,** *Z. Anorg. Allg. Chem.*, 384, 193, 1971.
12. **Markovskii, L. N., Soloviev, A. V., and Chermolovitch, Y. G.,** *Zh. Obshch. Khim.*, 50, 2184, 1980.
13. **Kukhar, V. P., Grichkun, E. V., and Rudavskii, V. P.,** *Zh. Obshch. Khim.*, 50, 1017, 1980.
14. **Nesterova, L. I. and Sinitsa, A. D.,** *Zh. Obshch. Khim.*, 55, 2624, 1985.
15. **Kukhar, V. P., Kacheva, T. N., and Iltchenko, A. I.,** *Zh. Obshch. Khim.*, 47, 476, 1977.
16a. **Becke-Goehring, M. and Wald, H. J.,** *Z. Anorg. Allg. Chem.*, 371, 88, 1969.
16b. **Becke-Goehring, M. and Schwind, H.,** *Z. Anorg. Allg. Chem.*, 372, 285, 1970.
17. **Roesky, H. W., Djarrah, H., Amirzadeh-Asl, D., and Scheldrick, W. S.,** *Chem. Ber.*, 114, 1554, 1981.
18. **Day, R. O., Schmidpeter, A., and Holmes, R. R.,** *Inorg. Chem.*, 22, 3696, 1983.
19. **Day, R. O., Schmidpeter, A., and Holmes, R. R.,** *Inorg. Chem.*, 21, 3916, 1982.
20. **Dennis, L. W., Bartuska, V. J., and Maciel, G. E.,** *J. Am. Chem. Soc.*, 104, 230, 1982.
21. **Gloede, J.,** *J. Prakt. Chem.*, 323, 621, 1981.
22. **Flück, E. and Vargas, M.,** *Z. Anorg. Allg. Chem.*, 437, 53, 1977.
23. **Storzer, W., Röschenthaler, G. V., Schmutzler, R., and Sheldrick, W. S.,** *Chem. Ber.*, 114, 3609, 1981.
24. **Budilova, I. Yu, Gusar, N. I., and Gololobov, Yu G.,** *Zh. Obshch. Khim.*, 54, 1985, 1984.
25. **Ebeling, J. and Schmidpeter, A.,** *Angew. Chem.*, 81, 707, 1969.
26. **Appel, R. and Schöler, H.,** *Chem. Ber.*, 110, 2382, 1977.
27. **Schmidpeter, A., Tautz, H., and Schreiber, F.,** *Z. Anorg. Allg. Chem.*, 475, 211, 1981.
28. **Schmidpeter, A., Luber, J., and Tautz, H.,** *Angew. Chem.*, 89, 554, 1977.
29. **Becke-Goehring, M. and Leichner, L.,** *Angew. Chem.*, 76, 686, 1964.
30. **Kolodka, T. V., Loktionova, R. A., and Gololobov, Yu G.,** *Zh. Obshch. Khim.*, 53, 2476, 1983.
31. **Gloede, J. and Gross, H.,** *Phosphorus Sulfur*, 7, 57, 1979.
32. **Schmidpeter, A. and Luber, J.,** *Chem. Ber.*, 108, 820, 1975.
33. **Flück, E. and Wachtler, D.,** *Liebigs Ann. Chem.*, p. 1651, 1980.
34. **Klein, H. A. and Latscha, H. P.,** *Z. Anorg. Allg. Chem.*, 406, 214, 1974.
35. **Schöning, G. and Glemser, O.,** *Chem. Ber.*, 110, 3231, 1977.
36. **Kaltchenko, V. I., Negrebetskii, V. V., Rudyii, R. B., Atamas, L. I., Pavolotskii, M. I., and Markovskii, L. N.,** *Zh. Obshch. Khim.*, 53, 932, 1983.
37. **Kaltchenko, V. I., Rudyi, R. B., Negrebetskii, V. V., and Povolotskii, M. I.,** *Zh. Obshch. Khim.*, 54, 2207, 1984.
38. **Dunmur, R. E. and Schmutzler, R.,** *J. Chem. Soc. A*, p. 1289, 1971.
39. **Hormuth, P. B. and Becke-Goehring, M.,** *Z. Anorg. Allg. Chem.*, 372, 280, 1970.
40. **Gloede, J., Pakulski, M., Skowronska, A., Gross, H., and Michalski, J.,** *Phosphorus Sulfur*, 13, 163, 1982.
41. **Holmes, R. R. and Gallagher, W. P.,** *Inorg. Chem.*, 2, 433, 1963.
42. **Dmitriev, V. I., Sergienko, L. M., Ratovskii, G. B., Timokhin, B. V., and Glukhikh, V. I.,** *Zh. Obshch. Khim.*, 47, 1665, 1977.
43. **Ramirez, F., Bigler, A. J., and Smith, C. P.,** *J. Am. Chem. Soc.*, 90, 3507, 1968.
44. **Röschenthaler, G. V., Storzer, W., and Schmutzler, R.,** *J. Fluorine Chem.*, 19, 579, 1982.
45. **Gibson, J. A. and Röschenthaler, G. V.,** *J. Chem. Soc. Chem. Commun.*, p. 694, 1974.
46. **Gibson, J. A., Röschenthaler, G. V., Schomburg, D., and Sheldrick, W. S.,** *Chem. Ber.*, 110, 1887, 1977.
47. **Harris, R. K., Wazeer, M. I. M., Schlak, O., and Schmutzler, R.,** *J. Chem. Soc. Dalton Trans.*, p. 306, 1976.
48. **Bartsch, R., Weiss, J. V., and Schmutzler, R.,** *Z. Anorg. Allg. Chem.*, 537, 53, 1986.
49. **Becke-Goering, M. and Wieber, H.,** *Z. Anorg. Allg. Chem.*, 365, 185, 1969.
50. **Richman, J. E. and Flay, R. B.,** *J. Am. Chem. Soc.*, 103, 5265, 1981.
51. **Denney, D. B., Denney, D. Z., and Hsu, Y. F.,** *Phosphorus*, 4, 215, 1974.
52. **Röschenthaler, G. V., Bohlen, R., Storzer, W., Sopchik, A. E., and Bentrude, W. G.,** *Z. Anorg. Allg. Chem.*, 507, 93, 1983.

53. **Eikmeier, H. B., Hodges, K. C., Stelzer, O., and Schmutzler, R.,** *Chem. Ber.*, 111, 2077, 1978.
54. **Eisenhut, M. and Schmutzler, R.,** *J. Chem. Soc. Chem. Commun.*, p. 1452, 1971.
55. **Röschenthaler, G. V., Gibson, J. A., and Schmutzler, R.,** *Chem. Ber.*, 110, 611, 1977.
56. **Gibson, J. A., Röschenthaler, G. V., Schmutzler, R., and Starke, R.,** *J. Chem. Soc. Dalton Trans.*, p. 450, 1977.
57. **Ramirez, F., Smith, C. P., and Meyerson, S.,** *Tetrahedron Lett.*, p. 3651, 1966.
58. **Kubjacek, M. and Utvary, K.,** *Monatsh. Chem.*, 112, 305, 1981.
59. **Ramirez, F., Gulati, A. S., and Smith, C. P.,** *J. Am. Chem. Soc.*, 89, 6283, 1967.
60. **Roesky, H. W. and Petersen, O.,** *Angew. Chem. Int. Ed.*, 12, 415, 1973.
61. **Bohlen, R., Heine, J., Kuhn, W., Offermann, W., Stelten, J., Röschenthaler, G. V., and Bentrude, W. G.,** *Phosphorus Sulfur*, 27, 313, 1986.
62. **De'Ath, N. J., Denney, D. Z., Denney, D. B., and Hall, C. D.,** *Phosphorus*, 3, 205, 1974.
63. **Peake, S. C., Fild, M., Hewson, M. J. C., and Schmutzler, R.,** *Inorg. Chem.*, 10, 2723, 1971.
64. **Denney, D. B., Denney, D. Z., and Hsu, Y. F.,** *Phosphorus*, 4, 213, 1974.
65. **Lustig, M. and Roesky, H. W.,** *Inorg. Chem.*, 9, 1289, 1970.
66. **Peake, S. C., Hewson, M. J. C., Schlak, O., Schmutzler, R., Harris, R. K., and Wazeers, M. I. M.,** *Phosphorus Sulfur*, 4, 67, 1978.
67. **Schmutzler, R.,** *J. Chem. Soc. Chem. Commun.*, 2, 19, 1965.
68. **Rankin, D. W. H. and Wright, J. G.,** *J. Chem. Soc. Dalton Trans.*, p. 1070, 1979.
69. **Goodrich-Haines, R. and Gilje, J. W.,** *Inorg. Chem.*, 15, 470, 1976.
70. **Cavell, R. G., Pirakitigoon, S., and Van de Griend, L.,** *Inorg. Chem.*, 22, 1378, 1983.
71. **Schmutzler, R.,** *J. Chem. Soc. Dalton Trans.*, p. 2687, 1973.
72. **Cowley, A. H., Brown, R. W., and Gilje, J. W.,** *J. Am. Chem. Soc.*, 97, 434, 1975.
73. **Hewson, J. C. and Schmutzler, R.,** *Phosphorus Sulfur*, 8, 9, 1980.
74. **Brown, D. H., Crosbie, K. D., Fraser, G. W., and Sharp, D. W. A.,** *J. Chem. Soc. A*, p. 872, 1969.
75. **Gombler, W.,** *Z. Anorg. Allg. Chem.*, 439, 207, 1978.
76. **Peake, S. C. and Schmutzler, R.,** *J. Chem. Soc. A*, p. 1049, 1970.
77. **Röschenthaler, G. V. and Schmutzler, R.,** *J. Inorg. Nucl. Chem.*, Suppl., 19, 1976.
78. **Maier, L. and Schmutzler, R.,** *J. Chem. Soc. Chem. Commun.*, p. 961, 1969.
79. **Cooper, D. and Trippett, S.,** *Tetrahedron Lett.*, p. 1725, 1979.
80. **Majoral, J. P., Kraemer, R., N'Gando M'Pondo, T., and Navech, J.,** *Tetrahedron Lett.*, 21, 1307, 1980.
81. **Denney, D. B., Denney, D. Z., and Liu, L. T.,** *Phosphorus Sulfur*, 22, 71, 1985.
82. **Denney, D. B., Denney, D. Z., Hammond, P. J., Huang, C., Liu, L. T., and Tseng, K. S.,** *Phosphorus Sulfur*, 15, 281, 1983.
83. **Schmidpeter, A. and Weinmaier, J. H.,** *Chem. Ber.*, 111, 2086, 1978.
84. **Carpenter, L. E., De Ruiter, B., Van Aken, D., Buck, H. M., and Verkade, J. G.,** *J. Am. Chem. Soc.*, 108, 4918, 1986.
85. **Denney, D. B., Denney, D. Z., Hammond, P. J., Huang, C., and Tseng, K. S.,** *J. Am. Chem. Soc.*, 102, 5073, 1980.
86. **Bone, S. A., Trippett, S., and Whittle, P. J.,** *J. Chem. Soc. Perkin Trans. 1*, p. 80, 1977.
87. **Antczak, S., Bone, S. A., Brierly, J., and Tripett, S.,** *J. Chem. Soc. Perkin Trans. 1*, p. 278, 1977.
88. **Carpenter, L. E. and Verkade, J. G.,** *J. Am. Chem. Soc.*, 107, 7084, 1985.
89. **Francke, R., Di Giacomo, R., Dakternieks, D., and Röschenthaler, G. V.,** *Z. Anorg. Allg. Chem.*, 519, 141, 1984.
90. **Duff, E., Trippett, S., and Whittle, P. J.,** *J. Chem. Soc. Perkin Trans. 1*, p. 972, 1973.
91. **Arbuzov, B. A., Poliejaeva, N. A., Vinogradova, V. C., and Samitov, Yu Yu.,** *Izv. Akad. Nauk SSSR Ser. Khim.*, p. 1605, 1967.
92. **Trippett, S. and Whittle, P. J.,** *J. Chem. Soc. Perkin Trans. 1*, p. 2302, 1973.
93. **Kukhar, V. P., Grishkun, E. V., Rudavskii, V. P., and Giliarov, V. A.,** *Zh. Obshch. Khim.*, 50, 1477, 1980.
94. **Von Criegern, T. and Schmidpeter, A.,** *Z. Naturforsch.*, 34B, 762, 1979.
95. **Tuptchienko, S. K., Dudtchenko, T. N., and Sinitsa, A. D.,** *Zh. Obshch. Khim.*, 55, 1191, 1985.
96. **Lopez, L., Fabas, C., and Barrans, J.,** *Phosphorus Sulfur*, 7, 81, 1979.
97. **Sanchez, M., Brazier, J. F., Houalla, D., and Wolf, R.,** *Nouv. J. Chim.*, 3, 775, 1979.
98. **Bernard, D. and Burgada, R.,** *Phosphorus*, 5, 285, 1975.
99. **Garrigues, B. and Munoz, A.,** *Can. J. Chem.*, 62, 2179, 1984.
100. **Schwarz, W., Flück, E., and Vargas, M.,** *Phosphorus Sulfur*, 5, 217, 1978.
101. **Bone, S. A., Trippett, S., and Whittle, P. J.,** *J. Chem. Soc. Perkin Trans. 1*, p. 2125, 1974.
102. **Willson, M., Burgada, R., and Mathis, F.,** *C. R. Acad. Sci. Paris*, 280, 225, 1975.
103. **Fauduet, H. and Burgada, R.,** *Nouv. J. Chim.*, 4, 113, 1980.

104. **Bone, S. A. and Trippett, S.,** *Tetrahedron Lett.*, 19, 1583, 1975.
105. **Bernard, D. and Burgada, R.,** *Tetrahedron*, 31, 797, 1975.
106. **Bone, S. A. and Trippett, S.,** *J. Chem. Soc. Perkin Trans. 1*, p. 156, 1976.
107. **Ramirez, F., Nagabhushnaman, M., and Smith, C. P.,** *Tetrahedron*, 24, 1785, 1968.
108. **Bernard, D. and Burgada, R.,** *C. R. Acad. Sci. Paris*, 277, 433, 1973.
109. **Bui Cong, C., Gence, G., Garrigues, B., Koenig, M., and Munoz, A.,** *Tetrahedron*, 35, 1825, 1979.
110. **Kudryatseva, T. N., Proskurnina, M. V., and Lutsenko, I. F.,** *Zh. Obshch. Khim.*, 51, 1664, 1981.
111. **Bernard, D. and Burgada, R.,** *C. R. Acad. Sci. Paris*, 279, 883, 1974.
112. **Negrebetskii, V. V., Kaltchenko, V. I., Rudyi, R. B., and Markovskii, L. N.,** *Zh. Obshch. Khim.*, 55, 1982, 1986.
113. **Konovalova, I. V., Ofitserova, E. K., Yudina, T. V., and Pudovik, A. N.,** *Zh. Obshch. Khim.*, 47, 476, 1977.
114. **El Manouni, D., Leroux, Y., and Burgada, R.,** *Phosphorus Sulfur*, 25, 319, 1985.
115. **Denney, D. B., Denney, D. Z., Hammond, P. J., and Tseng, K. S.,** *J. Am. Chem. Soc.*, 103, 2054, 1981.
116. **Denney, D. B., Denney, D. Z., and Pastor, S. D.,** *Phosphorus Sulfur*, 22, 191, 1985.
117. **Schmidpeter, A., Schomburg, D., Sheldrick, W. S., and Weinmaier, J. H.,** *Angew. Chem. Int. Ed.*, 15, 781, 1976.
118. **Hamilton, W. C., Ricci, J. S., Jr., Ramirez, F., Kramer, L., and Stern, P.,** *J. Am. Chem. Soc.*, 95, 6335, 1973.
119a. **Navech, J., Kraemer, R., and Majoral, J. P.,** *Tetrahedron Lett.*, 21, 1449, 1980.
119b. **Kyba, E. P. and Alexander, D. C.,** *J. Chem. Soc. Chem. Commun.*, p. 934, 1977.
120. **Baceiredo, A., Bertrand, G., Majoral, J. P., Wermuth, U., and Schmutzler, R.,** *J. Am. Chem. Soc.*, 106, 7065, 1984.
121. **Lamandé, L. and Munoz, A.,** *Phosphorus Sulfur*, 32, 1, 1987.
122. **Brierley, J., Dickstein, J. I., and Trippett, S.,** *Phosphorus Sulfur*, 7, 167, 1979.
123. **Roesky, H. W., Amirzadeh-Asl, D., and Sheldrick, W. S.,** *J. Am. Chem. Soc.*, 104, 2919, 1982.
124. **Dickstein, J. I. and Trippett, S.,** *Tetrahedron Lett.*, p. 2203, 1973.
125. **Schmidpeter, A., Luber, J., and Von Criegern, T.,** *Z. Naturforsch.*, 32B, 845, 1977.
126. **Konovalova, I. V., Cherkina, M. V., Burnaeva, L. A., Goldfarb, E. I., and Pudovik, A. N.,** *Zh. Obshch. Khim.*, 52, 1969, 1982.
127. **Schmidpeter, A., Zeiss, W., Schomburg, D., and Sheldrick, W. S.,** *Angew. Chem. Int. Ed.*, 19, 825, 1980.
128. **Gilyarov, V. A., Tikhonina, N. A., Andrianov, V. G., Strutchkov, Yu, T., and Kabachnik, M. I.,** *Zh. Obshch. Khim.*, 48, 732, 1978.
129. **Schmidpeter, A. and Luber, J.,** *Phosphorus*, 5, 55, 1974.
130. **Tikhonina, N. A., Gilyarov, V. A., and Kabachnik, M. I.,** *Zh. Obshch. Khim.*, 48, 44, 1978.
131. **Kabachnik, M. I., Gilyarov, V. A., Tikhonina, N. A., Kalinin, A. E., Andrianov, V. G., Strutchkov, Yu T., and Timofeeva, G. I.,** *Phosphorus Sulfur*, 5, 65, 1974.
132. **Kolodka, T. V. and Gololobov, Yu G.,** *Zh. Obshch. Khim.*, 53, 1013, 1983.
133. **Tikhonina, N. A., Timofeeva, G. I., Matrosov, E. I., Gilyarov, V. A., and Kabachnik, M. I.,** *Zh. Obshch. Khim.*, 45, 2414, 1975.
134. **Gonçalves, H., Dormoy, J. R., Chapleur, Y., Castro, B., Fauduet, H., and Burgada, R.,** *Phosphorus Sulfur*, 8, 147, 1980.
135. **Schmidpeter, A. and Weinmaier, J. H.,** *Angew. Chem.*, 89, 903, 1977.
136. **Allcock, H. R. and Kugel, R. L.,** *J. Chem. Soc. Chem. Commun.*, p. 1606, 1968.
137. **Sanchez, M., Brazier, J. F., Houalla, D., Munoz, A., and Wolf, R.,** *J. Chem. Soc. Chem. Commun.*, p. 730, 1976.
138. **Ramirez, F., Patwardhan, A. V., Kugler, H. J., and Smith, C. P.,** *J. Am. Chem. Soc.*, 89, 6276, 1967.
139. **Fauduet, H. and Burgada, R.,** *Nouv. J. Chim.*, 3, 555, 1979.
140. **Marre, M. R., Brazier, J. F., Wolf, R., and Klaebe, A.,** *Phosphorus Sulfur*, 11, 87, 1981.
141. **Bernard, D. and Burgada, R.,** *Phosphorus*, 3, 187, 1974.
142. **Mulliez, M. and Wolf, R.,** *Bull. Soc. Chim. Fr.*, p. 101, 1986.
143. **Konovalova, I. V., Cherkina, M. V., Burnaeva, L. A., and Pudovik, A. N.,** *Zh. Obshch. Khim.*, 52, 207, 1982.
144. **Chermolovitch, Yu G., Lantchenko, E. A., Soloviev, A. V., Tratchevskii, U. V., and Markovskii, L. N.,** *Zh. Obshch. Khim.*, 56, 560, 1986.
145. **Krawczyk, E., Michalski, J., Pakulski, M., and Skowronska, A.,** *Tetrahedron Lett.*, 23, 2019, 1977.
146. **Denney, D. B., Denney, D. Z., Chang, B. C., and Marsi, K. L.,** *J. Am. Chem. Soc.*, 91, 5243, 1969.
147. **Zeiss, W.,** *Angew. Chem. Int. Ed.*, 15, 555, 1976.
148. **Röschenthaler, G. V., Sauerbrey, K., and Schmutzler, R.,** *Chem. Ber.*, 111, 3105, 1978.

149. **Gololobov, Yu, Gusar, N. I., and Chaus, M. P.,** *Tetrahedron*, 41, 793, 1985.
150. **Bellan, J., Brazier, J. F., Zenati, N., and Sanchez, M.,** *C. R. Acad. Sci. Paris*, 289, 449, 1979.
151. **Mulliez, M.,** *Bull Soc. Chim. Fr.*, p. 1211, 1985.
152. **Bernard, D. and Burgada, R.,** *C. R. Acad. Sci. Paris*, 271, 418, 1970.
153. **Razhabov, A. A. and Yusupov, M. M.,** *Zh. Obshch. Khim.*, 52, 2209, 1982.
154. **Ramirez, F., Patwardhan, A. V., Kugler, H. J., and Smith, C. P.,** *Tetrahedron*, 24, 2275, 1968.
155. **Schmidpeter, A., Junius, M., Weinmaier, J. H., Barrans, J., and Charbonnel, Y.,** *Z. Naturforsch.*, 32B, 841, 1977.
156. **Denney, D. B., Denney, D. Z., Gavrilovic, D. M., Hammond, P. J., Huang, C., and Tseng, K. S.,** *J. Am. Chem. Soc.*, 102, 7072, 1980.
157. **Denney, D. B., Denney, D. Z., and Pastor, S. D.,** *Phosphorus Sulfur*, 22, 183, 1985.
158. **Tautz, H. and Schmidpeter, A.,** *Chem. Ber.*, 114, 825, 1981.
159. **Schomburg, D., Wermuth, U., and Schmutzler, R.,** *Z. Naturforsch.*, 41B, 207, 1986.
160. **Roesky, H. W. and Djarrah, H.,** *Inorg. Chem.*, 21, 844, 1982.
161. **Schiebel, H. M., Schmutzler, R., Schomburg, D., and Wermuth, U.,** *Z. Naturforsch.*, 38B, 702, 1983.
162. **Richman, J. E., Day, R. O., and Holmes, R. R.,** *J. Am. Chem. Soc.*, 102, 3955, 1980.
163. **Schmidpeter, A., Gross, J., Schrenk, E., and Sheldrick, W. S.,** *Phosphorus Sulfur*, 14, 49, 1982.
164. **Appel, R. and Halstenberg, M.,** *Angew. Chem. Int. Ed.*, 16, 263, 1977.
165. **Schomburg, D., Wermuth, U., and Schmutzler, R.,** *Phosphorus Sulfur*, 26, 193, 1986.
166. **Denney, D. B., Denney, D. Z., and Gavrilovic, D. M.,** *Phosphorus Sulfur*, 11, 1, 1981.
167. **De'Ath, N. J. and Denney, D. B.,** *J. Chem. Soc. Chem. Commun.*, p. 395, 1972.
168. **Hall, D.,** personal communication.
169. **Denney, D. B., Denney, D. Z., Hammond, P. J., and Wang, Y. P.,** *J. Am. Chem. Soc.*, 103, 1785, 1981.
170. **Denney, D. B., Denney, D. Z., Hammond, P. J., Liu, L. T., and Wang, Y. P.,** *J. Org. Chem.*, 48, 2159, 1983.
171. **Chermolovitch, Yu G., Kolesnik, N. P., Vasil'ev, V. V., Pashinnik, V. E., and Markovskii, L. N.,** *Zh. Obshch. Khim.*, 51, 542, 1981.
172. **Röschenthaler, G. V., Bohlen, R., and Schomburg, D.,** *Z. Naturforsch.*, 40B, 1589, 1985.
173. **Brierley, J., Trippett, S., and White, M. W.,** *J. Chem. Soc. Perkin Trans. 1*, p. 273, 1977.
174. **Lamandé, L., Munoz, A., and Boyer, D.,** *J. Chem. Soc. Chem. Commun.*, p. 225, 1984.
175. **Johnson, M. P. and Trippett, S.,** *J. Chem. Soc. Perkin Trans. 1*, p. 3074, 1981.
176. **Starke, U., Rösch, L., and Schmutzler, R.,** *Phosphorus Sulfur*, 27, 297, 1986.
177. **Chang, L. L., Denney, D. B., Denney, D. Z., and Kazior, R. J.,** *J. Am. Chem. Soc.*, 99, 2293, 1977.
178. **Denney, D. B. and Jones, D. H.,** *J. Am. Chem. Soc.*, 91, 5821, 1969.
179. **Van Aken, D., Paulissen, L. M. C., and Buck, H. M.,** *J. Org. Chem.*, 46, 3189, 1981.
180. **Markovskii, L. N., Soloviev, A. V., Pachinnik, V. E., and Chermolovitch, Yu G.,** *Zh. Obshch. Khim.*, 50, 807, 1980.
181. **Markovskii, L. N., Kolesnik, N. P., and Chermolovitch, Yu G.,** *Zh. Obshch. Khim.*, 49, 1764, 1979.
182. **Ramirez, F., Lee, S., Stern, P., Ugi, I., and Gillespie, P. D.,** *Phosphorus*, 4, 21, 1974.
183. **Denney, D. B., Denney, D. Z., and Liu, L. T.,** *Phosphorus Sulfur*, 13, 1, 1982.
184. **Ramirez, F., Tasaka, K., and Hershberg, R.,** *Phosphorus*, 2, 41, 1972.
185. **Thompson, Q. E.,** *J. Am. Chem. Soc.*, 83, 845, 1961.
186. **Castelijns, M. M. C. F., Schipper, P., Van Aken, D., and Buck, H. M.,** *J. Org. Chem.*, 46, 47, 1981.
187. **Bartlett, P. D., Baumstark, A. L., Landis, M. E., and Lerman, C. L.,** *J. Am. Chem. Soc.*, 96, 5267, 1974.
188. **Ramirez, F., Kugler, H. J., Patwardhan, A. V., and Smith, C. P.,** *J. Org. Chem.*, 33, 1185, 1968.
189. **Ramirez, F., Patwardhan, A. V., and Smith, C. P.,** *J. Org. Chem.*, 31, 474, 1966.
190. **Ramirez, F., Patwardhan, A. V., and Smith, C. P.,** *J. Org. Chem.*, 31, 3159, 1966.
191. **Ramirez, F., Bhatia, S. B., and Smith, C. P.,** *J. Am. Chem. Soc.*, 89, 3030, 1967.
192. **Marre, M. R., Sanchez, M., Brazier, J. F., Wolf, R., and Bellan, J.,** *Can. J. Chem.*, 60, 456, 1982.
193. **Ramirez, F., Bhatia, S. B., and Smith, C. P.,** *Tetrahedron*, 23, 2067, 1967.
194. **Ramirez, F. and Smith, C. P.,** *J. Chem. Soc. Chem. Commun.*, p. 662, 1967.
195. **Ramirez, F., Marecek, J., Ugi, I., and Marquarding, D.,** *Phosphorus*, 3, 91, 1973.
196. **Ramirez, F., Patwardhan, A. V., and Smith, C. P.,** *J. Org. Chem.*, 30, 2575, 1965.
197. **Ramirez, F. and Desai, N. B.,** *J. Am. Chem. Soc.*, 85, 3252, 1963.
198. **Chan, L. L. and Denney, D. B.,** *J. Chem. Soc. Chem. Commun.*, p. 84, 1974.
199. **Ramirez, F., Tasaka, K., Desai, N. B., and Smith, C. P.,** *J. Am. Chem. Soc.*, 90, 751, 1968.
200. **Ramirez, F., Bhatia, S. B., Patwardhan, A. V., Chen, E. H., and Smith, C. P.,** *J. Org. Chem.*, 33, 20, 1968.
201. **Aganov, A. V., Polezhaeva, N. A., Khayarov, A. I., and Arbuzov, B. A.,** *Phosphorus Sulfur*, 22, 303, 1985.

202. **Koole, L. H., Lanters, E. J., and Buck, H. M.,** *J. Am. Chem. Soc.*, 106, 5451, 1984.
203. **Hamilton, W. C., La Placa, S. J., Ramirez, F., and Smith, C. P.,** *J. Am. Chem. Soc.*, 89, 2268, 1967.
204. **Storzer, W. and Röschenthaler, G. V.,** *Z. Naturforsch.*, 33B, 305, 1978.
205. **Ramirez, F., Bigler, A. J., and Smith, C. P.,** *Tetrahedron*, 24, 5041, 1968.
206. **Clennan, E. L. and Heah, P. C.,** *J. Org. Chem.*, 47, 3329, 1982.
207. **Abdou, W. H., Denney, D. B., Denney, D. Z., and Pastor, S. D.,** *Phosphorus Sulfur*, 22, 99, 1985.
208. **El Khatib, F., Caminade, A. M., and Koenig, M.,** *Phosphorus Sulfur*, 20, 55, 1984.
209. **Koenig, M., El Khatib, F., Munoz, A., and Wolf, R.,** *Tetrahedron Lett.*, 23, 421, 1982.
210. **Ramirez, F., Nowakowski, M., and Marecek, J. F.,** *J. Am. Chem. Soc.*, 99, 4515, 1977.
211. **Ovchinikov, V. V., Eustaf'eva, Yu G., Cherkasov, R. A., and Pudovik, A. N.,** *Zh. Obshch. Khim.*, 53, 1923, 1983.
212. **Munoz, A., Lamandé, L., Koenig, M., and Wolf, R.,** *Phosphorus Sulfur*, 11, 71, 1981.
213. **Kudryatseva, T. N., Karlstedt, M. B., Proskurnina, M. V., Frolovskii, V. A., and Lutsenko, I. F.,** *Zh. Obshch. Khim.*, 54, 552, 1984.
214. **Laurenço, C. and Burgada, R.,** *Tetrahedron*, 32, 2253, 1976.
215. **Kimura, Y., Miyamoto, M., and Saegusa, T.,** *J. Org. Chem.*, 47, 916, 1982.
216. **Osman, F. H., El Hamouly, W. S., Abdel-Gawad, M. N., and Abbasi, M. M.,** *Phosphorus Sulfur*, 14, 1, 1982.
217. **Bernard, D. and Burgada, R.,** *Tetrahedron Lett.*, 36, 3455, 1973.
218. **Munoz, A., Garrigues, B., and Wolf, R.,** *Phosphorus Sulfur*, 4, 47, 1978.
219. **Ramirez, F., Prasad, V. A. V., and Marecek, J. F.,** *J. Am. Chem. Soc.*, 96, 7269, 1974.
220. **Schmidpeter, A., Von Criegern, T., Sheldrick, W. S., and Schomburg, D.,** *Tetrahedron Lett.*, 32, 2857, 1978.
221. **Bohlen, R., Hacklin, H., Heine, J., Offermann, W., and Röschenthaler, G. V.,** *Phosphorus Sulfur*, 27, 321, 1986.
222. **Kaltchenko, V. I., Atamas, L. I., Serguchev, Yu A., and Markovskii, L. N.,** *Zh. Obshch. Khim.*, 54, 1754, 1984.
223. **Ramirez, F., Pfohl, S., Tsolis, E. A., Pilot, J. F., Smith, C. P., Ugi, I., Marquarding, D., Gillespie, P., and Hoffmann, P.,** *Phosphorus*, 1, 1, 1971.
224. **Munoz, A., Garrigues, B., and Koenig, M.,** *Tetrahedron*, 36, 2467, 1980.
225. **Dubourg, A., Roques, R., Declerq, J. P., Boyer, D., Lamandé, L., Munoz, A., and Wolf, R.,** *Phosphorus Sulfur*, 17, 97, 1983.
226. **Treichel, P. M., Goodrich, R. A., and Pierce, S. B.,** *J. Am. Chem. Soc.*, 89, 2017, 1968.
227. **Holmes, R. R. and Storey, R. N.,** *Inorg. Chem.*, 5, 2146, 1966.
228. **Pudovik, M. A., Terentieva, S. A., Iliasov, A. V., Tchernov, A. N., Nafikova, A. A., and Pudovik, A. N.,** *Z. Obshch. Khim.*, 54, 2448, 1984.
229. **Milbrath, D. S. and Verkade, J. G.,** *J. Am. Chem. Soc.*, 99, 6607, 1977.
230. **Lafaille, L., Mathis, F., and Barrans, J.,** *C. R. Acad. Sci. Paris*, 285, 575, 1977.
231. **Lopez, L. and Barrans, J.,** *C. R. Acad. Sci. Paris*, 273, 1540, 1971.
232. **Wolf, R., Sanchez, M., Houalla, D., and Schmidpeter, A.,** *C. R. Acad. Sci. Paris*, 275, 151, 1972.
233. **Sanchez, M., Wolf, R., Burgada, R., and Mathis, F.,** *Bull. Soc. Chim. Fr.*, p. 773, 1968.
234. **Koenig, M., Munoz, A., Wolf, R., and Houalla, D.,** *Bull Soc. Chim. Fr.*, p. 1413, 1972.
235. **Burgada, R., Bon, M., and Mathis, F.,** *C. R. Acad. Sci. Paris*, 265, 1499, 1967.
236. **Garrigues, B., Munoz, A., Koenig, M., Sanchez, M., and Wolf, R.,** *Tetrahedron*, 33, 635, 1977.
237. **Munoz, A., Koenig, M., Garrigues, B., and Wolf, R.,** *C. R. Acad. Sci. Paris*, 274, 1413, 1972.
238. **Sanchez, M., Brazier, J. F., Houalla, D., and Wolf, R.,** *Bull Soc. Chim. Fr.*, p. 3930, 1967.
239. **Burgada, R. and Laurenço, C.,** *J. Organomet. Chem.*, 66, 255, 1974.
240. **Laurenço, C. and Burgada, R.,** *C. R. Acad. Sci. Paris*, 275, 237, 1972.
241. **Diallo, O. S., Brazier, J. F., Klaebe, A., and Wolf, R.,** *Phosphorus Sulfur*, 22, 93, 1985.
242. **Garrigues, B., Bui Cong, C., Munoz, A., and Klaebe, A.,** *J. Chem. Res. (S)* p. 172, 1979.
243. **Koenig, M., Munoz, A., Garrigues, B., and Wolf, R.,** *Phosphorus Sulfur*, 6, 435, 1979.
244. **Laurenço, C. and Burgada, R.,** *C. R. Acad. Sci. Paris*, 278, 291, 1974.
245. **Arduengo, A. J., III, Stewart, C. A., Davidson, F., Dixon, D. A., Becker, J. Y., Culley, S. A., and Mizen, M. B.,** *J. Am. Chem. Soc.*, 109, 627, 1987.
246. **Lopez, L., Majoral, J. P., Meriem, A., N'Gando M'Pondo, T., Navech, J., and Barrans, J.,** *J. Chem. Soc. Chem. Commun.*, p. 183, 1984.
247. **Charbonnel, Y. and Barrans, J.,** *C. R. Acad. Sci. Paris*, 277, 571, 1973.
248. **Garrigues, B., Munoz, A., and Mulliez, M.,** *Phosphorus Sulfur*, 9, 183, 1980.
249. **Laurenço, C., and Burgada, R.,** *C. R. Acad. Sci. Paris*, 278, 291, 1974.
250. **Bonningue, C., Houalla, D., Sanchez, M., and Wolf, R.,** *J. Chem. Soc. Perkin Trans. 2*, p. 19, 1981.
251. **Dupart, J. M., Le Borgne, G., Pace, S., and Riess, J. G.,** *J. Am. Chem. Soc.*, 107, 1202, 1985.
252. **Dupart, J. M., Pace, S., and Riess, J. G.,** *J. Am. Chem. Soc.*, 105, 1051, 1983.

253. **Lattman, M., Chopra, S. K., Cowley, A. H., and Arif, A. M.,** *Organometallics*, 5, 677, 1986.
254. **Richman, J. R. and Atkins, T. J.,** *Tetrahedron Lett.*, p. 4333, 1978.
255. **Nifantiev, E. E., Zavalishina, I., Sorokina, A. I., and Borisenko, A. A.,** *Zh. Obshch. Khim.*, 46, 471, 1976.
256. **Boisdon, M. T., Malavaud, C., Mathis, F., and Barrans, J.,** *Tetrahedron Lett.*, p. 3501, 1977.
257. **Germa, H., Willson, M., and Burgada, R.,** *C. R. Acad. Sci. Paris*, 270, 1474, 1970.
258. **Germa, H., Sanchez, M., Burgada, R., and Wolf, R.,** *Bull. Soc. Chim. Fr.*, p. 612, 1970.
259. **Burgada, R. and Bernard, D.,** *C. R. Acad. Sci. Paris*, 273, 164, 1971.
260. **Rankin, D. W. H. and Wright, J. G.,** *J. Chem. Soc. Dalton Trans.*, p. 2049, 1980.
261. **Bruzik, K., Stec, W., Houalla, D., and Wolf, R.,** *J. Chem. Res. (M)*, p. 4036, 1981.
262. **Stec, W., Uznanski, B., Houalla, D., and Wolf, R.,** *C. R. Acad. Sci. Paris*, 281, 727, 1975.
263. **Nixon, J. F. and Schmutzler, R.,** *Spectrochim. Acta*, 20, 1835, 1964.
264. **Fild, M., Kurpat, R., and Stankiewicz, T.,** *Z. Anorg. Allg. Chem.*, 439, 145, 1978.
265. **Francke, R., Röschenthaler, G. V., Di Giacomo, R., and Dakternieks, D.,** *Phosphorus Sulfur*, 20, 107, 1984.
266. **Schmutzler, R.,** *Halogen Chemistry, Vol. 2*, Gutman, V., Ed., Academic Pres, New York, 1967, 31.
267. **Burros, B. C., De'Ath, N. J., Denney, D. B., Denney, D. Z., and Kipnis, I. J.,** *J. Am. Chem. Soc.*, 100, 7300, 1978.
268. **Timokhin, B. V., Dmitriev, V. I., Glukhikh, V. I., and Korchevin, N. A.,** *Izv. Akad. Nauk SSSR Ser. Khim.*, 6, 1334, 1978.
269. **Kolodiajnyi, O. I.,** *Zh. Obshch. Khim.*, 57, 821, 1987.
270. **Bauer, G. and Hägele, G.,** *Angew. Chem. Int. Ed.*, 16, 477, 1977.
271. **Balitskii, Yu V., Gololobov, Yu G., Yurtchenko, V. M., Antinin, M. Yu, Strutchkov, Yu T., and Boldeskul, I. E.,** *Zh. Obshch. Khim.*, 50, 291, 1980.
272. **Svara, J. and Flück, E.,** *Phosphorus Sulfur*, 25, 129, 1985.
273. **Harris, R. K., Lewelly, M., Wazeer, M. I. M., Woplin, J. R., Dunmur, R. E., Hewson, M. J. C., and Schmutzler, R.,** *J. Chem. Soc. Dalton Trans.*, p. 61, 1975.
274. **Schlack, O., Schmutzler, R., Harris, R. K., McVicker, E. M., and Wazeer, M. I. M.,** *Phosphorus Sulfur*, 10, 87, 1981.
275. **Fild, M., Sheldrick, W. S., and Stankiewicz, T.,** *Z. Anorg. Allg. Chem.*, 415, 43, 1974.
276. **Flück, E., Svara, J., and Riffel, H.,** *Z. Anorg. Allg. Chem.*, 548, 63, 1987.
277. **Dakternieks, D., Röschenthaler, G. V. and Schmutzler, R.,** *J. Fluorine Chem.*, 12, 413, 1978.
278. **Robert, D., Demay, C., and Riess, J. G.,** *Inorg. Chem.*, 21, 1805, 1982.
279. **Schmutzler, R.,** *Angew. Chem. Int. Ed.*, 4, 496, 1965.
280. **Connelly, A. and Harris, R. K.,** *J. Chem. Soc. Dalton Trans.*, p. 1547, 1984.
281. **Gibson, J. A. and Schmutzler, R.,** *Z. Anorg. Allg. Chem.*, 416, 222, 1975.
282. **Schmutzler, R and Reddy, G. S.,** *Inorg. Chem.*, 4, 191, 1965.
283. **Negrebetskii, V. V., Kaltchenko, V. I., Rudyi, R. B., and Markovskii, L. N.,** *Zh. Obshch. Khim.*, 55, 271, 1985.
284. **Cavell, R. G., The, K. I., Gibson, J. A., and Yap, N. T.,** *Inorg. Chem.*, 18, 3400, 1979.
285. **Althoff, W., Fild, M., and Schmutzler, R.,** *Chem. Ber.*, 114, 1082, 1981.
286. **Althoff, W., Fild, M., Rieck, H. P., and Schmutzler, R.,** *Chem. Ber.*, 111, 1845, 1978.
287. **Weiss, J. V. and Schmutzler, R.,** *J. Chem. Soc. Chem. Commun.*, p. 643, 1976.
288. **Reddy, G. S. and Schmutzler, R.,** *Z. Naturforsch.*, 25B, 1199, 1970.
289. **Fild, M. and Schmutzler, R.,** *J. Chem. Soc. A*, p. 840, 1969.
290. **Schomburg, D., Weferling, N., and Schmutzler, R.,** *J. Chem. Soc. Chem. Commun.*, p. 609, 1981.
291. **Bentrude, W. G., Kawashima, T., Keys, B. A., Garroussian, M., Heide, W., and Wedegaertner, D. A.,** *J. Am. Chem. Soc.*, 109, 1227, 1987.
292. **Cadogan, J. I. G., Gould, R. O., and Tweddle, N. J.,** *J. Chem. Soc. Chem. Commun.*, p. 774, 1975.
293. **Buono, G. and Llinas, J. R.,** *J. Am. Chem. Soc.*, 103, 4532, 1981.
294. **Burgada, R., El Khoshnieh, Y. O., and Leroux, Y.,** *Phosphorus Sulfur*, 22, 225, 1985.
295. **Schmidpeter, A., Luber, J., Riedl, H., and Volz, M.,** *Phosphorus Sulfur*, 3, 171, 1977.
296. **Pudovik, M. A., Terentieva, S. A., and Pudovik, A. N.,** *Zh. Obshch. Khim.*, 55, 2461, 1985.
297. **Wieber, M., Mulfinger, O., and Wunderlich, H.,** *Z. Anorg. Allg. Chem.*, 477, 108, 1981.
298. **Gonçalves, H. and Majoral, J. P.,** *Phosphorus Sulfur*, 4, 343, 1978.
299. **Tarasova, R. I., Sinitsyna, N. I., Dvoinishnikova, T. A., Remizov, A. B., Alparova, M. V., Zykova, T. V., and Moskva, V. V.,** *Zh. Obshch. Khim.*, 54, 1489, 1984.
300. **Bagrov, F. V., Razumova, N. A., and Petrov, A. A.,** *Zh. Obshch. Khim.*, 49, 955, 1979.
301. **Konovalova, I. V., Burnaeva, L. A., Kashtanova, N. M., Gareev, R. D., and Pudovik, A. N.,** *Zh. Obshch. Khim.*, 54, 2445, 1984.
302. **Willson, M. and Burgada, R.,** *Phosphorus Sulfur*, 7, 115, 1979.

303. **Schmidpeter, A., Weinmaier, J. H., Sheldrick, W. S., and Schomburg, D.,** *Z. Naturforsch.*, 34B, 906, 1979.
304. **Korchin, E. E. and Mukhametov, F. S.,** *Zh. Obshch. Khim.*, 56, 961, 1986.
305. **Burger, K. and Penninger, S.,** *Synthesis*, p. 526, 1978.
306. **Cadogan, J. I. G., Strathdee, R. S., and Tweddle, N. J.,** *J. Chem. Soc. Chem. Commun.*, p. 891, 1976.
307. **Bentrude, W. G., Johnson, W. D., and Khan, W. A.,** *J. Am. Chem. Soc.*, 94, 923, 1972.
308. **Chaus, M. P., Gusar, N. I., and Gololobov, Yu G.,** *Zh. Obshch. Khim.*, 52, 24, 1982.
309. **Cadogan, J. I. G., Gosney, I., Henry, E., Naisby, T., Nay, B., Stewart, N. J., and Tweddle, N. J.,** *J. Chem. Soc. Chem. Commun.*, p. 189, 1979.
310. **Dahl, B. M., Dahl, O., and Trippett, S.,** *J. Chem. Soc. Perkin Trans. 1*, p. 2239, 1981.
311. **Konovalova, I. V., Burnaeva, L. A., Khusnutdinova, E. K., Khafizova, G. S., and Pudovik, A. N.,** *Zh. Obshch. Khim.*, 55, 2189, 1985.
312. **Malavaud, C., Charbonnel, Y., and Barrans, J.,** *Tetrahedron Lett.*, p. 497, 1975.
313. **Garrigues, B. and Munoz, A.,** *C. R. Acad. Sci. Paris*, 293, 677, 1981.
314. **Pudovik, M. A., Mikhailov, Yu B., and Pudovik, A. N.,** *Zh. Obshch. Khim.*, 53, 1950, 1983.
315. **Schmidpeter, A. and Von Criegern, T.,** *Chem. Ber.*, 112, 2762, 1979.
316. **Roesky, H. W., Ambrosius, K., Banck, M., and Sheldrick, W. S.,** *Chem. Ber.*, 113, 1847, 1980.
317. **Roesky, H. W., Ambrosius, K., and Sheldrick, W. S.,** *Chem. Ber.*, 112, 1365, 1979.
318. **Roesky, H. W., Mehrotra, S. K., Platte, C., Amirzadeh-Asl, D., and Roth, B.,** *Z. Naturforsch.*, 35B, 1130, 1980.
319. **Schmidpeter, A. and Von Criegern, T.,** *J. Chem. Soc. Chem. Commun.*, p. 470, 1978.
320. **Richman, J. E.,** *Tetrahedron Lett.*, p. 559, 1977.
321. **Sau, A. C. and Holmes, R. R.,** *J. Organomet. Chem.*, 156, 253, 1978.
322. **Kimura, Y., Kokura, T., and Saegusa, T.,** *J. Org. Chem.*, 48, 3815, 1983.
323. **Ragulin, V. V., Zakharov, V. I., Petrov, A. A., and Razumova, N. A.,** *Zh. Obshch. Khim.*, 51, 34, 1981.
324. **Burger, K., Ottlinger, R., Franck, A., and Schubert, U.,** *Angew. Chem. Int. Ed.*, 17, 774, 1978.
325. **Dmitriev, V. I., Koslov, E. S., Thimokhin, B. V., Dubenko, L. G., and Kalabina, A. V.,** *Zh. Obshch. Khim.*, 50, 2230, 1980.
326. **Röschenthaler, G. V.,** *Z. Naturforsch.*, 33B, 131, 1978.
327. **Scharf, D. J.,** *J. Org. Chem.*, 41, 28, 1976.
328. **Odinets, I. L., Novikova, Z. S., and Lutsenko, I. F.,** *Zh. Obshch. Khim.*, 55, 1196, 1985.
329. **Gazizov, T. Kh., Sudarev, Yu I., Goldfarb, E. I., and Pudovik, A. N.,** *Zh. Obshch. Khim.*, 46, 924, 1976.
330. **Denney, D. B. and Wagner, F. A.,** *Phosphorus*, 2, 281, 1973.
331. **Burger, K., Fehn, F., and Mole, E.,** *Chem. Ber.*, 104, 1826, 1971.
332. **Albanbauer, J., Burger, K., Burgis, E., Marquarding, D., Schabl, L., and Ugi, I.,** *Liebigs Ann. Chem.*, p. 36, 1976.
333. **Markovskii, L. N., Kolesnik, N. P., Bakhmutov, Yu L., Kudryatsev, A. A., and Chermolovitch, Yu G.,** *Zh. Obshch. Khim.*, 53, 1994, 1983.
334. **Labaudiniere, L. and Burgada, R.,** *Phosphorus Sulfur*, 24, 235, 1985.
335. **Cadogan, J. I. G., North, R. A., and Rowley, A. G.,** *J. Chem. Res. (M)*, p. 178, 1978.
336. **Laurenço, C. and Burgada, R.,** *Tetrahedron*, 32, 2089, 1976.
337. **Schomburg, D., Storzer, W., Bohlen, R., Kuhn, W., and Röschenthaler, G. V.,** *Chem. Ber.*, 116, 3301, 1983.
338. **Wieber, M. and Hoos, W. R.,** *Tetrahedron Lett.*, p. 5333, 1968.
339. **Burgada, R. and Mohri, A.,** *Phosphorus Sulfur*, 13, 85, 1982.
340. **Germa, H. and Burgada, R.,** *Bull. Soc. Chim. Fr.*, p. 2607, 1975.
341. **Richter, W. J.,** *Phosphorus Sulfur*, 10, 395, 1981.
342. **Savignac, P., Richard, B., Leroux, Y., and Burgada, R.,** *J. Organomet. Chem.*, 93, 331, 1975.
343. **Ofitsierova, E. Kh., Ivanova, O. E., Ofitsierova, E. N., Konovalova, I. V., and Pudovik, A. N.,** *Zh. Obshch. Khim.*, 51, 505, 1981.
344. **Wieber, M., Foroughi, K., and Klingl, H.,** *Chem. Ber.*, 107, 639, 1974.
345. **Sarma, R., Ramirez, F., and Marecek, J. F.,** *J. Org. Chem.*, 41, 473, 1976.
346. **Burgada, R., El Khoshnieh, Y. O., and Leroux, Y.,** *Tetrahedron*, 41, 1207, 1985.
347. **Lloyd, J. R., Lowther, N., and Hall, C. D.,** *J. Chem. Soc. Perkin Trans. 2*, p. 245, 1985.
348. **Bowman, D. A., Denney, D. B., and Denney, D. Z.,** *Phosphorus Sulfur*, 4, 229, 1978.
349. **Baldwin, J. E. and Swallow, J. C.,** *J. Org. Chem.*, 35, 3583, 1970.
350. **Weinmaier, J. H., Brunnhuber, G., and Schmidpeter, A.,** *Chem. Ber.*, 113, 2278, 1980.
351. **Janzen, A. F., Lemire, A. E., Marat, R. K., and Queen, A.,** *Can. J. Chem.*, 61, 2264, 1983.
352. **Griffiths, D. V. and Tebby, J. C.,** *J. Chem. Soc. Chem. Commun.*, p. 607, 1981.

353. **Burgada, R. and Mohri, A.,** *Phosphorus Sulfur*, 9, 285, 1981.
354. **McGall, G. H. and McClelland, R. A.,** *J. Chem. Soc. Chem. Commun.*, p. 1222, 1982.
355. **Johnson, A. W. and Trippett, S.,** *J. Chem. Soc. Perkin Trans. 1*, p. 191, 1982.
356. **McGall, G. H. and McClelland, R. A.,** *J. Am. Chem. Soc.*, 107, 5198, 1985.
357. **Appel, R. and Gilak, A.,** *Chem. Ber.*, 108, 2693, 1975.
358. **Seel, F., Gombler, F., and Rudolph, K. H.,** *Z. Naturforsch.*, 23B, 387, 1968.
359. **Falardeau, E. R., Morse, K. W., and Morse, J. G.,** *Inorg. Chem.*, 14, 1239, 1975.
360. **Houalla, D., Brazier, J. F., Sanchez, M., and Wolf, R.,** *Tetrahedron Lett.*, p. 2969, 1972.
361. **Houalla, D., Mouheich, T., Sanchez, M., and Wolf, R.,** *Phosphorus Sulfur*, 5, 229, 1979.
362. **Bonningue, C., Brazier, J. F., Houalla, D., and Osman, F. H.,** *Phosphorus Sulfur*, 5, 291, 1979.
363. **Malavaud, C. and Barrans, J.,** *Tetrahedron Lett.*, p. 3077, 1975.
364. **Jeanneaux, F., Riess, J. G., and Wachter, J.,** *Inorg. Chem.*, 23, 3036, 1984.
365. **Contreras, R., Murillo, A., Uribe, G., and Klaebe, A.,** *Heterocycles*, 23, 2187, 1985.
366. **Haddad, M., N'Gando M'Pondo, T., Malavaud, C., Lopez, L., and Barrans, J.,** *Phosphorus Sulfur*, 20, 333, 1984.
367. **Cavell, R. G., Gibson, J. A., and The, K. I.,** *J. Am. Chem. Soc.*, 99, 7841, 1977.
368. **Kolodyazhnyi, O. I.,** *Zh. Obshch. Khim.*, 54, 966, 1984..
369. **Von Allwörden, U., Tseggai, I., and Röschenthaler, G. V.,** *Phosphorus Sulfur*, 21, 177, 1984.
370. **Poulin, D. D. and Cavell, R. G.,** *Inorg. Chem.*, 13, 2324, 1974.
371. **Yap, N. T. and Cavell, R. G.,** *Inorg. Chem.*, 18, 1301, 1979.
372. **Stegmann, H. B., Dumm, H. V., and Scheffler, K.,** *Phosphorus Sulfur*, 5, 159, 1978.
373. **Harris, R. K., Wazeer, M. I. M., Schlak, O., Schmutzler, R., and Sheldrick, W. S.,** *J. Chem. Soc. Dalton Trans.*, p. 517, 1977.
374. **Röschenthaler, G. V.,** *Z. Naturforsch.*, 33B, 311, 1978.
375. **Gibson, J. A., Röschenthaler, G. V., and Schmutzler, R.,** *Z. Naturforsch.*, 32B, 599, 1977.
376. **Volkholz, M., Stelzer, O., and Schmutzler, R.,** *Chem. Ber.*, 111, 890, 1978.
377. **Peake, S. C., Hewson, M. J. C., and Schmutzler, R.,** *J. Chem. Soc. A*, p. 2364, 1970.
378. **Evangelidou-Tsolis, E., Ramirez, F., Pilot, J. F., and Smith, C. P.,** *Phosphorus*, 4, 109, 1974.
379. **Acher, F., Juge, S., and Wakselman, M.,** *Tetrahedron*, 43, 3721, 1987.
380. **Harper, S. D. and Arduengo, A. J., III,** *J. Am. Chem. Soc.*, 104, 2497, 1982.
381. **Cadogan, J. I. G., Husband, J. B., and McNab, H.,** *J. Chem. Soc. Perkin Trans. 1*, p. 1449, 1984.
382. **Stegmann, H. B., Dumm, H. V., Burmester, A., and Scheffler, K.,** *Phosphorus Sulfur*, 8, 59, 1980.
383. **Roesky, H. W., Pogatzki, V. W., Dhathathreyan, K. S., Thiel, A., Schmidt, H. G., Dyrbusch, M., Noltemeyer, M., and Scheldrick, G. M.,** *Chem. Ber.*, 119, 2687, 1986.
384. **Schmidpeter, A. and Luber, J.,** *Chem. Ber.*, 109, 3581, 1976.
385. **Baccolini, G., Spagnolo, P., and Todesco, P. E.,** *Phosphorus Sulfur*, 8, 127, 1980.
386. **Stegmann, H. B., Dumm, H. V., Burmester, A., and Scheffler, K.,** *Phosphorus Sulfur*, 9, 99, 1980.
387. **Röschenthaler, G. V., Von Allwörden, U., and Schmutzler, R.,** *Polyhedron*, 5, 1387, 1986.
388. **Bone, S. A., Trippett, S., and Whittle, P. J.,** *J. Chem. Soc. Perkin Trans. 1*, p. 437, 1977.
389. **Oram, R. K. and Trippett, S.,** *J. Chem. Soc. Perkin Trans. 1*, p. 1300, 1973.
390. **Oram, R. K. and Trippett, S.,** *J. Chem. Soc. Chem Commun.*, p. 554, 1972.
391. **Schomburg, D., Weferling, N., and Schmutzler, R.,** *J. Chem. Soc. Chem. Commun.*, p. 810, 1981.
392. **Ramirez, F., Pilot, J. F., Madan, O. P., and Smith, C. P.,** *J. Am. Chem. Soc.*, 90, 1275, 1968.
393. **Bentrude, W. G. and Johnson, W. D.,** *J. Am. Chem. Soc.*, 90, 341, 1968.
394. **Antczak, S. and Trippett, S.,** *J. Chem. Soc. Perkin Trans. 1*, p. 1326, 1978.
395. **Ragulin, V. V., Petrov, A. A., Zakharov, V. I., and Razumova, N. A.,** *Zh. Obshch. Khim.*, 52, 239, 1982.
396. **Koole, L. H., Van Kooyk, R. J. L., and Buck, H. M.,** *J. Am. Chem. Soc.*, 107, 4032, 1985.
397. **Granoth, I. and Martin, J. C.,** *J. Am. Chem. Soc.*, 100, 5229, 1978.
398. **Granoth, I. and Martin, J. C.,** *J. Am. Chem. Soc.*, 101, 4623, 1979.
399. **Perozzi, E. F., Michalak, R. S., Figuly, G. D., Stevenson, W. H., III, Dess, D. B., Ross, M. R., and Martin, J. C.,** *J. Org. Chem.*, 46, 1049, 1981.
400. **Segall, Y. and Granoth, I.,** *J. Am. Chem. Soc.*, 101, 3687, 1979.
401. **Segall, Y. and Granoth, I.,** *J. Am. Chem. Soc.*, 100, 5130, 1978.
402. **Granoth, I. and Martin, J. C.,** *J. Am. Chem. Soc.*, 101, 4618, 1979.
403. **Burgada, R., Leroux, Y., and El Khoshnieh, Y. O.,** *Tetrahedron Lett.*, p. 3533, 1981.
404. **Chermolovitch, Yu G., Kolesnik, N. P., and Markovskii, L. N.,** *Zh. Obshch. Khim.*, 56, 217, 1986.
405. **Röschenthaler, G. V., Bohlen, R., and Schomburg, D.,** *Z. Naturforsch.*, 40B, 1593, 1985.
406. **Krylova, V. N., Gornaeva, N. P., Chvets, V. I., and Evstigneeva, R. P.,** *Dokl. Akad. Nauk SSSR*, 246, 339, 1979.
407. **Ross, M. R. and Martin, J. C.,** *J. Am. Chem. Soc.*, 103, 1234, 1981.
408. **Granoth, I. and Martin, J. C.,** *J. Am. Chem. Soc.*, 100, 7434, 1978.

409. **Granoth, I. and Martin, J. C.,** *J. Am. Chem. Soc.*, 103, 2711, 1981.
410. **Kirschkel, H. and Röschenthaler, G. V.,** *Phosphorus Sulfur,* 27, 371, 1986.
411. **Kolodiajnyi, O. I.,** *Zh. Obshch. Khim.*, 56, 283, 1986.
412. **Granoth, I., Alkabets, R., and Segall, Y.,** *J. Chem. Soc. Chem. Commun.*, p. 622, 1981.
413. **Thimokin, B. V., Dmitriev, V. K., Dmitriev, V. I., Istomin, B. I., and Donshikh, V. I.,** *Zh. Obshch. Khim.*, 51, 1989, 1981.
414. **The, K. W. and Cavell, R. G.,** *Inorg. Chem.*, 15, 2518, 1976.
415. **Ramirez, F. and Tsolis, E. A.,** *J. Am. Chem. Soc.*, 92, 7553, 1970.
416. **Ruppert, I. and Bastian, V.,** *Angew. Chem. Int. Ed.*, 16, 718, 1977.
417. **Grosse, J. and Schmutzler, R.,** *Phosphorus,* 4, 49, 1974.
418. **De'Ath, N. J., Denney, D. B., Denney, D. Z., and Hsu, Y. F.,** *J. Am. Chem. Soc.*, 98, 768, 1976.
419. **Neumann, S., Schomburg, D., Richtarsky, G., and Schmutzler, R.,** *J. Chem. Soc. Chem. Commun.*, 946, 1978.
420. **Schmidpeter, A. and Von Criegern, T.,** *Angew. Chem.*, 90, 64, 1978.
421. **Schmidpeter, A. and Von Criegern, T.,** *Chem. Ber.*, 112, 3472, 1979.
422. **Scheffler, K., Burmester, A., Haller, R., and Stegmann, H. B.,** *Chem. Ber.*, 114, 23, 1981.
423. **Stegmann, H. B., Bauer, G., Breitmeier, E., Herrmann, E., and Scheffler, K.,** *Phosphorus Sulfur,* 5, 207, 1975.
424. **Stegmann, H. B., Müller, H., Ulmschneider, K. B., and Scheffler, K.,** *Chem. Ber.*, 112, 2444, 1979.
425. **Van de Griend, L. and Cavell R. G.,** *Inorg. Chem.*, 19, 2070, 1980.
426a. **Appel, R. and Halstenberg, M.,** *Angew. Chem. Int. Ed.*, 14, 768, 1975.
426b. **Sinyashin, O. G., Sinyashina, T. N., Batyeva, E. S., Pudovik, A. N., and Ofitserov, E. N.,** *Zh. Obshch. Khim.*, 52, 2438, 1982.
427. **Cavell, R. G., Leary, R. D., and Tomlinson, A. J.,** *Inorg. Chem.*, 11, 2578, 1972.
428. **Ul Haque, M., Caughlan, C. N., Ramirez, F., Pilot, J. F., and Smith, C. P.,** *J. Am. Chem. Soc.*, 93, 5229, 1971.
429. **Caster, K. C. and Quin, L. D.,** *Tetrahedron Lett.*, 24, 5831, 1983.
430. **Quin, L. D. and Spence, S. C.,** *Tetrahedron Lett.*, 23, 2529, 1982.
431. **Denney, D. Z., White, D. W., and Denney, D. B.,** *J. Am. Chem. Soc.*, 93, 2066, 1971.
432. **Denney, D. B., Denney, D. Z., Hall, C. D., and Marsi, K. L.,** *J. Am. Chem. Soc.*, 94, 245, 1972.
433. **Althoff, W., Day, R. O., Brown, R. K., and Holmes, R. R.,** *Inorg. Chem.*, 17, 3265, 1978.
434. **White, D. W., De'Ath, N. J., Denney, D. Z., and Denney, D. B.,** *Phosphorus,* 1, 91, 1971.
435. **Clennan, E. L. and Heah, P. C.,** *J. Org. Chem.*, 48, 2621, 1983.
436. **Aly, H. A. E., Barlow, J. H., Russel, D. R., Smith, D. J. H., Swindles, M., and Trippett, S.,** *J. Chem. Soc. Chem. Commun.*, p. 449, 1976.
437. **Evangelidou-Tsolis, E. and Ramirez, F.,** *Phosphorus,* 4, 121, 1974.
438. **Caesar, J. C., Vaughan Griffiths, D., and Tebby, J. C.,** *Phosphorus Sulfur,* 29, 123, 1987.
439. **Swank, D. D., Caughlan, C. N., Ramirez, F., and Pilot, J. F.,** *J. Am. Chem. Soc.*, 93, 5236, 1971.
440. **Gareev, R. D., Pudovik, A. N., and Schermergoorn, I. M.,** *Zh. Obshch. Khim.*, 53, 38, 1983.
441. **Segall, Y., Granoth, I., Kalir, A., and Bergmann, E. D.,** *J. Chem. Soc. Chem. Commun.*, p. 399, 1975.
442. **Von Itzstein, M. and Jenkins, I. D.,** *J. Chem. Soc. Chem. Commun.*, p. 164, 1983.
443. **Grochowski, E., Hilton, B. D., Kupper, R. J., and Michijda, C. J.,** *J. Am. Chem. Soc.*, 104, 6877, 1982.
444. **Von Itzstein, M. and Jenkins, I. D.,** *J. Chem. Soc. Perkin Trans. 1,* p. 437, 1986.
445. **Von Itzstein, M. and Jenkins, I. D.,** *Aust. J. Chem.*, 37, 2447, 1984.
446. **Robinson, P. L., Barry, C. N., Kelly, J. W., and Evans, S. A., Jr.,** *J. Am. Chem. Soc.*, 107, 5210, 1985.
447. **Pogatski, V. W. and Roesky, H. W.,** *Chem. Ber.*, 119, 771, 1986.
448. **Hellwinkel, D. and Krapp, W.,** *Chem. Ber.*, 111, 13, 1978.
449. **Landvatter, E. F. and Rauchfuss, T. B.,** *J. Chem. Soc. Chem. Commun.*, p. 1171, 1982.
450. **Granoth, I.,** *J. Chem. Soc. Perkin Trans. 1,* p. 735, 1982.
451. **Hellwinkel, D., Blaicher, W., Krapp, W., and Sheldrick, W. S.,** *Chem. Ber.*, 113, 1406, 1980.
452. **Granoth, I., Alkabets, R., and Shirin, E.,** *J. Chem. Soc. Chem. Commun.*, p. 981, 1981.
453. **The, K. I. and Cavell, R. G.,** *J. Chem. Soc. Chem. Commun.*, p. 279, 1975.
454. **Cavell, R. G. and The, K. I.,** *Inorg. Chem.*, 17, 355, 1978.
455. **Schmidbaur, H., Mitschke, K. H., Buchner, W., Stühler, H., and Weidlein, J.,** *Chem. Ber.*, 106, 1226, 1973.
456. **Schmidbaur, H. and Holl, P.,** *Z. Anorg. Allg. Chem.*, 458, 249, 1979.
457. **The, K. I. and Cavell, R. G.,** *Inorg. Chem.*, 16, 2887, 1977.
458. **Schmidbaur, H., Stühler, H., and Buchner, W.,** *Chem. Ber.*, 106, 1238, 1973.
459. **Schmidbaur, H. and Holl, P.,** *Chem. Ber.*, 109, 3151, 1976.

460. **Schmidbaur, H. and Holl, P.,** *Chem. Ber.*, 112, 501, 1979.
461. **Vedejs, E., Meier, G. P., and Snoble, K. A. J.,** *J. Am. Chem. Soc.*, 103, 2823, 1981.
462. **Vedejs, E. and Snoble, K. A. J.,** *J. Am. Chem. Soc.*, 95, 5778, 1973.
463. **Vedejs, E. and Meier, G. P.,** *Angew. Chem. Int. Ed.*, 22, 56, 1983.
464. **Reitz, A. B., Mutter, M. S., and Maryanoff, B. E.,** *J. Am. Chem. Soc.*, 106, 1873, 1984.
465. **Huisgen, R. and Wulff, J.,** *Chem. Ber.*, 102, 746, 1969.
466. **Daniel, H., Turcant, A., and Le Corre, M.,** *Phosphorus Sulfur*, 29, 211, 1987.
467. **Birum, G. H. and Matthews, C. N.,** *J. Org. Chem.*, 32, 3554, 1967.
468. **Garbers, C. F., Malherbe, J. S., and Schneider, D. F.,** *Tetrahedron Lett.*, p. 1421, 1972.
469. **Van de Griend, L. and Cavell, R. G.,** *Inorg. Chem.*, 22, 1817, 1983.
470. **Hellwinkel, D., Linder, W., and Wilfinger, H. J.,** *Chem. Ber.*, 107, 1428, 1974.
471. **Schlosser, M., Kadibelban, T., and Steinhoff, G.,** *Liebigs Ann. Chem.*, 743, 25, 1971.

Chapter 19

^{31}P NMR DATA OF SIX COORDINATE PHOSPHORUS COMPOUNDS

Compiled and presented by
Lydia Lamandé, Max Koenig, and Keith Dillon

The NMR data are presented in Table R with no subdivisions. The definitions of the ring systems which incorporate phosphorus follow that of the previous table, but differ from other sections in that all rings involving the phosphorus are given and those which are fused are identified by placing the letter 'f' immediately following the ring size.

TRENDS IN CHEMICAL SHIFTS AND COUPLING CONSTANTS

The ^{31}P NMR chemical shifts of hexacoordinated phosphorus compounds are all located to low frequency (high field) of 85% H_3PO_4. The values are spread between −57 ppm for the tris chelate CatGlcP*CH$_2$CH=CMeO*, and −441 ppm for $Br_2PCl(CN)_3$. The overall formal charge carried by the hexacoordinate species may be negative (the most common, e.g., PCl_6^-), positive (e.g., $PCl_4Pyr_2^+$), or neutral (Lewis acid-Lewis base adducts, e.g., PCl_5←Pyr). The effect of charge on the chemical shift is marked, and an increase in negative charge causes a displacement to lower frequency (higher field).

	$PCl_4Pyr_2^+$	PCl_5Pyr	PCl_6^-
δ_p	−188.2(neat)	−224.7(neat)	−298.5(neat)

In the table the formal charges are shown only where it is not readily apparent.

The nature of the substituents bound to phosphorus is a major factor determining the chemical shifts. Electronegativity and anisotropic effects have crucial roles.

←
Electronegativity

PF_6^-	$P(OR)^-$	$P(N_3)^-$	$P(NCS)^-$	PCl_6^-	$P(NCO)_6^-$
−140	−150	−180	−262	−298	−383

←

Incorporation of phosphorus in five-membered rings increases the chemical shift frequency. In the pyrocatechol derivatives for each additional ring the increment is about 20 ppm.

	$P(OPh)_6$	$Cat_2P(OPh)_2$	Cat_3P
δ_p	−153	−107	−82

The same trend is observed in the corresponding fluoro series.

	PF_6	$CatPF_4$	Cat_2PF_2(*cis*)	Cat_3P
δ_p	−145	−122	−102	−82 ppm

Care is needed in the interpretation of data where structural isomerism is possible, since the *trans* isomer usually has a lower field shift than the *cis* isomer.

The size and nature of the counterion do not affect greatly the values of the chemical shifts, e.g., PF_6^- with various cations is -146 ± 4.5 ppm for solutions and -145 ± 5 ppm for solid states.

The salts are naturally soluble in polar solvents. The polarity of the solvent affects only slightly the chemical shifts, whereas its basicity can play an important role if it affords a displacement of the equilibrium between P^V and P^{VI} species.

S

SH^+

S = CH_2Cl_2 δ = −29
S = DMF δ = −83

The coupling constant $^1J_{PX}$ (X = H, F, . . .) increases with the coordination around phosphorus for neutral compounds P^{III} to P^V, but always decreases from neutral P^V to the salts P^{VI}.

(StrucB)

$^1J_{PX}$ \ X	H	F
(P^V)	900 Hz	1000 Hz
(P^{VI})	860 Hz	757 Hz

TABLE R

P-bound atoms/rings	Connectivities	Structure	NMR (δ_p[solv, temp] J Hz)	Ref.
$BrClO_4$/5,5	C'_4	$(Cat)_2PBrCl$	−69.5(ne)	1
$BrN'O_4$/5,5	$C'_2;C'_4$	$BrP(Cat)_2Pyr$	−85.5(r)	1
$Br_2N'_2O_2$/5	$(C'_2)_2;C'_2$	$[(Cat)PBr_2(Pyr)_2]^+Br_3^-$	−217.7(ne)	1
$Br_2N'_2O_2$/5,5	$(C'_2)_2;C'_2$	$[(Cat)PBr_2(Bipyr)]_2{}^+Br_3^-,Br^-$	−198(r,n)	1
	$(C'_2)_2;C'_2$	$[(Cat)PBr_2(Bipyr)]^+BBr_4^-$	−193.5(n) −195.1(ne)	1
ClF_2N_2O/4	$(CC^+)_2;C$	$ClPF_2[N(Me)C(Ph)N(Me)]OC_4H_9$	−125.3(PhCl, −30°) J_{PF}885	2
$ClF_2N'_3$	C''_3	$ClPF_2(NCS)_3$	−204.5(r) J_{PF}655/887	3
$ClF_3N'_2$	N''_2	$ClPF_3(N_3)_2$	−132(r) J_{PF}861/714	3
	C''_2	$ClPF_3(NCO)_2$	−176(r) J_{PF}738 fac, −156.8(r) J_{PF}738 mer	4
	C''_2	$ClPF_3(NCS)_2$	−179(r) J_{PF}869/1039 fac	3
ClF_5		$ClPF_5$	−133.2 J_{PF}720/840, −138(m) J_{PF}725/860 138(m)	5, 6
$ClN'O_4$/5,5	$C'_2;C'_4$	$(Cat)_2PCl(Pyr)$	−84.5(c), −85.4(r)	7, 8
	$N'';C'_4$	$(Cat)_2PCl(N_3)$	−112.1(r)	9
	$N'';C'_4$	$(Cat)_2PCl(N_3Tms)$	−90($PrNO_2$, −100°)	10
ClN'_3O_2/5	$N''_3;C'_2$	$(Cat)PCl(N_3)_3$	−122.2(r)	10
	$N''_3;C'_2$	$(Cat)PCL(N_3)_2(N_3Tms)$	−124(EtCl, −75°)	10
ClN'_5	N''_4,C'_2	Pyr $PCl(N_3)_4$	−171.9(p)	9
	N''_5	$ClP(N_3)_5$	−167.7(r)	11
	C''_5	$ClP(NCO)_5$	−340(r)	12
	C''_5	$ClP(NCS)_5$	−264.4(r)	13
ClN'_5/5	$N''_3,(C'_2)_2$	$(Bipyr)PCl(N_3)_3$	−142.7(fe)	9
ClO_5	C''_5	$ClP(OCN)_5$	−156.4(m)	12
ClO_5/5,5	C'_4H	$(Cat)_2P(OH)Cl$	−83.1(r)	9
$Cl_2FN'_3$	N''_3	$Cl_2PF(N_3)_3$	−156.3(fe) J_{PF}882	3
$Cl_2F_2N'_2$	N''_2	$Cl_2PF_2(N_3)_2$	−149.3(r) J_{PF}901	3
	C''_2	$Cl_2PF_2(NCO)_2$	−162.9^T(r) J_{PF}748	4
	C''_2	$Cl_2PF_2(NCS)_2$	−194(r) J_{PF}757/1029	3
Cl_2F_3N'	N''	$Cl_2PF_3(N_3)$	−143.4(r) J_{PF}855/948	3
	C''	$Cl_2PF_3(NCO)$	−158.8(r) J_{PF}949/1004	4
	C''	$Cl_2PF_3(NCS)$	−168(r) J_{PF}836/996	3
Cl_2F_4		Cl_2PF_4	−141^C(m) J_{PF}870/1015	5
			−158^T J_{PF}970 *trans*	14
$Cl_2N'_2O_2$/5	$N''_2;C'_2$	$(Cat)PCl_2(N_3)_2$	−117.9(r), −132($PrNO_2$, −100°)	9, 10
	$N''_2;C'_2$	$(Cat)PCl_2(N_3Tms)(N_3)$	−127(ce, −90°)	10
	$(C'_2)_2;C'_2$	$(Cat)PCl_2(Pyr)_2$	−125.5(ne)	7
$Cl_2N'_2O_2$/5,5	$(C'_2)_2;C'_2$	$(Cat)PCl_2(Bipyr)$	−119(ne,r)	7
Cl_2N_4/4,4	$(CP^5)_2,(CC^+)_2$	**1**	−199	15
$Cl_2N'_4$	N''_4	$Cl_2P(N_3)_4$	−171.2^C(r)	11
	N''_3,C'_2	Pyr $PCl_2(N_3)_3$	−165.5(p)	9
	C''_4	$Cl_2P(NCO)_4$	−315.9^C(r)	12
	C''_4	$Cl_2P(NCS)_4$	−266.8^C(r)	13
$Cl_2N'_4$/5	$N''_2,(C'_2)_2$	$(Bipyr)PCl_2(N_3)_2$	−156.9(fe)	9
Cl_2O_4	C''_4	$Cl_2P(OCN)_4$	−172.5^C(m); −162^T(m)	12
Cl_2O_4/5,5	C'_4	$(Cat)_2PCl_2$	−69.5(n) −66.3(ne)	7
$Cl_3FN'_2$	N''_2	$Cl_3PF(N_3)_2$	−169.5(fe) J_{PF}911	3
Cl_3F_2N'	N''	$Cl_3PF_2(N_3)$	−161.7(r) J_{PF}910	3
Cl_3F_3		Cl_3PF_3	−156.8(r) J_{PF}956 fac	3
			−142.7(r) J_{PF}970/1081 mer	3
$Cl_3N'O_2$/5	$N'';C'_2$	$(Cat)PCl_3(N_3)$	−157(ne, −100°) −130.6(r)	9, 10
	$N'';C'_2$	$(Cat)PCl_3(N_3Tms)$	−140(ce, −100°)	10
	$C'_2;C'_2$	$(Cat)PCl_3(Pyr)$	−129.9 −136.9(ne)^I	7

TABLE R (continued)

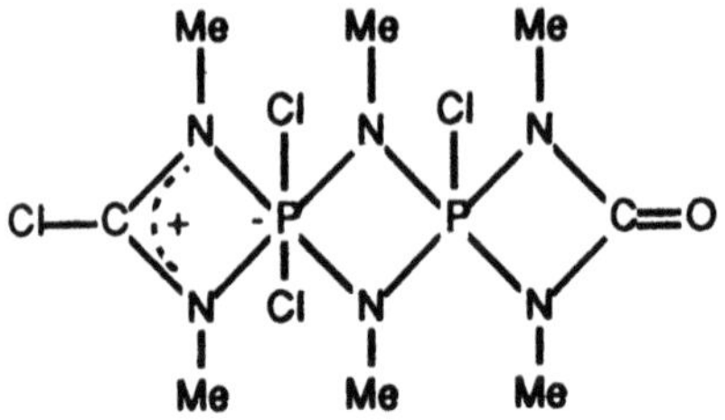

1

P-bound atoms/rings	Connectivities	Structure	NMR (δ_p[solv, temp] J Hz)	Ref.
$Cl_3N'_3$	N''_3	$Cl_3P(N_3)_3$	−183.4(r)fac	11
	N''_2,C'_2	Pyr-$PCl_3(N_3)_2$	−159.7(p)	9
	C''_3	$Cl_3P(NCO)_3$	−290(r)fac	12
	C''_3	$Cl_3P(NCS)_3$	−270.5(r)mer	13
$Cl_3N'_3/5$	$N'',(C'_2)_2$	(Bipyr)$PCl_3(N_3)$	−166.9(fe)	9
Cl_3O_3	C'_3	$Cl_3P(OCN)_3$	−195.1(r) −197.4(m)fac	12
Cl_4FN'	N''	$Cl_4PF(N_3)$	−193.9(fe) J_{PF}998	3
Cl_4F_2		Cl_4PF_2	−177.4(r) J_{PF}1022, −180(n) J_{PF}1010	3, 5
$Cl_4N_2/4$	$(CP^4)_2$	$Cl_4P[N(Me)PCl_2N(tBu)]$	−216	16
	$(CC^+)_2$	$Cl_4P[N(Me)C(Cl)N(Me)]$	−202	17
	$(CC^+)_2$	$Cl_4P[N(Me)C(CCl_3)N(Me)]$	−193.1(m)	18
	$(CC^+)_2$	$Cl_4P[N(Me)C(Tf)N(Me)]$	−198.5(m)	18
$Cl_4N'_2$	N''_2	$Cl_4P(N_3)_2$	-206.4^C(r)	11
	N'',C'_2	Pyr—$PCl_4(N_3)$	-199.6^C(p)	9
	$(C'_2)_2$	Cl_4P(Pyrazin)$_2$	−180.7(ne)	19
	$(C'_2)_2$	Cl_4P(3-I Pyr)$_2$	−188(ne) −189.5(n)	19
	$(C'_2)_2$	Cl_4PPyr_2	−188.2(n) −180.8(ne)	19
	$(C'_2)_2$	Cl_4P(2-CN Pyr)$_2$	−107.3(ne)	19
	$(C'_2)_2$	Cl_4P(3,5-Me_2 Pyr)$_2$	−176.4(n)	19
	C''_2	$Cl_4P(NCO)_2$	-274^C(r)	12
	C''_2	$Cl_4P(NCS)_2$	-276.1^T(r)	13
$Cl_4N'_2/5$	$(C'_2)_2$	Cl_4P(Bipyr)	−192.6(fe)	9
	$(C'_2)_2$	Cl_4P(Phenanthroline)	−192(ne) −199(n)	20
Cl_4O_2	C''_2	$Cl_4P(OCN)_2$	−203.6(m) -204.7^T(r) −214.4(m)*cis*	12
$Cl_4O_2/5$	C'_2	(Cat)PCl_4	−162(n), −157.3(r,ne)	7
Cl_5F		Cl_5PF	−224(m) J_{PF}1050/1070	6, 14
Cl_5N'	N''	$Cl_5P(N_3)$	−243.5(r)	11
	C'_2	PCl_5(3,5-Cl_2 Pyr)	−222.7(ne)	20
	C'_2	Pyr PCl_5	−230.6(p) −234(p) −224.7(n)	9, 21, 20
	C'_2	PCl_5(2-CN Pyr)	−170.8(2-CN Pyr)	20
	C'_2	PCl_5(3-Me Pyr)	−237(n) −228(3-Me Pyr)	20
	C''	$Cl_5P(NCO)$	−280.4(r) −282(fe)	12
	C''	$Cl_5P(NCS)$	−286.9(r)	13
Cl_5O'	P^4	$(nBu)_3P{=}O{-}PCl_5$	−296(r)	22
Cl_6		PCl_6	−282(n), −297.9(ne), −307(m, −40°)	23—25
$FNO_4/5,5,5f$	$C'_2;C_2C'_2$	**2**; R^1R^2 = Glc, R^3 = F	−94.1(m) J_{PF}695	26
	$C'_2;C'_4$	**2**; R^1R^2 = Cat, R^3 = F	−93(m) J_{PF}720	26

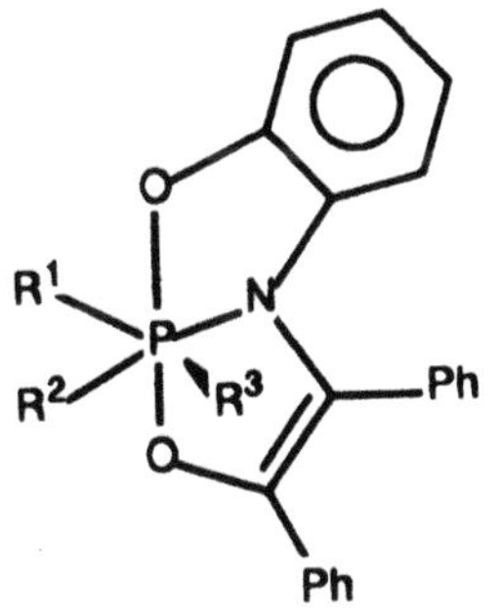

2

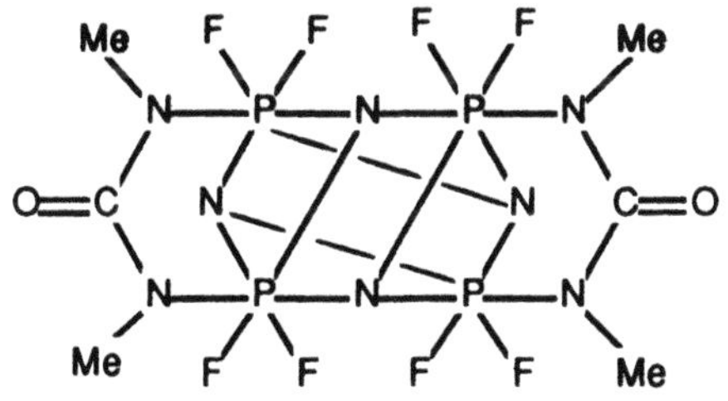

3

P-bound atoms/rings	Connectivities	Structure	NMR (δ_p[solv, temp] J Hz)	Ref.
FN'_5	N''_5	$FP(N_3)_5$	−167.3(r) J_{PF}884, −168.3(fe) J_{PF}870	3
	C''_5	$FP(NCS)_5$	−233.6(r) J_{PF}741	3
FO_5/5,5	$C_2C'_3$	(Glc)FP[OCTf=CTfO]O(4-FC_6H_4)	−102.5(k) J_{PF}762	27
F_2NO_3/5,5	C_2;CC'_2	[OCH_2CH_2N(Me)] PF_2(OCTf=CTfO)	−108.5(k) J_{PF}820	28
$F_2N'_2O_2$/4,5	$(CC')_2$;C'_2	(Cat)PF_2[N(Me)C(Ph)N(Me)]	−116.3(de) J_{PF}801/840/871	29
	$(CC')_2$;C'_2	(Cat)PF_2[N(Me)C(CCl_3)N(Me)]	−120.9(de) J_{PF}808/845/876	29
F_2N_4/4,4f,6f	$(P^6)_3$,CC'	**3**	−173.6	30
	$(CP^6)_2$,$(CP^5)_2$	**4**	−166.5	31
$F_2N'_4$	N''_4	$F_2P(N_3)_4$	−158.3^C(r) J_{PF}853	3
	C''_4	$F_2P(NCS)_4$	−209.5(r) J_{PF}760, −209.6(r) J_{PF}746	3
F_2O_3S/5,5	CC'_2;C	(OCTf=CTfO)PF_2(OCH_2CH_2S)	−89.5^C(m,k) J_{PF}864/888	28
			−88.1^C(m,k) J_{PF}882	28
F_2O_4/5,5	$C_2C'_2$	(Pnc)PF_2(Chloranyl)	−102.3^C J_{PF}735, −99.6^T J_{PF}768	27
	$C_2C'_2$	(Glc)PF_2(OCTf=CTfO)	−101^C J_{PF}728, −98.8^T J_{PF}756 (m,k)	27
	C'_4	$(Cat)_2PF_2$	−102^T J_{PF}757	32
F_3NO_2/5,5f	CC',C'_2	**5**	−113 J_{PF}782	33
F_3N_3/4,4f	$(CP^5)_3$	**6**	−166.5	31
$F_3N'_3$	N''_3	$F_3P(N_3)_3$	−151.2(r) J_{PF}732	3
	C''_3	$F_3P(NCS)_3$	−185.6(r)mer	3
$F_3O_2P^4$/5	C_2;C_3	(Pfp)$PF_3(PMe_3)$	−130(m) J_{PF}726/898	34
$F_4N'O$/5	C'_2;C'	F_4P(Quin)	−118.7(m) J_{PF}791	35
F_4N_2/4	CHP^6	[F_4P-NHMe]$_2$	−155	36
	$(CP^4)_2$	F_4P[N(Me)P^+ ($NMeCH_2CH_2NMe$)N(Me)]	−151.6(r) J_{PF}744/789	37
	$(CP^4)_2$	F_4P[N(tBu)PF(tBu)N(tBu)]	−139(r) J_{PF}771/822/843	37
	$(CC^+)_2$	F_4P[N(Me)C(Ph)N(Me)]	−137(c) J_{PF}816/836	29
	$(CC^+)_2$	F_4P[N(Me)C(Tf)N(iBu)]	−147.5(n) J_{PF}852/888	29
F_4N_2/6	$(C'H)_2$	F_4P[NHC(Tf)NC(Tf)NH]	−162.7(r) J_{PF}780	38
$F_4N'_2$	C''_2	$F_4P(NCO)_2$	−152^T(r) J_{PF}738	4
	C''_2	$F_4P(NCS)_2$	−174.2^C(r) J_{PF}735	3
F_4O_2/5	C_2	F_4P(Pfp)	−122.9 J_{PF}776	34
	C'_2	F_4P(Cat)	−122 J_{PF}710/760	32
F_4O_2/6	C'_2	F_4P(acac)	−148.2 J_{PF}762/823	39
F_5N	H_3	$PF_5 \cdot NH_3$	−146.5(m) J_{PF}762	40
	O_2	$(NO_2)PF_5$	−125(r, −95°) J_{PF}738/766	14
	C_2	F_5PNMe_2	−146.8(r) J_{PF}796	41

TABLE R (continued)

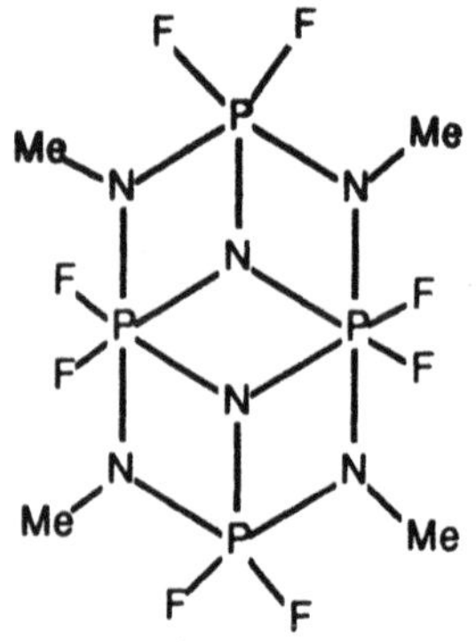

4 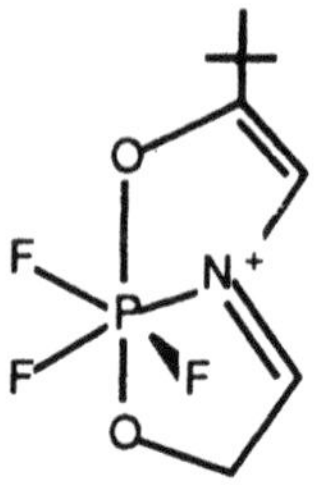 5 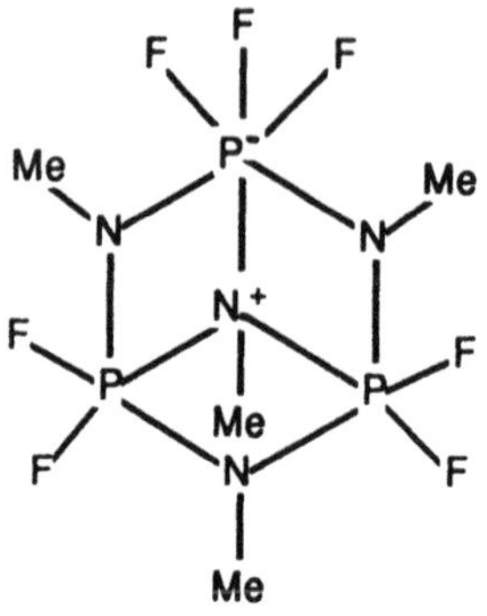 6

P-bound atoms/rings	Connectivities	Structure	NMR (δ_p[solv, temp] J Hz)	Ref.
F_5N'	C'_2	F_5P 3-MePyr	−145.5(r) J_{PF}761/803	35
	C″	$F_5P(NCO)$	−148.4(fe) J_{PF}745	4
	C″	$F_5P(NCS)$	−154.2(r) −156.4(m, −50°)	3, 42
	C″	F_5PMeCN	−146(m) J_{PF}787	43
F_5O'	P^4	$F_5P\text{-}O{=}PPh_3$	−145.9(m,r) J_{PF}740/760	42
	C′	$(CH_3COO)PF_5$	−153(SO_2) J_{PF}706/754	14
F_5P^4	C_2H	$Me_2(H)P\ PF_5$	−149(r, −10°) J_{PP}723 J_{PH}416, J_{PF}780/887	44
	C_3	$Me_3P\text{-}PF_5$	−120(m) J_{PP}720, J_{PF}780/890	44
	C'_3	$Ph_3P\text{-}PF_5$	−144.6(m,r) J_{PF}758/785	42
F_6		PF_6	−140(n), −146(m), −151 J_{PF}720	45, 6, 46
NO_5/4,5	$C'_2;C'_3O_2$	(Quin)P(OOO)$(OPh)_2$	−123	47
NO_5/5,5	$C_2;CC'_4$	(OCTf=CTfO)P $(NMeCH_2CH_2O)[O(4\text{-}FC_6H_4)]_2$	−99.6(k)	28
NO_5/5,5,5	$CH_2;CC'_4$	(Cat)(Chloranyl) $P[OC(O)CMe_2NH_2]$	−103(r)	48
	$CH_2;C'_5$	$(Cat)[O(2,4\text{-}tBu_2C_6H_2)O]$ $P[OC(O)CMe_2NH_2]$	−104(k)	48
	$CC'H;C'_5$	(Cat)(O-anthracenyl-O)P $[OC(O)CH_2NHPh]$	−89	48
	$CC'H;C'_5$	$(Cat)[O(2,4\text{-}tBu_2C_6H_2)O]$ $P[OC(O)CH_2NHPh]$	−90	48
	$CC'H;C_2C'_3$	$(Glc)[O(2,4\text{-}tBu_2C_6H_2)O]$ $P[OC(O)CH_2NHPh]$	−83	48
NO_5/5,5,5f	$C'_2;CC'_4$	**2**; R^1R^2 = Cat, R^3 = OMe	−91.6(m)	26
	$C'_2;C'_5$	**2**; R^1R^2 = Cat, R^3 = OPh	−95(m)	26
$N'O_5$/5,5	$C'_2;C'_5$	(OCTf=CTfO)P(Cat)(OPh)Pyr	−82	49
$N'O_5$/4,5,5	$C'_2;C'_3O_2$	**7**; X = H, Y = Cl	−102	47
	$C'_2;C'_3O_2$	**7**; X = Y = H	−102	47
	$C'_2;C'_3O_2$	**7**; X = H, Y = Me	−93	47
$N'O_5$/5,5,5	$C'_2;C_2C'_3$	(Glc)(Quin)P(OCPh=CPhO)	−85	50
	$C'_2;CC'_4$	**8**; R^1 = R^2 = H	−87(c)	51
	$C'_2;CC'_4$	**8**; R^1 = H, R^2 = $P(O)(OMe)_2$	−84.2(c)	52
	$C'_2;CC'_4$	**8**; R^1 = 2-Pyridil, R^2 = $P(O)(OMe)_2$	−87.6(c)	52
	$C'_2;CC'_4$	**8**; R^1 = Ph, R^2 = $P(O)(OMe)_2$	−86.1(c)	52

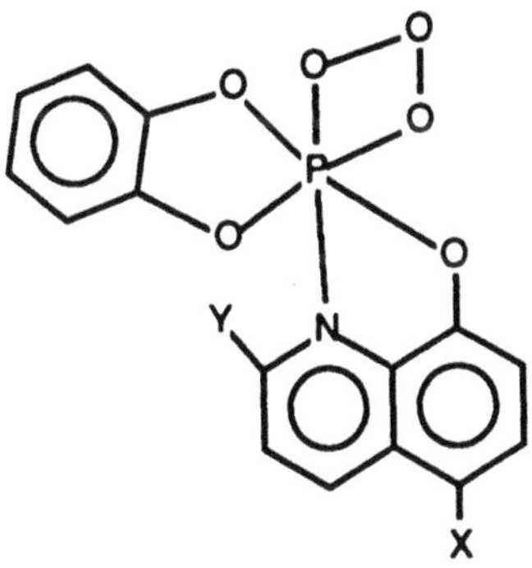

7

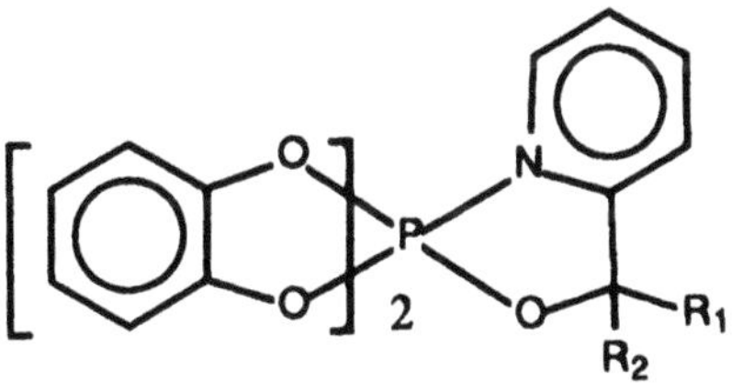

8

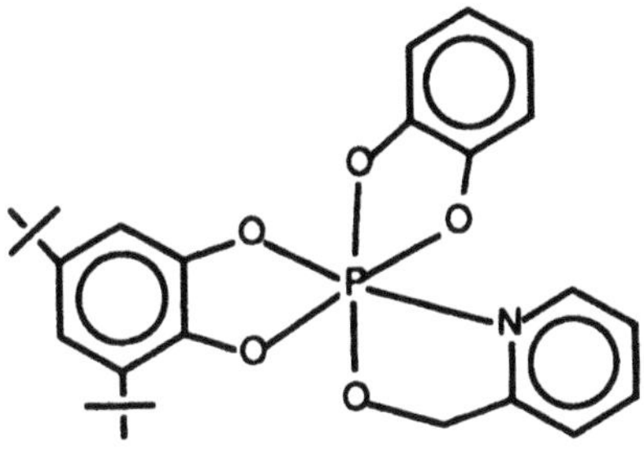

9

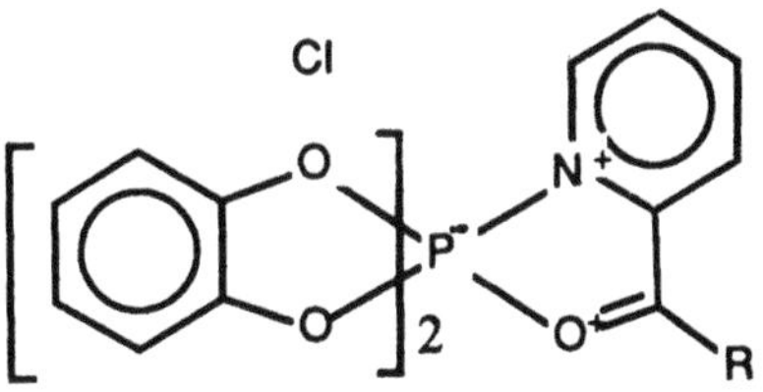

10

P-bound atoms/rings	Connectivities	Structure	NMR (δ_p[solv, temp] J Hz)	Ref.
	C'_2;CC'_4	**9**	−86	50
	C'_2;C'_5	[O(2-MeC_6H_4)O]$_2$P(7 Me-Quin)	−78	50
	C'_2;C'_5	(Cat)(Quin)P(OCPh=CPhO)	−94(r)	50
	C'_2;C'_5	(Cat)(Quin)P[O(2,4-$tBu_2C_6H_2$)O]	−88	50
	C'_2;C'_5	(Cat)(7 Me-Quin)P [O(2,4-$tBu_2C_6H_2$)O]	−77	50
$N'O_4O'$/5,5,5	C'_2;C'_4;C'	**10**; R = H	−90.2(c)	52
	C'_2;C'_4;C'	**10**; R = 2-Pyridil	−87.7(c)	52
	C'_2;C'_4;C'	**10**; R = Ph	−86.5(c)	52
N_2O_4/4,5,5	$(CC^+)_2$;C'_4	$(Cat)_2$P[N(Me)C(Ph)N(Me)]	−113.7(ne)	53
N_2O_4/5,5,5	$(C_2)_2$;C'_4	$(Cat)_2$P(Bipyr)	−95	8
N_2O_4/5,5,6	$(C_2)_2$;C'_4	$(Cat)_2$P(2,2′NC_5H_4-NH-C_5H_4N)	−112	8
$NN'O_4$/5,5,5f	C'_2;C'_2;C'_4	**2**; R^1R^2 = Cat, R^3 = Pyr	−105(m)	26
$NN'O_4$/5,5,6	C′N′;C′N;C'_4	**11**	−117	54
	C′N′;C′N;C'_4	**12**	−84(m)	54
N'_2O_4/5,5	N''_2;C'_4	$(Cat)_2P(N_3)_2$	−113($PrNO_2$)	10
	N''_2;C'_4	$(Cat)_2P(N_3)(N_3Tms)$	−99($PrNO_2$, −100°)	10
	$(C'_2)_2$;C'_4	$(Cat)_2P(Pyr)_2$	−100.2(c + p)	8
N'_2O_4/5,5,5	$(C'_2)_2$;C'_4	$(Cat)_2$P(Phenanthroline)	−91.2(ne)	7
N'_2O_4/5,5,6	$(C'_2)_2$;C'_4	$(Cat)_2$P(2,2′NC_5H_4=NC_5H_4N)	−120	8
N'_4O_2/5	N''_4;C'_2	$(Cat)P(N_3)_4$	−143.8(r)	9
	N''_4;C'_2	$(Cat)P(N_3)_3(N_3Tms)$	−120(ce)	10
N'_6	N''_6	$P(N_3)_6$	−180(r,m,fe), −184.1(m)	11, 55
	C'_6	$P(NCO)_6$	−388.4(r)	12
	C'_6	$P(NCS)_6$	−261.9(r)	13
N'_6/5	$(C'_2)_2$;N''_4	$(Bipyr)P(N_3)_4$	−150.9(fe)	9

TABLE R (continued)

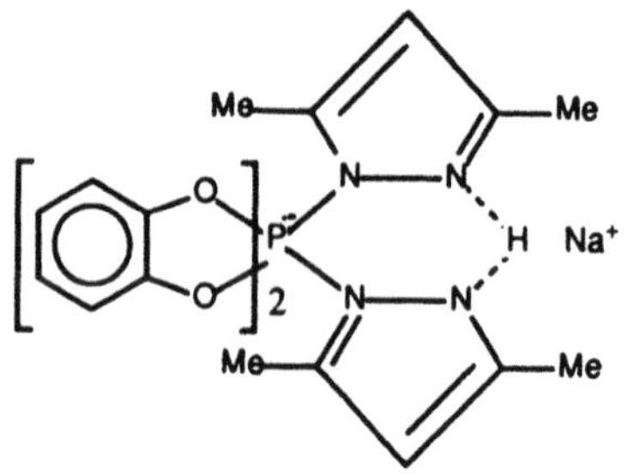

P-bound atoms/rings	Connectivities	Structure	NMR (δ_p[solv, temp] J Hz)	Ref.
O_5S/5,5	CC'_4;C	(OCTf=CTfO)P $(OCH_2CH_2S)[O(4\text{-}FC_6H_4)]_2$	−91.8 −96.8^I(k)	28
O_6	C_6	$P(OMe)_6$	−145(b)	56
	C'_6	$P(OPh)_6$	−153(v + m)	57
	C''_6	$P(OCN)_6$	−150.9(m)	12
O_6/5	$(P^4)_4,C'_2$	$(Cat)P[OP(O)(OEt)_2]_4$	−144.7	58
	C_6	$(Glc)P(OCH_2Tf)_4$	−128(b)	59
O_6/5,5	C_6	$(Glc)_2P(OCH_2Tf)_2$	−102.2 −105.9^I(b)	59
	$C_2C'_4$	(OCTf=CTfO)P $(Glc)[O(4\text{-}FC_6H_4)]_2$	−105.1 −105.5^I(k)	27
	$C_2C'_4$	(OCTf=CTfO)P $(Pnc)[O(4\text{-}FC_6H_4)]_2$	−110.1^C(k)	60
	$C_2C'_4$	$(Chloranyl)P(Pnc)[O(4\text{-}FC_6H_4)]_2$	−106.3^C(k)	60
	C'_6	$(OCTf{=}CTfO)P(Cat)(OPh)_2$	−109.5(r)	49
	C'_6	$(Cat)_2P(OPh)_2$	−107(v)	61
O_6/5,6	$C_2C'_4$	(OCTf=CTfO)P $(OCH_2CMe_2CH_2O)[O(4\text{-}FC_6H_4)]_2$	−128.5^C(k)	60
O_6/5,5,5	$C'_3N'_3$	$P[OC(Ph){=}NO]_3$	−77(m)	62
	C_6	$P(Glc)_3$	−89	63
	C_6	$P[OCH(CH_2OH)$ $CH(CHOHCH_2OH)O]_3$	−86	64
	C_6	$P[OCH(CO_2tBu)CH(CO_2tBu)O]_3$	−86	65
	$C_4C'_2$	$(Glc)_2P(OCPh{=}CPhO)$	−88	66
	$C_4C'_2$	$(Glc)_2P[O(2,4\text{-}tBu_2C_6H_2)O]$	−86	64
	$C_4C'_2$	(Pnc)(Glc)P(OCPh=CPhO)	−92	66
	$C_3C'_3$	$P[OC(O)CH_2O]_3$	−92.5 −94.5	64
	$C_3C'_3$	$P[OC(O)CMe_2O]_3$	−106 −108	64
	$C_3C'_3$	(Glc) $(OCPh{=}CPhO)P[OC(O)CMe_2O]$	−95	67
	$C_3C'_3$	(Glc) (OCPh=CPhO)P[OC(O)CHPhO]	−93	67
	$C_3C'_3$	(Pnc) $(OCPh{=}CPhO)P[OC(O)CMe_2O]$	−97	67
	$C_2C'_4$	**13**	−95(r)	68
	$C_2C'_4$	$(Cat)_2P(Glc)$	−85	52
	$C_2C'_4$	(Cat)P(OCPh=CPhO)(Glc)	−85	66
	$C_2C'_4$	(Cat)P(OCPh=CPhO)(Pnc)	−89	66
	CC'_5	$(Cat)_2P[OC(O)CMe_2O]$	−91	69
	CC'_5	$(Cat)_2P[OC(O)CHPhO]$	−90	69

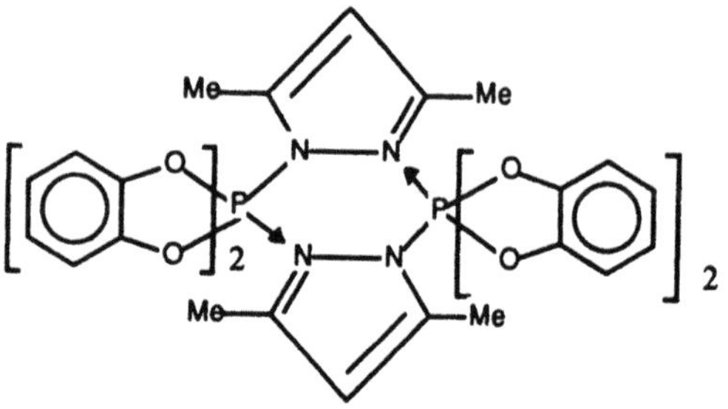

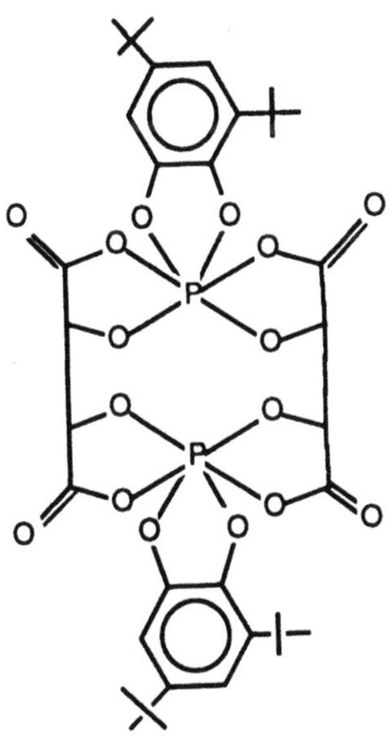

13

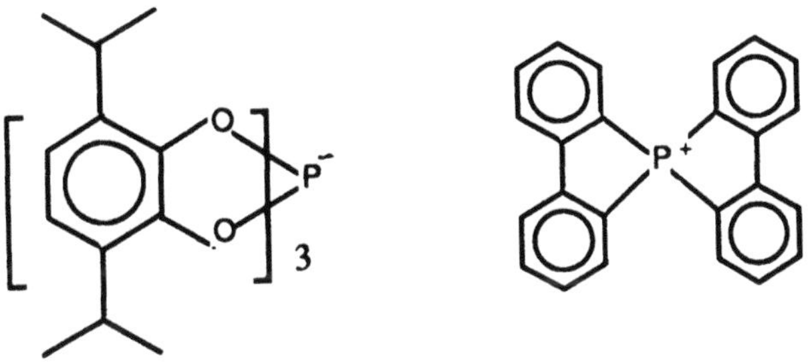

14

P-bound atoms/rings	Connectivities	Structure	NMR (δ_p[solv, temp] J Hz)	Ref.
	C'_6	$(Cat)_3P$	−82(v)	61
	C'_6	$(Cat)_2P(OCPh{=}CPhO)$	−86	66
	C'_6	$P[O(2\text{-}tBuC_6H_3)O]_3$	−84	70
	C'_6	**14**	−79	71
	C'_6	$P[O(2,4\text{-}tBu_2C_6H_2)O]_3$	−83.5 -83.7^{l}(m)	72
O_6/5,5,6	$C_2C'_4$	$(Cat)_2P(OCH_2CH_2CH_2O)$	−105	52
	C'_6	$(Cat)_2P(acac)$	−104	50
	C'_6	$(Cat)_2P[OC(O)DopO]$	−106	69
O_6/7,7,7	C'_6	P(O-Biphenyl-O)$_3$	−93.2	73
HF_5		HPF_5	−141.5 J_{PF}730/820, J_{PH}945; −123(r)	74, 14
HNO_4/5,5	$CH;CC'_3$	$(Cat)PH[OC(O)CMe_2NH]OMe$	−110 J_{PH}765	75
	$CH;CC'_3$	$(Cat)PH[OC(O)CMe_2NH]OtBu$	−112 J_{PH}740	75
$HN'O_4$/5,5	$C'_2;C'_4$	$[O(2\text{-}MeC_6H_3)O]_2PH(Pyr)$	−93 J_{PH}870	76
	$C'_2;C'_4$	$[O(4\text{-}MeC_6H_3)O]_2PH(Pyr)$	−99 J_{PH}865	76
	$C'_2;C'_4$	(Cat)PH(Quin)OPh	−100(r,v) J_{PH}860	77
HO_5	C_5	**15**	−95 J_{PH}785	65

TABLE R (continued)

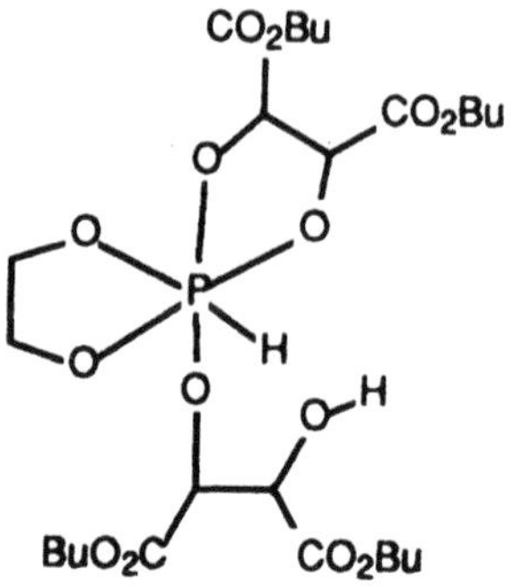

15

P-bound atoms/rings	Connectivities	Structure	NMR (δ_p[solv, temp] J Hz)	Ref.
HO_4/5,5	$C_3C'_2$	$[OC(O)CMe_2O]_2PH(OMe)$	−107 J_{PH}830	65
	CC'_4	$(Cat)_2PH(OMe)$	−97.2(m) J_{PH}800	78
	C'_5	$(Cat)_2PH[O(6\text{-}HOC_6H_4)]$	−99(m) J_{PH}802	79
$C''Cl_2F_3$	N″	$Cl_2PF_3(CN)$	−180 J_{PF}810/1010, −167.2 J_{PF}835/925 −168.5 J_{PF}940 (r)^{I}	80
$C''Cl_3F_2$	N″	$Cl_3PF_2(CN)$	−200.7(r) J_{PF}989 −201.3(r) J_{PF}752/1009^{I}	4
CCl_3N_2/5	$Cl_3;(C'_2)_2$	$[Cl_3CP(Bipyr)Cl_3]^+SbCl_6^-$	−131.9(ne)	81
	$H_3;(C'_2)_2$	$(Me)P(Bipyr)Cl_3$	−145.1 −150^{I}(ne)	82
	$H_3;(C'_2)_2$	$(Me)P(Phenanthroline)Cl_3$	−143.5 −148.4^{I}(ne)	82
	$CH_2;(C'_2)_2$	$(Et)P(Bipyr)Cl_3$	−127.5 −135^{I}(r)	82
	$CH_2;(C'_2)_2$	$(Et)P(Phenanthroline)Cl_3$	−127.5 −130.6^{I}(r)	82
$CCl_3N'_2$/5	$CCl_2;(C'_2)_2$	$C_2Cl_5P(Phenanthroline)Cl_3$	−138.5(ne)	81
$C'Cl_3N_2$/5	$C'_2;(C'_2)_2$	$[C_6F_5P(Bipyr)Cl_3]^+SbCl_6^-$	−180.6 −190.2^{I}(ne)	81
	$C'_2;(C'_2)_2$	$[(Ph)P(Bipyr)Cl_3]^+Cl^-$	−137.9 −153.2^{I}(ne)	83
	$C'_2;(C'_2)_2$	$[(Ph)P(Bipyr)Cl_3]^+SbCl_6^-$	−162.6(n)	83
	$C'_2;(C'_2)_2$	$[(Ph)P(Phenanthroline)Cl_3]^+Cl^-$	−135.4 −150.5^{I}(ne)	83
	$C'_2;(C'_2)_2$	$[(Ph)P(Phenanthroline)Cl_3]^+SbCl_6^-$	−165.9(n)	83
$C'Cl_3N'_2$	$C'_2;C''_2$	$C_6F_5PCl_3(NCS)_2$	−254.8(r)	81
$C'Cl_3N'_2$/5	$C'_2;(C'_2)_2$	$C_6F_5P(Phenanthroline)Cl_3$	−190.2(ne)	81
$C''Cl_4F$	N″	$Cl_4PF(CN)$	−244.4^{C}(r) J_{PF}964, −243.1^{T} J_{PF}1193	4
$C'Cl_4N'$	$C'_2;C'_2$	$C_6F_5PCl_4(Pyr)$	−192.8, −201^{I}(ne)	81
	$C'_2;C'_2$	$PhPCl_4(Pyr)$	−161.7(n)	83
	$C'_2;C'_2$	$PhPCl_4(3,5\text{-}Me_2Pyr)$	−182.6(n)	83
	$C'_2;C''$	$C_6F_5PCl_4(NCS)$	−260.2^{C}(r)	81
CCl_5	Cl_3	Cl_3CPCl_5	−196.6(ne)	84
	H_3	$MePCl_5$	−208(n) −205(ne)	82
	CCl_2	$C_2Cl_5PCl_5$	−168.9(r)	81
	CH_2	$EtPCl_5$	−138(ne) −160(r)	82, 84
$C'Cl_5$	C'_2	$C_6F_5PCl_5$	−240(r)	84
	C'_2	$PhPCl_5$	−223(n) −203.4(ne)	83
$C''Cl_5$	N″	$Cl_5P(CN)$	−309.5(r)	13
$CF_3N'O$/5	$H_3;C'_2;C'$	$(Quin)P(Me)F_3$	−105(m) J_{PF}883	35
	$CH_2;C'_2;C'$	$(Quin)P(Et)F_3$	−105(r) J_{PF}920	35
$C'F_3N'O$/5	$C'_2;C'_2;C'$	$(Quin)P(Ph)F_3$	−119.6(d) J_{PF}890	35
$C'F_3N_2$/4	$C'_2;(CC^+)_2$	$[N(Me)C(Ph)N(Me)]P(Ph)F_3$	−123.6 J_{PF}847/894	29
			−134.6 J_{PF}740/893(de)$_{I}$	29

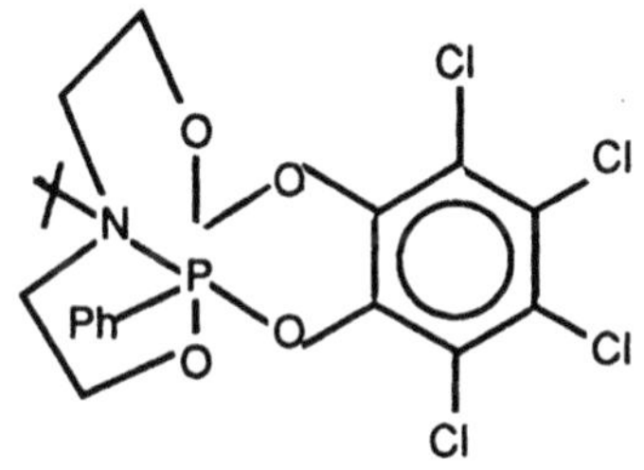

16

P-bound atoms/rings	Connectivities	Structure	NMR (δ_p[solv, temp] J Hz)	Ref.
$CF_3O_2/4$	F_3, $C^+{}_2$	$TfF_3P[OC(NMe_2)O]$	−133.9(−30°) J_{PF}922/936	85
			−145 J_{PF}831/954[I]	85
CF_5	H_3	$MePF_5$	−126.8(m) J_{PF}674/831	86
			−124.9(m) J_{PF}668/833	
	CH_2	$EtPF_5$	−126.5(m) J_{PF}665/855	87
$C'F_5$	C'_2	$PhPF_5$	−136(m) J_{PF}697/816	88
$C''F_5$	N″	$(CN)PF_5$	−157.7(r) J_{PF}744	80
$C'NO_4$/5,5,5f	C'_2;C_3;$C_2C'_2$	**16**	−96.2(b)	89
$C''NO_4$/5,5,5f	N″;C'_2;C'_4	**2**; R^1R^2 = Cat, R^3 = CN	−115(m)	26
CO_5/5,5,5	$C'H_2$,$C_2C'_3$	(Cat) $(OCH_2CH_2O)P(OCMe{=}CHCH_2)$	−57(d)	90
	$C'H_2$,$C_2C'_3$	(Cat)(OCHMeCHMeO)P $(OCMe{=}CHCH_2)$	−84.5(d)	90
CHO_4/5,5	H_3;C'_4	$(Cat)_2P(H)Me$	−113.5(m) J_{PH}620	91
$C'HO_4$/5,5	$C'_2C'_4$	$(Cat)_2P(H)C_6F_5$	−133.5(m) J_{PH}716	91
	$C'_2C'_4$	$(Cat)_2P(H)Ph$	−108.8(m) J_{PH}642	91
$C''HF_4$	N″	$(CN)P(H)F_4$	−157(r, −95°) J_{PF}695/739/847 J_{PH}832	14
$CC''Cl_4$	CH_2;N″	$EtPCl_4(CN)$	−206.5(r)*cis*	84
C''_2ClF_3	N''_2	$(CN)_2PClF_3$	−190.4(r) J_{PF}791/905	80
$C''_2Cl_2F_2$	N''_2	$(CN)_2PCl_2F_2$	−227.7(r) J_{PF}761/978	4
C''_2Cl_3F	N''_2	$(CN)_2PCl_3F$	−268.7 J_{PF}811, −269.3 J_{PF}1035; −258.7 J_{PF}789(r)[I]	4
$CC''Cl_4$	Cl_3;N″	$Cl_3CPCl_4(CN)$	−221.9[C](ne), −210.7[T](ne)	84
	H_3;N″	$MePCl_4(CN)$	−224[C], −217.7[T](r)	92
	C'_2;N″	$C_6F_5P(CN)Cl_4$	−277.3[C](r)	84
$C'C''Cl_4$	C'_2;N″	$PhPCl_4(CN)$	−225.9[C](r,ne)	92
C''_2Cl_4	N″	$PCl_4(CN)_2$	−331.2[C], −315[T](r)	13
$C_2F_2O_2/4$	$(F_3)_2$,$C^+{}_2$	$Tf_2F_2P[OC(NMe_2)O]$	−146(−30°) J_{PF}1057	85
C''_2F_4	N''_2	$PF_4(CN)_2$	−183.8[C](r) J_{PF}753; −172.6[T](r) J_{PF}741	80
$C_2F_4/5$	$(C'H_2)_2$	$F_4P(CH_2CMe{=}CMeCH_2)$	−109.9(k) J_{PF}754	60
C'_2O_4/5,5,5	$(C'_2)_2$;C'_4	$(Cat)_2P$(Biphenyl)	−106(v)	61
C''_3BrCl_2	N''_3	$BrPCl_2(CN)_3$	−398.9; −373.9(r)[I]	13
C''_3Br_2Cl	N''_3	$Br_2PCl(CN)_3$	−440.8; −417.3(r)[I]	13
C''_3ClF_2	N''_3	$ClPF_2(CN)_3$	−228.7 J_{PF}726, −234.4 J_{PF}666 −236.8 J_{PF}707/867[I](r)	4
C''_3Cl_2F	N''_3	$Cl_2PF(CN)_3$	−290 J_{PF}910, −284 J_{PF}697[I](t) −263.1 J_{PF}840	4
CC''_2Cl_3	Cl_3;N''_2	$Cl_3CPCl_3(CN)_2$	−241.2 −243.6 −249.3[I](ne)	84
	H_3;N''_2	$MePCl_3(CN)_2$	−267.5(r,ne,fe)	92
	CH_2;N''_2	$EtPCl_3(CN)_2$	−243.5; −246.6[I](r)	84
$C'C''_2Cl_3$	C'_2;N''_2	$C_6F_5PCl_3(CN)_2$	−303, −307; −315.9(r)[I]	84
	C'_2;N'_2	$PhPCl_3(CN)_2$	−259.5, −277.3; −285.3(r)[I]	92
C''_3Cl_3	N''_3	$PCl_3(CN)_3$	−351 fac −340(r)mer	13

TABLE R (continued)

P-bound atoms/rings	Connectivities	Structure	NMR (δ_p[solv, temp] J Hz)	Ref.
$C_3FO_2/4$	$(F_3)_3,C^+$	$Tf_3FP[OC(NMe)_2O]$	$-148.5(-50°)$ $J_{PF}997$	85
C''_3F_3	N''_3	$PF_3(CN)_3$	-225.5(r) $J_{PF}744$ fac -210.5 $J_{PF}684/780$ mer	80
CC''_3Cl_2	$Cl_3;N''_3$	$Cl_3CPCl_2(CN)_3$	-262.1; -258.1(n)[I] -262.9 -261.3; -258.9(ne)[I]	84
	$H_3;N''_3$	$MePCl_2(CN)_3$	-297.4(r,ne,fe)	92
	$CH_2;N''_3$	$EtPCl_2(CN)_3$	-275.7(r)	84
$C'C''_3Cl_2$	$C'_2;N''_3$	$C_6F_5PCl_2(CN)_3$	-338.6 -332; -328.8(r)[I]	84
	$C'_2;N''_3$	$PhPCl_2(CN)_3$	-299.8 -309.5; -314.2(r,ne)[I]	92
C''_4Cl_2	N''_4	$PCl_2(CN)_4$	-356.2[T](r)	13
C''_4F_2	N''_4	$PF_2(CN)_4$	-250.7[T](r) $J_{PF}853$	80
$C_4OS/4$	$(F_3)_3,H_3;C^+;C^+$	$Tf_3MeP[OC(NMe_2)S]$	-148.5; -157.5[I](f + r)	93
$C_4O_2/4$	$(F_3)_3$, H_3; C^+_2	$Tf_3MeP[OC(NMe_2)O]$	-138.5(f + r)	93
$C'_4O_2/5,5,5$	$(C'_2)_4;C'_2$	(Cat)P(Biphenyl)$_2$	-147(v)	61
$C'_4O_2/5,5,6$	$(C'_2)_4;C'_2$	(ONaphtylO)P(Biphenyl)$_2$	-168(v)	61
$C_4S_2/4$	$(F_3)_3,H_3;C^+_2$	$Tf_3MeP[SC(NMe_2)S]$	-185.5(f + r)	93
CC''_4Cl	$H_3;N''_4$	$MePCl(CN)_4$	-325.5; -315.9(r)[I]	92
	$CH_2;N''_4$	$EtPCl(CN)_4$	-296.6; -293.6(r)[I]	84
	$C'_2;N''_4$	$C_6F_5PCl(CN)_4$	-341(n), -340(r,ne)	84
CC''_5	$H_3;N''_5$	$MeP(CN)_5$	-333.7(r), -335.3(ne)	92
	$CH_2;N''_5$	$EtP(CN)_5$	-306.2(r,ne)	84
$C'_6/5,5,5$	$(C'_2)_6$	P(Biphenyl)$_3$	-181	94

REFERENCES

1. **Deng, R. M. K. and Dillon, K. B.,** *J. Chem. Soc. Dalton Trans.*, p. 1917, 1984.
2. **Markovskii, L. N., Sinitsa, A. D., Kaltchenko, V. I., Atamas, L. I., and Negrebetskii, V. V.,** *Zh. Obshch. Khim.*, 52, 445, 1982.
3. **Dillon, K. B. and Platt, A. W. G.,** *J. Chem. Soc. Dalton Trans.*, p. 1159, 1983.
4. **Platt, A. W. G.,** *Ph.D. thesis, University of Durham*, U.K., 1980.
5. **Buslaiev, Yu A., Il'in, E. G., and Sherbakova, M. N.,** *Dokl. Akad. Nauk SSSR*, 217, 337, 1974; 219, 1154, 1974.
6. **Il'in, E. G., Shcherbakova, M. N., and Buslaev, Yu A.,** *Koord. Khim.*, 1, 1179, 1975.
7. **Dillon, K. B., Reeve, R. N., and Waddington, T. C.,** *J. Chem. Soc. Dalton Trans.*, p. 1465, 1978.
8. **Schmidpeter, A., Von Criegern, T., and Blanck, K.,** *Z. Naturforsch.*, 31B, 1058, 1976.
9. **Dillon, K. B., Platt, A. W. G., and Waddington, T. C.,** *J. Chem. Soc. Dalton Trans.*, p. 2292, 1981.
10. **Skowronska, A., Pakulski, M., and Michalski, J.,** *J. Am. Chem. Soc.*, 101, 7412, 1979.
11. **Dillon, K. B., Platt, A. W. G., and Waddington, T. C.,** *J. Chem. Soc. Chem. Commun.*, p. 889, 1979.
12. **Dillon, K. B. and Platt, A. W. G.,** *Phosphorus Sulfur*, 19, 299, 1984.
13. **Dillon, K. B. and Platt, A. W. G.,** *J. Chem. Soc. Dalton Trans.*, p. 1199, 1982.
14. **Chevrier, P. J. and Brownstein, S.,** *J. Inorg. Nucl. Chem.*, 42, 1397, 1980.
15. **Becke-Goering, M. and Schwind, Z.,** *Z. Anorg. Allg. Chem.*, 372, 285, 1970.
16. **Filonenko, L. P., Povolotskii, M. I., and Pinchuk, A. M.,** *Zh. Obshch. Khim.*, 54, 216, 1984.
17. **Latscha, H. P. and Hormuth, P. B.,** *Angew. Chem. (Int. Ed.)*, 7, 299, 1968.
18. **Kaltchenko, V. I., Negrebetskii, V. V., Rudyi, R. B., Atamas, L. I., Povolotskii, M. I., and Markovskii, L. N.,** *Zh. Obshch. Khim.*, 53, 932, 1983.
19. **Dillon, K. B., Reeve, R. N., and Waddington, T. C.,** *J. Chem. Soc. Dalton Trans.*, p. 2382, 1977.
20. **Dillon, K. B., Reeve, R. N., and Waddington, T. C.,** *J. Chem. Soc. Dalton Trans.*, p. 1410, 1977.
21. **Latscha, H. P.,** *Z. Naturforsch.*, 23B, 139, 1968.
22. **Binder, H. and Flück, E.,** *Z. Anorg. Allg. Chem.*, 365, 166, 1969.

23. **Andrew, E. R. and Wynn, V. T.,** *Proc. R. Soc. London Ser. A,* 291, 257, 1966.
24. **Dillon, K. B., Lynch, R. J., Reeve, R. N., and Waddington, T. C.,** *J. Inorg. Nucl. Chem.*, 36, 815, 1974.
25. **Schmulbach, C. D. and Ahmed, I. Y.,** *J. Chem. Soc. A,* p. 3008, 1968.
26. **Schmidpeter, A. and Weinmaier, J. H.,** *Chem. Ber.*, 111, 2086, 1978.
27. **Font Freide, J. J. H. M. and Trippett, S.,** *J. Chem. Soc. Chem. Commun.*, p. 157, 1980.
28. **Font Freide, J. J. H. M. and Trippett, S.,** *J. Chem. Res. (S)* p. 218, 1981.
29. **Negrebetskii, V. V., Kaltchenko, V. I., Rudyi, R. B., and Markovskii, L. N.,** *Zh. Obshch. Khim.*, 55, 271, 1985.
30. **Utvary, K. and Kubjacek, M.,** *Monatsh. Chem.*, 110, 211, 1979.
31. **Charwath, M., Utvary, K., and Kanamueller, J. M.,** *Monatsh. Chem.*, 108, 1359, 1977.
32. **Koenig, M., Thesis,** *Toulouse n°886,* 1979.
33. **Arduengo, A. J., III, Stewart, C. A., Davidson, F., Dixon, D. A., Becker, J. Y., Culley, S. A., and Mizen, M. B.,** *J. Am. Chem. Soc.*, 109, 627, 1987.
34. **Gibson, J. A., Röschenthaler, G.-V., and Schmutzler, R.,** *J. Chem. Soc. Dalton Trans.*, p. 918, 1975.
35. **John, K. P., Schmutzler, R., and Sheldrick, W. S.,** *J. Chem. Soc. Dalton Trans.*, p. 1842 and 2466, 1974.
36. **Meindl, W. and Utvary, K.,** *Monatsh. Chem.*, 110, 129, 1979.
37. **Schlak, O., Schmutzler, R., Schiebel, H.-M., Wazeer, M. I. M., and Harris, R. K.,** *J. Chem. Soc. Dalton Trans.*, p. 2153, 1974.
38. **Schöning, G. and Glemser, O.,** *Z. Naturforsch.*, 32B, 117, 1977.
39. **Sheldrick, W. S. and Hewson, M. J. C.,** *Z. Naturforsch.*, 33B, 834, 1978.
40. **Storzer, W., Schomburg, D., Röschenthaler, G.-V., and Schmutzler, R.,** *Chem. Ber.*, 116, 367, 1983.
41. **Khabbas, N. D., Ph.D. thesis,** University of Durham, U.K., 1981.
42. **Il'in, E. G., Nazarov, A. P., and Sherbakova, M. N.,** *Dokl. Akad. Nauk SSSR* (Engl. Transl.), 250, 21, 1980.
43. **Wieker, W., Grimmer, A.-R., and Kolditz, L.,** *Z. Chem.*, 7, 434, 1967.
44. **Schultz, C. W. and Rudolph, R. W.,** *J. Am. Chem. Soc.*, 93, 1898, 1971.
45. **Wieker, W. and Grimmer, A.-R.,** *Z. Naturforsch.*, 21B, 1103, 1966.
46. **Kolditz, L., Lehmann, K., Wieker, W., and Grimmer, A. R.,** *Z. Anorg. Allg. Chem.*, 360, 259, 1968.
47. **Koenig, M., El Khatib, F., Munoz, A., and Wolf, R.,** *Tetrahedron Lett.*, 23, 421, 1982.
48. **Garrigues, B., Thesis,** *Toulouse n°920,* 1980.
49. **Ramirez, F., Prasad, V. A. V., and Marecek, J. F.,** *J. Am. Chem. Soc.*, 96, 7269, 1974.
50. **Bui Cong, C., Gence, G., Garrigues, B., Koenig, M., and Munoz, A.,** *Tetrahedron,* 35, 1825, 1979.
51. **Gence, G., Thesis,** *Toulouse n°1471,* 1975.
52. **Von Criegern, T. and Schmidpeter, A.,** *Phosphorus Sulfur,* 7, 305, 1979.
53. **Negrebetskii, V. V., Bogel'fer, L. Yu., Sinitsa, A. D., Krishtal, V. S., Kaltchenko, V. I., and Markovskii, L. N.,** *Zh. Obshch. Khim.*, 51, 956, 1981.
54. **Von Criegern, T. and Schmidpeter, A.,** *Z. Naturforsch.*, 34B, 762, 1979.
55. **Volgnandt, P. and Schmidt, A.,** *Z. Anorg. Allg. Chem.*, 425, 189, 1976.
56. **Denney, D. B., Denney, D. Z., and Ling, C.-F.,** *J. Am. Chem. Soc.*, 98, 6755, 1976.
57. **Lermann, C. L. and Westheimer, F. H.,** *J. Am. Chem. Soc.*, 98, 179, 1976.
58. **Skowronska, A., Burski, T., Krowczyk, E., and Pakulski, M.,** *Phosphorus Sulfur,* 27, 119, 1986.
59. **Denney, D. B., Denney, D. Z., Hammond, P. J., Wang, Y. P., and Liu, L. T.,** *J. Am. Chem. Soc.*, 103, 1785, 1981.
60. **Font Friede, J. J. H. M. and Trippett, S.,** *J. Chem. Soc. Chem. Commun.*, p. 934, 1980.
61. **Hellwinkel, D. and Wilfinger, H.-J.,** *Chem. Ber.*, 103, 1056, 1976.
62. **Flück, E. and Vargas, M.,** *Z. Anorg. Allg. Chem.*, 437, 53, 1977.
63. **Chang, B. C., Denney, D. B., Powell, R. L., and White, D. W.,** *Chem. Commun.*, p. 1070, 1971.
64. **Munoz, A.,** personal communication.
65. **Koenig, M., Munoz, A., Garrigues, B., and Wolf, R.,** *Phosphorus Sulfur,* 6, 435, 1979.
66. **Bernard, D. and Burgada, R.,** *C. R. Acad. Sci. Paris,* 279C, 883, 1974.
67. **Koenig, M., Munoz, A., Houalla, D., and Wolf, R.,** *J. Chem. Soc. Chem. Commun.*, p. 182, 1974.
68. **Munoz, A., Lamandé, L., Koenig, M., and Wolf, R.,** *Phosphorus Sulfur,* 11, 71, 1981.
69. **Munoz, A., Gence G., Koenig, M., and Wolf, R.,** *Bull. Soc. Chim. Fr.*, p. 1433, 1975.
70. **Gallagher, M., Munoz, A., Gence, G., and Koenig, M.,** *J. Chem. Soc. Chem. Commun.*, p. 321, 1976.
71. **Hellwinkel, D. and Krapp, W.,** *Phosphorus,* 6, 91, 1976.
72. **Kabachnik, M. I., Lobanov, D. I., and Petrovskii, P. V.,** *Izv. Akad. Nauk SSSR Ser. Khim.*, 10, 2398, 1979.
73. **Koenig, M., Klaebe, A., Munoz, A., and Wolf, R.,** *J. Chem. Soc. Perkin Trans.*, p. 40, 1979.
74. **Riesel, L. and Kant, M.,** *Z. Chem.*, 24, 382, 1984.

75. **Garrigues, B., Munoz, A., and Mulliez, M.,** *Phosphorus Sulfur,* 9, 183, 1980.
76. **Munoz, A., Koenig, M., Gence, G., and Wolf, R.,** *C. R. Acad. Sci. Paris,* 278C, 1353, 1974.
76b. **Munoz, A., Gence, G., Koenig, M., and Wolf, R.,** *C. R. Acad. Sci. Paris,* 280C, 395, 1975.
77. **Bui Cong, C., Thesis,** *Toulouse n°2004,* 1977.
78. **Burgada, R., Bernard, D., and Laurenço, C.,** *C. R. Acad. Sci. Paris,* 276C, 297, 1973.
79. **Lopez, L., Boisdon, M. T., and Barrans, J.,** *C. R. Acad. Sci. Paris,* 275C, 295, 1972.
80. **Dillon, K. B. and Platt, A. W. G.,** *J. Chem. Soc. Chem. Commun.,* p. 1089, 1983.
81. **Ali, R., Ph.D. thesis,** University of Durham, U.K., 1987.
82. **Deng, R. M. K. and Dillon, K. B.,** *J. Chem. Soc. Dalton Trans.,* p. 1911, 1984.
83. **Dillon, K. B., Reeve, R. N., and Waddington, T. C.,** *J. Chem. Soc. Dalton Trans.,* p. 1318, 1978.
84. **Ali, R. and Dillon, K. B.,** *J. Chem. Soc. Dalton Trans.,* p. 2077, 1988.
85. **Cavell, R. G. and Van de Griend, L.,** *Phosphorus Sulfur,* 18, 89, 1983.
86. **Reddy, G. S. and Schmutzler, R.,** *Inorg. Chem.,* 5, 164, 1966.
87. **Schmutzler, R.,** *Inorg. Chem.,* 7, 1327, 1968.
88. **Schmutzler, R.,** *J. Am. Chem. Soc.,* 86, 4500, 1964.
89. **Osman, F., Abdel Gawad, M. M., and Abbasi, M. M.,** *J. Chem. Soc. Perkin Trans.,* p. 1189, 1984.
90. **Voznenskaya, A. K., Razumova, N. A., Petrov, A. A., and Sheptienko, D. G.,** *Zh. Obshch. Khim.,* 47, 1432, 1977.
91. **Wieber, M., Foroughi, K., and Klingl, H.,** *Chem. Ber.,* 107, 639, 1974.
92. **Deng, R. M. and Dillon, K. B.,** *J. Chem. Soc. Dalton Trans.,* p. 1843, 1986.
93. **The, K. I., Van de Griend, L., Whitla, W. A., and Cavell, R. G.,** *J. Am. Chem. Soc.,* 99, 7379, 1977.
94. **Hellwinkel, D.,** *Chem. Ber.,* 98, 576, 1965.

INDEX

A

B

C

D

E

F

G

H

I

J

L

M

N

O

P

R

S

T

Y

Z

Milton Keynes UK
Ingram Content Group UK Ltd.
UKHW051929141024
449569UK00027B/1408